How to Evaluate, Simplify, and Solve

The most common directions in algebra are *evaluate, simplify,* and *solve.*

EVALUATE: To find the value (usually numerical) of an expression.
To replace variables and constants with numbers in a formula or expression.
To replace the function variable with numbers or another expression.

SIMPLIFY: To change an expression, making it less complicated.

The meaning of *simplify* expands as you progress through the book.

Chapter 1 — Do the operations shown, such as +, −, ×, and ÷.

Chapter 2 — Do the operations shown, taking into account the order of operations.
Do operations inside parentheses.
Do the operations shown with power expressions, using the properties of exponents.
Eliminate common factors in a fraction, leaving the fraction in lowest terms. It is not necessary to change from an improper fraction to a mixed number in algebra.
Use the distributive property to eliminate parentheses from an expression.
Eliminate common units (feet, inches, etc.) from the numerator and denominator of a fraction.
Do the operations shown using number properties, such as the associative properties of + and × or the commutative properties of + and ×.
Add (or subtract) like terms.

Chapter 3 — Do the operations needed to solve an equation.
Carry out any of the above steps on the expressions on either side of an equation before solving the equation.

Chapter 4 — Do the operations needed, after substituting numbers into an expression or function.

Chapter 5 — Eliminate common factors in a ratio, leaving the ratio in lowest terms.
Eliminate common units from a unit analysis and do the operations.

Chapter 6 — Add (or subtract) like terms in a polynomial.
Arrange terms in descending order of exponents on whichever variable is first alphabetically.
Do the operations shown with power expressions, leaving no zero or negative exponents.
Do the operations shown with power expressions, using the properties of exponents.

Chapter 8 — Do the operations shown in and with square root expressions.
Apply properties of square roots to variable expressions.

Chapter 9 — Factor rational expressions and eliminate common factors.
Apply the properties of fractions to adding, subtracting, multiplying, and dividing rational expressions.

SOLVE: To isolate the specified variable on just one side of an equation or inequality.
To draw on a number line the solution set to a linear inequality in one variable.
To isolate the specified variable on just one side of a formula.
To find the values of both variables in a system of two equations.
To find the values of three variables in a system of three equations.
To draw on rectangular coordinate axes the solution set to a linear inequality in two variables.

IF PACKAGE HAS BEEN OPENED, THIS CD-ROM CANNOT BE RETURNED FOR REFUND.

Autoplay

This CD-ROM has been programmed to launch automatically upon insertion into your computer. If the program does not start automatically, try the following steps:

1. Check to see if autoplay of CD-ROM has been disabled on your system. On a Macintosh®, this is done through the QuickTime® control panel, and on a PC it is done through the device control section of the system control panel. Enable autoplay if it is disabled.
2. Eject the CD-ROM and reinsert it into your CD-ROM drive.
3. Wait at least one minute. If nothing has happened, double-click the "Starthere" file on a Macintosh or the "Starthere.exe" file on a PC.
4. If you see the window with the Brooks/Cole-Thomson Learning logo but you do not subsequently see your browser launch, then the CD program is having difficulty locating your browser.
5. Or, if you see the browser launch but a window with the program does not open, the program is timing out after finding the browser but before it can open the correct file on the CD-ROM.
6. In either of these last two cases, open your browser program, and then from the browser open the "start_here.htm" file on the CD-ROM.

QuickTime

This program requires that QuickTime be installed on your system. It will automatically check to see if this software is installed, and if it is not, the correct installation program for either Macintosh or Windows® will be launched. Follow the on-screen instructions to add QuickTime to your system. You will need to restart your system if it is a Macintosh. If it is a PC, simply eject the CD-ROM and reinsert it.

2e

Introductory Algebra A Just-in-Time Approach

Alice Kaseberg

Lane Community College

Pacific Grove • Albany • Belmont • Boston • Cincinnati • Johannesburg • London • Madrid
Melbourne • Mexico City • New York • Scottsdale • Singapore • Tokyo • Toronto

Sponsoring Editor: *Bob Pirtle*
Marketing Team: *Leah Thomson, Samantha Cabaluna*
Marketing Assistant: *Debra Johnston*
Editorial Assistant: *Erin Wickersham*
Production Coordinator: *Keith Faivre*
Production Service: *Lifland et al., Bookmakers*
Manuscript Editor: *Sally Lifland*
Interior Design: *Vernon T. Boes*
Cover Design: *Christine Garrigan*
Cover Illustration: *Harry Briggs*
Art Coordinator: *Lisa Torri*
Interior Illustration: *Scientific Illustrators, Cyndie C. H. Wooley*
Print Buyer: *Vena Dyer*
Typesetting: *The Beacon Group*
Cover Printing: *Phoenix Color Corporation*
Printing and Binding: *World Color Corporation, Versailles*

For more information, contact:
BROOKS/COLE
511 Forest Lodge Road
Pacific Grove, CA 93950 USA
www.brookscole.com

Credits: page 124, Figure 1 adapted with permission from Michelle Hymen, "Partying for Profit," *The Register Guard*, May 10, 1994; page 267, Figure 1 by Kevork Djansezian, AP/Wide World Photos; page 337, Figure 1 reprinted with permission of Cornell University.

Printed in United States of America

10 9 8 7 6 5 4 3 2

Library of Congress Cataloging-in-Publication Data

Kaseberg, Alice.
Introductory algebra : a just-in-time approach/Alice Kaseberg. -- 2nd ed.
p. cm.
Includes index.
ISBN 0-534-35747-4 (hardcover)
1. Algebra. I. Title.
QA152.2.K37 1999
512.9--dc21
99-32648
CIP

Contents

3 Solving Equations and Inequalities in One Variable 124

Linear equations and inequalities are solved with tables, graphs, and symbols.

4 Formulas, Functions, Linear Equations, and Inequalities in Two Variables 187

Previous work with linear equations is formalized by studying formulas, functions, intercepts, slope, and equations of graphs. Inequalities are solved with graphs and symbols.

5 Ratios, Rates, and Proportional Reasoning 267

Proportional reasoning is one of the most common applications of linear equations. Ratios and proportions are examined in a variety of settings, including geometry. The chapter introduces work with averages and with the rational expressions that will be studied in Chapter 9.

Preface

Introductory Algebra: A Just-in-Time Approach is a nontraditional approach to algebra, based on

- the premise that concept development and understanding of mathematical thinking are facilitated by problem solving and discovery,
- agreement with the reforms advocated by organizations such as the National Council of Teachers of Mathematics (NCTM) and the American Mathematical Association of Two Year Colleges (AMATYC),
- the availability of technology and its considered use, and
- the mastery of certain basic skills.

The material is personalized by a 30-year career in mathematics, including experience at the community college, high school, and junior high levels; a curiosity about what mathematics is good for (which led me to degrees in business administration, mathematics, and engineering science); and, most importantly, an appreciation for problem solving, hands-on ways to present mathematics, and the beauty and wonder of mathematics inspired by the work of George Polya, W. W. Sawyer, and M. C. Escher.

I wrote the text because I want students to appreciate applications in mathematics, to understand rather than memorize skills, and to be prepared for in-depth function work in college algebra. I also want both the veteran and the novice instructor to think about new connections, new methods, and new ways of learning.

The backgrounds of developmental algebra students are diverse. There are those who have never had algebra, those who have been out of school for a while and need to relearn forgotten math skills, those who are fresh out of high school but need to review, and those who failed algebra in high school.

Because students' levels of optimism and energy are highest at the beginning of the term, Chapter 1 introduces problem solving and connections among numeric, visual, verbal, and symbolic information (see Figure A). Armed with the problem-solving background provided in Chapter 1, students will be interested and challenged by the more traditional skill review in Chapter 2.

From the beginning of this text, students are asked to make connections among tabular, graphical, verbal, and symbolic information; to integrate ideas from algebra and geometry (and, through projects, ideas from trigonometry, probability, and statistics); and to apply mathematics to real-world settings. There is more reading than in many math books, because the text includes the thinking

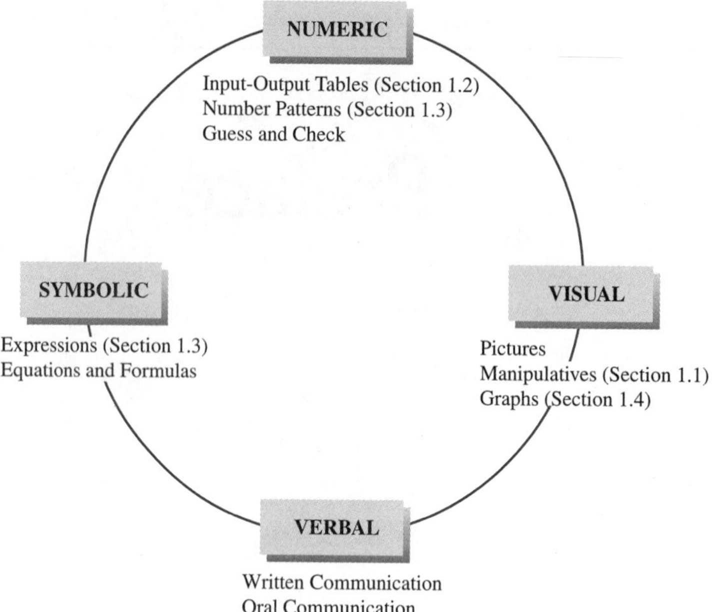

Figure A

that leads to a concept, the connections between the concept and other mathematics, and the applications for the mathematics. Where possible, material is presented through a discovery approach: exploration, question, summary, example. This approach takes longer than the traditional mode of statement and example.

The second edition includes considerably more skill building than the first edition did, as well as more explicit connections among objectives, examples, exercises, and tests. The text has been completely revised to provide a more accessible reading level and a friendlier appearance.

Pedagogy

Objectives

The learning outcomes for each section are listed at the beginning of the section. They serve as a summary for both students and instructors.

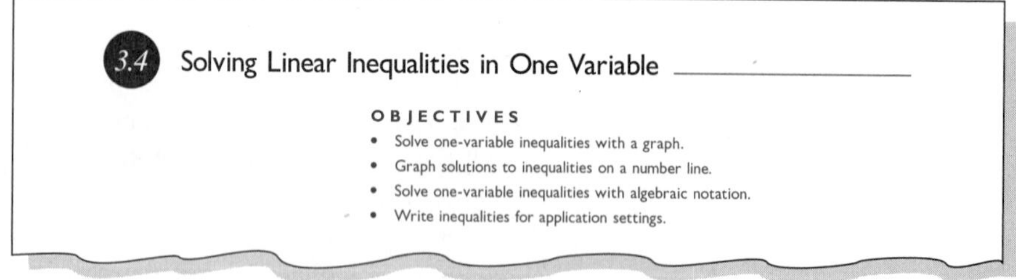

Projects

Projects are intended for group work or for individual effort. They may be more complicated problems related to the topic at hand, activity-based problems using manipulatives, or real-world applications that require research outside class. Projects suited to in-class group work are marked with an asterisk in the Index of Projects. The Index of Projects is repeated in the *Instructor's Resource Manual.*

92. ***Even and Odd Integer Sums***

a. Make a two-column listing, with the first two positive even integers in one column and their sum in the second column. Next list the first three even integers in the first column and their sum in the second column. Repeat for the first four even integers, and so forth up to 10 even integers. Your last entry in the first column should be $2 + 4 + 6 + 8 + 10 + 12 + 14 + 16 + 18 + 20$.

b. Find a pattern and write a rule for the sum of x positive even integers.

c. Repeat part a for the positive odd integers.

d. Find a rule for the sum of x positive odd integers.

93. ***Stacking Up Coins***

a. Arrange 25 coins into four stacks that fit the following conditions: The second stack is 3 times the first stack. The third stack is 1 less than the second. The fourth stack is 2 more than the first. How many coins are in each stack? Write an equation that would solve the same problem.

b. Arrange 28 coins into four stacks that fit the following conditions: The third stack is 3 more than the second stack. The first stack is twice the second stack. The fourth stack is 1 more than the first stack. How many coins are in each stack? Write an equation that would solve the same problem.

c. Describe a strategy to arrange the coins into the requested stacks.

94. ***Consecutive Integer Sums***

a. Add these consecutive integers and divide the sum by 3:

34, 35, 36

23, 24, 25

c. Add these consecutive even numbers and divide the sum by 3:

12, 14, 16

80, 82, 84

d. Explain any patterns you see in parts a, b, and c.

e. Add and divide the result by 3:

$x + (x + 1) + (x + 2)$

$x + (x + 2) + (x + 4)$

f. Explain how to find the three consecutive numbers that add to a given sum. Change your rule so it works for consecutive even numbers and for consecutive odd numbers.

g. Extend your rule in part f to finding other sets of consecutive multiples given their sum. *Hint:* Look for patterns in these sums:

$3 + 4 + 5 + 6$

$5 + 10 + 15 + 20$

$24 + 30 + 36 + 42 + 48$

$3 + 6 + 9 + 12 + 15$

h. Try your rules on these sums:

Four consecutive multiples of 10 that add to 220

Five consecutive multiples of 3 that add to 195

Four consecutive multiples of 7 that add to 210

Warm-ups

The Warm-up at the beginning of each section is designed to serve as a class opener, reviewing important concepts and linking prior and upcoming topics. Warm-ups tend to be skill-oriented; they generally connect to the algebra needed to solve text examples. The answers to the Warm-up appear in the Answer Box at the end of the section.

WARM-UP

In Exercises 1 to 4, place > or < between the two expressions.

1. a. $4 \_\_\_ 5$ **b.** $4 - 2 \_\_\_ 5 - 2$ **c.** $4 + (-5) \_\_\_ 5 + (-5)$

2. a. $-3 \_\_\_ -5$ **b.** $-3 - 4 \_\_\_ -5 - 4$ **c.** $-3 + 5 \_\_\_ -5 + 5$

3. a. $4 \_\_\_ -2$ **b.** $4(5) \_\_\_ -2(5)$ **c.** $4 \div 2 \_\_\_ -2 \div 2$

4. a. $-3 \_\_\_ 6$ **b.** $-3(4) \_\_\_ 6(4)$ **c.** $-3 \div 3 \_\_\_ 6 \div 3$

5. Does adding or subtracting a number change the inequality sign between two numbers or expressions? Does multiplying or dividing by a positive number change the inequality sign between two numbers or expressions?

6. Sketch a number-line graph for each of the following inequalities:
a. $x \geq 2$ **b.** $x < 0$ **c.** $x \leq 5$ **d.** $x \leq 0$

7. Write each inequality in Exercise 6 as an interval.

Small-Group Work

Some sections contain introductory questions or activities. These are intended to be done in class in small groups.

In Section 1.1, Exercise 17, on Numbers in Words, calls attention to the fact that students have different backgrounds and experiences. The exercise demonstrates how each student may contribute to the class and, in turn, learn from others. It is important to emphasize that students improve their own understanding by helping others.

17. ***Numbers in Words.*** This exercise reminds you to learn vocabulary, which is important in this course. What word best describes each of the phrases? Here's a hint to get you started: The prefixes commonly associated with "one" are *mono* and *uni.*
    a. One
    Legendary animal with one horn ________
    Ride it; one wheel; common in a circus ________
    One-lens eyepiece worn by the British ________
    Fancy letters printed on clothing or stationery ________
    b. Two
    Two singers ________
    Two-base hit in baseball ________
    Ride it; two wheels ________
    Able to speak two languages ________
    1,000,000,000 (number with 9 zeros) ________
    Muscle in the upper arm ________
    Sea shell with two parts ________
    Ocean animal with eight legs ________
    Full set of eight notes in music ________
    Halloween month ________
    Eight-sided geometric shape ________
    h. Ten
    Ten years ________
    Last month of calendar year ________
    Dot in our number system; the ________ point
    Track event with ten activities ________
    i. One hundred
    A hundred years ________
    One-hundredth of a dollar ________
    One-hundredth of a meter ________
    j. One thousand
    A thousand thousands ________
    A thousand years ________
18. Find words that have a prefix or root meaning six or nine.

Problem Solving

George Polya's four-step approach to problem solving—understanding the problem, making a plan, carrying out the plan, and checking the results—is introduced in Section 1.1 and revisited where appropriate. The text then focuses on ten important planning strategies. The strategies of *trying a simpler problem, using manipulatives,* and *drawing a picture* lay the foundation for Section 1.1. *Making a table of inputs and outputs* is introduced formally in Section 1.2. *Looking for a number pattern* starts in Section 1.3. *Making a graph* first appears in Section 1.4. *Working backwards* is the fundamental idea in solving equations and formulas in Sections 3.1, 4.1, and 8.3. *Choosing a test number or ordered pair and checking it* is used in drawing a line graph in Sections 2.6 and 3.4; in identifying half-planes for two-variable inequalities in Section 4.5; and in solving systems of inequalities in Section 7.5. *Making a systematic list* is an essential component of factoring in Sections 6.1 and 6.3. *Guessing and checking,* which is a natural extension of choosing a test number for inequalities, is essential in building and solving systems of equations in Section 7.2.

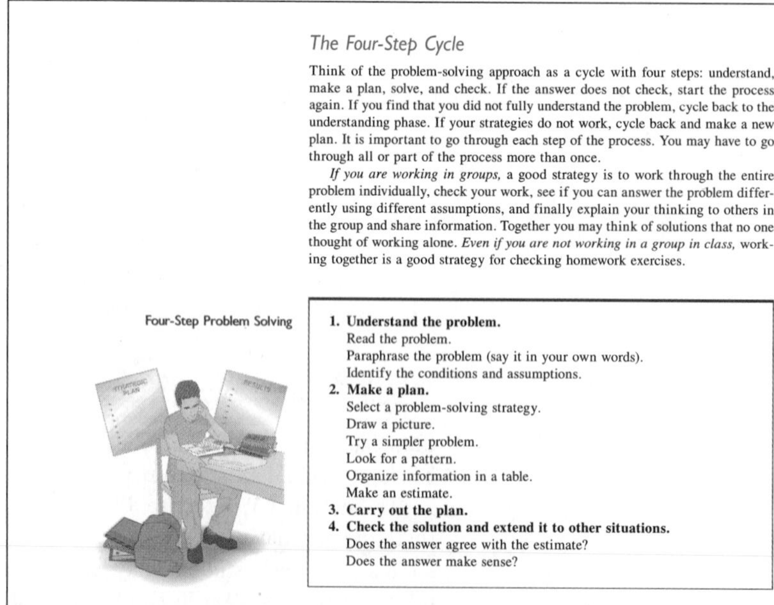

The Four-Step Cycle

Think of the problem-solving approach as a cycle with four steps: understand, make a plan, solve, and check. If the answer does not check, start the process again. If you find that you did not fully understand the problem, cycle back to the understanding phase. If your strategies do not work, cycle back and make a new plan. It is important to go through each step of the process. You may have to go through all or part of the process more than once.

If you are working in groups, a good strategy is to work through the entire problem individually, check your work, see if you can answer the problem differently using different assumptions, and finally explain your thinking to others in the group and share information. Together you may think of solutions that no one thought of working alone. *Even if you are not working in a group in class,* working together is a good strategy for checking homework exercises.

Four-Step Problem Solving

1. **Understand the problem.**
   Read the problem.
   Paraphrase the problem (say it in your own words).
   Identify the conditions and assumptions.
2. **Make a plan.**
   Select a problem-solving strategy.
   Draw a picture.
   Try a simpler problem.
   Look for a pattern.
   Organize information in a table.
   Make an estimate.
3. **Carry out the plan.**
4. **Check the solution and extend it to other situations.**
   Does the answer agree with the estimate?
   Does the answer make sense?

Explorations

Some examples are intended to be used in class for individual or group exploration. The solutions to these exploratory examples are included in the Answer Box at the end of the section.

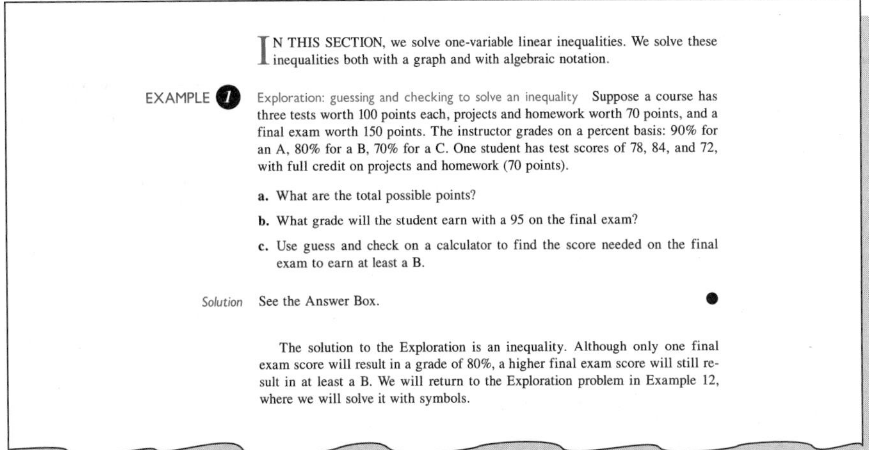

IN THIS SECTION, we solve one-variable linear inequalities. We solve these inequalities both with a graph and with algebraic notation.

EXAMPLE 1 Exploration: guessing and checking to solve an inequality Suppose a course has three tests worth 100 points each, projects and homework worth 70 points, and a final exam worth 150 points. The instructor grades on a percent basis: 90% for an A, 80% for a B, 70% for a C. One student has test scores of 78, 84, and 72, with full credit on projects and homework (70 points).

a. What are the total possible points?

b. What grade will the student earn with a 95 on the final exam?

c. Use guess and check on a calculator to find the score needed on the final exam to earn at least a B.

Solution See the Answer Box. ●

The solution to the Exploration is an inequality. Although only one final exam score will result in a grade of 80%, a higher final exam score will still result in at least a B. We will return to the Exploration problem in Example 12, where we will solve it with symbols.

Applications

To encourage creative thinking and depth in understanding, the text often poses a variety of questions about a single application setting. In addition, several applications, such as the credit card payment schedule, are repeated throughout the text so that students may observe the continuity and connections among topics.

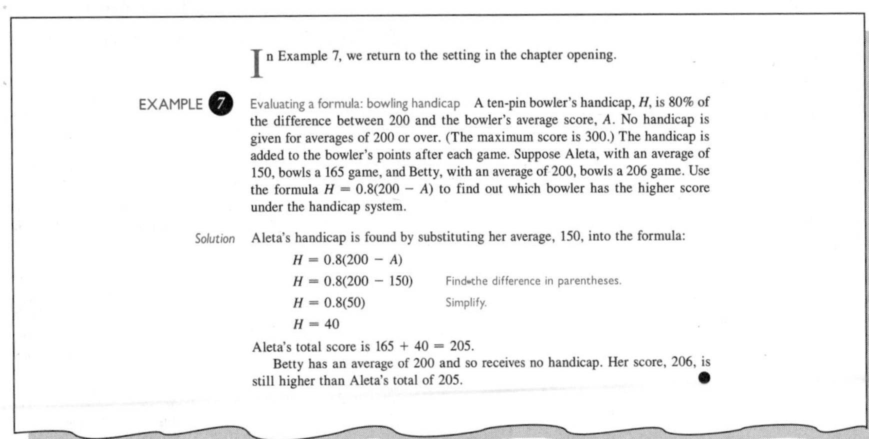

In Example 7, we return to the setting in the chapter opening.

EXAMPLE 7 Evaluating a formula: bowling handicap A ten-pin bowler's handicap, H, is 80% of the difference between 200 and the bowler's average score, A. No handicap is given for averages of 200 or over. (The maximum score is 300.) The handicap is added to the bowler's points after each game. Suppose Aleta, with an average of 150, bowls a 165 game, and Betty, with an average of 200, bowls a 206 game. Use the formula $H = 0.8(200 - A)$ to find out which bowler has the higher score under the handicap system.

Solution Aleta's handicap is found by substituting her average, 150, into the formula:

$H = 0.8(200 - A)$

$H = 0.8(200 - 150)$ Find the difference in parentheses.

$H = 0.8(50)$ Simplify.

$H = 40$

Aleta's total score is $165 + 40 = 205$.

Betty has an average of 200 and so receives no handicap. Her score, 206, is still higher than Aleta's total of 205. ●

Answer Boxes

Answers to the Warm-up, Explorations, and "Think about it" questions are placed in the Answer Box at the end of the section (just before the exercises). By providing answers as feedback, the Answer Box permits the text to be used in class or as a laboratory manual for group work or independent study.

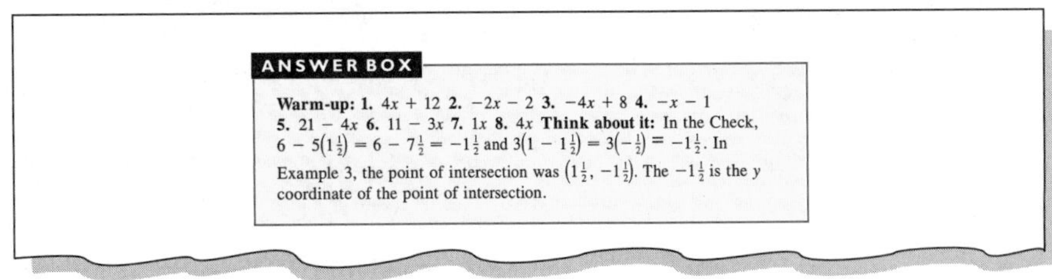

ANSWER BOX

Warm-up: 1. $4x + 12$ **2.** $-2x - 2$ **3.** $-4x + 8$ **4.** $-x - 1$ **5.** $21 - 4x$ **6.** $11 - 3x$ **7.** $1x$ **8.** $4x$ **Think about it:** In the Check, $6 - 5\left(1\frac{1}{2}\right) = 6 - 7\frac{1}{2} = -1\frac{1}{2}$ and $3\left(1 - 1\frac{1}{2}\right) = 3\left(-\frac{1}{2}\right) = -1\frac{1}{2}$. In Example 3, the point of intersection was $\left(1\frac{1}{2}, -1\frac{1}{2}\right)$. The $-1\frac{1}{2}$ is the y coordinate of the point of intersection.

Examples

Each example begins with a title, which states the purpose of the example. Usually these titles relate back to the objectives for the section.

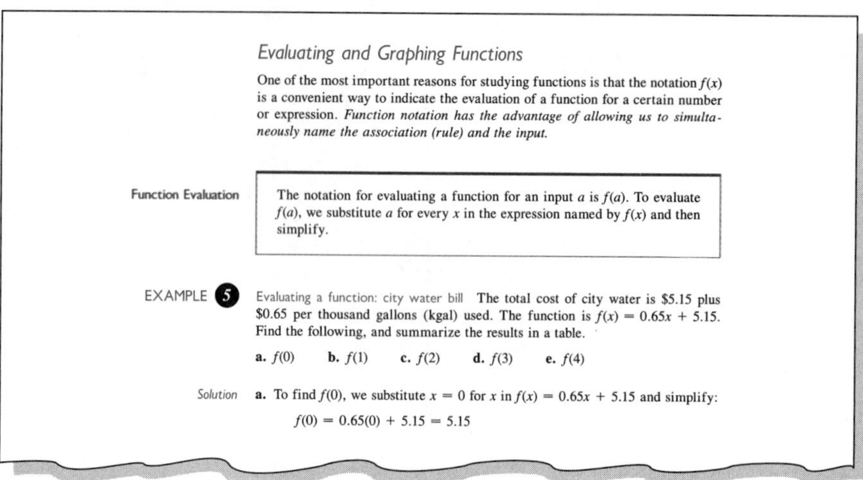

Evaluating and Graphing Functions

One of the most important reasons for studying functions is that the notation $f(x)$ is a convenient way to indicate the evaluation of a function for a certain number or expression. *Function notation has the advantage of allowing us to simultaneously name the association (rule) and the input.*

Function Evaluation

The notation for evaluating a function for an input a is $f(a)$. To evaluate $f(a)$, we substitute a for every x in the expression named by $f(x)$ and then simplify.

EXAMPLE **5** Evaluating a function: city water bill The total cost of city water is \$5.15 plus \$0.65 per thousand gallons (kgal) used. The function is $f(x) = 0.65x + 5.15$. Find the following, and summarize the results in a table.

a. $f(0)$ **b.** $f(1)$ **c.** $f(2)$ **d.** $f(3)$ **e.** $f(4)$

Solution **a.** To find $f(0)$, we substitute $x = 0$ for x in $f(x) = 0.65x + 5.15$ and simplify:

$$f(0) = 0.65(0) + 5.15 = 5.15$$

Tables and Graphs

Extensive use is made of data in tabular form. Tables encourage organization of information and promote observation of patterns. They also prepare students for spreadsheet technology. Where appropriate, a graph is related to the table, to underscore the connections among algebra, geometry, statistics, and the real world. Numbers and their corresponding equations and graphs are employed to emphasize the fact that algebra is the transition language between arithmetic and analysis.

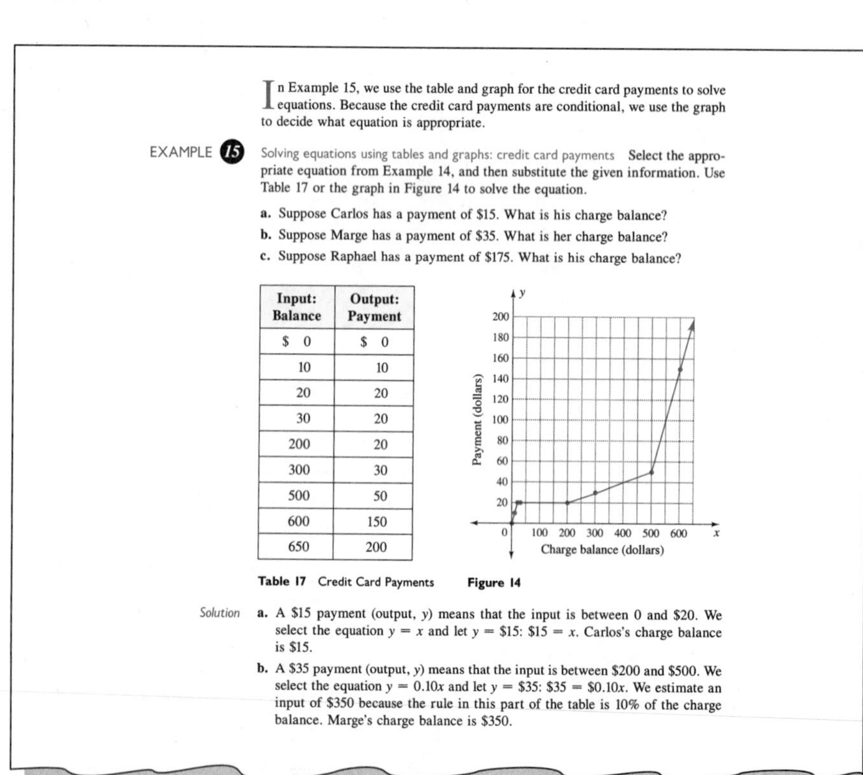

In Example 15, we use the table and graph for the credit card payments to solve equations. Because the credit card payments are conditional, we use the graph to decide what equation is appropriate.

EXAMPLE **15** Solving equations using tables and graphs: credit card payments Select the appropriate equation from Example 14, and then substitute the given information. Use Table 17 or the graph in Figure 14 to solve the equation.

a. Suppose Carlos has a payment of \$15. What is his charge balance?
b. Suppose Marge has a payment of \$35. What is her charge balance?
c. Suppose Raphael has a payment of \$175. What is his charge balance?

Input: Balance	Output: Payment
\$ 0	\$ 0
10	10
20	20
30	20
200	20
300	30
500	50
600	150
650	200

Table 17 Credit Card Payments **Figure 14**

Solution **a.** A \$15 payment (output, y) means that the input is between 0 and \$20. We select the equation $y = x$ and let $y =$ \$15: \$15 $= x$. Carlos's charge balance is \$15.

b. A \$35 payment (output, y) means that the input is between \$200 and \$500. We select the equation $y = 0.10x$ and let $y =$ \$35: \$35 $=$ \$0.10x. We estimate an input of \$350 because the rule in this part of the table is 10% of the charge balance. Marge's charge balance is \$350.

Hands-On Materials

The text supports use of an assortment of hands-on materials. Section 1.1 opens with examples in which toothpicks model pony-pen panels. Chapter 2 recommends use of a variety of hands-on materials, including colored plastic chips for integers and algebra tiles for adding like terms.

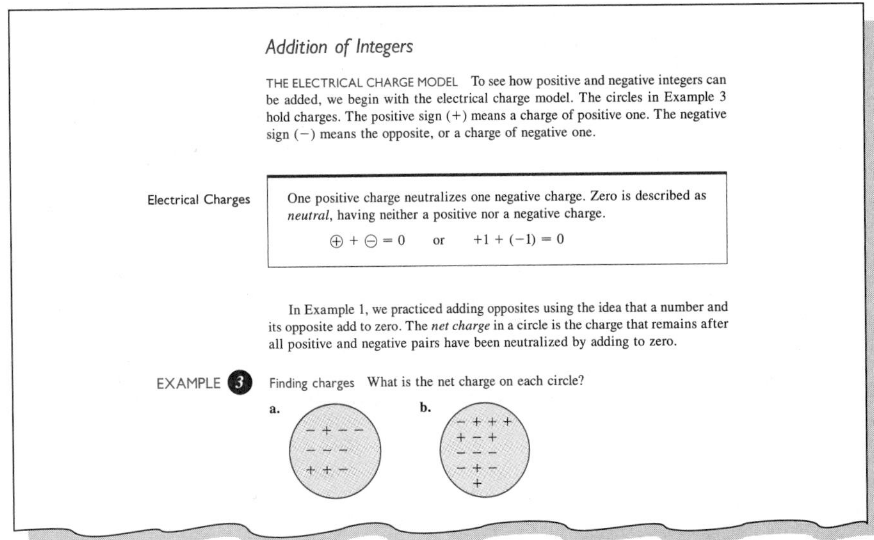

Addition of Integers

THE ELECTRICAL CHARGE MODEL To see how positive and negative integers can be added, we begin with the electrical charge model. The circles in Example 3 hold charges. The positive sign (+) means a charge of positive one. The negative sign (−) means the opposite, or a charge of negative one.

Electrical Charges

One positive charge neutralizes one negative charge. Zero is described as *neutral*, having neither a positive nor a negative charge.

$\oplus + \ominus = 0$ or $+1 + (-1) = 0$

In Example 1, we practiced adding opposites using the idea that a number and its opposite add to zero. The *net charge* in a circle is the charge that remains after all positive and negative pairs have been neutralized by adding to zero.

EXAMPLE 3 Finding charges What is the net charge on each circle?

Exercises

Tables and graphs in the exercises give students practice in skills such as solving equations. They also help students to learn graphing technology.

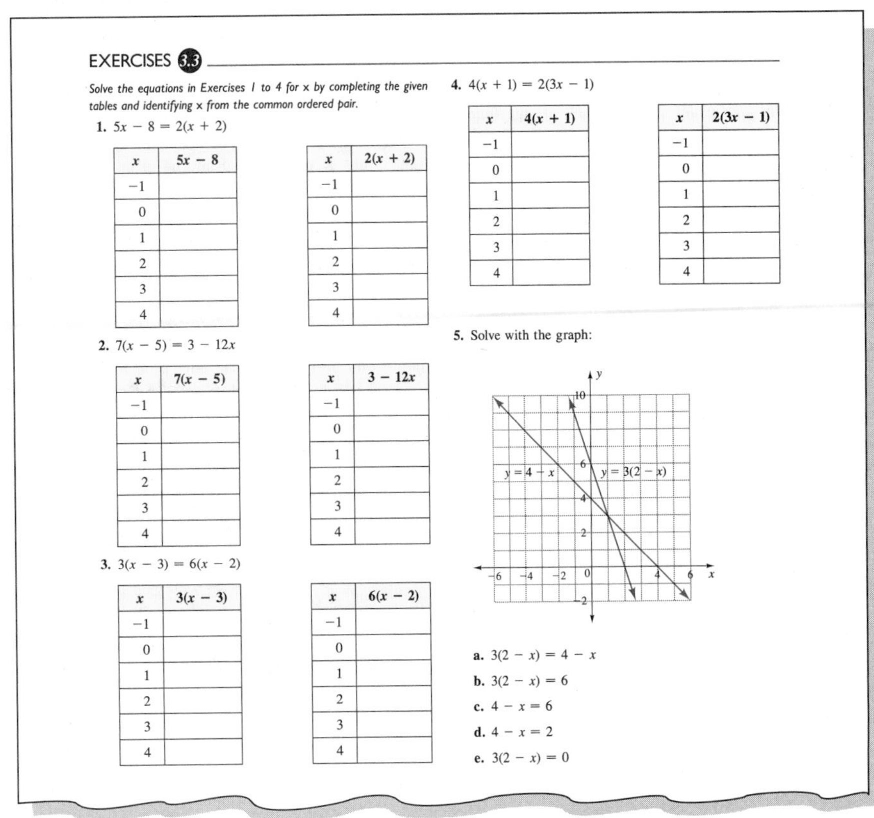

EXERCISES 3.3

Solve the equations in Exercises 1 to 4 for x by completing the given tables and identifying x from the common ordered pair.

1. $5x - 8 = 2(x + 2)$

x	$5x - 8$
−1	
0	
1	
2	
3	
4	

x	$2(x + 2)$
−1	
0	
1	
2	
3	
4	

2. $7(x - 5) = 3 - 12x$

x	$7(x - 5)$
−1	
0	
1	
2	
3	
4	

x	$3 - 12x$
−1	
0	
1	
2	
3	
4	

3. $3(x - 3) = 6(x - 2)$

x	$3(x - 3)$
−1	
0	
1	
2	
3	
4	

x	$6(x - 2)$
−1	
0	
1	
2	
3	
4	

4. $4(x + 1) = 2(3x - 1)$

x	$4(x + 1)$
−1	
0	
1	
2	
3	
4	

x	$2(3x - 1)$
−1	
0	
1	
2	
3	
4	

5. Solve with the graph:

a. $3(2 - x) = 4 - x$

b. $3(2 - x) = 6$

c. $4 - x = 6$

d. $4 - x = 2$

e. $3(2 - x) = 0$

Calculator Techniques

At the very least, a scientific calculator is required for this text. General keystrokes for scientific calculators are supplied where appropriate.

The use of graphing calculator technology—even if it is only by the instructor—enhances learning, as it gives students an understanding of basic concepts. Calculator suggestions are provided throughout the text in Graphing Calculator Technique boxes.

Graphing Calculator Technique: Solving an Equation by Graph and Table

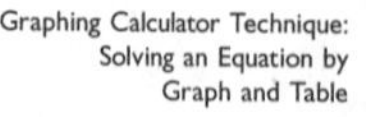

Enter the left side of the equation as Y_1 and the right side of the equation as Y_2.

To solve from a graph: Set up the viewing window with the endpoints of each axis and the scale for each axis. If a window is not near the origin, with both x and y in $[-10, 10]$, evaluate each side of the equation for the one or two inputs and use the resulting ordered pairs to estimate the window settings. Graph the equations. Trace to the point of intersection. Zoom and trace to increase the accuracy of the solution.

To solve from a table: Estimate a solution. Go to TABLE SET-UP. As a starting number for your table, choose a number near your estimate. As a change in x, or Δx, number for your table, choose a number similar to the one you used for the scale on the x-axis. Go to TABLE. You want Y_1 and Y_2 to be equal in the table. Continue to choose a new starting number, if needed, and a new change in x number until the outputs from Y_1 and Y_2 are equal. The x corresponding to the equal outputs is the solution to the original equation.

Reading Aids

A large capital letter extending down two lines (called a drop cap) signals a transition for the reader. The drop cap has been proven to aid readability. Here it leads the reader into an explanatory introduction to the next example.

The graphing calculator is useful when the common ordered pair, or intersection of two graphs, has inputs between two integers. With a graphing calculator, we can adjust the table to decimal inputs or we can trace and zoom in to a point of intersection. The solution to Example 3 is estimated with a table and graph. You may wish to solve Example 3 with a graphing calculator.

Cumulative Review

To help students maintain skills, a set of Cumulative Review Exercises is placed at the end of each even-numbered chapter. A Final Exam Review is included after the last chapter.

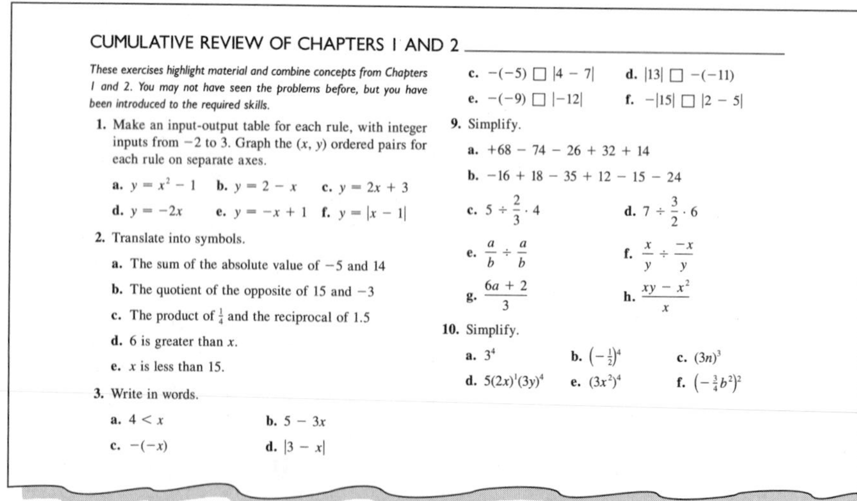

CUMULATIVE REVIEW OF CHAPTERS 1 AND 2

These exercises highlight material and combine concepts from Chapters 1 and 2. You may not have seen the problems before, but you have been introduced to the required skills.

1. Make an input-output table for each rule, with integer inputs from -2 to 3. Graph the (x, y) ordered pairs for each rule on separate axes.

a. $y = x^2 - 1$ **b.** $y = 2 - x$ **c.** $y = 2x + 3$
d. $y = -2x$ **e.** $y = -x + 1$ **f.** $y = |x - 1|$

2. Translate into symbols.

a. The sum of the absolute value of -5 and 14
b. The quotient of the opposite of 15 and -3
c. The product of $\frac{1}{4}$ and the reciprocal of 1.5
d. 6 is greater than x.
e. x is less than 15.

3. Write in words.

a. $4 < x$ **b.** $5 - 3x$
c. $-(-x)$ **d.** $|3 - x|$

c. $-(-5) \square |4 - 7|$ **d.** $|13| \square -(-11)$
e. $-(-9) \square |-12|$ **f.** $-|15| \square |2 - 5|$

9. Simplify.

a. $+68 - 74 - 26 + 32 + 14$
b. $-16 + 18 - 35 + 12 - 15 - 24$
c. $5 \div \frac{2}{3} \cdot 4$ **d.** $7 \div \frac{3}{2} \cdot 6$
e. $\frac{a}{b} \div \frac{a}{b}$ **f.** $\frac{x}{y} \div \frac{-x}{y}$
g. $\frac{6a + 2}{3}$ **h.** $\frac{xy - x^2}{x}$

10. Simplify.

a. 3^4 **b.** $\left(-\frac{1}{2}\right)^4$ **c.** $(3n)^3$
d. $5(2x)^1(3y)^4$ **e.** $(3x^2)^4$ **f.** $\left(-\frac{3}{4}b^2\right)^2$

Mid-Chapter Test

To keep students engaged and build their confidence, a Mid-Chapter Test is included in each chapter. This test gives students the opportunity to check their progress. All answers are included in the back of the book, in the Selected Answers section.

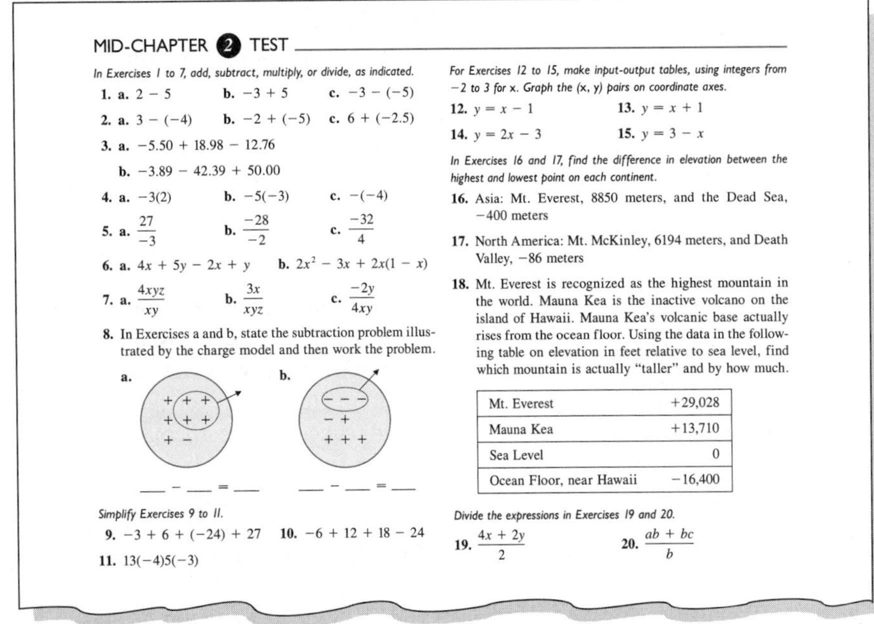

MID-CHAPTER 2 TEST

In Exercises 1 to 7, add, subtract, multiply, or divide, as indicated.

1. a. $2 - 5$ **b.** $-3 + 5$ **c.** $-3 - (-5)$

2. a. $3 - (-4)$ **b.** $-2 + (-5)$ **c.** $6 + (-2.5)$

3. a. $-5.50 + 18.98 - 12.76$

b. $-3.89 - 42.39 + 50.00$

4. a. $-3(2)$ **b.** $-5(-3)$ **c.** $-(-4)$

5. a. $\frac{27}{-3}$ **b.** $\frac{-28}{-2}$ **c.** $\frac{-32}{4}$

6. a. $4x + 5y - 2x + y$ **b.** $2x^2 - 3x + 2x(1 - x)$

7. a. $\frac{4xyz}{xy}$ **b.** $\frac{3x}{xyz}$ **c.** $\frac{-2y}{4xy}$

8. In Exercises a and b, state the subtraction problem illustrated by the charge model and then work the problem.

a. ___ − ___ = ___ **b.** ___ − ___ = ___

Simplify Exercises 9 to 11.

9. $-3 + 6 + (-24) + 27$ **10.** $-6 + 12 + 18 - 24$

11. $13(-4)5(-3)$

For Exercises 12 to 15, make input-output tables, using integers from −2 to 3 for x. Graph the (x, y) pairs on coordinate axes.

12. $y = x - 1$ **13.** $y = x + 1$

14. $y = 2x - 3$ **15.** $y = 3 - x$

In Exercises 16 and 17, find the difference in elevation between the highest and lowest point on each continent.

16. Asia: Mt. Everest, 8850 meters, and the Dead Sea, −400 meters

17. North America: Mt. McKinley, 6194 meters, and Death Valley, −86 meters

18. Mt. Everest is recognized as the highest mountain in the world. Mauna Kea is the inactive volcano on the island of Hawaii. Mauna Kea's volcanic base actually rises from the ocean floor. Using the data in the following table on elevation in feet relative to sea level, find which mountain is actually "taller" and by how much.

Mt. Everest	+29,028
Mauna Kea	+13,710
Sea Level	0
Ocean Floor, near Hawaii	−16,400

Divide the expressions in Exercises 19 and 20.

19. $\frac{4x + 2y}{2}$ **20.** $\frac{ab + bc}{b}$

Chapter Summary, Chapter Review Exercises, and Chapter Test

Every chapter ends with a Chapter Summary, Chapter Review Exercises, and a Chapter Test. The student is provided with answers to the odd-numbered Review Exercises and answers to all of the Chapter Test questions.

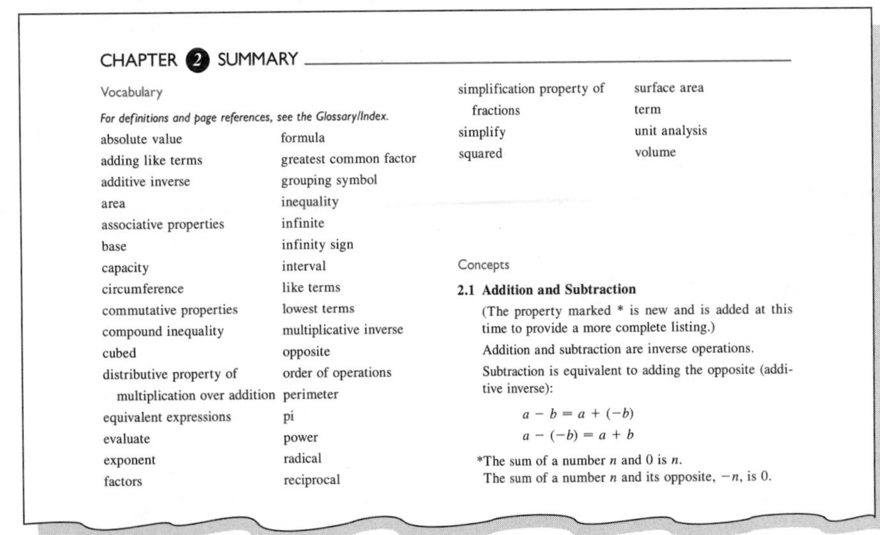

CHAPTER 2 SUMMARY

Vocabulary

For definitions and page references, see the Glossary/Index.

absolute value
adding like terms
additive inverse
area
associative properties
base
capacity
circumference
commutative properties
compound inequality
cubed
distributive property of multiplication over addition
equivalent expressions
evaluate
exponent
factors
formula
greatest common factor
grouping symbol
inequality
infinite
infinity sign
interval
like terms
lowest terms
multiplicative inverse
opposite
order of operations
perimeter
pi
power
radical
reciprocal
simplification property of fractions
simplify
squared
surface area
term
unit analysis
volume

Concepts

2.1 Addition and Subtraction

(The property marked * is new and is added at this time to provide a more complete listing.)

Addition and subtraction are inverse operations.

Subtraction is equivalent to adding the opposite (additive inverse):

$$a - b = a + (-b)$$
$$a - (-b) = a + b$$

*The sum of a number n and 0 is n.
The sum of a number n and its opposite, $-n$, is 0.

Glossary/Index

For the convenience of students and instructors, essential vocabulary is defined and referenced by page in the combined Glossary/Index. Vocabulary is considered essential if it is listed in the Chapter Summary. Nonessential terms are referenced only by page in the Glossary/Index.

Ancillaries to Accompany the Text

For Instructors

ANNOTATED INSTRUCTOR'S EDITION (AIE) For the instructor's convenience, the *Annotated Instructor's Edition* includes the answer to each exercise adjacent to that exercise; answers too long to fit in the available space are included in the Additional Answers section at the very end of the AIE. Annotations in the margin offer planning hints and teaching strategies, to supplement the more detailed planning and teaching information provided in the *Instructor's Resource Manual.*

INSTRUCTOR'S RESOURCE MANUAL (IRM) The *Instructor's Resource Manual* starts out with suggestions on planning and teaching a developmental algebra course. This extensive discussion is followed by section-by-section comments and lesson plans, plus key worksheets and overhead transparency masters. Included are fresh ideas for the experienced instructor, as well as extensive guidelines for successful mathematics instruction for the novice instructor. The IRM is designed to both provide inservice support and function as a daily reference.

The variety of instructional methods mentioned in the AIE and the IRM give you the permission and tools to be the teacher you want to be: lecturer, instructor/leader, coach, resource person, and/or facilitator of group work.

ASSESSMENT MATERIALS Test items, sample tests, and a project for each chapter are included in the assessment materials.

For Students

STUDENT'S SOLUTION MANUAL Complete worked-out solutions for all odd-numbered exercises are provided in the separate *Student's Solution Manual.* For Mid-Chapter Tests and Chapter Tests, complete worked-out solutions are given for all problems.

VIDEOTAPES A set of videotapes presents examples from each section as well as new examples. A videotape icon in the margin points out the examples covered in tapes.

Acknowledgments

I would like to thank the following reviewers and class-testers for their helpful comments and significant contributions, both to this edition and to the first edition:

Carol Achs
Mesa Community College

Rick Armstrong
Florissant Valley Community College

Linda Bastian
Portland Community College–Sylvania Campus

Paula Castagna
Fresno City College

Deann Christianson
University of the Pacific

Jennifer Dollar
Grand Rapids Community College

Dennis C. Ebersole
Northampton Community College

Grace P. Foster
Beaufort County Community College

Dave Gillette
Chemeketa Community College

Judith H. Hector
Walters State Community College

Tracey Hoy
College of Lake County

Donna L. Huck
North Central High School, Spokane

Charlotte Hutt
Southwest Oregon Community College

Virginia Lee
Brookdale Community College

Marveen McCready
Chemeketa Community College

Charles Miller
Albuquerque Technical Vocational Institute

Alice A. Mullaly
Southern Oregon State College

Susan D. Poston
Chemeketa Community College

Douglas Robertson
University of Minnesota

Gil Rodriguez
Los Medanos College

John Thickett
Southern Oregon State College

Susan M. White
DeKalb College

Tom Williams
Rowan-Cabarrus Community College

Robert Wynegar
University of Tennessee at Chattanooga

My deepest gratitude goes to my husband, Rob Bowie, and to Cosmo—the regular occupants of the empty chair in my home office. Thanks go to our parents for their patience and understanding over this past seven years and to the friends and colleagues at Lane Community College who gave emergency help: Cathy, Gayle, Jill, Jim, Karen, Phil, Tom, and Wendy. A special thank-you is due to Anne and Tami for getting me started, to Marveen and Charlotte for keeping me going, to Toni for her friendship and for the IRM, to Brigitta for her feedback on learning disabilities, and especially to Bob Pirtle and everyone at Brooks/Cole Publishing Company and Lifland et al., Bookmakers.

I appreciate the tremendous work done by everyone using the first edition. Fixing what was left undone has been a matter of pride. The second edition serves as my thank-you to those who provided feedback and encouragement.

Alice Kaseberg

To the Student

"Just in Time" is an industrial engineering term that describes a modern inventory management scheme in a manufacturing plant. Materials used in the manufacturing process are scheduled for purchase and delivery at the precise moment at which they are needed. The old inventory method of accumulating huge stockpiles of all manufacturing materials was costly, and during the recessions of recent decades many manufacturers either changed to the just-in-time inventory method or went out of business.

Introductory Algebra: A Just-in-Time Approach presents material for an introductory algebra course. "Just in time" describes this algebra in two ways. First, it refers to the book's novel approach to the study of algebra, in which you work on real-world problems, learning algebraic principles and procedures just in time as you need them. Second, it describes the book's new curriculum, which allows you to concentrate on what you need to know in this age of modern technology, just in time for the twenty-first century.

Introductory Algebra: A Just-in-Time Approach will help you to understand algebraic concepts, not memorize skills. Tables and graphs are used from the beginning to give numerical and visual meaning to algebra. Because of the new graphing and algebraic technology, the focus in algebra now is on when to use it and what it means. Estimating skills are more important than ever because you need to know whether your results are reasonable and whether they are meaningful. The text permits you to use calculator technology regularly.

Beginning the New Term

Problem Solving

Section 1.1 introduces problem-solving steps in a mathematical context. Right now, think about these four basic problem-solving steps in the context of your planning for the next few months:

1. *Understand the problem.* Your problem is that you don't have enough time to do everything you would like to do.
2. *Make a plan.* One time management strategy is to make a list of everything you have to do and when you have to do it. Make a chart showing each waking hour for the next seven days. For each hour, write in what you plan to do with that time. (There are many problem-solving strategies for planning. For a more complete listing, see Problem Solving in the Preface.)
3. *Carry out the plan.* Follow your plan (your schedule) for a week. Write notes on it to indicate when you varied from the plan and why.
4. *Check and refine.* After one week, review the plan. Did you get everything done that you needed to do? What do you need to change in your plan? Are your school load and work load reasonable? Redo the schedule to accommodate needed changes.

Time Management

Make sure your course load is sensible. Exceptionally few people can productively manage 60 hours per week of classes, study, and work. Check how sensible your plan for this term is:

> Multiply your number of credits hours by 3, and then add your number of hours of employment.

For most people, 40 to 45 hours is a reasonable commitment to school and employment.

Make or buy a calendar for the term, with spaces large enough for noting assignments, tests, appointments, and errands. Keep the calendar with you at all times.

Sticking to Your Plan

Successful students plan their time carefully. Because studying is most productively done in the daytime and in hour-long segments, plan to study between classes. Unless you have a pressing need to leave campus, stay an extra hour and study again after your last class. Not only will a schedule help you be more efficient; it will also remind you of your priorities and prevent you from avoiding tasks that need to be done.

Are You in the Right Course?

Each mathematics course has one or more prerequisite courses. Having passed a placement test does not ensure that you are prepared to succeed. If you have studied the background material recently, then usually with time, effort, confidence, and patience you will be able to learn the new material. If you have had a semester or a quarter or a summer break since your last math course, it is necessary to review. If the review provided in the text is not sufficient for you to recall, say, operations with fractions, you should immediately seek advice from your instructor or outside help. Use your prior book as a reference. If you took the prerequisite course more than a year ago, you may want to retake it before going on.

Beginning the Course

Here are a number of different issues for you to think about as you begin this course.

Getting a Good Start

To succeed, you need to attend class, read the book before class, and do the homework in a timely manner. Plan your study time. Some students set up their schedules to have the hour after math class free, to review notes and start the assignment.

Success also depends on being prepared with the proper equipment: an appropriate calculator, a six-inch ruler also marked in centimeters, and graph paper. Do all your graphs on graph paper.

Keep in mind that your first homework paper is a "grade application," just as a cover letter and resume are part of a job application. First impressions count. Neatness and completeness make a lasting impression on the instructor. So does having homework ready to turn in as you walk into class.

Taking Notes

Observe the five R's of note-taking (the Cornell system):

1. *Record.* Write down the ideas and concepts in a lecture. (Reading the text ahead of time will help identify these items.) Don't recopy notes.
2. *Reduce.* Summarize notes immediately after class (or as soon as possible); highlight important items.
3. *Recite.* Say out loud in your own words the main ideas and concepts.
4. *Reflect.* Think about how the material fits in with what you already know.
5. *Review.* Once a week, go over the ideas from each class so far in the term.

After the First Class

As you review your course syllabus, write test dates and other deadlines on your calendar. If you are working or taking classes at two schools, make sure there are no schedule conflicts with final exams. Talk with your instructor this week to resolve any scheduling problems.

Homework

Do the homework as one of the steps in your learning—not just as a requirement of the course. Work on assignments as soon as possible, right after class or early in the day or weekend. This gives you the option of going back later and spending more time on a difficult exercise. During long study periods, build in breaks to keep yourself fresh: Work for an hour, do another subject for an hour, and then come back.

Make the homework meaningful. Write notes to yourself on homework papers. Highlight exercises that were difficult and that you want to review again later. Highlight formulas or key steps. Re-read the objectives. Summarize the definitions and solution methods in your own words to be sure you understand. Describe how the current section fits in with prior sections.

Using the Answer Box and Answer Section Effectively

Practice working quickly. Do not work with the answers in front of you. Wait to check your answers until you have finished several exercises or half or more of the homework assignment. Let your own reasoning tell you whether something is correct.

Preparing for the Next Day's Class

Each of the following steps will get you progressively more prepared for your next class.

- Skim first. Read objectives. List unfamiliar words and identify new skills or concepts.
- Scan the section and look for definitions of vocabulary.
- Write vocabulary words and definitions on note cards.
- Outline the section, including summaries of skills and applications. (For your convenience in outlining, the objectives, headings, and example titles are in color in the text.)
- Read through the steps in several examples.
- Try the homework ahead of time.

What to Expect from This Course

Learning Styles

Because you and your classmates have different cultural backgrounds, with a wide variety of past and present life experiences, no single example or presentation will appeal to all of you. Consider how you best learn directions to a friend's house—in words over the phone (verbally), from a map (visually), or from having been there with someone else (kinesthetically). As a student of algebra, you may prefer words (a verbal approach), drawings, pictures, and graphs (a visual approach), getting up and moving around (a kinesthetic approach), or a combination of these approaches. To be successful in mathematics, you need to know your learning style and focus on those ways of learning information that best fit your style. It is advantageous, in the long run, to begin to learn in the other styles also. To help you achieve success with algebra, *Introductory Algebra: A Just-in-Time Approach* presents concepts in as many ways as space permits.

Independent Thinking

Although you will find the examples helpful, this text is designed to encourage your independent thinking. Look for patterns and relationships; discover concepts for yourself; seek out applications that are meaningful to you. Try to work through examples on your own first, before you look at the solution. The more involvement you have with the material, the more useful it will be and the longer you will remember it.

Groups

You are encouraged to work with others throughout this course. One of the most important benefits of working on mathematics with other people is that you clarify your own understanding when you explain an idea to someone else. This is especially true for the kinesthetic learner.

Alternative Approaches

Those of you who have had algebra before may find many familiar concepts in the text. Some concepts will appear just the way you learned them the first time; others will be presented quite differently. You are being asked to learn alternative approaches, not to discard your former skills.

Alternative approaches are important for several reasons. Your old way may not work in all situations. The new way may help introduce a later concept; it may be more efficient or give more useful results. Acknowledgment of alternative approaches validates your own discoveries. New and often better ways to do mathematics are being discovered all the time.

Now, get on with the course. Come back to the following suggestions if you run into difficulty at a later time.

Preventing Big Problems

Stuck on the Homework?

Suppose you took notes in class, read the section, and still are stumped by an exercise. If you understand the directions but can't get the problem to work, try it again on a clean sheet of paper. If you don't know how to do an exercise, summarize the relevant information and drawings and go on to another exercise. Be sure to read the exercise aloud before you give up. Sometimes we hear things that we miss when reading.

Sometimes we get too close to a problem and overlook the obvious. A fresh point of view may help. Come back later. If necessary, call another student. If you are off-campus, call your instructor during office hours to get a hint or suggested strategy. Many instructors also welcome e-mail questions.

Obtain help from your teacher or from the resource center as you need it. Don't wait until just before a test.

Falling Behind?

If you find yourself falling behind, let your instructor know that you are trying to catch up. Set up a plan that allows two to three days for each missed assignment. Most important, do current assignments first, even if you have to skip a few problems because you missed material. Work immediately after the class session. By doing the current assignment first, you will stay with the class. If you gradually complete missed work, within a reasonable amount of time you will be completely caught up. Do not skip class because you are behind or confused.

Forgetting Material?

Many students select one or two exercises from each section and write them on $3'' \times 5''$ cards, with complete solutions on the back. These "flash cards" may then be shuffled and practiced at any time for review. Cards provide an excellent way to study for tests and the final exam. Include vocabulary words in your card set.

Strategies for Taking and Learning from Tests

Prepare Yourself Academically

1. Attend class, and do the homework completely and regularly. If there are exercises on the homework that you do not know how to do, get help—from a classmate, the teacher, or another appropriate source.
2. Work under time pressure on a regular basis. Set yourself a limited amount of time to do portions of the homework. Use a time limit when doing review exercises or the practice tests at the middle and end of each chapter. Working in one- or two-hour blocks of time is usually more productive than spending all afternoon and evening on math one day a week.

Prepare Yourself Physically

3. Get a good night's sleep. Being rested helps you think clearly, even if you know less material.

Prepare Yourself Mentally

Psych yourself up! This is especially important if you have your test later in the day.

4. If you have a test at 8:00 A.M., use the last few minutes before bed to get everything ready for the next day. Make your lunch, set your books or pack on a chair by the door, set out your umbrella or appropriate weather gear, and make sure you have change for the bus or train or that your car's tires, battery, and gasoline level are okay.

5. Plan 10 or 15 minutes of quiet time before the test. Try to arrive early, if possible.

6. Mentally picture yourself taking the test.

- **a.** Imaging writing your name on the test.
- **b.** Imaging reading through the test completely to see where the instructor put various types of questions.
- **c.** Imagine writing notes, formulas, or reminders to yourself on the test.
- **d.** Imagine working your favorite type of problem first.

Take the Test Right

7. Arrive early. Be ready—pencil sharpened and homework papers ready to turn in.

8. Concentrate on doing the steps that you imagined in item 6 above.

9. Work quickly and carefully through those problems you know how to solve. Don't spend over two minutes on one problem until you have tried every problem.

10. Be confident that, having prepared for the test, you can succeed.

Learn from the Test

11. After you turned in the test, did you remember information that would have helped you on the test? Would reading through the test more thoroughly at the start have given you time to recall information you needed?

12. Before you forget, look up and write down anything that you needed to know for the test but did not know.

13. When you get the test back, look at each item you missed. Which ones did you know how to do, and which ones did you not know how to do? Figure out what caused you to miss the ones you knew how to do. Carefully re-work on paper the ones you did not know how to do, getting help as needed.

14. Write down what you will do differently in preparing for the next test.

1

Introduction to Expressions and the Coordinate Graph

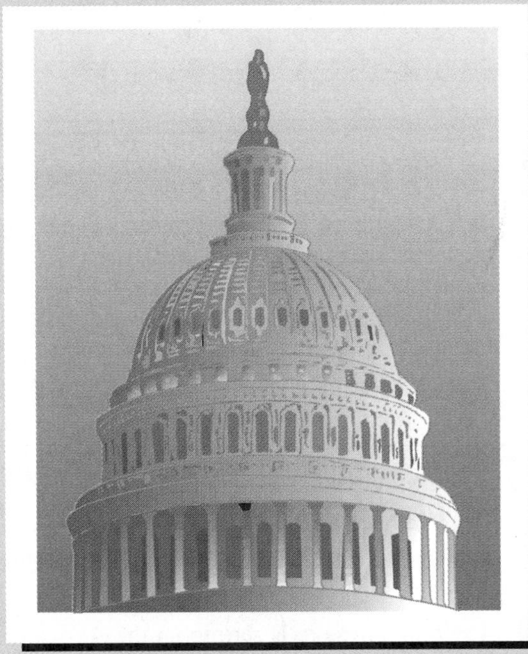

Figure 1 U.S. Capitol Building

The dome on the United States Capitol building, seen in Figure 1, is the center for street naming and numbering in Washington, D.C. The Mall and the three streets extending from the Capitol divide Washington, D.C. into four quadrants: N.E., N.W., S.E., and S.W. A similar quadrant division is introduced in Section 1.4 for graphing numerical data. Graphing is an important component of the approach to algebra presented in this textbook. The first section introduces problem solving and stresses thinking and verbal skills. Each remaining section in this chapter introduces one of the numeric, symbolic, or graphic approaches to algebra incorporated in the text. All sections include a verbal approach, with definitions and summaries.

1.1 Problem-Solving Steps

Student Note: Use the Objectives to guide your study. Ask yourself: What do I need to learn from this section? How do these ideas fit with what I already know? Study the section until you can answer yes to "Do I know the objectives?"

Student Note: Warm-ups review prior material or preview the current lesson. Answers to Warm-ups and selected examples are given in the Answer Box just before the Exercises.

OBJECTIVES

- Identify and apply Polya's four steps for problem solving.
- Identify conditions and assumptions in problem solving.

WARM-UP

Give the number most likely to be next in each pattern.

1. 2, 4, 6, 8, _

2. 3, 5, 7, 9, _

3. 4, 7, 10, 13, _

MATHEMATICS, ESPECIALLY ALGEBRA, was developed by people trying to solve real problems and to describe the world around them. New mathematics is being developed even now, and algebra is the language used to express these new ideas.

We will examine problem-solving techniques throughout the course. We start in this section by dividing problem solving into four steps. This four-step approach was first published by George Polya in *How To Solve It* in 1945.

Problem-Solving Step 1: Understand the Problem

The first step in solving a problem is to consider conditions in the problem and the assumptions we make about the problem.

EXAMPLE 

Exploration Jake is setting up the Agricultural Events Center for a pony show. He needs to build 20 pens for the ponies. For the sides of the pens he has a large number of straight, movable panels of one length, as shown in Figure 2. The panels fasten together only at the ends. How might he design and arrange the pens? How many panels will he need? These questions, like many in real life, have more than one correct answer and more than one correct way of getting the answer.

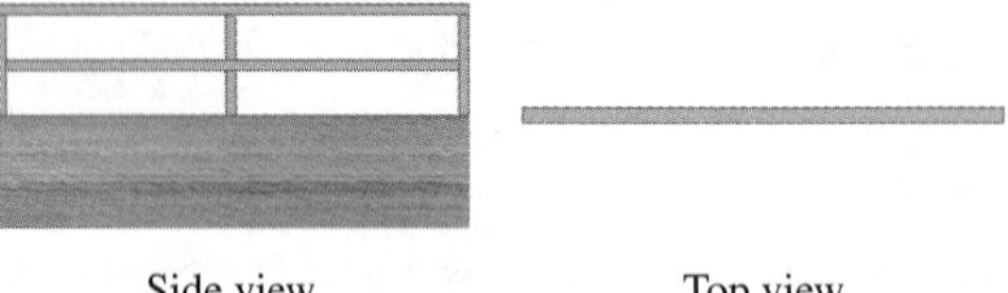

Figure 2

Solution You will find the solution to this exploration spread throughout the next several examples, or you may explore the problem yourself as an in-class activity. ●

Student Note: The end of an Example is marked by a ●.

In the exploration, Jake is given the task of building pens from panels. If he is new to the job and is given no instructions, he must first decide how to arrange the panels. Like us, he must first understand the problem.

Jake must understand the conditions and make some assumptions. A **condition** is a *requirement or restriction stated within a problem setting.* An **assumption** is *something not stated but taken as a fact.*

We all make assumptions and learn conditions on a daily basis. For example, you make a number of assumptions and learn various conditions during your first day of class. You might assume that your mathematics instructor will state the conditions for succeeding in the course. And when he or she does so, you learn what these conditions include, such as requirements for attendance, assignments, tests, and class participation. In turn, the instructor will assume that you have enrolled or intend to enroll in the course. Two conditions for enrollment may be to complete a registration process and to pay tuition.

We now look at the conditions and assumptions in Jake's problem.

EXAMPLE 2 Understanding the problem of building pens What conditions and assumptions must Jake consider?

Solution Reading the exploration statement a few times reveals that several conditions are stated in the problem: The panels are straight and movable. The panels are of one length and fasten only at the ends. It is not clear how the pens are to be arranged or even how many sides a pen should have. Jake must make assumptions in order to solve this problem. Jake might assume that each pen will hold only one pony, that the owners will need access to the pens, and that the pens should be arranged for public viewing. What other assumptions might he make? ●

Understanding the problem means understanding the questions, the given information (conditions), and any assumptions you have to make. Often you will have to read a problem several times to understand it clearly; then you will have to read it again to gather details. It is a good idea to paraphrase the problem—say it in your own words.

Problem-Solving Step 2: Make a Plan

Once you understand the problem, you need to plan how you will solve the problem. To make a plan, you think about your assumptions, develop a set of strategies for solving the problem, and search your memory for similar problems and the strategies you used to solve those problems.

EXAMPLE 3 Making a plan for building pens Suppose Jake assumes that the pens are to be connected and set out in one long line. Come up with a plan for predicting the number of panels needed, without actually building the pens.

Solution Many problem-solving strategies may be used to find a solution. Part of making a plan is deciding which strategy to use. *Drawing a picture* of a set of 20 pens and counting the panels would work. Another strategy would be to start by *trying a simpler problem,* perhaps drawing a picture of 1, 2, and 3 pens. We could then *look for a pattern* that would allow us to find the number of panels for any number of pens, without either drawing or building them all. We might *organize information by using a table.*

The final step in making a plan might be to ask, *Have we seen a problem like this before?* If so, what strategies did we learn that might help solve this problem? Can we look up a similar problem in the textbook? ●

Polya's research revealed that good problem solvers form at least one plan and think about a number of strategies before they start solving a problem. If the problem involves computation, *making an estimate* is also important.

Problem-Solving Step 3: Carry Out the Plan

Now, using what we understand about the problem, we can apply our strategies.

EXAMPLE 4 Drawing a picture of building pens Assuming that the pens will be square and in one long line, draw a picture and count the number of panels.

Solution The picture in Figure 3 shows the pens. If you count 10 panels for the first three pens, you are counting correctly. How many panels are there for 20 pens?

Top view

Figure 3

Drawing and counting panels for 20 pens may lead to error. By drawing a few of the pens, we can *do a simpler problem*. We then can organize our information in a table.

  Making a table about building pens If Jake first builds 1 pen, then attaches a second pen, and then adds a third, as shown in Figure 4, what is the pattern in the total number of panels used?

Solution Table 1 organizes the information. The headings in the table tell us to write the number of pens and the total number of panels used in building that number of pens.

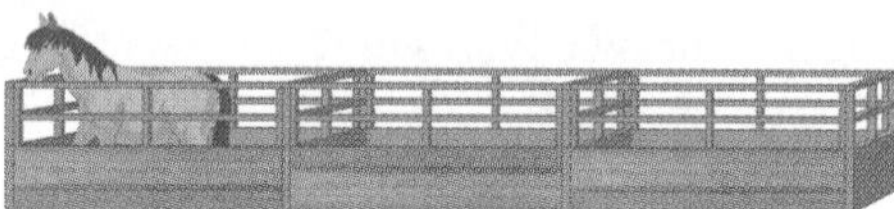

Side view

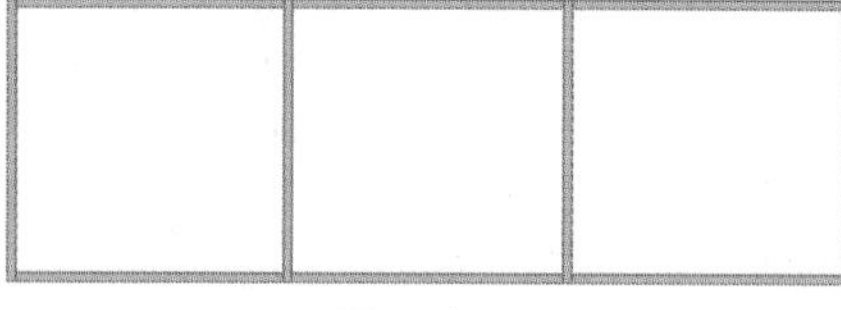

Top view

Figure 4

Number of Pens	Total Number of Panels
1	4
2	7
3	10

Table 1

From the figure and table we see that each new pen adds 3 more panels. How can we predict the total number of panels for 20 pens?

Problem-Solving Step 4: Check the Solution and Extend It to Other Situations

Go back to the exploration in Example 1. Does the number of panels for 20 pens make sense in the problem? Can we check the answer by solving the problem in a different way?

Example 4 used a picture, and Example 5 used a table of numbers. If the results are the same for two different solution methods, as they are in Examples 4 and 5, then we may be reasonably sure the answer is correct. Throughout this text, we will often use pictures (or graphs) and tables, as well as traditional algebraic methods, to solve the same problem.

Finally, we may extend our solution to other situations. What other assumptions and solutions are possible? Must we use square pens? Which pen designs use more panels? Which pen designs use fewer panels? Can we tie a pony to the panel and use fewer panels? Is there a space limitation? Will the design fit in the space? When reporting an answer, always list the assumptions you made in order to solve the problem.

The Four-Step Cycle

Think of the problem-solving approach as a cycle with four steps: understand, make a plan, solve, and check. If the answer does not check, start the process again. If you find that you did not fully understand the problem, cycle back to the understanding phase. If your strategies do not work, cycle back and make a new plan. It is important to go through each step of the process. You may have to go through all or part of the process more than once.

If you are working in groups, a good strategy is to work through the entire problem individually, check your work, see if you can answer the problem differently using different assumptions, and finally explain your thinking to others in the group and share information. Together you may think of solutions that no one thought of working alone. *Even if you are not working in a group in class,* working together is a good strategy for checking homework exercises.

Four-Step Problem Solving

1. **Understand the problem.**
   Read the problem.
   Paraphrase the problem (say it in your own words).
   Identify the conditions and assumptions.
2. **Make a plan.**
   Select a problem-solving strategy.
   Draw a picture.
   Try a simpler problem.
   Look for a pattern.
   Organize information in a table.
   Make an estimate.
3. **Carry out the plan.**
4. **Check the solution and extend it to other situations.**
   Does the answer agree with the estimate?
   Does the answer make sense?

For more information on the four-step plan of problem solving, find a copy of Polya's *How to Solve It.* The book describes how to use several problem-solving strategies, including those discussed in this lesson.

ANSWER BOX

Warm-up: 1. 10 **2.** 11 **3.** 16 **Example 4:** 61 panels **Example 5:** We add 3 panels for each new pen. The number of panels required for 20 pens in a straight line is 4 (for the first pen) plus 3 per pen times 19 additional pens; $4 + 3 \times 19$ is 61 panels. We assume that the expression $4 + 3 \times 19$ means to multiply 3 and 19 and then add the 4.

EXERCISES 1.1

In Exercises 1 to 4, suggest an assumption that could be made. Answers may vary.

1. You earn $48 each day of work for a month.
2. You write your first name but not your last name on your test.
3. You leave this message on your instructor's voice mail: "I'm Anna and I won't be in class today."
4. You allow your daughter to go to a movie.

Exercises 5 to 10 offer practice in using shapes to find number patterns and to solve problems. Think about the conditions and assumptions in each exercise.

5. Jake's assistant, Kelly, suggests using triangular pens for the ponies, as shown in the figure. How many panels would be needed to set up 20 triangular pens?

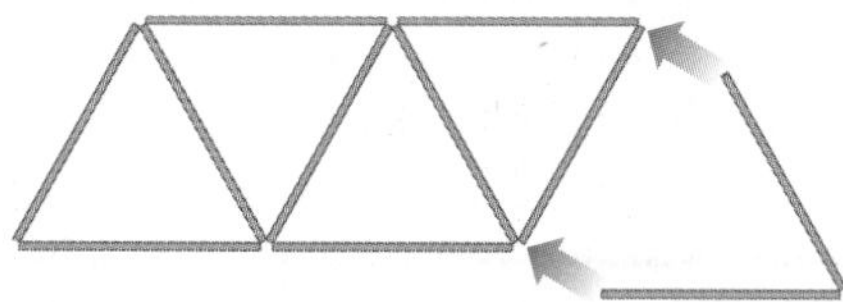

Top view

6. Suppose the pony pens are square but built with space between each pen, as shown in the figure. What is the total number of panels needed for 20 pens?

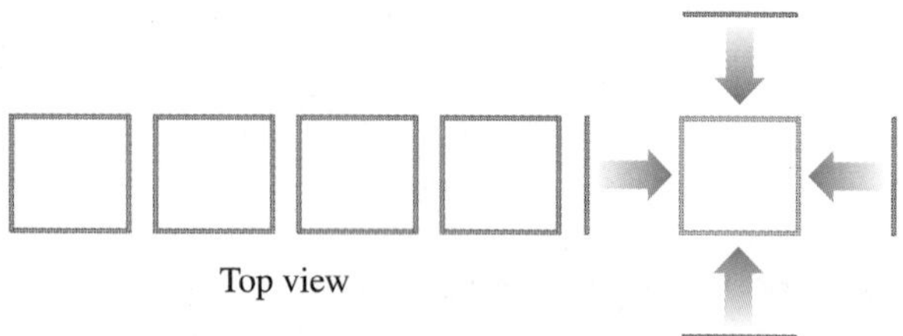

Top view

7. Jan is operations manager for a new conference center. He wants to order panels for the exhibitor display area. The design in the figure shows 6 exhibit displays built from 7 panels. How many panels will he need to build a row of 20 exhibit displays?

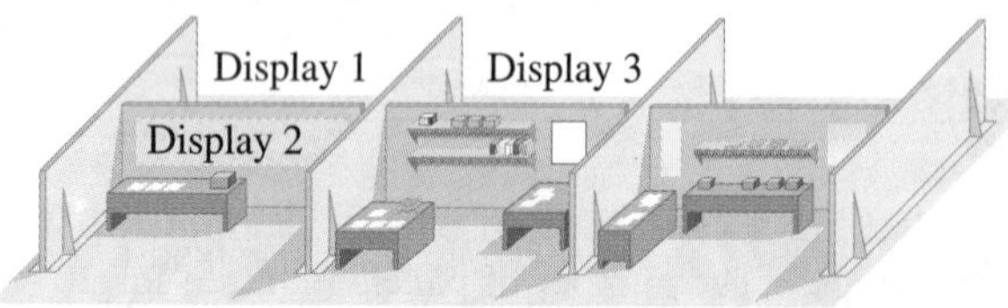

8. Cassandra is facilities manager for the Agricultural Events Center featured in the Warm-up. When Jake comes to ask for more money to buy additional panels, she suggests that he build the pens in a double row, as shown in the figure. How many panels will he now need for the 20 pens?

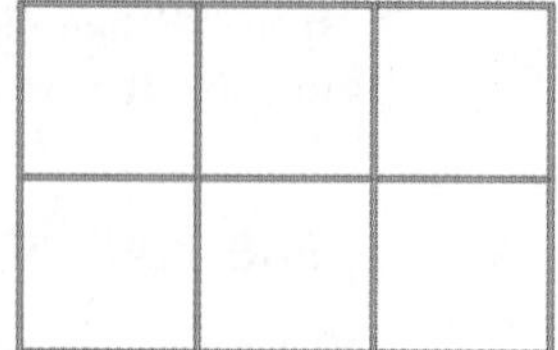

Top view

9. Mirielle hires Franck to redecorate her restaurant. He suggests buying trapezoid-shaped tables (see the figure). How many people may be seated at 1, 2, 3, and 4 of these tables if they are placed end to end with slanted sides touching?

10. Mirielle has opened a restaurant. She has square tables that may be rearranged as needed. The floor plan is sketched in the figure. Show how she may arrange some tables to seat 12 people together.

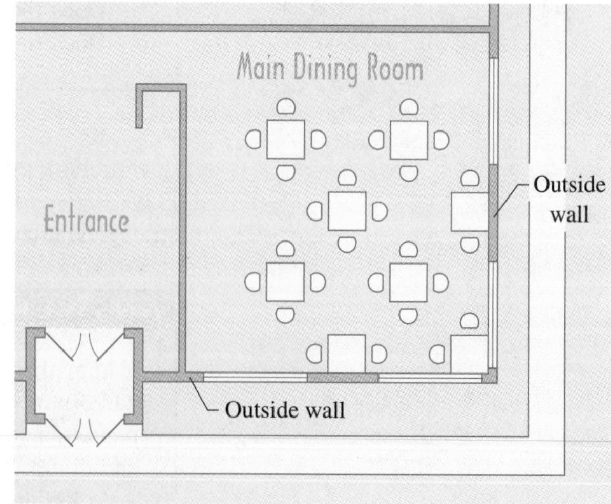

Top view

Answers to Exercises 11 to 14 may vary.

11. Write two conditions you face in your life that may affect your performance in this course.

12. Write two conditions for the course, as stated in your syllabus.

13. Write one assumption you have made about your course work.

14. Write one assumption your instructor might have made about you or your background as related to this course.

Exercises 15 and 16 are designed to trick you by making you focus on a wrong assumption.

15. A man and his son were in an auto accident. The father was killed, the son critically injured. The son arrives at the hospital and is wheeled immediately into surgery. The surgeon declares, "I cannot operate on this boy; he is my son." How is this possible?

16. A researcher is following a bear. The bear travels due south 1 mile, due east another mile, and due north 1 mile. The bear has now returned to its original position. What color is the bear?

Project

17. ***Numbers in Words.*** This exercise reminds you to learn vocabulary, which is important in this course. What word best describes each of the phrases? Here's a hint to get you started: The prefixes commonly associated with "one" are *mono* and *uni*.

a. One
Legendary animal with one horn ________
Ride it; one wheel; common in a circus ________
One-lens eyepiece worn by the British ________
Fancy letters printed on clothing or stationery ________

b. Two
Two singers ________
Two-base hit in baseball ________
Ride it; two wheels ________
Able to speak two languages ________
1,000,000,000 (number with 9 zeros) ________
Muscle in the upper arm ________
Sea shell with two parts ________
Glasses with two-part lenses ________
World War I aircraft with two wings, one above the other ________
Number system using only 0s and 1s ________

c. Three
Three babies at one birth ________
Three-sided geometric shape ________
Stable camera stand ________
Three-wheeled toy to ride ________

d. Four
Group of four singers ________
Four of these make a gallon in liquid measure ________
Four babies at one birth ________

e. Five
Five-sided geometric shape ________
Olympic event (fence, swim, run, shoot, ride horse) ________

f. Seven
Month when school typically begins ________
Track event with seven activities ________

g. Eight
Ocean animal with eight legs ________
Full set of eight notes in music ________
Halloween month ________
Eight-sided geometric shape ________

h. Ten
Ten years ________
Last month of calendar year ________
Dot in our number system; the ________ point
Track event with ten activities ________

i. One hundred
A hundred years ________
One-hundredth of a dollar ________
One-hundredth of a meter ________

j. One thousand
A thousand thousands ________
A thousand years ________

18. Find words that have a prefix or root meaning six or nine.

1.2 Sets of Numbers and Input-Output Tables

Student Note: To prepare for each class, read the Objectives, take notes on the vocabulary, look at subject headings, and read quickly through application examples.

OBJECTIVES

- Add, subtract, multiply, and divide numbers in fraction and mixed-number notation.
- Use the basic language for sets of numbers and for operations.
- Build an input-output table from a problem setting.
- Read and interpret data in single-rule and conditional-rule tables.

WARM-UP

We add, subtract, multiply, and divide $\frac{1}{3}$ and $\frac{1}{4}$ as follows. (The = symbol means "is equal to," and the dot commonly replaces a multiplication sign.)

$$\frac{1}{3} + \frac{1}{4} = \frac{4}{12} + \frac{3}{12} = \frac{7}{12}$$

$$\frac{1}{3} - \frac{1}{4} = \frac{4}{12} - \frac{3}{12} = \frac{1}{12}$$

$$\frac{1}{3} \cdot \frac{1}{4} = \frac{1}{12}$$

$$\frac{1}{3} \div \frac{1}{4} = \frac{1}{3} \cdot \frac{4}{1} = \frac{4}{3} = 1\frac{1}{3}$$

Now, add, subtract, multiply, and divide $\frac{1}{2}$ and $\frac{1}{5}$.
What assumptions did you make?

THIS SECTION HAS a numeric and verbal emphasis. After an initial overview of algebra as a language, the section summarizes several types of numbers and introduces input-output tables as a way of displaying number relationships or patterns.

Overview: Algebra as a Language

Algebra is often compared with a foreign language. It is the language used to describe numerical relationships. To learn a new language, you have to hear it in context, as well as read, speak, and write it. Hearing, reading, speaking, and writing algebra are important parts of our learning process. In order to think in terms of algebra, you need to use it.

In algebra, we use common words such as *sum, difference,* and *product* in special ways. We have words such as *quotient* that appear only in mathematics.

Definitions: Answers to Operations

A **sum** is the answer to an addition problem.
A **difference** is the answer to a subtraction problem.
A **product** is the answer to a multiplication problem.
A **quotient** is the answer to a division problem.

EXAMPLE 1 Practicing vocabulary Use the first three words in two sentences. In the first sentence, the word should have its mathematical meaning. In the second sentence, the meaning of the word should not be mathematical. Use the fourth word in a mathematical sentence only.

a. sum **b.** difference **c.** product **d.** quotient

Solution

a. The *sum* of three and four is seven.
Luisa has a *sum* of money.

b. The *difference* between twelve and nine is three.
Bill and George had a *difference* of opinion.

c. The *product* of six and five is thirty.
The manufacturer makes a quality *product.*

d. The *quotient* of 30 and six is five.

Notice that "the difference between twelve and nine" is 12 − 9, not 9 − 12. When using *difference* and *quotient,* we assume that the difference of *a* and *b* is written as $a - b$, *not* $b - a$, and the quotient of a and b is written as the fraction $\frac{a}{b}$, not $\frac{b}{a}$. We will apply this assumption in solutions to other examples and exercises. Many of the division problems in algebra involve fraction notation.

Division as a Fraction

> Fractions play an important role in algebra because almost all divisions are written in fraction notation, with the fraction bar representing the division.

Example 2 and the Warm-up review basic operations with fractions. Fraction skills are essential for the study of algebra. Practice these operations until you can do them correctly without using a calculator.

EXAMPLE 2 Calculating with fractions Write each phrase using the correct arithmetic symbol. Then do the calculation. Finally, describe each operation in a word setting.

a. The sum of $2\frac{3}{4}$ and $1\frac{2}{5}$ **b.** The difference between $2\frac{3}{4}$ and $1\frac{2}{5}$
c. The product of $2\frac{3}{4}$ and $1\frac{2}{5}$ **d.** The quotient of $2\frac{3}{4}$ and $1\frac{2}{5}$

Solution (The mixed numbers were changed to improper fractions before computation to show similarities among the operations.)

a. $2\frac{3}{4} + 1\frac{2}{5} = \frac{11}{4} + \frac{7}{5} = \frac{55}{20} + \frac{28}{20} = \frac{83}{20} = 4\frac{3}{20}$
If the Winters family eats $2\frac{3}{4}$ pizzas and the Summers family eats $1\frac{2}{5}$ pizzas (see Figure 5), the sum is what they eat together.

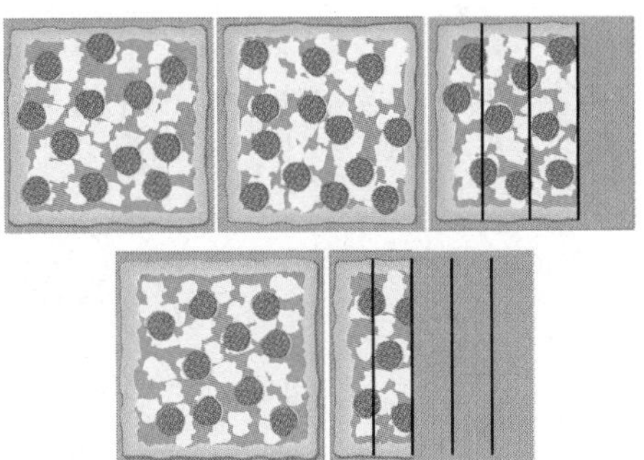

Figure 5

b. $2\frac{3}{4} - 1\frac{2}{5} = \frac{11}{4} - \frac{7}{5} = \frac{55}{20} - \frac{28}{20} = \frac{27}{20} = 1\frac{7}{20}$
The difference is how much more is eaten by the Winters family than by the Summers family.

c. $2\frac{3}{4} \cdot 1\frac{2}{5} = \frac{11}{4} \cdot \frac{7}{5} = \frac{77}{20} = 3\frac{17}{20}$
The product is the miles walked at $2\frac{3}{4}$ miles per hour in $1\frac{2}{5}$ hours after eating the pizza.

d. $2\frac{3}{4} \div 1\frac{2}{5} = \frac{11}{4} \div \frac{7}{5} = \frac{11}{4} \cdot \frac{5}{7} = \frac{55}{28} = 1\frac{27}{28}$
The quotient is how long it takes to walk $2\frac{3}{4}$ miles at $1\frac{2}{5}$ miles per hour. ●

Sets of Numbers and Notation

CHANGING NOTATION Think about the many ways we have to write numbers: fractions, decimals, percents, improper fractions, mixed numbers, and so forth. The number 1.5, which is in decimal notation, can also be written as $1\frac{1}{2}$, or $\frac{3}{2}$, or even 150%.

The **numerator** is *the top number in fraction notation,* and the **denominator** is *the bottom number.* "Numerator" is like the word *number,* saying *how much.* "Denominator" is like the word *denomination,* indicating *what kind* or *what size.* In the fraction $\frac{2}{3}$, the 3 shows the size, or kind, of the fraction—thirds. The 2 shows two parts, each of which is of size one-third. The $\frac{2}{3}$ also shows two units divided into three shares.

Within Example 2, we changed mixed numbers into improper fractions so that we could do the operations more easily. In Example 3, we change decimal notation to fraction and percent notation.

EXAMPLE 3 Changing decimals to fractions and percents Write 0.79 as a fraction and as a percent.

Solution There are two decimal places in 0.79, so the decimal is read "seventy-nine hundredths." We write 0.79 in fraction notation by placing the 79 over 100. We use 100 as the denominator because the 9 is in the hundredths place.

The word *cent* means 100, so *percent* means per hundred. We write

$$0.79 = \frac{79}{100} = 79\%$$

●

In Example 4, we change fraction notation to decimal and percent notation.

EXAMPLE 4 Changing fractions to decimals and percents Write $\frac{3}{8}$ as a decimal and as a percent.

Solution We change a fraction to a decimal by dividing the numerator by the denominator:

$$\tfrac{3}{8} = 0.375$$

On a calculator, we enter 3 [÷] 8 [=].

We change the decimal to a percent by changing it to a fraction with denominator 100:

$$\frac{0.375}{1} = \frac{0.375 \times 100}{1 \times 100} = \frac{37.5}{100} = 37.5\%$$

●

NAMING SETS Once mathematicians had a variety of numbers, they began to group them into types of numbers, or sets. A **set** is *a collection of objects or numbers.* A listing of the contents of a set is usually placed in braces, { }. For example, the **natural numbers** are *the numbers in the set {1, 2, 3, 4, . . .}.* The three dots indicate that the number list continues without end.

We build other sets of numbers using the natural numbers and other symbols, or notation. The **whole numbers** are *the set of natural numbers along with the number zero: {0, 1, 2, 3, . . .}.* The **integers** are *the numbers in the set {. . . , −3, −2, −1, 0, 1, 2, 3, . . .},* or *the set of whole numbers and their opposites.* We associate the integers with a number line, as shown in Figure 6.

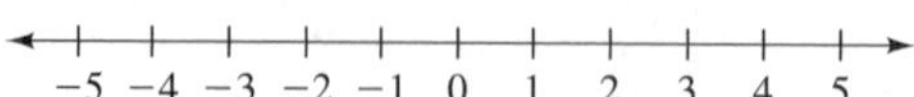

Figure 6

Student Note: To help you learn and remember new material, make 3-inch by 5-inch flash cards for notation, symbols, and vocabulary.

We use the number line to define opposites. Two different numbers are **opposites** if *they are on opposite sides of zero and the same distance from zero on a number line.* The opposite of 5 is −5; the opposite of −10 is 10. We will use integers on number lines in this chapter, learn operations with integers in Chapter 2, and use integers and their opposites to solve equations in Chapter 3.

We use integers and the fraction bar to build a rational number. **Rational numbers** are *the set of numbers that may be written as a quotient of integers,* $\frac{a}{b}$, $b \neq 0$.

To see why b cannot be zero, consider how multiplication and division facts are related: If $\frac{10}{2} = 5$, then $10 = 5 \cdot 2$. If the numerator a in $\frac{a}{b}$ is not zero and the denominator b is zero, as in $\frac{a}{0} = n$, then $a = n \cdot 0 = 0$. This is a false conclusion, so we say that division by zero is **undefined** (*has no meaning*).

Student Note: "Think about it" questions check your understanding, stress key ideas, or make connections between ideas. Possible answers are listed in the Answer Box.

Think about it: Examples 3 and 4 suggest that rational numbers may be written in several different notations, including decimals, fractions, and percents. Show that integers are also rational numbers.

All of the numbers mentioned above are members of the set of real numbers. **Real numbers** are *rational or irrational numbers* and can be shown on a number line. **Irrational numbers** are *those numbers whose decimal notation cannot be changed to a quotient of integers;* irrational numbers include $\sqrt{2}$, $\sqrt{3}$, π (pi), and $\sqrt{5}$. We will study irrational numbers in Chapter 8. Here we will focus on rational numbers. In Example 5, we think about the order of rational numbers on the number line.

EXAMPLE 5 Placing numbers in order on a number line Draw a number line with -5 on the left and 5 on the right. Show the locations for -3.5, $-\frac{1}{2}$, 1.5, $\sqrt{4}$, $3\frac{1}{4}$, and $\sqrt{25}$.

Solution The negative number -3.5 is between -4 and -3. The negative number $-\frac{1}{2}$ is between -1 and 0. Finding square roots, we have $\sqrt{4} = 2$ and $\sqrt{25} = 5$. The positions of the numbers are shown in Figure 7.

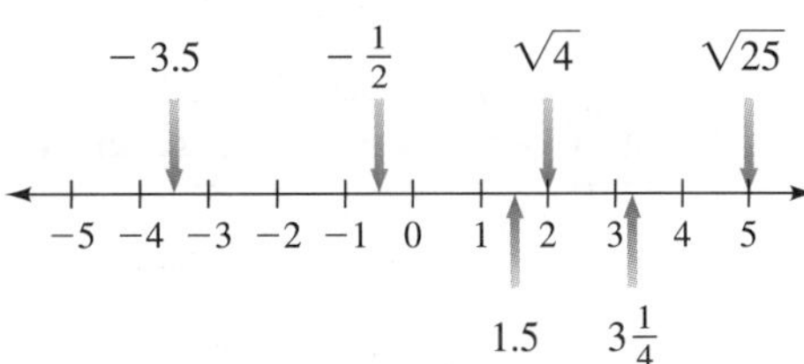

Figure 7

In Example 6, we practice identifying rational numbers with their names.

EXAMPLE 6 Matching numbers with names Write each number with every set to which it belongs: 5, 1.5, -3, $-\frac{7}{5}$, $\sqrt{9}$, -1.3, 0.2, $1\frac{1}{3}$, 0.

a. natural numbers

b. whole numbers

c. integers

d. rational numbers

Solution

a. The natural numbers include 5 and $\sqrt{9}$ (or 3).

b. The whole numbers are zero and the natural numbers, so they include 0, 5, and $\sqrt{9}$.

c. The integers are the whole numbers and their opposites, so they include -3, 0, 5, and $\sqrt{9}$.

d. All of the numbers are rational numbers.

Set of Real Numbers

Rational numbers can be written as the quotient of two integers. Sets of rational numbers include

- natural numbers $\{1, 2, 3, \dots\}$
- whole numbers $\{0, 1, 2, 3, \dots\}$
- integers $\{\dots, -3, -2, -1, 0, 1, 2, 3, \dots\}$

as well as numbers in special notation, such as

- fractions $\left(\frac{1}{2}, \frac{1}{4}, \dots\right)$
- square roots with exact decimal values $\{\sqrt{1}, \sqrt{4}, \sqrt{9}, \sqrt{16}, \dots\}$

Irrational numbers cannot be written as the quotient of two integers. They include

- square roots without exact decimal values $\{\sqrt{2}, \sqrt{3}, \sqrt{5}, \dots\}$
- pi, or π, which is equal to 3.14159265...

Input-Output Tables

We will use **input-output tables** throughout this course as *a tabular form for describing or summarizing numerical relationships.* Think of the left side, or input, of Table 2 as a source of numbers and the right side, or output, as the number matched with that input by some rule. The **input-output rule** tells us *what to do to the input to get the output.* In Section 1.1, we used such a rule to predict the number of partitions required for 20 pens.

Calculating college tuition provides an example of using a rule to match input to output. Table 2 shows the tuition at a low-cost college for 1 to 4 credit hours. Read across the table: 1 credit costs \$48, 2 credits cost \$96, and so on.

Input: Credit Hours	Output: Tuition Paid
1	\$ 48
2	\$ 96
3	\$144
4	\$192

Table 2 Tuition at Low-Cost College

EXAMPLE 7 Reading an input-output table on tuition Refer to Table 2:

a. What is the cost of a 3-hour writing course?

b. What is the cost of a 4-hour mathematics course?

c. What is the cost of a 5-hour chemistry course?

Solution **a.** From Table 2, a 3-hour writing course costs \$144.

b. From Table 2, a 4-hour mathematics course costs \$192.

c. We have at least two methods for answering the third question.

Method 1: The first and each additional credit costs \$48. If the fifth credit costs another \$48, the cost for 5 credits would be \$192 for the first 4 credits and \$48 for the fifth credit, for a total of \$240.

Method 2: If we divide each tuition amount by the number of credit hours, we find each credit to be worth \$48. The product of 5 credits at \$48 per credit is \$240.

The second method illustrates that the multiplication of the input (credit hours) by a cost per credit hour gives the output (tuition). Both methods assume that tuition rates continue in the same pattern for any possible number of credit hours. ●

EXAMPLE 8

Answering questions about tuition and fees based on an input-output table Suppose that, in addition to the tuition, there is a $20 fee charged to each student, regardless of the number of hours taken.

a. Predict how Table 2 will change.

b. What will 9 credits cost?

c. If the tuition and fees were $500, how many credit hours were taken?

Solution **a.** Each output should increase by $20, as shown in Table 3.

b. If we assume all credits cost the same amount, the total cost is $48 times the number of credit hours plus the $20 fee, or

$$\begin{aligned} \$48 \cdot 9 \text{ credits} + \$20 &= \$432 + \$20 \\ &= \$452 \end{aligned}$$

c. One way to think about this question is to note that the $500 cost is $48 more than the cost of 9 credits, so it represents 10 credits.

Input: Credit Hours	Output: Tuition Paid
1	$ 68
2	$116
3	$164
4	$212

Table 3 Tuition and Fees at Low-Cost College ●

EXAMPLE 9

Building an input-output table from a figure As you may recall from Section 1.1, Kelly, Jake's assistant, suggested using triangular pens, as shown in Figure 8. Make an input-output table in which the number of pens is the input and the total number of panels is the output.

Solution Kelly will need three panels for the first pen and two for each additional pen. The input-output table is shown in Table 4.

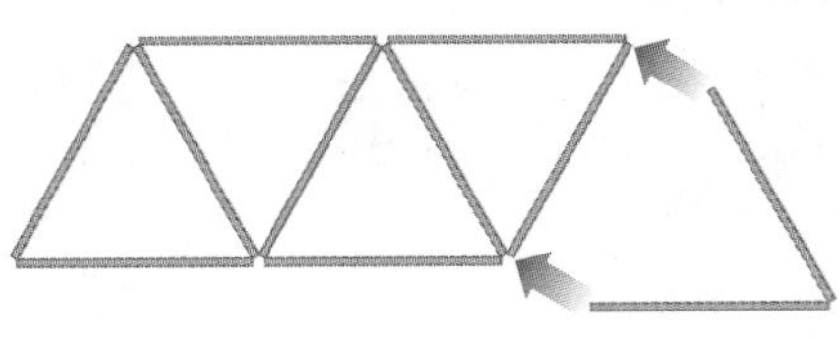

Figure 8

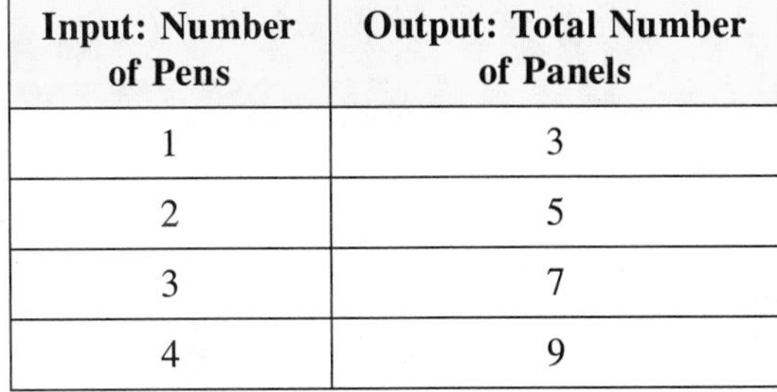

Input: Number of Pens	Output: Total Number of Panels
1	3
2	5
3	7
4	9

Table 4 ●

For each of the preceding three tables, one rule describes the entire table. But often rules change as inputs get larger. Consider Table 5, which describes the payment schedule for Chevron credit cards. The number inputs are grouped according to which output rule affects them. Such groups are called intervals. An **interval** is *a set containing all the numbers between its endpoints and including one endpoint, both endpoints, or neither endpoint.*

Input: Charge Balance	Output: Payment Due
\$0 to \$20	Full amount
\$20.01 to \$500	10% or \$20, whichever is greater
\$500.01 or more	\$50 plus the amount in excess of \$500

Table 5

EXAMPLE 10 Describing an interval Name the intervals listed in the input column in Table 5. Suppose the credit card limit is \$700.

Solution The intervals are \$0 to 20, \$20.01 to \$500, and \$500.01 to \$700. ●

The phrase **conditional rule** describes *a rule that changes for inputs in different intervals.* Table 5 gives the conditional rule for the payment due on a credit card balance at the end of the month. We use the table to calculate payment due in Example 11. Recall that we can find 10% of a number by dividing by 10.

EXAMPLE 11 Reading a payment schedule What is the payment due for each of these amounts?

a. \$15 **b.** \$25 **c.** \$200 **d.** \$400 **e.** \$550

Solution These solutions are only hints. The payments are listed in the Answer Box.

a. Pay the full amount.

b. Compare 10% of \$25 with \$20.

c. Compare 10% of \$200 with \$20.

d. Compare 10% of \$400 with \$20.

e. Add \$50 to the difference between \$550 and \$500. ●

The rule in Table 5 has three different parts. Its complexity explains why the rules are printed only on the folder containing the credit card and not on the monthly statement.

ANSWER BOX

Warm up: $\frac{7}{10}, \frac{3}{10}, \frac{1}{10}, \frac{5}{2}$. We assume that subtraction and division are done in the order listed: $\frac{1}{2} - \frac{1}{5}$ and $\frac{1}{2} \div \frac{1}{5}$. **Think about it:** An integer may be written with 1 as the denominator, thereby creating a rational number $\frac{a}{1}$. **Example 11: a.** \$15 **b.** \$20 **c.** \$20 **d.** \$40 **e.** \$100

EXERCISES 1.2

Write the phrases in Exercises 1 to 20 using the correct operation symbol (+, −, ·, ÷), and then do the operation. Write answers as fractions and decimals, where appropriate. Round decimals to the nearest hundredth. Assume the second number is subtracted from or divided into the first number.

1. The product of 4 and $\frac{1}{2}$
2. The quotient of 5 and $\frac{1}{2}$
3. The sum of $\frac{1}{3}$ and $\frac{1}{2}$
4. The difference between $\frac{3}{4}$ and $\frac{1}{3}$
5. The quotient of 5 and 15
6. The sum of 0.25 and $\frac{3}{4}$
7. The product of $\frac{3}{5}$ and $\frac{4}{9}$
8. The difference between $\frac{7}{5}$ and $\frac{3}{4}$
9. The difference between $2\frac{2}{3}$ and $1\frac{1}{4}$
10. The product of $2\frac{2}{3}$ and $1\frac{1}{4}$
11. The quotient of $2\frac{2}{3}$ and $1\frac{1}{4}$
12. The sum of $2\frac{2}{3}$ and $1\frac{1}{4}$
13. Find the difference between $\frac{4}{5}$ and $\frac{1}{3}$.
14. Find the product of $\frac{5}{6}$ and $\frac{3}{4}$.
15. Find the sum of $1\frac{1}{2}$ and $2\frac{1}{4}$.
16. Find the quotient of $1\frac{1}{2}$ and $2\frac{1}{4}$.
17. Find 10% of the sum of 1.5 and 2.5.
18. Find 10% of the product of 1.25 and 8.
19. Find 10% of the difference between 20 and 15.
20. Find 10% of the quotient of 80 and 5.
21. If the tuition cost in Example 7 remains \$48 per credit hour, what is the cost of 12 credit hours?
22. If the tuition cost in Example 7 was \$480, how many credit hours were taken?
23. In Example 8, a \$20 fee is added to the \$48 per credit hour tuition cost. How many credit hours were taken if the total cost was \$404?
24. In Example 8, we found how many credit hours were taken if the total of tuition and fees was \$500. What is another way to calculate the number of credit hours? Use complete sentences to describe your method. Try your method on a total cost of \$788.

In Exercises 25 to 28, use the following table.

Input: Shirt Size	Output: Fabric Needed
Small	$1\frac{1}{2}$ yards
Medium	$1\frac{3}{4}$ yards
Large	$1\frac{3}{4}$ yards
Extra Large	$2\frac{1}{8}$ yards

25. **a.** Find the total yards of fabric needed to make a small shirt and a large shirt.
    b. How many yards of fabric are needed to make 5 large shirts?
26. **a.** How many extra large shirts can be made with 17 yards of fabric?
    b. How many yards of fabric are needed for 2 medium shirts?
27. **a.** If you buy fabric for an extra large shirt and then make only a large shirt, how much fabric should you have remaining?
    b. How many medium shirts can be made with 21 yards of fabric?
28. **a.** Find the total yards of fabric needed to make a small shirt and an extra large shirt.
    b. How many more yards of fabric are needed to make an extra large shirt than a small shirt?

In Exercises 29 to 32, draw the next figure in each pattern, and explain your reasoning. Make an input-output table for the figure. Predict the output when the input is 10.

29. The input is the number of dots; the output is the number of non-overlapping segments (pieces of lines) separated by the dots.

30. The input is the number of rays (arrows); the output is the number of non-overlapping angles formed. There is one angle formed by the first ray—a full rotation of 360 degrees.

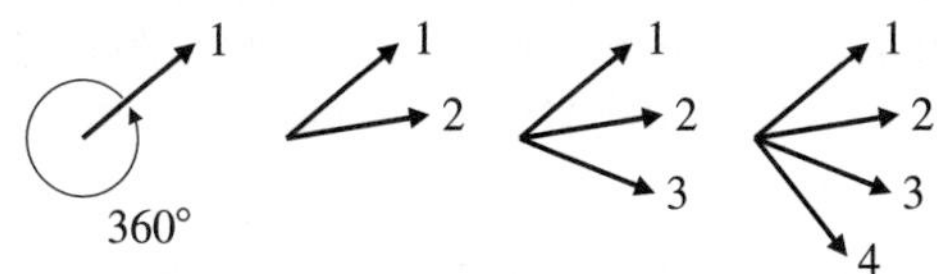

31. The input is the number of lines; the output is the number of regions separated (like pieces of pie) by the lines.

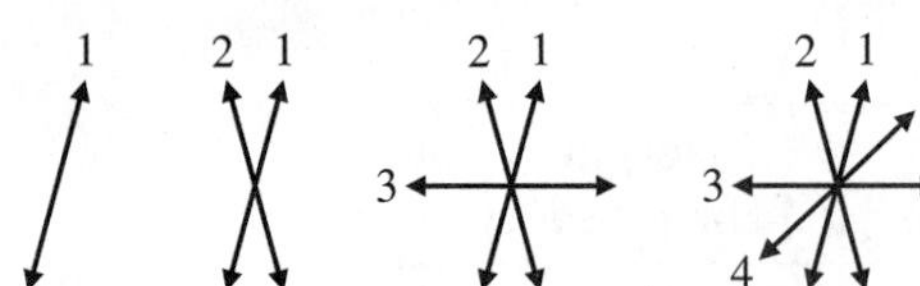

32. The input is the number of sides in each figure; the output is the number of vertices (corners) in each figure.

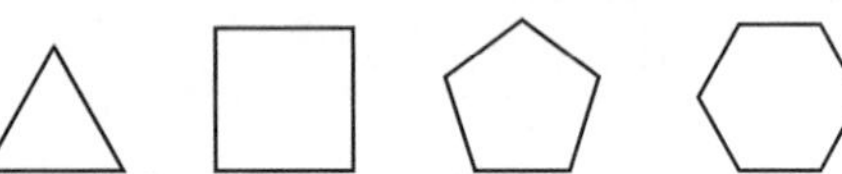

In Exercises 33 to 38, write a sentence that uses the word in its mathematical meaning. Then write a second sentence in which the word's meaning is not mathematical.

33. set **34.** interval

35. rational **36.** input

37. whole **38.** rule

For Exercises 39 to 42, use a ruler to draw a number line. Mark and label the location of the following sets of numbers.

39. $-3, 2\frac{1}{4}, -\frac{1}{2}, \frac{3}{4}, 1.5, -2\frac{1}{2}$

40. $2, -2\frac{1}{4}, \frac{1}{2}, -\frac{1}{4}, 2.5, -1.5$

41. $1.25, \frac{8}{7}, 1\frac{1}{6}$

42. $-2.5, -2\frac{4}{7}, -\frac{16}{7}$

In Exercises 43 to 46, use the credit card rule in Table 5.

43. Find the payment due if the charge balance is \$375.

44. If the payment is \$50, what was the charge balance?

45. If the payment is \$75, what was the charge balance?

46. Find the payment due if the charge balance is \$850.

For Exercises 47 and 48, make a conditional input-output table for the rule given. The **even numbers** *are the integers divisible by two. The* **odd numbers** *are the integers not divisible by two.*

47. The output is 5 if the input is an even number. The output is twice the input if the input is an odd number. Use integer inputs $\{0, 1, 2, \ldots, 8\}$.

48. The output is the sum of the input and 2 if the input is even. The output is the product of the input and 2 if the input is odd. Use integer inputs 0 through 8.

49. Make an input-output table showing the cost for 8 to 12 credit hours if the tuition is \$55 per credit hour for 1 to 10 credit hours and \$600 for more than 10 credit hours.

50. Make an input-output table for the following tuition model: Tuition is \$75 per credit hour for 1 to 11 hours; \$900 total for 12 to 18 hours; and an additional \$40 per credit hour for each hour above 18 hours. Let inputs be the even numbers from 2 to 20 hours and outputs be the total tuition cost.

51. Make a tuition and fees table for your school or a nearby college or university.

52. Make an input-output table in which the input is the temperature inside an office building and the output is whether the heat, air conditioning, or neither should be turned on.

53. a. Which set of real numbers includes positive integers but not negative integers or zero?

b. What kind of numbers might be used to write elevations below sea level?

c. Which set of real numbers includes $\frac{2}{3}$, $\frac{5}{3}$, and 3.03?

d. When we combine the even numbers and the odd numbers, we get which set of real numbers?

54. From the set $\left\{\frac{2}{3}, -2, 0, 2, \frac{6}{2}\right\}$, select a number that fits each statement.

a. Division by this number is undefined.

b. This rational number is not defined in the set of integers.

c. This integer is not defined in the set of natural numbers.

Problem-Solving Exercises: Guess and Check Skills

Use the following hint to do Exercises 55 and 56: Guess a pair of numbers that fits the first condition; see how closely your guess fits the second condition.

55. a. Find two numbers that add to 12 and multiply to 27.

b. Find two numbers that add to 10 and multiply to 21.

56. a. Find two numbers that multiply to 12 and add to 8.

b. Find two numbers that multiply to 24 and add to 10.

Computation Review

Exercises 57 and 58 contain modified input-output tables. Use the two input values to calculate the outputs, according to the rule given at the top of each output column. (The dot between a and b means to multiply.)

57.

Input a	Input b	Output $a + b$	Output $a - b$	Output $a \cdot b$	Output $a \div b$
15	5				
$\frac{3}{4}$	$\frac{2}{5}$				
0.36	0.06				
5.6	0.7				
2.25	1.5				

58.

Input a	Input b	Output $a+b$	Output $a-b$	Output $a \cdot b$	Output $a \div b$
9	6				
$\frac{2}{3}$	$\frac{1}{6}$				
0.5	0.25				
49	0.7				
6.25	2.5				

59. Change to a percent each number in the Input columns of Exercise 57.

60. Change to a percent each number in the Input columns of Exercise 58.

MID–CHAPTER  TEST

1. For the figure shown, make a table in which the number of sides is the input and the number of triangles inside is the output. Draw the next figure in the pattern. (Hint: It should contain 5 triangles.) Predict how many triangles would be inside a 20-sided figure.

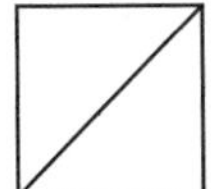
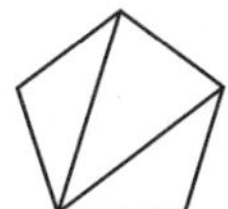

2. Tell whether each statement is true or false. If the statement is false, explain why.

a. A number may be rational and written as a percent.

b. A number may be both an integer and a rational number.

c. Zero is a natural number.

d. Dividing any two integers gives another integer.

3. a. Find the sum of $2\frac{1}{3}$ and $3\frac{3}{4}$.

b. Find the product of $1\frac{1}{2}$ and $1\frac{1}{4}$.

c. Find the difference between $3\frac{1}{4}$ and $1\frac{3}{4}$.

d. Find the quotient of $3\frac{1}{2}$ and $1\frac{3}{4}$.

4. Use the table to answer the following questions about long-distance calls.

a. What would it cost to talk for 3 minutes on Monday during the daytime?

b. What would it cost to talk for 3 minutes on Monday evening?

c. What would it cost to talk for 5 minutes on Saturday?

d. For how many minutes could we talk on the weekend for $5.25?

Input: Time of Day	Output: Charge
Daytime: Monday through Friday, 8 a.m. to 5 p.m.	$1.50 for the first minute $1.00 for each additional minute
Evening: Sunday through Thursday, 5 p.m. to 8 a.m.	$1.35 for the first minute $0.90 for each additional minute
Weekend: Friday, 5 p.m., to Sunday, 5 p.m.	$1.05 for the first minute $0.70 for each additional minute

1.3 Tables, Patterns, and Algebraic Expressions

OBJECTIVES

- Find the rules for input-output tables, and express them in words and in algebraic notation.
- Identify variables, constants, numerical coefficients, and expressions.
- Change word phrases into algebraic expressions.

WARM-UP

Complete the tables.

1.

Input	Output: Six Times Input
0	
1	
2	
3	
4	

2.

Input	Output: Six More Than Input
0	
1	
2	
3	
4	

IN THIS SECTION, we describe patterns with words and then write the word rules in algebraic notation. After algebraic notation and its vocabulary are introduced, we translate word statements into algebraic expressions.

Building a Table

We return to the pony pens of Section 1.1.

EXAMPLE 1 Exploration A year has passed; Kelly has taken over Jake's job at the Agricultural Events Center. She assumes she can build all the pony pens along a wall. Make an input-output table from her sketch, shown in Figure 9. Let the input be 1 to 4 pens and the output be the total number of panels.

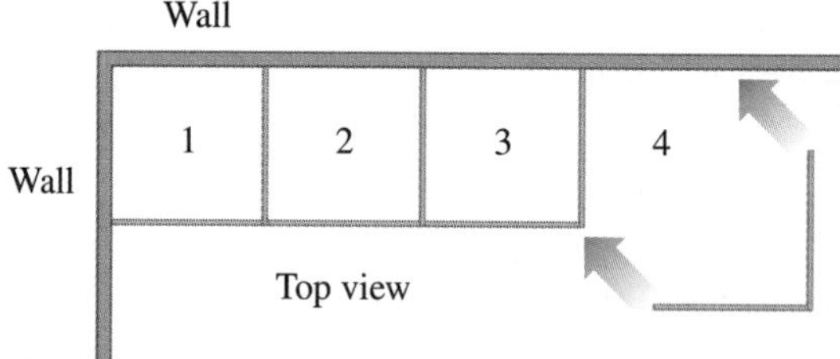

Figure 9

Solution The input-output table is shown in Table 6.

Input: Number of Pens	Output: Total Number of Panels
1	2
2	4
3	6
4	8

Table 6

Finding and Describing Rules Using Words and Algebraic Notation

To find a rule, look for a pattern. Follow these steps:

1. Match each input with an output.

2. Describe what we do to the input number to obtain the output.

EXAMPLE 2 Finding a pattern Continue Table 6 from Example 1, and use it to predict the number of panels needed for 10, 50, and 100 pony pens.

Solution In Table 6, one pen takes 2 panels, so the input 1 matches with the output 2. Two pens take 4 panels, so the input 2 matches with the output 4. Similarly, the input 3 matches with 6, and the input 4 with 8.

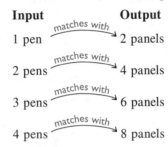

Input		**Output**
1 pen	matches with →	2 panels
2 pens	matches with →	4 panels
3 pens	matches with →	6 panels
4 pens	matches with →	8 panels

The numbers suggest that the output is two times the input. So, 10 pens take 20 panels, 50 pens take 100 panels, and 100 pens take 200 panels. ●

DESCRIBING A RULE IN WORDS To write a rule, look for one or two operations (addition, subtraction, multiplication, or division) that, when applied to each input, give the matching output.

EXAMPLE 3 Exploration What rule describes the input-output table in Example 1? Describe your rule in a sentence using *input* and *output*.

Solution All except one of the following statements describe the rule for the input table in Example 1. Find which statement does *not* state the rule.

1. Add the input to itself to get the output.

2. Multiply the input by itself to get the output.

3. The output is twice the input.

4. Double the input to get the output.

5. The output is two times the input.

See the Answer Box for the answer. ●

WRITING A RULE IN ALGEBRAIC NOTATION To change a rule into algebraic notation, we start by replacing the word *input* by a letter, such as n. We call n the input variable.

Definition of Variable

A **variable** is a letter or symbol that can represent any number from some set.

In Example 4, we write our rule using variables.

EXAMPLE 4 Writing rules in algebraic notation Describe the five choices in Example 3 in algebraic notation. Let n be the variable.

Solution Different symbols describe the various statements listed in Example 3. For example, $n + n$ describes statement 1, and $2 \times n$, or $2n$, describes statements 3, 4, and 5. In Section 2.3, you will see why the notation $n + n$ and the notation $2 \times n$ can both be used to express the same rule. Statement 2 is described by $n \times n$, which is not the same as $n + n$ or $2 \times n$ except when n equals 2. ●

Because the letter x is commonly used as a variable, mathematicians generally avoid the use of $\times$ as a multiplication sign. Multiplication can be written several ways.

> When a number is multiplied by a letter, as in 2 times n, we may write $2n$, $2 \cdot n$, $2(n)$, or $(2)(n)$; $2n$ is the most common form.
>
> When two different letters are multiplied together, as in a times b, we write $a \cdot b$ or, more commonly, ab.
>
> When the two letters are the same, as in n times n, we write n^2.

In Example 2, we associated the number of pens (input) with the total number of panels (output) needed to build the pens. In Example 5, we associate the position (input) of a design with the number of sides (output) in the design.

EXAMPLE 5 Finding a rule Look at the paper clip designs in Figure 10. The numbers below the designs show which is the first, second, third, and fourth position in the pattern.

a. Make an input-output table in which the position of the design is the input and the number of sides (paper clips) in the design is the output.

b. Describe the input and output matching needed to find a rule.

c. Write a sentence describing the rule for obtaining the output from the input.

d. Use the rule to predict outputs for inputs of 50 and 100.

e. Write the rule in symbols.

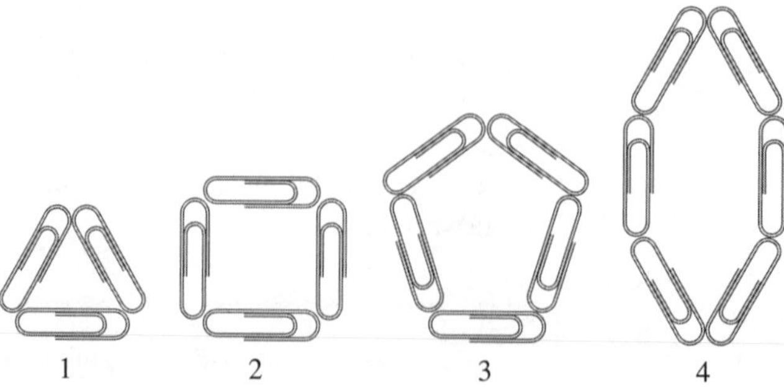

Figure 10

Solution **a.**

Input	Output
1	3
2	4
3	5
4	6
5	7

b. The first design has three sides, so 1 matches with 3. The second design has four sides, so 2 matches with 4. The third design has 5 sides, so 3 matches with 5. Similarly, 4 matches with 6, and 5 with 7.

c. Is your rule listed below? (Note: One of the sentences listed is not correct.)

1. Add 2 to the input to get the output.
2. The output is 2 greater than the input.
3. The output is 2 more than the input.
4. The input less 2 gives the output.
5. The input plus 2 gives the output.

The answers to parts c, d, and e are in the Answer Box. ●

As the answers to Example 5 suggest, $2 + n$ and $n + 2$ are the same whereas $n - 2$ is different. You will see why this is true in Section 2.3.

We now return to the triangular pony pens shown in Figure 8 (Example 9, Section 1.2) for an example of finding a rule by *comparing the outputs of one table with the outputs of another table.*

EXAMPLE 6 Finding a pattern by comparing outputs Kelly has decided to build triangular pony pens, in the arrangement shown in Figure 11. Table 7 shows the number of pens as input and the total number of panels as output. Suppose that, by listing more inputs, she finds that 20 pens require 41 panels.

a. Predict the number of panels needed for 50 and 100 pens.

b. Describe the pattern between inputs and outputs in words and in symbols.

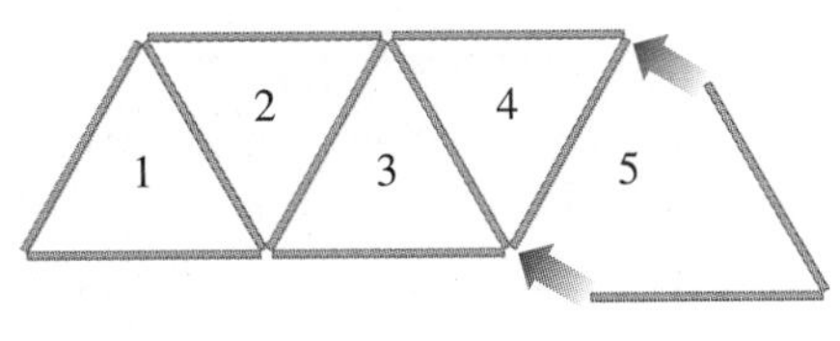

Figure 11

Input: Number of Pens	Output: Total Number of Panels
1	3
2	5
3	7
4	9
20	41
n	

Table 7

Solution *Hint:* Keep in mind that we want to match input with output and then compare the outputs in Table 7 with the outputs in Table 8 (based on Table 6 from the exploration in Example 1).

Input: Number of Pens	Output: Total Number of Panels
1	2
2	4
3	6
4	8
20	40
n	$2n$

Table 8

a. Each of the outputs in Table 7 is one more than the corresponding output in Table 8. The outputs in Table 8 are twice the inputs. Thus, we can predict that the output for the input 50 will be 2 times 50 plus 1, or 101. The output for the input 100 will be 2 times 100 plus 1, or 201.

b. To find the output, we can either multiply the input by 2 and then add 1 or find 1 more than twice the input. These rules are written $2n + 1$ and $1 + 2n$. Other rules may be possible. ●

In Example 6, for each change of 1 in the input, the change in output was 2, and the rule, $2n + 1$, contained a multiplication by 2. This was not a coincidence. Look for this pattern elsewhere. We will return to it in Section 4.3.

Writing Expressions

VOCABULARY We used variables to describe our inputs above. In $6 + x$, $5 - y$, and $2\pi r$, the x, y, and r are variables, like n. The numbers 6, 5, and 2π are constants.

Definition of Constant

> A **constant** is a number, letter, or symbol whose value is fixed.

The 2 in $2n$ is a constant for multiplication. The 3 in $n + 3$ is a constant for addition. The Greek letter pi, π, is also a constant.

When written together, as in $6 + x$, $5 - y$, and $2\pi r$, the variables, constants, numbers, and operations form expressions.

Definition of Expression

> An **expression** is any combination of numbers, constants, and variables with operations such as addition, subtraction, multiplication, or division.

There is one special type of constant: a numerical coefficient. An example of a numerical coefficient is the 2 in the expression $2n$.

Definition of Numerical Coefficient

> A **numerical coefficient** is the sign and number multiplying the variable or variables.

A number or numerical coefficient without a positive or negative sign is assumed to be positive. Thus, 2 means a positive 2. *A variable without a numerical coefficient is assumed to have a positive 1 coefficient.* Thus, n means positive $1n$.

EXAMPLE 

Practicing vocabulary Identify the numerical coefficients, variables, and constants in these expressions.

a. $2x + 1$ **b.** πr^2 **c.** x **d.** $-3x + 25$ **e.** $0.75n$

Solution

a. The 2 is the numerical coefficient of the variable x. Both 2 and 1 are constants.

b. The π (or pi) is both a numerical coefficient and a constant. The variable is r.

c. The variable x is assumed to have as a numerical coefficient the constant 1.

d. The -3 is the numerical coefficient of the variable x. Note that the negative sign on the 3 is part of the coefficient. Both -3 and 25 are constants.

e. The 0.75 is both a numerical coefficient and a constant. The letter n is the variable. ●

WORDS FOR OPERATIONS In Section 1.2, sum, difference, product, and quotient were defined as the answers in addition, subtraction, multiplication, and division, respectively. There are also several other words and phrases associated with these operations. You may want to keep a list and add your own words to those listed here.

Some words and phrases that indicate addition are *plus, added to, greater than, more than,* and *increased by.* Subtraction words include *less, less than, fewer, minus,* and *decreased by.* Generally, *if the word* than *appears in a subtraction setting, the order of the numbers is the reverse of the word order.*

EXAMPLE 

Writing expressions from words Write these subtractions as expressions.

a. The difference in age between 6 and 2

b. Fifteen dollars less ten dollars

c. Five less than twenty

d. Six less than n

e. x fewer than 4

f. x decreased by 5

Solution

In parts a, b, and f, write the numbers in the order in which they are stated. Parts c, d, and e contain *than,* so the order in which the numbers or variables should be written is the reverse of the word order. See the Answer Box for the answers. ●

Multiplication and division words may be harder to recognize than addition and subtraction words. The word *at* suggests multiplication; for example, "5 hamburgers at \$4.95 each" means to multiply 5 times \$4.95. *Double a number* and *twice a number* mean multiplication by 2, while *triple a number* means multiplication by 3. The word *of* means multiplication with fractions, decimals, and percents; for example, "$\frac{1}{2}$ of 4" means $\frac{1}{2}$ times 4. The word *half* may be troublesome. *Half of ten* means one-half times ten, or $10 \div 2$. *Divide 6 in half* means $6 \div 2 = 3$, whereas *divide 6 by a half* means $6 \div \frac{1}{2} = 6 \cdot 2 = 12$.

EXAMPLE  Writing expressions from words Write these phrases as expressions. Let n be the input number.

a. One less than twice the input

b. The difference between half the input and five

c. The quotient of three and double the input

d. 4 credit hours at n dollars per credit hour

e. The product of four and the input is then decreased by one

Solution Parts a, b, and e contain a subtraction and a multiplication. Part c contains a division and a multiplication, and part d contains a multiplication. See the Answer Box for the answers. ●

EXAMPLE  Writing expressions from word problems

a. The input is the number of payments. Each payment is \$35. Write an expression for the total value of x payments.

b. The input is the number of rides on the subway. Each ride costs \$1.25. Write an expression for the total cost of x rides.

c. The situation is the same as in part b, but you start with a prepaid ticket worth \$20. Write an expression for the value on the ticket after x rides on the subway.

Solution Parts a and b involve multiplication. Part c has a subtraction and a multiplication. See the Answer Box for the answers. ●

PERCENT A percent should be changed to a decimal before it is written with a variable or in an expression.

EXAMPLE  Writing percent expressions from words Write each percent phrase as an expression.

a. 20% of a number n **b.** 75% of a number n

c. $62\frac{1}{2}\%$ of a number n **d.** $\frac{1}{2}\%$ of a number n

Solution The numerator and denominator of the fraction are multiplied by 10 in parts c and d, to clear the decimal point in 62.5 and 0.5.

a. 20% of $n = \frac{20}{100}n$ or $0.20n$

b. 75% of $n = \frac{75}{100}n$ or $0.75n$

c. $62\frac{1}{2}\%$ of $n = 62.5\%$ of $n = \frac{62.5}{100}n = \frac{625}{1000}n$ or $0.625n$

d. $\frac{1}{2}\%$ of $n = 0.5\%$ of $n = \frac{0.5}{100}n = \frac{5}{1000}n$ or $0.005n$ ●

ANSWER BOX

Warm-up: 1. 0, 6, 12, 18, 24 **2.** 6, 7, 8, 9, 10 **Example 3:** Statement 2 is not correct. **Example 5: c.** Except for statement 4, all of the statements listed describe the rule. **d.** The output for 50 is 52 and for 100 is 102. **e.** $2 + n$ or $n + 2$ **Example 8: a.** $6 - 2$ **b.** \$15 − \$10 **c.** $20 - 5$ **d.** $n - 6$ **e.** $4 - x$ **f.** $x - 5$ **Example 9: a.** $2n - 1$ **b.** $\frac{1}{2}n - 5$ or $\frac{n}{2} - 5$ **c.** $3 \div 2n$ is correct, but fraction notation is preferred: $\frac{3}{2n}$ **d.** \$4n **e.** $4n - 1$ **Example 10: a.** \$35x **b.** \$1.25x **c.** \$20 − \$1.25x

EXERCISES 1.3

In Exercises 1 to 4, write an expression in symbols for each word phrase. Let n be the input. To find a difference or quotient, write the numbers in the order in which they are stated.

1. a. The product of three and the input

b. The quotient of eight and the input

c. The difference between the input and four

d. The quotient of the input and five

e. 15 pounds at n dollars per pound

2. a. The difference between four and the input

b. The quotient of the input and eight

c. The product of the input and three

d. The sum of three and the input

e. n credit hours at \$89 per credit

3. a. Three plus twice the input

b. The difference between four and triple the input

c. The product of the input and seven, increased by four

d. Multiply the input by itself

e. n ounces at \$0.79 per ounce

4. a. The quotient of the input and three, decreased by two

b. Three times the input, less two

c. Add two to twice the input

d. The product of seven and half the input

e. 2.5 yards at n dollars per yard

Identify the constants, numerical coefficients, and variables in the expressions in Exercises 5 and 6.

5. a. $2\pi r$

b. $1.5x$

c. $-4n + 3$

d. $x^2 - 9$

6. a. πd

b. $x/2$

c. $x^2 - 4$

d. $-2n - 1$

In Exercises 7 to 10, write a sentence that uses the word in its mathematical meaning. Write a second sentence in which the word's meaning is not mathematical.

7. expression **8.** constant

9. variable **10.** pattern

Write each percent statement in Exercises 11 and 12 as an expression.

11. a. 35% of n **b.** 10% of x

c. $87\frac{1}{2}\%$ of n **d.** $37\frac{1}{2}\%$ of x

e. $\frac{1}{2}\%$ of n **f.** 108% of x

12. a. 25% of x **b.** 15% of n

c. $6\frac{1}{2}\%$ of x **d.** 150% of n

e. $2\frac{1}{4}\%$ of x **f.** $12\frac{1}{2}\%$ of n

Make tables for the rules in Exercises 13 and 14. Use integer inputs from 1 to 5. Describe the output in algebraic notation.

13. The output is two more than three times the input.

14. The output is one more than the input times itself.

Match each input-output table in Exercises 15 to 18 with the appropriate set of figures in a to d. The input number, n, is shown below each figure. The output is the number of small squares in the figure. State the rule for the table in words and with an expression. Find the remaining outputs.

a.

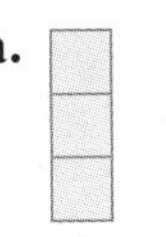
1

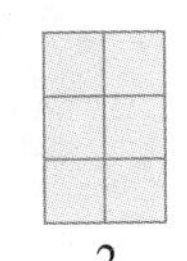
2

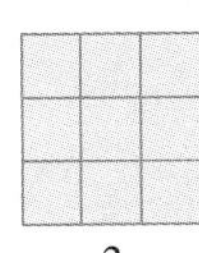
3

b.

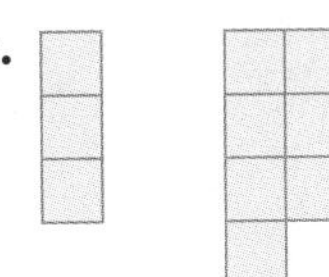
1 2

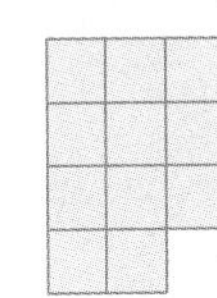
3

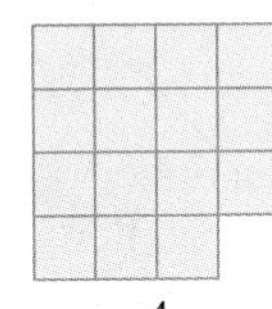
4

c.

1 2

3

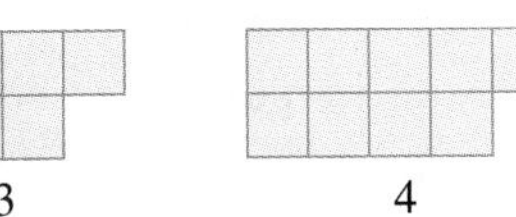
4

d.

1

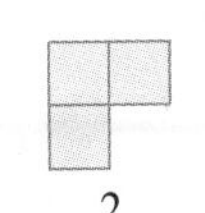
2

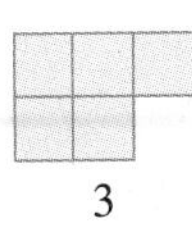
3

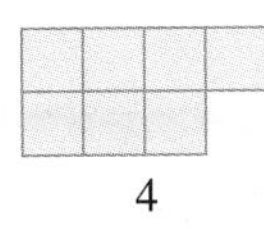
4

15.

Input	Output
1	3
2	5
3	7
4	
20	41
50	
100	
n	

16.

Input	Output
1	1
2	3
3	5
4	
20	39
50	
100	
n	

17.

Input	Output
1	3
2	7
3	11
4	
20	79
50	
100	
n	

18.

Input	Output
1	3
2	6
3	9
4	
20	60
50	
100	
n	

For Exercises 19 to 22, build an input-output table.

19. Input: number of hours, t
Output: distance traveled, D, in miles
Rule: Distance traveled is 55 miles per hour times the time t.
Use inputs $\{1, 2, 3, t\}$.

20. Input: number of cans, n
Output: total cost, C, in dollars
Rule: Cost is \$0.86 per can.
Use inputs $\{1, 2, 3, n\}$.

21. Input: number of sales made
Output: total income in dollars
Rule: Income is \$250 plus \$75 per sale.
Use inputs $\{1, 2, 3, n\}$.

22. Input: number of days on vacation
Output: total cost of vacation
Rule: Cost is \$1590 plus \$200 per day.
Use inputs $\{1, 2, 3, n\}$.

23. Is x a positive number? Explain.

24. Is $-x$ a negative number? Explain.

Conditional Rules

Make an input-output table for the conditional situations in Exercises 25 to 27.

25. The output is 0 if the input is a multiple of 3. The output is 2 otherwise. Use integer inputs $n = \{-6, -5, -4, \ldots, 6\}$.

26. The output is the input number if the input is positive or zero. The output is the opposite value if the input is negative (5 is the opposite of -5). Use integer inputs $n = \{-3, -2, -1, 0, 1, 2, 3\}$.

27. The output equals the input for numbers larger than 0. The output is 0 for inputs smaller than or equal to 0. Use integers -3 to 3 as inputs. (This is a rule used in mechanical engineering.)

28. Write each of the output rules for the Chevron credit card payment schedule, repeated in the table. Use n as the input amount. The output rule for \$20.01 to \$500 needs two descriptions separated by *or* and followed by *whichever is greater.*

Input: Charge Balance, n	Output: Payment Due	Output Rule (Use n as input.)
\$0 to \$20	Full amount	
\$20.01 to \$500	10% or \$20, whichever is greater	
\$500.01 or more	\$50 plus the amount in excess of \$500	

Complete the pattern in the tables in Exercises 29 and 30.

29.

Input	Output
1	2
2	3
3	6
4	5
5	10
6	7
7	14
8	
50	
101	
Even n	
Odd n	

30.

Input	Output
1	1
2	0
3	3
4	0
5	5
6	
7	
8	
50	
101	
Even n	
Odd n	

Problem Solving

31. ***Building Pens in Pairs.*** Cassandra suggests that Jake build pens in pairs, as shown in the figure. The pairs and panels pattern is shown in the following table. Describe the rule in words and with an expression. Use it to predict the number of panels needed for 10 and 25 pairs of pens.

Top view

Pairs of Pens	Panels
1	7
2	12
3	17
4	
x	
10	
25	

32. ***Building Pens.*** Jake's pens require the panels pattern shown in the following table. Three pens are shown in the figure. Describe the rule in words and with an expression. Use it to predict the number of panels needed for 20 and 50 pens.

Pens	Panels
1	4
2	7
3	10
4	
n	
20	
50	

Top view

33. ***Seating at Trapezoidal Tables.*** Three of Franck's trapezoid-shaped restaurant tables are shown in the figure. Eleven chairs may be placed at the 3 tables. Find the number of chairs for other table arrangements as you complete the following table. Include an expression for the number of seats at n tables in a long row.

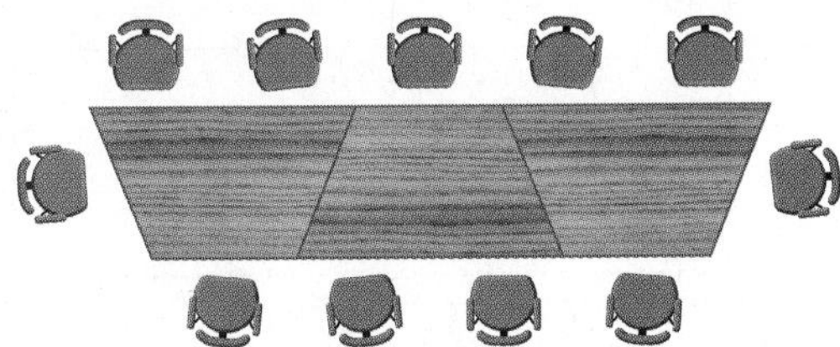

Tables	Chairs
1	5
2	
3	11
4	
n	
10	

34. ***Seating at Square Tables.*** The seating at 3 square tables placed next to each other at Mirielle's restaurant is shown in the figure. Eight chairs may be placed at the 3 tables. Find the number of chairs for other arrangements as you complete the following table. Include an expression for n square tables placed in a long row.

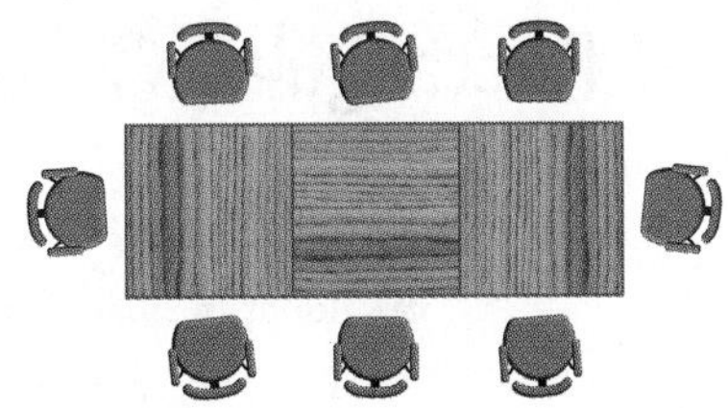

Tables	Chairs
1	4
2	6
3	8
4	
n	
20	

Computation Review

Exercises 35 and 36 contain modified input-output tables. Use the two input values to calculate the outputs according to the rule given at the top of each output column. You may need to find one or both inputs.

35.

Input x	Input y	Output xy	Output $x + y$
4	3	12	
		12	13
5	4		
	1	20	
		4	4
		18	9

36.

Input x	Input y	Output xy	Output $x + y$
2	6	12	
10	2		
		15	8
1	15		
	2	18	
4		4	

1.4 Tables and Rectangular Coordinate Graphs

OBJECTIVES

- Identify quadrants, axes, the origin, ordered pairs, and coordinates.
- Build a graph from an input-output table.
- Build a table and graph to fit a single rule or conditional rule.
- Read and interpret a graph.

WARM-UP

Complete these input-output tables.

1.

Input: x	Output: 3 plus 2 times the input
0	
0.5	
1.0	
1.5	
2.0	
2.5	
4.0	

2.

Input: x	Output: 3 plus the input times itself
0.0	
0.5	
1.0	
1.5	
2.0	
2.5	
4.0	

HAVE YOU EVER USED a map to find your way? In this section, we use graphs to describe position and to relate algebraic expressions to points and lines. The section's focus is more visual than numeric or algebraic. We look at how street numbering can help us locate buildings on a city map and then at how we locate points using the numbering of lines on a coordinate graph. We think about the steps that help us build graphs from rules or tables. Finally, we practice reading graphs.

Coordinate Graphs

A **coordinate graph** *identifies a position by two or more numbers or letters.* In 1791, the Frenchman Pierre Charles L'Enfant designed the national capital, Washington, D.C., as a coordinate graph so that a number and letter could identify each street intersection. His plan required that the city be divided into four quadrants: N.E., N.W., S.W., and S.E. The dome of the U.S. Capitol building, shown in Figure 12, is the center of the city.

Figure 12 U.S. Capitol Building

Andrew Ellicott surveyed the city, in which north and south streets are numbered from the Capitol building and east and west streets are lettered from the Capitol building. A portion of the city map is shown in Figure 13. To help you see the basic map design, the diagonal streets as well as the actual names assigned to the streets labeled A and B have been omitted.

In his book *What Do You Care What Other People Think?* (New York: Bantam Books, 1988), Richard Feynman relates his problem with the system. (Richard Feynman is the scientist who brought national attention to the o-ring problem leading to the 1986 Challenger space shuttle disaster.) When Feynman hired a taxi to take him from his Washington, D.C., hotel to 7th and B, which he thought was the address of the National Aeronautics and Space Administration (NASA), the taxi took him to an empty lot.

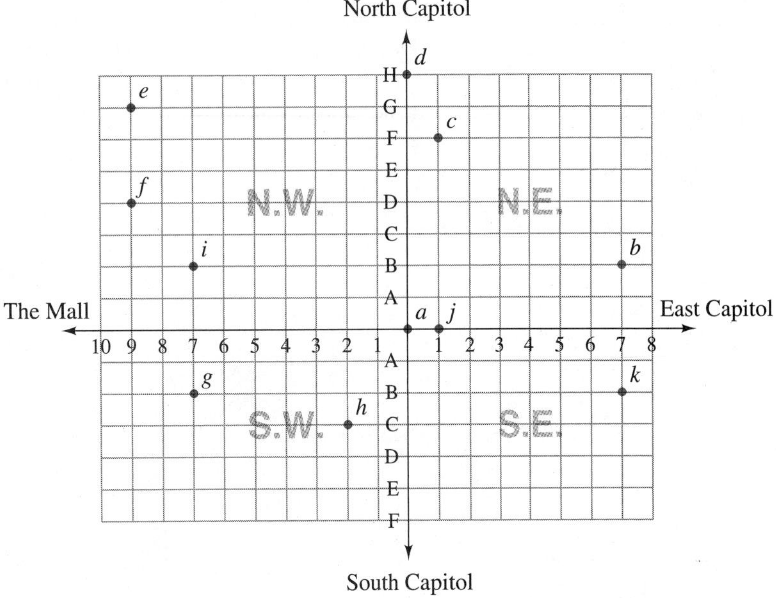

Figure 13

EXAMPLE 1 Exploration On the map in Figure 13, find the four points that mark possible locations of the NASA building, whose address is 7th and B.

Solution The answers are listed in the Answer Box.

The actual location of NASA is in the S.W. quadrant. As Dr. Feynman discovered, a major disadvantage of the Washington, D.C., street plan is that unless the quadrant is named, there are four possible locations for most addresses. ●

In the exploration in Example 2, we find buildings, given the names of the two streets that intersect nearby. Notice how important the quadrants are in locating the buildings.

EXAMPLE 2 Exploration On the map in Figure 13, find the points that mark the locations of the following buildings.

1. U.S. Capitol building, intersection of East Capitol and North Capitol
2. Union Station, 1st and F, N.E.
3. Martin Luther King Memorial Library, 9th and G, N.W.
4. Food and Drug Administration, 2nd and C, S.W.
5. Federal Bureau of Investigation, 9th and D, N.W.
6. Veterans Administration, North Capitol and H
7. Supreme Court, 1st and East Capitol

Solution See the Answer Box. ●

Locating Positions on a Graph

To identify position in mathematics, we use a plan similar to the Washington, D.C., map. Look for how this plan avoids using N.W., N.E., S.W., and S.E.

We place two number lines at right angles so that they cross at zero. The lines divide a flat surface into four *sections* called **quadrants**, as shown in Figure 14. The *flat surface* is the **coordinate plane**, and the *number lines* are the **axes**. The *number line that goes left to right* is the **horizontal axis**, and the *number line that runs up and down* is the **vertical axis**. The *horizontal axis* is commonly called the ***x*-axis**, and the *vertical axis* is called the ***y*-axis.** The *point where the number lines cross* is called the **origin**. The U.S. Capitol building is at the origin on the map in Figure 13.

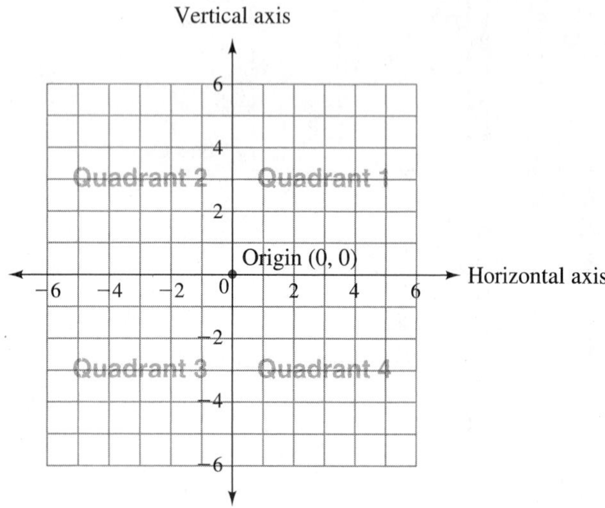

Figure 14

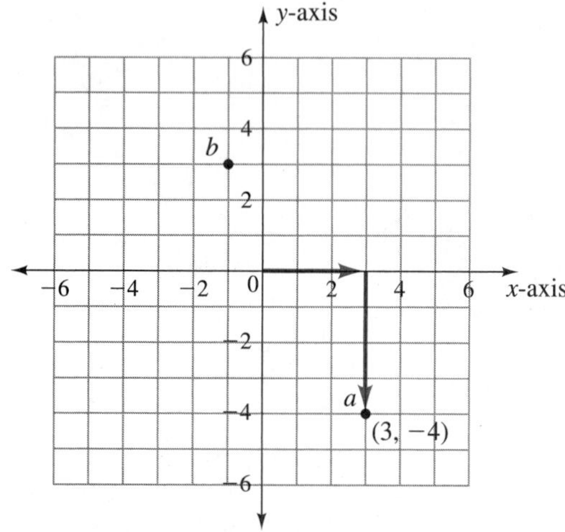

Figure 15

ORDERED PAIRS When locating a position on the coordinate plane, mathematicians use a pair of numbers. The *pair of numbers*, called an **ordered pair**, is placed in parentheses. For example, (0, 0) describes the origin.

Ordered Pairs

An ordered pair contains two real numbers (x, y) that describe a position on a coordinate plane. For each ordered pair:

- The first number, x, describes the horizontal distance from the origin, along the x-axis. Positive numbers are to the right; negative numbers are to the left.
- The second number, y, describes the vertical distance from the x-axis. Positive numbers are up; negative numbers are down.

The numbers in an ordered pair may be zero, positive, or negative. The ordered pair $(3, -4)$ describes the position found by counting 3 units to the right of the origin, then down 4 units, as shown in Figure 15 above. The word **unit** describes *a distance of length one in either the horizontal or the vertical direction.*

Think about it 1: What is the ordered pair describing the point labeled b in Figure 15?

EXAMPLE 3 Matching ordered pairs to points on a graph In Figure 16, find the lettered point that matches each of the ordered pairs $(4, 2)$, $(-2, 3)$, $(3, -2)$, $(-5, -2)$, $(0, 5)$, and $(2, 0)$. What is the letter shown at each location, and in which quadrant or on which axis is each point located?

Solution To find $(4, 2)$ go to 4 on the horizontal axis and count up 2 units in the vertical direction. The point $(4, 2)$, labeled e, is in the first quadrant.

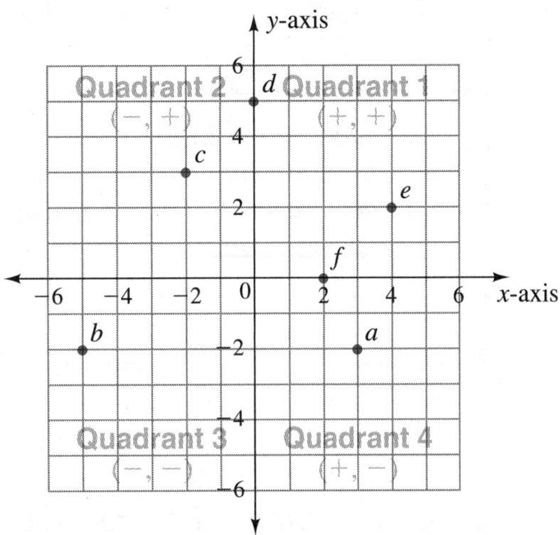

Figure 16

To find $(-2, 3)$, go to -2 on the horizontal axis and count up 3 units in the vertical direction, to the point labeled c in the second quadrant.

To find $(3, -2)$, go to 3 on the horizontal axis and count down 2 units in the vertical direction, to the point labeled a in the fourth quadrant. Observe that $(-2, 3)$ and $(3, -2)$ do not describe the same point.

To find $(-5, -2)$, go to -5 on the horizontal axis and count down 2 units, to the point labeled b in the third quadrant.

A zero in the ordered pair means no movement in one direction, so the point is on one of the axes. The point $(0, 5)$ is on the vertical axis at d, 5 units up from the origin. The point $(2, 0)$ is on the horizontal axis at f, 2 units to the right of the origin. ●

Ordered pairs are also called Cartesian coordinates or **coordinates**, after their French inventor René Descartes (1596–1650).

EXAMPLE 4 Graphing ordered pairs On a coordinate graph, locate the point described by each ordered pair. Name the quadrant in which each point is located.

a. $(3, -2)$ **b.** $(4, 2)$ **c.** $(-1, 2)$ **d.** $(-3, -5)$

Solution The points are shown in Figure 17.

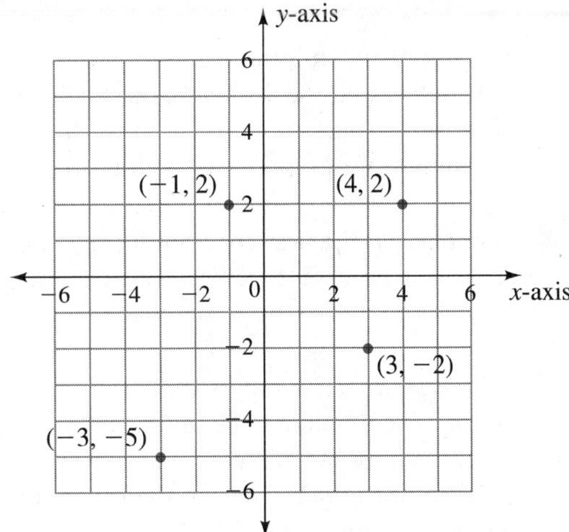

Figure 17

a. Quadrant 4 **b.** Quadrant 1 **c.** Quadrant 2 **d.** Quadrant 3 ●

When graphing ordered pairs, we label each point with its ordered pair.

> The word *graph* is used as both a noun and a verb. Used as a noun, **graph** refers to *the points that have been plotted* or *the line or curve drawn through those points.* As a verb, **graph** means *to locate points on the coordinate plane as described by ordered pairs.*

Think about it 2: Each of the quadrants in Figure 16 is labeled with a pair of signs, positive or negative. Compare the pairs of signs in the labels with the signs of the ordered pairs graphed in each quadrant.

Creating Graphs: The Four-Step Process

We now return to the four problem-solving steps, which we apply to building a graph.

The steps are listed below, along with some questions and instructions that might help you complete each problem-solving step.

1. **Understand the problem.**
   Which information is the input, to be labeled on the horizontal axis?
   Which information is the output, to be labeled on the vertical axis?
   Keep in mind that *the output depends on the input*; that is, the variable placed on the vertical axis depends on the variable placed on the horizontal axis.
2. **Plan the graph.**
   Which quadrants are implied or make sense in the problem situation?
   What numbers are needed on the axes?
   What units are needed on the axes?
3. **Sketch the graph.**
   Make a table if needed.
   Graph the ordered pairs.
   Look for a pattern, from left to right. What shape do the points make?
   Should the points be connected?
4. **Check the graph, and extend the results as needed.**
   Do all the data points lie on the graph?
   Do other points on the graph make sense in the problem situation?
   How might new data or information change the graph?

We begin building the graph by labeling the inputs on the horizontal axis and the outputs on the vertical axis. Each row of the table may then be written as an ordered pair (x, y), with x used for the input and y for the output. In the next two examples, we graph first-quadrant data from the tables in the Warm-up.

EXAMPLE 5 Graphing in the first quadrant from an input-output table based on a rule Use the four-step process to make a first-quadrant graph with the ordered pairs in Table 9. Should the points be connected?

Solution ***Understand:*** The inputs and outputs are given.
Plan: The graph is to be in the first quadrant. The numbers are reasonably small, so we number to 8 on the horizontal axis and to 12 on the vertical axis.
Sketch the graph: We plot the ordered pairs, as shown in Figure 18. The points lie in a straight line. Because other positive inputs make sense, we draw a line through the points and label the rule on the line.
Check and extend: On the line formed by all the ordered pairs, we select another point, such as $(x, y) = (3, 9)$. We find that $x = 3$, $y = 9$ makes a true statement

Input x	Output $y = 2x + 3$	Ordered pairs (x, y)
0	3	(0, 3)
0.5	4	(0.5, 4)
1.0	5	(1.0, 5)
1.5	6	(1.5, 6)
2.0	7	(2.0, 7)
2.5	8	(2.5, 8)
4.0	11	(4.0, 11)

Table 9

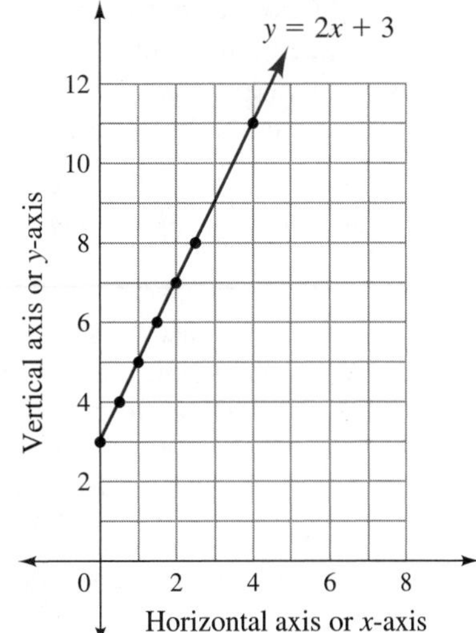

Figure 18

when we place the numbers in the rule $y = 2x + 3$; that is, $9 = 2(3) + 3$. The numbers $x = 3$ and $y = 9$ for the ordered pair (3, 9) could be listed in the table. ●

EXAMPLE  6 **Graphing in the first quadrant from an input-output table based on a rule** Use the four-step process to write each row of Table 10 as an ordered pair and to make a first-quadrant graph. Should the points be connected?

Solution ***Understand:*** The inputs and outputs are given. The ordered pairs (x, y) for the table values are (0, 3), (0.5, 3.25), (1.0, 4), (1.5, 5.25), (2.0, 7), (2.5, 9.25), and (4.0, 19).

Plan: The graph is to be in the first quadrant. The input numbers are small, so we can use zero and the positive integers from 1 to 7 on the horizontal axis. The output numbers go from 0 to 19, so we count by 2 on the vertical axis to fit the numbers 0 to 20.

Sketch the graph: We plot the ordered pairs in Figure 19. Other positive inputs make sense in the rule, so we connect the points with a curve.

Input x	Output $y = x^2 + 3$
0	3
0.5	3.25
1.0	4
1.5	5.25
2.0	7
2.5	9.25
4.0	19

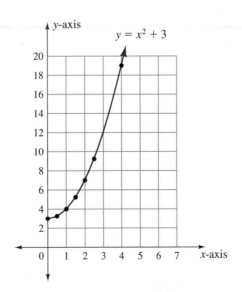

Table 10

Figure 19

Check and extend: On the curve formed by all the points, we select another ordered pair, such as $(x, y) = (3, 12)$. We find that placing $x = 3$ and $y = 12$ in $y = x^2 + 3$ gives $12 = 3^2 + 3$, a true statement. The numbers $x = 3$ and $y = 12$ could be listed in the table. ●

Finding the Scale

When we decide what numbers to count by on the axes, we are finding the **scale** for the axes. When numbers are very large or very small, we need to count by numbers other than 1 or 2.

One way to estimate the scale for an axis is to subtract the lowest input (or output) value in a table from the highest and divide by 10. Then round the quotient to an easy counting number, such as 5, 10, 20, 50, or 100.

$$\text{Estimated scale} = \frac{\text{Highest number} - \text{lowest number}}{10}$$

Then round to an easy counting number.

EXAMPLE 7 Finding the scale for the output axis The data in Table 11 show median incomes for households in the United States by age in 1993. The input data naturally fit counting by ten. What would be an appropriate scale for the output axis, for graphing ordered pairs based on Table 11? Graph the data from the table.

Solution To estimate the output scale, we subtract the lowest output from the highest and divide by 10. We then round to the nearest easy counting number.

$$\begin{aligned}\text{Estimated scale} &= \frac{\text{Highest output} - \text{lowest output}}{10}\\ &= \frac{46200 - 17800}{10}\\ &= 2840\\ &\approx 3000\end{aligned}$$

The symbol ≈ means "approximately equal to." We count by 3000 on the vertical axis. The graph is in Figure 20.

Age* x (years)	Median Income y (dollars)
20	19,300
30	31,300
40	40,900
50	46,200
60	33,500
70	17,800

*The ages are midpoints of intervals: 20 stands for ages 15 to 24, 30 stands for ages 25 to 34, . . . , 70 stands for ages 65 and older.
Source: Statistical Abstract

Table 11

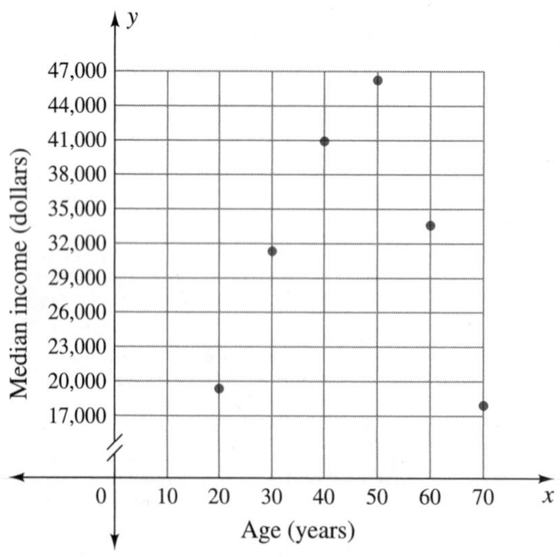

Figure 20

We do not connect the points in Figure 20, because the inputs represent a range of numbers and there is no rule for obtaining outputs for other inputs. ●

Think about it 3: We started the numbering on the vertical axis at $17,000 instead of zero. What are the advantages and disadvantages of this approach?

Student Note: Use the double slash only between zero and the next number on the axes.

Note that a *double slash*, //, is used on the vertical axis in Figure 20 to indicate that the space between zero and 17,000 is not the same as the space between each of the other numbers on the axis.

If the data have units (inches, dollars, liters, years, etc.), *the axes are labeled with the units.* In the conditional table in the next example, based on the credit card repayment example from Section 1.2, both the input and the output are in dollars.

Graphing with Conditional Rules

If the output rule has more than one part, it helps to find some ordered pairs before graphing.

EXAMPLE 8 Graphing a conditional rule about credit card payments Use Table 12 to build a table with inputs 0, 10, 20, 30, 200, 300, 500, 600, and 650. Then apply the four problem-solving steps to graph the conditional rule, using the data in the table.

Input: Charge Balance	**Output: Payment Due**
$0 to $20	Full amount
$20.01 to $500	10% or $20, whichever is greater
$500.01 or more	$50 plus the amount in excess of $500

Table 12 Credit Card Payment Schedule

Solution The values are shown in Table 13. Thus, the ordered pairs are (0, 0), (10, 10), (20, 20), (30, 20), (200, 20), (300, 30), (500, 50), (600, 150), and (650, 200).

Understand: Using the inputs and outputs given in the tables, we will assume that the payment due depends on the charge balance. Both axes will be labeled with dollars.

Plan: Because negative money values are not reasonable, this is a first-quadrant graph. To determine the horizontal scale, we observe that the inputs lie on the interval 0 to 650.

Input: Balance	**Output: Payment**
$ 0	$ 0
10	10
20	20
30	20
200	20
300	30
500	50
600	150
650	200

Table 13 Credit Card Payments

$$\text{Estimated scale} = \frac{\text{Highest input} - \text{lowest input}}{10}$$

$$= \frac{650 - 0}{10}$$

$$= 65$$

We round the 65 to 50 and use 50 as the scale for the horizontal axis. We find the scale for the vertical axis similarly.

$$\text{Estimated scale} = \frac{\text{Highest output} - \text{lowest output}}{10}$$

$$= \frac{200 - 0}{10}$$

$$= 20$$

Do the plan, graph: First we plot the points. Then, to add in the other values included by the rule, we connect the points. The graph, as shown in Figure 21, has four line segments.

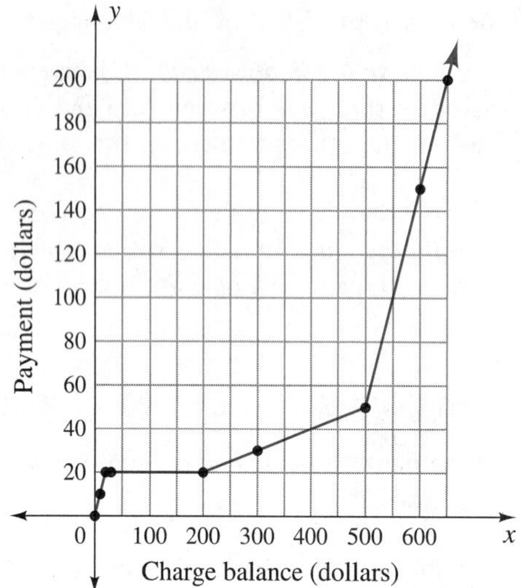

Figure 21

Check: The data fit the graph.

Think about it 4: There were only three parts in Table 12. Why are there four segments in the graph in Figure 21?

Reading Graphs

Tables and graphs help answer these important questions: What is the output value when the input is given? What is the input value when the output is given? Graphs are particularly useful in estimating numbers between the entries on a table and in observing patterns in tabular data.

Reading a Graph

- To find an output, locate the given input on the input axis, trace vertically to the graph, and then trace horizontally to the output axis.
- To find an input, locate the given output on the output axis, trace horizontally to the graph, and then trace vertically to the input axis.

EXAMPLE 9

Reading a graph about bulk food purchases Brown rice costs $0.89 per pound in the bulk food section of a grocery store. Table 14 and its graph (Figure 22) show the total cost in dollars (output) in terms of the number of pounds of brown rice purchased (input).

a. Which point (A, B, C, D, etc.) shows that 2 pounds of rice costs $1.78?

b. What fact from the table does point B represent?

c. Which point (A, B, C, D, etc.) would be used to determine the cost of 2.5 pounds of rice? Estimate the cost from the graph, calculate the exact cost, and compare your results.

d. Which point shows how much rice could be purchased for $4.00? Estimate the weight from the graph, calculate the exact weight, and compare your results.

Input: Pounds	Output: Cost	Coordinate Point (x, y)
1	\$0.89	(1, 0.89)
2	1.78	(2, 1.78)
3	2.67	(3, 2.67)
4	3.56	(4, 3.56)

Table 14 Cost of Brown Rice

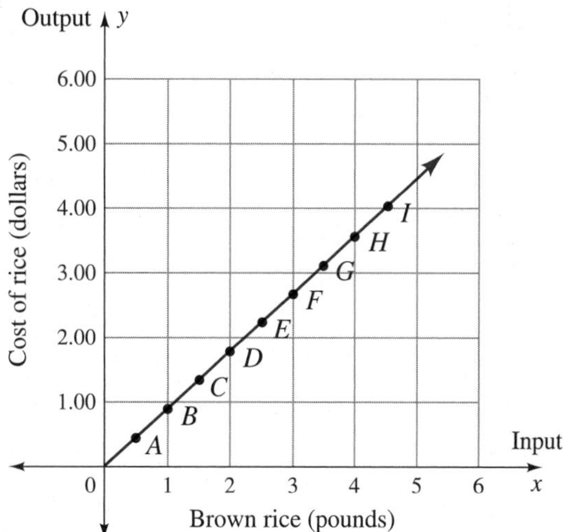

Figure 22

Solution

a. Point D shows that 2 pounds of rice costs \$1.78.

b. Point B represents the \$0.89 cost of 1 pound of rice.

c. Point E would be used. Reading to the left of E, we see that the cost is between \$2.00 and \$2.50. To calculate exactly, we multiply 2.5 pounds times \$0.89 per pound and round to the nearest cent: \$2.23.

d. Point I shows that about 4.5 pounds could be purchased for \$4.00. To calculate exactly, we divide \$4.00 by \$0.89 per pound.

$$\$4.00 \div \frac{\$0.89}{1 \text{ pound}}$$ Change the division to multiplication by the reciprocal

$$= \frac{\$4.00}{1} \cdot \frac{1 \text{ pound}}{\$0.89}$$ Divide \$4.00 by \$0.89

$$\approx 4.49 \text{ pound}$$

For \$4.00, we can buy 4.49 pounds.

HISTORICAL NOTE

The abbreviation for pound is lb and is from the Latin word *libra*, an ancient Roman unit of weight.

ANSWER BOX

Warm-up: 1. 3, 4, 5, 6, 7, 8, 11 **2.** 3, $3\frac{1}{4}$, 4, $5\frac{1}{4}$, 7, $9\frac{1}{4}$, 19 **Example 1:** *b, g, i, k* **Example 2: 1.** *a* **2.** *c* **3.** *e* **4.** *h* **5.** *f* **6.** *d* **7.** *j* **Think about it 1:** $(-1, 3)$ **Think about it 2:** All the ordered pairs in a quadrant have the same combination of positive and negative signs. **Think about it 4:** The middle condition (10% of the input or \$20, whichever is greater) is really two rules. For \$20.01 to \$200, \$20 is greater than 10% of the input. Between \$200 and \$500, 10% of the input is greater than \$20.

EXERCISES 

Write ordered pairs for points A to I in Exercises 1 and 2.

1.

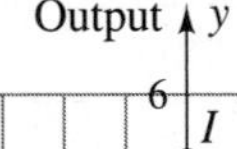

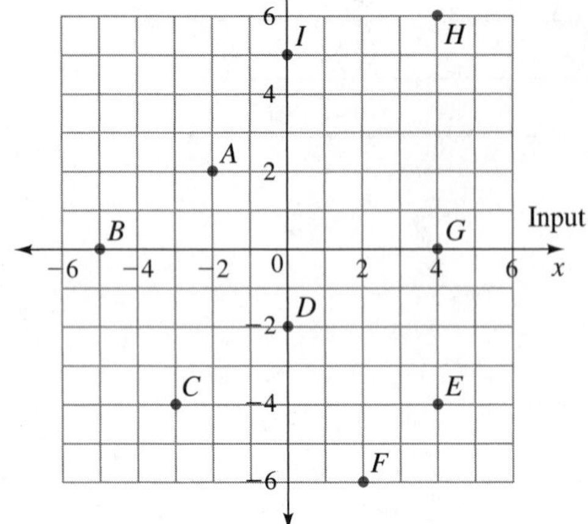

2.

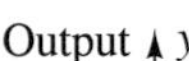

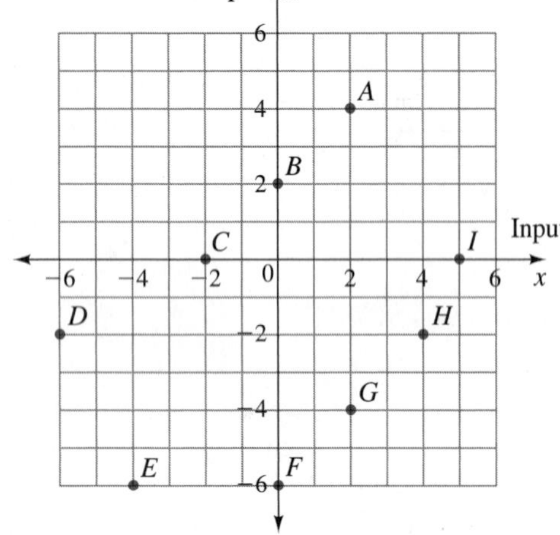

For Exercises 3 and 4, name the quadrant in which each point is located.

3. a. $(-4, 3)$ **b.** $(3, -4)$ **c.** $(-2, -4)$ **d.** $(-3, -2)$

4. a. $(4, -3)$ **b.** $(-3, 2)$ **c.** $(-2, -3)$ **d.** $(2, -3)$

For Exercises 5 and 6, name the axis on which each point is located.

5. a. $(0, -4)$ **b.** $(0, -2)$ **c.** $(3, 0)$

6. a. $(0, 4)$ **b.** $(-2, 0)$ **c.** $(-3, 0)$

For Exercises 7 and 8, use the graph in the figure to complete the table.

7.

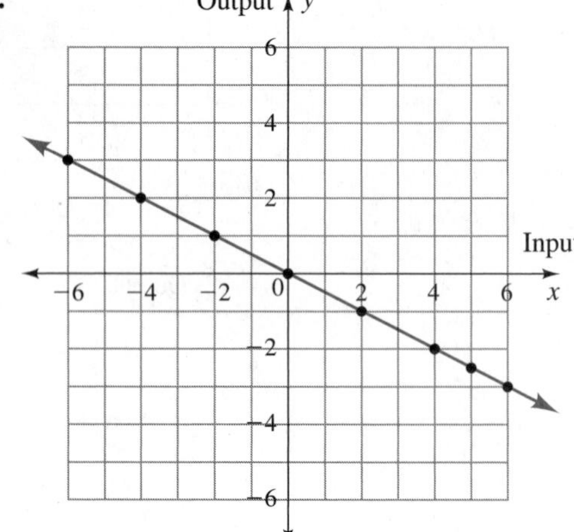

Input x	Output y
0	
	1
-4	
	-1
	-2
	3
5	

8.

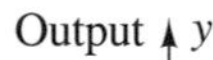

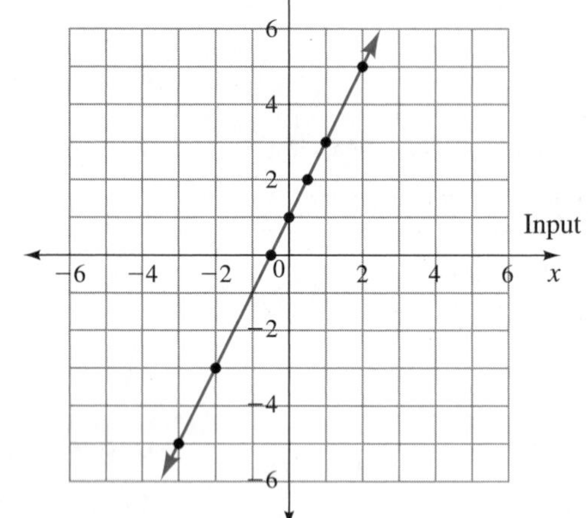

Input x	Output y
0	
	0
2	
	−5
	3
−2	
	2

In Exercises 9 and 10, make a table and a graph for each rule. Use integer inputs of 0 to 5.

9. The output is 5 more than twice the input.

10. The output is 2 more than four times the input.

In Exercises 11 to 14, draw a graph from the input-output table.

11.

Input x	Output y
−3	5
−2	3
−1	1
0	−1
1	−3

12.

Input x	Output y
−2	5
−1	2
0	−1
1	−4
2	−7

13.

Input x	Output y
−3	9
−2	7
−1	5
0	3
1	1

14.

Input x	Output y
−2	−11
−1	−8
0	−5
1	−2
2	1

In Exercises 15 to 22, graph the two ordered pairs. Then, for each exercise, draw a straight line through the two points and write ordered pairs for two other points on the line.

15. (1, 2), (3, 6)

16. (3, 3), (5, 5)

17. (2, −1), (8, −4)

18. (3, 1), (9, 3)

19. (8, 4), (2, 1)

20. (4, −4), (2, −2)

21. (1, 3), (2, 6)

22. (3, 5), (5, 7)

23. ***Bulk Food Purchases.*** Mixed nuts in the bulk foods department cost \$6.50 per pound.

a. Use the four-step approach to build a table and graph for inputs of 0 to 4 pounds. Let outputs be total cost. Label your steps.

b. Use your graph to estimate the cost of these purchases: $2\frac{1}{2}$ pounds, $1\frac{3}{4}$ pounds, $3\frac{1}{4}$ pounds.

c. One and a half pounds of the same nut mixture is available in packages for \$9.49. Plot this information as a data point. Which is a better buy, bulk or packaged nuts?

24. ***Bulk Food Purchases, Continued.*** In the bulk foods department, candy costs \$1.29 per pound.

a. Use the four-step approach to build a table and graph for inputs of 0 to 4 pounds. Let outputs be total cost. Label your steps.

b. Use your graph to estimate the total cost of these candy purchases: $2\frac{1}{2}$ pounds, $1\frac{3}{4}$ pounds, $3\frac{1}{4}$ pounds.

c. Two and a half pounds of the same candy is available in packages for \$3.98. Plot this information as a data point. Which is a better buy, bulk candy or packaged candy?

25. ***Telephone Card.*** You have a prepaid \$24 telephone card. Calls cost \$1.50 per minute.

a. Use the four-step approach to build a table and graph for inputs of 0 to 16 minutes, counting by 4. Let outputs be the value remaining on the telephone card. Label your steps.

b. Circle the point on your graph that matches the value remaining after 8 minutes have been used.

c. What is the meaning of the ordered pair (0, 24) on your graph?

d. What is the meaning of the ordered pair (16, 0) on your graph?

26. ***Gift Certificate.*** You have been given a \$36 gift certificate to Koffee Klatch. Each cup of coffee costs \$3.

a. Use the four-step approach to build a table and graph for inputs of 0 to 12 cups of coffee, counting by 2. Let outputs be the value remaining on the gift certificate. Label your steps.

b. Circle the point on your graph that matches the value remaining after buying 4 cups of coffee.

c. What is the meaning of the ordered pair (0, 36) on your graph?

d. What is the meaning of the ordered pair (12, 0) on your graph?

In Exercises 27 and 28, graph the stock prices in the table, using a scale like the one shown in the coordinate plane below. The tables are displayed in a horizontal format. Horizontal tables have the inputs on the top row and the outputs on the bottom row.

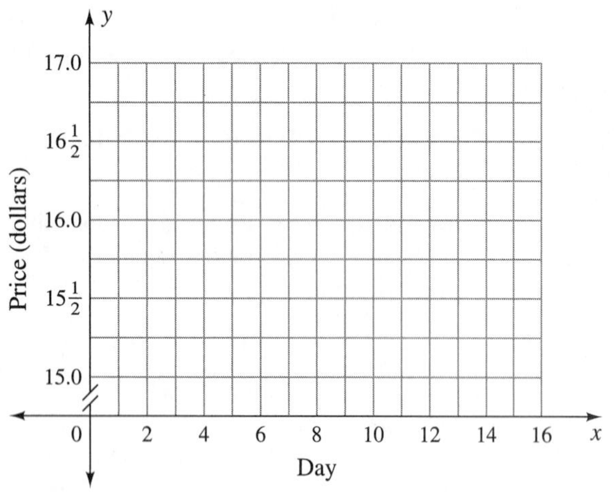

27. ***Stock Prices***

a. Use the four steps to graph the stock prices from the table.

b. What do you observe about the stock prices over this 15-day period? State your observation in a complete sentence.

Day	1	2	3	4	5	6	7
Price	$15\frac{1}{2}$	$15\frac{3}{4}$	$15\frac{1}{4}$	$15\frac{1}{4}$	15	$15\frac{1}{4}$	16

Day	8	9	10	11	12	13	14	15
Price	$15\frac{1}{2}$	$15\frac{3}{4}$	16	$15\frac{3}{4}$	16	$16\frac{1}{2}$	$16\frac{3}{4}$	$16\frac{1}{2}$

28. ***Stock Prices, Continued***

a. Use the four steps to graph the stock prices from the table.

b. What do you observe about the stock prices over this 15-day period? State your observation in a complete sentence.

Day	1	2	3	4	5	6	7
Price	$8\frac{1}{2}$	9	$8\frac{1}{2}$	$8\frac{1}{4}$	$8\frac{1}{4}$	$8\frac{1}{2}$	8

Day	8	9	10	11	12	13	14	15
Price	$7\frac{3}{4}$	$8\frac{1}{2}$	$8\frac{1}{4}$	8	$7\frac{1}{2}$	$7\frac{3}{4}$	8	$7\frac{1}{2}$

Use the following graph to find or estimate the answers in Exercises 29 to 32.

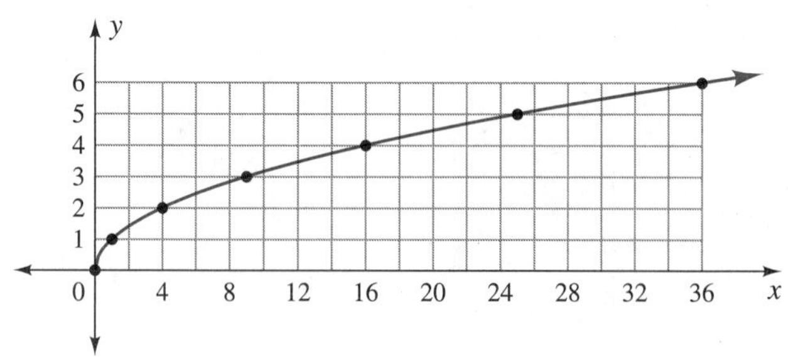

29. Describe the scale used on the horizontal axis.

30. What is the output for each of these inputs?

a. 4 b. 20 c. 25 d. 36

31. What is the input for each of these outputs? Estimate to the nearest whole number.

a. 3 b. 3.5 c. 4 d. 5.5

32. Predict the output for an input of 49.

Use the following graph to find or estimate the answers in Exercises 33 to 40.

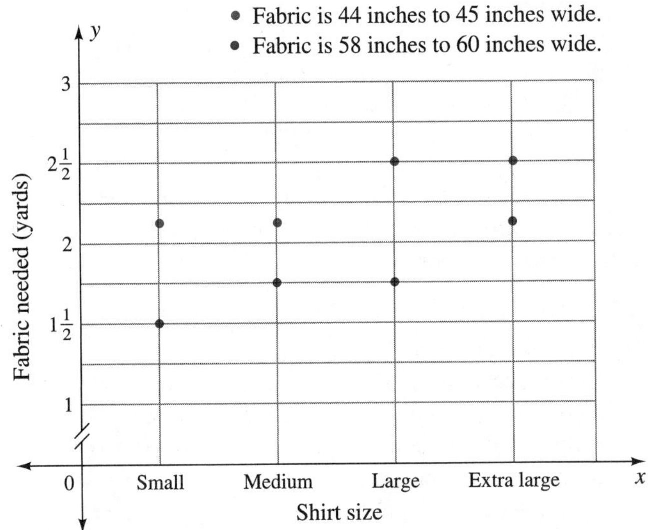

33. How much 45-inch-wide fabric is needed for an extra large shirt?

34. How much 60-inch-wide fabric is needed for a medium shirt?

35. Why is the double slash needed between 0 and 1?

36. What size shirt can be made with $1\frac{1}{2}$ yards of 60-inch-wide fabric?

37. How much more fabric is needed for a large shirt with 44-inch-wide fabric than with 58-inch-wide fabric?

38. How much more fabric is needed for a small shirt with 44-inch-wide fabric than with 58-inch-wide fabric?

39. If 36-inch-wide fabric were available, predict where the output values for the graph would be located.

40. If 72-inch-wide fabric were available, predict where the output values for the graph would be located.

In Exercises 41 to 46, write answers in complete sentences.

41. Why is (x, y) called an ordered pair?

42. How do you find the point described by an ordered pair?

43. How do you find the ordered pair describing a point?

44. How do you find in which quadrant an ordered pair is located?

45. How do you find on which axis an ordered pair $(a, 0)$ or $(0, b)$ is located?

46. How do you use a graph to build a table?

Refer to Example 8 for Exercises 47 to 51.

47. Why is the graph in Figure 21 level for inputs of \$20 to \$200?

48. Why is there a change in the graph at \$200?

49. What is the payment (output) for these charge balances?

a. \$50 **b.** \$100 **c.** \$400 **d.** \$700

50. What is the charge balance (input) for these payments?

a. \$12 **b.** \$25 **c.** \$45 **d.** \$60

51. If the payment (output) was \$20, can we find the charge balance?

Refer to Example 9 for Exercises 52 and 53.

52. Suppose the bulk food department offers storage containers for \$1.00. If the total cost rises by \$1.00 for all inputs, how will the graph change?

53. How would a higher price per pound for rice change the graph? Write your answer as a complete sentence.

Problem Solving

In Exercises 54 to 57, find the ordered pairs for points A and B from these portions of the coordinate plane. Each space equals one. The axes are hidden, so you must reason relative positions from the given point(s).

54.

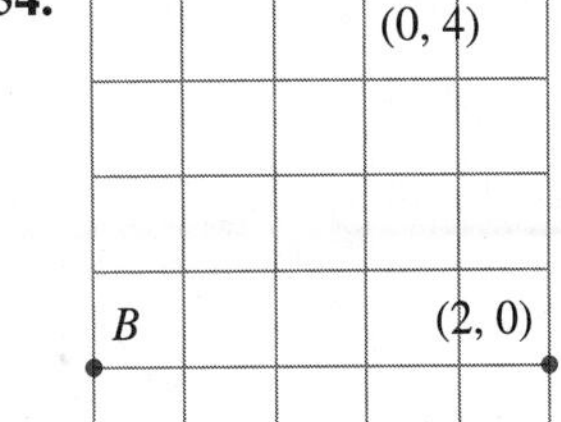

55.

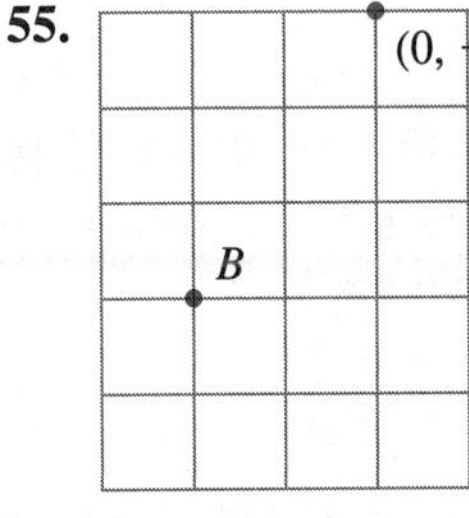

56.

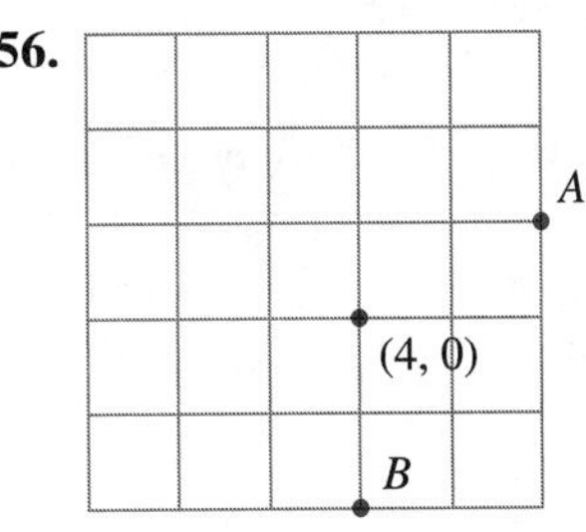

57.

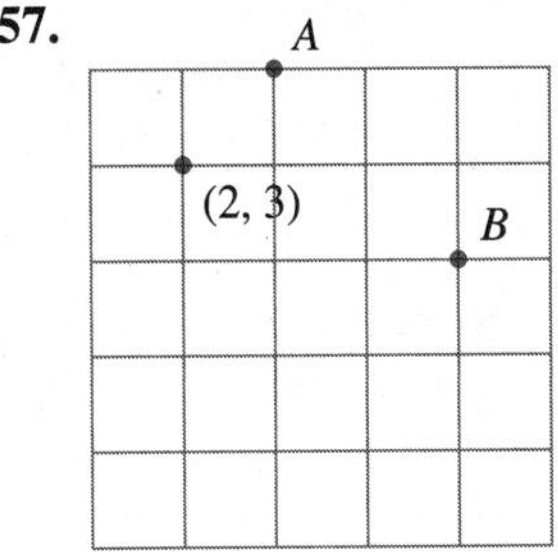

CHAPTER 1 SUMMARY

Vocabulary

For definitions and page references, see the Glossary/Index.

assumption	numerical coefficient
axes	opposites
condition	ordered pair
conditional rule	origin
constant	product
coordinate graph	quadrant
coordinate plane	quotient
coordinates	rational numbers
denominator	real numbers
difference	set
expression	scale
graph	sum
horizontal axis	undefined
input-output rule	unit
input-output table	variable
integers	vertical axis
interval	whole numbers
irrational numbers	x-axis
natural numbers	y-axis
numerator	

Concepts

Algebra is presented numerically with number patterns and tables; visually with charts, pictures, and graphs; and symbolically with expressions and equations in algebraic notation.

1.1 Problem-Solving Steps

The four problem-solving steps are to understand the problem, make a plan, carry out the plan, and check the solution. We find conditions and make assumptions when we solve problems.

Problem-solving strategies include drawing a picture, trying a simpler problem, looking for a pattern, organizing information by using a table, and making an estimate.

1.2 Sets of Numbers and Input-Output Tables

The language of algebra permits us to describe and use sets of numbers and to describe and solve problems. Input-output tables display number relationships and patterns.

1.3 Tables, Patterns, and Algebraic Expressions

We find a rule from an input-output table by matching each input with its output. The rule describes what operations we do to the input number to obtain the output. Comparing the outputs of one table with those of another table is another strategy for finding a rule.

Rules in algebraic notation are built with expressions containing numbers, constants, numerical coefficients, and variables. The multiplication of 2 times n may be written as $2n$, $2 \cdot n$, $2(n)$, or $(2)(n)$. The division of a by b is usually written in fraction notation as a/b.

1.4 Tables and Rectangular Coordinate Graphs

An ordered pair of real numbers describes a position on a set of axes. We apply the problem-solving steps to graphing by considering the questions and following the instructions on page 32.

The scale markings on the axes should be equally spaced and clearly numbered. When the scale does not start at zero, we make a double slash between the origin and the first mark on the axis to indicate a break in the numbering. The axes should be labeled with units.

To estimate scale, find the difference between the highest and lowest value, divide by 10, and round to a convenient number.

CHAPTER 1 REVIEW EXERCISES

1. State an assumption you might make about each of these sentences:

   a. You want to catch the 3:15 p.m. bus home from campus.

   b. You want to subtract 8 and 5.

   c. You want to divide 12 and 8.

   d. You want to ride your bicycle to school.

2. State a condition given in each of these settings:

   a. Calculating the area of a rectangle where the length is twice the width.

   b. Measuring the angles in a triangle where the sum of the angles is 180°.

   c. You want to go to a movie that starts at 4:45 p.m.

   d. You are building a fence from 8-foot sections of prefabricated material.

3. The figure shows a set of 28 dominos, each marked with 0 to 6 dots on a side. There is only one domino that has zero dots on each side.

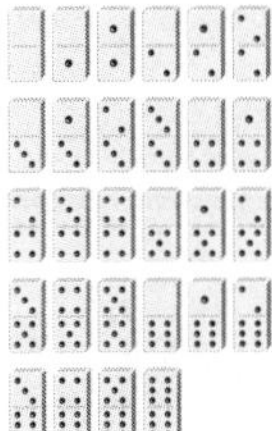

a. How many dominos have, at most, 1 dot on each side?

b. How many dominos have, at most, 2 dots on each side?

c. Use your results in parts a and b to start a table. Let the number of dots in question be the input, and let the total number of dominos with, at most, that number of dots on each side be the output.

d. Continue counting until you find a pattern with which to extend the table to 6 dots, at most, on each side. *Hint:* Your table will contain the ordered pairs (3, 10) and (6, 28).

e. Now consider a set of dominos in which each side is marked with 0 to 9 dots. Using your pattern, extend the table to predict the number of dominos with, at most, 9 dots on each side.

4. Find the words in the following list that match the given definitions: bifocal, quadrille, quintet, binomial, bimetal, triad, biathlon, tricolor, quadrilateral.

a. Two thin metal strips fastened together and used to detect small changes in temperature

b. Set of three musical notes

c. Olympic Games event involving skiing and shooting

d. Flag with three colors in stripes, such as the French flag

e. Eyeglasses with two parts (top and bottom)

f. Square dance for four couples

5. Add, subtract, multiply, and divide the two numbers. In doing subtraction or division, subtract or divide in the order listed.

a. $\frac{5}{6}$ and $\frac{3}{8}$

b. $2\frac{1}{6}$ and $1\frac{3}{4}$

6. Each word describes the answer to an arithmetic operation. Name the operation.

a. product b. quotient c. difference d. sum

7. For each of the following numbers, name all of the number sets to which it belongs: real numbers, rational numbers, integers, whole numbers.

a. 1.5 b. 6 c. 0.75

d. $\frac{1}{2}$ e. $\frac{1}{3}$ f. -5

Make an input-output table for Exercises 8 and 9.

8. The input, h, is the number of kilowatt hours of electricity used. The output is \$5.00 plus \$0.04 per kilowatt hour used.

9. The input, n, is the number of pounds of rice purchased. The rice costs \$0.89 per pound. The output is the total cost, including a \$0.10 "store coupon" refund.

In Exercises 10 to 13, make a table for whole-number inputs on the interval 0 to 6.

10. The output is half of the input.

11. The output is 4 more than the input.

12. The output is double the input if the input is even. The output is 4 if the input is odd.

13. The output is 5 if the input is even. The output is 2 less than the input if the input is odd.

14. Use the following table to find what discount is received on each of these purchases.

Total Purchase	Discount
\$0 to \$149.99	5% of the purchase
\$150 to \$499.99	6% of the purchase
\$500 and over	\$50 or 8% of the purchase, whichever is greater

a. \$65 b. \$145 c. \$250

d. \$500 e. \$550 f. \$700

15. Write each expression in words.

a. $3x - 4$ b. $x^2 + 3$ c. $x \div 3$

16. Write each phrase with expressions. Let x be the input.

a. Four more than the product of two and the input

b. The difference between five and the square of the input

c. Fifteen percent of the input

d. 5% of the input

e. $8\frac{1}{2}$% of the input

17. Use the following words as column headings:

Addition Subtraction Multiplication Division

List each of these words, phrases, or symbols under the appropriate heading: decreased by, increased by, product, sum, more than, per, half (write it in two different ways), twice, of, fewer than, longer than, less than, difference, quotient, farther, slower than, for each, times, loses, increases, altogether, plus, combined, one third (two ways), faster than, multiplied by, a less b, a greater than b, a diminished by b, b subtracted from a, a decreased by b, b bigger than a, a exceeds b by 3, the fraction bar, $a \cdot b$, $a - b$, $(a)(b)$, a/b, ab, $a \div b$, b/a, $a + b$, $a(b)$, $a \times b$

18. In the expression $3x - 4$, which number is the numerical coefficient?

19. Identify the constant term in the expression $x^2 + 3$.

20. What is the variable in the expression $x \div 3$?

In Exercises 21 and 22,
a. Describe how the data in the table match the design in the figure.
b. Predict the outputs for inputs of 50 and 100.
c. Write a rule for the table in words and in symbols.

21. 

1 2 3 4

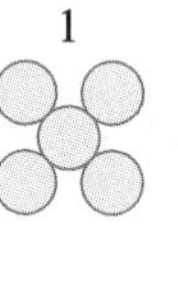

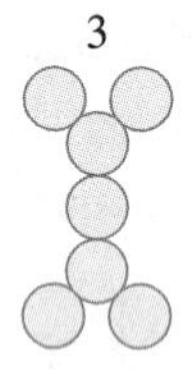
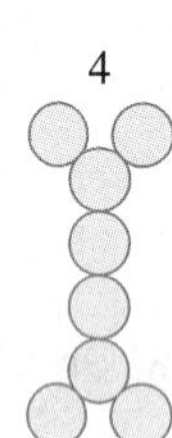

Input	Output
1	5
2	6
3	7
4	8
5	9
19	23
50	
100	
n	

22.

1 2 3 4

Input	Output
1	1
2	5
3	9
4	13
5	17
19	73
50	
100	
n	

23. a. Pinto beans cost $0.37 per pound in the bulk foods department. Use the four-step process (Section 1.4) to make a table and graph. Let weight in pounds be the input and total cost in dollars be the output.

b. What is the rule describing the bulk purchase graph? Let y be the total cost of x pounds.

c. Several sizes of packaged pinto beans are also available: 1 pound at $0.49, 2 pounds at $0.74, 4 pounds at $1.66, and 8 pounds at $2.88. Graph the weight and cost for the packaged choices as individual ordered pairs.

d. Discuss which might be the best buy and under what circumstances.

24. What words describe the parts of the following graph labeled a, b, and c? What ordered pairs describe the locations labeled d, e, f, and g?

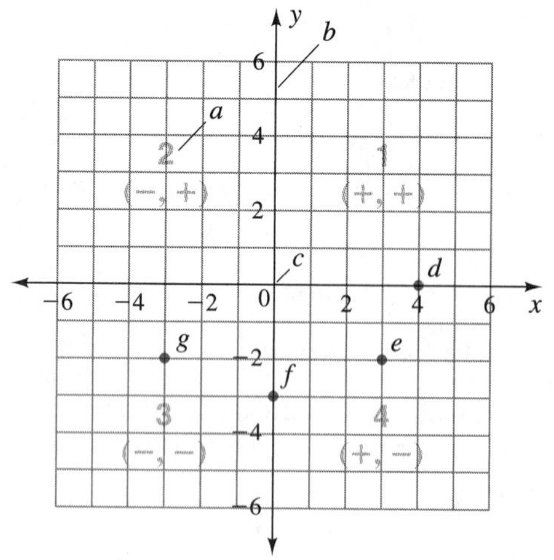

25. Graph these ordered pairs:

a. $(-3, -2)$ **b.** $(-2, 1)$ **c.** $(2, 3)$ **d.** $(3, -1)$
e. $(1, -2)$ **f.** $(0, -4)$ **g.** $(-1, 0)$ **h.** $(0, 3)$
i. $(1, 0)$ **j.** $(0, 0)$

26. The following graphs represent three different dieting experiences. Describe each experience in words. Use complete sentences.

a.

y
16
14
12
10
8
6
4
2
0
5 10 15 20 25 30 35 40 45 x
Total weight loss (pounds)
Days after diet starts

b.

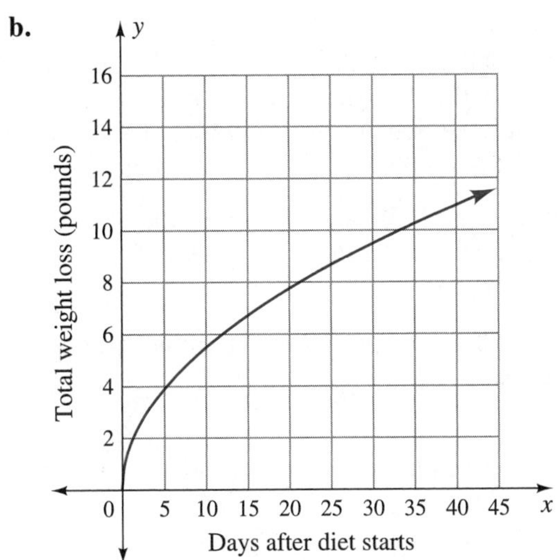

c.

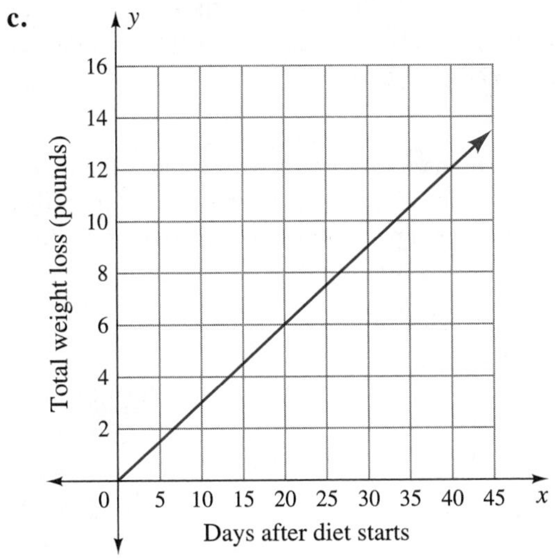

For Exercises 27 and 28, assume that the coordinate axes are temporarily invisible and the scale on each graph is one space equals one unit. Give the coordinates of A and B in each.

27.

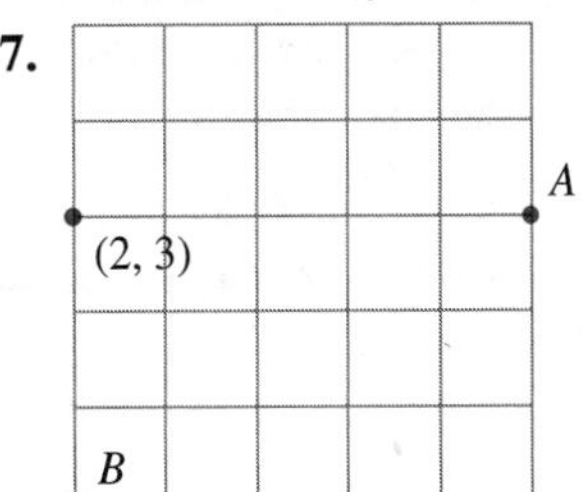

28.

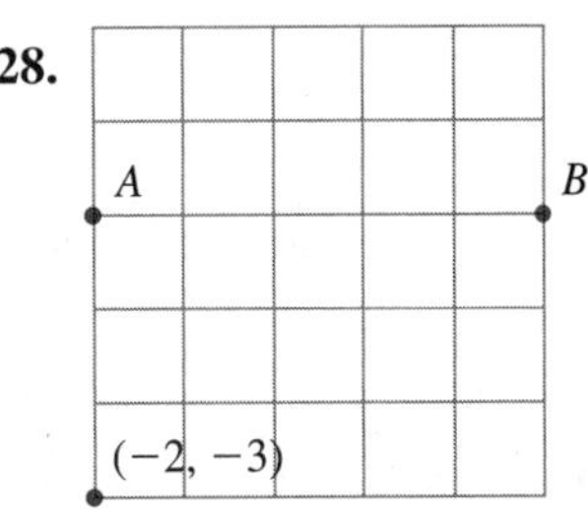

29. Plot these ordered pairs and connect them in the order listed. Do not connect ordered pairs separated by the word "lift"; lift your pencil.
(9, 5), (3, 2), (6, 8), (7, 2), (2, 5), (9, 5), lift,
(−7, 6), (−3, 2), (−5, 8), (−6, 2), (−1, 6), (−7, 6), lift,
(−3, −2), (−8, −6), (−3, −6), (−8, −2), (−6, −10), (−3, −2), lift,
(3, −8), (8, −4), (2, −4), (6, −8), (6, −2), (3, −8)

30. Plot these ordered pairs and connect them in the order listed. Do not connect ordered pairs separated by the word "lift"; lift your pencil.
(7, −1), (6, 1), (7, 1), (7, 3), (6, 3), lift,
(−5, 3), (−4, 2), (−2, 2), (−1, 3), lift,
(−6, 3), (−7, 3), (−7, 1), (−6, 1), (−7, −1), lift,
(1, 3), (2, 2), (4, 2), (5, 3), lift,
(−3, −2), (−2, −3), (2, −3), (3, −2)

31. Plot the sets of ordered pairs in parts a, b, and c on one graph and the sets of ordered pairs in parts d, e, and f on another.

a. (1, 2), (2, 3), (3, 4), (4, 5)

b. (1, 1), (2, 0), (3, −1), (4, −2)

c. (−4, 5), (−3, 4), (−2, 3), (−1, 2)

d. (1, 5), (2, 10), (3, 15), (4, 20)

e. (2, 1), (4, 2), (6, 3), (8, 4)

f. (1, 3), (2, 5), (3, 7), (4, 9)

32. What ordered pair $(0, y)$ would fit each pattern in Exercise 31? Describe how you found the ordered pair.

33. For the graphs in Exercise 31, write a description of the output in words or symbols, using x as the input.

In Exercises 34 to 36, make an input-output table for the situation described, and graph the ordered pairs from the table.

34. Decaffeinated coffee beans cost \$14 per pound. Let the input be the number of pounds used and the output be the total cost of coffee. Show the cost for 0 to 8 pounds.

35. The first eight uses of an automatic teller machine (ATM) card are free; each additional use costs \$0.75. Let the input be the number of uses and the output be the total cost for a month. Show the cost for 0 to 12 uses of the card.

36. A 30-minute long-distance telephone card costs \$9.99.

a. What is the cost per minute?

b. Let the input be the number of minutes talked and the output be the value remaining on the card. Show the value of the card for inputs of 0 to 30 minutes in steps of 5 minutes.

37. Why would we not connect points in Exercise 35?

38. Why would we connect points in Exercises 34 and 36?

CHAPTER 1 TEST

1. List several assumptions that a student might make about a chapter test.

2. List several conditions that a teacher might place on a chapter test.

Fill in the missing word in Exercises 3 to 5.

3. The answer to a subtraction problem is the ________ .

4. A listing of numbers in braces, { }, or a collection of objects is a ________ .

5. The numbers described by $\{\ldots, -3, -2, -1, 0, 1, 2, 3, \ldots\}$ are the ________ .

For Exercises 6 and 7, refer to the graph below.

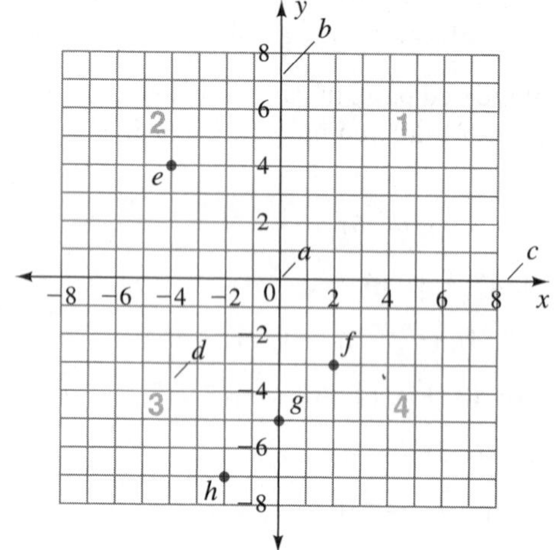

6. What words describe the parts of the graph labeled *a*, *b*, *c*, and *d*?

7. What ordered pairs describe the locations labeled *e*, *f*, *g*, and *h*?

8. Make an input-output table with integer inputs of 0 to 5 for the following rule: Output is twice the input plus 3. Graph the data.

9. A car rental costs \$150 for one week plus \$0.10 for each mile. Use the four-step process to make a table and graph for one week's rental. Let input be miles, counting by 100 to 600. Let output be total cost.

a. ***Understand:*** What units go on the axes? What is an expression for the original rental?

b. ***Plan:*** Explain how you determine quadrants and scales.

c. ***Do the graph:*** Make a table, and draw a graph from the table.

d. ***Check and extend:*** Another company offers the same car for \$200 a week with no per-mile charge. At what number of miles per week does the second car become a better deal?

10. Make an input-output table with integer inputs of 0 to 8 for this rule: If the input is an even number, the output is half the input; if the input is an odd number, the output is double the input.

11. Find the sum of $\frac{5}{8}$ and $\frac{1}{6}$.

12. Find the product of $1\frac{1}{9}$ and $1\frac{5}{12}$.

13. Find 9% of 35.

14. Complete the table and write the rule in words and in algebraic notation.

Input	Output
1	4
2	8
3	12
4	16
5	
100	
n	

15. The vertical and horizontal axes have been omitted from the figure below. Use the given point to name the ordered pairs for points A and B.

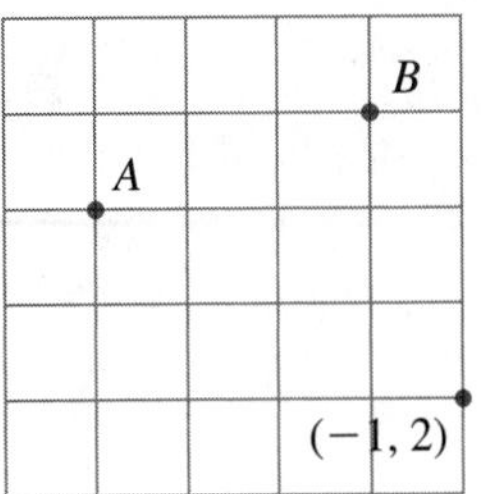

2

Operations with Real Numbers and Expressions

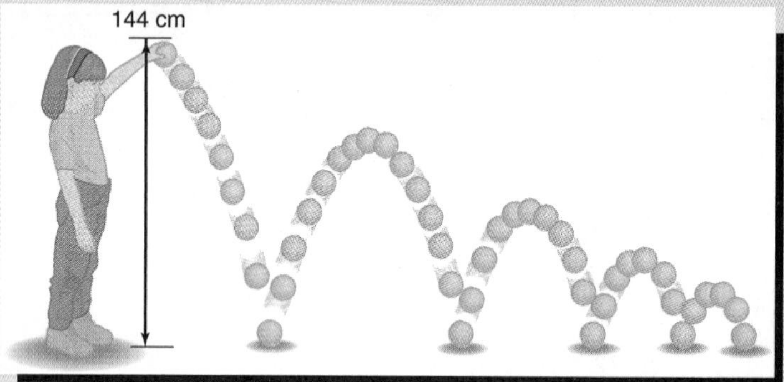

Figure 1

A ball is dropped from a height of 144 centimeters. After each time it hits the floor, the ball bounces two-thirds as high as it did on the last bounce (see Figure 1). What is the height of the ball after it hits the floor ten times? We solve this problem in Section 2.4.

In this chapter, we do operations with positive and negative numbers, exponents, and units of measure. We look at how numbers behave (their *properties*). We use rules that tell us which operations to do first in a problem (the *order of operations*). At the end of the chapter, we name sets of numbers with *inequalities* and *intervals*.

2.1 Addition and Subtraction with Integers

OBJECTIVES

- Find the opposite (called the additive inverse) and the absolute value of a number.
- Use opposites and absolute value to add and subtract integers.
- Use models—the electrical charge model and the distance model—to add and subtract integers.

WARM-UP

The horizontal scale and the vertical scale on the following grid are both 1 unit. The distance between a and b is 3 units. Count to find the distance between the indicated points.

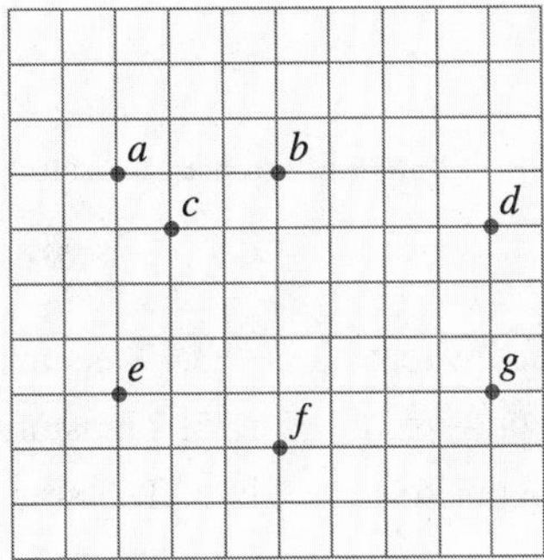

1. c and d **2.** e and g **3.** a and e **4.** b and f **5.** d and g

IN THIS AND THE NEXT SECTION, we will add, subtract, multiply, and divide with integers. We will find rules for these operations that apply to all real numbers, but in these first two sections will use them only for integers.

Integers and Opposites

INTEGERS In Section 1.2, we defined the *integers* as the set of whole numbers and their opposites: $\{\ldots, -3, -2, -1, 0, 1, 2, 3, \ldots\}$. Figure 2 shows a number line with the integers from -5 to 5.

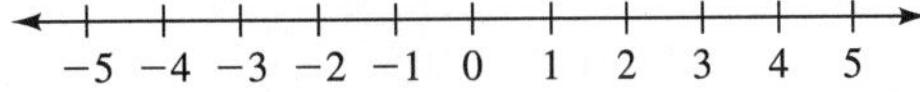

Figure 2

Integers describe the distance to a number from the reference point, zero. In terms of the reference point (zero), a *positive number* is greater than zero and a *negative number* is less than zero.

In applications, we use integers to describe such measurements as elevation in reference to sea level, which is elevation zero. Mt. McKinley in Alaska is $+20{,}320$ feet, and the ocean floor near Hawaii is $-16{,}400$ feet. Integers describe temperature in reference to the freezing point of water at 0°C (zero degrees Celsius). Water boils at +100°C. Dry ice (the solid carbon dioxide that provides "fog" at carnivals) "freezes" at approximately −78°C.

Integers may also describe *change*, such as change in value (+5 dollars, −10 dollars) or change in position (+6 yards, −3 yards). In change, a positive number indicates a gain or increase while a negative number indicates a loss or decrease.

OPPOSITES Numbers are opposites if they are the same distance from zero in opposite directions. The formal name for opposites is *additive inverse,* where *inverse* means reversed or opposite in position or direction. The additive inverse is the number that, *added* to its opposite, equals zero.

Definition of Opposites

The **opposite**, or **additive inverse**, of a number is *the number that is the same distance from zero on the number line but on a different side of zero.*

The opposite of a is written $-a$.

A number and its opposite add to zero:

$$a + -a = 0$$

EXAMPLE 1 Finding opposites Use the number line in Figure 2 to find the opposite of each expression. Then add each expression to its opposite.

a. 3 **b.** −5 **c.** $-\frac{1}{2}$ **d.** x

Solution

a. The opposite of 3 is −3. The sum of 3 and −3 is 0.

b. The opposite of −5 is 5. The sum of −5 and 5 is 0.

c. The opposite of $-\frac{1}{2}$ is $\frac{1}{2}$. The sum of $-\frac{1}{2}$ and $\frac{1}{2}$ is 0.

d. The opposite of x is $-x$. The sum of x and $-x$ is 0.

We now consider the opposite of the opposite. Suppose we place a dot at 3 on a number line and draw a line to its opposite, −3 (see Figure 3). Where is the opposite of −3?

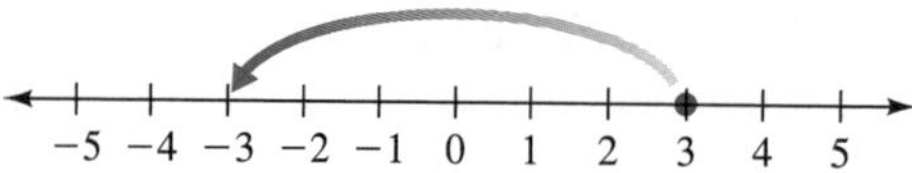

Figure 3

If you said 3, the original dot, you guessed right (see Figure 4).

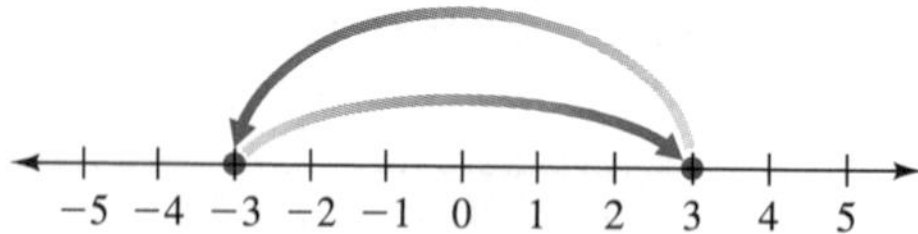

Figure 4

This leads to an important property of opposites: When you take the opposite of an opposite, you get the original number.

Opposite of an Opposite

The opposite of the opposite of a is a:

$$-(-a) = a$$

We say that a and $-(-a)$ are equivalent expressions. **Equivalent expressions** are *expressions that have the same value for all replacements of the variables.* Numbers can also be equivalent.

EXAMPLE 2 Finding an equivalent expression Write each expression as an equivalent number.

a. $-(-4)$ **b.** $-(-12)$ **c.** $-(-a)$, where $a = -5$

Solution Here's a hint: $-(-a)$ means the opposite of the opposite of a. See the Answer Box for the answers.

Addition of Integers

THE ELECTRICAL CHARGE MODEL To see how positive and negative integers can be added, we begin with the electrical charge model. The circles in Example 3 hold charges. The positive sign (+) means a charge of positive one. The negative sign (−) means the opposite, or a charge of negative one.

Electrical Charges

> One positive charge neutralizes one negative charge. Zero is described as *neutral*, having neither a positive nor a negative charge.
>
> $\oplus + \ominus = 0$ or $+1 + (-1) = 0$

In Example 1, we practiced adding opposites using the idea that a number and its opposite add to zero. The *net charge* in a circle is the charge that remains after all positive and negative pairs have been neutralized by adding to zero.

EXAMPLE 3 Finding charges What is the net charge on each circle?

a.

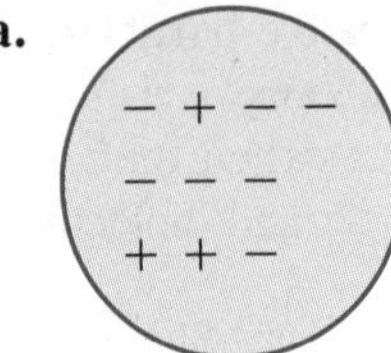

b.

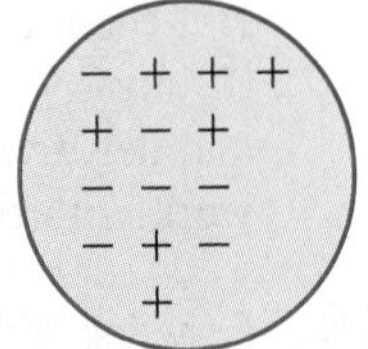

c.

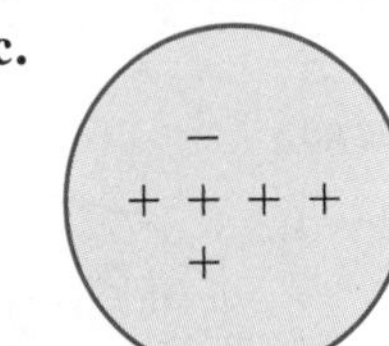

d.

Solution Possible solutions are shown below.

a. Net charge = −4

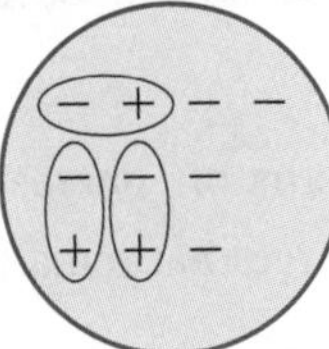

b. Net charge = 0

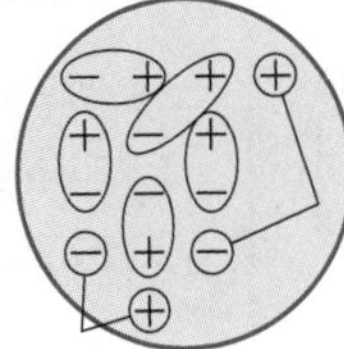

c. Net charge = +4 **d.** Net charge = −1

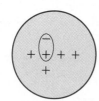

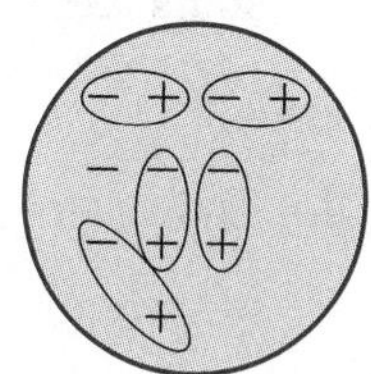

To find the net charges in Example 3, you may have matched pairs of positive and negative charges. Or you may have counted the set of all positive charges and then the set of all negative charges.

If you counted the charges in sets, you were thinking of the positive charges as a positive number and the negative charges as a negative number. When we group the charges in sets of positives and negatives, we can write the charges as *addition of integers*.

EXAMPLE 4 Identifying addition of integers Guess an answer to each problem, and then match each pair of addition problems with one of the charge circles in Example 3.

a. $+5 + (-1)$
$-1 + (+5)$

b. $+5 + (-6)$
$-6 + (+5)$

c. $-7 + (+3)$
$+3 + (-7)$

d. $+7 + (-7)$
$-7 + (+7)$

Solution See the Answer Box.

Check your solutions to Example 4 with a scientific calculator. Enter a negative number into the calculator with the [±] or [(−)] key. Do not use the subtraction operation for a negative sign. Some calculators need the [±] before the number; others need it after the number. Practice until your answers agree with the solutions.

If there are more negative particles, the answer is negative. If there are more positive particles, the answer is positive. This way of thinking about positive and negative numbers is called the *charge model* because it is closely related to work with charged atoms and molecules in chemistry.

EXAMPLE 5 Drawing addition of integers Show these additions with charges.

a. $-9 + (+3)$ **b.** $+4 + (-7)$

Solution Answers vary. Here is one.

a.

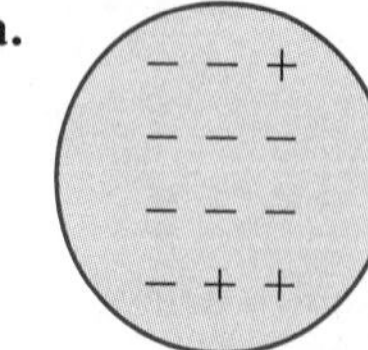

b.

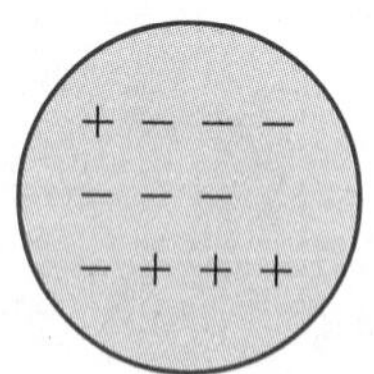

Adding Integers with the Charge Model

To add charges:

1. Eliminate the pairs of charges that add to zero.
2. Find the net charge that remains.

EXAMPLE 6 Adding integers Do these additions, referring to the charge circles in the solution to Example 5.

a. $-9 + (+3)$ **b.** $+4 + (-7)$

Solution Use these hints; then see the Answer Box.

a. Hint: In $-9 + (+3)$, which integer has more charges and by how many?

b. Hint: In $+4 + (-7)$, which integer has more charges and by how many? ●

ABSOLUTE VALUE When you think about which integer has more charges, -9 or $+3$, you are starting to think in terms of absolute value. In the Warm-up, to find the distance we counted spaces, because no numbers were given. The distances were positive numbers.

The **absolute value** of a number is *the distance it is from zero.* We consider distance to be positive, so the absolute value of a number is positive. The symbol for absolute value is two vertical lines placed around a number; for example, the absolute value of three is written $|3|$.

EXAMPLE  Finding absolute value Find the absolute value and show it on a number line.

a. -3 **b.** 4

Solution **a.** The absolute value of -3 is $|-3| = 3$, because -3 is a distance of three units from zero.

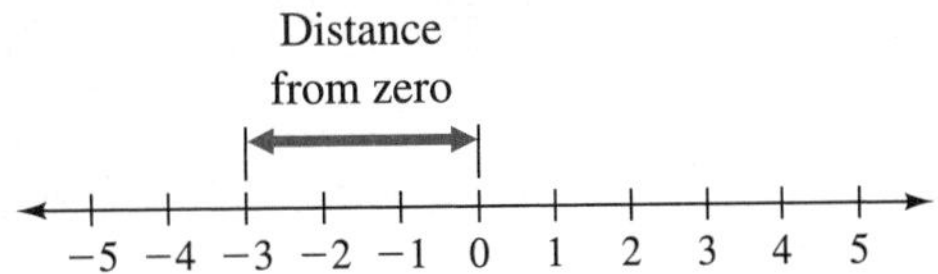

b. The absolute value of $+4$ is $|+4| = 4$, because $+4$ is a distance of four units from zero.

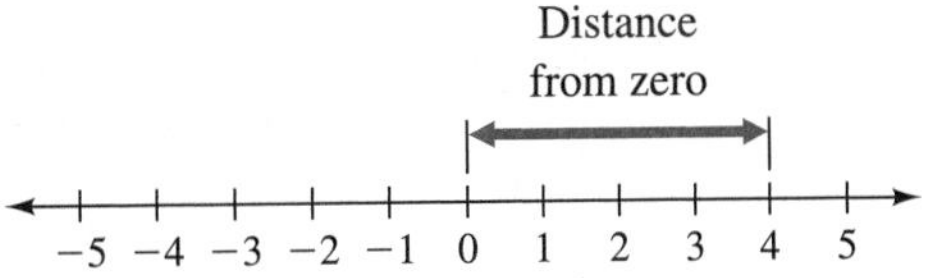

●

Think about it 1: Which number is further from zero, -3 or 4? Which number has more charges, -3 or 4?

ABSOLUTE VALUE AND ADDITION In our charge model, the absolute value is the number of charged particles, regardless of sign.

EXAMPLE 8 Adding with absolute value Which of the numbers in each addition has the greater absolute value? When the numbers are added, which number controls the sign of the answer? Predict the sign on these sums; then do the additions.

a. $-8 + 2$ **b.** $9 + (-3)$

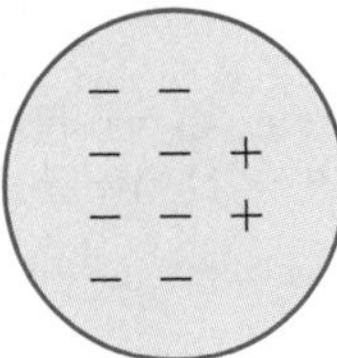

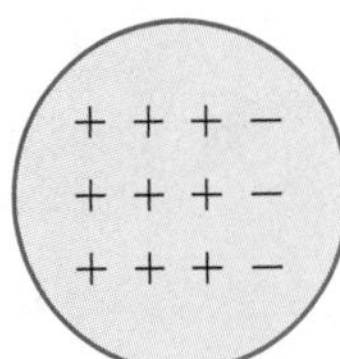

c. $-4 + (-3)$ **d.** $+5 + (+2)$

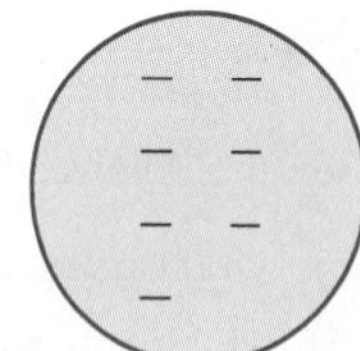

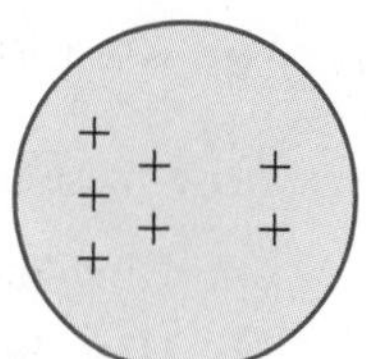

Solution **a.** In $-8 + 2$, the -8 has more charges and the greater absolute value. The sign on the sum is negative.

$$-8 + 2 = -6$$

b. In $9 + (-3)$, the 9 has more charges and the greater absolute value. The sign on the sum is positive.

$$9 + (-3) = +6$$

c. In $-4 + (-3)$, the -4 has greater absolute value. Both are negative. The sign on the sum is negative.

$$-4 + (-3) = -7$$

d. In $+5 + (+2)$, the $+5$ has the greater absolute value. Both are positive. The sign on the sum is positive.

$$+5 + (+2) = +7$$

Absolute value lets us write a formal description of adding positive and negative numbers.

Addition of Positive and Negative Numbers Using Absolute Value

- To add numbers with the same sign, add their absolute values and place the common sign on the answer.
- To add numbers with opposite signs, subtract their absolute values and place the sign from the number with the greater absolute value on the answer.

EXAMPLE 9 Adding integers Find these sums. Explain your answer in terms of absolute value.

a. $+13 + (+15)$ **b.** $+15 + (-26)$

c. $-5 + (+8)$ **d.** $-8 + (-7)$

e. $-3 + 15 + (-18)$ **f.** $16 + (-4) + (-21)$

Solution In most cases, if the sign is positive, we drop the sign.

a. The signs are alike, so we add the absolute values.

$$13 + 15 = 28$$

b. The signs are different, so we subtract the absolute values. We mentally do the subtraction $26 - 15 = 11$. We write

$$+15 + (-26) = -11$$

c. The signs are different, so we subtract the absolute values. We mentally do the subtraction $8 - 5 = 3$. We write

$$-5 + (+8) = 3$$

d. The signs are alike, so we add the absolute values. We mentally add $8 + 7$ and write

$$-8 + (-7) = -15$$

e. The signs are different. We add the -3 and -18 to get -21, and then we subtract 21 and 15. We write

$$-3 + 15 + (-18) = -6$$

f. The signs are different. We add the -4 and -21 to get -25, and then we subtract 25 and 16. We write

$$16 + (-4) + (-21) = -9$$

Subtraction of Integers

In the charge model, shown in Figure 5, subtraction means removing objects from a set.

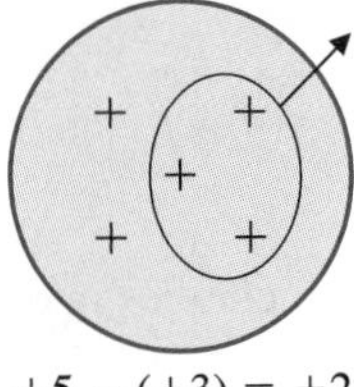

$+5 - (+3) = +2$

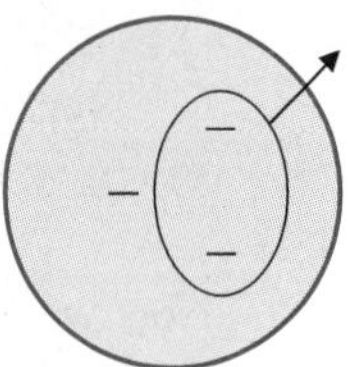

$-3 - (-2) = -1$

Figure 5

The charge model may not seem to work for subtraction when there are no charges available to be removed. However, *because a pair of opposite charges, one positive and one negative, adds to zero, we can add any number of pairs of opposites to the charge circle without changing its net charge.*

We use these steps.

Subtracting Integers with the Charge Model

To subtract $x - y$:

1. Start with x charges.
2. Add in pairs of opposite charges until there are y charges available to subtract.
3. Subtract.
4. Count the net charge.

EXAMPLE **10** Subtracting with the charge model Subtract by adding in pairs of opposites.

a. $3 - (+5)$ **b.** $-2 - (-3)$ **c.** $+3 - (-2)$

Solution **a.** For $+3 - (+5)$, the charge circle starts with 3 positive charges. We need a total of 5 positive charges for the subtraction, so we add 2 (or more) pairs of opposites to the circle. (See Figure 6.)

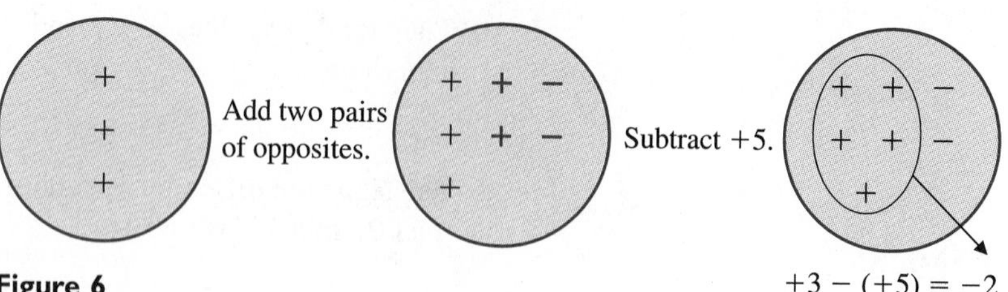

Figure 6

Now we have 5 positive charges and can remove them. The net charge is -2.

$$3 - (+5) = -2$$

b. For $-2 - (-3)$, the charge circle starts with 2 negative charges. We need a total of 3 negative charges for the subtraction, so we add 1 pair of opposites (or more) to the circle. (See Figure 7.)

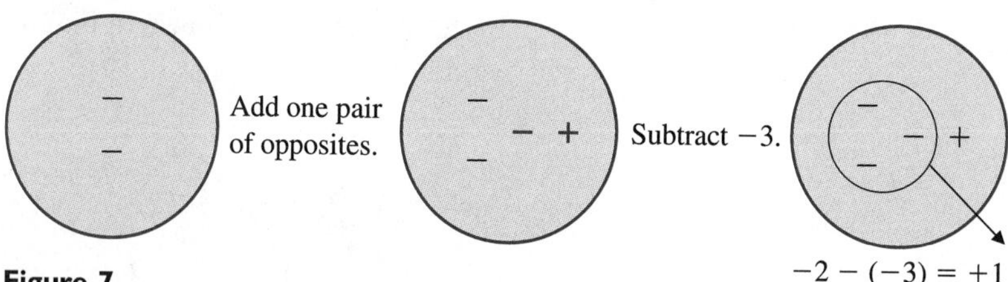

Figure 7

Now we have 3 negative charges and can remove them. The net charge is $+1$.

$$-2 - (-3) = +1$$

c. For $+3 - (-2)$, the charge circle starts with 3 positive charges. We need 2 negative charges for the subtraction, so we add 2 (or more) pairs of opposites to the circle. (See Figure 8.)

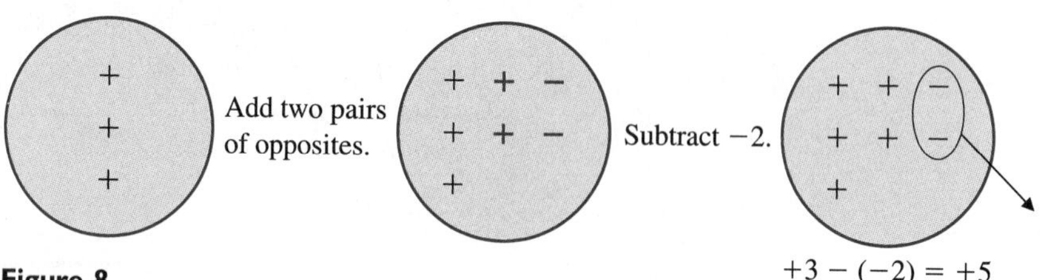

Figure 8

Now we have 2 negative charges and can remove them. The net charge is $+5$.

$$+3 - (-2) = +5$$

●

Think about it 2: Complete these sentences with the word *increase* or *decrease:*

a. Subtracting a positive results in a(n) ________ in the net charge.

b. Subtracting a negative results in a(n) ________ in the net charge.

EXAMPLE **11** Subtracting integers Subtract these integers, using a charge model.

a. $6 - 5$ **b.** $2 - 4$ **c.** $-1 - (+2)$

d. $2 - (+3)$ **e.** $2 - (-2)$ **f.** $-5 - (-3)$

Solution **a.** $6 - 5 = 1$ **b.** $2 - 4 = -2$ **c.** $-1 - (+2) = -3$

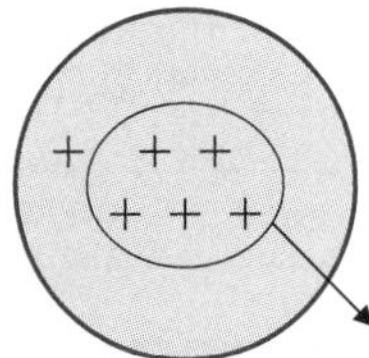

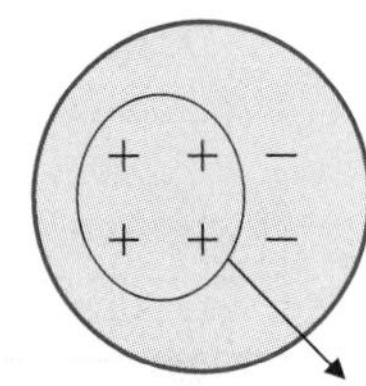

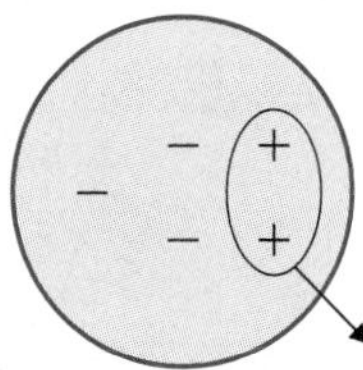

d. $2 - (+3) = -1$ **e.** $2 - (-2) = 4$ **f.** $-5 - (-3) = -2$

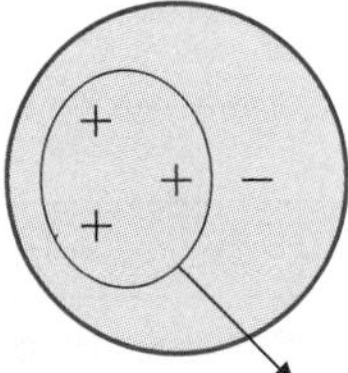

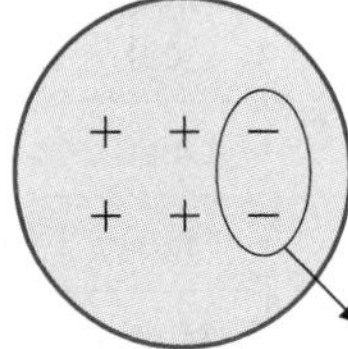

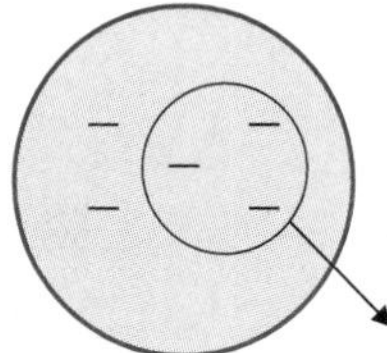

Think about it 3: If you are doing these subtractions mentally, you are probably—without realizing it—using absolute value to get your answer. Explain how.

When numbers are small, do subtractions mentally. When numbers become more complicated, use a calculator. Try working Example 11 with a calculator. Practice until your answers agree with the solutions.

To subtract $x - y$ with the charge model, we could add y pairs of opposites. Those pairs would contain both y original charges and y opposite charges. In mathematics, we say call this process *subtraction by adding the opposite.*

Subtraction of Integers

To subtract two numbers, change the subtraction to addition of the opposite number:

$$x - y = x + (-y)$$

Examples: $3 - 6 = 3 + (-6) = -3$ $4 - (-3) = 4 + (+3) = 7$

EXAMPLE 12 Subtracting by adding the opposite Rework Example 11, changing each subtraction to addition of the opposite number.

Solution See the Answer Box.

Application: Elevations

We use the word *elevation* when we are looking at vertical distance. We describe a *difference* in elevation with a positive number. A *change* in elevation, as when a plane is landing, can be negative. *To find the difference in elevation, we subtract the lesser height from the greater height.*

EXAMPLE 13 Finding differences in elevation Use the data in Table 1 to find the difference in elevation between the two places given.

a. Mt. McKinley in Alaska and Pike's Peak in Colorado

b. Mt. Everest and the Mariana Trench in the Pacific Ocean (south of Japan and east of the Philippines)

c. Death Valley in California and the Dead Sea between Israel and Jordan

Mt. Everest	+29,028
Mt. McKinley	+20,320
Pike's Peak	+14,110
Mauna Kea	+13,710
Sea Level	0
Death Valley	−282
Dead Sea	−1,312
Ocean Floor, near Hawaii	−16,400
Mariana Trench	−35,840

Table 1 Elevation in Feet Relative to Sea Level

Solution **a.** The elevation of Mt. McKinley minus the elevation of Pike's Peak is

$$20{,}320 - 14{,}110 = 6{,}210 \text{ ft}$$

b. The elevation of Mt. Everest minus the elevation of Mariana Trench is

$$29{,}028 - (-35{,}840) = 29{,}028 + (+35{,}840) = 64{,}868 \text{ ft}$$

c. The elevation of Death Valley minus the elevation of the Dead Sea is

$$-282 - (-1312) = -282 + (+1312) = 1030 \text{ ft}$$

●

ANSWER BOX

Warm-up: 1. 6 **2.** 7 **3.** 4 **4.** 5 **5.** 3 **Example 2: a.** 4 **b.** 12 **c.** −5 **Example 4: a.** 4, c **b.** −1, d **c.** −4, a **d.** 0, b **Example 6: a.** −6 **b.** −3 **Think about it 1:** 4 is further from zero and has more charges. **Think about it 2: a.** decrease **b.** increase **Example 12:** **a.** $6 + (-5) = 1$ **b.** $2 + (-4) = -2$ **c.** $-1 + (-2) = -3$ **d.** $2 + (-3) = -1$ **e.** $2 + (+2) = 4$ **f.** $-5 + (+3) = -2$

EXERCISES 2.1

In Exercises 1 and 2, give the opposite of each number or expression.

1. a. 5 **b.** $-\frac{1}{2}$ **c.** 0.4 **d.** x **e.** $-2x$

2. a. −5 **b.** $\frac{2}{3}$ **c.** 2.5 **d.** $3x$ **e.** $-ab$

In Exercises 3 to 8, calculate the expression.

3. a. $|4|$ **b.** $|-6|$ **c.** $-(-5)$ **d.** $-(-2)$

4. a. $|-5|$ **b.** $-(-4)$ **c.** $-|-3|$ **d.** $|7|$

5. a. $-|7|$ **b.** $-|-8|$ **c.** $-(-3)$ **d.** $|-7|$

6. a. $-(-6)$ **b.** $-|4|$ **c.** $|-(-2)|$ **d.** $-|-5|$

7. a. $|-4|$ **b.** $|5|$ **c.** $|4 - 9|$ **d.** $-|2 + 5|$

8. a. $|7|$ **b.** $|-6|$ **c.** $|3 - 9|$ **d.** $-|3 + 4|$

9. What is $-(-x)$ if

a. $x = +4$ **b.** $x = -4$

10. What is $|x|$ if

a. $x = -3$ **b.** $x = 3$

11. What is $-|x|$ if

a. $x = 6$ **b.** $x = -6$

12. What is $-x$ if

a. $x = -5$ **b.** $x = 5$

13. Write in words the expression $-|x|$.

14. Write in words the expression $-(-x)$.

What addition fact describes the circles in Exercises 15 to 20? What is the net charge?

15.

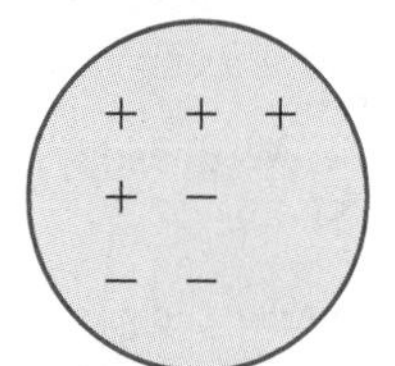

___ + ___ = ___
Net charge = ___

16.

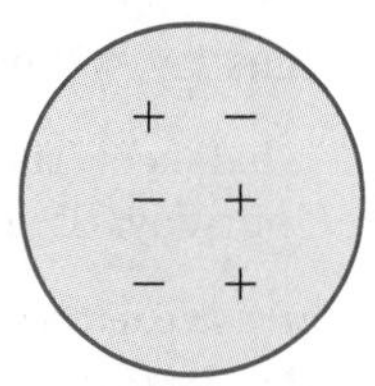

___ + ___ = ___
Net charge = ___

17.

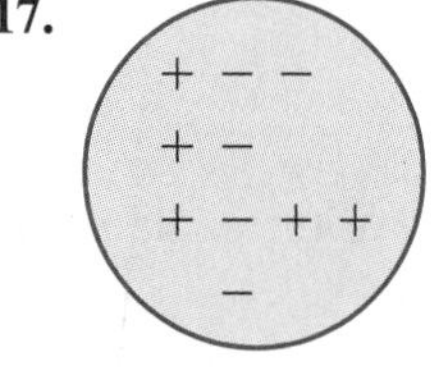

___ + ___ = ___
Net charge = ___

18.

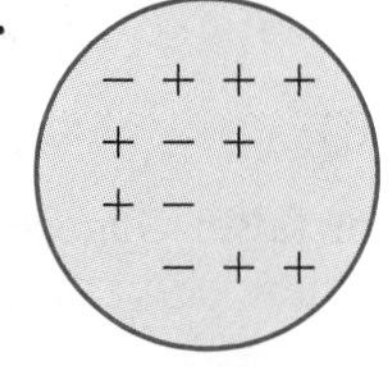

___ + ___ = ___
Net charge = ___

19.

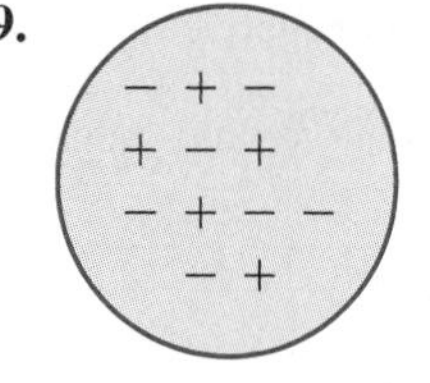

___ + ___ = ___
Net charge = ___

20.

___ + ___ = ___
Net charge = ___

For Exercises 21 to 24, draw a charge circle for the problem, and then add. Circles may vary.

21. $-3 + (+2) =$ ___ **22.** $+6 + (-4) =$ ___

23. $+3 + (-5) =$ ___ **24.** $-4 + (-3) =$ ___

In Exercises 25–32, find the sum and then check with a calculator.

25. a. $-8 + 3$ **b.** $4 + (-7)$ **c.** $+4 + (-4)$

26. a. $-5 + (-2)$ **b.** $+3 + (-3)$ **c.** $5 + (-12)$

27. a. $-3 + (+3)$ **b.** $-4 + (-7)$ **c.** $-12 + (+8)$

28. a. $5 + (+8)$ **b.** $-9 + (+9)$ **c.** $15 + (-8)$

29. a. $-3 + (-4) + (+5)$ **b.** $+7 + (-8) + (-3)$

30. a. $4 + (-6) + (-1)$ **b.** $-5 + (-4) + (+4)$

31. a. $2 + (-5) + (-4)$ **b.** $-6 + (-3) + (+7)$

32. a. $-4 + (-6) + (+1)$ **b.** $4 + (-7) + (-9)$

Check your answers in Exercises 33 to 36 with a calculator.

33.
$$\begin{array}{r} -12 \\ +\ 6 \\ -15 \\ \underline{+\ 7} \end{array}$$

34.
$$\begin{array}{r} +15 \\ -16 \\ +17 \\ \underline{-\ 9} \end{array}$$

35. $-6 + (-7) + (-8) + 20$

36. $-18 + 7 + (-9) + (-12)$

State the subtraction problem illustrated by each of the charge circles in Exercises 37 and 38, and then work the problem.

37. a.

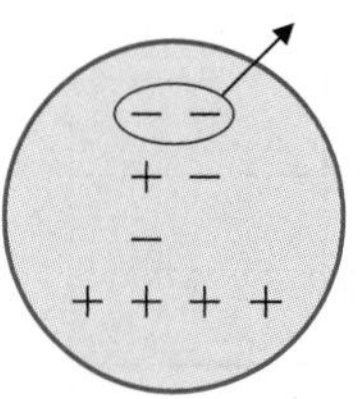

___ − ___ = ___

b.

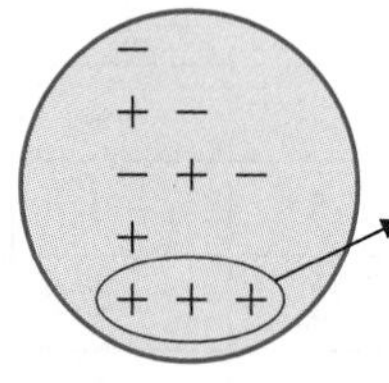

___ − ___ = ___

38. a.

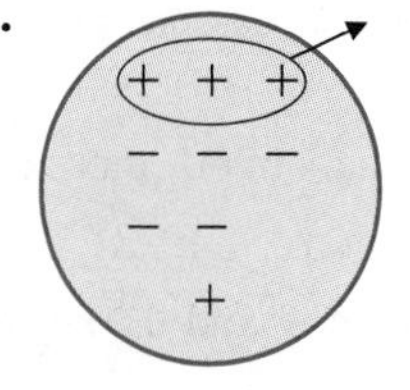

___ − ___ = ___

b.

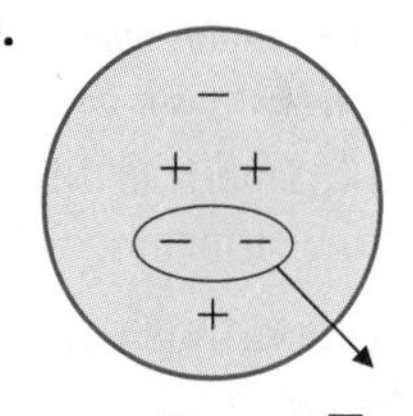

___ − ___ = ___

Make a charge circle that shows each problem in Exercises 39 and 40, and solve the problem. Circles may vary.

39. a. $-4 - (-3)$ **b.** $-1 - (-3)$

40. a. $-3 - (+2)$ **b.** $-8 - 4$

In Exercises 41 to 48, find the difference and then check with a calculator.

41. a. $8 - 11$ **b.** $-8 - (-11)$ **c.** $-16 - 3$
d. $0 - 5$ **e.** $-17 - 4$

42. a. $12 - 7$ **b.** $-12 - 7$ **c.** $-17 - (-4)$
d. $17 - 4$ **e.** $16 - (-3)$

43. a. $-4 - (-4)$ **b.** $14 - (10)$ **c.** $0 - (-5)$
d. $8 - 9$ **e.** $-4 - 5$

44. a. $10 - 14$ **b.** $-8 - (-9)$ **c.** $-7 - 6$
d. $-3 - (-3)$ **e.** $4 - (-5)$

45. a. $2 - 7$ **b.** $2 - 5$ **c.** $2 - (-5)$
d. $-2 - (-5)$ **e.** $-7 - (-2)$

46. a. $-7 - 5$ **b.** $-5 - 7$ **c.** $-5 - (-7)$
d. $-7 - (-5)$ **e.** $5 - 7$

47. a. $8 - (+3) - (-4)$ **b.** $-5 - (-7) + (-1)$

48. a. $6 - (-7) - (+2)$ **b.** $-4 - (+8) - (-5)$

Complete the tables in Exercises 49 and 50.

49.

Input x	Input y	Output $x + y$	Output $x - y$
4	5		
5	−4		
−4	5		
−4	−5		

50.

Input x	Input y	Output $x + y$	Output $x - y$
5	4		
−5	−4		
−5	4		
4	−5		

In Exercises 51 to 54, what is the difference in elevation between the highest and lowest point on each continent?

51. South America: Mt. Aconcagua, 6960 meters, and Valdes Peninsula, −40 meters

52. Africa: Kilimanjaro, 5896 meters, and Lake Assal, −156 meters

53. Australia: Mt. Kosciusko, 2228 meters, and Lake Eyre, −16 meters

54. The Mariana Trench (Pacific Ocean) is 10,930 meters below sea level. The Puerto Rico Trench (Atlantic Ocean) is 8600 meters below sea level. Write an appropriate subtraction, and determine how much deeper the Mariana Trench is than the Puerto Rico Trench.

Checkers at grocery stores must balance their cash drawer each shift. An error in either direction, cash-over (too much money) or cash-under (too little), is recorded. For Exercises 55 to 58, write an expression using absolute value to describe the total error due to cash-over/cash-under.

55. Monday, \$15 under; Tuesday, \$10 under; Wednesday, balance; Thursday, \$5 under; Friday, \$10 over

56. Monday, \$5 under; Tuesday, \$5 over; Wednesday, \$15 over; Thursday, \$5 under; Friday, balance

57. Monday, \$5 over; Tuesday, \$12 over; Wednesday, balance; Thursday, \$7 under; Friday, \$10 under

58. Monday, balance; Tuesday, \$13 over; Wednesday, \$20 under; Thursday, \$5 under; Friday, \$8 over

For the rules listed in Exercises 59 to 64, build a table and graph with the integers −3 to 3 as inputs. Do operations inside the absolute value signs first.

59. $y = |x|$ **60.** $y = -x$

61. $y = |x + 1|$ **62.** $y = |x - 2|$

63. $y = |x| - 2$ **64.** $y = |x| + 1$

65. True or false: Opposite numbers have the same absolute value.

66. True or false: $-|x| = |-x|$ for all x.

67. Explain what absolute value means.

68. Explain how to add with positive and negative numbers.

69. Explain how to subtract with positive and negative numbers.

2.2 Multiplication and Division with Integers

OBJECTIVES

- Use the web site sales model to multiply integers.
- Find the reciprocal (called the multiplicative inverse) of a real number.
- Multiply and divide integers.

WARM-UP

What are the missing numbers? Guess and check. Some answers may be fractions.

1. $-3 + \_\_\_ = 0$ **2.** $\_\_\_ + 5 = 1$ **3.** $\_\_\_ + (-2) = 0$

4. $4 + \_\_\_ = 1$ **5.** $\_\_\_ \cdot 3 = 1$ **6.** $6 + \_\_\_ = 0$

7. $5 \cdot \_\_\_ = 1$ **8.** $\frac{3}{4} \cdot \_\_\_ = 0$ **9.** $\frac{1}{2} \cdot \_\_\_ = 1$

IN THIS SECTION, we will multiply and divide with integers, rewriting division problems as multiplication problems. The rules we find here will apply to all real numbers.

Multiplication of Integers

WEB SALES MODEL Thinking in terms of a web site sales model gives us a way of figuring out what sign is needed when we multiply positive and negative integers.*

Suppose you open a sales site, www.sale.com, on the web. All sales and all bills are delivered by your local server. A delivery is recorded as a positive number. Sales are written as positive numbers, and bills are written as negative numbers.

The server delivers three sales for \$20 each. You write $+3(+20)$. Your net worth is increased, so $+3(+20) = +60$.

The server delivers two bills for \$10 each. You write $+2(-10)$. Your net worth is decreased, so $+2(-10) = -20$.

The name of your site is close to that of another site, www.sell.com. You sometimes receive sales and bills for the other site. When the server takes away items delivered by mistake because of a wrong address, the removal is recorded as a negative number.

The server takes away four sales for \$5 each. You write $-4(+5)$. Your net worth is decreased, so $-4(+5) = -20$.

EXAMPLE 1 Exploration: investigating the web sales model Write each transaction in positive or negative numbers and find how the net worth changes.

a. The server brings two sales for \$35 each.

b. The server brings three bills for \$50 each.

c. The server takes away four sales for \$30 each.

d. The server takes away five bills for \$8 each.

Solution See the Answer Box. ●

When the server takes away bills, you do not need to pay them, and so your net worth is increased.

EXAMPLE  Finding net worth Use the web sales model to write each of these transactions as a multiplication of positive and negative numbers. Find the change in net worth.

a. The server brings three sales for \$50 each.

b. The server takes away four sales for \$10 each.

c. The server takes away four bills for \$25 each.

d. The server brings two bills for \$75 each.

e. The server takes away three bills for \$20 each.

f. The server takes away two sales for \$25 each.

Solution The dollar signs are omitted in the computation step.

a. $+3(+50) = +150$; net worth increases \$150.

b. $-4(+10) = -40$; net worth decreases \$40.

c. $-4(-25) = +100$; net worth increases \$100.

d. $+2(-75) = -150$; net worth decreases \$150.

*Thanks to Dr. Judith H. Hector, Walters State Community College, Morristown, Tennessee, for her delivery person model that led to this model.

e. $-3(-20) = +60$; net worth increases \$60.

f. $-2(+25) = -50$; net worth decreases \$50.

Think about it: Complete these sentences based on the example.

a. A positive number multiplied by a positive number is _______.

b. A positive number multiplied by a negative number is _______.

c. A negative number multiplied by a positive number is _______.

d. A negative number multiplied by a negative number is _______.

PATTERNS IN THE PRODUCTS OF INTEGERS The web sales model suggests that we can multiply positive and negative numbers and have results that make sense. We now develop rules for multiplication of positive and negative integers by looking at patterns in tables.

EXAMPLE 3 Multiplying a positive times a negative Complete the input-output table in Table 2. What happens to the output as the input, x, gets smaller? What might we conclude about a negative number multiplied by a positive number?

Input x	**Output** $+2x$
+2	$2(2) = 4$
+1	$2(1) = 2$
0	
−1	
−2	
−3	

Table 2

Solution

Input x	**Output** $+2x$
+2	$+2(2) = 4$
+1	$+2(1) = 2$
0	$+2(0) = 0$
−1	$+2(-1) = -2$
−2	$+2(-2) = -4$
−3	$+2(-3) = -6$

As the input decreases by 1, the output decreases by 2.

Use a calculator to practice multiplying the positive and negative numbers in Example 3 until your answers agree with the solution.

Example 3 suggests the following rule.

Multiplying a Positive Integer and a Negative Integer

> A negative number multiplied by a positive number is a negative number.

In Example 4, we explore what happens when we multiply a negative by a negative. We make a table and look for a pattern, as we did in Example 3.

EXAMPLE 4 Multiplying a negative times a negative Use the first three rows of Table 3 to find a pattern in the outputs. Then use the pattern to complete the table. What does the pattern say about a negative number times a negative number?

Input x	**Output** $-3x$
+2	$-3(2) =$ ____
+1	$-3(1) =$ ____
0	$-3(0) =$ ____
−1	$-3(-1) =$ ____
−2	$-3(-2) =$ ____
−3	$-3(-3) =$ ____

Table 3

Solution

$$-3(2) = -6$$
$$-3(1) = -3$$
$$-3(0) = 0$$

As the input decreases by 1, the output increases by 3. In order for this pattern to continue, the next three outputs must be 3, 6, and 9. Thus, the pattern indicates that

$$-3(-1) = 3$$
$$-3(-2) = 6$$
$$-3(-3) = 9$$

●

Check your results in Example 4 with a calculator.

Example 4 suggests the following rule.

Multiplying a Negative Integer and a Negative Integer

> A negative number multiplied by a negative number is a positive number.

EXAMPLE  Multiplying integers Multiply the integers in parts a–h. Then look for patterns in the answers and complete the sentences in parts i–l.

a. $3(-4)$ **b.** $-3(-4)$

c. $3(4)$ **d.** $-3(4)$

e. 3(0) **f.** 0(4)

g. 3(1) **h.** 1(4)

i. When the signs on two integers are alike, their product is ______.

j. When the signs on two integers are unlike, their product is ______.

k. When one of the numbers is zero, the product is ______.

l. When a number n is multiplied by 1, the product is ______.

Solution **a.** −12 **b.** 12 **c.** 12 **d.** −12

e. 0 **f.** 0 **g.** 3 **h.** 4

i. positive **j.** negative **k.** zero **l.** n ●

The multiplication rules are summarized in these four statements.

Multiplication

If two numbers have like signs, their product is positive.

$8(9) = 72 \qquad -6(-7) = 42$

If two numbers have unlike signs, their product is negative.

$8(-7) = -56 \qquad -9(6) = -54$

The product of zero and a number is zero.

$0 \cdot 4 = 0$

The product of 1 and a number n is n.

$1 \cdot 6 = 6$

EXAMPLE 6 **Multiplying integers** Find the products without a calculator.

a. −7(3) **b.** 4(−8) **c.** 7(8) **d.** 9(−6)(5)

e. −3(−4)(5) **f.** −8(−9)(−2) **g.** 8(0) **h.** −4(1)

Solution See the Answer Box. ●

Practice doing the multiplications in Example 6 with a calculator.

Division of Integers

RECIPROCAL OR MULTIPLICATIVE INVERSE The Warm-up had these exercises:

1. $-3 + \_\_\_ = 0$ **2.** $\_\_\_ + 5 = 1$ **3.** $\_\_\_ + (-2) = 0$

4. $4 + \_\_\_ = 1$ **5.** $\_\_\_ \cdot 3 = 1$ **6.** $6 + \_\_\_ = 0$

7. $5 \cdot \_\_\_ = 1$ **8.** $\frac{3}{4} \cdot \_\_\_ = 0$ **9.** $\frac{1}{2} \cdot \_\_\_ = 1$

The exercises that add to zero are based on adding opposites, also called additive inverses.

$$-3 + 3 = 0 \qquad 2 + -2 = 0 \qquad 6 + -6 = 0$$

The exercises that multiply to 1 are based on multiplying reciprocals, or multiplicative inverses.

$$\frac{1}{3} \cdot 3 = 1 \qquad 5 \cdot \frac{1}{5} = 1 \qquad \frac{1}{2} \cdot 2 = 1$$

Definition of Reciprocals

> The **reciprocal**, or **multiplicative inverse**, of a number n is *the number that, when multiplied by* n, *gives* 1. Thus, for all real numbers except zero,
>
> $$n \cdot \frac{1}{n} = 1$$

Because the product of a number and its reciprocal is positive one, the number and its reciprocal have the same sign; either both are positive or both are negative.

EXAMPLE 7 Finding a reciprocal Find the reciprocal of each number. Check by multiplying the number and its reciprocal.

a. 8 **b.** -6 **c.** $-\frac{2}{3}$ **d.** $1\frac{1}{4}$ **e.** 0.75

Solution

a. The reciprocal of 8 is $\frac{1}{8}$.
Check: $8 \cdot \frac{1}{8} = \frac{8}{8} = 1$ ✓

b. The reciprocal of -6 is $-\frac{1}{6}$.
Check: $-6 \cdot (-\frac{1}{6}) = +\frac{6}{6} = 1$ ✓

c. The reciprocal of $-\frac{2}{3}$ is $-\frac{3}{2}$ or $-1\frac{1}{2}$.
Check: $-\frac{2}{3} \cdot (-\frac{3}{2}) = +\frac{6}{6} = 1$ ✓

d. Because $1\frac{1}{4} = \frac{5}{4}$, its reciprocal is $\frac{4}{5}$.
Check: $\frac{5}{4} \cdot \frac{4}{5} = \frac{20}{20} = 1$ ✓

e. Because $0.75 = \frac{3}{4}$, its reciprocal is $\frac{4}{3}$ or $1\frac{1}{3}$.
Check: $\frac{3}{4} \cdot \frac{4}{3} = \frac{12}{12} = 1$ ✓ ●

The calculator reciprocal key is [1/x] or [x^{-1}]. Practice with this key by entering each number in Example 7 and finding its reciprocal.

CHANGING DIVISION TO MULTIPLICATION In Section 2.1, Example 12, page 57, we subtracted integers by adding the opposite: $a - b = a + (-b)$. In fraction notation, we divide by multiplying by the reciprocal:

$$\frac{a}{b} \div \frac{c}{d} = \frac{a}{b} \cdot \frac{d}{c}$$

We use the idea of inverses (opposites) to change subtraction to addition. Similarly, we use the idea of inverses (reciprocals) to change division to multiplication.

EXAMPLE 8 Dividing by integers Divide by multiplying by the reciprocal.

a. $24 \div 3$ **b.** $63 \div (-9)$

c. $-36 \div 6$ **d.** $-45 \div (-9)$

Solution

a. $24 \div 3 = 24 \cdot \frac{1}{3} = 8$

b. $63 \div (-9) = 63 \cdot (-\frac{1}{9}) = -7$
We are multiplying a positive number by a negative number. The signs are different, and the answer is negative.

c. $-36 \div 6 = -36 \cdot \frac{1}{6} = -6$
We are multiplying a negative number by a positive number. The signs are different, and the answer is negative.

d. $-45 \div (-9) = -45 \cdot (-\frac{1}{9}) = 5$
We are multiplying a negative number by a negative number. The signs are alike, and the answer is positive. ●

Because the negative sign does not change when we find the reciprocal of a negative number, the rules for multiplication of positive and negative numbers apply to division.

Multiplication and Division

In multiplication and division of two real numbers:

- If the signs are alike, the answer is positive.

$$\frac{21}{3} = 7 \qquad \frac{-14}{-2} = +7$$

- If the signs are different, the answer is negative.

$$\frac{28}{-4} = -7 \qquad \frac{-35}{7} = -5$$

SIGNS ON FRACTIONS In part b of Example 8, the reciprocal of -9 was written $-\frac{1}{9}$. It might have been more natural to write $\frac{1}{-9}$. The two fractions $-\frac{1}{9}$ and $\frac{1}{-9}$ are equivalent expressions; they have the same value.

$$-\frac{1}{9} = \frac{-1}{9} = \frac{1}{-9}$$

The placement of signs on fractions can be confusing. In general, we place the negative sign in front of the fraction in writing an answer. However, it is also correct—and more convenient when doing operations such as multiplication—to place the negative sign in the numerator. We rarely, if ever, leave the one negative sign in the denominator.

EXAMPLE 9 Placing signs on fractions Write each as a fraction in lowest terms.

a. $9 \div -45$ **b.** $-9 \div 45$ **c.** $-(9 \div 45)$

Solution **a.** $9 \div -45 = \dfrac{9}{-45} = -\dfrac{1}{5}$ **b.** $-9 \div 45 = \dfrac{-9}{45} = -\dfrac{1}{5}$

c. $-(9 \div 45) = \dfrac{-9}{45} = -\dfrac{1}{5}$ ●

If a fraction has one negative sign, the value of that fraction is the same regardless of where the negative sign appears.

Signs on Fractions

For all real numbers, b not zero,

$$-\frac{a}{b} = \frac{-a}{b} = \frac{a}{-b}$$

ANSWER BOX

Warm-up: 1. 3 **2.** -4 **3.** 2 **4.** -3 **5.** $\frac{1}{3}$ **6.** -6 **7.** $\frac{1}{5}$ **8.** 0 **9.** 2
Example 1: a. $+2(+35) = +70$ **b.** $+3(-50) = -150$
c. $-4(+30) = -120$ **d.** $-5(-8) = +40$ **Think about it: a.** positive
b. negative **c.** negative **d.** positive **Example 6: a.** -21 **b.** -32 **c.** 56
d. -270 **e.** 60 **f.** -144 **g.** 0 **h.** -4

EXERCISES 2.2

In Exercises 1 to 8, write an expression and find the change in net worth for your web site business.

1. The server takes away two sales for \$150 each.
2. The server brings four bills for \$20 each.
3. The server brings two sales for \$400 each.
4. The server takes away three bills for \$70 each.
5. The server takes away eight bills for \$40 each.
6. The server takes away two sales for \$300 each.
7. The server brings three bills for \$90 each.
8. The server brings four sales for \$125 each.

Complete the tables in Exercises 9 to 12. Then graph the ordered pairs.

9.

Input x	Output $3x$
2	
1	
0	
−1	
−2	

10.

Input x	Output $4x$
2	
1	
0	
−1	
−2	

11.

Input x	Output $-2x$
2	
1	
0	
−1	
−2	

12.

Input x	Output $-5x$
2	
1	
0	
−1	
−2	

Find the products in Exercises 13 to 20. Do not use a calculator.

13. a. $-7(-6)$ b. $6(-8)$ c. $3(-15)$ d. $-5(-7)$
14. a. $4(-7)$ b. $-8(-4)$ c. $-6(9)$ d. $-7(9)$
15. a. $12(-4)$ b. $8(-8)$ c. $15(4)$ d. $-9(-9)$
16. a. $5(-12)$ b. $-7(7)$ c. $-5(-15)$ d. $-11(11)$
17. a. $6(-8)(-1)$ b. $-5(5)(-1)$ c. $4(3)(-2)$
18. a. $5(4)(-3)$ b. $-9(-8)(-1)$ c. $-4(2)(-8)$
19. a. $-5(-6)(-7)$ b. $-6(9)(2)$ c. $-3(0)(-7)$
20. a. $-5(-8)(-8)$ b. $-2(-3)(0)$ c. $5(-2)(-9)$

In Exercises 21 to 24, write the reciprocal of each number or expression. Assume the variables do not equal zero.

21. a. 4 b. -2 c. $\frac{1}{2}$ d. $-\frac{3}{4}$ e. 0.5
22. a. -3 b. 6 c. $-\frac{1}{3}$ d. $\frac{2}{3}$ e. 0.25
23. a. $3\frac{1}{3}$ b. 6.5 c. x d. $\frac{a}{b}$ e. $-x$
24. a. $2\frac{3}{4}$ b. 8.2 c. n d. $\frac{x}{y}$ e. $-y$
25. Multiply these expressions.
    a. $-3(\frac{1}{3})$ b. $-3(-\frac{1}{3})$ c. $2(-\frac{1}{2})$ d. $-\frac{1}{4}(-4)$
26. Multiply these expressions.
    a. $-2(-\frac{1}{2})$ b. $3(-\frac{1}{3})$ c. $-4(\frac{1}{4})$ d. $\frac{1}{2}(-2)$

In Exercises 27 and 28, divide using multiplication by the reciprocal.

27. a. $15 \div (-5)$ b. $-45 \div 9$ c. $42 \div (-6)$
28. a. $-18 \div 6$ b. $-18 \div (-9)$ c. $-36 \div 2$

In Exercises 29 to 32, divide and then check with a calculator.

29. a. $\frac{56}{-8}$ b. $\frac{-56}{-8}$ c. $-\frac{56}{8}$
30. a. $\frac{49}{-7}$ b. $\frac{-72}{-8}$ c. $\frac{-72}{-6}$
31. a. $\frac{-55}{-11}$ b. $\frac{-28}{4}$ c. $\frac{-27}{9}$
32. a. $\frac{36}{-4}$ b. $\frac{-36}{12}$ c. $\frac{-48}{-24}$

Complete the tables in Exercises 33 and 34. Write a sentence about patterns you observe.

33.

a	b	$-\left(\frac{b}{a}\right)$	$\frac{-b}{a}$	$\frac{b}{-a}$
5	35			
−27	3			

34.

a	b	$-\left(\frac{b}{a}\right)$	$\frac{-b}{a}$	$\frac{b}{-a}$
−2	8			
−20	−4			

35. What two expressions are correctly described by "the opposite of b divided by a"?
36. Match the symbols $-\frac{x}{y}$, $\frac{-x}{y}$, and $\frac{x}{-y}$ with the word descriptions.
    a. The opposite of the quotient of x and y
    b. The quotient of x and the opposite of y
    c. The quotient of the opposite of x and y

In Exercises 37 and 38, make an input-output table for these rules, with the integers −3 to 3 as inputs.

37. If the input is zero or positive, the output equals the input. If the input is negative, the output is negative one times the input (a rule used in mathematics).

38. The output is negative one times the input if the input is even. The output equals the input if the input is odd.

Computation Review

In Exercises 39 and 40, do the divisions.

39. **a.** $15 \div \frac{1}{3}$ **b.** $\frac{2}{3} \div \frac{4}{5}$ **c.** $\frac{5}{8} \div \frac{2}{3}$ **d.** $\frac{3}{4} \div \frac{1}{4}$

40. **a.** $16 \div \frac{8}{5}$ **b.** $24 \div \frac{3}{4}$ **c.** $\frac{3}{4} \div \frac{6}{7}$ **d.** $\frac{2}{3} \div \frac{5}{6}$

In Exercises 41 and 42, complete the tables.

41.

x	y	$x \cdot y$	$x + y$
2	−2		
3		−6	
−4		−12	
	3		0
2			−1

42.

x	y	$x \cdot y$	$x + y$
−4		12	
		−16	0
		−15	2
−5			−8
−1		6	

43. Explain how to multiply two negative numbers.

44. Explain how to find the reciprocal of a number in decimal notation, such as 2.5.

45. One student says that the reciprocal of $\frac{2}{3}$ is $\frac{3}{2}$. Another student says that the reciprocal is $\frac{1}{\frac{2}{3}}$. Show that the answers are equivalent.

Projects

46. ***Greatest Product***

a. List six pairs of numbers that add to 13.

b. Find the product of each pair.

c. What is the greatest product of two whole numbers that add to 13?

d. What fractional or decimal values give a larger product than the whole numbers do?

e. Repeat parts a to d for numbers that add to 15.

f. Describe how the largest product is related to the sum of the numbers.

47. ***Stock Prices.*** Using the Internet or your library's collection of back issues of newspapers, obtain three weeks' closing prices for one share of a common stock of your choice.

a. Make a table with three headings: date, closing price, change.

b. Record the date and the closing price of the stock for 15 consecutive trading days.

c. In the third column, write the change in closing price since the preceding day.

d. Make a graph of the dates and closing prices.

e. Comment, in complete sentences, on any trend you observe.

2.3 Properties of Real Numbers, Simplifying Fractions, and Adding Like Terms

OBJECTIVES

- Identify terms and factors.
- Use the associative and commutative properties of real numbers to add and multiply.
- Use the distributive property to do multiplication over addition.
- Identify the greatest common factor and divide expressions.
- Simplify fractions containing variables.
- Add like terms.

WARM-UP

Describe ways to make these problems easy to work without a calculator. Then do the problems.

1. $60 + (-20) + 30 + 40 + 70 + 20$ **2.** $5 \cdot 27 \cdot 2$

3. $3\frac{1}{4} + 2\frac{1}{2} + 1\frac{3}{4}$ **4.** $\$4.75 + \$8.98 + \$6.25$

5. $0.04(4)(\$25)$ **6.** $7(\$19.98)$

THIS SECTION EMPHASIZES algebraic notation. We practice using positive and negative numbers in fraction and decimal notation, and we explore the associative, commutative, and distributive properties. We use the properties with algebraic expressions to simplify expressions in fractional notation and to add like terms.

Terms and Factors

A **term** is *a number, variable, or expression being added or subtracted.* In the expression $x + y$, the x and y are terms.

EXAMPLE 1 Counting terms How many terms are in each of these expressions?

a. $x + y - z$ **b.** xyz **c.** $-2ab$ **d.** $a - 2b$ **e.** $\dfrac{bh}{2}$

Solution **a.** 3 terms **b.** 1 term

The remaining answers are in the Answer Box. ●

Factors are *numbers, variables, or expressions being multiplied.* In the expression $x \cdot y$, the x and y are factors. The expression $x(y + z)$ has two factors: x and $(y + z)$. Placing parentheses around two (or more) terms creates a single factor. In the one-term expression $2\pi r$, the 2, π, and r are factors.

Because the number 1 is a factor in every expression, it is not counted in the number of factors.

EXAMPLE 2 Counting factors How many factors are in each of these expressions?

a. $(x + y - z)$ **b.** xyz **c.** $\dfrac{bh}{2}$

d. $-2ab$ **e.** $(x + y)(x - y)$ **f.** $\dfrac{h}{2}(a + b)$

Solution **a.** 1 factor **b.** 3 factors

c. 3 factors; because division by 2 is the same as multiplication by $\frac{1}{2}$, the three factors are $\frac{1}{2}$, b, and h.

The remaining answers are in the Answer Box. ●

Properties of Real Numbers

The properties of real numbers give us a powerful tool for creating shortcuts to solve problems. You may already use these shortcuts without knowing they are mathematical properties.

SHORTCUTS USING PROPERTIES OF REAL NUMBERS In the Warm-up, you were asked to describe ways to make the problems easy to work. Example 3 shows ways the Warm-ups might have been done.

EXAMPLE 3 Finding shortcuts in operations Add or multiply the following. Explain any shortcuts you use.

a. $60 + (-20) + 30 + 40 + 70 + 20$ **b.** $5 \cdot 27 \cdot 2$

c. $3\frac{1}{4} + 2\frac{1}{2} + 1\frac{3}{4}$ **d.** \$4.75 + \$8.98 + \$6.25

e. 0.04(4) (\$25) **f.** 7(\$19.98)

Solution Here are some shortcuts you may have used.

a. Change the order of the numbers so that pairs of numbers that add to 100 or add to zero are next to each other. Add these pairs first.

$$60 + (-20) + 30 + 40 + 70 + 20$$
$$= (60 + 40) + ((-20) + 20) + (30 + 70)$$
$$= 100 + 0 + 100 = 200$$

b. Change the order of the numbers so that pairs of numbers that multiply to 10 are next to each other. Multiply these pairs first.

$$5 \cdot 27 \cdot 2 = (5 \cdot 2) \cdot 27$$
$$= 10 \cdot 27 = 270$$

c. Consider the fractions separate from the whole numbers. First add the whole numbers. Next add the fractions that add to 1 (that is, $\frac{1}{4}$ and $\frac{3}{4}$). Add the $\frac{1}{2}$ last.

$$3\frac{1}{4} + 2\frac{1}{2} + 1\frac{3}{4} = (3 + 2 + 1) + \left(\frac{1}{4} + \frac{3}{4}\right) + \frac{1}{2}$$
$$= 6 + 1 + \frac{1}{2} = 7\frac{1}{2}$$

d. Consider the dollars separate from the cents; look for dollars that add to 10 dollars and for cents that add to 1 dollar. Add the 4, 6, and 8 in that order, then the 0.75 and 0.25, and finally the 0.98.

$$\$4.75 + \$8.98 + \$6.25$$
$$= (\$4 + \$6) + \$8 + (\$0.75 + \$0.25) + \$0.98$$
$$= \$10 + \$8 + \$1 + \$0.98$$
$$= \$19.98$$

e. Look for numbers that multiply to 100. Multiply the 4 and \$25 first.

$$0.04(4)(\$25) = 0.04(\$100) = \$4$$

f. \$19.98 is almost \$20. First estimate the total as 7 times \$20 = \$140. Then find the exact answer by subtracting 7 times \$0.02 from \$140 (see Figure 9).

$$7(\$19.98) = 7(\$20 - \$0.02) = \$140 - \$0.14 = \$139.86$$

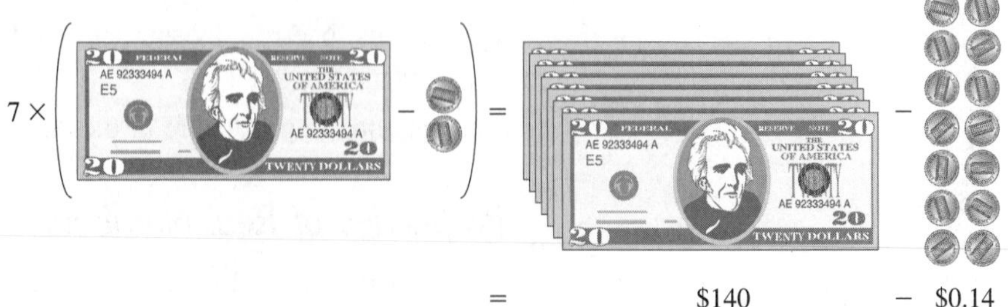

Figure 9

Think about it 1: If you used different shortcuts, how similar are they to the ones used here?

Example 3 illustrates several shortcuts:

- Change the order of the numbers to be added or multiplied.
- Find and add pairs of numbers whose sums are 0, 10, or 100.
- Find and multiply pairs of numbers whose products are 1, 10, or 100.
- Rewrite a number as the sum or difference of two numbers.

We now look at five real number properties that allow us to use these shortcuts.

ASSOCIATIVE PROPERTIES OF ADDITION AND MULTIPLICATION When we added $60 + 40 + (-20) + 20 + 30 + 70$, the **associative property**, which rules *the way expressions are grouped,* allowed us to add the -20 and 20 separately from the other numbers. The term *associative property* is based on the word *associate,* meaning "to select groups" or "to choose (business or social) connections" (see Figure 10).

Figure 10 Business associates associate

The Associative Properties of Addition and Multiplication

$a + (b + c) = (a + b) + c$ for all real numbers

$a \cdot (b \cdot c) = (a \cdot b) \cdot c$ for all real numbers

The associative property says we can group numbers in any way that is convenient to add or multiply the numbers.

EXAMPLE 4 Grouping numbers conveniently Using the associative property, group the numbers to take advantage of sums of 0 or 10 or products of 10 or 100. Insert parentheses to show the groups.

a. $8 + 2 + 4 + 6 + 7 + 9 + 1$

b. $4 + (-3) + 3 + 9 + (-9) + 2$

c. $35 \cdot 5 \cdot 2$

d. $7 \cdot (-2) \cdot (-50)$

Solution

a. $(8 + 2) + (4 + 6) + 7 + (9 + 1) = 10 + 10 + 7 + 10$
$= 37$

b. $4 + [(-3) + 3] + [9 + (-9)] + 2 = 4 + 0 + 0 + 2$
$= 6$

c. $35 \cdot (5 \cdot 2) = 35 \cdot 10 = 350$

d. $7 \cdot [(-2) \cdot (-50)] = 7 \cdot 100$
$= 700$

COMMUTATIVE PROPERTIES OF ADDITION AND MULTIPLICATION When we multiplied $5 \cdot 27 \cdot 2$, we changed the order of 2 and 27, to get $5 \cdot 2 \cdot 27$. This change, which made the problem easier, was permitted by the commutative property of multiplication. The **commutative property** refers to *the order in which actions or operations are done.* The word *commutative* comes from the word *commute,* which means "to change position or order" (see Figure 11).

Figure 11 Workers commute

The Commutative Properties of Addition and Multiplication

$a + b = b + a$ for all real numbers
$a \cdot b = b \cdot a$ for all real numbers

The commutative properties let us change the order of the numbers in addition and multiplication problems.

EXAMPLE 5 Changing number positions Show how a change of order with the commutative property makes it possible to do the problems mentally.

a. $5 \cdot 13 \cdot 4$ **b.** $-4 \cdot 3 \cdot (-25)$
c. $30 + 88 + 70$ **d.** $-6 + 5 + (-4)$

Solution In each problem, we change the order (commutative property) and then group two numbers (associative property) to multiply or add.

a. $5 \cdot 13 \cdot 4 = (5 \cdot 4) \cdot 13$
$= 20 \cdot 13 = 260$

b. $-4 \cdot 3 \cdot (-25) = (-4 \cdot (-25)) \cdot 3$
$= 100 \cdot 3 = 300$

c. $30 + 88 + 70 = (30 + 70) + 88$
$= 100 + 88 = 188$

d. $-6 + 5 + (-4) = (-6 + (-4)) + 5$
$= -10 + 5 = -5$ ●

Not all real-world and mathematical operations are associative and commutative. If you put on your socks and then your shoes, the outcome is entirely different than if you put on your shoes followed by your socks. Finding whether subtraction and division are commutative is left to the exercises.

DISTRIBUTIVE PROPERTY OF MULTIPLICATION OVER ADDITION When we multiplied 7($19.98), we changed $19.98 into the sum $20 + (−$0.02), or $20 − $0.02, and then multiplied 7($20 − $0.02). This change made use of the **distributive property of multiplication over addition,** which says that *multiplying a sum is equivalent to multiplying each term separately and then adding.*

Distributive Property of Multiplication Over Addition

> For all real numbers a, b, and c,
>
> $$a(b + c) = ab + ac$$

One example of the distributive property is $2(x + 5) = 2x + 10$. Other examples are given in Examples 6 to 9.

The name *distributive* may have been chosen because multiplying each term in the parentheses by a is like dealing cards to each person in a game or serving cake to each guest at a party. Both dealing and serving are *distributive* actions (see Figure 12). Caution: The multiplication sign is omitted in $a(b + c)$, as well as in ab and ac.

Figure 12 Card players distribute

SUBTRACTION WITHIN THE DISTRIBUTIVE PROPERTY Because we can write subtraction as addition of the opposite, we can distribute multiplication over subtraction:

$$a(b - c) = ab - ac$$

EXAMPLE 

Using the distributive property Multiply these expressions.

a. $5(x + 3)$ **b.** $0.27(x - 175)$

c. $-3(x - 4)$ **d.** $6(a + b - c)$

Solution In some problems we use the multiplication dot, and in others we use parentheses. Use whichever seems most natural.

a. $5 \cdot x + 5 \cdot 3 = 5x + 15$

b. $0.27(x) - 0.27(175) = 0.27x - 47.25$; parentheses are used here because the multiplication dot could be confused with the decimal point.

c. $-3(x) - 3(-4) = -3x - (-12) = -3x + 12$; parentheses are used here to help separate the numbers with negative signs.

d. $6 \cdot a + 6 \cdot b - 6 \cdot c = 6a + 6b - 6c$ ●

In part c of Example 6, we distributed a negative number. Look carefully at how this process changes the signs inside the parentheses: $-3(x - 4) = -3x + 12$. The signs of both terms in the parentheses change.

EXAMPLE 

Using the distributive property Rewrite the numbers so that the multiplication can be done mentally.

a. 5($4.03) **b.** 8($11.96)

Solution **a.** 5($4.00 + $0.03) = 5($4.00) + 5($0.03)
= $20.00 + $0.15 = $20.15

b. $8(\$12.00 - \$0.04) = 8(\$12.00) - 8(\$0.04)$
$= \$96.00 - \$0.32 = \$95.68$ ●

The distributive property also applies when the number being distributed does not show. We assume there to be a 1 in front of the parentheses in $-(x - 12)$. Thus,

$$-(x - 12) = -1(x - 12) = (-1)(x) - (-1)(12) = -x + 12$$

EXAMPLE 8 Using the distributive property to multiply by a negative Use the assumed 1 and the distributive property to do these multiplications.

a. $-(x + 4)$ **b.** $-(x - 3)$

Solution Start by placing a 1 between the negative sign and the parentheses.

a. $-1(x + 4) = (-1)(x) + (-1)(4) = -1x - 4 = -x - 4$

b. $-1(x - 3) = (-1)(x) - (-1)(3) = -1x + 3 = -x + 3$

Note that both signs inside the parentheses change when we multiply by a negative. ●

DIVISION WITHIN THE DISTRIBUTIVE PROPERTY Because we can write division as multiplication by the reciprocal, we can distribute division over addition or subtraction:

$$\frac{b+c}{a} = \frac{b}{a} + \frac{c}{a} \qquad \frac{b-c}{a} = \frac{b}{a} - \frac{c}{a}$$

EXAMPLE  Using the distributive property Divide these.

a. $\dfrac{24x + 16}{8}$ **b.** $\dfrac{3x - xy}{x}$

Solution **a.** $\dfrac{24x + 16}{8} = \dfrac{24x}{8} + \dfrac{16}{8}$ **b.** $\dfrac{3x - xy}{x} = \dfrac{3x}{x} - \dfrac{xy}{x}$ ●

Later, in Example 11, we will change the fractions in parts a and b to lowest terms.

SIMPLIFICATION PROPERTY OF FRACTIONS We say that a fraction is in **lowest terms** when *the numerator and denominator have no common factors*. The fraction $\frac{5}{7}$ is in lowest terms because there is no number other than 1 that divides evenly into both 5 and 7. To change a fraction to lowest terms, we use the fact that $a/a = 1$, stated in the simplification property of fractions.

Simplification Property of Fractions

> For all real numbers, a not zero and c not zero,
>
> $$\frac{ab}{ac} = \frac{a \cdot b}{a \cdot c} = \frac{a}{a} \cdot \frac{b}{c} = 1 \cdot \frac{b}{c} = \frac{b}{c}$$

The associative property allows us to regroup the a/a from ab/ac. The factor a is called the *common factor* because it is a factor that appears in both the numerator and the denominator. The **greatest common factor** is *the largest possible factor.*

In words, the **simplification property of fractions** says that *if the numerator and denominator of a fraction contain the same factor (a common factor), those factors can be eliminated.* In short, to change fractions to lowest terms, *eliminate common factors.* The fraction $\frac{6}{8}$ can be changed to lowest terms by eliminating the common factor of 2:

$$\frac{6}{8} = \frac{\boxed{2} \cdot 3}{\boxed{2} \cdot 4} = \frac{3}{4}$$

In the following examples, the word *simplify* will be used as a shorter way to write instructions. **Simplify** means *to use the simplification property of fractions to eliminate common factors as well as to do the given operations.*

EXAMPLE 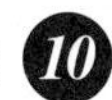 Simplifying fractions to lowest terms Identify the common factors in the numerator and denominator, and simplify to lowest terms.

a. $\frac{6x}{4x}$ **b.** $\frac{12ac}{15bc}$

Solution **a.** The common factors are 2 and x.

$$\frac{6x}{4x} = \frac{\boxed{2} \cdot 3 \cdot \boxed{x}}{\boxed{2} \cdot 2 \cdot \boxed{x}} = \frac{3}{2}$$

b. The common factors are 3 and c.

$$\frac{12ac}{15bc} = \frac{2 \cdot 2 \cdot \boxed{3} \cdot a \cdot \boxed{c}}{\boxed{3} \cdot 5 \cdot b \cdot \boxed{c}} = \frac{4a}{5b}$$

In Example 11, we simplify the fractions from Example 9.

EXAMPLE  Simplifying fractions Simplify these fractions by eliminating common factors.

a. $\frac{24x}{8} + \frac{16}{8}$ **b.** $\frac{3x}{x} - \frac{xy}{x}$

Solution **a.** $\frac{24x}{8} + \frac{16}{8} = \frac{3 \cdot \boxed{8}x}{\boxed{8}} + \frac{2 \cdot \boxed{8}}{\boxed{8}} = 3x + 2$

b. $\frac{3\boxed{x}}{\boxed{x}} - \frac{\boxed{x}y}{\boxed{x}} = 3 - y$

Dividing Expressions

> To divide $a \div b$:
>
> 1. Write the terms as a fraction, $\frac{a}{b}$.
> 2. Factor the numerator and denominator.
> 3. Simplify the fraction to lowest terms by eliminating common factors.
>
> To divide $(a + b) \div c$:
>
> 1. Write the expression as a fraction, $\frac{a + b}{c}$.
> 2. Change to $\frac{a}{c} + \frac{b}{c}$ with the distributive property.
> 3. Simplify each fraction to lowest terms.

Adding Like Terms

THE TILE MODEL Like terms are added using the commutative and distributive properties. We can model the addition of like terms with algebra tiles.

In Example 12, let a be the side of the large square, let b be the side of the small square, and let a and b be the sides of the rectangle.

EXAMPLE 12 Finding area Explain how to find the area of each tile in Figure 13.

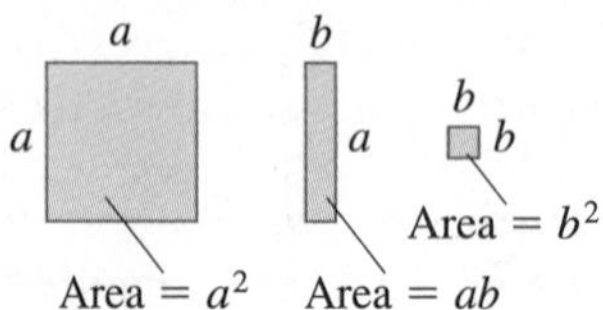

Figure 13

Solution The large square has area $a \cdot a = a^2$. The small square has area $b \cdot b = b^2$. The rectangle has area ab. ●

EXAMPLE 13 Adding area Using the dimensions given in Figure 13, write an expression for the total area in each set.

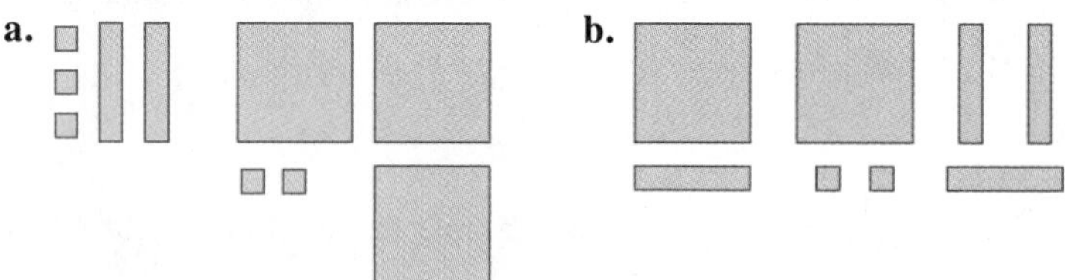

Solution **a.** $a^2 + a^2 + a^2 + ab + ab + b^2 + b^2 + b^2 + b^2 + b^2 =$
$3a^2 + 2ab + 5b^2$

b. $a^2 + a^2 + ab + ab + ab + ab + b^2 + b^2 =$
$2a^2 + 4ab + 2b^2$ ●

Student Note: Many examples in this book will be like Example 14. Each line of symbols is followed on the right by instructions on what to do to get to the next line. A good reading strategy is to cover up the next line with a piece of paper and use the directions to write the next line yourself. Move the paper down one line, and check your work. Repeat the process until you reach the end.

In writing the total area for each set of tiles in Example 13, it is sensible to count the a^2 terms together because they describe the same tile shape. The ab tiles have a different shape so they cannot be counted with either the a^2 or the b^2 tiles.

LIKE TERMS We say that the a^2 terms are *like terms.* The b^2 terms are *like terms,* and the ab terms are *like terms.* **Like terms** *have identical variable factors.* When we *count the like terms and write a single term to describe the total count,* we are **adding like terms**.

EXAMPLE 14 Adding like terms What is the sum of all 18 tiles in parts a and b of Example 13?

Solution

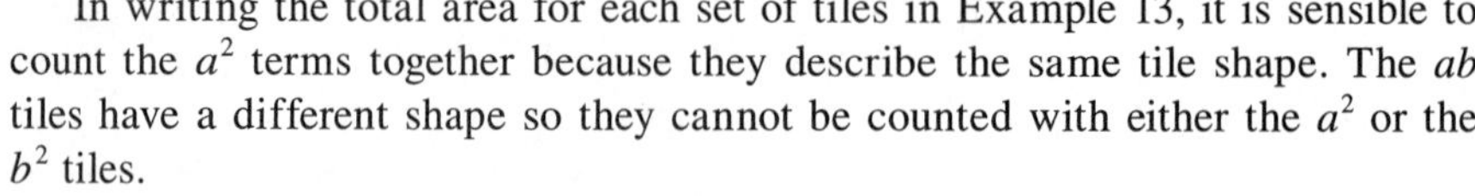

$3a^2 + 2ab + 5b^2 + 2a^2 + 4ab + 2b^2$ Arrange by like terms.
$= 3a^2 + 2a^2 + 2ab + 4ab + 5b^2 + 2b^2$ Add the like terms.
$= 5a^2 + 6ab + 7b^2$ ●

Think about it 2: How did the number properties help us to add the terms in Example 14?

In Example 15, we use new dimensions on the sides of the tiles. Let the side of the large square be x, and let the side of the small square be 1. The area of the large square is $x \cdot x = x^2$ square units, and the area of the small square is $1 \cdot 1 = 1$ square unit. The area of the rectangle is $x \cdot 1 = x$ square units. The areas are marked on the tiles.

EXAMPLE 15 Adding like terms Find the total area of each set of tiles by adding like terms.

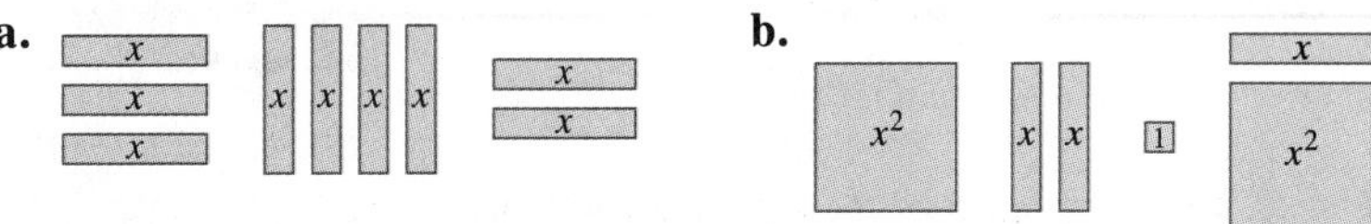

Solution **a.** The tiles are all like tiles: $3x + 4x + 2x = x(3 + 4 + 2) = 9x$

b. See the Answer Box.

The tiles suggest the following rule.

Adding Like Terms

> To add like terms, add the numerical coefficients of terms with identical variables and exponents.

EXAMPLE 16 Adding like terms Add the like terms, if any.

a. $4a^2 + 5a^2$ **b.** $6a^2 - 5a^2$ **c.** $4a^2 - 5a^2$

d. $4x + 4y$ **e.** $3x^2 + 15x^2 - 4x^2$ **f.** $7x + 7x^2$

g. $1 + 4x + 5x^2 + 2x - x^2 + 2$ **h.** $3ab + 4ba$

Solution **a.** $9a^2$ **b.** $1a^2$ or just a^2 **c.** $-1a^2$ or $-a^2$

d. No like terms; the expressions cannot be added. **e.** $14x^2$

f. No like terms; the expressions cannot be added.

g. $4x^2 + 6x + 3$ **h.** $ab = ba$, so the sum is $7ab$.

ANSWER BOX

Warm-up: 1. 200 **2.** 270 **3.** $7\frac{1}{2}$ **4.** \$19.98 **5.** 4 **6.** \$139.86
Example 1: c. 1 term **d.** 2 terms **e.** 1 term
Example 2: d. 3 factors: -2, a, and b **e.** 2 factors, each containing 2 terms **f.** 3 factors: $\frac{1}{2}$, h, and $(a + b)$ **Think about it 2:** The commutative property of addition let us change the order so that like terms were next to each other. The associative property of addition let us pair up the like terms. **Example 15: b.** $2x^2 + 3x + 1$

EXERCISES 2.3

1. How many terms are in each expression?

a. $2c + d$ **b.** $3x + 3y - 1$

c. xy **d.** $1 - 2a$

e. $-2x$ **f.** $x^2 - y^2$

2. How many terms are in each expression?

a. xxx **b.** $w + x + y + 2z$

c. $x^4 + 4x^3 + 6x^2 + 4x + 1$ **d.** $5y^5$

e. $ax^2 + bx + c$ **f.** $2L + 2W$

3. How many factors are in each expression?

a. $\frac{1}{2}xy$ **b.** $(x + 2)(x - 2)$

c. $2x(x + 3)$ **d.** $4abc$ **e.** $\dfrac{a + b + c}{3}$

4. How many factors are in each expression?

a. $4(a + b)(c + d)$ **b.** $x(x + y)$

c. $-0.5xy$ **d.** $ab(c + d)$ **e.** $\frac{1}{2}bh$

In Exercises 5 to 10, use shortcuts to find the sums and products.

5. $2\frac{1}{2} + 1\frac{1}{3} + 3\frac{2}{3} + 5\frac{1}{2}$

6. $6\frac{1}{4} + 3\frac{1}{2} + 5\frac{3}{4} + 2\frac{1}{2}$

7. $\frac{1}{4} \cdot 25 \cdot 8 \cdot 4$

8. $\frac{1}{2} \cdot 15 \cdot 4 \cdot 3$

9. $4.25 + 2.98 + 1.75$

10. $2.75 + 6.15 + 1.85$

For Exercises 11 to 18, find the value of each side. Do the operations in parentheses first. If the statement is true, write which property it uses.

11. $-3(4)(-5) = -3(-5)(4)$

12. $-3(4 + 5) = -3(4) - 3(5)$

13. $-3 + 4 + 5 = 4 + (-3) + 5$

14. $-3(4 \cdot 5) = (-3 \cdot 4) \cdot 5$

15. $3 + (4 + 5) = (3 + 4) + 5$

16. $3 + (4 \cdot 5) = (3 + 4) \cdot 5$

17. $3(4 \cdot 5) = (3 \cdot 4) + (3 \cdot 5)$

18. $3(4 \cdot 5) = 3 \cdot 4 \cdot 3 \cdot 5$

19. In parts a–d, what is the value of each expression if you do the operation in parentheses first? Are any of the expressions equal?

a. $8 - (5 - 3)$ **b.** $(8 - 5) - 3$

c. $16 \div (4 \div 2)$ **d.** $(16 \div 4) \div 2$

e. Is subtraction associative?

f. Is division associative?

20. In parts a–d, tell whether the statement is true or false.

a. $2 - 3 = 3 - 2$ **b.** $4 \div 5 = 5 \div 4$

c. $8 \div 2 = 2 \div 8$ **d.** $6 - 5 = 5 - 6$

e. Is subtraction commutative?

f. Is division commutative?

In Exercises 21 to 24, explain how you would find the products mentally, using the distributive property.

21. 4 times \$4.97 **22.** 7 times \$5.99

23. 3 times \$10.98 **24.** 6 times \$7.96

In Exercises 25 to 30, simplify with the distributive property.

25. **a.** $6(x + 2)$ **b.** $-3(x - 3)$ **c.** $-6(x + 4)$

26. **a.** $-(2 - x)$ **b.** $x(x - 3)$ **c.** $-(3 + x)$

27. **a.** $-3(x + y - 5)$ **b.** $-(x - y - z)$

28. **a.** $4(x - 4)$ **b.** $-4(x + 2)$ **c.** $-5(x - 3)$

29. **a.** $-(x - 3)$ **b.** $y(4 + y)$ **c.** $-(2 - y)$

30. **a.** $-5(2x - 4 + y)$ **b.** $-(x + y - z)$

Simplify the expressions in Exercises 31 to 36. Assume that no variable is zero.

31. **a.** $\dfrac{-2x}{-6x}$ **b.** $\dfrac{-14a}{21a}$ **c.** $\dfrac{6x}{15xyz}$

32. **a.** $\dfrac{2xy}{-10xyz}$ **b.** $\dfrac{-15ab}{9ac}$ **c.** $\dfrac{-6ac}{-27cx}$

33. **a.** $\dfrac{15xy}{21y}$ **b.** $\dfrac{39abc}{13acd}$ **c.** $\dfrac{-12xy}{48xz}$

34. **a.** $\dfrac{9ab}{45bc}$ **b.** $\dfrac{72xz}{18xy}$ **c.** $\dfrac{16ab}{48ac}$

35. **a.** $\dfrac{ab}{bc}$ **b.** $\dfrac{ab}{ac}$ **c.** $\dfrac{ay}{by}$

36. **a.** $\dfrac{ac}{bc}$ **b.** $\dfrac{xy}{xyz}$ **c.** $\dfrac{bx}{xy}$

In Exercises 37 and 38, divide the expressions. Simplify to lowest terms.

37. **a.** $\dfrac{3x + 4}{4}$ **b.** $\dfrac{4x + 8}{4}$

c. $\dfrac{x^2 + xy}{x}$ **d.** $\dfrac{ab - bc}{b}$

38. **a.** $\dfrac{2x + 6}{2}$ **b.** $\dfrac{5x + 2}{5}$

c. $\dfrac{x^2 + xy}{y}$ **d.** $\dfrac{ab - ac}{a}$

In Exercises 39 to 42, write an expression for the total area.

39.

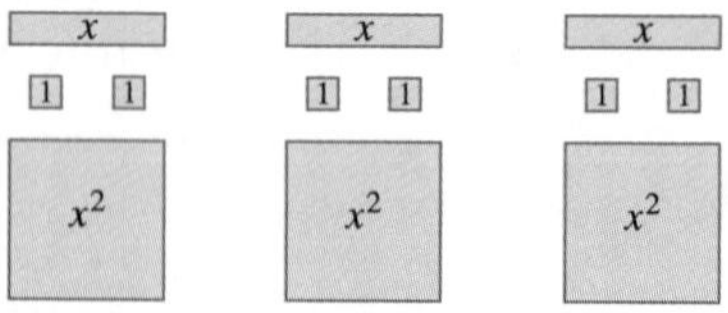

x 1 1 x^2 | x 1 1 x^2 | x 1 1 x^2

40.

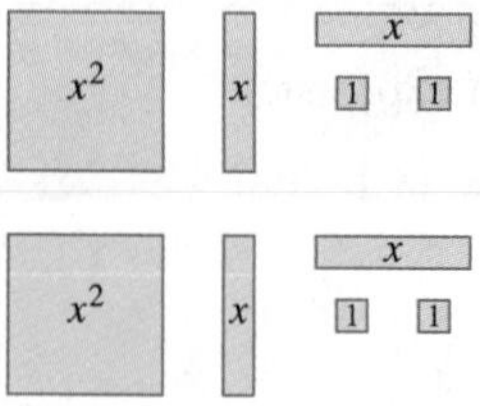

x^2 x x 1 1

x^2 x x 1 1

41.

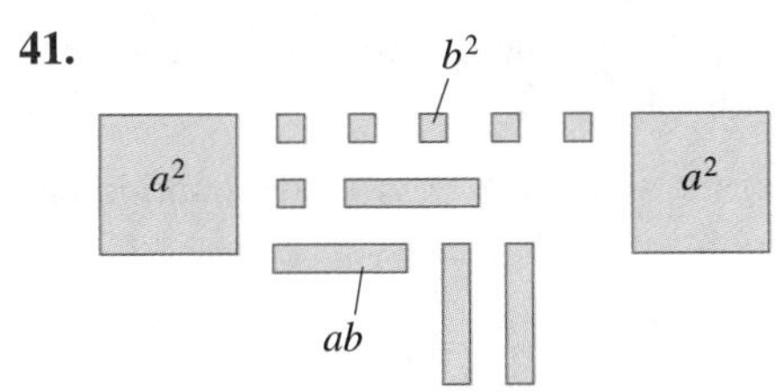

42.

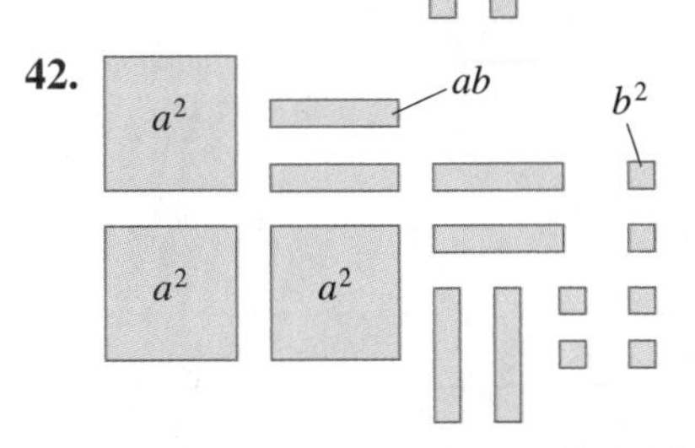

Identify the numerical coefficient in each of the terms in Exercises 43 and 44.

43. **a.** $-4x$ **b.** x **c.** $-x$

44. **a.** $-3x^2$ **b.** $5x$ **c.** y

In Exercises 45 to 48, add like terms. Multiply out parentheses first.

45. **a.** $-3x + 4x - 6x + 8x$

b. $-6y^2 + 8y^2 - 10y^2 - 3y^2$

c. $2a + 6 + 3a + 9$

d. $3(x + 1) - 2(x - 1)$

e. $3(x - 4) + 4(4 - x)$

f. $2x - 3y + 2x - 3y$

46. **a.** $-4y - 6y + 8y - 10y$

b. $+7x^2 - 12x^2 - 5x^2 + 8x^2$

c. $5x - 15 + 4x - 12$

d. $4(x + 1) - 3(x - 1)$

e. $7(x - 3) + 5(3 - x)$

f. $2x + 3y - 2x - 3y$

47. **a.** $\frac{1}{2}x + \frac{1}{4}y + \frac{1}{2}y - \frac{1}{4}x$

b. $0.5a + 0.75b - 0.5b + 1.5a$

c. $-2x + 3y + (-4x) + (-6x)$

d. $2(b + c) - 2(b - c)$

48. **a.** $\frac{1}{2}a - \frac{1}{4}b + \frac{1}{2}a - \frac{1}{2}b$

b. $0.25x + 0.5y + 0.5y - 0.75x$

c. $-4a + 8b - 6b + 9a$

d. $3(x - y) - 3(x + y)$

49. Write the commutative property of multiplication using variables x and y.

50. Write the associative property of addition using variables x, y, and z.

51. Explain how the distributive property changes a product into a sum.

52. Explain how to change a fraction to lowest terms.

53. How do you recognize like terms?

54. How do you find the greatest common factor?

55. How do you distinguish factors from terms?

56. Is $a(bc)$ an example of the distributive property? (*Hint:* Compare $4(5 \cdot 6)$ with $4(5) + 4(6)$ and $4(5) \cdot 4(6)$.)

57. In dividing $2x + 2y + 2$ by 2 a student writes $x + y$. Explain what is wrong, and provide the correct answer.

Prime numbers *have no integer factors except 1 and the number itself. The first prime number is 2.* **Composite numbers** *have factors other than themselves and 1. The number 1 is neither prime nor composite. In Exercises 58 and 59, find the prime factors for the composite numbers.*

58. **a.** 39 **b.** 28 **c.** 51 **d.** 61

59. **a.** 45 **b.** 59 **c.** 72 **d.** 111

60. Which of these activities are commutative? Give situations that might change your answers.

a. get dressed, eat breakfast

b. start car, fasten seatbelt

c. put key in ignition, start car

d. turn right, walk five steps forward

Projects

61. ***Percent and the Distributive Property.*** Writing expressions involving percent often requires adding like terms and applying the distributive property. In each set of expressions, find the one expression that does not reflect the situation described. Explain why it is different.

a. An 8% sales tax is added to the price, n:
$n + 0.08n$, $1n + 0.08n$,
a number plus 8% of the number, $n + 8\%$

b. A 7% sales tax is added to the price, p:
$1.00p + 0.07p$, $p(1 + 0.07)$,
$1.07p$, $1 + 0.07p$

c. The original price, x, is discounted by 5%:
$x - 0.05x$, $x - 5\%$,
$0.95x$, $1x - 0.05x$

d. The original price, n, is discounted by 20%:
the number minus 20% of the number,
$n - \frac{n}{5}$, $n - \frac{1}{5}$, $n(1 - \frac{1}{5})$

62. ***Distinguishing Terms from Factors***

a. Make a two-column list. Name one column *Expressions with Two or More Terms* and the other *Expressions with One Term and Multiple Factors.*

b. Record each of these expressions under the appropriate heading:

$12xy$	$3x + 2y - 7z$	$2x - y$	$2 + \sqrt{3}$
abc	$a + b + c$	$\dfrac{a + b}{c}$	$\dfrac{a}{c} + \dfrac{b}{c}$
$\dfrac{12xy}{4x^2}$	$\dfrac{15ab}{abc}$	$\dfrac{-b}{2a}$	$\frac{1}{2}h(a + b)$
$\dfrac{a}{b}$	$\frac{1}{2}ah + \frac{1}{2}bh$	$\dfrac{-b}{2a} + \dfrac{\sqrt{b^2 - 4ac}}{2a}$	

c. Circle the expressions that are equivalent, and connect them with a line.

d. Write each of the following after its description below: $a(b + c)$, $a + b$, $xy + wz$, $(x + y)(x - y)$

2 terms, each with 2 factors

2 factors, each with 2 terms

1 term containing 2 factors

2 terms, each with 1 factor

MID-CHAPTER 2 TEST

In Exercises 1 to 7, add, subtract, multiply, or divide, as indicated.

1. **a.** $2 - 5$ **b.** $-3 + 5$ **c.** $-3 - (-5)$

2. **a.** $3 - (-4)$ **b.** $-2 + (-5)$ **c.** $6 + (-2.5)$

3. **a.** $-5.50 + 18.98 - 12.76$

b. $-3.89 - 42.39 + 50.00$

4. **a.** $-3(2)$ **b.** $-5(-3)$ **c.** $-(-4)$

5. **a.** $\dfrac{27}{-3}$ **b.** $\dfrac{-28}{-2}$ **c.** $\dfrac{-32}{4}$

6. **a.** $4x + 5y - 2x + y$ **b.** $2x^2 - 3x + 2x(1 - x)$

7. **a.** $\dfrac{4xyz}{xy}$ **b.** $\dfrac{3x}{xyz}$ **c.** $\dfrac{-2y}{4xy}$

8. In Exercises a and b, state the subtraction problem illustrated by the charge model and then work the problem.

a.

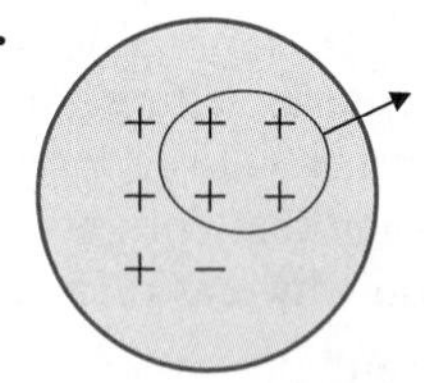

___ − ___ = ___

b.

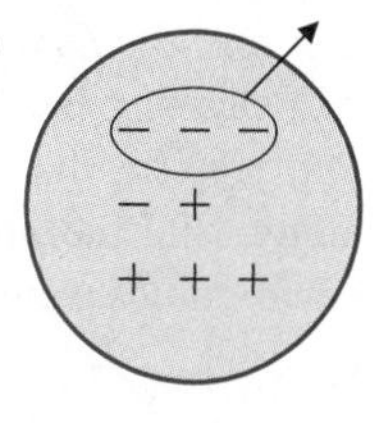

___ − ___ = ___

Simplify Exercises 9 to 11.

9. $-3 + 6 + (-24) + 27$ **10.** $-6 + 12 + 18 - 24$

11. $13(-4)5(-3)$

For Exercises 12 to 15, make input-output tables, using integers from −2 to 3 for x. Graph the (x, y) pairs on coordinate axes.

12. $y = x - 1$ **13.** $y = x + 1$

14. $y = 2x - 3$ **15.** $y = 3 - x$

In Exercises 16 and 17, find the difference in elevation between the highest and lowest point on each continent.

16. Asia: Mt. Everest, 8850 meters, and the Dead Sea, −400 meters

17. North America: Mt. McKinley, 6194 meters, and Death Valley, −86 meters

18. Mt. Everest is recognized as the highest mountain in the world. Mauna Kea is the inactive volcano on the island of Hawaii. Mauna Kea's volcanic base actually rises from the ocean floor. Using the data in the following table on elevation in feet relative to sea level, find which mountain is actually "taller" and by how much.

Mt. Everest	+29,028
Mauna Kea	+13,710
Sea Level	0
Ocean Floor, near Hawaii	−16,400

Divide the expressions in Exercises 19 and 20.

19. $\dfrac{4x + 2y}{2}$ **20.** $\dfrac{ab + bc}{b}$

2.4 Exponents and Order of Operations

OBJECTIVES

- Identify the base and exponent of an expression.
- Write expressions with and without exponents.
- Simplify expressions containing exponents.
- Identify grouping symbols.
- Simplify expressions with the order of operations.

> WARM-UP
> Simplify.
> **1.** $5 \cdot 5 \cdot 5$ **2.** $(-2)(-2)(-2)(-2)$ **3.** $(-2) \cdot 2 \cdot 2 \cdot 2$

IN THIS SECTION, we examine properties of expressions with exponents and agreements about simplifying expressions containing two or more operations.

Bases and Exponents

Let's return to the problem posed at the beginning of the chapter, on page 48.

EXAMPLE  Exploration of a bouncing ball pattern A ball is dropped from a height of 144 centimeters. After each time it hits the floor, the ball bounces $\frac{2}{3}$ as high as it did on the last bounce. What is the height of the ball after it hits the floor ten times?

Solution Use your calculator, and record your work on paper. Start with an input-output table. Let the input be the number of times the ball has hit the floor, with zero as the starting number. Let the output be the height of the ball, with 144 centimeters as the starting height.

Number of times ball has hit floor	Height (cm)
0	144
1	$144(\frac{2}{3}) = 96$
2	$144(\frac{2}{3})(\frac{2}{3}) = 64$

Table 4

Table 4 shows the first three entries; complete the table for inputs 3 to 10. See the Answer Box for outputs. *Hint:* After the sixth time the ball hits the floor, the ball's height is

$$144(\tfrac{2}{3})(\tfrac{2}{3})(\tfrac{2}{3})(\tfrac{2}{3})(\tfrac{2}{3})(\tfrac{2}{3}) \approx 12.6 \text{ cm}$$ ●

IDENTIFYING BASES AND EXPONENTS An exponent provides a shorter way to write an expression for the height of the ball.

$$144(\tfrac{2}{3})(\tfrac{2}{3})(\tfrac{2}{3})(\tfrac{2}{3})(\tfrac{2}{3})(\tfrac{2}{3}) = 144(\tfrac{2}{3})^6$$

The $\frac{2}{3}$ is the base. The **base** is *the number repeated in the multiplication. The small raised number to the right of the base* is an **exponent**. The exponent tells us the number of times the base is repeated. *The base and exponent together* are called a **power**.

> In the power expression x^n, the *positive integer exponent* n indicates the number of factors of the base x. That is,
>
> $$x^n = x \cdot x \cdot x \cdot x \cdot \cdots \cdot x$$
>
> with n factors of x.

EXAMPLE 2 Finding bases and applying exponents Identify each base; then write the expression as factors and multiply.

a. 5^3 **b.** $(-2)^4$ **c.** -2^4

Solution **a.** The base is 5; $5^3 = 5 \cdot 5 \cdot 5 = 125$.

b. The base is -2; $(-2)^4 = (-2)(-2)(-2)(-2) = 16$.

c. The base is 2, not -2; $-(2^4) = -(2 \cdot 2 \cdot 2 \cdot 2) = -16$. ●

Mathematicians have agreed to place bases with negative signs within parentheses, as in part b of Example 2. Thus, in part c, the negative sign on -2^4 is not part of the base. The expression -2^4 means *the opposite of* 2^4. The base 2 is raised to the exponent 4 first, and then the negative sign is applied: $-2^4 = -(2^4)$, *not* $(-2)^4$.

Negative Signs and Bases

Place a negative base in parentheses when applying an exponent to it:

$$(-4)^2 = (-4)(-4) = 16$$

A negative sign in front of the base means "opposite of," not a negative base:

$$-4^2 = \text{opposite of } 4^2 = -(4 \cdot 4) = -16$$

EXAMPLE 3 Finding bases and applying exponents Identify each base; then write the expression as factors and multiply.

a. $144(\frac{2}{3})^{10}$ **b.** $(-3)^4$ **c.** -3^4 **d.** $(-6)^3(\frac{1}{2})^2$

Solution **a.** The base is $\frac{2}{3}$; 144 is the numerical coefficient, which is not repeated and is not part of the base:

$$144(\tfrac{2}{3})^{10} = 144(\tfrac{2}{3})(\tfrac{2}{3})(\tfrac{2}{3})(\tfrac{2}{3})(\tfrac{2}{3})(\tfrac{2}{3})(\tfrac{2}{3})(\tfrac{2}{3})(\tfrac{2}{3})(\tfrac{2}{3}) \approx 2.5$$

b. The base is -3; $(-3)^4 = (-3)(-3)(-3)(-3) = 81$.

c. The base is 3, not -3; $-(3^4) = -(3 \cdot 3 \cdot 3 \cdot 3) = -81$.

d. The bases are -6 and $\frac{1}{2}$; $(-6)(-6)(-6)(\frac{1}{2})(\frac{1}{2}) = -\frac{216}{4} = -54$. ●

Squares and Cubes

When *2 is used as an exponent,* we say the base is **squared**. When *3 is an exponent,* the base is **cubed**. These words come from the measurement of area and volume, to be discussed in Section 2.5.

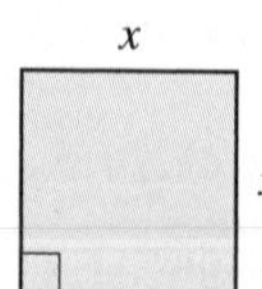

Square with side x

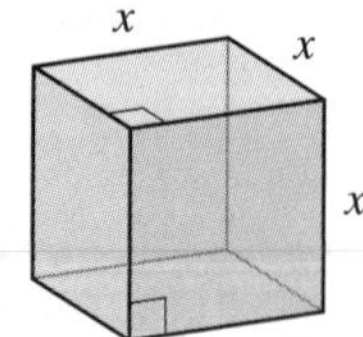

Cube with side x

THE BASE PROPERTY When we write

$$\left(\frac{2}{3}\right)^2 = \left(\frac{2}{3}\right) \cdot \left(\frac{2}{3}\right) = \left(\frac{2^2}{3^2}\right) = \frac{4}{9}$$

we are using our first property of exponents and bases:

Base Property

> An exponent outside the parentheses applies to all parts of a product or quotient inside the parentheses:
>
> $$(x \cdot y)^a = x^a \cdot y^a \qquad \left(\frac{x}{y}\right)^a = \frac{x^a}{y^a}$$

We use the name *base property* because we repeat the base a times, which creates a factors of each part of the base.

EXAMPLE 4 Using the base property of exponents Write these powers containing squares and cubes without parentheses.

a. $\left(\frac{2}{3}\right)^3$ **b.** $(-3x)^2$ **c.** $\left(\frac{x}{4}\right)^2$ **d.** $2(3a)^3$

Solution

a. $\left(\frac{2}{3}\right)^3 = \frac{2^3}{3^3} = \frac{8}{27}$

b. $(-3x)^2 = (-3)^2 \cdot x^2 = 9x^2$

c. $\left(\frac{x}{4}\right)^2 = \frac{x^2}{4^2} = \frac{x^2}{16}$

d. $2(3a)^3 = 2 \cdot 3^3 \cdot a^3 = 54a^3$ ●

In Example 4, we used the base property to simplify the expressions. We get the same results if we use the definition of exponents instead of the base property. For example, in part d,

$$2(3a)^3 = 2(3a)(3a)(3a) = 2 \cdot 3 \cdot 3 \cdot 3 \cdot a \cdot a \cdot a = 54a^3$$

MULTIPLICATION AND DIVISION OF LIKE BASES Look for patterns in the multiplications and divisions in the next two examples, to see whether you can predict the rules.

EXAMPLE 5 Finding patterns in expressions with exponents Write these expressions with a single base and exponent:

a. $x^4 \cdot x^3$ **b.** $x^1 \cdot x^6$

Solution **a.** $x^4 \cdot x^3 = xxxx \cdot xxx = x^7$ **b.** $x^1 \cdot x^6 = x \cdot xxxxxx = x^7$ ●

Think about it 1: What other exponent expressions would make x^7? What happens to the exponents when we multiply?

EXAMPLE 6 Simplifying expressions Write these expressions with a single base and exponent. (*Hint:* Use the simplification property of fractions.)

a. $\frac{x^5}{x^3}$ **b.** $\frac{x^7}{x^5}$ **c.** $\frac{x^4}{x^2}$

Solution **a.** $\frac{x^5}{x^3} = \frac{xxxxx}{xxx} = \frac{x}{x} \cdot \frac{x}{x} \cdot \frac{x}{x} \cdot x \cdot x = 1 \cdot 1 \cdot 1 \cdot x \cdot x = x^2$

b. $\frac{x^7}{x^5} = \frac{xxxxxxx}{xxxxx} = \frac{x}{x} \cdot \frac{x}{x} \cdot \frac{x}{x} \cdot \frac{x}{x} \cdot \frac{x}{x} \cdot x \cdot x = 1 \cdot 1 \cdot 1 \cdot 1 \cdot 1 \cdot x \cdot x = x^2$

c. $\frac{x^4}{x^2} = \frac{xxxx}{xx} = \frac{x}{x} \cdot \frac{x}{x} \cdot x \cdot x = 1 \cdot 1 \cdot x \cdot x = x^2$ ●

Think about it 2: What other exponent expressions would make x^2? What happens to the exponents when we divide?

Expressions have *like bases* when their bases are the same. Examples 5 and 6 suggest these properties:

Multiplication and Division Properties with Like Bases

> To multiply expressions with like bases, keep the base and add the exponents:
>
> $x^a \cdot x^b = x^{a+b}$
>
> To divide expressions with like bases, keep the base and subtract the exponents:
>
> $\frac{x^a}{x^b} = x^{a-b}$

THE POWER PROPERTY In Example 7, we explore what happens when the base already has an exponent.

EXAMPLE 7 Simplifying expressions using the definition of exponents Use the definition of positive integer exponents to write each of the expressions without parentheses.

a. $(x^3)^2$ **b.** $(2x^2)^3$ **c.** $(0.5y^4)^2$

Solution **a.** $(x^3)^2 = (x^3)(x^3) = x^{3+3} = x^6$

b. $(2x^2)^3 = (2x^2)(2x^2)(2x^2) = 2 \cdot 2 \cdot 2 \cdot x^{2+2+2} = 8x^6$

c. $(0.5y^4)^2 = (0.5y^4)(0.5y^4) = (0.5)(0.5)y^{4+4} = 0.25y^8$ ●

Example 7 shows that a base x^a in $(x^a)^b$ is multiplied as a factor b times, giving us the power property.

Power Property

> To apply an exponent to a power expression, multiply exponents:
>
> $(x^a)^b = x^{a \cdot b}$

In Example 8, we extend our definition of **simplify** to mean *combine expressions with like bases and use the properties of exponents to do the indicated operations.*

EXAMPLE **8** Simplifying expressions Simplify.

a. $\frac{a^8}{a^5}$ **b.** a^6a^7 **c.** $(b^6)^2$ **d.** x^4y^3 **e.** $(a^2b^3)^3$ **f.** $\frac{x^2y^3}{xy}$

g. $\left(\frac{a^2b^3}{ab}\right)^3$ **h.** $\left(\frac{-3x^5y^6}{xy^2}\right)^2$

Solution

a. $\frac{a^8}{a^5} = a^{8-5} = a^3$ **b.** $a^6a^7 = a^{6+7} = a^{13}$ **c.** $(b^6)^2 = b^{6\cdot 2} = b^{12}$

d. The bases are not alike, so the expression cannot be simplified.

e. $(a^2b^3)^3 = a^{2\cdot 3}b^{3\cdot 3} = a^6b^9$

f. $\frac{x^2y^3}{xy} = x^{2-1}y^{3-1} = x^1y^2 = xy^2$

Note that a base with no exponent shown has exponent 1.

g. $\left(\frac{a^2b^3}{ab}\right)^3 = (a^{2-1}b^{3-1})^3 = (ab^2)^3 = a^{1\cdot 3}b^{2\cdot 3} = a^3b^6$

We assume that we simplify inside the parentheses first.

h. $\left(\frac{-3x^5y^6}{xy^2}\right)^2 = (-3x^{5-1}y^{6-2})^2 = (-3x^4y^4)^2 = 9x^8y^8$ ●

Although applying the properties is generally more convenient, it is always possible to use the definition of exponents to simplify expressions. For example, in part g of Example 8,

$$\left(\frac{a^2b^3}{ab}\right)^3 = \left(\frac{aabbb}{ab}\right)^3 \qquad \frac{a}{a} = 1, \frac{b}{b} = 1.$$

$$= (ab^2)^3 \qquad \text{Write the base 3 times.}$$

$$= (ab^2)(ab^2)(ab^2) \qquad \text{Multiply.}$$

$$= a^3b^6$$

We now summarize the operations with exponents.

Operations with Exponents

Base Property. An exponent outside the parentheses applies to all parts of a product or quotient inside the parentheses:

$$(x \cdot y)^a = x^a \cdot y^a \qquad \left(\frac{x}{y}\right)^a = \frac{x^a}{y^a}$$

Multiplication Property. To multiply expressions with like bases, keep the base and add the exponents:

$$x^a \cdot x^b = x^{a+b}$$

Division Property. To divide expressions with like bases, keep the base and subtract the exponents:

$$\frac{x^a}{x^b} = x^{a-b}$$

Power Property. To apply an exponent to a power expression, multiply exponents:

$$(x^a)^b = x^{a\cdot b}$$

Order of Operations

Simplifying expressions with more than one operation leads us to ask the question "How do we know which operation to do first, second, and so on?" The answer is "Use the order of operations."

The **order of operations** is *an agreed-upon order in which we do mathematics so that everyone obtains the same answers.* Note the word *agreement*—this is an order mathematicians have adopted, not a property. As Example 9 shows, this order is not even programmed into all calculators.

EXAMPLE 9 Trying different orders of operations Find the value of $1 - 9 + 9 \cdot 9$ in the following ways:

a. By calculating from left to right

b. With a four-function, credit-card-size, or business calculator

c. With a scientific calculator

d. By doing the multiplication first and then the subtraction and addition, in order from left to right.

Solution **a.** $1 - 9 + 9 \cdot 9 = 9$. When we calculate from left to right, $1 - 9 = -8$. The -8 and $+9$ add to 1, leaving $1 \cdot 9 = 9$.

b. Generally, the order of operations for a four-function calculator is the order in which the numbers and operations are entered. If we enter $1 - 9 + 9 \cdot 9$, we obtain 9, as in part a.

c. With a scientific calculator, we enter $1 - 9 + 9 \cdot 9$ to obtain 73.

d.

$1 - 9 + 9 \cdot 9$	Do the multiplication first.
$1 - 9 + 81$	Do the subtraction next.
$-8 + 81$	Do the addition last.
73	

The correct answer in algebra is the answer found with the scientific calculator, which follows the order described in part d. This order is the one that has been agreed upon in the order of operations.

The Order of Operations

1. Calculate expressions within parentheses and other grouping symbols.
2. Calculate exponents and square roots.
3. Do the remaining multiplication and division in the order of appearance, left to right.
4. Do the remaining addition and subtraction in the order of appearance, left to right.

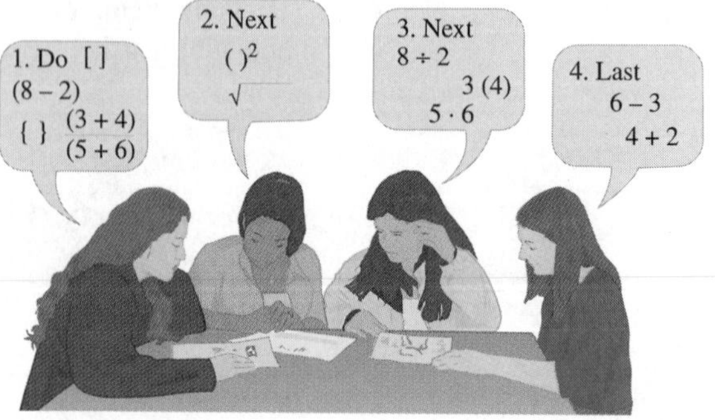

EXAMPLE 10 Using the order of operations with exponents Which step in the order of operations explains why these expressions are equal?

a. $144\left(\frac{2}{3}\right)^{10} = 144 \cdot \frac{2^{10}}{3^{10}}$ **b.** $\left(\frac{a^2b^3}{ab}\right)^3 = (ab^2)^3$

Solution **a.** We apply the exponent before doing the multiplication between the 144 and the expression in parentheses.

b. We simplify the expression in the parentheses before applying the outside exponent.

In Example 11, we use the order of operations to find the value of expressions that are used in formulas elsewhere in this book. Again we expand the meaning of the word *simplify*. Here, **simplify** means *to use the order of operations, to calculate the value of an expression, to apply the properties of exponents, to add like terms, to change fractions to lowest terms, or to use the number properties* (from Section 2.3).

EXAMPLE 11 Using the order of operations to simplify expressions Simplify these expressions, commonly used for the indicated applications.

a. Surface area of a box: $2 \cdot 5 \cdot 6 + 2 \cdot 5 \cdot 8 + 2 \cdot 6 \cdot 8$

b. Area of a trapezoid: $\frac{1}{2} \cdot 7(8 + 10)$

c. Surface area of a cylinder: $2(3.14) \cdot 3 \cdot 10 + 2(3.14) \cdot 3^2$

d. Height of an object: $-\frac{1}{2}(32)(3)^2 + 64(3) + 50$

e. General term of a sequence: $9 - 4(n - 1)$

Solution **a.**

$2 \cdot 5 \cdot 6 + 2 \cdot 5 \cdot 8 + 2 \cdot 6 \cdot 8$	Do multiplication left to right.
$= 60 + 80 + 96$	Do addition left to right.
$= 236$	

b.

$\frac{1}{2} \cdot 7(8 + 10)$	Do operations in parentheses.
$= \frac{1}{2} \cdot 7 \cdot 18$	Do multiplication left to right.
$= 63$	

The commutative property of multiplication (and other properties from Section 2.3) may be applied before any step in the order of operations. Reversing the 7 and 18 allows us to do the problem mentally: $\frac{1}{2} \cdot 18 \cdot 7 = 9 \cdot 7 = 63$.

c.

$2(3.14) \cdot 3 \cdot 10 + 2(3.14) \cdot 3^2$	Apply the exponent first.
$= 2(3.14) \cdot 3 \cdot 10 + 2(3.14) \cdot 9$	Do multiplication left to right.
$= 188.4 + 56.52$	Do addition.
$= 244.92$	

d.

$-\frac{1}{2}(32)(3)^2 + 64(3) + 50$	Apply the exponent first.
$= -\frac{1}{2}(32) \cdot 9 + 64(3) + 50$	Do multiplication left to right.
$= -144 + 192 + 50$	Do addition left to right.
$= 98$	

e.

$9 - 4(n - 1)$	The expression in the parentheses cannot be simplified. The multiplication with the distributive property must be done before the subtraction.
$= 9 - 4n + 4$	Add like terms.
$= 13 - 4n$	

Caution: In part f of Example 11, a common error is to subtract 9 − 4 first. Look for problems like this one in the exercises for this section and again in Chapter 3.

Grouping Symbols

When parentheses are needed within parentheses, we sometimes use different symbols, such as brackets or braces. *The preferred way to write multiple grouping symbols is* {[()]}, with *parentheses* on the inside, the square-shaped *brackets* used next, and *braces* (like little wires) used outermost. If the context is not confusing, double parentheses, (( )), are acceptable.

EXAMPLE 12 Using grouping symbols Simplify the following expressions. Work the innermost parentheses, ( ), first; then the brackets, [ ]; and finally the braces, { }.

a. $\{-1 + [(9 - 9) + 8]\}^2$

b. $-1 + [9 - (9 + 8)]$

Solution **a.** $\{-1 + [(9 - 9) + 8]\}^2 = \{-1 + [0 + 8]\}^2 = \{7\}^2 = 49$

b. $-1 + [9 - (9 + 8)] = -1 + [9 - 17] = -1 + [-8] = -9$ ●

Parentheses, brackets, and braces are just one type of **grouping symbol**, *used to place terms together.* Other grouping symbols include the absolute value symbol, the square root symbol, and the horizontal fraction bar.

The *absolute value* symbol may act as a grouping symbol if it contains an expression, such as $|1 - 9|$. First calculate the expression inside the absolute value, and then find the absolute value.

The *square root* symbol, or **radical** sign, is a grouping symbol if it contains an expression rather than a single number, as in $\sqrt{8 + 1}$. The expression inside should be calculated before taking the square root. In writing an expression under a radical sign, it is important to draw the overbar over the whole expression.

The *horizontal fraction bar* acts as a grouping symbol. The numerator and denominator of the fraction are calculated separately. (See part a in Example 13.)

EXAMPLE 13 Using special grouping symbols and the order of operations Simplify these expressions.

a. Slope of a line: $\dfrac{-3 - 6}{10 - 7}$

b. Distance formula: $\sqrt{(10 - (-2))^2 + (-3 - 6)^2}$

c. Mean absolute deviation: $\dfrac{|1.3 - 1.5| + |1.4 - 1.5| + |1.8 - 1.5|}{3}$

d. Part of the quadratic formula: $\dfrac{-1 + \sqrt{49}}{2}$

Solution

a. $\dfrac{-3 - 6}{10 - 7}$ Do numerators and denominators first.

$= \dfrac{-9}{3}$ Divide.

$= -3$

b. $\sqrt{(10 - (-2))^2 + (-3 - 6)^2}$ Do parenthetical expressions.

$= \sqrt{12^2 + (-9)^2}$ Apply exponents.

$= \sqrt{144 + 81}$ Add the expression under the radical sign.

$= \sqrt{225}$ Take the square root.

$= 15$

c. $\dfrac{|1.3 - 1.5| + |1.4 - 1.5| + |1.8 - 1.5|}{3}$ Do subtractions inside absolute value.

$= \dfrac{|-0.2| + |-0.1| + |0.3|}{3}$ Find the absolute value.

$= \dfrac{0.2 + 0.1 + 0.3}{3}$ Add the numerator.

$= \dfrac{0.6}{3}$ Divide.

$= 0.2$

d. $\dfrac{-1 + \sqrt{49}}{2}$ Find the square root.

$= \dfrac{-1 + 7}{2}$ Add the numerator.

$= \dfrac{6}{2}$ Divide.

$= 3$

The order of operations is often memorized by means of an acronym such as PEMDAS (parentheses, exponents, multiplication, division, addition, subtraction) or a saying such as "Please Excuse My Dear Aunt Sally." Neither of these stresses the left-to-right order, though. Ask your instructor for other favorite aids to memorizing the order of operations.

When using a calculator, place parentheses around the expression inside a square root sign and around numerator and denominator expressions in a fraction. The square root expression $\sqrt{}(5^2 + 12^2)$ is written as it should be entered into the calculator, as is the expression $(-4 + 2) \div (3 - 5)$ for the fraction $\dfrac{-4 + 2}{3 - 5}$.

Most calculators have a squaring key, $\boxed{x^2}$, but squaring may also be done with the exponent key, $\boxed{y^x}$, $\boxed{x^y}$, or $\boxed{\wedge}$.

ANSWER BOX

Warm-up: 1. 125 **2.** 16 **3.** -16 **Example 1:** The heights, rounded to the nearest tenth, are $144(\frac{2}{3})^3 \approx 42.7$, $144(\frac{2}{3})^4 \approx 28.4$, 19.0, 12.6, 8.4, 5.6, 3.7, and $144(\frac{2}{3})^{10} \approx 2.5$. **Think about it 1:** x^2x^5 or any other expression in which the exponents add to 7; when we multiply expressions with like bases, we add the exponents. **Think about it 2:** x^6/x^4 or any other expression in which the exponents subtract to 2; when we divide expressions with like bases, we subtract the exponents.

EXERCISES 2.4

In Exercises 1 and 2, identify the base for the exponent 2, write the expression using factors, and write the expression in words.

1. a. $3x^2$ **b.** $-3x^2$ **c.** $(-3x)^2$ **d.** ax^2 **e.** $-x^2$ **f.** $(-x)^2$

2. a. $-b^2$ **b.** $(-b)^2$ **c.** ab^2 **d.** mn^2 **e.** $(ab)^2$ **f.** $-2x^2$

Simplify the expressions in Exercises 3 to 8.

3. a. 3^5 **b.** 2^6 **c.** $(-2)^2$ **d.** $(-3)^3$

4. a. $(\frac{2}{3})^2$ **b.** $(-\frac{1}{2})^2$ **c.** $(-\frac{1}{3})^3$ **d.** $-(\frac{1}{2})^2$

5. a. $(\frac{1}{3})^3$ **b.** $(\frac{4}{5})^3$ **c.** $(-\frac{2}{3})^3$ **d.** $-(-\frac{1}{3})^2$

6. a. 4^4 **b.** 5^3 **c.** $(-2)^3$ **d.** $(-3)^2$

7. a. $2 \cdot 4^2$ **b.** -2^2 **c.** $3(-2)^2$ **d.** $-4 \cdot 3^2$

8. a. $3 \cdot 5^2$ **b.** -4^2 **c.** $-3 \cdot 3^3$ **d.** $4(-3)^2$

Simplify Exercises 9 and 10. Check with a calculator.

9. a. $2^2 2^3$ **b.** $(3^4)^2$ **c.** $4^3 4^2$ **d.** $\dfrac{3^7}{3^2}$

10. a. $3^4 3^5$ **b.** $(3^2)^3$ **c.** $2^5 2^3$ **d.** $\dfrac{4^9}{4^3}$

Simplify the expressions in Exercises 11 to 14. Assume no variable is zero.

11. a. $m^3 m^5$ **b.** $n^4 n^4$ **c.** $a^6 a^2$ **d.** $a^7 a^1$

12. a. $m^3 m^3$ **b.** $m^1 m^5$ **c.** $b^2 b^4$ **d.** $n^2 n^3$

13. a. $\dfrac{x^5}{x^2}$ **b.** $\dfrac{a^8}{a^5}$ **c.** $\left(\dfrac{x}{y}\right)^2$ **d.** $\left(\dfrac{2x}{y}\right)^3$

14. a. $\dfrac{x^6}{x^3}$ **b.** $\dfrac{b^8}{b^4}$ **c.** $\left(\dfrac{x}{3y}\right)^2$ **d.** $\left(\dfrac{2a}{3b}\right)^3$

15. What exponent n makes $(\frac{2}{5})^n = \frac{8}{125}$?

16. What exponents n make $(-1)^n$ a positive number?

17. What pairs of exponents make $x^{\square}x^{\circ} = x^{12}$?

18. What pairs of exponents make $x^{\square}x^{\circ} = x^{10}$?

Simplify the expressions in Exercises 19 to 28. Assume no variable is zero.

19. a. $(x^2)^3$ **b.** $(xy)^2$ **c.** $(x^2y^3)^2$

20. a. $(a^3)^2$ **b.** $(a^2b)^2$ **c.** $(a^3b^2)^3$

21. a. $(2a)^3(2a)^6$ **b.** $(2a)^2(3b)^3$ **c.** $4(3a)^3(-1)^1$

22. a. $(2b)^2(2b)^4$ **b.** $(3b)^3(2b)^2$ **c.** $6(3a)^2(-1)^2$

23. a. $\dfrac{a}{a}$ **b.** $\dfrac{-4ab}{6a}$ **c.** $\dfrac{a^3b^2}{ab}$ **d.** $\dfrac{-4x^4y^3}{-10x^2y}$

24. a. $\dfrac{xy}{x}$ **b.** $\dfrac{-6b}{9b}$ **c.** $\dfrac{x^5y^2}{xy^2}$ **d.** $\dfrac{-2a^7b^3}{-8ab^2}$

25. a. $(x^2)^2$ **b.** $(2x^2)^3$ **c.** $(-\frac{1}{4}a^2)^2$

26. a. $5(x^2)^4(2y^2)^1$ **b.** $10(x^2)^2(-2y^2)^3$ **c.** $\left(\dfrac{a^2b}{a}\right)^2$

27. a. $(y^2)^3$ **b.** $(3y^3)^2$ **c.** $(\frac{1}{3}y^3)^2$

28. a. $10(x^2)^3(2y^2)^2$ **b.** $5(x^2)(2y^2)^4$ **c.** $\left(\dfrac{xy^2}{y}\right)^3$

Simplify the expressions in Exercises 29 to 60 without a calculator.

29. a. $1 + 9 \cdot 9 - 8$ **b.** $1 + \sqrt{9} \cdot 9 - 8$

c. $(1 + \sqrt{9}) \cdot 9 - 8$

30. a. $1 - \sqrt{9} \cdot (9 - 8)$ **b.** $1 + 9 \cdot (\sqrt{9} - 8)$

c. $(1 + 9) \cdot \sqrt{9} - 8$

31. $2 \cdot 3 \cdot 4 + 2 \cdot 4 \cdot 5 + 2 \cdot 3 \cdot 5$

32. $2 \cdot 2 \cdot 3 + 2 \cdot 3 \cdot 6 + 2 \cdot 2 \cdot 6$

33. $\frac{1}{2} \cdot 5 \cdot (6 + 8)$ **34.** $\frac{1}{2} \cdot 15 \cdot (4 + 6)$

35. $6^2 + 8^2$ **36.** $9^2 + 12^2$

37. $5^2 + 12^2$ **38.** $8^2 + 15^2$

39. $7 - 2(4 - 1)$ **40.** $5 - 2(6 + 2)$

41. $(7 - 2)(4 - 1)$ **42.** $(5 - 2)(6 + 2)$

43. $6 - 3(x - 4)$ **44.** $7 - 5(4 - x)$

45. $(6 - 3)(x - 4)$ **46.** $(7 - 5)(4 - x)$

47. $3[8 - 2(3 - 5)]$ **48.** $4[9 - 6(5 - 3)]$

49. $|4 - 6| + |6 - 4|$ **50.** $|3 - 7| - |7 - 3|$

51. $|5 - 2| + |2 - 5|$ **52.** $|3 - 8| - |8 - 5|$

53. a. $\dfrac{4 - 7}{2 - (-3)}$ **b.** $\dfrac{-2 - 3}{-2 - (-3)}$ **c.** $\dfrac{5 - 2}{3 - (-4)}$

54. a. $\dfrac{2 - 5}{-3 - 4}$ **b.** $\dfrac{7 - 2}{6 - (-3)}$ **c.** $\dfrac{-7 - 2}{-6 - 3}$

55. a. $\dfrac{3 + \sqrt{25}}{4}$ **b.** $\dfrac{9 - \sqrt{36}}{2}$ **c.** $\dfrac{12 + \sqrt{64}}{5}$

56. a. $\dfrac{5 + \sqrt{81}}{2}$ **b.** $\dfrac{14 - \sqrt{16}}{5}$ **c.** $\dfrac{15 + \sqrt{121}}{4}$

57. $\sqrt{(-6 - 6)^2 + (12 - 3)^2}$

58. $\sqrt{(1 - 4)^2 + (3 - 7)^2}$

59. $\sqrt{(4 - (-2))^2 + (4 - (-4))^2}$

60. $\sqrt{(3 - (-9))^2 + (-3 - 2)^2}$

Simplify the expressions in Exercises 61 to 70. Use a calculator as needed. Round to the nearest hundredth.

61. $\dfrac{|3 - 7.5| + |9 - 7.5| + |10 - 7.5|}{3}$

62. $\dfrac{|2 - 2.75| + |2.25 - 2.75| + |4 - 2.75|}{3}$

63. $3.14(2)^2 \cdot 6$ **64.** $3.14(4)^2 \cdot 6$

65. $\frac{4}{3}(3.14) \cdot 2^3$ **66.** $\frac{4}{3}(3.14) \cdot 4^3$

67. $2(3.14) \cdot 3 \cdot 4 + 2(3.14)(3^2)$

68. $2(3.14) \cdot 6 \cdot 4 + 2(3.14)(6^2)$

69. $-\frac{1}{2}(32)(-1)^2 + 16(-1) + 50$

70. $-\frac{1}{2}(32)(-2)^2 + 16(-2) + 50$

71. What could it mean to *simplify* an expression?

72. If n is a positive integer, what does x^n mean?

73. The property $\frac{x^a}{x^b} = x^{a-b}$ can be written "When dividing expressions with like bases, keep the base and subtract the exponents." Write in words the following properties.

a. $x^a \cdot x^b = x^{a+b}$

b. $(x^a)^b = x^{ab}$

c. $(xy)^a = x^a y^a$

d. $\left(\frac{x}{y}\right)^a = \frac{x^a}{y^a}$

74. Write the rule for the order of operations.

Error Analysis

75. Which student has the correct answer? Write a sentence or two to explain to the other students what they did wrong.

Student #1 has $(3x^3)^2 = 9x^9$.
Student #2 has $(3x^3)^2 = 6x^6$.
Student #3 has $(3x^3)^2 = 9x^6$.

76. Which student has the correct answer? Write a sentence or two to explain to the other students what they did wrong.

Student #1 has $(-0.2x^3)^2 = 0.4x^9$.
Student #2 has $(-0.2x^3)^2 = -0.4x^6$.
Student #3 has $(-0.2x^3)^2 = 0.04x^9$.
Student #4 has $(-0.2x^3)^2 = 0.04x^6$.

77. a. Check these problems. One is not correct. Change one operation sign to make it true.

$1 \cdot 9 + 9 - 8 = 10$
$1 - 9 + 9 + 8 = 9$
$1 \cdot [9 - (9 - 8)] = 8$
$-1 + 9 - 9 - 8 = 7$
$-1 - 9 \div 9 + 8 = 6$

b. Check these problems. One is not correct. Change one operation sign to make it true.

$-1 + (9 + 9) \div 9 = 1$
$1 \cdot (9 + 9) \div 9 = 2$
$1 + (9 + 9) \div 9 = 3$
$1 \cdot (9 \div 9) + \sqrt{9} = 4$
$1 + (9 \div 9) - \sqrt{9} = 5$

Explain the errors made in Exercises 78 and 79.

78. Simplifying $7 - 4(x - 3)$ as $3(x - 3)$

79. Simplifying $5 - 3(x - 1)$ as $5 - 3x - 3$

Problem Solving

In Exercises 80 to 83, begin by building a table.

80. A bouncing ball bounces half the previous height with each bounce. What is the height of the fifth bounce if the ball starts from 96 centimeters?

81. A basketball tournament starts with 64 teams. The teams play on Wednesdays, Saturdays, and Sundays. How many game days are needed to identify the two teams for the final game if play starts on Wednesday and half the teams are eliminated each day? On which game day will the final game be played?

82. A bacteria population doubles every 2 hours. If there are 1,000,000 bacteria to start with, how many will there be in 24 hours?

83. Suppose a gambling machine returns an average of 75% of the coins placed in it. In the first round, someone starts with $20 in nickels, plays all the money, and sets aside the money returned. In each of the following rounds, the person plays all the remaining money and sets aside the money returned. How many rounds will the person play before all the money will be gone? Assume there will be no fractions of coins.

Projects

84. ***Bouncing Balls.*** The rules of tennis require that a tennis ball bounce between 4 feet 5 inches and 4 feet 10 inches when dropped 8 feet 4 inches onto a concrete base.

a. What decimal compares a bounce of 4 feet 5 inches to the original height of 8 feet 4 inches? What percent of the original height is the bounce?

b. What decimal compares a bounce of 4 feet 10 inches to the original height of 8 feet 4 inches? What percent of the original height is the bounce?

c. Why might the height 8 feet 4 inches have been chosen as the drop height?

d. Suppose that on the first bounce the ball reaches the larger height (4 feet 10 inches). If the ball is permitted to continue bouncing, what height, in inches, will it reach on the second bounce? the third bounce? the sixth bounce? Round answers to the nearest tenth of an inch.

Using any ball and a yardstick or meterstick, plan an experiment to find the height of the ball after one bounce, the height of the ball after three bounces, and the percent rebound. A 36-inch or 1-meter starting height is recommended. State any assumptions.

e. Gather data for both the first and the third bounce. Repeat five times. Average the heights for the first bounce. Average the heights for the third bounce.

f. Calculate what percent of the original height the first bounce is. Use this percent to predict the height of the third bounce. Compare your data with your prediction.

g. Use your results to predict the height for the sixth bounce.

h. Summarize your findings.

85. ***Birth Year.*** Use the four digits in the year of your birth, together with the four basic operations and the square root, to find 15 whole-number values between 1 and 25, as was done in Exercise 77. Show all parentheses needed to create the values.

86. ***Exponent Puzzle.*** The variables A, B, C, and D in the equation $A^B C^D = ABCD$ represent numbers from the set $\{0, 1, 2, 3, 4, 5, 6, 7, 8, 9\}$. The expression $ABCD$ in the equation represents a number with four digits, not a product of the numbers A and B and C and D. Use a calculator to guess the numbers. (For example: $7^3 \cdot 8^3$ gives the six-digit answer 175,616, not 7383.) Digits may be used twice.

87. ***Words and Symbols.*** Change the word phrases to symbols and the symbols to word phrases. Tell which expressions have the same value.

a. The sum of 3 and 4 is divided by the difference between 5 and 8.

b. Three plus the quotient of 4 and 5 is decreased by 8.

c. The sum of 3 and 4 is divided by 5 and decreased by 8.

d. Three is added to the quotient of 4 and the difference between 5 and 8.

e. $3 + 4 \div 5 - 8$

f. $(3 + 4) \div 5 - 8$

g. $3 + 4 \div (5 - 8)$

h. $(3 + 4) \div (5 - 8)$

i. $3 + \dfrac{4}{5 - 8}$

j. $\dfrac{3 + 4}{5 - 8}$

k. $3 + \dfrac{4}{5} - 8$

l. $\dfrac{3 + 4}{5} - 8$

88. ***Exponents That Are Not Positive Integers***

a. Find the pattern, and complete the blanks in both directions:

$\{\_\_, \_\_, \_\_, 2, 4, 8, 16, \_\_, \_\_\}$

$\{\_\_, \_\_, \_\_, 2^1, 2^2, 2^3, 2^4, \_\_, \_\_\}$

$\{\_\_, \_\_, \_\_, 10, 100, 1000, 10000, \_\_, \_\_\}$

$\{\_\_, \_\_, \_\_, 10^1, 10^2, 10^3, 10^4, \_\_, \_\_\}$

$\{\_\_, \_\_, \_\_, 3, 9, 27, \_\_, \_\_\}$

$\{\_\_, \_\_, \_\_, 3^1, 3^2, 3^3, \_\_, \_\_\}$

b. Use your calculator to explore the effect of 0 as an exponent. Look for a pattern. Look for an exception.

115^0 $\quad 4^0$ $\quad 1^0$

0^0 $\quad (-1)^0$ $\quad (-2)^0$

2.5^0 $\quad \pi^0$ $\quad 3^0$

Put negative numbers, such as -2, into the calculator in parentheses.

Which number above produced an error message when entered into the calculator?

Complete the statement: $x^0 = \_\_$ for all numbers x except $\_\_$.

c. Use your calculator to explore the effect of -1 as an exponent. The calculator will give decimals. Write the decimals as fractions in lowest terms to see the effect of a -1 exponent more clearly.

5^{-1} $\quad (-4)^{-1}$ $\quad 2^{-1}$

$(-25)^{-1}$ $\quad 20^{-1}$ $\quad (-50)^{-1}$

$(\frac{1}{2})^{-1}$ $\quad (\frac{1}{4})^{-1}$ $\quad (-\frac{1}{5})^{-1}$

Describe the pattern in a sentence or two.

d. Use your calculator to explore the meaning of -2 as an exponent.

$(\frac{1}{2})^{-2}$ $\quad 2^{-2}$ $\quad 3^{-2}$

$(\frac{1}{3})^{-2}$ $\quad (-\frac{1}{4})^{-2}$ $\quad (\frac{1}{5})^{-2}$

Write a sentence or two to describe the pattern.

e. Investigate x^{-3}.

f. In general, what does a negative exponent do?

g. We are not limited to integers as exponents. Explore the answers to these problems:

$1^{0.5}$ $\quad 2^{0.5}$ $\quad 3^{0.5}$

$4^{0.5}$ $\quad 9^{0.5}$ $\quad 16^{0.5}$

$49^{0.5}$ $\quad 6.25^{0.5}$ $\quad 0.01^{0.5}$

Describe the pattern that emerges. What is the meaning of 0.5, or 1/2, as an exponent? We return to the 0.5 exponent in Section 8.2.

2.5 Unit Analysis and Formulas

OBJECTIVES

- Change from one unit of measure to another using unit analysis.
- Use formulas to calculate perimeter, area, surface area, and volume.

WARM-UP

Use the simplification property of fractions on both the numbers and the variables; then multiply these fractions.

1. $\frac{1{,}000{,}000m}{1} \cdot \frac{h}{60m} \cdot \frac{d}{24h} \cdot \frac{y}{365d}$
2. $\frac{1000L}{1} \cdot \frac{1.0567q}{L} \cdot \frac{g}{4q}$
3. $\frac{f^3}{1} \cdot \frac{12i}{f} \cdot \frac{12i}{f} \cdot \frac{12i}{f}$
4. $\frac{c}{1} \cdot \frac{q}{4c} \cdot \frac{g}{4q} \cdot \frac{231i^3}{g} \cdot \frac{1}{0.0625i}$

IN THIS SECTION, we examine unit analysis, a process that helps you work with units of measure, organize your thinking, and present your answers clearly. The second half of the section provides practice using a variety of geometric formulas. Unit analysis and order of operations are especially important in formula work.

Using Unit Analysis

Unit analysis is *a method for changing from one unit of measure to another or from one rate to another.* Unit analysis is a clear step-by-step process that helps us solve complicated problems quickly.

Unit analysis is a powerful tool for critical thinking. News broadcasts, magazines, and newspapers are filled with "facts" containing numbers and units of measure. Many of these "facts" are misleading. Unit analysis lets us examine the facts and decide whether they are reasonable.

Technology, science, and engineering students use unit analysis to get correct answers without having to memorize processes and formulas. You may already use unit analysis when you do mental math.

Unit analysis is presented in two parts. In this section, we change from one unit of measure to another. Later, in Chapter 5, Section 5.1, we will change from one rate to another.

In unit analysis, we use $\frac{a}{a} = 1$ to eliminate unwanted units, writing facts such as 1 minute = 60 seconds as $\frac{1 \text{ minute}}{60 \text{ seconds}} = 1$. We can also use the fact that $a = \frac{a}{1}$ to write facts as fractions and eliminate units.

EXAMPLE 1 Eliminating like units Simplify these expressions.

a. $\frac{\text{liters} \cdot \text{quarts} \cdot \text{gallons}}{\text{liters} \cdot \text{quarts}}$ **b.** $\frac{\text{inches}^3}{\text{inches}}$

Solution **a.** $\frac{\text{liters} \cdot \text{quarts} \cdot \text{gallons}}{\text{liters} \cdot \text{quarts}} = \text{gallons}$

b. $\frac{\text{inches}^3}{\text{inches}} = \frac{\text{inches} \cdot \text{inches} \cdot \text{inches}}{\text{inches}} = \text{inches}^2$ ●

Example 2 shows how unit analysis describes the thinking many of us do without realizing it. We will organize our solution with the four problem-solving steps: understand, plan, do, and check.

EXAMPLE 2 Using equivalent units in unit analysis Change 1,000,000 minutes into years.

Solution ***Understand:*** We start with minutes and want to end with years, so we list the units-of-measure facts that move step by step from minutes to years:

60 minutes = 1 hour

24 hours = 1 day

365 days ≈ 1 year (an assumption)

Plan: Starting with minutes, we arrange the facts as fractions so that units (minutes, hours, and days) are eliminated and only years remain.

Do:

$$\frac{1{,}000{,}000 \text{ minutes}}{1} \cdot \frac{1 \text{ hour}}{60 \text{ minutes}} \cdot \frac{1 \text{ day}}{24 \text{ hours}} \cdot \frac{1 \text{ year}}{365 \text{ days}}$$

$$= \frac{1{,}000{,}000}{60 \cdot 24 \cdot 365} \text{ years}$$

$$\approx 1.9 \text{ years}$$

Check: We make sure that the facts are correctly written, all units except years are eliminated, and the answer is reasonable. ●

Think about it 1: Adding parentheses may be important if you do Example 2 on a calculator. Try these different sequences of keystrokes and compare answers. Which is correct?

a. 1 000 000 [÷] 60 [×] 24 [×] 365 [=]

b. 1 000 000 [÷] 60 [÷] 24 [÷] 365 [=]

c. 1 000 000 [÷] [(] 60 [×] 24 [×] 365 [)] [=]

A unit analysis shows that you have all the facts you need, that every unit is accounted for, and that you have multiplied or divided appropriately.

In the next few examples and exercises, you will be given the facts. Later, you will need to make your own list of facts. Go back to Example 2, and look through the facts carefully. Is there a clear path in units of time between minutes and years? Trace a path from the starting unit to the ending unit. Some students like to use a pencil to connect the units in their list of facts before changing the list into a product of fractions.

The following examples are intended to show the variety of ways we can apply unit analysis, changing systems of measure (metric to American standard), using facts more than once, and deriving surprising information from the facts.

EXAMPLE 3 Changing systems of measure How many gallons are in the 1000-liter container shown in Figure 14? How much water is 1000 liters—enough for a lemonade glass, a child's plastic wading pool, or a swimming pool?

Figure 14

Solution ***Understand:*** We list facts, starting with liters and ending with gallons.

$$1 \text{ liter} = 1.0567 \text{ quarts}$$
$$4 \text{ quarts} = 1 \text{ gallon}$$

Plan: Starting with liters, we arrange the facts so that liters and quarts are eliminated.

Do:
$$\frac{1000 \text{ liters}}{1} \cdot \frac{1.0567 \text{ quarts}}{1 \text{ liter}} \cdot \frac{1 \text{ gallon}}{4 \text{ quarts}}$$
$$= \frac{1000(1.0567)}{4} \text{ gallons}$$
$$\approx 264 \text{ gallons}$$

Check: Many sodas are sold in 2-liter plastic bottles. The 1000-liter container would fill 500 plastic bottles or a child's large wading pool. ●

The next example uses foot and cubic foot, or $(\text{foot})^3$. If there is more than one foot in a problem, we write *feet*. You may use $\frac{\text{foot}}{\text{feet}} = 1$, but remember that $\frac{\text{foot}}{(\text{foot})^3} = \frac{1}{(\text{foot})^2}$.

EXAMPLE 4 Using facts more than once How many cubic inches are there in 1 cubic foot?

Facts: 1 foot = 12 inches

Solution ***Understand:*** The term *cubic foot* means $(\text{foot})^3$.

Plan: Our fact contains only the foot, not the cubic foot. To obtain the cubic foot, or $(\text{foot})^3$, we need to use foot as a factor three times.

Do:
$$\frac{1 \text{ cubic foot}}{1} \cdot \frac{12 \text{ inches}}{1 \text{ foot}} \cdot \frac{12 \text{ inches}}{1 \text{ foot}} \cdot \frac{12 \text{ inches}}{1 \text{ foot}}$$
$$= \frac{12^3 \text{ inches}^3}{1}$$
$$= 1728 \text{ cubic inches}$$

Check: Our work contains foot · foot · foot and inches · inches · inches. The number part of the answer, 12^3, has 3 factors, like the units. ●

Example 5 is a setting familiar to parents.

EXAMPLE 5 Confirming expectations Suppose a glass contains 1 cup of milk. How many square inches will the milk cover when spilled and spread to a depth of $\frac{1}{16}$ inch?

Facts: 1 gallon = 231 cubic inches
4 quarts = 1 gallon
1 quart = 4 cups

Solution ***Understand:*** We want to start with the 1 cup of milk and change it into cubic inches. We will then need to get square inches, as requested.

Plan: First we look for a "path" through the facts from cups to cubic inches. The next step is to decide what to do with $\frac{1}{16}$ inch. To obtain square inches, or $(\text{inches})^2$, we need to place $\frac{1}{16}$ inch in the denominator and let the inch eliminate one of the inches from cubic inches.

Do:

$$\frac{1 \text{ cup}}{1 \text{ spill}} \cdot \frac{1 \text{ quart}}{4 \text{ cups}} \cdot \frac{1 \text{ gallon}}{4 \text{ quarts}} \cdot \frac{231 \text{ cubic inches}}{1 \text{ gallon}} \cdot \frac{1}{\frac{1}{16} \text{ inch}}$$

$$= \frac{231}{4 \cdot 4 \cdot \frac{1}{16}} \frac{\text{square inches}}{1 \text{ spill}}$$

$$= 231 \text{ square inches per spill}$$

Note that $4 \cdot 4 \cdot \frac{1}{16} = 1$.

Check: All units except square inches per spill are eliminated. ●

Think about it 2: If we had placed $\frac{1}{16}$ inch in the numerator, what units would we have had?

Unit Analysis: Step by Step

1. Identify the unit of measure to be changed, and identify the unit of measure for the answer. Write both as fractions.
2. List facts containing the unit to be changed and all units needed to get to the unit for the answer.
3. Write the unit to be changed (or a fact containing the unit) as the first fraction. Set up a product of fractions, using the list of facts. Arrange fractions so that the units to be eliminated appear in both a numerator and a denominator.
4. Eliminate units of measure where possible, and calculate the numbers.

Using Geometric Formulas

A **formula** is *a rule or principle written in mathematical language.* In geometric formulas, we use units of measure.

The following examples involve the formulas for perimeter, area, volume, and surface area. We begin with a definition of perimeter and area and an example of a practical use. Later, we consider volume and surface area.

PERIMETER AND AREA Table 5 summarizes the formulas for two-dimensional, or flat, objects. In all the formulas, the base and height (or length and width) refer to lines that are perpendicular. *Perpendicular lines* form square corners where they cross. The square corner is a 90 degree angle and is labeled with a small square, ∟.

Definitions of Perimeter and Area

Perimeter is *the distance around the outside of a flat object. The perimeter of a circle* is called the **circumference**. Perimeter is measured by units of length such as centimeters, meters, and kilometers (in the metric system) or inches, yards, and miles (in the American standard system). To buy fencing, you need to know the perimeter of the garden.

Area is *the measure of the surface enclosed within a flat object.* Area is measured in square units, such as square meters or square feet. To buy wall-to-wall carpet, you need to know the area of the floor.

Triangle	Area $= \frac{1}{2}$ base $\cdot$ height $A = \frac{1}{2}bh$	h, b
Square	Perimeter $= 4 \cdot$ side, $P = 4s$ Area $=$ side $\cdot$ side, $A = s^2$	s, s
Rectangle	Perimeter, $P = 2l + 2w$ Area $=$ length $\cdot$ width, $A = lw$	w, l
Parallelogram	Area $=$ base $\cdot$ height $A = bh$	h, b
Trapezoid	Area $= \frac{1}{2}$height $\cdot$ (sum of parallel sides) $A = \frac{1}{2}h(a + b)$	a, h, b
Circle	Circumference, $C = 2\pi r = \pi d$ Diameter, $d = 2r$ Area, $A = \pi r^2$, where $r =$ radius	r

Table 5 Selected Geometric Formulas for Two-Dimensional Figures

In earlier sections, we shortened instructions by using the word *simplify*. In this section we will use another instruction: *evaluate*. **Evaluate** formulas and expressions by *substituting numbers in place of the variables and simplifying the result.*

In Example 6, we evaluate a formula for perimeter. The perimeter is the distance around the outside of a flat object.

EXAMPLE  **Finding the perimeter of a rectangle: guarding a hazardous waste site** Suppose a guard walks around a rectangular waste site (see Figure 15). Which formula in Table 5 finds the distance once around? Evaluate the formula.

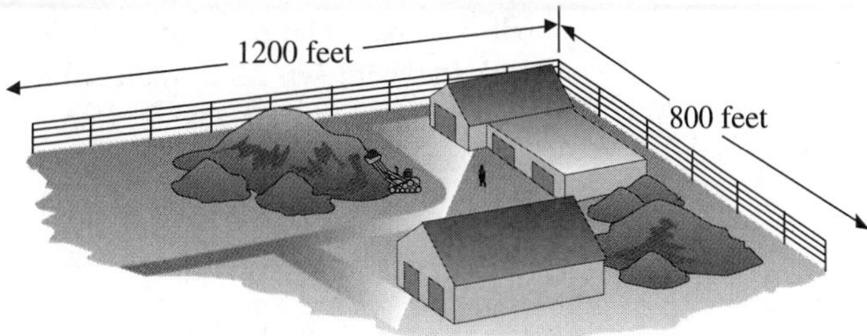

Figure 15

Solution The distance around is 2 lengths of 1200 feet and 2 widths of 800 feet, which is the same as the perimeter formula for a rectangle:

$$\begin{aligned} P &= 2l + 2w \\ P &= 2(1200 \text{ ft}) + 2(800 \text{ ft}) \\ &= 2400 \text{ ft} + 1600 \text{ ft} \\ &= 4000 \text{ ft} \end{aligned}$$

Using a Formula

To use a formula:

1. State the formula to be used.
2. List the facts, with units.
3. Substitute the facts into the formula.
4. Calculate, either by hand or by entering the entire expression into the calculator, using parentheses as needed.

Following the steps as outlined will help you learn the formulas as well as do the exercises.

Several formulas, including the one for the circumference of a circle, contain the constant pi. (As mentioned earlier, the perimeter of a circle is called the circumference.) **Pi** is *the number found by dividing the circumference of any circle by its diameter.* The symbol for pi is π. Use either the $\boxed{\pi}$ key on your calculator or 3.14 to approximate pi. The calculator key will give slightly more accurate results.

Example 7 is typical of many problems in that it requires calculating parts of the perimeter separately and then adding the results.

EXAMPLE 7 Summing sections to find the perimeter of a track around a soccer field What is the perimeter of a track placed around a rectangular soccer pitch (playing field), as shown in Figure 16? Assume that the curved parts are half-circles.

Solution This perimeter is the sum of the circumference of the curved parts (totaling a complete circle), πd, and the two straight sides, $2l$. The perimeter is

$$\text{Perimeter} = \pi d + 2l \qquad d = 73 \text{ m} \quad l = 100 \text{ m} \quad \pi \approx 3.14$$

$$P \approx 3.14(73 \text{ m}) + 2(100 \text{ m})$$

$$\approx (229.22 \text{ m}) + (200 \text{ m})$$

$$\approx 429.22 \text{ m}$$

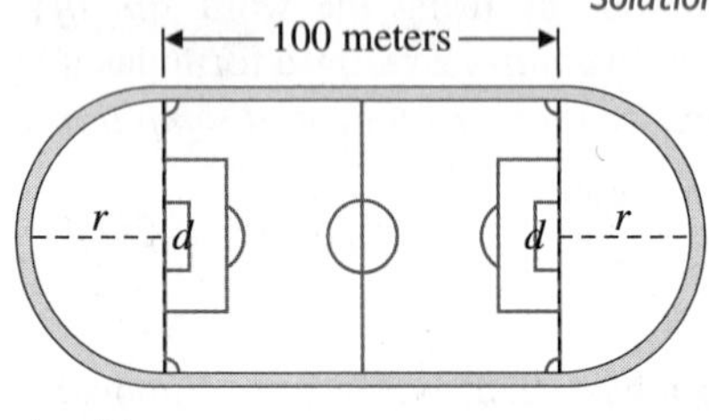

$d = 73$ meters

Figure 16

Spotting the Difference

Abbreviations for units—such as m for meters—are generally written without periods. (The one exception is inches, written in. to avoid confusion with the word *in*.) As a result, 73 m may be confused with $73m$, the product of 73 and the variable m. Here is how you spot the difference: If the m or other letter has a space in front of it, it refers to a unit of measurement. If there is no space in front of the letter, it is the variable in an expression.

The area of a figure is how much surface it covers. Area is described in square units such as square inches, square feet, square centimeters, and square meters. In choosing measurements for calculating area, make sure the base and height (or length and width) are perpendicular. Look for a small square in the corner indicating perpendicular lines.

EXAMPLE 8 Finding the area of a trapezoid Ray has a trapezoidal area, shown in Figure 17, of a hooked wall hanging to complete with yarn. What is the area in square inches?

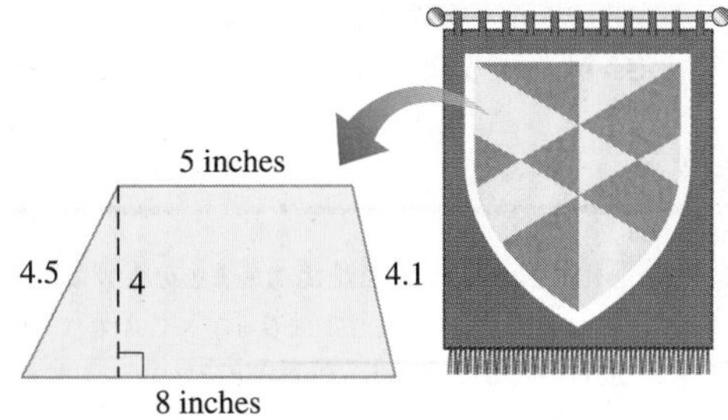

Figure 17

Solution The parallel sides are $a = 5$ in. and $b = 8$ in.

$$\begin{aligned} \text{Area} &= \tfrac{1}{2}h(a + b) \qquad h = 4 \text{ in.} \\ A &= (\tfrac{1}{2} \cdot 4 \text{ in.})(5 \text{ in.} + 8 \text{ in.}) \\ &= (2 \text{ in.})(13 \text{ in.}) \\ &= 26 \text{ in}^2 \end{aligned}$$

●

EXAMPLE 9 Finding the area of circles: pizza sizes One pizza has a radius of 5 inches. Another has a radius of 7 inches. How many times as large as the area of the 5-inch pizza is the area of the 7-inch pizza?

Solution We assume the pizzas are round and of the same thickness.

$$\begin{aligned} \text{Area of 7-inch pizza: } A &= \pi r^2 \qquad r = 7 \text{ in.} \quad \pi \approx 3.14 \\ A &= 3.14(7 \text{ in.})^2 \\ A &\approx 154 \text{ in}^2 \\ \text{Area of 5-inch pizza: } A &= \pi r^2 \qquad r = 5 \text{ in.} \\ A &= 3.14(5 \text{ in.})^2 \\ A &\approx 79 \text{ in}^2 \end{aligned}$$

To compare, divide 154 by 79, obtaining 1.9. The area of the 7-inch pizza is nearly two times as large as the area of the 5-inch pizza. ●

Graphing Calculator Technique: Evaluating an Expression

Method 1: Evaluating an Expression on the Computation Screen

Write the expression on the screen, substituting the numbers for the variables. In Example 9, to evaluate $A = \pi r^2$ for $r = 7$, write $\pi \times 7^2$ and press ENTER. To evaluate the same expression for $r = 5$, use the replay option to obtain $\pi \times 7^2$, replace 7 with 5, and press ENTER. Repeat as needed for other numbers.

This method works for an expression with any number of variables.

Method 2: Evaluating a One-Variable Expression with TABLE

To evaluate $A = \pi r^2$, enter $y_1 = \pi x^2$. Go to the table setup options. Choose the option ASK for the independent variable. Go to TABLE. Enter the number 7 for x and then 5 for x. Your answers should agree with those in the example.

To prevent confusion later, be sure to return to table setup and change the table back to automatic.

SURFACE AREA AND VOLUME Table 6 summarizes the formulas for surface area and volume of common three-dimensional, or solid, objects.

Definitions of Surface Area and Volume

Surface area is *the area needed to cover a three-dimensional object.* Surface area is measured in square units, just like area. When you wrap a gift, you need to estimate the surface area in order to cut enough paper.

Volume is *the space taken up by a three-dimensional object.* Volume is measured in cubic units, such as cubic meters or cubic feet. The amount of space cooled by a window air conditioner is given in cubic feet. **Capacity** is *the amount (especially of liquids) a container holds.* We change units of capacity, such as gallons or quarts, to units of volume by describing them as cubic units. For example, 1 gallon = 231 cubic inches.

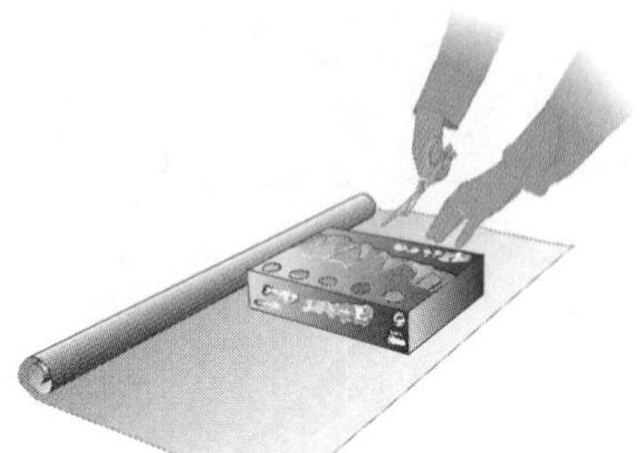

Surface area is the amount of area needed to cover the outside of a three-dimensional object. For example, because its thickness is negligible, we call the metal forming a home fuel tank or a can of food the surface area of a cylinder. The soap film forming a bubble floating in the air, and the leather forming a basketball are examples of the surface area for a sphere.

Rectangular prism (box)	Surface area, $S = 2lw + 2hl + 2hw$ Volume, $V = lwh$
Cylinder	Surface area, $S = 2\pi r^2 + 2\pi rh$ Volume, $V = \pi r^2 h$, where r = radius, h = height
Sphere	Surface area, $S = 4\pi r^2$ Volume, $V = (\frac{4}{3})\pi r^3$

Table 6 Selected Geometric Formulas for Three-Dimensional Figures

EXAMPLE 10 Finding surface area of a storage tank If we wanted to paint a cylindrical storage tank, we would need to know the surface area to determine the amount of paint needed. Find the surface area of the tank shown in Figure 18. Describe the order of operations needed to evaluate the formula.

Solution Note that the height need not be shown in the vertical position. The "height" of this cylinder, 10 feet, is the distance between the circular ends. The radius is 2 feet.

Surface area	$= 2\pi r^2 + 2\pi rh$	$r = 2$ ft $h = 10$ ft $\pi \approx 3.14$
S	$\approx 2(3.14)(2 \text{ ft})^2$	Simplify powers and units
	$+ 2(3.14)(2 \text{ ft})(10 \text{ ft})$	
	$+ 2(3.14)(2 \text{ ft})(10 \text{ ft})$	
	$\approx 2(3.14)(4 \text{ ft}^2) + 2(3.14)2(10 \text{ ft}^2)$	Do multiplications.
	$\approx (25.12 \text{ ft}^2) + (125.6 \text{ ft}^2)$	Do addition.
	$\approx 150.72 \text{ ft}^2$	

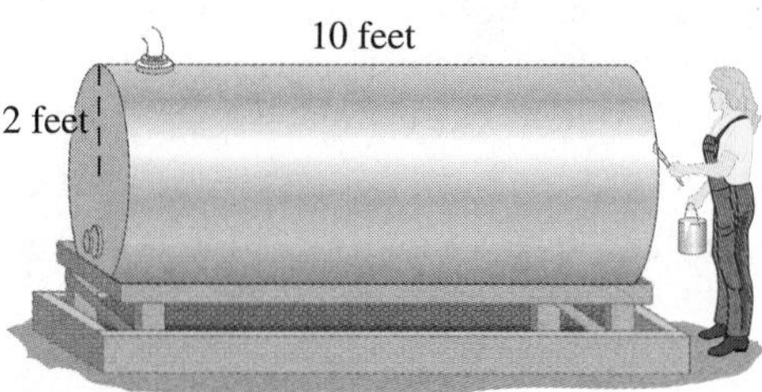

Figure 18

Volume describes how much space an object takes up. Volume is measured in cubic units such as cubic inches, cubic millimeters, and cubic meters. We talk about the volume of air inside a balloon. A doctor prescribes an injection with cc's (cubic centimeters). We order cubic yards of gravel and concrete. Bales of peat moss are labeled with cubic feet. Three common volume formulas are shown in Table 6.

EXAMPLE 11 **Finding the volume of a sphere: inflating balloons** A nonsmoker in excellent physical condition, using deep inhalation and forced exhalation, might exhale 250 to 275 cubic inches. (Some air must remain or the lungs will collapse.) Suppose a balloon approximates a sphere (see Figure 19). Could someone inflate

a. an empty balloon to a radius of 3 inches in one breath?

b. an empty balloon to a radius of 4 inches in one breath?

c. a balloon from a radius of 4 inches to a radius of 5 inches?

Figure 19

Solution We need to compare volumes by evaluating the formula with several radii (plural of *radius*). We will *round* our answers to the nearest whole number, and we will use π from the calculator. The volume of a sphere is given by $V = \frac{4}{3}\pi r^3$.

For radius $r = 3$ in., $V = \frac{4}{3}\pi(3 \text{ in.})^3 \approx 113 \text{ in}^3$

For radius $r = 4$ in., $V = \frac{4}{3}\pi(4 \text{ in.})^3 \approx 268 \text{ in}^3$

For radius $r = 5$ in., $V = \frac{4}{3}\pi(5 \text{ in.})^3 \approx 524 \text{ in}^3$

a. Reaching a radius of 3 inches is likely.

b. Only a fit person could reach a radius of 4 inches.

c. The *change in volume* is $524 - 268 = 256$ cubic inches. A fit person could inflate from a radius of 4 inches to a radius of 5 inches.

Caution: The numbers in Example 11 do not take into account the resistance of the balloon against air going into it. You may want to seek medical advice before trying Example 11 as an experiment.

USING A CALCULATOR TO EVALUATE FORMULAS When a portion of a formula is used repeatedly, using the calculator memory saves time and improves accuracy. A number is stored in memory with the key [STO], [M in], or [$x \to$M]. With multiple storage locations, either [$x \to$M] or [STO▶] is followed by a letter or number. To recall the stored number, use [RCL], [M out], or [M$\to x$]. With multiple memory locations, a letter or number is entered after the recall command.

EXAMPLE 12 Using calculator memory for repeated calculation In Example 11, $\frac{4}{3}\pi$ appeared in each volume expression. List the keystrokes needed to store $\frac{4}{3}\pi$ and to recall it for each volume calculation.

Solution Following is a sample solution for scientific calculators:

To store $\frac{4}{3}\pi$, enter

4 [÷] 3 [=] [×] [π] [=] [STO]

To calculate volume, enter

3 [y^x] 3 [×] [RCL] [=]

4 [y^x] 3 [×] [RCL] [=]

5 [y^x] 3 [×] [RCL] [=] ●

When adding to or subtracting from memory, use the [SUM], [M+], [+M], or [M−] key. Using both [±] and [SUM] keys will subtract from memory if the calculator does not have an [M−] key.

Graphing Calculator Technique

On most graphing calculators, you enter the entire expression before evaluating. If you make an error in keying in the expression, pressing [2nd] [ENTER] or [replay] will restore the expression to the display for editing. This feature also permits you to recall and change the formula for a new radius, thus replacing the need for memory, as was used in Example 12.

ANSWER BOX

Warm-up: 1. $1.9y$ **2.** $264g$ **3.** $1728i^3$ **4.** $231i^2$ **Think about it 1:** Both b and c are correct. **Think about it 2:** (inches)^4 per spill

EXERCISES 2.5

Simplify the expressions in Exercises 1 to 8. Assume $\frac{\text{feet}}{\text{foot}} = 1$.

1. $\dfrac{8 \text{ ounces} \cdot 1 \text{ pound}}{16 \text{ ounces}}$

2. $\dfrac{8 \text{ fluid ounces} \cdot 4 \text{ cups}}{1 \text{ cup} \cdot 1 \text{ quart}}$

3. $\dfrac{100 \text{ yards} \cdot 3 \text{ feet} \cdot 12 \text{ inches}}{1 \text{ yard} \cdot 1 \text{ foot}}$

4. $\dfrac{55 \text{ mm} \cdot 1 \text{ cm} \cdot 1 \text{ in.}}{10 \text{ mm} \cdot 2.54 \text{ cm}}$

5. $\dfrac{\text{feet}^3}{\text{foot}}$

6. $\dfrac{\text{meters}^3}{\text{meters}^2}$

7. $\dfrac{\text{inches}^2}{\text{inches}^3}$

8. $\dfrac{\text{cm}}{\text{cm}^3}$

Use unit analysis to change the units in Exercises 9 to 12. Round to the nearest whole number.

9. Change 1 kilometer into feet.
   12 inches = 1 foot
   39.37 inches = 1 meter
   1000 meters = 1 kilometer

10. How many ounces are in 1 ton?
    1 ton = 2000 pounds
    16 ounces = 1 pound

11. How many grams are in 120 pounds?
    2.2 pounds = 1 kilogram
    1000 grams = 1 kilogram

12. How many cups are in a 5-gallon water can?
    1 gallon = 4 quarts
    4 cups = 1 quart

In Exercises 13 to 26, list the facts needed and write a unit analysis to solve the problem. Unless otherwise noted, the facts you need are in the examples or earlier exercises. Round to the nearest tenth.

13. How many seconds are in 1 day?

14. Change 1,000,000 ounces into tons.

15. Change 1,000,000 seconds into days.

16. How many minutes are in 1 year?

17. Change 72 years into seconds.

18. Change 1 square meter into square centimeters.
    100 centimeters = 1 meter

19. Change 1 cubic foot into cubic inches.

20. Change 1 square mile into square feet.
    5280 feet = 1 mile

21. Change 1 square yard into square inches.

22. Change 1 cubic yard into cubic inches.

23. How many square feet are in 1200 square inches? Round the answer to the nearest tenth of a square foot.

24. How many square inches are in 12 square feet?

25. The difference in elevation between Mt. Everest and the Mariana Trench is 19.77 kilometers. Write this elevated difference in feet. Round to the nearest 1000.

26. We need to replace the fence around the hazardous waste site in Example 6. Suppose the fencing manufacturer uses metric dimensions. Change 4000 feet to meters.

In Exercises 27 to 30, round to the nearest tenth. Use the $\boxed{\pi}$ *key.*

27. Find the perimeter and area of each shape.

a.

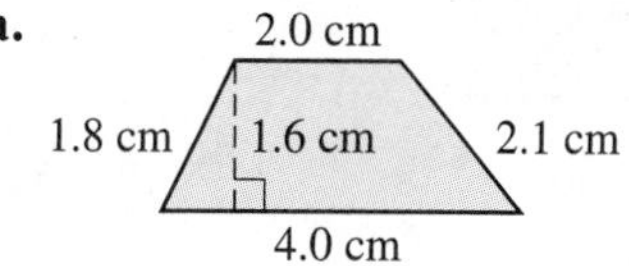

b.

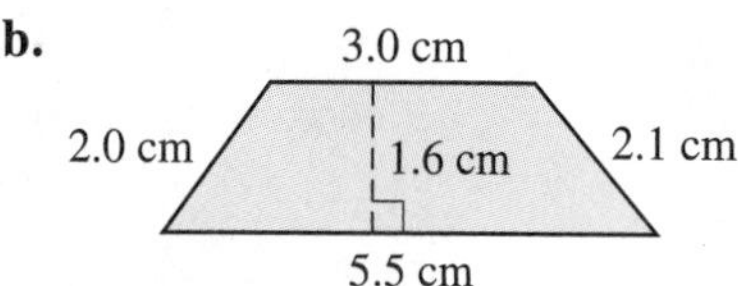

c.

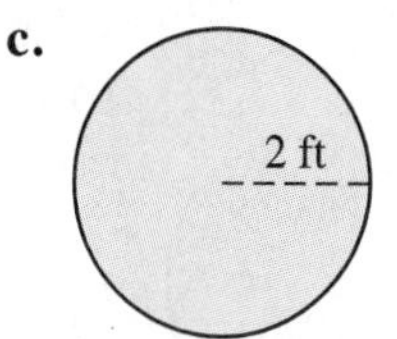

d.

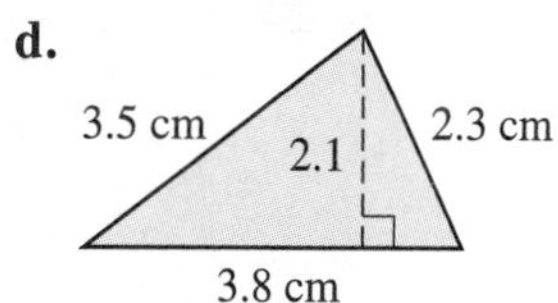

28. Find the perimeter and area of each shape.

a.

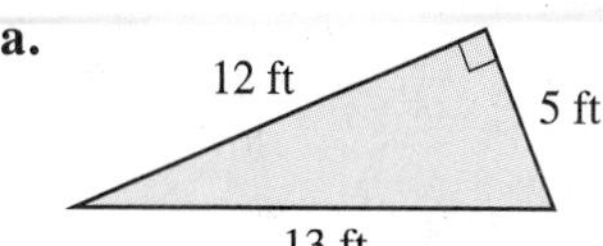

b.

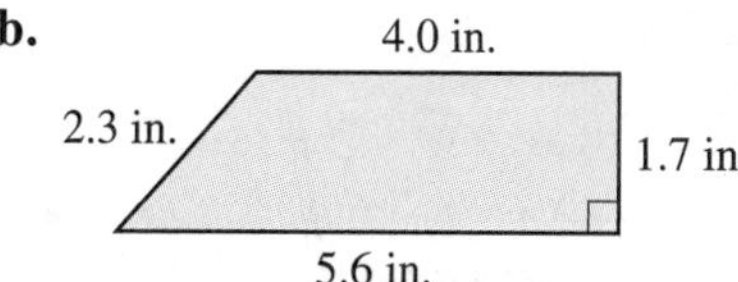

c.

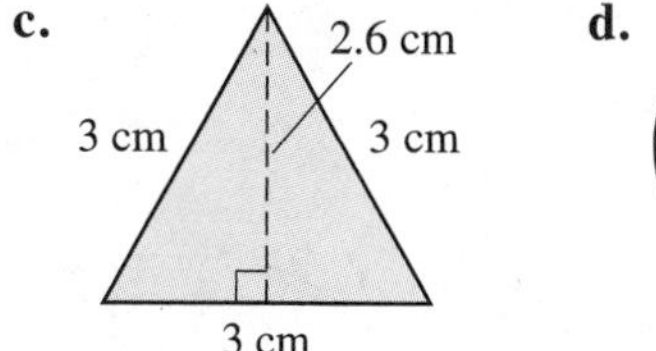

d.

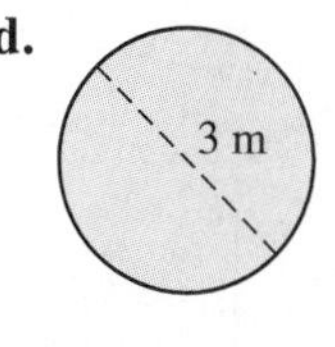

29. Find the perimeter and area of each shape.

a.

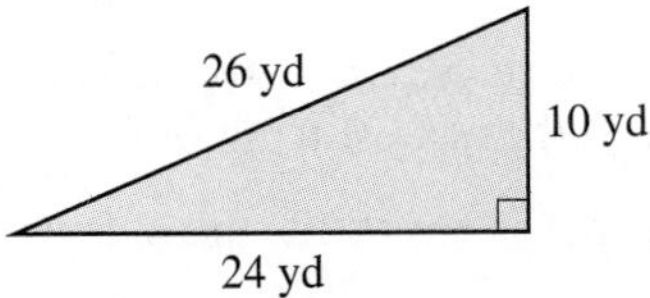

b.

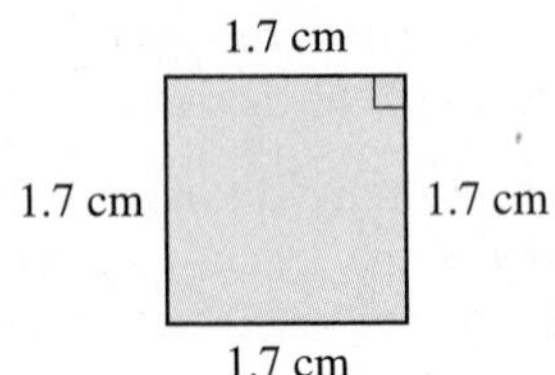

c.

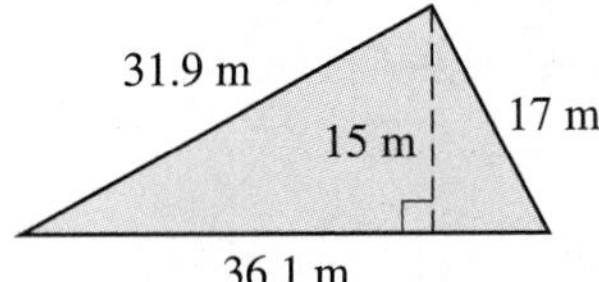

d.

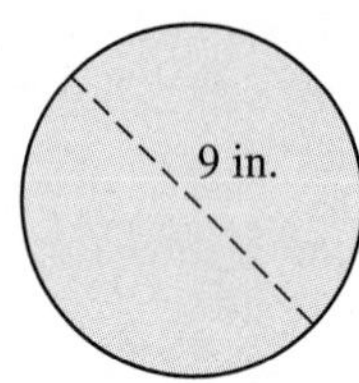

30. Find the perimeter and area of each shape.

a.

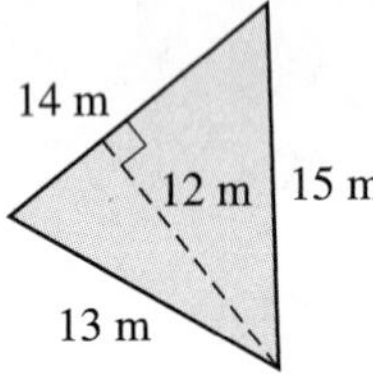

b.

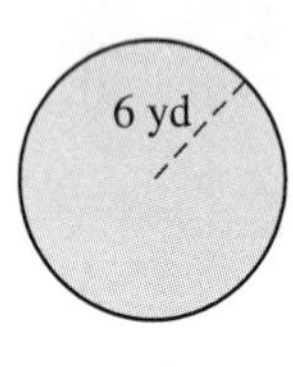

c.

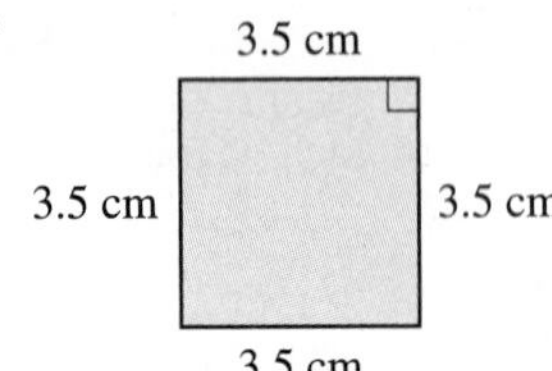

d.

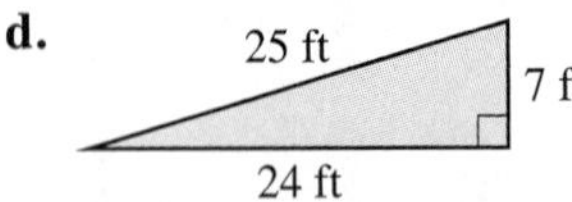

What are the surface area and volume of each shape in Exercises 31 to 36? Use $\pi \approx 3.14$. Round to the nearest tenth.

31. a.

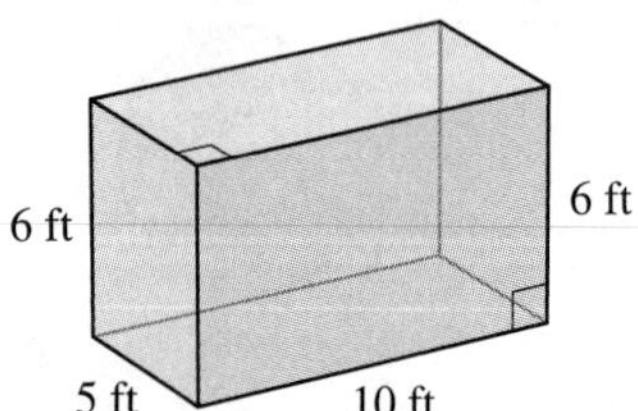

b.

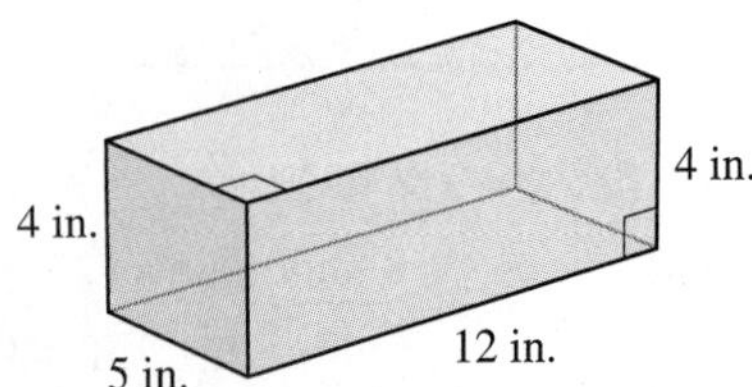

32. a.

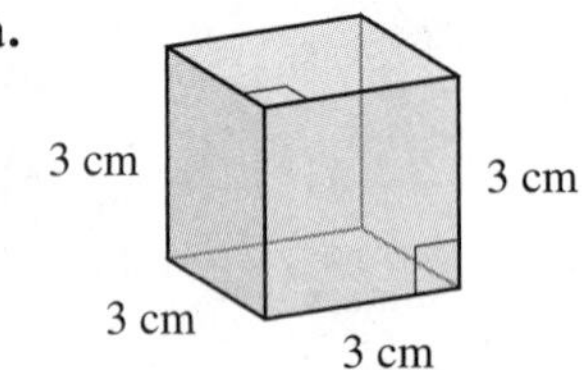

b.

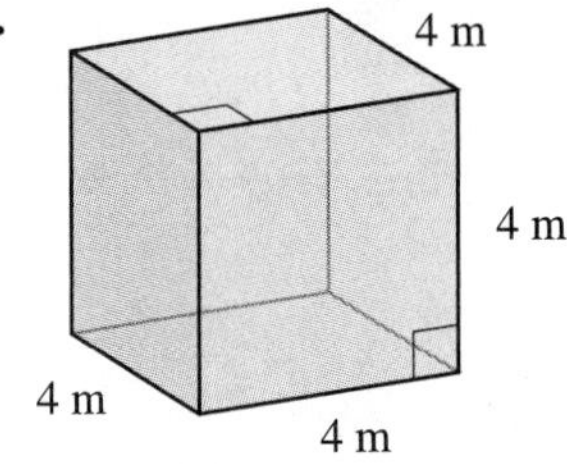

33. a.

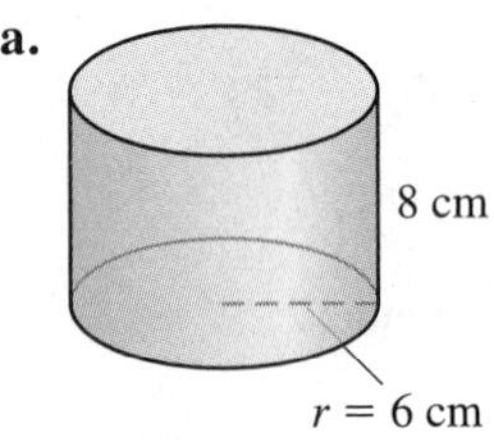

b.

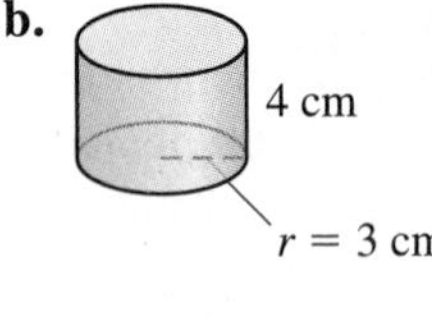

34. a.

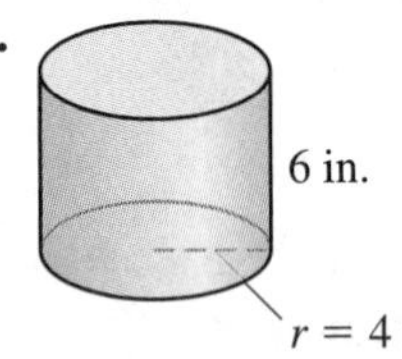

b.

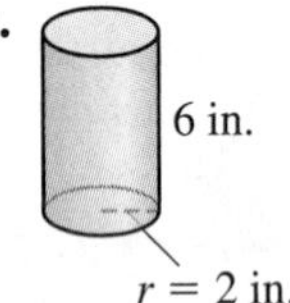

35. a.

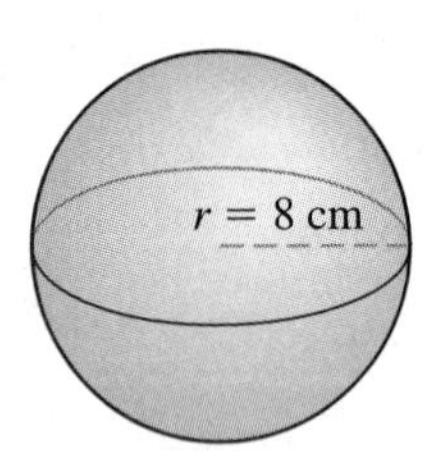

b.

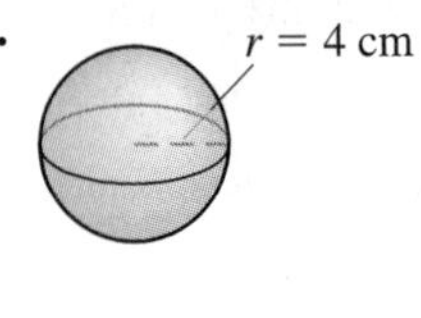

36. a.

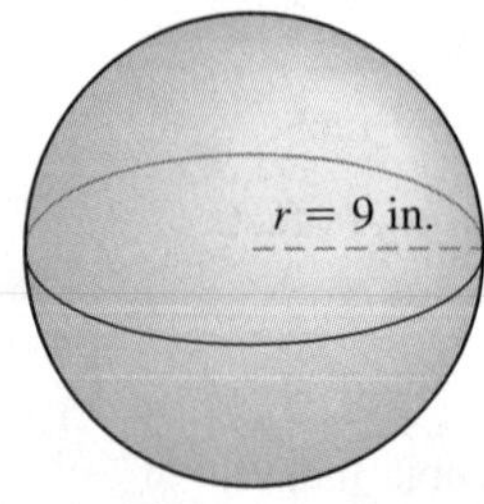

b.

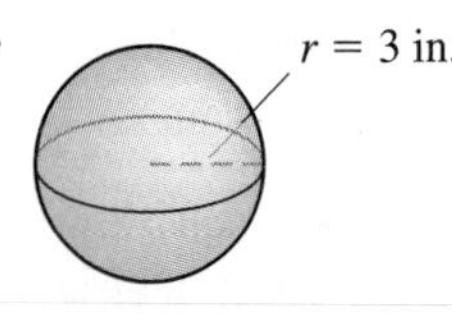

Problem Solving

For Exercises 37 and 38, assume the pizzas are round and of the same thickness.

37. How many times as large in area as a pizza with a 6-inch radius is a pizza with a 9-inch radius?

38. How many times as large in area as a pizza with a radius of x inches is a pizza with a radius of $2x$ inches?

39. What is the volume of an athletic shoe box $4\frac{1}{2}$ inches in height, 7 inches wide, and $13\frac{1}{2}$ inches long? How many cubic inches would it take to store 1000 such boxes? How many cubic feet?

40. What is the volume of a cereal box with depth 7 centimeters, width 21 centimeters, and height 30.5 centimeters? How many cubic centimeters are needed to store 1000 such boxes? How many cubic meters?

Use unit analysis in Exercises 41 to 46.

41. Suppose a 12-fluid-ounce can of soda spills and spreads to a thickness of $\frac{1}{16}$ inch. How many square inches will the spill cover? 1 cup = 8 fluid ounces.

42. How long a sidewalk can be built with 4 cubic yards of concrete? Suppose the sidewalk is 4 inches thick and 4 feet wide.

43. For a reception, Robin needs 200 servings of punch at 12 fluid ounces per serving. How many gallons does she need to prepare?

44. If a paper clip weighs 3 grams, how many paper clips are in a pound?

45. The Cheez-It® snack crackers box weighs 16 ounces. The serving size is 12 crackers. There are 140 calories in each ounce. A box contains 32 servings. Find the number of calories per cracker.

46. A 15-ounce bag of Diane's Tortilla Chips® contains 15 servings. There are 80 milligrams (mg) of sodium in each serving. How many milligrams of sodium are consumed in eating a bag of chips?

Exercises 47–50 concern an accident in Boston on January 15, 1919. A tank holding 2,500,000 gallons of molasses broke open, and a 6-foot-high flood poured out. Use unit analysis to answer the questions. Facts: 231 cubic inches = 1 gallon, 640 acres = 1 square mile.

47. If the molasses ran down a 40-foot-wide street, how long would it reach if it maintained the 6-foot thickness?

48. Suppose the molasses was stored in a cylindrical tank 50 feet high and 100 feet in diameter. What is the volume of such a tank in gallons? Round to the nearest ten thousand.

49. How many square feet would the molasses cover if it spread out to a depth of 1 inch? Would it cover a square mile?

50. How many acres would the molasses cover if it spread out to a depth of 1 inch?

51. Find the perimeter and area of the squares in parts a and b.

a.

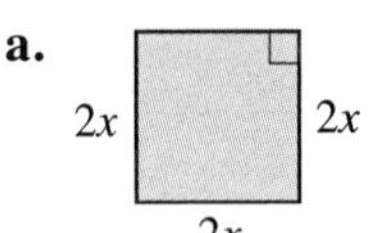

b.

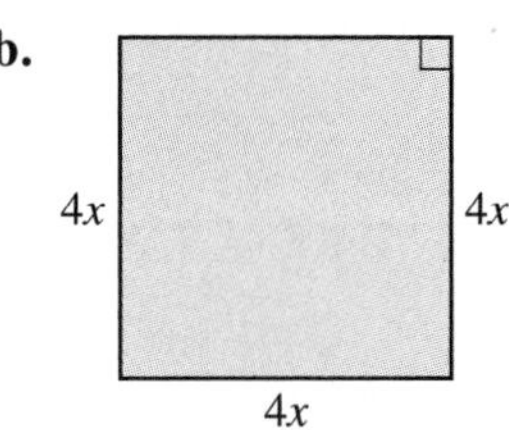

c. Using a fraction, compare the perimeter of the square in part a to that of the square in part b.

d. Using a fraction, compare the area of the square in part a to that of the square in part b.

52. Find the surface area and volume of the cubes in parts a and b.

a.

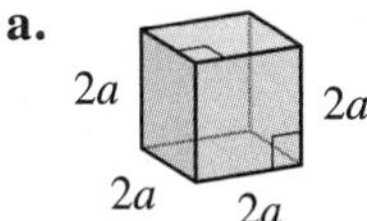

b.

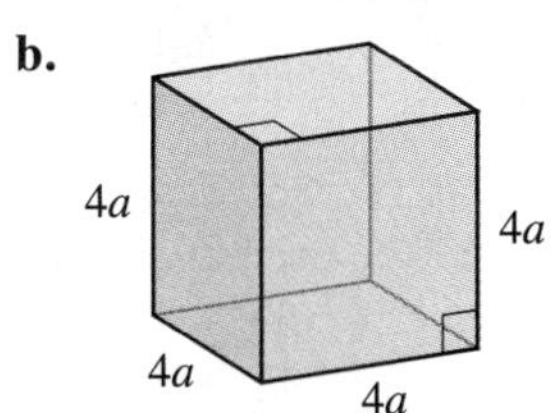

c. Using a fraction, compare the surface area of the cube in part a to that of the cube in part b.

d. Using a fraction, compare the volume of the cube in part a to that of the cube in part b.

53. Explain how to find the area of a triangle given height and base.

54. Explain how to find the area of a trapezoid given lengths of parallel sides and height.

55. Explain how to find the volume of a sphere given radius.

56. Explain how to find the volume of a rectangular box given length, width, and height.

57. How are area and surface area the same?

58. How are area and surface area different?

59. Explain how finding the perimeter of a rectangle is an example of adding like terms.

60. Explain how finding the surface area of a rectangular prism (box) is an example of adding like terms.

In Exercises 61 to 66, use reasoning to find missing information. Use $\pi \approx 3.14$, and round to the nearest tenth.

61. Find the area of the blue part.

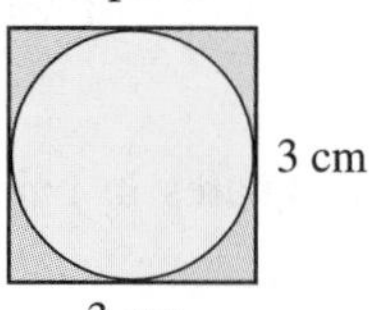

62. Find the area of the blue part.

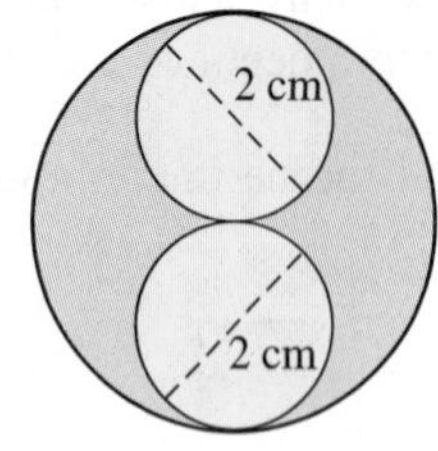

63. Find the area inside the oval track.

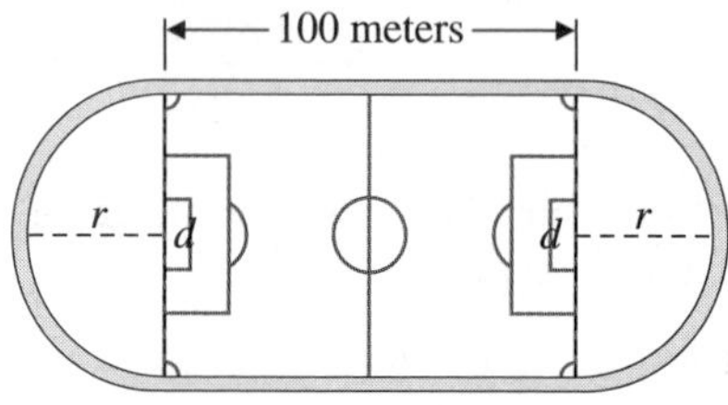

$d = 73$ meters

64. Find the area of the blue part.

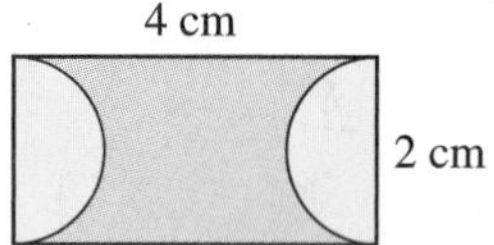

65. Find the area of the shaded part.

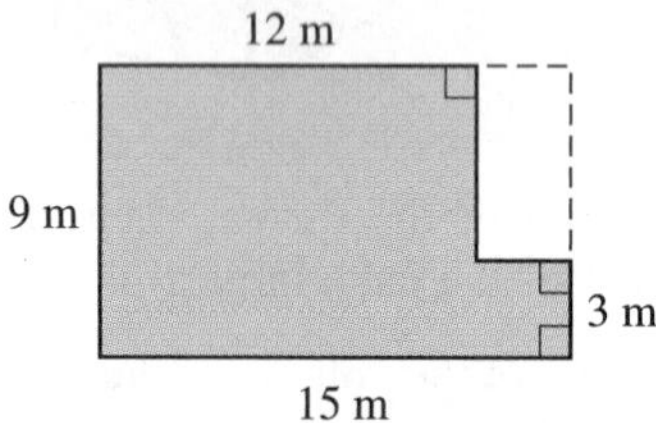

66. Find the area of the shaded part.

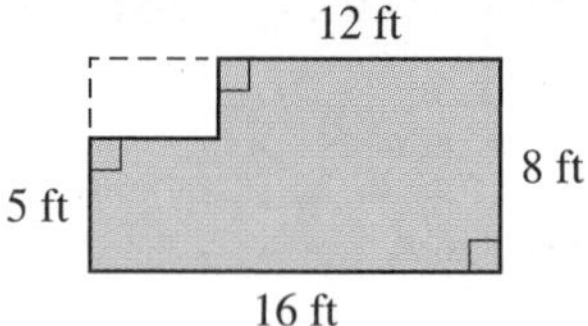

Projects

67. ***Soup Cans.*** The marketing manager for the Bell Soup Company wanted more visibility on grocery shelves for the company's soup. He reasoned that doubling the radius and the height of the soup can would double the shelf space and make the product easier to see. The original can was a cylinder with diameter 7.5 centimeters and height 10.5 centimeters. Use the $\boxed{\pi}$ key.

a. What is the volume of the original can?

b. Find the diameter and height of the new can.

c. What is the volume of the new can?

d. How many times as large in volume as the original can is the new can?

e. The marketing manager recommended that the price also be doubled. Was he correct?

f. If the old can sold for \$0.69, what should be the price of the new can? Justify your answer.

g. Do you think that the new can, priced as in part f, will sell as well as the old can?

68. ***Measuring Items.*** Copy the following list of units of measure on your paper.

cups
gallons or liters
square yards or square meters
cubic feet
feet or meters
pounds
square feet
yards or meters
cubic yards or cubic meters

a. Identify each as a measure of length, area, volume, weight, or capacity.

b. Write each of the following items beside the appropriate unit(s) of measure: fabric area, fabric when purchased, concrete, paint when purchased, paint when applied, rope, peat moss, carpeting, butter when purchased, butter when measured for cooking.

c. Write three other items that are measured with each unit of measure.

69. ***Fixed Perimeter and Area.*** Cut a piece of string 30 inches long. Form it into a rectangle with width 1 inch.

a. Measure the length of the rectangle, and record it in the table below. Calculate the area of the rectangle.

Width	Length	Area
1 in.		
2 in.		
3 in.		
4 in.		
5 in.		
6 in.		

b. Form another rectangle with width 2 inches. Measure the length of the new rectangle, and record it. Calculate the area.

c. Repeat this procedure to complete the table. Add more rows until you reach a largest possible area.

d. On one set of axes, graph the length and width pairs, using length as x and width as y.

e. On another set of axes, graph the length and area pairs, using length as x and area as y.

f. Are there fractional widths and lengths that give an area larger than 56 square inches? How is this number related to the length of the string (30 inches)?

g. Where on the length and area graph is the point representing the largest area?

70. ***Volume of Box.*** Cut a piece of paper to 8 by $10\frac{1}{2}$ inches. Cut a $\frac{1}{2}$-inch square from each corner, as shown in the figure. Fold the paper on the dotted lines to make a box without a top.

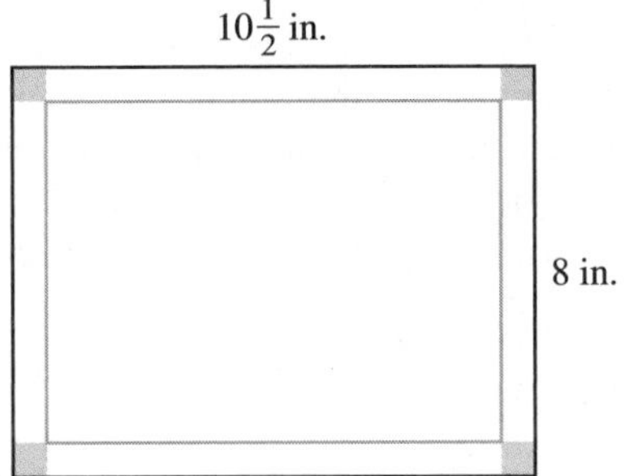

a. Measure the length, width, and height of the box. Place the data in the first row of the table, and calculate the volume of the box.

Corner Square	Length (in.)	Width (in.)	Height (in.)	Volume (in^3)
$\frac{1}{2}$				
1				
$1\frac{1}{2}$				
2				
$2\frac{1}{2}$				
3				
$3\frac{1}{2}$				

b. Repeat the calculations for a 1-inch square cut from each corner, then a $1\frac{1}{2}$-inch square, and so forth, until the table is complete.

c. What is the largest volume of the box?

d. Draw a graph with the corner square's side length on the x-axis and the volume on the y-axis.

e. Is it possible to choose a different corner square that would yield a larger volume?

71. ***Water Usage.*** Water charges at Everybody's Utility include a basic \$5.15 fee and \$0.65 for each thousand gallons used. An average American family uses 8000 gallons each month.

a. Make a table and graph for the total cost of 0 to 8000 gallons.

b. How many gallons does your household use each month? Assume 30 days, and estimate usage. Use the following table, which gives domestic water use in the United States in gallons. If your water is metered, how close is your estimate to the metered use?

Washing car with hose running	180
Watering lawn for 10 minutes	75
Washing machine at top level, 1 load	60
Ten-minute shower	25–50
Average bath	36
Hand-washing dishes with water running	30
Dripping faucet, per day	25–30
Shaving with water running	20
Automatic dishwasher	10
Toilet flush	5–7
Brushing teeth with water running	2

Source: Data reprinted with permission from G. Tyler Miller, Jr., *Living in the Environment,* Wadsworth Publishing Co., © 1992, p. 356.

c. A gallon is 231 cubic inches. How big a container would be needed to hold 8000 gallons? Describe such a container's shape and dimensions (if rectangular, give length, width, and height; if cylindrical, give radius of base and height) in suitable units of measurement (inches, feet, yards, miles, etc.).

2.6 Inequalities and Intervals

OBJECTIVES

- Compare two numbers using inequality signs.
- Write sets of numbers using inequalities.
- Write sets of numbers using intervals.
- Graph intervals and inequalities on a number line.
- Use inequalities and intervals to describe inputs in conditional situations.

WARM-UP

The positions for 0 and 1 are marked on the number line.

a b c d e f g h i j k l

0 1

1. Match each number with a letter on the number line.

$-0.25, 0.75, -2, 1\frac{3}{4}, 2.25, -1\frac{1}{4}, -1.5$

2. What is the number marked by each of these letters on the number line: *d, e, g, i, k*?

INEQUALITIES AND INTERVALS are special ways of writing sets of numbers. They are useful in describing input and output sets, and consequently they take on considerable importance in work with calculators and computers. We will return to the credit card payment schedule in Chapter 1 because its inputs are sets of numbers. We start with inequalities, go on to intervals, and then relate the two with line graphs.

Inequalities

An **inequality** is *a statement that one quantity is greater than or less than another quantity.* Inequality symbols are used to compare the relative positions of two numbers on a number line or the relative sizes of two or more expressions. The symbol $>$ in $4 > -5$ indicates that 4 is greater than -5. The number 4 is to the right of -5 on the number line in Figure 20.

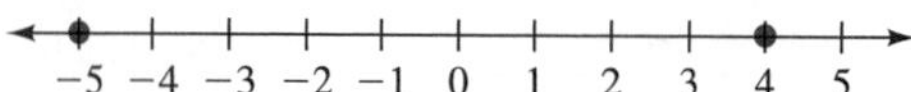

Figure 20

An easy way to read inequalities is to remember that the point of the symbol always faces the lesser number and the wide part always faces the greater number.

There are four inequality symbols:

Inequality Symbol	Meaning
$<$	is less than (to the left of another number on the number line)
$>$	is greater than (to the right of another number on the number line)
$\le$	is less than or equal to
$\ge$	is greater than or equal to

EXAMPLE  Writing inequalities Place the correct inequality signs between these statements.

a. $3 \square -5$ **b.** $-5 \square 2$ **c.** $-2 \square -2$ **d.** $\frac{1}{2} \square \frac{1}{4}$ **e.** $-\frac{1}{2} \square -\frac{1}{4}$

Solution There are two solutions to each part, because the *or* in the meanings of $\le$ and $\ge$ always makes at least one of these symbols appropriate. Thus, whenever $=$ is correct, $\le$ and $\ge$ are correct also; whenever $<$ is correct, $\le$ is correct also; and whenever $>$ is correct, $\ge$ is correct also.

a. 3 is to the right of -5 on the number line, so 3 is greater than -5. Either $3 > -5$ or $3 \ge -5$.

b. -5 is to the left of 2 on the number line, so -5 is less than 2. Either $-5 < 2$ or $-5 \le 2$.

c. -2 is equal to itself. Thus, any symbol that includes equality is correct. Either $-2 \le -2$ or $-2 \ge -2$.

d. $\frac{1}{2}$ is to the right of $\frac{1}{4}$, so $\frac{1}{2}$ is greater than $\frac{1}{4}$. Either $\frac{1}{2} > \frac{1}{4}$ or $\frac{1}{2} \ge \frac{1}{4}$.

e. $-\frac{1}{2}$ is to the left of $-\frac{1}{4}$, so $-\frac{1}{2}$ is less than $-\frac{1}{4}$. Either $-\frac{1}{2} < -\frac{1}{4}$ or $-\frac{1}{2} \le -\frac{1}{4}$.

●

The set of inputs from \$0 to \$20 in the credit card example in Chapter 1 may be described with a variable and two inequalities. For the set of numbers greater than or equal to zero, we write $x \ge 0$. For the set of numbers less than or equal to 20, we write $x \le 20$.

A **compound inequality** is *two inequalities in one statement.* We write $x \ge 0$ and $x \le 20$ together as $0 \le x \le 20$. We read this as

x is the set of numbers between 0 and 20, including 0 and 20.

EXAMPLE 2 Writing compound inequalities Write each pair of inequalities as a compound inequality. Use only $<$ or $\le$.

a. $x < 0$ and $x > -3$ **b.** $x \le 5$ and $x > -2$

c. $x > 0$ and $x \le 3$ **d.** $x < -\frac{1}{2}$ and $x \ge -\frac{3}{4}$

Solution To place the x and the numbers in order, remember that the point on the inequality faces the smaller number and the x goes between the two numbers.

a. $-3 < x < 0$ **b.** $-2 < x \le 5$

c. $0 < x \le 3$ **d.** $-\frac{3}{4} \le x < -\frac{1}{2}$

It is a good habit to place the smaller number to the left. ●

EXAMPLE  Writing inequalities: credit card payments Use inequalities or compound inequalities to describe the inputs of the credit card payment schedule in Table 7. Write each of the inequalities in words.

Input: Charge Balance	Output: Payment Due
\$0 to \$20	Full amount
\$20.01 to \$500	10% or \$20, whichever is greater
\$500.01 or more	\$50 plus the amount in excess of \$500

Table 7 Credit Card Payment Schedule

Solution Table 8 shows inequalities and word descriptions for the inputs in Table 7.

Inequalities	Words
$0 \le x \le 20$	The set of numbers between 0 and 20, including 0 and 20
$20.01 \le x \le 500$	The set of numbers between 20.01 and 500, including 20.01 and 500
$x \ge 500.01$	The set of numbers greater than or equal to 500.01

Table 8 Credit Card Payment Schedule

The inequalities in Table 8 are satisfactory in business and finance, but not in mathematics. There are number gaps between the given input sets. Between \$20 and \$20.01 are all the fraction and decimal portions of one cent: $\$20.00\frac{1}{2}$, \$20.005, and even \$20.009. There is another gap between \$500 and \$500.01. In mathematics, we prefer to include all numbers in our intervals, leaving no gaps.

EXAMPLE  Removing gaps: credit card payments, continued Use inequalities to describe the inputs of the credit card payment schedule in Table 8 without any gaps between the input sets. Describe each set in words.

Solution The first inequality, $0 \le x \le 20$, remains the same.

In the second inequality, we start at 20 but do not wish to include 20. We want to include 500. The continued inequality becomes $20 < x \le 500$, which excludes 20 but includes 500.

In the third inequality, we eliminate the gap between 500 and 500.01 with $x > 500$, which excludes the endpoint 500. The results are summarized in Table 9.

Inequalities Without Gaps	Words
$0 \le x \le 20$	The set of numbers greater than or equal to 0 and less than or equal to 20
$20 < x \le 500$	The set of numbers greater than 20 and less than or equal to 500
$x > 500$	The set of numbers greater than 500

Table 9

Intervals

We may also state inequalities as intervals. We used intervals when we first described sets of inputs in Section 1.2. Recall that an **interval** is *a set containing all the numbers between its endpoints as well as one endpoint, both endpoints, or neither endpoint.* Intervals indicate the inclusion or exclusion of endpoints through the use of brackets or parentheses. Brackets, [ ], indicate that the endpoints are included in the set. Parentheses, ( ), are used when the endpoints are excluded from the set. We may mix brackets and parentheses in one interval, as shown in Table 10.

Inequality	Interval	Words
$3 \le x \le 7$	[3, 7]	The set of numbers between 3 and 7, including 3 and 7
$3 \le x < 7$	[3, 7)	The set of numbers between 3 and 7, including 3
$3 < x \le 7$	(3, 7]	The set of numbers between 3 and 7, including 7
$3 < x < 7$	(3, 7)	The set of numbers between 3 and 7

Table 10

EXAMPLE 5 Writing intervals: credit card payments revisited Use intervals to describe the inputs of the credit card payment schedule: $0 \le x \le 20$, $20 < x \le 500$, $x > 500$.

Solution

Inequalities	Intervals
$0 \le x \le 20$	x in [0, 20]
$20 < x \le 500$	x in (20, 500]
$x > 500$	x in (500, $+\infty$)

To *include* both 0 and 20, we write x *in* [0, 20]. The brackets, [ ], include the endpoints.

To describe the set of numbers from 20 to 500 but not including 20, we write x *in* (20, 500]. We *exclude* the endpoint 20 by using a parenthesis.

The last interval, x *in* (500, $+\infty$), describes all numbers larger than 500. There is no greatest number, so we need a symbol to say that *the numbers get large without bound.* We use an **infinity sign**, ∞. Realistically, the credit card company would object to this much spending and put a limit, say $1000, on the card. ●

Infinite means *without bound.* Infinity describes a concept, not a number. Number lines and axes on coordinate graphs all have arrows on their ends because the lines go on without bound. The numbers on the number line go to the right and left forever. Placing a positive sign before the infinity sign means infinite to the right on the number line; a negative sign indicates infinite to the left (see Figure 21).

We generally write inequalities with the smaller number on the left. Using the greater than symbol places the larger number on the left and sometimes makes the inequality confusing or hard to read. We *always* write intervals with the smaller number on the left. For either $5 \ge x$ or $x \le 5$, we write $(-\infty, 5]$ instead of $[5, -\infty)$.

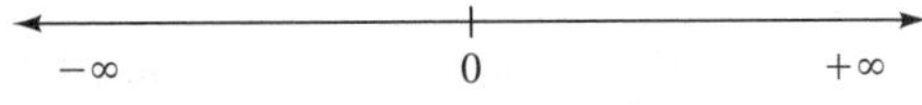

Figure 21

Line Graphs

It is sometimes useful to draw a picture of an inequality or interval. We use a line graph to do so. The graph of the inequality $1 \le x \le 4$ or interval [1, 4] is a number line with dots (closed circles) at 1 and 4 and a line segment connecting them. (Using brackets on the number line instead of dots is also acceptable.)

Inequality: $1 \leq x \leq 4$
Interval: $[1, 4]$
Words: The set of numbers between 1 and 4, including 1 and 4
Line Graph:

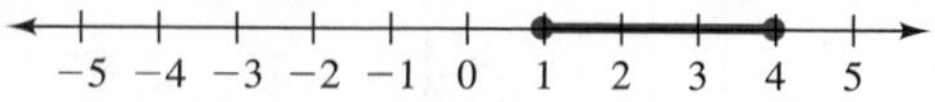

−5 −4 −3 −2 −1 0 1 2 3 4 5

The graph of $1 < x < 4$ or its interval $(1, 4)$ has *small (open) circles* as endpoints at 1 and 4 and a line segment connecting them. (Using parentheses instead of the small circles is also acceptable.)

Inequality: $1 < x < 4$
Interval: $(1, 4)$
Words: The set of numbers between 1 and 4
Line Graph:

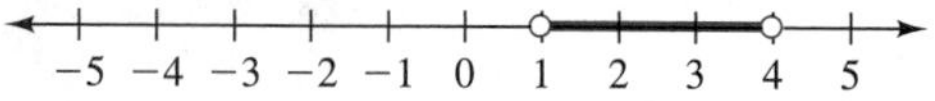

−5 −4 −3 −2 −1 0 1 2 3 4 5

EXAMPLE 6 Drawing line graphs of inequalities Write each inequality as an interval, express it in words, and make a line graph.

a. $-2 < x \leq 4$ **b.** $-3 \leq x < 4$

Solution Both endpoint forms are shown in the graphs.

a. Inequality: $-2 < x \leq 4$
Interval: $(-2, 4]$
Words: The set of numbers between −2 and 4, including 4
Line Graph:

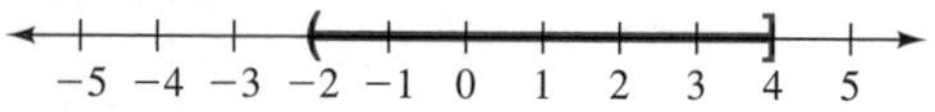

−5 −4 −3 −2 −1 0 1 2 3 4 5

OR

−5 −4 −3 −2 −1 0 1 2 3 4 5

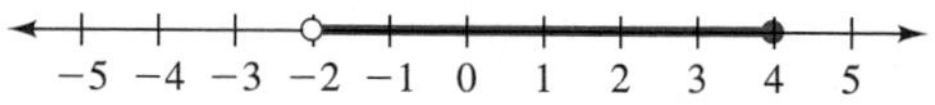

b. Inequality: $-3 \leq x < 4$
Interval: $[-3, 4)$
Words: The set of numbers between −3 and 4, including −3
Line Graph:

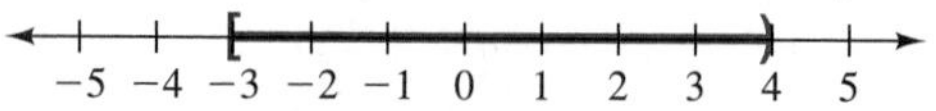

−5 −4 −3 −2 −1 0 1 2 3 4 5

OR

−5 −4 −3 −2 −1 0 1 2 3 4 5

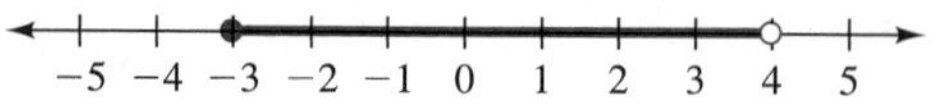

Think about it: Write the word descriptions in Example 6 in another way.

Expressions such as $-2 < x$ may be read either as "x is greater than −2" or as "−2 is less than x." It is not always obvious whether the graph is to the left or right of the number. To be sure that we have the line graph drawn correctly, we select a *test point* on the number line. If the test point makes the inequality true, then the graph should pass through the point. If the test point makes the inequality false, then the graph goes in the opposite direction. *Zero is a convenient test point for a line graph.*

EXAMPLE 7 Drawing line graphs of inequalities Write each inequality as an interval, express it in words, and make a line graph. Use a test point to check the graph.

a. $x < 4$ **b.** $x \geq -2$

Solution **a.** The inequality $x < 4$ is the set of all numbers less than 4 and describes the interval $(-\infty, 4)$. The test point 0 gives $0 < 4$, which is true. Thus, the graph goes through 0. Either the parenthesis or the open circle on the 4 excludes 4 from the graph.

Inequality: $x < 4$

Interval: $(-\infty, 4)$

Words: The set of numbers less than 4

Line Graph:

−5 −4 −3 −2 −1 0 1 2 3 4 5

OR

−5 −4 −3 −2 −1 0 1 2 3 4 5

b. The inequality $x \geq -2$ is the set of all numbers greater than or equal to -2, and the interval is $[-2, +\infty)$. The test point 0 gives $0 \geq -2$, which is true. The graph goes through 0. The bracket or dot at -2 shows the inclusion of -2 in the set.

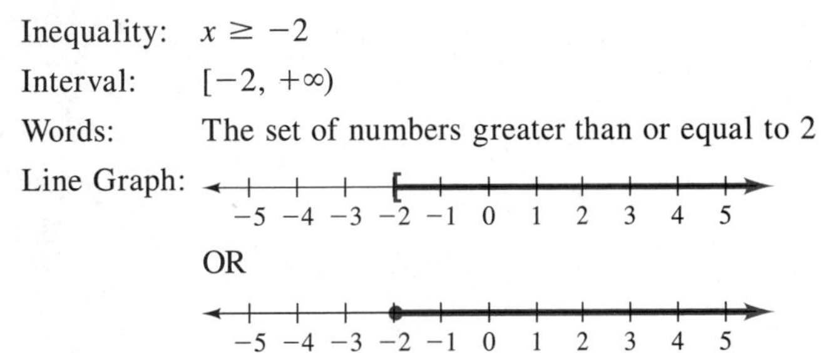

In reading the summary chart in Table 11, observe the use of parentheses with $<$, $>$, or the infinity sign (∞) and the use of brackets with $\leq$ or $\geq$.

Inequality Symbol	Interval Notation	Word Meaning	Line Graph Notation
$<$	(,)	is less than	open circle or (,)
$>$	(,)	is greater than	open circle or (,)
$=$		is equal to	dot
$\leq$	[,]	is less than or equal to	dot or [,]
$\geq$	[,]	is greater than or equal to	dot or [,]
$+\infty$	, $+\infty$)	positive infinity	$\rightarrow$
$-\infty$	($-\infty$,	negative infinity	$\leftarrow$

Table 11 Symbols Used in Inequalities and Intervals

Caution: Interval notation may look like the coordinates of a point. Read carefully when you come across (a, b) to see whether the reference is to a coordinate point (a, b) or an interval (a, b) describing the set $a < x < b$. We will use the word *interval* to introduce interval notation in this text.

ANSWER BOX

Warm-up: 1. *f, h, a, j, l, c, b* **2.** $-1, -\frac{1}{2}, \frac{1}{4}, 1\frac{1}{4}, 2$ **Think about it:** **a.** x is greater than -2 and less than or equal to 4. **b.** x is greater than or equal to -3 and less than 4.

EXERCISES 2.6

In Exercises 1 and 2, draw a number line from −2 to 2, marked in eighths, and write the numbers where they belong on the line.

1. $-\frac{7}{8}, 0.5, -\frac{5}{4}, -1\frac{3}{4}, \frac{1}{4}, \frac{3}{2}, -\frac{1}{8}, 0.75$

2. $-\frac{6}{8}, -0.25, -1.5, 1\frac{5}{8}, \frac{1}{2}, \frac{7}{4}, -1\frac{3}{8}, \frac{3}{4}$

In Exercises 3 to 6, copy the statement and fill in the correct sign, <, =, or >.

3. a. $-8 \square -3$ **b.** $+4 \square -9$

c. $(-3)^2 \square 3^2$ **d.** $0.5 \square 0.5^2$

e. $6 \square -5$ **f.** $-2(6) \square -2(-5)$

g. $-6 \square -5$ **h.** $-2(-6) \square -2(-5)$

4. a. $-7 \square 5$ **b.** $2(-7) \square 2(5)$

c. $-2(-7) \square -2(5)$ **d.** $0.2^2 \square 0.2$

e. $1.5 \square 1.5^2$ **f.** $\frac{3}{4} \square \left(\frac{3}{4}\right)^2$

g. $3(-7) \square 3(-6)$ **g.** $-3(-7) \square -3(-6)$

5. a. $-3.75 \square -3.25$ **b.** $3(-2) \square -3(2)$

c. $\frac{1}{2} \square -\frac{1}{2}$ **d.** $|-4| \square |2|$

e. $-2(-3) \square 2(-4)$ **f.** $\left(\frac{1}{2}\right)^2 \square \left(-\frac{1}{2}\right)^2$

g. $-2.5 \square -3$ **h.** $\frac{22}{7} \square \pi$

6. a. $0.5 \square 0.25$ **b.** $-0.5 \square -0.75$

c. $3(6) \square 3(-2)$ **d.** $|-3| \square -3$

e. $3(-5) \square 4^2$ **f.** $(-2)^2 \square 2^3$

g. $-2(-3)(-4) \square -4(-3)(-2)$ **h.** $\pi \square 3.14$

In Exercises 7 and 8, write each pair of inequalities as a compound inequality. Use only < or ≤.

7. a. $x < 4$ and $x > 0$ **b.** $x \le -2$ and $-5 < x$

8. a. $x > -2$ and $0 \ge x$ **b.** $x \ge -\frac{1}{2}$ and $\frac{1}{4} > x$

In Exercises 9 and 10, write each compound inequality as two separate inequalities. More than one answer is possible.

9. a. $3 < x < 8$ **b.** $-3 < x \le -1$

c. $-2 < x < 1$

10. a. $-5 \le x < 0$ **b.** $-\frac{1}{4} \le x < \frac{1}{2}$

c. $-3 < x < 0$

Choose one listed inequality and one listed interval that describe each line graph in Exercises 11 to 16.

Inequality	Interval
a. $-3 < x < 3$	**p.** $(-\infty, -3)$
b. $-3 \le x \le 3$	**q.** $(-\infty, 3]$
c. $x \ge -3$	**r.** $[-3, 3]$
d. $x > -3$	**s.** $(-\infty, 3)$
e. $x \ge 3$	**t.** $[-3, +\infty)$
f. $x < 3$	**u.** $(-3, 3)$
g. $x > 3$	**v.** $(-3, +\infty)$
h. $x \le 3$	**w.** $[3, +\infty)$

11. −5 −4 −3 −2 −1 0 1 2 3 4 5

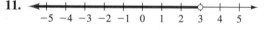

12. −5 −4 −3 −2 −1 0 1 2 3 4 5

13. −5 −4 −3 −2 −1 0 1 2 3 4 5

14. −5 −4 −3 −2 −1 0 1 2 3 4 5

15. −5 −4 −3 −2 −1 0 1 2 3 4 5

16. −5 −4 −3 −2 −1 0 1 2 3 4 5

In Exercises 17 and 18, complete the tables.

17.

	Inequality	Interval	Words	Line Graph
a.	$-1 \le x < 3$			
b.		$(-4, -1]$		
c.			x is between -3 and 5, including -3.	
d.	$x < -4$			
e.				-2 0 5
f.	$x > -2$			
g.			x is greater than -4 and less than or equal to 2.	
h.		$[-3, +\infty)$		

18.

	Inequality	Interval	Words	Line Graph
a.	$-4 < x < -1$			
b.		$[3, 8)$		
c.			x is greater than or equal to -2 and less than 3.	
d.	$x > -3$			
e.				-3 0 5
f.	$x < 2$			
g.			x is between -1 and 3, including -1.	
h.		$(-\infty, 4]$		

19. What is an inequality?

20. Explain the difference between $<$ and $\le$. Write a number statement containing each.

21. What is an interval?

22. Explain the meaning of the ∞ in $(4, +\infty)$.

23. Why is $4 < x < 2$ not a true statement?

24. Why is $3 > x > 5$ not a true statement?

25. Why is $2 \le 2$ a true statement?

26. Why is $3 \ge 3$ a true statement?

Conditional Rules

In Exercises 27 to 32, write inequalities to describe the situations. Do not leave any gaps between your inequalities.

27. A tax schedule. The tax, x, is

a. not over 2000

b. over 2000 but not over 5000

c. Over 5000

28. A purchase discount schedule. The number of items, x, is

a. 10 or less

b. greater than 10 and less than 50

c. 50 or more

29. x is

a. -5 or less

b. greater than -5 and less than 5

c. 5 or larger

30. A pattern in squaring numbers. x is

a. less than -1

b. -1 to 1

c. larger than $+1$

31. x is

a. less than five

b. five to fifty

c. larger than fifty

32. Age categories. The age, x, is

a. less than eighteen

b. eighteen and above but less than twenty-one

c. twenty-one and above

Computation Review

33. Complete the table by finding n percent of each number or expression in the top row.

***n* percent**	**\$1.00**	**\$5.00**	**\$10.00**	x
6%				
10%				
25%				
100%				
150%				

Projects

34. ***Utility Payments.*** Read your utility bill (electricity, natural gas, heating oil, or water) or call the billing department of your local utility to get a current cost schedule. Find out whether the cost is conditional. If it is conditional, write input conditions in both interval and inequality form.

35. ***Size of Numbers.*** Make a table with the following three headings: x^2 is less than x, x^2 is equal to x, x^2 is greater than x.

a. Square each number x listed below. Compare x^2 with x and place the original number, x, under the appropriate heading in your table.

$\frac{2}{3}$, 1.5, $(-\frac{1}{2})$, $\frac{1}{3}$, 1, 2.5, (-1), 0.5
0.1, 0, (-0.1), 2, (-2), (-2.5)

(*Hint:* The negative numbers in the listing above are placed in parentheses to remind you to use parentheses when squaring on the calculator.)

b. In words, summarize the relationship between the size of the square of a number and the position of the original number on a number line.

For parts c and d, find a number that makes the statement true and another that makes the statement false.

c. $-x > x$ **d.** $\frac{1}{x} > x$

CHAPTER 2 SUMMARY

Vocabulary

For definitions and page references, see the Glossary/Index.

absolute value
adding like terms
additive inverse
area
associative properties
base
capacity
circumference
commutative properties
compound inequality
cubed
distributive property of multiplication over addition
equivalent expressions
evaluate
exponent
factors
formula
greatest common factor
grouping symbol
inequality
infinite
infinity sign
interval
like terms
lowest terms
multiplicative inverse
opposite
order of operations
perimeter
pi
power
radical
reciprocal
simplification property of fractions
simplify
squared
surface area
term
unit analysis
volume

Concepts

2.1 Addition and Subtraction

(The property marked * is new and is added at this time to provide a more complete listing.)

Addition and subtraction are inverse operations.

Subtraction is equivalent to adding the opposite (additive inverse):

$$a - b = a + (-b)$$
$$a - (-b) = a + b$$

*The sum of a number n and 0 is n.

The sum of a number n and its opposite, $-n$, is 0.

If the signs are alike, add the numbers and place the common sign on the answer.

If the signs are different, subtract the number portion and place the sign of the number farthest from zero on the answer.

2.2 Multiplication and Division

Multiplication and division are inverse operations.

Division by n is equivalent to multiplication by the reciprocal, $\frac{1}{n}$:

$$a \div n = a \cdot \frac{1}{n}$$

The product of a number n and 1 is n.

The product of a number n and 0 is 0.

In multiplication and division of two real numbers, if the signs are alike, the answer is positive. If the signs are different, the answer is negative.

2.3 Properties of Operations with Real Numbers

Associative property of addition:
$(a + b) + c = a + (b + c)$

Associative property of multiplication:
$(a \cdot b) \cdot c = a \cdot (b \cdot c)$

Commutative property of addition: $a + b = b + a$

Commutative property of multiplication:
$a \cdot b = b \cdot a$

Distributive property of multiplication over addition: $a(b + c) = ab + ac$

Simplification property of fractions: For all real numbers, a not zero and c not zero,

$$\frac{ab}{ac} = \frac{a}{a} \cdot \frac{b}{c} = 1 \cdot \frac{b}{c} = \frac{b}{c}$$

2.4 Operations with Exponents

Base property: An exponent outside the parentheses applies to all parts of a product or quotient inside the parentheses:

$$(x \cdot y)^a = x^a \cdot y^a \qquad \left(\frac{x}{y}\right)^a = \frac{x^a}{y^a}$$

Multiplication property: To multiply numbers with like bases, add the exponents:

$$x^a \cdot x^b = x^{a+b}$$

Division property: To divide numbers with like bases, subtract the exponents:

$$\frac{x^a}{x^b} = x^{a-b}$$

Power property: To apply an exponent to a power expression, multiply exponents:

$$(x^a)^b = x^{a \cdot b}$$

Order of operations: The order of operations for algebra and scientific calculators:

1. Calculate expressions within parentheses and other grouping symbols.
2. Calculate exponents and square roots.
3. Do remaining multiplication and division in the order of appearance, left to right.
4. Do remaining addition and subtraction in the order of appearance, left to right.

2.5 Unit Analysis

1. Identify the unit of measure to be changed, and identify the unit of measure for the answer. Write both as fractions.
2. List facts containing the unit to be changed and all units needed to get to the unit for the answer.
3. Write the unit to be changed (or a fact containing the unit) as the first fraction. Set up a product of fractions using the list of facts. Arrange fractions so that the units to be eliminated appear in both a numerator and a denominator.
4. Eliminate units of measure where possible, and calculate the numbers.

See Table 5 on page 97 and Table 6 on page 100 for selected geometric formulas.

2.6 Symbols for Inequalities and Intervals

See Table 11 on page 113.

CHAPTER 2 REVIEW EXERCISES

1. Add or subtract, as indicated.

a. $-3 + -5$ **b.** $3 + (-8)$ **c.** $4 - 17$

d. $-5 - (-18)$ **e.** $-21 + 7$ **f.** $-26 + 19$

g. $14 - (-28)$ **h.** $12 - 36$ **i.** $-32 - (-16)$

j. $-4 - (-16)$ **k.** $-11 + 22$ **l.** $8 - (-5)$

2. Add or subtract, as indicated.

a. $-4 + 9.5$ **b.** $-5 + 1.1$ **c.** $2.5 - (-6)$

d. $-3 - (-0.7)$ **e.** $-1.0 + 0.6$ **f.** $-1.3 + 0.8$

g. $2.6 - (-1.3)$ **h.** $-1.9 - 4.7$ **i.** $-0.3 + (-0.4)$

j. $\frac{5}{6} - \frac{7}{9}$ k. $-\frac{1}{4} + \left(-\frac{3}{4}\right)$ l. $\frac{5}{3} + \left(-\frac{2}{3}\right)$

m. $-\frac{2}{3} - \left(-\frac{3}{4}\right)$ n. $-\frac{4}{5} + \frac{5}{3}$ o. $-\frac{5}{6} + \left(-\frac{3}{4}\right)$

3. Multiply or divide, as indicated.

a. $(-9)(6)$ b. $(-9)(-6)$ c. $(-18)(-3)$

d. $(-18)(3)$ e. $(-8)(-7)$ f. $8(-7)$

g. $4 \cdot (-14)$ h. $(-4)(-14)$ i. $(-48) \div (-24)$

j. $(-48) \div (12)$ k. $48 \div (-6)$ l. $(-48) \div (-6)$

m. $-48 \div (-3)$ n. $-48 \div 3$ o. $48 \div (-8)$

4. Multiply or divide, as indicated.

a. $-1.0(0.6)$ b. $(-1.3)(-0.8)$

c. $(2.6) \div (-1.3)$ d. $(-1.7) \div (5.1)$

e. $-0.3(-0.4)$ f. $(0.7) \div (-1.4)$

g. $\left(-\frac{1}{4}\right)\left(-\frac{3}{4}\right)$ h. $\left(\frac{5}{3}\right)\left(-\frac{2}{3}\right)$

i. $\left(-\frac{2}{3}\right) \div \left(-\frac{4}{3}\right)$ j. $\left(-\frac{4}{5}\right) \div \left(\frac{5}{3}\right)$

5. Simplify.

a. $-(-2)^2$ b. $4 - (-2) + (-2)^2$

c. $5 - (-3) + (-2)^2$ d. $-(-3)^2$

e. $\sqrt{3^2 + 4^2}$ f. $\sqrt{8^2 + 6^2}$

g. $\sqrt{25^2 - 20^2}$ h. $\sqrt{15^2 - 12^2}$

i. $\sqrt{1.5^2 + 2^2}$ j. $\sqrt{10^2 - 6^2}$

k. $-|-4|$ l. $|-6 - (-5)|$

6. Simplify. Leave expressions without parentheses.

a. $4(x - 3)$ b. $3(x + y - 5)$

c. $-2(x - y + 3)$ d. $5 - 3(x - 4)$

e. $2 - (x - 3)$ f. $-2(x - 4)$

g. $4 - 2(x - 4)$ h. $3 - (x - 4)$

7. Complete these statements.

a. 2(______) $= 2x + 2y$ b. $ac + ab =$ ____$(c + b)$

c. 4(__________) $= 4x^2 - 8x + 12$

d. $3xy + 4x^2y = xy($______$)$

e. ____$(2x + 4y - 5) = 6x + 12y - 15$

f. $15a^2bc + 5ab^2 + 10abc = 5ab($__________$)$

8. Add like terms.

a. $12 - 3x + 5 + 6x$ b. $-4 - (2 - 3x)$

c. $4(2a^2 + 4) - (6 - 3a)$

d. $3a + 3(a - 2) - (2a + 5)$

e. $2a^2 + 3a + 2 - 4a^2 - a - 5$

f. $x^2 - (-3x) + 2 + 6x^2 - 2x - 1$

g. $4x^2 - 3xy - 2y^2 - 6x^2 + 6xy - y^2$

h. $a^2b + a^2b^2 + a^2 - 2a^2b^2$

i.

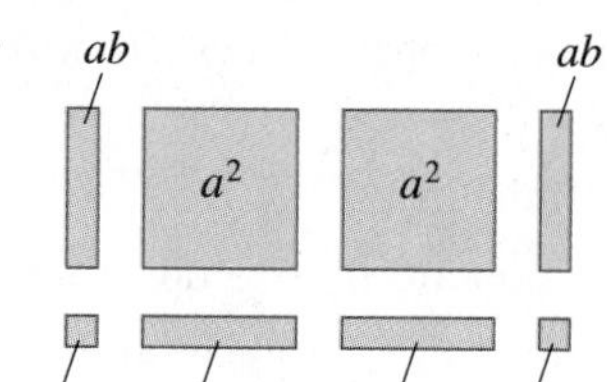

j.

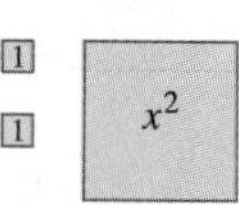

9. Simplify these expressions. Answers should contain no parentheses.

a. $\dfrac{abc}{bcd}$ b. $\dfrac{4xy}{6xz}$ c. $\dfrac{-21cd}{14ad}$

d. $(-2x)^2$ e. $(-3y)^3$ f. $(-2y)^4$

g. $(-ab)^2$ h. $(ab)^2$ i. m^4m^5

j. m^2m^7 k. $m^5 \div m^2$ l. $m^7 \div m^4$

m. $\dfrac{3x + 6y}{3}$ n. $\dfrac{mn + n^2}{n}$ o. $\dfrac{2a + 4b}{4}$

10. Simplify these expressions. Answers should contain no parentheses.

a. $\left(\dfrac{4x}{y}\right)^2$ b. $\left(\dfrac{x}{3y}\right)^2$ c. $(-3ab^2)^2$

d. $\dfrac{3x^2}{9x}$ e. $\dfrac{4ab^2}{a^2b}$ f. $\dfrac{xy^2z}{x^2yz^2}$

g. $\dfrac{(-x)^2}{-x^2}$ h. $\dfrac{-3x^2}{(9x)^2}$ i. $\dfrac{-4x^2}{(-4x)^2}$

11. Simplify these expressions.

a. $4 - 3(3 - 5)$ b. $(4 - 3)(3 - 5)$

c. $\sqrt{5^2 - 4(2)(-12)}$ d. $\dfrac{4 - \sqrt{49}}{4}$

e. $|6 - 1| - |3 - 9|$ f. $(7 - 2)^2 + (4 - 1)^2$

g. $\dfrac{-3 - (-5)}{-6 - 4}$ h. $\frac{1}{2} \cdot 11(5 + 7)$

12. Simplify with a calculator, as needed. Round to a whole number.

a. $-\frac{1}{2}(32)(4)^2 + 64(4) + 60$

b. $1000\left[\left(1 + \dfrac{0.08}{2}\right)^2\right]^5$

13. Evaluate these formulas. Round to the nearest tenth.

a. $A = \pi r^2$, $r = 2.5$ ft

b. $A = \frac{1}{2}bh$, $b = 5$ yd, $h = 4$ yd

c. $V = \frac{4}{3}\pi r^3$, $r = 3$ m **d.** $V = s^3$, $s = 1.5$ cm

14. Arrange the facts listed after each problem into a unit analysis that solves the problem. Round to the nearest tenth.

a. 140 pounds is how many grams (g)?
2.2 pounds = 1 kilogram
1000 grams = 1 kilogram

b. How many yards is 100 meters?
3 feet = 1 yard
12 inches = 1 foot
39.37 inches = 1 meter

c. A polo ground is 300 yards by 200 yards. How many square feet are in this rectangular field?
1 yard = 3 feet

d. Change 1 cubic foot into cubic inches.
1 foot = 12 inches

15. Find the perimeters and areas of these figures. Round to the nearest tenth. Use $\pi \approx 3.14$.

a.

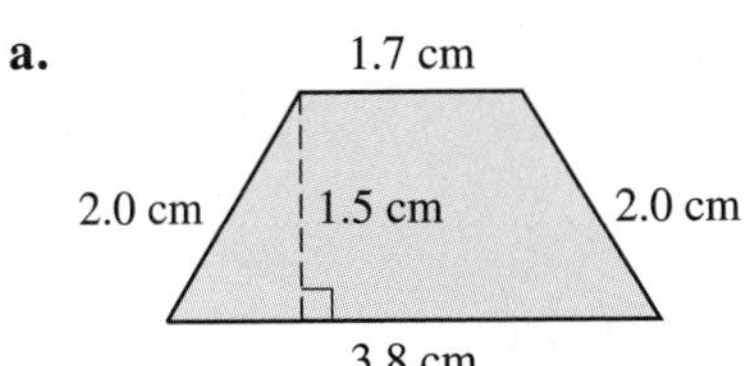

b.

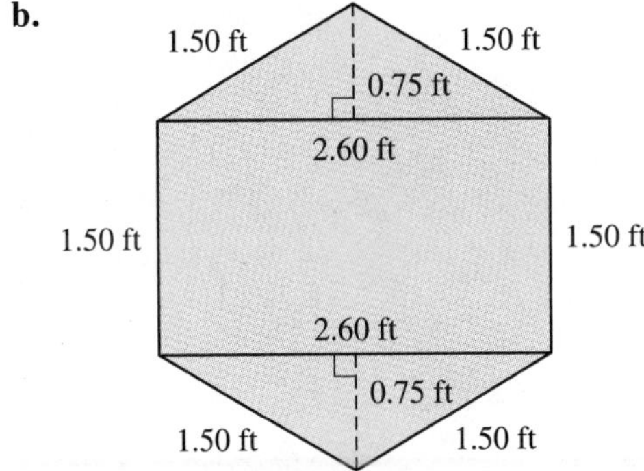

c.

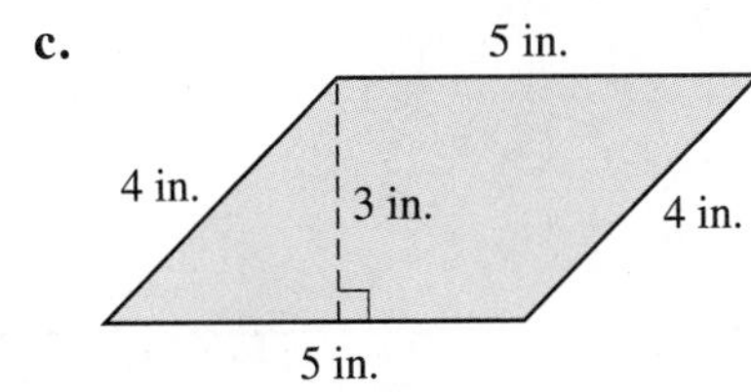

d.

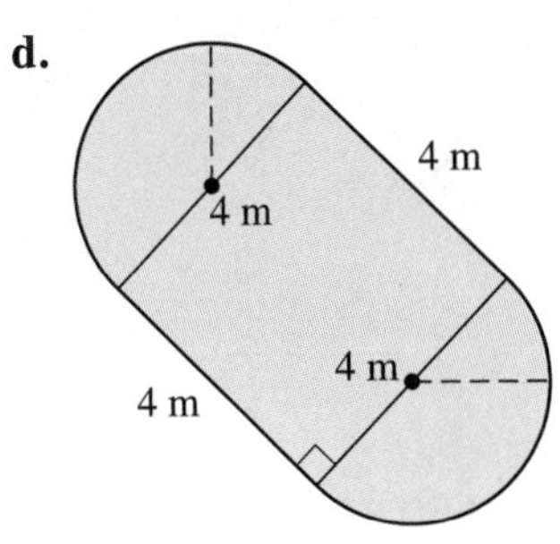

16. Find the surface areas and volumes of these figures. Round to a whole number. Use $\pi \approx 3.14$.

a. **b.**

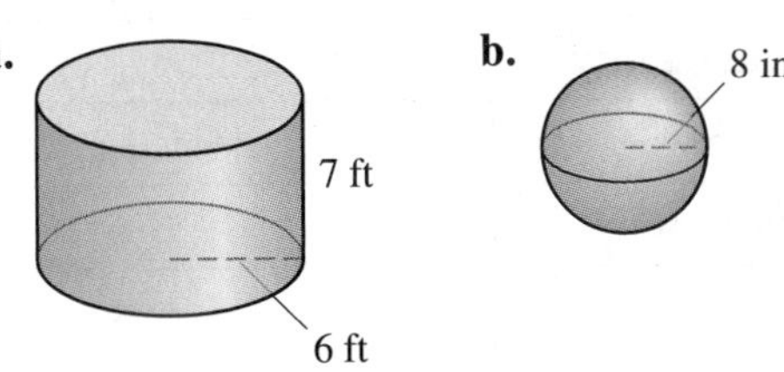

c.

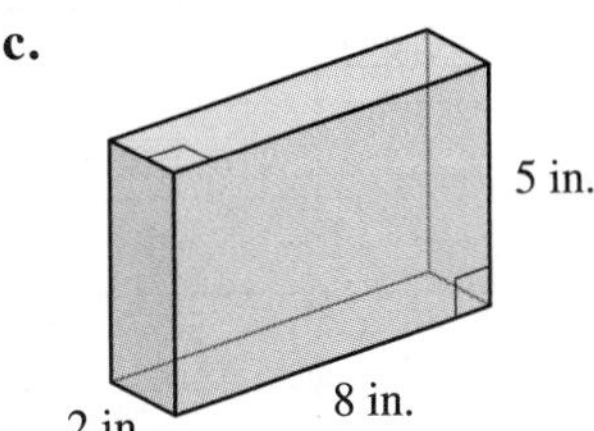

17. Give an example to show that three numbers can be regrouped to add.

18. Give an example to show that three numbers can be rearranged to multiply.

19. Explain when to use the small circle and when to use the dot in a line graph.

20. Explain why $3x^2$ does not equal $(3x)^2$.

21. Explain why $-x^2$ is not the same as $(-x)^2$.

22. Explain how to find the reciprocal of 3.5.

23. Explain how to find the reciprocal of $2\frac{1}{4}$.

24. Explain why $|5|$ and $|-5|$ are equal.

25. Explain the role of inverse operations in subtraction of negatives and division of fractions.

26. Write three sentences that illustrate the difference in meaning among *base of triangle, base of exponent,* and any other use of *base.*

27. Write five sentences that illustrate the difference in meaning among *set of points, set of numbers, set of rules, set of line segments,* and *set of axes.*

28. Find the missing numbers.

a. $-8 + \square = 0$ **b.** $5 \cdot \square = 1$

c. $\frac{3}{4} \cdot \square = 1$ **d.** $\square + (-3) = 0$

e. $\square \cdot \left(-\frac{1}{2}\right) = 1$ **f.** $5 + \square = 0$

29. Place the correct sign, <, =, or >, between these expressions.

a. $4 \square -3$ **b.** $2(-3) \square -2(-3)$

c. $(-2)^2 \square -2^2$ **d.** $|-4| \square |4|$

e. $-2^3 \square (-2)^3$ **f.** $|-5| \square -|5|$

g. $-\frac{1}{4} \square -\frac{1}{2}$ **h.** $-1.3 \square -1.5$

30. Each row in the table below contains equivalent statements. Fill in the blanks for each row.

	Inequality	Interval	Words	Line Graph
a.	$x \le 5$		x is less than or equal to 5	
b.	$-3 \le x < 4$			
c.		$[-3, 5]$		
d.		$(2, 4)$		
e.		$(-5, 0]$		
f.			x is greater than -5	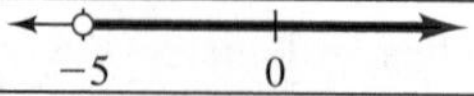
g.				
h.				
i.			x is zero or positive	
j.			x is negative	

31. Complete this table for a conditional setting. Let x be the cost of the item.

Input: Cost of Item	Inequality	Interval
Less than \$50		
\$50 to \$500		
Over \$500		

32. Complete each table, graph the points (x, y), and then connect them.

a.

Input x	Output $y = -x - 2$
-2	
-1	
0	
1	
2	

b.

Input x	Output $y = -2x$
-2	
-1	
0	
1	
2	

CHAPTER 2 TEST

1. Complete the table. Graph the points (x, y), and connect them.

x	$y = 3 - x$
-2	
0	
2	
4	

2. Simplify.

a. $\frac{3}{4} + \frac{5}{6}$ **b.** $\frac{3}{4} - \frac{5}{6}$

c. $\frac{3}{4} \cdot \frac{-5}{6}$ **d.** $\frac{-3}{4} \div \frac{5}{6}$

3. What property allows us to change $1 + 4 + 9 + 16 + 25$ to $1 + 9 + 4 + 16 + 25$?

4. What property allows us to add the problem in Exercise 3 as $(1 + 9) + (4 + 16) + 25$ instead of following the usual left-to-right order in addition?

5. Simplify.

a. $-5 + 9$ **b.** $-1.4 + 2.5 - 3.6$

c. $-4 - (-3)$ **d.** $(-3)(4)(-5)$

e. $8 - (-3)^2$ **f.** $\sqrt{26^2 - 24^2}$

g. m^2m^9 **h.** $m^7 \div m^3$

i. $36 \div 2 \cdot 2 - 3 + (3^2 - 5)$ **j.** $\dfrac{ace}{aft}$

k. $\dfrac{6x^2}{9x}$ **l.** $\dfrac{-x^3}{(-x)^2}$

m. $(a^2b^3)^3$ **n.** $\left(\dfrac{a^3}{2b^2}\right)^3$

o. $\dfrac{3x - 9}{3}$ **p.** $\dfrac{x^2 + 2x}{x}$

6. Add like terms. Remove parentheses as necessary.

a. $3x + 2y - 2x - 3y + 4x - 4y$

b. $x^3 - 3x^2 + x - 2x^2 + 6x - 2$

c. $2(x - 2) + 3(x - 1)$

d. $12(x - 1) - 5(x - 1)$

e.

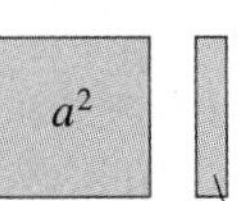

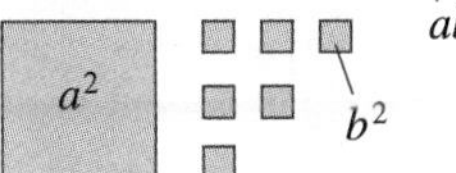

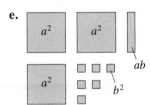

7. Complete each statement with the distributive property.

a. $3(2x + 9y)$ **b.** $3(2a + 9b)$

c. $2(3x^2 + 4x - 2)$ **d.** $ab(b - ab + a^2)$

8. Are $(-x)^2$ and $-x^2$ the same? Explain why or why not.

9. The formula for the area of a circle is $A = \pi r^2$. Let $\pi \approx 3.14$. Round to the nearest hundredth.

a. Find A if $r = 2$ ft. **b.** Find A if $r = 20$ ft.

c. Divide the areas to find how many times as large as the area in part a the area in part b is. Complete this statement: When the radius is multiplied by 10, the area of a circle is multiplied by ____ .

10. The record for airplane flight duration is almost 5,616,000 seconds. Change the seconds into days. (The record is 2 hours less, but the above number comes out to an even number of days. The flight was refueled in the air.)

11. The area of the world's largest mural (in Long Beach Arena, CA) is 18,446,400 square inches. (12 inches = 1 foot, 1 yard = 3 feet.)

a. Change the area into square feet.

b. If the height of the rectangular mural is 35 yards, what is the length?

c. What is the perimeter of the mural in yards?

12. What is the perimeter of each figure? Let $\pi \approx 3.14$. Round to the nearest hundredth.

a.

1.5 ft
0.5π ft
0.5π ft
1.5 ft

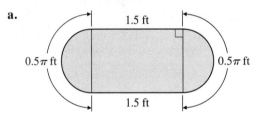

b.

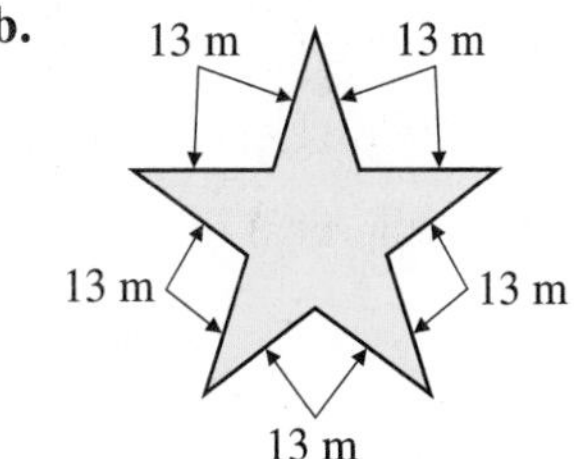

13. What are the volume and surface area of each figure? Round to the nearest whole number. Let $\pi \approx 3.14$.

a.

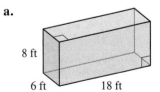

8 ft
6 ft
18 ft

b.

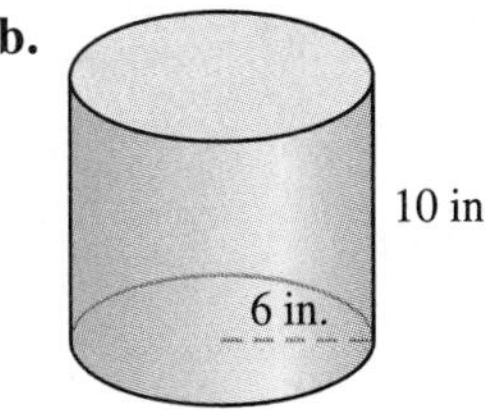

In Exercises 14 and 15, complete the table.

14.

	Inequality	Interval	Line Graph
0 to 20			
More than 20 and less than 50	$20 < x < 50$	(20, 50)	
50 or greater			

15.

x	y	$x + y$	$x - y$	$x \cdot y$	$x \div y$
−8	4				
−6		−8			
			9	−18	

CUMULATIVE REVIEW OF CHAPTERS 1 AND 2

These exercises highlight material and combine concepts from Chapters 1 and 2. You may not have seen the problems before, but you have been introduced to the required skills.

1. Make an input-output table for each rule, with integer inputs from −2 to 3. Graph the (x, y) ordered pairs for each rule on separate axes.

a. $y = x^2 - 1$ **b.** $y = 2 - x$ **c.** $y = 2x + 3$

d. $y = -2x$ **e.** $y = -x + 1$ **f.** $y = |x - 1|$

2. Translate into symbols.

a. The sum of the absolute value of −5 and 14

b. The quotient of the opposite of 15 and −3

c. The product of $\frac{1}{4}$ and the reciprocal of 1.5

d. 6 is greater than x.

e. x is less than 15.

3. Write in words.

a. $4 < x$ **b.** $5 - 3x$

c. $-(-x)$ **d.** $|3 - x|$

In Exercises 4 and 5, add like terms. Simplify expressions to remove parentheses as necessary.

4. a. $-2x + 3y - 4x + 2x - 5y$

b. $x^3 + 2x^2 - x - 3x^2 - 6x + 3$

5. a. $3(x + 1) + 5(x - 2)$ **b.** $8 - 2(x - 1)$

c. $8(x + 2) - 3(x - 2)$ **d.** $5 - 3(2 - x)$

6. Change to like units and add. (16 ounces = 1 pound.)

a. 3 feet + 24 inches + 2 yards

b. 2.5 pounds + 8 ounces + 1.5 pounds

7. Complete the table.

x	y	$x + y$	$x - y$	$x \cdot y$	$x \div y$
−4		2			
5				−15	
$-\frac{2}{3}$	$\frac{3}{4}$				
$\frac{7}{8}$	$-\frac{7}{10}$				
1.44	1.8				
0.25	−0.5				

8. Write the correct symbol between these expressions. Choose from <, =, and >.

a. $|-4| \square |3|$ **b.** $-|-6| \square -(-3)$

c. $-(-5) \square |4 - 7|$ **d.** $|13| \square -(-11)$

e. $-(-9) \square |-12|$ **f.** $-|15| \square |2 - 5|$

9. Simplify.

a. $+68 - 74 - 26 + 32 + 14$

b. $-16 + 18 - 35 + 12 - 15 - 24$

c. $5 \div \frac{2}{3} \cdot 4$ **d.** $7 \div \frac{3}{2} \cdot 6$

e. $\frac{a}{b} \div \frac{a}{b}$ **f.** $\frac{x}{y} \div \frac{-x}{y}$

g. $\frac{6a + 2}{3}$ **h.** $\frac{xy - x^2}{x}$

10. Simplify.

a. 3^4 **b.** $\left(-\frac{1}{2}\right)^4$ **c.** $(3n)^3$

d. $5(2x)^1(3y)^4$ **e.** $(3x^2)^4$ **f.** $\left(-\frac{3}{4}b^2\right)^2$

g. $\left(\frac{2x}{3x^2}\right)^2$ **h.** $\left(-\frac{0.2x}{y^3}\right)^3$

11. Concrete is sold by the cubic yard. We wish to make a sidewalk 40 inches wide, 8 inches thick, and 20 feet long. How many cubic yards of concrete will be needed?

12. The oval shape shown is called an *ellipse*. The formula for the area of an ellipse is $A = \pi rR$. Use $\pi \approx 3.14$.

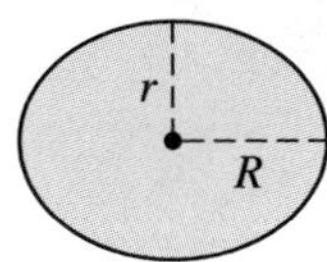

a. Find the area of an ellipse with $R = 4$ and $r = 3$.

b. If $r = R$, what shape is created?

13. A portion of a mileage chart for Pennsylvania is shown below.

	Erie	**Pittsburgh**	**Reading**	**Scranton**
Erie	—	135	325	300
Pittsburgh		—	255	272
Reading			—	99
Scranton				—

a. What assumption about the distances between two cities makes it possible to complete the table?

b. Complete the table.

c. Which number property is similar to our assumption about the table?

14. Adam Moving Company decides to double the dimensions (length, width, and height) of its book boxes so that movers can put more books in each box. The original book boxes were cubes, 1.5 feet on each side.

a. What is the volume of the old box?

b. Find the new length, width, and height.

c. What is the volume of the new "double" box?

d. How many times as large in volume is the new box?

e. If the original box held 30 pounds of books, what would the new box, full of books, weigh?

f. What happened to the employee who thought of this idea?

3

Solving Equations and Inequalities in One Variable

PARTY PACKAGES

- **American Gymnastics Training Center:** $70 up to 15 kids, $3 each additional. Includes instructor, gym activities, party room, supplies, balloons, treat bags.
- **Fairfield Lanes:** $5.50 per child includes bowling, party with balloons, food, drinks.
- **Farrell's Ice Cream Parlour & Restaurant:** Four packages, $3.95 to $5.50 per child. Includes hats, balloons, drinks, ice cream, and/or meal.
- **Fun Base One:** $30 for eight kids, $3.50 each additional. Includes games, party room with supplies.
- **Grand Slam U.S.A.:** Four packages, $5.95 to $10.95 per child. Includes games, ice cream, and/or meal, drinks, party favors.
- **Lane County Ice:** $85 for 10 kids, $4.75 each additional. Includes skate rental, lesson, and party with cake, ice cream, punch, hats, balloons.
- **Papa's Pizza:** $3 per child includes pizza, drinks, ice cream, candy, hats, balloons. Playground.
- **Pietro's Pizza Restaurants:** $3 per child includes pizza, drinks, toys, candy, and decorations. Playgrounds; one location has merry-go-round.
- **Skate World:** $40 to $55 for 10 skaters, admission plus 50 cents each additional. Includes skating, party room, ice cream, and drinks. Add $10 for balloons, favors, hats.
- **The Little Gym:** $75 for up to eight children, $100 for nine to 14, then $6 each additional. Includes gym, party room, supplies, juice.
- **Willamalane Wave Pool:** Admission ($2 each in district, $3 outside) plus $10 an hour for party room.

Figure 1

Whether you are planning a birthday party (with options as shown in Figure 1), a wedding reception, or a grand opening of a new business, you are faced with a number of questions. How much will the event cost? What choices do you have within your given budget? For how many people will two different choices have the same cost? Equations and inequalities can help you answer these questions.

In this chapter, we solve equations and inequalities in one variable. We solve equations in three ways: with symbols, tables, and graphs. We focus on graphic and symbolic solutions of inequalities.

3.1 Solving Equations with Algebraic Notation

OBJECTIVES

- Identify identities, conditional equations, and equivalent equations.
- Translate word sentences into equations in one variable.
- Solve equations in one variable.
- Use the addition property of equations.
- Use the multiplication property of equations.

WARM-UP

Rewrite each exercise using an inverse operation and then simplify.
Example: $3 - (-4) = 3 + (+4) = 7$

1. $3 - 4$ **2.** $-3 - (-4)$
3. $-3 - 4$ **4.** $6 \div \frac{1}{2}$
5. $-4 \div \frac{2}{3}$ **6.** $9 \div \frac{3}{4}$

Evaluate each expression. Describe in a complete sentence the order of operations you use to evaluate each. (*Hint:* Start with "Take the number replacing x, . . .")

7. $3x - 2$ for $x = 1$ **8.** $2 - 6x$ for $x = -\frac{1}{2}$
9. $4x + 1$ for $x = 2.25$ **10.** $\frac{2}{3}x + 4$ for $x = 27$

IN THIS SECTION, we review several uses of the equal sign and then define an equation and an identity. By writing equations from word statements, we practice recognizing the operations contained in an equation. Then we look at where to start and what steps to follow in using algebraic notation to solve an equation.

Equations

THE EQUAL SIGN The **equal sign**, =, placed between two numbers *says that the numbers are equal.* Thus, $\frac{1}{2} = 0.5$ indicates that the fraction notation $\frac{1}{2}$ is equal in value to the decimal notation 0.5. *We use the equal sign between steps in changing notation or expressions*—for example, when we simplify fractions to lowest terms or use the order of operations to simplify a numerical expression.

EXAMPLE  Using the equal sign Tell how the equal sign is used in each of the following.

a. $0.75 = 75\%$ **b.** $\dfrac{15}{18} = \dfrac{5 \cdot 3}{6 \cdot 3} = \dfrac{5}{6}$

c. $\dfrac{13}{5} = \dfrac{10 + 3}{5} = \dfrac{10}{5} + \dfrac{3}{5} = 2 + \dfrac{3}{5} = 2\dfrac{3}{5}$

d. $8 - 3(2 - 5) = 8 - 3(-3) = 8 + 9 = 17$

Solution The equal sign says that the numbers or expressions on each side are equal

a. when we change from decimal to percent notation.

b. when we simplify a fraction to lowest terms.

c. when we change an improper fraction to a mixed number.

d. when we apply the order of operations step by step.

CONDITIONAL EQUATIONS Most of the equations we write in algebra are conditional equations.

Definition of Conditional Equation

An equation is a statement of equality between two expressions. A **conditional equation** is an equation that is true for only certain values of the variable(s) in the equation.

The conditional equation $3x = 6$ is true for $x = 2$ and no other values of x. When you see the word *equation,* it usually means conditional equation.

IDENTITIES An *identity* is a special type of equation that is true for all numbers. When we added like terms in Section 2.3, in problems like $5x + 6x = 11x$, we used the equal sign. Here the equal signs says that $5x + 6x$ has the same value as $11x$ for all values of x. Because the expressions on the left and right sides are identical in value, $5x + 6x = 11x$ is called an identity.

Definition of Identity

An **identity** is formed when the expression on the left side of the equal sign is equal to the expression on the right side for all values of the variable(s).

The statement of the associative property of multiplication, $a(b \cdot c) = (a \cdot b)c$, is an identity because it is true for all real numbers a, b, and c.

VARIABLES The statements $n = 4$, $2n = 8$, and $n + 1 = 5$ are examples of **equations in one variable** as *these statements contain only a single variable* (in this case, n).

Equations can have more than one variable. "The output is 2 less than 3 times the input" is written as $y = 3x - 2$, a *two-variable equation.* The formula $V = l \cdot w \cdot h$ is an equation in *four variables,* V, l, w, and h.

EXAMPLE 2 Identifying types of equations What kind of equation is each of the following—an identity or a conditional equation? State how many variables each conditional equation contains.

a. $x - 3 = 15$ **b.** $y = x - 3$

c. $a(b + c) = ab + ac$ **d.** $C = 2\pi r$

e. $A = \frac{h}{2}(a + b)$ **f.** $b + c = c + b$

Solution

a. Conditional equation, 1 variable

b. Conditional equation, 2 variables

c. Identity. This equation states the distributive property of multiplication over addition, which is true for all real numbers.

d. Conditional equation, 2 variables. The letter π is a constant.

e. Conditional equation, 4 variables

f. Identity. This equation states the commutative property of addition, which is true for all real numbers.

Writing Equations in One Variable

We will now review writing equations in algebraic notation. Look carefully at how the operations (addition, subtraction, multiplication, and division) are written in the equations. In order to solve equations, you want to be in the habit of reading the algebraic notation as numbers and operations.

When writing equations from sentences, we use the equal sign to replace the word *is*. Words such as *gives* or *to get* may also mean *equals.* In Section 1.2, we looked at words such as *sum, difference, product,* and *quotient,* which describe the answer to an operation. You may want to review that section if any of the words in Example 3 are not familiar.

Writing Equations

> In writing equations, always define the variables, as in "let x be the input number," "let t be time," or "let n be the missing number."

EXAMPLE 3

Writing sentences as equations in one variable Write an equation for each sentence. Let x be the unknown number.

a. The product of 6 and a number is 24.

b. Half of a number is 24.

c. A number subtracted from 15 is 10.

d. The sum of a number and 6 is -3.

e. A number is multiplied by two and then added to three to get five.

f. Thirty less than four times a number is 6.

g. The difference between 14 and twice a number is a negative eight.

h. Subtracting three times a number from five gives eight.

Solution

a. *Product* means that 24 is the answer to a multiplication: $6x = 24$.

b. Here we think of *half* as multiplication by $\frac{1}{2}$: $\frac{1}{2}x = 24$.

We may also think of *half* as division by 2: $\dfrac{x}{2} = 24$.

c. The *from 15* tells us to write the 15 first: $15 - x = 10$.

d. *Sum* means that x and 6 are added to obtain the answer of -3: $x + 6 = -3$.

e. Here *to get* means the equal sign: $2x + 3 = 5$.

f. Here *less than* means that 30 is subtracted from $4x$: $4x - 30 = 6$.

g. We assume that 14 is written first in the equation because it is written first in the sentence: $14 - 2x = -8$.

h. The word *from* tells us that 5 is written first. The word *gives* locates the equal sign: $5 - 3x = 8$. ●

In Example 4, we write word statements from equations. Get in the habit of saying the operations as you read equations. This will help you understand solving equations.

EXAMPLE 4

Writing sentences from equations Write each equation in words. Start each sentence with "A number. . . ."

a. $x + 4 = -3$ **b.** $3x = 75$ **c.** $6 - x = 10$

Solution Answers may vary.

a. A number added to 4 gives negative 3.

b. A number is multiplied by 3 to give 75.

c. A number subtracted from 6 gives 10. ●

Some equations can be written as sentences in many ways. The task in Example 5 is to find a sentence that does not describe a given equation.

EXAMPLE 5 Translating sentences into equations Which sentence does not describe the equation $6 - 2x = 10$?

a. A number is multiplied by -2 and then added to 6 to get 10.

b. The difference between six and twice a number is 10.

c. Six less twice a number is 10.

d. A number is multiplied by 2 and then subtracted from 6 to get 10.

e. Six less than twice a number is 10.

Solution See the Answer Box. ●

Solving Equations

When we find the numbers that make a conditional equation true, we solve the equation.

Useful Terms

> A **solution** to an equation is a value of the variable that makes the equation true.
> The **solution set** is the set of all solutions to an equation.
> **Solving an equation** is the process of finding the values of the variable for the solution set.

One of the most effective ways to solve an equation is to work backwards.

WORKING BACKWARDS Working backwards from the answer is a common problem-solving strategy. (How often have you not known what to do with a problem and been helped by looking at the answer in the back of the book?)

EXAMPLE 6 Exploration: working backwards I am thinking of a number. When I multiply it by 3 and subtract 2, I get 10. What is my number? (See Figure 2.)

Figure 2

Solution The strategy is to work backwards. If the result was 10 and it was found by subtracting 2, then the prior number must have been $10 + 2$, or 12. If the answer after multiplying the original number by 3 is 12, then the original number must be $12 \div 3$, or 4.

Check: Multiplying 4 by 3 and subtracting 2 gives 10. We can write the check step as an equation, $3x - 2 = 10$, and substitute 4 for x: $3(4) - 2 \stackrel{?}{=} 10$. ✔

The symbol $\stackrel{?}{=}$ in the Check statement asks the question "Does it equal?" The symbol ✔ says "Yes, it does." Checking is an important step in solving equations.

Example 7 contains four one-operation exercises to work backwards.

EXAMPLE 7 Working backwards What is my number?

a. When I add 5 to my number I get 11. What is my number?

b. When I subtract 7 from my number I get 8. What is my number?

c. When I divide my number by 3, I get 6. What is my number?

d. When I multiply my number by 2, I get 15. What is my number?

Solution

a. If the result is 11 after addition of 5, then the number must be 11 take away 5, or $11 - 5 = 6$.

Check: $5 + 6 \stackrel{?}{=} 11$ ✔

b. If the result is 8 after subtraction of 7, then the number was 8 plus 7, or $8 + 7 = 15$.

Check: $15 - 7 \stackrel{?}{=} 8$ ✔

c. If the result is 6 after division by 3, then the number was 6 times 3, or $6 \cdot 3 = 18$.

Check: $18 \div 3 \stackrel{?}{=} 6$ ✔

d. If the result is 15 after multiplication by 2, then the number was 15 divided by 2, or $15 \div 2 = 7.5$.

Check: $7.5 \cdot 2 \stackrel{?}{=} 15$ ✔

INVERSE OPERATIONS WITH ONE OPERATION In algebra, we use the term *inverse operations* to describe working backwards to solve an equation. Solving equations using inverse operations usually requires writing one or more equivalent equations.

Definition of Equivalent Equations

> **Equivalent equations** are two or more equations that have the same solution set.

We say that $x + 6 = 9$ and $x = 9 - 6$ are equivalent equations because, for each equation, the solution 3 makes the equation true.

Two properties of equations—the addition property of equations and the multiplication property of equations—allow us to use inverse operations to write equivalent equations.

Properties of Equations

The **addition property of equations** states that adding the same number to both sides of an equation produces an equivalent equation.

If $a = b$, then $a + c = b + c$

The **multiplication property of equations** states that multiplying both sides of an equation by the same nonzero number produces an equivalent equation.

If $a = b$ and $c \neq 0$, then $ac = bc$

We omit 0 in the multiplication property because we will be including division by c in some equations.

In Example 7, we used working backwards, or inverse operations, to solve the problems. We now use the inverse operations along with the addition and multiplication properties of equations to find equivalent equations and solve the equation.

EXAMPLE 

Solving equations with inverse operations and properties of equations Solve each equation and check your solution. State the inverse operation and the property of equations used.

a. $x - 2 = -5$ **b.** $\frac{x}{5} = 2.5$

Solution **a.** In $x - 2 = -5$, we subtract 2 from x. The inverse operation is to add 2.

Student Note: The comments to the right tell what was done at each step.

$x - 2 = -5$ State the equation.

$\quad + 2 \quad +2$ Add 2 on each side (addition property).

$x \quad = -3$

Check: $-3 - 2 \stackrel{?}{=} -5$ ✔

b. In $\frac{x}{5} = 2.5$, we divide x by 5. The inverse operation is to multiply by 5.

$\frac{x}{5} = 2.5$ State the equation.

$\frac{5}{1} \cdot \frac{x}{5} = 5(2.5)$ Multiply each side by 5 (multiplication property). We write $\frac{5}{1}$ instead of 5 on the left to remember to multiply the numerators and the denominators of the fractions.

$x = 12.5$

Check: $\frac{12.5}{5} \stackrel{?}{=} 2.5$ ✔

The addition and multiplication properties also apply to subtraction and division. Recall that in Sections 2.1 and 2.2 we changed subtraction to *addition of the opposite* and changed division to *multiplication by the reciprocal.* Exercises 1 to 6 in the Warm-up for this chapter reviewed these operations.

EXAMPLE 

Solving equations with inverses State the inverse operation and solve the equation.

a. $4 + x = 11$ **b.** $-2x = 12$

Solution **a.** In $4 + x = 11$, we add 4 to x. The inverse operation is to subtract 4.

$$\begin{aligned} 4 + x &= 11 && \text{State the equation.} \\ -4 \quad &\quad -4 && \text{Subtract 4 from each side (addition property).} \\ x &= 7 \end{aligned}$$

Check: $4 + 7 \stackrel{?}{=} 11$ ✓

b. In $-2x = 12$, we multiply x by -2. The inverse operation is to divide by -2.

$$\begin{aligned} -2x &= 12 && \text{State the equation.} \\ \frac{-2x}{-2} &= \frac{12}{-2} && \text{Divide each side by } -2 \text{ (multiplication property).} \\ x &= -6 \end{aligned}$$

Check: $(-2)(-6) \stackrel{?}{=} 12$ ✓ ●

INVERSE OPERATIONS WITH TWO OPERATIONS When an equation has more than one operation, such as the equation $5x - 1 = 9$, we need to know the answer to the question "Which operation is done first in solving the equation for x?" To find the answer, we look at the order of operations on x and then reverse that order and do inverse operations. In Warm-up Exercises 7 to 10, we wrote a sentence to describe the order of operations on x. Writing such a sentence is the first step in our plan in Example 10.

EXAMPLE **10** Solving an equation Solve the equation $3x - 1 = 12.5$, using the following four-step process.

Understand: Make an estimate.

Plan: Write the order of operations on x, and then list the order backwards with the inverse operations.

Carry out the plan.

Check.

Solution ***Understand:*** Because 3(4) is 12, x is close to 4.

Plan: In the equation, we multiply x by 3 and then we subtract 1. The reverse order of operations with inverses is to add 1 and then divide by 3.

Carry out the plan:

$$\begin{aligned} 3x - 1 &= 12.5 && \text{Write the equation.} \\ +1 \quad &\quad +1 && \text{Add 1 to each side.} \\ 3x \quad &= 13.5 \\ \frac{3x}{3} &= \frac{13.5}{3} && \text{Divide by 3 on each side.} \\ x &= 4.5 \end{aligned}$$

Check: $3(4.5) - 1 \stackrel{?}{=} 12.5$ ✓ ●

Summary of Solving Equations

> The solution to an equation requires the opposite, or inverse, of each operation in the reverse order of operations on x.

In the next few problems, we will focus on the plan. It is a good idea to estimate the solution; if you do not estimate, be sure to check your answer.

EXAMPLE 11 Stating a plan Solve $3x - 2 = 1$.

Solution ***Plan:*** In the equation, we multiply x by 3 and subtract 2. To solve, we will add 2 and then divide by 3.

$$3x - 2 = 1 \quad \text{Write the equation.}$$
$$3x - 2 + 2 = 1 + 2 \quad \text{Add 2 to each side.}$$
$$3x = 3$$
$$\frac{3x}{3} = \frac{3}{3} \quad \text{Divide each side by 3.}$$
$$x = 1$$

Check: $3(1) - 2 \stackrel{?}{=} 1$ ✔

In Example 11, the addition step is on the same line as the equation. In Example 10, the addition step is below the equation. Use whichever method you or your instructor prefers.

In Example 12, we change a subtraction to addition of the opposite in order to write the order of operations.

EXAMPLE 12 Solving equations with subtraction of the variable term Write a plan and then solve the equation $5 = 2 - 6x$.

Solution ***Plan:*** The equation may be written as $5 = 2 + (-6)x$. We multiply x by -6 and then add 2. To solve, we will subtract 2 and then divide by -6.

Student Note:
The comments to the right tell what to do in the next step. Try writing the next step before reading it.

$$5 = 2 - 6x \quad \text{Change the subtraction to addition of the opposite.}$$
$$5 = 2 + (-6)x \quad \text{Subtract 2 from each side.}$$
$$5 - 2 = 2 + (-6)x - 2$$
$$3 = (-6)x \quad \text{Divide each side by } -6.$$
$$\frac{3}{-6} = \frac{(-6)x}{-6} \quad \text{Simplify.}$$
$$-\tfrac{1}{2} = x$$

Check: $5 \stackrel{?}{=} 2 - 6\left(-\tfrac{1}{2}\right)$ ✔

In Examples 13 and 14, we have fractions as coefficients of the variable. The examples show two different ways to solve an equation containing fractions. In Example 13, we use the fact that division by $\frac{2}{3}$ is the same as multiplication by $\frac{3}{2}$.

EXAMPLE 13 Solving fractional equations Solve $\frac{2}{3}x + 4 = 22$.

Solution ***Plan:*** Because $\frac{2}{3}$ is multiplied by x before 4 is added, we will subtract 4 and then divide by $\frac{2}{3}$.

$$\tfrac{2}{3}x + 4 = 22 \quad \text{Subtract 4 from each side.}$$
$$\tfrac{2}{3}x + 4 - 4 = 22 - 4 \quad \text{Add like terms.}$$
$$\tfrac{2}{3}x = 18 \quad \text{Divide each side by } \tfrac{2}{3}.$$
$$\tfrac{2}{3}x \div \tfrac{2}{3} = 18 \div \tfrac{2}{3} \quad \text{Change } \div \tfrac{2}{3} \text{ to } \cdot \tfrac{3}{2}.$$
$$\tfrac{2}{3}x \cdot \tfrac{3}{2} = 18 \cdot \tfrac{3}{2} \quad \text{Simplify.}$$
$$x = 27$$

Check: $\frac{2}{3}(27) + 4 \stackrel{?}{=} 22$ ✔

The inverse nature of multiplication and division means we can multiply by a reciprocal rather than divide by a fraction.

EXAMPLE 14 Using an alternative solution method for fractional equations Solve $\frac{2}{3}x + 4 = 22$.

Solution ***Plan:*** Because $\frac{2}{3}$ is multiplied by x before 4 is added, we will subtract 4 and then multiply by the reciprocal of $\frac{2}{3}$.

$\frac{2}{3}x + 4 = 22$	Subtract 4 from each side.
$\frac{2}{3}x + 4 - 4 = 22 - 4$	Simplify.
$\frac{2}{3}x = 18$	Multiply by the reciprocal, $\frac{3}{2}$, on each side.
$\frac{3}{2} \cdot \frac{2}{3}x = \frac{18}{1} \cdot \frac{3}{2}$	Simplify.
$1x = \frac{54}{2}$	Simplify.
$x = 27$	

Check: $\frac{2}{3}(27) + 4 \stackrel{?}{=} 22$ ✔ ●

Graphing Calculator Technique: Checking a Solution

The =, >, <, ≥, ≤ options under the test key may be used to check equations and inequalities on the calculation screen. If you apply one of these options to any statement, the calculator will give 1 if the statement is true or 0 if the statement is false.

To check Example 12, enter $5 = 2 - 6(-1/2)$. Be sure to use a negative sign, not a subtraction sign, in front of the 1/2. The calculator will give 1 for true.

To check Example 13, enter $(2/3) \times 27 + 4 = 22$. The calculator will give 1 for true.

ANSWER BOX

Warm-up: **1.** $3 + (-4) = -1$ **2.** $-3 + (+4) = 1$ **3.** $-3 + (-4) = -7$ **4.** $6 \times \frac{2}{1} = 12$ **5.** $-4 \times \frac{3}{2} = -6$ **6.** $9 \times \frac{4}{3} = 12$ **7.** 1. Take the number replacing x, multiply by 3, and subtract 2. **8.** 5. Take the number replacing x, multiply by a negative 6, and add 2. **9.** 10. Take the number replacing x, multiply by 4, and add 1. **10.** 22. Take the number replacing x, multiply by $\frac{2}{3}$, and add 4. **Example 5:** The sentence in part e means $2x - 6 = 10$.

EXERCISES 3.1

In Exercises 1 to 8, copy the problem and do the indicated operation. Use equal signs in your work.

1. Simplify to lowest terms: $\frac{28}{35}$.

2. Change 0.8 to a percent.

3. Change $\frac{45}{6}$ to a mixed number.

4. Simplify to lowest terms: $\frac{32}{40}$.

5. Change 0.3 to a percent.

6. Simplify $4 - 2(3 - 6)$.

7. Simplify $8 - 3(2 - 8)$.

8. Change $\frac{24}{9}$ to a mixed number.

In Exercises 9 and 10, describe the equation as an identity or a conditional equation. For conditional equations, state the number of variables in the equation.

9. a. $x + 4 = y$ **b.** $x + 4 = -7$

c. $A = \frac{1}{2}bh$ **d.** $a(b \cdot c) = (ab) \cdot c$

e. $3x + 4x = 7x$ **f.** $n + 0.03n = 1.03n$

10. a. $y = 2x + 3$ **b.** $\frac{ab}{c} = a \cdot \frac{b}{c}$

c. $\frac{-a}{b} = -\frac{a}{b}$ **d.** $5x - 2x = 3x$

e. $2x = 7$ **f.** $x + 0.05x = 1.05x$

In Exercises 11 to 22, write equations. Let x be the unknown number.

11. Half of a number is 8.

12. Twice a number is 2.5.

13. Six plus a number is 4.

14. Eight subtracted from a number is 12.

15. Fifteen less than twice a number is -9.

16. Seven is two more than three times a number.

17. Nineteen is the sum of three and twice a number.

18. Twelve is 5 less than twice a number.

19. Four less than 3 times a number is 17.

20. Eight more than 5 times a number is 43.

21. The difference between 26 and 4 times a number is 2.

22. Subtracting three times a number from 7 is -2.

In Exercises 23 to 30, write each equation as a sentence.

23. $x - 4 = 6$ **24.** $5 - 3x = -4$

25. $3x - 5 = 16$ **26.** $\frac{3}{4}x = 27$

27. $\frac{2}{3}x = 24$ **28.** $4 - x = 15$

29. $6 - 2x = 10$ **30.** $x + 3 = -10$

In Exercises 31 to 34, check these "solutions" to equations by substituting the numbers into the equation. Note any wrong "solutions." Explain how to solve the equation correctly.

31. a. $2n = 8, n = 4$

b. $x + 1 = -5, x = -6$

c. $n - 3 = 15, n = 12$

32. a. $3n = 6, n = 3$

b. $1 - x = 4, x = 3$

c. $\frac{1}{2}n = 15, n = 30$

33. a. $55t = 440, t = 20$

b. $\frac{1}{2}x = 10, x = 5$

c. $1.05x = 42, x = 40$

34. a. $5r = 320, r = 64$

b. $2x - 3 = 13, n = 8$

c. $0.9x = 7.2, x = 8$

Copy the equations in Exercises 35 to 52. Show the inverse operation. Write the equivalent equation that solves the original equation.

35. $x - 5 = 8$ **36.** $x + 4 = 7$

37. $9 = x + 12$ **38.** $x - 6 = -3$

39. $x - 6 = -10$ **40.** $17 = x - 5$

41. $2x = 26$ **42.** $3x = 36$

43. $3 = 8x$ **44.** $2 = 7x$

45. $\dfrac{x}{4} = 16$ **46.** $\dfrac{x}{3} = 12$

47. $\dfrac{-x}{12} = 4$ **48.** $\dfrac{-x}{10} = 5$

49. $-4 = \frac{1}{2}x$ **50.** $-8 = \frac{1}{4}x$

51. $-\frac{3}{4}x = 12$ **52.** $-\frac{2}{3}x = 6$

Match each equation in Exercises 53 to 58 with one of sentences a to g. Copy the equation and its sentence, and then solve the equation.

a. x times -4 plus 2 gives 3.

b. x times $\frac{1}{4}$ subtract 2 gives 3.

c. x times $\frac{1}{2}$ subtract 4 gives 3.

d. x times 4 plus 2 gives 3.

e. x times -2 subtract 4 gives 3.

f. x times -2 plus 4 gives 3.

g. x times 2 plus 4 gives 3.

53. $4 - 2x = 3$ **54.** $2 - 4x = 3$ **55.** $\frac{1}{2}x - 4 = 3$

56. $2x + 4 = 3$ **57.** $\frac{1}{4}x - 2 = 3$ **58.** $4x + 2 = 3$

Solve for x in each of the equations in Exercises 59 to 80.

59. $4x + 3 = 23$ **60.** $5x - 2 = 23$

61. $3x - 2 = 43$ **62.** $6x + 4 = 52$

63. $10 - x = -2$ **64.** $4 - 2x = 10$

65. $3 = 10 - x$ **66.** $-3 = 4 - 2x$

67. $0 = 3 - 3x$ **68.** $6 = 2 - 4x$

69. $3 - 3x = 9$ **70.** $2 - 4x = -1$

71. $\frac{1}{2}x + 3 = -2$ **72.** $\frac{1}{2}x - 2 = -4$

73. $0 = \frac{3}{2}x + 3$ **74.** $0 = \frac{3}{2}x - 2$

75. $-4 = \frac{2}{3}x - 2$ **76.** $12 = \frac{2}{3}x - 2$

77. $4.2x - 3 = -9.3$ **78.** $2.5x + 2 = -3.5$

79. $7.5 = 4.2x - 3$ **80.** $5 = 2.5x + 2$

In Exercises 81 to 88, tell what was done to the first equation to get to the second equation.

81. $2x = 10$; $x = 5$

82. $x + 4 = 6$; $x = 2$

83. $2x + 5 = 11$; $2x = 6$

84. $2x - 5 = 11$; $2x = 16$

85. $2x - 4 = 10$; $2x = 14$

86. $x - 7 = -3$; $x = 4$

87. $\frac{1}{2}x = 15$; $x = 30$

88. $\frac{1}{2}x = 10$; $x = 20$

Applications

The equations in Exercises 89 to 92 have letters other than x as variables. Solve for the variable.

89. $110 = 55t$

90. $150 = 3r$

91. $212 = \frac{9}{5}C + 32$

92. $32 = \frac{9}{5}C + 32$

In Exercises 93 to 102, place the given number in the formula and solve for the remaining letter.

93. $D = 55t$; if $D = 200$ solve for t

94. $D = 55t$; if $D = 450$ solve for t

95. $D = 3r$; if $D = 200$ solve for r

96. $D = 3r$; if $D = 450$ solve for r

97. The 6% tax on a purchase, p, is $T = 0.06p$. If $T = \$0.10$, solve for p.

98. The 6% tax on a purchase, p, is $T = 0.06p$. If $T = \$0.25$, solve for p.

99. Electricity cost is $C = \$5.00 + \$0.03715x$, where x is the number of kilowatt hours used. If $C = \$75$, solve for x.

100. Electricity cost is $C = \$5.00 + \$0.03715x$, where x is the number of kilowatt hours used. If $C = \$50$, solve for x.

101. The area of a triangle is $A = \frac{1}{2}bh$. If $A = 15$ and $b = 3$, solve for h.

102. The area of a triangle is $A = \frac{1}{2}bh$. If $A = 28$ and $h = 7$, solve for b.

103. Explain why we can use the addition property of equations to subtract the same number on both sides of an equation.

104. Explain how to solve $ax = b$ for x.

105. Explain how to solve $x - a = b$ for x.

106. Explain how to solve $ax + b = 0$ for x.

107. Explain the difference between an identity and a conditional equation.

Projects

108. ***Equivalent Equations.*** Which equation, if any, in each set is not equivalent to the other three? Show clearly how you made your choice.

a. $2x = 6$, $6 \div 2 = x$, $2 \div x = 6$ (for x not zero), $6 = 2x$

b. $x + 3 = -2$, $x = -6$, $x = -2 \cdot 3$, $x \div (-2) = 3$

c. $5x + 4 = 24$, $5x = 20$, $24 = 5x + 4$, $4 - 24 = 5x$

d. $4 + x = 9$, $9 = x + 4$, $9 - 4 = x$, $9 - x = 4$

e. $x - 6 = -3$, $x + 3 = 6$, $x - 3 = 0$, $x = -9$

f. $2x + 3 = 15$, $2x = 12$, $3 - 15 = 2x$, $15 = 3 + 2x$

109. ***Inequality or Equation.*** When we place variables in inequalities, we obtain a *conditional inequality* if the inequality is not true for some value of the variable(s). In the following review of the names of algebraic statements, x is the unknown number.

"Four is less than five" is an inequality: $4 < 5$.

"Four is less than five times a number" is a conditional inequality: $4 < 5x$.

"Four less than a number is 5" is a conditional equation: $x - 4 = 5$.

"Four less than nine is five" is an identity: $9 - 4 = 5$.

Write each statement below in algebraic notation. Explain why each is an inequality, a conditional inequality, a conditional equation, or an identity. Let x be the unknown number.

a. Six is less than eight.

b. Six is two less than eight.

c. Four greater than negative three is one.

d. Four is greater than negative three.

e. Four greater than a number is negative three.

f. Four times negative three is greater than negative thirty.

g. Negative three is greater than a number.

h. Negative three is less than three.

i. Negative three less three is negative six.

j. Negative three is greater than negative six.

k. Negative six divided by a number is two.

l. Negative six divided by negative three is greater than zero.

3.2 Solving Equations with Tables and Graphs

OBJECTIVES

- Identify linear and nonlinear equations.
- Write sentences as two-variable equations, and write equations as sentences.
- Identify independent and dependent variables.
- Find the solution to an equation from a table and a graph.
- Interpret the meaning of the intersection of a graph with the x and y axes.

WARM-UP

Complete the tables for the given inputs.

1.

Input x	Output $3x - 2$
-1	
0	
1	
2	

2.

Input x	Output $2 - 6x$
-1	
0	
1	
2	

3.

Input x	Output x^2
-2	
-1	
0	
1	
2	

4.

Input x	Output $\sqrt{x}$
0	
1	
4	
9	

5. What is the rule describing the pattern below? Let x be the number of the figure in the pattern sequence and y be the number of squares.

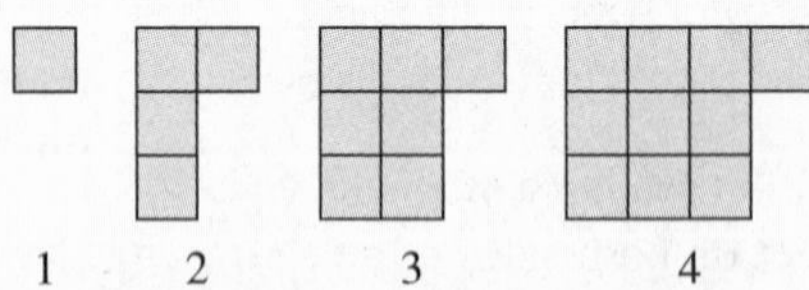

IN THIS SECTION, we return to the rules that tell us how to get from the input to the output. We state the rules with two-variable equations. We practice writing equations from sentences and changing sentences into equations. We use tables and graphs to solve equations.

Linear Equations

EXAMPLE 1 Reviewing graphing Graph the ordered pairs from the four Warm-up tables. Let y be the output.

a. $y = 3x - 2$ **b.** $y = 2 - 6x$ **c.** $y = x^2$ **d.** $y = \sqrt{x}$

Solution

a.

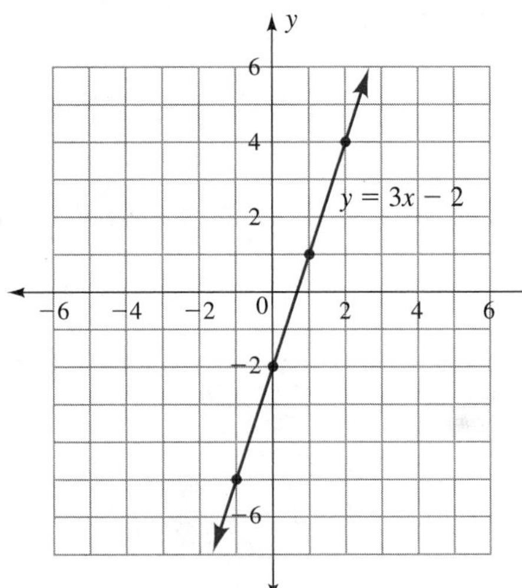

b.

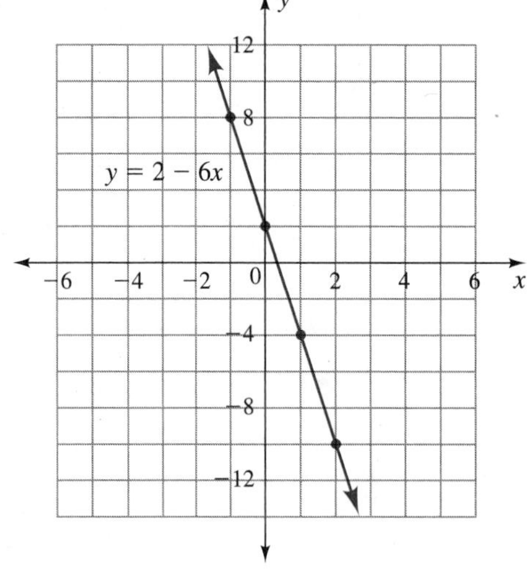

c.

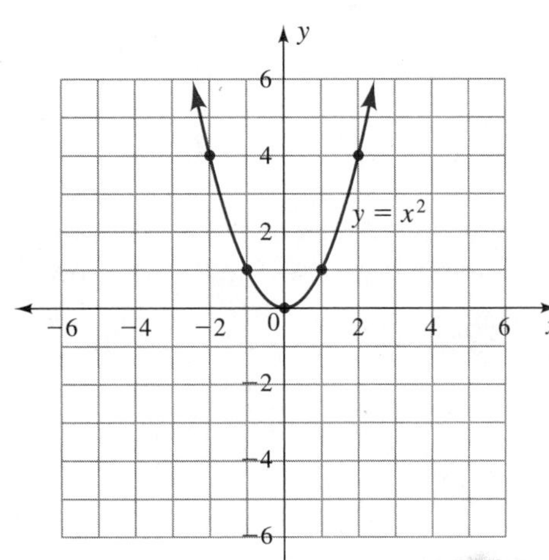

d.

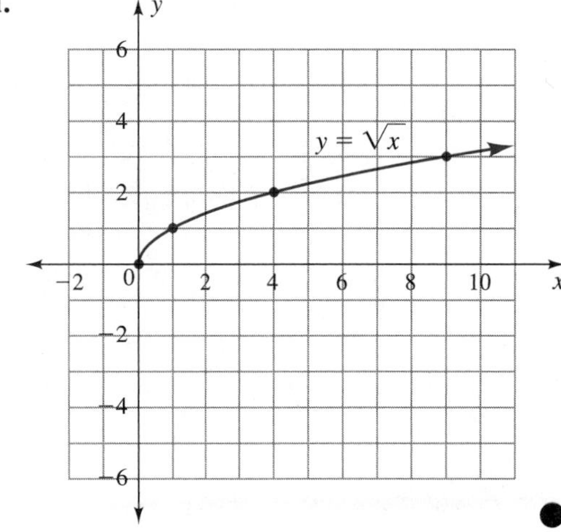

A **linear equation in two variables** describes *the set of input-output pairs, or ordered pairs* (x, y), *whose graph makes a straight line.* For example, the rule "the output is two less than three times the input" has the linear equation $y = 3x - 2$. This is the rule for Exercises 1 and 5 of the Warm-up and part a of Example 1. The rule for part b of Example 1 is also a linear equation: $y = 2 - 6x$.

A **nonlinear equation** is the name given to *equations whose graphs do not form straight lines.* The graph $y = x^2$, in part c of Example 1, and the square root curve $\sqrt{x}$, in part d of Example 1, are two examples of nonlinear equations.

One way to identify linear equations is to arrange the equations into a certain form, such as that of $y = 3x - 2$ and $y = -6x + 2$. *Each nonvertical linear equation has the input multiplied by a constant, then a constant added or subtracted.* The constant multiplied may be any real number, including zero. The constant added may also be zero.

Linear Equations

Student Note:
In this text, we solve equations in one variable for inputs when given the outputs. Equations in two variables usually are used to describe rules for tables or graphs.

A **linear equation in one variable** may be written in the form

$$mx + b = 0$$

where x is the variable and m and b are real-number constants.

A **linear equation in two variables** may be written in the form

$$y = mx + b$$

where x and y are variables and m and b are real-number constants.

Think of the constant m as *multiplying* the input x, and think of the letter b as *being added.*

EXAMPLE 

Writing linear equations Write equations in parts a to c in the form $mx + b = 0$, and write equations in parts d and e in the form $y = mx + b$.

a. $x = 4$

b. $x - 5 = 6$

c. $5 + x = 3$

d. $y - x + 2 = 0$

e. $x - y = 3$

Solution See the Answer Box. ●

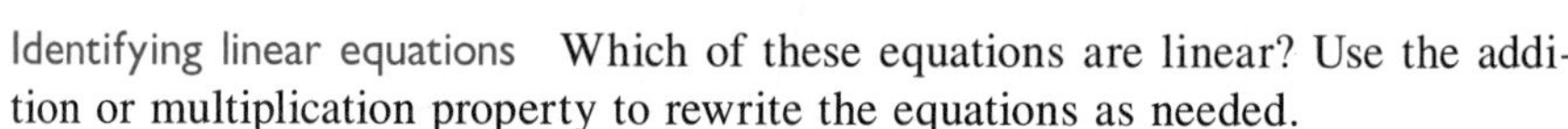

EXAMPLE 3 Identifying linear equations Which of these equations are linear? Use the addition or multiplication property to rewrite the equations as needed.

a. $x + y = 10$ **b.** $x \cdot y = 10$ **c.** $x + 10 = 5$ **d.** $x^2 = 10$

Solution In these solutions, the operation is stated to the right of the step.

a.
$$\begin{aligned} x + y &= 10 \\ -x \quad & \quad -x \\ y &= -x + 10 \end{aligned}$$
Subtract x from both sides.

The equation is in the form $y = mx + b$ with $m = -1$ and $b = 10$. It is a two-variable linear equation.

b.
$$\frac{x \cdot y}{x} = \frac{10}{x}$$
Divide both sides by x.
$$y = \frac{10}{x}$$

The equation is not in the form $y = mx + b$ because of the division by x. The equation is not linear.

c.
$$\begin{aligned} x + 10 &= 5 \\ -5 \quad & \quad -5 \\ x + 5 &= 0 \end{aligned}$$
Subtract 5 from both sides.

The equation is in the form $mx + b = 0$ with $m = 1$ and $b = 5$. It is a one-variable linear equation.

JUST FOR FUN

When something is filled halfway, some will say that it is half full and others will say that it is half empty. Equating these two ideas, we have

$\frac{1}{2}$ full $= \frac{1}{2}$ empty

The multiplication property of equations states that we can multiply both sides of an equation by the same number. Multiplying both sides by two gives

$2 \cdot \frac{1}{2}$ full $= 2 \cdot \frac{1}{2}$ empty

full = empty!

d. $x^2 = 10$

$-10 \quad -10$ Subtract 10 from both sides.

$x^2 - 10 = 0$

The equation is not in the form $mx + b = 0$ because of the x^2. The equation is not linear.

Writing Equations in Two Variables

We now review writing two-variable equations in words (Example 4) and writing rules as equations (Example 5).

EXAMPLE 4 Writing equations in words Which sentence does not describe the equation $y = -6x + 2$? The input is x, and the output is y.

a. The output is two added to the opposite of six times the input.

b. The output is the sum of two and the opposite of six times the input.

c. The product of negative six with the input is then added to two to get the output.

d. The output is the sum of the input and the opposite of six added to two.

e. The input is multiplied by negative six and then added to two to get the output.

Solution Recall that a *sum* is the answer to an addition and a *product* is the answer to a multiplication. See the Answer Box.

In equations, the letter x is usually the input variable, called the **independent variable**. In graphs, we place the independent variable on the horizontal axis. The letter y is usually the output, called the **dependent variable**. We place the dependent variable on the vertical axis. We say *y depends on x*.

Dependent Variables

> In applications, look for the output, y, depending on the input, x. Write equations so that *y depends on x*.

EXAMPLE 5 Writing rules as equations Write an equation that describes each of these problem situations. Define the input and output variables.

a. What is the total cost of bulk rice at \$0.89 per pound with a \$0.50 deposit on a reusable container?

b. What is the total cost of tuition at \$64 per credit hour plus \$10 in fees?

c. What is the sales tax on a purchase in a city where taxes are 6% of the price?

d. What is the value remaining on a prepaid copy machine card that costs \$5.00 and is charged \$0.05 per copy as the copy machine is used?

Solution In each equation, x is the input and y is the output.

a. Total cost depends on the number of pounds purchased, so x is the number of pounds and y is the total cost. The total cost is $y = 0.89x + 0.50$ if we use dollars or $y = 89x + 50$ if we use cents.

b. Total cost depends on the number of credits taken, so x is the number of credits and y is the total cost. The total cost, in dollars, is $y = 64x + 10$.

c. Tax depends on price, so x is the price and y is the tax. Tax, in dollars, is $y = 0.06x$. The phrase *of the price* reminds us to multiply the 6% by the input x.

d. The value of the card depends on the number of copies made, so x is the number of copies and y is the value remaining on the card. The value remaining, in dollars, is $y = -0.05x + 5.00$. ●

Solving Equations from Tables

As mentioned earlier, the two-variable equation $y = 3x - 2$ describes the table shown in Warm-up Exercise 1 and the pattern shown in Warm-up Exercise 5 (and repeated in Figure 3).

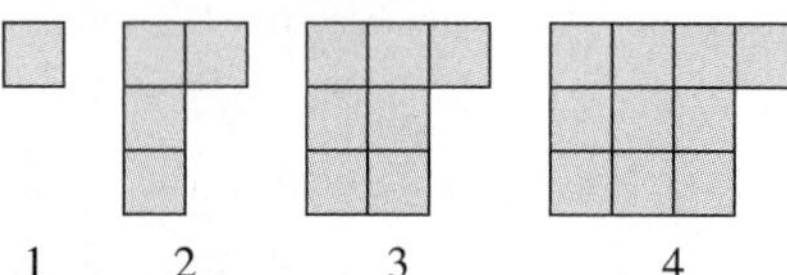

Figure 3

> When either an input or an output number is given, the two-variable equation becomes a one-variable equation that can be solved for the other variable.

When $x = 5$, the equation $y = 3x - 2$ becomes $y = 3(5) - 2$. Solving this equation means simplifying the right side to obtain $y = 13$. A fifth set of squares in Figure 3 would contain 13 squares.

In Example 6, we replace the output y with a number.

EXAMPLE 6 Solving an equation from a table

a. What equation is formed when $y = 4$ in $y = 3x - 2$?

b. Solve the equation in part a from Table 1.

Input x	**Output** $y = 3x - 2$
-1	-5
0	-2
1	1
2	4

Table 1

Solution **a.** If $y = 4$, then $y = 3x - 2$ becomes $4 = 3x - 2$.

b. To solve $4 = 3x - 2$, we look down the output column, find 4, and then look in the input column for the solution: $x = 2$. Thus, $4 = 3x - 2$ when $x = 2$. To check, we replace x in $4 = 3x - 2$ with the input 2. We write the checking step as follows:

Check: $4 \stackrel{?}{=} 3(2) - 2$ ✓ ●

Because a table shows many input-output pairs, we can solve many different equations with the same table. We can use the table to estimate solutions when the input-output pairs are between entries in the table. To find still other solutions, we can extend the table using patterns.

EXAMPLE 7

Finding and estimating solutions from tables Solve these equations from Table 2. Extend the table and estimate solutions as needed.

a. $3x - 2 = -5$

b. $3x - 2 = 2$

c. $3x - 2 = 9$

Solution **a.** To solve $3x - 2 = -5$, we look down the output column for -5. Looking in the input column, we find $x = -1$.

Check: $3(-1) - 2 \stackrel{?}{=} -5$ ✓

Input x	Output $y = 3x - 2$
−1	−5
0	−2
1	1
2	4

Table 2

Input x	Output $y = 3x - 2$
1	1
$1\frac{1}{3}$	2
$1\frac{2}{3}$	3
2	4

Table 3

b. To solve $3x - 2 = 2$, we look down the output column for 2. The number 2 does not appear, but it would be between the outputs 1 and 4. In fact, 2 is $\frac{1}{3}$ of the distance between 1 and 4. For this linear equation, the input that goes with 2 will be $\frac{1}{3}$ of the distance from 1 to 2, or $1\frac{1}{3}$ (see Table 3).

Check: $3\left(1\frac{1}{3}\right) - 2 \stackrel{?}{=} 2$ ✓

c. To solve $3x - 2 = 9$, we extend the table. We can place other inputs into the equation $y = 3x - 2$, or we can look at number patterns for the input change, Δx, and the output change, Δy. As Table 4 shows, the output increases by 3 for every increase of 1 in the input, so we can continue the pattern. An output of 9 will have an input x between 3 and 4. Because 9 is $\frac{2}{3}$ of the distance between 7 and 10, the solution will be $\frac{2}{3}$ of the distance between 3 and 4, or $3\frac{2}{3}$ (see Table 5).

Δx	Input x	Output $y = 3x - 2$	Δy
	−1	−5	
+1			+3
	0	−2	
+1			+3
	1	1	
+1			+3
	2	4	
+1			+3
	3	7	
+1			+3
	4	10	

Table 4 Change in x and in y

Input x	Output $y = 3x - 2$
3	7
$3\frac{1}{3}$	8
$3\frac{2}{3}$	9
4	10

Table 5

Check: $3\left(3\frac{2}{3}\right) - 2 \stackrel{?}{=} 9$ ✓

The small triangle in part c of Example 7 is the Greek letter delta. The letter **delta** (Δ) before a variable means *the change in that variable.*

If it is easy to estimate solutions from a table, we do so; otherwise, we find solutions with another method.

Summary: Solving an Equation from a Table

To solve the equation $n = mx + b$ from a table for $y = mx + b$, look for n in the output column. The solution to the equation $n = mx + b$ is the input x for output n.

If n is between numbers in the output column, estimate x. If n is above or below the numbers in the output column, extend the table.

This solution method also works for other equations written in $y =$ form.

Graphing Calculator Technique: Solving Equations from a Table

Repeat Example 7 using the table feature of your graphing calculator. Enter the equation $y = 3x - 2$ into [Y =]. Go to TABLE SET-UP to enter the starting number and the input change number. Many calculators indicate the change in inputs with Δx. You should be able to set Δx first as 1 to solve part a and then as $\frac{1}{3}$ to solve parts b and c.

Solving Equations from Graphs

The next two examples illustrate the use of graphs to solve equations. The process is the same as finding input-output pairs in Chapter 1 because we are assuming x is the input and the number on the right side of the equation is y, or the output.

EXAMPLE 8 Solving equations from a graph Use the graph in Figure 4 to solve these equations.

a. $3x - 2 = 4$ **b.** $3x - 2 = -5$ **c.** $3x - 2 = 9$

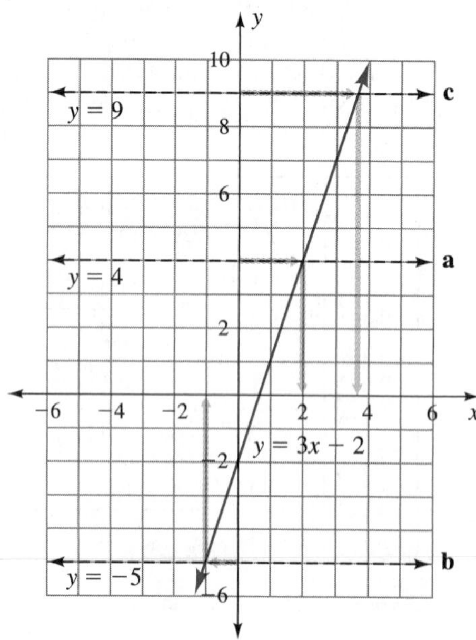

Figure 4

Solution **a.** To solve $3x - 2 = 4$, we trace along the horizontal line $y = 4$ to where it intersects $y = 3x - 2$. We look below this intersection on the x-axis. The input $x = 2$ lies directly below the intersection. Thus, $3x - 2 = 4$ at the point where $x = 2$.

Check: $3(2) - 2 \stackrel{?}{=} 4$ ✓

b. To solve $3x - 2 = -5$, we trace along the horizontal line $y = -5$ to where it intersects $y = 3x - 2$. The value of x directly above that intersection is $x = -1$. Thus, $3x - 2 = -5$ at the point where $x = -1$.

Check: $3(-1) - 2 \stackrel{?}{=} -5$ ✓

c. To solve $3x - 2 = 9$, we trace along the horizontal line $y = 9$. The lines $y = 3x - 2$ and $y = 9$ cross between $x = 3$ and $x = 4$, so the solution is between 3 and 4, in the interval $(3, 4)$.

Check: See part c of Example 7. ✓

Graphing Calculator Technique: Solving an Equation from a Graph

One way to solve an equation on the graphing calculator is to graph each side as a separate equation and trace to the x-value of the intersection of the two graphs.

For example, solve $3x - 2 = 9$ as follows. Enter $Y_1 = 3X - 2$ and $Y_2 = 9$. Set the viewing window using the interval $[-10, 10]$ for x and $[-10, 10]$ for y. Graph, and trace on $y = 3x - 2$ to $y = 9$. Zoom in. Trace on $3x - 2$ again to $y = 9$, zoom in, and trace to $y = 9$ again. Read the corresponding x value, $x \approx 3.67$ (which may be recognized as $3\frac{2}{3}$). Check by storing 3.67 in x and evaluating $y = 3x - 2$.

The equations in Example 8 had either one solution or a solution on an interval. In Example 9, we have a curved graph, illustrating that some equations may have more than one solution or no solution at all.

EXAMPLE 9 Solving equations from a graph Use the graph of $y = x^2$ in Figure 5 to find the solutions to the equations. Use sets to describe the solutions.

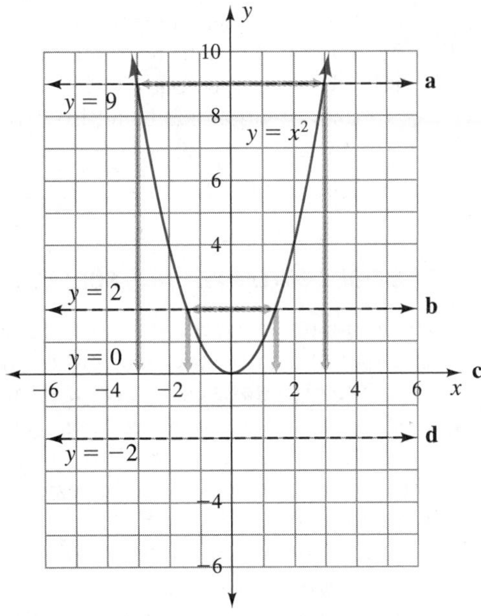

Figure 5

a. $x^2 = 9$

b. $x^2 = 2$

c. $x^2 = 0$. What is special about this point on the graph of $y = x^2$?

d. $x^2 = -2$

Solution **a.** We look for the inputs that give $y = 9$. For $x^2 = 9$, there are two solutions: $x = 3$ and $x = -3$. Thus, the solution set is $\{-3, 3\}$.

Check: $(-3)^2 \stackrel{?}{=} 9$ and $3^2 \stackrel{?}{=} 9$ ✓

b. We look for the inputs that give $y = 2$. For $x^2 = 2$, there are two solutions: $x \approx 1.4$ and $x \approx -1.4$. Thus, the approximate solution set is $\{-1.4, 1.4\}$.

Check: $(-1.4)^2 \stackrel{?}{=} 2$ and $(1.4)^2 \stackrel{?}{=} 2$ if answers are rounded to the nearest tenth. ✓

c. We look for the inputs that give $y = 0$. For $x^2 = 0$, there is only one solution: $x = 0$. The solution set is $\{0\}$. Note that the ordered pair (0, 0) gives the lowest point on the curve $y = x^2$.

Check: $0^2 \stackrel{?}{=} 0$ ✓

d. We look for the inputs that give $y = -2$. There are no solutions to $x^2 = -2$, because the curve never goes below the x-axis. We say that $x^2 = -2$ has no real-number solution. ●

When an equation such as that in Example 9 has no real-number solution, we say the solution set is empty and write the symbol { } or ϕ. The symbol { } shows *a set with nothing in it,* the **empty set**. The other symbol, ϕ, is also common. Choose either symbol.

Caution: Compare parts c and d of Example 9 carefully; {0} and { } are not the same result.

Solving an Equation from a Graph

> To solve the equation $n = mx + b$ from a graph of $y = mx + b$, look for the horizontal line where $y = n$. The solution to the equation $n = mx + b$ is the input x for the point of intersection of $y = n$ with $y = mx + b$.
>
> If n is between numbers on the output axis, estimate y and the corresponding x. It may be necessary to extend the graph vertically or horizontally to find the output n or its corresponding input x.
>
> This solution method also works for other equations written in $y =$ form.

Special Solutions: Intersections with the Axes

When a graph intersects one or both of the axes, the ordered pairs $(a, 0)$ and $(0, b)$ have special significance. We consider these pairs now.

The **horizontal axis intercept point**, defined by the pair $(a, 0)$, is *the point where a graph crosses the horizontal axis.* Because we commonly label the horizontal, or input, axis with x, $(a, 0)$ is also called the ***x*-intercept point**. The ***x*-intercept** is *the number a*. Remember, the x-intercept point is where the output is zero: $y = 0$ (see Figure 6).

The **vertical axis intercept point**, defined by the pair $(0, b)$, is *the point where a graph crosses the vertical axis.* Because we commonly label the vertical, or output, axis with y, $(0, b)$ is also called the ***y*-intercept point**. The ***y*-intercept** is *the number b*. Remember, the y-intercept is where $x = 0$ (see Figure 6).

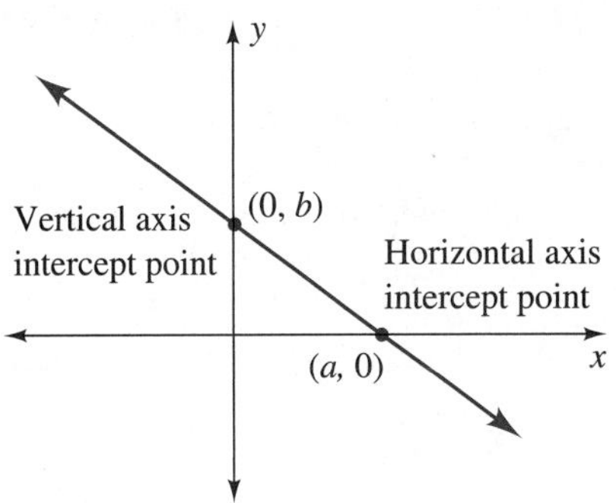

Figure 6

EXAMPLE **10** Solving equations for horizontal and vertical intercepts

a. Make a table and graph for $3x + 2y = 12$.

b. Solve $3x + 2y = 12$ for its horizontal axis intercept, using the table and graph. Check by solving with algebraic notation.

c. Solve $3x + 2y = 12$ for its vertical axis intercept, using the table and graph. Check by solving with algebraic notation.

Solution **a.** The table appears in Table 6 and the graph in Figure 7.

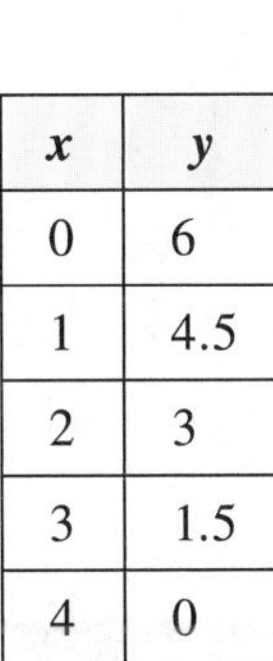

x	y
0	6
1	4.5
2	3
3	1.5
4	0

Table 6 $3x + 2y = 12$

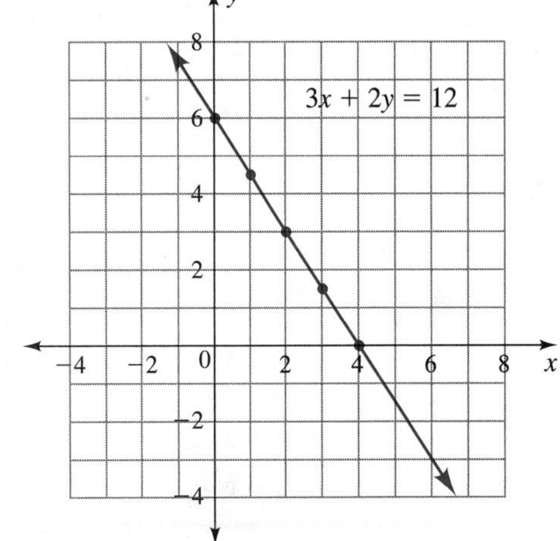

Figure 7

b. *To find the horizontal axis intercept,* we need the ordered pair $(a, 0)$. Looking in the table for $y = 0$, we find $x = 4$. On the graph, we look for the point of intersection of the graph of $3x + 2y = 12$ with the horizontal axis, or $x =$ axis. The intersection is $(4, 0)$. Again, if $y = 0$, then $x = 4$.

With algebraic notation, we have

Student Note:
The comments tell what to do next.

$$3x + 2y = 12 \qquad \text{Let } y = 0.$$

$$3x + 2 \cdot 0 = 12 \qquad \text{Divide both sides by 3.}$$

$$\frac{3x}{3} = \frac{12}{3} \qquad \text{Simplify.}$$

$$x = 4$$

Check: $3(4) + 2(0) \stackrel{?}{=} 12$ ✓

c. *To find the vertical axis intercept,* we need the ordered pair $(0, b)$. Looking in the table for $x = 0$, we find $y = 6$. On the graph, we look for the point of intersection of the graph of $3x + 2y = 12$ with the vertical axis. If $x = 0$, then $y = 6$.

With algebraic notation, we have

Student Note:
The comments tell what to do next.

$3x + 2y = 12$	Let $x = 0$.
$3(0) + 2y = 12$	Divide both sides by 2.
$\frac{2y}{2} = \frac{12}{2}$	Simplify.
$y = 6$	

Check: $3(0) + 2(6) \stackrel{?}{=} 12$ ✓

Summary of Horizontal and Vertical Intercepts

To find a horizontal axis intercept, let $y = 0$ in the table or equation and find the corresponding x. To find the horizontal axis intercept on the graph, find the intersection of the graph with the horizontal axis.

To find a vertical axis intercept, let $x = 0$ in the table or equation and find the corresponding y. To find the vertical axis intercept on the graph, find the intersection of the graph with the vertical axis.

	x	y
x-intercept	a	0
y-intercept	0	b

Applications

In Examples 11 to 13, we return to applications. Example 11 is about the prepaid copy machine card from part d of Example 5. In the exercises, you will solve the same equations with algebraic notation.

EXAMPLE  Finding and interpreting intercepts: photocopies Suppose photocopies cost \$0.05 each. After x copies have been made on a \$5 prepaid copy machine card, the value remaining on the card may be described by the equation $y = -0.05x + 5.00$.

a. Solve the equation $3.50 = -0.05x + 5.00$, using Table 7 and the graph in Figure 8. What do the equation and solution mean in this problem situation?

b. Find the vertical axis intercept and tell what it means.

c. Find the horizontal axis intercept and tell what it means.

Solution **a.** In the table, \$3.50 is halfway between \$3 and \$4. The input x should be halfway between 20 and 40. Thus, from the table and the graph, when the value y of the card is \$3.50, $x = 30$. The equation asks for the number of copies that have been made when \$3.50 remains on the copy card. The solution is 30 copies.

b. The vertical axis intercept, or y-intercept, has $x = 0$. From the table and the graph, $y = 5.00$. At the vertical axis intercept, no copies have been made: $(0, 5)$. The intercept shows the original value of the copy card.

Number of Copies x	Value Remaining on Card (dollars) $y = -0.05x + 5.00$
0	5.00
20	4.00
40	3.00
60	2.00
80	1.00
100	0

Table 7

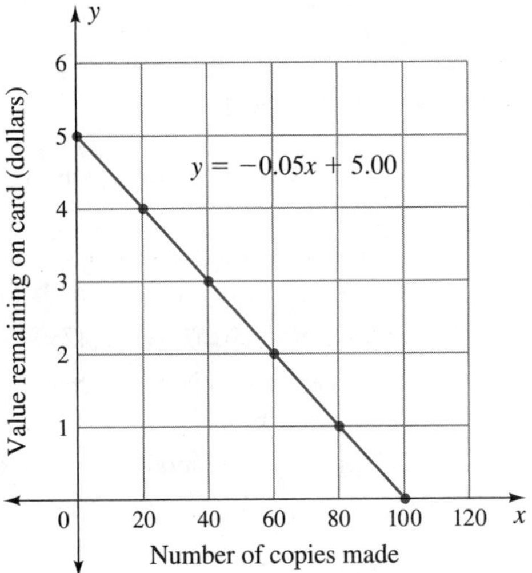

Figure 8

c. The horizontal axis intercept, or x-intercept, has $y = 0$. From the table and the graph, $x = 100$ at $y = 0$. At the horizontal axis intercept, no value remains on the card. With 100 copies, the \$5 prepaid value is used up. ●

In Example 11, the output (value remaining on the card) decreases as the input (number of copies) increases. In Example 12, the output increases as the input increases.

EXAMPLE 12 Finding and interpreting intercepts: electric bill West Coast Electric charges a basic monthly fee of \$5.00 plus \$0.04 per kilowatt hour of electricity used.

a. Make a table for the cost each month of 0 to 4000 kilowatt hours. Graph the data from the table.

b. What is the vertical axis intercept and what does it mean?

c. What is the cost if 1500 kilowatt hours are used?

d. How much electricity is used if the total cost is \$75?

e. Write an equation for the monthly cost. Let x be the number of kilowatt hours used and y be the total monthly cost.

f. What is the horizontal intercept and what does it mean?

Solution **a.** We build Table 8 and Figure 9.

Kilowatt Hours Used	Monthly Cost (dollars)
0	5
1000	45
2000	85
3000	125
4000	165

Table 8 Monthly Electric Cost

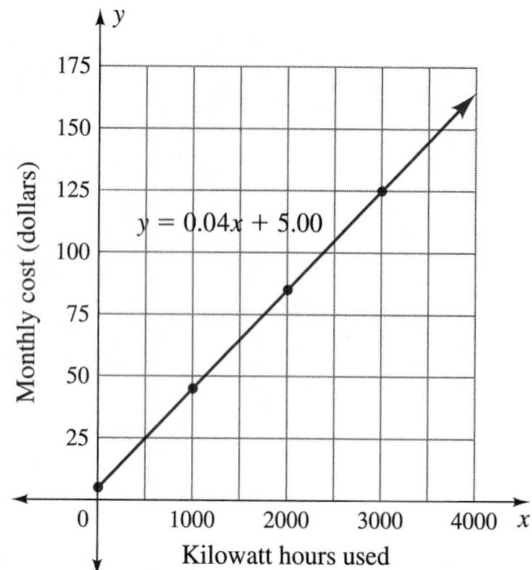

Figure 9

b. The vertical axis intercept is (0, 5); the cost is \$5.00 even if no electricity is used.

c. In the table, 1500 kilowatt hours is halfway between 1000 and 2000, so the cost is \$65, which is halfway between \$45 and \$85.

d. From the graph, \$75 buys 1750 kilowatt hours.

e. The cost (output) is 0.04 times the kilowatt hours (input) plus the \$5.00 fee, or $y = 0.04x + 5.00$.

f. To find the horizontal axis intercept, we let $y = 0$ and solve for x.

Student Note:
The comments tell what to do next.

$y = 0.04x + 5$	Let $y = 0$.
$0 = 0.04x + 5$	Subtract 5 from both sides.
$0 - 5 = 0.04x + 5 - 5$	Simplify.
$-5 = 0.04x$	Divide both sides by 0.04.
$\frac{-5}{0.04} = \frac{0.04x}{0.04}$	Simplify.
$-125 = x$	

The horizontal intercept is negative. Unless we can sell electricity back to the electric company from a home generator, a negative number of kilowatt hours used has no meaning in the problem situation. ●

Think about it: What is the same and what is different about the equations and graphs in Examples 11 and 12?

ANSWER BOX

Warm-up: 1. −5, −2, 1, 4 **2.** 8, 2, −4, −10 **3.** 4, 1, 0, 1, 4 **4.** 0, 1, 2, 3 **5.** We multiply the number of the figure (input) by 3 and then subtract 2 to get the number of squares (output): $y = 3x - 2$. **Example 2: a.** $x - 4 = 0$ **b.** $x - 11 = 0$ **c.** $x + 2 = 0$ **d.** $y = x - 2$ **e.** $y = x - 3$ **Example 4:** d **Think about it:** The y-intercepts are both \$5.00. The number multiplying x is negative in Example 11 and positive in Example 12.

EXERCISES 3.2

In Exercises 1 and 2, tell whether or not the equations are linear and why.

1. **a.** $x = y$ **b.** $6 = x$ **c.** $xy = 5$
d. $y = 2 + x$ **e.** $y = x^2$ **f.** $x + 3 = 5$

2. **a.** $x = 4$ **b.** $y = 3$ **c.** $\frac{1}{x} = y$
d. $3 + x^2 = y$ **e.** $3 + x = y$ **f.** $y = x$

Write each sentence in Exercises 3 to 8 as an equation with x as the input, or independent, variable and y as the output, or dependent, variable.

3. The output is five more than twice the input.

4. The input is multiplied by negative two to get the output.

5. Six less than triple the input gives the output.

6. The output is three less than four times the input.

7. Five less than half the input is the output.

8. The product of negative three and the input is added to 4 to give the output.

In Exercises 9 to 20, define variables and write equations so that y depends on x. (Hint: Distance traveled is the product of the rate and time.)

9. Find the tip if a tip is usually 15% of the cost of a meal.

10. Find the amount of tax for an 8.5% sales tax on a purchase.

11. Find the Medicare payment at 1.45% of wages.

12. Find the Social Security payment at 6.2% of wages.

13. Find the total cost for any quantity of bulk candy at \$1.49 per pound.

14. Find the distance traveled at 35 miles per hour for a given number of hours.

15. Find the distance traveled in 3 hours at various speeds.

16. Find the cost of x thousand gallons of water at \$1.15 per thousand gallons.

17. Find the total cost of tuition and fees for x credit hours taken. Tuition is \$75 per credit hour, and fees are \$32.

18. Find the value remaining on a \$20 prepaid mass transit ticket when each ride costs \$1.85.

19. Find the value remaining on a prepaid \$26 coffee card when your usual beverage costs \$3.25.

20. Find the value remaining on a \$25 theater gift certificate when each ticket costs \$3.

21. Solve these equations from the table.

x	$y = 10 - x$
1	9
2	8
3	7
4	6

a. $10 - x = 8$ **b.** $10 - x = 6$
c. $10 - x = 3$ **d.** $10 - x = -2$

22. Solve these equations from the table.

x	$y = 4 - 2x$
1	2
2	0
3	−2
4	−4

a. $4 - 2x = -4$ **b.** $4 - 2x = 0$
c. $4 - 2x = -10$ **d.** $4 - 2x = 6$

23. Solve these equations from the table.

x	$y = 2 - 4x$
1	−2
2	−6
3	−10
4	−14

a. $2 - 4x = -2$ **b.** $2 - 4x = -6$
c. $2 - 4x = 6$ **d.** $2 - 4x = -8$

24. Solve these equations from the table.

x	$y = 3 - 3x$
1	0
2	−3
3	−6
4	−9

a. $3 - 3x = -6$ **b.** $3 - 3x = 3$
c. $3 - 3x = -15$ **d.** $3 - 3x = -5$

In Exercises 25 to 28, complete the tables. Your inputs will contain fractions or decimals.

25.

x	$y = 2x + 1$
3	7
	8
4	9
	10
5	11
	12

26.

x	$y = 1 - 4x$
−3	13
	12
	11
	10
−2	9
	8

27.

x	$y = 5x - 4$
3	11
	12
	13
	14
	15
4	16

28.

x	$y = 5 - 3x$
0	5
	4
	3
1	2
	1
	0

29. Solve these equations from the graph.

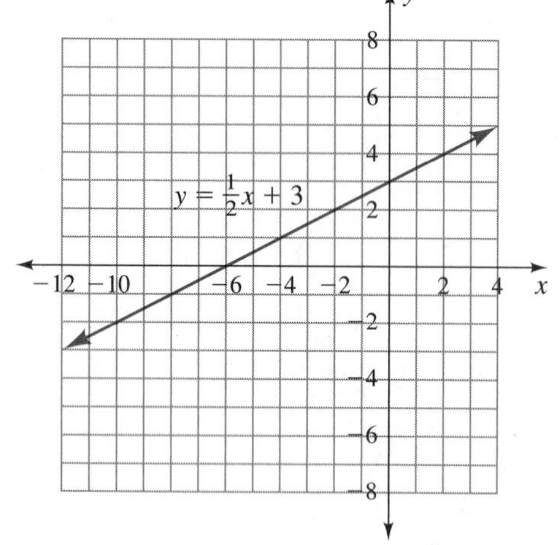

a. $\frac{1}{2}x + 3 = 1$

b. $\frac{1}{2}x + 3 = 0$

c. $\frac{1}{2}x + 3 = -2$

30. Solve these equations from the graph.

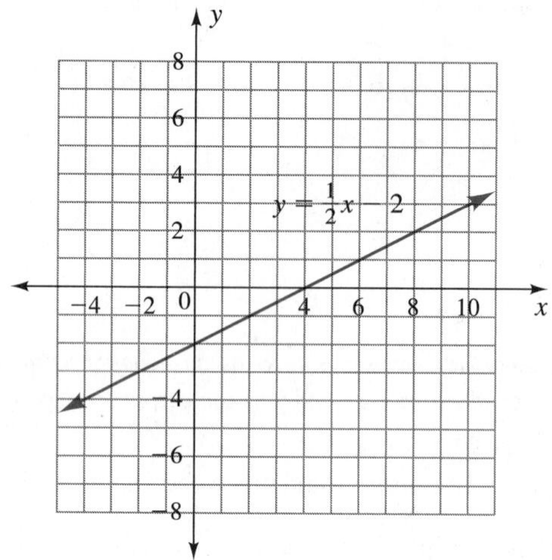

a. $\frac{1}{2}x - 2 = 3$

b. $\frac{1}{2}x - 2 = 0$

c. $\frac{1}{2}x - 2 = -4$

31. Solve these equations from the graph.

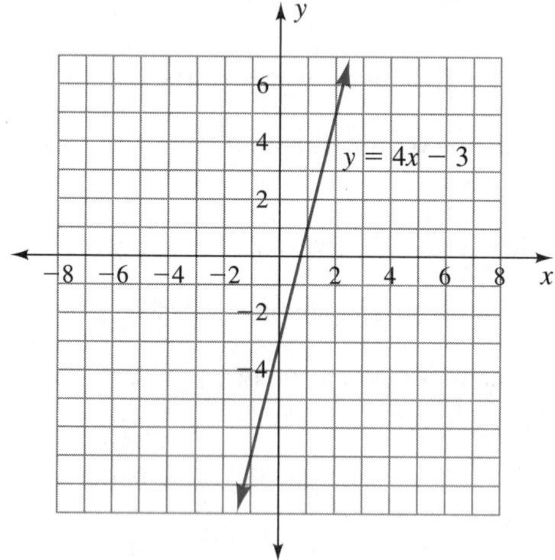

a. $4x - 3 = 1$

b. $4x - 3 = -7$

c. $4x - 3 = 5$

32. Solve these equations from the graph.

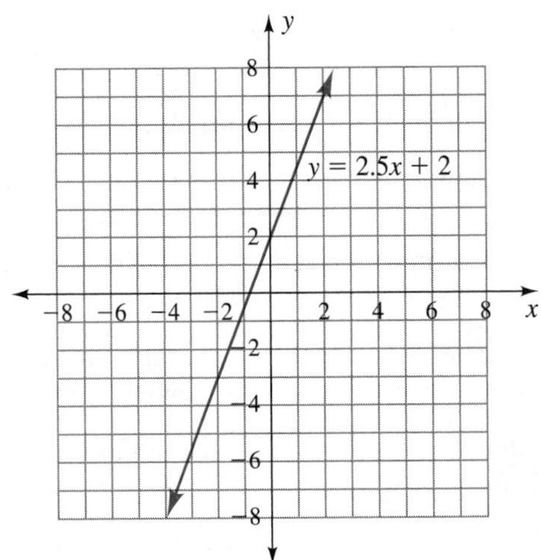

a. $2.5x + 2 = 7$

b. $2.5x + 2 = -3$

c. $2.5x + 2 = -8$

Use the graphs to find the solutions to the equations in Exercises 33 and 34. Give answers in sets.

33.

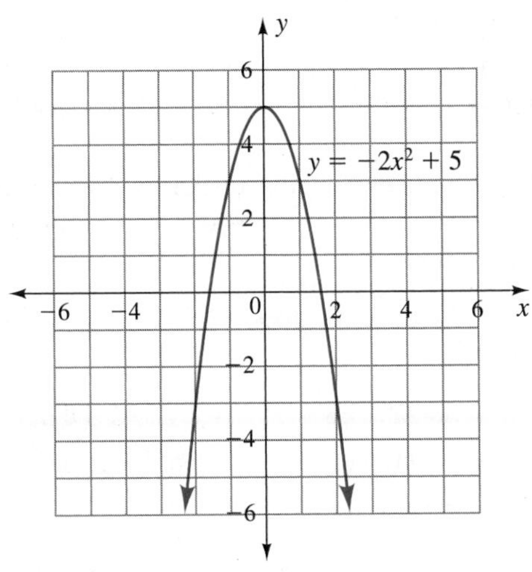

a. $-2x^2 + 5 = -3$

b. $-2x^2 + 5 = 3$

c. $-2x^2 + 5 = 5$

d. $-2x^2 + 5 = 6$

34.

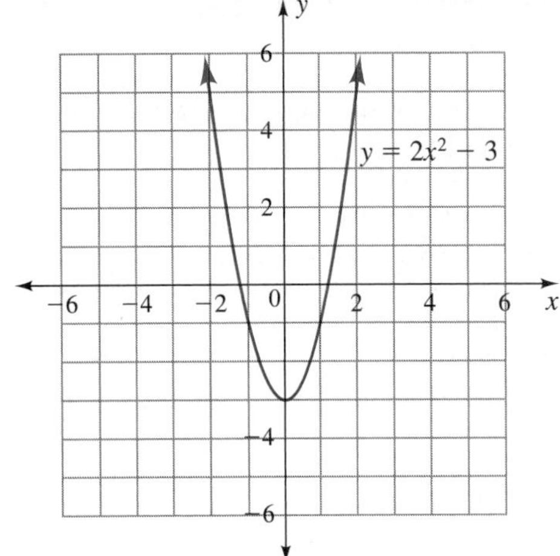

a. $2x^2 - 3 = -3$ **b.** $2x^2 - 3 = -1$

c. $2x^2 - 3 = 5$ **d.** $2x^2 - 3 = -5$

In Exercises 35 to 38, find the x-intercept and y-intercept for each equation. Sketch a graph showing the axes, the intercepts, and the line passing through the intercepts.

35. a. $3x - 4y = -12$ **b.** $2x + 5y = 10$

36. a. $2x - 3y = 6$ **b.** $3x + 5y = 15$

37. a. $y = \frac{2}{3}x - 4$ **b.** $y = \frac{3}{4}x + 3$

38. a. $y = \frac{3}{4}x - 6$ **b.** $y = \frac{2}{3}x + 8$

In Exercises 39 and 40, show whether the given values make the equation true.

39. a. $y = 3x - 2$; $x = 0, y = -2$

b. $2 + 5x = y$; $x = -2, y = -8$

c. $y = \frac{1}{2}x$; $x = 2, y = 4$

d. $x + y = 8$; $x = 3, y = 5$

e. $\frac{1}{2}bh = 36$; $b = 8, h = 9$

40. a. $y = 3x - 2$; $x = 1, y = 1$

b. $2 + 5x = y$; $x = 2, y = 10$

c. $y = \frac{1}{2}x$; $x = 6, y = -3$

d. $x + y = 8$; $x = -3, y = 11$

e. $\frac{1}{2}bh = 36$; $b = 18, h = 2$

41. How do we identify a linear equation (either one-variable or two-variable)?

42. How can we identify the independent (or input) variable in an application?

43. Explain how to solve $mx + b = n$ for x from a table for $y = mx + b$. Assume that the number n appears in the table.

44. Explain how to solve $mx + b = n$ for x from a graph of $y = mx + b$. Assume that the number n appears on the vertical axis.

45. Explain how to find the y-intercept of the graph of an equation.

46. Explain how to find the x-intercept of the graph of an equation.

47. Explain the difference in meaning between the symbols $\{\ \}$ and $\{0\}$.

48. Is $C = 2\pi r$ a linear equation? Explain.

49. True or false: If the numbers in (x, y) make a true statement when substituted into an equation, then (x, y) lies on the graph of the equation.

50. True or false: All ordered pairs for points on a graph make a true statement when substituted into the equation for the graph.

51. True or false: If an ordered pair (a, b) is on a graph, then the ordered pair (b, a) is also on the graph.

52. True or false: When we let $x = 0$ in an equation, we are finding the x-intercept.

The equations in Exercises 53 and 54 are related to the applications in this section. Some were solved from tables or graphs. Solve them with algebraic notation.

53. **a.** $3.50 = -0.05x + 5.00$

b. $y = -0.05(0) + 5.00$

c. $0 = -0.05x + 5.00$

54. **a.** $y = 0.04(0) + 5.00$

b. $75.00 = 0.04x + 5.00$

c. $0.04x + 5.00 = 0$

55. The value (in dollars) remaining on the photocopy card in Example 11 after x copies is $y = -0.05x + 5.00$.

a. How many copies have been made if the card value is \$2.25?

b. Which equation in Exercise 53 finds the x-intercept?

c. Which equation in Exercise 53 finds the y-intercept?

56. The total cost of x kilowatt hours of electricity in Example 12 is $y = 0.04x + 5.00$.

a. How many kilowatt hours were used if the electric bill is \$100?

b. Which equation in Exercise 54 finds the x-intercept?

c. Which equation in Exercise 54 finds the y-intercept?

57. You purchase a prepaid telephone card. The card costs \$12. Each minute costs \$0.30.

a. Make a table for the value remaining on the card after 0, 10, 20, 30, 40, and 50 minutes.

b. From the table, what is the value remaining after 30 minutes?

c. How many minutes were used if \$9 remains on the card?

d. How many minutes were used if \$2 remains on the card?

e. What is the x-intercept of a graph of the table?

f. What is the meaning, if any, of the x-intercept?

g. What is the y-intercept of a graph of the table?

h. What is the meaning, if any, of the y-intercept?

i. What equation describes the value remaining on the card after x minutes?

In Exercises 58 to 59, use D = rt as the relationship among distance, rate, and time.

58. **a.** Build a table showing the distances traveled at a constant rate of 55 miles per hour for times of 0, 1, 2, 3, and 4 hours.

b. From the table, how long will it take to travel 27.5 miles?

c. How many miles would be driven in $2\frac{1}{2}$ hours?

d. What is the x-intercept of a graph of the table? What is its meaning?

59. **a.** Build a table showing the distances traveled in 3 hours at speeds of 0, 10, 20, 30, and 40 miles per hour.

b. Karen-Louise, a marathon runner, traveled 15 miles. From the table, how fast did she run?

c. From the table, at what speed would Fabio, a bicyclist, travel in covering a distance of 75 miles?

d. What is the y-intercept of a graph of the table? What is its meaning?

60. How might Exercise 56 (Example 12) be changed so that the graph passed through the origin, (0, 0)?

Projects

61. *Equations and Graphs.* The following equations contain only x, y, 2, =, and one operation, yet they create very different graphs.

$$y = x + 2 \quad (1)$$

$$y = 2 - x \quad (2)$$

$$y = x - 2 \quad (3)$$

$$y = x^2 \quad (4)$$

$$y = \frac{2}{x} \quad (5)$$

$$y = 2x \quad (6)$$

$$y = \frac{x}{2} \quad (7)$$

$$x = y^2 \quad (8)$$

For each equation (1) to (8), do the following:

a. Make a table of at least six ordered pairs.

b. Plot the ordered pairs and sketch a line or curve through the points.

c. State any assumptions that you made in drawing lines (or curves) through the sets of points.

d. Identify the graphs as linear or nonlinear.

e. Which equation has the same graph as $y = 2 + x$? Why? Which other equation can be written in two ways to give the same graph?

f. Why is the graph of $y = 2 - x$ different from the graph of $y = x - 2$?

g. Why is the graph of $y = 2/x$ different from the graph of $y = x/2$?

62. *Number of Solutions.* Equations that make nonlinear graphs often have multiple solutions. Enter $y = -x^4 - 1.5x^3 + 2x^2 + x + 1$ into a graphing calculator. Set a window of -3 to 3 for the x-axis. Set a window of -6 to 6 for the y-axis. Make a sketch of the graph. Answer the questions below for your graph. Estimate to the nearest whole number.

a. For what number y can you draw a horizontal line that passes through the curve once? Draw the horizontal line and label it y = [your number]. This number, when placed into the equation for y, creates an equation with one solution.

b. For what number y will the equation have two solutions?

c. For what y will the equation have three solutions?

d. For what y will the equation have four solutions?

e. For what y will the equation have five solutions?

f. For what y will the equation have no solutions?

g. Repeat parts a to f for $y = x^3 - 3x^2$.

h. Draw a graph that will have five solutions for some number y.

MID-CHAPTER 3 TEST

1. Say whether each equation is an identity or a conditional equation.

a. $2x + 4 = -1$ **b.** $2x + 3x = 5x$

c. $4(x - 2) = 4x - 8$ **d.** $2x = x$

2. Which pairs of equations are not equivalent?

a. $4x + 5 = 29$, $4x = 24$

b. $\frac{1}{2}x = 16$, $x = 8$

c. $2x + 3x = 10$, $5x^2 = 10$

d. $3x - 2 = 8$, $2x = 10$

In Exercises 3 to 6, is the given input x a solution to the equation?

3. $3x - 4 = -16$, $x = -4$

4. $4x - 3 = -9$, $x = -3$

5. $-8x + 5 = 1$, $x = \frac{1}{2}$

6. $-6 - 10x = -11$, $x = \frac{1}{2}$

In Exercises 7 to 12, solve for x.

7. $x - 4 = 3$ **8.** $\frac{2}{3}x = 24$

9. $2x + 3 = -7$ **10.** $\frac{1}{2}x - 8 = -1$

11. $3x = \frac{1}{2}$ **12.** $3 - 2x = 8$

In Exercises 13 to 16, write equations. Let x be the unknown number.

13. Five more than twice a number is 10.

14. Five is 6 less than half a number.

15. How many credit hours can be taken for $545 if tuition is $85 per credit hour and fees are $35?

16. How many minutes were used if there is $7.20 left on a $20 telephone card and each minute costs $0.40?

In Exercises 17 and 18, write each equation in words.

17. $5 = 3x - 4$

18. $4 + 2x = -3$

19. Use the table to solve the following equations.

x	$y = \frac{1}{2}x + 4$
2	5
4	6
6	7
8	8

a. $\frac{1}{2}x + 4 = 5$

b. $\frac{1}{2}x + 4 = 8$

c. $\frac{1}{2}x + 4 = 4$

d. $\frac{1}{2}x + 4 = 6.5$

Solve the equations in Exercises 20 and 21 from the graphs.

20.

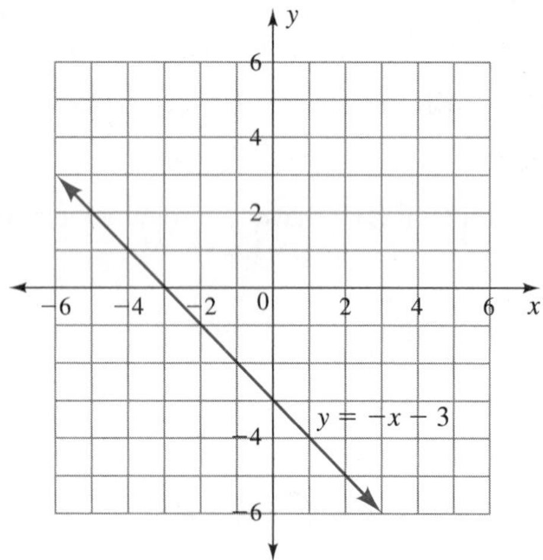

a. $-x - 3 = -5$

b. $-x - 3 = 1$

c. $-x - 3 = -2$

21.

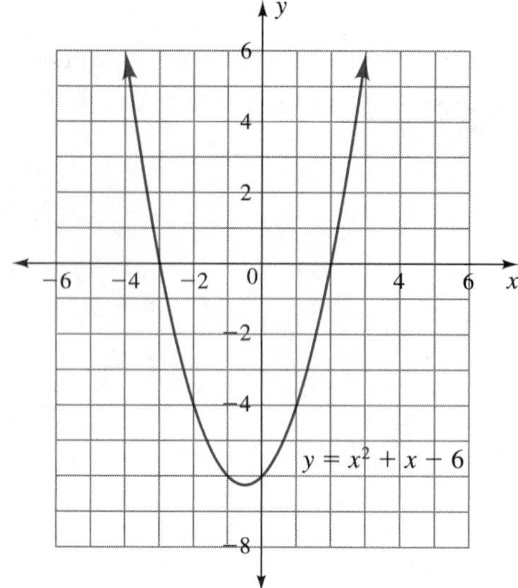

a. $x^2 + x - 6 = 6$

b. $x^2 + x - 6 = 0$

c. $x^2 + x - 6 = -6$

22. A student solves an equation by writing $2x = 10 = x = 5$. Why is this incorrect?

23. The American Gymnastics Training Center charges \$70 for a birthday party for up to 15 children, plus \$3 each for additional children.

a. Make a table and graph to show the output (total cost) for an input of x children. Count by 5 children in the table. Be sure to label your graph.

b. What is the cost for 10 children?

c. What is the cost for 20 children?

d. If the points on the graph were connected, where would the graph cross the vertical axis? Is there any meaning to this point? Why?

e. Write an equation that can be solved to find the number of children to invite for a total cost of \$100.

3.3 Solving Linear Equations with Tables, Graphs, and Algebraic Notation

OBJECTIVES

- Solve equations with expressions on the left and right sides.
- Solve equations containing parentheses.
- Set up and solve equations for consecutive integers and other applications.

WARM-UP

Simplify:

1. $4(x + 3)$ **2.** $-2(x + 1)$ **3.** $-4(x - 2)$

4. $-(x + 1)$ **5.** $13 - 4(x - 2)$ **6.** $8 - 3(x - 1)$

7. $2x + 3x - 4x$ **8.** $5x - 2x + x$

IN THIS SECTION, we use tables and graphs to solve equations with inputs x on both sides. We extend our work with equation solving by first simplifying each side with the distributive property and/or addition of like terms and then using the addition and multiplication properties of equations to solve the equations. We also set up and solve equations from applications.

Solving Equations: Tables and Graphs

In the following examples, we use tables and graphs to solve equations with variables on both sides.

Solving a Linear Equation Using a Table and a Graph

1. Let $y =$ the left side of the equation. Build a table and plot a graph.
2. Let $y =$ the right side of the equation. Build a table and plot a graph on the same axes used for the left side.
3. Look for the ordered pair that appears in both tables; it is also the ordered pair at the point of intersection. The x in the ordered pair is the solution to the equation. The y in the ordered pair is the value obtained when x is substituted into both sides of the original equation.

EXAMPLE 

Using tables and graphs to solve an equation Solve $3(x - 1) = x + 1$ using Tables 9 and 10 and the graph in Figure 10.

x	The Left Side: $3(x - 1)$
-1	-6
0	-3
1	0
2	3
3	6

Table 9

x	The Right Side: $x + 1$
-1	0
0	1
1	2
2	3
3	4

Table 10

Solution To solve from the tables, we find the ordered pair (2, 3), appearing in both tables. The input $x = 2$ makes the two expressions $3(x - 1)$ and $x + 1$ equal to 3. Thus, $x = 2$ solves the equation $3(x - 1) = x + 1$.

Check: $3(2 - 1) \stackrel{?}{=} 2 + 1$ ✓

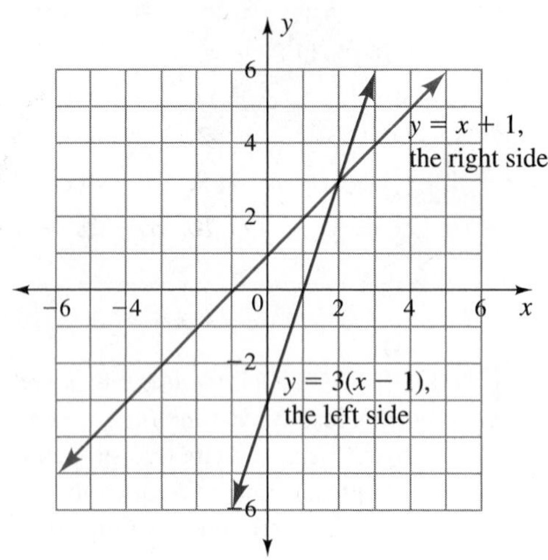

Figure 10

To solve $3(x - 1) = x + 1$ from the graph, we graph $y =$ the left side and $y =$ the right side. We look for the point of intersection of the two graphs $y = 3(x - 1)$ and $y = x + 1$. The graphs intersect at (2, 3). This means that the input $x = 2$ makes both $3(x - 1)$ and $x + 1$ equal to 3. Thus, $x = 2$ solves the equation.

Check: $3(2 - 1) \stackrel{?}{=} 2 + 1$ ✔ ●

Example 2 shows an equation for which longer tables are needed to find a common ordered pair.

EXAMPLE 2 Using tables and graphs to solve an equation Solve $4(x - 2) = 2(x + 1)$ with tables and a graph.

Solution First we build Tables 11 and 12, extending the tables until a common ordered pair is found. Both tables contain the ordered pair (5, 12). At $x = 5$, $4(x - 2) = 12$ and $2(x + 1) = 12$. Thus, $x = 5$ solves the equation $4(x - 2) = 2(x + 1)$.

x	$4(x - 2)$
−2	−16
−1	−12
0	−8
1	−4
2	0
3	4
4	8
5	12

Table 11

x	$2(x + 1)$
−2	−2
−1	0
0	2
1	4
2	6
3	8
4	10
5	12

Table 12

Check: $4(5 - 2) \stackrel{?}{=} 2(5 + 1)$ ✔

Next we plot the graph of each side of the equation, as shown in Figure 11. The ordered pair (5, 12) locates the point of intersection of the two graphs $y = 4(x - 2)$ and $y = 2(x + 1)$. Thus, $x = 5$ as an input to both $4(x - 2)$ and $2(x + 1)$ gives 12 as an output, so $x = 5$ solves the equation $4(x - 2) = 2(x + 1)$.

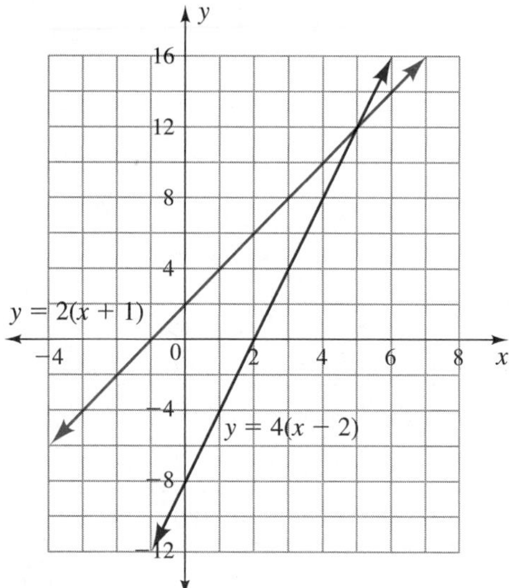

Figure 11

Check: $4(5 - 2) \stackrel{?}{=} 2(5 + 1)$ ✓

Graphing Calculator Technique: Solving an Equation by Graph and Table

Enter the left side of the equation as Y_1 and the right side of the equation as Y_2.

To solve from a graph: Set up the viewing window with the endpoints of each axis and the scale for each axis. If a window is not near the origin, with both x and y in $[-10, 10]$, evaluate each side of the equation for the one or two inputs and use the resulting ordered pairs to estimate the window settings. Graph the equations. Trace to the point of intersection. Zoom and trace to increase the accuracy of the solution.

To solve from a table: Estimate a solution. Go to TABLE SET-UP. As a starting number for your table, choose a number near your estimate. As a change in x, or Δx, number for your table, choose a number similar to the one you used for the scale on the x-axis. Go to TABLE. You want Y_1 and Y_2 to be equal in the table. Continue to choose a new starting number, if needed, and a new change in x number until the outputs from Y_1 and Y_2 are equal. The x corresponding to the equal outputs is the solution to the original equation.

The graphing calculator is useful when the common ordered pair, or intersection of two graphs, has inputs between two integers. With a graphing calculator, we can adjust the table to decimal inputs or we can trace and zoom in to a point of intersection. The solution to Example 3 is estimated with a table and graph. You may wish to solve Example 3 with a graphing calculator.

EXAMPLE 3 Solving with tables and a graph Solve $6 - 5x = 3(1 - x)$ with tables and a graph.

Solution Tables 13 and 14 have no coordinate point in common. The coordinates are closest between $x = 1$ and $x = 2$.

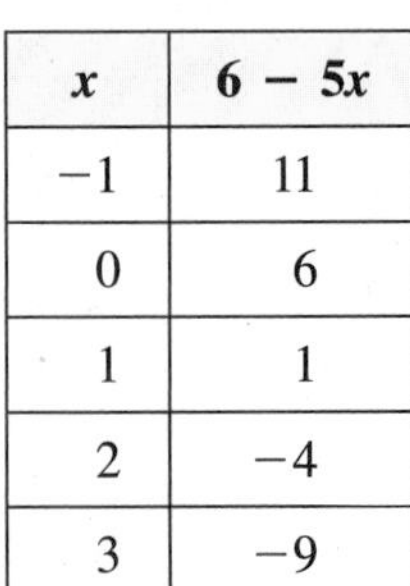

x	$6 - 5x$
−1	11
0	6
1	1
2	−4
3	−9

Table 13

x	$3(1 - x)$
−1	6
0	3
1	0
2	−3
3	−6

Table 14

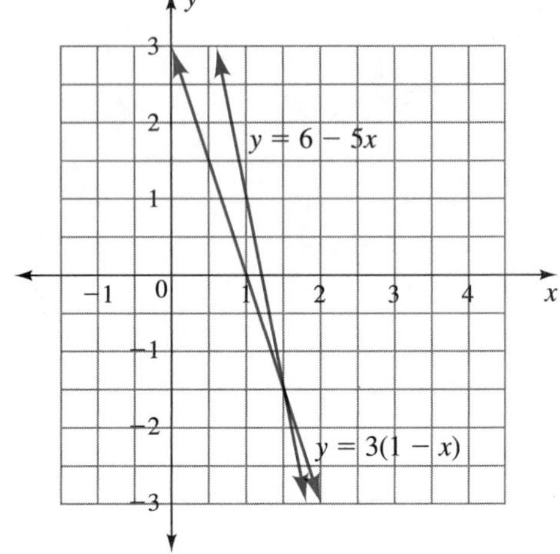

Figure 12

The graph in Figure 12 shows the point of intersection to be a fraction for both x and y. We guess $\left(1\frac{1}{2}, -1\frac{1}{2}\right)$ and check it.

Check: $6 - 5\left(1\frac{1}{2}\right) \stackrel{?}{=} 3\left(1 - 1\frac{1}{2}\right)$ ✓ ●

Solving Equations with Algebraic Notation

In Example 4, we will solve the equation from Example 3 with symbols. For many equations like the one in Example 3, symbolic solutions are easier than tables or graphs. If the symbolic method becomes difficult or even impossible, we can turn to graphing calculators or computers for a solution.

Solving Linear Equations with Algebraic Notation

1. Multiply on both sides by the least common denominator to remove fractions or decimals.
2. Apply the distributive property to expressions containing parentheses.
3. Collect like terms on each side.
4. Using the addition property of equations, get all the terms with the variable to be solved for onto one side of the equation. Get the constant terms to the other side of the equation.
5. Collect like terms again, if necessary.
6. Solve for the variable, using the multiplication property of equations.

Caution: When an equation has a variable on both sides, you will *make fewer errors if you keep the variable's coefficient positive when you add variable terms to each side* (step 4).

EXAMPLE  Solving equations using algebraic notation Solve for x: $6 - 5x = 3(1 - x)$.

Solution

$6 - 5x = 3(1 - x)$	Distributive property
$6 - 5x = 3 - 3x$	Add $5x$ (addition property) to keep the coefficient positive.
$6 - 5x + 5x = 3 - 3x + 5x$	Add like terms.
$6 = 3 + 2x$	Subtract 3 (addition property).
$6 - 3 = 3 + 2x - 3$	Add like terms.
$3 = 2x$	Divide both sides by 2 (multiplication property).
$\frac{3}{2} = \frac{2x}{2}$	Simplify.
$1\frac{1}{2} = x$	

Student Note:
The comments tell what to do next.

Check: $6 - 5(1\frac{1}{2}) \stackrel{?}{=} 3(1 - 1\frac{1}{2})$ ✓

Think about it: What number is obtained on both sides of the Check in Example 4? What is the meaning of this number in Example 3?

In Example 5, we use the distributive property and addition of like terms to simplify the left side before solving the equation.

EXAMPLE  Solving equations using algebraic notation Solve for x: $13 - 4(x - 2) = 1$.

Solution

$13 - 4(x - 2) = 1$	Use the distributive property.
$13 - 4x + 8 = 1$	Add like terms.
$21 - 4x = 1$	Subtract 21 from both sides (addition property).
$21 - 4x - 21 = 1 - 21$	
$-4x = -20$	Divide both sides by -4 (multiplication property).
$\frac{-4x}{-4} = \frac{-20}{-4}$	Simplify.
$x = 5$	

Check: $13 - 4(5 - 2) \stackrel{?}{=} 1$ ✓

In Example 6, we multiply both sides of the equation by the denominator to eliminate the fraction notation.

EXAMPLE  Solving equations containing fractions Solve $\frac{5}{6}x = x + 4$.

Solution

$\frac{5}{6}x = x + 4$	Multiply both sides by 6.
$\frac{6}{1} \cdot \frac{5}{6}x = 6(x + 4)$	Apply the distributive property.
$5x = 6x + 24$	Subtract $5x$ from both sides.
$-5x \quad -5x$	
$0 = x + 24$	Subtract 24 from both sides.
$-24 \quad -24$	
$-24 = x$	

Check: $\frac{5}{6}(-24) \stackrel{?}{=} (-24) + 4$ ✓

In Example 7, we eliminate the fraction and use the distributive property in solving the equation.

EXAMPLE 7 Solving equations containing fractions Solve $2(x - 7) = \frac{1}{2}x - 2$.

Solution

$2(x - 7) = \frac{1}{2}x - 2$	Multiply both sides by $\frac{2}{1}$.
$\frac{2}{1} \cdot 2(x - 7) = \frac{2}{1}(\frac{1}{2}x - 2)$	Apply the distributive property.
$4x - 28 = x - 4$	Subtract x from both sides.
$-x \qquad -x$	
$3x - 28 = -4$	Add 28 to both sides.
$+28 \qquad +28$	
$3x = 24$	Divide by 3 on both sides.
$\frac{3x}{3} = \frac{24}{3}$	Simplify.
$x = 8$	

Check: $2(8 - 7) \stackrel{?}{=} \frac{1}{2}(8) - 2$ ✔

In Example 8, we apply the order of operations, doing a multiplication before a subtraction. A common error is to subtract the 6 and 4 first.

EXAMPLE 8 Solving equations containing parentheses Solve $6 - 4(x - 2) = 11$.

Solution

$6 - 4(x - 2) = 11$	Do the multiplication on the left side first.
$6 - 4x + 8 = 11$	Add like terms.
$14 - 4x = 11$	Subtract 14 from both sides.
$-14 \qquad -14$	
$-4x = -3$	Divide by -4 on both sides.
$\frac{-4x}{-4} = \frac{-3}{-4}$	Simplify.
$x = \frac{3}{4}$	

Check: $6 - 4\left(\frac{3}{4} - 2\right) \stackrel{?}{=} 11$ ✔

Applications

Starting with word descriptions or problem settings, we now practice writing and solving equations containing parentheses.

COMMON PHRASES REQUIRING PARENTHESES Grouping symbols such as parentheses are needed in writing equations from sentences containing the following phrases:

for each additional minute
for each additional day
(an operation is done to) the sum of a and b
(an operation is done to) the difference between a and b

EXAMPLE 9 Writing equations Write each sentence as a one- or two-variable equation containing parentheses. Define your variables.

a. What is the total cost of a telephone call if the charge is \$0.75 for the first minute and \$0.10 for each additional minute?

b. What is the total cost of a pressure-washer rental at \$40 for the first day plus \$25 for each additional day?

c. Take a number, subtract 8, and then multiply the difference by 2. The result is -24.

d. Half the sum of the input and three gives the output.

Solution Parts a, b, and d are two-variable equations. Part c is a one-variable equation.

a. Let $x =$ the total number of minutes used and $y =$ the total cost: $y = 0.75 + 0.10(x - 1)$. (Recall that the cost of the first minute is \$0.75.)

b. Let $x =$ the total number of rental days and $y =$ the total cost: $y = 40 + 25(x - 1)$.

c. Let x be the input number: $2(x - 8) = -24$.

d. Let $x =$ input and $y =$ output: $y = \frac{1}{2}(x + 3)$. ●

CONSECUTIVE INTEGERS If you go shopping on three consecutive days or make five consecutive shots in basketball, you are engaging in events that occur one after another without interruption. Appointments on four consecutive Mondays have the interruption of the intervening days, but the Mondays themselves are in a row (see Figure 13).

Figure 13

Consecutive integers are *integers that follow one after another without interruption.* A set of consecutive integers such as 4, 5, and 6 may be described in terms of the smallest number: x, $x + 1$, and $x + 2$. In Example 11, we describe the set in terms of the other two numbers.

EXAMPLE 10 Writing expressions Write a consecutive integer description for 4, 5, and 6 with x as the largest number. Repeat with x as the middle number.

Solution

	4	**5**	**6**
Let $x = 6$	$x - 2$	$x - 1$	x
Let $x = 5$	$x - 1$	x	$x + 1$

●

Recall that the numbers including 0, 2, 4, 6, . . . are the *even* integers and the numbers including 1, 3, 5, 7, . . . are the *odd* integers. There is a surprising result in the description of these integers.

EXAMPLE 11 Evaluating expressions Evaluate x, $x + 2$, and $x + 4$ for $x = 11$ and for $x = 12$. Identify the resulting numbers as consecutive even integers, consecutive odd integers, or neither.

Solution

	x	$x + 2$	$x + 4$	
$x = 11$	11	13	15	Consecutive odd integers
$x = 12$	12	14	16	Consecutive even integers

Example 11 suggests that the symbolic description x, $x + 2$, $x + 4$ for consecutive even integers is identical to the symbolic description for consecutive odd integers. It is a common error to write consecutive odd numbers as x, $x + 1$, and $x + 3$.

Examples 12 and 13 illustrate the use of consecutive integers in number puzzle problems. Such puzzles are important only because they provide practice in writing and solving equations.

EXAMPLE 12 Writing equations The sum of three consecutive odd integers is 447. Set up an equation that shows this relationship, and solve it.

Solution The equation is $x + (x + 2) + (x + 4) = 447$. Solving for x, we find

$$x + (x + 2) + (x + 4) = 447$$
$$3x + 6 = 447$$
$$3x = 441$$
$$x = 147$$

Thus, the numbers are $x = 147$, $x + 2 = 149$, and $x + 4 = 151$.

EXAMPLE 13 Writing expressions Write expressions for these years: 1990, 1995, 2000, 2005, 2010. Let $x =$ the first year, 1990.

Solution If $x = 1990$, then

$$1995 = x + 5$$
$$2000 = x + 10$$
$$2005 = x + 15$$
$$2010 = x + 20$$

CREDIT CARD PAYMENTS We now return to the credit card payment schedule.

EXAMPLE 14 Finding equations for a conditional table: credit card payments Use the output descriptions in Table 15 to write equations. Let x be the charge balance in dollars and y be the payment due in dollars.

Input, x: Charge Balance	Output, y: Payment Due
[0, 20]	Full amount
(20, 500]	10% or \$20, whichever is greater
(500, $+\infty$)	\$50 plus the amount in excess of \$500

Table 15 Credit Card Payment Schedule

Solution The equations are shown in Table 16.

The phrase *full amount* means the payment due exactly equals the charge balance. Because output equals input, the equation for the first set of inputs is $y = x$.

Ten percent means 0.10 times the input, or $y = 0.10x$. Twenty dollars as output means $y = \$20$. The two outputs need to be compared and the highest output paid.

The rule for the last set of inputs has the phrase "the amount in excess of \$500." We subtract \$500 from the input before adding \$50. The equation is $y = (x - \$500) + \$50$.

Input, x: Charge Balance	Output, y: Payment Due
[0, 20]	$y = x$
(20, 500]	$y = 0.10x$ or $y = \$20$, whichever is greater
$(500, +\infty)$	$y = (x - \$500) + \$50$

Table 16 Credit Card Payment Schedule

Three items in Example 14 are noteworthy. First, we always define variables. Second, we can write percents in decimal notation in equations as well as in expressions. Third, although the parentheses are not needed in the third equation, they serve to remind us to find the *excess* amount first.

In Example 15, we use the table and graph for the credit card payments to solve equations. Because the credit card payments are conditional, we use the graph to decide what equation is appropriate.

EXAMPLE  Solving equations using tables and graphs: credit card payments Select the appropriate equation from Example 14, and then substitute the given information. Use Table 17 or the graph in Figure 14 to solve the equation.

a. Suppose Carlos has a payment of \$15. What is his charge balance?

b. Suppose Marge has a payment of \$35. What is her charge balance?

c. Suppose Raphael has a payment of \$175. What is his charge balance?

Input: Balance	Output: Payment
\$ 0	\$ 0
10	10
20	20
30	20
200	20
300	30
500	50
600	150
650	200

Table 17 Credit Card Payments

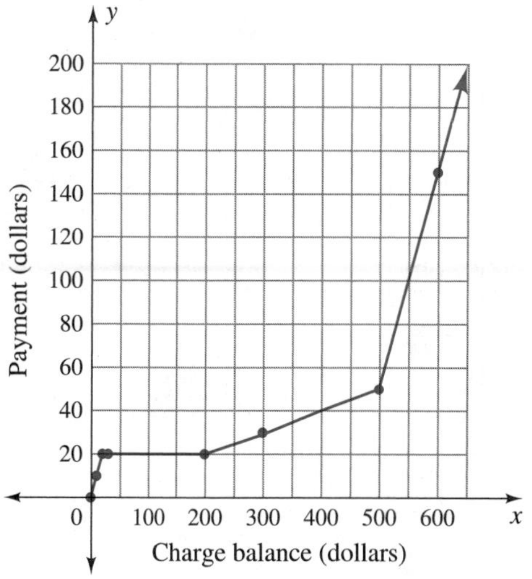

Figure 14

Solution

a. A \$15 payment (output, y) means that the input is between 0 and \$20. We select the equation $y = x$ and let $y = \$15$: $\$15 = x$. Carlos's charge balance is \$15.

b. A \$35 payment (output, y) means that the input is between \$200 and \$500. We select the equation $y = 0.10x$ and let $y = \$35$: $\$35 = \$0.10x$. We estimate an input of \$350 because the rule in this part of the table is 10% of the charge balance. Marge's charge balance is \$350.

c. A \$175 payment means that the input is greater than \$500. We select the equation $y = (x - 500) + 50$ and let $y = \$175$: $\$175 = (x - \$500) + \$50$. From the table, \$175 is halfway between \$150 and \$200, so its input would be \$625. Raphael's charge balance is \$625. ●

ANSWER BOX

Warm-up: 1. $4x + 12$ **2.** $-2x - 2$ **3.** $-4x + 8$ **4.** $-x - 1$ **5.** $21 - 4x$ **6.** $11 - 3x$ **7.** $1x$ **8.** $4x$ **Think about it:** In the Check, $6 - 5(1\frac{1}{2}) = 6 - 7\frac{1}{2} = -1\frac{1}{2}$ and $3(1 - 1\frac{1}{2}) = 3(-\frac{1}{2}) = -1\frac{1}{2}$. In Example 3, the point of intersection was $(1\frac{1}{2}, -1\frac{1}{2})$. The $-1\frac{1}{2}$ is the y coordinate of the point of intersection.

EXERCISES 3.3

Solve the equations in Exercises 1 to 4 for x by completing the given tables and identifying x from the common ordered pair.

1. $5x - 8 = 2(x + 2)$

x	$5x - 8$
−1	
0	
1	
2	
3	
4	

x	$2(x + 2)$
−1	
0	
1	
2	
3	
4	

2. $7(x - 5) = 3 - 12x$

x	$7(x - 5)$
−1	
0	
1	
2	
3	
4	

x	$3 - 12x$
−1	
0	
1	
2	
3	
4	

3. $3(x - 3) = 6(x - 2)$

x	$3(x - 3)$
−1	
0	
1	
2	
3	
4	

x	$6(x - 2)$
−1	
0	
1	
2	
3	
4	

4. $4(x + 1) = 2(3x - 1)$

x	$4(x + 1)$
−1	
0	
1	
2	
3	
4	

x	$2(3x - 1)$
−1	
0	
1	
2	
3	
4	

5. Solve with the graph:

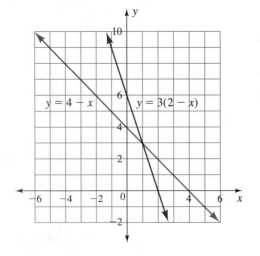

a. $3(2 - x) = 4 - x$

b. $3(2 - x) = 6$

c. $4 - x = 6$

d. $4 - x = 2$

e. $3(2 - x) = 0$

6. Solve with the graph:

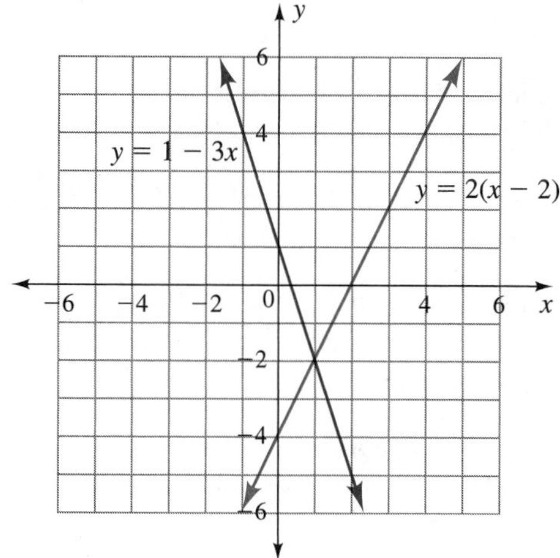

a. $1 - 3x = 2(x - 2)$

b. $2(x - 2) = 2$

c. $1 - 3x = 4$

d. $1 - 3x = -5$

e. $2(x - 2) = 0$

7. Solve with the graph:

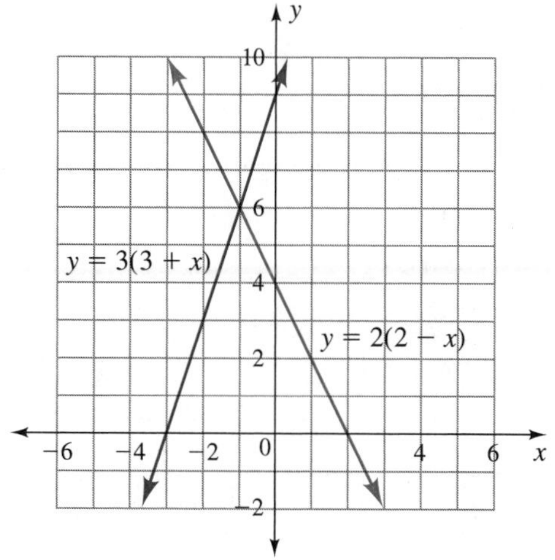

a. $3(3 + x) = 2(2 - x)$

b. $3(3 + x) = 0$

c. $2(2 - x) = 0$

d. $2(2 - x) = 4$

e. $3(3 + x) = 3$

8. Solve with the graph:

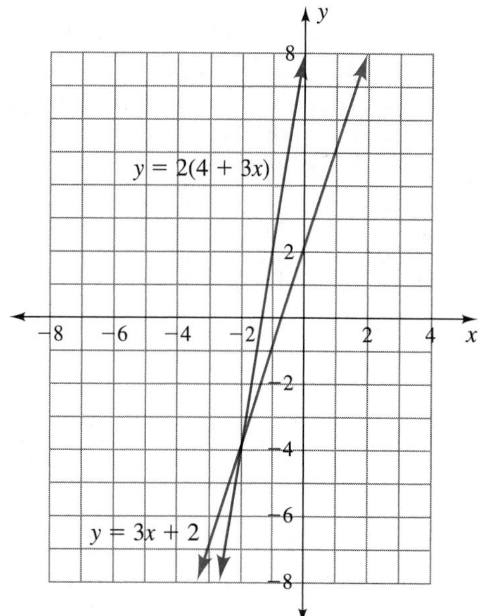

a. $2(4 + 3x) = 3x + 2$

b. $2(4 + 3x) = 8$

c. $3x + 2 = 8$

d. $3x + 2 = -1$

e. $2(4 + 3x) = 2$

9. a. Make a table and graph for $y = 8 - 3(x + 2)$.

b. Solve $8 - 3(x + 2) = 11$ from the table and with algebraic symbols.

c. Solve $8 - 3(x + 2) = 0$ from the graph and with algebraic symbols.

d. Solve $8 - 3(x + 2) = -4$ with any of the three methods.

10. a. Make a table and graph for $y = 8 - 3(x + 4)$.

b. Solve $8 - 3(x + 4) = 2$ from the table and with algebraic symbols.

c. Solve $8 - 3(x + 4) = 0$ from the graph and with algebraic symbols.

d. Solve $8 - 3(x + 4) = -4$ with any of the three methods.

Solve the equations in Exercises 11 to 40 using any of the three methods.

11. $2(x - 3) = 0$

12. $3(x - 1) = 2$

13. $-2(x + 1) = 5$

14. $-3(x - 2) = 0$

15. $2(3x + 1) = 5 - 3x$

16. $3(3x + 1) = 13 - 6x$

17. $4x + 6 = 2(1 + 3x) + 1$

18. $7x + 2 = 3(1 + x) + 2$

19. $7(x + 1) = 11 + x$ **20.** $3(x + 2) = 4 - x$

21. $2(x + 3) = 3x - 2$ **22.** $2(2 + x) = 13 - 4x$

23. $\frac{3}{5}x = 15$ **24.** $\frac{5}{8}x = 20$

25. $\frac{1}{3}x = x - 12$ **26.** $\frac{2}{3}x + 4 = x + 7$

27. $\frac{3}{4}x = x + 3$ **28.** $\frac{4}{7}x - 12 = x$

29. $\frac{1}{2}(x + 5) = x - 4$ **30.** $\frac{3}{5}(2 - x) = x + 6$

31. $6 - 4(x - 2) = 22$ **32.** $5 - 3(x - 2) = 9$

33. $2(8 - x) = 1 + 4x$

34. $3(x + 1) - 2x = 3 - 4x$

35. $2 - 5(x + 1) = 4 - 3x$

36. $5x - 2(x - 1) = 4x - 3$

37. $7 - 2(x - 1) = 5 - 3x$

38. $4 - x + 3 = -2 + 4x$

39. $5 - 2(x - 3) = 3(x - 3)$

40. $7 - 2(x - 1) = 5(x - 1)$

In Exercises 41 to 46, tell what was done to the first equation to obtain the second equation. Which property of equations or real numbers lets us do each step?

41. $2(2x + 3) = 2 + 6x + 1$
$4x + 6 = 2 + 6x + 1$

42. $7x + 2 = 3 + 3x + 2$
$4x + 2 = 3 + 2$

43. $7x + 7 = 11 + x$
$7x = 4 + x$

44. $3(x + 2) = 4 - x$
$3x + 6 = 4 - x$

45. $3x + 6 = 4 - x$
$4x + 6 = 4$

46. $4x + 6 = 6x + 2 + 1$
$4x + 6 = 6x + 3$

Write each sentence in Exercises 47 to 60 as a one- or two-variable equation containing parentheses. Define your variables so that y depends on x. Solve the one-variable equations.

47. Take a number, add 5, and then multiply the sum by 2. The result is 14.

48. Take a number, subtract 6, and then multiply the difference by 3. The result is 15.

49. Take a number, subtract 4, and then multiply the difference by negative 2. The result is 6.

50. Take a number, add 3, and then multiply the sum by 7. The result is −21.

51. Half the difference between the input and five gives the output.

52. The output is the product of negative three and four more than the input.

53. The output is the product of fifteen and the difference between three and the input.

54. Twice the sum of 6 and a number gives the output.

55. A bowling handicap is 80% of the difference between 200 and the bowling average.

56. One recommended exercise pulse rate is 60% of the difference between 220 and your age.

57. A rental car costs a basic fee of \$65 for the first 100 miles and \$0.15 per mile over 100 miles. How many miles, x, may be driven for a total of \$100?

58. A pressure washer costs \$30 for the first 4 hours and \$10 per additional hour. What is the total hours rented, x, if the total cost is \$70?

59. A transcript costs \$2.00 for the first copy and \$1.50 for each additional copy. If the total cost is \$6.50, what is the number of copies made, x?

60. At an outdoor concert the staff adds 3 seats to each row of x seats. There are 25 rows altogether, and there are 1200 seats altogether. How many seats, x, were in the original rows?

Complete the tables in Exercises 61 to 64. Write each number in terms of x, as defined in the left column.

61. ***Consecutive Integers***

	4	**5**	**6**
Let $x = 4$	x	$x + 1$	
Let $x = 5$		x	
Let $x = 6$		$x - 1$	x

62. ***Consecutive Even Integers***

	6	**8**	**10**
Let $x = 6$	x		
Let $x = 8$		x	
Let $x = 10$		$x - 2$	x

63. ***Consecutive Odd Integers***

	5	**7**	**9**
Let $x = 5$	x		
Let $x = 7$		x	
Let $x = 9$		$x - 2$	x

64. ***Consecutive Integers***

	−7	−6	−5
Let $x = -5$			x
Let $x = -6$		x	
Let $x = -7$	x		

65. The sum of three consecutive integers is 42. Write an equation and solve it to find the three integers.

66. The sum of three consecutive integers is 288. Write an equation and solve it to find the three integers.

67. The sum of three consecutive odd integers is 177. Write an equation and solve it to find the three integers.

68. The sum of three consecutive odd integers is 429. Write an equation and solve it to find the three integers.

69. Draw a number line with equally spaced numbers from 1 to 16. Circle the odd numbers, and connect each with the next odd number using a curved line above the number. Place a small square around the even numbers, and connect each to the next even number using a curved line below the number line. How does this show that one description of both even and odd numbers is x, $x + 2$, $x + 4$?

70. If $2005 = x$, describe these years (consecutive 5-year intervals) in terms of x: 1990, 1995, 2000, 2005, 2010.

71. The U.S. presidential elections, the summer Olympics, and leap years all occur every 4 years. Furthermore, these years are evenly divisible by 4. Which year—2010 or 2012—is one of these special years? If x represents the first year, give a description for the next three leap years in terms of x.

72. The concept of events' taking place at regular intervals even appears in the U.S. Constitution: "The actual Enumeration shall be made within three Years after the first Meeting of the Congress of the United States, and within every subsequent term of ten Years, . . ." (Article One, Section 2.3). What important activity does this article describe?

73. A series of books is published at 7-year intervals. When the seventh book is issued, the sum of the publication years is 13,741. When was the first book published?

74. A series of books is published at 4-year intervals. When the fifth book is issued, the sum of the publication years is 10,020. When was the first book published?

The equations in Exercises 75 to 80 are related to the credit card applications in this section. Solve them with algebraic notation.

75. $40 = 0.10x$

76. $15 = 0.10x$

77. $35 = 0.10x$

78. $100 = 50 + (x - 500)$

79. $550 = 50 + (x - 500)$

80. $175 = 50 + (x - 500)$

81. Describe how to find the solution to an equation of the form $ax + b = cx + d$ from tables.

82. Describe how to find the solution to an equation of the form $ax + b = cx + d$ from a graph.

83. What algebraic steps do we use in solving an equation with variables on both sides (for example, $ax + b = cx + d$)?

84. What algebraic steps do we use in solving an equation containing parentheses (for example, $a(x + c) = d$)?

85. What algebraic steps do we use in solving an equation containing a fraction (for example, $\frac{ax}{b} + c = d$)?

86. What phrases in a word problem tell us that its equation may contain parentheses?

87. Find the values of the expressions x, $x + 1$, and $x + 3$ for $x = 1$. Repeat for $x = 2$. Why might someone incorrectly use these expressions to describe a set of odd numbers?

88. Compare the algebraic descriptions for consecutive even integers and consecutive odd integers.

89. What restrictions are there on the inputs, x, in Exercises 55 to 60?

Error Analysis

90. Explain what is wrong with this equation-solving step:

$$\begin{array}{l} 5 + 2x - 7 = 15 \\ \underline{+7 \qquad\quad +7} \\ 12 + 2x \qquad = 15 \end{array}$$

Projects

91. ***Equations and Identities.*** Use guess and check or algebraic notation to solve each equation. Write one of the following statements for each equation:

Has only $x = 0$ as a solution.

Has no real-number solutions.

Is an identity (true for all real numbers).

Has both $x = 0$ and $x = 1$ as a solution.

Has one solution, $x = 4$.

a. $-(x - 5) = 5 - x$ **b.** $3 + 3x = 3 - 3x$

c. $3x = 5x$ **d.** $-3(x - 2) = -3x - 2$

e. $3x + 6 = 3(x + 2)$ **f.** $-2(x - 4) = -2x - 8$

g. $2x + 3x = 5x^2$ **h.** $2x + 3x = 5x$

i. $5 - 2(x - 3) = 3(x - 3)$

j. $5 - 2(x - 3) = 11 - 2x$

92. ***Even and Odd Integer Sums***

a. Make a two-column listing, with the first two positive even integers in one column and their sum in the second column. Next list the first three even integers in the first column and their sum in the second column. Repeat for the first four even integers, and so forth up to 10 even integers. Your last entry in the first column should be $2 + 4 + 6 + 8 + 10 + 12 + 14 + 16 + 18 + 20$.

b. Find a pattern and write a rule for the sum of x positive even integers.

c. Repeat part a for the positive odd integers.

d. Find a rule for the sum of x positive odd integers.

93. ***Stacking Up Coins***

a. Arrange 25 coins into four stacks that fit the following conditions: The second stack is 3 times the first stack. The third stack is 1 less than the second. The fourth stack is 2 more than the first. How many coins are in each stack? Write an equation that would solve the same problem.

b. Arrange 28 coins into four stacks that fit the following conditions: The third stack is 3 more than the second stack. The first stack is twice the second stack. The fourth stack is 1 more than the first stack. How many coins are in each stack? Write an equation that would solve the same problem.

c. Describe a strategy to arrange the coins into the requested stacks.

94. ***Consecutive Integer Sums***

a. Add these consecutive integers and divide the sum by 3:

34, 35, 36

23, 24, 25

b. Add these consecutive odd integers and divide the sum by 3:

33, 35, 37

27, 29, 31

c. Add these consecutive even numbers and divide the sum by 3:

12, 14, 16

80, 82, 84

d. Explain any patterns you see in parts a, b, and c.

e. Add and divide the result by 3:

$x + (x + 1) + (x + 2)$

$x + (x + 2) + (x + 4)$

f. Explain how to find the three consecutive numbers that add to a given sum. Change your rule so it works for consecutive even numbers and for consecutive odd numbers.

g. Extend your rule in part f to finding other sets of consecutive multiples given their sum. *Hint:* Look for patterns in these sums:

$3 + 4 + 5 + 6$

$5 + 10 + 15 + 20$

$24 + 30 + 36 + 42 + 48$

$3 + 6 + 9 + 12 + 15$

h. Try your rules on these sums:

Four consecutive multiples of 10 that add to 220

Five consecutive multiples of 3 that add to 195

Four consecutive multiples of 7 that add to 210

3.4 Solving Linear Inequalities in One Variable

OBJECTIVES

- Solve one-variable inequalities with a graph.
- Graph solutions to inequalities on a number line.
- Solve one-variable inequalities with algebraic notation.
- Write inequalities for application settings.

WARM-UP

In Exercises 1 to 4, place > or < between the two expressions.

1. a. $4 \_\_\_ 5$ b. $4 - 2 \_\_\_ 5 - 2$ c. $4 + (-5) \_\_\_ 5 + (-5)$
2. a. $-3 \_\_\_ -5$ b. $-3 - 4 \_\_\_ -5 - 4$ c. $-3 + 5 \_\_\_ -5 + 5$
3. a. $4 \_\_\_ -2$ b. $4(5) \_\_\_ -2(5)$ c. $4 \div 2 \_\_\_ -2 \div 2$
4. a. $-3 \_\_\_ 6$ b. $-3(4) \_\_\_ 6(4)$ c. $-3 \div 3 \_\_\_ 6 \div 3$
5. Does adding or subtracting a number change the inequality sign between two numbers or expressions? Does multiplying or dividing by a positive number change the inequality sign between two numbers or expressions?
6. Sketch a number-line graph for each of the following inequalities:
   a. $x \geq 2$ b. $x < 0$ c. $x \leq 5$ d. $x \leq 0$
7. Write each inequality in Exercise 6 as an interval.

IN THIS SECTION, we solve one-variable linear inequalities. We solve these inequalities both with a graph and with algebraic notation.

EXAMPLE  Exploration: guessing and checking to solve an inequality Suppose a course has three tests worth 100 points each, projects and homework worth 70 points, and a final exam worth 150 points. The instructor grades on a percent basis: 90% for an A, 80% for a B, 70% for a C. One student has test scores of 78, 84, and 72, with full credit on projects and homework (70 points).

a. What are the total possible points?

b. What grade will the student earn with a 95 on the final exam?

c. Use guess and check on a calculator to find the score needed on the final exam to earn at least a B.

Solution See the Answer Box. ●

The solution to the Exploration is an inequality. Although only one final exam score will result in a grade of 80%, a higher final exam score will still result in at least a B. We will return to the Exploration problem in Example 12, where we will solve it with symbols.

One-Variable Linear Inequalities

Definition of One-Variable Linear Inequality

A **linear inequality in one variable** can be written $ax + b < c$, where a, b, and c are real numbers and a is not 0.

The above definition and the properties and definitions that follow are true also for the other inequality symbols, $>$, $\geq$, and $\leq$.

The **solution set of a one-variable inequality** is *the set of values of the input variable that make the inequality a true statement.*

SOLVING ONE-VARIABLE INEQUALITIES FROM A COORDINATE GRAPH We commonly refer to the solution set of an inequality as the *set of all solutions.* We show the set of solutions to a one-variable inequality on a number-line graph (see Warm-up Exercise 6). Return to Section 2.6 to review number-line graphs.

When we solve the equation $3x - 2 = -5$ with a graph, we graph both the left side of the equation, $y = 3x - 2$, and the right side, $y = -5$ (see Figure 15). The ordered pair (x, y) at the point of intersection gives us both the solution to the equation, $x = -1$, and the value of each side of the equation when x is substituted into the equation, $y = -5$.

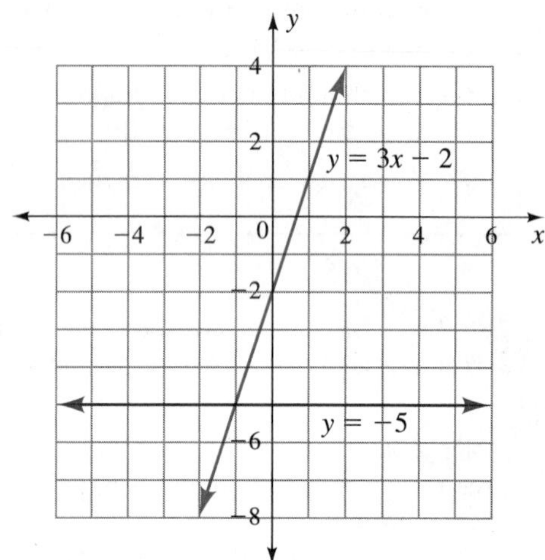

Figure 15

To find the graphical solution to an inequality, we follow a process similar to the one we used to find the graphical solution to an equation. We again graph the left and right sides as separate equations and then find the point of intersection (x, y). The additional steps are to mark the x on a number line and then use test numbers to find which other values on the number line satisfy the inequality.

EXAMPLE 2 Solving an inequality with a graph Solve $3x - 2 > -5$ with the graph in Figure 15.

Solution The lines $y = 3x - 2$ and $y = -5$ intersect at $(-1, -5)$ in Figure 15. We mark $x = -1$ on the number line, using a small circle because of the $>$ inequality sign. To find other points in the solution set, we select a test point—say, $x = 2$. We then substitute the test point into $3x - 2 > -5$:

$$3(2) - 2 > -5$$
$$6 - 2 > -5$$
$$4 > -5$$

The inequality is true, so we draw an arrow from $x = -1$ through the point $x = 2$. The solution set is $x > -1$.

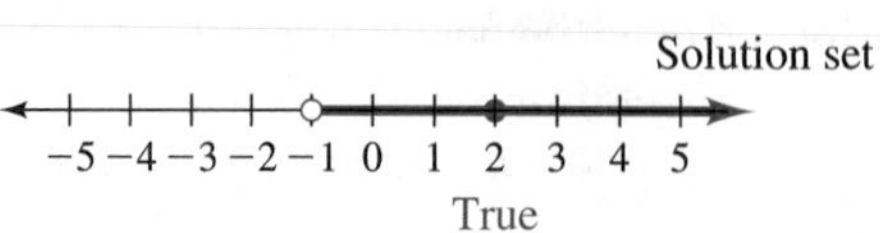

Solving One-Variable Inequalities with a Graph

1. Graph $y =$ the left side of the inequality and $y =$ the right side.
2. Find the point of intersection (x, y). Mark the x on a number line. Use a dot if the inequality is $\geq$ or $\leq$; use a small circle if the inequality is $>$ or $<$.
3. Choose a test number on either side of the x on the number line. Substitute the test number into the inequality. If the number makes the statement true, draw an arrow on the number line from x through the test number. If the number makes the statement false, draw the arrow from x in the opposite direction from the test number.
4. Write an inequality for the solution set shown on the number line.

In each of the following examples, we use a coordinate graph to find the point of intersection and then graph the set of solutions on a number line below the graph.

EXAMPLE 3 Solving an inequality with a graph Solve $6 - 2x \leq 12$ with a graph. Show the solution set as a line graph, an inequality, and an interval.

Solution The graphs of $y = 6 - 2x$ and $y = 12$ are shown in Figure 16. The point of intersection is $(-3, 12)$.

We mark $x = -3$ on a number line, using a dot because of the $\leq$ inequality sign. We then select $x = 0$ as a test point:

$$6 - 2x \leq 12 \qquad \text{Substitute } x = 0.$$
$$6 - 2(0) \leq 12$$
$$6 \leq 12$$

The inequality is true, so $x = 0$ is in the solution set. We mark 0 on the number line and draw an arrow from -3 through 0. The solution set is $x \geq -3$, or $[-3, +\infty)$.

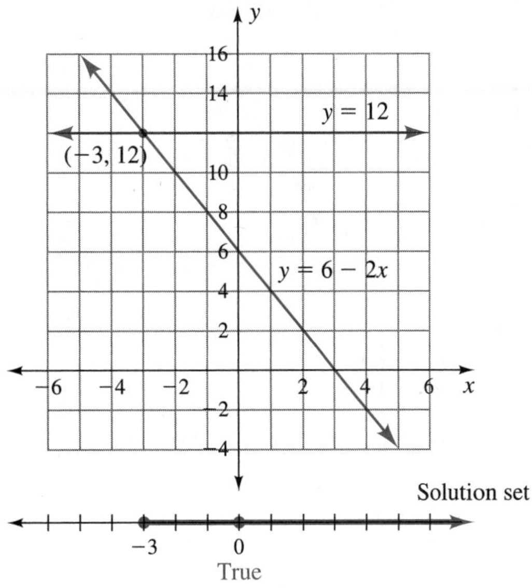

Figure 16

Graphing Calculator Solution The graphing calculator will provide a line graph solution to an inequality. If an input x makes the inequality true, the calculator will output $y = 1$. If an input x makes the inequality false, the calculator will output $y = 0$. The calculator will make a graph of these ones and zeros instead of a graph of the function.

Enter $Y_1 = 6 - 2X \leq 12$.

Set a window similar to that shown in Figure 16. Graph.

A horizontal line will appear at $y = 1$ for $x \geq -3$. Although a horizontal line will also appear at $y = 0$ for $x < -3$, it will not show unless you choose to shut off the axes, because it lies on the x-axis.

TRACE will show $y = 0$ for $x < -3$ and $y = 1$ for $x \geq -3$.

If the graphical results are unclear, look at the zeros and ones in TABLE. ●

Solving an inequality with a graph Suppose Audrey decides to have a two-hour party at Willamalane Wave Pool. The total cost of \$3 per person plus \$20 party room rental is given by $y = 3x + 20$, where x is the number of people. Her budget is \$65.

a. Graph her cost equation and budget equation. Name the point of intersection.

b. Write an inequality that describes the number of people Audrey can have at the wave pool.

c. Solve the inequality. Show the solution set as a line graph, an inequality, and an interval.

Solution **a.** The graphs, shown in Figure 17, intersect at (15, 65).

b. The number of people at the party must lie where the cost is less than or equal to the budget: $3x + 20 \leq 65$.

c. The intersection is at (15, 65), so we mark $x = 15$ on a number line. We choose a test number—say, $x = 20$—and substitute it into the inequality:

$$3(20) + 20 \leq 65$$
$$80 \leq 65$$

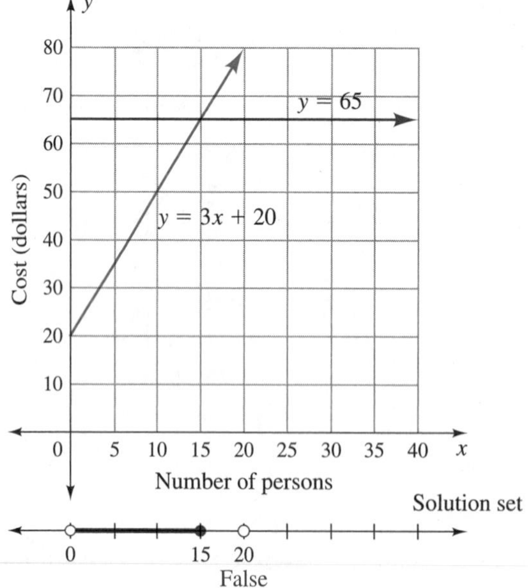

Figure 17

The test number gives a false statement, so we draw an arrow from $x = 15$ in the opposite direction. The solution to $3x + 20 \leq 65$ is $x \leq 15$. Audrey can have 15 or fewer people. The number of people must be positive, so the solution set is $0 < x \leq 15$, or $(0, 15]$. ●

Example 5 has variables on both sides of the inequality. The solution process is the same as in the other examples.

EXAMPLE 5 Solving an inequality with a graph Solve $2x + 3 < -3x - 2$ by graphing.

Solution The graphs of $y = 2x + 3$ and $y = -3x - 2$ are in Figure 18. The lines intersect at $(-1, 1)$. We mark $x = -1$ on a number line, using a small circle. We then choose a test number—say, $x = 0$. Next we substitute $x = 0$ into the inequality:

$$2x + 3 < -3x - 2$$
$$2(0) + 3 < -3(0) - 2$$
$$3 < -2$$

The test number gives a false statement, so we draw an arrow from $x = -1$ in the opposite direction from zero. The solution set is $x < -1$.

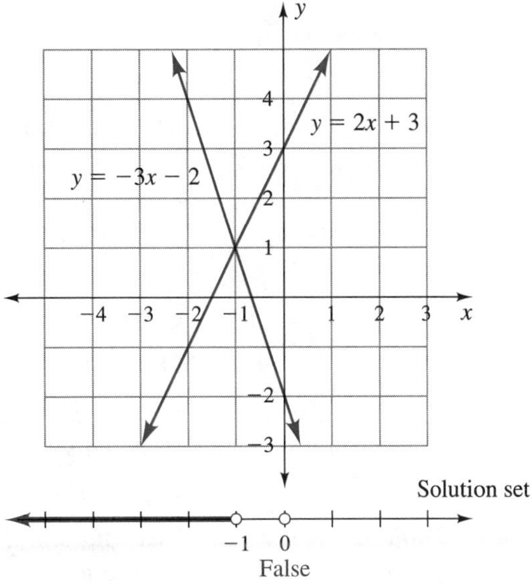

Figure 18 ●

Think about it 1: For what values of x is the graph of $y = 2x + 3$ below the graph of $y = -3x - 2$? Could the positions of the graphs be used to solve the inequality $2x + 3 < -3x - 2$?

EXAMPLE 6 Solving an inequality with a graph Suppose Julian wishes to find the number of persons for which a two-hour rental at Willamalane (see Example 4) will be cheaper than going to Farrell's Ice Cream at \$3.95 per person.

a. State and graph the cost equations.

b. Write an inequality describing the cost comparison.

c. Solve the inequality from the graph.

Solution **a.** The cost equations are $y = 3x + 20$ and $y = 3.95x$. They are graphed in Figure 19.

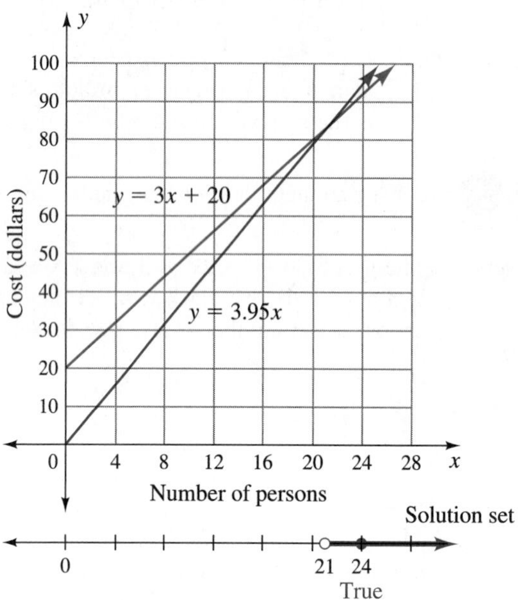

Figure 19

b. Willamalane will be cheaper when $3x + 20 < 3.95x$.

c. The point of intersection of the two graphs is approximately (21, 83). We mark $x = 21$ on a number line, using a small circle. We then choose a test number—say, $x = 24$. Next we substitute $x = 24$ into the inequality:

$$3x + 20 < 3.95x$$
$$3(24) + 20 < 3.95(24)$$
$$92 < 94.80$$

The test number gives a true statement, so we draw a line from $x = 21$ through $x = 24$. The solution set is $x > 21$. ●

Think about it 2: For what values of x is the graph of $y = 3x + 20$ below the graph of $y = 3.95x$? Could the positions of the graphs be used to solve the inequality $3x + 20 < 3.95x$?

SOLVING ONE-VARIABLE INEQUALITIES USING ALGEBRAIC NOTATION When we solved equations with algebraic notation, we created our equivalent equations at each step in the solution. Similarly, the inequalities in each step of a solution are said to be equivalent. *Equivalent inequalities* have the same solution set. The following exploration shows a surprising property of operations on inequalities, which you must remember to ensure that you produce equivalent inequalities.

EXAMPLE  7 Exploration Perform the indicated operation on each true statement. Is the resulting statement true or false?

a. $-6 < 7$ Multiply both sides by 3.

b. $-5 < 4$ Add -6 to both sides.

c. $-3 < 5$ Subtract 10 from both sides.

d. $-2 < 1$ Multiply both sides by -3.

e. $-4 < -2$ Divide both sides by -2.

Solution See the Answer Box.

To get an equivalent inequality when we multiply or divide by a negative number, we reverse the direction of the inequality sign.

Multiplication (by a *Negative* Number) Property of Inequalities

> If the same *negative* number is used to multiply (or divide) on both sides of an inequality, the direction of the inequality sign is changed.
>
> If $a < b$, then $ac > bc$
>
> where a and b are real numbers and $c < 0$.

It is always a good idea to predict one number that will make an inequality true before solving it. Predicting is like choosing a test number, as we did for the line graphs in earlier examples. The prediction helps us make sure that our solution is correct.

EXAMPLE 8 Solving an inequality with algebraic notation

a. Predict a number that will satisfy $-3x < 15$.

b. Solve the inequality and make sure that the predicted number satisfies the resulting inequality.

c. Graph the solution set.

Solution **a.** The product of x and -3 must be smaller than 15. Because $(-3)(-2) = 6$, we predict that $x = -2$ will satisfy the inequality and will be in the solution set.

b. $-3x < 15$ Divide both sides by -3 and reverse the inequality sign.

$\frac{-3x}{-3} > \frac{15}{-3}$ Simplify.

$x > -5$

Check: $x = -2$ satisfies $x > -5$.

c. Solution set:

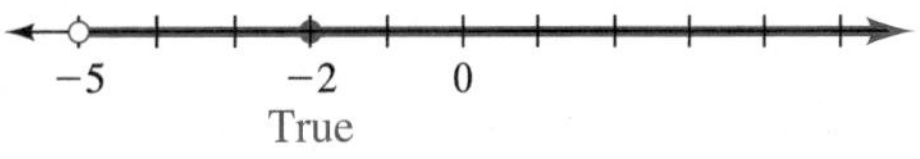

The next two inequality properties state that multiplication and division by a positive number leave the inequality sign unchanged, as do all additions and subtractions.

Multiplication (by a *Positive* Number) Property of Inequalities

> If the same *positive* number is used to multiply (or divide) on both sides of an inequality, the direction of the inequality sign is not changed.
>
> If $a < b$, then $ac < bc$
>
> where a and b are real numbers and $c > 0$.

Addition Property of Inequalities

> If the same number is added (or subtracted) on both sides of an inequality, the direction of the inequality sign is not changed.
>
> If $a < b$, then $a + c < b + c$
>
> for all real numbers a, b, and c.

In Example 9, we return to Example 3 and solve the inequality with symbols.

EXAMPLE 9 Solving an inequality Solve $6 - 2x \leq 12$. Write the solution set as an inequality and an interval, and graph the solution set.

Solution We predict $x = 0$ because $6 - 2(0) \leq 12$ is true.

$$6 - 2x \leq 12 \qquad \text{Subtract 6 from each side.}$$

$$-2x \leq 6 \qquad \text{Divide by } -2 \text{ and reverse the inequality sign.}$$

$$\frac{-2x}{-2} \geq \frac{6}{-2} \qquad \text{Simplify.}$$

$$x \geq -3$$

Solution set: $x \geq -3$, $[-3, +\infty)$

Check: Our predicted number, $x = 0$, is in the solution set $x \geq -3$. ●

In many cases, we can avoid multiplication and division by a negative number if we keep the coefficient on the variable positive. In Example 10, it is natural to keep the variable positive.

EXAMPLE 10 Keeping the coefficient positive Solve for x: $20 + 3x < 3.95x$. Compare the solution with that in Example 6.

Solution

$$20 + 3x < 3.95x \qquad \text{Subtract } 3x.$$

$$20 + 3x - 3x < 3.95x - 3x$$

$$20 < 0.95x \qquad \text{Divide by 0.95.}$$

$$\frac{20}{0.95} < \frac{0.95x}{0.95}$$

$$21.05 < x$$

Solving the inequality gives the same result we obtained by graphing. The inequality may be read either as "21.05 is less than x" ($21.05 < x$) or as "x is greater than 21.05" ($x > 21.05$). In this situation we round to the nearest whole number and read it as $x > 21$ people, which is more natural than saying "21 people is less than x." Willamalane is cheaper for more than 21 people. ●

We will solve the inequality in Example 11 two ways, to illustrate division by a negative.

EXAMPLE 11 Solving an inequality with division by a negative Solve for x in two different ways: $2x + 3 \leq -3x - 2$. Compare the solutions with that found in Example 5.

Solution First we will use division by a negative.

$$\begin{aligned} 2x + 3 &\le -3x - 2 && \text{Subtract } 2x. \\ 3 &\le -5x - 2 && \text{Add 2.} \\ 5 &\le -5x && \text{Divide by } -5 \text{ and reverse the inequality sign.} \\ \frac{5}{-5} &\ge \frac{-5x}{-5} \\ -1 &\ge x \end{aligned}$$

Now we will solve the equation again, keeping the coefficient on the variable positive.

$$\begin{aligned} 2x + 3 &\le -3x - 2 && \text{Add } 3x. \\ 5x + 3 &\le -2 && \text{Subtract 3.} \\ 5x &\le -5 && \text{Divide by 5.} \\ \frac{5x}{5} &\le \frac{-5}{5} && \text{The inequality sign is not changed.} \\ x &\le -1 \end{aligned}$$

The pointed end of the inequality faces x in both solutions. Thus, $-1 \ge x$ and $x \le -1$ represent the same set of numbers. It is useful to be able to recognize an inequality in either direction. ●

Keeping Coefficients Positive

> Keeping the variable's coefficient positive has the distinct advantage of eliminating the need to change the inequality sign. It is possible to keep the variable positive in solving most inequalities.

In Example 12, we return to the exploration in Example 1 to write an inequality and solve it.

EXAMPLE 12 Writing and solving inequalities: grades Suppose a course has three tests worth 100 points each, projects and homework worth 70 points, and a final exam worth 150 points. The instructor grades on a percent basis: 90% for an A, 80% for a B, 70% for a C. One student has test scores of 78, 84, and 72, with full credit on projects and homework (70 points). Write an inequality showing the points earned, the points possible, and the final exam score needed to earn at least a B. Solve the inequality.

Solution The grade is based on points earned relative to total points. We add the points earned using a variable to represent the last test, and place this sum over the total possible points to obtain a percent. Because any percent larger than 80% will give a B, we write an inequality using ≥ 0.80.

$$\begin{aligned} \frac{78 + 84 + 72 + 70 + x}{100 + 100 + 100 + 70 + 150} &\ge 0.80 && \text{Simplify.} \\ \frac{304 + x}{520} &\ge 0.80 && \text{Multiply by 520.} \\ (520)\frac{304 + x}{520} &\ge 520(0.80) \\ 304 + x &\ge 416 && \text{Subtract 304.} \\ 304 + x - 304 &\ge 416 - 304 \\ x &\ge 112 \end{aligned}$$

The student had a C+ on tests: $(78 + 84 + 72) \div 3 = 78$ average. The student needs $\frac{112}{150} = 75\%$ on the final for a B in the course. The homework helped! ●

Solving One-Variable Linear Inequalities with Algebraic Notation

1. As with linear equations, use addition, subtraction, multiplication, and division to isolate the variable.
2. When you multiply or divide by a negative number, reverse the direction of the inequality sign.
3. Try to keep the variable's coefficient positive to avoid the mistakes that tend to arise when you multiply or divide by a negative number and reverse the direction of the inequality sign.
4. Remember that the solution set for an inequality is an inequality.

ANSWER BOX

Warm-up: **1.** all < **2.** all > **3.** all > **4.** all < **5.** no, no

6. a.

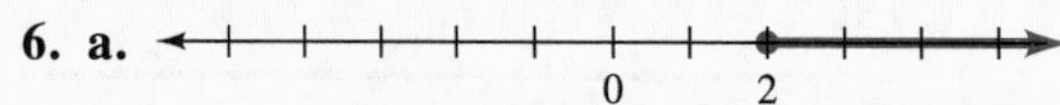

b. −2 0

c. 0 5

d. −2 0

7. a. $[2, +\infty)$ **b.** $(-\infty, 0)$ **c.** $(-\infty, 5]$ **d.** $(-\infty, 0]$ **Example 1: a.** 520 points **b.** The percent is $399/520 \approx 77\%$. The student will earn a C. **c.** A score of 112 is needed on the final exam for a B. **Think about it 1:** The graph of $y = 2x + 3$ is below that of $y = -3x - 2$ for $x < -1$. This is the same solution we arrived at with a test point. **Think about it 2:** The graph of $3x + 20$ is below that of $3.95x$ for $x > 21$. This is the same solution we arrived at with the test point. **Example 7: a.** $-18 < 21$ is true. **b.** $-11 < -2$ is true. **c.** $-13 < -5$ is true. **d.** $6 < -3$ is false. **e.** $2 < 1$ is false. We must reverse the direction of the inequality signs in parts d and e to obtain a true statement.

EXERCISES 3.4

In Exercises 1 to 8, graph each solution set on a number line and write each as an interval.

1. $x > 5$

2. $x < 3$

3. $x \le -2$

4. $x \ge -3$

5. $x \ge 0$

6. $x < 0$

7. $-1 > x$

8. $-4 < x$

Use the graph to solve the inequalities in Exercises 9 to 12. Graph the solution set on a number line.

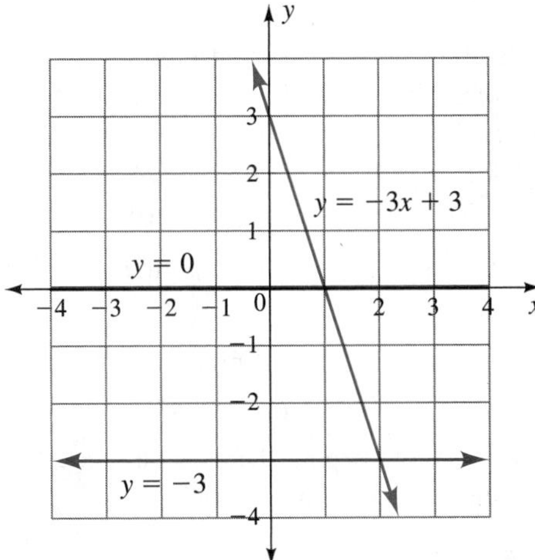

9. $-3 > -3x + 3$ **10.** $-3 < -3x + 3$

11. $0 < -3x + 3$ **12.** $0 > -3x + 3$

Use the graph to solve the inequalities in Exercises 13 to 16. Graph the solution set on a number line.

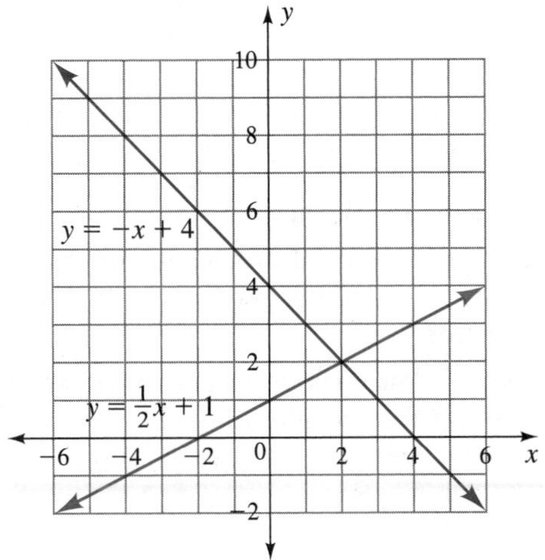

13. $-x + 4 < \frac{1}{2}x + 1$ **14.** $-x + 4 > \frac{1}{2}x + 1$

15. $-x + 4 > 0$ **16.** $\frac{1}{2}x + 1 > 0$

In Exercises 17 to 24, solve each inequality with a coordinate graph. Graph the solution set on a number line.

17. $1 < x + 3$ **18.** $2 > x - 1$

19. $3x - 5 \geq 7$ **20.** $2 - 3x \leq 5$

21. $-x < -3x + 2$ **22.** $x > 2x - 1$

23. $2 - 2x < 3 - 3x$ **24.** $2 - 2x > 3 - 3x$

In Exercises 25 to 56, solve each inequality using algebraic notation. Write the solution set as an inequality and an interval.

25. $-3 > -3x + 3$ **26.** $-3 < -3x + 3$

27. $0 < -3x + 3$ **28.** $0 > -3x + 3$

29. $-x + 4 < \frac{1}{2}x + 1$ **30.** $-x + 4 > \frac{1}{2}x + 1$

31. $-x + 4 > 0$ **32.** $\frac{1}{2}x + 1 > 0$

33. $-1 < -3x + 2$ **34.** $3 < 2x - 1$

35. $-2 \leq 2 - 2x$ **36.** $0 \leq 3 - 3x$

37. $3 - 2x \leq -5$ **38.** $7 + 2x > -3$

39. $-4 > x + 5$ **40.** $-2 < x - 4$

41. $1 - 4x < -4$ **42.** $2x - 7 > -2$

43. $-2 > 2x + 1$ **44.** $3 \leq 4 - 2x$

45. $2x < x + 5$ **46.** $x > 3x - 4$

47. $4 - 2x \geq x - 2$ **48.** $4 - 3x \leq 8 + x$

49. $3x - 4 < -2x + 1$ **50.** $3x - 3 > -x + 1$

51. $2(x + 3) < 5x$ **52.** $3(x - 1) \leq 4x$

53. $\frac{1}{2}x > 4 + x$ **54.** $\frac{1}{2}x < x + 1$

55. $-x > -\frac{1}{2}x + 1$ **56.** $-\frac{1}{2}x > 1 - x$

Exercises 57 to 60 relate to Mrs. Kay's math class, where there are 150 points on the final exam and 520 total points for the course. Just prior to the final exam, several students are thinking about their grades. For each exercise, write an inequality and solve it.

57. A student has earned 78, 84, and 72 points on tests and only 5 points on projects and homework. Is it possible for the student to earn a B or better (80%)?

58. Is it possible for the student in Exercise 57 to earn a C (70%)?

59. A student has earned the full 70 points on homework and projects and expects 135/150 on the final. She wants to know at least how many total points she must have had on the three tests to get an A (90%) in the course.

60. A student missed one test, has no homework, and has test scores of 74 and 84 on the other two tests. Is it possible for the student to get a D (60%)?

61. Test scores are 88 out of 100, 84 out of 100, and 89 out of 100. Homework is 70 out of 70. What final exam score (200 points possible) is needed to get 90% or better?

62. Test scores are 92 out of 100, 88 out of 100, and 91 out of 100. Homework is 25 out of 70. What final exam score (200 points possible) is needed to get 90% or better?

63. Fairfield Lanes charges \$5.50 per child for a bowling party. Seth has to buy a cake for \$17 and has a \$160 budget. Write and solve an inequality that shows the number of children who can attend on Seth's budget.

64. Papa's Pizza offers a party at \$3 per child with a \$40 fee for the cake and playground supervisor. Write and solve an inequality that shows the number of children who can attend on a \$76 budget.

65. For a wedding reception, the Country Inn charges a \$350 service fee plus \$8.50 per person for food. Write and solve an inequality that shows the number of people who can attend on a \$1030 budget.

66. The Valley Inn charges \$17.50 per person and a \$100 service fee. Write an inequality that shows the number of persons who can attend a wedding reception on a \$1500 budget.

67. Using the information in Exercises 65 and 66, write an inequality that shows for how many people will it cost less to use the Country Inn.

68. Using the information in Exercises 63 and 64, write an inequality that shows for how many children it will cost less to go to Papa's Pizza.

69. Rewrite the following statement so that it correctly describes division of an inequality by a negative number: If $a < b$, then $ac < bc$, where a and b are real numbers and $c > 0$.

70. Describe the effect on $-3 < 4$ of multiplying by -2.

71. Describe the effect on $-8 < -6$ of dividing by -2.

72. What is the advantage of keeping the variable's coefficient positive when solving an inequality?

Project

73. ***Party Costs.*** Call four locations for children's parties or wedding receptions and obtain cost information. Write four inequality problems from your data and solve them. Include at least two coordinate graphs in your solutions.

CHAPTER 3 SUMMARY

Vocabulary

For definitions and page references, see the Glossary/Index.

addition property of equations
conditional equation
consecutive integers
delta (Δ)
dependent variable
empty set
equal sign
equation in one variable
equivalent equations
horizontal axis intercept point
identity
independent variable
linear inequality in one variable
linear equation in one variable
linear equation in two variables
multiplication property of equations
multiplication (by a negative number) property of inequalities
multiplication (by a positive number) property of inequalities
nonlinear equation
solution
solution set
solution set of a one-variable inequality
solving an equation
vertical axis intercept point
x-intercept
x-intercept point
y-intercept
y-intercept point

Concepts

3.1 Solving Equations with Algebraic Notation

Solutions to equations may be found by guessing or observation, through step-by-step algebraic procedures, from an input-output table, or from a graph.

Changing sentences into equations or equations into sentences helps us recognize the operations in a problem. We use the inverse operations to solve an equation using algebraic notation. Solutions may make use of any or all of these steps:

- Reverse order of operations.
- Addition property of equations: Adding the same number to both sides of an equation produces an equivalent equation.
- Multiplication property of equations: Multiplying both sides of an equation by the same nonzero number produces an equivalent equation.

The addition and multiplication properties permit us to subtract or divide both sides of an equation by the same nonzero number. We may multiply both sides of an equation by the reciprocal of a fraction instead of dividing by that fraction.

3.2 Solving Equations with Tables and Graphs

The table solution to an equation is the input entry in the table that makes the left and right sides of the equation equal.

The graphical solution to an equation is the first number in the ordered pair at the point of intersection of the graphs of the left and right sides of the equation. The second number in the ordered pair is the value obtained on each side when the first number is substituted.

All ordered pairs (x, y) that make an equation true belong on the input-output table or graph for that equation. Likewise, all ordered pairs on the table or graph make the equation true.

3.3 Solving Linear Equations

Solving a linear equation using a table and a graph:

1. Let $y =$ the left side of the equation. Build a table and plot a graph.
2. Let $y =$ the right side of the equation. Build a table and plot a graph on the same axes used for the left side.
3. Look for the ordered pair that appears in both tables; it is also the ordered pair at the point of intersection. The x in the ordered pair is the solution to the equation. The y in the ordered pair is the value obtained when x is substituted into both sides of the original equation.

Solving a linear equation with algebraic notation:

1. Multiply on both sides by the least common denominator to remove fractions or decimals.
2. Apply the distributive property to expressions containing parentheses.
3. Collect like terms on each side.
4. Using the addition property of equations, get all the terms with the variable to be solved for onto one side of the equation. Get the constant terms to the other side of the equation.
5. Collect like terms again, if necessary.
6. Solve for the variable, using the multiplication property of equations.

3.4 Solving Linear Inequalities

The solution set for an inequality is an inequality. Solving a one-variable inequality with a graph:

1. Graph $y =$ the left side of the inequality and $y =$ the right side.
2. Find the point of intersection (x, y). Mark the x on a number line. Use a dot if the inequality is $\geq$ or $\leq$; use a small circle if the inequality is $>$ or $<$.
3. Choose a test number on either side of the x on the number line. Substitute the test number into the inequality. If the number makes the statement true, draw an arrow on the number line from x through the test number. If the number makes the statement false, draw the arrow from x in the opposite direction from the test number.
4. Write an inequality for the solution set shown on the number line.

Solving one-variable linear inequalities with algebraic notation:

1. As with linear equations, use addition, subtraction, multiplication, and division to isolate the variable.
2. When you multiply or divide by a negative number, reverse the direction of the inequality sign.
3. Try to keep the variable's coefficient positive to avoid the mistakes that tend to arise when you multiply or divide by a negative number and reverse the direction of the inequality sign.

Solutions using algebraic notation to inequalities may require any or all of these steps:

- Reverse order of operations.
- Multiplication (by a negative number) property of inequalities: Multiplying both sides of an inequality by the same negative number makes an equivalent inequality if the direction of the inequality is reversed. (Note: If you keep the coefficient of the variable term positive, the direction of the inequality will not need to be reversed.)
- Multiplication (by a positive number) property of inequalities: Multiplying both sides of an inequality by the same nonzero number produces an equivalent inequality.
- Addition property of inequalities: Adding the same number to both sides of an inequality produces an equivalent inequality.
- Distributive property and adding like terms.

CHAPTER 3 REVIEW EXERCISES

In Exercises 1 to 5, fill in the blank with one or more of the following: identity, conditional equation, equivalent equations, two-variable equation, nonlinear equation.

1. $3x + 4 = 7$ and $3x = 3$ are ______________.
2. $x + 3 = 3 + x$ is a(n) ______________.
3. $y = 3x + 4$ is a(n) ______________.
4. $y = x^2 + x$ is a(n) ______________.
5. $3x - 5 = -2$ is a(n) ______________.

In Exercises 6 to 10, is the given input x a solution to the equation?

6. $-4x + 5 = -3, x = -2$
7. $-5x + 4 = 9, x = -1$

8. $2.5x - 3 = -1, x = 0.8$

9. $8 - 7.5x = 5, x = 0.4$

10. $2x + 1 = 4x - 3, x = 2$

In Exercises 11 to 28, solve for the variable.

11. $x + 3 = -4$

12. $x - 4 = 8$

13. $3x = 27$

14. $4x = -12$

15. $\frac{1}{2}x = 12$

16. $\frac{1}{4}x = 20$

17. $4x - 2 = 22$

18. $3x - 5 = 34$

19. $-2x + 3 = 9$

20. $-4x + 5 = 21$

21. $-2(x - 4) = 18$

22. $-6(x - 5) = 42$

23. $5 - 2(x + 1) = -11$

24. $7 - 3(x + 2) = -8$

25. $3x - 1 = x + 1$

26. $5 - 3(x - 4) = x + 9$

27. $-2(x - 3) = \frac{1}{2}x + 3$

28. $3 - (x - 4) = x + 3$

In Exercises 29 to 36, write equations. Let x be the unknown number.

29. Six less than three times a number is −15.

30. Four is five times a number less eleven.

31. The product of a number and negative seven is 21.

32. The quotient of a number and six is 12.

33. Three times a number is four more than twice the number.

34. Five less than twice a number is the number.

35. Six times the sum of two and a number is negative six.

36. Twice the difference between seven and a number is −4.

In Exercises 37 to 40, write each equation in words.

37. $\frac{x}{5} = 15$

38. $4x - 3 = 29$

39. $3(x - 4) = -18$

40. $5 - 2(x - 1) = 7 + x$

Exercises 41 to 48 use the algebraic description of consecutive integers.

41. If $x = 12$, what are $x - 1$ and $x + 1$?

42. If $x = 8$, what are $x - 1$ and $x + 1$?

43. If $x = 15$, what are $x + 3$ and $x + 6$?

44. If $x = -11$, what are $x + 3$ and $x + 6$?

45. Let $x = 10$; write 9 and 11 in terms of x.

46. Let $x = -5$; write −4 and −3 in terms of x.

47. Let $x = -3$; write −5 and −1 in terms of x.

48. Let $x = 12$; write 11 and 10 in terms of x.

In Exercises 49 and 50, write an equation and solve it.

49. An integer and twice the next consecutive integer add to 17.

50. An integer and half the next consecutive integer add to 47.

51. Use the table to solve the following equations.

x	$y = 3x - 2$
2	4
3	7
4	10
5	13

a. $3x - 2 = 4$ **b.** $3x - 2 = 13$

c. $3x - 2 = 8$ **d.** $3x - 2 = 16$

52. Solve these equations using the graph.

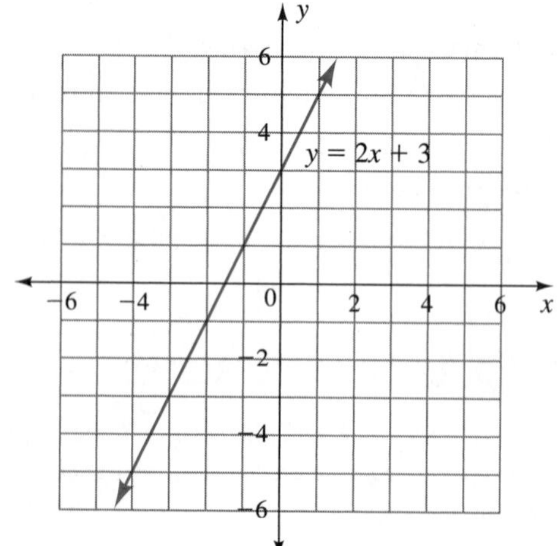

a. $2x + 3 = 5$ **b.** $2x + 3 = 1$

c. $2x + 3 = -2$

53. Solve these equations using the graph.

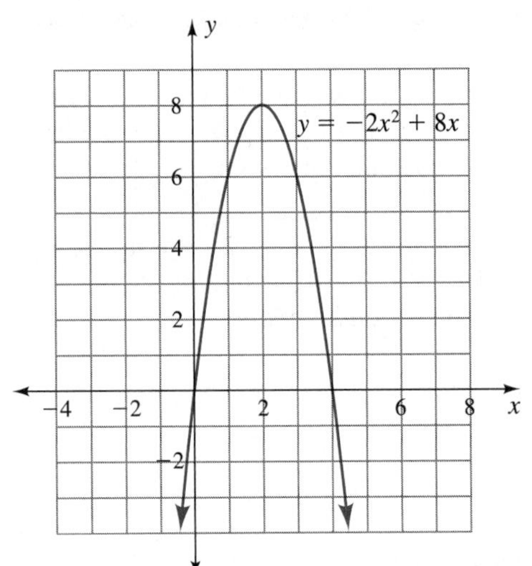

a. $8 = -2x^2 + 8x$

b. $6 = -2x^2 + 8x$

c. $0 = -2x^2 + 8x$

54. Use the graph and the equation $x + 1 = 3 - x$ to do the following.

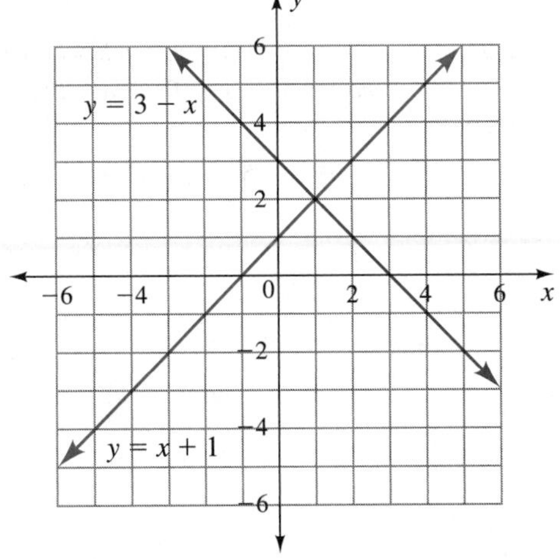

a. Find the point of intersection of the two lines.

b. Substitute the intersection point into each equation shown on the graph.

c. Solve the equation $x + 1 = 3 - x$.

d. Describe how the equation $x + 1 = 3 - x$ relates to the graph.

55. Use the graph to solve the following equations.

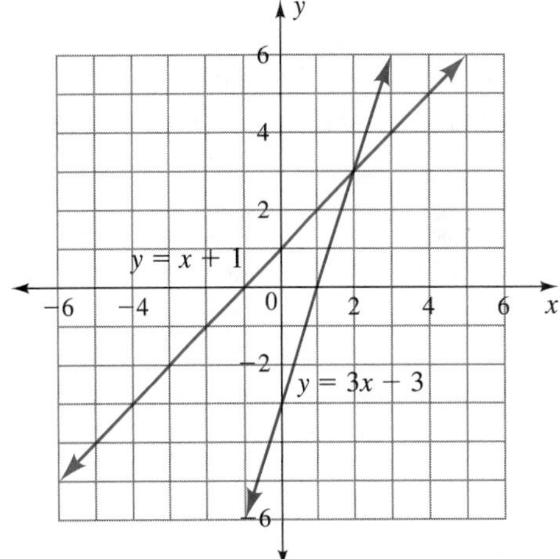

a. $3x - 3 = x + 1$

b. $x + 1 = -2$

c. $3x - 3 = -6$

56. Use the graph to solve the following equations.

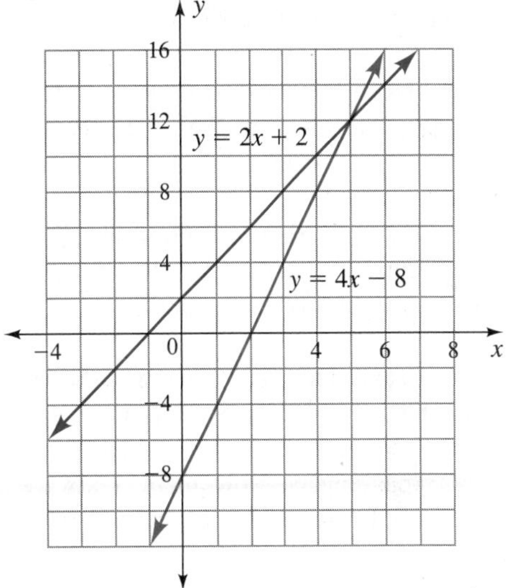

a. $2x + 2 = 4x - 8$

b. $2x + 2 = -4$

c. $4x - 8 = 8$

In Exercises 57 to 64, define variables and write equations so that y depends on x. Find and explain the meaning of the horizontal and vertical intercepts for each problem.

57. Find the distance traveled in 4 hours at various speeds.

58. Find the distance traveled at 65 miles per hour for a given number of hours.

59. Find the $7\frac{1}{2}\%$ tax on a meal.

60. Find the 0.2% property tax on the value of a house.

61. Find the total cost for a given number of credits at a college that charges \$300 per credit tuition and \$150 in fees.

62. Find the total cost for a given number of cubic feet of natural gas at \$0.04 per cubic foot plus a \$5.00 service charge.

63. Find the value remaining after a given number of visits on a \$520 health club deposit when \$10 is deducted for each visit.

64. Find the value remaining after a given number of years on a \$20,000 car that drops in value \$2000 each year.

To answer the questions in Exercises 65 to 68, write an equation and solve it.

65. To make room for a larger than expected crowd at an outdoor wedding, the chair rental company places 3 additional seats at the end of each row of x seats. There are 20 rows of seats and 440 seats altogether. How many seats were in each original row?

66. The perimeter of a right triangle is 12 inches. The sides are three consecutive integers. What are the sides of the triangle?

67. The perimeter of a right triangle is 60 inches. The sides are three consecutive multiples of 5 (that is, x, $x + 5$, etc.). What are the sides of the triangle?

68. By making the seats and aisle narrower, Squeez-um Airlines was able to add 2 more seats in each of the 47 rows on its economy flights. The economy flight now holds 423 passengers. How many seats were in each of the original rows? (Assume that all rows have the same number of seats.)

69. Why would the notation x, $x + 1$, $x + 2$ be preferable to a, b, c for describing three consecutive numbers?

70. After each step below, write what was done to obtain the next step:

$5 < x$ __________

$5 - x < 0$ __________

$-x < 0 - 5$ __________

$x > 5$

What do these steps tell you about all statements of the form $a < b$ and $b > a$?

The two students in Exercises 71 and 72 both want to earn at least a B (80%). If the final exam is worth 150 points, is it possible for each student to earn at least a B? Write an inequality for each student and solve the inequality.

71. Student 1 has earned 82 and 72 on tests (100 points each); 20, 0, 20, 20, and 18 on quizzes (20 points each); and 12 of 70 points on homework.

72. Student 2 has earned 82 and 72 on tests (100 points each); 15, 15, 15, 16, and 16 on quizzes (20 points each); and 70 of 70 points on homework.

73. Which is a solution set for $2 < x - 3$? Draw the solution set on a number line.

a. $x > 5$ **b.** $x < 5$ **c.** $x > 1$ **d.** $x < 1$

74. Which is a solution set for $5 - x \geq 2$? Draw the solution set on a number line.

a. $x \geq 3$ **b.** $-3 \leq x$ **c.** $3 \geq x$ **d.** $x \leq 7$

In Exercises 75 to 82, solve for x. Write the solution set as an inequality and an interval.

75. $1 < x - 4$

76. $5 > 2 - x$

77. $-\frac{1}{2}x > 8$

78. $4 \geq -\frac{1}{3}x$

79. $5 - 3x > 13$

80. $6 - 5x \leq 31$

81. $13 \leq 7 - \dfrac{x}{3}$

82. $5 < -10 - \dfrac{x}{3}$

In Exercises 83 and 84, sketch a copy of the graph and label each line with its equation. Solve the indicated inequality first with the graph and then using symbols.

83. $y = 15 - 2x$ and $y = x - 6$; $15 - 2x < x - 6$

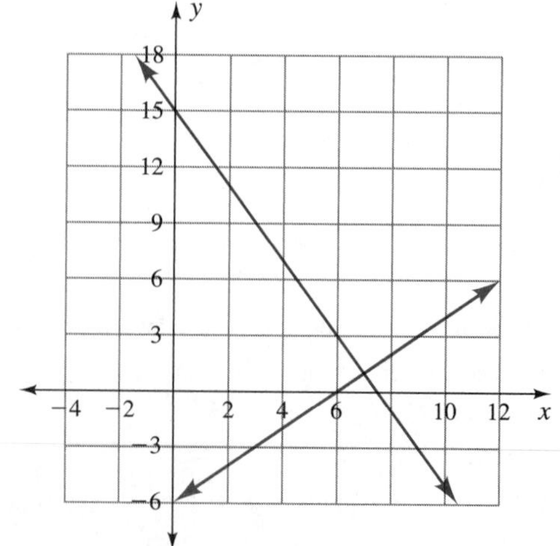

84. $y = x - 1$ and $y = 5 - x$; $5 - x < x - 1$

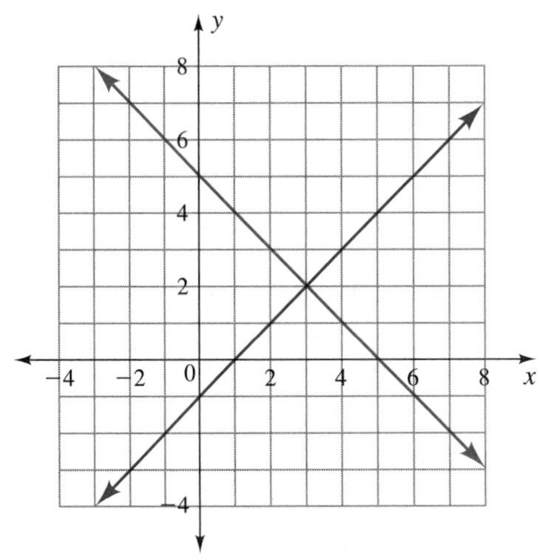

In Exercises 85 and 86, solve with a graph and with symbols.

85. $3x - 5 \le 3 - x$

86. $x - 4 > 2 - x$

In Exercises 87 to 96, solve for x. Write the solution set as an inequality and an interval.

87. $-2x < 1 - x$

88. $-x > x + 2$

89. $x > 2x + 3$

90. $2x - 2 < 1 - x$

91. $x + 1 \le 3 - x$

92. $x + 5 \ge 1 - 3x$

93. $-2x + 2 > -2 - x$

94. $2x + 4 < 2 - x$

95. $3x + 2 \ge 3 - 2x$

96. $-x - 1 \le 2x - 3$

In Exercises 97 to 100, define variables, write an equation, and solve the equation.

97. A wedding cake costs $550, and the catering price per person is $25. Up to how many persons may attend on a $3500 budget?

98. An anniversary reception is planned for a location that charges a $100 service fee and $8 per person. How many people can attend on a budget of $2100?

99. One sidewalk repair bid is $250 plus $12 per linear foot of sidewalk. A second bid is $320 plus $8 per linear foot. For what lengths of sidewalk will the first bid be cheaper?

100. One plumber charges a $30 travel fee and $45 per hour. A second plumber charges a $60 travel fee and $30 per hour. For what numbers of hours will the first plumber be cheaper?

CHAPTER 3 TEST

In Exercises 1 to 10, solve for x using algebraic notation.

1. $x + 8 = -3$

2. $4 - x = 5$

3. $\frac{2}{5}x = 30$

4. $-6x = 3$

5. $5 - 2x = 3$

6. $\frac{1}{2}x + 5 = -3$

7. $2(x - 2) = -x - 1$

8. $4(x - 3) = 6$

9. $4 - 2(x - 4) = 2x$

10. $-2(x - 3) = -0.5(x - 6)$

11. Circle the solution to the equation $5 - 2x = 3$ on the table and on the graph.

Input: x	Output: $y = 5 - 2x$
−1	7
0	5
1	3
2	1
3	−1

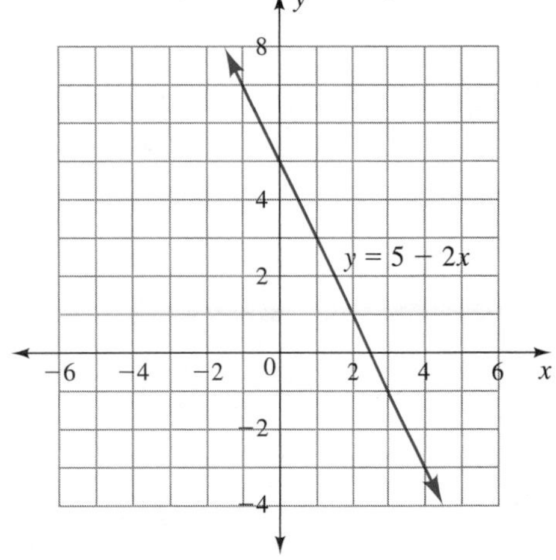

12. Explain how to estimate the solution to $5 - 2x = 0$ from the table in Exercise 11.

13. Explain how to estimate the solution to $5 - 2x = 0$ using the graph in Exercise 11.

14. The graph in the figure may be used to answer the following.

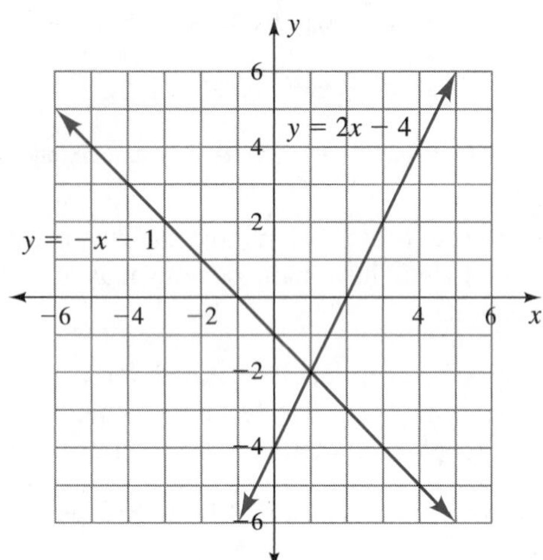

a. Find the point of intersection of the two lines.

b. Substitute the intersection point into each equation shown on the graph.

c. Solve the equation $2x - 4 = -x - 1$.

d. What does the graph indicate about the equation $2x - 4 = -x - 1$?

In Exercises 15 to 19, define variables and write the sentence in symbols. Solve the one-variable equations.

15. Six more than half a number is 15.

16. Three times a number less 7 is -31.

17. The output is a third of the input.

18. The output is two less than twice the input.

19. Four consecutive integers add to -74.

In Exercises 20 to 22, solve the inequality. Show the solution set on a line graph, with an inequality, and with an interval.

20. $2x + 5 < -3$

21. $3 - 2x > 11$

22. $2x + 8 \geq \frac{1}{2}(x + 1)$

23. Tell how the intersection of the graphs of $y = 15 - 2x$ and $y = x - 6$ is used to find the solution to the inequality $15 - 2x > x - 6$. Solve the inequality with symbols and then make a line graph of the solution set.

24. NW Natural Gas Company charges home users a \$4 monthly fee plus \$0.615 per therm (1 therm = 100 cubic feet).

a. Make a table for the total monthly cost of 0 to 100 therms. Count by 20.

b. Graph the data from part a.

c. Write an equation to find the total cost for a month's use of gas with the input in therms.

d. Use your equation to find the amount of gas (in therms) used by a well-insulated energy-efficient home with a bill of \$47.05.

e. What is the horizontal axis intercept? What does it mean?

f. What is the vertical axis intercept? What does it mean?

4

Formulas, Functions, Linear Equations, and Inequalities in Two Variables

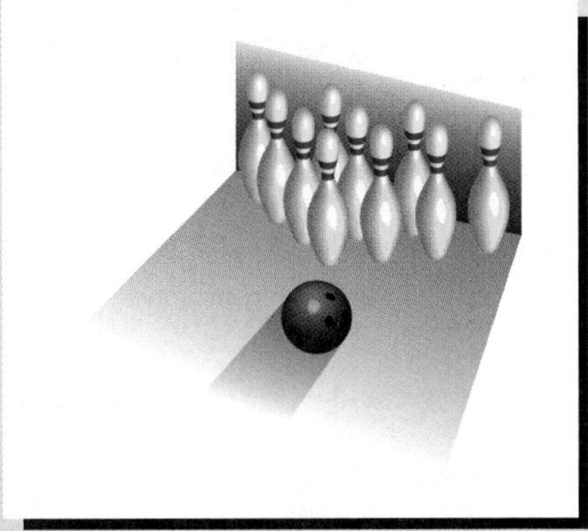

Figure 1

In amateur bowling, the members of each team are given a bonus score called a handicap. The handicap permits teams of differing ability to be compared. In league play, the average and handicap are recalculated each weak.

The bowler's average, A, is used in the formula $H = 0.8(200 - A)$ to find the handicap, H. No handicap is given for averages of 200 or over. (The maximum score is 300.) The handicap is added to the bowler's points after each game. Suppose Bowler A, with an average of 150, bowls a 165 game, and Bowler B, with an average of 200, bowls a 206 game. Which bowler has the higher score under the handicap system?

This opening example makes use of a formula. Section 4.1 on formulas focuses on extending equation-solving skills. We then return to graphing equations and finding intercepts as we consider functions in Section 4.2. Section 4.3 develops the concept of slope. In Section 4.4, we apply slope and intercepts to writing the equation of a line from data. The chapter closes with the graphing of inequalities in two variables.

4.1 Formulas

OBJECTIVES

- Describe relationships using the phrase *in terms of.*
- Solve a formula for one variable.
- Evaluate formulas that include variables with subscripts.

WARM-UP

Use the order of operations to simplify each of the following.

1. a. $\dfrac{6 + 10}{2}$ **b.** $\dfrac{6}{2} + \dfrac{10}{2}$

2. a. $\dfrac{6 - 9}{3}$ **b.** $\dfrac{6}{3} - \dfrac{9}{3}$

3. a. $\dfrac{12 + 7}{6}$ **b.** $\dfrac{12}{6} + \dfrac{7}{6}$

4. What property permits us to write the expressions in Exercises 1 to 3 in the two different ways shown in parts a and b?

5. How could $\dfrac{P - 2l}{2}$ be written in a different way?

IN THIS SECTION, we practice solving and evaluating formulas and dealing with subscripted variables.

EXAMPLE 

Exploration The international airport is 55 miles from your home, 30 miles of which is freeway. You need to be at the airport at 8:30 a.m. for a 9:30 a.m. flight. What time do you need to leave home? Discuss what assumptions you make about type of streets traveled and rate of travel.

Solution To solve this problem, you need to know $d = rt$, the distance, rate, and time formula. You need to guess how fast you can drive the 55 miles. You might assume that you can drive the speed limit—say, 25 miles per hour on city streets and 60 miles per hour on the freeway. There are 25 miles of city streets and 30 miles of freeway.

Writing the miles per hour as a fraction, $\dfrac{\text{miles}}{\text{hours}}$, we find the time for the freeway travel using $d = rt$:

$$30 \text{ miles} = \frac{60 \text{ miles}}{1 \text{ hour}} \cdot t \qquad \text{Multiply by 1 hour.}$$

$$30 \text{ miles} \cdot 1 \text{ hour} = 60 \text{ miles} \cdot t \qquad \text{Divide by 60 miles.}$$

$$\frac{30 \text{ miles} \cdot 1 \text{ hour}}{60 \text{ miles}} = t \qquad \text{Simplify the left side.}$$

$$\frac{1}{2} \text{ hour} = t$$

We then find the time for the city street travel using $d = rt$:

$$25 \text{ miles} = \frac{25 \text{ miles}}{1 \text{ hour}} \cdot t \qquad \text{Multiply by 1 hour.}$$

$$25 \text{ miles} \cdot 1 \text{ hour} = 25 \text{ miles} \cdot t \qquad \text{Divide by 25 miles.}$$

$$\frac{25 \text{ miles} \cdot 1 \text{ hour}}{25 \text{ miles}} = t \qquad \text{Simplify the left side.}$$

$$1 \text{ hour} = t$$

We travel $\frac{1}{2}$ hour on the freeway and 1 hour on the city streets, for a total of $1\frac{1}{2}$ hours. We need to leave $1\frac{1}{2}$ hours before 8:30 a.m., or at 6:00 a.m. ●

Think about it 1: How long will the trip take if the freeway is congested and traffic slows to an average of 30 miles per hour?

Using the Phrase "in terms of"

When we solve a formula, we apply addition and multiplication properties to get one variable by itself on one side of the equation. We say that the variable is **in terms of** other variable when *the variable is by itself on one side of the equal sign and terms containing other variables, numbers, and operations are on the other side.*

EXAMPLE 2 Using *in terms of* Describe each equation using the phrase *in terms of.* If you recognize the formula, use words instead of variables.

a. $d = rt$ **b.** $t = \frac{d}{r}$ **c.** $A = \frac{1}{2}bh$ **d.** $H = 0.8(200 - A)$

e. $h = -\frac{1}{2}gt^2 + v_0t + h_0$

Solution

a. Distance is in terms of rate and time.

b. Time is in terms of distance and rate.

c. Area (of a triangle) is in terms of base and height.

d. A bowling handicap is in terms of a bowling average.

e. The variable h is in terms of g, t, v_0, and h_0. This is the formula for the height (h) of an object thrown straight up into the air in terms of gravity (g), time (t), initial velocity (v_0), and initial height (h_0). ●

Figure 2

Subscripts

The small zeros to the right of the variables in part e of Example 2 are subscripts. **Subscripts** are *a common way to label a particular item from a group of similar items.* In Example 2, the subscripts represent the velocity and height when time, t, is zero. In the nutrition field, vitamins are identified by subscripts: vitamin B_1, vitamin B_2, vitamin B_6 (see Figure 2). Vitamin B_1 is not the same as vitamin B_2. In music, middle C is C_4, and a chord might be described as G_4-C_5-E_5. In Section 4.3, we will use subscripts in naming coordinate points.

Solving Formulas for Repeated Use

In Example 1, we solved the equation $D = rt$ for the time, t, at each different rate traveled. If you plan to make repeated use of formulas, it is easier to first solve the formula for the variable you want.

EXAMPLE 3 Solving formulas Solve $d = rt$ for t in terms of d and r.

Solution

$$d = r\overset{\downarrow}{t}$$ Mark the variable t with an arrow. Divide both sides by r.

$$\frac{d}{r} = \frac{rt}{r}$$ The value of $\frac{r}{r}$ is 1.

$$\frac{d}{r} = t$$

Formula Solving versus Equation Solving

One difference between the equation solving we did in other sections and the formula solving here is that here we have variables remaining on both sides of the answer. Equations in one variable solve to a numerical answer, whereas formulas remain in expression form. Expect the solutions to formulas to look different from the solutions to equations.

A formula is *solved for a variable* when the variable appears by itself on one side and the other side no longer contains the variable. In Example 4, we find out whether formulas are solved for a variable.

EXAMPLE 4 Solving formulas Which of these formulas are solved for the variable on the left?

a. $A = \frac{1}{2}h(a + b)$ **b.** $n = \frac{n + 1}{2}$ **c.** $r = \frac{A - 2\pi r^2}{2\pi h}$

Solution

a. The formula is correctly solved for A. The variable a on the right is the length of one of the parallel sides of a trapezoid. The variable A on the left is the area of the trapezoid. *In copying a formula, do not change the form of any letter, A to a or a to A.*

b. The formula is not solved for n. The variable n appears on both sides.

c. The formula is not solved for r. The variable r appears on both sides.

Steps in Solving Formulas

Because the solutions to formulas contain variables and operations, it is important to apply the equation-solving steps carefully.

Solving Formulas

- Keep track of the selected variable. Mark the variable for which you are solving with an arrow, ↓.
- Make a plan. Observe the order of operations on the selected variable. Write a plan that uses the inverse operations in the reverse order.

EXAMPLE 5 Solving and evaluating formulas

a. Solve $A = \frac{1}{2}bh$ for b in terms of A and h.

b. If the height of the triangle is 6 feet and the area is 30 square feet, what is the base?

Solution

a. ***Plan:*** Note that multiplication by $\frac{1}{2}$ is the same as division by 2. Thus, b is multiplied by h and the result is divided by 2. We will use the inverse operations in the reverse order: multiply by 2 and divide by h.

$$A = \frac{1}{2}\overset{\downarrow}{b}h$$ Mark the letter with an arrow. Multiply both sides by 2 to eliminate the $\frac{1}{2}$ on the right.

$$2 \cdot A = 2 \cdot \frac{1}{2}bh$$ Divide both sides by h to eliminate the h on the right.

$$\frac{2 \cdot A}{h} = \frac{bh}{h}$$ Simplify.

$$\frac{2A}{h} = b$$

b. $b = \dfrac{2A}{h} = \dfrac{2(30 \text{ square feet})}{6 \text{ feet}} = 10 \text{ feet}$ ●

EXAMPLE 6 Solving and evaluating formulas

a. Solve $P = 2l + 2w$ for w in terms of P and l.

b. Find the width of a rectangle whose perimeter is 90 inches and length is 29 inches.

Solution **a.** ***Plan:*** w is multiplied by 2 and then is added to $2l$. We must subtract $2l$ and then divide by 2.

$$P = 2l + 2\overset{\downarrow}{w}$$ Subtract $2l$ from both sides.

$$P - 2l = 2w$$ Divide both sides by 2.

$$\frac{P - 2l}{2} = w$$

b. $w = \dfrac{P - 2l}{2}$

$$w = \frac{90 \text{ inches} - 2(29 \text{ inches})}{2}$$

$$w = 16 \text{ inches}$$ ●

Think about it 2: What do we obtain when we divide by 2 in $\dfrac{P - 2l}{2}$?

n Example 7, we return to the setting in the chapter opening.

EXAMPLE 7 Evaluating a formula: bowling handicap A ten-pin bowler's handicap, H, is 80% of the difference between 200 and the bowler's average score, A. No handicap is given for averages of 200 or over. (The maximum score is 300.) The handicap is added to the bowler's points after each game. Suppose Aleta, with an average of 150, bowls a 165 game, and Betty, with an average of 200, bowls a 206 game. Use the formula $H = 0.8(200 - A)$ to find out which bowler has the higher score under the handicap system.

Solution Aleta's handicap is found by substituting her average, 150, into the formula:

$$H = 0.8(200 - A)$$

$$H = 0.8(200 - 150)$$ Find the difference in parentheses.

$$H = 0.8(50)$$ Simplify.

$$H = 40$$

Aleta's total score is $165 + 40 = 205$.

Betty has an average of 200 and so receives no handicap. Her score, 206, is still higher than Aleta's total of 205. ●

In Example 7, we used the bowling formula in its original form. In Example 8, we solve the formula for A.

EXAMPLE 8 Solving a formula: bowling averages His computer's hard drive crashed, and Scott lost the team's bowling averages. He has last week's handicaps for the following bowlers and wants to find their averages:

Fran, 60 Miguel, 20 Trina, 16 Scott, 0

a. Solve the bowling handicap formula, $H = 0.8(200 - A)$, for the average, A.

b. Use the formula to find the average for each member of Scott's team.

Solution **a.**

$H = 0.8(200 - \overset{\downarrow}{A})$	Multiply using the distributive property.
$H = 160 - 0.8A$	Add $0.8A$ to both sides.
$H + 0.8A = 160$	Subtract H from both sides.
$0.8A = 160 - H$	Divide by 0.8 on both sides.
$A = \dfrac{160 - H}{0.8}$	

b. For Fran, with $H = 60$,

$$A = \frac{160 - H}{0.8}$$

$$A = \frac{160 - 60}{0.8} = 125$$

For Miguel, with $H = 20$,

$$A = \frac{160 - H}{0.8}$$

$$A = \frac{160 - 20}{0.8} = 175$$

For Trina, with $H = 16$,

$$A = \frac{160 - H}{0.8}$$

$$A = \frac{160 - 16}{0.8} = 180$$

Because Scott's handicap is 0, his average score could be any number 200 or larger. ●

Because formulas can often be solved in several ways, answers may not always look the same. Students in higher mathematics classes often use their algebra skills to make their answers look like those in the answer section of the textbook. Examples 9 and 10 show two different ways to solve the area of a trapezoid formula for one of its variables. The answers will not look alike.

EXAMPLE 9 Solving a formula With a reverse order of operations, solve $A = \frac{1}{2}h(a + b)$ for a.

Solution ***Plan:*** Observe that b is added to a, then the sum is multiplied by h and divided by 2. The reverse order, with opposite operations, is to multiply by 2, divide by h, and subtract b.

$A = \frac{1}{2}h(\overset{\downarrow}{a} + b)$	Mark a with an arrow. Multiply by 2.
$2 \cdot A = 2 \cdot \frac{1}{2}h(a + b)$	Simplify.

$$2A = h(a + b) \qquad \text{Divide by } h.$$

$$\frac{2A}{h} = \frac{h(a + b)}{h} \qquad \text{Simplify.}$$

$$\frac{2A}{h} = a + b \qquad \text{Subtract } b.$$

$$\frac{2A}{h} - b = a + b - b \qquad \text{Simplify.}$$

$$\frac{2A}{h} - b = a$$

●

EXAMPLE 10

Solving a formula Clearing the formula of fractions and using the distributive property to remove the parentheses, solve for a in $A = \frac{1}{2}h(a + b)$.

Solution

$$A = \tfrac{1}{2}h(a + b) \qquad \text{Multiply by 2 to clear the fraction.}$$

$$2 \cdot A = 2 \cdot \tfrac{1}{2}h(a + b)$$

$$2A = h(a + b) \qquad \text{Multiply } h \text{ times } a \text{ and } b.$$

$$2A = ha + hb \qquad \text{Subtract } hb \text{ from both sides.}$$

$$2A - hb = ha \qquad \text{Divide by } h \text{ on both sides.}$$

$$\frac{2A - hb}{h} = \frac{ha}{h} \qquad \text{Simplify.}$$

$$\frac{2A - hb}{h} = a$$

●

Think about it 3: Write $\dfrac{2A - hb}{h}$ in another way.

EXAMPLE 

Solving equations containing subscripts Solve $v = gt + v_0$ for g in terms of v, t, and v_0.

Solution The subscript on v_0 indicates that it stands for velocity at time $t = 0$. The two variables v_0 and v are not the same and cannot be combined; v and v_0 must be treated as different variables.

$$v = gt + v_0 \qquad \text{Subtract } v_0 \text{ from both sides.}$$

$$v - v_0 = gt \qquad \text{Divide both sides by } t.$$

$$\frac{v - v_0}{t} = g$$

●

ANSWER BOX

Warm-up: 1. both 8 **2.** both −1 **3.** both $3\frac{1}{6}$ **4.** the distributive property of multiplication over addition, used here as division over addition or subtraction **5.** $\dfrac{P}{2} - \dfrac{2l}{2}$ or $\dfrac{P}{2} - l$. **Think about it 1:** If freeway traffic is moving at 30 miles per hour, it will take 1 hour to travel the 30 miles, so the trip will take a total of 2 hours. **Think about it 2:** When we divide each term by 2, we get $\dfrac{P}{2} - l$. **Think about it 3:** $\dfrac{2A}{h} - \dfrac{hb}{h} = \dfrac{2A}{h} - b$

EXERCISES 4.1

In Exercises 1 to 6, use the phrase in terms of *to describe each formula or equation. If you recognize the formula, use words; otherwise, use the variables themselves.*

1. $I = prt$

2. $P = 2l + 2w$

3. $r = \dfrac{d}{t}$

4. $C = 2\pi r$

5. $G = \dfrac{T_1 + T_2 + T_3 + H + E}{P}$, where G = percent earned, T = test, H = homework, E = final exam, and P = total points possible

6. The volume of a sphere: $V = \dfrac{4}{3}\pi r^3$

What formula is described by each statement in Exercises 7 to 10? You may need to look up the correct formula.

7. The area of a rectangle in terms of length and width

8. The time needed for a trip in terms of distance and rate

9. The area of a circle in terms of radius

10. The circumference of a circle in terms of diameter

In Exercises 11 to 16, write the statements as formulas. Some formulas may contain parentheses.

11. The area, A, is half the product of b and h.

12. The perimeter, P, is twice the sum of l and w.

13. The perimeter, P, is the sum of twice l and twice w.

14. The area, A, is half of h multiplied by the sum of a and b.

15. The number of calories used, C, is the product of the number of calories per minute, f, and the number of minutes of exercise, m.

16. The last term, L, is the sum of f and the product of d with the difference between n and 1.

Solve each formula in Exercises 17 to 48 for the indicated letter.

17. $p = 5n$ for n

18. $q = 4d$ for d

19. $A - P = H$ for A

20. $A - P = H$ for P

21. $C = 2\pi r$ for r

22. $A = lw$ for w

23. $A = bh$ for h

24. $d = rt$ for r

25. $I = prt$ for t

26. $I = prt$ for r

27. $C = \pi d$ for d

28. $A = \pi r^2$ for r^2

29. $P = R - C$ for C

30. $P = R - C$ for R

31. $PV = nRT$ for n

32. $PV = nRT$ for R

33. $C_1V_1 = C_2V_2$ for V_1

34. $C_1V_1 = C_2V_2$ for C_2

35. $P = a + b + c$ for c

36. $P = 2l + 2w$ for l

37. $A = \frac{1}{2}h(a + b)$ for h

38. $A = \frac{1}{2}h(a + b)$ for b

39. $V = \frac{1}{3}\pi r^2 h$ for r^2

40. $V = \frac{1}{3}\pi r^2 h$ for h

41. $x = \dfrac{-b}{2a}$ for b

42. $x = \dfrac{-b}{2a}$ for a

43. $y = mx + b$ for b

44. $y = mx + b$ for m

45. $d^2 = \dfrac{3h}{2}$ for h

46. $S = \dfrac{a}{1 - r}$ for a

47. $t^2 = \dfrac{2d}{g}$ for g

48. $t^2 = \dfrac{2d}{g}$ for d

Exercises 49 to 56 provide practice in solving for the variable y, a skill needed to make a table or graph. Solve for y in terms of x.

49. $xy = -4$

50. $xy = -6$

51. $3x - y = 10$

52. $2x - y = 3$

53. $x - 2y = -5$

54. $2x - 3y - 4 = 0$

55. $3x - 2y = 6$

56. $2x + 3y = 9$

In Exercises 57 and 58, A = amount, P = principal, t = time in years, and r = percent interest.

57. a. Solve $A = P + Prt$ for r.

b. An amount of \$11,050 is received on a two-year time certificate with a \$10,000 principal. What is the rate of interest, r?

58. a. Solve $A = P + Prt$ for t.

b. An amount of \$60,125 is received on a time certificate at 6.75% interest on a \$50,000 principal. What is the number of years, t, on the certificate?

59. a. Solve $C = \frac{5}{9}(F - 32)$ for F.

b. What is the Fahrenheit temperature, F, corresponding to a C of 37° Celsius?

60. a. Solve $K = C + 273$ for C.

b. What is the Celsius temperature corresponding to absolute zero, 0 K?

c. Use the answer to part b and the answer to part a of Exercise 59 to obtain the Fahrenheit temperature corresponding to absolute zero.

61. The bowling handicap, H, in terms of bowling average, A, is $H = 0.8(200 - A)$.

a. What is the bowling handicap for a bowler with a 140 average?

b. Solve the formula for A.

c. What is the bowling average if the handicap is 24?

62. An aerobic heart rate for exercise is $R = 0.7(220 - A)$, where A is age in years.

a. What is the heart rate for a 20-year-old?

b. Solve the formula for A.

c. What is the age for someone with a predicted aerobic heart rate of 119?

63. One equation for a straight line is $y = mx + b$.

a. Solve the equation for b.

b. Find b if $x = 3$, $y = 4$, and $m = 2$.

c. Find b if $x = 3$, $y = 4$, and $m = -2$.

d. Find b if $x = 3$, $y = 4$, and $m = \frac{1}{2}$.

e. Find b if $x = 3$, $y = 4$, and $m = -\frac{1}{2}$.

Projects

64. ***Velocity of Falling Objects.*** The formula $v = gt + v_0$ gives the velocity at which something is falling after t seconds; g is the gravitational constant of 32.2 feet per second squared, and v_0 is the starting velocity.

a. Solve $v = gt + v_0$ for t in terms of v, g, and v_0.

b. If we drop a rock ($v_0 = 0$) from the top of a high cliff, how long will it be until the rock reaches a velocity of 66 feet per second?

c. If we throw the rock downward with an initial velocity $v_0 = 22$ feet per second, how long will it be until the rock reaches a velocity of 66 feet per second?

d. Solve $v = gt + v_0$ for v_0 in terms of v, g, and t.

e. What is the initial velocity if the velocity after 7 seconds is 300 feet per second?

f. What is the initial velocity if the velocity after 10 seconds is 322 feet per second?

65. ***Equivalent Formulas.*** Which formula, if any, in each set is not equivalent to the first one? Explain how each equivalent formula was obtained from the first formula.

a.
$$P = R - C$$
$$P - C = R$$
$$P + C = R$$
$$P - R = -C$$

b.
$$C = p + 0.08p$$
$$C = 1.08p$$
$$C = p(1 + 0.08)$$
$$\frac{C}{1.08} = p$$

c.
$$P = 2l + 2w$$
$$P - 2l = 2w$$
$$2P = l + w$$
$$\frac{P}{2} = l + w$$

d.
$$d = rt$$
$$\frac{d}{r} = t$$
$$\frac{d}{t} = r$$
$$dr = t$$

e.
$$I = prt$$
$$\frac{I}{pr} = t$$
$$\frac{I}{pt} = r$$
$$\frac{I}{rt} = p$$

f.
$$A = \frac{a + b + c}{3}$$
$$\frac{A}{3} = a + b + c$$
$$3A = a + b + c$$
$$A = \frac{a}{3} + \frac{b}{3} + \frac{c}{3}$$

g.
$$y = mx + b$$
$$mx + b = y$$
$$y = b + mx$$
$$y + mx = b$$

66. ***Subscript Research.*** Subscripts are used in genetics to describe generations after a parent generation. The letter P represents the parent generation; F_1 is the next, or first filial, generation; F_2 is the second filial generation; F_3 is the third; and so forth. Research an application of subscripts in a subject of interest to you. Give several examples and some detail about the application. Explain why subscripts are necessary in the application.

67. ***The Mouse and the Earth's Circumference.*** Assume a 12-inch section is added into a steel cable that formerly fitted the Earth snugly on the equator. If the cable were now held a uniform distance from the Earth's surface, could a mouse go under it? Guess and then investigate by calculating the new radius for spheres of various radii.

First, perform steps a to d for each of the three radii

$r = 6$ inches, $r = 3600$ inches,
$r = 1{,}000{,}000$ inches

a. Calculate the circumference with $C = 2\pi r$. Use calculator π, not 3.14.

b. Add 12 inches to the circumference and find the resulting new radius, r_n, by setting the new circumference equal to $2\pi r_n$ and solving for r_n.

c. Subtract the beginning radius r from the new radius, r_n.

d. Could a mouse go under this cable?

Finally, to prove the result, carry out steps e to h.

e. Solve the circumference formula, $C = 2\pi r$, for r.

f. Solve the formula for the new circumference, $C + 12 \text{ in.} = 2\pi r_n$, for r_n.

g. Subtract r from r_n.

h. Calculate the difference in step g with a calculator.

4.2 Functions and Graphs

OBJECTIVES

- Define a function in terms of input and output.
- Find the domain and range of a function.
- Evaluate expressions written in function notation.
- Graph functions.
- Use the vertical-line test.
- Find horizontal and vertical axis intercepts of functions.

WARM-UP

In Exercises 1 to 3, complete the table for the equation. The background settings for the equations will be given in the examples in this section.

1.

Number of Trips, x	Remaining Value, y (dollars)
0	
2	
4	
6	
8	

Value Remaining on Prepaid Transit Ticket, $y = 20 - 2.50x$

2.

Amount of Water, x (1000 gallons)	Cost, y (dollars)
0	
1	
2	
3	
4	

Cost of City Water, $y = 0.65x + 5.15$

3.

Input, x	Output, y
	−2
	−1
	0
	1
	2

An Absolute Value Equation, $x = |y|$

THIS SECTION introduces functions and function notation. We review graphing as we evaluate functions and draw graphs of functions. We look at graphs to find out whether an equation is a function. We find intercept points of functions with the axes and describe the intercepts in function notation.

Defining Functions

After almost every personal tragedy, major illness, or natural disaster, we speculate on what could have been done to prevent, predict, or reduce the impact of the event. In our search to find answers, we look for evidence that an outcome is *a function of* a particular event (input) or set of events (set of inputs). We might ask: Is the increased use of guns in school violence a function of uncontrolled gun sales? Is lung cancer a function of second-hand smoking? Is river flooding a function of logging water sheds, straightening river channels, and filling wetlands for development? These are issues about which people have very strong opinions. Compromise—and progress toward solutions—is more likely to occur when both sides have reached agreement on the "is a function of" relationship between events and their outcomes.

In mathematics, we have a similar goal: finding relationships or associations that connect an input and an output. If the relationship or association is such that, for each input x, only one outcome y can result, then we have a function.

Definition of Function

A **function** is a relationship or association where for each input x there is exactly one output y.

To describe a function, we can use the same ordered pair (x, y) as in coordinate graphing, except we say *y is a function of x*. Example 1 shows that not all sets of ordered pairs are functions.

EXAMPLE 

Finding whether ordered pairs are functions Which sets of ordered pairs are functions?

a. (5, 25), (6, 36), (7, 49)

b. (5, 6), (5, 7), (5, 8)

c. (25, 5), (25, −5), (9, 3), (9, −3)

d. (5, 6), (6, 6), (7, 6)

e. (Cuba, Maria Conchita Alonso), (Cuba, Orlando Hernandez), (Cuba, Desi Arnaz)

f. (Roberto Clemente, Puerto Rico), (Jose Feliciano, Puerto Rico), (Rita Moreno, Puerto Rico)

Solution

a. Function. Each input is associated with exactly one output.

b. Not a function. For the input 5, there are three different outputs: 6, 7, and 8.

c. Not a function. For the input 25, there are two outputs: 5 and −5. For the input 9, there are two outputs: 3 and −3.

d. Function. Each input is associated with exactly one output. This example shows that the outputs do not need to be different numbers.

e. Not a function. For the input Cuba, there are different outputs.

f. Function. For each input name, there is one output. ●

The ordered pairs in part a of Example 1 are from the equation $y = x^2$. Because each input x has only one square ($x \cdot x$, or x^2), we say that y is the *squaring function.*

Think about it 1: What rule describes each of the other parts of Example 1?

Sets of Inputs (Domain) and Sets of Outputs (Range)

Finding the set of inputs and the set of outputs for a function helps in describing the function. It also helps us decide what numbers to use on the axes when we draw a graph of the function.

EXAMPLE 2 Finding sets of inputs and sets of outputs What is the set of inputs for the given ordered pairs? What is the set of outputs?

a. (5, 25), (6, 36), (7, 49) **b.** (5, 6), (6, 6), (7, 6)

Solution **a.** The set of inputs is {5, 6, 7}. The set of outputs is {25, 36, 49}.

b. The set of inputs is {5, 6, 7}. The output is a single number, {6}. ●

In Example 3, we look in applications for input and output sets that are functions.

EXAMPLE  Finding the input and output sets Explain why each sentence describes a function. What are the input and output phrases in each sentence? What sets describe the inputs and outputs?

a. The remaining value of a prepaid transit ticket is a function of the number of identical trips taken.

b. The total cost of a city water bill is a function of the number of thousands of gallons of water used each month.

Solution **a.** For each *number of trips taken* (input), there is exactly one *remaining value of the transit ticket* (output). Each identical trip should cause the same cost to be subtracted from the ticket. The inputs are the set of positive integers; the outputs are positive real numbers in dollars and cents.

b. For every *number of thousands of gallons used* (input), there is exactly one *cost* (output). Each thousand gallons of water usage should have the same cost. The inputs are positive real numbers in thousands of gallons, and the outputs are positive real numbers in dollars and cents. ●

If we want to be more formal in describing the sets of inputs and outputs, we use the words *domain*, as in "domestic" or "home," and *range*, as in "to go out."

Definition of Domain and Range

> The **domain** is the set of inputs to a function.
> The **range** is the set of outputs from a function.

Function Notation

We can use **function notation** *to write functions in symbols:* If V is the remaining value on a transit ticket and x is the number of trips, then $V = f(x)$ says that the value is a function of the number of trips.

Function Notation

> The notation $f(x)$ is read "function of x" or "f of x." The f and x in $f(x)$ are not being multiplied.

EXAMPLE 

Writing function notation Define variables and write each sentence in function notation.

a. The total cost of city water is a function of the thousands of gallons used.

b. The monthly payment due on a credit card is a function of the charge balance.

c. The area of a circle is a function of the radius of the circle.

Solution

a. Let C = total cost and g = number of thousands of gallons of water; then $C = f(g)$.

b. Let p = monthly payment due and b = charge balance; then $p = f(b)$.

c. Let A = area and r = radius; then $A = f(r)$. The number π in $A = \pi r^2$ is a constant and is left out of the function notation. ●

Note: As mentioned earlier, the parentheses in function notation such as $f(g)$, $f(b)$, and $f(r)$ do *not* imply a multiplication.

Think about it 2: How is the product of two variables such as f and g usually written?

We can write functions with letters that represent variables in the problem, or we can use x for the input and $f(x)$ for the output. For some variables, we commonly use capital letters—for example, A for area. For others, we more often use lowercase letters—for example, r for radius. Your instructor will indicate any preferences with respect to the style of letters used.

Evaluating and Graphing Functions

One of the most important reasons for studying functions is that the notation $f(x)$ is a convenient way to indicate the evaluation of a function for a certain number or expression. *Function notation has the advantage of allowing us to simultaneously name the association (rule) and the input.*

Function Evaluation

> The notation for evaluating a function for an input a is $f(a)$. To evaluate $f(a)$, we substitute a for every x in the expression named by $f(x)$ and then simplify.

EXAMPLE 

Evaluating a function: city water bill The total cost of city water is \$5.15 plus \$0.65 per thousand gallons (kgal) used. The function is $f(x) = 0.65x + 5.15$. Find the following, and summarize the results in a table.

a. $f(0)$ **b.** $f(1)$ **c.** $f(2)$ **d.** $f(3)$ **e.** $f(4)$

Solution

a. To find $f(0)$, we substitute $x = 0$ for x in $f(x) = 0.65x + 5.15$ and simplify:

$$f(0) = 0.65(0) + 5.15 = 5.15$$

b. To find $f(1)$, we substitute $x = 1$ for x in $f(x) = 0.65x + 5.15$ and simplify:

$$f(1) = 0.65(1) + 5.15 = 5.80$$

c. To find $f(2)$, we let $x = 2$ in $f(x)$ and simplify:

$$f(2) = 0.65(2) + 5.15 = 6.45$$

d. We let $x = 3$ in $f(x)$:

$$f(3) = 0.65(3) + 5.15 = 7.10$$

e. We let $x = 4$ in $f(x)$:

$$f(4) = 0.65(4) + 5.15 = 7.75$$

The evaluations are summarized in Table 1.

Amount of Water, x (kgal)	Cost, $f(x) = 0.65x + 5.15$ (dollars)
0	$f(0) = 0.65(0) + 5.15 = 5.15$
1	$f(1) = 0.65(1) + 5.15 = 5.80$
2	$f(2) = 0.65(2) + 5.15 = 6.45$
3	$f(3) = 0.65(3) + 5.15 = 7.10$
4	$f(4) = 0.65(4) + 5.15 = 7.75$

Table 1 Cost of City Water

We can take the pairs of numbers in Table 1 and build a graph. To graph a function, we graph the ordered pairs $(x, f(x))$. Because $y = f(x)$ for a function, it is sensible to place x on the horizontal axis and $f(x)$ on the vertical axis.

In Example 6, we use the problem-solving steps—understand, plan, do, and check—to build a graph. Review the steps as needed (Section 1.4, page 32).

EXAMPLE 6 Graphing functions What are the ordered pairs in Example 5? Use the problem-solving steps to build a graph from these ordered pairs.

Solution The ordered pairs are (0, 5.15), (1, 5.80), (2, 6.45), (3, 7.10), and (4, 7.75).

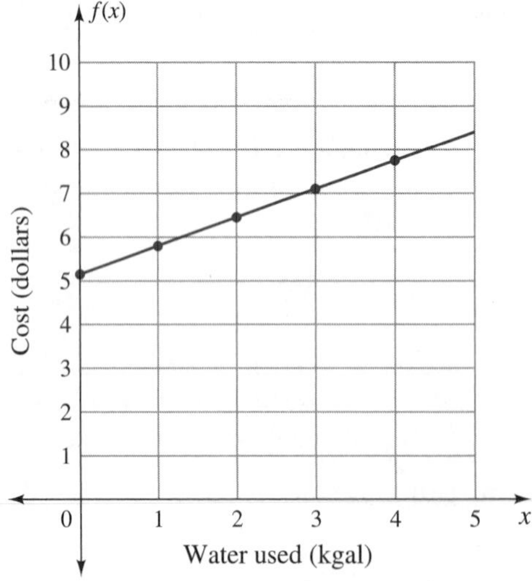

Figure 3

Understand: The input is the amount of water used. We place "Water used (kgal)" on the horizontal axis. The output is the cost. We place "Cost (dollars)" on the vertical axis.

Plan: This is a first-quadrant graph, as negative gallons and negative cost make no sense. If the water meter could be read very accurately, it might be possible to have water measured in fraction or decimal notation, so we can connect the points formed by the ordered pairs. The horizontal axis should start at zero and count by ones to 5. The vertical axis might be marked in ones up to 10.

Do: We plot the ordered pairs in Figure 3. The points form a straight line.

Check: The points all lie on the line. ●

We evaluate a function in Example 7 and then graph it in Example 8.

EXAMPLE  Evaluating a function: prepaid transit ticket After x trips at $2.50 each, a $20 prepaid transit ticket is worth $-\$2.50x + \$20$. The function describing the value remaining on the ticket is $f(x) = -2.50x + 20$. Find the following, and summarize the results in a table.

a. $f(0)$ **b.** $f(2)$ **c.** $f(4)$ **d.** $f(6)$ **e.** $f(8)$

Solution **a.** To find $f(0)$, we substitute $x = 0$ for x in $f(x) = -2.50x + 20$ and simplify:

$$f(0) = -2.50(0) + 20 = 20$$

b. To find $f(2)$, we substitute $x = 2$ for x and simplify:

$$f(2) = -2.50(2) + 20 = 15$$

c. $f(4) = -2.50(4) + 20 = 10$

d. $f(6) = -2.50(6) + 20 = 5$

e. $f(8) = -2.50(8) + 20 = 0$

The evaluations are summarized in Table 2.

Number of Trips, x	**Remaining Value, $f(x) = -2.50x + 20$ (dollars)**
0	$f(0) = -2.50(0) + 20 = 20$
2	$f(2) = -2.50(2) + 20 = 15$
4	$f(4) = -2.50(4) + 20 = 10$
6	$f(6) = -2.50(6) + 20 = 5$
8	$f(8) = -2.50(8) + 20 = 0$

Table 2 Value Remaining on Prepaid Transit Ticket after x Trips ●

EXAMPLE  Graphing a function Plot the ordered pairs $(x, f(x))$ to graph the function in Example 7. Use the problem-solving steps to plan the graph.

Solution ***Understand:*** The ordered pairs are $(x, f(x))$.

Plan: The number of trips is the input and is placed on the horizontal axis. The remaining value in dollars is the output and is placed on the vertical axis.

Do: The points, shown in Figure 4, lie in a line that goes down from left to right.

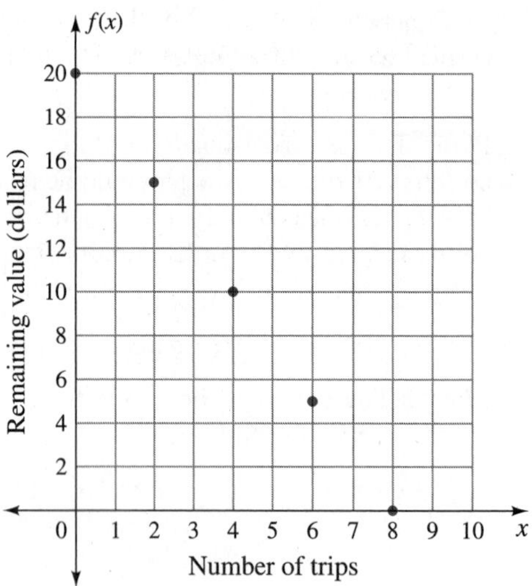

Figure 4

Check: Although the points all lie in a line, we do not draw a solid line because fractions of trips do not make sense. ●

Intercepts

In Section 3.2, we looked at the intercept points where graphs cross the horizontal and vertical axes. In Example 6, the cost of water graph crossed the vertical axis at the point (0, 5.15). In this application, the ordered pair (0, 5.15) indicates that if we use no water, we still pay $5.15. The definitions of the intercept points and intercepts are restated here for your convenience:

Definitions of Intercept Points and Intercepts

> The **horizontal axis intercept point** has the ordered pair $(a, 0)$.
> The **horizontal axis intercept, or x-intercept,** is the number a in $(a, 0)$.
> The **vertical axis intercept point** has the ordered pair $(0, b)$.
> The **vertical axis intercept, or y-intercept,** is the number b in $(0, b)$.

If a problem asks for a point, give the ordered pair. If a problem asks for just the intercept, give the number a or b.

In Example 9, we return to the prepaid transit ticket application to look at the meaning of the intersection of the graph with each axis.

EXAMPLE 9 Finding the meaning of intercept points Refer to the transit ticket graph in Figure 4.

a. What is the meaning of the vertical axis intercept point?

b. What is the meaning of the horizontal axis intercept point?

Solution

a. The prepaid transit ticket graph crosses the vertical axis at the point (0, 20). The intercept point tells us the value of the ticket after zero trips.

b. The prepaid transit ticket graph crosses the horizontal axis at (8, 0). The intercept point tells us that after 8 trips, the card has no remaining value. ●

Function notation gives us a way to describe how to find the horizontal and vertical axis intercepts.

> The horizontal, or x-axis, intercept, a, is where $y = 0$. To find a, solve $f(x) = 0$.
>
> The vertical, or y-axis, intercept, b, is where $x = 0$. To find b, let $b = f(0)$.

To find the horizontal axis intercept, we solve for x in $f(x) = 0$; that is, we let $f(x) = 0$ (or $y = 0$) and solve the resulting equation for the number a. To find the vertical axis intercept, we find $f(0)$; that is, we substitute $x = 0$ into the function and find the number b. Thus, finding intercepts provides practice in both solving equations and evaluating functions.

EXAMPLE 10 Finding intercepts Use algebra to find the vertical and horizontal axis intercepts for the transit ticket function, $f(x) = -2.50x + 20$.

Solution The vertical axis intercept is at $f(0)$: $f(0) = -2.50(0) + 20 = 20$, the initial value of the ticket. The horizontal intercept is where $f(x) = 0$:

$$0 = -2.50x + 20 \quad \text{Add } 2.50x \text{ to both sides.}$$
$$2.50x = 20 \quad \text{Divide both sides by 2.50.}$$
$$\frac{2.50x}{2.50} = \frac{20}{2.50} \quad \text{Simplify.}$$
$$x = 8$$

It takes 8 trips to use up the initial \$20 value of the transit ticket. ●

EXAMPLE 11 Finding intercepts Use algebra to find the vertical and horizontal axis intercepts for the city water function, $f(x) = 0.65x + 5.15$.

Solution The vertical axis intercept is where $x = 0$, at $f(0)$: $f(0) = 0.65(0) + 5.15 = 5.15$. The horizontal axis intercept is where $y = 0$, or $f(x) = 0$:

$$0 = 0.65x + 5.15 \quad \text{Subtract 5.15 from both sides.}$$
$$-5.15 = 0.65x \quad \text{Divide both sides by 0.65.}$$
$$\frac{-5.15}{0.65} = \frac{0.65x}{0.65} \quad \text{Simplify.}$$
$$-7.923 \approx x$$

The horizontal axis intercept is negative. Because a negative amount of water is not sensible in the problem setting, we say that this intercept has no meaning. ●

Think about it 3: Does the result $x = -7.923$ when $y = 0$ agree with the graph in Figure 3?

Vertical-Line Test

Not all associations or relations are functions. Relationships in which there are two (or more) possible outcomes for each input are not functions. Think about the money received from buying a lottery ticket. Whether you spend \$1 or \$100 when you purchase a lottery ticket, you probably will win nothing. However, you might win some money or, very rarely, a large amount of money. Because there is more than one possible outcome, the money received is not a function of the money spent. In Example 12, we look at an equation that is not a function.

EXAMPLE 12 Identifying when an equation is not a function

a. Complete Table 3 for $x = |y|$.
b. Graph the ordered pairs (x, y).
c. If we consider x to be the input, why is the equation $x = |y|$ not a function?

$x = \|y\|$	y
	-2
	-1
	0
	1
	2

Table 3

Solution **a.** The ordered pairs for the table are $(2, -2)$, $(1, -1)$, $(0, 0)$, $(1, 1)$, and $(2, 2)$.

b. These five points are plotted in Figure 5.

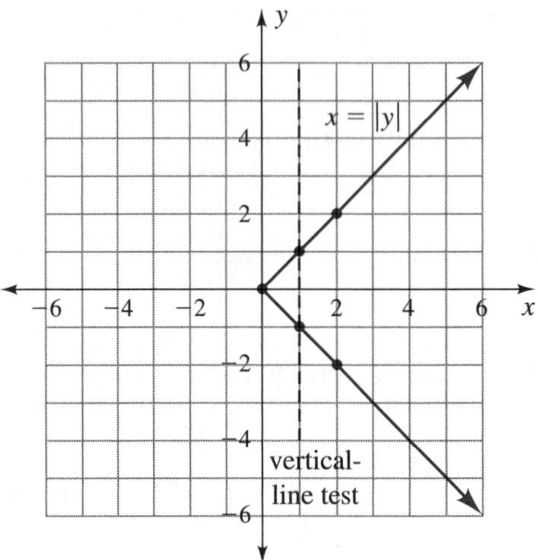

Figure 5

c. The equation $x = |y|$ is not a function because for $x = 1$, y is either 1 or -1. Similarly, for $x = 2$, $y = 2$ or -2. Thus, y is not a function of x.

Looking at a graph is an easy way to find out if an equation is a function. In Figure 5, the points $(1, 1)$ and $(1, -1)$ are on the same vertical line. A graph having two points on the same vertical line is not a function.

Vertical-Line Test

> The graph of a function has at most one output for each input. A vertical line will intersect a function in at most one point.

EXAMPLE 13 Applying the vertical-line test Which of these graphs represent functions? If the graph does not represent a function, name two ordered pairs that show that it does not.

a.

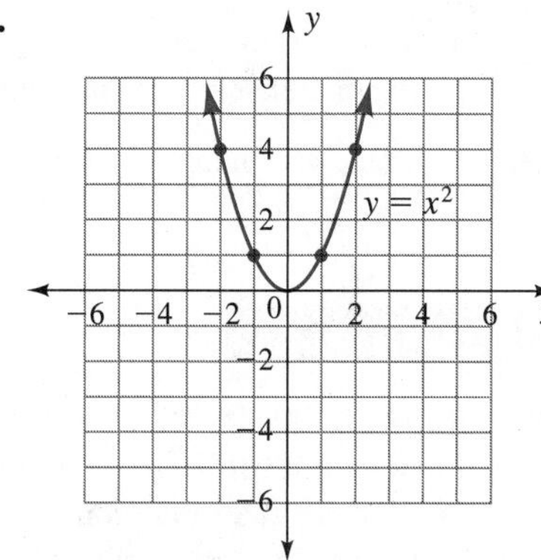

b.

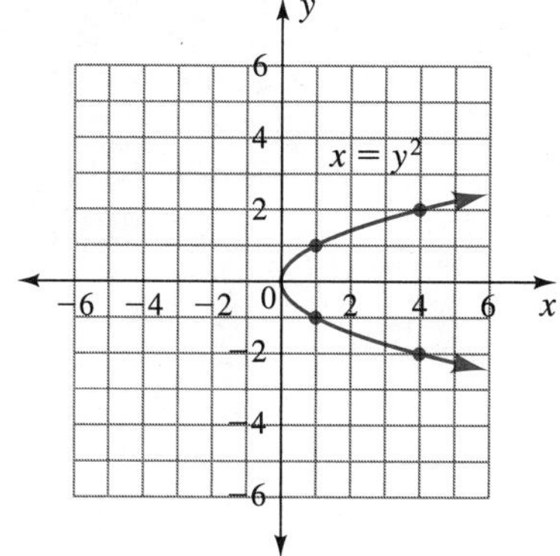

c.

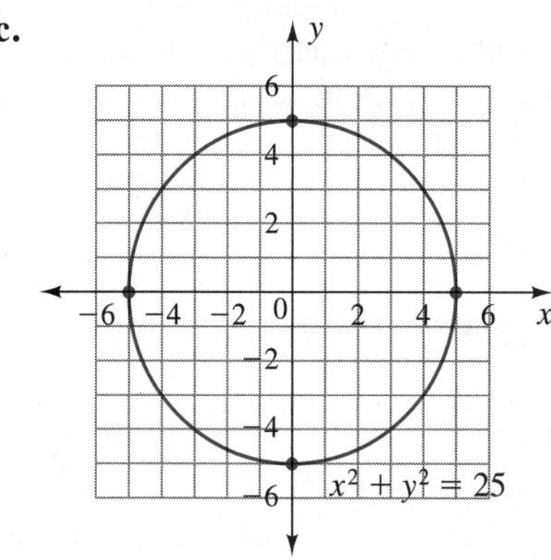

Solution **a.** Function

b. Not a function. For example, (4, 2) and (4, −2) are on the same vertical line.

c. Not a function. For example, (0, 5) and (0, −5) are on the same vertical line.

Think about it 4: What two ordered pairs in part b of Example 13 have $x = 1$? What two ordered pairs in part c have $x = 3$? What two ordered pairs in part c have $x = -3$?

Caution: In Section 4.1 we used the phrase *in terms of.* The *in terms of* phrase describes a relationship between variables, as does a function. However, *in terms of* is more general than is *a function of* and may be used to describe all types of equations, either functions or not functions. In Example 13, equations b and c are in terms of x and y but neither one is a function.

Functions in Two or More Variables

We close the section with a note about functions containing more than one variable. Many formulas contain multiple variables and are functions.

An example of a two-variable function is $A = lw$. The area of a rectangle is a function of length and width and may be written $A = f(l, w)$.

An example of a three-variable function is $V = lwh$. The volume of a box is a function of length, width, and height and may be written $V = f(l, w, h)$.

EXAMPLE  Writing functions in two or more variables Define variables and write each sentence using function notation.

a. The perimeter of a rectangle is a function of its length and width.

b. The area of a trapezoid is a function of its height and the lengths of its two parallel sides.

Solution **a.** Let p = perimeter, l = length, and w = width; $p = f(l, w)$.

b. Let A = area, h = height, a and b be the parallel sides; $A = f(h, a, b)$.

ANSWER BOX

Warm-up: 1. 20, 15, 10, 5, 0 **2.** 5.15, 5.80, 6.45, 7.10, 7.75 **3.** 2, 1, 0, 1, 2 **Think about it 1: b.** $x = 5$ **c.** $y^2 = x$ **d.** $y = 6$ **e.** The output is a person born in the input country. **f.** The output is the place of birth of the person named as input. **Think about it 2:** The product of f and g is usually written $f \cdot g$ or just fg. This prevents confusion with the function notation $f(g)$. **Think about it 3:** In Figure 3, if we extended the graph to the left, it would intersect the horizontal axis to the left of zero. The negative result for x is reasonable. **Think about it 4:** (1, −1) and (1, 1); (3, 4) and (3, −4); (−3, 4) and (−3, −4)

EXERCISES 4.2

In Exercises 1 to 8, which sets of ordered pairs are functions? For each set that is not a function, explain why. For each function, name the set of inputs (domain) and the set of outputs (range).

1. a. (5, 5), (−5, 5), (6, 6), (−6, 6)

b. (5, −5), (5, 5), (6, 6), (6, −6)

2. a. (5, 5), (5, 6), (5, 7)

b. (5, 6), (6, 7), (7, 8)

3. a. $\left(2, \frac{1}{2}\right), \left(3, \frac{1}{3}\right), \left(4, \frac{1}{4}\right)$

b. $\left(2, \frac{1}{2}\right), \left(2, \frac{1}{3}\right), \left(2, \frac{1}{4}\right)$

4. a. (1, 11), (1, 111), (1, 1111)

b. (−2, 2), (−3, 2), (−4, 2)

5. (Eden, Barbara), (Tuckman, Barbara), (McClintock, Barbara)

6. (Jordan, Barbara), (Jordan, Michael), (Jordan, Vernon)

7. (CA, Amy Tan), (CA, Maxine Hong Kingston), (CA, Ursula Le Guin)

8. (Haing Ngor, Academy Award), (Aung San Suu Kyi, Nobel Peace Prize), (Ieoh Ming Pei, Gold Medal of the American Institute of Architects)

In Exercises 9 to 12, give a reason why each relationship is true. Name the set of inputs (domain) and the set of outputs (range).

9. The cost of a long distance telephone call is a function of how long the parties talk.

10. The cost of the roof of a house is a function of the area of the roof.

11. The hours of sunlight on a clear day in December is a function of the distance from the equator.

12. For an hourly worker, the amount earned is a function of the number of hours worked.

In Exercises 13 to 18, give three examples to show why each statement does not describe a function. Answers will vary.

13. Name a city starting with the letter A.

14. Name a city in Texas.

15. Name a rational number x such that $4 < x < 6$.

16. Name an odd number.

17. Name a real number smaller than zero.

18. Name a negative number greater than −5.

In Exercises 19 to 22, define variables and write each sentence in function notation.

19. Auto registration cost is a function of the value of a car.

20. The capital of a state is a function of the name of the state.

21. The circumference of a circle is a function of the diameter.

22. The volume of a sphere is a function of the radius.

Find f(−2), f(−1), f(0), f(1), and f(2) for the functions in Exercises 23 to 28. Sketch a graph of each function using the ordered pairs (x, f(x)).

23. $f(x) = 2x - 1$ **24.** $f(x) = 1 - 2x$

25. $f(x) = 2 - 3x$ **26.** $f(x) = \frac{1}{2}x - 1$

27. $f(x) = \frac{1}{4}x + 1$ **28.** $f(x) = 3 - 2x$

Let f(x) = 0 for each function in Exercises 29 to 34; these are the same functions as in Exercises 23 to 28. Solve the resulting equation for x, and mark the point (x, 0) on the graph you sketched earlier. Explain how you could have found x without solving the equation.

29. $f(x) = 2x - 1$ **30.** $f(x) = 1 - 2x$

31. $f(x) = 2 - 3x$ **32.** $f(x) = \frac{1}{2}x - 1$

33. $f(x) = \frac{1}{4}x + 1$ **34.** $f(x) = 3 - 2x$

In Exercises 35 to 44, in which equations or inequalities is y a function of x? Explain your reasoning.

35. $y = x + 2$ **36.** $y = 1 - x$

37. $x + y = 3$ **38.** $x - y = 2$

39. $y = x^2$ **40.** $x = y^2$

41. $y^2 + 1 = x$ **42.** $y = x^2 + 1$

43. $y < x$ **44.** $y > x + 2$

In Exercises 45 to 52, the functions are labeled with the letters h, H, g, and G instead of f. Find h(4) and h(−4) in Exercises 45 and 46.

45. $h(x) = \left|\frac{1}{2}x\right|$ **46.** $h(x) = |2x|$

Find H(4) and H(−4) in Exercises 47 and 48.

47. $H(x) = x - x^2$ **48.** $H(x) = x^2 - 2x$

Find g(−2) and g(1) in Exercises 49 and 50.

49. $g(x) = x^2 + 1$ **50.** $g(x) = 1 - x^2$

Find G(−2) and G(1) in Exercises 51 and 52.

51. $G(x) = 2 - x^2$ **52.** $G(x) = x^2 + x - 1$

53. Without graphing the functions in Exercises 45, 47, 49, and 51, find the vertical axis intercept for each function.

54. Without graphing the functions in Exercises 46, 48, 50, and 52, find the vertical axis intercept for each function.

Find x if f(x) = 0 for the functions in Exercises 55 to 60.

55. $f(x) = 3(x - 4) + x$

56. $f(x) = 5(x - 8) - x$

57. $f(x) = 9 - 3(x - 1)$

58. $f(x) = 8 - 2(x + 4)$

59. $f(x) = 2x + 3(x - 5)$

60. $f(x) = 5x - 2(x - 5)$

For the functions in Exercises 61 to 64, find the horizontal axis and vertical axis intercepts, and explain their meaning, if any.

61. Total tuition in dollars, where x = number of credit hours taken: $f(x) = 115x + 48$

62. Value remaining (in dollars) on a Coffee Corner gift certificate, where x = number of lattes: $f(x) = 15 - 3x$

63. Value remaining (in dollars) on a phone card, where x = number of minutes used: $f(x) = 19.50 - 0.75x$

64. Total cost of transcripts, where x = number of transcripts ordered: $f(x) = 5 + 2(x - 1)$

Which of the graphs in Exercises 65 to 70 represent functions?

65.

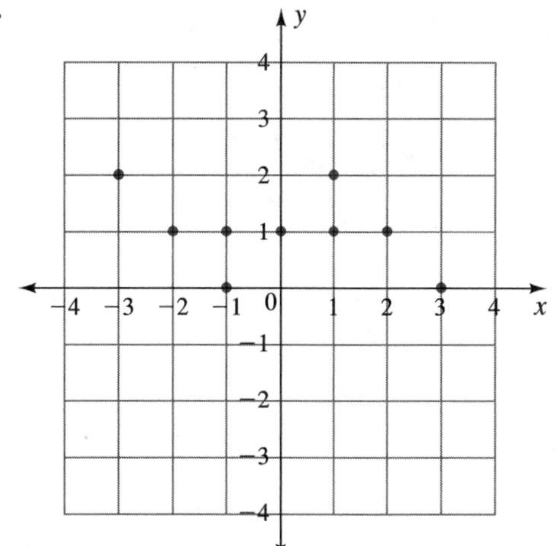

66.

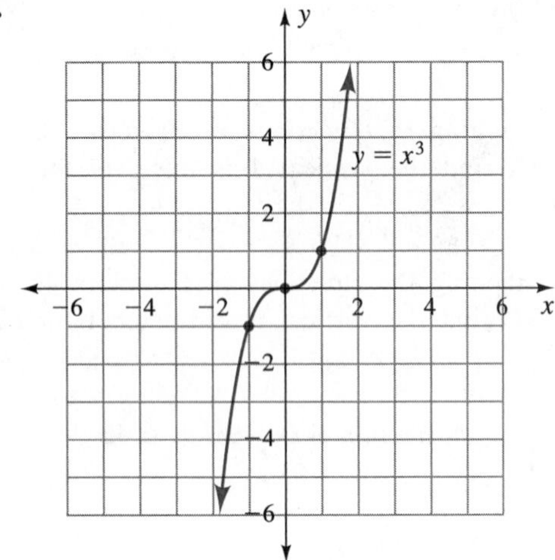

67.

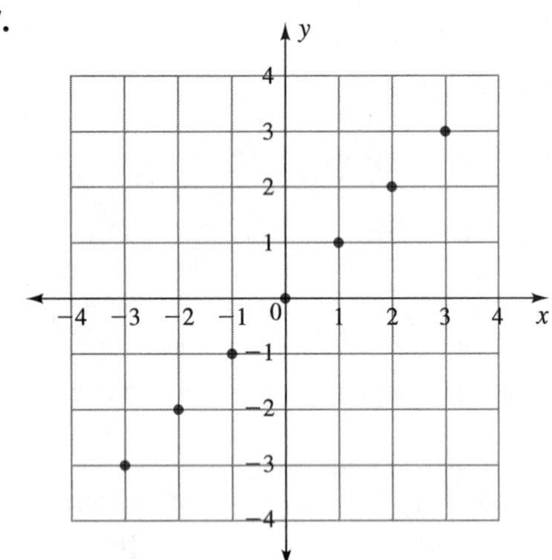

68.

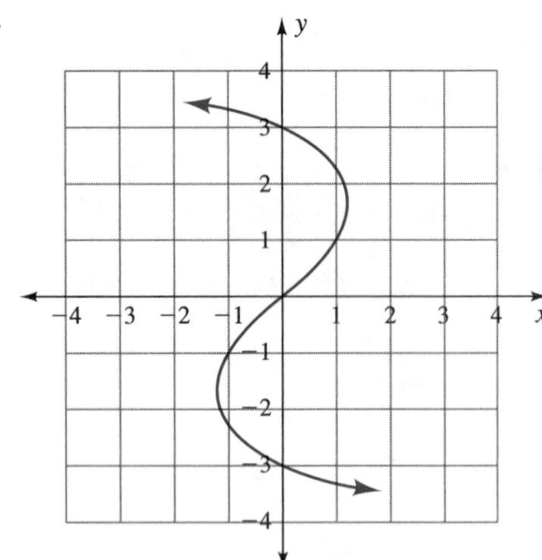

69.

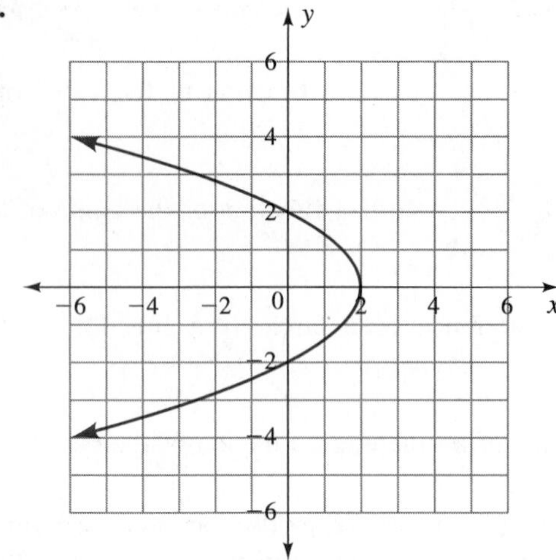

70.

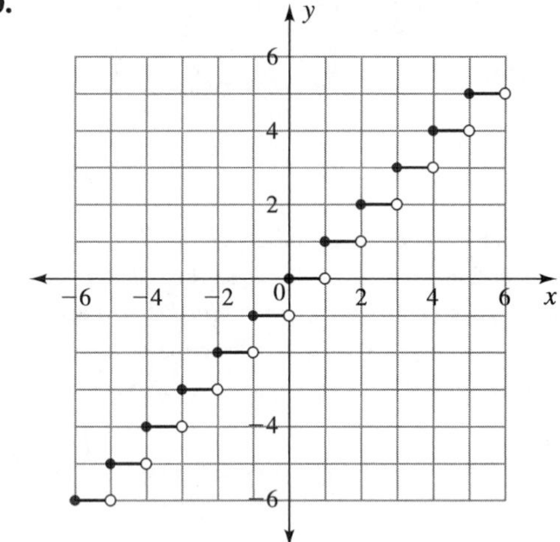

71. Explain how to find $f(a)$ for a function $y = f(x)$.

72. Explain how to find a horizontal axis intercept.

73. Explain how to find a vertical axis intercept.

74. Explain how the vertical-line test identifies a graph of a function.

75. How would you explain to someone that $f(x)$ does not mean multiplication?

In Exercises 76 to 80, define two or more variables and write each sentence in function notation.

76. A child's height is a function of age, genetics, and nutrition.

77. Distance traveled is a function of rate and time.

78. Annual interest earned is a function of the amount invested (principal) and the rate of interest.

79. The volume of a cylinder is a function of the radius and height.

80. The area of a parallelogram is a function of the base and height.

Projects

81. ***Birthday Candle.*** Use a ballpoint pen to mark centimeters from the top of a birthday candle. Place the candle in a holder, and then mount the candle holder on a piece of styrofoam covered with aluminum foil. Make sure that the styrofoam is large enough that it will not tip over.

Predict the length of time required for one centimeter of the candle to burn. Repeat for two centimeters. Light the candle and record the time it takes for the candle to burn 1, 2, 3, and 4 centimeters. Is the distance burned a function of time? Use the data to predict the length of time required to burn 6 centimeters on a candle.

82. ***Probability.*** Functions are important because they guarantee one output for each input. The field of probability deals with multiple outputs for a given input.

a. If the event (input) is to toss four coins and read the top faces, we have 16 different possible outcomes (outputs). If H = head on top and T = tail (or back of the coin) on top, one of the outcomes is HHTH. Another outcome is HTHH. Both HHTH and HTHH have 3 heads and 1 tail. List the 16 different outcomes and group them according to numbers of heads and tails. Which grouping has the most outcomes?

b. If the event (input) is to roll two six-sided dice and read the numbers of dots on the top faces of the dice, there are 36 different outcomes (outputs). If the numbers are 2 and 1, we have a sum of 3. We also have a sum of 3 if the numbers are 1 and 2. List the 36 possible outcomes and their sums. What is the most common sum?

c. Make up your own problem. The problem might be related to the birth order of children, the outcomes for a well-known basketball player shooting a free throw, or any other event or activity with multiple possible outcomes.

83. ***Utility Bills.*** Look for examples of functions on your monthly utility bills. Record the function or data that you think give a function. Be sure to include units.

84. ***Domain and Range.*** Make a chart with these four headings: Function, Sketch of graph with inputs -5 to $+5$, Domain, Range. List these functions in the left-hand column, and then complete the chart.

a. The squaring function, $f(x) = x^2$

b. The absolute value function, $f(x) = |x|$

c. $f(x) = x$

d. $f(x) = -x$

e. $f(x) = 2$

Hint: To find the domain, ask if every real number can be an input. To find the range, ask if every real number can be an output.

85. ***Absolute Value.*** How is the graph of $y = |x|$ the same as the graph of $x = |y|$? How is it different?

86. ***Visualizing Functions.*** Discuss which of the following pictures show functions. The "rule" for each part is the line connecting a number in the input set to a number in the output set. Suggest a rule for each part.

a.

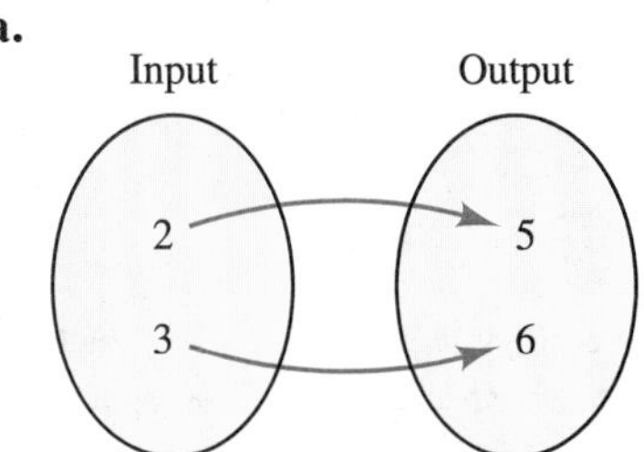

b.

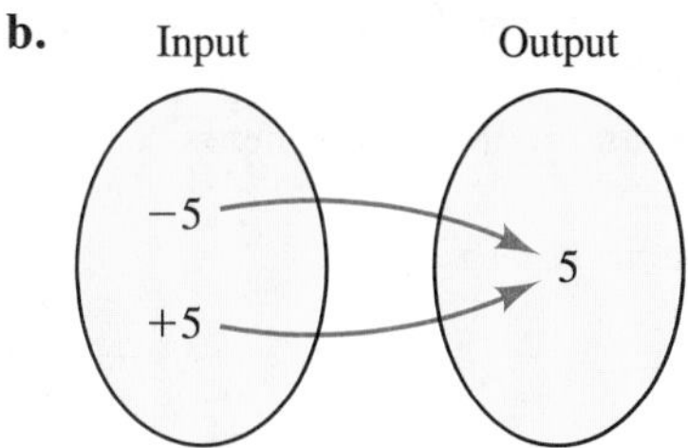

c.

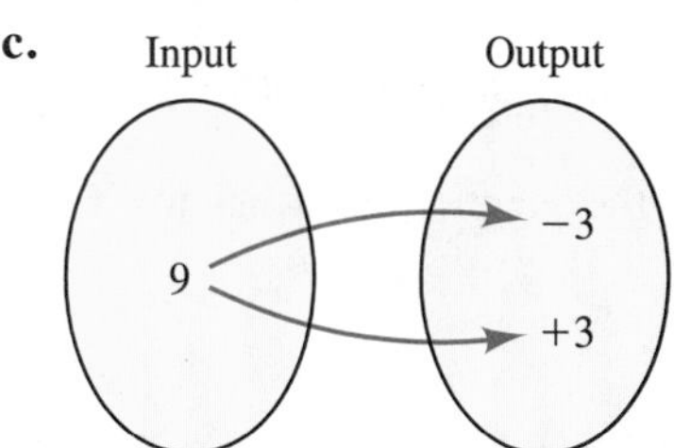

d.

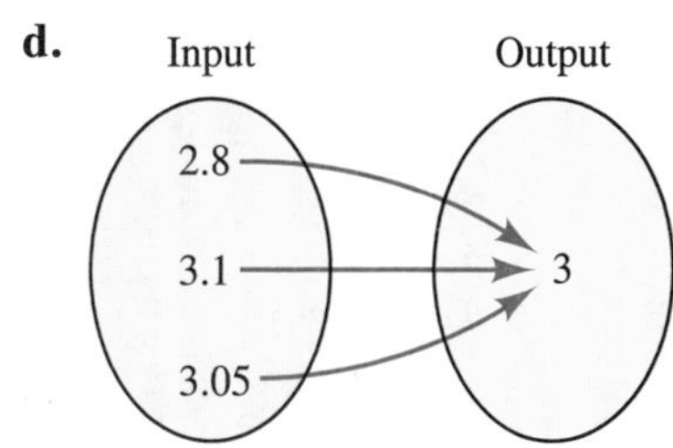

87. ***Credit Card Payment Schedule***

Charge Balance	Payment Due
\$0 to \$20	Full amount
\$20.01 to \$500	10% or \$20, whichever is greater
\$500.01 or more	\$50 plus the amount in excess of \$500

If $f(x)$ is the credit card repayment schedule, find

a. $f(0)$ **b.** $f(20)$

c. $f(35)$ **d.** $f(195)$

e. $f(200)$ **f.** $f(300)$

g. $f(500)$ **h.** $f(625)$

Find the value of x for

i. $f(x) = 0$ **j.** $f(x) = 15$

k. $f(x) = 20$ **l.** $f(x) = 25$

m. $f(x) = 35$ **n.** $f(x) = 55$

o. $f(x) = 100$ **p.** $f(x) = 150$

88. ***Functions and Inequalities.*** Graph $f(x)$ and $g(x)$ on one set of axes. Let $f(x) = -x + 4$ and $g(x) = \frac{1}{2}x + 1$. Solve the equations in parts a–h and complete the sentences in parts i–j.

a. $f(x) = g(x)$ **b.** $f(x) \geq g(x)$

c. $f(x) \leq g(x)$ **d.** $f(x) \leq 0$

e. $0 \geq g(x)$ **f.** $f(x) \geq 4$

g. $g(x) \leq 3$ **h.** $g(x) > f(x)$

i. If $f(x) > g(x)$, then the graph of $f(x)$ is ________ the graph of $g(x)$ for inputs in the solution set.

j. If $f(x) = g(x)$, then the graph of $f(x)$ ________ the graph of $g(x)$.

MID-CHAPTER 4 TEST

In Exercises 1 and 2, write each formula in words, using the phrase in terms of. *The meanings of the variables are given following the formula.*

1. $C = \frac{5}{9}(F - 32)$, Celsius and Fahrenheit temperatures

2. $V = \frac{4}{3}\pi r^3$, volume and radius of a sphere

In Exercises 3 and 4, solve the equations for b.

3. $3 = 4(-2) + b$ **4.** $-4 = \frac{2}{3}(-6) + b$

In Exercises 5 to 7, solve the formulas for the indicated variable.

5. Temperature scales: $C = K - 273$ for K

6. Distance seen on the moon from a height of h feet: $d^2 = \dfrac{3h}{8}$ for h

7. Last term in a number sequence: $l = a + (n - 1)d$ for d

8. Find $f(6)$ for

a. $f(x) = 3x$ **b.** $f(x) = \frac{1}{2}x + 2$

c. $f(x) = \frac{1}{2}(x + 2)$ **d.** $f(x) = x^2 - 2$

9. Graph $f(x) = 3 - 4x$.

10. Answer the following questions for the set of ordered pairs (2, 3), (3, 3), (4, 3), (5, 4), (6, 4), (7, 4).

a. Does the set describe a function?

b. What is the set of inputs (domain)?

c. What is the set of outputs (range)?

11. a. Sketch a graph of the set of ordered pairs (2, 3), (2, 4), (2, 5).

b. Complete this statement: A relationship is a function if for each input there is ____ output.

c. Explain, in terms of inputs and outputs, why the set of ordered pairs in part a is not a function.

d. Show whether or not the set of ordered pairs in part a passes the vertical-line test.

12. Find $f(0)$ for $f(x) = 3x - 5$.

13. Suppose $f(x) = \frac{1}{2}x + 2$. Solve for x in $f(x) = 0$.

14. What is $f(0)$ on the graph of a function?

4.3 Linear Functions: Slope and Rate of Change

OBJECTIVES

- Find the slope of a line from its graph.
- Find the slope of a line from ordered pairs.
- Find the slope of a line from a table.
- Find the slopes of horizontal and vertical lines.
- Write the meaning of slope in terms of units on the axes.
- Draw the graph of a line with a given slope through a given point.
- Recognize the slope concept in a variety of situations.

WARM-UP

Simplify these expressions.

1. $\dfrac{250 - 100}{55 - 50}$ **2.** $\dfrac{-3 - 1}{0 - (-5)}$ **3.** $\dfrac{5333 - 4477}{1990 - 1980}$

4. $\dfrac{4.00 - 4.50}{20 - 10}$ **5.** $\dfrac{725 - 550}{1750 - 1650}$ **6.** $\dfrac{5 - (-1)}{6 - (-3)}$

IN THIS SECTION and the next, we focus on linear functions. We define the linear function and its slope, or rate of change. We look at slope numerically with tables, visually with graphs, and symbolically with a formula involving ordered pairs.

EXAMPLE 

Exploration Angie spends most of her free time outdoors. She has planned a cross country ski trip for her winter vacation and a bicycling trip for her summer vacation. On her cross country ski trip, Angie plans to make a map that shows the elevation of the trail. Angie wants to describe the steepness, or slope, of the trail as she skis from left to right. Figure 6 shows some line segments that could appear on her map, as well as two segments that could not appear on her map. The four types of slope are shown in Figure 7.

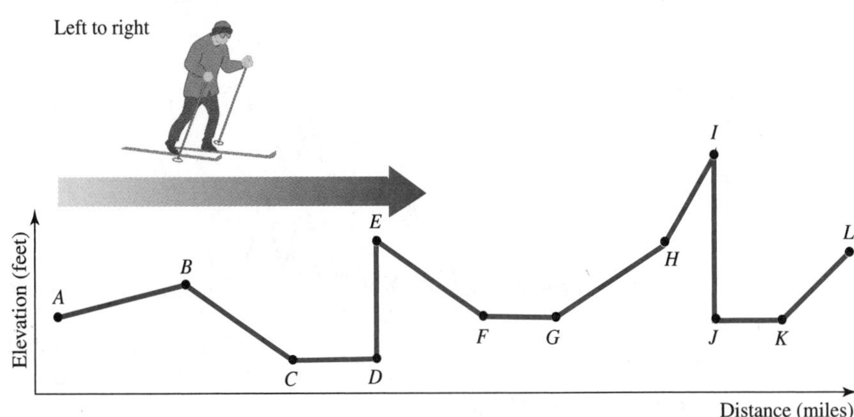

Figure 6 Elevation of ski trip

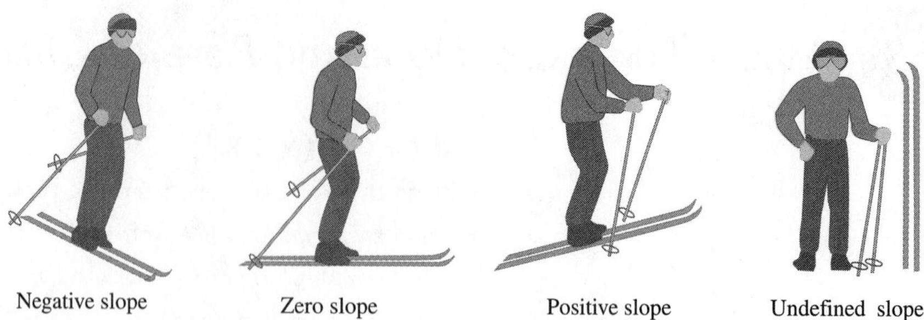

Figure 7 Type of slope

a. Use the term *positive*, *negative*, *zero*, or *undefined* to label the type of slope represented by the elevation of each of the following parts of the "trail" in Figure 6: *AB, BC, CD, DE, EF, FG, GH, HI, IJ, JK, KL*.

b. Which two segments in Figure 6 are not likely to appear on a cross country ski trail? Why?

c. Label the slope of each of the following pieces of the trail as *positive, negative, zero*, or *undefined*:

For her summer vacation, Angie plans to bicycle the Oregon Coast. The Oregon Department of Transportation provided her with a map of the first 100 miles, shown in Figure 8. Left to right on the map is north to south. The graph below the map shows the elevation in feet of the road, where 0 is sea level.

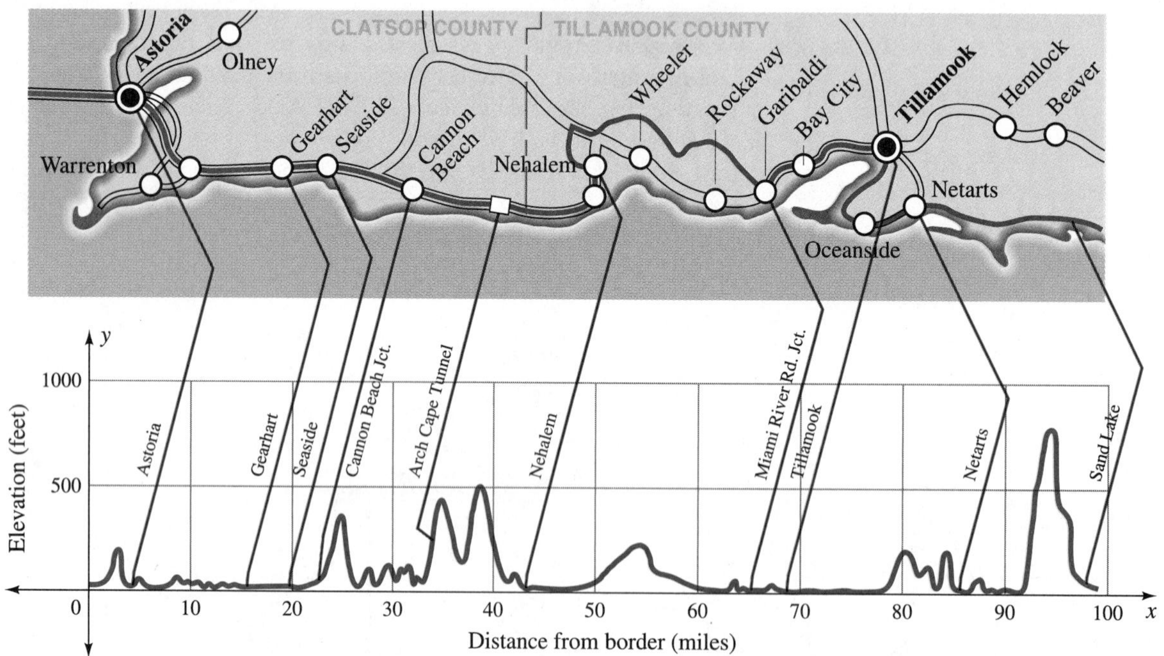

Figure 8 Elevation of bicycle trip

d. In what ten-mile interval is the road flattest?

e. In what ten-mile interval is the road steepest?

f. Describe the type of slope between Gearhart (mile 15) and Seaside (mile 20).

g. Describe the type of slope between miles 45 and 55.

h. Describe the type of slope between miles 55 and 62.

i. Describe the type of slope between miles 92 and 95.

j. In traveling from left to right, where would you rather ride—where the slope is positive or where it is negative?

k. Where would it be possible to ride—where the slope is zero or where it is undefined?

l. Are there any undefined slopes on the map? Why?

Solution See the Answer Box. ●

Finding Slope from a Graph

As we saw in Example 1, the slope, or rate of change, is one of the most important features of the graph of a line. After looking at slope visually (with a graph), we will look at it verbally (with a formula) and then numerically (with a table), so as to be sure to find the way that best suits your learning style.

Slope describes the steepness and direction of a line. As we trace from left to right along a graph, *the vertical change relative to the horizontal change* defines the **slope** of the line.

Finding Slope from a Graph

> To find slope from a graph, divide the vertical change between two points by the horizontal change between the same points.
>
> - A line with positive slope rises from left to right.
> - A line with negative slope falls from left to right.

The definition lets us find a number to describe slope.

The map in Figure 8 shows a vertical change of 750 feet in the horizontal distance of 3 miles between miles 92 and 95. The slope is

$$\frac{\text{Vertical change}}{\text{Horizontal change}} = \frac{750\text{ ft} - 0\text{ ft}}{95\text{ mi} - 92\text{ mi}} = \frac{750\text{ ft}}{3\text{ mi}} = 250\text{ ft per mi}$$

The units, feet per mile, are important. Look for units in all application problems.

EXAMPLE 2 Finding the slope from a graph Return to Figure 8.

a. Estimate the height of the graph at mile 50.

b. Estimate the height of the graph at mile 55.

c. Estimate the slope between miles 50 and 55.

Solution **a.** The height is approximately 100 feet.

b. The height is approximately 250 feet.

c. The slope is

$$\frac{\text{Vertical change}}{\text{Horizontal change}} = \frac{250\text{ ft} - 100\text{ ft}}{55\text{ mi} - 50\text{ mi}} = \frac{150\text{ ft}}{5\text{ mi}} = 30\text{ ft per mi}$$ ●

Think about it 1: Which is a steeper slope: 250 feet per mile or 30 feet per mile?

We will return to applications in Example 11. For the next several examples, we will use numbers only. In Example 3, we use two letters to name a line segment. Look for this notation in other examples and exercises.

EXAMPLE 3 Finding the slope from a graph Use the graph in Figure 9 to find these slopes.

a. The slope of BC, the line segment from B to C

b. The slope of BD, the line segment from B to D

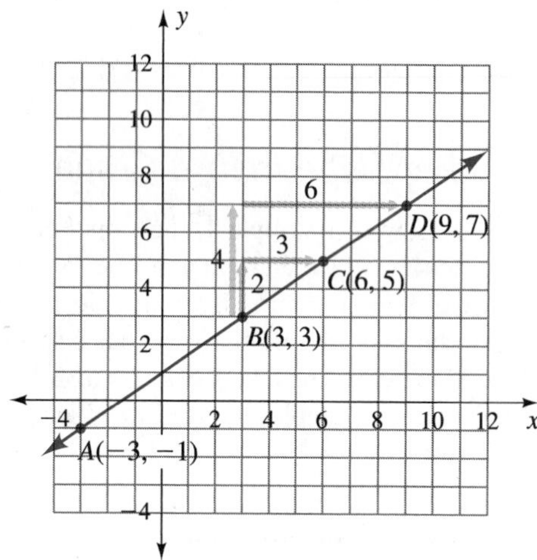

Figure 9

Solution Both BC and BD have positive slope. To find the vertical change, we count the vertical distance between the points. To find the horizontal change, we count the horizontal distance between the points. The triangles drawn on the line show the vertical and horizontal change for each pair of points.

a. Segment BC rises 2 units for 3 units from left to right. The slope is

$$\frac{\text{Vertical change}}{\text{Horizontal change}} = \frac{2}{3}$$

b. Segment BD rises 4 units for 6 units from left to right. The slope is

$$\frac{\text{Vertical change}}{\text{Horizontal change}} = \frac{4}{6} = \frac{2}{3}$$

The slopes for the two segments are equal. ●

Example 3 suggests that we can describe slope as *rise over run*. Think about rise over run in Example 4.

EXAMPLE 4 Finding the slope from a graph Find the slope of EF from the graph in Figure 10.

Solution Line EF has a negative slope. The vertical change is 4 units, from 1 to -3, as shown in Figure 11. Note that here the "rise" is down. The *horizontal change*, or **run**, is 5 units, from -5 to 0. The slope is

$$\frac{\text{Vertical change}}{\text{Horizontal change}} = \frac{\text{rise}}{\text{run}} = -\frac{4}{5}$$

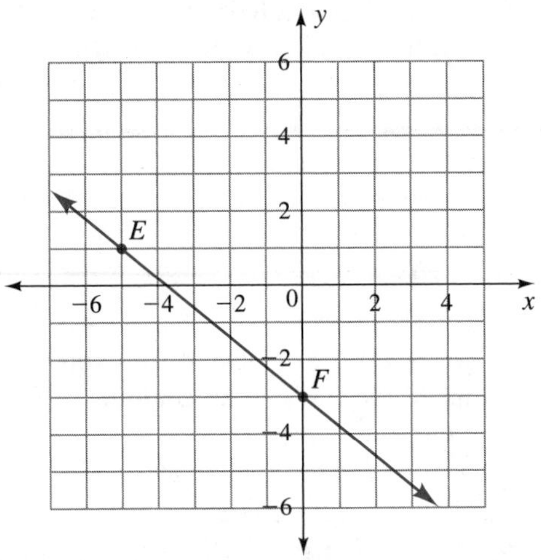

Figure 10

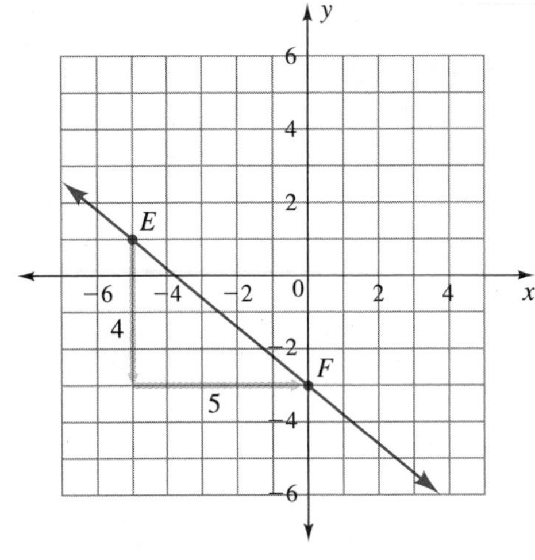

Figure 11

For a negatively sloped line, the *vertical change*, or **rise**, is down. We associate negative slopes with decreasing values as we move from left to right.

Negative Slope

> A function with a negative slope decreases as we trace its graph from left to right.

Finding Slope from Ordered Pairs

In Example 5, we check our results from Example 4 with a second method, using a formula to find the slope from the ordered pairs for E and F.

Slope Formula

> For ordered pairs (x_1, y_1) and (x_2, y_2),
>
> $$\text{Slope} = \frac{y_2 - y_1}{x_2 - x_1}$$

Generally, we let the ordered pair (x_1, y_1) be the point on the left and the ordered pair (x_2, y_2) be the point on the right.

EXAMPLE 5 Finding slope from ordered pairs Find the slope of EF, using the ordered pairs $E(-5, 1)$ and $F(0, -3)$.

Solution We let $(x_1, y_1) = (-5, 1)$ and $(x_2, y_2) = (0, -3)$.

$$\text{Slope } AB = \frac{y_2 - y_1}{x_2 - x_1} = \frac{-3 - 1}{0 - (-5)} = \frac{-4}{5}$$

Think about it 2: Does the slope depend on the order of the points? Repeat Example 5 but let $(x_1, y_1) = (0, -3)$ and let $(x_2, y_2) = (-5, 1)$.

Finding Slope from a Table

A third method of finding slope is from a table. In Example 6, we find slope by first calculating the change, or difference, between the numbers in each column of the table and then dividing to obtain slope. The results of the subtractions are listed in new columns headed Δx for "change in x" and Δy for "change in y."

> In a table, slope $= \dfrac{\Delta y}{\Delta x}$.

EXAMPLE 6 **Finding slope from a table** Table 4 contains several ordered pairs from an equation. Complete the Δx and Δy columns in the table and find the slope for each pair of changes in x and y.

Δx	Input, x	Output, y	Δy	Slope
	−3	−1		
	3	3		
	6	5		
	9	7		

Table 4

Solution We complete the Δx and Δy columns by subtracting to find the change between entries in the table, as shown in Table 5.

Δx	Input, x	Output, y	Δy	Slope
	−3	−1		
6			4	$\frac{4}{6} = \frac{2}{3}$
	3	3		
3			2	$\frac{2}{3}$
	6	5		
3			2	$\frac{2}{3}$
	9	7		

Table 5

The slope of the line containing all the ordered pairs in the table is $\frac{2}{3}$. ●

As mentioned earlier, the small triangle used in the table is the capital Greek letter *delta*. **Delta**, Δ, indicates *a change in the value of the variable that follows it.*

Finding Slopes of Horizontal and Vertical Lines

In Example 7, we use ordered pairs to find the slopes of horizontal and vertical lines.

EXAMPLE  **Finding slopes of horizontal and vertical lines** Use the graph in Figure 12 to find these slopes.

a. The slope of the vertical line containing A and B

b. The slope of the horizontal line containing B and C

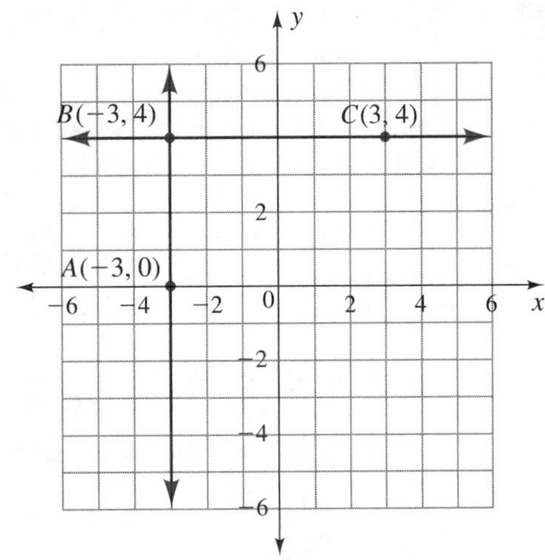

Figure 12

Solution **a.** Segment AB connects $(-3, 0)$ and $(-3, 4)$.

$$\text{Slope } AB = \frac{y_2 - y_1}{x_2 - x_1} = \frac{4 - 0}{(-3) - (-3)} = \frac{4}{0}$$

The zero in the denominator means division by zero. Since division by zero is not defined, we say that a vertical line has **undefined slope**.

b. Segment BC connects $(-3, 4)$ and $(3, 4)$.

$$\text{Slope } BC = \frac{y_2 - y_1}{x_2 - x_1} = \frac{4 - 4}{(3) - (-3)} = \frac{0}{6} = 0$$

A horizontal line has **zero slope**. ●

Slope and Linear Functions

Nonvertical straight lines may be described as linear functions.

Linear Functions

> A linear function is any association between x and y that can be written
>
> $$y = mx + b$$
>
> We may write a linear function as $f(x) = mx + b$, where $y = f(x)$.

Think about it 3: Why are vertical lines not functions?

In Example 3 and Example 6, different ordered pairs on the same line gave the same slope. This suggests two conclusions: (1) the slope of a line is constant (the same number) between any two points on the line and (2) slope can be used to show that three points lie in a straight line.

> - The slope of a linear function is constant.
> - If the slope between points A and B is the same as the slope between points A and C, then points A, B, and C all lie on the same line.

EXAMPLE 8 Showing that three points lie on a line Do $A(-3, -1)$, $B(3, 3)$, and $C(6, 5)$ lie on the same line?

Solution We must find the slope of segments AB and AC.

Let $(x_1, y_1) = (-3, -1)$ and $(x_2, y_2) = (3, 3)$.

$$\text{Slope } AB = \frac{y_2 - y_1}{x_2 - x_1} = \frac{3 - (-1)}{3 - (-3)} = \frac{4}{6} = \frac{2}{3}$$

Let $(x_1, y_1) = (-3, -1)$ and $(x_2, y_2) = (6, 5)$.

$$\text{Slope } AC = \frac{y_2 - y_1}{x_2 - x_1} = \frac{5 - (-1)}{6 - (-3)} = \frac{6}{9} = \frac{2}{3}$$

The slopes of AB and AC are the same, so A, B, and C lie on a straight line.

Using Slope in Graphing Lines

In many applications, we are given the slope and need to draw a line.

> To draw a line with a given slope, choose a first point and use the slope to find a second point.

EXAMPLE 9 Drawing lines given the slope

a. Draw a line with a $\frac{3}{4}$ slope.

b. Draw a line with a $-\frac{5}{2}$ slope.

c. Draw a line through $(-1, 3)$ with a -2 slope.

Solution

a. To draw a $\frac{3}{4}$ slope, mark a first point A anywhere on a piece of graph paper. Count 3 units up from A and then count 4 units to the right to mark point B. The line through A and B, shown in Figure 13, will have a $\frac{3}{4}$ slope.

b. To draw a $-\frac{5}{2}$ slope, mark a first point A anywhere. Count 5 units down from A and then count 2 units to the right to mark point B. The line through A and B, shown in Figure 14, will have a $-\frac{5}{2}$ slope.

c. A slope of -2 is the same as $-\frac{2}{1}$. To draw a line through $(-1, 3)$ with slope $-\frac{2}{1}$, place point A at $(-1, 3)$. Count 2 units down from A and then count 1 unit to the right to mark point B. The line through A and B, shown in Figure 15, will have a -2 slope and will pass through $(-1, 3)$.

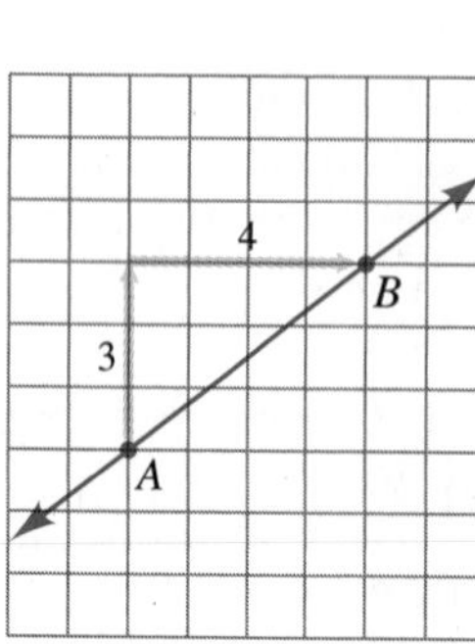

Figure 13

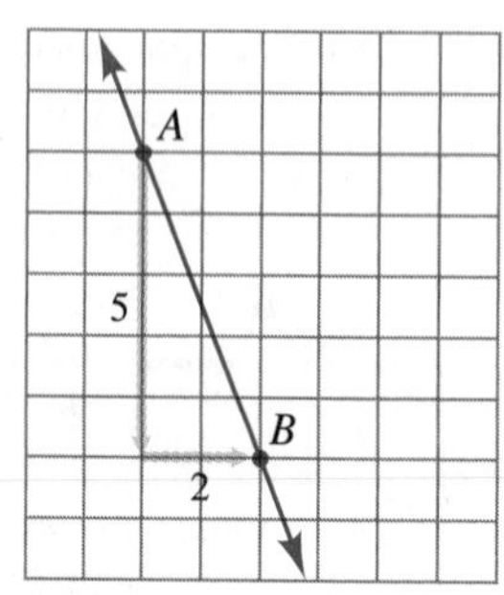

Figure 14

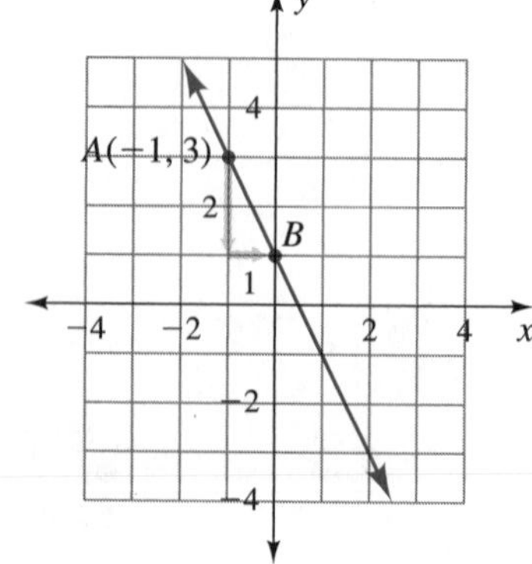

Figure 15

Example 9 suggests that we need to know both the slope and a point to locate a specific line. The slope alone will not tell us where to draw the line.

> To locate a line, we need to know both its slope and a point on the line.

EXAMPLE 10 Finding the graph of a line Match a line graph in Figure 16 with each slope and point given.

a. slope $= -\frac{2}{3}$, point $= (-6, 2)$

b. slope $= 0$, point $= (-6, 6)$

c. slope is undefined, point $= (1, 5)$

d. slope $= \frac{1}{3}$, point $= (-5, -3)$

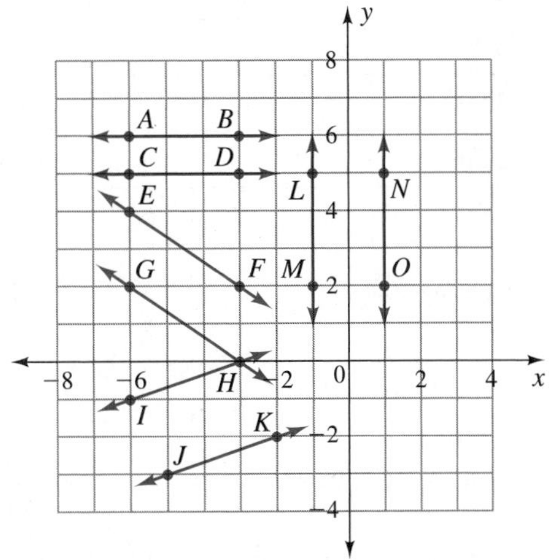

Figure 16

Solution Only one of each pair of parallel lines in the figure is used. See the Answer Box.

Using Slope to Graph Lines

> To draw a line given a slope, start with a point anywhere on the graph. Find a second point on the line by first counting the vertical change (up for a positive slope or down for a negative slope) and then counting the horizontal change to the right. Draw the line through the two points.
>
> To draw a line given a point (a, b) and a slope, start by graphing (a, b). Find a second point by first counting the vertical change (up or down) and then counting the horizontal change to the right. Draw the line through the two points.

Applications

Because slope is the change in y divided by the change in x, *the meaning of slope in applications is the units for y divided by the units for x.*

In Example 11, we return to the prepaid photocopy machine card.

EXAMPLE 11 Finding the meaning of slope A prepaid photocopy machine card costs \$5.00. Each photocopy is \$0.05. Table 6 lists the value remaining on the card after x copies are made. The letters in the table match positions on the graph in Figure 17.

a. Find the slope between points A and B.

b. Find the slope between points C and D.

c. What is the meaning of the slope? Why is it negative?

Copies, x	Card Value, y	Points on Graph
0	\$5.00	
10	4.50	A
20	4.00	B
30	3.50	C
40	3.00	
50	2.50	
60	2.00	D

Table 6 Photocopy Card Value

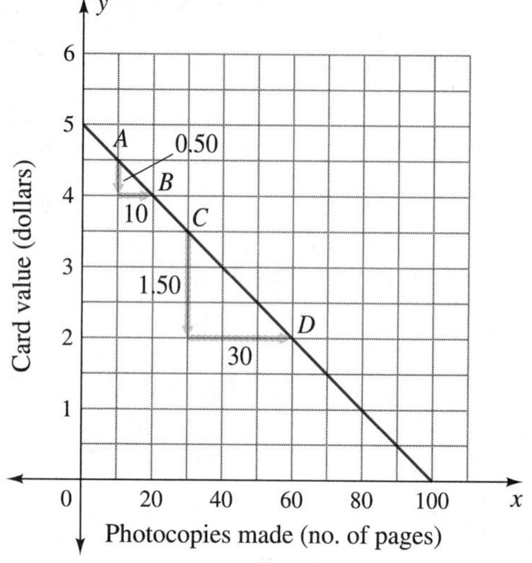

Figure 17

Solution **a.** Slope $AB = \dfrac{y_2 - y_1}{x_2 - x_1} = \dfrac{\$4.00 - \$4.50}{(20 - 10)\text{ copies}} = \dfrac{-\$0.50}{10\text{ copies}} = -\$0.05$ per copy

b. Slope $CD = \dfrac{y_2 - y_1}{x_2 - x_1} = \dfrac{\$2.00 - \$3.50}{(60 - 30)\text{ copies}} = \dfrac{-\$1.50}{30\text{ copies}} = -\$0.05$ per copy

c. The vertical axis is labeled with card value in dollars. The horizontal axis is labeled with the number of photocopies. The units for slope are $\dfrac{\text{dollars}}{\text{number of copies}}$, or $-\$0.05$ per copy. The slope is negative because the vertical axis is labeled with the value of the card and the card value decreases with each additional photocopy made. ●

Think about it 4: Here are some questions that draw on what you learned in Section 4.1:

a. Why is the value of the photocopy card a function?

b. Where does the graph in Figure 17 cross the vertical axis? What is the meaning of this point?

c. Write a function for the value of the card after x photocopies.

d. Where does the graph cross the horizontal axis? What is the meaning of this point?

In Example 12, we use slope to describe the steepness of a roof. The A-frame house in Figure 18 is popular in climates with lots of snow. The slope of the roof is quite large so that snow does not pile up on the roof. In hot climates, where there is no snow, the roof is relatively flat (Figure 19). Because we are given the measurements of the roof instead of ordered pairs, *rise over run* is the appropriate way to find the slope of a roof. The *rise* is the vertical distance; the *run* is half the horizontal distance across the bottom of the roof.

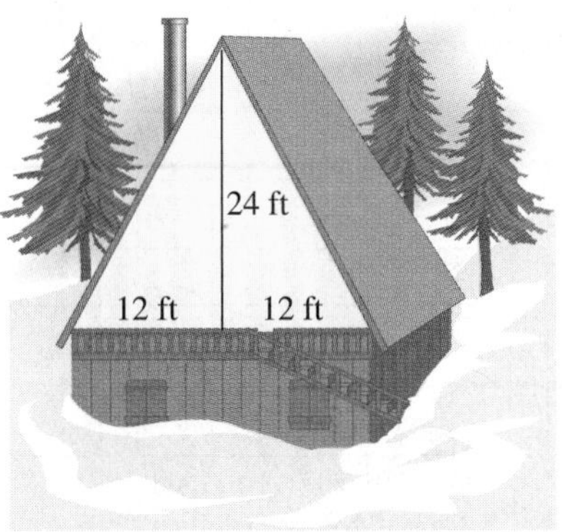

Figure 18

Figure 19

EXAMPLE 12 Finding the meaning of slope: building roofs

a. What is the slope of the roof in Figure 18?

b. What is the slope of the roof in Figure 19?

c. What units describe the slope?

d. Is a positive or negative slope appropriate for a roof?

Solution **a.** $\text{Slope} = \dfrac{\text{rise}}{\text{run}} = \dfrac{24 \text{ ft}}{12 \text{ ft}} = \dfrac{2}{1}$

b. $\text{Slope} = \dfrac{\text{rise}}{\text{run}} = \dfrac{2 \text{ ft}}{12 \text{ ft}} = \dfrac{1}{6}$

c. The units for the slope of a roof are feet over feet.

d. Because one side of the roof has a positive slope and the other a negative slope, we usually omit the positive and negative signs. ●

In Example 13, we return to the credit card payment schedule, repeated in Table 7.

Charge Balance	Payment Due
\$0 to \$20	Full amount
\$20.01 to \$500	10% or \$20, whichever is greater
\$500.01 or more	\$50 plus the amount in excess of \$500

Table 7 Credit Card Payment Schedule

EXAMPLE  Finding the meaning of slope: credit card payments The graph in Figure 20 shows the following ordered pairs from the credit card payment schedule: $A(0, 0)$, $B(20, 20)$, $C(200, 20)$, $D(500, 50)$, and $E(600, 150)$. Find the slopes of the following segments and explain what the slopes mean.

a. AB **b.** BC **c.** CD **d.** DE

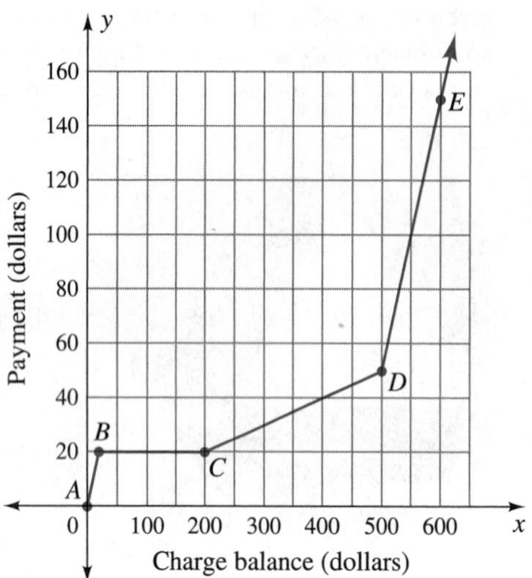

Figure 20

Solution The units of slope are $\dfrac{\text{dollars of payment due}}{\text{dollars of charge balance}}$.

a. Slope $AB = \dfrac{y_2 - y_1}{x_2 - x_1} = \dfrac{20 - 0}{20 - 0} = \dfrac{20}{20} = 1$

The slope is 1 dollar of payment for each dollar of charge balance, because for a charge balance of \$20 or less, the payment due is the full charge balance.

b. Slope $BC = \dfrac{y_2 - y_1}{x_2 - x_1} = \dfrac{20 - 20}{200 - 20} = \dfrac{0}{180} = 0$

The slope is zero, because the graph is horizontal. Between a \$20 and a \$200 charge balance, the payment is \$20. There is no increase in the payment due for each dollar increase in the charge balance.

c. Slope $CD = \dfrac{y_2 - y_1}{x_2 - x_1} = \dfrac{50 - 20}{500 - 200} = \dfrac{30}{300} = \dfrac{1}{10}$

The slope is $\frac{1}{10}$, or 10% in percent notation. Between a \$200 and a \$500 charge balance, the payment due is 10% of the balance.

d. Slope $DE = \dfrac{y_2 - y_1}{x_2 - x_1} = \dfrac{150 - 50}{600 - 500} = \dfrac{100}{100} = 1$

The slope is 1 dollar of payment for each dollar of charge balance over \$500, because the payment due includes all of the charge balance over \$500. ●

Our final example concerns world population. Here the slope, or rate of change, gives us a comparison over time.

EXAMPLE **14** Comparing slopes: world population from 1650 to 1990 Use the data in Table 8 to answer the following questions.

a. What is the rate of change in population between 1650 and 1750?

b. What is the rate of change in population between 1980 and 1990?

c. Use division to compare the result in part b with that in part a.

d. Does the graph in Figure 21 confirm your results in parts a and b?

1650	550,000,000
1750	725,000,000
1850	1,175,000,000
1900	1,600,000,000
1950	2,565,000,000
1980	4,477,000,000
1990	5,333,000,000

Table 8 Estimated World Population
Data from *The World Almanac and Book of Facts 1992* (New York: Pharos Books, 1991), p. 822.

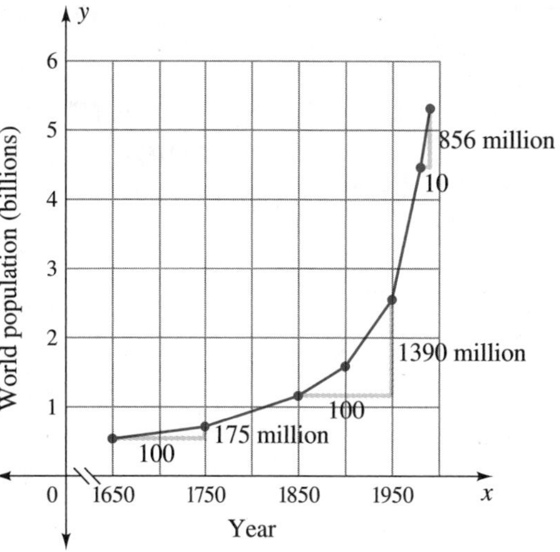

Figure 21

Solution **a.** Rate of change $= \dfrac{\text{change in population}}{\text{change in time}}$

$$= \frac{(725 - 550)\text{ million}}{(1750 - 1650)\text{ yr}} = \frac{175\text{ million}}{100\text{ yr}}$$

$$= 1.75\text{ million per yr}$$

b. Rate of change $= \dfrac{\text{change in population}}{\text{change in time}}$

$$= \frac{(5333 - 4477)\text{ million}}{(1990 - 1980)\text{ yr}} = \frac{856\text{ million}}{10\text{ yr}}$$

$$= 85.6\text{ million per yr}$$

c. The rate of change from 1650 to 1750 is 1.75 million per year. The rate of change from 1980 to 1990 is 85.6 million per year.

If we divide 85.6 million by 1.75 million, the quotient is about 50. Thus, the rate of change in population in recent years is about 50 times that of the period from 1650 to 1750.

d. The lines drawn on the graph give the same results. ●

Think about it 5: Use the graph in Figure 21 to find the rate of change between 1850 and 1950. Use division to compare the result in part b with your result.

Example 14 suggest that when the rate of change, or slope, is different for various parts of a graph, the graph is not a linear function.

Curved graphs have slopes that change as we move along the graph.

Here is what we have found so far about slope:

Summary

- Slope $= \dfrac{\text{vertical change}}{\text{horizontal change}} = \dfrac{\text{output change}}{\text{input change}} = \dfrac{\text{rise}}{\text{run}} = \dfrac{\Delta y}{\Delta x}$
- Slope formula: slope $= \dfrac{y_2 - y_1}{x_2 - x_1}$
- Lines with positive slope rise, or increase, from left to right.
- Lines with negative slope fall, or decrease, from left to right.
- Horizontal lines have zero slope.
- Vertical lines have undefined slope.
- The units for slope are the units on the vertical axis divided by the units on the horizontal axis.

Here is what slope tells us about linear functions:

- All nonvertical straight lines are linear functions.
- The slope of a linear function is constant.
- If the slope between points A and B is the same as the slope between points A and C, then points A, B, and C all lie on the same line.

ANSWER BOX

Warm-up: 1. 30 **2.** $-\frac{4}{5}$ **3.** 85.6 **4.** -0.05, or $-\frac{1}{20}$ **5.** 1.75 **6.** $\frac{2}{3}$ **Example 1: a.** positive, negative, zero, undefined, negative, zero, positive, positive, undefined, zero, positive **b.** *DE* and *IJ*; they are cliffs **c.** zero, positive, negative, undefined, negative, positive **d.** miles 10 to 20 **e.** miles 90 to 100 **f.** zero **g.** positive **h.** negative **i.** positive (almost undefined) **j.** negative, as it is downhill **k.** zero, as it is flat **l.** No; an undefined slope would be like a cliff, straight up and down. **Think about it 1:** 250 feet per mile is a steeper slope. **Think about it 2:** No; $\dfrac{y_2 - y_1}{x_2 - x_1} = \dfrac{1 - (-3)}{(-5) - 0} = \dfrac{4}{-5} = -\dfrac{4}{5}$. **Think about it 3:** A vertical line does not pass the vertical-line test; for each x, there are many y values. **Example 10: a.** *GH* **b.** *AB* **c.** *NO* **d.** *JK*. **Think about it 4: a.** For each number of copies made, there is exactly one value of the card. **b.** The vertical axis intercept is at \$5.00, the initial value of the card. **c.** $f(x) = -0.05x + 5.00$ **d.** The horizontal axis intercept is at 100, the number of photocopies that can be made for \$5.00 at \$0.05 per copy. **Think about it 5:** slope $= \dfrac{1390 \text{ million}}{100 \text{ yr}} =$ 13.9 million per yr. Dividing 85.6 by 13.9 tells us that the rate of change in population between 1980 and 1990 is 6 times the rate between 1850 and 1950.

EXERCISES 

In Exercises 1 to 6, tell whether the slope of each line will be positive, negative, zero, or undefined. Use the rise and run in the graph to find the slope of each line.

1.

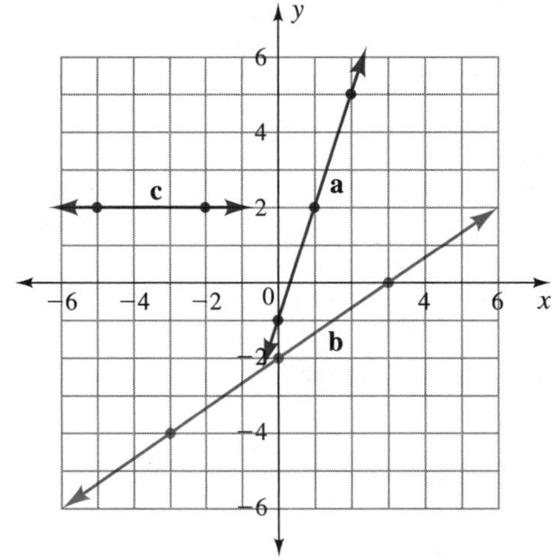

2.

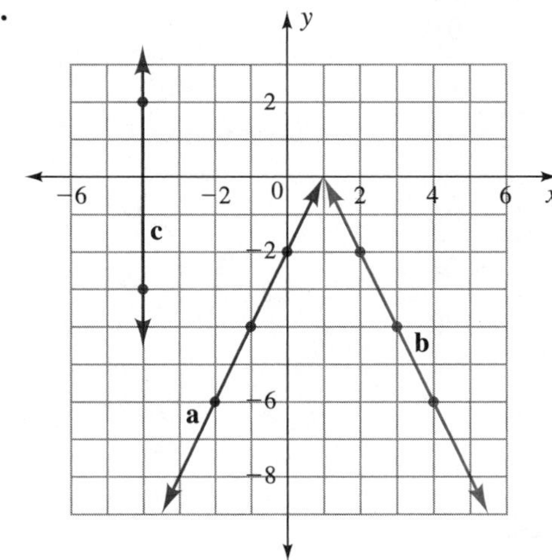

3.

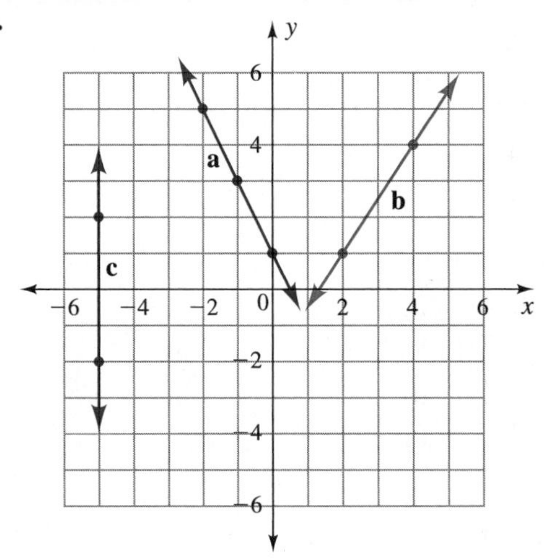

4.

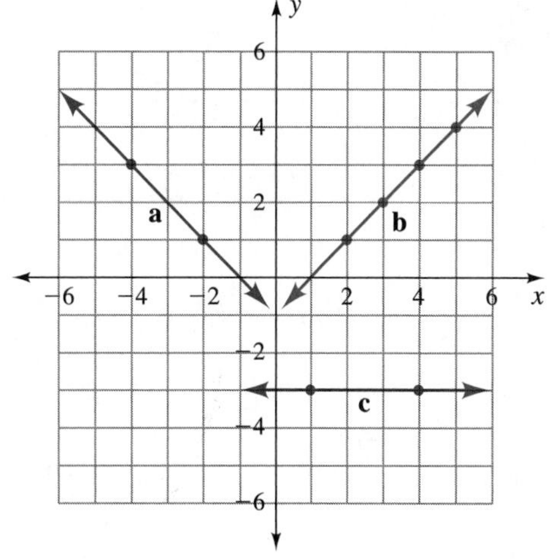

5.

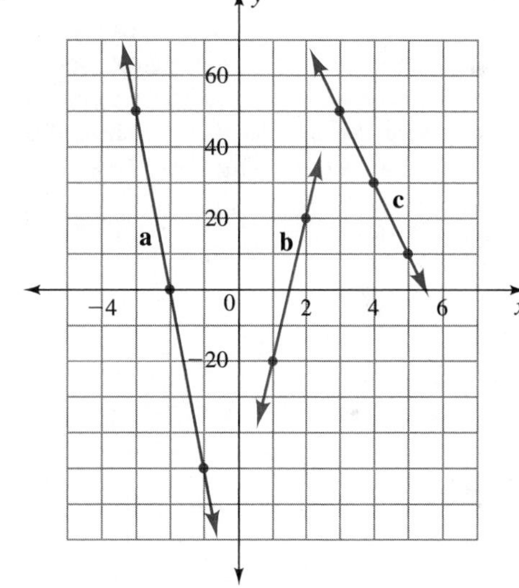

6.

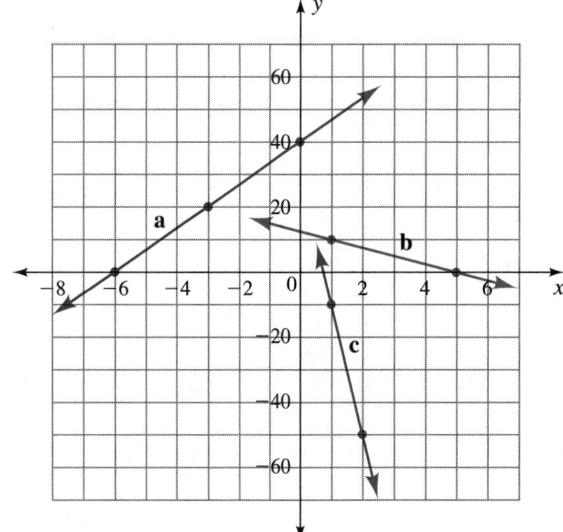

In Exercises 7 to 22, tell whether the slope of the line through the ordered pairs will be positive, negative, zero, or undefined. Use the slope formula to find the slope. (Hint: To learn the formula, write the complete formula each time you use it.)

7. (0, 2) and (4, 3)

8. (2, 0) and (4, 3)

9. (−2, 3) and (0, −4)

10. (−3, 2) and (4, 0)

11. (4, 3) and (4, 4)

12. (4, 3) and (3, 3)

13. (0, 2) and (−2, 2)

14. (4, 4) and (−4, 4)

15. (−2, 3) and (4, −1)

16. (3, −4) and (−5, 2)

17. (2, −3) and (4, −3)

18. (−3, −2) and (−3, 5)

19. (0, 4) and (5, 0)

20. (−3, 0) and (0, 2)

21. (3, −4) and (3, 4)

22. (3, −2) and (−3, −2)

In Exercises 23 to 38, tell whether the slope of the graph formed from the table will be positive, negative, zero, or undefined. State whether the table represents a linear function. For linear functions, find the slope using Δx and Δy. State the units on the slope, if any.

23.

x	y
0	−3
1	−6
2	−9

24.

x	y
0	−9
1	−4
2	1

25.

x	y
0	−5
1	2
2	9

26.

x	y
0	2
1	8
2	14

27.

Hours	Earnings
2	$18
4	36
6	54
8	72

28.

Cookies	Calories
12	900
18	1350
24	1800
30	2250

29.

Kilogram x	Cost y
1	$0.50
2	1.00
3	1.50

30.

Gallons	Cost
2	$3.00
3	4.50
4	6.00
5	7.50

31.

Pounds x	Cost y
1	$0.32
3	0.96
5	1.60

32.

Credit	Cost
1	$ 24
5	120
8	192
10	240

33.

Time (sec)	Distance (ft)
0	0
1	16
2	64
3	144

34.

Length (ft)	Width (ft)
5	10
8	7
9	6
12	3

35.

Copies	Value
0	$15.00
10	12.50
20	10.00
25	8.75

36.

Rides	Value
0	$20
2	16.50
4	13.00
10	2.50

37.

Time (hr)	Distance (mi)
1	40
2	70
3	90
4	100

38.

Radius (ft)	Area (sq ft)
1	3.14
2	12.57
3	28.27
4	50.27

Use the slope formula to find the slope in Exercises 39 to 44.

39. (a, b) and (c, d)

40. $(0, 0)$ and (m, n)

41. $(a, 0)$ and $(0, b)$

42. (m, n) and (p, q)

43. (a, b) and (a, c)

44. (a, b) and (c, b)

In Exercises 45 to 54, draw a line with the given slope.

45. $\frac{4}{3}$

46. $-\frac{3}{1}$

47. $\frac{3}{4}$

48. $-\frac{1}{3}$

49. $-\frac{1}{2}$

50. $\frac{2}{5}$

51. $-\frac{2}{1}$

52. $\frac{2}{3}$

53. 0

54. undefined

A slope and an ordered pair are given in Exercises 55 to 64. Plot the point, and draw a line with the given slope through the point.

55. $\frac{1}{2}$ and $(3, -4)$

56. $-\frac{4}{3}$ and $(1, 2)$

57. -4 and $(2, 3)$

58. $\frac{2}{3}$ and $(4, -1)$

59. $-\frac{3}{2}$ and $(2, 0)$

60. 2 and $(-1, 3)$

61. 0 and $(0, 4)$

62. undefined slope and $(3, 0)$

63. undefined slope and $(-2, 0)$

64. 0 and $(-2, -2)$

In Exercises 65 to 70, arrange the slopes from flattest to steepest.

65. $\frac{4}{3}, \frac{3}{4}, 1$

66. $-1, -\frac{3}{1}, -\frac{1}{3}$

67. $-\frac{1}{2}, 0, -\frac{2}{1}$

68. $\frac{2}{5}, \frac{6}{5}, \frac{1}{2}$

69. $1, \frac{3}{4}, \frac{3}{2}, \frac{3}{5}$

70. $-2, -\frac{1}{2}, -\frac{2}{3}, -\frac{3}{2}$

For Exercises 71 to 74, find the slope of the roof or roof support.

71.

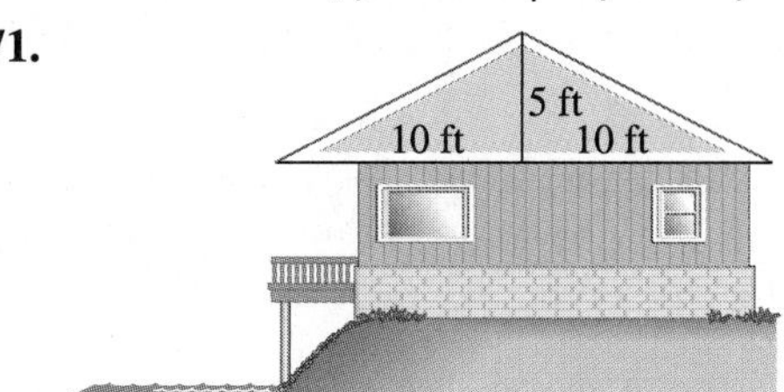

72.

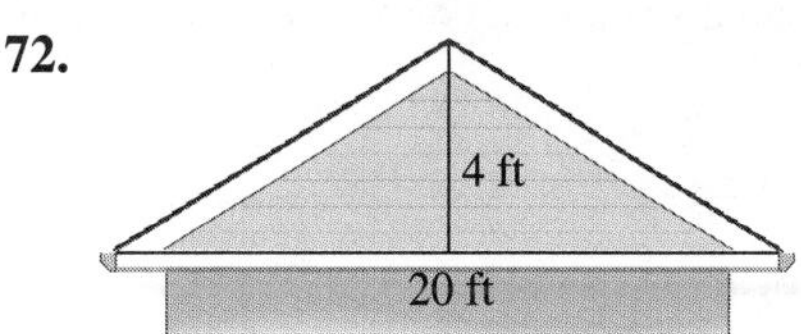

73.

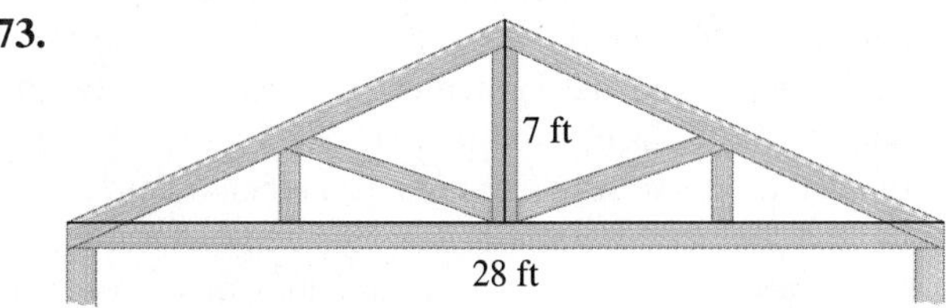

74.

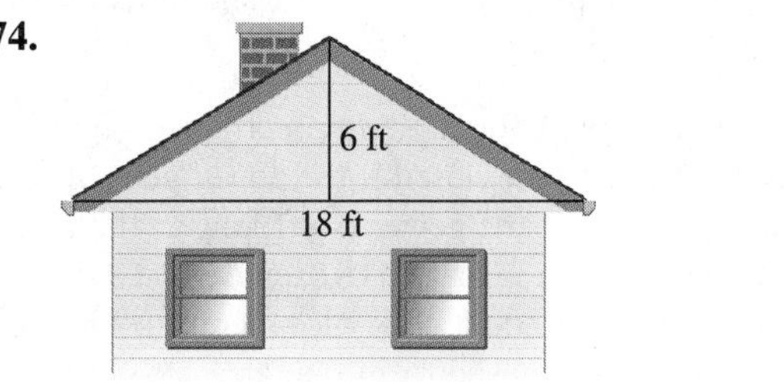

For the applications in Exercises 75 to 84, do the following:

a. *Make a table for inputs 0, 1, and 2.*

b. *Find the slope and write the meaning of the slope.*

c. *Write an equation describing the output as a function of the input. Define input and output variables as needed.*

75. The output is the total cost of g gallons of gasoline at \$1.55 per gallon.

76. The output is the total calories in n cookies at 65 calories per cookie.

77. The output is the total earnings for h hours at \$6.25 per hour.

78. The output is the total miles traveled in h hours at 55 miles per hour.

79. The output is the total kilometers traveled in h hours at 80 kilometers per hour.

80. The output is the total cost for x credit hours at \$100 per credit hour.

81. The output is the total cost per semester of x hours on the computer. Each semester, students are charged a \$3 fee plus \$1 for each hour.

82. The output is the total cost of a taxi ride of x miles. The taxi driver charges a \$2 fee plus \$3 for each mile traveled.

83. The output is the total cost of a meal, where x is the price of the meal and a tip of 15% of the price of the meal is left for the serving person.

84. The output is the total cost of a shirt, where x is the price of the shirt and a sales tax of 7% of the price is added.

85. Why is $f(0) \neq 0$ in Exercises 81 and 82?

86. Why is $f(0) = 0$ in Exercises 83 and 84?

87. Water freezes at 0° Celsius and 32° Fahrenheit. Water boils at 100°C and 212°F. Let the Celsius temperature C be the input and the Fahrenheit temperature F be the output. Write an ordered pair (C, F) for the temperature at which water freezes and another ordered pair for the temperature at which water boils. Assume the temperature scales are linear functions of each other. Using the ordered pairs, find the slope and units for the slope.

88. The owner's manual for a Kenmore freezer suggests leaving the freezer door closed if the power goes off for less than 24 hours. If the power is off longer, the manual recommends placing dry ice (frozen carbon dioxide) in the freezer. The amount of dry ice recommended for each 24-hour period is shown in the table. Assume the dry ice requirement is a linear function of the size of freezer. From the table, find the slope and units for the slope.

Size of Freezer, x	Dry Ice, y
5 ft^3	20 lb
15 ft^3	40 lb

89. What does the slope tell you about the graph of a line?

90. Explain how to tell if a line has a positive slope.

91. Explain how to tell if a line drawn from a table will have a negative slope.

92. Explain how to find slope given two ordered pairs.

93. Explain how to find slope from a graph.

94. Explain how the units on the axes give meaning to the slope of a line.

95. Explain how to draw a line for slope a/b.

96. ***Error Analysis.*** Explain what is wrong with using this expression to find the slope between (a, b) and (c, d): $\dfrac{b - a}{d - c}$.

97. ***Error Analysis.*** Explain what is wrong with using this expression to find the slope between (a, b) and (c, d): $\dfrac{c - a}{d - b}$.

98. In calculating the slope between (8, 2) and (4, 5), one student started with $(5 - 2)$ divided by $(4 - 8)$. Another student started with $(2 - 5)$ divided by $(8 - 4)$. Will they both obtain the correct slope? Explain why or why not.

99. Jane purchased a pick-up for \$30,000. The value of the pick-up over several years is shown in the table.

Year, x	Value, y
0	\$30,000
1	24,000
2	21,600
3	1,000

a. Draw a graph of the data.

b. What is the slope from year 0 to year 1?

c. What is the slope from year 1 to year 2?

d. What would be the meaning of a line drawn between (2, 21,600) and (2, 1000)?

e. What is the slope of the line from (2, 1000) to (3, 1000)?

f. Describe a likely story about Jane's pick-up.

100. ***Marriage Ages***

Year	Men	Women
1890	26.1	22.0
1900	25.9	21.9
1910	25.1	21.6
1920	24.6	21.2
1930	24.3	21.3
1940	24.3	21.5
1950	22.8	20.3
1960	22.8	20.3
1970	23.2	20.8
1980	24.7	22.0
1990	26.1	23.9

Median Age at First Marriage

Reprinted with permission from *The World Almanac and Book of Facts 1992.* Copyright © 1991. All rights reserved. The World Almanac is an imprint of Funk & Wagnalls Corporation.

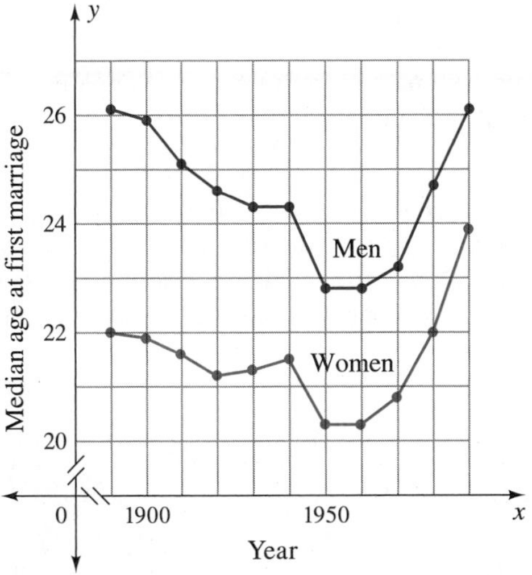

a. What is the meaning of the vertical distance between the graph of men's ages and the graph of women's ages? Where is it greatest? Where is it smallest?

b. What are some possible causes for the changes in age at first marriage?

c. In which decade (ten-year period) was there zero change for women?

d. In which decades did the marriage age decrease for men?

e. For which decades does the women's age graph have a positive slope?

f. For which decade does the women's age graph have the steepest negative slope? Find the slope.

g. For which decade does the men's age graph have the steepest positive slope? Find the slope.

h. What is the meaning of the slope in these graphs?

Projects

101. ***Slope and Scale on Axes.*** Although the graphs below appear to have the same slope, the three lines actually have different slopes because of the different scales on the axes. List the slope for each graph. Tell which graph—graph 1, 2, or 3—is most appropriate for each of the following input-output situations.

Graph 1:

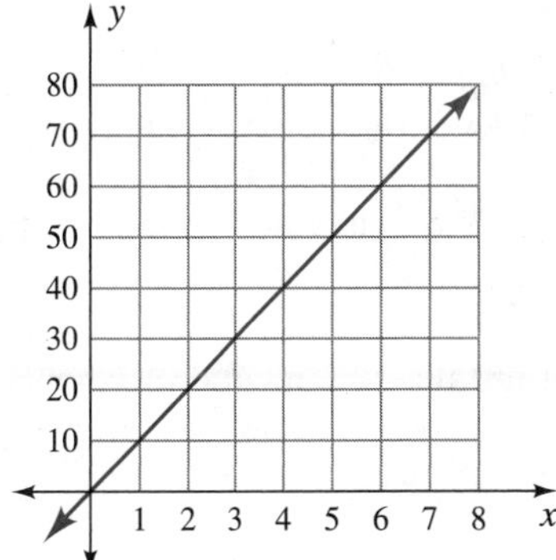

Graph 2:

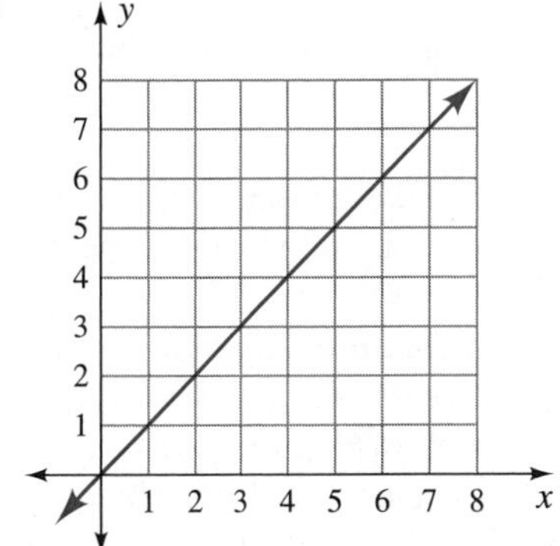

Graph 3:

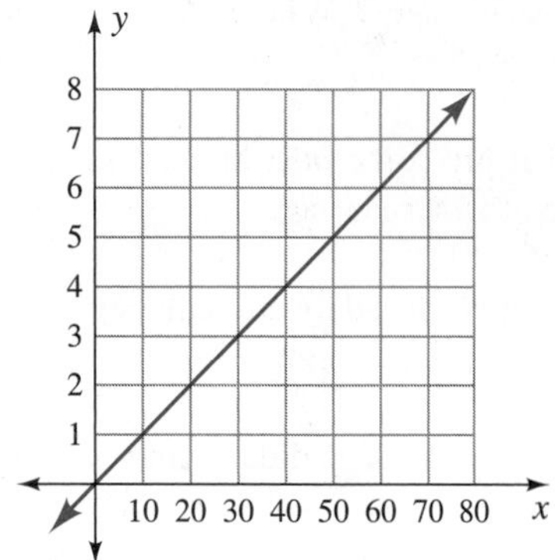

For parts a to c, assume the scale on the y-axis is in feet.

a. The height, y, of a wheelchair ramp of length x feet

b. The height, y, climbed up a rock cliff for x horizontal feet traveled

c. The height, y, climbed up a steep staircase for x horizontal feet traveled

For parts d to f, assume the scale on the vertical axis is in dollars.

d. The total cost, y, of x items at the dollar store

e. The total cost, y, of x pieces of candy at 10 cents each

f. The total cost, y, of x movie tickets at \$10 each

For parts g to i, assume the scale on the vertical axis is in miles.

g. The distance, y, traveled in x hours by the NASA space shuttle between the assembly building and the launch pad

h. The distance, y, traveled by bicycle in x hours

i. The distance, y, paved by a highway construction crew in x hours

Extension: Make up one description for each graph, stating your assumptions about the units on the axes.

102. ***Graphing Calculator Exploration: Comparing Slope and y-intercepts.*** For parts a to d, enter each set of equations into $\boxed{\text{Y} =}$, set a viewing window with both x and y on the interval $[-5, 5]$, and graph. Then list the equations in order from flattest graph to steepest graph. For parts c and d, predict your answers before graphing.

a. $y = 2x$, $y = 4x$, $y = \frac{1}{2}x$

b. $y = -3x$, $y = -x$, $y = -\frac{1}{3}x$

c. $y = 3x$, $y = 1.5x$, $y = x$, $y = 2.5x$

d. $y = -0.25x$, $y = -4x$, $y = -0.5x$, $y = -x$

For parts e to g, tell which two equations have the same slope and which two equations have the same y-intercept.

e. $y = 2x + 1$, $y = 2x - 1$, $y = -2x + 1$

f. $y = 3x + 3$, $y = -3x - 3$, $y = -3x + 3$

g. $y = -\frac{1}{2}x - 1$, $y = \frac{1}{2}x + 1$, $y = -\frac{1}{2}x + 1$

Parts h and i refer to equations written in $y = mx + b$ form.

h. Which part of the equation controls the slope?

i. Which part of the equation controls the y-intercept?

103. ***Staircase Slope.*** On a staircase, the riser is the vertical distance (rise) on each step and the tread is the horizontal distance (run) between two risers. Some steps have a slight overhang, which must be ignored in measuring the tread.

a. From memory, estimate the rise to run ratio for steps in a nearby staircase. Look at a ruler in making your estimates. Record your estimate first as a fraction and then as a decimal.

b. Measure and record the riser and the tread.

c. Write the slope for the stairs first as a fraction and then as a decimal.

d. Use your results in part b to estimate the total rise and run of the staircase.

e. Comment on how your estimate compares with the actual measurements.

Extension: Find five other staircases, in locations such as an office building, a tourist attraction, and a concert hall. Measure the riser and the tread, and calculate the slope of the stairs. Discuss the relationship between the steepness of the stairs and the purpose of the stairs.

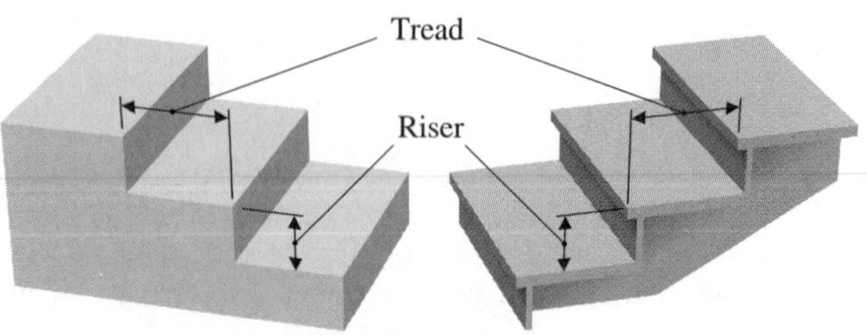

4.4 Linear Equations

OBJECTIVES

- Find a linear equation from a graph or ordered pairs.
- Find the slope and vertical axis intercept from a linear equation.
- Write a linear equation from a slope and vertical axis intercept.
- Write equations for parallel and perpendicular lines.
- Draw the graphs of horizontal, vertical, parallel, and perpendicular lines.

WARM-UP

Divide and simplify.

1. $\dfrac{-3x + 6}{2}$ **2.** $\dfrac{4x - 12}{3}$

3. $\dfrac{2x + 3}{3}$ **4.** $\dfrac{x - 2}{6}$

IN THIS SECTION, we return to linear functions. Thus far, we have been given equations in algebraic notation or in word statements. Now, our focus will be on writing equations when we are given the slope and vertical axis intercept from a graph, data from a table, or two ordered pairs. We will graph a variety of lines, including horizontal and vertical lines, parallel and perpendicular lines. We will identify the appropriate domain and range in applications of linear functions.

Building Linear Equations from Graphs

Given a slope and vertical axis intercept, we can build a linear equation. If the slope and intercept are not given, we can find them from a graph of the equation. We begin by exploring slope and y-intercepts. In Example 1, we multiply x by three different numbers and look at the effects on the graph.

EXAMPLE 1 Exploring equations and graphs

a. Which line in Figure 22 is the graph of each of the following: $y = x$, $y = 2x$, and $y = 5x$?

b. What is the slope of each graph?

c. Compare the steepness of the graphs.

d. What is the y-intercept point for each graph?

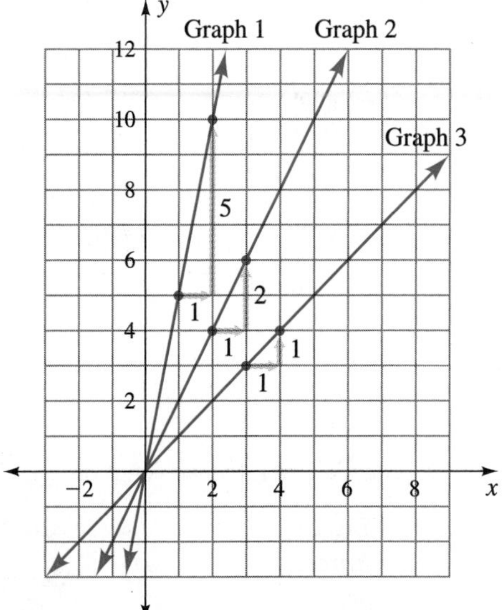

Figure 22

Solution **a.** $y = x$ is graph 3, $y = 2x$ is graph 2, and $y = 5x$ is graph 1.

The answers to b, c, and d are in the Answer Box. ●

In every linear function, the slope constant, $\Delta y/\Delta x$, multiplies x. Back in Section 1.3, we were finding slope in every table, but with $\Delta x = 1$.

Slope and the Linear Function

> The linear function always contains a multiplication of the input variable by the slope. The slope is represented by the letter m in $y = mx + b$.

Once we know the slope, m, we need one other number, b, to write the equation for a linear function. The source of the number b is most easily seen by comparing graphs of several linear equations with the same slope.

EXAMPLE 2 Exploring equations and graphs

a. Which line in Figure 23 is the graph of each of the following: $y = 2x$, $y = 2x + 1$, and $y = 2x - 3$?

b. What is the y-intercept of each equation?

c. What is the source of the number b in $y = mx + b$?

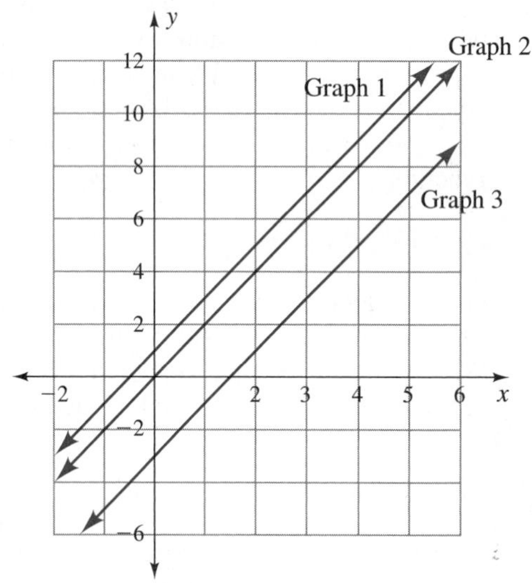

Figure 23

Solution **a.** $y = 2x$ is graph 2, $y = 2x + 1$ is graph 1, and $y = 2x - 3$ is graph 3.

See the Answer Box for the other answers. ●

We can summarize our findings about a linear equation as follows:

Equation of a Linear Function

> The equation for a linear function is $y = mx + b$, where the number m is the slope and the number b is the vertical axis intercept.

Because the vertical axis is frequently labeled y, we usually call b the y-intercept rather than the vertical axis intercept.

EXAMPLE 3

Finding an equation from a graph The data in Table 9, showing the value remaining on a prepaid commuter train ticket, are graphed in Figure 24.

a. What is the slope? What is its meaning?

b. What is the y-intercept point? What is its meaning?

c. What is the equation of the line?

Trips, x	Remaining Value, y
0	\$20.00
4	16.80
8	13.60
12	10.40
16	7.20
20	4.00

Table 9 Prepaid Transit Ticket

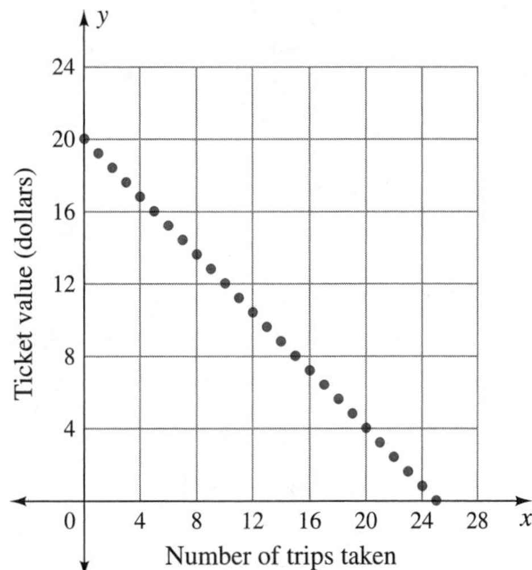

Figure 24

Solution **a.** The slope can be calculated from any two data points. We will use the two points (0, 20) and (20, 4):

$$\text{Slope} = \frac{y_2 - y_1}{x_2 - x_1} = \frac{\$4 - \$20}{20 - 0} = \frac{-\$16}{20} = -\$0.80 \text{ per trip}$$

The slope represents the fare for one trip. It is negative because the value remaining on the prepaid ticket decreases with each additional trip made.

b. The y-intercept point is (0, 20); \$20 is the initial cost of the ticket.

c. To find the equation, we substitute the slope and y-intercept into $y = mx + b$:

$$y = mx + b$$
$$y = -\$0.80x + \$20$$

Think about it 1: What are the domain (set of inputs) and range (set of outputs) for Example 3? How would the graph change if the prepaid ticket were for \$25 instead of \$20? How would the graph change if the cost of a trip changed from \$0.80 to \$0.90?

Finding Slope and y-intercept from Linear Equations

Some equations must first be solved for y in order to find the slope and y-intercept. In Example 4, we read slope and y-intercept from equations in $y = mx + b$ form.

EXAMPLE  Finding slope and y-intercept from an equation Find the slope and y-intercept for each equation, labeling your answers with m and b. First write the equation in $y = mx + b$ form, if it is not already in this form.

a. $y = -2x + 4$ **b.** $y = 2 - 4x$ **c.** $2y + 3x = 6$

d. $4x - 3y = 12$ **e.** $y = \$5.00 - \$0.05x$ **f.** $y = 40 + 25(x - 1)$

Solution **a.** $m = -2$, $b = 4$

b. $y = -4x + 2$; $m = -4$, $b = 2$

c. $2y + 3x = 6$ Subtract $3x$ from each side.

$2y = -3x + 6$ Divide each side by 2.

$y = \dfrac{-3x + 6}{2}$ Change the right side to $mx + b$.

$y = -\frac{3}{2}x + 3$

$m = -\frac{3}{2}$, $b = 3$

d. $4x - 3y = 12$ Subtract 12 from each side.

$4x - 12 - 3y = 0$ Add $3y$ to each side.

$4x - 12 = 3y$ Divide each side by 3.

$\dfrac{4x - 12}{3} = y$ Change the left side to $mx + b$.

$\frac{4}{3}x - 4 = y$

$m = \frac{4}{3}$, $b = -4$

e. $y = -\$0.05x + \$5.00$; $m = -\$0.05$, $b = \$5.00$

f. $y = 40 + 25(x - 1)$ Apply the distributive property.

$y = 40 + 25x - 25$ Combine like terms.

$y = 25x + 15$

$m = 25$, $b = 15$ ●

Even if linear equations do not contain x and y, we can still find the slope and vertical axis intercept. If the output is not identified, look for a variable that is a function of or depends on another variable. The output will equal the slope times the input variable plus the vertical axis intercept.

EXAMPLE  Identifying slope and y-intercept from an equation Find the slope and vertical axis intercept for each formula.

a. $C = 500n + 2000$ (The total cost depends on the number of items manufactured plus fixed costs.)

b. $c = 60 + 50h$ (The plumber's bill includes a fee to show up and a charge per hour.)
c. $C = \pi d$ (The circumference of a circle depends on the diameter.)
d. $h \approx 1.732s$ (The height of an equilateral triangle depends on the length of the side.)

Solution **a.** The slope is 500; the vertical axis intercept is 2000.
b. The slope is 50; the vertical axis intercept is 60.
c. The slope is π; the intercept is 0.
d. The slope is 1.732; the intercept is 0. ●

EXAMPLE 6 Finding an equation and graphing, given the slope and y-intercept For the slope and y-intercept listed, write a linear equation and then graph the line.

a. $m = -2, b = 3$ **b.** $m = \frac{1}{4}, b = -3$

Solution **a.** $y = -2x + 3$. To graph, we start with the intercept, $b = 3$, and plot the point (0, 3). Using the slope to find a second point on the graph, we move 2 units down and 1 unit to the right from the intercept. We draw a line through the two points.
b. $y = \frac{1}{4}x - 3$. To graph, we start with the intercept, $b = -3$, and plot the point (0, −3). Using the slope to find a second point on the graph, we move 1 unit up and 4 units to the right from the intercept. We draw a line through the two points.

The graphs of the lines are shown in Figure 25.

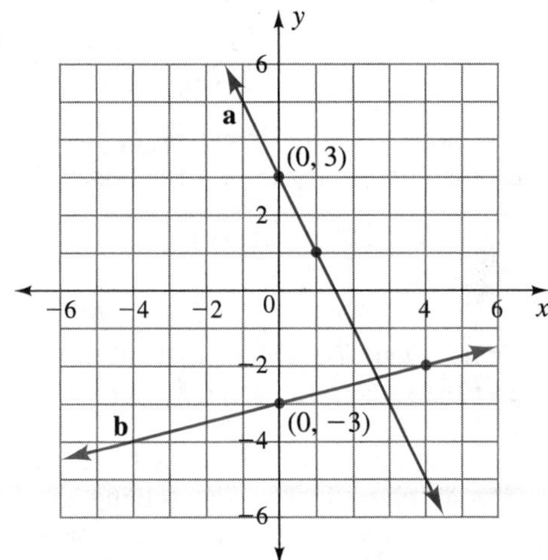

Figure 25 ●

Finding Linear Equations from Ordered Pairs

In Example 7, we find the equation of a line, given two ordered pairs. To write the linear equation $y = mx + b$, we need both m and b. We use the slope formula to find m from two ordered pairs. To find b, we substitute one ordered pair and the slope into $y = mx + b$ and then solve for b. We then substitute m and b into $y = mx + b$ to build an equation.

EXAMPLE 7 Finding an equation from ordered pairs The ordered pairs (−4, 2) and (3, −1) lie on a line.

a. Find the slope, m, of the line.
b. Find the y-intercept, b, of the line.
c. Write an equation for the line.

Algebraic Solution

a. Slope $= \dfrac{y_2 - y_1}{x_2 - x_1} = \dfrac{-1 - 2}{3 - (-4)} = \dfrac{-3}{7}$

b. To find the y-intercept, we substitute the slope and one of the ordered pairs into $y = mx + b$.

$$y = mx + b \quad \text{Substitute } m = -\tfrac{3}{7} \text{ and } (x, y) = (-4, 2).$$
$$2 = -\tfrac{3}{7}(-4) + b \quad \text{Simplify.}$$
$$2 = \tfrac{12}{7} + b \quad \text{Subtract } \tfrac{12}{7} \text{ from both sides.}$$
$$2 - \tfrac{12}{7} = b \quad \text{Simplify.}$$
$$\tfrac{14}{7} - \tfrac{12}{7} = b$$
$$b = \tfrac{2}{7}$$

c. We substitute the slope and y-intercept into $y = mx + b$:

$$y = mx + b \quad \text{Substitute } m = -\tfrac{3}{7} \text{ and } b = \tfrac{2}{7}.$$
$$y = -\tfrac{3}{7}x + \tfrac{2}{7}$$

Graphing Calculator Solution

We can place the ordered pairs in the lists provided under the statistics function and then find the equation for the straight line with the linear regression option. Here's how it's done:

To start, clear old data as needed.

Let x be in list 1 and y in list 2. Enter the data points $(-4, 2)$ and $(3, -1)$.

Choose the linear regression option under the statistical calculations.

The regression gives the slope and the y-intercept:

$$a \approx -0.43 \qquad \text{and} \qquad b \approx 0.29$$

The regression equation must be written in the form $y = ax + b$:

$$y = -0.43x + 0.29$$

The slope, a, and y-intercept, b, are listed under statistical variables and may be recalled and changed into fractions, to obtain

$$y = \frac{-3x}{7} + \frac{2}{7}$$

●

Think about it 2: Show that substituting $(3, -1)$ in part b of Example 7 would give the same y-intercept, $\tfrac{2}{7}$, as $(-4, 2)$ did.

In Example 7, we substituted one ordered pair and the slope into $y = mx + b$ and solved for b. The solving step may be avoided if we first solve $y = mx + b$ for b to create a formula for the y-intercept, as shown in Example 8. (Of course, it is always an option to use linear regression on the graphing calculator.)

EXAMPLE 8 Finding a shortcut for calculating b Make a formula for the y-intercept.

Solution We can solve $y = mx + b$ directly for b:

$$y = mx + b \quad \text{Subtract } mx \text{ from both sides.}$$
$$y - mx = b$$

●

The y-intercept Formula

The vertical axis intercept, b, for a linear equation is

$$b = y - mx$$

where m is the slope and (x, y) is any point on the line.

We now find a linear equation from application data.

EXAMPLE 9 Finding an equation from ordered pairs The owner's manual for a Kenmore freezer recommends placing dry ice in the freezer if the power is off for longer than 24 hours. The data are shown in Table 10. Assume the data are linear.

Size of Freezer, x	Dry Ice, y
5 ft^3	20 lb
15 ft^3	40 lb

Table 10 Emergency Cooling

a. Find a linear equation to calculate the amount of dry ice needed.

b. Graph the data.

Solution **a.** First we find the slope:

$$\text{Slope} = m = \frac{(40 - 20)\text{ lb}}{(15 - 5)\text{ ft}^3} = \frac{20\text{ lb}}{10\text{ ft}^3} = 2\text{ lb per ft}^3$$

Then we substitute $m = 2$ and (5, 20) into the equation:

$$b = y - mx$$
$$b = 20 - 2(5)$$
$$b = 10$$

Next we substitute $b = 10$ into $y = 2x + b$ to get the equation relating x cubic feet to y pounds of dry ice needed for each 24-hour period: $y = 2x + 10$.

b. Figure 26 shows that the graph of the line through (5, 20) and (15, 40) passes through the y-intercept point, (0, 10), which confirms our symbolic work.

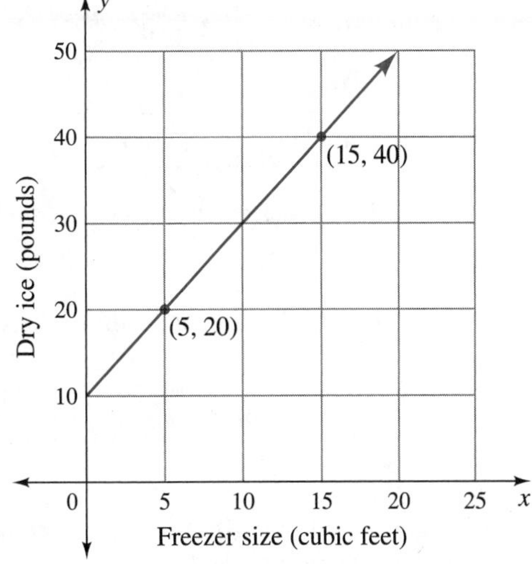

Figure 26

Think about it 3: In Example 9, what is the x-intercept? Does it have any meaning? What is the y-intercept? Does it have any meaning?

EXAMPLE *10* Finding an equation from ordered pairs Find the equation of a line passing through (0, 8) and (3, 5).

Solution Slope:

$$m = \frac{y_2 - y_1}{x_2 - x_1} = \frac{5 - 8}{3 - 0} = \frac{-3}{3} = -1$$

y-intercept:

$$b = y - mx \qquad \text{Let } m = -1 \text{ and } (x, y) = (3, 5).$$

$$b = 5 - (-1)3 = 8$$

Equation:

$$y = mx + b \qquad \text{Let } m = -1 \text{ and } b = 8.$$

$$y = -1x + 8$$

●

Think about it 4: Could we have written the equation immediately after finding slope?

Always look to see if the vertical axis intercept is given in the problem. This will save you time in finding a linear equation.

EXAMPLE  Finding an equation from ordered pairs: temperature conversion revisited Table 11 gives the two coordinate points relating Celsius and Fahrenheit temperatures. Using $y = mx + b$, with $x = C$ for Celsius and $y = F$ for Fahrenheit, find the equation of the line $F = mC + b$.

Student Note: In cases where there is no clear dependence between variables, assign variables alphabetically.

	Input (°C)	**Output (°F)**
Water freezes	0	32
Water boils	100	212

Table 11 Celsius and Fahrenheit

The slope of the line containing these data is

$$\text{Slope} = m = \frac{212 - 32}{100 - 0} = \frac{180}{100} = \frac{9}{5}$$

Solution The ordered pair (0, 32) is the vertical axis intercept point. We substitute slope $m = \frac{9}{5}$ and intercept $b = 32$ into the equation $F = mC + b$, to obtain $F = \frac{9}{5}C + 32$.

●

In summary:

Building a Linear Equation

To find the equation of a line from two ordered pairs (x_1, y_1) and (x_2, y_2):

1. Find the slope, using
$$m = \frac{y_2 - y_1}{x_2 - x_1}$$
2. Find the y-intercept, using $b = y - mx$, the slope, and either ordered pair.
3. Substitute the numbers for m and b into $y = mx + b$.
4. Use another ordered pair, if available, to check.

Horizontal and Vertical Lines

As Figure 27 reminds us,

- All horizontal lines have constant outputs.
- All vertical lines have constant inputs.

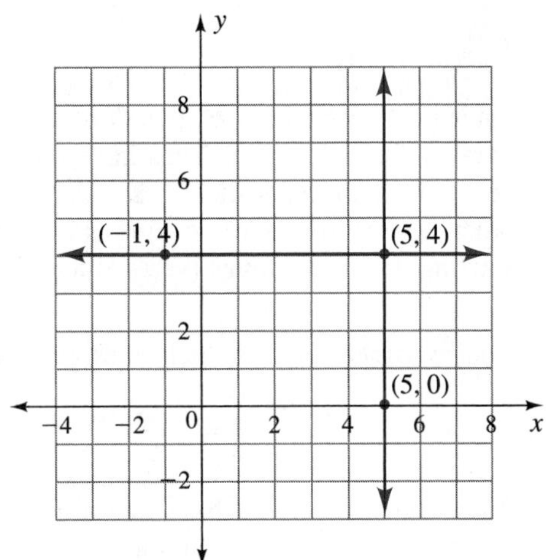

Figure 27

In Example 12, we find the equation of a horizontal line. We will then use the results from Example 12 to state the general equations for both horizontal and vertical lines.

EXAMPLE **12** Finding the equation of a horizontal line Find the equation of the line through $(-1, 4)$ and $(5, 4)$, shown in Figure 27.

Solution To find the equation, we first find the slope and y-intercept. The slope is

$$m = \frac{y_2 - y_1}{x_2 - x_1} = \frac{4 - 4}{5 - (-1)} = \frac{0}{6} = 0$$

From the graph in Figure 27, the y-intercept is $b = 4$.

Next we find the equation:

$y = mx + b$ Substitute $m = 0$ and $b = 4$.

$y = 0x + 4$ Simplify.

$y = 4$ ●

Example 12 suggests that the equation of a horizontal line is found by setting y equal to the second number in any of the ordered pairs forming the line.

Because the slope of a vertical line is undefined, we are unable to use the linear equation $y = mx + b$ to find its equation. However, the ordered pairs on the graph of vertical line all have the same first number. In Figure 27, the ordered pairs have $x = 5$. The equation for the vertical line passing through (5, 0) on the horizontal axis is $x = 5$.

Equations of Horizontal and Vertical Lines

> The equation of a horizontal line is $y = b$, where (x, b) is any point on the line.
>
> The equation of a vertical line is $x = a$, where (a, y) is any point on the line.

Parallel Lines

Knowing the position of the slope number in an equation helps us to identify parallel lines. **Parallel lines** have *the same slope but different y-intercepts.* In geometry, we use parallel lines to determine the nature or properties of shapes. In other applications, parallel lines indicate the same rate of change but different starting or initial values (vertical axis intercepts).

EXAMPLE  Finding parallel lines Which lines are parallel?

a. $2x + y = 4$ **b.** $y - 2x = 3$ **c.** $\frac{1}{2}y - x = 5$ **d.** $\frac{1}{2}y + x = 1$

Solution The parallel lines will have the same slope. To compare the slopes of the lines, we must change each equation into an equation of the form $y = mx + b$.

a. $2x + y = 4$ Subtract $2x$ from both sides.

$y = -2x + 4$ The slope is -2.

b. $y - 2x = 3$ Add $2x$ to each side.

$y = 2x + 3$ The slope is 2.

c. $\frac{1}{2}y - x = 5$ Add x on each side.

$\frac{1}{2}y = x + 5$ Multiply each side by 2.

$y = 2x + 10$ The slope is 2.

d. $\frac{1}{2}y + x = 1$ Subtract x from each side.

$\frac{1}{2}y = -x + 1$ Multiply each side by 2.

$y = -2x + 2$ The slope is -2.

The lines in parts a and d have slope -2 and are parallel. The lines in parts b and c have slope 2 and are parallel. ●

EXAMPLE 14 Applying parallel lines: photocopy costs revisited In Example 11 of Section 4.3, the prepaid photocopy machine card cost \$5.00 and each photocopy cost \$0.05. Figure 28 shows a graph for the card's value after photocopies have been made. Suppose the card value changes from \$5.00 to \$10.00 but the copies remain the same price.

a. What is the new equation for the card's value?

b. Why does the graph (see Figure 29) change?

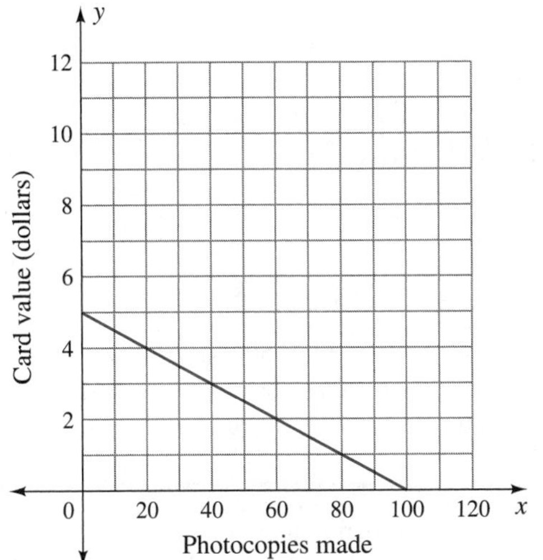

Figure 28

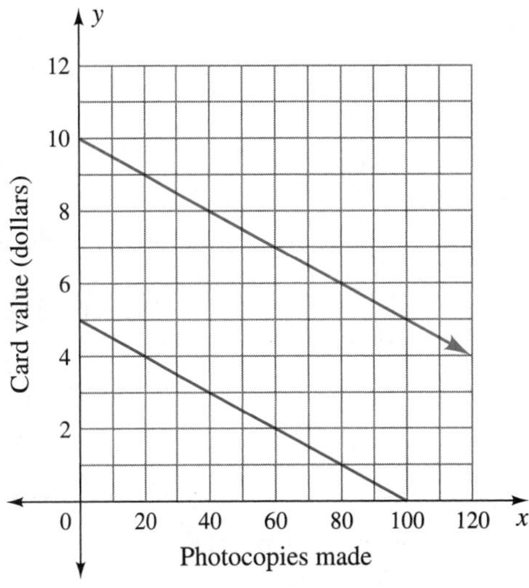

Figure 29

Solution **a.** The slope, or rate, remains \$0.05 per copy. The initial cost is \$10.00 instead of \$5.00. The equation is therefore $y = -0.05x + 10.00$.

b. The graph, in Figure 29, shifts up because the y-intercept is now \$10.00. The new graph is parallel to the old graph. ●

Perpendicular Lines

Two lines that cross at a right angle are **perpendicular lines.** In designing computer graphics and in finding areas of rectangles and triangles, we use perpendicular lines to show that right angles are formed. Every pair of horizontal and vertical lines is perpendicular. Example 15 suggests the numerical relationship between the slopes of perpendicular lines.

EXAMPLE 15 Exploring perpendicular lines Lines AO and BO in Figure 30 (page 242) intersect at the origin.

a. What is the slope of AO?

b. What is the slope of BO?

c. On a separate piece of paper, trace the axes and lines AO and BO. Place the traced figure over the book so that it matches the original figure. Press with a pencil point at O while turning the x-axis 90° (a right angle) counterclockwise. Describe the new positions of the positive x-axis and AO.

d. What can you conclude about the lines AO and BO?

e. What can you conclude about the slopes of perpendicular lines?

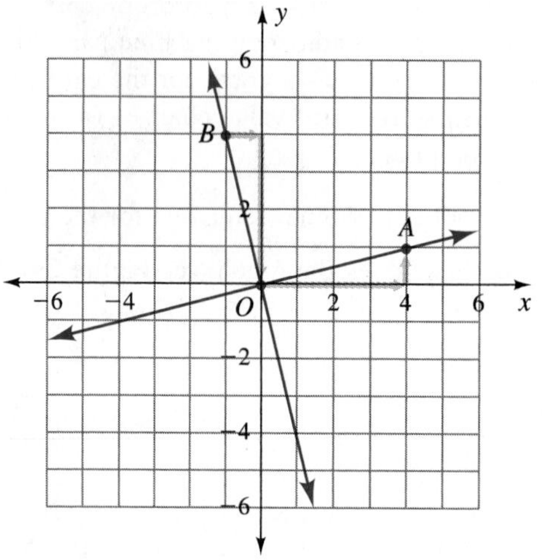

Figure 30

Solution The answers are in the Answer Box.

Our exploration suggests that the lines in Figure 30 are perpendicular. The slopes of the lines were opposite reciprocals. Recall that reciprocals multiply to 1. These ideas suggest the following conclusion.

Perpendicular Lines

> Two lines are perpendicular if
>
> **1.** their slopes multiply to -1 (that is, their slopes are opposite reciprocals) or
>
> **2.** their slopes are the same as those of the horizontal and vertical axes.

EXAMPLE 16 Finding slopes of lines perpendicular to a given line What would be the slope of a line perpendicular to each of the following?

a. $y = 3x + 2$ **b.** $y = \frac{1}{2}x - 3$ **c.** $y = -1.5x$ **d.** $y = 3$

Solution

a. The slope of $y = 3x + 2$ is 3. The slope of a perpendicular line would be the opposite reciprocal of 3: $-\frac{1}{3}$.

b. The slope of $y = \frac{1}{2}x - 3$ is $\frac{1}{2}$. The slope of a perpendicular line would be the opposite reciprocal of $\frac{1}{2}$: $-\frac{2}{1}$, or -2.

c. The slope of $y = -1.5x$ is -1.5, or $-\frac{3}{2}$. The opposite reciprocal is $\frac{2}{3}$.

d. The line $y = 3$ is a horizontal line with slope 0. Any vertical line $x = a$ would be perpendicular. The slopes of all vertical lines are undefined. ●

EXAMPLE 17 Using slope to discover geometric properties A square is placed on the coordinate axes (see Figure 31), with one corner at the origin and two sides along the horizontal and vertical axes. The length of each side is n, as indicated by the coordinates at points A, B, C, and D. Two *diagonal* lines connect the corners: AC and BD.

a. Find the slopes of the diagonals.

b. What can you conclude about the diagonals of a square?

Solution

a. Slope of $AC = \dfrac{y_2 - y_1}{x_2 - x_1} = \dfrac{n - 0}{n - 0} = \dfrac{n}{n} = 1$

Slope of $BD = \dfrac{y_2 - y_1}{x_2 - x_1} = \dfrac{0 - n}{n - 0} = \dfrac{-n}{n} = -1$

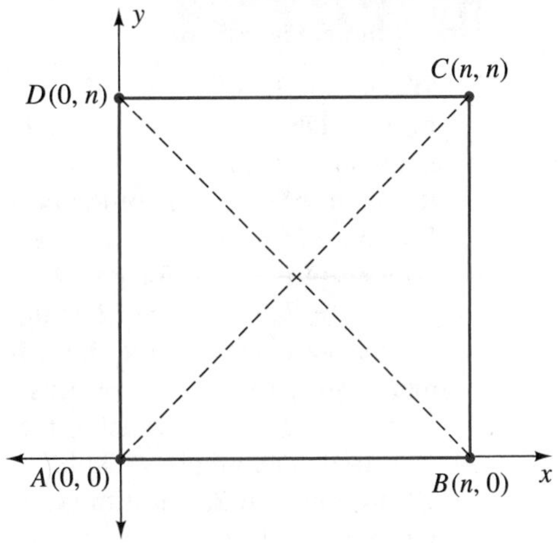

Figure 31

b. The slopes are opposite reciprocals and multiply to -1. The diagonals are perpendicular. Because n is a variable, we may conclude that *the diagonals of any square are perpendicular.*

We now combine several ideas by finding equations of lines given information about slopes and/or intercepts. In some problems, we use the given information to find the slope or y-intercept. In other problems, the required equation is that for a vertical line in the form $x = a$.

EXAMPLE 18 Writing linear equations Find the equation of a line with the given characteristics.

a. slope 0 and y-intercept at -1

b. parallel to $y = 3x - 4$ and y-intercept at -2

c. parallel to $y = -\frac{1}{2}x + 5$ and passing through the origin

d. slope is undefined and x-intercept at -1

e. perpendicular to $y = 3x - 4$ and containing $(0, 5)$

f. perpendicular to $y = -\frac{2}{5}x + 2$ and with y-intercept point $(0, -3)$

Solution **a.** Let $m = 0$ and $b = -1$ in $y = mx + b$; $y = 0x - 1$; $y = -1$.

b. Because the line is parallel, the slope is the same: $m = 3$. The y-intercept is $b = -2$. The line is $y = 3x - 2$.

c. Because the line is parallel, the slope is the same: $m = -\frac{1}{2}$. The origin lies on the y-axis, and so $b = 0$. The line is $y = -\frac{1}{2}x + 0$, or $y = -\frac{1}{2}x$.

d. Because the slope is undefined, the line is vertical. If the line has an x-intercept at -1, then $x = -1$ at that point and the equation is $x = -1$.

e. A line perpendicular to $y = 3x - 4$ has slope equal to the opposite reciprocal of 3, which is $-\frac{1}{3}$. The point $(0, 5)$ is on the y-axis, so $b = 5$. The equation is $y = -\frac{1}{3}x + 5$.

f. A line perpendicular to $y = -\frac{2}{5}x + 2$ has a slope of $\frac{5}{2}$. The y-intercept point is $(0, -3)$, so $b = -3$. The line is $y = \frac{5}{2}x - 3$.

ANSWER BOX

Warm-up: 1. $-\frac{3}{2}x + 3$ **2.** $\frac{4}{3}x - 4$ **3.** $\frac{2}{3}x + 1$ **4.** $\frac{1}{6}x - \frac{1}{3}$ **Example 1: b.** The slopes are 1 for $y = x$, 2 for $y = 2x$, and 5 for $y = 5x$. **c.** The graph of $y = x$ is the flattest, and the graph of $y = 5x$ is the steepest. **d.** All three graphs have the origin (0, 0) as the y-intercept point. **Example 2: b.** For $y = 2x$, the y-intercept is 0; for $y = 2x + 1$, the y-intercept is 1; and for $y = 2x - 3$, the y-intercept is -3. **c.** In $y = mx + b$, the number b is the y-intercept and $(0, b)$ is the y-intercept point. **Think about it 1:** The domain is positive integers representing the number of trips. The range is multiples of \$0.80. The graph would move parallel to the old graph if the prepaid value increased. The graph would be steeper if the cost per trip increased. **Think about it 2:** Substituting $m = -\frac{3}{7}$ and $(x, y) = (3, -1)$ into $y = mx + b$ gives $-1 = -\frac{3}{7}(3) + b$. Solving for b and simplifying, we get $b = \frac{2}{7}$. **Think about it 3:** Neither the x-intercept (-5 cubic feet) nor the y-intercept (10 pounds) has any meaning; clearly a freezer of zero cubic feet would require no dry ice. **Think about it 4:** Yes; solving for the y-intercept is not necessary because (0, 8) is the y-intercept point. **Example 15: a.** $\frac{1}{4}$ **b.** $-\frac{4}{1}$ **c.** The positive x-axis should lie over the original positive y-axis. The line AO now matches BO. **d.** Because AO and BO matched when we turned our traced figure 90°, the exploration suggests that the lines cross at a right angle and so are perpendicular. **e.** The slopes, $\frac{1}{4}$ and $-\frac{4}{1}$, are both opposite in sign and reciprocals of each other.

EXERCISES 4.4

In Exercises 1 to 10, find the slope and y-intercept for each graph and write an equation. Note the labeling on the axes in selecting your variables.

1.

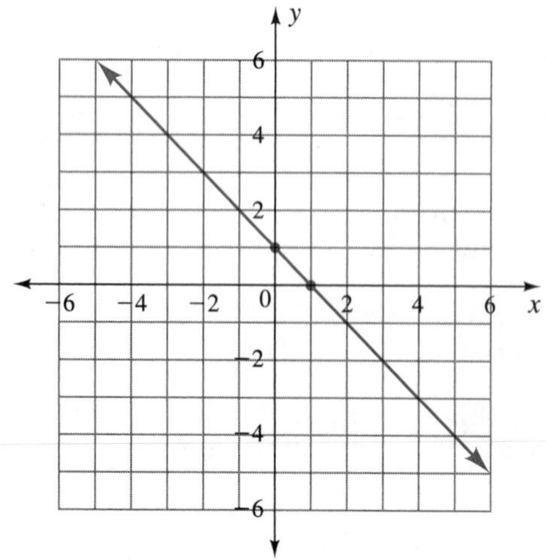

2.

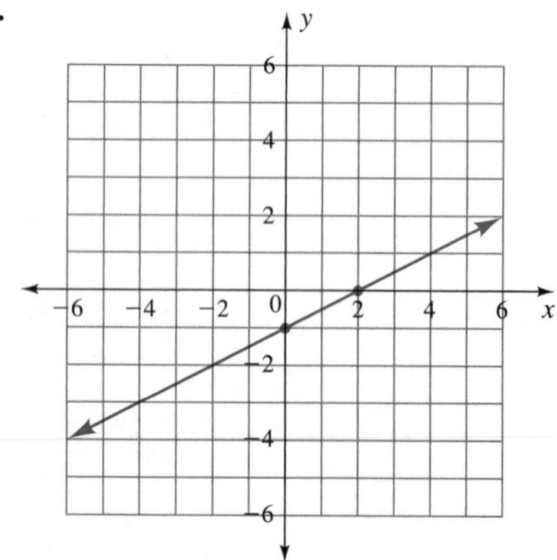

3.

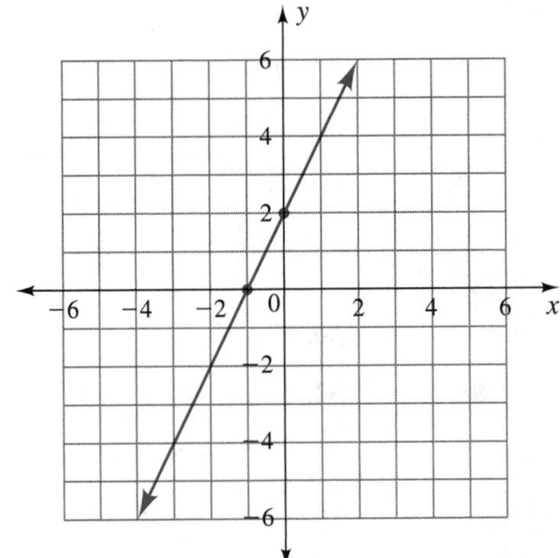
y
x
6
4
2
−2
−4
−6
−6
−4
−2
0
2
4
6

4.

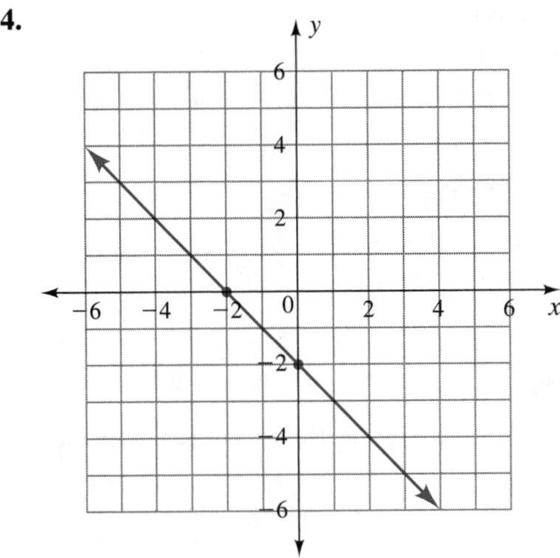
y
x
6
4
2
−2
−4
−6
−6
−4
−2
0
2
4
6

5.

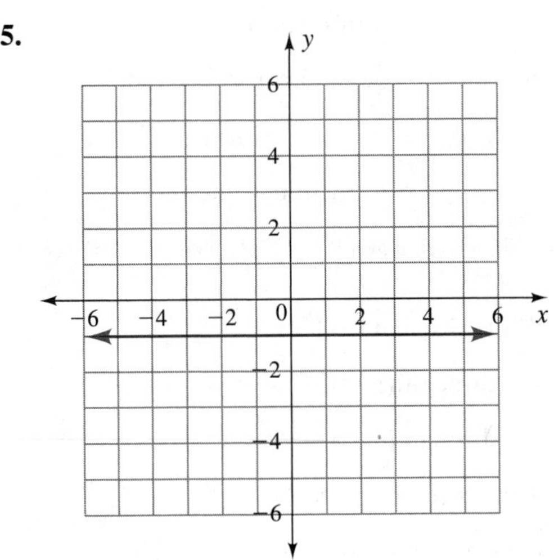
y
x
6
4
2
−2
−4
−6
−6
−4
−2
0
2
4
6

6.

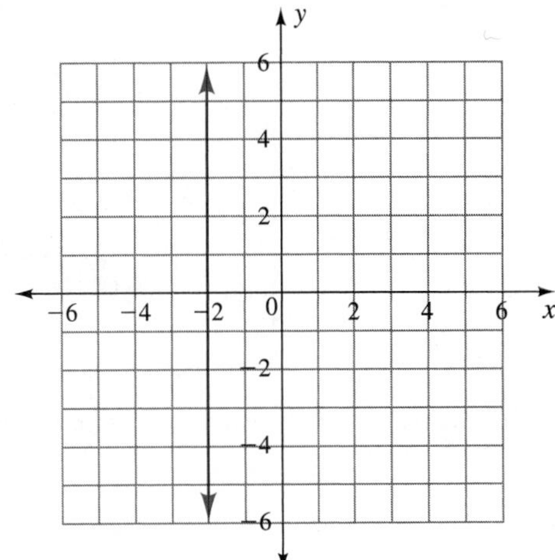
y
x
6
4
2
−2
−4
−6
−6
−4
−2
0
2
4
6

7.

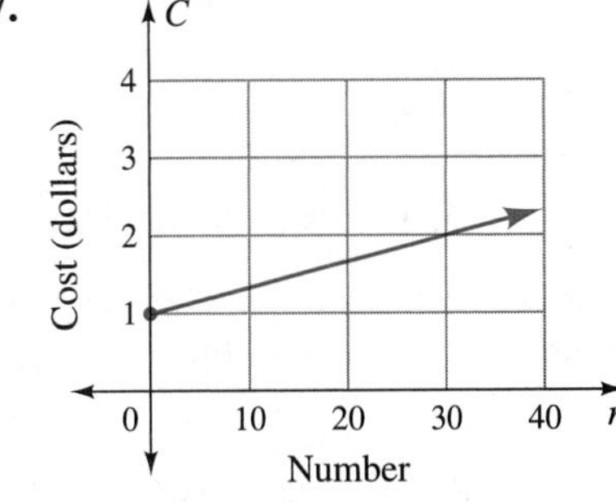
C
Cost (dollars)
4
3
2
1
0
10
20
30
40
n
Number

8.

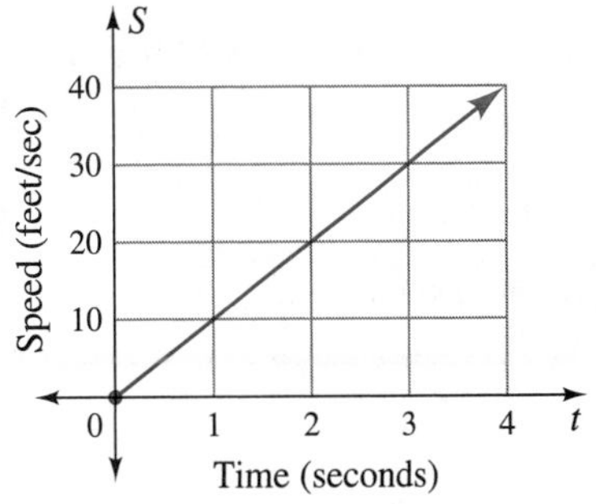

S
Speed (feet/sec)
40
30
20
10
0
1
2
3
4
t
Time (seconds)

9.

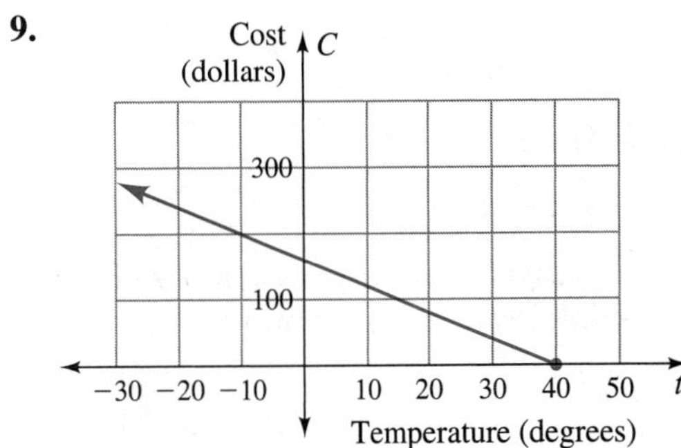
Cost (dollars)
C
300
100
−30
−20
−10
10
20
30
40
50
t
Temperature (degrees)

10.

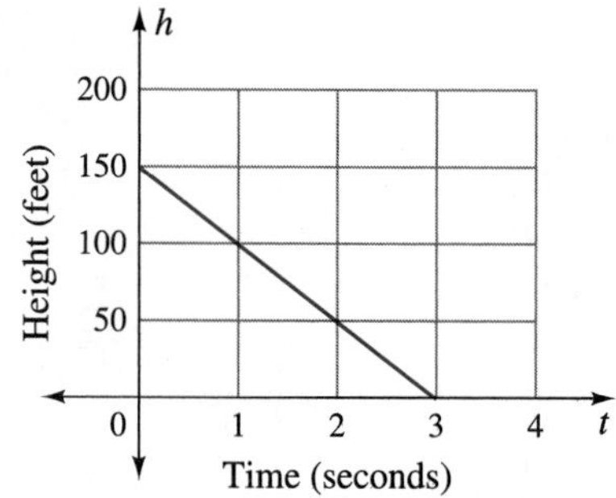

In Exercises 11 to 26, find the slope and y-intercept for each equation.

11. $y = 2x - \frac{1}{2}$ **12.** $y = 3x - \frac{4}{3}$

13. $y = 15 - 4x$ **14.** $y = 4 - 3x$

15. $y = -\frac{3}{4}x$ **16.** $y = \frac{1}{2}x$

17. $2x = y + 4$ **18.** $3x = y - 6$

19. $2x + 3y = 12$ **20.** $3y + 5x = 15$

21. $5y - 2x = 10$ **22.** $x + 3y = 6$

23. $x - 4y = 4$ **24.** $4x - 3y = 24$

25. $y = 12 - 0.30x$ **26.** $y = (x - 500) + 50$

In Exercises 27 to 38, find the slope and vertical axis intercept for each equation. Assume that the output variable is on the left.

27. $D = 55t$ **28.** $C = 2\pi r$

29. $C = 8 + 2\pi r$ **30.** $D = 225 + 45t$

31. $p = 2.98n + 0.50$ **32.** $c = 0.08p + 25$

33. $V = 50 - 0.29n$ **34.** $V = 20 - 0.19n$

35. $H = 0.8(200 - A)$ **36.** $R = 0.6(220 - A)$

37. $C = 65 + 0.15(d - 100)$

38. $c = 30 + 10(h - 4)$

In Exercises 39 to 46, write an equation for the given data.

39. $m = \frac{1}{2}$, $b = 3$

40. $m = \frac{3}{2}$, $b = 4$

41. slope $= \frac{2}{3}$, y-intercept $= -2$

42. slope $= -4$, y-intercept $= \frac{1}{2}$

43. slope $= 5$, y-intercept $= \frac{1}{4}$

44. slope $= \frac{3}{4}$, y-intercept $= -3$

45. $m = -\frac{3}{2}$, $(0, 1)$

46. $m = \frac{4}{5}$, $(0, -2)$

In Exercises 47 to 58, a slope and ordered pair are given. Write a linear equation having the slope and containing the given point.

47. $m = 4$, $(3, -1)$ **48.** $m = -2$, $(2, -3)$

49. $m = -1$, $(4, -2)$ **50.** $m = 5$, $(1, 3)$

51. $m = \frac{1}{2}$, $(2, 4)$ **52.** $m = \frac{2}{3}$, $(6, -2)$

53. $m = \frac{4}{5}$, $(-10, 3)$ **54.** $m = \frac{1}{3}$, $(-9, 1)$

55. $m = \frac{5}{3}$, $(-3, 1)$ **56.** $m = \frac{5}{2}$, $(-4, 2)$

57. $m = -2$, $(1.5, 3)$ **58.** $m = 2$, $(-2.5, 4)$

In Exercises 59 to 70, write an equation for the line passing through the two given ordered pairs.

59. $(1, 1)$, $(3, 9)$ **60.** $(2, 3)$, $(4, 7)$

61. $(2, -2)$, $(5, -8)$ **62.** $(-2, -4)$, $(1, 2)$

63. $(-3, 1)$, $(0, -1)$ **64.** $(-4, -2)$, $(0, 5)$

65. $(13, 6)$, $(10, 0)$ **66.** $(3, 6)$, $(0, 0)$

67. $(-5, 6)$, $(-4, -2)$ **68.** $(3, -4)$, $(-1, 4)$

69. $(5, 2)$, $(3, 3)$ **70.** $(2, 3)$, $(7, 1)$

In Exercises 71 to 74, four equations are shown. Write each equation in $y = mx + b$ form if it is not already in this form. Which lines are parallel? Which are perpendicular?

71. a. $y = 2x + 3$ **b.** $y = \frac{1}{2}x + 3$
c. $y = -\frac{1}{2}x + 3$ **d.** $2y = x + 4$

72. a. $y = -\frac{1}{3}x + 2$ **b.** $y = 3x + 4$
c. $y = -3x + 2$ **d.** $3x + y = 4$

73. a. $y = \frac{1}{3}x - 4$ **b.** $x - \frac{1}{3}y = 6$
c. $3y - x = 2$ **d.** $y + \frac{1}{3}x = 4$

74. a. $2y + x = 4$ **b.** $1 + y = \frac{1}{2}x$
c. $3 = \frac{1}{2}x - y$ **d.** $-2x = 1 - y$

In Exercises 75 to 82, sketch a graph and find an equation of each line, as described.

75. parallel to $y = 4x + 1$ through the origin

76. perpendicular to $y = 4x + 1$ through $(0, 5)$

77. perpendicular to $y = 2x - 3$ through the origin

78. parallel to $y = 2x - 3$ through $(0, -1)$

79. parallel to $y = \frac{1}{3}x - 5$ through $(0, 4)$

80. perpendicular to $y = \frac{1}{3}x - 5$ through the origin

81. perpendicular to $y = -\frac{3}{4}x + 2$ through $(0, -2)$

82. parallel to $y = -\frac{3}{4}x + 2$ through the origin

In Exercises 83 to 90, sketch a graph and find an equation of each line, as described.

83. horizontal line through $(-2, 4)$

84. vertical line through $(3, 5)$

85. vertical line through $(4, 3)$

86. horizontal line through $(-5, 2)$

87. vertical line through $(3, 0)$

88. horizontal line through $(0, -3)$

89. horizontal line through $(4, 0)$

90. vertical line through $(0, -4)$

In Exercises 91 to 94, answer the following questions.

a. *Which fact gives the slope?*

b. *Which fact gives the y-intercept?*

c. *Write the equation using* $y = mx + b$.

91. Hwang prepays $50 on racquetball court rental of $2 per hour. The equation describes the prepaid amount that remains after x hours of rental time.

92. Yolanda's $500 monthly expense account is set up through an automatic teller machine (ATM). She withdraws funds, using the $40 Fast Cash Option. The equation describes the amount that remains in her account after x withdrawals during the month.

93. Carmen rents a Cessna 152 for $42 per hour plus a $28 insurance fee. The equation describes the total cost of x hours flying time.

94. Alberto earns a weekly salary of $250 plus 10% of his sales volume. The equation describes the total weekly earnings for x dollars in sales.

95. What words identify the slope in Exercises 91 to 94?

96. What words identify the y-intercept in Exercises 91 to 94?

In Exercises 97 to 102, will the described change give a parallel line or a steeper line? Write the new equation.

97. The total rental cost on a car is $C = \$0.25n + \$35$, where n is in miles. The fixed cost increases from $35 to $45.

98. The monthly cost of water is $C = \$0.65g + \$5.15$, where g is in thousands of gallons. The basic charge for water service rises $2, but the cost per thousand gallons remains $0.65.

99. The total cost of gasoline is $C = \$1.50g$, where g is in gallons. The gasoline rises in price by $0.10 per gallon.

100. The value remaining on a transit ticket is $V = \$20 - \$0.80n$, where n is the number of rides. Rosa buys a $30 ticket instead of a $20 ticket.

101. The total monthly cost of a loan is $C = 0.01x + \$3$, where x is the number of dollars borrowed. The service fee rises from $3 to $5.

102. The distance traveled at 40 miles per hour is $d = 40t$, where t is in hours. Duane increases his speed by 10 miles per hour.

103. In the late afternoon, a 7-minute call to Sweden costs $6.01. On another afternoon, an 8-minute call costs $6.79. Assume that the cost of a call is linear and x is the time in minutes. Find a linear equation that gives the cost of an afternoon call.

104. One Sunday, a 22-minute call to Sweden cost $23.71. On another Sunday, a 31-minute call cost $33.16. Assume that the cost of a call is linear and x is the time in minutes. Find a linear equation that gives the cost of a Sunday call.

105. ***Credit Card Payment Schedule.*** The following points mark the end of line segments in a credit card payment graph:

$A(0, 0)$, $B(20, 20)$, $C(200, 20)$, $D(500, 50)$, $E(600, 150)$

For each segment AB, BC, CD, and DE, do the following.

a. Find the slope.

b. Find the vertical axis intercept. Does it have meaning in the problem setting?

c. Find the equation.

d. What is the domain of the segment?

e. What is the range of the segment? Are any of the segments parallel? Why?

106. Experiments show that a heat pump has an output of 36,000 Btu/hr at 48°F outside temperature. The same heat pump has an output of 15,000 Btu/hr at 18°F outside temperature. Assume that the output in Btu's/hr is linear for input temperatures between 18°F and 48°F.

a. Find an equation that gives the Btu/hr output at any temperature output.

b. Why might there be limitations on the inputs for this equation?

c. What is the y-intercept, and does it have any meaning?

d. What is the x-intercept, and does it have any meaning?

For Exercises 107 to 112, explain how to find the given expression(s).

107. a linear equation from two ordered pairs

108. a linear equation from its graph

109. the slope and y-intercept from $y = cx + d$

110. the slope and y-intercept from $ax + cy = d$

111. the slope of a line perpendicular to $y = \frac{ax}{b} + d$

112. the equation of a vertical line

113. ***Slope and Intercepts***

a. Use the slope formula to find the slope of the line through $(a, 0)$ and $(0, b)$.

b. Explain how to find the slope of a line, given its x- and y-intercept points.

c. Try your rule on $(-5, 0)$ and $(0, -4)$. Check the slope another way.

Projects

114. ***Calculator Regression***

Eyelets (pairs)	Length (inches)
2	21
4	24
6	27
8	30

Kiwi Brand Dress-Shoe Lace Chart

Eyelets (pairs)	Length (inches)
6	40
7	45
8	54
9	63
10	63
11 & over	72

Kiwi Brand Boot Lace Chart

a. Is the dress-shoe lace length a function of the pairs of eyelets?

b. Fit a linear equation to the dress-shoe lace chart.

c. Is the boot lace length a linear function?

d. Fit a linear equation to the boot lace chart.

e. What is the meaning of the slope?

f. What is the meaning of the vertical axis intercept?

g. What other factors besides eyelets might control the length of the lace?

115. ***Slope and Right Triangles.*** The sides forming the right angle in a right triangle are perpendicular. Use the slopes of AB, AC, and BC to show that each set of ordered pairs forms a right triangle. Name the perpendicular sides. (Hint: By showing slope on a carefully drawn graph you can avoid use of the slope formula.)

a. $A(0, 2)$, $B(3, 6)$, $C(7, 3)$

b. $A(-2, -1)$, $B(3, -3)$, $C(5, 2)$

c. $A(-2, 3)$, $B(0, -1)$, $C(2, 5)$

d. $A(-3, -4)$, $B(-1, -7)$, $C(0, -2)$

4.5 Inequalities in Two Variables

OBJECTIVES

- Find out if an ordered pair is a solution of a two-variable inequality.
- Find solutions to a two-variable inequality from a graph.
- Graph a two-variable linear inequality.

WARM-UP

These exercises provide practice with the properties of inequalities (Section 3.4).

1. Solve each of these for y in terms of x.

a. $2x - y < 4$ **b.** $3x + 2y \le 6$

c. $2x - 3y > 6$ **d.** $2y - x \ge 4$

2. Change each of these into a form matching $ax + by < c$. For this exercise, the coefficient of x may be a fraction or negative number.

a. $y < 3x - 1$ **b.** $y \ge -4x + 2$

c. $y > \frac{1}{2}x + 3$ **d.** $y \le \frac{1}{3}x - 3$

IN THIS SECTION, we focus on graphical representation of the solution sets to inequalities in two variables. We practice operations with inequalities, find out which ordered pairs are solutions to inequalities, and graph two-variable linear inequalities.

Definition of Two-Variable Linear Inequality

> A **linear inequality in two variables** can be written $ax + by < c$, where a, b, and c are real numbers and a and b are not both 0.

The above definition and the properties and definitions that follow are true for the other inequality symbols, $>$, $\geq$, and $\leq$.

In the Warm-up exercises, you practiced changing two-variable inequalities into their two most common forms. To use a graphing calculator, we must solve an inequality for y. The inequality form $ax + by < c$ is generally for non-calculator use and most commonly has a positive integer for a, the coefficient on x. Example 1 shows how to change from one inequality form to another.

EXAMPLE  Changing the form of an inequality

a. Solve for y: $2x - y < 4$.

b. Change to $ax + by$ form: $y \leq \frac{3x}{2} + 3$

Solution **a.**

$2x - y < 4$ — Subtract $2x$ on both sides.

$-y < -2x + 4$ — Multiply by -1 on both sides and reverse the direction of the inequality sign.

$-1(-y) > -1(-2x + 4)$ — Simplify.

$y > 2x - 4$

b.

$y \leq \frac{3x}{2} + 3$ — Multiply both sides by 2.

$2y \leq 3x + 6$ — Subtract $3x$ on both sides.

$-3x + 2y \leq 6$ — Multiply both sides by -1 and reverse the direction of the inequality sign.

$-1(-3x + 2y) \geq -1(6)$ — Simplify.

$3x - 2y \geq -6$ ●

Because we can solve $ax + by = c$ for the intercepts by letting $x = 0$ and $y = 0$, the form $ax + by < c$ is useful for graphing by hand. In Example 2, we review finding the intercepts of equations.

EXAMPLE 2 Finding intercept points What are the x-intercept point and y-intercept point for each of these equations?

a. $y - x = 2$ **b.** $y - x = 0$ **c.** $40x + 20y = 160$

Solution The x-intercept point has $y = 0$. The y-intercept point has $x = 0$.

a.

$y - x = 2$ — Substitute $y = 0$.

$0 - x = 2$

$x = -2$ — The x-intercept point is $(-2, 0)$.

$y - x = 2$ — Substitute $x = 0$.

$y - 0 = 2$

$y = 2$ — The y-intercept point is $(0, 2)$.

b. $y - x = 0$ Substitute $y = 0$.

$0 - x = 0$

$x = 0$ This line passes through the origin. The ordered pair (0, 0) is both the x-intercept point and the y-intercept point.

c. $40x + 20y = 160$ Substitute $y = 0$.

$40x + 20(0) = 160$

$x = 4$ The x-intercept point is (4, 0).

$40x + 20y = 160$ Substitute $x = 0$.

$40(0) + 20y = 160$

$y = 8$ The y-intercept point is (0, 8).

Solving a Two-Variable Inequality

CHECKING SOLUTIONS A solution of a two-variable inequality is an ordered pair that makes the inequality a true statement. In Example 3, we test ordered pairs (x, y) to see if they make an inequality true.

EXAMPLE 3 Finding out if ordered pairs are in a solution set Substitute each ordered pair into the inequality $y \geq x + 2$ to see if it makes the inequality true. Plot the ordered pairs, and compare the positions of the points to that of the line $y = x + 2$.

a. (0, 0) **b.** (2, 1) **c.** (−2, 0) **d.** (0, 2) **e.** (−3, 2)

f. (0, 4) **g.** (1, 3) **h.** (−3, −1)

Solution **a.** The point (0, 0) makes a false inequality, $0 \geq 0 + 2$.

b. The point (2, 1) makes a false inequality, $1 \geq 2 + 2$.

c. The point (−2, 0) makes a true inequality, $0 \geq -2 + 2$.

d. The point (0, 2) makes a true inequality, $2 \geq 0 + 2$.

e. The point (−3, 2) makes a true inequality, $2 \geq -3 + 2$.

f. The point (0, 4) makes a true inequality, $4 \geq 0 + 2$.

g. The point (1, 3) makes a true inequality, $3 \geq 1 + 2$.

h. The point (−3, −1) makes a true inequality, $-1 \geq -3 + 2$.

The "false" points (0, 0) and (2, 1) are both to the right of the line. The "true" points (−2, 0), (0, 2), (1, 3), and (−3, −1) lie on the line $y = x + 2$ (see Figure 32). The points (−3, 2) and (0, 4) are also true and lie to the left of the line.

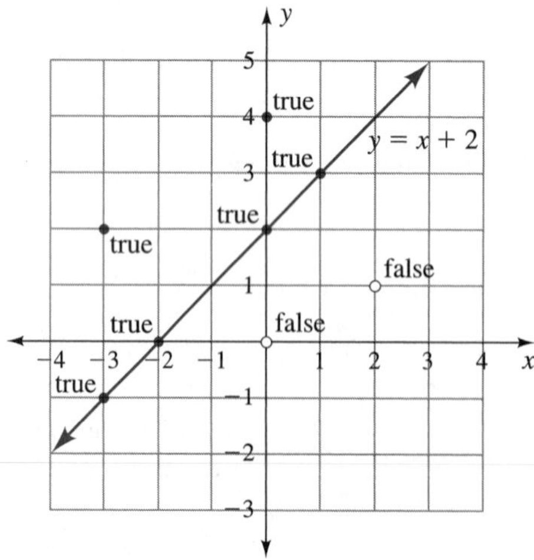

Figure 32

GRAPHING SOLUTIONS Example 3 suggests that the solutions to two-variable inequalities will be regions on the coordinate plane. In fact, the solution set to a two-variable inequality is a half-plane. A **half-plane** is *the region on one side of a line.* The graph of every line determines two half-planes. Only one of these half-planes will make an inequality true. *The line between the half-planes* is called the **boundary line.** Both the half-plane and the boundary line must be considered in graphing the solution set.

To find which half-plane or region is in the solution set, we use a test point. Recall that when we solved one-variable inequalities in Section 3.4, we used a test number to find out which way to draw the arrow on a line graph. In this section, we use an ordered pair as a test point to find and then shade the correct half-plane for our solution set.

The inequality sign tells us whether the boundary line is part of the solution set.

Boundary Lines

If the inequality contains $>$ or $<$, the boundary line on the half-plane is not in the solution set and we draw a dashed line.

If the inequality contains $\geq$ or $\leq$, the boundary line on the half-plane is in the solution set and we draw a solid line.

EXAMPLE 

Finding a solution set Graph $y \geq x + 2$ on coordinate axes. Use a test point to find the region described by the inequality.

Solution The graph of $y \geq x + 2$ combines the graph of a boundary line, $y = x + 2$, with the half-plane given by $y > x + 2$.

We graph the boundary line $y = x + 2$ in Figure 33. Because the inequality contains $\geq$, we use a solid line for the boundary. Ordered pairs on the boundary line make the inequality true.

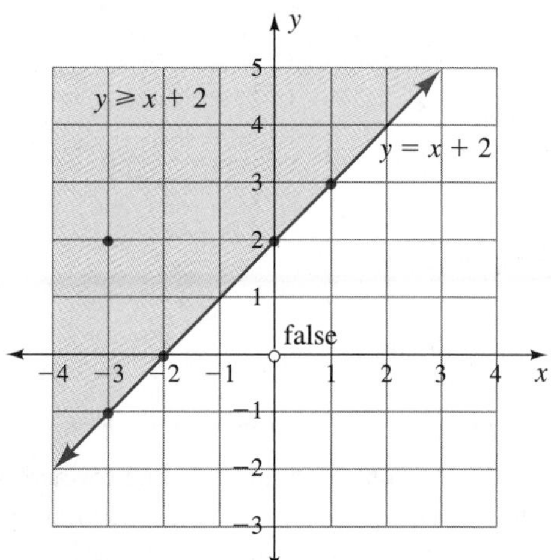

Figure 33

We next select a test point—say, (0, 0)—and substitute it into the inequality.

$y \geq x + 2$ Let $x = 0$, $y = 0$.

$0 \geq 0 + 2$ Simplify.

$0 \geq 2$

The inequality is false. We shade the region on the opposite side from (0, 0). The solution set is the boundary line and the half-plane to its left. ●

There are two special ideas to remember about test points:

First, always substitute the test point into the original inequality.

Second, the easiest test point to check in any inequality is the origin, (0, 0). The origin is a good test point unless it is on the boundary line.

EXAMPLE 5 Graphing a solution set where the boundary line passes through the origin Graph the inequality $y - x \geq 0$.

Solution To find the boundary line, we replace the inequality with an equal sign and solve for y.

$$y - x = 0$$
$$y = x$$

We graph the boundary line $y = x$ in Figure 34.

The boundary line passes through the origin, so we choose another test point—say, (0, 5).

$$y - x \geq 0$$
$$5 - 0 \geq 0$$
$$5 \geq 0$$

The inequality is true, so we shade the (0, 5) side of the boundary line.

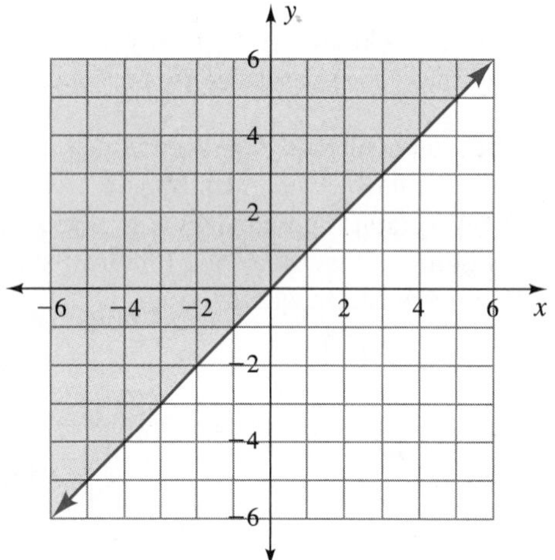

Figure 34

In summary:

Graphing a Two-Variable Linear Inequality

1. Graph the boundary line formed by replacing the inequality sign with an equal sign. Use a dashed line if the inequality is $<$ or $>$. Use a solid line if the inequality is $\leq$ or $\geq$. The dashed line indicates that the boundary is not included in the solution set. The solid line indicates that the boundary is included in the solution set.
2. Select a test point not on the boundary line. Substitute the ordered pair for the test point into the inequality.
3. If the test point in step 2 creates a true statement, shade the half-plane that contains the test point. If the test point creates a false statement, shade the half-plane that does not contain the test point.

Applications

Example 6 illustrates an application where the boundary line is a horizontal or vertical line and where we must think about the domain and range of the problem situation in drawing our graph.

EXAMPLE  Finding a solution set with limited domain and range Audrey has \$65 to spend on a birthday party. Let x be the number of people who attend and y be the total cost of the birthday party.

a. Write Audrey's budget as an inequality.

b. What is one condition placed on the input set (domain)? What is one condition placed on the output set (range)?

c. Draw a graph for Audrey's budget inequality. Include appropriate domain and range information on the graph.

Solution **a.** Audrey's budget is $y \leq \$65$. Her budget inequality does not contain x. The boundary line at \$65 is solid because Audrey can spend up to or exactly \$65.

b. The number of people who attend must be positive: $x > 0$. The number of dollars spent on the party must be positive: $y > 0$.

c. The conditions suggest a first-quadrant graph. The boundary line is $y = 65$. When we substitute the test point (0, 0) into $y \leq 65$, we get

$$0 \leq 65$$

The inequality is true, so we shade below the budget line, as shown in Figure 35.

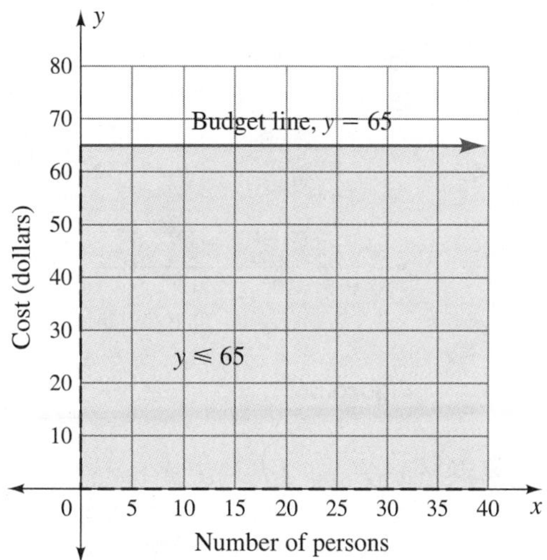

Figure 35

Think about it: If Audrey spends the whole \$65, what will happen to the money spent per person as the number of people who attend increases?

EXAMPLE  Writing and graphing an inequality Smart-Mart keeps the currency in its overnight cash drawer limited to \$100. Suppose the cash drawer has x one-dollar bills and y five-dollar bills and no larger bills.

a. What ordered pair describes the number of one-dollar bills that can be held with 10 fives? no fives?

b. What ordered pair describes the number of fives that can be held with 10 ones? no ones?

c. Write an expression for how much money there is in x ones.

d. Write an expression for how much money there is in y fives.

e. What inequality describes the possible numbers of the two types of bills that can be held with the \$100 limit?

f. What is one limit on the domain for this application? on the range? Graph the inequality from part e.

Solution

a. With 10 fives, there can be up to 50 ones: (50, 10). With no fives, there can be up to 100 ones: (100, 0).

b. With 10 ones, there can be up to 18 fives: (10, 18). With no ones, there can be up to 20 fives: (0, 20).

c. $1x$

d. $5y$

e. $1x + 5y \le 100$

f. The number of ones and the number of fives must be zero or positive, so $x \ge 0$ and $y \ge 0$. The graph is in Figure 36. The origin makes the inequality true.

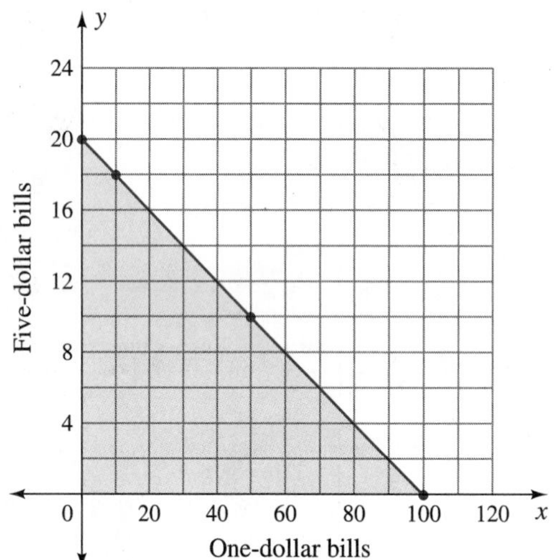

Figure 36

EXAMPLE 8

Writing and graphing an inequality Sens-a-diet allows a daily maximum of 160 calories in snacks. Caramel candies have 40 calories each, and ginger snaps have 20 calories each. Let x be the number of caramels and y be the number of ginger snaps.

a. What ordered pair describes the number of ginger snaps that can be eaten with 3 caramels? 0 caramels?

b. What ordered pair describes the number of caramels that can be eaten with 4 ginger snaps? 0 ginger snaps?

c. Write an expression for the number of calories in x caramels.

d. Write an expression for the number of calories in y ginger snaps.

e. What inequality describes the numbers of the two types of snacks that can be eaten with the 160-calorie maximum?

f. What is one limit on the domain for this application? on the range? Graph the inequality from part e.

Solution **a.** (3, 2), (0, 8)

b. (2, 4), (4, 0)

c. $40x$

d. $20y$

e. $40x + 20y \leq 160$

f. *To graph the inequality by hand*, we first determine the domain and range. The number of candies and the number of ginger snaps must be zero or positive, so the domain is $x \geq 0$ and the range is $y \geq 0$. The solution set is in the first quadrant and includes the axes. We then locate the boundary line $40x + 20y = 160$ by plotting the four ordered pairs from parts a and b. The line is shown in Figure 37. Then we choose a test point—say, (3, 1).

$$40x + 20y \leq 160 \quad \text{Substitute (3, 1).}$$
$$40(3) + 20(1) \leq 160 \quad \text{Simplify.}$$
$$140 \leq 160$$

The inequality is true, so we shade on the (3, 1) side of the boundary line.

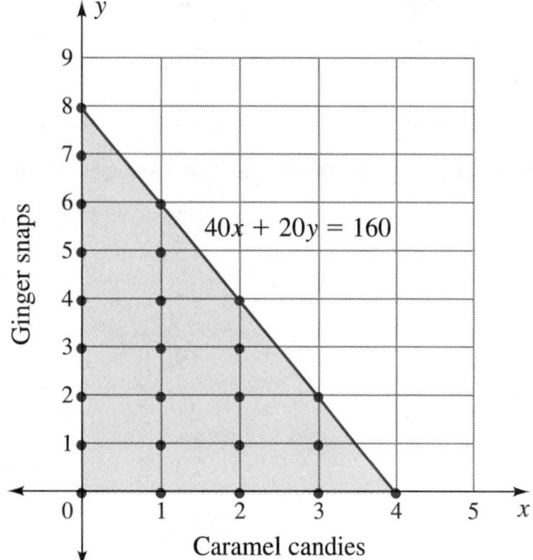

Figure 37

To graph the inequality on a calculator, we first solve the inequality for y:

$$40x + 20y \leq 160 \quad \text{Subtract } 40x \text{ on both sides.}$$
$$20y \leq -40x + 160 \quad \text{Divide by 20 on both sides.}$$
$$y \leq -2x + 8$$

Then we enter $Y_1 = -2X + 8$, using a window that shows the first quadrant and the intercepts in parts a and b. Next we select an ordered pair as a test point—say, (3, 1).

$$y \leq -2x + 8 \quad \text{Substitute } x = 3, y = 1.$$
$$1 \leq -2(3) + 8$$
$$1 \leq 2$$

The point is true, so we shade the half-plane containing (3, 1), which is below the line. ●

Graphing Calculator Technique

Some calculators have a shading option with the [Y =] key. Enter $Y_1 = -2X + 8$. Place the cursor to the left of Y_1 and press [ENTER] until you see the shading below the line option, ◣. Set an appropriate viewing window. Graph. This graph represents $y \leq -2x + 8$. To graph $y \geq -2x + 8$, place the cursor to the left of Y_1 and press [ENTER] until you see the shading above option, ◥. Reset the viewing window as needed. Graph.

ANSWER BOX

Warm-up: 1. a. $y > 2x - 4$ **b.** $y \leq -\frac{3x}{2} + 3$ **c.** $y < \frac{2x}{3} - 2$ **d.** $y \geq \frac{1}{2}x + 2$ **2. a.** $-3x + y < -1$ **b.** $4x + y \geq 2$ **c.** $-\frac{1}{2}x + y > 3$ **d.** $-\frac{1}{3}x + y \leq -3$ (The answers to Exercise 2 are commonly written without fractions or negative integers on the x term: **2. a.** $3x - y > 1$ **b.** $4x + y \geq 2$ **c.** $x - 2y < -6$ **d.** $x - 3y \geq 9$) **Think about it:** The spending per person will decrease as the number of people who attend increases.

EXERCISES 4.5

In Exercises 1 to 6, identify and carry out the steps needed to change the first inequality into the second.

1. $-3x + y < -1$ to $3x - y > 1$

2. $-2x - y < 2$ to $2x + y > -2$

3. $-\frac{1}{2}x + y > 3$ to $x - 2y < -6$

4. $-\frac{1}{3}x + y \leq -3$ to $x - 3y \geq 9$

5. $2x - 3y > 6$ to $y < \frac{2x}{3} - 2$

6. $2y - x \geq 4$ to $y \geq \frac{1}{2}x + 2$

In Exercises 7 to 14, which ordered pairs are solutions to the inequality?

7. $x + y > 3$; $(-2, 3)$, $(4, 0)$, $(1, 4)$

8. $x - y < 2$; $(4, 1)$, $(1, 4)$, $(-2, 1)$

9. $\frac{1}{2}x + y \geq 2$; $(-2, 3)$, $(4, -2)$, $(6, -1)$

10. $y - \frac{1}{2}x \leq 3$; $(3, 2)$, $(0, 4)$, $(5, 2)$

11. $2x - 3y > 4$; $(0, 0)$, $(1, -1)$, $(3, 0)$

12. $2x + y \leq 1$; $(0, 1)$, $(1, 0)$, $(-1, 1)$

13. $y \leq -2$; $(-2, 3)$, $(-2, 0)$, $(0, -2)$

14. $x < 4$; $(3, 5)$, $(5, 4)$, $(-5, 0)$

In Exercises 15 to 28, write the intercepts for the boundary line and then graph the inequality on the coordinate plane.

15. $y < -3x + 2$ **16.** $y < 2x - 11$

17. $y \geq 4x - 1$ **18.** $y \geq x + 3$

19. $y \leq 2 - 2x$ **20.** $y \leq 3 - 3x$

21. $2x + y > 5$ **22.** $2x - y < 4$

23. $2 - 2y < 4x$ **24.** $6 + 3y \geq 2x$

25. $x > 4$ **26.** $y \geq -3$

27. $y \leq 4$ **28.** $x < -2$

In Exercises 29 to 36, write the equation of the boundary line and then write the inequality represented by the shaded coordinate plane.

29.

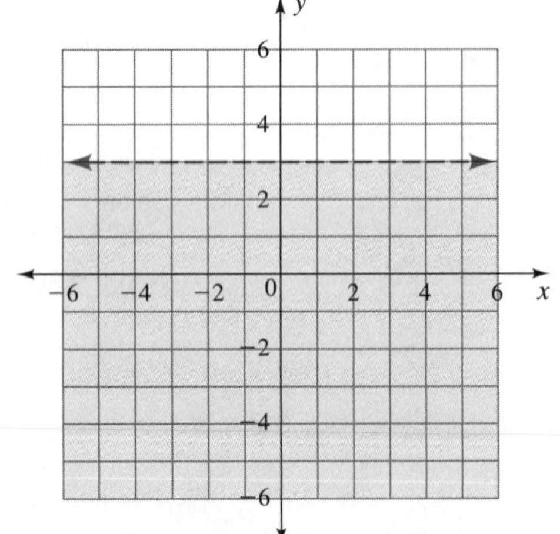

30.

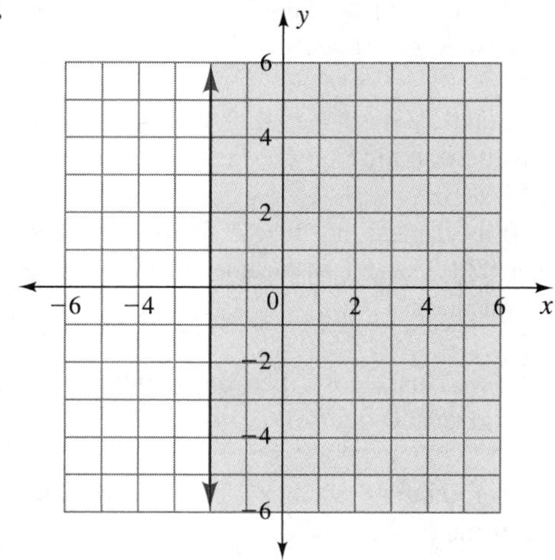

31.

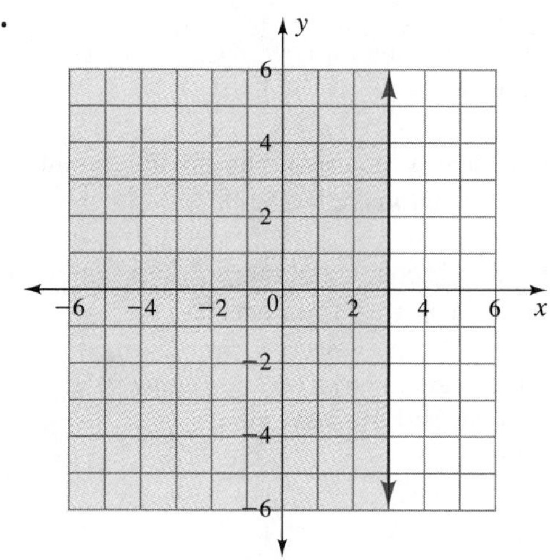

32.

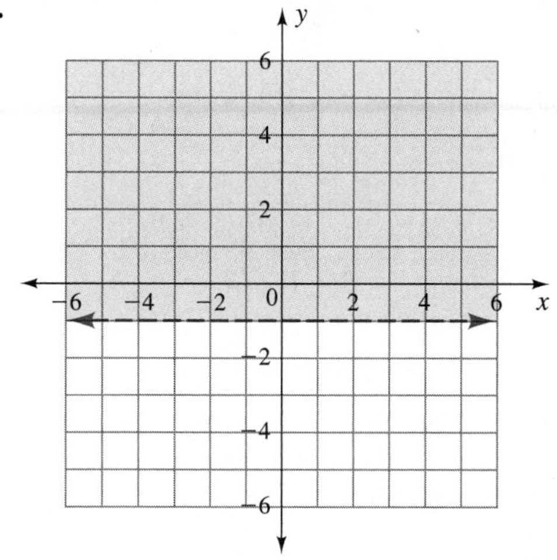

33.

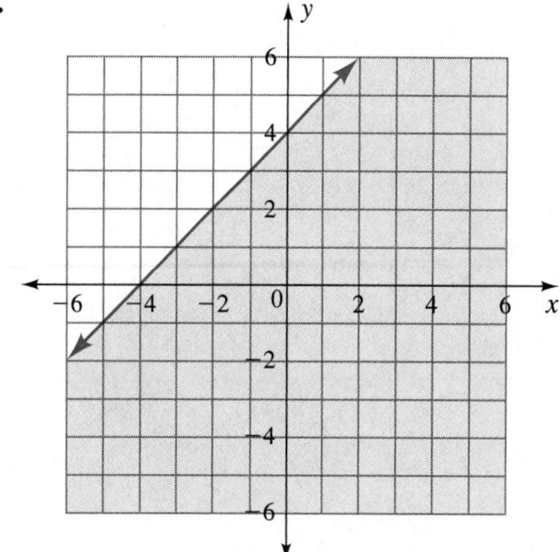

34.

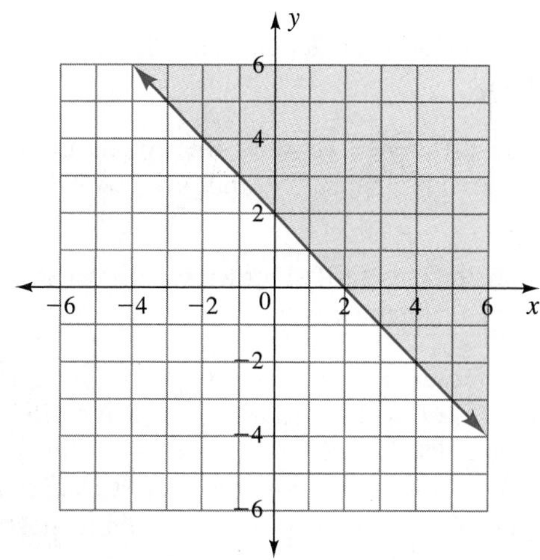

35.

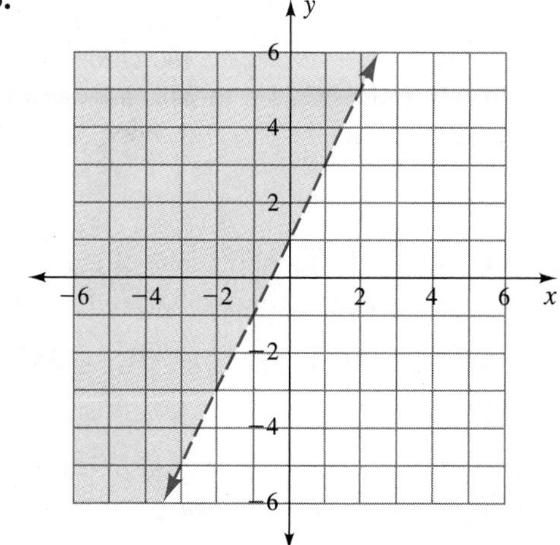

36.

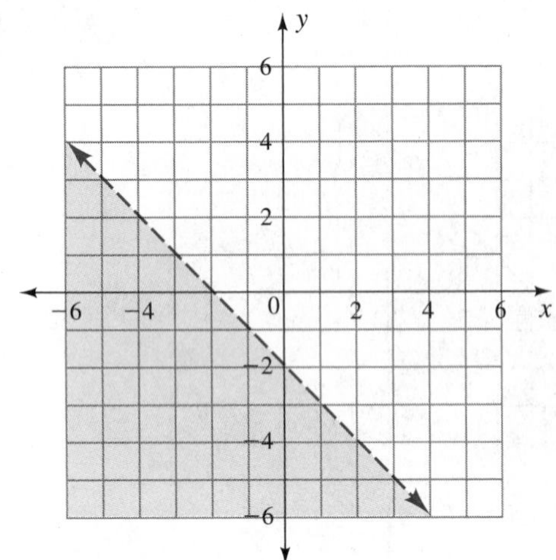

37. Explain how to find which side of a boundary line to shade when graphing the solution set to an inequality.

38. Explain why some boundary lines are dashed and others are solid.

39. Explain the difference between the graphs of $x \geq 3$ as a one-variable inequality and as a two-variable inequality.

40. Explain how to find out if an ordered pair is a solution of a two-variable inequality.

41. To meet expenses, a local theater group has a ticket sales goal of \$2400. Regular tickets sell for \$16, and student/senior tickets sell for \$12. Let x = number of regular tickets sold and y = number of student/senior tickets sold. Write an inequality describing possible combinations of ticket sales that would meet the goal. Draw a graph showing all possible combinations of ticket sales that would meet the goal.

42. In Example 8, we used the inequality $40x + 20y \leq 160$ to describe the snacks allowed with a 160-calorie maximum. If we increase the calories to 240, how will the graph change? (*Hint:* Find the new intecepts.)

43. A dieter is allowed 140 calories for a snack.

a. What are three possible combinations of apricots at 20 calories each and tangerines at 35 calories each?

b. Plot the boundary line, and describe the region that shows sensible solutions.

c. What inequality describes the snacks? What is one limit on the domain? on the range?

44. A dieter limits a snack to 60 calories.

a. What are three possible combinations of small carrots at 20 calories each and medium celery stalks at 3 calories each?

b. Plot the boundary line, and describe the region that shows sensible solutions.

c. What inequality describes the snacks? What is one limit on the domain? on the range?

45. Sesha has only dimes and quarters in her pocket. She has at most \$2.00. Let x = number of dimes and y = number of quarters.

a. What ordered pair describes the number of dimes with 4 quarters? 0 quarters?

b. What ordered pair describes the number of quarters with 15 dimes? 0 dimes?

c. Write an expression for how much money there is in x dimes.

d. Write an expression for how much money there is in y quarters.

e. What inequality describes the possible number of coins that could make the \$2.00?

f. What are the domain and range in this application? Graph the inequality from part e.

46. Describe the region shown in the figure. What party scenario might the figure describe?

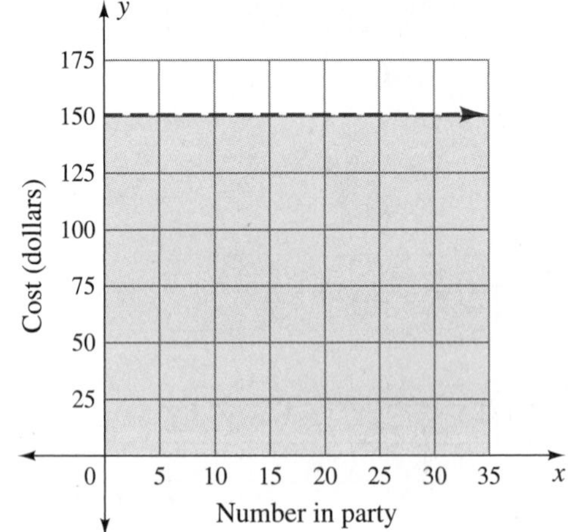

For Exercises 47 to 49, identify the regions described by the inequalities. (Hint: $x \geq 0$ and $y \geq 0$ describe points in the first quadrant, or on the positive portions of the axes.)

47. $x > 0$ and $y \leq 0$

48. $x < 0$ and $y \geq 0$

49. $x < 0$ and $y < 0$

What inequalities describe the regions in Exercises 50 to 52?

50. The second quadrant without axes

51. The fourth quadrant without axes

52. The third quadrant together with the negative x-axis and the negative y-axis

Project

53. ***Pricing to Undersell the Competition.*** The Fun Corporation is building a bowling alley to compete with Fairfield Lanes. Fairfield charges \$5.50 per person. Fun Corporation wants to attract parties with budgets of \$100 or less and to keep the per-person cost below that charged by Fairfield.

a. Let the number of people be the input, and let total cost be the output. Draw a line representing the \$100 budget limit. Draw a line showing the cost at Fairfield Lanes. Shade the pricing option region for Fun Corporation.

b. If a party were held at Fairfield Lanes for \$100, how many people could attend?

c. Suppose Fun Corporation charged \$5 per person. Mark and identify the points that show how many people could attend for \$50 and for \$100.

d. What is the Fun Corporation charge per person if (10, 35) is a point on its cost line?

CHAPTER 4 SUMMARY

Vocabulary

For definitions and page references, see the Glossary/Index.

boundary line	range
domain	rise
function	run
function notation	slope
half-plane	slope formula
linear inequality in two variables	subscripts
in terms of	undefined slope
parallel lines	vertical-line test
perpendicular lines	y-intercept formula
	zero slope

Concepts

4.1 Formulas

When solving formulas, observe the order of operations on the selected variable and write a plan that uses the inverse operations in the reverse order.

Solving formulas results in answers in expression form. Solving one-variable equations results in numerical answers.

4.2 Functions

The function notation $f(x)$ is read "function of x" or "f of x." The f and x in $f(x)$ are not being multiplied. The notation for evaluating a function for an input a is $f(a)$. To evaluate $f(a)$, substitute a for every x in the expression and then simplify.

The horizontal axis intercept, a, is where the output is zero or $f(x) = 0$.

The vertical axis intercept, b, is where the input is zero, or $f(0)$.

4.3 Slope

Because the scales on the axes affect the appearance of steepness of a line, we use a number, *slope*, to describe the steepness. The slope describes a rate of change.

$$\text{Slope} = \frac{\text{vertical change}}{\text{horizontal change}} = \frac{\text{rise}}{\text{run}} = \frac{\text{change in output}}{\text{change in input}} = \frac{\Delta y}{\Delta x}$$

$$m = \frac{y_2 - y_1}{x_2 - x_1} \quad \text{slope formula}$$

Lines that rise from left to right have *positive slope.*

Lines that drop from left to right have *negative slope.*

A horizontal line has *zero slope*, because the change in y is zero.

A vertical line has *undefined slope*, because the change in x is zero.

Straight lines have constant slope; that is, the slope between any two points on a straight line is the same as that between any other two points on the same line.

Curved graphs have slopes that change as we move along the graph.

To draw a line with a given slope, choose a first point and then use the slope to find a second point.

To locate a line, we need to know both its slope and a point on the line.

The units on slope are the units on the vertical axis divided by the units on the horizontal axis.

4.4 Linear Equations

The equation for a linear function is $y = mx + b$, where the number m is the slope and the number b is the vertical axis intercept, or y-intercept. Find m using the definition or slope formula. Find b using $b = y - mx$, where (x, y) is any point on the line.

The equation of a vertical line is $x = a$, where (a, y) is any point on the line.

The equation of a horizontal line is $y = b$, where (x, b) is any point on the line.

Parallel lines have the same slope but different y-intercepts.

Two lines are perpendicular

- if their slopes multiply to -1 (that is, their slopes are opposite reciprocals) or
- if their slopes are the same as those of the horizontal and vertical axes.

4.5 Inequalities

To solve $y > ax + b$, graph $y = ax + b$ as a boundary line and then substitute in a test point to find out which side of the boundary line to shade. Use a solid boundary line for inequalities containing $\geq$ or $\leq$. Use a dashed boundary line for inequalities containing $>$ or $<$.

CHAPTER 4 REVIEW EXERCISES

In Exercises 1 to 4, solve the equations for the indicated variable.

1. $6 = 4(-1) + b$ for b

2. $-5 = \frac{1}{2}(-6) + b$ for b

3. $37 = \frac{5}{9}(F - 32)$ for F

4. $100 = \frac{5}{9}(F - 32)$ for F

5. Describe the formula in Exercise 9 below, using the phrase *in terms of.*

6. Describe the formula in Exercise 10 below, using the phrase *in terms of.*

In Exercises 7 to 22, solve the formulas for the indicated variable.

7. $W = hp$ for h

8. $d = rt$ for t

9. $A = \dfrac{bh}{2}$ for b

10. $A = \dfrac{a + b + c}{3}$ for b

11. $I = \dfrac{AH}{T}$ for T

12. $R = \dfrac{E^2}{W}$ for W

13. $PV = nRT$ for T

14. $W = I^2R$ for R

15. $ax + by = c$ for x

16. $A = 2l + 2w$ for l

17. $C = 35 + 5(k - 100)$ for k

18. $C = FW - 10(A - 35)$ for A

19. $P_1V_1 = P_2V_2$ for P_2

20. $y_2 - y_1 = m(x_2 - x_1)$ for m

21. $A = \frac{1}{2}h(b_1 + b_2)$ for b_1

22. $\dfrac{P_1V_1}{T_1} = \dfrac{P_2V_2}{T_2}$ for V_2

In Exercises 23 and 24, a formula and values are given. Solve for the remaining variable.

23. If $A = \frac{1}{2}h(a + b)$ with $A = 27$, $h = 3$, and $a = 9$, what is b?

24. If $A = \dfrac{a + b + c}{3}$ with $A = 80$, $a = 75$, and $b = 78$, what is c?

In Exercises 25 to 28, find f(0), f(3), f(−5), and f(a) for each function.

25. $f(x) = 7 - 2x$

26. $f(x) = 3x^2$

27. $f(x) = x^2 + x$

28. $f(x) = -3x - 4$

In Exercises 29 to 34, solve for x where f(x) = 0.

29. $f(x) = 2x - 5$

30. $f(x) = -2x + 7$

31. $f(x) = -3x - 4$

32. $f(x) = 3x + 9$

33. $f(x) = \frac{1}{2}x - 6$

34. $f(x) = \frac{1}{4}x + 8$

35. What is $f(0)$ on the graph of a function?

36. What is $f(x) = 0$ on the graph of a function?

37. This function refers to slang names for money, with the inputs in "bits":

(2, 25), (4, 50), (6, 75), (8, 100)

a. What is the set of inputs, or domain?

b. What is the set of outputs, or range?

c. Extra: What are the outputs?

38. This function refers to the card game cribbage:

(5, 10), (6, 9), (7, 8), (8, 7), (9, 6), (10, 5)

a. What is the domain?

b. What is the range?

c. Extra: What is the rule?

39. This function is common in describing nails:

(1, 2), (2, 6), (3, 10), (4, 20), (5, 40), (6, 60)

The input number is the length of the nail.

a. What is the domain?

b. What is the range?

c. Extra: What is the output?

40. This function is common in metric measurement:

(deka, 10), (hecto, 10^2), (kilo, 10^3), (mega, 10^6), (giga, 10^9), (tera, 10^{12})

The input is the prefix name.

a. What is the domain?

b. What is the range?

c. Extra: What is the output?

In Exercises 41 to 46, which rules are functions? The parentheses show possible input-output ordered pairs.

41. What writing prize did the person win? (Gwendolyn Brooks in 1950, Pulitzer Prize), (Alice Walker in 1983, Pulitzer Prize), (Charles Gordone in 1970, Pulitzer Prize)

42. Who was born in the given year? (1918, Paul Harvey), (1918, Ann Landers), (1918, Abigail Van Buren)

43. What was the person's year of birth? (Delores Huerta, 1930), (Cesar Chavez, 1927), (Philip Vera Cruz, 1905)

44. In what country was this person a political leader? (Emiliano Zapata, Mexico), (Simon Bolivar, Venezuela), (Salvador Allende Gossens, Chile)

45. What is the last name of a composer whose first initial is G? (G, Bizet), (G, Donizetti), (G, Gershwin), (G, Handel)

46. What is each person's given name? (Adams, John), (Adams, John Quincy), (Adams, Abigail)

Matching: Complete each sentence in Exercises 47 to 50 by choosing one of the following:

a. *let* $x = a$.

b. *substitute* $y = 0$ *into the equation and solve for* x.

c. *let* $y = b$.

d. *let* $x = b$.

e. *substitute* $x = 0$ *into the equation and simplify.*

47. To find the horizontal intercept for $y = mx + b$,

48. To find the vertical axis intercept for $y = mx + b$,

49. To find the equation of a vertical line through $(a, 0)$,

50. To find the equation of a horizontal line through $(0, b)$,

51. Simplify these expressions.

a. $\dfrac{4 - (-2)}{6 - 4}$ **b.** $\dfrac{3 - (-4)}{-6 - 4}$

c. $\dfrac{0 - (-2)}{6 - (-4)}$ **d.** $\dfrac{1 - 3}{-6 - 9}$

52. Find the slope of the line passing through the two points.

a. (2, −3) and (5, −1) **b.** (0, −4) and (−5, −4)

c. (2, 3) and (2, −4) **d.** (−3, −2) and (4, 1)

53. Draw a line with a $-\frac{5}{3}$ slope.

54. Draw a line with a $\frac{2}{5}$ slope.

55. Draw a line with a $\frac{2}{3}$ slope passing through (−1, 4).

56. Draw a line with a $\frac{4}{3}$ slope passing through (2, −3).

57. Explain how to tell if a line has negative slope.

58. Explain how to tell from a table if the slope will be positive.

59. Explain how to draw a line with slope a/b through a point (x, y).

60. Choose two ordered pairs, graph them, predict the slope of the line passing through them, and then calculate the slope.

In Exercises 61 to 64, find the slope, y-intercept, and equation for each line.

61.

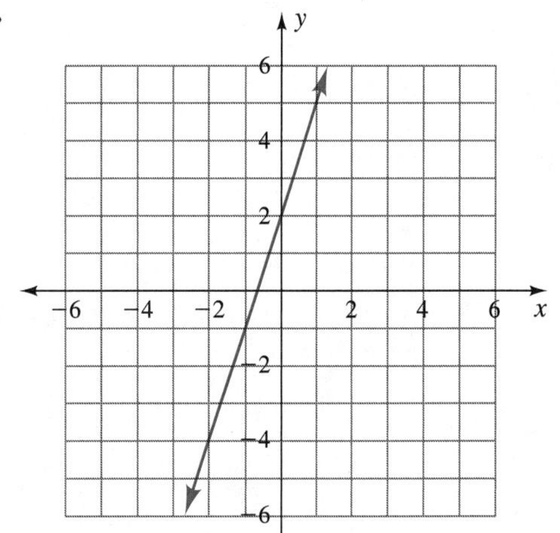

62.

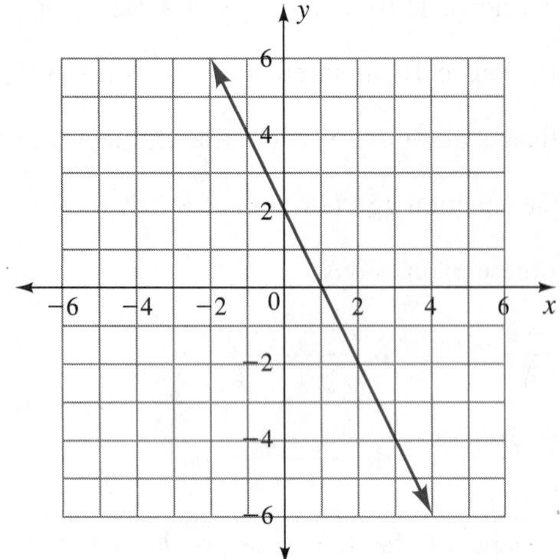

63.

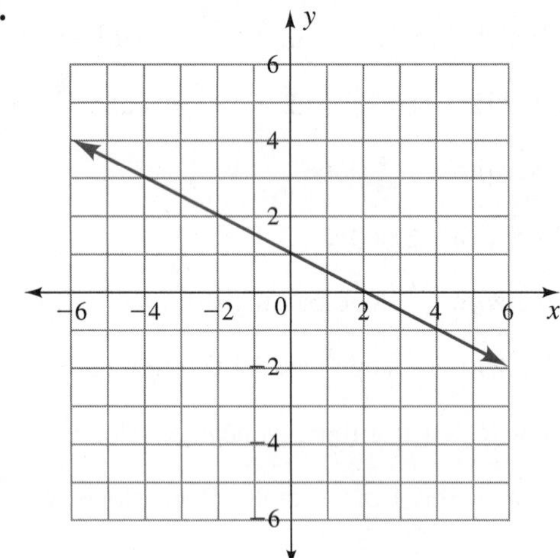

64.

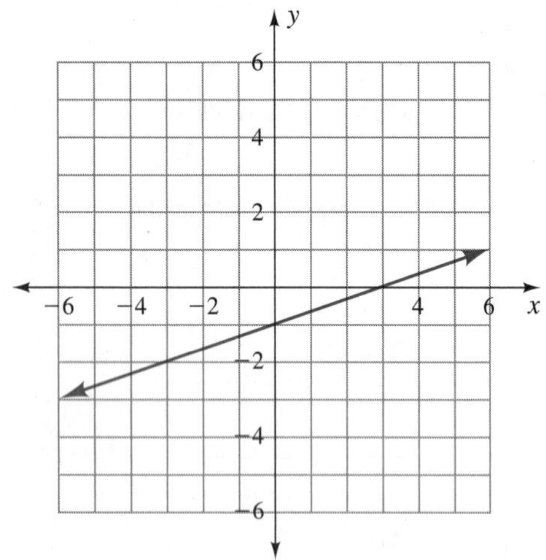

For Exercises 65 to 68, determine whether the data in the table are linear. If they are linear, give the slope.

65.

x	y
1	5
3	2
5	−1

66.

x	y
1	4
3	7
5	11

67.

x	y
1	5
3	8
5	13

68.

x	y
1	6
3	1
5	−4

For Exercises 69 to 72, answer the following questions.

a. *Write each problem situation using data points. Select the inputs appropriately. Assume a linear relationship.*

b. *What is the slope of the line through these points?*

c. *What is the meaning of the slope?*

69. 14 minutes at \$4.44 and 5 minutes at \$2.10 (long-distance phone calls)

70. 2 minutes at \$0.26 and 7 minutes at \$0.91 (local toll phone calls)

71. 12 feet takes 12 hours and 32 feet takes 17 hours (side-walk repair)

72. 6 dozen in 2 hours and 12 dozen in 3 hours (cookie baking)

73. Which one of these equations is not equivalent to the others?

a. $y = mx + b$ **b.** $y = b + mx$

c. $mx + b = y$ **d.** $y + b = mx$

e. $b + mx = y$

74. Write this equation in three other ways: $5x + 4 = y$.

In Exercises 75 to 84, change the equation into $y = mx + b$ *form, if possible, and find the slope,* m*, and the constant term (y-intercept),* b.

75. $3x + 5y = 15$ **76.** $3y - 9x = -6$

77. $5x - 2y = 10$ **78.** $4x - y = -10$

79. $4y - 3x = 8$ **80.** $3y + 2x = 9$

81. $y + 3 = 0$ **82.** $x + 3 = 0$

83. $x = 3$ **84.** $2 - y = 0$

Find the slope and vertical axis intercept for each formula in Exercises 85 and 86. Assume the output variable is on the left.

85. $c = 2 + 1.5(n - 1)$ **86.** $S = 25(c + 3)$

In Exercises 87 to 94, use the slope and y-intercept to build an equation. Find an equivalent equation containing no fractions. Sketch a graph.

87. $m = 2, b = 4$

88. $m = 0, b = 3$

89. $m = 3, b = 0$

90. $m = -2, b = -3$

91. $m = \frac{1}{2}, b = 2$

92. $m = -3, b = \frac{1}{2}$

93. $m = -3, b = \frac{1}{4}$

94. $m = \frac{1}{4}, b = -1$

In Exercises 95 to 106, find the equation and then graph the line. For Exercises 95 to 102, you may put multiple graphs on the same axes.

95. Vertical line through $(4, -2)$

96. Horizontal line through $(3, -1)$

97. Vertical line through $(-2, 1)$

98. Horizontal line through $(-1, 3)$

99. Horizontal line through $(0, -2)$

100. Vertical line through $(3, 0)$

101. Horizontal line through $(2, 0)$

102. Vertical line through $(0, -1)$

103. Line parallel to $y = 3x + 2$ through the origin

104. Line parallel to $y = \frac{1}{2}x - 4$ through $(0, 5)$

105. Line perpendicular to $y = -\frac{2}{3}x + 3$ through $(0, -2)$

106. Line perpendicular to $y = -5x - 2$ through the origin

107. Do the ordered pairs $(-2, 4)$, $(0, 3)$, and $(4, 1)$ lie on a straight line? If so, what is the equation of the line?

For Exercises 108 to 110, write an equation based on the following information: A chemistry book suggests using this procedure to estimate the number of calories you need each day. Multiply your weight in pounds by an activity factor and then subtract 10 calories for each year over the age of 35 (to a maximum of 400 calories). The activity factors are 10 calories per pound for physically inactive; 15 calories per pound for moderately active; and 20 calories per pound for very active.

108. The calories needed, in terms of weight, by a moderately active 40-year-old person

109. The calories needed, in terms of weight, by an inactive 25-year-old

110. The calories needed, in terms of weight, by a very active 50-year-old

111. In the early 1980s, the federal government gave away surplus cheese, butter, and powdered milk. Eligibility guidelines for monthly family income were as follows: one person, \$507; two, \$682; three, \$857; four, \$1032; five, \$1207; six, \$1382; seven, \$1557; eight, \$1,732. Are the data linear? Explain why or why not. If they are linear, fit a linear equation for maximum income in terms of number of people in the household.

112. The Ajax Car Rental cost schedule for a compact car is shown in the table. The cost is also shown in the figure.

a. What is the input, or independent variable?

b. What is the output, or dependent variable?

c. What is the cost of driving 125 miles?

d. How many miles may be driven for \$57?

e. What is the y-intercept? What does it mean?

f. What is the slope of the line? What does it mean?

g. What is the equation of the line?

h. What is the x-intercept? What does it mean?

Miles, x	Cost, y
50	\$39
100	43
150	47
200	51
250	55

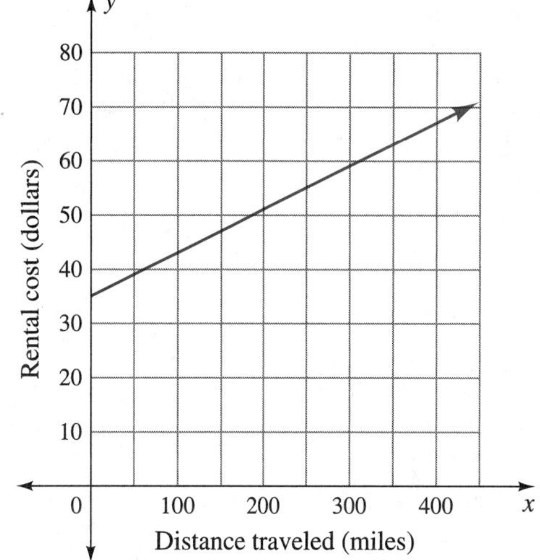

113. Which ordered pair or pairs make $y \leq x - 3$ true?

a. $(4, 2)$ **b.** $(-1, -6)$

c. $(3, 8)$ **d.** $(0, 0)$

114. Which ordered pair or pairs make $y > 3x - 6$ true?

a. $(0, -6)$ **b.** $(-2, -12)$

c. $(2, 1)$ **d.** $(3, 4)$

115. ***Matching.*** Which inequality, $x \geq 0$, $x \leq 0$, $x > 0$, $x < 0$, $y \geq 0$, $y \leq 0$, $y > 0$, or $y < 0$, describes each of the following half-planes?

a. All points on the x-axis or in Quadrant 1 or 2

b. All points on the y-axis or in Quadrant 1 or 4

c. All points to the left of the y-axis in Quadrants 2 and 3

d. All points below the x-axis in Quadrants 3 and 4

e. All points to the right of the y-axis in Quadrants 1 and 4

f. All points on the x-axis or in Quadrant 3 or 4

116. Write a description similar to those in parts a to f of Exercise 115.

a. $x \leq 0$

b. $y > 0$

In Exercises 117 to 122, graph the inequalities. Note that the boundary equations appeared in Exercises 75 to 80.

117. $3x + 5y > 15$ **118.** $3y - 9x \geq -6$

119. $5x - 2y < 10$ **120.** $4x - y \leq -10$

121. $4y - 3x \leq 8$ **122.** $3y + 2x > 9$

123. A diet has a 400-calorie snack limit. Vincente chooses to eat olives for his snack. Green olives contain 20 calories each, and ripe (black) olives contain 25 calories each. Define the variables. What are four different combinations of olives with fewer than 400 calories? What are four different combinations with exactly 400 calories? Plot your solutions on a graph. Shade the region of the graph that shows all combinations that satisfy the diet.

CHAPTER 4 TEST

1. Solve $5 = \frac{1}{2}(-8) + b$ for b.

2. Write a formula describing your course grade in terms of tests, homework, and final exam.

3. Solve each formula for the indicated variable.

a. $C = \pi d$ for d

b. $A = \frac{1}{2}bh$ for h

c. $y = mx + b$ for b

d. $P_1 V_1 = P_2 V_2$ for V_2

4. Which are functions?

a. To what Native American nation did the person belong? (Sitting Bull, Sioux), (Crazy Horse, Sioux), (Cochise, Apache), (Geronimo, Apache), (Captain Jack, Modoc)

b. What person played the indicated instrument? (trumpet, Davis), (trumpet, Armstrong), (trumpet, Gillespie)

c. What is the first initial for each composer? (Holst, G), (Mahler, G), (Puccini, G), (Rossini, G), (Verdi, G)

d. (2, 3), (3, 4), (4, 5)

e. (2, 3), (2, 4), (2, 6)

f. (3, 2), (4, 2), (5, 2)

5. Match each function expression or equation with one of the following descriptions: $f(x) = 0$, $f(a)$, or $f(0)$.

a. the x-intercept of the graph of a function

b. the y-intercept (or vertical axis intercept) of the graph of a function

c. the function $f(x)$ evaluated at $x = a$

6. a. Find the slope for each line in the figure.

b. Which line—graph 1, 2, 3, or 4—shows the distance traveled by a bicyclist averaging 10 miles per hour?

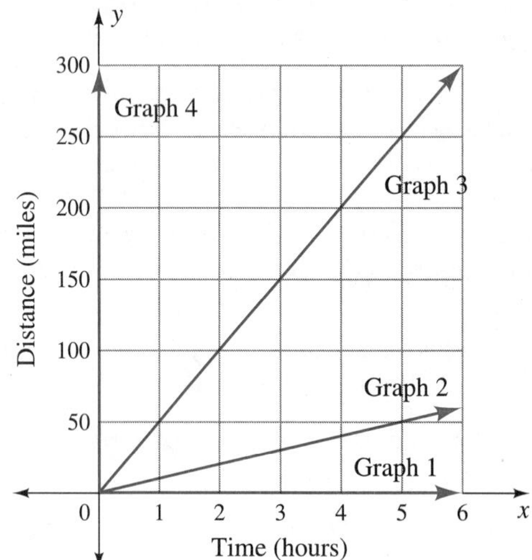

7. Find the slope of the line passing through the two points.

a. (5, −3) and (2, −1)

b. (1, −4) and (0, −4)

c. (1, 3) and (2, −4)

d. (4, −3) and (4, 1)

8. Draw a line with a $-\frac{3}{7}$ slope.

9. Draw a line with a $\frac{5}{3}$ slope passing through (2, −5).

10. What are the slope and y-intercept for each equation?

a. $y = 5 - 2x$

b. $6y + 3 = 2x$

11. State whether the pair of equations describes parallel lines, perpendicular lines, or neither.

a. $y = 3x + 2$ and $y = -3x + 2$

b. $y = 2x + 3$ and $y = 2x - \frac{1}{3}$

c. $y = -\frac{1}{2}x + 3$ and $y = 2x + 3$

d. $y = 2x - 3$ and $y = \frac{1}{2}x + 3$

12. Explain how to find a linear equation from a slope and one ordered pair.

13. Explain how to find the equation of a line passing through $(0, k)$ that is parallel to $y = ax + d$.

14. Explain how to find the equation of a horizontal line through (j, k).

In Exercises 15 and 16, write a linear equation for the given data. Write an equivalent equation containing no fractions.

15. $m = 5$, $b = -1$

16. $m = \frac{1}{3}$, $b = 2$

17. Graph $y \geq 2x + 6$.

18. Graph $2x + 3y < 12$.

19. The Help-U Company rents a compact car for the costs shown in the table. The costs are graphed in the figure.

Miles: x	Cost: y
50	$35
100	40
150	45
200	50
250	55

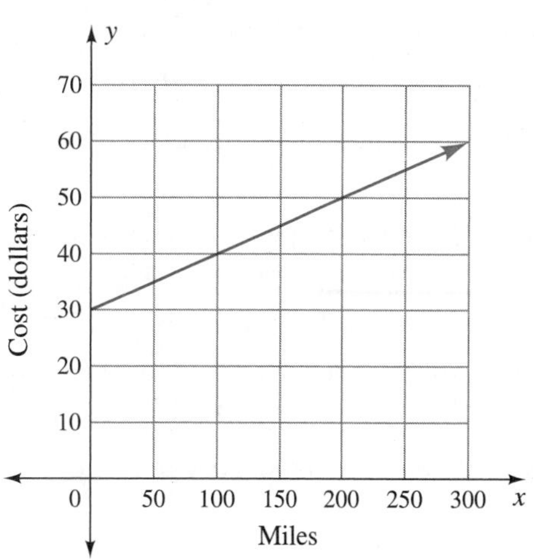

a. What is the input, or independent variable?

b. What is the output, or dependent variable?

c. What is the cost for driving 125 miles?

d. How many miles can be driven for $57?

e. What is the y-intercept? What does it mean?

f. What is the slope of the line? What does it mean?

g. What is the equation of the line?

h. What is the x-intercept? What does it mean?

CUMULATIVE REVIEW OF CHAPTERS 1 TO 4

For Exercises 1 to 5, follow the form of the example.
Example: Add, subtract, multiply, and divide $2\frac{2}{3}$ and $1\frac{1}{4}$.

$$2\tfrac{2}{3} + 1\tfrac{1}{4} = \tfrac{8}{3} + \tfrac{5}{4} = \tfrac{32}{12} + \tfrac{15}{12} = \tfrac{47}{12} = 3\tfrac{11}{12}$$
$$2\tfrac{2}{3} - 1\tfrac{1}{4} = \tfrac{8}{3} - \tfrac{5}{4} = \tfrac{32}{12} - \tfrac{15}{12} = \tfrac{17}{12} = 1\tfrac{5}{12}$$
$$2\tfrac{2}{3} \cdot 1\tfrac{1}{4} = \tfrac{8}{3} \cdot \tfrac{5}{4} = \tfrac{2}{3} \cdot \tfrac{5}{1} = \tfrac{10}{3} = 3\tfrac{1}{3}$$
$$2\tfrac{2}{3} \div 1\tfrac{1}{4} = \tfrac{8}{3} \div \tfrac{5}{4} = \tfrac{8}{3} \cdot \tfrac{4}{5} = \tfrac{32}{15} = 2\tfrac{2}{15}$$

1. Add, subtract, multiply, and divide $\frac{1}{3}$ and $\frac{3}{7}$.

2. Add, subtract, multiply, and divide $3\frac{1}{5}$ and $1\frac{3}{4}$.

3. Add, subtract, multiply, and divide $+4.8$ and -6.4.

4. Add, subtract, multiply, and divide $8x$ and $-2x$.

5. Add, subtract, multiply, and divide $4x^2$ and $6x$.

6. a. Make a table and graph showing the total cost of a meal with a 22% total tax and tip. Use the input set $\{0, 10, 20, 30\}$.

b. What is the vertical axis intercept point? What does this point mean?

c. What equation describes the relationship in part a?

d. What is the meaning of the slope in this setting?

7. Unit analysis: How many days are there in 1,000,000 minutes?

For Exercises 8 and 9, use the formula for the area of a trapezoid, $A = \frac{1}{2}h(a + b)$, *where a and b are the parallel sides.*

8. Find the area if $h = 7$ ft, $a = 12$ ft, and $b = 21$ ft.

9. The area is 91 square inches, and the sum of the parallel sides is 13 inches. Find the height, h.

For Exercises 10 to 13, solve for x in each equation.

10. $\frac{2}{3}x = 64$

11. $2(x + 5) = x + 3$

12. $10 - x = 3x - 6$

13. $9 - 2(x - 3) = 21$

14. Solve for y: $3x + 2y = 18$.

15. a. What is the slope of the graph of the equation in Exercise 14?

b. What is the vertical axis intercept for this graph?

16. Solve for x: $5 - 3x \geq -4$. Show the solution set on a line graph.

17. Draw a coordinate graph of the solution set of $y \leq -x + 5$.

18. Find $f(-2)$ if $f(x) = x + 6$.

19. Find $f(-2)$ if $f(x) = x^2$.

20. Solve $s = \dfrac{a}{1 - r}$ for a.

21. Find the slope, y-intercept, and equation of the line containing the points $(-4, 0)$ and $(2, -6)$.

5

Ratios, Rates, and Proportional Reasoning

Figure 1 Federal Express DC-10 at Los Angeles International Airport, August 21, 1991

The airplane in Figure 1 was out of balance and became an embarrassment when it tipped. The balance point in an airplane is called the center of gravity. In this case, loading or unloading the plane caused the center of gravity to shift behind the main landing gear, making the plane tilt.

In this chapter, we study ratios, percents, and proportions and relate them to various applications, including rates, similar triangles, and word problems. We extend proportional reasoning to averages in statistics and geometry.

5.1 Ratios, Percents, and Rates

OBJECTIVES

- Write ratios.
- Find equivalent ratios.
- Simplify ratios containing expressions.
- Apply continued ratios.
- Write percents as ratios.
- Find percent change.
- Simplify ratios using unit analysis.
- Simplify rates and solve problems using unit analysis.

WARM-UP

1. Identify the greatest common factors and simplify to lowest terms. Leave as improper fractions.

   a. $\frac{10}{25}$ **b.** $\frac{42}{14}$ **c.** $\frac{128}{360}$

2. Simplify these exponential expressions.

   a. $\frac{x^7}{x^3}$ **b.** $\frac{x^5}{x}$ **c.** $\frac{x}{x^5}$

   d. $\frac{a^2b}{ab^2}$ **e.** $\frac{a^3b^3}{ab^2}$ **f.** $\frac{a^2b^2}{a^3b}$

3. Multiply and simplify to lowest terms.

   a. $\frac{7.5(10)}{10.5(10)}$ **b.** $\frac{6.75(100)}{10.75(100)}$

IN THIS SECTION, we examine ratios, percents, and rates. We return to unit analysis, using it to simplify ratios and to change from one rate to another.

Ratios

Our lives are filled with numbers, and we are continually comparing them: how many times as fast, how many times as tall, and so forth. A ratio is one of the ways we have to compare two (or more) numbers.

Definition of Ratio

> A **ratio** is a comparison of two (or more) like or unlike quantities.

There are three common ways to write a ratio: as a fraction, with the word *to*, or with a colon.

EXAMPLE  Writing ratios Write each ratio in two other ways.

a. $\frac{3}{4}$ **b.** 5 to 2 **c.** 4:5

Solution **a.** 3 to 4, 3:4 **b.** 5:2, $\frac{5}{2}$ **c.** $\frac{4}{5}$, 4 to 5 ●

If ratios have the same decimal notation or simplify to the same fraction, they are equivalent.

Equivalent Ratios

Two ratios are **equivalent ratios** if they simplify to the same number.

EXAMPLE 2 Finding equivalent ratios Show that $\frac{4}{6}$ and $\frac{6}{9}$ are equivalent ratios.

Solution

$$\frac{4}{6} = \frac{2 \cdot 2}{3 \cdot 2} = \frac{2}{3} \quad \text{and} \quad \frac{6}{9} = \frac{2 \cdot 3}{3 \cdot 3} = \frac{2}{3}$$

Each ratio contains a factor common to the numerator and denominator. These common factors simplify to $\frac{1}{1}$ and can be removed, leaving the same ratio: $\frac{2}{3}$. ●

SIMPLIFYING RATIOS Simplifying ratios that contain expressions or units of measure is exactly like simplifying fractions: We cross out the greatest common factor in the numerator and denominator. The ratios in Example 3 contain numbers and expressions. Beginning with Example 13, we will consider ratios containing units.

EXAMPLE 3 Simplifying ratios to lowest terms Write each ratio as a fraction. Then find and cross out the greatest common factor.

a. $3x^2:15x$ **b.** $24x^2:16x^3$

c. $14ab^3:8a^3b^2$ **d.** $(a + b)(a - b)$ to $(a - b)$

Solution **a.** $3x$ is the greatest common factor.

$$\frac{3x^2}{15x} = \frac{3x \cdot x}{3x \cdot 5} = \frac{x}{5}$$

b. $8x^2$ is the greatest common factor.

$$\frac{24x^2}{16x^3} = \frac{8x^2 \cdot 3}{8x^2 \cdot 2x} = \frac{3}{2x}$$

c. $2ab^2$ is the greatest common factor.

$$\frac{14ab^3}{8a^3b^2} = \frac{2ab^2 \cdot 7b}{2ab^2 \cdot 4a^2} = \frac{7b}{4a^2}$$

d. $(a - b)$ is the greatest common factor.

$$\frac{(a + b)(a - b)}{(a - b)} = \frac{(a + b)}{1}$$ ●

APPLICATIONS OF RATIOS TO SLOPE We return to the topic of slope, introduced in Section 4.3. A slope is a ratio written as a fraction.

$$\text{Slope} = \frac{\text{rise}}{\text{run}} = \frac{\text{change in output}}{\text{change in input}} = \frac{\Delta y}{\Delta x} = \frac{y_2 - y_1}{x_2 - x_1}$$

Recall that a slope of -2 is written as $-2/1$. The denominator, 1, completes the ratio of the change in output to the change in input, or rise to run.

EXAMPLE 4 Finding slope from a table and from a graph Find the slope of the graph of $y = -2x$, and write this slope as a ratio.

Solution *Note*: We could read the slope directly, as the coefficient of x in the equation, but instead we will review finding the slope from the data in Table 1 and from the coordinates on the graph in Figure 2.

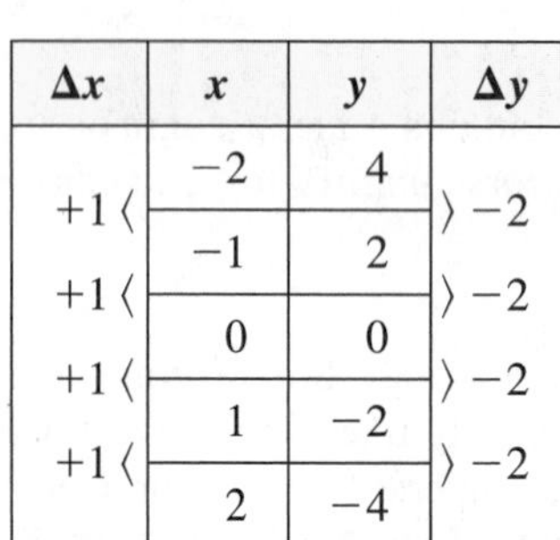

Δx	x	y	Δy
	-2	4	
$+1$			-2
	-1	2	
$+1$			-2
	0	0	
$+1$			-2
	1	-2	
$+1$			-2
	2	-4	

Table 1

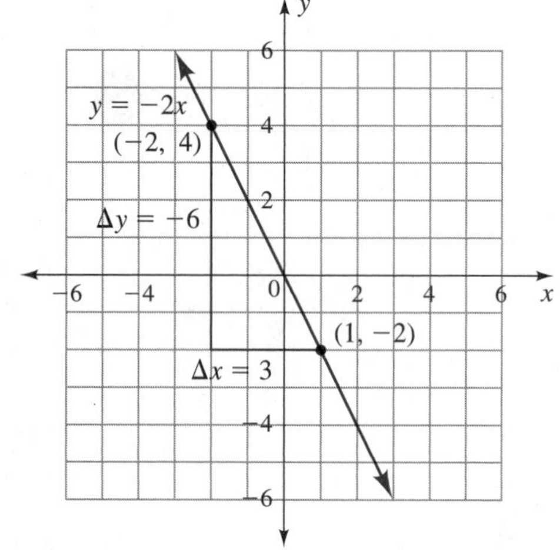

Figure 2

Table 1 shows a 2-unit decrease in output for a 1-unit increase in input. Thus, $y = -2x$ has a slope of $-2/1$, or -2. The slope between two arbitrarily chosen coordinates, $(-2, 4)$ and $(1, -2)$, is

$$\frac{y_2 - y_1}{x_2 - x_1} = \frac{-2 - 4}{1 - (-2)} = \frac{-6}{3} = \frac{-2}{1}$$

The slope is the ratio -2 to 1. The line drops 2 units for each 1-unit change to the right.

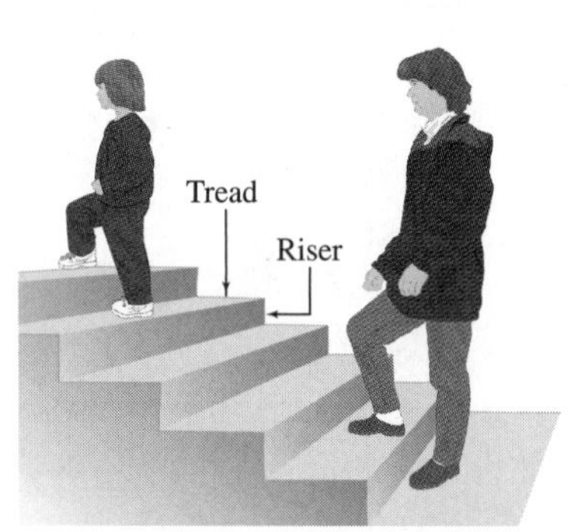

Figure 3

Slopes are common in construction. We looked at the slopes of roofs in Section 4.3. Here we simplify the slopes of staircases. The slope of a staircase is the ratio formed by dividing the height of the *riser* by the width of the *tread* (see Figure 3). The challenge in simplifying actual staircase ratios is that the measurements are in fractions of an inch, but the measurements in Example 5 have already been changed from fraction notation to decimal notation. Look for how the decimals are removed before the slope ratios are simplified.

EXAMPLE 5 Simplifying staircase ratios Write each ratio of rise to run in fraction notation. Simplify to lowest terms, if possible.

a. 5 to 12 **b.** 7.5 to 10.5 **c.** 6.75 to 10.75

Solution **a.** $\frac{5}{12}$

b. First multiply the numerator and denominator by 10 to clear the decimals.

$$\frac{7.5}{10.5} = \frac{7.5(10)}{10.5(10)} = \frac{75}{105} = \frac{3 \cdot 5 \cdot 5}{3 \cdot 5 \cdot 7} = \frac{5}{7}$$

Factor the numerator and denominator, and eliminate common factors.

c. First multiply the numerator and denominator by 100 to clear the decimals.

$$\frac{6.75}{10.75} = \frac{6.75(100)}{10.75(100)} = \frac{675}{1075} = \frac{25 \cdot 27}{25 \cdot 43} = \frac{27}{43}$$

Factor the numerator and denominator, and eliminate common factors. ●

APPLICATION OF RATIOS TO BATTING AVERAGES A baseball batting average is the ratio of hits to times at bat. Many ratios, such as batting averages, are changed from fraction to decimal notation to make them easier to compare.

$$\text{Batting average} = \frac{\text{number of hits}}{\text{number of times at bat}}$$

EXAMPLE 6 **Comparing ratios: batting averages** In 1894, Hugh Duffy (Boston, National League) had 236 hits in 539 at-bats. In 1924, Rogers Hornsby (St. Louis, National League) had 227 hits in 536 at-bats. Write each batting average as a ratio in fraction notation, and find the equivalent decimal, rounded to the nearest thousandth. Who had the higher batting average?

Solution Duffy had an average of $\frac{236}{539} \approx 0.438$, while Hornsby had $\frac{227}{536} \approx 0.424$. Duffy's was higher. ●

Batting averages are often written as a ratio of 1000. The averages for the players in Example 6 are 438 to 1000 and 424 to 1000. Baseball fans drop the reference to 1000 and just say a 438 or 424 batting average.

CONTINUED RATIOS In Example 6, we might say that Duffy and Hornsby had batting averages of 438 and 424, respectively.

Respectively means "in the same order as listed" or "as given." The use of the word *respectively* saves repeating information in a paragraph. We use the word *respectively* with our next form of ratio: a continued ratio.

A **continued ratio** is *a ratio of three or more quantities.* In business, a continued ratio might represent various individuals' interests in an estate or in profit sharing. Such a ratio might be used to describe the quantities of ingredients in food manufacturing or to allocate time spent by a specialist on a number of different projects. Continued ratios are also the basis for writing formulas for chemical compounds.

Finding Shares from a Continued Ratio

To find the shares from a continued ratio $a:b:c$ numerically:

1. Add the numbers in the ratio: $a + b + c$.
2. Divide the amount to be shared by the total of the numbers in the ratio, to get the amount per share.
3. Multiply each number in the continued ratio by the amount per share.
4. Check by adding the individual shares.

To find the shares from a continued ratio $a:b:c$ with algebra:

1. Let $x =$ the amount per share.
2. Write the continued ratio as an expression: $ax + bx + cx$.
3. Set the expression equal to amount to be shared.
4. Solve the equation and check.

EXAMPLE  Applying continued ratios to find shares: profit sharing The owners of a business have contributed funds in the continued ratio 5:3:1. If there is a $90,000 profit, how much will each receive?

Numeric Solution The ratio 5:3:1 means that for $5 + 3 + 1 = 9$ total parts, 5 go to the first owner, 3 to the second owner, and 1 to the third owner. If the profit is $90,000, one part will be

$$\$90{,}000 \div 9 = \$10{,}000$$

The three owners will receive $10,000 times their shares, or $50,000, $30,000, and $10,000, respectively.

Algebraic Solution Let $x =$ the value of one share of the profit. The continued ratio gives the expression $5x + 3x + 1x$, and the sum of the shares must be $90,000.

$$\begin{aligned} 5x + 3x + 1x &= 90{,}000 \\ 9x &= 90{,}000 \\ x &= 10{,}000 \end{aligned}$$

Each share is $10,000. The three owners will receive $50,000, $30,000, and $10,000, respectively. ●

EXAMPLE 8 Applying continued ratios to find shares: time on the job During one day, an accounting consultant spent 4 hours on project A, 3 hours on project B, and 1 hour on project C.

a. Write a continued ratio to describe the time spent on the three projects.

b. Find the amount charged to each project if the consultant's daily salary is $400.

Solution **a.** The ratio 4:3:1 describes the time spent on projects A, B, and C.

b. The continued ratio 4:3:1 means that for $4 + 3 + 1 = 8$ total parts, 4 parts apply to project A, 3 parts to project B, and 1 part to project C. If the total value of the day's work is $400, one part will be

$$\$400 \div 8 = \$50$$

If each hour is valued at $50, the projects will be charged $200, $150, and $50, respectively. (Exercise 23 asks you to solve this problem algebraically.) ●

A chemical formula is a continued ratio—the ratio of atoms in a molecule. The water molecule, with formula H_2O, has 2 atoms of hydrogen for each atom of oxygen. A subscript 1 is assumed on oxygen: $O = O_1$. Nicotine, $C_{10}H_{14}N_2$, is composed of 10 atoms of carbon, 14 atoms of hydrogen, and 2 atoms of nitrogen.

Three molecules of water would be written $3H_2O$, to represent 6 atoms of hydrogen combined with 3 of oxygen. The 3 in front is a coefficient of the entire formula and is applied to each chemical element. It does not change the ratio of the elements in the formula.

EXAMPLE 9 Interpreting chemical formulas Two molecules of caffeine is written $2C_8H_{10}N_4O_2$. How many atoms of each element are contained in these two molecules?

Solution The formula $2C_8H_{10}N_4O_2$ represents 16 atoms of carbon, 20 atoms of hydrogen, 8 atoms of nitrogen, and 4 atoms of oxygen. ●

Ratios and Percents

Because both ratios and percents can be expressed as fractions, we can rewrite percents as ratios, and vice versa.

FINDING RATIOS FROM PERCENTS Recall that a percent n is *a ratio of n to 100.* The quantity 15 percent means 15 per hundred and is written $\frac{15}{100}$ as a fraction or ratio.

In highway construction, *slope* is specified in terms of **percent grade**. To change a percent grade to a slope ratio, rewrite the percent as a fraction and simplify to lowest terms.

EXAMPLE  Finding ratios from percents: slope of a road grade Highway signs, like the one in Figure 4, are placed near the tops of mountain passes to warn truckers and motorists about the steepness of the downhill grade. A 6% grade indicates that there is a drop of 6 vertical feet for every 100 feet traveled horizontally. Write the average slope of the road as a ratio.

Figure 4

Solution The average grade is 6%:

$$\frac{6}{100} = \frac{3}{50}$$

Although we can write ratios as percents, sometimes the wording or emphasis changes. Suppose that, in a recent local election, the ratio of votes in favor of building a swimming pool to votes against is 2 to 1. As in a continued ratio, the votes make $2 + 1 = 3$ total parts. The ratio of votes for the pool to total votes is 2 to 3. The fraction of voters supporting the pool is $\frac{2}{3}$. The percent of voters supporting the pool is $66\frac{2}{3}\%$.

Finding Percent

To find the percent, we divide the part by the total.

Example 11 illustrates how we usually compare parts to a total in working with a percent.

EXAMPLE  Finding percents Here is how Monique spent her time during one week. What percent of her week was spent on each activity?

Classes:	12 hours
Study:	30 hours
Work:	20 hours
Sleep:	56 hours (8 hours each night)
Transportation:	7 hours
Other:	remaining time (walking between classes, with family, eating, etc.)

Solution 24 hours per day × 7 days per week = 168 hours per week

Classes:	12 ÷ 168 = 0.071 = 7.1%
Study:	30 ÷ 168 = 0.179 = 17.9%
Work:	20 ÷ 168 = 0.119 = 11.9%
Sleep:	56 ÷ 168 = 0.333 = 33.3%
Transportation:	7 ÷ 168 = 0.042 = 4.2%
Other:	168 − (12 + 30 + 20 + 56 + 7) = 43 hr remaining
	43 ÷ 168 = 0.256 = 25.6%

Check: The percents add to 100%. ✓ ●

FINDING PERCENT CHANGE To find the **percent change**, we *subtract the original number from the new number and divide the difference by the original number.* If the answer is positive, the percent change is an increase. If the answer is negative, the percent change is a decrease.

$$\text{Percent change} = \frac{\text{new number} - \text{original number}}{\text{original number}}$$

In Example 12, we compare prices in 1986 and 1995. (The project in Exercise 80 asks you to compare prices in 1995 and the current year.)

EXAMPLE 12 Finding percent change The following prices are from a general-merchandise drug store. Estimate the percent change in prices. Then find the percent change to the nearest tenth of a percent.

a. T-120 videotape: 1986, \$4; 1995, \$1.89

b. Three-piece bathroom-rug set: 1986, \$8; 1995, \$25.97

Solution **a.** The difference in prices is about \$2, so the change in price is near 50% of the 1986 price.

$$\frac{\$1.89 - \$4.00}{\$4.00} = \frac{-2.11}{4.00} = -0.5275$$

The 1995 price is 52.8% lower than the 1986 price.

b. The difference in prices is about \$18, or a little more than twice \$8. The percent change will be about 200%.

$$\frac{\$25.97 - \$8.00}{\$8.00} = \frac{17.97}{8.00} = 2.24625$$

The 1995 price is 224.6% higher than the 1986 price. ●

Think about it 1: If you estimated 300% in part b of Example 12, you compared the 1995 price to the 1986 price, instead of comparing the change in price to the 1986 price. Although \$25.97 is about three times \$8, the change, \$17.97, is only slightly more than twice \$8. Calculate the percent formed by dividing the new price by the old price in each part of Example 12.

Ratios and Unit Analysis

The ratios in the rest of this section contain units of measure. In some problems, the units are the same. In other problems, the units can be changed to like units through unit analysis (Section 2.5).

In Example 13, the units are alike and drop out when we build ratios. The ratios are from dosage computation—an important topic for nurses, medical record clerks, and pharmacy students.

EXAMPLE 

Finding ratios with like units Adult dosage applies to a person who weighs at least 150 pounds, is at least 150 months of age, or has a body surface area of at least 1.7 square meters. (Body surface area estimates are important in burn treatment.) Suppose a 60-pound 8-year-old has a body surface area of 1 square meter.

a. What is the ratio of the child's weight to adult weight?

b. What is the ratio of the child's age to adult age?

c. What is the ratio of the child's body surface area to adult body surface area?

d. With decimal notation, compare the ratios from parts a, b, and c.

Solution

a. $\dfrac{60 \text{ lb}}{150 \text{ lb}} = \dfrac{2 \cdot 30}{5 \cdot 30} = \dfrac{2}{5}$

b. Using unit analysis, we have $8 \text{ yr} \cdot \dfrac{12 \text{ mo}}{1 \text{ yr}} = 96 \text{ mo}$.

$$\frac{96 \text{ mo}}{150 \text{ mo}} = \frac{6 \cdot 16}{6 \cdot 25} = \frac{16}{25}$$

c. $\dfrac{1 \text{ m}^2 \text{ body surface area}}{1.7 \text{ m}^2 \text{ body surface area}} = \dfrac{1(10)}{1.7(10)} = \dfrac{10}{17}$

d. We use division to change the ratios to decimals. The weight ratio is 0.4 to 1. The age ratio is 0.64 to 1. The body surface area ratio is 0.59 to 1. The weight ratio would give the lowest dosage, and the age ratio would give the highest dosage. ●

In part b of Example 13, we used unit analysis when we multiplied by 12 mo/1 yr. That step changed years into months. Recall that *unit analysis* is a method for changing from one unit of measure or rate to another. Here are the steps in unit analysis, from Section 2.5.

Unit Analysis: Step by Step

1. Identify the unit of measure to be changed, and identify the unit of measure for the answer. Write both as fractions.
2. List facts containing the unit to be changed and all units needed to get to the unit for the answer.
3. Write the unit to be changed (or a fact containing the unit) as the first fraction. Set up a product of fractions, using the list of facts. Arrange fractions so that the units to be eliminated appear in both a numerator and a denominator.
4. Eliminate units of measure where possible, and calculate the numbers.

When we simplify a ratio, we find out how many times as large as the bottom of a ratio the top is. If the units in the parts of a ratio are different, they must be made the same in order to compare the parts. In Example 14, we change to like units with unit analysis in order to simplify ratios and compare their parts.

EXAMPLE 

Finding ratios with unlike units Use unit analysis to change the units on each ratio to like units. Simplify the ratio. Use the simplified ratio to compare the measures.

a. 100 centimeters: 2 meters **b.** 1500 grams to 1 kilogram

c. Ostrich height to smallest hummingbird length, 9 feet to $2\frac{1}{4}$ inches

Solution In each part below, the useful fact contains both units.

a. A useful fact is 100 cm = 1 m.

$$\frac{100 \text{ cm}}{2 \text{ m}} \cdot \frac{1 \text{ m}}{100 \text{ cm}} = \frac{100}{200} = \frac{1}{2}$$

The units can all be eliminated. We see that 100 centimeters is half of 2 meters.

b. A useful fact is 1000 g = 1 kg.

$$\frac{1500 \text{ g}}{1 \text{ kg}} \cdot \frac{1 \text{ kg}}{1000 \text{ g}} = \frac{1500}{1000} = \frac{3}{2}$$

The units can all be eliminated. We see that 1500 grams is $\frac{3}{2}$, or $1\frac{1}{2}$, times as large as 1 kilogram.

c. A useful fact is 12 in. = 1 ft.

$$\frac{9 \text{ ft}}{2\frac{1}{4} \text{ in.}} \cdot \frac{12 \text{ in.}}{1 \text{ ft}} = \frac{108}{2.25} = \frac{48}{1}$$

The units can all be eliminated. We divide 108 by 2.25 with a calculator, to get 48. The ostrich's height is 48 times the length of the hummingbird. ●

Rates and Unit Analysis

As the final topic of this section, we look at **rates**, *special ratios for comparing quantities with different units.* A rate often contains the word *per,* which means "for each" or "for every."

Some rates are abbreviated with just a single letter for each word, such as mpg and rpm in Example 15. When we change a rate into a ratio, we place the word that follows *per* in the denominator.

EXAMPLE 15 Writing rates as fractions Write these rates in fraction notation.

a. mph (miles per hour)

b. liters per minute

c. mpg (miles per gallon)

d. rpm (revolutions per minute)

e. dollars per hour

Solution Observe the abbreviations used within the fraction form.

a. $\frac{\text{mi}}{\text{hr}}$ **b.** $\frac{\text{L}}{\text{min}}$ **c.** $\frac{\text{mi}}{\text{gal}}$ **d.** $\frac{\text{rev}}{\text{min}}$ **e.** $\frac{\text{dollars}}{\text{hr}}$ ●

To change from one rate to another, we use unit analysis. We usually have to change two different kinds of units. In Example 16, we change miles per hour into feet per minute.

EXAMPLE 16 Changing a rate with unit analysis Change 100 miles per hour to feet per minute.

Solution We use the four steps in a unit analysis.

Step 1: We need to change miles to feet and hours to minutes.

Step 2: Useful facts:

5280 ft = 1 mi

60 min = 1 hr

Step 3: We start with 100 miles per hour as a fraction.

$$\frac{100 \text{ mi}}{1 \text{ hr}} \cdot \frac{5280 \text{ ft}}{1 \text{ mi}} \cdot \frac{1 \text{ hr}}{60 \text{ min}} = \frac{100 \cdot 5280 \cdot 1 \text{ ft}}{1 \cdot 1 \cdot 60 \text{ min}}$$
$$= \frac{8800 \text{ ft}}{1 \text{ min}}$$

Step 4: We see that 100 miles per hour is 8800 feet per minute. ●

Think about it 2: Is 8800 feet per minute faster or slower than one mile per minute?

In Step 3 of Example 16, the units that we eliminated were not next to each other. If you want the units to be next to each other in a rate problem, place the rate to be changed in the center and write the facts as fractions on each side.

$$\frac{5280 \text{ ft}}{1 \text{ mi}} \cdot \frac{100 \text{ mi}}{1 \text{ hr}} \cdot \frac{1 \text{ hr}}{60 \text{ min}}$$

We will use this suggestion in Example 17, where we change drops per second into cups per hour.

EXAMPLE 17 Using unit analysis to change rates Suppose a faucet leaks 1 drop per second. Find out how many cups per hour are wasted.

Solution Suppose an experiment indicates that there are 16 drops per teaspoon.

Step 1: We need to change drops into cups and seconds into hours.

Step 2: Useful facts:

16 drops = 1 teaspoon (tsp)
3 tsp = 1 tablespoon (tbsp)
16 tbsp = 1 cup
60 sec = 1 min
60 min = 1 hr

Step 3: We place the 1 drop per second in the center of the paper and write the other facts as fractions on each side of the 1 drop per second fraction.

Change time ← Start here → Change quantity

$$\frac{60 \text{ min}}{1 \text{ hr}} \cdot \frac{60 \text{ sec}}{1 \text{ min}} \cdot \frac{1 \text{ drop}}{1 \text{ sec}} \cdot \frac{1 \text{ tsp}}{16 \text{ drops}} \cdot \frac{1 \text{ tbsp}}{3 \text{ tsp}} \cdot \frac{1 \text{ cup}}{16 \text{ tbsp}}$$
$$= \frac{60 \cdot 60 \text{ cup}}{16 \cdot 3 \cdot 16 \text{ hr}}$$
$$\approx 4.7 \text{ cups per hour}$$ ●

Calculator note: If you do Example 17 on a calculator, you need to be careful about grouping. Use these keystrokes, and compare answers. Which is correct?

a. 60 [×] 60 [÷] 16 [×] 3 [×] 16 [=]
b. 60 [×] 60 [÷] 16 [÷] 3 [÷] 16 [=]
c. 60 [×] 60 [÷] [(] 16 [×] 3 [×] 16 [)] [=]

Recall that the fraction bar is a grouping symbol. Option a illustrates why the numerator and denominator must often be entered within parentheses. Because of the order of operations built into the scientific calculator, the keystrokes in option a will not give the correct solution. Both option b and option c are correct.

ANSWER BOX

Warm-up: 1. a. 5, $\frac{2}{5}$ **b.** 14, $\frac{3}{1}$ **c.** 8, $\frac{16}{45}$ **2. a.** x^4 **b.** x^4 **c.** $1/x^4$ **d.** a/b **e.** a^2b **f.** b/a **3. a.** $\frac{5}{7}$ **b.** $\frac{27}{43}$ **Think about it 1: a.** $1.89/4.00 \approx 47.3\%$ **b.** $25.97/8 \approx 324.6\%$

Think about it 2: Faster, $\dfrac{8800 \text{ ft}}{1 \text{ min}} \cdot \dfrac{1 \text{ mi}}{5280 \text{ ft}} \approx 1.67$ mi per min

EXERCISES 

In Exercises 1 and 2, write each ratio in two other ways.

1. a. 5 to 3 **b.** $\frac{3}{2}$ **c.** 4:9

2. a. $\frac{4}{5}$ **b.** 8:5 **c.** 7 to 11

In Exercises 3 to 8, identify common factors and simplify to lowest terms.

3. a. $\frac{15}{35}$ **b.** $\frac{48}{16}$ **c.** $\frac{5280}{3600}$

4. a. $\frac{84}{32}$ **b.** $\frac{52}{26}$ **c.** $\frac{1024}{288}$

5. a. $12x:4x^2$ **b.** $2xy$ to $6x^2y$ **c.** $x:3x^4$

6. a. $18x:6x^3$ **b.** $6x^2y$ to $9xy^2$ **c.** $x:4x^3$

7. a. abc to ace **b.** $a(b + c)$ to ac

c. $(a + b):a(a + b)$

8. a. def to dig **b.** $b(c + d)$ to bc

c. $(a - b):b(a - b)$

For Exercises 9 to 18, find the slope ratio for each equation.

9. $y = 3x$ **10.** $y = 4 + 2x$ **11.** $y = 2 - 4x$

12. $y = -5x$ **13.** $y = \frac{1}{2}x$ **14.** $y = -\frac{1}{4}x$

15. $y = 2.98x + 5.00$ **16.** $y = 20 - 1.50x$

17. $y = 5 - 0.05x$ **18.** $y = 0.65x + 5.15$

For Exercises 19 and 20, write the staircase slopes in fraction notation and simplify to lowest terms. All measures are in inches, so units may be disregarded.

19. a. 3 to 15 **b.** 7.5 to 10

c. 6.75 to 10.5 **d.** 7.25 to 9.75

20. a. 6 to 12 **b.** 7 to 10.5

c. 7.75 to 10.25 **d.** 6.75 to 10.25

Find the batting average, in decimal notation, for each of the players in Exercises 21 and 22. Round to the nearest thousandth.

21. a. Roberto Clemente: 209 hits in 585 times at bat

b. Juan Gonzalez: 152 hits in 584 times at bat (The hits included 43 home runs.)

22. a. Ken Griffey, Jr.: 174 hits in 565 times at bat

b. Ivan Calderon: 141 hits in 470 times at bat

23. Write an algebraic solution for Example 8, where the consultant works 4 hours, 3 hours, and 1 hour on three projects, A, B, and C. His daily charge is $400.

24. A $1 million estate is to be divided among Goodwill, Habitat for Humanity, and United Way in the ratio 10:10:5. How much does each receive?

25. Cosmo, Timo, and Sven earn a profit of 2500 crowns (Swedish currency) on a dance. They split the profits 4:3:3. How much will each receive?

26. Virginia, James, and Ursula receive a $5000 advance on royalties. All three authors are writing the book, but Ursula is contributing another full share by keying the book into the computer. They agree to split the $5000 in a 1:1:2 ratio. How much should each earn?

What is the ratio of atoms in the chemical formulas in Exercises 27 to 30? (Na is sodium and Ca is calcium; the other elements were listed in the text.)

27. $C_9H_8O_4$ (the active ingredient in aspirin)

28. $NaHCO_3$ (baking soda)

29. $CaCO_3$ (chalk)

30. $C_{12}H_{22}O_{11}$ (sugar)

31. Wizard Island in Crater Lake, Oregon is a small cinder cone. The sides of the cone form a 62.5% grade. What is the average slope in fraction notation?

32. Lookout Mountain in Chattanooga, Tennessee has an inclined railway with a grade near the top of 72.7%. What is the slope in fraction notation? Is the grade closest to $\frac{3}{4}$, $\frac{5}{6}$, or $\frac{6}{7}$?

What percent grade is each of the ramps described in Exercises 33 to 36?

33. A wheelchair access ramp with a slope of 1 to 12

34. A loading ramp with a slope of $\frac{1}{4}$

35. An access ramp with a slope of $\frac{1}{8}$

36. A freeway ramp with a slope of 1 to 20

37. Marguerite spends the following each month: rent, \$360; utilities, \$132; food, \$240; clothes, \$84; travel (including bus pass), \$96; entertainment, \$48; renters' insurance, \$60; books and magazines, \$24; savings, \$156. What percent of her total monthly income does she spend on each item?

38. Jasmin spends the following each month: rent, \$400; utilities, \$140; food, \$320; clothes, \$130; car payment, \$300; car insurance, \$150; gasoline, \$60; savings, \$20; other, \$80. What percent of her total monthly income does she spend on each item?

In Exercises 39 to 42, estimate the percent change. Then find the percent change to the nearest tenth of a percent. Finally, find the percent the 1995 price is of the 1986 price.

39. Conair folding hair dryer: 1986, \$12; 1995, \$16

40. S&W tomatoes, 28-oz can: 1986, \$1; 1995, \$0.79

41. True Temper plastic rake: 1986, \$5; 1995, \$2.99

42. Comet powdered cleaner: 1986, \$0.34; 1995, \$0.50

43. Tom has a 20% rent increase.

a. If his old rent is \$450, what is his new rent?

b. If his new rent is \$450, what is his old rent? Guess and check. Write an equation using the percent change formula. Solve your equation.

44. Alex has a 25% salary increase.

a. If her old salary is \$3200, what is her new salary?

b. If her new salary is \$3200, what is her old salary? Guess and check. Write an equation using the percent change formula. Solve your equation.

In Exercises 45 and 46, change the units to like units with a unit analysis multiplication, and simplify the ratio.

Useful facts:
1000 milliliters = 1 liter
1000 grams = 1 kilogram
1000 meters = 1 kilometer
100 centimeters = 1 meter
16 ounces = 1 pound
1 yard = 36 inches
1 yard = 3 feet
1 mile = 5280 feet
4 quarts = 1 gallon
2 pints = 1 quart

45. **a.** $\frac{1}{2}$ foot to 1 inch

b. 3000 grams to 1 kilogram

c. 32 ounces to 5 pounds

d. $2\frac{2}{3}$ yards to 2 feet

e. 200 milliliters to 20 liters

f. 2 years to 150 months

g. 20 minutes to $\frac{1}{4}$ hour

46. **a.** 1 foot to 3 inches

b. 2 meters to 1200 centimeters

c. 1500 meters to 1 kilometer

d. 2 quarts to 4 gallons

e. 2 liters to 2000 milliliters

f. 4 years to 150 months

g. 120 minutes to $\frac{1}{2}$ hour

Change the slopes in Exercises 47 and 48 (from the bicycling tour in Section 4.3) into percent grades. The units must be alike to change to percents.

47. 30 feet per mile

48. 250 feet per mile

In Exercises 49 and 50, write a sentence using each of the rates in an appropriate setting.

49. **a.** stitches per inch **b.** feet per second

c. miles per gallon **d.** revolutions per hour

50. **a.** revolutions per minute **b.** gallons per hour

c. kilometers per hour **d.** calories per day

For Exercises 51 to 58, write a ratio from the sentence and simplify it, if possible.

51. Sergei adds 3 cans of water to each can of frozen orange juice.

52. Frantisek adds 2 cans of milk to a can of soup.

53. Stephanie pays $480 tuition for 12 credits.

54. Jocelyn earns $250 in a 40-hour week.

55. Mikael adds a half pint of oil to 1 gallon of gasoline.

56. Sylvia adds 1 tablespoon of plant fertilizer to a gallon of water.

57. Betty orders an intravenous feeding of 240 cc in 16 hours.

58. Johann fills an intravenous feeding order for 250 mL in 12 hours.

In Exercises 59 to 62, change miles per hour to feet per second. Start with a fraction showing miles per hour, and then set up the unit analysis product of fractions.

Useful facts: 5280 feet = 1 mile
60 seconds = 1 minute
60 minutes = 1 hour

59. 30 miles per hour **60.** 55 miles per hour

61. 80 miles per hour **62.** 500 miles per hour

In Exercises 63 to 66, change feet per second to miles per hour (mph).

63. 88 feet per second **64.** 66 feet per second

65. 220 feet per second **66.** 4.4 feet per second

67. Use unit analysis to find the number of feet you have driven in a 4-second reaction time at 60 miles per hour. Guess before you start: Will it be more than, equal to, or less than a residential city block of length 334 feet?

68. A lawn mower cuts 100 feet in 30 seconds. How many miles per hour is this?

69. The oil in a lawn mower needs to be changed every 25 hours. If the mower travels 100 feet in 30 seconds and shuts off when stopped, how many miles will it travel between oil changes?

70. How far is it possible to bicycle in 10 minutes at 20 miles per hour?

71. How far is it possible to walk in 30 minutes at 6 miles per hour?

72. 1 gallon for five miles is how many dollars per day of driving?

1 hr to travel 55 mi

1 gal is $1.25

1 driving day is 10 hr

73. 250 mL in 12 hours is how many microdrops per minute?

60 microdrops is 1 mL

1 hr is 60 min

74. An infant of 1 year should receive how many milligrams (mg)?

Adult dosage is 500 mg

150 mo = 1 adult dosage

1 yr = 12 mo

75. An airplane is traveling 300 miles per hour. The pilot announces that the plane is 40 miles from the airport. In how many minutes will the flight be at the airport?

In Exercises 76 and 77, a mole is a chemical unit of measure, not a small, furry animal. Round answers to the nearest tenth.

76. Apply unit analysis to 25.0 grams CH_4 to find grams HCN. (It is assumed that the environment has an ample source of nitrogen, N.) *Hint:* Start with 25.0 grams CH_4 over 1.

1 mole CH_4 is 16.0 grams CH_4

1 mole HCN is 27.0 grams HCN

2 moles CH_4 is equated with 2 moles HCN

77. Apply unit analysis to 7.0 grams Na to find grams NaCl. (It is assumed that the environment has an ample source of chlorine, Cl.) *Hint:* Start with 7.0 grams Na over 1.

1 mole NaCl is 58.5 grams NaCl

2 moles Na is equated with 2 moles NaCl

1 mole Na is 23.0 grams Na

Projects

78. ***Circumference and Diameter.*** Use a ruler and string to measure the diameter and circumference of ten circular objects, and record your results in a table. Measure in centimeters.

a. Graph your results, with diameter as input and circumference as output.

b. Draw a straight line that seems to pass through most of the points.

c. Where should your line cross the y-axis? the x-axis?

d. Pick two points and calculate the slope of the line.

e. What should the slope be? Why?

79. ***Staircase Rule.*** Carpenters have a "rule of thumb" for designing staircases. See if you can discover the rule. All measurements in Example 5 and Exercises 19 and 20 are consistent with the carpenter's rule. It is possible to study the data and predict the rule. Linear regression on a graphing calculator will give you a hint, but the rule is much simpler than the calculator results imply. To use a calculator, do the following:

a. Make a table, with width of tread (run) as input and height of riser (rise) as output. Enter the data, with inputs in L1 and outputs in L2.

b. Graph the data. Imagine a line through the data.

c. Fit an equation to the data with linear regression.

d. Graph the equation from the data. To discover the carpenter's rule, look for nearby lines with equations containing whole numbers.

e. Check that your rule works with the data.

Your written report should include a table, a graph, and a summary of your thinking, whether you used reasoning or a calculator.

80. ***Price Changes.*** Example 12 and Exercises 39 to 42 used 1986 and 1995 prices for selected items. Find prices of comparable products today, and calculate the percent change since 1995. Discuss why some items got cheaper and others more expensive.

81. ***Circle Graphs.*** A protractor can be used to divide a half circle into 180 degrees (or 180°) and a full circle into 360°. Find the part of a circle needed to show each of the percents in Example 11 on a circle graph, or pie chart. (*Hint:* Multiply each percent by 360°.) Draw the circle graph.

82. ***Slopes and Angles.*** The figure shows a protractor drawn on graph paper, with 0 to 10 marked on horizontal and vertical lines. The graph is not a coordinate graph. The angle between line AB and the horizontal is about 26.5° because line AB passes between 26° and 27°. From the horizontal and vertical lines, line AB has a slope (rise over run) of $\frac{5}{10} = \frac{1}{2} = 0.5$. Thus, we can say that a line with a slope of $\frac{1}{2}$ makes a 26.5° angle with the horizontal. By using a ruler to connect A and points on the vertical scale, we can read other angle measures and slopes.

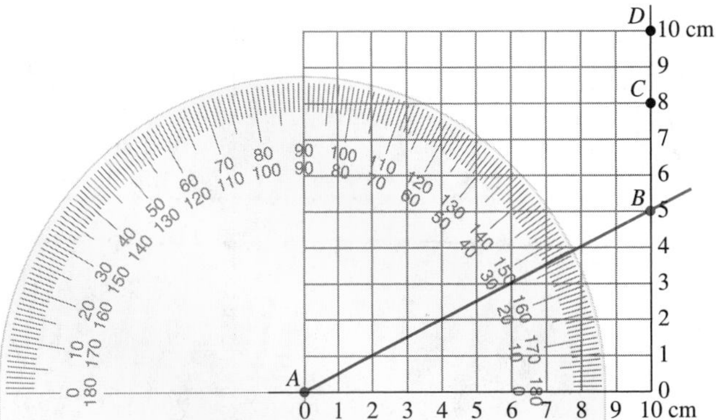

a. What angle does line AC make with the horizontal?

What is the slope of AC?

b. What angle does line AD make with the horizontal?

What is the slope of AD?

Estimate the slopes for the angles in parts c to e:

c. 10° **d.** 20° **e.** 30°

Estimate the angles for the slopes in parts f to h:

f. 1/10 **g.** 3/10 **h.** 7/10

83. ***Angles, Slope, and Percent Grade.*** For selected values of the angle a line makes with the positive horizontal axis, the table shows the slope of the line, both as a decimal and as a percent (percent grade).

Angle with Horizontal	Slope as Decimal	Slope as Percent (percent grade)
0°	0	0
10°	0.176	17.6
30°	0.577	57.7
60°	1.732	173.2

a. Use $\Delta y/\Delta x$ to find three slopes from the first two columns of the table.

b. Is the relationship between angle and slope a linear function?

5.2 Proportions and Proportional Reasoning

OBJECTIVES

- Write equivalent ratios as proportions.
- Check proportions using cross multiplication.
- Solve percent problems and equations using proportions.
- Set up proportions from applications.
- Solve proportions and equations using cross multiplication.

WARM-UP

These Warm-up exercises provide practice with the distributive property. We will use the distributive property in Example 7 to solve equations. Multiply, using the distributive property.

1. $6(x + 4)$ **2.** $4(x + 5)$ **3.** $8(x + 8)$

4. $20(x - 1)$ **5.** $6(x - 6)$ **6.** $-4(x - 3)$

7. $-4(x + 1)$ **8.** $-5(x - 1)$ **9.** $-2(x + 3)$

10. $-3(x - 3)$

IN THIS SECTION, we look at proportions and apply them to solving application problems and equations. The emphasis is on numeric and algebraic work.

Proportions and Proportional Statements

Safety guidelines for home and professional ladders warn people to set the top of a straight ladder four times as high as the distance the foot of the ladder is from the wall (see Figure 5). The height-to-base ratio is 4 to 1. This 4:1 ratio is also the slope of the ladder.

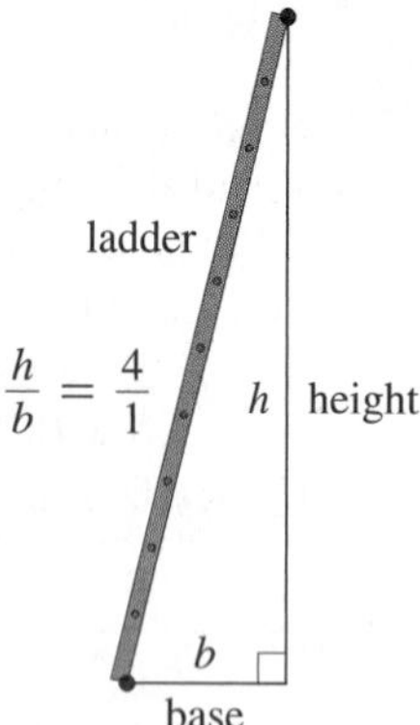

Figure 5

EXAMPLE 1 Exploration: safe-ladder ratio Each of the following ladders reaches 12 feet up a wall. Write a simplified height-to-base ratio for each ladder position. Compare the simplified ratio with 4:1. How is an accident likely to happen if the ratio is not safe?

a.

Unsafe or safe ratio?

b.

12 ft

2 ft

Unsafe or safe ratio?

c.

Unsafe or safe ratio?

Solution

a. 12:3 = 4:1, a safe ratio

b. 12:2 = 6:1, not a safe ratio; the ladder may tip over backwards because the ratio is larger than 4:1.

c. 12:6 = 2:1, not a safe ratio; the bottom of the ladder may slide away from the wall because the ratio is smaller than 4:1. ●

Proportions and Cross Multiplication

In part a of Example 1, the safe-ladder ratios form a proportion:

$$12:3 = 4:1$$

or, in fraction notation,

$$\frac{12}{3} = \frac{4}{1}$$

This leads to the definition of a proportion.

Definition of Proportion

Two equal ratios form a **proportion**:

$$\frac{a}{b} = \frac{c}{d}$$

where $b \neq 0$ and $d \neq 0$. By the cross multiplication property, the proportion

$$\frac{a}{b} = \frac{c}{d} \quad \text{implies} \quad a \cdot d = b \cdot c.$$

Once we know the safe-ladder ratio, we can use cross multiplication to check any ladder position to see whether it is safe. If a ladder reaches 20 feet up a wall and is set 5 feet from the base of the wall, its height-to-base ratio is 20 to 5. Either we can write $\frac{20}{5}$ in lowest terms as $\frac{4}{1}$ or we can check with a proportion:

$$\frac{20}{5} = \frac{4}{1}$$

The numbers on the two diagonals (20 and 1, 5 and 4) multiply to the same number: 20. **Cross multiplication** *sets these diagonal products equal,* with the equation $a \cdot d = b \cdot c$. The process is called cross multiplication because lines drawn between the parts being multiplied form an X-shaped cross.

EXAMPLE 2 Finding out if two ratios form a proportion Use cross multiplication to find out if each of the following is a proportion.

a. $\frac{9}{12} \stackrel{?}{=} \frac{6}{8}$ **b.** $\frac{16}{15} \stackrel{?}{=} \frac{15}{14}$ **c.** $\frac{25}{26} \stackrel{?}{=} \frac{24}{25}$

Solution **a.** Yes, this is a proportion. Both diagonals multiply to 72.

72 72

$\frac{9}{12} \times \frac{6}{8}$

b. No, this is not a proportion. The diagonals multiply to different numbers.

224 225

$\frac{16}{15} \times \frac{15}{14}$

c. No, this is not a proportion. The diagonals multiply to different numbers.

625 624

$\frac{25}{26} \times \frac{24}{25}$

Think about it 1: How would simplifying each ratio show whether a proportion was formed?

Later in this section, in Example 9, we will use a proportion to find the distance the ladder needs to be from the wall. In Chapter 8, we will find the length of the ladder.

Example 3 shows that proportions written in the form $a:b = c:d$ also can be written $a \cdot d = b \cdot c$. This form emphasizes the fact that equal products come from the multiplication of the two inside numbers and the multiplication of the two outside numbers. We call this way of writing a proportion the *inner-outer product method.* The **inner-outer product** is *an alternative to writing a proportion with fractions.*

EXAMPLE 3 Finding the inner-outer product Verify that $8:12 = 12:18$, using the inner-outer product and cross-multiplication methods.

Solution ***Inner-outer product method***:

Inner product = 144

$8:12 = 12:18$

Outer product = 144

Cross-multiplication method:

144 144

$\frac{8}{12} \times \frac{12}{18}$

Example 4 shows that a set of four numbers may be arranged into a proportion in several ways.

EXAMPLE 4 Building proportions Use the set of numbers 6, 8, 15, and 20 to make four proportions that appear to be different.

Solution We check the proportions with cross multiplication.

$$\frac{6}{15} = \frac{8}{20} \quad 6 \cdot 20 = 15 \cdot 8, \text{ a proportion}$$

$$\frac{15}{6} = \frac{20}{8} \quad 15 \cdot 8 = 6 \cdot 20, \text{ a proportion}$$

$$\frac{6}{8} = \frac{15}{20} \quad 6 \cdot 20 = 8 \cdot 15, \text{ a proportion}$$

$$\frac{8}{6} = \frac{20}{15} \quad 8 \cdot 15 = 6 \cdot 20, \text{ a proportion}$$

Not all pairs of fractions with these numbers are proportions.

$$\frac{6}{20} \stackrel{?}{=} \frac{8}{15} \quad 6 \cdot 15 = 90,\ 20 \cdot 8 = 160, \text{ not a proportion}$$

Example 5 repeats the idea from Example 4, but with numbers and units.

EXAMPLE  Building proportions: medication dosages Make proportions showing that a child of age 96 months should receive a dosage of 160 milligrams if 250 milligrams is the dosage for persons 150 months or older.

Solution Some proportions are

$$\frac{96 \text{ mo}}{150 \text{ mo}} = \frac{160 \text{ mg}}{250 \text{ mg}}$$

$$\frac{96 \text{ mo}}{160 \text{ mg}} = \frac{150 \text{ mo}}{250 \text{ mg}}$$

$$\frac{250 \text{ mg}}{150 \text{ mo}} = \frac{160 \text{ mg}}{96 \text{ mo}}$$

Think about it 2: Write a sentence to describe each proportion in Example 5.

The units in Example 5 draw attention to the importance of matching the units in a proportion.

Matching Units in Proportions

When proportions contain units, the units must match—either side to side or within numerator and denominator pairs.

In either case, the product of the units will be identical for the two multiplications when we cross multiply.

Solving Problems Using Proportions

We now turn to using variables in proportions and application settings. We will start with proportions without units, as we solve percent problems with proportions. We will return to proportions with units in Example 8.

SOLVING PERCENT PROBLEMS AS PROPORTIONS Many percent problems are phrased in sentences, as shown on the left. The equation for each sentence is shown on the right.

Sentence	**Equation**
15% of what number is 45?	$0.15x = 45$
72 is what percent of 80?	$72 = \frac{x}{100} \cdot 80$
15 is what percent of 5?	$15 = \frac{x}{100} \cdot 5$
What number is 150% of 18?	$x = 1.50(18)$

Writing these sentences as equations requires that we remember to change percents into decimals. Some students like to use equations. Other students prefer to write the sentences as proportions. The above problems all fit this form:

$$\text{If } a \text{ is } n\% \text{ of } b, \text{ then } \frac{n}{100} = \frac{a}{b}.$$

The symbol $n\%$ is equal to the ratio $n/100$, which forms the left side of the proportion. The right side is the percent quantity over the base quantity. Note that *n is out of 100* and *the percent quantity is out of the base, b.*

EXAMPLE 6 Writing percent problems as proportions Change each percent problem to a proportion and solve.

a. 15% of what number is 45?

b. 72 is what percent of 80?

c. 15 is what percent of 5?

d. What number is 150% of 18?

Solution **a.** Write 15% as 15/100. The percent quantity is 45. The base is x.

$$\frac{15}{100} = \frac{45}{x} \qquad \text{Cross multiply.}$$

$$15x = (100)(45) \qquad \text{Divide each side by 15.}$$

$$x = \frac{4500}{15} \qquad \text{Simplify.}$$

$$x = 300$$

Check: 15% of 300 is 45. ✓

b. Write "what percent" as $x/100$. The percent quantity is 72. The base is 80.

$$\frac{x}{100} = \frac{72}{80} \qquad \text{Cross multiply.}$$

$$80x = (100)(72) \qquad \text{Divide each side by 80.}$$

$$x = \frac{7200}{80} \qquad \text{Simplify.}$$

$$x = 90$$

Check: 72 is 90% of 80. ✓

c. Write "what percent" as $x/100$. The percent quantity is 15. The base is 5.

$$\frac{x}{100} = \frac{15}{5} \quad \text{Cross multiply.}$$
$$5x = (100)(15) \quad \text{Divide each side by 5.}$$
$$x = \frac{1500}{5} \quad \text{Simplify.}$$
$$x = 300$$

Check: 15 is 300% of 5. ✓

d. Write 150% as 150/100. The percent quantity is x. The base is 18.

$$\frac{150}{100} = \frac{x}{18} \quad \text{Cross multiply.}$$
$$100x = (150)(18) \quad \text{Divide each side by 100.}$$
$$x = \frac{2700}{100} \quad \text{Simplify.}$$
$$x = 27$$

Check: 27 is 150% of 18. ✓ ●

Parts c and d of Example 6 have a percent larger than 100. Some of the price changes we found when we calculated percent change in Section 5.1 were also larger than 100%. Percents larger than 100 are common—and very desirable when they reflect the percent of change in the prices of things people own and want to sell later, such as Beanie Babies or stocks.

You should know how to solve percent problems at least one way—with either equations or proportions.

SOLVING EQUATIONS AS PROPORTIONS Before going on to applications, we will use proportions to solve equations with denominators on both sides. The cross multiplication property of a proportion applies to any equation that is in the form $\frac{a}{b} = \frac{c}{d}$. In this section, we will study only equations with numbers in the denominator. We will delay study of those with variables in the denominator until Chapter 9.

Example 7 shows how to solve equations with fractions on both sides. Look for use of the distributive property in the solution.

EXAMPLE 7 Solving equations by cross multiplication. Solve each equation for x.

a. $\frac{x-1}{4} = \frac{x+1}{5}$ **b.** $\frac{2x}{9} = \frac{2x+4}{12}$

Solution **a.**

$$\frac{x-1}{4} = \frac{x+1}{5} \quad \text{Cross multiply.}$$
$$5(x-1) = 4(x+1) \quad \text{Use the distributive property.}$$
$$5x - 5 = 4x + 4 \quad \text{Subtract } 4x \text{ on both sides.}$$
$$x - 5 = 4 \quad \text{Add 5 on both sides.}$$
$$x = 9$$

Check: $\frac{9-1}{4} \stackrel{?}{=} \frac{9+1}{5}$ ✓

JUST FOR FUN

What four consecutive letters of the alphabet form a word? You may scramble the order of the letters. (*Note:* "Figh" is not a fruit, and "d'cab" is not what you call to get a ride to d'airport!) Find a proportion that gives the word upon cross multiplication.

b. $\frac{2x}{9} = \frac{2x + 4}{12}$ Cross multiply.

$12(2x) = 9(2x + 4)$ Use the distributive property.

$24x = 18x + 36$ Subtract 18x on both sides.

$6x = 36$ Divide by 6 on both sides.

$x = 6$

Check: $\frac{2 \cdot 6}{9} \stackrel{?}{=} \frac{2 \cdot 6 + 4}{12}$ ✓

Note that parentheses were placed around the numerators in the cross multiplication step to assure proper use of the distributive property.

SOLVING APPLICATION PROBLEMS CONTAINING UNITS In Example 8, we return to problems involving units. We set up a proportion with a variable and cross multiply to solve for the variable.

EXAMPLE 8 Setting up and solving a proportion: sewing budget Sam needs more ribbon for a sewing project. He has \$30 remaining in the project's budget. If 7 yards of ribbon cost \$80, how many yards can he buy for \$30?

Solution

$\frac{7 \text{ yd}}{\$80} = \frac{x \text{ yd}}{\$30}$ Set up the proportion and cross multiply.

$210 = 80x$ Divide by 80.

$x = 2.625$

Check: $7 \text{ yd} \cdot \$30 \stackrel{?}{=} 2.625 \text{ yd} \cdot \$80$ ✓

Sam can buy 2.625, or $2\frac{5}{8}$, yards.

Note that in Example 8 the units in each ratio were the same: yards over dollars. In the safe-ladder ratio problems, the 4-to-1 ratio was height over base.

EXAMPLE  Setting up and solving a proportion: safe-ladder ratio A ladder needs to reach 22 feet up a wall. How far from the wall must the ladder be in order to meet the 4-to-1 safe-ladder ratio?

Solution We set up the proportion as two ratios with height over base.

$\frac{\text{height}}{\text{base}} = \frac{\text{height}}{\text{base}}$

$\frac{22 \text{ ft}}{x \text{ ft}} = \frac{4}{1}$ Cross multiply.

$4x = 22$ Divide by 4.

$x = 5.5$

The ladder must be 5.5 feet from the wall.

In Example 10, we return to slope. Recall that slope may be defined as rise over run or the vertical change over the horizontal change. Rise over run is like height over base in Example 9.

EXAMPLE 10 Setting up and solving a proportion: slopes and ramps The slope of a wheelchair access ramp should be 1 to 10. The ramp needs to rise 20 inches. What horizontal length is needed to build the ramp?

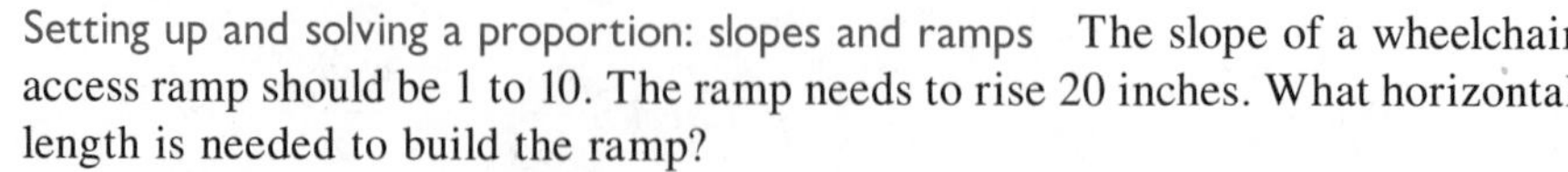

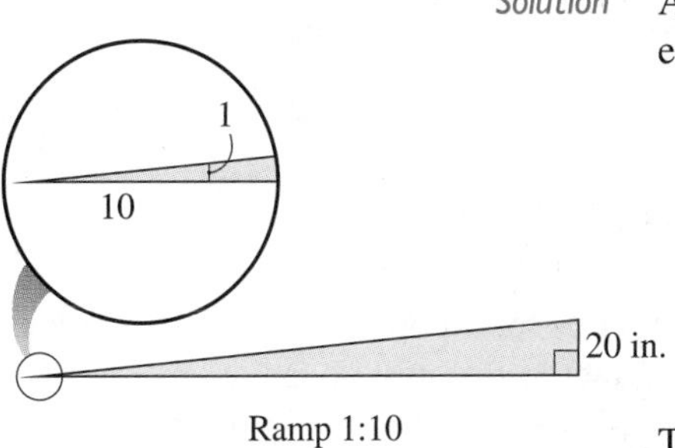

Figure 6

Solution A sketch like Figure 6 may be helpful. We start with a line of slope 1 to 10, and extend it an unknown distance to a vertical line of 20 inches.

$$\frac{\text{rise}}{\text{run}} = \frac{\text{rise}}{\text{run}}$$

$$\frac{1}{10} = \frac{20 \text{ in.}}{x} \qquad \text{Cross multiply.}$$

$$1x = 200$$

The horizontal length is 200 inches. Using unit analysis to change inches to feet, we have

$$200 \text{ in.} \cdot \frac{1 \text{ ft}}{12 \text{ in.}} = 16\tfrac{2}{3} \text{ ft}$$

The horizontal length of the ramp is $16\frac{2}{3}$ feet. ●

Think about it 3: Use a proportion to change 200 inches into feet.

PROPORTIONS AND STATISTICS Statistics is a field of mathematics where we gather information and then draw conclusions or make predictions based on that information. How the information is gathered (random surveys or random samples) is an entire course by itself, but as the next three examples show, proportional reasoning plays an important role in the field. (A project is included in the exercise set for those who want to research the meaning of *random survey* and *random sample* and think about the assumptions and conclusions in the examples given here.)

Example 11 applies proportions to TV ratings.

EXAMPLE 11 Setting up and solving a proportion: TV ratings Suppose 500 American households are surveyed randomly and 6 are found to be watching NBC's 8 p.m. program. Estimate the total number of households watching NBC nationwide if there are 93 million households.

Solution We have two kinds of data: data on those who are watching NBC and data on totals. We set up the proportion with this in mind.

$$\frac{6 \text{ NBC households}}{500 \text{ households surveyed}} = \frac{x \text{ NBC households}}{93 \text{ million households}} \qquad \text{Cross multiply.}$$

$$6(93 \text{ million}) = 500x \qquad \text{Divide by 500.}$$

$$x = 1.116 \text{ million}$$

About 1 million households appear to be watching the NBC program. ●

In Example 11, we assumed that the households surveyed were typical of the nation as a whole. In wildlife population sampling, it is assumed that when randomly tagged animals are released, they will mix back into the general population. Any subsequent random sample will have tagged animals in the same proportion as the total population.

EXAMPLE 12 Setting up and solving a proportion: fish population At random locations in a certain lake, fish management personnel catch and tag 50 fish. They return the fish unharmed to the lake. They come back to the same locations two weeks later and catch 100 fish. If 4 of the fish are tagged and 96 are not tagged, estimate the total population of fish. What assumptions are made?

Solution

$$\frac{50 \text{ initial tagged}}{x \text{ total population}} = \frac{4 \text{ caught with tag}}{100 \text{ total caught}} \quad \text{Cross multiply.}$$

$$50 \cdot 100 = 4 \cdot x \quad \text{Divide by 4.}$$

$$x = 1250$$

Check: $500(100) \stackrel{?}{=} 4(1250)$ ✓

This solution uses a ratio of tagged fish to total population. Other solutions are possible, using tagged to untagged or untagged to total population.

We assume that no tagged fish died and that the tagged fish returned to the general fish population and mixed sufficiently so they and the other fish were equally likely to be caught the second time. ●

In Example 13, we solve a problem first with proportions and then again with a unit analysis.

EXAMPLE 13 Setting up and solving a proportion: bat and mosquito population Researchers sample several locations on the ceiling of a cave and find an average of 18 adult bats per square foot. They estimate the dimensions of the ceiling to be 30 feet by 75 feet. Use a proportion to estimate the number of adult bats, x, in the cave. Then estimate the number of mosquitoes, m, eaten in 30 nights if each bat consumes 500 mosquitoes per night.

Solution with Proportions We first find the number of bats, x.

$$\frac{18 \text{ bats}}{1 \text{ ft}^2} = \frac{x \text{ bats}}{(30 \text{ ft})(75 \text{ ft})} \quad \text{Cross multiply.}$$

$$x = 40{,}500 \text{ bats}$$

To obtain the number of mosquitoes eaten each night, we set up the proportion

$$\frac{1 \text{ bat}}{500 \text{ mosquitoes}} = \frac{40{,}500 \text{ bats}}{m \text{ mosquitoes}}$$

Cross multiplying gives $m = 500(40{,}500) = 20{,}250{,}000$ mosquitoes.

To obtain the number of mosquitoes eaten in 30 nights, we set up the proportion

$$\frac{1 \text{ night}}{20{,}250{,}000 \text{ mosquitoes}} = \frac{30 \text{ nights}}{n \text{ mosquitoes}}$$

Cross multiplying gives $n = 30(20{,}250{,}000) = 607{,}500{,}000$ mosquitoes.

Solution with Unit Analysis We use the four steps:

Step 1: We need to change the number of bats into the number of mosquitoes.

Step 2: Useful facts: Area of cave ceiling is 30 ft by 75 ft
18 adult bats in 1 ft^2
1 bat eats 500 mosquitoes per night
30 nights

Step 3: Suppose we start with the area of the cave's ceiling.

$$\frac{(30 \text{ ft})(75 \text{ ft})}{1} \cdot \frac{18 \text{ adult bats}}{1 \text{ ft}^2} \cdot \frac{500 \text{ mosquitoes per night}}{1 \text{ adult bat}} \cdot \frac{30 \text{ nights}}{1}$$

$$= 30 \cdot 75 \cdot 18 \cdot 500 \cdot 30 \text{ mosquitoes}$$

$$= 607{,}500{,}000 \text{ mosquitoes}$$

Step 4: The results are the same as with the proportion. ●

Unit analysis is handy when there are several units involved. Proportions work well when there are just two units and you know the fact relating them. Try to learn at least one of the two methods.

ANSWER BOX

Warm-up: 1. $6x + 24$ **2.** $4x + 20$ **3.** $8x + 64$ **4.** $20x - 20$ **5.** $6x - 36$ **6.** $-4x + 12$ **7.** $-4x - 4$ **8.** $-5x + 5$ **9.** $-2x - 6$ **10.** $-3x + 9$ **Think about it 1:** Equivalent ratios would simplify to the same number or divide to the same decimal. **Think about it 2:** The ratio of the child's age, 96 months, to the adult age, 150 months, is the same as the ratio of the child's dosage, 160 mg, to the adult dosage, 250 mg. The ratio of the child's age, 96 months, to the child's dosage, 160 mg, is the same as the ratio of the adult age, 150 months, to the adult dosage, 250 mg. The ratio of the adult dosage, 250 mg, to the adult age, 150 months, is the same as the ratio of the child's dosage, 160 mg, to the child's age, 96 months. **Think about it 3:** Our fact is 12 in. = 1 ft, so 200 in. = x ft. One proportion is $\frac{12 \text{ in.}}{200 \text{ in.}} = \frac{1 \text{ ft}}{x \text{ ft}}$. Other proportions are possible. Cross multiplication gives $12x = 200$ or $x = 16\frac{2}{3}$ ft.

EXERCISES 5.2

In Exercises 1 to 6, simplify each ratio, compare it to 4:1, and state whether it is a safe-ladder ratio. If the ratio represents an unsafe ladder position, guess whether the ladder will slip or tip.

1. 15 feet to 5 feet

2. 18 feet to 5 feet

3. 18 feet to 4 feet

4. 20 feet to 4 feet

5. 20 feet to 5 feet

6. 18 feet to 4.5 feet

In Exercises 7 to 12, which are proportions and which are false statements? You may want to write each in fraction notation first.

7. $6:8 = 15:20$

8. $8:10 = 12:15$

9. $4:6 = 6:9$

10. $9:12 = 15:18$

11. $9:21 = 21:35$

12. $9:6 = 15:10$

13. Make four different proportions with the numbers 2, 4, 5, and 10. Use cross multiplication to check.

14. Make four different proportions with the numbers 2, 3, 4, and 6. Use cross multiplication to check.

Solve the proportions in Exercises 15 to 22.

15. $\frac{3}{4} = \frac{x}{15}$

16. $\frac{5}{3} = \frac{27}{x}$

17. $\frac{x}{5} = \frac{2}{3}$

18. $\frac{5}{x} = \frac{3}{8}$

19. $\frac{7}{3} = \frac{5}{x}$

20. $\frac{2}{5} = \frac{x}{12}$

21. $\frac{4}{x} = \frac{3}{7}$

22. $\frac{x}{6} = \frac{7}{5}$

In Exercises 23 to 38, write as equations or proportions and solve. Round to the nearest tenth or tenth of a percent.

23. 45% of what number is 36?

24. 65% of what number is 39?

25. 56 is what percent of 84?

26. 125 is what percent of 225?

27. What number is 60% of 42?

28. What number is 35% of 70?

29. 56 is what percent of 40?

30. 72 is what percent of 60?

31. What number is 18% of 25?

32. What number is 90% of 45?

33. 64% of what number is 16?

34. 32% of what number is 24?

35. 104 is what percent of 80?

36. 45 is what percent of 75?

37. 28 is what percent of 40?

38. 112 is what percent of 64?

In Exercises 39 to 52, solve for x.

39. $\frac{x-1}{2}=\frac{x+3}{3}$

40. $\frac{x-4}{3}=\frac{x-1}{4}$

41. $\frac{x+4}{15}=\frac{x+2}{10}$

42. $\frac{x-6}{10}=\frac{x-4}{6}$

43. $\frac{x+7}{9}=\frac{x+3}{6}$

44. $\frac{x-1}{8}=\frac{x+2}{12}$

45. $\frac{3x}{2}=\frac{6x+3}{5}$

46. $\frac{4x}{3}=\frac{3x-1}{2}$

47. $\frac{2x-3}{7}=\frac{x+2}{4}$

48. $\frac{5x-1}{9}=\frac{x+7}{3}$

49. $\frac{4x+2}{5}=\frac{x-1}{2}$

50. $\frac{x-3}{4}=\frac{3x+1}{7}$

51. $\frac{x+2}{2}=\frac{3x+1}{5}$

52. $\frac{x}{3}=\frac{3x-5}{8}$

Solve the unit problems in Exercises 53 to 58 using the following conversion information. Use either a proportion or unit analysis. Round to the nearest hundredth.

1 acre = 43,560 square feet
1 liter = 1.0567 quarts
1 kilogram = 2.2 pounds
1 kilometer = 1000 meters
1 square mile = 640 acres
1 meter = 39.37 inches
1 kilometer = 0.621 mile

53. How many meters are in 65 inches?

54. How many kilograms are in 160 pounds?

55. How many kilometers are in 15 miles?

56. How many liters are in 16 quarts?

57. How many square feet are in 640 acres?

58. How many square miles are in 10,000 acres?

In Exercises 59 to 72, set up a proportion and solve. Round to the nearest hundredth unless otherwise noted.

59. The base of a ladder can sit securely $6\frac{1}{2}$ feet from a wall. How far up the wall can the ladder reach and still satisfy the 4:1 safe-ladder ratio?

60. How far from a wall must the base of a ladder be set in order to safely reach 21 feet up a wall?

61. An access ramp needs to rise 54 inches. If the ramp is to have a slope ratio of 1:8, what horizontal distance is needed to build the ramp? Give the answer in feet.

62. An access ramp needs to rise 18 inches. If the ramp is to have a slope ratio of 1:8, what horizontal distance is needed to build the ramp? Give the answer in feet.

63. A staircase has a slope of 7 to 11. What horizontal distance is needed for an 8-foot vertical distance?

64. A staircase has a slope of 6.5 to 11.5. What horizontal distance is needed for an 8.5-foot vertical distance?

65. If a 10-foot storm surge (high wave of water) from Hurricane Andrew hits the Louisiana coast and comes inland 13 miles, what is the average slope of the coastal region at that location? (*Hint:* The units must be the same for a sensible answer.) Leave the answer as a fraction.

66. If a 10-foot storm surge hits the Oregon coast and comes inland $\frac{1}{8}$ mile, what is the average slope of the coastline at that location? (*Hint:* The units must be the same.) Leave the answer as a fraction.

67. A 40-pound child should receive how many units of penicillin if the dosage is 500,000 units for a 150-pound adult? Round to the nearest ten thousand.

68. How many milligrams (mg) of atrophine sulfate should be given to a 6-month infant if the dosage is 0.4 mg for a patient of 150 months?

69. A ski lift goes up a 62.5% grade. If the ski lift covers a 3000-foot horizontal distance, what is the change in elevation?

70. If a ski lift rises 9000 vertical feet, find the horizontal distance covered if the average slope is 62.5%.

71. A narrow mountain road has a 9% grade. If this grade covers a 2-mile horizontal distance, what is the rise in elevation to the nearest foot?

72. A highway sign indicates a 6% downhill grade. If this 6% grade is 7 miles of horizontal distance, what is the change in elevation, to the nearest foot?

73. ***Bird Population.*** A wildlife management team traps pheasants in nets and tags them at randomly located areas in a fire-damaged setting. They tag 75 birds altogether. Two weeks later they trap again, and they capture 10 tagged birds and 55 untagged birds. Use a proportion to estimate the population. Assume that the birds didn't learn to avoid the nets after being caught the first time.

74. ***Fish Population.*** Suppose 15,000 hatchery fish are released in a river. At maturity, these fish return to the river along with the native fish. Assume that the hatchery fish have a 5% survival and return rate. Of 85 mature fish caught by people fishing along the river, 82 have the clipped fin of the hatchery fish. Use a proportion to estimate the number of native fish in the river.

75. Explain how the units help us place numbers into a proportion.

76. Prove the cross multiplication property by multiplying both sides of $\frac{a}{b} = \frac{c}{d}$ by bd. Explain what happens to the denominators.

Projects

77. ***Population Estimates.*** Place a large number (at least 100, but uncounted) of like objects (dry beans, marbles, coins) in a paper bag.

a. Remove a handful, count, and "tag" them with some sort of label. Return the tagged objects to the bag and mix thoroughly.

b. Remove another handful, and count the tagged objects.

c. Set up a proportion to estimate the total number of objects in the bag. State your assumptions.

d. How could you improve your estimate? What other ways are there to count the objects? Count the objects or use another method to estimate the total objects.

78. ***Clock Angles.*** Examine the clock faces below. In answering the questions, explain your reasoning carefully and completely.

a. What is the measure of the angle formed by the hands of a clock at 7:30? *Hint:* A circle contains 360°.

b. What is the measure of the angle at 4:35?

c. What is the rate in degrees per hour for the minute hand? the hour hand? the second hand?

79. ***Chemistry Formulas.*** Chemistry and physics have several laws relating pressure, volume, and temperature of gases, as well as concentration and volume. All contain subscripted variables. We solved several of these formulas in Section 4.1 for certain variables. In parts a to d, use cross multiplication to transform the proportions into formulas without denominators.

a. $\frac{V_1}{T_1} = \frac{V_2}{T_2}$ **b.** $\frac{V_1}{C_2} = \frac{V_2}{C_1}$

c. $\frac{P_1}{V_2} = \frac{P_2}{V_1}$ **d.** $\frac{P_1 V_1}{T_1} = \frac{P_2 V_2}{T_2}$

In parts e to h, change the chemistry formulas back into proportion form.

e. $C_1 V_1 = C_2 V_2$; make a proportion with the *V*s on top.

f. $P_1 V_1 = P_2 V_2$; make a proportion with the *P*s on top.

g. $V_1 T_2 = V_2 T_1$; make a proportion with the *T*s on top.

h. $P_1 V_1 = P_2 V_2$; make a proportion with the *V*s on top.

80. ***Random Survey.*** Research the meanings of *random survey* and *random sample* (perhaps in an elementary statistics textbook). Why is it important that a survey of TV viewers be random? What assumptions might have been made in the TV survey in Example 11? What assumptions did we make about the behavior of the tagged fish in Example 12 or the counting of bats in Example 13? Why are these assumptions important in our conclusions?

81. ***Safe-Ladder Experiment.*** Your task is to find out for what decimal ratios of height to base a ladder—modeled with meter sticks—is safe. You will need three stiff, smoothly finished meter sticks and tape to secure one stick to a wall.

a. Fasten the first meter stick vertically to a wall. The lower end of the meter stick should be a measured distance—say, 20 centimeters—above the floor. Secure the meter stick with tape at the upper end and lower end only. Lay the second meter stick on the floor so that it is perpendicular to the wall and directly below the first stick. Lean the third meter stick (the ladder) so that it forms a steep triangle with the other two meter sticks. Press gently inward and downward on the midpoint of the meter stick. If the meter stick (ladder) does not move, the ladder is in a relatively "safe" position. If the base of the ladder moves toward the wall or away from the wall, the ladder is unsafe.

Set up a table with four headings: Height of triangle, Base of triangle, Safe or unsafe, Ratio of height to base. For the position in which you have put the ladder, record on the table the height of the triangle, the base of the triangle, and whether the ratio is safe or unsafe. Complete the last column with a calculator. Repeat for 10 or more other positions of the ladder.

b. Describe the type of accident modeled when the base of the ladder slides toward the wall.

c. Describe the type of accident modeled when the base slides away from the wall.

82. ***Slope and Trigonometry.*** In trigonometry, we relate angle measures and ratios of the sides of a right triangle. One of the ratios, a to b, is the same as the slope of a line, $\Delta y/\Delta x$. The relationship is described by the tangent equation

$$\tan A = \frac{a}{b}$$

This equation says: For the right triangle ABC, the tangent of angle A (measured in degrees) is the ratio of the length of side a to the length of side b.

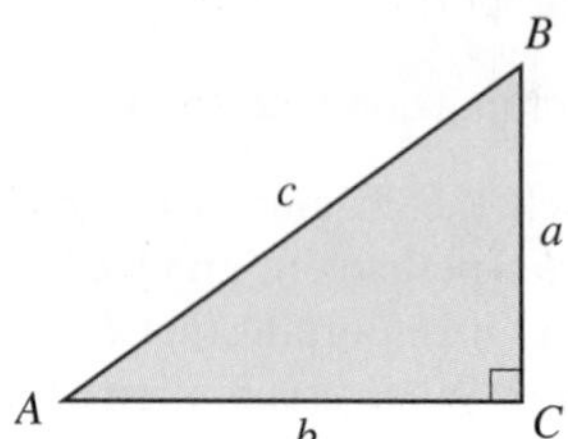

The tangent function is abbreviated *tan* and is built into calculators with the [TAN] key. To change an angle measure into a slope, set the calculator to degree measure and use A [TAN] or [TAN] A. The equation $\tan 26.6° \approx 0.5$ means that a line making a 26.6° angle with the horizontal has a slope of 0.5, or $\frac{1}{2}$.

What is the slope (tangent) of each angle in parts a to f? Round to the nearest tenth.

a. 22° **b.** 45° **c.** 76°

d. 10° **e.** 20° **f.** 30°

To change slopes back into angle measures, we use the shifted tangent key, [2nd] [TAN] or [INV] [TAN]. To change a slope of 1/2 to its angle with the horizontal axis, we enter [2nd] [TAN] (1/2). The answer is 26.6°.

What is the angle for each slope in parts g to l?

g. $\frac{1}{10}$ **h.** $\frac{3}{10}$ **i.** $\frac{7}{10}$

j. 1 **k.** 1.5 **l.** 2

m. What percent grade is a 45° angle?

n. What are the slope and angle with the horizontal for a mountain highway with a 6% grade?

o. What are the slope and angle for the 62.5% grade on the sides of Wizard Island in Crater Lake, Oregon?

p. What are the slope and angle for the 72.7% grade on the Lookout Mountain Incline Railway in Chattanooga, Tennessee?

5.3 Proportions in Similar Figures and Similar Triangles

OBJECTIVES

- Identify similar figures.
- Identify corresponding parts of similar figures.
- Write and solve proportions to find unknown lengths in similar triangles.

WARM-UP

Solve Exercises 1 to 6 for x.

1. $\frac{20}{12} = \frac{x}{7.5}$ **2.** $\frac{6}{8} = \frac{x}{20}$ **3.** $\frac{5.5}{3} = \frac{x}{34}$

4. $\frac{10 - x}{4} = \frac{10}{9}$ **5.** $\frac{4}{3} = \frac{x + 4}{5}$ **6.** $\frac{n}{3} = \frac{n + 3}{8}$

In Exercises 7 to 10, use this line segment:

A ———— B ———— C

Hint: If $AB = 20$ and AC is 25, then $BC = 25 - 20$.

7. If $AB = x$ and $AC = 8$, what expression describes BC?

8. If $AC = x$ and $AB = 8$, what expression describes BC?

9. If $AB = x$ and $BC = 8$, what expression describes AC?

10. If $AC = x$ and $BC = 8$, what expression describes AB?

IN THIS SECTION, we relate proportions to geometric figures. We use slope and similar triangles in applications in photography and forestry.

Similar Figures

EXAMPLE 

Exploring the height of a rectangle A magazine layout designer wishes to enlarge a photograph to fit the bottom of a page. The photograph's size is 12 centimeters (base) by 7.5 centimeters (height). The enlargement is to have a base of 20 centimeters. To finish the rest of the page layout, the designer needs to know the height of the photograph after the enlargement.

a. Estimate the height of the enlarged photo.

b. Find the height, using Figure 7 and the concept of similarity. In Figure 7, a diagonal is drawn from the lower left to the upper right corner of a rectangle with the same shape as the photograph. The diagonal is then extended until it crosses a vertical line at width 20 centimeters. The height of the vertical line at width 20 is the height of the enlarged photograph.

c. What is the slope of the diagonal line in Figure 7?

d. Let x be the missing height. Use the slope to write a proportion to find the missing height. Solve the proportion.

Figure 7

Solution See the Answer Box. ●

In Section 4.3, we found that a straight line has a constant slope. The fact that there is a constant slope ratio all along the diagonal line helps explain why we can apply a proportion in part d of Example 1. In Example 2, we check that the constant slope controls the shape of the rectangle as it is enlarged (or reduced). In part c, two letters are used to identify line segments.

EXAMPLE 

Exploring ratios of height to base

a. What is the slope of the diagonal line on the grid in Figure 8?

b. Which horizontal and vertical lines on the grid match the original photo in Figure 7?

c. Find these ratios: BF to AB, CG to AC, DH to AD, EI to AE.

d. What do you observe about the ratios? (*Hint:* If necessary, change the ratios to decimal notation.)

e. Complete this sentence about the rectangles drawn on the grid: For each 4-unit increase in the (horizontal) base, the rectangle increases ____ units in (vertical) height.

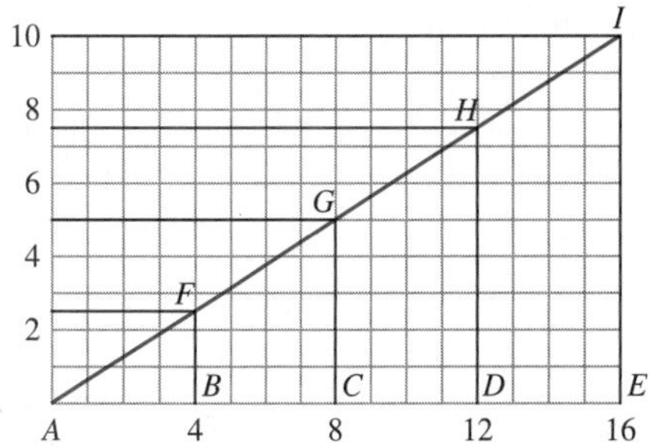

Figure 8

Solution See the Answer Box. ●

Example 2 suggests that enlargement (or reduction) of a photograph will make a rectangle that is the same shape as the original but of a different size. There are two important ideas here.

First, we say that the heights of the two rectangles are corresponding sides. **Corresponding** means *in the same relative position.* The bases are also corresponding sides. Second, the right angles forming the corners of the rectangle remain square corners during the enlargement. These two ideas give us the basis for a definition of similar figures.

Definition of Similar Figures

> **Similar figures** have corresponding sides that are proportional and corresponding angles that are equal.

Enlargements or reductions are similar to the original rectangles because the ratio of any height-to-base pair is constant.

If we are told that two figures are similar, we can assume that the corresponding sides are proportional.

EXAMPLE 3 Finding similar figures A photograph is $3\frac{1}{2}$ inches by 5 inches. Find the length of the shorter side if the longer side is

a. enlarged to 8 inches

b. reduced to 3 inches

Solution Because enlargement or reduction creates a similar figure, we can use a proportion. We set up a proportion, with short side to long side in each ratio.

a. $\dfrac{3.5}{5} = \dfrac{x}{8}$, $x = 5.6$ in.

b. $\dfrac{3.5}{5} = \dfrac{x}{3}$, $x = 2.1$ in. ●

The corresponding sides may not always be obvious. In Example 4, we look at relative sizes of sides (in part a) and relative sizes of angles (in part b) to find the corresponding sides. The sides are identified by two letters. The angles are identified by the letter at the **vertex**, or *corner of the triangle*.

EXAMPLE 4 Identifying corresponding sides

a. Name the corresponding sides of the triangles. Give the ratios of the corresponding sides. Indicate why the triangles are similar.

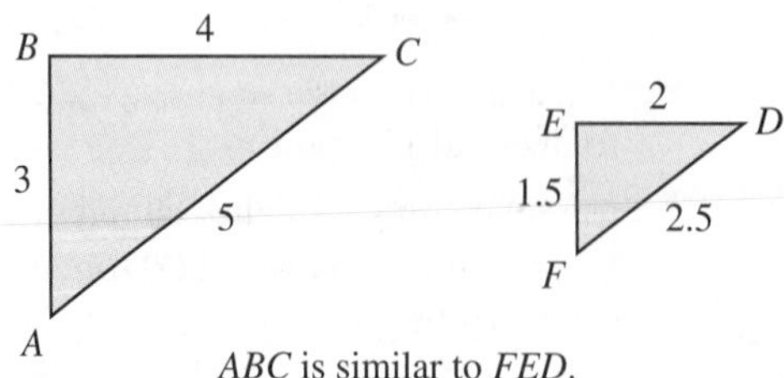

ABC is similar to *FED*.

b. Name the corresponding angles of the triangles, and indicate why the triangles are similar.

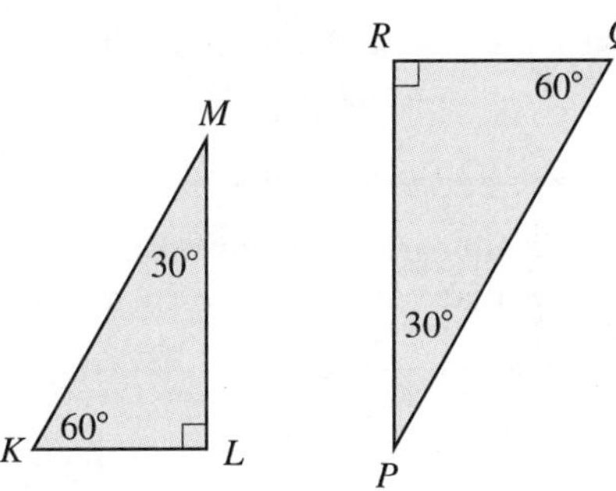

KLM is similar to *QRP*.

Solution **a.** The longest sides in both triangles are corresponding, so *AC* corresponds with *FD*. The middle-length sides are the next correspondence: *BC* and *ED*. Finally, the shortest sides give the third correspondence: *AB* and *FE*.

The ratio *AC* to *FD* is 5 to 2.5; the ratio simplifies to 2 to 1. The ratio *BC* to *ED* is 4 to 2; the ratio simplifies to 2 to 1. The ratio *AB* to *FE* is 3 to 1.5; the ratio simplifies to 2 to 1.

The triangles are similar because the ratios of corresponding sides are the same: 2 to 1.

b. Angle *L* corresponds with angle *R* because they are both right angles. Angle *M* corresponds with angle *P* because they are both 30°. Angle *K* corresponds with angle *Q* because they are both 60°.

The two triangles are similar because their corresponding angles are equal. ●

In Example 5, we use ratios to test for similarity between figures.

EXAMPLE 5 Identifying corresponding parts of figures Write ratios of corresponding lengths, and find out if the two figures in the pair are similar.

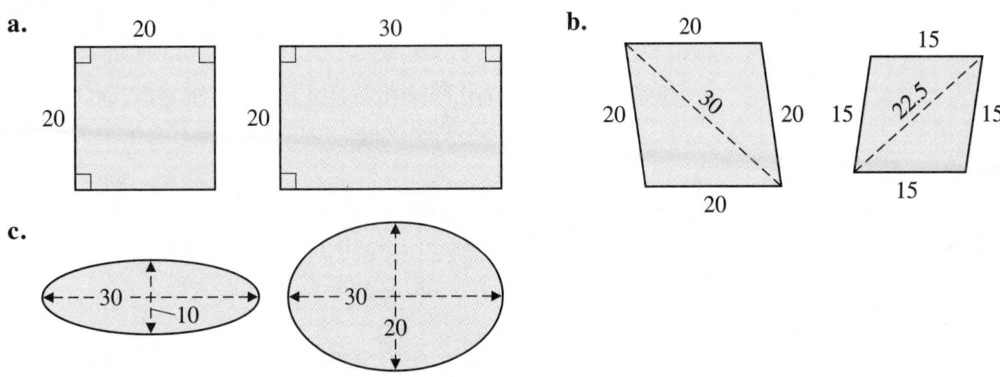

Solution **a.** The figures are both rectangles, but the ratios of their sides, 20/20 and 20/30, are not equal. The rectangles are not similar.

b. The outer sides are in a ratio of 20 to 15. The longest diagonals are in a ratio of 30 to 22.5. To find out whether the ratios are equal, we cross multiply

$$\frac{20}{15} \stackrel{?}{=} \frac{30}{22.5}$$

and get 450 in both directions. The figures are similar.

c. These two figures are ellipses. Within the figures, the horizontal distances are equal and the vertical distances are different. They are not similar. ●

We restate the relationships of similar figures in terms of the relationships of similar triangles:

Similar Triangles

> Similar triangles have corresponding angles that are equal.
> Similar triangles have corresponding sides that are proportional.

EXAMPLE 6 Identifying similar triangles Explain why the pairs of triangles are not similar.

a.

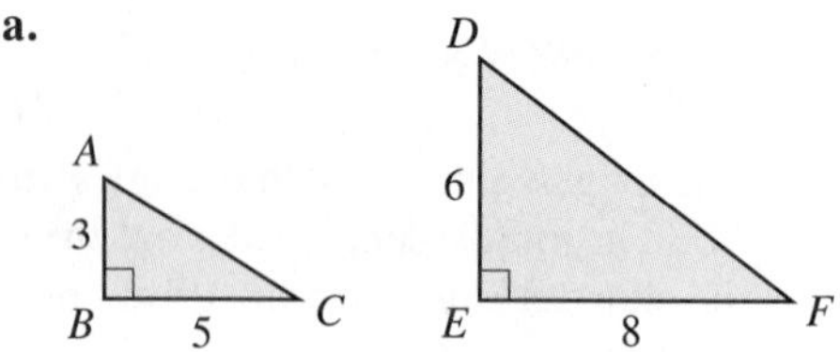

b.

L
100°
40°
40°
K
M
Q
120°
30°
30°
P
R

Solution **a.** The ratio of the shortest sides is $AB:DE = 3:6$, which simplifies to $1:2$. The ratio of the middle-length sides is $BC:EF = 5:8$. The ratios of corresponding sides are different, so the triangles are not similar.

b. Angle L and angle Q are the largest angles in the two triangles; but they are not equal, so the triangles are not similar. ●

In Example 7, we use proportions and the relationships of similar triangles to find missing sides.

EXAMPLE 7 Using proportions to find missing sides In part a of Example 6, if EF remains 8, what length would DE have to be for triangle DEF to be similar to triangle ABC?

Solution The ratio of base to height for triangle ABC is 3 to 5. Let x be the unknown side, DE. The proportion is

$$\frac{3}{5} = \frac{x}{8} \quad \text{Cross multiply.}$$

$$5x = 24 \quad \text{Divide both sides by 5.}$$

$$x = 4.8$$ ●

Similar Triangles and Indirect Measurement

The proportionality of similar triangles gives us a powerful tool for finding unknown lengths. The following examples involve *indirect measurement*—finding the measure of objects we are able to see but not actually measure.

The proportional relationship between heights and shadows in Example 8 was known to the ancient Greeks (see Exercise 43).

EXAMPLE 8 Finding an unknown length with a proportion: trees and shadows If a tree casts a 34-foot shadow along the ground while a $5\frac{1}{2}$-foot-tall person casts a shadow 3 feet long, how tall is the tree? (Because of space limitations, the triangles shown in Figure 9 are similar but are not to scale with each other.)

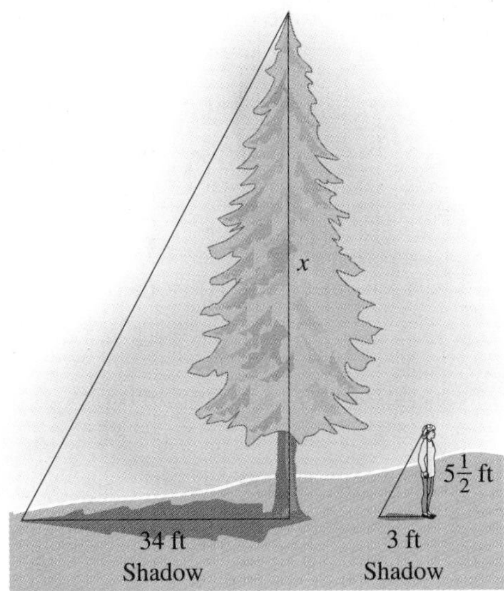

Figure 9

Geometry fact: The sun's rays form equal angles at the top of each triangle. It is assumed that the tree and the person are at right angles to the ground. This information is sufficient to create similar triangles.

Solution We form a proportion with the heights and shadows:

$$\frac{\text{tree height}}{\text{tree shadow}} = \frac{\text{person height}}{\text{person shadow}}$$

$$\frac{x}{34 \text{ ft}} = \frac{5.5 \text{ ft}}{3 \text{ ft}}$$

$$x(3 \text{ ft}) = (34 \text{ ft})(5.5 \text{ ft})$$

$$x \approx 62.3 \text{ ft}$$

The tree is approximately 62.3 feet tall. ●

In Example 9, the triangles overlap. The large triangle has the streetlight as its height, and its base extends to the tip of the shadow on the ground. The base is the sum 12 ft + 8 ft. The small triangle has the person as the height and the shadow as its base.

EXAMPLE 

Finding an unknown length with a proportion: streetlight and shadows A 6-foot-tall person has an 8-foot shadow formed by a streetlight. The person is standing 12 feet from the streetlight. Figure 10 shows a side view of the situation. How high is the streetlight?

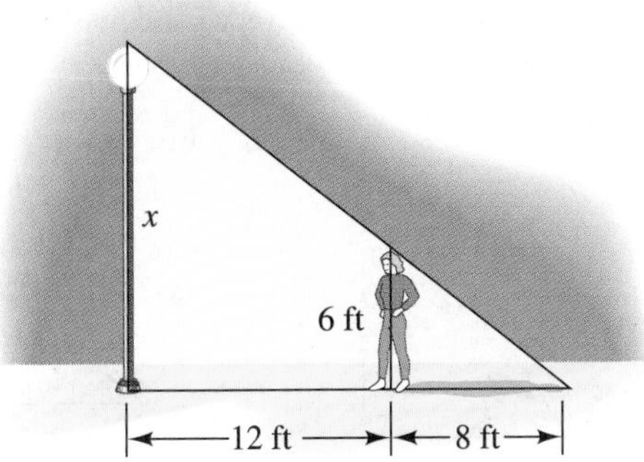

Figure 10

Solution The length of the base of the larger triangle is the sum of the 12 feet between the lamppost and the person and the 8-foot shadow.

$$\frac{\text{height of light}}{\text{base of triangle}} = \frac{\text{person height}}{\text{person shadow}}$$

$$\frac{x}{(12 + 8)\text{ ft}} = \frac{6\text{ ft}}{8\text{ ft}}$$

$$x(8\text{ ft}) = (6\text{ ft})(20\text{ ft})$$

$$x = 15\text{ ft}$$

The streetlight is 15 feet high. ●

Not all similar triangles are right triangles. In Example 10, we are given overlapping similar triangles and must find a missing height.

EXAMPLE 10 Finding an unknown length with a proportion Set up a proportion and solve for the indicated height in Figure 11.

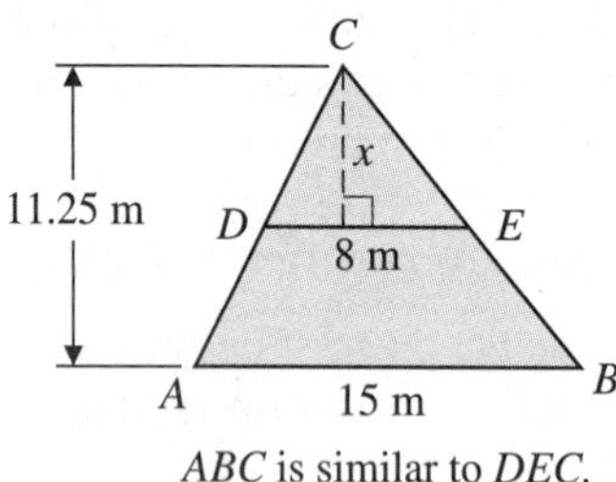

Figure 11

Solution The height-to-base ratio of the large triangle is 11.25 meters to 15 meters. The base of the small triangle is 8 meters.

$$\frac{11.25}{15} = \frac{x}{8} \qquad \text{Cross multiply.}$$

$$90 = 15x \qquad \text{Divide by 15.}$$

$$x = 6$$

The height of the small triangle is 6 meters. ●

In Example 11, the unknown length is part of one side of overlapping similar triangles.

EXAMPLE 11 Finding unknown lengths Find length x in each figure. Round answers to the nearest tenth.

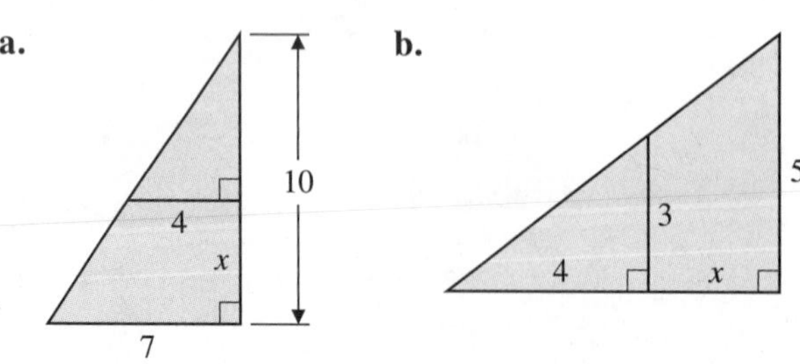

Solution **a.** The height of the small triangle is $10 - x$.

$$\frac{10}{7} = \frac{10 - x}{4} \qquad \text{Cross multiply.}$$

$$7(10 - x) = 40 \qquad \text{Apply the distributive property.}$$

$$70 - 7x = 40 \qquad \text{Subtract 70 on both sides.}$$

$$-7x = -30 \qquad \text{Divide by } -7.$$

$$x \approx 4.3$$

b. The base of the large triangle is $x + 4$.

$$\frac{5}{x + 4} = \frac{3}{4} \qquad \text{Cross multiply.}$$

$$3(x + 4) = 20 \qquad \text{Apply the distributive property.}$$

$$3x + 12 = 20 \qquad \text{Subtract 12 on both sides.}$$

$$3x = 8 \qquad \text{Divide by 3.}$$

$$x \approx 2.7$$

●

In Example 12, information about the lengths of the sides is contained in the ordered pairs that label the corners of the triangles. We solve for the missing information in two ways: first using similar triangles and second using slope.

EXAMPLE 12 Using proportions to find ordered pairs Find the ordered pairs that describe points A and B in Figure 12.

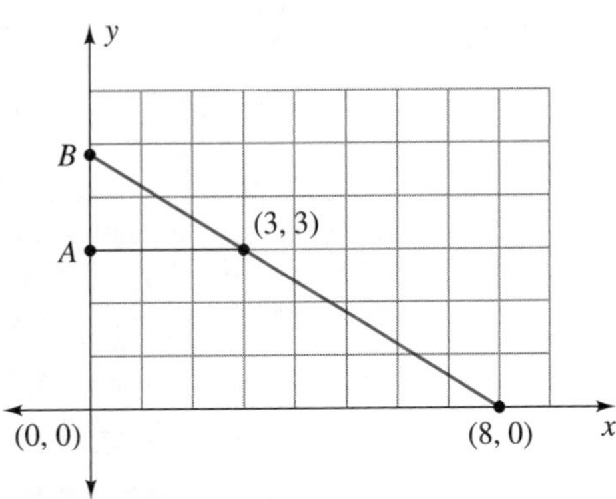

Figure 12

Solution Using Similar Triangles Point A is (0, 3) because it is on the vertical axis passing through (0, 0) and it is on the same horizontal line as (3, 3). Point B is also on the vertical axis.

Let n be the distance from A to B. From A to the origin is 3. The height of the large triangle is $n + 3$. The base of the small triangle on top is 3. The base of the large triangle (resting on the x-axis) is 8.

$$\frac{n}{n + 3} = \frac{3}{8} \qquad \text{Cross multiply.}$$

$$8n = 3(n + 3) \qquad \text{Apply the distributive property.}$$

$$8n = 3n + 9 \qquad \text{Subtract } 3n \text{ on both sides.}$$

$$5n = 9 \qquad \text{Divide by 5.}$$

$$n = 1.8$$

The height of the large triangle is $1.8 + 3 = 4.8$. B is at (0, 4.8).

Solution Using Slope As mentioned in the prior solution, point A is (0, 3). Because a straight line has constant slope, we can use the slope formula twice to obtain two different ratio expressions and then set them equal to make a proportion to find point B.

Between (3, 3) and (8, 0), the slope is

$$\frac{y_2 - y_1}{x_2 - x_1} = \frac{0 - 3}{8 - 3} = \frac{-3}{5}$$

Between point B, $(0, y)$, and (3, 3), the slope is

$$\frac{y_2 - y_1}{x_2 - x_1} = \frac{3 - y}{3 - 0} = \frac{3 - y}{3}$$

We set the two slope ratios equal and solve:

$$\frac{-3}{5} = \frac{3 - y}{3} \quad \text{Cross multiply.}$$

$$5(3 - y) = -9 \quad \text{Use the distributive property.}$$

$$15 - 5y = -9 \quad \text{Subtract 15 on both sides.}$$

$$-5y = -24 \quad \text{Divide by } -5 \text{ on both sides.}$$

$$y = 4.8$$

B is at (0, 4.8).

ANSWER BOX

Warm-up: 1. $x = 12.5$ **2.** $x = 15$ **3.** $x \approx 62.3$ **4.** $x \approx 5.6$ **5.** $x \approx 2.7$ **6.** $n = 1.8$ **7.** $8 - x$ **8.** $x - 8$ **9.** $x + 8$ **10.** $x - 8$ **Example 1: a.** Under 15 cm; the length does not quite double between 12 cm and 20 cm, so the height can be no more than twice 7.5 cm. **b.** Slightly more than 12 cm; measuring the picture shows that the height is only slightly more than the original width. **c.** 7.5/12 **d.** $\frac{7.5}{12} = \frac{x}{20}$, $x = 12.5$ cm **Example 2: a.** 5/8 **b.** AD and DH **c.** 2.5/4, 5/8, 7.5/12, 10/16 **d.** All are 0.625. **e.** 2.5

EXERCISES 5.3

1. If the designer in Example 1 reduced the 7.5-centimeter by 12-centimeter photograph to an 8-centimeter base, estimate what the new height would be. Find the height with a proportion.

2. If a designer enlarged a $3\frac{1}{2}$-inch by 5-inch photograph so that its shortest side was 14 inches, estimate what the new base would be. Find the base with a proportion.

For Exercises 3 to 6, trace the rectangles, and use the designer's diagonal method to enlarge or reduce them to the given base. Measure and label the heights.

3. Enlarge to a 2.5-inch base.

4. Reduce the rectangle in Exercise 3 to a 1-inch base.

5. Reduce the rectangle to a 1-inch base.

6. Enlarge the rectangle in Exercise 5 to a 3-inch base.

In Exercises 7 to 10, use ratios to find out if the triangles are similar. If they are similar, name the corresponding sides of the triangles.

7.

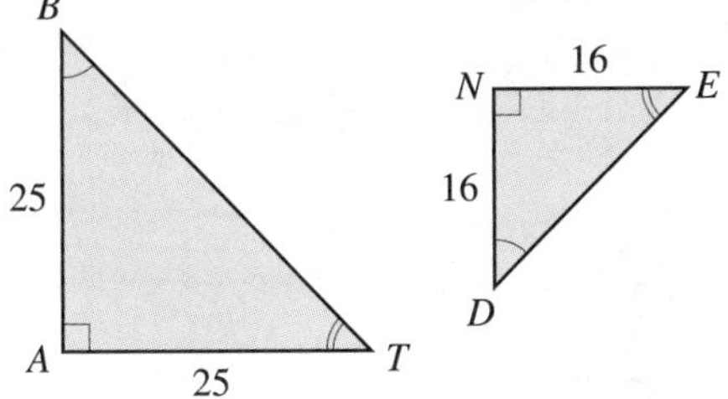

8.

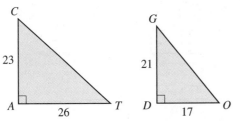

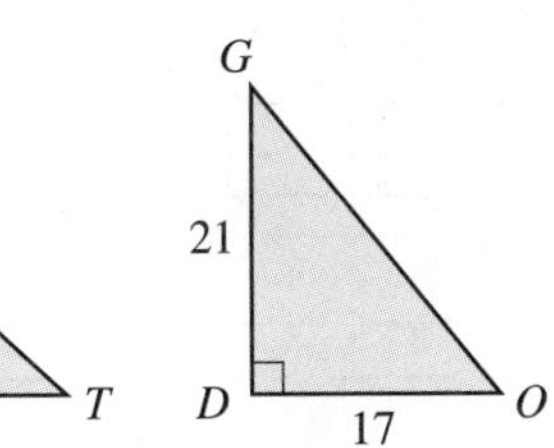

9.

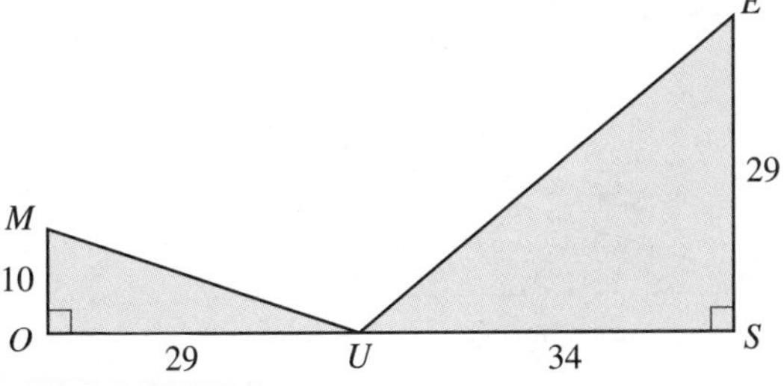

10.

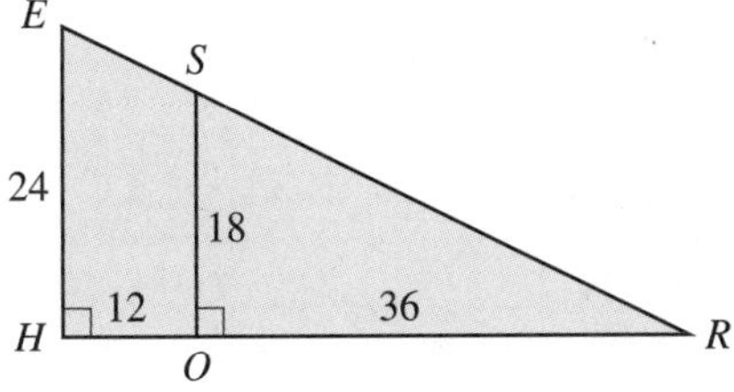

In Exercises 11 to 14, use ratios to show whether the figures are similar. Name two pairs of corresponding line segments for each pair of similar figures.

11.

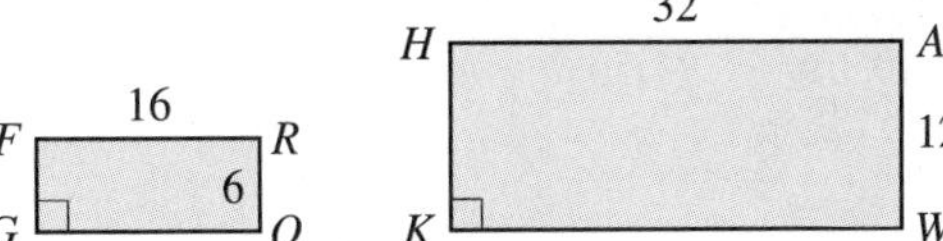

12.

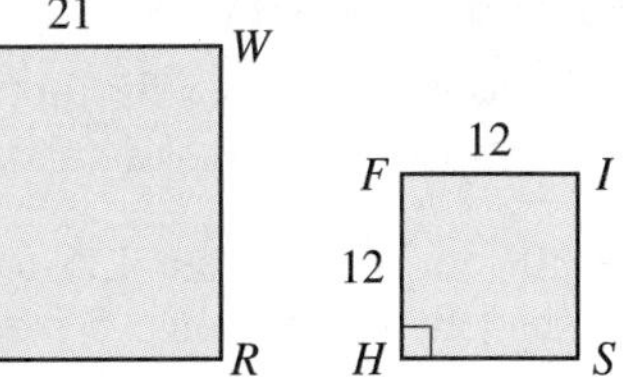

13.

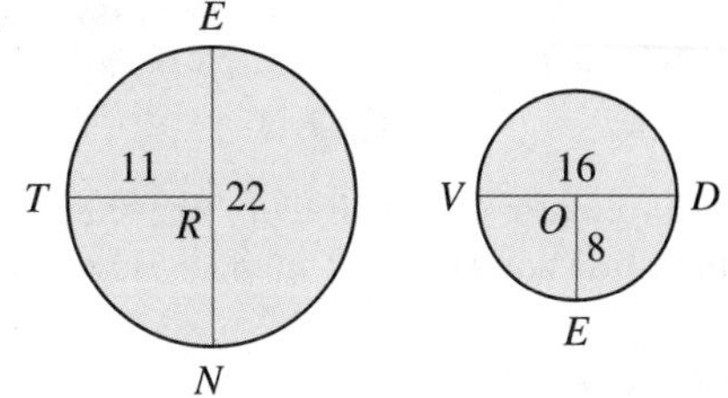

14.

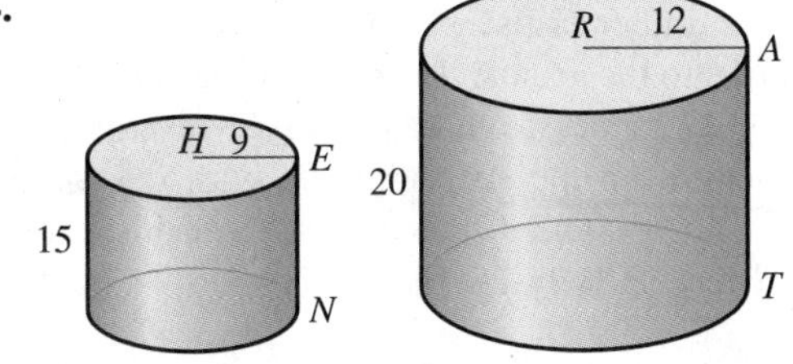

In Exercises 15 to 18, find the unknown side, n, in each pair of similar figures.

15.

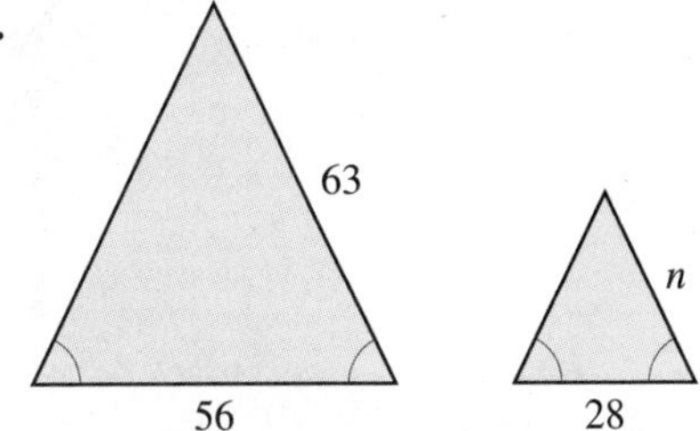

16.

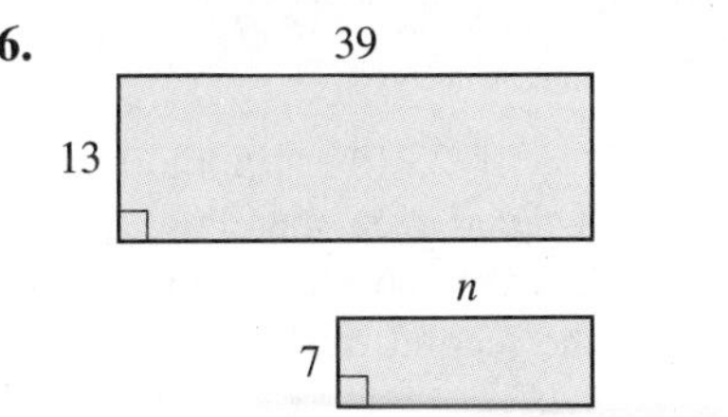

17.

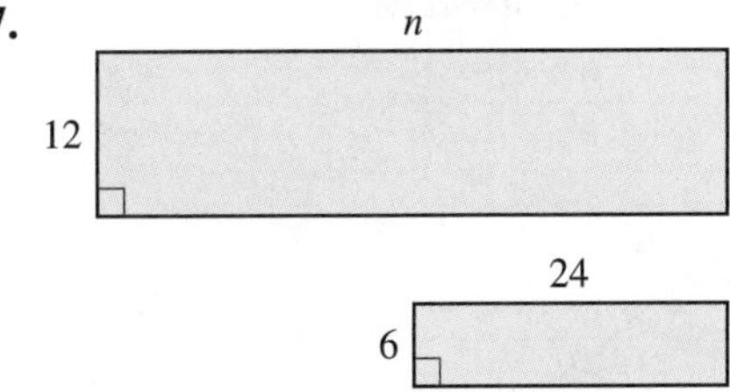

18.

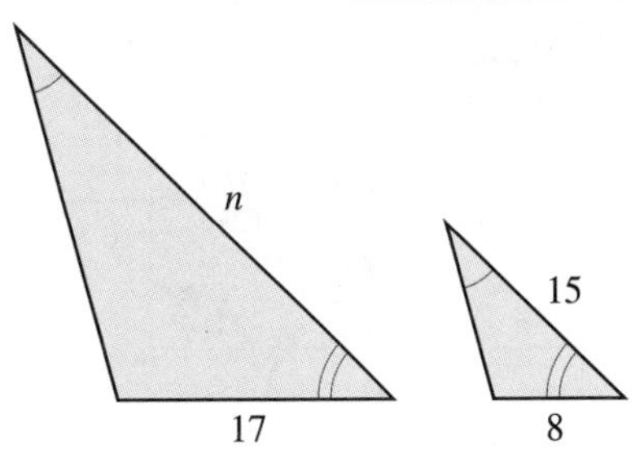

In Exercises 19 to 24, round to the nearest tenth.

19. At 3:00 p.m., a 30-foot tree casts a 35-foot shadow. A person 4 feet tall will cast how long a shadow?

20. At 5:00 p.m., a 30-foot tree casts a 125-foot shadow. How long is the shadow of a person 5.5 feet tall?

21. The shadow of a flagpole at 10 a.m. is 4 feet. The shadow of a 7-foot person is 1.4 feet. How tall is the flagpole?

22. The shadow of a tree at 2 p.m. is 8 feet. The shadow of a nearby 10-inch-tall squirrel is 2 inches. How tall is the tree?

23. The person who estimates the amount of wood in a tree is called a timber cruiser. A timber cruiser holds her arm parallel to the ground. In her hand she holds a stick, vertically, 27 inches from her eye (see the figure). A 14-inch length on the stick lines up with the top and bottom of a tree. The distance from the cruiser to the tree is 78 feet. How tall is the tree?

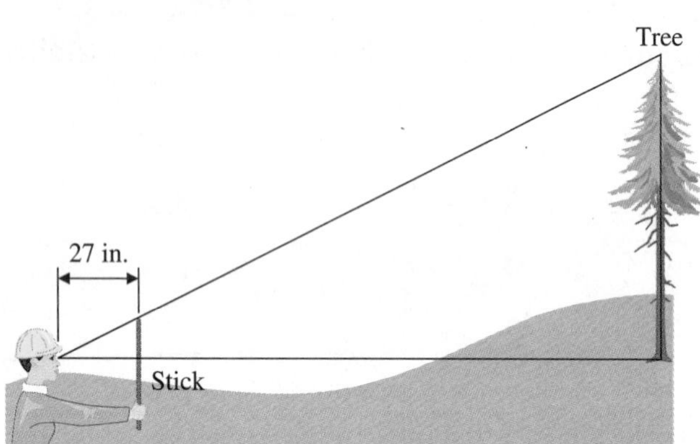

24. The timber cruiser in Exercise 23 lines up another tree in the same way. The second tree matches up with 30 inches on the stick when she is 60 feet from the base of the tree. How tall is the second tree?

Use proportions to find x in Exercises 25 to 30.

25.

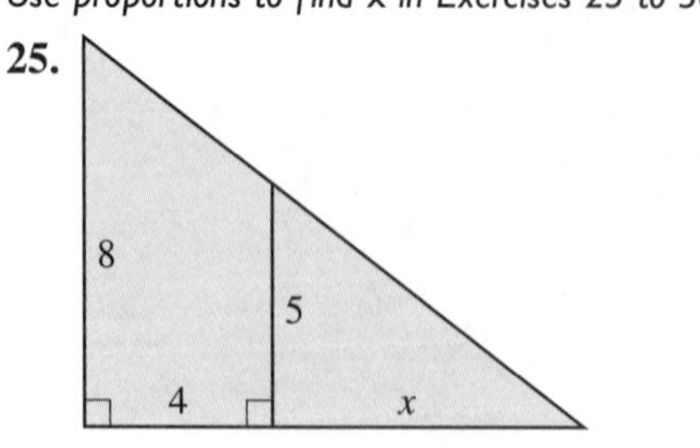

26.

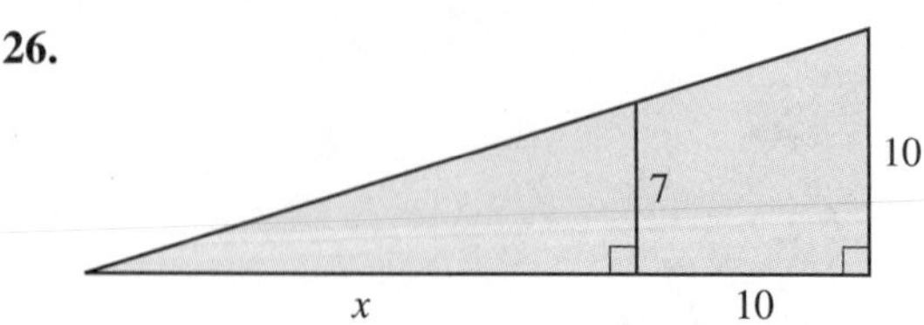

27.

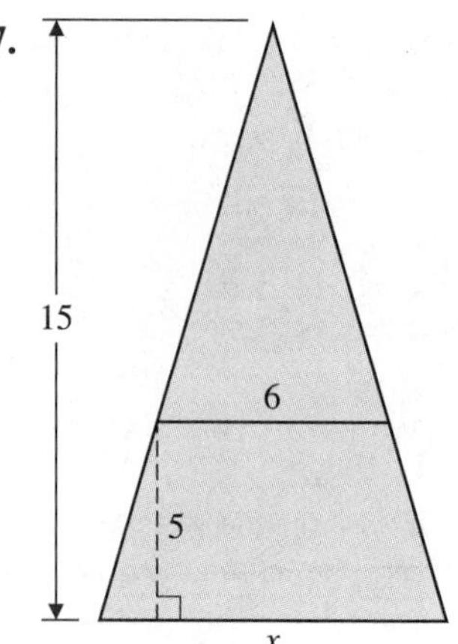

28.

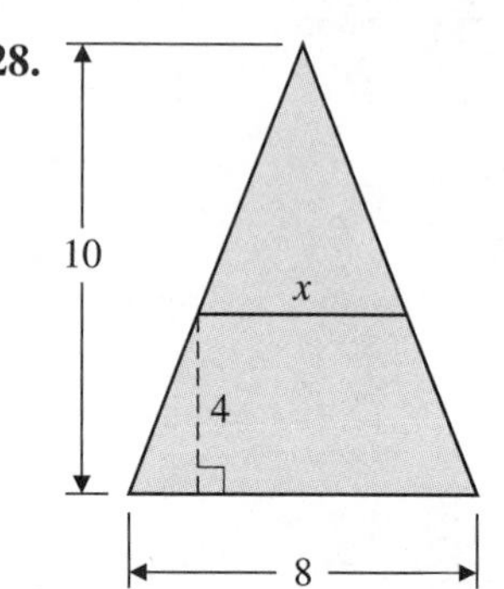

29.

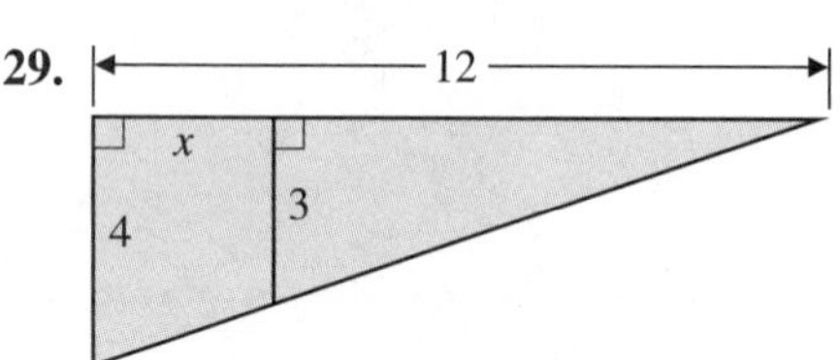

30.

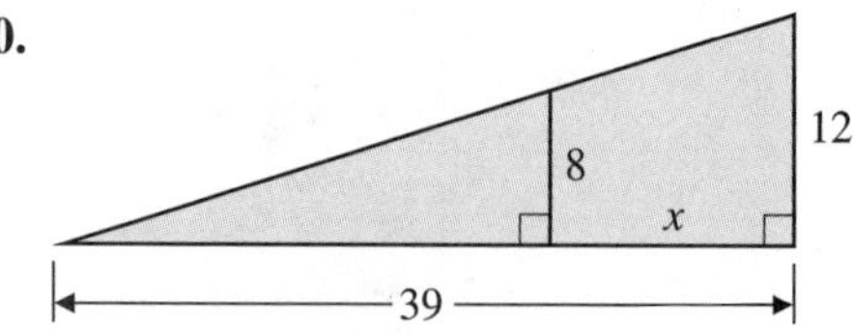

Identify the coordinates labeled A and B in Exercises 31 to 34. Use properties of similar triangles as needed.

31.

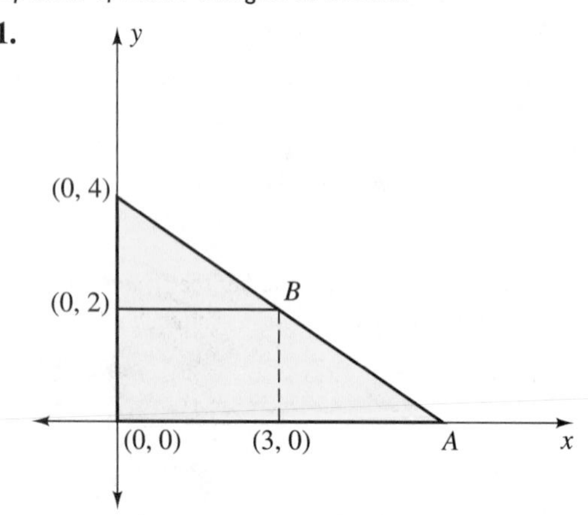

32.

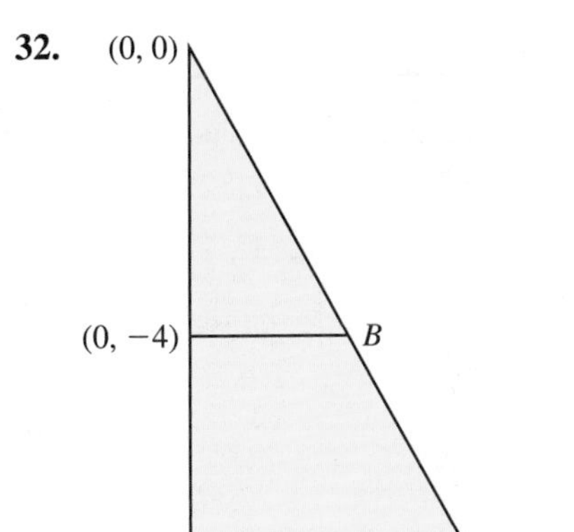

33.

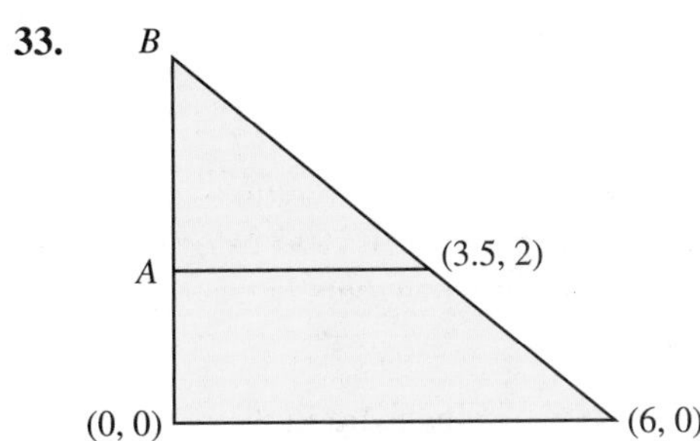

34.

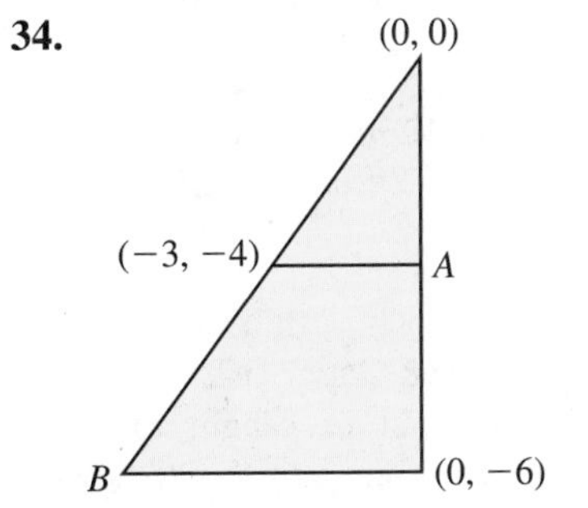

For a hint for Exercises 35 and 36, see Exercises 37 to 40.

35. A 5-foot parking meter has a 4-foot shadow from a nearby streetlight. The streetlight is 15 feet tall. How far is the streetlight from the parking meter?

36. Late one evening, a 6-foot person is standing 4 feet from a streetlight. The streetlight is 18 feet tall. How long is the person's shadow?

For Exercises 37 to 40, use this line segment:

A B C

37. If segment AC is 10 units and BC is x units, write an expression for AB.

38. If segment AB is 10 units and BC is x units, write an expression for AC.

39. If segment AC is x units and AB is 10 units, write an expression for BC.

40. If segment AB is x units and AC is 10 units, write an expression for BC.

41. Name three geometric figures that are always similar.

42. Explain why rectangles are not all similar.

Projects

43. ***Eratosthenes.*** Research how the ancient Greek mathematician Eratosthenes estimated the circumference of Earth. Include drawings in your report.

44. ***Parallelograms.*** The opposite sides of a parallelogram are parallel and have the same slope.

a. Three of the four vertices (corners) of a parallelogram are given: (0, 0), (3, 4), and (5, 0). Plot them. Find a fourth vertex. There are three different possible fourth vertices. Find all three, and plot them. Find the area of the triangle formed by the three possible fourth vertices. Compare that area with the area formed by the original three vertices. Are the triangles similar?

b. Repeat for (1, 1), (5, 3), and (1, 8) and then for three points of your choice.

45. ***Similar Rectangles and the Graphing Calculator Window.*** The standard window setting for some graphing calculators is a horizontal setting of $-10 \le x \le 10$ and a vertical setting of $-10 \le y \le 10$. The height of the standard window is 20 units, and the width is also 20 units.

a. Why are the graphs on a standard window distorted?

A window setting with a horizontal to vertical ratio of 3 to 2 will produce a graph with little distortion. An example is a horizontal setting of $-15 \le x \le 15$ and a vertical setting of $-10 \le y \le 10$, which has a width of 30 units and a height of 20 units.

$$\frac{\text{Horizontal width}}{\text{Vertical height}} = \frac{30 \text{ units}}{20 \text{ units}} = \frac{3}{2}$$

In parts b to i, use a horizontal to vertical ratio of 3 to 2 to find the missing window size or setting.

b. Horizontal is 9 units wide. Find the height.

c. Vertical is 50 units tall. Find the width.

d. Vertical is 40 units tall. Find the width.

e. Horizontal is 18 units wide. Find the height.

f. $-8 \le x \le 16, -5 \le y \le$ ___

g. $-10 \le x \le 11, -6 \le y \le$ ___

h. $-20 \le x \le 20, -10 \le y \le$ ___

i. $0 \le x \le 100, 0 \le y \le$ ___

MID–CHAPTER  TEST

For Exercises 1 and 2, write the ratio in simplified form.

1. $12xy$ to $15y^2z$
2. $\dfrac{(x-2)(x+2)}{x(x+2)}$

Change the ratios in Exercises 3 and 4 to like units, and simplify.

3. 18 inches : 6 feet
4. 3 meters to 75 centimeters

In Exercises 5 and 6, use either proportions or equations to find the missing number.

5. 16% of what number is 11.2?
6. 360 is 250% of what number?
7. What percent increase is a salary change from \$24,000 to \$30,000?
8. What percent decrease is a CD price change from \$15.99 to \$13.99?

In Exercises 9 and 10, describe in a complete sentence a setting for each rate.

9. gallons per mile
10. pounds per week

Solve the equations in Exercises 11 to 15.

11. $\dfrac{5}{x} = 500{,}000$
12. $\dfrac{3}{5} = \dfrac{16}{x}$
13. $\dfrac{x}{12} = \dfrac{10}{8}$
14. $\dfrac{x+5}{8} = \dfrac{2x-1}{12}$
15. $\dfrac{5x-2}{3x} = \dfrac{4}{3}$
16. Solve for b: $\dfrac{a}{b} = \dfrac{c}{d}$.
17. Use similar triangles and proportions to find x in the figure.

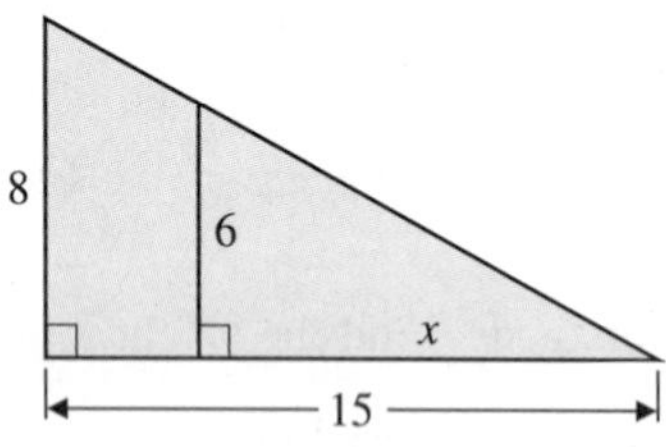

18. Use similar triangles and proportions as needed to find coordinates A and B in the figure.

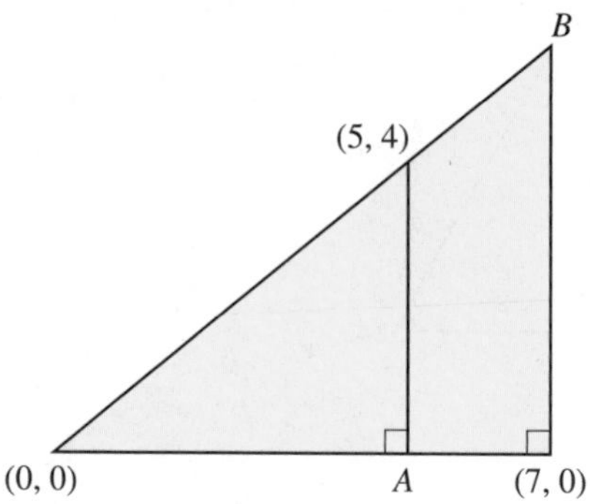

19. Calories in a diet are distributed in a ratio of 2:3:5 for fat, carbohydrate, and protein. If 1500 calories are to be consumed, how many may be allocated to each source?
20. The angle measures in a triangle are in the ratio 1:1:2. The sum of the angles is 180°. What is the measure of each angle?
21. An access ramp needs to rise 48 inches. If the ramp is to have a slope ratio of 1:12, what horizontal distance is needed to build the ramp?
22. Ten thousand hatchery trout, each with a clipped fin, are released in a lake. Two weeks later, fishing season starts. Within two days, 260 trout are caught, and 250 have a clipped fin. Estimate the number of native trout (not clipped) originally in the lake.
23. The lawsuit against Exxon, after the Valdez spill in Alaska, resulted in a \$5 billion judgment. A newspaper report indicated that the 5.9% annual interest on the judgment during the appeal process was \$9.40 per second.
    a. What is 5.9% of \$5 billion?
    b. Use unit analysis to show whether the newspaper statement was correct.

5.4 Averages

OBJECTIVES

- Find the mean, median, and mode of a set of data.
- Find the weighted average of a set of data.
- Find the midpoint of a line segment, given its endpoints as ordered pairs.
- Find the centroid of a figure, given its corners as ordered pairs.

WARM-UP

1. Solve for x:

$$\frac{0.78 + 0.70 + 0.90 + x}{4} = 0.80$$

2. Solve for x:

$$\frac{1.93 + 1.5x}{4} = 0.80$$

In Exercises 3 to 6, find the average for the four scores. The scores are for two tests, homework, and the final exam, each worth 100 points.

3. 95, 70, 20, 70

4. 70, 70, 90, 70

5. 75, 75, 95, 75

6. 85, 75, 100, 100

IN THIS SECTION, we look at the concept of average. We work with numbers and practice writing and solving equations. We relate averages both to geometry and to the proportions we studied in Sections 5.1 to 5.3.

Averages

An average describes the "center," or middle, of a set of numbers. In some settings, we think of average as being "normal." In other settings, an average helps us compare an individual with a group or compare one group with another group. In geometry, the average is a description of middle.

We now define three important ways of describing the middle of a set of numbers—with the mean, the median, and the mode.

MEAN Usually, people say they are finding the "average" when they add numbers together and divide by how many numbers were added. This average is called the mean or mean average, to distinguish it from the other forms of averages (see median and mode, below).

Finding the Mean

The **mean** of a set of numbers is the sum of the numbers divided by the number of numbers in the set.

The mean average is closely related to ratios and proportions. Recall (from Section 5.1) that continued ratios show shares of a quantity. Finding each person's contribution when three people share the total cost of a month's food purchases in the continued ratio 1:1:1 is the same as finding the average spent per person.

If researchers wanted to make more reliable predictions of wildlife population or TV viewing (Section 5.2), they might repeat the survey and average the results. The desire for improved accuracy must be balanced against the cost of repeating the survey.

In Example 1, we find a mean average in the context of travel.

EXAMPLE 1

Finding the mean Suppose the sets of numbers below reflect the weights of suitcases belonging to different families. For domestic baggage, the airlines have a 70-pound limit. If a family is allowed to average the weights, which sets will be accepted by the airlines?

a. 95, 58, 52, 88 **b.** 72, 47, 90 **c.** 98, 55, 58, 75, 63

Solution **a.** $(95 + 58 + 52 + 88) \div 4 = 73.25$, over the 70-pound average

b. $(72 + 47 + 90) \div 3 = 69.7$, under the 70-pound average

c. $(98 + 55 + 58 + 75 + 63) \div 5 = 69.8$, under the 70-pound average ●

In Section 3.4, we calculated the score needed on a final exam to earn a specific grade. In Example 2, each test and the total homework are worth the same number of points, so we can find the mean average of the percent scores.

EXAMPLE 2 Finding the mean: class average Suppose an instructor gives equal weight to homework, each of two midterms, and the final exam. A student has 78% and 70% on the midterms and 90% on homework. What percent does this student need on the final to have an 80% mean average for the class?

Solution

$$\frac{0.78 + 0.70 + 0.90 + x}{4} = 0.80 \qquad \text{Add the numbers in the numerator.}$$

$$\frac{2.38 + x}{4} = 0.80 \qquad \text{Multiply both sides by 4.}$$

$$2.38 + x = 3.20 \qquad \text{Subtract 2.38 from both sides.}$$

$$x = 0.82$$

The student needs 82% on the final exam. ●

Note that we could add the percents in Example 2 because the instructor gives equal importance, or weight, to each. In general, *do not average percents unless they are percents of the same number.*

The mean does not always give a good description of a set of numbers. In Example 1, we found the mean of 95, 58, 52, and 88 to be 73.25. To the airport baggage handlers, the mean doesn't matter because they have to lift each bag. In Example 3, the mean would not give a good description of the income of each family.

EXAMPLE 3 Finding the mean: mean income Suppose the annual incomes of five families are \$4000, \$8000, \$9000, \$9000, and \$100,000, respectively. Find the mean, and explain why the mean does not give a good description of this set of families.

Solution The mean is

$$(\$4000 + \$8000 + \$9000 + \$9000 + \$100{,}000) \div 5 = \$26{,}000$$

The mean appears to indicate that all the families have an income well above the 1990 poverty level for families of four persons, \$13,359. Yet in reality four of the five families are below the poverty level. ●

MEDIAN Averaging salaries is one of several settings in which the mean does not provide a good description of a set of numbers. A very high or low number, as in Example 3, can cause the mean not to be close to most of the numbers.

Because the mean does not always give a good description, statisticians invented other averages, such as the median and the mode.

Finding the Median

> The **median** is found by selecting the middle number when the numbers are arranged in numerical order (from smallest to largest or largest to smallest). If there is no single middle number, the median is the mean of the two middle numbers.

Finding the median: income Find the median of each set of incomes.

a. \$4000, \$8000, \$9000, \$9000, \$100,000

b. \$50,000, \$20,000, \$8000, \$16,000

Solution **a.** The median of \$4000, \$8000, \$9000, \$9000, and \$100,000 is \$9000, because the numbers are already arranged in order and \$9000 is the middle number.

b. The second set of numbers must be rearranged into order: \$8000, \$16,000, \$20,000, and \$50,000. The set has no middle number. In this case, we find the mean of the two middle numbers and use it as the median. The median is (\$16,000 + \$20,000) ÷ 2 = \$18,000. ●

As suggested by Example 4, the median is more useful than the mean when one number in the set of data is some distance from the rest of the numbers.

MODE The mode is a third form of average. The mode is useful when the average needs to be one of the numbers—the number that occurs more often than any other number.

Finding the Mode

> The **mode** of a set is the number that occurs most often.

In the set of numbers {2, 2, 2, 2, 3, 4}, the mode is 2 because 2 appears most often.

Table 2 shows a tally by states of the age at which one is allowed unrestricted operation of private passenger cars. Many states with an age of 18 allow the applicant to drive at a lower age if he or she has taken a driver's education course.

Driver's Age, Years	States (including the District of Columbia)
15	//
16	卌 卌 卌 卌
16.5	/
17	///
18	卌 卌 卌 卌 ////
19	/

Table 2 Age for Unrestricted Operation of Private Passenger Cars
Data from *The World Almanac and Book of Facts 1992* (New York: Pharos Books, 1991), p. 678.

Finding median and mode from a table What are the median and mode for the data on driving age in Table 2?

Solution The tally places the data in order by age, so we count to the middle tally mark. The middle of the 51 tally marks is the 26th tally mark, because it has 25 marks before it and 25 after it. The 26th tally mark is the last mark for age 17, so age 17 is the median. The mode is the number that is tallied most often: age 18. ●

In Example 5, the minimum age for driving is about evenly split between age 16 and age 18. If the tally showed the same number of states for ages 16 and 18, the data would be **bimodal**—that is, *having two modes.*

Weighted Averages

There are times when some numbers in a set are more important than others. A final exam may have the same number of points as a midterm but be worth twice as much in calculating a grade for the course. The number of points a basketball player scores by making a ball go into the basket depends on when and from where the player throws the ball. The fine for a traffic ticket in a construction or school zone is double (or triple) the fine in other places.

To *give importance to numbers,* we assign a **weight.** A **weighted average** is *an average found by multiplying each number by its weight, adding the products, and dividing by the total weight* (rather than the total number of numbers). Here is a formula:

$$\text{Weighted average} = \frac{\text{sum of the product of each number and its weight}}{\text{total weight}}$$

A table can help organize the weights and the products. In the table in Example 6, we multiply the numbers in the rows (across the table) and then add the numbers in the last column to get the total.

EXAMPLE  Finding weighted grade averages Suppose the instructor in Example 2 gives homework half the weight of the midterm and gives the final one and a half times the weight of the midterm. Find the weighted average by completing Table 3. The weighted average, which is placed below the Score column, is the sum of the $w \cdot s$ column divided by the sum of the weights. Use the weighted average formula to set up and solve an equation to find what score the student needs on the final exam to average 80%.

	Weight, w	**Score, s**	$w \cdot s$
Midterm	1	0.78	
Midterm	1	0.70	
Homework	0.5	0.90	
Final	1.5	x	
Total		( )	

Table 3

Solution The completed table is shown in Table 4.

	Weight, w	**Score, s**	$w \cdot s$
Midterm	1	0.78	0.78
Midterm	1	0.70	0.70
Homework	0.5	0.90	0.45
Final	1.5	x	$1.5x$
Total	4.0	(0.80)	$1.93 + 1.5x$

↑ Weighted average

Table 4

The weighted average formula gives the same information:

$$\frac{1(0.78) + 1(0.70) + 0.5(0.90) + 1.5x}{1 + 1 + 0.5 + 1.5} = 0.80$$ Add the numbers in the numerator and denominator.

$$\frac{1.93 + 1.5x}{4} = 0.80$$ Multiply both sides by 4.

$$1.93 + 1.5x = 4(0.80)$$ Subtract 1.93 from both sides.

$$1.5x = 1.27$$ Divide by 1.5 on both sides.

$$x \approx 0.85$$

The student needs an 85 on the final for an 80% average. ●

EXAMPLE 7 Finding a weighted average Suppose the instructor in Example 2 makes the homework worth 10% of the grade, each midterm 20%, and the final 50%. Complete Table 5 and set up a weighted average equation to find what score is needed on the final exam to average 80%.

	Weight, w	Score, s	$w \cdot s$
Midterm	0.20	0.78	
Midterm	0.20	0.70	
Homework	0.10	0.90	
Final	0.50	x	
Total			

Table 5

Solution The completed table is shown in Table 6.

	Weight, w	Score, s	$w \cdot s$
Midterm	0.20	0.78	0.156
Midterm	0.20	0.70	0.140
Homework	0.10	0.90	0.090
Final	0.50	x	$0.5x$
Total	1.00	0.80	$0.386 + 0.5x$

↑ Weighted average

Table 6

The weighted average formula gives the same information:

$$\frac{0.20(0.78) + 0.20(0.70) + 0.10(0.90) + 0.50x}{0.20 + 0.20 + 0.10 + 0.50} = 0.80$$ Add the numbers in the numerator and denominator.

$$\frac{0.386 + 0.5x}{1} = 0.80$$ Multiply both sides by 1.

$$0.386 + 0.5x = 1(0.80)$$ Subtract 0.386 from both sides.

$$0.5x = 0.414$$ Divide by 0.5 on both sides.

$$x \approx 0.83$$

The student needs an 83 on the final for an 80% average. ●

Examples 6 and 7 show that weights can be either numbers or percents.

Think about it 1: Why was the score needed by the student in Example 7 lower than that needed by the student in Example 6?

In a weighted average, each item's influence on the average depends on the quantity of that item present. We return to these tables in Section 5.5.

Statistics and Geometry

The mean average appears in two important applications in geometry: midpoint and centroid. These topics are presented here, to give more practice with ordered pairs and finding the mean.

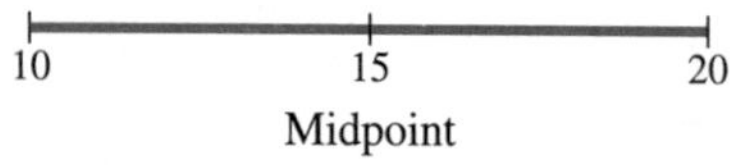

10 15 20
Midpoint

Figure 13

MIDPOINT The **midpoint of a line** is *the center, or the point halfway between its endpoints.* Figure 13 shows a line of length 10 units, with a midpoint 5 units from both ends. The point labeled 15 is halfway between a point labeled 10 and another labeled 20. The midpoint is the average of the coordinates of the endpoints; we can find the midpoint of any line segment by calculating this average.

The **midpoint on a coordinate graph** is *the mean in both the x and the y direction.* Thus, the midpoint of (x_1, y_1) and (x_2, y_2) is the average of x_1 and x_2 and the average of y_1 and y_2. Figure 14 shows the midpoint of the line connecting (x_1, y_1) and (x_2, y_2). The formula for finding midpoints is

$$\text{Midpoint} = \left(\frac{x_1 + x_2}{2}, \frac{y_1 + y_2}{2}\right)$$

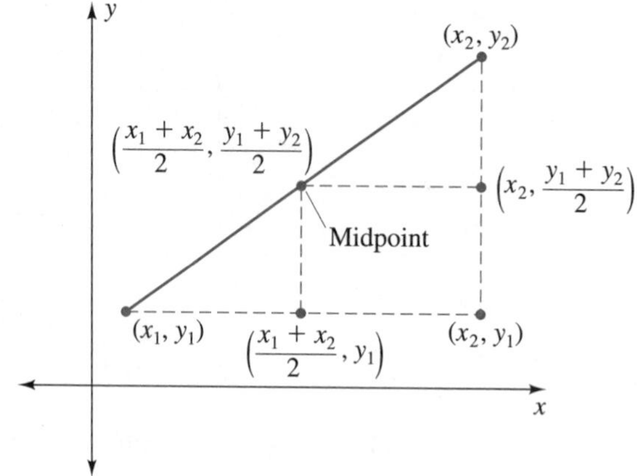

Figure 14

EXAMPLE 8 Finding midpoints Find the midpoints for the segments formed by the following coordinates. First use the formula. Then check the answer by sketching the line segment on a graph and plotting the midpoint.

a. (0, 2) and (5, 0) **b.** (0, 0) and (5, 2) **c.** (7, 3) and (9, −1)

Solution **a.** For (0, 2) and (5, 0), the midpoint is

$$\left(\frac{x_1 + x_2}{2}, \frac{y_1 + y_2}{2}\right) = \left(\frac{0 + 5}{2}, \frac{2 + 0}{2}\right) = \left(\frac{5}{2}, 1\right) = (2.5, 1)$$

b. For (0, 0) and (5, 2), the midpoint is

$$\left(\frac{x_1 + x_2}{2}, \frac{y_1 + y_2}{2}\right) = \left(\frac{0 + 5}{2}, \frac{0 + 2}{2}\right) = \left(\frac{5}{2}, 1\right) = (2.5, 1)$$

c. For (7, 3) and (9, −1), the midpoint is

$$\left(\frac{x_1 + x_2}{2}, \frac{y_1 + y_2}{2}\right) = \left(\frac{7 + 9}{2}, \frac{3 + (-1)}{2}\right) = (8, 1)$$

The midpoint for the first two segments is the same, (2.5, 1). It is labeled M_a and M_b in Figure 15. The midpoint for the third segment is (8, 1) and is labeled M_c.

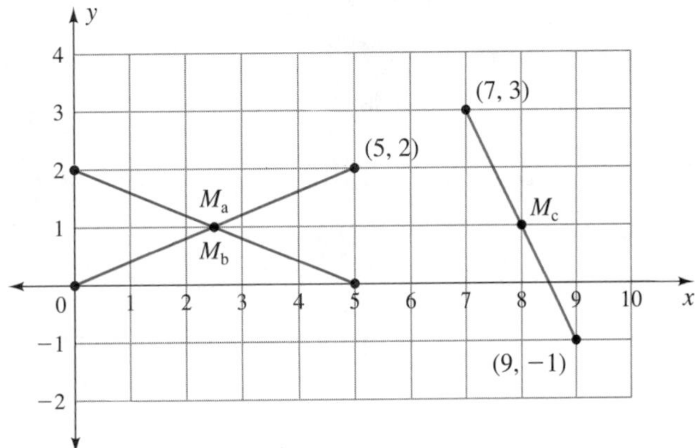

Figure 15

Think about it 2: Geometry includes many applications of averages. How can the formula for the area of a trapezoid be rewritten to show a mean average?

CENTROID The midpoint locates the middle of a line, whereas the **centroid** is *the center of a flat or solid geometric shape*. In more complicated structures such as the human body, bicycles, and airplanes, the center is called the *center of mass* or *center of gravity*.

The airplane in the photograph on the chapter opener became so out of balance that it tipped. Not only was the situation embarrassing; it was also a graphic illustration of how important the location of the center of gravity is. The *center of gravity is the average position of weight.* The pilot determines the center of gravity from the weight of the luggage and cargo, the weight of the fuel, and an average weight assigned to each passenger. Additional fuel, as needed, is ordered by weight and loaded into the various fuel tanks so as to maintain an appropriate center of gravity. If the suitcases and cargo are loaded into the back of the plane before the plane is refueled, the loading may cause the plane to tip as shown.

Finding the Centroid

> The centroid of a rectangle, square, or triangle is the average of the coordinates of its corners (vertices).

EXAMPLE 9 Finding the centroid Find the position of the centroid for the triangle in Figure 16.

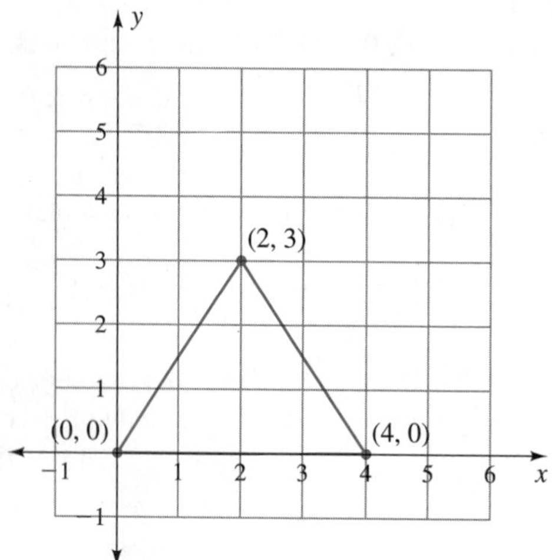

Figure 16

Solution The average of the x-coordinates of the vertices is

$$x_c = \frac{0 + 4 + 2}{3} = \frac{6}{3} = 2$$

The average of the y-coordinates of the vertices is

$$y_c = \frac{0 + 0 + 3}{3} = 1$$

The coordinates of the centroid are $(x_c, y_c) = (2, 1)$. ●

Centroids (and centers of gravity) are important because the behavior of an object in motion is dependent on where the centroid is located. Thus, the flight of an airplane, the rotation of a tire, the spinning of a washing machine, and the wild motion of a carnival ride depend on a correct centering of weight. Many sports such as ice skating, platform diving, pole vaulting, gymnastics, and ski jumping require careful control of the body's center to achieve top performance.

ANSWER BOX

Warm-up: 1. $x = 0.82$ **2.** $x \approx 0.847$ **3.** ≈ 64 **4.** 75 **5.** 80 **6.** 90 **Think about it 1:** The final exam has a larger weight in Example 7 than in Example 6, so the 83, although lower than the 85, is worth more in the grade calculation. **Think about it 2:** The area of a trapezoid is $A = \frac{1}{2}h(a + b)$, where h is the height and a and b are the parallel sides. We can write this formula as

$$A = h\left(\frac{a + b}{2}\right)$$

This formula for area can be read as the height multiplied by the average of the two parallel sides.

EXERCISES 5.4

In Exercises 1 to 4, calculate the mean, median, and mode.

1. Age at inauguration of Presidents of the United States.

a. 57, 61, 57, 57, 58
(Washington, Adams, Jefferson, Madison, Monroe)

b. 57, 61, 54, 68, 51
(Adams, Jackson, Van Buren, Harrison, Tyler)

c. 61, 52, 69, 64, 46
(Ford, Carter, Reagan, Bush, Clinton)

d. 65, 52, 56, 46, 54
(Buchanan, Lincoln, Johnson, Grant, Hayes)

2. Advertised puppy prices, in dollars

a. Pomeranian male:
225, 350, 200

b. Pomeranian female:
325, 225

c. Weimaraner male:
500, 600, 600, 500

d. Weimaraner female:
600, 650, 600, 500

3. Advertised rent, in dollars, for one-bedroom apartments

a. In a city with population 50,000:
315, 410, 440, 375, 415, 430, 415, 365, 395, 395, 340, 350, 385, 365

b. In a city with population 150,000:
550, 550, 410, 375, 515, 285, 400, 405, 325, 515, 510, 410, 435, 435

4. Cost of used sport utility vehicles, in dollars

a. 1997 Chevy Suburban:
27500, 27500, 27988, 27200, 32500, 25600, 26995

b. 1997 Ford Explorer:
21900, 22500, 19466, 24988, 24988, 22495, 19995

5. If the mean is the same for two sets of data, does this imply that the numbers in the sets are the same? Explain.

6. How does one large piece of data affect the mean? the mode? the median?

7. Comment, using complete sentences, on the effect of one low grade or one high grade on the mean of the test scores.

8. Choose one: The mean is influenced by (every, most, few) measurements in the set. Explain.

9. Choose one: The median (is, is not) influenced by one large or small measurement. Explain.

10. If a fly ball or strikeout is worth 0 and a hit is worth 1, is a baseball batting average (hits divided by times at bat) a median, mode, or mean?

11. When might the median provide a better description of the average than the mean?

12. When might the mean provide a better description of the average than the median?

13. Is it possible to have $\frac{3}{4}$ of the students above the median test score? Explain.

14. What can you conclude about a set of data if the mean is larger than the median?

15. What can you conclude about a set of data if the median is larger than the mean?

16. List a set of numbers for which the mean is larger than the median. List a set of numbers for which the mean is smaller than the median. Explain how you found your list.

17. *Sampling* involves selecting a number of objects from a set. Explain why a test must sample what a student knows.

18. Why might the *median of a set of numbers* and the *median of a triangle* be given such similar names?

19. The guard rail or strip of ground between the traffic lanes of a freeway is called the *median*. Explain why this is an appropriate word.

20. *Mode* has the same root as *modern*, *model*, and *a la mode*, which are associated with current, fashionable, or most popular styles. Explain how *mode* as an average fits with these other words.

Exercises 21 and 22 provide practice in calculating with weighted numbers.

21. Three colleges are competing in a track and field meet. A first-place finish is worth 5 points; second place, 3 points; third place, 1 point. GRCC has 4 first places, 3 seconds, and 2 thirds. LCC has 5 first places and 4 thirds. BCC has 1 first, 7 seconds, and 4 thirds. What is each team's total score?

22. A basketball team has 10 free throws at 1 point each, 25 field goals at 2 points each, and 6 three-point shots. The opposing team has 16 free throws, 20 field goals, and 4 three-point shots. What is each team's total score?

The student scores in Exercises 23 to 26 are from a course with three midterms and a final exam. The final is worth twice as much as each midterm. Estimate an average for each student. Find a weighted average for each student.

23. Tests: 0.85, 0.70, 0.80; final: 0.95

24. Tests: 0.75, 0.70, 0.75; final: 0.68

25. Tests: 0.80, 0.80, 0.70; final: 0.60

26. Tests: 0.90, 0.80, 0.60; final: 0.85

The students in Exercises 27 to 30 are in the same course as those in Exercises 23 to 26. What final exam score does each student need to earn a 0.80 weighted average? A perfect final exam score is 1.00.

27. Tests: 0.76, 0.81, 0.72

28. Tests: 0.60, 0.65, 0.70

29. Tests: 0.50, 0.70, 0.70

30. Tests: 0.60, 0.70, 0.75

Find the midpoints of the line segments connecting the sets of points given in Exercises 31 and 32. (Hint: A sketch on coordinate axes may confirm that the midpoints are reasonable.)

31. a. (0, 4), (5, 2) **b.** (−1, 3), (3, −3)

32. a. (2, 3), (8, 12) **b.** (2, −5), (−4, 3)

Find the midpoints of the line segments connecting the sets of points given in Exercises 33 and 34. Assume that the variables a and b are positive numbers.

33. a. $(a, 0), (0, b)$ **b.** $(a, a), (0, 0)$

c. $(0, b), (0, 0)$

34. a. $(a, 0), (0, a)$ **b.** $(0, 0), (a, 0)$

c. $(a, b), (0, 0)$

Find the midpoint of each side of the triangles in Exercises 35 to 38. Find the centroid of each triangle.

35.

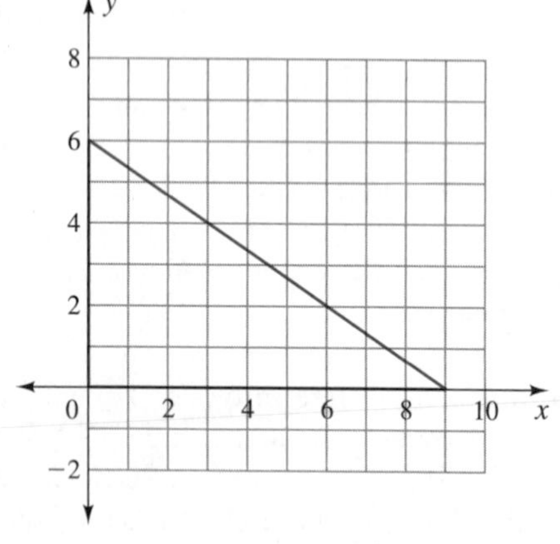

36.

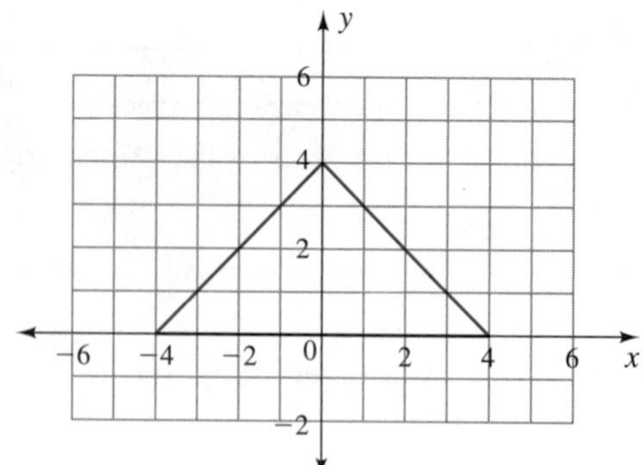

37.

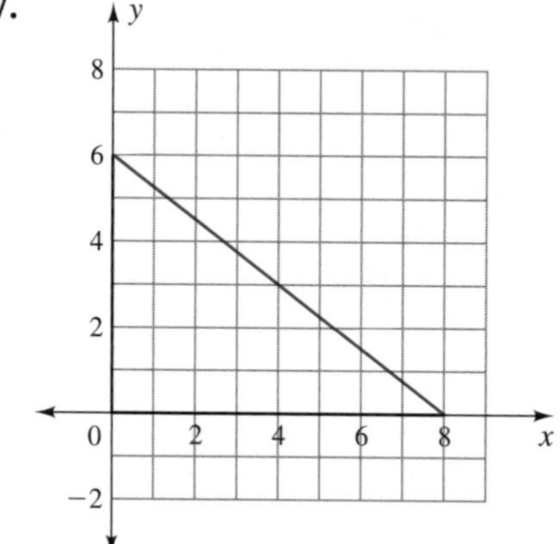

38.

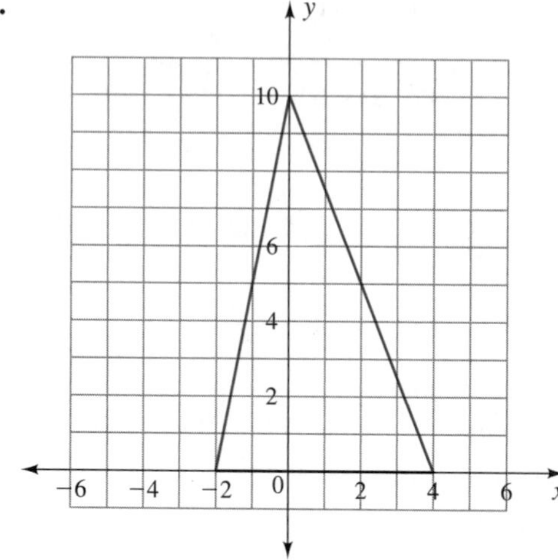

Projects

39. *Bookstore Discount.* A local bookstore offers a plan to "Buy 12 books, get one free." The store records each purchase on a card. After you buy 12 books, the store averages the purchases and you get to spend the average on your free book.

a. What is the value of the free book for a card with 55, 6, 15, 40, 8, 35, 42, 6, 20, 15, 10, and 15? A card with 10, 18, 40, 34, 12, 8, 50, 30, 45, 20, 10, and 5?

b. What is the total value of the two free books from part a?

c. Suppose you were able to sort out the purchases and place the 12 largest on one card and the others on the second card. Could you increase the total dollar value of the two free books?

d. Describe a way, if any, to sort the purchases to your advantage.

40. ***Center of Population.*** The U.S. Center of Population is like a centroid of population. It would be the balance point for the United States if each person (assume equal weight) were standing at her or his home location on a rigid plate the size of the country.

a. Trace a small map of the United States, and guess where the population center started and how it moved during the period from 1790 to 1990.

b. The center is recalculated after each ten-year census. An almanac or other general census reference will give the center, decade by decade. Plot the 21 centers of population for 1790 to 1990, and connect them.

41. ***Centroid Exploration***

a. Cut a large triangle from a stiff piece of notebook-size cardboard. Hold a pencil in a vertical position, and move the triangle around on the eraser until the triangle balances. Mark this point on the triangle. The balance point is the centroid of the triangle.

b. Measure the sides of the triangle, and mark the midpoint of each side. Connect each midpoint with the vertex opposite that side. The intersection of these three lines is the centroid. Does this centroid match that found in part a?

c. Use a paper punch to put a hole near each corner of the cardboard triangle. Place a pencil point through one of the holes, and hang a string with a weight tied at the bottom from the pencil point. The string and the triangle must be able to turn freely. Mark the point where the string crosses the opposite side of the triangle. Connect the hole and the mark with a line. Repeat for each of the other two holes. The intersection of the lines is the centroid. How close is this experimental centroid to the balance point? Where does the string cross the opposite side?

d. Place the triangle on a piece of graph paper. Find an ordered pair for each corner and use the averaging method to find the centroid.

e. How do your results in parts a to d agree?

5.5 Writing and Solving Equations with Quantity-Value Tables

OBJECTIVES

- Distinguish between quantity and value in problem situations.
- Use quantity-value tables to summarize word problems and set up equations.
- Explain when quantity sums and average values are meaningful.
- Estimate the average value, given the quantities and values.
- Estimate a quantity (or value) needed to find a given average value.

WARM-UP

Solve for x.

1. $200(55) + x(60) = 29{,}000$
2. $0.07x + 0.12(8000 - x) = 800$
3. $9(35) + 140x = (9 + x)(103)$
4. $24(1.875) + 3x = (24 + x)(2.50)$
5. $24(1.875) + 3x = (24 + x)(3.00)$

IN THIS SECTION, we look at quantity and value in a variety of applications. The section is in three parts. In the first part, we consider the difference between quantity and value. In the second, we build tables and find total quantities and average values. In the third, we set up and solve equations with the help of tables.

This section may seem out of place in a chapter on proportions, but being able to estimate outcomes for these problems requires an understanding of proportional thinking.

Distinguishing Between Quantity and Value

The poem below illustrates the fact that quantity and value are two different concepts. A young child learns quantity—how many—with counting. Value—what something is worth—is learned much later.

SMART
Shel Silverstein

My dad gave me one dollar bill
'Cause I'm his smartest son,
And I swapped it for two shiny quarters
'Cause two is more than one!

And then I took the quarters
And traded them to Lou
For three dimes—I guess he don't know
That three is more than two!

Just then, along came old blind Bates
And just 'cause he can't see
He gave me four nickels for my three dimes,
And four is more than three!

And I took the nickels to Hiram Coombs
Down at the seed-feed store,
And the fool gave me five pennies for them,
And five is more than four!

And then I went and showed my dad,
And he got red in the cheeks
And closed his eyes and shook his head—
Too proud of me to speak!

In the poem, confusing value and quantity caused a decrease in what the child's money was worth. The child measured "worth" in terms of the quantity of coins. **Quantity** *answers the question "How many?" or "How much?"*

His father measured "worth" in terms of value. There are many types of value. **Value**, as in the poem, may be *monetary worth*: cost per item, 10 cents per dime. Value may be a *percent*: an investment yielding 8.5%. Value may be a *rate*: miles per hour, pounds per square inch. The word *per* is a clue to recognizing a rate, but it is not the only word indicating a value or rate.

Value may also be *numbers given to specific results*: points assigned to letter grades to find a grade point average, different orders in which runners finish a track meet, and types of scores in a basketball game.

EXAMPLE 1 Distinguishing between quantity and value Which of these describe quantities? Which describe values or rates?

a. 5 points for first place, 3 points for second place

b. 5 cents per nickel
c. 8 meters
d. 12 fluid ounces
e. 1 point for a free throw
f. 15 free throws
g. $1.49 per pound
h. 8% interest

Solution Parts c, d, and f describe quantities. The rest are values. ●

A key idea in this section is that quantity and value (or rate) take many different forms, so problems that seem different may really be the same.

Building Quantity-Value Tables

When we place the quantity and value (or rate) from a problem setting into a table, called a **quantity-value table**, we find that the product, quantity times value, has meaning.

EXAMPLE 2 Completing a quantity-value table Use the information in the poem "SMART" to fill in Table 7. The first two rows have been completed.

Item	Quantity	Value (of each)	Quantity · Value
Dollar bill	1	$1.00	$1.00
Quarters	2	$0.25	$0.50
Dimes			
Nickels			
Pennies			

Table 7

Solution There are 3 dimes at $0.10 each, for a total of $0.30.
There are 4 nickels at $0.05 each, for a total of $0.20.
There are 5 pennies at $0.01 each, for a total of $0.05. ●

Table 7 illustrates the pattern of multiplying the quantity and the value to obtain a total item value.

> The last column of the quantity-value table is the product of the quantity of an item and the item's value (rate).

For most tables, we will head the last column $Q \cdot V$.

The sum of the quantity column has no meaning in the "SMART" table, nor does the sum of the quantity · value column. Look for the column sums in the next several examples to have progressively more meaning.

Cash register receipts from many large grocery stores give the name of the item, the quantity purchased, the item's unit value, the item's total cost, and the total cost for all items. Example 3 contains information from one shopping trip.

EXAMPLE 3 Building a quantity-value table: grocery bill Make a table showing the quantity and value for each item in the purchase, and calculate the item's total value (quantity · value) and the total cost. The purchase is 0.13 pound of garlic at \$2.88 per pound, 1.98 pounds of bananas at \$0.58 per pound, 4 cans of pears at \$0.78 per can, and 2 cans of peaches at \$1.05 per can.

Solution The completed table is shown in Table 8.

Item	Quantity	Value (rate)	Quantity · Value $Q \cdot V$
Garlic	0.13 lb	\$2.88/lb	\$0.38
Bananas	1.98 lb	0.58/lb	1.15
Pears	4 cans	0.78/can	3.12
Peaches	2 cans	1.05/can	2.10
Total			\$6.75

Table 8

Adding the entries in the last column gives the total cost of the purchase. ●

Think about it 1: Can the items in the quantity column be added? Would it be useful to know the sum if we knew the weight of the cans? Can the items in the value column be added?

In Example 4, both the sum of the quantities and the sum of the $Q \cdot V$ column have meaning.

EXAMPLE 4 Building a quantity-value table: lunch for four Make a table for this meal at Jaime's Hamburgers, identifying quantity and value for each item ordered: 4 hamburgers at \$3.85 each, 2 orders of fries at \$1.65 each, 2 coffees at \$0.95 each, and 2 milks at \$1.10 each. Calculate the item total and the total cost of the meal.

Solution The completed table is shown in Table 9.

Item	Quantity	Value (rate)	Quantity · Value $Q \cdot V$
Hamburger	4	\$3.85	\$15.40
Fries	2	1.65	3.30
Coffee	2	0.95	1.90
Milk	2	1.10	2.20
Total	10 items		\$22.80

Table 9

The sum of the $Q \cdot V$ column, \$22.80, is the cost of the meal. ●

Think about it 2: What is the meaning of the total of the quantity column?

In Example 4, dividing \$22.80 by 10 gives the average cost, \$2.28, of the 10 items. This average cost is meaningful but not particularly useful. The average cost may be written as the last entry in the value column. The value column is never added.

Finding Average Value

The **average value** may be included as the last entry in the value column. To find the average value, divide the sum of the $Q \cdot V$ column by the sum of the quantity column.

In Examples 5 and 6, we divide total value by total quantity to get an average that is both meaningful and useful. To organize our thinking, we return to the four-step problem-solving process.

EXAMPLE 5 Building a quantity-value table and finding average value: investment earnings

Silvia invests \$2000 at 8% interest and \$2500 at 6% interest. Build and complete a quantity-value table. Estimate her total earnings and the average interest rate earned (value). Include the estimation as part of your four-step process for finding the total earnings and average value. What is the meaning of the average value?

Solution ***Understand:*** To estimate total earnings, we look for easy mental multiplications. Noting that 8% is close to 10% and 6% is close to 5%, we calculate

$$(\$2000 \cdot 10\%) + (\$2500 \cdot 5\%) = \$200 + \$125 = \$325$$

The amounts invested at the two rates are similar, so the average interest rate will be near 7%; but because more money is invested at 6%, the rate will be slightly closer to 6% than to 8%.

We must also consider formulas. The interest earned in one year is the product of the investment and the interest rate written as a decimal.

Plan: Set up a quantity-value table, as shown in Table 10.

Quantity (money invested)	Value (interest rate)	$Q \cdot V$ (interest earned)
\$2000	0.08	\$160.00
\$2500	0.06	\$150.00
Total: \$4500		\$310.00

Table 10 Investment

Carry out the plan: The total earnings are \$310.

Check and extend: Is the sum of the quantity column meaningful? Is dividing the total interest earned by the sum of the quantity column meaningful? The quantity column sum gives the total money invested, \$4500. The quotient of the total interest earned and the total money invested (\$310 ÷ \$4500) gives the average interest rate earned on the total invested, approximately 6.9%. Thus, both calculations are meaningful. ●

Example 6 may be helpful in your own school career.

EXAMPLE 6 Building a quantity-value table: calculating a grade point average (GPA) To calculate grade point averages, some schools set these rates: 4 points per A, 3 points per B, 2 points per C, and 1 point per D. Make a quantity-value (rate) table for the following grade report. Calculate the grade point average.

French:	4 credit hours	C
Algebra:	4 credit hours	A
Writing:	3 credit hours	B
Tennis:	1 credit hour	A

Solution ***Understand:*** To estimate the grade point average, we observe that the student has slightly more credit hours of As than of Cs, so the GPA is likely to be just over 3.00. The credit hours are quantities. The points per letter grade are rates and belong in the value column.

Plan: We make a quantity-value table, as shown in Table 11.

Subject and Grade	Quantity (credit hours)	Value or Rate (grade points)	$Q \cdot V$
French, C	4 hr	2 pts per hr	8 pts
Algebra, A	4 hr	4 pts per hr	16 pts
Writing, B	3 hr	3 pts per hr	9 pts
Tennis, A	1 hr	4 pts per hr	4 pts
Total	12 hr		37 pts

Table 11

Carry out the plan: The grade point average is 37 points divided by 12 hours, or approximately 3.08. The average, 3.08, is placed in the last row of the value or rate column.

Check and extend: Are the sum of the quantity column and the sum of the $Q \cdot V$ column meaningful? Is the answer reasonable? The sum of the quantity column gives the total number of credits. The sum of the $Q \cdot V$ column gives the total points earned. The average is close to our estimate—a little larger than 3.00. ●

Think about it 3: What would this student's GPA be if all courses were assigned the same credit? (*Hint*: Suppose the quantity column changed to all 3s.)

Solving Equations with the Help of Quantity-Value Tables

What makes the quantity-value approach important is that it is not limited to purchases, investments, and grade point averages. Quantities and values appear in medicine, business, science, engineering, transportation, and other fields.

Avoid the temptation to learn this material application by application. Try to see the product quantity · value in each problem. The remaining examples are grouped in three categories: applications with *money as value*, applications with *percent as value*, and applications with *unusual units as value*. You have seen all these categories in Examples 3 to 6. The grocery store receipt and restaurant meal problems had money in the value column. The investment problem had percents in the value column. The grade point average problem had points per credit hour in the value column. For lack of another name, we consider points per credit hour to be a value with unusual units.

MONEY AS VALUE In Example 5, the money invested was the quantity because we were multiplying by an interest rate to find the interest earned. In Example 7, the money invested is in the last, or $Q \cdot V$, column because it is the product of number of shares (quantity) and cost per share (rate or value).

EXAMPLE 7 Finding a missing quantity Inez has \$29,000 to invest in the stock market. Texaco stock costs \$60 per share. GTE stock costs \$55 per share. Estimate how many shares of Texaco she can buy if she already has decided to buy 200 shares of GTE. Use a quantity-value table to build an equation, and solve for the number of shares. What is the average value and what does it mean?

Solution ***Understand:*** Inez will spend a little over \$10,000 on GTE. The prices of the two stocks are close, so she should be able to purchase more than 200 shares of Texaco. The number of shares is a quantity, and the price per share is a value. Let x be the number of shares of Texaco.

Plan: We set up a quantity-value table, as shown in Table 12.

Stock	Quantity (shares)	Value (price per share)	$Q \cdot V$
GTE	200	\$55	200(55)
Texaco	x	\$60	$x(60)$
Total	$200 + x$	Average purchase price per share	\$29,000

Table 12

Carry out the plan: We find the equation by adding the $Q \cdot V$ products in the last column and setting the resulting expression equal to the total investment, \$29,000.

$$200(55) + x(60) = 29{,}000$$
$$11{,}000 + 60x = 29{,}000 \quad \text{Subtract 11,000 from each side.}$$
$$60x = 18{,}000 \quad \text{Divide by 60.}$$
$$x = 300$$

The average value is \$29,000/500 = \$58 average cost per share.

Check: 300 shares at \$60 plus 200 shares at \$55 gives \$29,000. ●

Think about it 4: Which word tells us that price per share is the value for Table 12 in Example 7?

In prior examples, we have had only one unknown quantity. In the next two examples, the quantities of two items are unknown. *To write two different quantities in terms of one variable, we need a fact relating them*, such as the fact that their sum is 60. If we let one quantity be x, then our fact tells us that the other is $60 - x$. We used the same idea with similar triangles in Section 5.3; given the total length and one part of a segment, we wrote an expression for the other part.

In Example 8, we are given the average value and the sum of the two quantities and asked to find the quantity of each of the two items.

EXAMPLE 8 Finding two quantities, given their sum and the average value Janelle has two kinds of chocolates with which to fill one-pound boxes. The chocolate truffles sell for \$36 per pound, and the chocolate creams sell for \$20 per pound. She wants to make 60 one-pound boxes that sell for \$24 per pound. How many pounds of each chocolate should she use?

Solution ***Understand*:** We will guess that she will use mostly chocolate creams because they are cheaper—the desired price per pound is closer to \$20 than to \$36. The two unknown quantities add to 60, so we let x be the first quantity and $60 - x$ be the second.

***Plan*:** We set up a quantity-value table, as shown in Table 13, and use it to find an equation.

Item	Quantity (pounds)	Value (dollars per pound)	$Q \cdot V$ (dollars)
Chocolate truffles	x	\$36	$36x$
Chocolate creams	$60 - x$	\$20	$1200 - 20x$
Total	60	\$24	1440

Table 13

***Carry out the plan*:** The equation comes from the fact that the sum of the $Q \cdot V$ column is 1440.

$$36x + 1200 - 20x = 1440 \quad \text{Add like terms; subtract 1200.}$$
$$16x = 240 \quad \text{Divide by 16.}$$
$$x = 15$$

Janelle should use 15 pounds of chocolate truffles and $60 - 15$, or 45, pounds of chocolate creams.

***Check*:** $15(36) + 45(20) \stackrel{?}{=} 60(24)$ ✓ ●

In this section, we write the two unknown quantities with one variable.

When two quantities add to a number S, let one be x and the other be $S - x$.

In Chapter 7, we will write the two unknown quantities with two variables.

PERCENT AS VALUE We now return to problems with percent in the value column. When using percent as value, always change the percent to a decimal. In Example 9, we are given the average value and asked to find the quantity of each ingredient. Look for how the two quantities are written with one variable. In Example 9, as in Examples 7 and 8, we build the equation from the lower right corner.

Building an Equation from a Quantity-Value Table

Finding two expressions for the lower right corner of the quantity-value table is the key to building an equation. The sum of the $Q \cdot V$ column must equal the product across the total row.

EXAMPLE 

Finding two quantities, given an average value and the total quantity The Healthy Options Company is mixing 8000 pounds of cat food from two ingredients. The first ingredient has 7% protein, and the second has 12% protein. The mixture needs to have 10% protein. Use a quantity-value table to find out how many pounds of each ingredient are needed.

Solution ***Understand:*** Because 10% is between 7% and 12%, we start with a guess of 4000 pounds of each ingredient. Writing each percent as a decimal, we come up with Table 14.

Item	Quantity (pounds)	Value (protein per pound)	$Q \cdot V$ (pounds of protein)
Ingredient 1	4000	0.07	280
Ingredient 2	4000	0.12	480
Total	8000	$760/8000 = 0.095$	760

Table 14

An equal amount of each ingredient gives 9.5% protein, so we need a little more of the 12% ingredient than of the 7% ingredient.

Plan: If the two ingredients add to 8000 and one is x, then the other is the difference, $8000 - x$. We build a table with x and $8000 - x$ as the quantities and 10% as the average value, as shown in Table 15. (Percents are written as decimals.)

Item	Quantity (pounds)	Value (protein per pound)	$Q \cdot V$ (pounds of protein)
Ingredient 1	x	0.07	$0.07x$
Ingredient 2	$8000 - x$	0.12	$0.12(8000 - x)$
Total	8000	0.10	800

Table 15

Carry out the plan: The sum of the $Q \cdot V$ column is $0.07x + 0.12(8000 - x)$. The product across the last row is $8000(0.10) = 800$. We set these equal to build the equation.

$0.07x + 0.12(8000 - x) = 800$	Apply the distributive property.
$0.07x + 960 - 0.12x = 800$	Add like terms.
$-0.05x + 960 = 800$	Add $0.05x$ and subtract 800 on both sides.
$160 = 0.05x$	Divide by 0.05.
$x = 3200$	

The company needs to mix 3200 pounds of the 7% ingredient with 8000 − 3200, or 4800, pounds of the 12% ingredient.

Check: Our estimate was that more of the 12% ingredient would be needed. $3200(0.07) + 4800(0.12) \stackrel{?}{=} 8000(0.10)$ ✓ ●

Our guess-and-check step in Example 9 produces an important result:

> When equal quantities of two items are used, the average value is the average of the item values.

We can conclude that if the desired average value is less than halfway between item values, we should use more of the smaller value. If the desired average value is more than halfway between item values, we should use more of the larger value.

UNUSUAL UNITS AS VALUE In Example 10, we blend water of two different temperatures. Although the example can be modeled quite well in the classroom with small quantities of water and inexpensive thermometers (see Exercise 69), the results are not accurate enough for scientific purposes. The laws of heat transfer in the field of thermodynamics are far too complex to be studied or modeled here.

The units on the values in Example 10 are degrees, so when we multiply by gallons we get degree · gallons, a very unusual unit.

EXAMPLE  Finding a missing quantity, given an average value: a hot bath One cold winter evening, you plan to take a leisurely bath, but you get called away after you turn the water on. When you get back, you discover that only the cold water was turned on. The tub is one-fourth full (9 gallons). The cold water temperature in winter is about 35°. The hot water heater is set at 140°. You would rather not waste water by draining the tub and starting over. Use a quantity-value table to find how much hot water must be added to correct the temperature to the desired 103°.

Solution

Item	Quantity	Value (temperature)	$Q \cdot V$
Cold water	9 gal	35°	9(35)
Hot water	x gal	140°	$140x$
Total	$9 + x$	103°	(Two expressions)

Table 16 Bath Water

The two expressions for the lower right corner of Table 16 are *the total down*, $9(35) + 140x$, and *the product across*, $(9 + x)(103°)$. We set them equal and solve for x:

$$9(35) + 140x = (9 + x)(103) \quad \text{Use the distributive property.}$$

$$315 + 140x = 927 + 103x \quad \text{Subtract 315 and } 103x \text{ from both sides.}$$

$$37x = 612 \quad \text{Divide by 37.}$$

$$x \approx 16.5 \text{ gal}$$

You need to add approximately 16.5 gallons of hot (140°) water. Will the added water raise the water level too high (either before or after you step into the tub)? ●

As in the previous examples, look carefully at how the two expressions for the lower right corner are used to build the equation.

EXAMPLE 11 Finding a missing quantity, given an average value: GPAs, continued A student with a 1.875 GPA has 24 credit hours. Using the grade rates from Example 6, find how many credit hours of Bs he needs to raise his GPA to a 2.5. (*Hint*: Let x be the number of credit hours of Bs.)

Solution ***Plan*:** The desired GPA, 2.50, is placed in the last row of the value column. The sum of the last column (the total points from current and future grades) gives one expression for the lower right corner. The product of the total quantity of credit hours and the desired GPA gives a second expression for the lower right corner.

Item	Quantity (credit hours)	Value (GPA)	$Q \cdot V$ (grade points)
Current GPA status	24	1.875 points	24(1.875)
Future B grades	x	3.00 points	$3x$
Total	$24 + x$	2.50 average	(Two expressions: the sum down and the product across)

Table 17 Grade Point Average

***Carry out the plan*:** The two expressions for the lower right corner of Table 17 are *the total down*, $24(1.875) + 3x$, and *the product across*, $(24 + x)(2.50)$. We set them equal and solve for x:

$$24(1.875) + 3x = (24 + x)(2.50)$$

$$45 + 3x = 60 + 2.50x$$

$$0.5x = 15$$

$$x = 30 \text{ hr}$$

***Check*:** $24(1.875) + 3(30) \stackrel{?}{=} (24 + 30)(2.50)$ ✓ ●

Finding two expressions for the lower right corner is the key to building an equation.

EXAMPLE 12 Finding a missing quantity, given an average value: more GPAs Suppose the student in Example 11 wants to raise his GPA to a 3.00. How many credit hours of Bs does he need to raise his GPA to a 3.00? Use Table 18 to write the two expressions that describe the lower right corner.

Item	Quantity (credit hours)	Value (GPA)	$Q \cdot V$
Current GPA	24	1.875 points	24(1.875)
Desired grades	x	3.00 points	$3x$
Total	$24 + x$	3.00 average	(Write two expressions)

Table 18 Grade Point Average

Solution The two expressions for the lower right corner are *the total down*, $24(1.875) + 3x$, and *the product across*, $(24 + x)(3.00)$. We set them equal and solve for x:

$$24(1.875) + 3x = (24 + x)(3.00)$$

$$45 + 3x = 72 + 3.00x$$

$$0 = 27 \qquad \text{No real-number solution}$$

There is no real number x that makes the equation true, so the solution set is empty, { }.

Think about it 5: What does "No real-number solution" mean in terms of the problem setting?

ANSWER BOX

Warm-up: 1. $x = 300$ **2.** $x = 3200$ **3.** $x \approx 16.5$ **4.** $x = 30$ **5.** No real-number solution **Think about it 1:** The items in the quantity column are unlike and cannot be added. It would be useful to know the weight of the cans if we were walking home. The items in the value column cannot be added. **Think about it 2:** The quantity column contains the number ordered of each item. The total is the number of items ordered. **Think about it 3:** 3.25 **Think about it 4:** *Per* suggests a rate or value. **Think about it 5:** "No real-number solution" means that it is impossible for the student to raise his GPA to a 3.00 by earning only Bs. Some As will be needed.

EXERCISES 5.5

No algebraic expressions or equations are needed in Exercises 1 to 28. In Exercises 1 to 10, which phrases describe quantities? Which describe values?

1. 25 cents per quarter
2. $85 per share
3. 10 liters
4. 12 coins
5. 3 points per credit hour of Bs
6. 6% interest rate
7. A molarity value of 12
8. $2.88 per pound
9. 40% boric acid solution
10. $2,000

For Exercises 11 to 28, do the following:

a. *Set up a quantity-value table.*

b. *Write the meaning, if any, of the sum of the quantity column.*

c. *Find the average value for the table by dividing the sum of the Q · V column by the sum of the quantity column.*

d. *Write the meaning of the sum of the Q · V column.*

11. Demi's snack shop has 5 kilograms of peanuts at $8.80 per kilogram and 2 kilograms of cashews at $24.20 per kilogram.
12. Reuel's café has 100 pounds of coffee at $9.00 per pound and 100 pounds of coffee at $10.80 per pound.
13. Abraham buys 3 pounds of grapes at $0.98 per pound, 5 pounds of potatoes at $0.49 per pound, and 2 pounds of broccoli at $0.89 per pound. What is the total cost of his purchase?
14. Ludvina buys 15 airletters at $0.50, 50 first-class postage stamps at $0.33, and 10 postcards at $0.20. What is the total cost of her purchase?
15. Serena has 15 dimes and 20 quarters.
16. Andrzej has 12 half-dollars and 30 nickels.
17. Erin has $1500 invested at 9% interest and $1500 at 6% interest.
18. Kim has $2500 invested at 8% interest and $3000 at 4% interest.
19. Amel, a veterinarian, has 100 pounds of dog food containing 12% protein and 50 pounds of dog food containing 15% protein.
20. La Deane, a veterinarian, has 30 kilograms of cat food containing 8% fat and 40 kilograms of cat food containing 14% fat.
21. JuLeah, a horse trainer, has 150 pounds of alfalfa at 10% protein and 25 pounds of straw at 0% protein.
22. Ward, a horse trainer, has 1 gallon of coat conditioner in a 5% solution to be added to 20 gallons of water (0% solution).
23. Lenny has 5 hours of Ds (1 point per credit hour), 4 hours of Cs (2 points per credit hour), and 3 hours of Bs (3 points per credit hour).
24. Lenny has 4 hours at a 2.00 GPA from fall term, 4 hours at a 1.75 GPA from winter term, and 12 hours at a 1.83 GPA from spring term.
25. Loki drives 3 hours at 80 kilometers per hour and 2 hours at 30 kilometers per hour.
26. Kana drives 4 hours at 50 miles per hour and 3 hours at 18 miles per hour.
27. Li, a chemist, has 150 milliliters of sulfuric acid with a molarity value of 18 and 100 milliliters of sulfuric acid with a molarity value of 3. (*Molarity* is a chemical term; the larger the molarity value, the more concentrated the acid.)
28. Ingrid, a chemist, has 200 milliliters of nitric acid with a molarity value of 16 and 500 milliliters of nitric acid with a molarity value of 6.

In Exercises 29 to 34, each quantity is the sum of two numbers. The variable x is one of the two numbers. What is an expression for the other number?

29. 300 pounds
30. 50 kilograms
31. $15,000
32. 18 hours
33. 16 liters
34. 500 shares

In Exercises 35 to 42, solve by setting up a quantity-value table and building an equation from the table.

35. Georgia purchases 200 shares of Boeing stock at $54 per share. How many shares of Nike stock can she purchase at $71 per share if she has a total of $25,000 to invest?
36. Mikhail has $22,000 to invest. How many shares of Ford stock at $37 per share can he buy if he also buys 300 shares of General Motors at $24 per share?
37. One week, Neva works 30 hours on one job at $5.80 per hour. Her second job pays $7.20 per hour. If she needs to earn an average of $6.36 per hour, estimate and then find how many hours she must work at the second job.

38. Florence works 36 hours one week at $7.50 per hour. She is offered overtime at $11 per hour. Estimate and then find how many hours she must work overtime to average $8 per hour.

39. Estimate and then find how many hours Neva (Exercise 37) needs to work at the second job to earn an average of $7 per hour. Is this a reasonable expectation?

40. Estimate and then find how many hours Florence (Exercise 38) needs to work overtime to average $9.10 per hour.

41. Alan blends two types of coffee beans to sell at a price of $8.35 per pound. Colombian coffee beans sell for $7.25 per pound, and Sumatran beans sell for $10.00 per pound. If he wishes to make 300 pounds of blend, how many pounds of each are needed?

42. Bridget's blend of coffee combines Colombian beans at $7.25 per pound with Sumatran beans at $10.00 per pound. She wishes to blend 300 pounds to sell at $9.45 per pound. How many pounds of each does she need?

Bill and Pat's Nut Shop mixes cashews and peanuts. Cashews sell for $24 a kilogram, and peanuts sell for $10 a kilogram. Estimate how many kilograms of each are needed to make a 50-kilogram mixture to sell for the prices listed in Exercises 43 to 46. Use a quantity-value table to set up and solve an equation to find the number of kilograms of each. Round to the nearest tenth.

43. $12 per kilogram

44. $16 per kilogram

45. $18 per kilogram

46. $20 per kilogram

Maria has $15,000 to invest for one year. Investment A pays 5% interest for the year, and Investment B pays 8% interest. In Exercises 47 and 48, set up a quantity-value table, find how much she earns, and predict and find the average rate of return (average value). No equation is needed.

47. a. Maria invests all at 8% and nothing at 5%.

    b. Maria invests half the money at 8% and half at 5%.

48. a. Maria invests all the money at 5% and nothing at 8%.

    b. Maria invests $\frac{1}{3}$ of the money at 5% and $\frac{2}{3}$ of the money at 8%.

Maria has $15,000 to invest for one year. Investment A pays 5% interest for the year, and Investment B pays 8% interest. In Exercises 49 and 50, estimate the investments in A and in B. Then set up a quantity-value table and an equation to find how much is invested in each.

49. a. If her total earnings are $1060, how much is invested in each?

    b. If her total earnings are $825, how much is invested in each?

50. a. If her total earnings are $1005, how much is invested in each?

    b. If her total earnings are $1140, how much is invested in each?

51. How many credit hours of Bs (at 3 points per credit hour) does Lenny need to raise his GPA to a 2.00? He has 20 credit hours at a 1.85 GPA.

52. How many credit hours of Bs (at 3 points per credit hour) does Alex need to raise 40 credit hours at a 1.85 GPA to a 2.25 GPA?

53. How many gallons of cold water (summer temperature 60°) would need to be added to a bathtub containing 12 gallons of hot water at 140° to lower the temperature to 103°?

54. How many gallons of cold water (winter temperature 35°) would need to be added to a bathtub containing 12 gallons of hot water at 140° to lower the temperature to 103°?

55. How many gallons of hot water at 140° would need to be added to a bathtub containing 15 gallons of cold water at 35° to raise the temperature to 103°?

56. How many gallons of hot water at 140° would need to be added to a bathtub containing 15 gallons of cold water at 60° to raise the temperature to 103°?

57. Predict the temperature if equal parts of hot (140°) and winter cold (35°) water are mixed. Is this true for any equal quantities?

58. Predict the temperature if equal parts of hot (140°) and summer cold (60°) water are mixed. Is this true for any equal quantities?

59. Abdulla needs to produce a sulfuric acid solution with a molarity value of 12 by blending an unknown quantity of sulfuric acid with a molarity value of 3 and 300 milliliters of sulfuric acid with a molarity value of 18. How many milliliters of the acid with a molarity value of 3 will he need?

60. Tsuki needs to make hydrochloric acid with a molarity value of 8. She wants to add an unknown quantity of hydrochloric acid with a molarity value of 12 to 500 milliliters of hydrochloric acid with a molarity value of 6. How many milliliters of the acid with a molarity value of 12 will she need?

61. Explain why total money invested appears in different places in the quantity-value tables in Example 5 (interest rate) and Example 7 (stock purchase).

62. Explain how to use one variable to describe two items when the total quantity of the two items is known.

63. Explain how to estimate the quantities of two items, given the value of each and the average value.

64. Explain how to estimate the average wage, given two different wages and the number of hours worked at each.

65. Explain how to estimate whether a quantity will have a large effect on an average value.

66. Two students both have a 3.00 GPA. This semester, each earns a 4.00 GPA on 12 credit hours. Which student will see more improvement in overall GPA: the second-year full-time student or the fourth-year full-time student? State any assumptions you make.

67. ***Review Problem.*** In the poem "SMART," the value decreased from \$1.00 to \$0.50 to \$0.30 to \$0.20 to \$0.05. Estimate which change reflected the largest percent decrease and which reflected the smallest percent decrease. What is the percent decrease with each change?

68. This exercise proves that when we have equal quantities of two items, the average value for the table is the average of the values for the two items.

a. Explain how you know that the table below shows equal quantities of the two items.

b. Copy and complete the table.

c. Set up an equation containing A.

d. Solve for A and explain the result.

Item	Quantity	Value	$Q \cdot V$
First item	x	m	
Second item	x	n	
Total		A	Two expressions set equal.

Projects

69. ***Water-Temperature Experiment.*** Model the bathtub water temperature problem in Example 10. Take two plastic containers marked in either fluid ounces or tenths of a liter. Place a known quantity of hot tap water in one container. Place a known quantity of cold water without ice (say, from a refrigerated drinking fountain) in the other container. Leave enough room in one container to pour in the water from the other container. Place a thermometer in each container. Set up a quantity-value table for the amounts of water and temperatures. With the table, predict the average temperature that will result from mixing the water. "Check" the table results by putting the water, together with both thermometers, into one container. Read the temperature when the two thermometers agree.

70. ***Spreadsheet.*** Create a computer spreadsheet similar to the table below to model the quantity-value tables. Use it to do the relevant exercises in this section. To run the spreadsheet, enter numbers in the positions A2, A3, B2, and B3. If A2, A3, B2, or B3 information is missing in the problem, solve the problem by guessing and then use the spreadsheet to check.

Row	Column A	Column B	Column C
1	Quantity	Value	$Q \cdot V$
2	A2	B2	+A2 * B2
3	A3	B3	+A3 * B3
4	+A2 + A3	+(C2 + C3)/A4	+C2 + C3
5	"Calculated Average Value =" +B4		
6	"Product Across =" +A4 * B4		
7	"$Q \cdot V$ Sum Down =" +C2 + C3		

71. ***Flipping a Coin.*** Flip a coin 20 times. Record each outcome—heads or tails—in the first column of a table.

a. Name the second and third columns "Cumulative Heads Percent" and "Cumulative Tails Percent," respectively. Calculate the percentages indicated. (The cumulative heads percent is calculated by dividing the total number of times the coins have come up heads by the total number of heads and tails together.) A sample table is shown below.

Outcome	Cumulative Heads Percent	Cumulative Tails Percent
T	$\frac{0 \text{ heads}}{1 \text{ total}} = 0\%$	$\frac{1 \text{ tail}}{1 \text{ total}} = 100\%$
H	$\frac{1 \text{ heads}}{2 \text{ total}} = 50\%$	$\frac{1 \text{ tail}}{2 \text{ total}} = 50\%$
H	$\frac{2 \text{ heads}}{3 \text{ total}} = 66.7\%$	$\frac{1 \text{ tail}}{3 \text{ total}} = 33.3\%$
H	$\frac{3 \text{ heads}}{4 \text{ total}} = 75\%$	$\frac{1 \text{ tail}}{4 \text{ total}} = 25\%$

b. Draw a set of axes, with numbers of flips on the x-axis and percents on the y-axis. Plot two graphs, one for percent heads and one for percent tails.

c. What do you observe about the two graphs? In what way should your graph be similar to the graphs in other people's projects? Why?

d. Compare the effect of the twentieth outcome on the percents to that of the second outcome.

CHAPTER 5 SUMMARY

Vocabulary

For definitions and page references, see the Glossary/Index.

bimodal
centroid
continued ratio
corresponding
cross multiplication
equivalent ratios
inner-outer product
mean
median
midpoint of a line
midpoint on a coordinate graph
mode
percent
percent change
percent grade
proportion
quantity
quantity-value table
rate
ratio
similar figures
similar triangles
value
vertex (vertices)
weight
weighted average

Concepts

5.1 Ratios, Percents, and Rates

Two ratios are equivalent if they simplify to the same number or if they divide to the same decimal value.

Ratios containing variables or units are simplified in the same way as fractions—by eliminating common variables or units where

$$\frac{a}{a} = 1 \quad \text{or} \quad \frac{\text{inches}}{\text{inches}} = 1$$

To find the shares from a continued ratio $a{:}b{:}c$ with algebra, let $x =$ the amount to be received by one share. Set the continued ratio expression $ax + bx + cx$ equal to amount to be shared, and solve for x.

To find percent change, calculate the difference between the new number and the original number and then divide by the original number.

Applying unit analysis to rates may require changing both the units in the numerator and the units in the denominator.

5.2 Proportions and Proportional Reasoning

If $\frac{a}{b} = \frac{c}{d}$, then $ad = bc$.

When a proportion contains units, the units must match either side to side or within numerator and denominator pairs.

In percents, if a is $n\%$ of base b, then

$$\frac{n}{100} = \frac{a}{b}$$

5.3 Similar Figures and Similar Triangles

Apply proportions to find the lengths of the sides in similar figures. When you know the total length or quantity of something that is separated into two parts, use one variable to describe the two parts. If the whole is 10 and one part is x, the other part is $10 - x$.

The sum of the angle measures in a triangle is 180°.

5.4 Averages

To find the mean of a set of numbers, add the numbers and divide by the number of numbers in the set.

To find the median, select the middle number when the numbers are arranged in numerical order (from smallest to largest or largest to smallest). If there is no single middle number, the median is the mean of the two middle numbers.

To find the mode, select the number appearing the highest number of times.

The ordered pair giving the midpoint of a line segment is the average of the x-coordinates and of the y-coordinates of the endpoints.

The ordered pair giving the centroid of a square, rectangle, or triangle is the average of the x-coordinates and of the y-coordinates of the vertices.

5.5 Quantity-Value Tables

Finding two expressions for the lower right corner of the quantity-value table is the key to building an equation. The sum of the $Q \cdot V$ column must equal the product across the total row (the sum of the quantities multiplied by the average value).

When two quantities add to a number S, let one be x and the other be $S - x$.

When equal quantities of two items are used, the average value is the average of the item values.

CHAPTER 5 REVIEW EXERCISES

For Exercises 1 and 2, simplify the ratios.

1. $16x^2$ to $4x^4$

2. $\dfrac{x(x - 3)}{(x - 3)(x + 3)}$

Simplify the ratios in Exercises 3 and 4.

3. 5 feet : 18 inches

4. 50 centimeters to 2 meters

5. 75% of what number is 108?

6. 108% of 75 is what number?

In Exercises 7 and 8, describe in a complete sentence an appropriate situation for each rate.

7. hours per revolution

8. dollars per day

9. Sales of beverages are in a 5:5:1 ratio for Pepsi products, Coca Cola products, and other brands. How many of each brand were sold if a total of 121,000 cases were sold?

10. The angle measures in a triangle are in the ratio 1:2:3. The sum of the angle measures is 180°. What is the measure of each angle?

In Exercises 11 and 12, use the fact that nutrition experts recommend that no more than 30% of a person's daily calorie intake come from fat. The remaining calories should come from carbohydrates and protein.

11. In a 1500-calorie diet, how many calories should come from fat? At 9 calories per gram of fat, how many grams of fat would this be?

12. In a 2000-calorie diet, how many calories should come from carbohydrates and protein together? At 4 calories per gram of protein and carbohydrate, how many grams of protein and carbohydrate would this be?

13. What is the percent decrease when a price drops from \$49.99 to \$42.49?

14. What is the percent increase when a wage rises from \$5.40 per hour to \$6.30 per hour?

15. On a trip to Canada, you pay \$0.60 Canadian per liter for gasoline. How much is this in U.S. dollars per gallon? Estimate first.

    1 Canadian dollar $\approx$ 0.70 U.S. dollar
    1.0567 quarts = 1 liter
    4 quarts = 1 gallon

16. How many miles per hour is 100 meters in 9 seconds?

    12 inches = 1 foot
    39.37 inches = 1 meter
    5280 feet = 1 mile
    3600 seconds = 1 hour

17. An access ramp needs to rise 30 inches. If the ramp is to have a slope ratio of 1:8, what horizontal distance is needed to build the ramp? Give the answer in feet.

18. A staircase needs to have a slope of 6.5 to 11. What horizontal distance is needed for an 8.5-foot vertical distance between floors?

19. Orange juice is a mixture of water and frozen orange juice concentrate. The ratio of water to concentrate is 3 to 1. How much water is needed to mix with 1.5 gallons of concentrate?

For Exercises 20 to 24, solve the equations.

20. $\frac{2}{3} = \frac{x}{17}$

21. $\frac{2}{6} = \frac{3}{x}$

22. $\frac{2}{x} = 2{,}000{,}000$

23. $\frac{x + 1}{2} = \frac{4x - 1}{6}$

24. $\frac{2x + 5}{7} = \frac{2x - 1}{5}$

For Exercises 25 and 26, solve the formulas.

25. $\frac{a}{b} = \frac{c}{d}$ for d

26. $\frac{V_1}{C_2} = \frac{V_2}{C_1}$ for V_2

27. Find a value or expression for x.

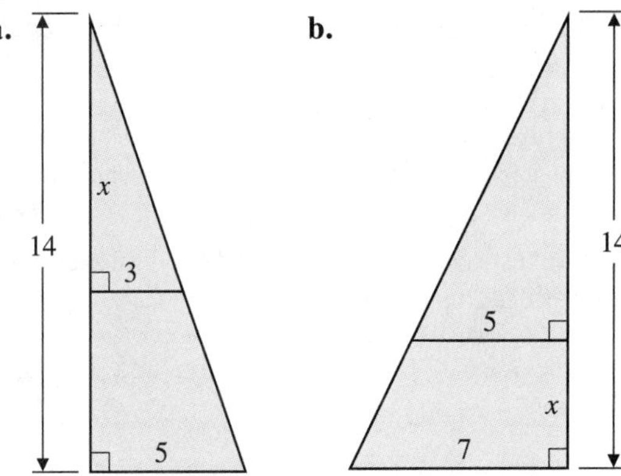

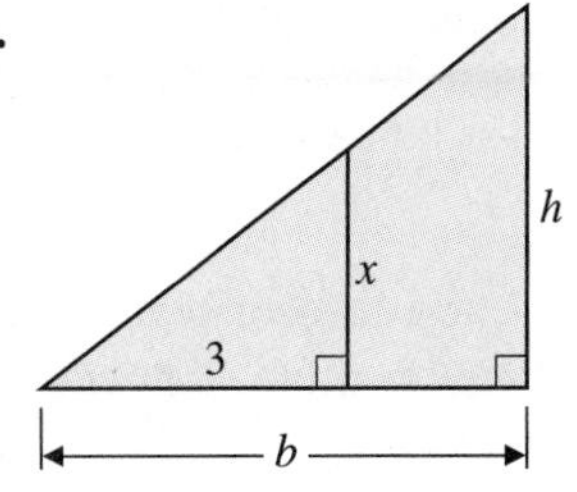

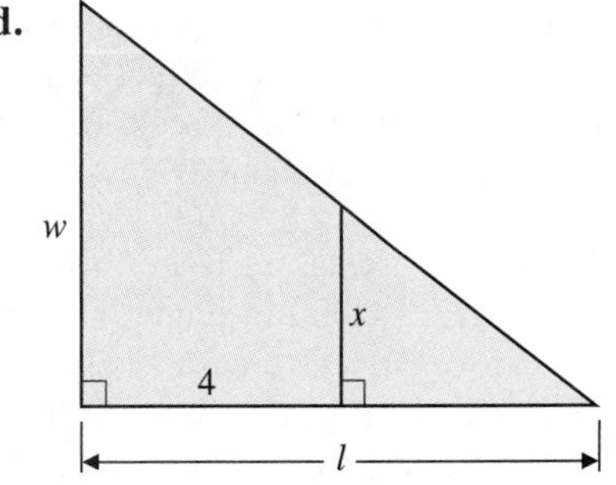

For Exercises 28 and 29, find the coordinates of A and B on the graph. Use slope, proportions, or counting, as necessary.

28.

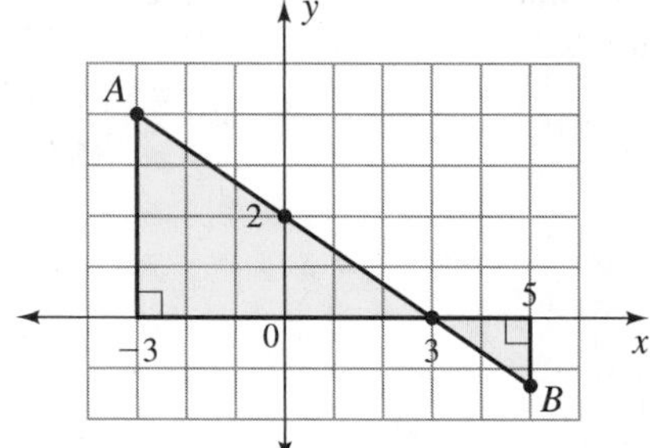

29.

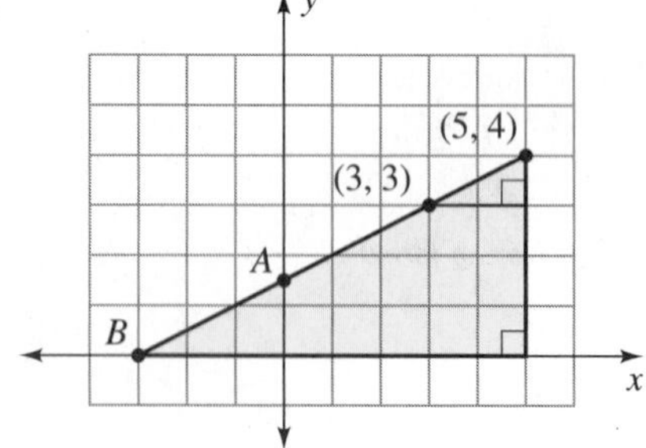

30. Kei's research team catches and tags 250 bats. The next week, they catch 300 bats from the same cave, and only 15 are tagged. Estimate the total bat population from this cave.

Make a quantity-value table for each problem situation in Exercises 31 to 35. Find appropriate totals and averages. Indicate units and whether any of the totals are meaningful. Suggest questions that could be asked.

31. Debi invests \$10,000 in a tax-free bond at 8% and \$5000 in savings earning 3.5%.

32. A plane flies 4 hours at 125 mph and 7.5 hours at 200 mph.

33. An aquarium is to be filled with 8 liters of hot water at 90°C and 50 liters of cold water at 5°C.

34. Max buys 6 gallons of 88-octane fuel and 10 gallons of 92-octane fuel.

35. Arlan earns an A on a 12-credit course and a B on a 4-credit course. Assume an A is 4 points per credit and a B is 3 points per credit.

36. Because of heat loss in uninsulated pipes, hot water at the bathtub faucet is 130°. How many gallons of hot water need to be added to 9 gallons of cold water (35°) to raise the average temperature to 103°?

37. Rafael wants to earn an average of 6% on his investments. He needs to keep \$10,000 in liquid investments (investments that can be turned to cash easily without penalty). Liquid assets pay only 3%. How much money does he need to invest at 8% to reach his goal?

Find the mean, median, and mode of each of the measurements in Exercises 38 to 41. Round to the nearest hundredth.

38. 4.2 grams, 4.3 grams, 4.3 grams

39. 31.7 cm, 31.8 cm, 31.6 cm, 31.8 cm

40. 6.9 miles, 6.9 miles, 6.8 miles, 7.0 miles, 6.8 miles, 6.8 miles

41. 45.0 mL, 44.0 mL, 45.5 mL, 45.0 mL, 44.5 mL

42. Refer to the figure.

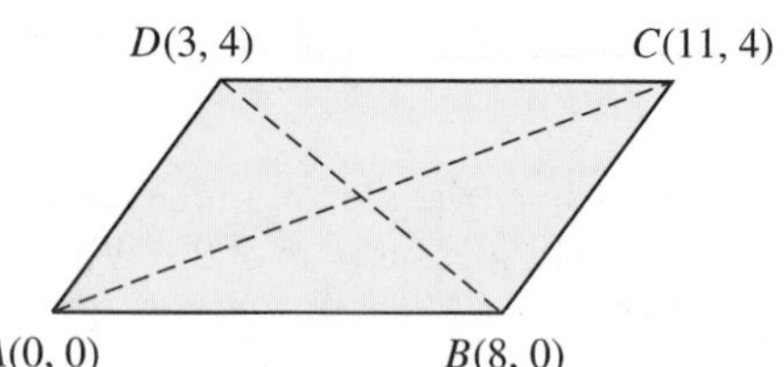

a. Find the midpoint of each line: AD, BC, AB, CD, AC, and BD.

b. Find the slope of each line: AD, AB, CD, and BC.

c. What is the name of the shape $ABCD$?

43. Refer to the figure.

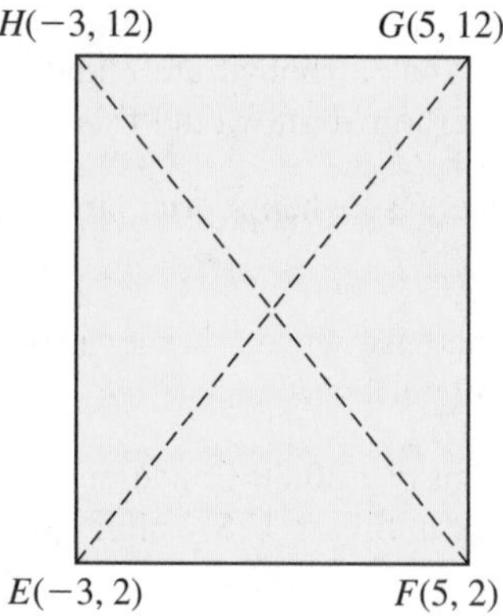

a. Find the midpoint of each line: EG, EF, HE, and HF.

b. Find the slope of each line: HF and EG.

c. What concept explains why the slope of HE is undefined? (*Hint*: Calculate the slope.)

d. What is the centroid of figure $EFGH$?

For Exercises 44 and 45, find the centroid.

44. $I(-5, 5)$ $J(2, 5)$

$K(-5, -1)$ $L(2, -1)$

45.

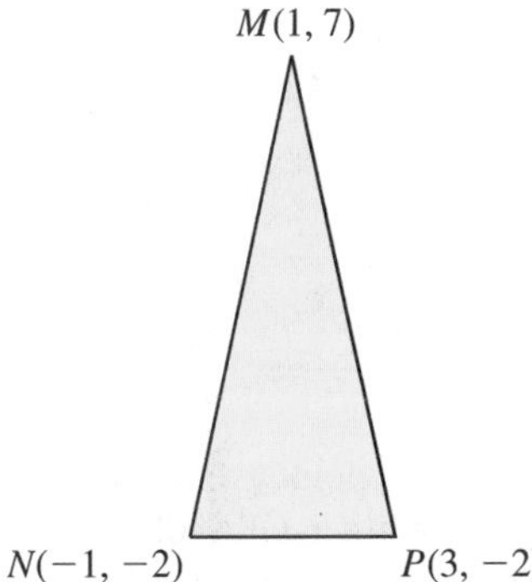

46. A pair's final score in a pairs figure skating competition is 0.4 times the pair's rank in the short program plus 1.0 times the pair's rank in the long program. The lowest score wins. What is each pair's score and which pair wins?

a. Pair A ranks 5 in the short program and 1 in the long program.

b. Pair B ranks 1 in the short program and 3 in the long program.

c. Pair C ranks 2 in the short program and 2 in the long program.

47. Langche's math instructor weights the term project and the final exam each as 2 and the two midterms each as 1. He gets 100 on the project, 85 on the final exam, and 95 and 75 on the midterms. What is his course average?

48. The table shows several examples of quantities and values. Fill in the missing units.

Quantity	Value or Rate	$Q \cdot V$
Number of coins	Value of coin, cents per coin	
Time in hours		Distance in miles
Dollars invested	Percent interest	
Credit hours	Point value of grade per credit hour	
Number of cans	Cost per can	
Number of pounds	Cost per pound	
	Wages per hour	Wages
Liters of acid	Molar value, moles per liter	

CHAPTER  TEST

For Exercises 1 and 2, give three ratios that are equivalent to these ratios.

1. $\frac{6}{14}$ **2.** $\frac{7.5}{100}$

Simplify the ratios in Exercises 3 to 5.

3. $\frac{7ab^3}{28a^2b^2}$ **4.** $\frac{(a + b)}{(a + b)(a - b)}$

5. $\frac{10 \text{ ft}}{24 \text{ in.}}$

6. 125% of what number is 105?

7. Arrange these facts into a unit analysis to find how many dollars per year are spent on 2 packs per day.

20 packs per carton
$35 per carton
365 days per year

8. A $3\frac{1}{2}$-inch by 5-inch photo is to be enlarged into a poster so that its longer side is 2 feet. How long will the other side be?

In Exercises 9 and 10, solve the equations for x.

9. $\frac{2}{5} = \frac{13}{x}$ **10.** $\frac{x + 1}{4} = \frac{x - 3}{3}$

11. Rosa wants a 90% average in math. She has 85% and 91% on her first two of three equal-value midterms. What does she need on the third test to obtain the 90?

What are the median and the mean of each set of data in Exercises 12 and 13?

12. Five ballpoint pens: $0.29, $0.29, $0.29, $0.29, $1.79

13. Five ballpoint pens: $0.69, $0.29, $0.19, $0.69, $1.09

14. Make some observations about the medians and means in Exercises 12 and 13 and what might have caused these results. Under what circumstances would each be a good description of the average cost of the pens?

For Exercises 15 and 16, refer to the figure.

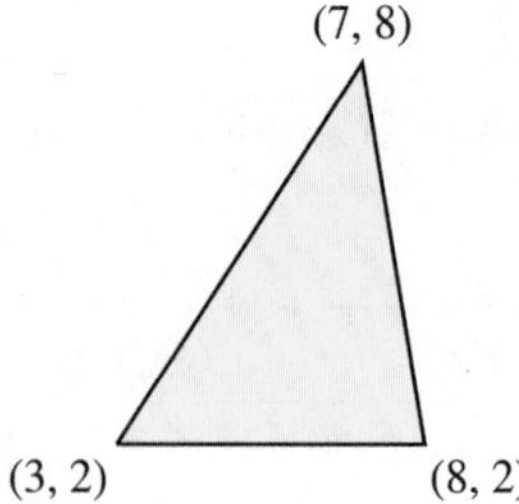

15. What is the midpoint of each side of the triangle?

16. What is the centroid of the triangle?

17. Eugene borrows $17,000 for a car at 5.8% interest and has a $2000 credit card balance at 14.9%. Make a quantity-value table. Explain whether the totals are meaningful, and if so, why. Is there a meaningful average value?

18. What is the length of the line labeled x in the triangle?

a.

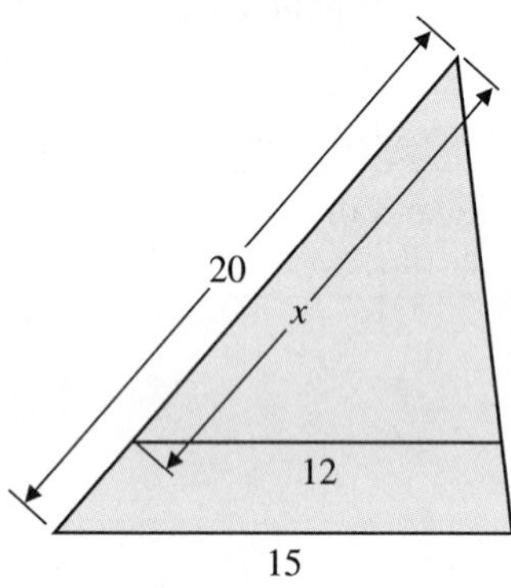

b.

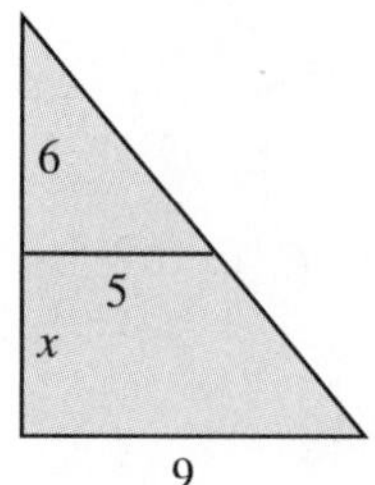

19. Pavel wants to raise his grade point average (GPA). He currently has 60 credits, with a GPA of 3.40. How many credits of As, at 4.00 points per credit, does he need to raise his GPA to a 3.50?

20. The base unit of a portable telephone must be connected to an electrical outlet. The small box plugged into the outlet is a transformer that changes 120-volt (V_1) electricity into 12-volt (V_2) electricity. The output current (I_2) from the transformer is 100 mA. What is the input current (I_1) when the voltage and current are related by the formula $\frac{V_1}{V_2} = \frac{I_2}{I_1}$? Is this formula the same as $\frac{V_1}{V_2} = \frac{I_1}{I_2}$?

6

Polynomial Expressions

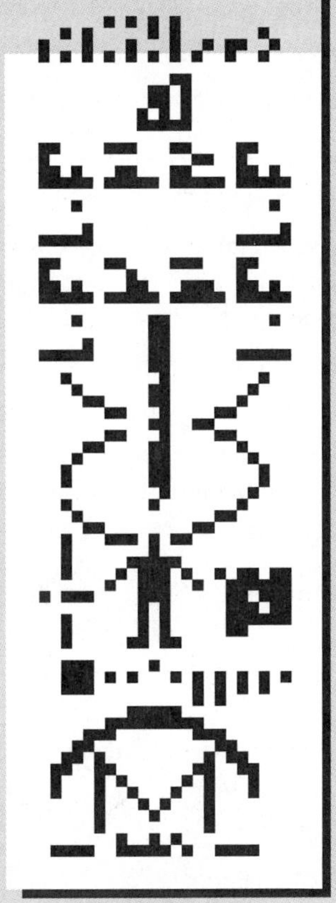

Figure 1

In 1974, the coded message in Figure 1 was sent by the transmitter at Arecibo, Puerto Rico, toward the star cluster M13, located 27,000 light-years from earth. The coded message is a set of 1679 ones and zeros. The number 1679 is factorable into only two primes, 23 and 73. When the ones and zeros are arranged as black and white squares in a rectangular grid of dimensions 23 by 73, the message appears.

In this chapter, we turn our attention to computation with algebraic notation. In Sections 6.1 and 6.3, factoring (such as that needed by distant life forms) is reviewed and extended to expressions called polynomials. Use of the basic operations—addition, subtraction, multiplication, and division—with polynomials is the main focus of the early sections. In Section 6.4, we investigate how many miles the coded message will travel in 27,000 years. Among the skills covered is working with exponents and with large and small numbers written in scientific notation.

6.1 Operations on Polynomials

OBJECTIVES

- Identify monomials, binomials, and trinomials.
- Arrange terms in descending order of exponents.
- Add and subtract polynomials.
- Multiply polynomials.
- Factor the greatest common monomial factor from an expression.
- Distinguish between terms and factors.

WARM-UP

These exercises review ideas in Section 2.3. In Exercises 1 to 4, use the distributive property to multiply.

1. $7(x + 5)$
2. $-2(x - 3)$
3. $-6a(a - b - 2c)$
4. Complete the table by multiplying each expression on the top by the expression on the left.

Multiply	$2a$	$-2b$	$+c$
$5a$			

In Exercises 5 to 7, divide the expressions.

5. $\dfrac{6a + 3b}{3}$

6. $\dfrac{8x^3 + 4x^2 - 6x}{2x}$

7. $\dfrac{a^2b - 2ab^2 + b^3}{ab}$

Add or subtract like terms or like units in Exercises 8 to 12.

8. 6 cm + 5 cm − 2 cm
9. 5 kg + 11 kg − 2 kg
10. $\frac{1}{2}x^2 + 3x^2 + 0.5x^2$
11. $-2ab + 3ab + \frac{1}{2}ab$
12. 5 mL + 12 mL − 3 mL

THIS SECTION RETURNS to the algebra tile model and skills introduced in Section 2.3—applying the distributive property, adding like terms, dividing, and factoring. We use like terms and factoring with special types of algebraic expressions called polynomials.

Polynomials

EXAMPLE 1 Exploring with tiles: width, length, perimeter, and area The shapes below represent algebra tiles. Arrange each set into a rectangle, and sketch the rectangle in the first column of Table 1. Then complete the rest of the table for each rectangle, counting the tiles as needed. Let a = the side of the large square, let b = the side of the small square, and let a and b be the sides of the rectangles.

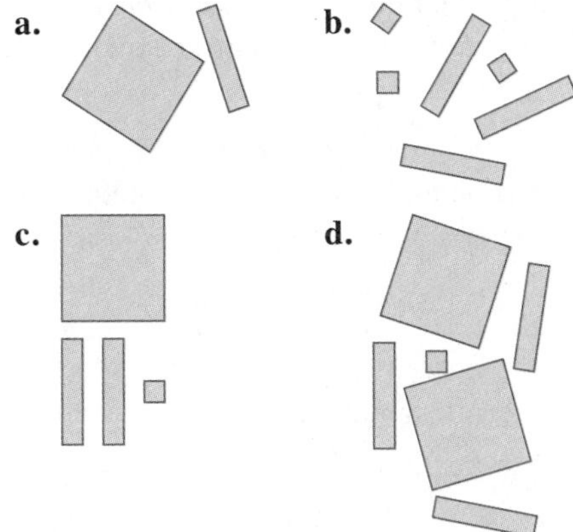

Sketch of Rectangle	Width (height)	Length (base)	Perimeter	Area
a.				$a^2 + ab$
b.	$3b$			
c.		$a + b$		
d.			$6a + 4b$	

Table 1

Solution Sketches of the rectangles for parts a and b appear in parts a and b of Example 8. Although the pieces within the rectangles can be arranged in different ways, all possible rectangles for part a will have the same dimensions, perimeter, and area. In part b, we could arrange the tiles into a long narrow rectangle or into the rectangle shown in part b of Example 8. The answers in the Answer Box are for the rectangle in Example 8. ●

Expressions such as $a^2 + ab$, $3ab + 3b^2$, $a^2 + 2ab + b^2$, and $2a^2 + 3ab + b^2$ (the areas in Example 1) are called polynomials. Polynomials may be used to describe many things: consecutive numbers, the path of a rocket, the height of the rocket given the time after launch, the surface area of a cylinder, the resistance to bending by a structural beam, and the amount of deflection of a beam when loaded.

Definition of Polynomial

> A **polynomial** is an expression containing one or more terms being added or subtracted. The exponents on the variables in each term must be positive integers.

SPECIAL NAMES The prefix *poly* means "many." *Polynomial* means "many terms." A *polynomial expression containing one, two, or three terms* is called, respectively, a **monomial**, **binomial**, or **trinomial**.

Number of terms	Example	Name
one	$2a^2b$	monomial
two	$x^2 + xy$	binomial
three	$x^2 + 2x + 1$	trinomial

No special names are given to expressions containing four, five, or more terms—all are polynomials. The expression $1x^5 + 2x^4 + 3x^3 + 4x^2 + 5x + 6$ is a six-term polynomial. The words *monomial*, *binomial*, *trinomial*, and *polynomial* are often used to describe patterns and to give directions for exercises.

EXAMPLE 2 Identifying monomials, binomials, and trinomials How many terms are in each expression and what is the polynomial name for each?

a. $4 + 3x + x^2$ **b.** $3a + a^2$ **c.** x^2
d. $2b + 3b^2$ **e.** $3 - 2x - x^2$ **f.** $-\frac{1}{2}gt^2$

Solution **a.** 3 terms; trinomial **b.** 2 terms; binomial **c.** 1 term; monomial
d. 2 terms; binomial **e.** 3 terms; trinomial **f.** 1 term; monomial ●

ORDER OF TERMS To make it easy to compare polynomials (for example, to compare your answers to the exercises with those in the back of the book), we write them in a special order. The expressions in Example 2 are written in **ascending order**; that is, the terms are listed with *the exponents from lowest to highest.* When we list polynomials with *the highest exponents first,* we have **descending order**. The expression $1x^5 + 2x^4 + 3x^3 + 4x^2 + 5x + 6$ is in descending order of exponents on x. Unless told to do otherwise, write polynomials in descending order.

EXAMPLE 3 Arranging terms Write the terms in descending order of exponents.

a. $4 + 3x + x^2$ **b.** $3a + a^2$ **c.** $2b + 3b^2$
d. $3 - 2x - x^2$ **e.** $3x^2 + x^4 - x^3 + 5 - x$

Solution **a.** $x^2 + 3x + 4$ **b.** $a^2 + 3a$ **c.** $3b^2 + 2b$
d. $-x^2 - 2x + 3$ **e.** $x^4 - x^3 + 3x^2 - x + 5$ ●

If there are two variables, list the variables in each term alphabetically.

EXAMPLE 4 Arranging terms Write the terms in descending order of exponents for the indicated variable.

a. $x^2y^2 + x^3y + xy^3$ for x **b.** $ab^2 - a^3b^2 + a^2b^2$ for a

Solution **a.** $x^3y + x^2y^2 + xy^3$ **b.** $-a^3b^2 + a^2b^2 + ab^2$ ●

Arranging Terms in Polynomials

- Unless told to do otherwise, arrange the variables in a term first in alphabetical order and then in descending order of the exponent on the first variable:

$$x^5 + x^4y + x^2y^2 - xy$$

- In expressions such as $x^2 + xy + y^2 + yz + z^2 + xz$, where the terms cannot be arranged both in alphabetical and in descending order, arrange them alphabetically first.

Adding and Subtracting Polynomials

To add or subtract polynomials, combine like terms. When all additions and subtractions of like terms within a polynomial have been completed and the terms are arranged in descending order of exponents, the expression is said to be *simplified.*

EXAMPLE 5 Adding and subtracting polynomials Simplify by adding or subtracting like terms.

a. $3a^2 + 4b^2 - 2a^2 - 8b^2 + 4a^2$

b. $(x^2 - 2x + 1) - (x^2 + 2x + 1)$

c. $(x^2 + 2x - 1) - 2(x^2 - 2x + 1)$

d. $(-8a^2 + 2ab + 5b^2) - (-9a^2 - 5ab + 6b^2)$

e. $(x^3 - 6x^2 + 9x) + (-3x^2 + 18x - 27)$

Solution **a.** $3a^2 + 4b^2 - 2a^2 - 8b^2 + 4a^2$ Arrange the terms in alphabetical order.

$= 3a^2 - 2a^2 + 4a^2 + 4b^2 - 8b^2$ Add like terms.

$= 5a^2 - 4b^2$

b. $(x^2 - 2x + 1) - (x^2 + 2x + 1)$ Change the subtraction to addition of the opposite.

$= (x^2 - 2x + 1) + (-1)(x^2 + 2x + 1)$ Apply the distributive property.

$= x^2 - 2x + 1 - x^2 - 2x - 1$ Add like terms.

$= -4x$

c. $(x^2 + 2x - 1) - 2(x^2 - 2x + 1)$ Change the subtraction to addition of the opposite.

$= x^2 + 2x - 1 + (-2)(x^2 - 2x + 1)$ Apply the distributive property.

$= x^2 + 2x - 1 - 2x^2 + 4x - 2$ Add like terms.

$= -x^2 + 6x - 3$

d. $(-8a^2 + 2ab + 5b^2) - (-9a^2 - 5ab + 6b^2)$

Change the subtraction to addition of the opposite.

$= (-8a^2 + 2ab + 5b^2) + (-1)(-9a^2 - 5ab + 6b^2)$

Apply the distributive property.

$= -8a^2 + 2ab + 5b^2 + 9a^2 + 5ab - 6b^2$ Add like terms.

$= a^2 + 7ab - b^2$

e. $(x^3 - 6x^2 + 9x) + (-3x^2 + 18x - 27)$ The distributive property is not needed.

$= x^3 - 6x^2 + 9x - 3x^2 + 18x - 27$ Add like terms.

$= x^3 - 9x^2 + 27x - 27$ ●

Think about it 1: Give the polynomial name of each answer in Example 5.

When we found the perimeter of the algebra tile designs in Example 1, we were adding polynomials. In Example 6, we find perimeters of other figures by adding like terms.

EXAMPLE 6 Finding perimeters Write a polynomial to describe the perimeter of each figure.

a. Rectangle:

$2a$

$3b$ $3b$

b. Equilateral triangle:

$6x$

$6x$

c.

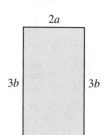

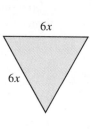

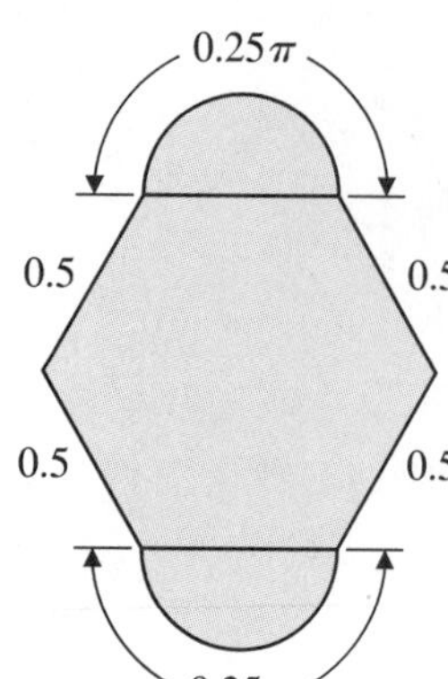

Solution **a.** Perimeter $= 2a + 3b + 2a + 3b = 4a + 6b$ units

b. Perimeter $= 6x + 6x + 6x = 18x$ units

c. Perimeter $= 0.5 + 0.5 + 0.25\pi + 0.5 + 0.5 + 0.25\pi = 2 + 0.5\pi$ units

(Pi may be approximated with 3.14. The perimeter is approximately 3.57 units.) ●

Multiplying Polynomials

EXAMPLE  Exploring with tiles The length and width of six rectangles are shown. Sketch in the tiles needed to complete the rectangles, and write the total area for each part. In parts a, b, and c, the side of the large square tile is a, and the side of the small square tile is b; the rectangle is a by b. In parts d, e, and f, the side of the large square tile is x, and the side of the small square tile is 1; the rectangle is x by 1.

a. $a(a + b)$

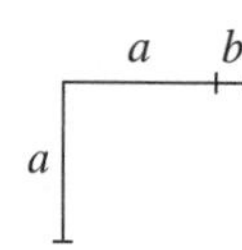

b. $2a(a + 2b)$

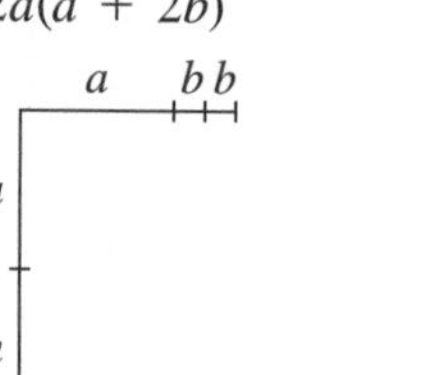

c. $b(a + b)$

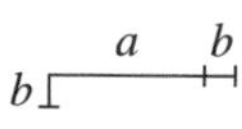

d. $2x(x + 1)$

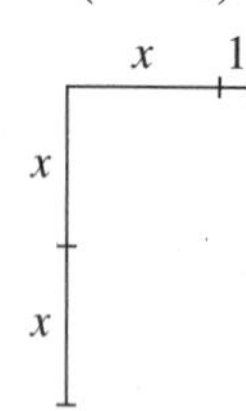

e. $3(x + 1)$

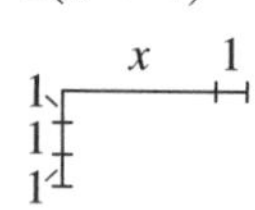

f. $x(x + 3)$

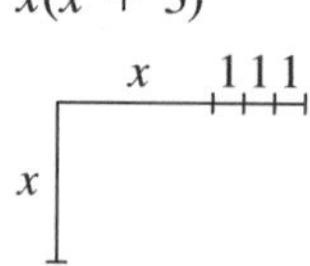

Solution See the Answer Box. ●

In Example 7, we found that the total area of the tiles is the product of the length and width of the rectangle. We have seen this multiplication before in the distributive property: $a(b + c) = ab + ac$.

MULTIPLYING A MONOMIAL AND A POLYNOMIAL The distributive property is the basis for all multiplication of polynomials. In Example 8, we multiply a monomial and a binomial using the distributive property and observe how the product relates to the area of the algebra tiles.

EXAMPLE 

Multiplying a monomial and a binomial Apply the distributive property to multiply these expressions. Explain how the algebra tile figures show the same product.

a. $a(a + b)$ 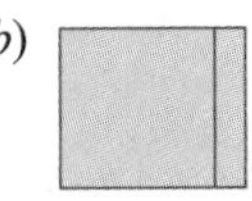

b. $3b(a + b)$ 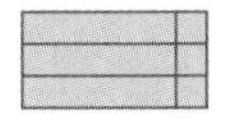

c. $a(a + 3b)$ 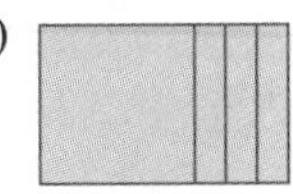

d. $b(a + 3b)$

Solution

a. $a(a + b) = a \cdot a + a \cdot b = a^2 + ab$
The rectangle formed by the tiles has a length of $a + b$. The width of the rectangle is a. The area of the tiles is a large square plus a rectangle: $a^2 + ab$. The area agrees with the product from the distributive property.

b. $3b(a + b) = 3b \cdot a + 3b \cdot b = 3ab + 3b^2$
The rectangle formed by the tiles has a length of $a + b$. The width of the rectangle is $3b$. The area of the tiles is three rectangles plus three small squares: $3ab + 3b^2$. This area agrees with the product from the distributive property.

c. $a(a + 3b) = a \cdot a + a \cdot 3b = a^2 + 3ab$
We have one large square and three rectangles: $a^2 + 3ab$. This area agrees with the product.

d. $b(a + 3b) = b \cdot a + b \cdot 3b = ab + 3b^2$
We have one rectangle and three small squares: $ab + 3b^2$. This area agrees with the product. ●

Think about it 2: Let $x =$ the side of the large algebra tile square, and let $1 =$ the side of the small square. The sides of the rectangle are x and 1.

a. What is the area of the large square?

b. What is the area of the small square?

c. What is the area of the rectangle?

d. Write each of the products in Example 8 using x and 1 for the sides of the large and small square tiles, respectively. Use the distributive property to do the multiplication. *Hint*: The first product is $x(x + 1)$.

Example 9 shows that the distributive property may be applied to three or more terms.

EXAMPLE 

Multiplying a monomial and a trinomial or other polynomial Use the distributive property on these expressions.

a. $6a(a^2 + 2a + 1)$ **b.** $3x^2(x + 2x^2 - 4x^3)$

c. $-4ab(a^2 - 2ab + b^2)$ **d.** $x(x^4 + 2x^3 + 3x + 1)$

Solution

a. $6a(a^2 + 2a + 1) = 6a^3 + 12a^2 + 6a$

b. $3x^2(x + 2x^2 - 4x^3) = 3x^3 + 6x^4 - 12x^5$

c. $-4ab(a^2 - 2ab + b^2) = -4a^3b + 8a^2b^2 - 4ab^3$

d. $x(x^4 + 2x^3 + 3x + 1) = x^5 + 2x^4 + 3x^2 + x$ ●

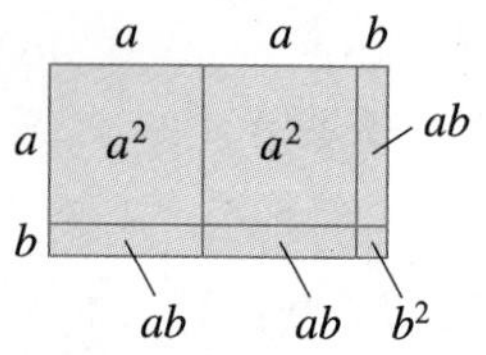

Figure 2

MULTIPLYING TWO BINOMIALS To give you a head start in thinking about the next section, Figure 2 shows multiplication of two binomials with tiles. The left side of the rectangle is $a + b$. The top of the rectangle is $2a + b$. To find the product $(a + b)(2a + b)$, we count the tiles. The area of the rectangle is $2a^2 + 3ab + b^2$.

EXAMPLE 10 Multiplying binomials with tiles Set up rectangles with tiles to show these multiplications.

a. $(a + b)(a + b)$ **b.** $(a + b)(a + 2b)$

Solution **a.** $(a + b)(a + b) = a^2 + 2ab + b^2$

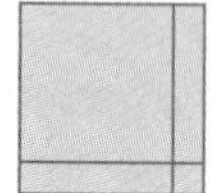

b. $(a + b)(a + 2b) = a^2 + 3ab + 2b^2$

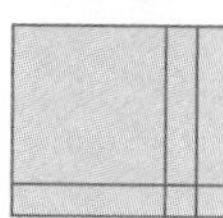

Factoring

When we write the distributive property in reverse, we are said to be factoring the sum of two terms, $ab + ac$, into a product $a(b + c)$.

Factoring with the Distributive Property

> For real numbers a, b, c,
>
> $$ab + ac = a(b + c)$$

The factor a is called the **common factor** because it is *a factor that appears in each term* of the right side. The **greatest common factor (gcf)** is *the largest possible common factor of all terms.*

EXAMPLE  Finding the greatest common factor, or gcf What is the greatest common factor for each expression?

a. $3x + 3$ **b.** $ab + b^2$ **c.** $a^2b - a^2b^2 + a^2b^3$ **d.** $x^2 - 1$

Solution **a.** gcf $= 3$ **b.** gcf $= b$ **c.** gcf $= a^2b$
d. gcf $= 1$; there is no other common factor.

In Example 12, we are given the common factor and asked to find the remaining factor.

EXAMPLE 12 Factoring, given the gcf The expression $3ab + 3b^2$ has $3b$ as the greatest common factor. Use division or tiles as needed to find the other factor.

Solution We can factor $3ab + 3b^2$ in three ways.

First, we can divide by the greatest common factor to find the other factor:

$$\frac{3ab + 3b^2}{3b} = \frac{3ab}{3b} + \frac{3b^2}{3b} = a + b$$

We then have

$$3ab + 3b^2 = 3b(a + b)$$

Second, we can circle the greatest common factor in each term and look at what remains to find the other factor:

$$3ab + 3b^2 = \textcircled{3}a\textcircled{b} + \textcircled{3b}b = 3b(a + b)$$

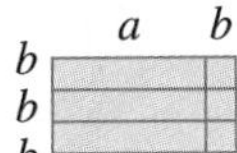

Figure 3

Third, we can use tiles. As shown in Figure 3, we build a rectangle with the greatest common factor, $3b$, as the width. Again, we have

$$3ab + 3b^2 = 3b(a + b)$$

Practice using any of the three methods suggested in Example 12 to factor the expressions in Example 13.

EXAMPLE **13** Factoring, given the gcf

a. Factor $a^2 + 3ab + a$, given that gcf $= a$.

b. Factor $x^2y - 4y^3$, given that gcf $= y$.

c. Factor $4x^2y^2 - 4x^3y^3 + 2x^4y^4$, given that gcf $= 2x^2y^2$.

Solution **a.** $a^2 + 3ab + a = a(a + 3b + 1)$

b. $x^2y - 4y^3 = y(x^2 - 4y^2)$

c. $4x^2y^2 - 4x^3y^3 + 2x^4y^4 = 2x^2y^2(2 - 2xy + x^2y^2)$

To check, multiply using the distributive property. ●

When the terms are all negative, the gcf is negative. In Example 14, we find both the gcf and the other factor.

EXAMPLE **14** Finding the greatest common factor and factoring the expression Identify the greatest common factor, or gcf, and then apply the distributive property to factor each sum into a product.

a. $4x + 6$ **b.** $-4x - 8$ **c.** $x^2 - x$

d. $2ab - 8a^2b + 6ab^2$ **e.** $-10x^2 - 10x - 15$ **f.** $x^2y^2 + 2xy^3 - y^3$

In parts g and h, the gcf is on the left, and the second factor is on the top of the table.

g.

Factor	
	$8a^3 + 12a^2 - 16a$

h.

Factor	
	$x^2 + 3x$

Solution **a.** gcf $= 2$; $4x + 6 = 2(2x + 3)$

b. gcf $= -4$; $-4x - 8 = -4(x + 2)$

c. gcf $= x$; $x^2 - x = x(x - 1)$

d. gcf $= 2ab$; $2ab - 8a^2b + 6ab^2 = 2ab(1 - 4a + 3b)$

e. gcf $= -5$; $-10x^2 - 10x - 15 = -5(2x^2 + 2x + 3)$

f. gcf $= y^2$; $x^2y^2 + 2xy^3 - y^3 = y^2(x^2 + 2xy - y)$

g. gcf $= 4a$

Factor	$2a^2 + 3a - 4$
$4a$	$8a^3 + 12a^2 - 16a$

h. gcf $= x$

Factor	$x + 3$
x	$x^2 + 3x$

●

Summary of Operations and Area

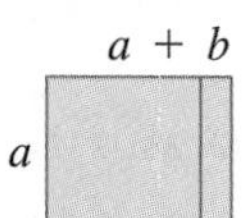

Figure 4

MULTIPLICATION When we are given the width and length and asked to find the area, as in Figure 4, we multiply using the distributive property.

$$\text{width} \cdot \text{length} = \text{area}$$
$$a(a + b) = a^2 + ab$$

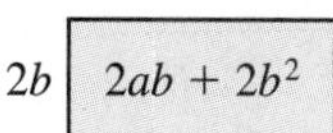

Figure 5

DIVISION When we are given one side and the area, as in Figure 5, we divide to find the other side of the rectangle. This "division" may consist of guessing an expression that we can multiply by the first side to get the area.

$$\text{width} \cdot \text{length} = \text{area}$$
$$2b(?) = 2ab + 2b^2$$
$$2b(a + b) = 2ab + 2b^2$$

If we cannot guess the other factor, $a + b$, we can write the division as a fraction:

$$\frac{2ab + 2b^2}{2b} = \frac{2ab}{2b} + \frac{2b^2}{2b} = a + b$$

FACTORING When we factor, we have the area and must find both the width and the length of the rectangle. Although there is only one greatest common factor, many different widths and lengths may be possible for a given area, as Example 15 reminds us.

EXAMPLE 15 Factoring Factor $6xy - 4xz$. Give three sets of possible widths and lengths that give an area of $6xy - 4xz$.

Solution The greatest common factor is $2x$, so $6xy - 4xz = 2x(3y - 2z)$. Other possible widths and lengths are $2(3xy - 2xz)$ and $x(6y - 4z)$. ●

In general, to avoid multiple answers, use the greatest common factor.

Factors versus Terms

As mentioned in Chapter 3, confusing the words *terms* and *factors* is a common error. Factors are the numbers, variables, or expressions being multiplied in a product. Terms are the expressions being added or subtracted. The product $(a + b)(a + b)(a - b)$ has three factors, and each factor is a two-term expression.

In Example 16, we identify terms, and in Example 17, we identify factors.

EXAMPLE 16 Identifying terms How many terms are there in these expressions?

a. $x^2 + 4x + 2$ **b.** $-b + d$ **c.** $3 - 2x$

d. $1 + 3y + 3y^2 + y^3$ **e.** $x^2 - y^2$

Solution See the Answer Box. ●

EXAMPLE 17 Identifying factors How many factors are there in each of these products?

a. $wxyz$ **b.** $(x + y)(x - y)$ **c.** $x(x + 1)(x + 2)$

d. $n(n - 1)$ **e.** $(x - 1)(x^2 + x + 1)$ **f.** πr^2

Solution See the Answer Box. ●

ANSWER BOX

Warm-up: **1.** $7x + 35$ **2.** $-2x + 6$ **3.** $-6a^2 + 6ab + 12ac$ **4.** $10a^2 - 10ab + 5ac$ **5.** $2a + b$ **6.** $4x^2 + 2x - 3$ **7.** $a - 2b + \frac{b^2}{a}$ **8.** 9 cm **9.** 14 kg **10.** $4x^2$ **11.** $1.5ab$ **12.** 14 mL

Example 1:

	Sketch of Rectangle	Width (height)	Length (base)	Perimeter	Area
a.	$a + b$; a	a	$a + b$	$4a + 2b$	$a^2 + ab$
b.	$a + b$; $3b$	$3b$	$a + b$	$2a + 8b$	$3ab + 3b^2$
c.	$a + b$; a, b	$a + b$	$a + b$	$4a + 4b$	$a^2 + 2ab + b^2$
d.	$2a + b$; a, b	$a + b$	$2a + b$	$6a + 4b$	$2a^2 + 3ab + b^2$

Think about it 1: **a.** binomial **b.** monomial **c.** trinomial **d.** trinomial **e.** polynomial

Example 7:

a. $a(a + b) = a^2 + ab$

a b; a

b. $2a(a + 2b) = 2a^2 + 4ab$

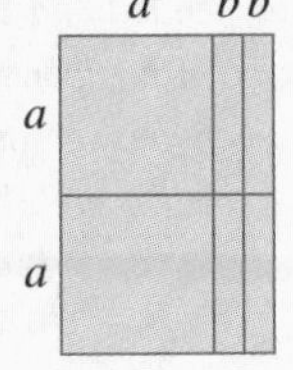

c. $b(a + b) = ab + b^2$

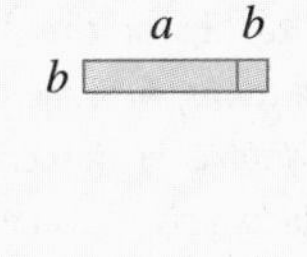

d. $2x(x + 1) = 2x^2 + 2x$

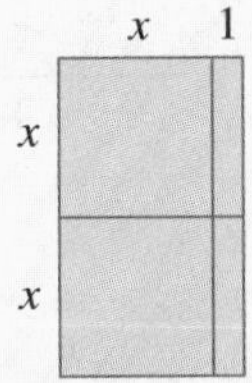

e. $3(x + 1) = 3x + 3$

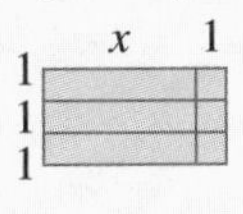

f. $x(x + 3) = x^2 + 3x$

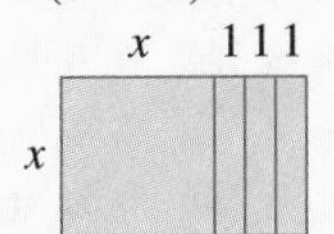

Think about it 2: **a.** $x \cdot x = x^2$ square units **b.** $1 \cdot 1 = 1$ square unit **c.** $x \cdot 1 = x$ square units **d.** $x(x + 1) = x^2 + x$; $3 \cdot 1(x + 1) = 3x + 3$; $x(x + 3 \cdot 1) = x^2 + 3x$; $1(x + 3 \cdot 1) = x + 3$

Example 16: **a.** 3 terms **b.** 2 terms **c.** 2 terms **d.** 4 terms **e.** 2 terms

Example 17: **a.** 4 factors **b.** 2 factors **c.** 3 factors **d.** 2 factors **e.** 2 factors **f.** 3 factors

EXERCISES 

Exercises 1 and 2 review integer addition and subtraction.

1. a. $-9 + 4$ **b.** $-8 + 3$ **c.** $-5 - 8$

d. $-7 - 3$ **e.** $-9 - (-5)$ **f.** $3 - (-4)$

2. a. $-4 + 6$ **b.** $-5 + 1$ **c.** $-4 - 3$

d. $-3 - 9$ **e.** $9 - (-8)$ **f.** $-5 - (-1)$

In Exercises 3 and 4, add or subtract like terms. Identify each answer as a monomial, a binomial, or a trinomial.

3. a. $5a + 3b - 2c - 4a - 6c + 9b$

b. $6m + 2n - 6p - 3m - 3n + 6p$

c. $8y + 5y + 5y - 5y + 8y$

d. $(x^2 + 3x) - (4x + 12)$

e. $(x^2 - 2x + 3) - (2x^2 - 4x + 6)$

4. a. $a - 8c - d + a + 4d - 6c$

b. $5m - 4p + 8n - 9p + 8m - 7n$

c. $9y + 7y + 7y + 8y - 2y$

d. $(2x^2 + 4x) - (3x - 4)$

e. $(y^2 + 3y - 2) - (3y^2 + 9y - 6)$

In Exercises 5 to 8, arrange the terms in descending order of exponents on x.

5. $5 - 3x^2 + 5x - x^3$ **6.** $4x + 3x^3 - 3 + 2x^2$

7. $x^2 - x^4 + 1 - x$ **8.** $-3 + x^5 - 4x^2 + x^3$

In Exercises 9 to 12, use the distributive property as needed and add or subtract like terms. Arrange the terms alphabetically, with exponents on the first variable in descending order. Identify each answer as a monomial, a binomial, a trinomial, or a polynomial with more than three terms.

9. a. $x^2 + 2x + 3x + 6$

b. $3x^2 + 6x + x + 2$

c. $5 - 4(x - 3)$

d. $6x^2 + 2x + 3x + 1$

e. $a^2 - ab + ab - b^2$

10. a. $x^2 + x + 2x + 2$

b. $2x^2 + 3x + 4x + 6$

c. $3 + 4(x - 5)$

d. $2x^2 - 2x + 3x - 3$

e. $a^2 + b^2 + ab + ab$

11. a. $2y(-x + 2y) + x(x - 2y)$

b. $x^2 - 4y^2 - 2xy + 2xy$

c. $x^3 + 2x^2 + x + x^2 + 2x + 1$

d. $x(4 + 4x + x^2) - 2(4 + 4x + x^2)$

e. $b^3 - ab^2 + a^2b + a^3 - a^2b + ab^2$

12. a. $3y(3y + x) - x(x + 3y)$

b. $-9y^2 + 3xy + x^2 - 3xy$

c. $-x^2 + 2x - 1 + x^3 - 2x^2 + x$

d. $2(4 - 2x + x^2) + x(4 - 2x + x^2)$

e. $a^3 - a^2b + ab^2 - b^3 - ab^2 + a^2b$

Exercises 13 to 16 show rectangles made from algebra tiles. Find the length, width, perimeter, and area of each rectangle.

13.

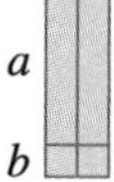

14.

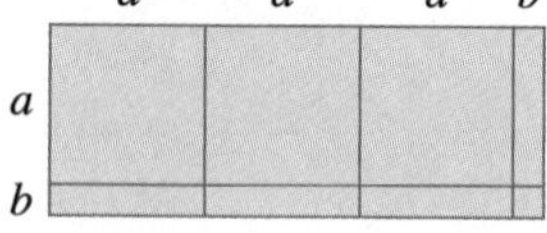

15.

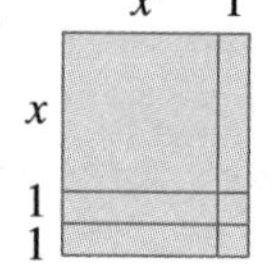

16.

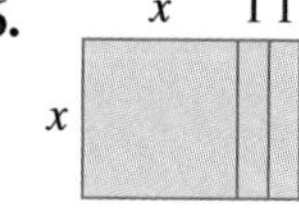

17. Find the perimeter of each shape.

a.

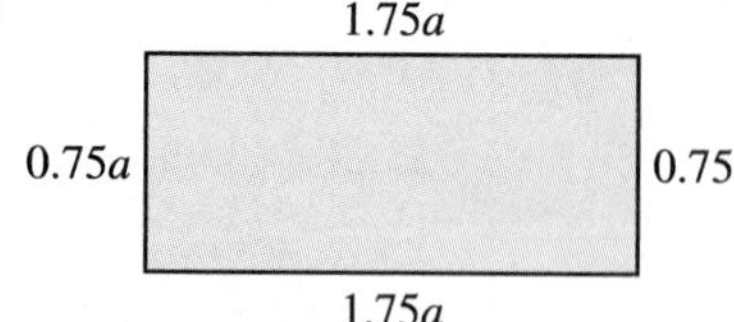

b.

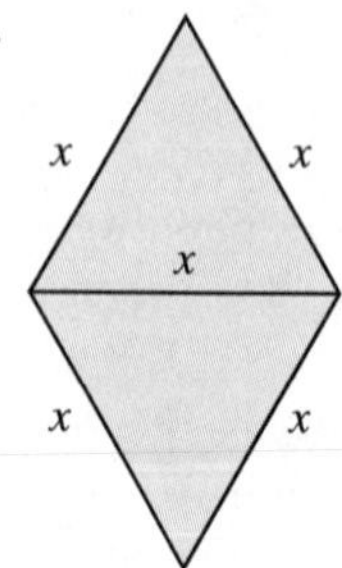

c.

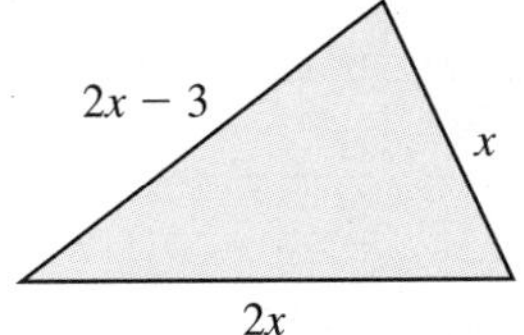

d.

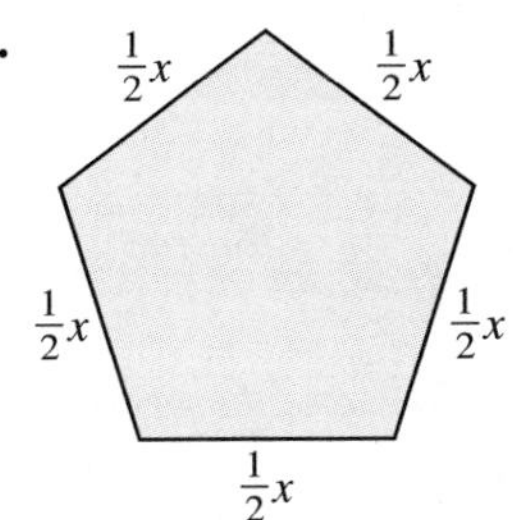

e.

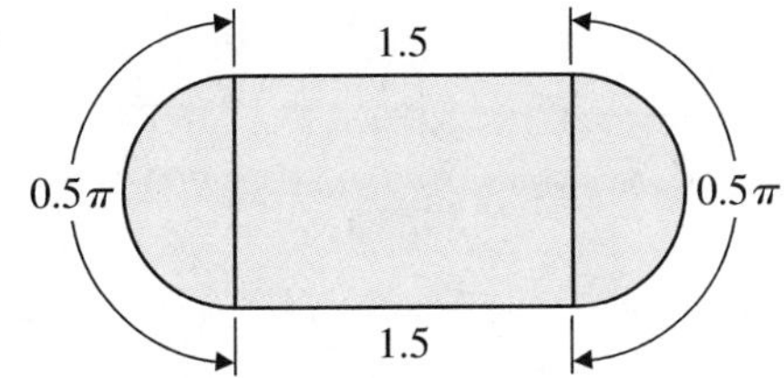

18. Find the perimeter of each shape.

a.

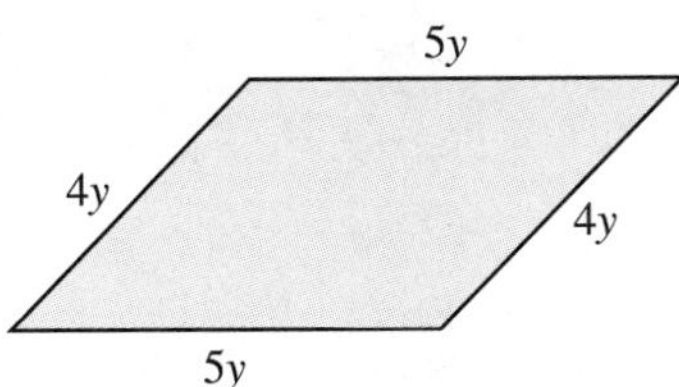

b.

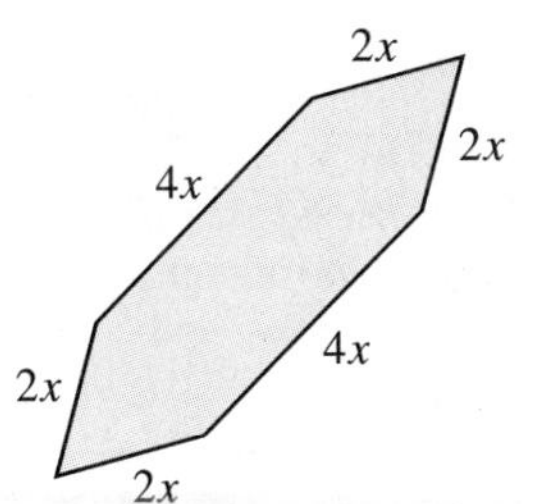

c.

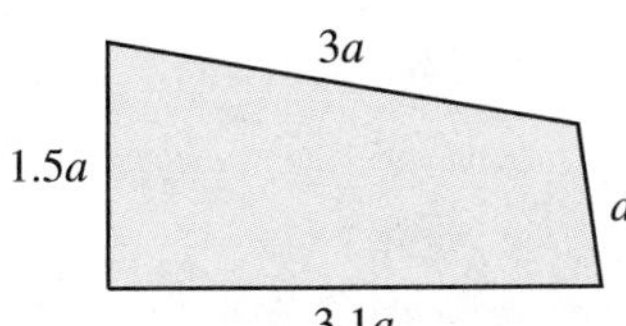

d.

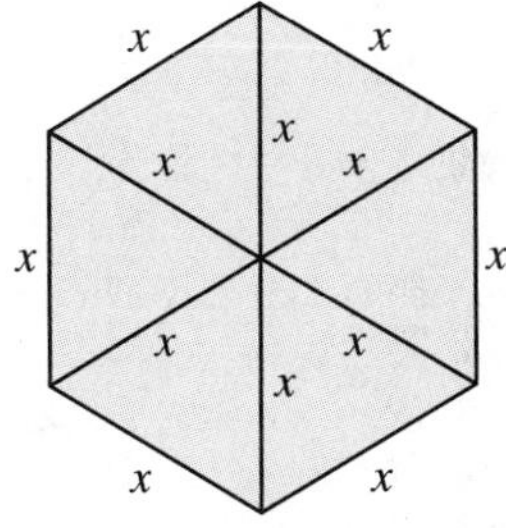

e.

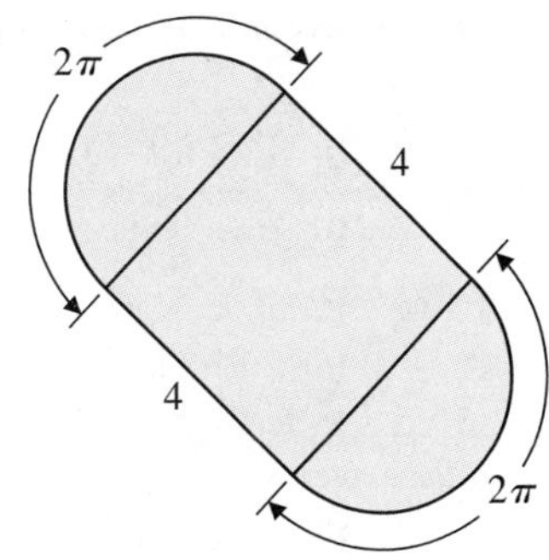

In Exercises 19 and 20, use the given perimeter, P, to find the missing side in each figure. Then find the area of each figure.

19. a.

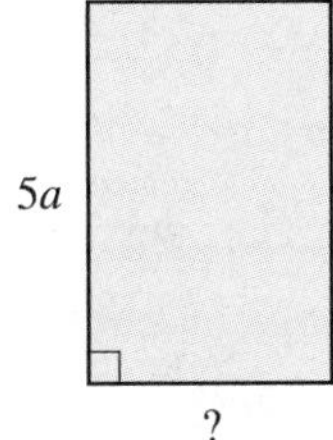

$P = 10a + 6c$

b.

$P = 3x + 5y$

c.

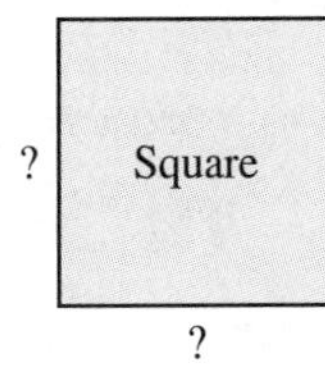

$P = 20y$

20. a.

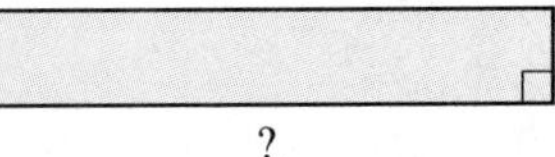

$P = 2x + 8y$

b.

y ?

$P = 10y$

c.

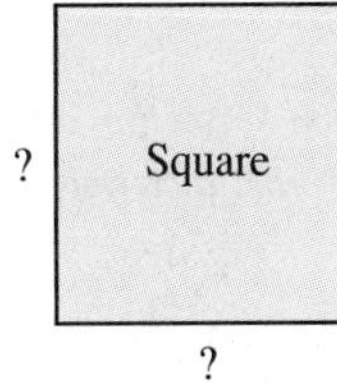

$P = 24x$

In Exercises 21 and 22, multiply the expressions. The exercises can be done with algebra tiles.

21. a. $2a(a + 2b)$ **b.** $2b(2a + b)$

c. $x(x + 3)$ **d.** $x(2x + 1)$

22. a. $a(a + 3b)$ **b.** $b(3a + b)$

c. $x(3x + 1)$ **d.** $2x(x + 2)$

In Exercises 23 and 24, multiply the expressions.

23. a. $2x(x^2 + 3x)$ **b.** $x^2(x - 1)$

c. $x^2(x^2 + 2x + 1)$ **d.** $ab(b^2 - 1)$

e. $b^2(a - b)$ **f.** $a^2(1 + b - b^2)$

24. a. $3x(x - 3x^2)$ **b.** $x(x^2 + 4)$

c. $x^2(1 - x - x^2)$ **d.** $b^2(a - a^2)$

e. $ab(b - a)$ **f.** $a^2(1 - b + b^2)$

Simplify the expressions in Exercises 25 to 30.

25. $5x - 3x(1 - x)$ **26.** $4a + 2a(3 - a)$

27. $8b + 2b(b + 1)$ **28.** $7a - 2a(4 - a)$

29. $4b - 2b(5 - b)$ **30.** $6x - x(x + 1)$

Sketch tiles to show each multiplication in Exercises 31 to 34.

31. $(x + 2)(x + 1)$ **32.** $(2x + 3)(x + 1)$

33. $(a + b)(a + 3b)$ **34.** $(a + 2b)(a + 2b)$

In Exercises 35 and 36, use guess and check on a calculator to factor the numbers into prime factors.

35. a. 111 **b.** 91

36. a. 1001 **b.** 129

What is the greatest common factor of the numerator and denominator in each of the expressions in Exercises 37 and 38? Simplify by eliminating common factors.

37. a. $\frac{36}{99}$ **b.** $\frac{66}{990}$ **c.** $\frac{185}{999}$

d. $\frac{mn}{mp}$ **e.** $\frac{4np}{24mn}$

38. a. $\frac{407}{999}$ **b.** $\frac{108}{999}$ **c.** $\frac{39}{1001}$

d. $\frac{xz}{xy}$ **e.** $\frac{3xy}{9yz}$

In Exercises 39 to 42, the greatest common factor for the expression inside the table is on the left. Write the other factor on the top of the table.

39.

Factor			
y	y^3	$+2xy^2$	$+x^2y$

40.

Factor			
x^2y^2	x^2y^3	$-x^3y^2$	$+x^2y^2$

41.

Factor			
$2ab$	$2a^2b$	$-4ab^2$	$+6ab^3$

42.

Factor			
$4ab^2$	$4ab^3$	$-8a^2b^3$	$-16a^2b^2$

For the expressions in Exercises 43 to 52, name the greatest common factor and then factor the expression.

43. $x^3 + 4x^2 + 4x$ **44.** $x^2y + xy^2 + y^3$

45. $a^2b + ab^2 + b^3$ **46.** $a^3 - a^2b + ab^2$

47. $6x^2 + 2x$ **48.** $12y^2 + 6y$

49. $15y^2 - 3y$ **50.** $8x^2 - 4x$

51. $15x^2y + 10xy^2$ **52.** $9x^2y^2 + 18xy^2 - 24y^2$

Factor the expressions in Exercises 53 to 60 in two ways. First, use a positive common factor. Second, use a negative common factor.

53. $-4x - 12$ **54.** $-12y - 3$

55. $-2xy + 4y^2$ **56.** $-6x^2 + 18xy$

57. $-12x^2 - 8x - 8$

58. $-x^2y - xy^2 - y^3$

59. $-y^2 + 4y^3 - 8y^4$

60. $-10x^2 + 15xy - 20y^2$

The rectangles in Exercises 61 and 62 show areas. Write the greatest common factor on the left, and then write the other factor above the area.

61. a. $2a - 4ab$ **b.** $x^2y + xy^2$

62. a. $3xy - 6y^2$ **b.** $ab^2 - a^2b^2$

How many terms are in each expression in Exercises 63 to 68?

63. $x^2 + 5xy$ **64.** $2xyz$

65. $\frac{1}{2}gt^2$ **66.** mc^2

67. $4x^2 + 8x + 8$ **68.** $2\pi rh + 2\pi r^2$

How many factors are in each expression in Exercises 69 to 74?

69. $4(x + 2)(x + 2)$ **70.** $x(x + 5y)$

71. $2\pi r(h + r)$ **72.** mc^2

73. $2xyz$ **74.** $\frac{1}{2}gt^2$

75. Why are the expressions in Exercises 63 and 70 equal?

76. Why are the expressions in Exercises 68 and 71 equal?

77. Complete the sentence: The distributive property, $a(b + c) = ab + ac$, changes the product of two ______ into the ____ of two ______.

78. Complete the sentence: Factoring $ab + ac = a(b + c)$ changes the sum of ____ terms into the ________ of two ______.

79. Describe how to tell if two terms are "like" terms.

80. There is a saying "You cannot add apples and oranges." How does this saying apply to adding like terms?

81. Describe how to find the greatest common factor.

Projects

82. ***Factors.*** The smallest number with just one factor is 1. The smallest number with exactly two factors is 2: $1 \cdot 2$.
The smallest number with exactly three factors is 4: $1 \cdot 4$, $2 \cdot 2$.
The smallest number with exactly four factors is 6: $1 \cdot 6$, $2 \cdot 3$.
Find the smallest number with exactly

a. five factors **b.** six factors

c. seven factors **d.** eight factors

e. nine factors **f.** ten factors

What do all numbers with an odd number of factors have in common?

83. ***Dessert Distribution.*** At a party, an unknown number of people are seated at a table of unknown size. A platter of small desserts is passed around the table. Each person takes one dessert and passes the platter on to the next person. The platter keeps getting passed around the table until all the desserts are gone. The platter originally contained 62 desserts. Gloria took the first and the next to last dessert (and possibly others as well). How many people were seated at the table? There are several answers. Find as many answers as possible. For each of your answers, find how many desserts Gloria took.

84. ***Coded Message.*** Scientists assume that intelligent extraterrestrials will factor 1679 into $23 \cdot 73$ and find the message in Figure 1 on the chapter-opening page.

a. Using zeros and ones to transmit data is common on Earth. Dot matrix printers, an early form of printer for personal computers, relied on a rectangular array of dots to form each letter or character. Suppose a dot matrix printer head has a rectangular array, 8 squares wide by 9 squares tall. Arrange the following message into an 8 by 9 array, and then decode the message by shading each 1 and leaving each zero unshaded.

001111000100001001000010001111000100
001001000010001111000000000000
00000

b. Suppose the following 15-digit coded message, similar to the one on the opening page of this chapter, was sent. What are the factors of 15? Into what rectangular shapes could the data be arranged? Only one of the possible rectangles will make sense.

101011110110101

c. Repeat part b for the following 50-digit message.

1100010100010001010111000101110100010001
1101010001

d. Why was the coded message on the chapter opener sent in symbols instead of words?

6.2 Multiplication of Binomials and Special Products

OBJECTIVES

- Multiply binomials with the tile model and the table method.
- Identify patterns in the products of binomial factors.
- Multiply binomials mentally.
- Identify binomials that multiply to perfect square trinomials and differences of squares.

WARM-UP

Complete the table.

m	n	Sum $m + n$	Product $m \cdot n$
3	4		
		8	12
		8	15
1	15		
4	6		
		11	24
2	12		
−4	6		
−4	−6		
2	−12		
−2	−12		
		5	−24
		−11	24
		−4	−12
		−8	15

IN THIS SECTION, we focus on the multiplication of two binomials called binomial factors. We start with the tile model, which leads to multiplication in a table. We then look at examples that can be multiplied mentally and at special products called perfect square trinomials and differences of squares.

Binomial Multiplication

EXAMPLE 1 Exploring with tiles Let the large square algebra tile have side x, the rectangle have sides x and 1, and the small square tile have side 1. Arrange the sets of tiles into rectangles that show these products.

a. $(x + 2)(x + 3)$ **b.** $(2x + 1)(x + 2)$

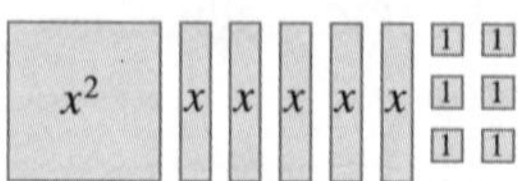

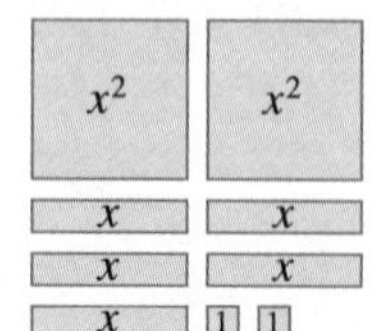

Solution One possible rectangle showing the first product appears in Example 2. The rectangle for part b appears in Example 3. ●

USING TILES TO BUILD TABLES We can arrange the tiles into a rectangular shape that matches the products in a multiplication table. To make this special arrangement, we start with the large squares in the upper left corner. We put some of the rectangles in the lower left corner and some in the upper right corner. Then we group all of the small squares in the lower right corner.

EXAMPLE 2 Showing multiplication with a table Multiply $(x + 2)(x + 3)$.

Solution We count the tiles in part (a) and find that the area is the sum of 1 large square, 5 rectangles, and 6 small squares, or $x^2 + 5x + 6$. Thus,

$$(x + 2)(x + 3) = x^2 + 5x + 6$$

Part (b) shows tiles in a table. The multiplication table on the right shows the product as $x^2 + 3x + 2x + 6$. By adding the like terms $2x$ and $3x$, we get $5x$. Thus,

$$(x + 2)(x + 3) = x^2 + 5x + 6$$

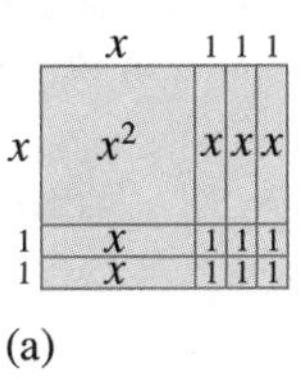

(a)

Multiply	x	3
x	x^2	x x x
2	x / x	1 1 1 / 1 1 1

(b)

Multiply	x	$+3$
x	x^2	$3x$
$+2$	$2x$	6

●

Example 3 shows the multiplication in part b of Example 1 with tiles and a table.

EXAMPLE 3 Showing multiplication with a table Show the product $(2x + 1)(x + 2)$ with tiles and a table.

Solution

Multiply	x	2
$2x$	x^2 / x^2	x x / x x
1	x	1 1

Multiply	x	$+2$
$2x$	$2x^2$	$+4x$
$+1$	$+1x$	$+2$

From the tiles, the product is the sum of 2 large squares, 5 rectangles, and 2 small squares, or $2x^2 + 5x + 2$. Thus,

$$(2x + 1)(x + 2) = 2x^2 + 5x + 2$$

In the table, we use the distributive property twice to get the product of $2x(x + 2)$ in the upper row and $1(x + 2)$ in the lower row. We add like terms in the product $2x^2 + 4x + 1x + 2$, to get $2x^2 + 5x + 2$.

Both the tiles and the table give the same product. ●

USING TABLES AND THE TRADITIONAL METHOD One advantage of the table method is that we can use terms with either positive or negative coefficients. See Exercise 78 for a project in which negative tiles are defined. Example 4 shows a table solution along with a traditional multiplication.

EXAMPLE 

Multiplying binomials containing negative coefficients Multiply $(x - 3)(2x - 1)$ with a table and with traditional multiplication.

Solution **Table method:**

Multiply	$2x$	-1
x	$2x^2$	$-x$
-3	$-6x$	$+3$

Traditional method:

$$\begin{array}{rrl} & 2x - 1 & \\ & \underline{x - 3} & \\ & -6x + 3 & \text{Multiply } -3(2x - 1). \\ \underline{2x^2 - 1x} & & \text{Multiply } x(2x - 1). \\ 2x^2 - 7x + 3 & & \text{Add the products.} \end{array}$$

Using both methods, we have $(x - 3)(2x - 1) = 2x^2 - 7x + 3$. ●

In the next example, we repeat both methods—table and traditional.

EXAMPLE 

Multiplying binomials containing negative coefficients Multiply $(3x + 1)(2x - 3)$ with a table and with traditional multiplication.

Solution **Table method:**

Multiply	$2x$	-3
$3x$	$6x^2$	$-9x$
$+1$	$+2x$	-3

Traditional method:

$$\begin{array}{rrl} & 2x - 3 & \\ & \underline{3x + 1} & \\ & 2x - 3 & \text{Multiply } 1(2x - 3). \\ \underline{6x^2 - 9x} & & \text{Multiply } 3x(2x - 3). \\ 6x^2 - 7x - 3 & & \text{Add the products.} \end{array}$$

Using both methods, we have $(3x + 1)(2x - 3) = 6x^2 - 7x - 3$. ●

Many students like the table method because it helps them remember that multiplication is like finding area. As we see again in the next section, factoring is like finding the width and length given the area. You may have a favorite way to multiply already. If your method works, don't switch. Just use the table method to help you understand the meanings of the operations and identify some useful patterns in this and the next section.

FINDING PATTERNS IN THE POSITIONS OF LIKE TERMS In Example 6, we multiply binomials and observe the positions of the like terms.

EXAMPLE 

Multiplying binomials Multiply these binomials with a table. Where are the like terms in the products?

a. $(5x - 3)(2x + 3)$ **b.** $(ax + b)(cx + d)$

Solution **a.**

Multiply	$2x$	$+3$
$5x$	$10x^2$	$+15x$
-3	$-6x$	-9

$(5x - 3)(2x + 3)$
$= 10x^2 + 9x - 9$

b.

Multiply	cx	$+d$
ax	acx^2	$+adx$
$+b$	$+bcx$	$+bd$

$(ax + b)(cx + d)$
$= acx^2 + (ad + bc)x + bd$

In part a, the like terms are on the diagonal in the table from lower left to upper right. The binomial in part b has letters instead of numbers. It shows that, no matter what the coefficients on x, the like terms will be on the diagonal from lower left to upper right. ●

Patterns in Like Terms

> Within a table for multiplication of binomials, the like terms will be on the diagonal from lower left to upper right.

FINDING PATTERNS IN THE PRODUCTS OF DIAGONALS The next pattern gives us a way to check whether the multiplication is correct. The pattern also gives us information that we will need in the next section, on factoring trinomials.

In Example 7, we return to the tables from Example 6.

EXAMPLE  Comparing diagonal products Multiply the two diagonal terms in each table. What pattern do you observe?

a.

Multiply	$2x$	$+3$
$5x$	$10x^2$	$+15x$
-3	$-6x$	-9

b.

Multiply	cx	$+d$
ax	acx^2	$+adx$
$+b$	$+bcx$	$+bd$

Solution **a.** $(10x^2)(-9) = -90x^2$

$(-6x)(+15x) = -90x^2$

The products of the diagonals are equal.

b. $(acx^2)(bd) = abcdx^2$

$(bcx)(adx) = abcdx^2$

Because a, b, c, and d represent any numbers, this pair of products shows that the products of the diagonals are equal for all tables. ●

Patterns in Diagonal Products

> In a table showing multiplication of binomials, the products of the diagonal terms are equal.

Student Note:
Confirm the diagonal property by checking prior examples.

In Example 8, we again look for patterns in the positions of like terms and products of diagonals, this time with the product of a binomial and a trinomial.

EXAMPLE 8 Finding the product of a binomial and a trinomial

a. Complete this table showing the multiplication of the binomial $x - 2$ and the trinomial $x^2 + 2x + 4$.

Multiply	x^2	$+2x$	$+4$
x			
-2			

b. Write the product from the table.

c. Are the like terms on a diagonal?

d. Are the diagonal products equal?

Solution **a.**

Multiply	x^2	$+2x$	$+4$
x	x^3	$+2x^2$	$+4x$
-2	$-2x^2$	$-4x$	-8

b. $(x - 2)(x^2 + 2x + 4) = x^3 - 8$

c. There are two sets of like terms. The like terms $-2x^2$ and $+2x^2$ are on a diagonal from lower left to adjacent upper right, as are the like terms $-4x$ and $+4x$. Both sets add to zero.

d. There are two sets of diagonal products:

$$(x^3)(-4x) = -4x^4, \qquad (-2x^2)(+2x^2) = -4x^4$$
$$(2x^2)(-8) = -16x^2, \qquad (-4x)(+4x) = -16x^2$$

Both sets of products are equal. ●

The patterns we observed in binomial and trinomial products apply to other products as well.

MENTAL MULTIPLICATION You are encouraged to multiply easy products mentally. In earlier examples, each time we multiplied two binomials, the product had four terms and two of the terms could be added. To mentally multiply a product such as $(x + 5)(x - 7)$, follow three steps:

1. Find the product of the first terms: $x \cdot x$.
2. Find the products that form like terms and add them: $-7x + 5x$.
3. Find the product of the last terms: $5(-7)$.

$$(x + 5)(x - 7) = \underbrace{x^2}_{1} + \underbrace{-7x + 5x}_{2} \underbrace{- 35}_{3} = x^2 - 2x - 35$$

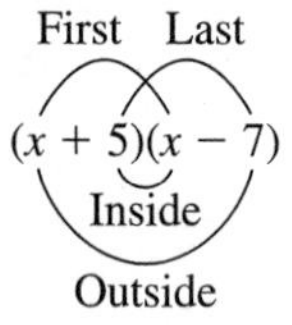

First Last
(x + 5)(x − 7)
Inside
Outside

Figure 6

This method is often called FOIL because we multiply the first terms, then outside terms, inside terms, and last terms. Some people use the visual memory aid shown in Figure 6.

Remember: FOIL is limited to multiplying binomials; it does not work for monomials, trinomials, or other polynomials. FOIL is a memory device and a shortcut, not a general procedure.

Special Products

After some practice, you should be able to recognize and do mentally the multiplications leading to two special types of products: perfect square trinomials and differences of squares.

PERFECT SQUARE TRINOMIALS Look for patterns in the first and last terms of the products in Example 9.

EXAMPLE 

Finding patterns in products Multiply these binomials. Describe patterns in the first and last terms. Describe patterns in the sum of the like terms.

a. $(x - 1)(x - 1)$ **b.** $(x + 3)(x + 3)$

c. $(2x + y)(2x + y)$ **d.** $(2ax + b)(2ax + b)$

Solution **a.** $(x - 1)(x - 1) = x^2 - 2x + 1$

b. $(x + 3)(x + 3) = x^2 + 6x + 9$

c. $(2x + y)(2x + y) = 4x^2 + 4xy + y^2$

d. $(2ax + b)(2ax + b) = 4a^2x^2 + 4abx + b^2$

In each problem, we are multiplying a binomial by itself. The answer always starts and ends with a perfect square. The middle term is twice the product of the terms in one factor. ●

The products in Example 9 are perfect square trinomials.

Definition of Perfect Square Trinomial

> The expression $a^2 + 2ab + b^2$ is a **perfect square trinomial** because its terms are the square of a, twice the product of a and b, and the square of b.

A perfect square trinomial is created when we multiply any binomial times itself—that is, *square* a binomial:

$$(a + b)(a + b) = (a + b)^2 = a^2 + 2ab + b^2$$

In Example 10, we multiply $(a + b)^2$ with both the tile and the table method.

EXAMPLE 10 Showing a perfect square trinomial Multiply $(a + b)^2$ with tiles and a table.

Solution $(a + b)^2 = (a + b)(a + b)$

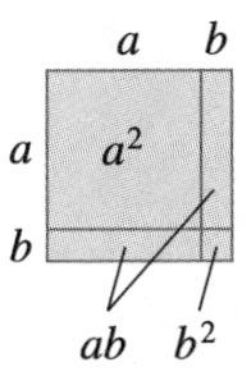

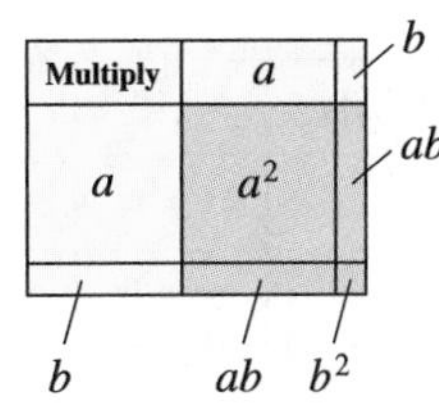

Multiply	a	$+b$
a	a^2	$+ab$
$+b$	$+ab$	$+b^2$

With both the tiles and table,

$$(a + b)(a + b) = a^2 + ab + ab + b^2 = a^2 + 2ab + b^2$$

The tiles form a square with $a + b$ on each side. ●

The phrase *perfect square trinomial* describes the answer or product. The phrase *square of binomial* describes the original multiplication, $(a + b)^2$ or $(x + 2)^2$. Both phrases describe square tile designs, as shown in Example 10. In Example 11, we look for perfect square trinomials and find other products that lead to our next pattern.

EXAMPLE 11 Identifying patterns in binomial multiplication Multiply these binomials. Which answers are perfect square trinomials? What patterns can you find in the other products?

a. $(x - 1)(x + 1)$ **b.** $(a + 3)(a + 3)$

c. $(a + 3)(a - 3)$ **d.** $(2 - b)(2 - b)$

Solution **a.** $(x - 1)(x + 1) = x^2 + 1x - 1x - 1 = x^2 - 1$

b. $(a + 3)(a + 3) = a^2 + 3a + 3a + 9 = a^2 + 6a + 9$, a perfect square trinomial

c. $(a + 3)(a - 3) = a^2 - 3a + 3a - 9 = a^2 - 9$

d. $(2 - b)(2 - b) = 4 - 2b - 2b + b^2 = 4 - 4b + b^2$, a perfect square trinomial

Parts a and c have only two terms in their products. The terms are both squared expressions and are separated by a subtraction. ●

DIFFERENCES OF SQUARES Expressions such as $x^2 - 1$ and $a^2 - 9$ in Example 11 are called differences of squares. The phrase *difference of squares* describes the answer.

Definition of Difference of Squares

> The expression $a^2 - b^2$ is a **difference of squares** because its two terms are the square of a and the square of b.

The original binomials contain the same variables but have opposite operations between the variables:

$$(a + b)(a - b) = a^2 - b^2$$

In Example 12, we look at why multiplying these binomials gives a difference of squares.

EXAMPLE 12 Finding differences of squares Multiply $(a + b)(a - b)$ with a table. What happens to the like terms?

Solution

Multiply	a	$-b$
a	a^2	$-ab$
$+b$	$+ab$	$-b^2$

Adding terms in the table gives $a^2 - ab + ab - b^2 = a^2 - b^2$. The like terms $-ab$ and $+ab$ add to zero. The remaining terms a^2 and b^2 are squares separated by a subtraction—hence the name *difference of squares*. ●

Forming a tile view of the difference of squares is left as a project (see Exercise 78). In the interests of time and space, negative tiles are not considered in the text. Tables permit us to multiply any polynomial with less introduction.

In Example 13, we practice identifying special products.

EXAMPLE 13 Identifying expressions that multiply to special products Predict whether the product from each of these expressions will be a perfect square trinomial, a difference of squares, or neither.

a. $(y - 6)(y + 6)$ **b.** $(y - 6)(y - 6)$

c. $(2x - 3y)(2x - 3y)$ **d.** $(2x + 3y)(2x - 3y)$

e. $(a + b)(-a - b)$

Solution **a.** The variables are alike, but the operations between the terms are different; the answer will be a difference of squares.

$$(y - 6)(y + 6) = y^2 + 6y - 6y - 36 = y^2 - 36$$

b. The variables and operations within the binomials are alike. The expression is the square of a binomial; the answer will be a perfect square trinomial.

$$(y - 6)(y - 6) = y^2 - 6y - 6y + 36 = y^2 - 12y + 36$$

c. The variables and operations within the binomials are alike. The expression is the square of a binomial; the answer will be a perfect square trinomial.

$$(2x - 3y)(2x - 3y) = 4x^2 - 6xy - 6xy + 9y^2 = 4x^2 - 12xy + 9y^2$$

d. The variables are alike, but the operations are different; the answer will be a difference of squares.

$$(2x + 3y)(2x - 3y) = 4x^2 - 6xy + 6xy - 9y^2 = 4x^2 - 9y^2$$

e. The variables are alike, but both signs in the second factor are different.

$$(a + b)(-a - b) = -a^2 - ab - ab - b^2 = -a^2 - 2ab - b^2$$

The square terms in the answer are negative, so the expression is not a perfect square trinomial. The like terms do not add to zero, so the expression is not a difference of squares. ●

Recognizing perfect square trinomials and differences of squares is useful in mental multiplication, factoring, and applications.

ANSWER BOX

Warm-up:

m	n	**Sum** $m + n$	**Product** $m \cdot n$
3	4	7	12
2	6	8	12
3	5	8	15
1	15	16	15
4	6	10	24
3	8	11	24
2	12	14	24
−4	6	2	−24
−4	−6	−10	24
2	−12	−10	−24
−2	−12	−14	24
−3	8	5	−24
−3	−8	−11	24
2	−6	−4	−12
−3	−5	−8	15

EXERCISES 

In Exercises 1 to 6, what binomial factors are shown in the figures? What is the product of each pair of factors?

1.

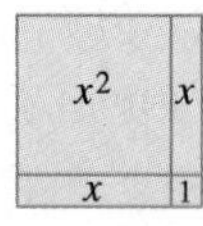

2.

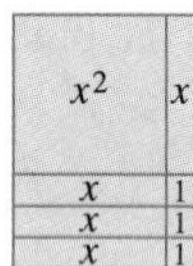

3.

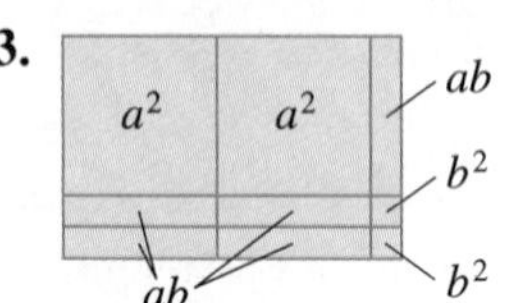

4.

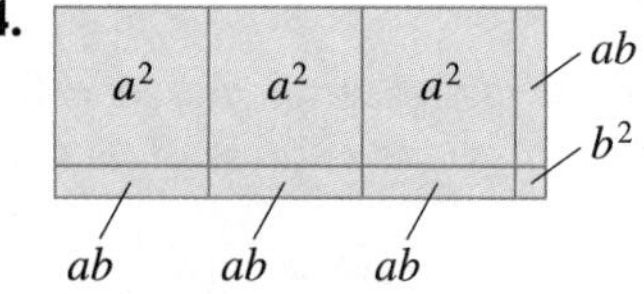

5.

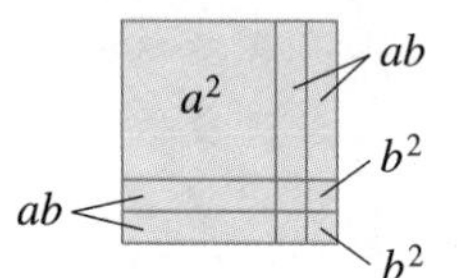

6.

x^2 x^2 x
x^2 x^2 x
x x 1

In Exercises 7 to 14, complete the tables and write the product.

7.

Multiply	$2x$	$+5$
x		
-4		

$(x - 4)(2x + 5) =$

8.

Multiply	x	-1
$3x$		
$+1$		

$(3x + 1)(x - 1) =$

9.

Multiply	x	-2
$2x$		
$+3$		

$(2x + 3)(x - 2) =$

10.

Multiply	$2x$	-3
x		
$+3$		

$(x + 3)(2x - 3) =$

11.

Multiply	x	-2
$2x$		
-1		

$(2x - 1)(x - 2) =$

12.

Multiply	$4x$	-1
x		
-5		

$(x - 5)(4x - 1) =$

13.

Multiply	x	-4
$5x$		
-1		

$(5x - 1)(x - 4) =$

14.

Multiply	x	-3
$3x$		
-1		

$(3x - 1)(x - 3) =$

Mentally multiply the expressions in Exercises 15 to 26.

15. a. $(x - 2)(x - 2)$ **b.** $(x + 2)(x - 2)$

16. a. $(x - 1)(x - 1)$ **b.** $(x - 1)(x + 1)$

17. a. $(a + 5)(a + 5)$ **b.** $(b + 5)(b - 5)$

18. a. $(b - 4)(b + 4)$ **b.** $(a + 4)(a + 4)$

19. a. $(a + b)(a - b)$ **b.** $(a - b)(a - b)$

20. a. $(a - b)(a + b)$ **b.** $(a + b)(a + b)$

21. a. $(x + 1)(x + 7)$ **b.** $(x + 1)(x - 7)$

22. a. $(x - 1)(x + 7)$ **b.** $(x - 1)(x - 7)$

23. a. $(b + 7)(b + 7)$ **b.** $(a + 7)(a - 7)$

24. a. $(a - 7)(a + 7)$ **b.** $(b - 7)(b - 7)$

25. a. $(x + y)(x + y)$ **b.** $(x - y)(x - y)$

26. a. $(x + y)(x - y)$ **b.** $(x - y)(x + y)$

27. Identify ten of the Exercises 15 to 26 that multiply to a perfect square trinomial.

28. Identify ten of the Exercises 15 to 26 that multiply to a difference of squares.

In Exercises 29 to 38, predict whether the expressions will multiply to a perfect square trinomial (pst), a difference of squares (ds), or neither. Next, find the products.

29. $(2x + 3)(2x + 3)$ **30.** $(2x - 3)(2x - 3)$

31. $(2x - 3)(2x + 3)$ **32.** $(2x + 3)(2x - 3)$

33. $(2x - 3)(3 - 2x)$ **34.** $(3x - 2)(2 - 3x)$

35. $(3x - 2)(3x - 2)$ **36.** $(3x + 2)(3x + 2)$

37. $(3x + 2)(3x - 2)$ **38.** $(3x - 2)(3x + 2)$

In Exercises 39 to 46, multiply the expressions.

39. $(x + 5)^2$ **40.** $(x + 7)^2$

41. $(a - 6)^2$ **42.** $(a - 8)^2$

43. $2(x + 3)^2$ **44.** $3(x - 2)^2$

45. $3(x - 5)^2$ **46.** $2(x + 6)^2$

47. Explain what is wrong:

$$(x - a)^2 = x^2 - a^2$$

48. Explain what is wrong:

$$3(a - b)^2 = 9a^2 - 18ab + 9b^2$$

Finish Exercises 49 to 54 with a mental multiplication.

Binomial factors: Trinomial product:

49. $(x - 1)(x + 12) = x^2 +$ ____ $x - 12$

50. $(x - 2)(x + 6) = x^2 +$ ____ $x - 12$

51. $(x - 3)(x + 4) = x^2 +$ ____ $x - 12$

52. $(x + 3)(x - 4) = x^2 -$ ____ $x - 12$

53. $(x + 2)(x - 6) = x^2 -$ ____ $x - 12$

54. $(x + 1)(x - 12) = x^2 -$ ____ $x - 12$

Multiply the factors in Exercises 55 to 66. Look for patterns.

55. a. $(x + 1)(x + 8)$ **b.** $(x + 1)(x - 8)$

56. a. $(x + 8)(x - 1)$ **b.** $(x - 8)(x - 1)$

57. a. $(x + 2)(x + 4)$ **b.** $(x + 2)(x - 4)$

58. a. $(x + 4)(x - 2)$ **b.** $(x - 4)(x - 2)$

59. a. $(2x - 3)(3x - 2)$ **b.** $(2x + 3)(3x - 2)$

60. a. $(2x - 3)(3x + 2)$ **b.** $(2x + 3)(3x + 2)$

61. a. $(6x + 1)(x + 6)$ **b.** $(6x - 1)(x + 6)$

62. a. $(6x - 1)(x - 6)$ **b.** $(6x + 1)(x - 6)$

63. a. $(2x + 5)(x + 1)$ **b.** $(2x + 5)(x - 1)$

64. a. $(2x - 5)(x + 1)$ **b.** $(2x - 5)(x - 1)$

65. a. $(2x - 1)(x + 5)$ **b.** $(2x + 1)(x + 5)$

66. a. $(2x - 1)(x - 5)$ **b.** $(2x + 1)(x - 5)$

67. All of the products in Exercises 49 to 54 end with -12. What are the six sets of binomial factors that multiply to $x^2 \pm$ ____ $+ 12$? Assume the factors contain only whole numbers.

68. Look again at Exercises 49 to 54. Are there any other binomial factors that multiply to $x^2 \pm$ ____ $- 12$? Assume the factors contain only whole numbers.

69. What are the six sets of binomial factors that multiply to $x^2 \pm$ ____ $+ 20$?

70. What are the six sets of binomial factors that multiply to $x^2 \pm$ ____ $- 20$?

71. What are the possible numbers n in $x^2 \pm nx + 24$?

72. What are the possible numbers n in $x^2 \pm nx - 24$?

73. Explain how we obtain the coefficient -2 on the middle term of the product in $(x + 3)(x - 5) = x^2 - 2x - 15$.

74. Explain how we obtain the coefficient -8 on the middle term of the product in $(x - 4)(x - 4) = x^2 - 8x + 16$.

Projects

75. ***Multiplying Binomials and Trinomials.*** Multiply these expressions. Use of a table is recommended.

a. $(x - 1)(x^2 - 2x + 1)$

b. $(x - 2)(x^2 + 2x + 4)$

c. $(a - b)(a^2 + ab + b^2)$

d. $(x + y)(x^2 - xy + y^2)$

e. $(x + y)(x^2 + 2xy + y^2)$

f. $(a + b)(a^2 + 2ab + b^2)$

76. ***Number Patterns.*** Use a calculator to evaluate the expressions in parts a to d.

a. $17^2 - 14^2 - (16^2 - 13^2)$

b. $16^2 - 13^2 - (15^2 - 12^2)$

c. $15^2 - 12^2 - (14^2 - 11^2)$

d. $14^2 - 11^2 - (13^2 - 10^2)$

e. What do you notice about the values of the expressions?

f. Write a symbolic expression for the number pattern in part a. Let $x = 17$, $x - 3 = 14$, $x - 1 = 16$, and $x - 4 = 13$. Does the same expression describe parts b, c, and d?

g. By simplifying your expression, prove that the number pattern works for all inputs x. (*Hint*: Multiply the squared terms and combine like terms.)

h. Write a description of the following pattern in symbols, and simplify to prove that the pattern always works.

$$1 \cdot 3 - 2 \cdot 2 = -1$$
$$2 \cdot 4 - 3 \cdot 3 = -1$$
$$3 \cdot 5 - 4 \cdot 4 = -1$$
$$4 \cdot 6 - 5 \cdot 5 = -1$$
$$5 \cdot 7 - 6 \cdot 6 = -1$$
$$\vdots$$
$$49 \cdot 51 - 50 \cdot 50 = -1$$
$$\vdots$$

(*Hint*: Let x be the first number.)

77. ***Tiles Representing Negative Terms.*** The expression $a - b$ can be thought of as either "a subtract b" or "a add the opposite of b." Both ways of thinking can be used to build rectangles for binomial products.

Subtraction Model: To subtract, we place tiles of a second color on top of the positive color tiles. The product $(a - b)(a + b)$ is shown in the figure.

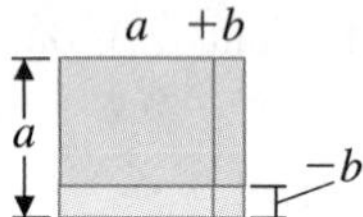

A small square $-b^2$ on top of the vertical rectangle ab comes from the product $b(-b)$ and is needed to complete the rectangular area, $a^2 - ab + ab - b^2$.

Addition of the Opposite Model: The product $(a + (-b))(a + b)$ has an addition of the opposite in the second factor. To add the opposite, we place a different-colored rectangular tile of width b and length a below the a^2 tile. With tiles, $(a + (-b))(a + b)$ is shown as follows:

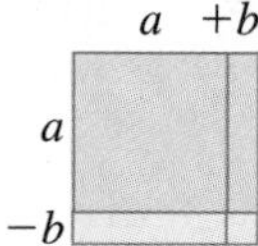

Again, a small square $-b^2$ comes from the product $b(-b)$ and is needed to complete the rectangular area, $a^2 - ab + ab - b^2$.

Build rectangles showing these products. Use either of the models described above.

a. $(a + b)(a - 2b)$

b. $(2a - b)(a - 3b)$ (*Hint*: $-b(-3b) = +3b^2$)

c. $(x + 2)(x - 1)$

d. $(x - 2)(x - 3)$

e. $(x - 2)(x + 3)$

f. $(2x - 3)(3x + 1)$

MID-CHAPTER 6 TEST

1. Simplify each of the following expressions. Arrange the terms alphabetically, with exponents on the first letter in descending order. Indicate whether the result is a monomial, binomial, trinomial, or other polynomial.

a. $7b - 8c + 3a + 4b - 5c - 6a$

b. $x^2 - 4x + x^2 - 6$

c. $4xy^2 + x^3y^2 - 3x^2y + x^3y^2 - 2xy^2$

d. $(5a - 3b - 2c) - (3a + 4b - 6c)$

e. $x(x^2 + 5x + 25) - 5(x^2 + 5x + 25)$

f. $9 - 3x(x + 3)$

g. $8 - 4(4 - x)$

2. Name the greatest common factor for each expression, and then factor the expression.

a. $6x^2 - 2x + 8$

b. $2abc - 3ac + 4ab$

3. Find the length, width, perimeter, and area of each rectangle.

a.

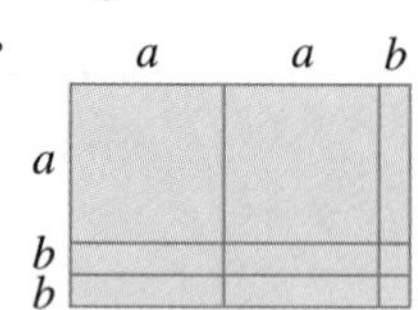

b.

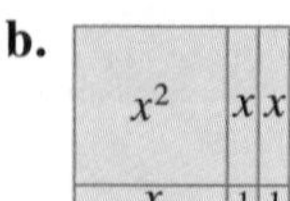

4. Explain the difference between terms and factors. Use expressions like ab and $a + b$ in your explanation.

5. Arrange these tiles into a rectangle. What multiplication is shown?

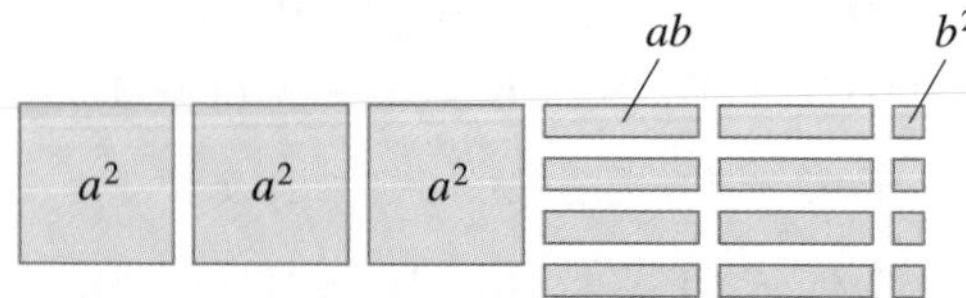

6. Complete the table. Write the binomial factors and trinomial product.

Multiply	$2x$	-5
$3x$		
$+2$		

7. What is the diagonal product in the table in Exercise 6?

8. Complete the table. Write the factors and the resulting product.

Multiply	$3x^2$	$-2x$	$+1$
x			
-2			

9. What are the like terms in the table in Exercise 8?

In Exercises 10 to 14, find the products.

10. $(x + 3)(x + 5)$ **11.** $(x - 4)(3x + 5)$

12. $(2x + 7)(3x - 1)$ **13.** $(x - 3)(x + 3)$

14. $(2x - 5)^2$

15. Identify each expression as a perfect square trinomial (pst), a difference of squares (ds), or neither.

a. $x^2 - 2x - 1$ **b.** $x^2 + 9$

c. $x^2 - 49$ **d.** $x^2 + 6x + 9$

16. What are all the possible numbers that could complete this statement?

$$(x \pm \_\_\_)(x \pm \_\_\_) = x^2 \pm \_\_\_ \pm 10$$

6.3 Factoring Binomials and Trinomials

OBJECTIVES

- Factor with algebra tiles.
- Factor with a table.
- Factor with guess and check.
- Factor special products.
- Factor out the greatest common factor.

> **WARM-UP**
>
> The whole-number factor pairs of 28 are $1 \cdot 28$, $2 \cdot 14$, and $4 \cdot 7$. Find all the whole-number factor pairs for these numbers:
>
> **1.** 45
> **2.** 60
> **3.** 36
> **4.** 24
> **5.** 56

THIS SECTION FOCUSES ON factoring, the reverse of the multiplication process in Section 6.2. We find binomial factors, given the trinomial or binomial product. We again use the tile and table models. We summarize a guess-and-check approach and factor the special products introduced in the last section: perfect square trinomials and differences of squares. The greatest common factor continues to be important.

The Tile Model of Factoring

In Section 6.1, we found the height and base from the area of a rectangle. We called this process *factoring*. In addition to being a basic algebraic process, factoring changes everyday formulas into more usable forms; see the project in Exercise 87.

We have factored expressions such as $3ab + 3b^2$ into $3b(a + b)$, the product of a monomial and a binomial. We now factor binomials and trinomials into two binomials. We begin with the tile model to review how length, width, and area are related to multiplication and factoring.

EXAMPLE  Exploration: finding width and length, given area Arrange these sets of tiles into rectangles. Describe the width and length of each rectangle. Write the binomial factors that give each set of tiles as a product.

a. $x^2 + 3x + 2$ **b.** $2x^2 + 7x + 3$

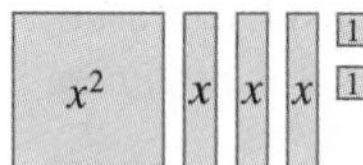

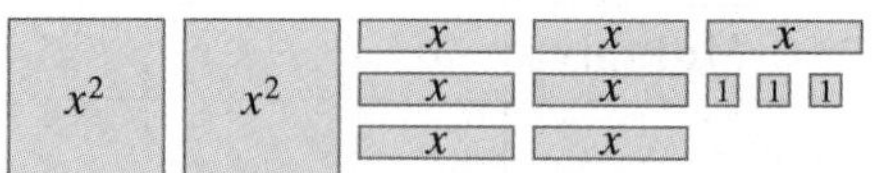

Solution For sketches of the tile rectangles, see Examples 2 and 3.

a. For $x^2 + 3x + 2$, the width and length are $x + 1$ and $x + 2$. Using binomial factors, we have

$$(x + 1)(x + 2) = x^2 + 3x + 2$$

b. For $2x^2 + 7x + 3$, the width and length are $2x + 1$ and $x + 3$. Using binomial factors, we have

$$(2x + 1)(x + 3) = 2x^2 + 7x + 3$$ ●

Example 2 reviews the multiplication patterns from Section 2.2.

EXAMPLE  Multiplying two binomial factors

a. Find the product of $(x + 1)(x + 2)$ using a table.

b. Compare the result to the product found with tiles.

c. In the table, find the sum of the like terms and the diagonal product.

d. Where in the table is the product of the first terms?

e. Where in the table is the product of the last terms?

Solution **a.**

Multiply	x	$+2$
x	x^2	$+2x$
$+1$	$+1x$	$+2$

b.

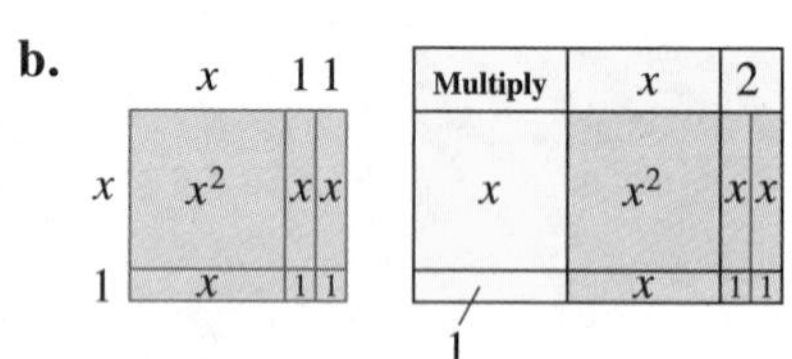

width · length = area

$$(x + 1)(x + 2) = x^2 + 3x + 2$$

c. The like terms on the diagonal from lower left to upper right add to $3x$. The terms on the diagonals multiply to the same value:

$$(x^2)(2) = 2x^2 \quad \text{and} \quad (2x)(1x) = 2x^2$$

d. The product of the first terms, x and x, is in the upper left corner.

e. The product of the last terms, 1 and 2, is in the lower right corner. ●

The Table Model of Factoring

To find the two binomial factors of a trinomial, we place the terms of the trinomial in a table. The position of each term helps with the factoring.

EXAMPLE 3 Factoring by table Use a table to factor $2x^2 + 7x + 3$.

Solution ***Step 1:*** We write the first term, $2x^2$, in the upper left corner. We write the last term, 3, in the lower right corner. We multiply the first and last terms to get the diagonal product, $6x^2$.

Factor		
	$2x^2$	
		$+3$

Diagonal product, $2x^2 \cdot 3 = 6x^2$

Step 2: From the diagonal product, $6x^2$, we list all the factors of the coefficient, 6.

$1 \cdot 6$

$2 \cdot 3$

From this list, we find the factors that add to the coefficient of the middle term, 7. The factors are $1x \cdot 6x$. We write the factors in the other diagonal, from lower left to upper right.

Factor		
	$2x^2$	$+6x$
	$+1x$	$+3$

Sum of like terms, $6x + 1x = 7x$

Step 3: To the left of $2x^2$, we write the greatest common factor of the first row:

gcf $= 2x$

Factor		
$2x$	$2x^2$	$+6x$
	$+1x$	$+3$

Step 4: We use the gcf to find the remaining factors. If the gcf is $2x$, the factor in the top left must be x.

Factors, $2x \cdot x = 2x^2$

Factor	x	
$2x$	$2x^2$	$+6x$
	$+1x$	$+3$

If the gcf is $2x$, the factor in the top right must be $+3$.

Factor	x	$+3$
$2x$	$2x^2$	$+6x$
	$+1x$	$+3$

Factors, $2x \cdot 3 = 6x$

If the factor in the top left is x, the factor in the bottom left must be $+1$.

Factors, $1 \cdot x = x$

Factor	x	$+3$
$2x$	$2x^2$	$+6x$
$+1$	$+1x$	$+3$

We use the last term, $+3$, to check the bottom left and top right factors.

Factor	x	$+3$
$2x$	$2x^2$	$+6x$
$+1$	$+1x$	$+3$

Factors, $1 \cdot 3 = 3$

Step 5: The table shows the binomial factors $(2x + 1)$ and $(x + 3)$. We multiply the binomials to check the answer:

$(2x + 1)(x + 3) = 2x^2 + 7x + 3$ ●

When we multiply the factors from Example 3 with the traditional method to find the product, the results match those from the tile model.

Traditional method:

$$\begin{array}{r} x + 3 \\ 2x + 1 \\ \hline x + 3 \\ 2x^2 + 6x \quad \\ \hline 2x^2 + 7x + 3 \end{array}$$

Tile model:

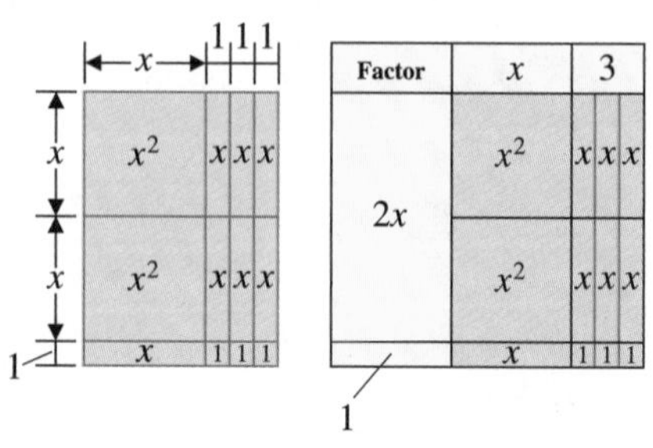

Think about it 1: If we switch the positions of $6x$ and x in the table, does that change the factors? Sketch the tile design for this set of factors.

Following is a summary of factoring by table, using the generic trinomial $ax^2 + bx + c$. You will see this trinomial again in Chapter 8 in the quadratic equation and the quadratic formula.

Summary: Factoring by Table

To find the binomial factors of $ax^2 + bx + c$ by table:

1. **a.** Write the first term of the trinomial, ax^2, in the upper left corner of the table.
   b. Write the last term of the trinomial, c, in the lower right corner.
   c. Multiply the first and last terms to get the diagonal product, acx^2.
2. **a.** From the diagonal product, list the factors of the coefficient, ac.
   b. Find the factors that add to the coefficient of the middle term of the trinomial, b.
   c. Write the factors adding to bx in the diagonal from lower left to upper right.
3. Write the greatest common factor of the first row to the left of ax^2.
4. **a.** Use the gcf to find the remaining factors at the top and left of the table.
   b. Check that the top right and bottom left factors multiply to the last term, c.
5. Multiply the binomial factors from the table to check the factoring.

Factor		
Greatest common factor (gcf) of this row	First term of trinomial, ax^2	____
	____	Last term of trinomial, c

Sum of like terms equals middle term of trinomial, bx

Diagonal product, acx^2

We again factor by table in Example 4.

EXAMPLE 4

Factoring by table Use a table to factor $3x^2 + 13x + 12$. Sketch a tile model that shows the factors.

Solution ***Step 1:*** We write the first term, $3x^2$, in the upper left corner. We write the last term, 12, in the lower right corner. We multiply the first and last terms to get the diagonal product, $36x^2$.

Factor		
	$3x^2$	
		$+12$

Diagonal product, $3x^2 \cdot 12 = 36x^2$

Step 2: From the diagonal product, $36x^2$, we list all the factors of the coefficient, 36.

$1 \cdot 36$

$2 \cdot 18$

$3 \cdot 12$

$4 \cdot 9$

$6 \cdot 6$

From this list, we find the factors that add to the coefficient of the middle term, 13. The factors are $4x \cdot 9x$. We write the factors in the diagonal from lower left to upper right.

Factor		
	$3x^2$	$+9x$
	$+4x$	$+12$

Sum of like terms, $9x + 4x = 13x$

Step 3: To the left of $3x^2$, we write the greatest common factor of the first row:

gcf = $3x$

Factor		
$3x$	$3x^2$	$+9x$
	$+4x$	$+12$

Step 4: We use the gcf to find the remaining factors. If the gcf is $3x$, the factor in the top left must be x. If the gcf is $3x$, the factor in the top right must be $+3$. If the factor in the top left is x, the factor in the bottom left must be $+4$. We use the last term, $+12$, to check the bottom left and top right factors.

Factor	x	$+3$
$3x$	$3x^2$	$+9x$
$+4$	$+4x$	$+12$

Factors, $4 \cdot 3 = 12$

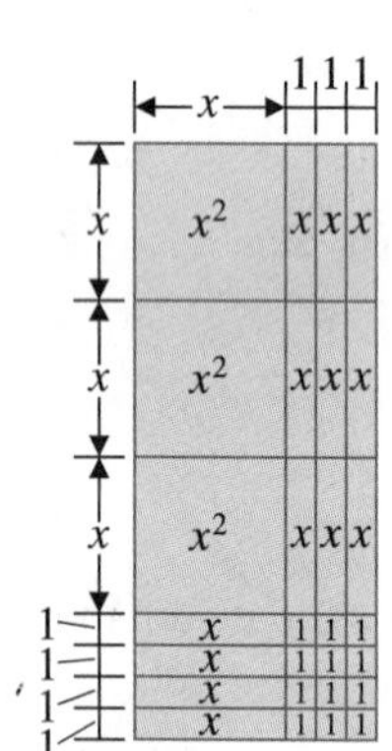

Figure 7

Step 5: The table shows the binomial factors $(3x + 4)$ and $(x + 3)$. We multiply the binomials to check the answer:

$(3x + 4)(x + 3) = 3x^2 + 13x + 12$

Figure 7 shows the tile model of $(3x + 4)(x + 3)$. ●

In Example 5, we factor a trinomial containing negative terms.

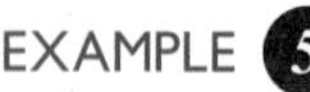

EXAMPLE 5 Factoring trinomials with negative terms Use a table to factor $6x^2 - 11x - 10$.

Solution *Step 1:* We write the first term, $6x^2$, in the upper left corner. We write the last term, -10, in the lower right corner. We multiply the first and last terms to get the diagonal product, $-60x^2$.

Factor		
	$6x^2$	
		-10

Diagonal product, $6x^2 \cdot (-10) = -60x^2$

Student Note:
You may find that only one list of factors of 60 is sufficient to find a pair that sum to -11.

Step 2: From the diagonal product, $-60x^2$, we list all the factors of the coefficient, -60. One of the two numbers will be negative.

$-1 \cdot 60$ $\quad 1 \cdot (-60)$
$-2 \cdot 30$ $\quad 2 \cdot (-30)$
$-3 \cdot 20$ $\quad 3 \cdot (-20)$
$-4 \cdot 15$ $\quad 4 \cdot (-15)$
$-5 \cdot 12$ $\quad 5 \cdot (-12)$
$-6 \cdot 10$ $\quad 6 \cdot (-10)$

From this list, we find the factors that add to the coefficient of the middle term, -11. The factors are $4x \cdot (-15x)$. We write the factors in the diagonal from lower left to upper right.

Factor		
	$6x^2$	$-15x$
	$+4x$	-10

Sum of like terms, $4x + (-15x) = -11x$

Step 3: To the left of $3x^2$, we write the greatest common factor of the first row:

gcf $= 3x$

Factor		
$3x$	$6x^2$	$-15x$
	$+4x$	-10

Step 4: We use the gcf to find the remaining factors. If the gcf is $3x$, the factor in the top left must be $2x$.

Factors, $3x \cdot 2x = 6x^2$

Factor	$2x$	
$3x$	$6x^2$	$-15x$
	$+4x$	-10

If the gcf is $3x$, the factor in the top right must be -5.

Factor	$2x$	-5
$3x$	$6x^2$	$-15x$
	$+4x$	-10

Factors, $3x \cdot (-5) = -15x$

If the factor in the top left is $2x$, the factor in the bottom left must be $+2$.

Factor	$2x$	-5
$3x$	$6x^2$	$-15x$
$+2$	$+4x$	-10

Factors, $2 \cdot 2x = 4x$

We use the last term, -10, to check the bottom left and top right factors.

Factor	$2x$	-5
$3x$	$6x^2$	$-15x$
$+2$	$+4x$	-10

Factors, $2 \cdot (-5) = -10$

Step 5: The table shows the binomial factors $(3x + 2)$ and $(2x - 5)$. We multiply the binomials to check the answer:

$(3x + 2)(2x - 5) = 6x^2 - 11x - 10$ ●

In the next example, along with the table method, we use the traditional method of factoring trinomials (called the factoring-by-grouping or *ac*-product method).

EXAMPLE 6 Factoring by table and the traditional method Use a table to factor $6x^2 - 11x + 3$. Compare the steps with those in the traditional method.

Table Solution ***Step 1:*** We write the first and last terms, $6x^2$ and $+3$, in the upper left and lower right corners. We multiply the terms to get the diagonal product, $18x^2$.

Factor		
	$6x^2$	
		$+3$

Student Note:
You may find that only one list of factors of 18 is sufficient.

Step 2: We list the factors of the coefficient of the diagonal product:

$1 \cdot 18$ $-1 \cdot (-18)$
$2 \cdot 9$ $-2 \cdot (-9)$
$3 \cdot 6$ $-3 \cdot (-6)$

We write the factors that add to $-11x$ in the diagonal from lower left to upper right.

Factor		
	$6x^2$	$-2x$
	$-9x$	$+3$

Step 3: To the left of $6x^2$, we write the gcf, $2x$, of the first row.

Step 4: We use the gcf to find the remaining factors at the top and left. We use the last term, $+3$, to check the bottom left and top right factors.

Factor	$3x$	-1
$2x$	$6x^2$	$-2x$
-3	$-9x$	$+3$

Step 5: We multiply the binomial factors to check the answer:

$$(3x - 1)(2x - 3) = 6x^2 - 11x + 3$$

●

Traditional Solution

The following traditional solution is called the *ac-product* or *factoring-by-grouping* method. The steps are numbered to show how they match those in the table method.

Student Note:
The comments to the right tell what was done to get to that step.

Step 1:	$6x^2 - 11x + 3$	
	$a = 6$	Find a and c in $ax^2 + bx + c$.
	$c = 3$	
	$6 \cdot 3 = 18$	Multiply a and c.
Step 2:	$1 \cdot 18, 2 \cdot 9, 3 \cdot 6$	List all the factors of ac.
	2 and 9 add to 11	Find the factors that add to bx.
	$2x + 9x = 11x$	
	$6x^2 - (2x + 9x) + 3$	Write $ax^2 + bx + c$ as a four-term expression by replacing bx with the two factors $2x$ and $9x$.
	$6x^2 - 2x - 9x + 3$	Apply the distributive property to $-(2x + 9x)$.
Step 3:	$2x(3x - 1) - 9x + 3$	Find the gcf and factor the first two terms.
Step 4:	$2x(3x - 1) - 3(3x - 1)$	Find the gcf and factor the last two terms.
	$(3x - 1)(2x - 3)$	Factor the $3x - 1$ from the two expressions.
Step 5:	$(3x - 1)(2x - 3) = 6x^2 - 11x + 3$	Multiply the binomial factors to check the factoring.

●

The advantage of the table method is that we do the work in a table that looks identical to the multiplication table. The traditional factoring method does not *look* like the traditional multiplication method. However, if you look closely, you will see that the traditional method is the same as the table method. The differences lie in how and where we write the information.

In the next example, we try to factor an expression that does not have whole numbers in the factors. We need to be able to recognize these cases.

EXAMPLE 7 Identifying a trinomial that does not factor Using both tiles and a table, show that $x^2 + x + 1$ does not factor.

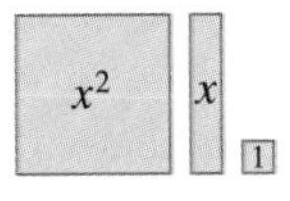

Figure 8

Solution With tiles, we see in Figure 8 that we have one large square, a rectangle, and one small square. There is no way to build a rectangle with these tiles.

With a table, we go through the following steps:

Step 1: We write x^2 and 1 in the diagonal. We multiply x^2 and 1 to obtain the diagonal product, $1x^2$.

Factor		
	x^2	
		$+1$

Step 2: From the diagonal product, $+1x^2$, we list the factors of the coefficient, 1.

$1 \cdot 1$

$-1 \cdot (-1)$

We try adding the factors of the diagonal product:

$1x + 1x = 2x$

$-1x + (-1x) = -2x$

We cannot get $1x$ from whole-number factors of the diagonal product. The trinomial $x^2 + x + 1$ does not factor. ●

Think about it 2: Some students prefer to use guess and check to try to factor $x^2 + x + 1$. Multiply these guesses to see if any give the product $x^2 + x + 1$.

a. $(x - 1)(x + 1)$ **b.** $(x - 1)(x - 1)$ **c.** $(x - 1)(-x - 1)$

d. $(x + 1)(x + 1)$ **e.** Try any other factors.

Do any of the binomial factors make special products?

Factoring with Guess and Check

With the availability of algebraic calculators and alternative traditional procedures (see the quadratic formula in Chapter 8), it is not as important to be able to factor complicated expressions. We do need to factor some expressions, though, and so we now turn to factoring mentally.

Students in higher mathematics courses find that most factoring can be done mentally. You are encouraged at this time to practice factoring expressions mentally. The easiest expressions to factor are the special products (perfect square trinomials and differences of squares) and trinomials with a first term of x^2.

The Warm-up to Section 6.2 (page 352) provides practice in finding two numbers, given their sum and product. If you found that you could do that Warm-up, then you have all the skills you need to factor mentally. If you had problems, go back and try the Warm-up again.

Factoring by Guess and Check

When the leading coefficient on the trinomial $ax^2 + bx + c$ is $a = 1$, the way to guess the factors is to look at the numbers b and c. The numbers in the blanks must multiply to c and add to b:

$$x^2 + bx + c = (x \pm \_\_\_\_)(x \pm \_\_\_\_)$$

The ± sign indicates *addition or subtraction* because the operations depend on the signs of b and c. Make a guess; then multiply, check, and guess again as needed.

In Example 8, we practice factoring mentally and find two trinomials that do not factor. In all examples, we consider factors with whole numbers only.

EXAMPLE 8 Factoring mentally Find the missing numbers in these statements. The addition and subtraction signs have been placed in the factors.

a. $x^2 + 8x + 12 = (x + \_\_\_\_)(x + \_\_\_\_)$

b. $x^2 + x + 12 = (x + \_\_\_\_)(x + \_\_\_\_)$

c. $x^2 + 8x + 15 = (x + \_\_\_\_)(x + \_\_\_\_)$

d. $x^2 + 10x + 15 = (x + \_\_\_\_)(x + \_\_\_\_)$

e. $x^2 + 5x - 24 = (x - \_\_\_\_)(x + \_\_\_\_)$

f. $x^2 - 11x + 24 = (x - \_\_\_\_)(x - \_\_\_\_)$

Solution **a.** The numbers that multiply to 12 are $1 \cdot 12$, $2 \cdot 6$, and $3 \cdot 4$. Only 2 and 6 add to 8.

$$x^2 + 8x + 12 = (x + 2)(x + 6)$$

b. The numbers that multiply to 12 are $1 \cdot 12$, $2 \cdot 6$, and $3 \cdot 4$ or the same pairs of numbers with both numbers negative. None of the pairs add to 1, so the trinomial $x^2 + x + 12$ cannot be factored with whole numbers.

c. The numbers that multiply to 15 are $1 \cdot 15$ and $3 \cdot 5$. Only 3 and 5 add to 8.

$$x^2 + 8x + 15 = (x + 3)(x + 5)$$

d. The numbers that multiply to 15 are $1 \cdot 15$ and $3 \cdot 5$ or the same pairs of numbers with both numbers negative. None of the pairs add to 10, so the trinomial $x^2 + 10x + 15$ cannot be factored with whole numbers.

e. The numbers that multiply to 24 are $1 \cdot 24$, $2 \cdot 12$, $3 \cdot 8$, and $4 \cdot 6$. Because the product is -24, one number must be negative. The sum is $+5$, so the numbers are -3 and 8.

$$x^2 + 5x - 24 = (x - 3)(x + 8)$$

f. Using the same factors of 24 as in part e, we get a sum of -11 with -3 and -8.

$$x^2 - 11x + 24 = (x - 3)(x - 8)$$

●

We now turn to other problems that can be factored mentally.

Factoring Special Products

We now factor special products. In Section 6.2, we multiplied identical binomial factors and got perfect square trinomials.

$$(a + b)(a + b) = a^2 + 2ab + b^2 \qquad \text{Perfect square trinomial}$$

Perfect square trinomials form squares when we use tiles.

A difference of squares has binomial factors with the same numbers and variables, but one factor contains addition and the other subtraction.

$$(a + b)(a - b) = a^2 - b^2 \qquad \text{Difference of squares}$$

EXAMPLE 9 Identifying and factoring special products Tell whether each of the following expressions is a perfect square trinomial (pst), difference of squares (ds), or neither. Factor all that can be factored.

a. $x^2 - 2x + 1$	**b.** $y^2 - 4y + 4$	**c.** $x^2 - 5x + 25$
d. $a^2 - 49$	**e.** $x^2 + 4x + 8$	**f.** $9x^2 + 25$
g. $y^2 + 2yz + z^2$	**h.** $4x^2 - 25$	

Solution **a.** $x^2 - 2x + 1 = (x - 1)(x - 1) = (x - 1)^2$; pst

b. $y^2 - 4y + 4 = (y - 2)(y - 2) = (y - 2)^2$; pst

c. $x^2 - 5x + 25$; neither. The middle term would need to be $-10x$ for a pst.

d. $a^2 - 49 = (a - 7)(a + 7)$; ds

e. $x^2 + 4x + 8$; neither. The last term would need to be 4 for a pst.

f. $9x^2 + 25$; neither. The operation would need to be subtraction for a ds.
g. $y^2 + 2yz + z^2 = (y + z)(y + z) = (y + z)^2$; pst
h. $4x^2 - 25 = (2x - 5)(2x + 5)$; ds ●

We now summarize factoring by adding one step to all the types of problems in this section.

Factoring Out the Greatest Common Factor

Before proceeding with traditional factoring processes, we must first factor out any greatest common factor from the binomial or trinomial. The next two examples include two solutions, one in which we first factor out the greatest common factor and the second in which we find a common factor later in the process.

EXAMPLE 10 Factoring out the greatest common factor Factor $2x^2 + 4x + 2$ by guess and check and with tiles.

Guess-and-Check Solution $2x^2 + 4x + 2 = 2(x^2 + 2x + 1)$ Remove the greatest common factor.

$= 2(x + \_\_\_)(x + \_\_\_)$ The numbers in the blanks must multiply to 1 and add to 2.

$= 2(x + 1)(x + 1)$ The numbers are 1 and 1.

You should recognize that $x^2 + 2x + 1$ is a perfect square trinomial and is equal to $(x + 1)^2$.

Tiles Solution The rectangle shown in Figure 9(a) represents $2x^2 + 4x + 2$. Observe that the tiles can be split into two sets to form two rectangles, as shown in part (b). This shows that

$$2x^2 + 4x + 2 = (x + 1)(2x + 2) = 2(x + 1)(x + 1) = 2(x + 1)^2$$

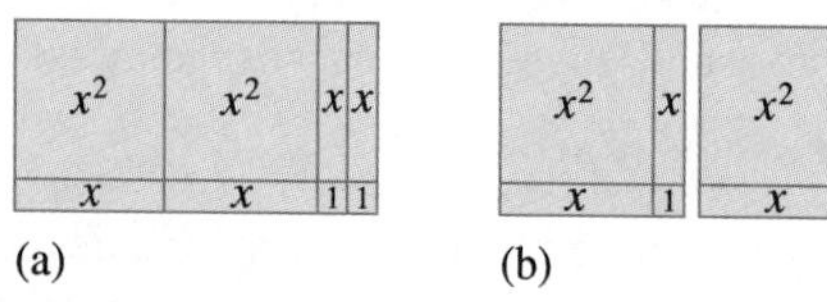

Figure 9 ●

In Example 11, a guess-and-check process requires more than finding numbers that add to b and multiply to c, because the coefficient a is not equal to 1, as it was in earlier examples.

EXAMPLE 11 Factoring out the greatest common factor Factor $4x^2 - 10x - 6$ by guess and check and with a table.

Guess-and-Check Solution $4x^2 - 10x - 6 = 2(2x^2 - 5x - 3)$ Remove the greatest common factor.

We cannot assume that this trinomial does not factor simply because there are no numbers that multiply to -3 and add to -5. The 2 on the x^2 term means that the guess-and-check process used earlier won't work. We need to try other guesses (or go on the table or traditional method).

The only way to get $2x^2$ is with $2x$ and $1x$. Thus, we have

$$2(2x \pm \_\_\_)(x \pm \_\_\_)$$

We must have 1 and 3 as last terms to have the product -3, so the expression must be of the form

$$2(2x \pm 3)(x \pm 1) \qquad \text{or} \qquad 2(2x \pm 1)(x \pm 3)$$

We try multiplying each possible option until a pair of factors works. Then we finish finding the factors. ●

Table Solution

Step 1: We write $4x^2$ and -6 in the diagonal. We multiply $4x^2$ and -6 to obtain the diagonal product, $-24x^2$.

Factor		
	$4x^2$	
		-6

Step 2: We list all the factors of the coefficient, -24:

$-1 \cdot 24 \qquad 1 \cdot (-24)$

$-2 \cdot 12 \qquad 2 \cdot (-12)$

$-3 \cdot 8 \qquad 3 \cdot (-8)$

$-4 \cdot 6 \qquad 4 \cdot (-6)$

The middle term is $-10x$. After finding the factors that add to -10, we place $+2x$ and $-12x$ in the other diagonal.

Factor		
	$4x^2$	$-12x$
	$+2x$	-6

Step 3: The greatest common factor of $4x^2$ and $-12x$ is $4x$, so we write $4x$ to the left of $4x^2$.

Step 4: We find the remaining table factors: $x - 3$ on top and $4x + 2$ on the left.

Factor	x	-3
$4x$	$4x^2$	$-12x$
$+2$	$+2x$	-6

Step 5: The factors $4x + 2$ and $x - 3$ multiply to $4x^2 - 10x - 6$. We can now remove the common factor from $4x + 2$ and complete the factoring:

$$4x^2 - 10x - 6 = 2(2x + 1)(x - 3)$$ ●

As a last example, we use the greatest common factor with another special product to factor a binomial.

EXAMPLE 12 Factoring with the greatest common factor Factor $3x^2 - 3$ by observing patterns.

Solution $3x^2 - 3 = 3(x^2 - 1)$ When we remove the greatest common factor, the parentheses contain a difference of squares.

$$= 3(x - 1)(x + 1)$$

●

ANSWER BOX

Warm-up: 1. $1 \cdot 45, 3 \cdot 15, 5 \cdot 9$ **2.** $1 \cdot 60, 2 \cdot 30, 3 \cdot 20, 4 \cdot 15, 5 \cdot 12, 6 \cdot 10$ **3.** $1 \cdot 36, 2 \cdot 18, 3 \cdot 12, 4 \cdot 9, 6 \cdot 6$ **4.** $1 \cdot 24, 2 \cdot 12, 3 \cdot 8, 4 \cdot 6$ **5.** $1 \cdot 56, 2 \cdot 28, 4 \cdot 14, 7 \cdot 8$ **Think about it 1:** The factor $(x + 3)$ would be on the left, and the factor $(2x + 1)$ would be on the top of the table.

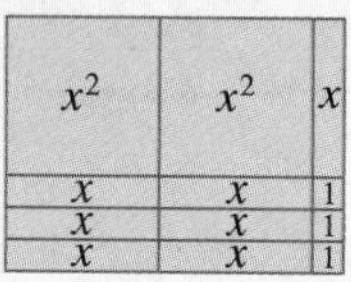

Think about it 2: a. This is a difference of squares, $x^2 - 1$. **b.** This is a square of a binomial and gives a perfect square trinomial, $x^2 - 2x + 1$. **c.** $-x^2 + 1$ or $1 - x^2$, a difference of squares. **d.** This is a square of a binomial and gives a perfect square trinomial, $x^2 + 2x + 1$.

EXERCISES 

In Exercises 1 to 4, arrange the tiles into a rectangle and state the area, which is the product of width and length.

1. x^2 x^2 x x x x x 1 1

2. x^2 x x x x 1 1 1 1

3. x^2 x^2 x^2 x^2 1 x x x x

4. x^2 x^2 x^2 x x x x x 1 1

In Exercises 5 to 10, complete the table and state the problem and factors described by the table.

5.

Factor		
	x^2	$+5x$
	$+4x$	$+20$

6.

Factor		
	x^2	$+5x$
	$-4x$	-20

7.

Factor		
	x^2	$+2x$
	$-10x$	-20

8.

Factor		
	x^2	$-2x$
	$+10x$	-20

9.

Factor		
	$6x^2$	$+x$
	$-18x$	-3

10.

Factor		
	$6x^2$	$-2x$
	$+9x$	-3

11. Describe why we get the coefficient -2 on the middle term of the product in $(x + 3)(x - 5) = x^2 - 2x - 15$.

12. Describe why we get the coefficient $+1$ on the middle term of the product in $(x - 4)(x + 5) = x^2 + x - 20$.

Mentally factor the binomials and trinomials in Exercises 13 to 24. Two of the expressions cannot be factored.

13. $x^2 + 8x + 12$ **14.** $x^2 - 8x + 12$

15. $x^2 - 13x + 12$ **16.** $x^2 - 7x + 12$

17. $x^2 + x + 12$ **18.** $x^2 + 13x + 12$

19. $x^2 + 7x + 12$ **20.** $x^2 - 4x + 12$

21. $x^2 + x - 12$ **22.** $x^2 - 4x - 12$

23. $x^2 - 11x - 12$ **24.** $x^2 - x - 12$

In Exercises 25 to 36, factor with any of the methods in this section.

25. $x^2 + 6x + 9$ **26.** $x^2 + 10x + 25$

27. $x^2 + 11x + 30$ **28.** $x^2 + 17x + 30$

29. $x^2 + 13x - 30$ **30.** $x^2 + x - 30$

31. $x^2 - 6x - 16$ **32.** $x^2 + 6x - 16$

33. $x^2 + 15x - 16$ **34.** $x^2 - 15x - 16$

35. $x^2 - 25$ **36.** $x^2 - 36$

Look for special products and greatest common factors in Exercises 37 to 48. Continue to factor mentally. Two of the expressions cannot be factored.

37. $x^2 - 4$ **38.** $x^2 + 36$

39. $4x^2 - 16$ **40.** $4x^2 - 36$

41. $x^2 + 12x + 36$ **42.** $x^2 - 12x + 36$

43. $4x^2 + 8x + 4$ **44.** $9x^2 + 36x + 36$

45. $x^2 + 4$ **46.** $x^2 - 8x + 16$

47. $x^2 - 6x + 9$ **48.** $5x^2 - 20$

In Exercises 49 to 86, factor using any of the methods in this section. Some of the expressions cannot be factored.

49. $2x^2 + 11x + 12$ **50.** $3x^2 + 13x + 12$

51. $2x^2 - 3x - 9$ **52.** $2n^2 + 3n - 9$

53. $2n^2 + n - 3$ **54.** $2x^2 - 5x - 3$

55. $3x^2 + 5x - 2$ **56.** $3a^2 + a - 2$

57. $3a^2 - 11a - 4$ **58.** $3x^2 - 4x - 4$

59. $3a^2 + 5a - 4$ **60.** $3n^2 - 9n - 4$

61. $3x^2 + 10x - 5$ **62.** $3x^2 - 5x - 5$

63. $9x^2 - 49$ **64.** $9x^2 - 25$

65. $16x^2 - 9$ **66.** $16x^2 - 81$

67. $6x^2 + x - 2$ **68.** $6x^2 - 7x - 5$

69. $6x^2 + 5x - 6$ **70.** $6x^2 - 13x + 5$

71. $2n^2 + 9n - 5$ **72.** $2n^2 + 11n - 6$

73. $25x^2 - 36$ **74.** $100x^2 - 49$

75. $3x^2 + 12x + 9$ **76.** $2x^2 + 6x + 4$

77. $3x^2 - 27$ **78.** $2x^2 - 18$

79. $5x^2 - 10x + 5$ **80.** $3x^2 - 18x + 27$

81. $18x^2 - 50$ **82.** $5x^2 - 80$

83. $3x^2 - 30x + 75$ **84.** $12x^2 + 4x - 60$

85. $3x^2 + 6x + 12$ **86.** $10x^2 - 20x + 40$

Projects

87. ***CD Earnings.**** Eli and Shana are calculating the value of a certificate of deposit (CD) at the end of years 1 to 4. The rate of interest, r, is 4%. The starting amount of the CD, P, is \$500. Eli's formulas are shown in the first table. Shana's formulas are shown in the second table.

a. Find the amount of money after each year using the two formulas.

b. What do you observe about the two formulas? Can you prove your observation with multiplication?

c. Which table (set of formulas) would you rather use?

d. Use Shana's table to create a formula for t years.

Years, n	Rule	Amount, \$
1	$P + Pr$	
2	$P + 2Pr + Pr^2$	
3	$P + 3Pr + 3Pr^2 + Pr^3$	
4	$P + 4Pr + 6Pr^2 + 4Pr^3 + Pr^4$	

Eli's Formulas

Years, n	Rule	Amount, \$
1	$P(1 + r)$	
2	$P(1 + r)^2$	
3	$P(1 + r)^3$	
4	$P(1 + r)^4$	

Shana's Formulas

88. ***Cubic Polynomials.*** Polynomials with x^3 or a^3 as the highest term are called cubic expressions. Multiply the expressions in parts a to f to obtain cubic expressions, copying the original factors and then writing the result of the multiplication.

a. $(x + 1)(x^2 + 2x + 1)$

b. $(x - 2)(x^2 + 2x + 4)$

c. $(x + 3)(x^2 + 6x + 9)$

*This project was suggested by Charlotte Hutt, Southwestern Oregon Community College.

d. $(a + b)(a^2 + 2ab + b^2)$

e. $(a - 3)(a^2 + 3a + 9)$

f. $(a - b)(a^2 + ab + b^2)$

Complete the tables in parts a to j to factor the expressions in the tables. Write the factors as well as the original expression in a statement below each table.

g.

Factor			
	x^3	$-2x^2$	$+x$
	$-x^2$	$+2x$	-1

__________ = (   )(   )

h.

Factor			
	a^3	$-2a^2b$	$+ab^2$
	$-a^2b$	$+2ab^2$	$-b^3$

__________ = (   )(   )

i.

Factor			
	a^3	$-a^2b$	$+ab^2$
	a^2b	$-ab^2$	$+b^3$

__________ = (   )(   )

j.

Factor			
	x^3	$-2x^2$	$+4x$
	$+2x^2$	$-4x$	$+8$

__________ = (   )(   )

k. When we multiply $(x + 1)(x^2 + 2x + 1)$, our answer is like the answer to $(a - b)(a^2 - 2ab + b^2)$. What other three problems, a to j, have the same pattern? Use the pattern to factor $n^3 + 3n^2 + 3n + 1$ and $y^3 - 3y^2 + 3y - 1$.

l. When we multiply $(x - 2)(x^2 + 2x + 4)$, our answer is like the answer to $(a + b)(a^2 - ab + b^2)$. What other three problems, a to j, have the same pattern? Use the pattern to factor $x^3 - 1$, $x^3 + 64$, and $x^3 - 27$.

89. ***Binomial Powers.*** Multiply the binomials, using a table as needed.

a. $(a + b)^2 = (a + b)(a + b) =$

b. $(a + b)^3$

c. $(a + b)^4$

d. $(a + b)^5$

To check, use the fact that

$$(a + b)^6 = a^6 + 6a^5b + 15a^4b^2 + 20a^3b^3 + 15a^2b^4 + 6ab^5 + 1b^6$$

Multiply the binomials, using a table as needed.

e. $(x + 1)^2 = (x + 1)(x + 1) =$

f. $(x + 1)^3$

g. $(x + 1)^4$

h. $(x + 1)^5$

To check, use the fact that

$$(x + 1)^6 = x^6 + 6x^5 + 15x^4 + 20x^3 + 15x^2 + 6x + 1$$

i. Predict $(x + y)^4$, $(x + 2)^4$, and $(x - 1)^4$.

90. ***Pascal's Triangle Research.*** Research Pascal's triangle and how it fits the powers of binomials in Exercises 87 and 89.

91. ***Hardy-Weinberg Research.*** Research the Hardy-Weinberg principle in genetics or biology books. Related information may also be found under the *Punnett-square method.*

92. ***Polynomials and Volume.*** The volume of a cube with side x is x^3. The volume of a cube with side $(x + y)$ is $(x + y)^3$. The parts making up the volume may also be shown geometrically with stacks of boxes.

a. The figure below illustrates $(x + y)^3$. It is split into a front stack of boxes and a back stack of boxes, shown separately. Write out the volume for $(x + y)^3$.

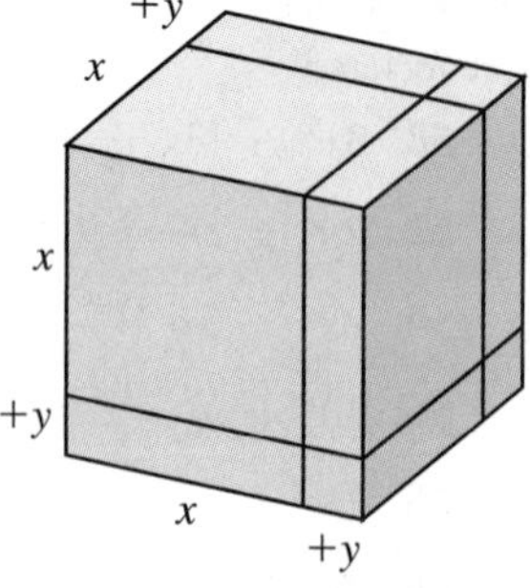

Front stack

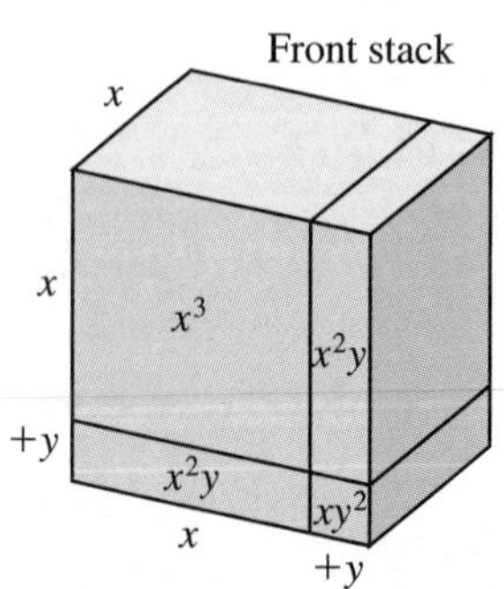

b. Sketch two stacks of boxes (front and back) that show $(x + 2)^3$. Multiply out $(x + 2)^3$ to find another expression for the volume.

c. Sketch three stacks of boxes that show $(x + y + z)^3$. Multiply out $(x + y + z)^3$ to find another expression for the volume.

6.4 Exponents and Scientific Notation

OBJECTIVES

- Simplify expressions containing zero and negative exponents.
- Simplify multiplication, division, and powers of exponential expressions.
- Change numbers between decimal notation and scientific notation.
- Estimate and use a calculator to perform operations in scientific notation.

WARM-UP

The following problems review exponent operations from Section 2.4. A summary of these operations appears on page 381. First, simplify and evaluate (if possible) the problems without a calculator; then check with a calculator.

1. $5^2 \cdot 5^3$	**2.** $2^6 \cdot 2^4$	**3.** $a^3 \cdot a^4$
4. $\dfrac{2^8}{2^3}$	**5.** $\dfrac{5^{10}}{5^8}$	**6.** $\dfrac{n^7}{n^2}$
7. $(3^3)^2$	**8.** $(4^3)^2$	**9.** $(b^5)^4$

IN SECTION 6.2, we used squares of binomials, $(x + y)^2$, to get perfect square trinomials, $x^2 + 2xy + y^2$. All polynomial expressions contain only positive integer exponents. We return to exponents in this section to extend the definition of exponents from positive integer exponents to zero and negative exponents. The last part of the section introduces scientific notation, a common application of negative exponents.

We examined the following definition of exponents in Section 2.4.

Positive Integer Exponents

When n is a positive integer exponent, n is the number of times the base, x, is used as a factor. Thus, $x^n = x \cdot x \cdot x \cdot x \cdot \cdots \cdot x$ has n factors of x.

Example 1 suggests the possibility of exponents other than positive integers.

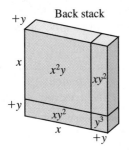

EXAMPLE 1 Exploring exponents Complete Table 2. The exponent on each number is 1 less than it was in the prior row. Look for a pattern between rows in the value column.

Exponent Expression	Value	Exponent Expression	Value	Exponent Expression	Value
10^3	1000	5^3	125	4^3	64
10^2	100	5^2		4^2	
10^1		5^1		4^1	
$10^{\square}$		$5^{\square}$		$4^{\square}$	
$10^{\square}$		$5^{\square}$		$4^{\square}$	
$10^{\square}$		$5^{\square}$		$4^{\square}$	
$10^{\square}$		$5^{\square}$		$4^{\square}$	

Table 2

Solution *Hint:* By what number do we divide 1000 to get 100? Does the pattern continue? See the Answer Box. ●

In Example 2, we use a calculator to explore the meaning of zero and negative exponents. This example suggests the possibility of fractions as exponents.

EXAMPLE 2 Exploring exponents Use a scientific calculator to complete Table 3. The exponent keys are [^], [y^x], and [x^y].

Expression	Value on Calculator
5^0	
$(-2)^0$	
$(1.5)^0$	
0^0	
4^{-1}	
3^{-1}	
2^{-1}	

Table 3

Solution See the Answer Box. ●

The calculator readily gives an answer for all the table entries except 0^0. We say that zero to the zero power is *undefined.*

We will use the following properties of exponents as we explore why zero and negative numbers can be used as exponents.

Operations with Exponents

To multiply numbers with like bases, add the exponents:

$x^a \cdot x^b = x^{a+b}$ Multiplication property

To divide numbers with like bases, subtract the exponents:

$\frac{x^a}{x^b} = x^{a-b}$ for $x \neq 0$ Division property

To apply an exponent to a power expression, multiply the exponents:

$(x^a)^b = x^{a \cdot b}$ Power property

An exponent outside the parentheses applies to all parts of a product or quotient inside the parentheses:

$(x \cdot y)^a = x^a \cdot y^a \quad \left(\frac{x}{y}\right)^a = \frac{x^a}{y^a}$ for $y \neq 0$ Base properties

Zero Exponents

For zero exponents,

$x^0 = 1$

The expression 0^0 is not defined.

EXAMPLE 3 Find the value of an expression with a zero exponent First simplify each expression to a number by using properties of fractions. Then simplify to an exponent expression by using the appropriate property.

a. $\frac{5^3}{5^3}$ **b.** $\frac{x^n}{x^n}$, where $x \neq 0$

Solution **a.** $\frac{5^3}{5^3} = 1$ by the simplification property of fractions.

$\frac{5^3}{5^3} = 5^{3-3} = 5^0$ by the division property.

The value of $5^0 = 1$.

b. $\frac{x^n}{x^n} = 1$, where $x \neq 0$, by the simplification property of fractions.

$\frac{x^n}{x^n} = x^{n-n} = x^0$, where $x \neq 0$, by the division property.

The value of $x^0 = 1$, where $x \neq 0$. ●

Example 4 reminds us that a zero exponent may be applied to any expression.

EXAMPLE 4 Simplifying with zero exponents Simplify these exponential expressions.

a. 4^0 **b.** $\left(\frac{1}{2}\right)^0$ **c.** $(x + y)^0$ **d.** x^5y^0 **e.** $(3x^2y^2 + 5x^{20})^0$

Solution Because 0^0 is undefined, we must avoid that expression.

a. $4^0 = 1$

b. $\left(\frac{1}{2}\right)^0 = 1$

c. If x and y do not add to zero, then $(x + y)^0 = 1$.

d. If $y \neq 0$, then $x^5y^0 = x^5$.

e. If the expression inside the parentheses does not equal zero, then $(3x^2y^2 + 5x^{20})^0 = 1$. ●

Negative Exponents

Negative Exponents

For negative one exponents,

$$x^{-1} = \frac{1}{x}$$

the reciprocal of x.

For negative two exponents,

$$x^{-2} = \frac{1}{x^2}$$

The square of the reciprocal of x^2.

In Example 5, we look at how the properties of exponents show the meaning of negative exponents.

EXAMPLE 5 Finding the value of negative exponents First simplify each expression to a number by using properties of fractions. Then simplify to an exponent expression by using the appropriate property.

a. $\dfrac{5^3}{5^4}$ **b.** $\dfrac{4^3}{4^5}$ **c.** $\dfrac{x^5}{x^6}, x \neq 0$ **d.** $\dfrac{x^5}{x^7}, x \neq 0$

Solution **a.** $\dfrac{5^3}{5^4} = \dfrac{1}{5}$ by the simplification property of fractions.

$\dfrac{5^3}{5^4} = 5^{3-4} = 5^{-1}$ by the division property.

Thus, $5^{-1} = \dfrac{1}{5}$.

b. $\dfrac{4^3}{4^5} = \dfrac{1}{4^2}$ by the simplification property of fractions.

$\dfrac{4^3}{4^5} = 4^{3-5} = 4^{-2}$ by the division property.

Thus, $4^{-2} = \dfrac{1}{4^2}$.

c. $\dfrac{x^5}{x^6} = \dfrac{1}{x}$ by the simplification property of fractions.

$\dfrac{x^5}{x^6} = x^{5-6} = x^{-1}$ by the division property.

Thus, $x^{-1} = \dfrac{1}{x}$, $x \neq 0$.

d. $\dfrac{x^5}{x^7} = \dfrac{1}{x^2}$ by the simplification property of fractions.

$\dfrac{x^5}{x^7} = x^{5-7} = x^{-2}$ by the division property.

Thus, $x^{-2} = \dfrac{1}{x^2}$, $x \neq 0$. ●

Example 6 ties the negative one exponent with the zero exponent.

EXAMPLE 6 Simplifying with negative exponents Simplify these exponential expressions as directed.

a. $5^1 \cdot 5^{-1}$, by adding exponents.

b. $5^1 \cdot 5^{-1}$, by changing to fractions.

c. $5^1 \cdot 5^{-1}$, with a calculator.

Solution **a.** $5^1 \cdot 5^{-1} = 5^{1+(-1)} = 5^0 = 1$

b. $5^1 \cdot 5^{-1} = 5 \cdot \dfrac{1}{5} = \dfrac{5}{5} = 1$

c. On a scientific calculator,

5 [y^x] 1 · 5 [y^x] 1 [±] = 1

On a graphing calculator,

5 [^] 1 · 5 [^] [±] 1 = 1 ●

Example 6 reminds us that the product of a number and its reciprocal is 1 and again shows that a number raised to the zero power equals one.

In Example 7, we practice using a negative one exponent on a variety of expressions.

EXAMPLE 7 Simplifying with a negative one exponent Simplify these expressions without a calculator. Write the answers as fractions or decimals, where appropriate.

a. 25^{-1} **b.** 9^{-1} **c.** $\left(\frac{2}{3}\right)^{-1}$ **d.** $(0.25)^{-1}$ **e.** 6.25^{-1}

f. 10^{-1} **g.** x^{-1} **h.** $\dfrac{1}{x^{-1}}$ **i.** $\left(\dfrac{a}{b}\right)^{-1}$ **j.** $(x + y)^{-1}$

Solution **a.** $25^{-1} = \frac{1}{25} = 0.04$

b. $9^{-1} = \frac{1}{9} = 0.1111\ldots$

c. $\left(\frac{2}{3}\right)^{-1} = \left(\frac{3}{2}\right) = 1.5$

d. $(0.25)^{-1} = \left(\frac{1}{4}\right)^{-1} = \frac{4}{1} = 4$

e. $6.25^{-1} = \left(\frac{25}{4}\right)^{-1} = \frac{4}{25} = 0.16$

f. $10^{-1} = \frac{1}{10} = 0.1$

g. $x^{-1} = \dfrac{1}{x}$

h. $\dfrac{1}{x^{-1}} = \dfrac{1}{\frac{1}{x}} = 1 \div \dfrac{1}{x} = 1 \cdot \dfrac{x}{1} = x$

i. $\left(\dfrac{a}{b}\right)^{-1} = \dfrac{b}{a}$

j. $(x + y)^{-1} = \dfrac{1}{x + y}$ ●

Calculator Note: The expression x^{-1}, or $1/x$, is the formal mathematical expression for the reciprocal of x. Scientific calculators have a key for reciprocals, $\boxed{x^{-1}}$ or $\boxed{1/x}$. To practice using the reciprocal key, repeat Example 7, parts a to f, with a calculator.

In Example 8, we look at numbers related by a negative two exponent.

EXAMPLE 8 Exploring exponents

a. With a calculator, verify the equality of the two expressions:

$$\left(\frac{25}{4}\right)^{-1} \text{ and } \left(\frac{4}{25}\right)$$

b. Write the numbers as exponential expressions containing 2 as an exponent.

c. What do the numbers suggest about a negative two exponent?

Solution **a.** Both expressions are equivalent to 0.16.

b. $$\left(\frac{25}{4}\right)^{-1} = \left(\frac{5^2}{2^2}\right)^{-1} = \frac{5^{-2}}{2^{-2}} = \left(\frac{5}{2}\right)^{-2}$$

$$\left(\frac{4}{25}\right) = \frac{2^2}{5^2} = \left(\frac{2}{5}\right)^2$$

c. The equality of the original expressions indicates that $\left(\frac{5}{2}\right)^{-2} = \left(\frac{2}{5}\right)^2$. The negative 2 exponent on $\frac{5}{2}$ gives the reciprocal of $\frac{5}{2}$ squared. ●

Example 8 shows how a negative two exponent combines both the reciprocal and the squaring function.

We can extend our definition of negative exponents to any number, located in the numerator or denominator.

Negative-Number Exponents

For any negative real-number exponents,

$$x^{-n} = \frac{1}{x^n}$$

If x is in fraction notation, then

$$\left(\frac{a}{b}\right)^{-n} = \left(\frac{b}{a}\right)^n$$

For any negative real-number exponent in the denominator,

$$\frac{1}{x^{-n}} = x^n$$

In Example 9, we simplify expressions containing -2 and -3 as exponents.

EXAMPLE 9 Simplifying exponent expressions Simplify these expressions by replacing negative exponent expressions with equivalent positive expressions.

a. $\left(\frac{2x}{3}\right)^{-2}$ **b.** $\left(\frac{4x^2}{y^3}\right)^{-2}$ **c.** $\left(\frac{1}{a}\right)^{-2}$

d. $\frac{1}{a^{-2}}$ **e.** $\frac{b}{a^{-3}}$ **f.** $\left(\frac{b}{a}\right)^{-3}$

Solution

a. $\left(\dfrac{2x}{3}\right)^{-2} = \left(\dfrac{3}{2x}\right)^{2} = \dfrac{3^2}{(2x)^2} = \dfrac{9}{4x^2}$

b. $\left(\dfrac{4x^2}{y^3}\right)^{-2} = \left(\dfrac{y^3}{4x^2}\right)^{2} = \dfrac{(y^3)^2}{(4x^2)^2} = \dfrac{y^6}{16x^4}$

c. $\left(\dfrac{1}{a}\right)^{-2} = \left(\dfrac{a}{1}\right)^{2} = a^2$

d. $\dfrac{1}{a^{-2}} = 1 \div a^{-2} = 1 \div \dfrac{1}{a^2} = 1 \cdot \dfrac{a^2}{1} = a^2$

e. $\dfrac{b}{a^{-3}} = b \div a^{-3} = b \div \dfrac{1}{a^3} = b \cdot a^3 = a^3b$

f. $\left(\dfrac{b}{a}\right)^{-3} = \left(\dfrac{a}{b}\right)^{3} = \dfrac{a^3}{b^3}$ ●

In Example 10, we perform a variety of operations with negative exponents.

EXAMPLE  10

Simplifying exponent expressions Use the appropriate property to simplify these expressions.

a. $x^{-4} \cdot x^7$ **b.** $\dfrac{x}{x^{-2}}$ **c.** $\dfrac{b^3}{b^{-3}}$

d. $\dfrac{xy^{-3}}{x^2y}$ **e.** $\dfrac{a^3b^{-1}}{a^{-2}b^3}$ **f.** $(x^{-5})^2$

Solution

a. $x^{-4} \cdot x^7 = x^{-4+7} = x^3$

b. $\dfrac{x}{x^{-2}} = x^{1-(-2)} = x^3$

c. $\dfrac{b^3}{b^{-3}} = b^{3-(-3)} = b^6$

d. $\dfrac{xy^{-3}}{x^2y} = x^{1-2}y^{-3-1} = x^{-1}y^{-4} = \dfrac{1}{xy^4}$

e. $\dfrac{a^3b^{-1}}{a^{-2}b^3} = a^{3-(-2)}b^{-1-3} = a^5b^{-4} = \dfrac{a^5}{b^4}$

f. $(x^{-5})^2 = x^{-5\cdot 2} = x^{-10} = \dfrac{1}{x^{10}}$ ●

When we simplify exponent expressions, we leave answers with positive exponents.

Scientific Notation

One of the most common uses of negative exponents is in scientific notation. **Scientific notation** is *a short way of writing large and small numbers,* such as the distance to a star or the size of a virus. Calculators automatically change into scientific notation when the number is too large or too small to fit the answer display, as in Example 11.

EXAMPLE 11

Exploring large and small numbers Use a scientific calculator to evaluate these numbers.

a. $(3600)^4$ **b.** $(0.00063)^5$

Solution The calculator answers are in an unexpected form. They are written in scientific notation. There is some variation among calculators, but the scientific notation display will probably appear in one of these two forms:

a. 1.6796 14 or 1.6796E14

b. 9.9244 −17 or 9.9244E−17 ●

Scientific notation is based on the powers of 10. Numbers written in scientific notation have two parts, as shown below.

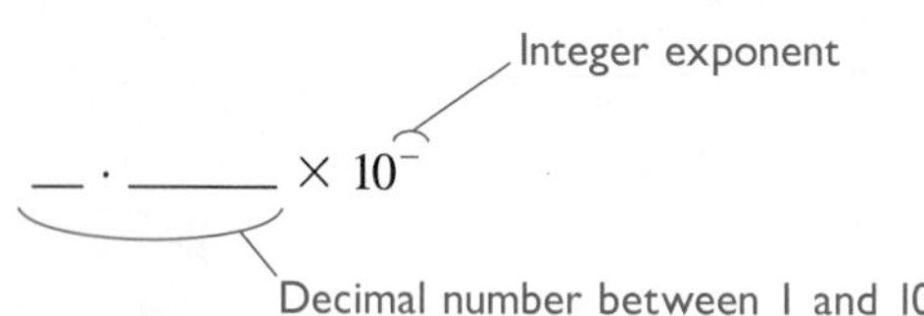

The first part is a decimal with one nonzero number to the left of the decimal point. The second part is a power of 10. The decimal number is multiplied by the power of 10. Calculator displays show the decimal part as well as the exponent but omit the symbol for multiplication and the base 10.

Note: We use the × sign for multiplication in scientific notation. The × is not a variable.

EXAMPLE 12 Exploring products and quotients with powers of ten Predict how many decimal places to the left or right the decimal point will move in each case. Do the operations without a calculator.

a. 34.6 × 10 **b.** 34.6 × 100 **c.** 2.78(1000)

d. 2.78(0.0001) **e.** 1.6(100,000) **f.** 1.6(0.000 001)

g. 219.1 ÷ 10 **h.** 219.1 ÷ 1000 **i.** 57.8 ÷ 0.01

j. 57.8 ÷ 0.0001

Solution

a. 1 place right; 346

b. 2 places right; 3460

c. 3 places right; 2780

d. 4 places left; 0.000 278

e. 5 places right; 160,000

f. 6 places left; 0.000 001 6

g. 1 place left; 21.91

h. 3 places left; 0.2191

i. 2 places right; 5780

j. 4 places right; 578,000 ●

Table 4 gives examples of scientific notation in a variety of settings.

We need to be able to change between decimal notation and scientific notation in order to interpret answers from a calculator and enter large and small numbers into a calculator.

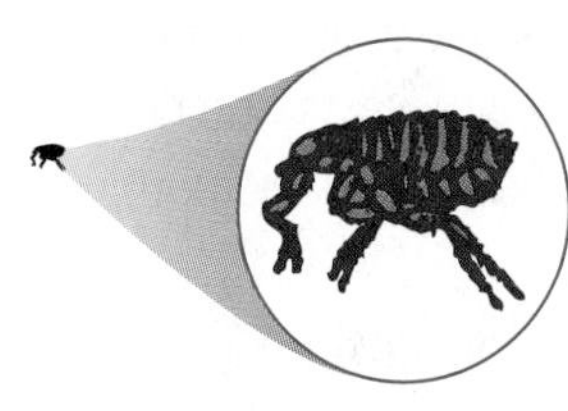

Power of 10	Value	Scientific Notation	Application
10^9	1,000,000,000 (billion)	4.6×10^9 years ago	Estimated formation of Earth
10^8	100,000,000	9.83×10^8 feet per second	Speed of light in a vacuum
10^7	10,000,000	2.46×10^7 years	Half-life of ^{236}U, radioactive uranium
10^6	1,000,000	6.38×10^6 meters	Radius of Earth
10^5	100,000	5.256×10^5 minutes	Minutes in a 365-day year
10^4	10,000	8.64×10^4 seconds	Seconds in one day
10^3	1,000	5.73×10^3 years	Half-life of ^{14}C, radioactive carbon
10^2	100	1×10^2 meters	Length of 100-meter dash
10^1	10		
10^0	1	8.04×10^0 days	Half-life of ^{131}I, radioactive iodine
10^{-1}	$1/10 = 0.1$		
10^{-2}	$1/100 = 0.01$		
10^{-3}	$1/1000 = 0.001$	3×10^{-3} meter	Size of a flea (3 mm)
10^{-7}	0.000 000 1	1×10^{-7} meter	Length of HIV virus
10^{-10}		1×10^{-10} meter	1 angstrom, measure of distance between atoms
10^{-27}		1.66×10^{-27} kilogram	Unit of atomic mass

Table 4 Applications of Scientific Notation

DECIMAL NOTATION TO SCIENTIFIC NOTATION Here are some things to keep in mind.

Changing Decimal Notation to Scientific Notation

To change to scientific notation:

1. Place the decimal point after the first nonzero digit.
2. Count the number of places the decimal point has moved, and use that number as the number exponent on ten.
3. a. If the original number is greater than 10, the exponent on ten is positive.
   b. If the original number is between 10 and 1, the exponent on ten is zero.
   c. If the original number is between 0 and 1, the exponent on ten is negative.

In Example 13, we apply the rules for changing to scientific notation. Example 14 shows more realistic applications of scientific notation.

EXAMPLE 13 Changing decimal numbers to scientific notation Change each number to scientific notation.

a. 0.5 **b.** 1.5 **c.** 10.5 **d.** 105

Solution **a.** $0.5 = 5.0 \times 10^{-1}$

The first nonzero digit is 5. The decimal moves 1 place, so we place a 1 in the exponent on ten. The number is between 0 and 1, so the exponent on ten is negative.

b. $1.5 = 1.5 \times 10^{0}$

The first nonzero digit is 1. The decimal does not move, so we place a 0 in the exponent on ten. The number is between 1 and 10, which provides a check that the exponent on ten is zero.

c. $10.5 = 1.05 \times 10^{1}$

The first nonzero digit is 1. The decimal moves 1 place, so we place a 1 in the exponent on ten. The number is greater than 10, so the exponent on ten is positive.

d. $105 = 1.05 \times 10^{2}$

The first nonzero digit is 1. The decimal moves 2 places, so we place a 2 in the exponent on ten. The number is greater than 10, so the exponent on ten is positive. ●

EXAMPLE 14 Changing numbers to scientific notation Predict the exponent on ten needed to change each number to scientific notation. Then change the number.

a. The minimum distance from the sun to Mercury is 28,600,000 miles.

b. The maximum distance from the sun to Mercury is 43,400,000 miles.

c. The maximum distance from Earth to Pluto is 4,644,000,000 miles.

d. The mass of an electron is

0.000 000 000 000 000 000 000 000 000 000 910 9 kilogram

e. The mass of a proton is

0.000 000 000 000 000 000 000 000 001 672 6 kilogram

Solution **a.** The decimal point will move 7 places. The number is larger than ten. The exponent on ten will be positive.

2.86×10^{7} miles

b. The decimal point will move 7 places. The number is larger than ten. The exponent on ten will be positive.

4.34×10^{7} miles

c. The decimal point will move 9 places. The number is larger than ten. The exponent on ten will be positive.

4.644×10^{9} miles

d. The decimal point will move 31 places. The number is between 0 and 1. The exponent on ten will be negative.

9.109×10^{-31} kilogram

e. The decimal point will move 27 places. The number is between 0 and 1. The exponent on ten will be negative.

1.6726×10^{-27} kilogram ●

In Example 15, we look at some common errors in writing scientific notation.

EXAMPLE  Finding errors in scientific notation Explain why each of the following is not in standard scientific notation. Guess what number is intended by each expression, and restate it in correct scientific notation.

a. $4.82\ 10^{11}$ **b.** 5.23^5 **c.** 34.6×10^{12} **d.** 0.465×10^{-2}

Solution

a. The multiplication sign was left out. The correct form is 4.82×10^{11}.

b. The $\times$ 10 was left out. Because many calculators display scientific notation answers in this form, this is a common error. The correct form is 5.23×10^5.

c. Correct scientific notation has only one nonzero digit to the left of the decimal point. The correct form is 3.46×10^{13}.

d. Correct scientific notation has one nonzero digit to the left of the decimal point. The correct form is 4.65×10^{-3}. ●

SCIENTIFIC NOTATION TO DECIMAL NOTATION Here are some things to keep in mind.

Changing Scientific Notation to Decimal Notation

To change from scientific notation to decimal notation:

1. Look at the exponent on ten to find the number of decimal places and the direction in which to move the decimal point.
2. **a.** If the exponent is negative, move the decimal point left to make a number between 0 and 1.
   b. If the exponent is zero, do not change the decimal point.
   c. If the exponent is positive, move the decimal point right to make a number greater than 10.

In Example 16, we change scientific notation to decimal notation.

EXAMPLE  Changing scientific notation to decimal notation Predict the direction in which the decimal point will move and then change each number to decimal notation.

a. The minimum distance from the sun to Earth is 9.14×10^7 miles.

b. In chemistry, Avogadro's number measures the number of molecules in a mole: 6.022×10^{23}.

c. The estimated human population of the world in 1990 was 5.333×10^9.

d. The human population on Earth in 1650 is estimated at 5.5×10^8.

e. The mass of a house spider is about 1×10^{-4} kilogram.

Solution *Remember*: Positive exponents on ten go with numbers greater than 10, and negative exponents on ten go with numbers between 0 and 1.

a. 91,400,000 miles

b. 602,200,000,000,000,000,000,000 molecules per mole

c. 5,333,000,000 people

d. 550,000,000 people

e. 0.0001 kilogram ●

UNIT ANALYSIS RESULTING IN SCIENTIFIC NOTATION In Example 17, unit analysis yields numbers in scientific notation.

EXAMPLE 17 Using scientific notation with unit analysis Apply unit analysis to the following problems. Write answers in scientific notation.

a. The speed of light is 186,000 miles per second. How many miles does light travel in one year? This distance is called a light-year.

b. If a giant sequoia (redwood) is 360 feet tall and estimated to be 6000 years old, what is its average growth rate in miles per hour?

Solution **a.** First we list facts:

speed of light = 186,000 mi per sec
60 sec = 1 min
60 min = 1 hr
24 hr = 1 day
365 days = 1 yr (rounded)

We make a unit analysis, starting with the speed of light in seconds and changing to miles per year.

$$\frac{186{,}000 \text{ mi}}{1 \text{ sec}} \cdot \frac{60 \text{ sec}}{1 \text{ min}} \cdot \frac{60 \text{ min}}{1 \text{ hr}} \cdot \frac{24 \text{ hr}}{1 \text{ day}} \cdot \frac{365 \text{ days}}{1 \text{ yr}}$$

A calculator shows that a light-year is 5.87×10^{12} miles.

b. First we list facts:

1 mi = 5280 ft
24 hr = 1 day
365 days = 1 yr (rounded)

We make a unit analysis, starting with 360 feet in 6000 years. On the left, we change feet to miles, and on the right, we change years to hours.

$$\frac{1 \text{ mi}}{5280 \text{ ft}} \cdot \frac{360 \text{ ft}}{6000 \text{ yr}} \cdot \frac{1 \text{ yr}}{365 \text{ days}} \cdot \frac{1 \text{ day}}{24 \text{ hr}}$$

A calculator shows that the average growth rate is about 1.30×10^{-9} mph.

●

OPERATIONS WITH SCIENTIFIC NOTATION It is important to know how to multiply and divide numbers in scientific notation both by estimating and by using a calculator. Both are important skills, as estimation is often easier with scientific notation and the data for many application problems are written in scientific notation.

In estimating, we perform operations on the two parts of the scientific notation separately.

Estimating Using Scientific Notation

1. **a.** To estimate the product of two numbers in scientific notation, multiply the decimals and then add the exponents on ten.
   b. To estimate the quotient of two numbers in scientific notation, divide the decimals and then subtract the exponents on ten.
2. If the product or the quotient of the decimals is greater than 10 or between 0 and 1, adjust the decimal point and the exponent on ten.

EXAMPLE 18 **Performing operations with scientific notation mentally** Use estimation techniques to do these operations mentally.

a. $(1.2 \times 10^3) \cdot (3.0 \times 10^2)$ **b.** $(1.5 \times 10^{-2}) \cdot (4.0 \times 10^5)$

c. $(1.5 \times 10^{-4}) \div (5.0 \times 10^2)$ **d.** $(2.4 \times 10^8) \div (3.0 \times 10^{-3})$

e. $\dfrac{1.6 \times 10^{-4}}{(2.0 \times 10^{12})(5.0 \times 10^{-8})}$

Solution

a. 1.2 times 3.0 is 3.6, and 10^3 times 10^2 is 10^5. The answer is 3.6×10^5, or 360,000.

b. 1.5 times 4.0 is 6.0, and 10^{-2} times 10^5 is 10^3. The answer is 6.0×10^3, or 6000.

c. 1.5 divided by 5.0 is 0.3, and 10^{-4} divided by 10^2 is 10^{-6}. The answer is

$$0.3 \times 10^{-6} = 3.0 \times 10^{-1} \times 10^{-6} = 3 \times 10^{-7}$$

d. 2.4 divided by 3.0 is 0.8, and 10^8 divided by 10^{-3} is 10^{11}. The answer is

$$0.8 \times 10^{11} = 8.0 \times 10^{-1} \times 10^{11} = 8.0 \times 10^{10}$$

e. 1.6 divided by the product of 2 and 5 is 0.16. 10^{-4} divided by the product of 10^{12} and 10^{-8} is $10^{-4-12-(-8)} = 10^{-8}$. The answer is

$$0.16 \times 10^{-8} = 1.6 \times 10^{-1} \times 10^{-8} = 1.6 \times 10^{-9}$$ ●

Calculators are designed to do operations with scientific notation. We use the keys [EE], [EXP], or [EEX]—but not [e^x]—to enter a number in scientific notation. The letters EE represent "*e*nter the *e*xponent on 10." Negative exponents may need to have the negative sign, [+/−], entered after the exponent. Many calculators permit a negative, [(−)], before the exponent. Experiment with your calculator.

EXAMPLE  **Using a calculator for scientific notation operations** Repeat the operations from Example 18 with a calculator.

Solution

a. 1.2 [EE] 3 [×] 3.0 [EE] 2 [=] 3.6×10^5
Some calculators give 360,000.

b. 1.5 [EE] 2 [+/−] [×] 4.0 [EE] 5 [=] 6.0×10^3
The calculator answer may be 6000.

c. 1.5 [EE] 4 [+/−] [÷] 5.0 [EE] 2 [=] 3.0×10^{-7}
The calculator answer may be 0.000 000 3.

d. 2.4 [EE] 8 [÷] 3.0 [EE] 3 [+/−] [=] 8.0×10^{10}
This is 80,000,000,000, but it is almost always shown in scientific notation.

e. 1.6 [EE] 4 [±] [÷] (2.0 [EE] 12 [×] 5.0 [EE] 8 [±]) [=] 1.6×10^{-9}
The calculator answer may be 0.000 000 001 6. ●

Think about it: When a number is placed into a calculator in scientific notation using the [EE] or [EXP] key, the entire number is considered to be one number, as in parts c and d of Example 19. Repeat these two parts using multiplication by 10 with the [^] or [y^x] key and no parentheses around the denominator. Explain the results.

ANSWER BOX

Warm-up: 1. 5^5, 3125 **2.** 2^{10}, 1024 **3.** a^7 **4.** 2^5, 32 **5.** 5^2, 25 **6.** n^5 **7.** 3^6, 729 **8.** 4^6, 4096 **9.** b^{20}

Example 1:

Exponent Expression	Value	Exponent Expression	Value	Exponent Expression	Value
10^3	1000	5^3	125	4^3	64
10^2	100	5^2	25	4^2	16
10^1	10	5^1	5	4^1	4
10^0	1	5^0	1	4^0	1
10^{-1}	1/10	5^{-1}	1/5	4^{-1}	1/4
10^{-2}	1/100	5^{-2}	1/25	4^{-2}	1/16
10^{-3}	1/1000	5^{-3}	1/125	4^{-3}	1/64

Example 2: $5^0 = 1$; $(-2)^0 = 1$; $(1.5)^0 = 1$; 0^0 is undefined; $4^{-1} = 0.25$; $3^{-1} = 0.333...$; $2^{-1} = 0.5$ **Think about it:** Incorrect answers are obtained. For c, we get 0.003; for d, 80,000. Without any parentheses around the denominator, the last multiplication is performed as though it were in the numerator.

EXERCISES 

Complete the input-output tables in Exercises 1 to 4. Look for a pattern down the output column. Change decimals to fractions.

1.

Input: x	Output: 3^x
2	
1	
0	
−1	
−2	
−3	

2.

Input: x	Output: 2^x
2	
1	
0	
−1	
−2	
−3	

3.

Input: x	Output: 4^x
2	
1	
0	
−1	
−2	
−3	

4.

Input: x	Output: 5^x
2	
1	
0	
−1	
−2	
−3	

Simplify the expressions in Exercises 5 to 10 without a calculator.

5. a. 2^{-2} **b.** 2^{-1} **c.** 2^0

6. a. $\left(\frac{1}{2}\right)^{-1}$ **b.** $\left(\frac{1}{2}\right)^{0}$ **c.** $\left(\frac{1}{2}\right)^{-2}$

7. a. $\left(\frac{1}{4}\right)^{0}$ **b.** $\left(\frac{1}{4}\right)^{-2}$ **c.** $\left(\frac{1}{4}\right)^{-1}$

8. a. $\left(\frac{3}{4}\right)^{0}$ **b.** $\left(\frac{3}{4}\right)^{-1}$ **c.** $\left(\frac{3}{4}\right)^{-2}$

9. a. $(0.5)^{-1}$ **b.** $(0.5)^0$ **c.** $(0.5)^{-2}$

10. a. $(0.25)^{-2}$ **b.** $(0.25)^{-1}$ **c.** $(0.25)^0$

11. Which exponent gives the reciprocal of a number?

12. Which exponent gives the square of the reciprocal of a number?

13. Which exponent gives 1 unless the base is zero?

14. Name a number with a reciprocal smaller than the original number. (*Hint*: Look at the reciprocal problems in Exercises 5 to 10.)

15. Name a number with a reciprocal larger than the original number.

16. Name a number with a reciprocal equal to the original number.

Simplify Exercises 17 to 24 using the appropriate property and then check with a calculator. Leave answers without denominators.

17. **a.** $5^3 \cdot 5^{-7}$ **b.** $6^5 \cdot 6^{-2}$ **c.** $10^3 \cdot 10^{-6}$ **d.** $10^2 \cdot 10^{-7}$

18. **a.** $10^5 \cdot 10^{-4}$ **b.** $10^{-4} \cdot 10^6$ **c.** $2^6 \cdot 2^{-3}$ **d.** $2^{-4} \cdot 2^2$

19. **a.** $10^{-15} \cdot 10^{-15}$ **b.** $10^{-28} \cdot 10^{19}$ **c.** $2^{24} \cdot 2^{-16}$ **d.** $2^{13} \cdot 2^{-5}$

20. **a.** $5^{-14} \cdot 5^6$ **b.** $6^8 \cdot 6^{-15}$ **c.** $10^{-22} \cdot 10^{14}$ **d.** $10^{12} \cdot 10^{-21}$

21. **a.** $\frac{3^5}{3^{-2}}$ **b.** $\frac{10^{-5}}{10^{-2}}$ **c.** $\frac{10^{-4}}{10^{12}}$ **d.** $\frac{1}{10^2}$

22. **a.** $\frac{10^{15}}{10^{-8}}$ **b.** $\frac{2^{-3}}{2^{-4}}$ **c.** $\frac{10^{-6}}{10^7}$ **d.** $\frac{1}{5^2}$

23. **a.** $(2^3)^4$ **b.** $(2^3)^{-4}$ **c.** $(10^5)^2$ **d.** $(10^{-6})^3$

24. **a.** $(2^{-3})^4$ **b.** $(2^{-3})^{-4}$ **c.** $(10^{-5})^{-1}$ **d.** $(10^4)^{-5}$

Simplify Exercises 25 to 32 using the appropriate property. Leave answers without denominators.

25. **a.** $a^5 \cdot a^{-12}$ **b.** $x^{-5} \cdot x^{13}$ **c.** $n^{-6} \cdot n^2$

26. **a.** $x^{-2} \cdot x^{-8}$ **b.** $n^{-9} \cdot n^{15}$ **c.** $a^{12} \cdot a^{-7}$

27. **a.** $\frac{1}{x^2}$ **b.** $\frac{1}{a^{-1}}$ **c.** $\frac{b}{b^{-2}}$

28. **a.** $\frac{x}{x^2}$ **b.** $\frac{a}{a^{-1}}$ **c.** $\frac{1}{b^{-3}}$

29. **a.** $\frac{a^3}{a^{-6}}$ **b.** $\frac{a^{-6}}{a^2}$ **c.** $\frac{x^4}{x^{-2}}$

30. **a.** $\frac{x^{-2}}{x^{-3}}$ **b.** $\frac{a^6}{a^{-3}}$ **c.** $\frac{a^{-5}}{a^3}$

31. **a.** $(x^2)^{-4}$ **b.** $(x^{-4})^{-3}$ **c.** $(b^{-2})^3$

32. **a.** $(b^{-4})^{-1}$ **b.** $(x^{-2})^5$ **c.** $(x^6)^{-3}$

In Exercises 33 to 36, simplify the expressions to remove negative and zero exponents.

33. **a.** x^{-1} **b.** $\left(\frac{x}{y}\right)^{-1}$ **c.** $\left(\frac{y}{x}\right)^{-1}$ **d.** $\left(\frac{a}{b}\right)^0$ **e.** $\left(\frac{a}{c}\right)^0$ **f.** $\left(\frac{a}{bc}\right)^{-1}$

34. **a.** y^{-1} **b.** $\left(\frac{b}{c}\right)^{-1}$ **c.** $\left(\frac{c}{b}\right)^{-1}$ **d.** $\left(\frac{c}{b}\right)^0$ **e.** $\left(\frac{x}{y}\right)^0$ **f.** $\left(\frac{c}{ab}\right)^{-1}$

35. **a.** y^{-3} **b.** $\left(\frac{y}{x}\right)^{-2}$ **c.** $\frac{1}{b^{-3}}$ **d.** $\left(\frac{a}{b}\right)^{-3}$ **e.** $\left(\frac{4a^2}{c}\right)^{-2}$ **f.** $\left(\frac{a}{b^2}\right)^{-3}$

36. **a.** x^{-3} **b.** $\left(\frac{c}{b}\right)^{-2}$ **c.** $\frac{1}{x^{-2}}$ **d.** $\left(\frac{c}{b}\right)^{-3}$ **e.** $\left(\frac{2x^2}{y}\right)^{-2}$ **f.** $\left(\frac{2c}{a^2b}\right)^{-3}$

Simplify Exercises 37 to 40 using the appropriate property. Leave answers with no zero or negative exponents.

37. **a.** $\frac{xy^{-2}}{x^2y^3}$ **b.** $\frac{x^{-1}y}{x^{-1}y^{-2}}$ **c.** $\frac{a^{-2}b^2}{a^3b^{-1}}$

38. **a.** $\frac{a^{-3}b^2}{ab^4}$ **b.** $\frac{x^2y^{-2}}{x^{-3}y^{-2}}$ **c.** $\frac{x^3y^{-2}}{x^{-1}y^3}$

39. **a.** $1(2x)^2(-3)^0 + 2(2x)^1(-3)^1 + 1(2x)^0(-3)^2$

b. $1(2x)^3(3)^0 + 3(2x)^2(3)^1 + 3(2x)^1(3)^2 + 1(2x)^0(3)^3$

c. $1(3x)^3(-2)^0 + 3(3x)^2(-2)^1 + 3(3x)^1(-2)^2 + 1(3x)^0(-2)^3$

40. **a.** $1(3x)^4(-2)^0 + 4(3x)^3(-2)^1 + 6(3x)^2(-2)^2 + 4(3x)^1(-2)^3 + 1(3x)^0(-2)^4$

b. $1(x)^3(\tfrac{1}{2})^0 + 3(x)^2(\tfrac{1}{2})^1 + 3(x)^1(\tfrac{1}{2})^2 + 1(x)^0(\tfrac{1}{2})^3$

c. $1(x)^3(-\tfrac{1}{2})^0 + 3(x)^2(-\tfrac{1}{2})^1 + 3(x)^1(-\tfrac{1}{2})^2 + 1(x)^0(-\tfrac{1}{2})^3$

41. Simplify these expressions. What do you observe?

a. $(x^{-1})^{-1}$, by multiplying exponents

b. $\frac{1}{x^{-1}}$

c. $\frac{1}{\frac{1}{x}}$, by division of fractions

42. Multiply these expressions. Describe the pattern.

a. $\frac{3}{4} \cdot \frac{4}{3}$ **b.** $\frac{8}{5} \cdot \frac{5}{8}$ **c.** $\frac{12}{7} \cdot \frac{7}{12}$

d. $a^{-1} \cdot a^1$ **e.** $b^2 \cdot \left(\frac{1}{b^2}\right)$ **f.** $(ab)^{-2} \cdot (ab)^2$

43. Complete the table.

x	10^x as fraction	10^x as decimal
0		
−1		
−2		
−3		
−4		
−5		

In Exercises 44 to 46, use a calculator and record answers in scientific notation. Round the decimal part to two decimal places.

44. a. 0.0012^2 **b.** $500{,}000^2$

45. a. 5280^3 **b.** 0.00008^3

46. a. 0.0054^2 **b.** $123{,}456^2$

Change the "calculator outputs" in Exercises 47 and 48 from scientific notation to decimals.

47. a. 2.34 −02 **b.** 3.14 03 **c.** 6.28 07

48. a. 9.01 −03 **b.** 4.56 08 **c.** 2.41 12

In Exercises 49 to 54, change each number to scientific notation (Hint: 1 million = 1,000,000.) Round the decimal part to two decimal places.

49. The diameter of the sun is 1,391,400 kilometers.

50. The maximum distance from the sun to Earth is 94.6 million miles.

51. The minimum distance from the sun to Pluto is 2756.4 million miles.

52. The mass of a bacteria is 0.000 000 000 000 1 kilogram.

53. The mass of a chicken is 1800 grams.

54. The mass of an average polar bear is 322 kilograms.

The numbers in Exercises 55 to 58 are not in standard scientific notation. Correct them.

55. a. 34×10^3 **b.** 560×10^{-2}

56. a. 450×10^{-1} **b.** 6700×10^4

57. a. 0.432×10^4 **b.** 0.567×10^{-5}

58. a. 0.025×10^{-3} **b.** 0.0042×10^4

In Exercises 59 to 66, change from scientific notation to decimal notation.

59. The mass of the sun is 1.99×10^{30} kilograms.

60. The mass of Earth is 5.98×10^{24} kilograms.

61. An electron charge is -1.602×10^{-19} coulomb.

62. The gravitational constant is 6.672×10^{-11} N · m^2/kg^2.

63. Dinosaurs first appeared on Earth about 2.0×10^8 years ago.

64. Dinosaurs were extinct by 6.5×10^7 years ago.

65. The mass of a neutron is 1.6750×10^{-27} kilogram.

66. The projected human population of Earth for 2025 is 8.17×10^9.

Place an inequality sign between the numbers in Exercises 67 to 72 to make a true statement.

67. a. (2×10^5)___(3×10^4)

b. (4×10^3)___(3×10^4)

68. a. (2.5×10^5)___(3.5×10^5)

b. (4.6×10^5)___(3.6×10^3)

69. a. (2×10^{-5})___(3×10^{-4})

b. (4×10^{-3})___(3×10^{-4})

70. a. (2×10^{-15})___(2×10^{-14})

b. (3×10^{-13})___(3×10^{-14})

71. a. (3.2×10^{-14})___(2.8×10^{-14})

b. (4.3×10^{-13})___(5.3×10^{-13})

72. a. (1.2×10^{-19})___(2.4×10^{-19})

b. (1.8×10^{-18})___(1.2×10^{-18})

73. Which has a smaller mass, an electron at 9.101×10^{-31} kilogram or a proton at 1.6726×10^{-27} kilogram?

74. Which is closer to the sun, Neptune at 1.860×10^9 miles or Saturn at 9.375×10^8 miles?

In Exercises 75 to 82, solve with a unit analysis. Assume 365 days per year. Round the decimal part to two decimal places.

75. How many miles does light travel in the 27,000 light-years to the M13 star cluster mentioned in the chapter opener?

76. How fast in feet per second is a silver birch tree growing if it is 30 feet tall after 15 years of growth?

77. The Ohio class submarine displaces 18,700 tons of water. How many quarts of water is this? (*Hint*: 2000 lb = 1 ton, 1 gal ≈ 8.3 lb of water. List other facts as needed.)

78. A basketball court is 26 meters by 14 meters. How many basketball courts would it take to cover the entire surface of the Earth at 510,070,000 square kilometers?

79. In 1993, Georgia farmers produced 1,360,000,000 pounds of peanuts. If there are 16 ounces per pound and 160 calories per ounce for dry-roasted unsalted peanuts, how many calories were in this peanut crop? How many calories per person did the crop represent if the population of the world at that time was 5.6 billion?

80. Suppose a 20-foot record-height tomato plant took 6 months to grow. What was the tomato's growth rate in inches per minute? Assume 30 days per month.

81. If an average boy grows from 20 inches to 5 feet 9 inches in 18 years, what is the average growth rate in miles per hour?

82. If an average girl grows from 19 inches to 5 feet 4 inches in 17 years, what is the average growth rate in miles per hour?

In Exercises 83 to 90, estimate the products and quotients without a calculator.

83. a. $(2 \times 10^{15})(3 \times 10^{12})$

b. $(4 \times 10^{13})(3 \times 10^{18})$

84. a. $(6 \times 10^{14})(2 \times 10^{-19})$

b. $(8 \times 10^{-12})(5 \times 10^{17})$

85. a. $(5 \times 10^{-16})(6 \times 10^{-11})$

b. $(8 \times 10^{-12})(6 \times 10^{-13})$

86. a. $(8 \times 10^{14}) \div (2 \times 10^{11})$

b. $(2.0 \times 10^{12}) \div (5 \times 10^{17})$

87. a. $(2.8 \times 10^{-14}) \div (7 \times 10^{-18})$

b. $(9 \times 10^{-25}) \div (4.5 \times 10^{-12})$

88. a. $(6 \times 10^{13}) \div (3 \times 10^{-18})$

b. $(1.5 \times 10^{-12}) \div (3 \times 10^{17})$

89. $\dfrac{4.8 \times 10^{-7}}{(1.6 \times 10^{8})(3.0 \times 10^{-18})}$

90. $\dfrac{5.6 \times 10^{15}}{(1.4 \times 10^{-8})(4.0 \times 10^{4})}$

Projects

91. ***Black Holes and Scientific Notation.**** No light escapes from a black hole. Thus, the formula relating the mass and radius of a black hole and the speed of light is the same as the formula for a rocket ship leaving Earth. The radius, r_B, of a black hole depends on its mass, m, according to the formula

$$r_B = \frac{2Gm}{c^2}$$

*Astronomy information from William K. Hartman, *The Cosmic Voyage Through Time and Space* (Belmont, CA: Wadsworth Publishing Company, 1992).

G is the gravitational constant 6.672×10^{-11} N · m²/kg², where N = kg · m/sec². The mass is in kilograms. The letter c is the speed of light, 2.998×10^{8} m/sec.

a. Substitute just the units into the formula to see what units describe the radius.

For parts b to e, find the radius to which each of these objects in our solar system would have to be compressed in order to be a black hole.

b. Sun: mass = 1.99×10^{30} kg

c. Jupiter: mass = 1.90×10^{27} kg

d. Earth: mass = 5.98×10^{24} kg

e. Moon: mass = 7.35×10^{22} kg

f. The diameters for the four objects in parts b through e are, respectively, 1,391,400 kilometers, 142,800 kilometers, 12,756 kilometers, and 3476 kilometers. To find a number describing the number of times each object is compressed, find the actual radius and divide it by the black hole radius.

92. ***Interest on CDs.*** The interest on certificates of deposit (CDs) at many banks and credit unions is calculated with annual compounding. If you want A dollars n years in the future, you need to invest P dollars now at interest rate r (expressed as a decimal).

a. Use the formula $P = A(1 + r)^{-n}$ in the table to calculate how much, P, you need to save now to have \$50,000 in the future. Round your answers to the nearest dollar.

Interest Rate	$P = 50000(1 + r)^{-10}$ (n = 10 years)	$P = 50000(1 + r)^{-20}$ (n = 20 years)
$r = 0.03$		
$r = 0.04$		
$r = 0.05$		
$r = 0.06$		
$r = 0.07$		
$r = 0.08$		

b. Use the completed table to predict the interest rate needed to reach \$50,000 in 10 years if you start with an investment of \$25,000.

c. Use the completed table to predict the interest rate needed to reach \$50,000 in 20 years if you start with an investment of \$25,000.

CHAPTER  SUMMARY

Vocabulary

For definitions and page references, see the Glossary/Index.

ascending order
binomial
common factor
descending order
difference of squares
greatest common factor
monomial
perfect square trinomial
polynomial
scientific notation
trinomial

Concepts

6.1 Operations on Polynomials

Multiplying two expressions is like finding the area of a rectangle given its base and height. Arrange answers in descending order of exponents.

Dividing two expressions is like finding the base of a rectangle given its height and area.

Factoring an expression is like finding the base and height of a rectangle given its area.

The first step in any factoring is to factor the greatest common factor.

6.2 Multiplication of Binomials

In multiplication of binomials by table, there are two patterns: The terms on the diagonal from lower left to upper right add to the middle term of the trinomial answer, and the terms on the two diagonals multiply to the same product.

A perfect square trinomial describes $(a + b)^2$. The product $(a + b)(a + b)$ has three terms: the square of a, twice the product of a and b, and the square of b.

$$(a + b)(a + b) = a^2 + 2ab + b^2$$

A difference of squares describes $a^2 - b^2$. The factors of $a^2 - b^2$ are $(a - b)(a + b)$.

6.3 Factoring Binomials and Trinomials

To factor $ax^2 + bx + c$ by table:

1. a. Write the first term of the trinomial, ax^2, in the upper left corner of the table.
   b. Write the last term of the trinomial, c, in the lower right corner.
   c. Multiply the first and last terms to get the diagonal product, acx^2.
2. a. From the diagonal product, list the factors of the coefficient, ac.
   b. Find the factors that add to the coefficient of the middle term of the trinomial, b.
   c. Write the factors adding to bx in the diagonal from lower left to upper right.
3. Write the greatest common factor of the first row to the left of ax^2.
4. a. Use the gcf to find the remaining factors at the top and left of the table.
   b. Check that the top right and bottom left factors multiply to the last term, c.
5. Multiply the binomial factors from the table to check the factoring.

Factor			
Greatest common factor (gcf) of this row	First term of trinomial, ax^2	____	Sum of like terms equals middle term of trinomial
	____	Last term of trinomial, c	Diagonal product, acx^2

Remember: The first step in any factoring is to factor the greatest common factor. If you forget to factor the gcf first, look for a gcf in the factors from the table as your last step.

6.4 Exponents and Scientific Notation

For zero exponents, $x^0 = 1$. The expression 0^0 is not defined.

For negative one exponents, $x^{-1} = \dfrac{1}{x}$, the reciprocal of x.

For negative two exponents, $x^{-2} = \dfrac{1}{x^2}$, the square of the reciprocal of x.

In general, $\left(\dfrac{a}{b}\right)^{-n} = \left(\dfrac{b}{a}\right)^n$.

To change decimal notation to scientific notation:

1. Place the decimal point after the first nonzero digit.
2. Count the number of places the decimal point has moved, and use that number as the number exponent on ten.
3. a. If the original number is greater than 10, the exponent on ten is positive.
   b. If the original number is between 10 and 1, the exponent on ten is zero.
   c. If the original number is between 0 and 1, the exponent on ten is negative.

To change from scientific notation to decimal notation:

1. Look at the exponent on ten to find the number of decimal places and the direction in which to move the decimal point.
2. a. If the exponent is negative, move the decimal point left to make a number between 0 and 1.
   b. If the exponent is zero, do not change the decimal point.
   c. If the exponent is positive, move the decimal point right to make a number greater than 10.

CHAPTER 6 REVIEW EXERCISES

1. Simplify each of the following expressions. Arrange the terms alphabetically, with exponents on the first letter in descending order. Indicate whether the result is a monomial, binomial, trinomial, or other polynomial.

a. $3a^2 - 5ab + 4a^2 - 3b^2 + 2ab - 7b^2$

b. $3 + 4x^2 + 5x - 2(x - 1)$

c. $x(x^2 + 4x + 16) - 4(x^2 + 4x + 16)$

d. $11x - 4x(1 - x)$

e. $9 - 4(x - 3)$

2. Name the greatest common factor for each expression, and then factor the expression.

a. $6x^2 + 9x + 6$

b. $14x^2 + 7xh + 49h^2$

c. $32xy - 24xy^2 + 16x^2y^2$

d. $25abc - 15ac + 35bc$

3. Find the perimeter and area of each shape.

a.

$2x$

$3x + 1$

b.

$x + 2$

$3x$

c.

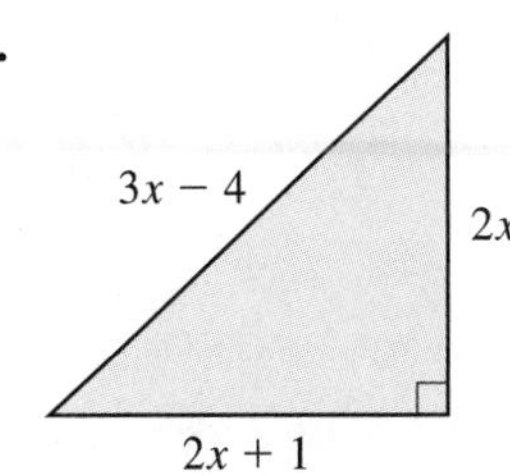

4. Find the missing side of each shape.

a.

$x - 3$

?

Area = $4x^2 - 12x$

b.

$3x + 2$

?

Area = $9x^2 + 6x$

c.

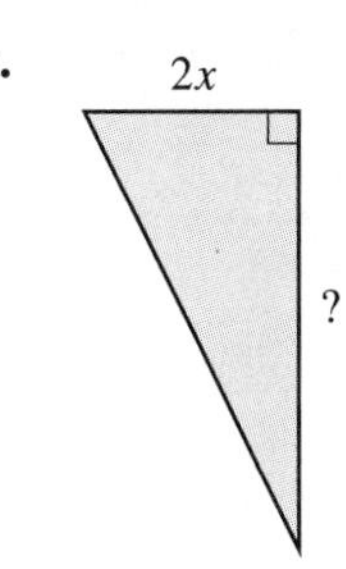

Area = $x^2 + 4x$

5. Arrange these tiles into a rectangle. What multiplication is shown?

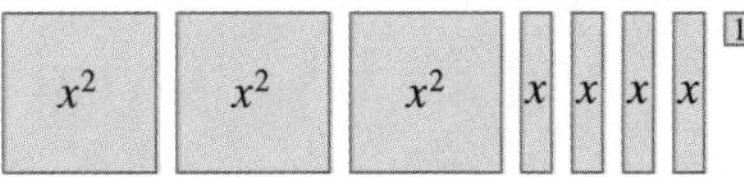

6. Complete these tables. Write the multiplication problem each table describes.

a.

Multiply	$2x$	$+5$
$3x$		

b.

Multiply		
	$3a^2$	-9

In Exercises 7 to 16, find the products.

7. $(x + 4)(x + 3)$

8. $(x + 4)(x - 3)$

9. $(2x - 5)(2x - 5)$

10. $(2x - 5)(2x + 5)$

11. $(3x + 5)(3x - 2)$

12. $(2x - 3)(3x - 2)$

13. $(2x - 3)(3x + 2)$

14. $(a + b)(a - b)$

15. $(a - b)(a - b)$

16. $(a - b)(b - a)$

17. Which answers in Exercises 7 to 16 are perfect square trinomials?

18. Which answers in Exercises 7 to 16 are differences of squares?

19. Complete these tables. Write the multiplication problem each table describes.

a.

Factor		
	x^2	$+7x$
	$-2x$	-14

b.

Factor		
	$6x^2$	$+10x$
	$+9x$	$+15$

20. Find the missing numbers and signs in these factorizations.

a. $x^2 + 2x - 24 = (x\_\_\_)(x\_\_\_)$

b. $x^2 - 10x + 24 = (x\_\_\_)(x\_\_\_)$

c. $x^2 - 10x - 24 = (x\_\_\_)(x\_\_\_)$

d. $x^2 - 14x + 24 = (x\_\_\_)(x\_\_\_)$

21. Complete these tables. Then write the original trinomial and its factors.

a. Factor $x^2 - 9x + 14$.

Factor		
	x^2	
		$+14$

b. Factor $2x^2 + 11x - 21$.

Factor		
	$2x^2$	
		-21

c. Factor $12x^2 - x - 1$.

Factor		
	$12x^2$	
		-1

d. Factor $20x^2 + 24x - 9$.

Factor		

Factor each expression in Exercises 22 to 47. Identify it as a difference of squares (ds), a perfect square trinomial (pst), or neither.

22. $x^2 - 16$ **23.** $x^2 - 3x + 2$

24. $x^2 - 6x + 9$ **25.** $9x^2 - 16$

26. $4x^2 + 3x - 1$ **27.** $2x^2 + x - 6$

28. $7x^2 - 5x - 2$ **29.** $9x^2 + 3x - 2$

30. $4x^2 - 4x + 1$ **31.** $x^2 + 6x + 8$

32. $x^2 - 8x + 16$ **33.** $x^2 - 11x + 10$

34. $y^2 + 12y + 36$ **35.** $25 - 9x^2$

36. $4x^2 - 1$ **37.** $y^2 + 8y + 12$

38. $2x^2 + 10x - 12$ **39.** $2x^2 - 3x - 35$

40. $4x^2 - 4x - 35$ **41.** $4x^2 - 8x + 4$

42. $4x^2 - 4$ **43.** $x^3 + 4x^2 + 4x$

44. $x^3 + 6x^2 + 9x$ **45.** $3x^2 - 27$

46. $5x^2 - 20x + 15$ **47.** $x^3 - 7x^2 + 10x$

Write the expressions in Exercises 48 to 63 without zero or negative exponents.

48. a. 5^0 **b.** 5^{-1} **c.** 5^{-2}

49. a. 3^{-1} **b.** 3^0 **c.** 3^{-2}

50. a. $\left(\frac{1}{3}\right)^{-2}$ **b.** $\left(\frac{1}{3}\right)^{-1}$ **c.** $\left(\frac{1}{3}\right)^0$

51. a. $\left(\frac{2}{3}\right)^0$ **b.** $\left(\frac{2}{3}\right)^{-1}$ **c.** $\left(\frac{2}{3}\right)^{-2}$

52. a. b^4b^{-3} **b.** $b^{-4}b^{-7}$ **c.** $x^{-5}x^5$

53. a. x^7x^{-2} **b.** x^3x^{-3} **c.** $b^{-5}b^{-5}$

54. a. $\dfrac{n^4}{n^5}$ **b.** $\dfrac{n^{-4}}{n^5}$

55. a. $\dfrac{n^4}{n^{-5}}$ **b.** $\dfrac{n^{-4}}{n^{-5}}$

56. a. $(x^{-2})^3$ **b.** $(b^{-1})^{-2}$ **c.** $(x^{-4})^0$

57. a. $(b^2)^{-4}$ **b.** $(x^{-2})^{-3}$ **c.** $(b^0)^{-2}$

58. a. $\left(\dfrac{a}{b}\right)^{-2}$ **b.** $\left(\dfrac{a}{b^2}\right)^0$

59. a. $\left(\dfrac{a^2}{b^2}\right)^{-1}$ **b.** $\left(\dfrac{2a}{b^2c}\right)^{-3}$

60. a. $\dfrac{x^3y^{-2}}{x^2y^{-1}}$ **b.** $\dfrac{a^4b^{-5}}{a^{-2}b^3}$

61. a. $\dfrac{x^{-3}y^{-3}}{x^5y^{-2}}$ **b.** $\dfrac{a^3b^{-6}}{a^6b^{-4}}$

62. a. $\dfrac{a^0}{a^{-1}}$ **b.** $\dfrac{x^{-4}}{x^0}$

63. a. $\dfrac{x^3x^0}{x^0}$ **b.** $\dfrac{a^{-2}a^2}{a^0}$

64. Change each number in the national debt and population columns of the table to scientific notation. Calculate the last column (divide national debt by population).

Year	National Debt	Population	Debt per Person
1900	1,200,000,000	76,200,000	
1920	24,200,000,000	106,000,000	
1950	256,100,000,000	151,300,000	
1990	3,233,300,000,000	248,700,000	

65. Complete the table by changing each number to decimal notation.

Chemical	Symbol for Isotope	Half-life	Half-life as Decimal
Potassium	^{40}K	1.4×10^9 years	
Calcium	^{41}Ca	1.2×10^5 years	
Radon	^{219}Rn	1.243×10^{-7} year	
Polonium	^{212}Po	3.0×10^{-7} second	

66. Use unit analysis to change the radon half-life to seconds.

67. Use unit analysis to change the polonium half-life to years.

In Exercises 68 to 71, find the product or quotient mentally and then check with a calculator.

68. **a.** $(1.5 \times 10^{-4})(3.0 \times 10^9)$

b. $(2.5 \times 10^{-9})(4.0 \times 10^3)$

69. **a.** $(6.0 \times 10^8)(7.0 \times 10^{-2})$

b. $(8.0 \times 10^7)(4.0 \times 10^{-3})$

70. **a.** $\dfrac{6.4 \times 10^{-12}}{1.6 \times 10^{-3}}$ **b.** $\dfrac{7.5 \times 10^{-13}}{2.5 \times 10^{-6}}$

71. **a.** $\dfrac{7.2 \times 10^9}{8.0 \times 10^{-8}}$ **b.** $\dfrac{5.4 \times 10^9}{9.0 \times 10^{-6}}$

In Exercises 72 to 73, change the number to standard scientific notation.

72. **a.** 38.5×10^{-3} **b.** 0.48×10^{-2}

73. **a.** 234×10^5 **b.** 0.0436×10^8

74. The moon travels about 1.5×10^6 miles as it orbits Earth. It completes one orbit in about 27.3 days. What is its speed to the nearest ten miles per hour?

75. A snail crawls 2 inches around a fishbowl in 15 minutes. Using scientific notation, write its speed in miles per hour.

CHAPTER 6 TEST

1. Choose from *factors* or *terms*: The expression $3x^2 + 4x + 1$ has three ________.

2. The $2ab$ that divides evenly into the expression $4a^2b - 8ab + 10ab^2$ is called the ________.

3. Choose from *dividing*, *factoring*, or *multiplying*: When we find the base and height of a rectangle with area $x^2 + 3x + 2$, we are ________.

4. Which exponent gives the reciprocal of a number?

5. What is the value of the exponent in $4^x = 1$?

6. What is the value of the exponent in $4^x = \frac{1}{16}$?

7. Write 3.482×10^{-4} in decimal notation.

8. Write 45,000,000,000 in scientific notation.

In Exercises 9 to 11, do the operation and indicate whether the result is a monomial, binomial, trinomial, or other polynomial.

9. $(a + 3b - 3c - d) + (3a - 5b + 8c - d)$

10. $x(x^2 + 3x + 9) - 3(x^2 + 3x + 9)$

11. $-4x(x - 5)$

12. Factor $14xy + 6x^2y - 18y^2$.

13.

Factor		
	$6x^2$	$+8x$
	$-15x$	-20

14. Find the perimeter and area of the rectangle in the figure.

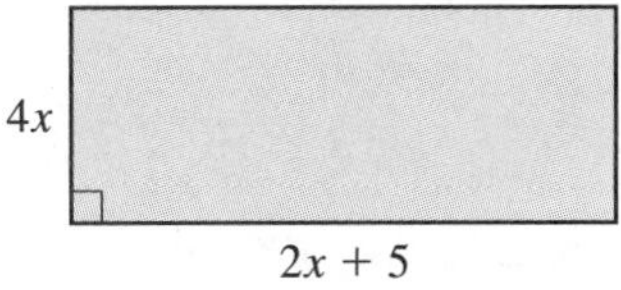

15. Find the width of the rectangle in the figure.

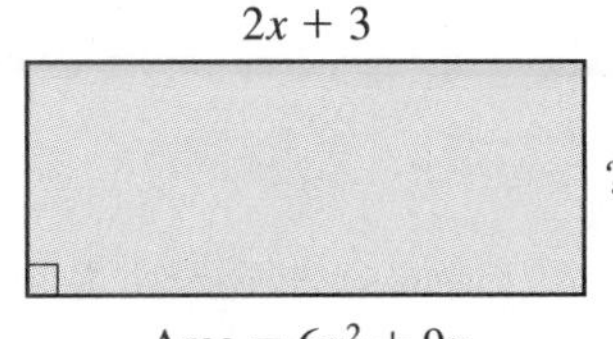

Multiply the expressions in Exercises 16 to 19.

16. $(x - 4)(x + 7)$ **17.** $(x - 7)(x - 7)$

18. $(2x - 7)(2x + 7)$ **19.** $2(x - 4)(x - 4)$

Factor the expressions in Exercises 20 to 24.

20. $x^2 - 9x + 20$ **21.** $2x^2 - 3x - 2$

22. $2x^2 - 8$ **23.** $x^2 - 8x + 16$

24. $9x^2 + 6x + 1$

25. Name one of Exercises 16 to 24 that has a difference of squares as a problem or answer.

Simplify each expression in Exercises 26 to 28. Assume none of the variables equal zero.

26. **a.** $b^{-2}b^3$ **b.** $(x^3)^{-2}$

27. **a.** $\dfrac{b^3}{b^{-2}}$ **b.** $\dfrac{ab^0}{a^2b^{-1}}$

28. **a.** $\left(\dfrac{a}{2b}\right)^0$ **b.** $\left(\dfrac{9x^2}{25y^2}\right)^{-2}$

29. Multiply $(2.5 \times 10^{-15})(4.0 \times 10^2)$. Write the answer in decimal notation.

30. Divide $\dfrac{1.25 \times 10^{-8}}{2.5 \times 10^{-4}}$. Write the answer in scientific notation.

Choose one of Exercises 31 to 33 to answer.

31. What are the possible sets of factors (containing whole numbers) and trinomials that could complete this statement?

$$(x \pm \_\_\_)(x \pm \_\_\_) = x^2 \pm \_\_\_ \pm 21$$

Explain how you know that your listing is complete.

32. Explain how the table multiplication model shows that $(a + b)^2$ cannot equal $a^2 + b^2$.

33. Why does entering 10 [EE] 3 on a calculator give 10,000, whereas entering 10 [y^x] 3 gives 1000? (*Note*: Some calculators have [EXP] or [EEX] instead of [EE], and some have [x^y] or [^] instead of [y^x].)

CUMULATIVE REVIEW OF CHAPTERS 1 TO 6

1. For each phone call, Save You Money company charges \$0.25 plus \$0.25 per minute. Make an input-output table and a graph. Use inputs of 0 to 30 minutes, in steps of 5.

Simplify the expressions in Exercises 2 to 6.

2. $-5 - (-12) - 7$

3. $5 - 2(4 - x)$

4. $3^3 - 2 \cdot 5 + 4(7 - 2)$

5. $x^3 \cdot x^8$

6. $\dfrac{a^9}{a^4}$

7. Evaluate $R = \dfrac{ab}{a + b}$ if $a = 30$ and $b = 20$.

8. Write in symbols: A number plus 3 equals six less than twice the number.

9. Solve for x: $3x - 13 = 13 - 2x$.

10. Solve for x: $9 - 4(x - 3) = -2 - 3x$.

11. Solve for the solution set: $x + 1 \geq 9 - 3x$. Sketch a line graph of the solution set.

12. Write in symbols: The output is a third of the input.

13. Solve for y: $3x + 5y = 15$.

14. Solve for y: $5x - 3y = 8$.

15. Solve for a: $\dfrac{a + b + c}{3} = m$.

16. If $f(x) = 3x - 5$, what are $f(2)$ and $f(5)$?

17. On a rectangular coordinate grid, first sketch a line (anywhere) with slope $\frac{5}{3}$ and then sketch a line with slope $-\frac{2}{3}$ that passes through (3, 5).

18. What is the equation of a line with slope 4 and y-intercept -5?

19. What is the equation of a line parallel to $3x + 5y = 15$ (from Exercise 13) through the origin?

20. What is the simplified ratio of 15 minutes to 3 hours?

21. Quarts of motor oil cost \$0.92 in 1986 and \$1.19 in 1995. What is the percent change in price between the two years?

22. Set up a proportion or equation and solve:

a. 15% of what number is 5.4?

b. 42 is what percent of 48?

c. What is 40% of 65?

23. Solve for x: $\dfrac{x - 1}{2} = \dfrac{x + 3}{3}$.

24. Suppose you earned \$6.50 per hour from age 15 to age 65. At this rate, how many dollars would you earn? State the facts needed to set up a unit analysis. State all your assumptions.

25. Multiply, then add like terms: $x(x^2 - 2x + 1) - (x^2 - 2x + 1)$.

26. Factor $8xy - 4xyz + 12yz$.

27. Multiply $(2x + 1)(2x + 3)$.

28. Factor $x^2 - 6x + 9$. What special product is this?

29. Factor $4x^2 - 25$. What special product is this?

30. Factor $6x^2 - 2x - 8$.

7

Systems of Equations and Inequalities

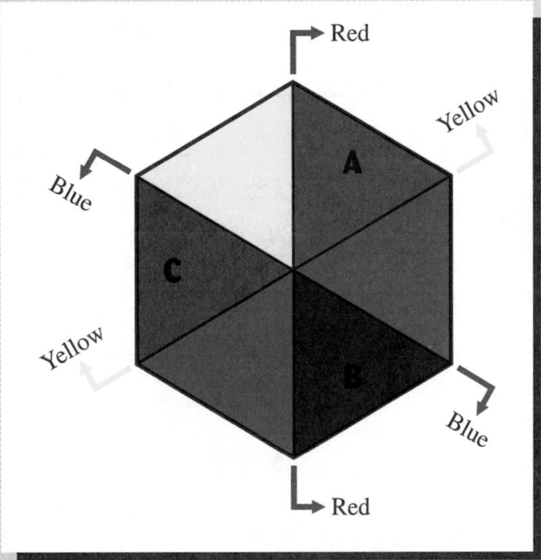

Figure 1

Figure 1 shows a color chart with the three primary colors—red, yellow, and blue—each shading half the hexagon. What colors appear in the regions where the colors overlap? We will look at overlapping regions in mathematics as we study graphs of systems of inequalities in Section 7.5.

This chapter is about sets (or systems) of equations and inequalities. There are four ways to solve a system of equations: with graphs and tables (Sections 7.1 and 7.2), by substitution (Section 7.3), and by elimination (Section 7.4). There are three different possible outcomes for a system of linear equations: no solution, one solution, and an infinite number of solutions. These outcomes are associated with distinctive algebraic and graphical results. The chapter closes with a discussion of graphical solutions to systems of linear inequalities in Section 7.5.

7.1 Solving Systems of Equations with Graphs

OBJECTIVES

- Graph a system of equations.
- Solve a system of equations by graphing.
- Give the geometric meaning for a system of equations having no solution, one solution, or infinitely many solutions.

WARM-UP

For each of these equations, name the slope ratio and y-intercept point and sketch a graph.

1. $y = 3x + 2$ **2.** $y = -x - 2$ **3.** $y = 3x - 2$

4. $y = -2x + 4$ **5.** $2y = 6 - 4x$ **6.** $y = \frac{1}{2}x - 2$

THIS SECTION INTRODUCES systems of equations and types of solutions. We start with systems of two linear equations and solve systems with one, no, and infinitely many solutions.

Systems of Linear Equations

Definition of System of Equations

A **system of equations** is a set of two or more equations that are to be solved for the values of the variables making all the equations true.

Taken together, the equations $x + y = 4$ and $x - y = 6$ form a system of two linear equations.

Definition of Solution to a System

A **solution to a system** of two linear equations is the ordered pair that makes both equations true.

The ordered pair $(5, -1)$ makes both $x + y = 4$ and $x - y = 6$ true.

We will find solutions in this section by graphing.

Definition of Point of Intersection

> The **point of intersection** of two graphs, where the graphs coincide, identifies a solution to the two equations graphed. Because every point on the graph of an equation makes the equation true, the point of intersection makes both equations true.

A system of two or more linear equations may have zero, one, or infinitely many solutions. The Exploration in Example 1 shows how these three outcomes might occur.

EXAMPLE  Exploring numbers of solutions Look at the lines in Figure 2.

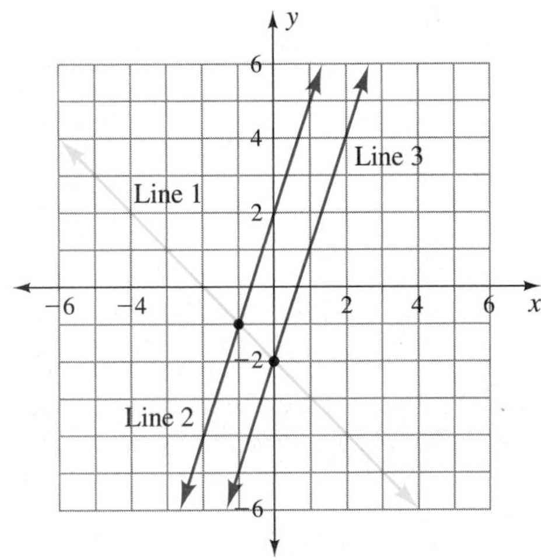

Figure 2

a. Which line has equation $y = 3x + 2$?

b. Which line has equation $y = -x - 2$?

c. Which line has equation $y = 3x - 2$?

d. What is the point of intersection of lines 1 and 2?

e. Some lines have one point of intersection. Substitute the ordered pair from part d into $y = 3x + 2$, and substitute the ordered pair from part d into $y = -x - 2$. What do you observe?

f. Other lines have no point of intersection. What is the point of intersection of lines 2 and 3? What can we say about these lines?

g. Still other lines have an infinite number of points of intersection. What is the point of intersection between line 2 and the graph of $y - 3x - 2 = 0$? (*Hint:* Use the equation $y - 3x - 2 = 0$ to find ordered pairs for $x = -1$ and $x = 0$. Plot the ordered pairs.) What do you observe? Prove your observation by solving $y - 3x - 2 = 0$ for y.

Solution See the Answer Box. ●

The Exploration in Example 1 illustrates the three types of solution sets we obtain from graphing a system of two linear equations: one solution, no solution, or an infinite number of solutions. Here is a summary of the three types of solution sets:

Solution Sets for a System of Two Linear Equations

In graphing a system of two linear equations, we have three possible outcomes:

1. *One solution:* The equations describe lines intersecting at exactly one point (point A). This system of equations has exactly one solution—the ordered pair at the point of intersection.

2. *No solution:* The equations describe parallel lines (lines BC and DE). The graphs have no point of intersection; hence, the system has no real-number solution.

3. *Infinite number of solutions:* The equations describe the same line (line FG). We say the graphs are *coincident.* The system has an infinite number of solutions because the coordinates of every point on the graph of the first equation satisfy the second equation.

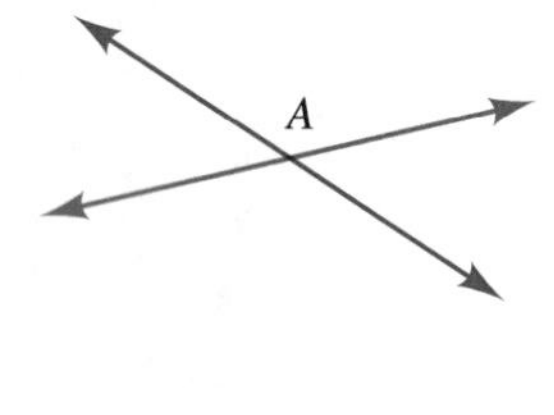

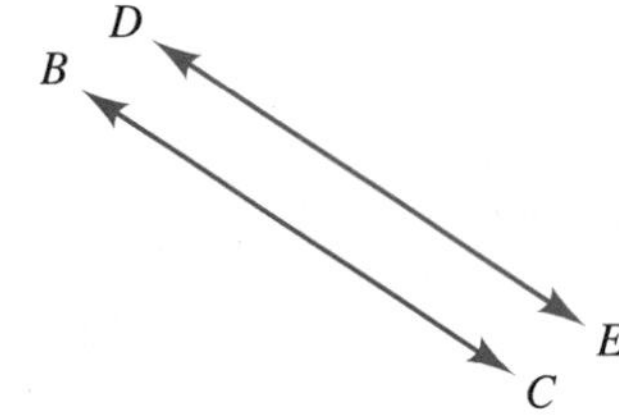

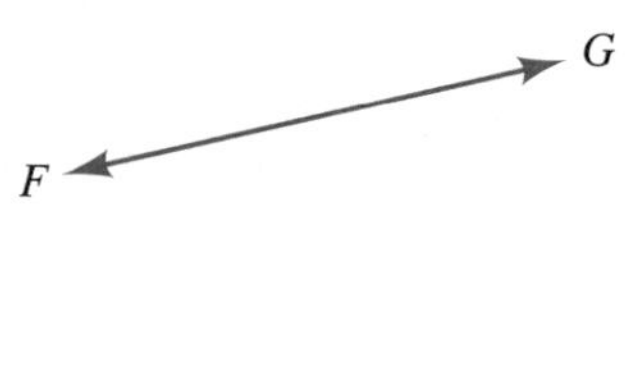

We now look at how the three types of solution sets are obtained and applications that result in each type of set.

Systems with Exactly One Solution

In Example 2, we graph two equations and find their intersection.

EXAMPLE 2 Solving a system with one solution Solve $2x + 3y = 6$ and $2y - x = -10$ by graphing.

Solution We solve each equation for y. After plotting the y-intercept point, we use the slope to find a second point on each line.

For the first equation, we have

$$2x + 3y = 6 \qquad \text{Subtract } 2x \text{ on each side.}$$
$$3y = -2x + 6 \qquad \text{Divide by 3 on each side.}$$
$$y = -\tfrac{2}{3}x + 2$$

The slope is $-\frac{2}{3}$, and the y-intercept point is $(0, 2)$. We plot $(0, 2)$ and then count down 2 units and to the right 3 units to find a second point, $(3, 0)$, as shown in Figure 3. We draw a line through $(0, 2)$ and $(3, 0)$.

For the second equation, we have

$$2y - x = -10 \qquad \text{Add } x \text{ to each side.}$$
$$2y = x - 10 \qquad \text{Divide by 2 on each side.}$$
$$y = \tfrac{1}{2}x - 5$$

The slope is $\frac{1}{2}$, and the y-intercept point is $(0, -5)$. We plot $(0, -5)$ and then count up 1 unit and to the right 2 units to find a second point, $(2, -4)$, as shown in Figure 3. We draw a line through $(0, -5)$ and $(2, -4)$.

The lines intersect at $(6, -2)$. The solution to the system of equations is the point of intersection, $(6, -2)$.

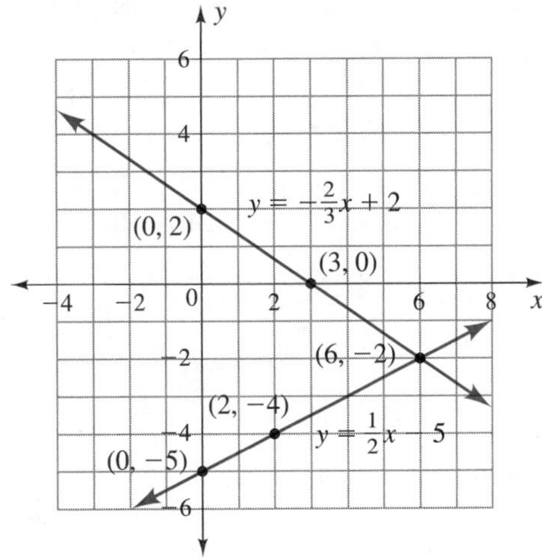

Figure 3

Check: We substitute $x = 6$ and $y = -2$ into each equation.

$$2(6) + 3(-2) \stackrel{?}{=} 6 \checkmark$$
$$2(-2) - (6) \stackrel{?}{=} -10 \checkmark$$

The ordered pair $(6, -2)$ makes both equations true and is the solution. ●

In Example 3, we write and solve a system of equations and use intervals and inequalities to describe the results.

EXAMPLE 3

Finding the solution to a system of equations and interpreting the meaning in an application setting Two competing car rental agencies offer similar cars. Ava's Rent-a-Car charges \$40 plus \$0.10 per mile. Herr's Rent-a-Car charges \$20 plus \$0.20 per mile.

a. Write a system of equations to describe the total rental costs, where x is the number of miles driven.

b. Match each of your equations to its graph in Figure 4. Use the graph to determine the solution to the system.

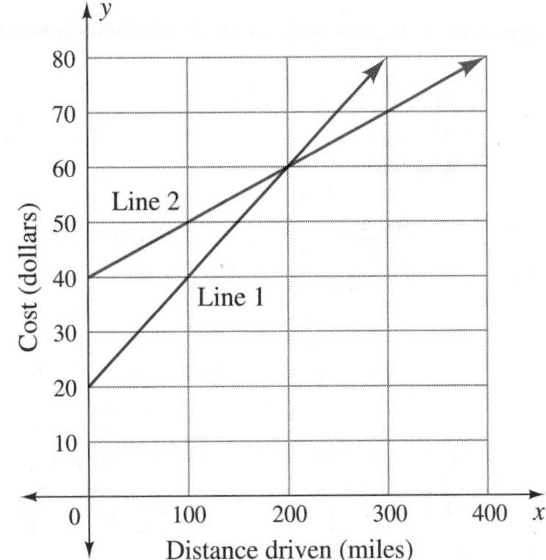

Figure 4

c. What is the meaning of the point of intersection in terms of the problem situation?

d. For what mileage will Ava's car be cheaper? Describe the x inputs with an interval and with an inequality.

e. For what mileage will Herr's car be cheaper? Describe the x inputs with an interval and with an inequality.

Solution **a.** The two rental costs are

Ava: $y = 0.10x + 40$

Herr: $y = 0.20x + 20$

b. Line 2 is Ava's graph. Line 1 is Herr's graph. The graph has miles on its horizontal axis and total dollar cost on its vertical axis. The solution to the system is the point of intersection of the graphs, (200, 60).

c. The point of intersection (200, 60) gives the number of miles, $x = 200$, and the total dollar cost, $y = 60$, at which the two agencies' charges are the same.

d. Ava's will be better for longer trips, with mileage in the interval $(200, +\infty)$, or $x > 200$. Of course, infinity $(+\infty)$ is not a reasonable mileage because there is a limit on the number of miles that can be driven.

e. Herr's will be better for short trips, with mileage in the interval $[0, 200)$, or $0 \leq x < 200$.

Graphing Calculator Solution

Let x = miles traveled and y = total dollar cost. Enter $Y_1 = 0.20X + 20$ and $Y_2 = 0.10X + 40$. Set the window with x in the interval [0, 400], scale 100, and y in the interval [0, 80], scale 10. Graph. Trace to the point of intersection, which shows $x \approx 200$ and $y \approx 60$. ●

Think about it 1: What is the meaning of the y-intercept of each equation in Example 3? What is the meaning of the slope of each line?

In Example 4, the equations include small decimal numbers, and a short table of values is helpful in graphing the equations. The solution is not easy to read from the graph, and so a graphing calculator zoom or intersection feature is helpful in solving the system.

EXAMPLE 

Finding a solution and interpreting its meaning in a problem situation: wage options
The Appliance Works is reviewing employee wages. Management is considering two plans, both based on a fixed salary plus a percent commission on retail sales. Plan A is \$300 salary plus 6% of sales, and Plan B is \$150 salary plus 10% of sales.

a. Write a system of equations for Plans A and B.

b. Which category, total income or sales, should go on the horizontal axis when the equations are graphed? Why?

c. To better estimate the scales on the axes, make a table for each plan, with inputs as sales dollars from 0 to 5000, counting by 1000.

d. Graph the system of equations.

e. For what level of sales will the plans be equal?

f. Under what conditions would an employee prefer Plan A?

g. What type of employee would prefer Plan B?

Solution

a. Plan A: $y = 300 + 0.06x$
Plan B: $y = 150 + 0.10x$

b. Because total income depends on the amount of sales, sales should be on the horizontal axis, or x-axis.

c. Using inputs 0 to 5000 in steps of 1000, we create Tables 1 and 2 for Plans A and B, respectively.

Sales (dollars)	Income (dollars)
0	300
1000	360
2000	420
3000	480
4000	540
5000	600

Table 1 Plan A

Sales (dollars)	Income (dollars)
0	150
1000	250
2000	350
3000	450
4000	550
5000	650

Table 2 Plan B

d. The equations are graphed in Figure 5.

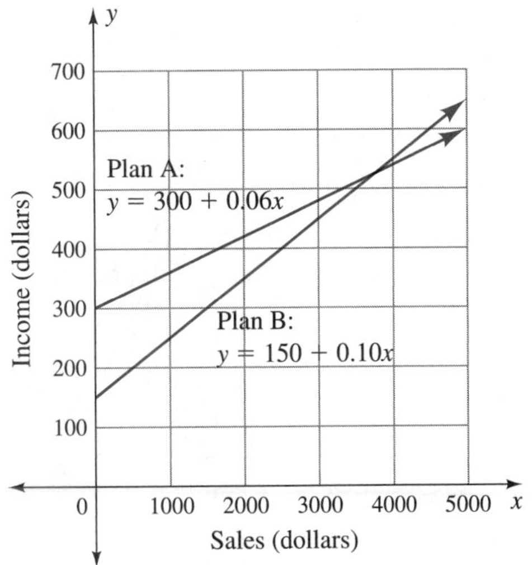

Figure 5

e. The lines intersect where the plans give the same total income, about \$500 to \$550 on sales of between \$3500 and \$4000.

f. Plan A yields more money with low sales. If business is slow or a salesperson is new, Plan A might be better.

g. A good salesperson would do well with Plan B.

Graphing Calculator Solution

Enter $Y_1 = 300 + 0.06X$ and $Y_2 = 150 + 0.10X$. Use the table feature to find sales levels that yield similar outputs for the two equations. To duplicate Figure 5, the window should have x in the interval [0, 5000], scale 1000, and y in the interval [0, 700], scale 100. Trace and zoom as needed to find the point of intersection, (3750, 525), or use the calculator's intersection feature.

Think about it 2:

a. What equation describes the point of intersection in Example 4?
b. For what levels of sales would Plan A produce more income?
c. For what levels of sales would Plan B produce more income?
d. Explain the meaning of the slopes of the lines.
e. Explain the meaning of the y-intercept of each line.

Systems without a Solution

We now consider examples in which there are no solutions to the system. In Example 5, we graph equations to solve the system.

EXAMPLE 5 Solving a system of equations by graphing Graph the equations $2x + y = 4$ and $2y = 6 - 4x$.

Solution First we solve the equations for $y = mx + b$, and then we graph them.

$$2x + y = 4 \qquad \text{Subtract } 2x \text{ on both sides of the first equation.}$$
$$y = -2x + 4$$
$$2y = 6 - 4x \qquad \text{Divide by 2 on both sides of the second equation.}$$
$$y = 3 - 2x \qquad \text{Rearrange the right side.}$$
$$y = -2x + 3$$

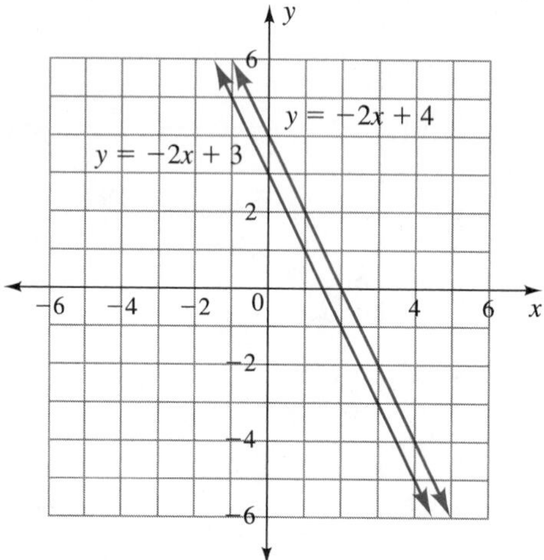

Figure 6

The lines have the same slope but different y-intercepts, as shown in Figure 6. They are parallel lines and have no point of intersection. ●

When a system of equations contains parallel lines, there is no point of intersection and no ordered pair makes both equations true. In this case, we say *there is no solution to the system of equations.*

Systems of Equations with No Solution

- A system containing parallel lines has no solution.
- When there is no solution, the solution set is empty, { } or ∅.

That parallel lines have no solution is perhaps easier to understand in application settings. In Example 6, we extend the application from Example 4 on wage options.

EXAMPLE 6 Interpreting solutions for parallel lines: more wage options Suppose Wage Plan A is given by $y = 300 + 0.06x$ and Wage Plan C is given by $y = 150 + 0.06x$.

a. Graph the wage plans.

b. Find the point of intersection.

c. Compare the two plans.

Solution **a.** The equations are graphed in Figure 7.

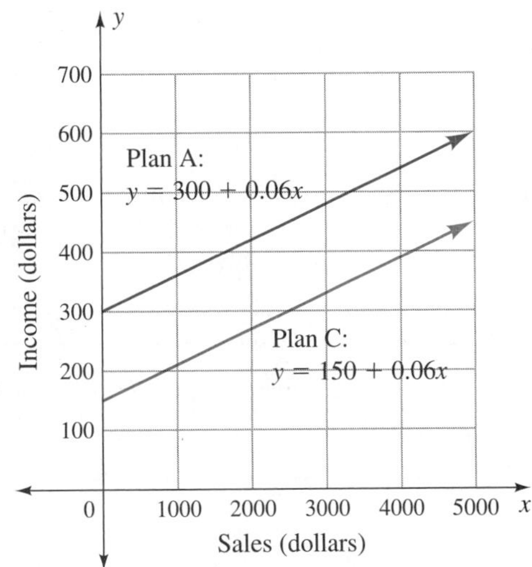

Figure 7

b. The two plans give parallel lines with no point of intersection; hence, there is no real-number solution to the system. The solution set is empty: { }.

c. The \$150 salary plus 6% of sales would never be better than the \$300 salary plus 6% of sales. ●

Systems with an Infinite Number of Solutions

EXAMPLE  Solving a system of equations by graphing Graph the system of equations below and describe the solution set.

$$y = \tfrac{1}{2}x - 2$$
$$2y + 4 = x$$

Solution The equation $y = \frac{1}{2}x - 2$ has a slope of $\frac{1}{2}$ and a y-intercept point at $(0, -2)$. The graph of $y = \frac{1}{2}x - 2$ is shown in Figure 8. To graph $2y + 4 = x$, we solve for y.

$$2y + 4 = x \qquad \text{Subtract 4 on each side.}$$
$$2y = x - 4 \qquad \text{Divide by 2 on each side.}$$
$$y = \tfrac{1}{2}x - 2$$

The equation $2y + 4 = x$ is the same as the first equation. The graphs will coincide. Every ordered pair that makes $y = \frac{1}{2}x - 2$ true will also make $2y + 4 = x$ true.

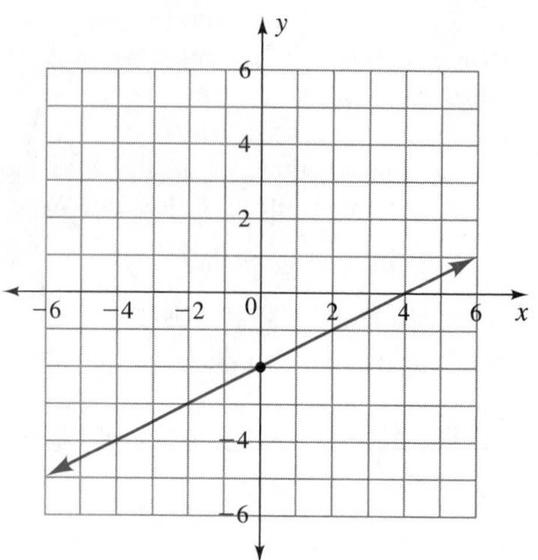

Figure 8 ●

Two lines that have the same equation are **coincident lines**. Because coincident lines have all points in common, we say that as a system *they have an infinite number of solutions.*

Systems of Equations with Coincident Lines

A system whose graphs are coincident lines has an infinite number of solutions.

EXAMPLE  8

Interpreting solutions for coincident lines Suppose management of the Appliance Works in Example 4 offers employees a wage package described by $2y = 600 + 0.12x$, instead of $y = 300 + 0.06x$. How should the employees react?

Solution We solve the new offer for y:

$$2y = 600 + 0.12x \qquad \text{Divide by 2.}$$
$$y = 300 + 0.06x$$

The equations are the same. The graphs will coincide, and there will be an infinite number of points in common. The new wage offer is identical to the old offer. The employees should be insulted by the meaningless new offer. ●

Predicting Solutions to Systems of Equations

To predict the number of solutions to a system of two linear equations, solve each equation for $y = mx + b$. Find the slope m and the y-intercept b.

- If the slopes are different, there is exactly one solution.
- If the slopes are the same and the y-intercepts are different, the lines are parallel and there is no solution to the system.
- If the slopes are the same and the y-intercepts are the same, the lines are the same (coincident), and there are an infinite number of solutions to the system.

ANSWER BOX

Chapter Opener: When the pigments are mixed, region A is orange, region B is violet (or purple), and region C is green.
Warm-up: 1. $m = \frac{3}{1}$, (0, 2) **2.** $m = -1$, (0, −2) **3.** $m = \frac{3}{1}$, (0, −2) **4.** $m = -\frac{2}{1}$, (0, 4) **5.** $m = -\frac{2}{1}$, (0, 3) **6.** $m = \frac{1}{2}$, (0, −2). For graphs, see Figures 2, 6, and 8. **Example 1: a.** line 2 **b.** line 1 **c.** line 3 **d.** (−1, −1) **e.** $-1 = 3(-1) + 2$ and $-1 = -(-1) - 2$ both are true, so (−1, −1) makes both equations true and is the solution set. **f.** There is no point of intersection; lines 2 and 3 are parallel. **g.** The ordered pairs (−1, −1) and (0, 2) lie on line 2. This suggests that $y - 3x - 2 = 0$ and $y = 3x + 2$ are coincident lines having the same equation. Solving $y - 3x - 2 = 0$ for y gives $y = 3x + 2$.
Think about it 1: The fee for the rental car; the cost per mile for the rental car **Think about it 2: a.** $300 + 0.06x = 150 + 0.10x$ **b.** Sales between 0 and \$3750 **c.** Sales greater than \$3750 **d.** The slope is the percent of sales received as commission. **e.** The y-intercept is the wage received if no sales are made.

EXERCISES 7.1

Find the slope and y-intercept for each equation in the systems listed in Exercises 1 to 16. Sketch graphs for each system, and find the point of intersection. To show that the point of intersection is a solution to the system, substitute it into each equation of the system.

1. $y = -2x$, $y = 1 - x$

2. $y = -x$, $y = x + 2$

3. $y = x$, $y = 2x + 3$

4. $y = 2x - 2$, $y = 1 - x$

5. $y = 3$, $x = -1$

6. $x = -3$, $y = -1$

7. $x + y = 5$, $y = -x + 5$

8. $x - y = 4$, $y = x - 8$

9. $x - y = 3$, $y = x - 6$

10. $y = x + 2$, $x - y = -2$

11. $y = x - 4$, $3x + y = 4$

12. $4x - y = -10$, $y + 2x = 4$

13. $4x - 2y = 2$, $4x + 3 = y$

14. $3y - 9x = -6$, $y + 2x = 13$

15. $2x + 3y = 12$, $x + y = 5$

16. $x + 2y = 10$, $x + y = 4$

Which of the pairs of equations in Exercises 17 to 22 form parallel lines? It is not necessary to graph them.

17. $2x + 2y = 100$, $y = 20 - x$

18. $y = x + 55$, $y = x + 25$

19. $y = 55x$, $y = 25x$

20. $y = 10 - x$, $y = 5 + x$

21. $y = 4x$, $y = \frac{1}{4}(12 + 16x)$

22. $y = 300 - 60x$, $y = 60 - 300x$

In Exercises 23 to 28, which of the pairs of equations form coincident lines?

23. $y + 60x = 300$, $y = 60 - 300x$

24. $y + x = 20$, $2x = 40 - 2y$

25. $y = x + 0.15x$, $y = 1.15x$

26. $y = 150 + 0.10x$, $y = 150x + 0.10$

27. $2x + y = 10$, $2y + x = 10$

28. $y = x + 0.08x$, $y = 1.08x$

In Exercises 29 to 32, what system of equations is represented by each graph? Write the equations in y = mx + b form. Find the point of intersection, and show that it is a solution to the system.

29.

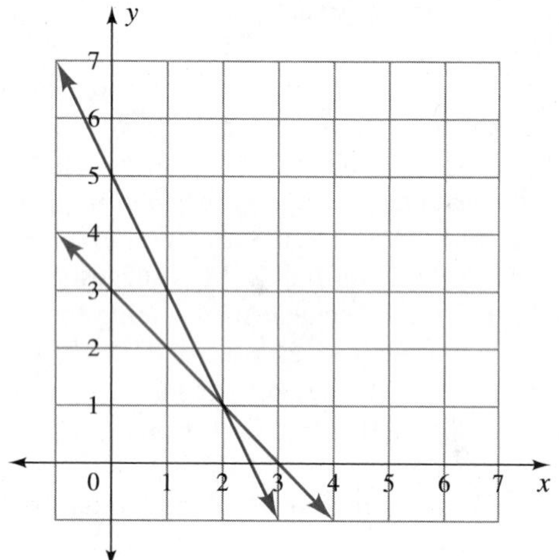

30.

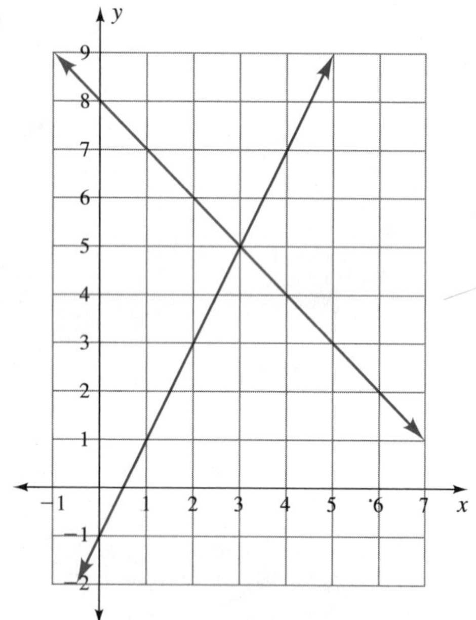

31.

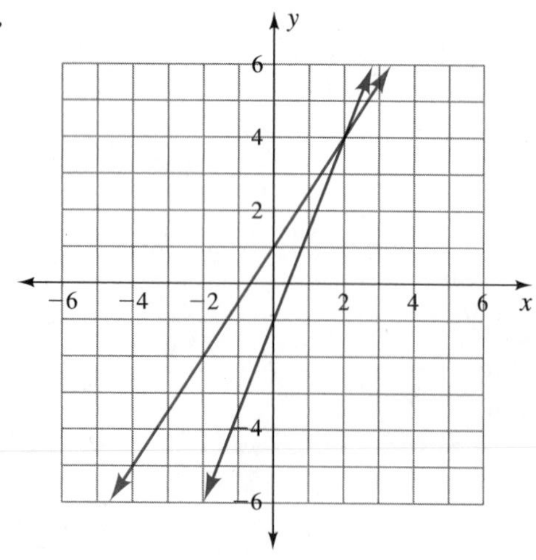

32.

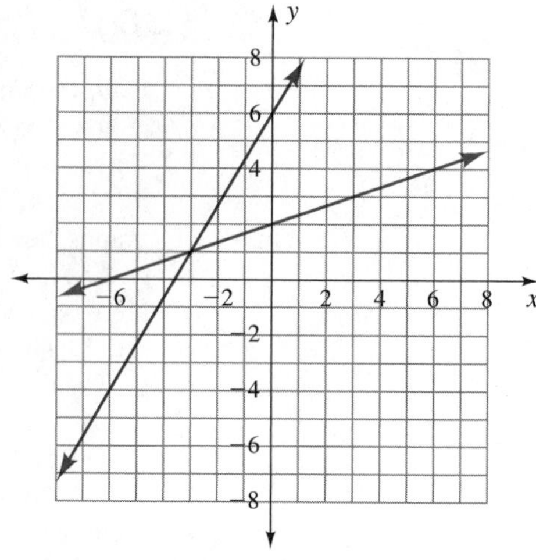

33. Sense Rent-a-Car charges \$50 plus \$0.05 per mile. Herr's Rent-a-Car charges \$20 plus \$0.20 per mile.

a. Write a system of equations to describe total rental costs y, where x is the number of miles driven.

b. Graph the equations.

c. Find the point of intersection, and describe its meaning in terms of the problem situation.

d. Which company has the steeper graph? Why?

e. What is the meaning of the y-intercept of each graph?

f. For what distances is Sense Rent-a-Car the better deal?

34. Ava's Rent-a-Car currently charges \$40 plus \$0.10 per mile. Ava's wants to compete with Herr's.

a. How will Ava's graph change if its new charge is \$40 plus \$0.08 per mile?

b. How will Ava's graph change if its new charge is \$30 plus \$0.10 per mile?

c. Which change (part a or part b) will be a better deal for customers than Herr's Rent-a-Car's charge of \$20 plus \$0.20 per mile? Explain.

d. What would Ava's charge be if the company halved its original rate per mile?

35. The Appliance Works wants to increase income, y, for people with large sales, x.

a. Which plan rewards big sales: $y = 350 + 0.04x$ or $y = 50 + 0.12x$?

b. Graph the two plans in part a.

c. What is the point of intersection of the graphs of the equations in part a?

d. Under what conditions would an employee prefer $y = 350 + 0.04x$?

e. What are the new equations if the employer doubles the rate per dollar of sales?

Exercises 36 to 40 refer to the figure below. The figure shows two lines: revenue and cost. Revenue is from registration fees at a student career conference in 1985. Costs are from printing the promotional flyer and ordering lunches for x participants. In business, the point of intersection of a cost graph and a revenue graph is the break-even point, B.E. Profit is equal to revenue minus cost.

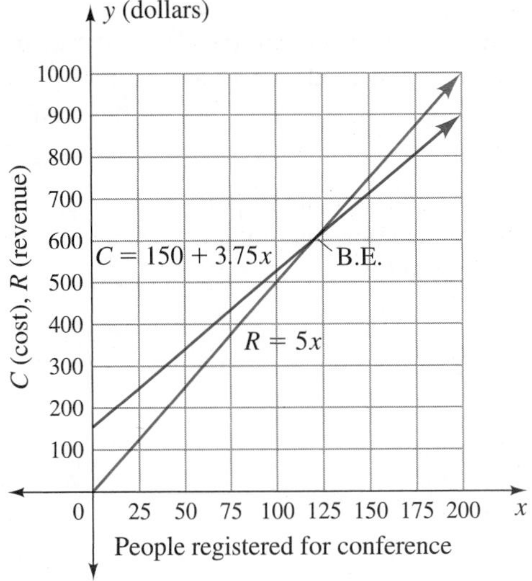

36. a. What is the registration fee?

b. What is the printing cost?

c. What is the cost of each lunch?

d. Estimate the ordered pair for the break-even point.

37. Suppose 100 students attend the conference.

a. What is the total cost for the people putting on the conference?

b. What is the total revenue to the conference?

c. What is the profit?

d. If the total registration is below the break-even point, which graph is on top?

38. Suppose 150 students attend the conference.

a. What is the total cost for the people putting on the conference?

b. What is the total revenue to the conference?

c. What is the profit?

d. If the total registration is above the break-even point, which graph is on top?

39. When this same conference was held again in 1999, a local sponsor gave \$200 and the charge for registration was \$10. The planning group paid \$250 for printing and \$8.50 for each lunch.

a. What is the equation for total cost?

b. What is the equation for total revenue?

c. Draw the graph for the new data.

d. What is the new break-even point?

40. Repeat Exercise 39 with an \$8.50 fee for registration instead of a \$10 fee.

41. Marti leaves home at 7 a.m. on her bicycle and travels 20 miles per hour. Jan leaves from the same house two hours later by car and travels 45 miles per hour along Marti's route. Let the input, t, be the number of hours after 7 a.m. Let the output, D, be in miles, with $D = rt$.

a. Draw a graph for each person that shows the distance in miles each travels between 7 a.m. and noon.

b. What equation describes Marti's distance from home?

c. What equation describes Jan's distance from home?

d. What is the intersection of the two graphs?

e. What does the intersection mean?

f. Who has the steeper graph, Marti or Jan? Why?

g. Why does Jan's graph start at (2, 0) and not at the origin?

Projects

42. ***Racing Times***

a. In a 500-mile Indy-car street race, Lyn St. James averaged 150 miles per hour for the entire race. Her main competitor, Bobby Rahal, averaged 160 miles per hour for the first 250 miles but, because of a slight brush with the safety wall, averaged 140 miles per hour for the second 250 miles of the race. Guess which driver won the race.

b. Draw a graph for Lyn with $D = 150t$. Place time on the horizontal axis and distance on the vertical axis. Suppose t_1 = time for Bobby to drive the first 250 miles at 160 miles per hour. Calculate t_1. Draw a conditional graph for Bobby with $D = 160t$ for $t < t_1$ and $D = 140(t - t_1) + 250$ for $t > t_1$.

c. Assume each driver completed the 500 miles. How long did it take Lyn to finish the race? How long did it take Bobby to finish the race?

d. What is the difference in elapsed time, if any, between the drivers? Use unit analysis to change the time to seconds.

e. At what time and distance after the race started did the winner take the lead?

43. ***Mt. St. Helens Eruption.*** Photographer Gary Rosenquist was 11 miles from Mt. St. Helens when it erupted on May 18, 1980. He began taking photographs and then realized that the pyroclastic flow from the mountain was traveling in his direction. He later learned that the flow traveled 120 miles per hour, had temperatures of up to 300°F, and flattened trees up to 15 miles away. Suppose that, to be safe, he had to travel 9 miles over graveled roads in a direction away from the mountain. How long could he take pictures and still get away safely? Use these suggestions and questions to guide your thinking.

a. How many miles per minute is 120 mph? 90 mph? 60 mph?

b. Draw axes with time after the eruption, in minutes, on the horizontal axis and distance from the mountain, in miles, on the vertical axis. Plot (and connect with a line) the position of the pyroclastic flow for each minute. Plot Rosenquist's original position as a horizontal line. Plot the safe position as another horizontal line.

c. In how many minutes does the pyroclastic flow reach the 11-mile position? In how many minutes could the pyroclastic flow reach the safe position?

d. Suppose Rosenquist leaves immediately without taking any pictures and travels 60 miles per hour. Draw a line showing his position each elapsed minute, and show that he reaches safety before the pyroclastic flow arrives.

e. Show how long Rosenquist can take pictures and still make it to safety if he can travel an average of 60 mph on the forest roads. State any assumptions you make. How long can he stay if he travels 90 mph after leaving?

f. Fit equations to your graphs in parts b, d, and e by selecting two points on each graph (and doing linear regression if needed).

7.2 Setting Up Systems of Equations

OBJECTIVES

- Use a quantity-value table to set up a system of equations.
- Use guess and check to set up a system of equations.

> **WARM-UP**
>
> Sort these units into quantities and values (rates):
>
> | | |
> |---|---|
> | 5 pounds per bag | 100 bags |
> | 60 nickels | \$0.25 per quarter |
> | 30 hours | \$7.50 per hour |
> | 20 quarters | \$0.05 per nickel |
> | 10 pounds per bag | 20 hours |
> | \$9.75 per hour | 400 bags |

IN THIS SECTION, we examine some strategies for setting up the system of equations necessary to solve a problem. We return to the quantity and value concept of Section 5.4. We use tables to organize a guess-and-check process for building equations.

Problem-Solving Steps in Setting Up Equations

To review the process for setting up equations from word problems, we return to the four problem-solving steps in Section 1.1:

1. Move toward *understanding* by reading carefully.

2. *Plan* by considering what might be reasonable inputs and preparing tables, charts, or pictures to organize the information. Guessing an input is a good strategy.
3. *Carry out the plan* by working through the problem with the chosen input.
4. *Check* by comparing the result with the conditions or requirements of the original problem.

Building Systems of Equations with Quantity-Value Tables

For our exploration, we use a puzzle problem, because it is easier to think about than more realistic applications.

EXAMPLE  Solving a quantity and value problem with guess and check To understand the problem, read it carefully: Robert likes to keep a number of nickels and quarters in the ashtray of his car, to use for tolls and parking meters. His 18-year-old daughter borrows the car and notices that there are no coins in the ashtray, so she gets \$10 in change. His 4-year-old daughter proudly tells him that there are now 80 coins in the ashtray. How many of each coin is in the ashtray?

To make a plan, consider these questions:

What are the inputs (the variables)?

What information is known?

What information is not known?

What information is assumed?

As we guess and check, what facts (the constants) do not change?

Can we estimate an answer?

Can we use a quantity-value table?

To carry out the plan, first check your understanding by explaining why 60 nickels might be a reasonable guess. Then complete Table 3 for 60 nickels. Make another guess for the number of nickels, and complete the table again.

Item	Quantity	Value	$Q \cdot V$
Nickels			
Quarters			
Total			

Table 3

Solution The inputs are the two types of coins: nickels and quarters. The number of each type of coin is unknown. The total number of coins, as well as the total value of the coins, is known. The problem assumes that we know that nickels have a value of \$0.05 and quarters have a value of \$0.25. The total number of coins, their values, and the total value are constants.

If we divide the total value, \$10, by the number of coins, 80, we get the average value of the coins, \$0.125. This average is somewhat closer to \$0.05 than to \$0.25, so we might guess that there are more nickels than quarters. If there are 80 coins and more are nickels than quarters, 60 nickels is a reasonable starting guess.

Because we have both the number (quantity) of coins and the value of the coins to consider, a quantity-value table is appropriate for the problem. The tables are in the Answer Box. ●

Think about it 1: How can the completed quantity-value tables for Example 1 be used to predict the numbers of coins? Find the number of each type of coin.

In Example 2, we use two variables to set up equations.

EXAMPLE  Finding equations in two variables from a quantity-value table Set up a system of two equations for the problem in Example 1 by letting $x =$ the number of nickels in the quantity-value table and $y =$ the number of quarters. Place a dollar sign in the column headings where appropriate.

Solution We write x, y, and 80 (the total number of coins) in the quantity column in Table 4. Next we write the values of the coins in the value column. The total value of the money is \$10, and so we put 10 in the lower right corner, as the total of the $Q \cdot V$ column.

Item	Quantity	Value (\$/coin)	$Q \cdot V$ (\$)
Nickels	x	0.05	$0.05x$
Quarters	y	0.25	$0.25y$
Total	80		10

Table 4

As in Section 5.4, we multiply across the table (Quantity · Value, or $Q \cdot V$) to obtain the total value of the nickels, $0.05x$, and the total value of the quarters, $0.25y$. From the columns, we write a system of two equations in two variables. The quantity column gives $x + y = 80$. The $Q \cdot V$ column gives $0.05x + 0.25y = 10$. ●

Think about it 2: Use one variable to describe the quantities of nickels and quarters in Example 2. (*Hint:* See Section 5.5, page 324.)

Note that the dollar signs are missing on the last equation. *You may find it convenient when working with numbers to use dollar signs. However, when working with algebraic notation, you may prefer to place the dollar sign in column headings and on numerical answers.*

As you work through the remaining examples, do extra guess-and-check steps. Only experience can show you the importance of guess and check in reading problems, setting up tables, observing patterns, and then writing equations.

In Example 2, the unknowns were the numbers of coins. In Example 3, the unknowns are the numbers of hours.

EXAMPLE 3 Writing equations in a quantity-value setting Jaynine works two different jobs for one employer. Last week she worked 50 hours altogether and earned a total of \$435.75. When she works in the shipping department, she earns \$7.50 per hour When she is filling orders, she earns \$9.75 per hour. She wants to know how many hours she was paid for at each job.

a. List the constants and variables.

b. Find the average wage, and use it to come up with a guess for a quantity-value table.

c. Make another table with two variables, and write a system of two equations that can be solved to find the number of hours for which she was paid at each job.

Solution **a.** The constants are the total hours of work, 50; the total earnings, \$435.75; and the hourly wages, \$7.50 and \$9.75, which reflect the value of each hour worked. The variables are the number of hours worked in shipping and the number of hours spent filling orders.

b. We find the average wage by dividing earnings by hours: \$435.75 ÷ 50 = \$8.715 per hour. This is slightly closer to \$9.75 than to \$7.50, so we estimate that the number of hours spent filling orders will be slightly higher than the number spent shipping. In Table 5, we try 30 hours for filling orders and 20 hours for shipping.

Item	Quantity (hr)	Value (\$/hr)	$Q \cdot V$ (\$)
Shipping	20	7.50	150
Filling orders	30	9.75	292.50
Total	50		442.50

Table 5

c. If we let x = the number of hours worked in shipping and y = the number of hours spent filling orders, then the table will look like Table 6.

Item	Quantity (hr)	Value (\$/hr)	$Q \cdot V$ (\$)
Shipping	x	7.50	$7.50x$
Filling orders	y	9.75	$9.75y$
Total	50		435.75

Table 6

The system of equations comes from the columns:

$$x + y = 50 \quad \text{Sum of the quantity column}$$
$$7.50x + 9.75y = 435.75 \quad \text{Sum of the } Q \cdot V \text{ column}$$

Think about it 3: Continue guessing to find the number of hours spent shipping and filling orders in Example 3.

Example 4 has a business decision setting. This time, the "value" is the weight of onion bags instead of a money value.

EXAMPLE  Writing equations from a quantity-value table: onion bags A produce distributor supplies fresh fruit and vegetables to grocery stores. The distributor receives a shipment of 4050 pounds of onions and needs to bag them for retail sale. Past history shows that stores sell four times as many 5-pound bags as 10-pound bags.

Set up a quantity-value table for a guess of one hundred 10-pound bags. Set up a second quantity-value table with two variables. Write a system of equations to describe the problem setting.

Solution The variables are the numbers of the two sizes of bags. The weights of the bags and the total weight of the onions are the constants.

Setting up the quantity-value table, as shown in Table 7, helps us read the problem carefully and translate the phrase "four times as many" into a number of

5-pound bags. The total number of pounds we come up with is about 1000 less than the required number of pounds, so we must raise the number of 10-pound bags if we continue to guess.

Item	Quantity (bags)	Value (lb/bag)	$Q \cdot V$ (lb)
5-pound bags	400	5	2000
10-pound bags	100	10	1000
Total	500		3000

Table 7

Using variables, we have Table 8.

Item	Quantity (bags)	Value (lb/bag)	$Q \cdot V$ (lb)
5-pound bags	x	5	$5x$
10-pound bags	y	10	$10y$
Total	???		4050

Table 8

We get one equation from the total weight:

$$5x + 10y = 4050$$

The second equation comes from the store sales:

$$x = 4y$$

Think about it 4: First explain why $x = 4y$ describes the number of 5-pound bags. Then use guess and check to solve Example 4.

Building Systems of Equations with Guess-and-Check Tables

In the remaining examples, we will use guess and check to help us understand the problem and then write the system of equations. These tables have *a column for each unknown and a row for each guess.*

EXAMPLE 

Using guess and check to understand the problem and write a system of equations: job sharing Cal and Joe share a job that takes a total of 243 hours each month. Cal works 17 hours more than Joe. Use a guess-and-check table to write a system of equations to find out how many hours each works.

Solution The unknowns are the numbers of hours Cal and Joe work, so we have a column for Cal's hours, another for Joe's hours, and a third for total hours. For a first guess of Cal's hours, we pick 150, a number that is more than half of 243. Because Cal works 17 more hours than Joe, our guess for Joe's hours is 133. Table 9 shows the total hours to be 283, which is too high.

Cal's Hours	Joe's Hours	Total Hours
150	$150 - 17 = 133$	$150 + 133 = 283$
140	$140 - 17 = 123$	$140 + 123 = 263$
x	y	243

Table 9

Our next guess is slightly lower: 140. Subtracting 17, we get 123 for Joe's hours. We add the hours to get 263 in total. In the third row, we let x = Cal's hours and y = Joe's hours. We then write two equations:

$y = x - 17$ The difference in their work hours

$x + y = 243$ The total hours worked ●

Think about it 5: Use guess and check to solve for the number of hours in Example 5.

In Example 6, we return to the party packages in the Chapter 3 opener, on page 124. Again we show the calculations in a guess-and-check table so that we can use these calculations to write equations.

EXAMPLE 6 Using guess and check to find patterns and write equations: party packages An evening party at Skate World costs \$55 for up to 10 people and \$6 for each additional person. At Lane County Ice, the cost is \$85 for up to 10 people and \$4.75 for each additional person. Use guess and check to estimate the number of people for which the cost will be the same for parties at these two locations. Write a system of equations.

Solution Table 10 has three column headings: the number of people and the total cost at each location. The steps in finding the total cost are shown in the columns.

Number of People	Skate World Cost (\$)	Lane County Ice Cost (\$)
10	55	85
20	$55 + 6(20 - 10) = 115$	$85 + 4.75(20 - 10) = 132.50$
30	$55 + 6(30 - 10) = 175$	$85 + 4.75(30 - 10) = 180$
40	$55 + 6(40 - 10) = 235$	$85 + 4.75(40 - 10) = 227.50$

Table 10

For 10 people, the costs at the two locations are clearly not the same. Lane County Ice costs \$30 more. We guess 20 people and complete the second row. The costs are still different but are less than \$20 apart. At 30 persons, the costs are within \$5. At 40 people, Skate World becomes more expensive by slightly more than \$5, so the answer is between 30 and 40 people and slightly closer to 30 people.

Let x = the number of people and y = the total cost in dollars. For Skate World, if $x \le 10$, $y = 55$. If $x > 10$,

$$y = 55 + 6(x - 10)$$

For Lane County Ice, if $x \leq 10$, $y = 85$. If $x > 10$,

$$y = 85 + 4.75(x - 10)$$

Think about it 6: Why do we subtract 10 from the number of people before multiplying by \$4.75 or \$6.00? Continue to guess and check until you find a solution and then compare the solution with the point of intersection of the equations on a graphing calculator graph (see the Graphing Calculator Box).

Graphing Calculator Technique: Graphing Conditional Equations

Graph the equations in Example 6 and find their point of intersection. For Skate World,

$$Y_1 = 55(X \leq 10) + (55 + 6(X - 10))(X > 10)$$

For Lane County Ice,

$$Y_2 = 85(X \leq 10) + (85 + 4.75(X - 10))(X > 10)$$

The conditions are written as inequalities and placed immediately after a set of parentheses containing the expression.

When one or more of the equations in a system is nonlinear (not a straight line), the possible number of solutions is hard to predict. In Example 7, one equation is linear and the other is a conditional equation in two line segments.

EXAMPLE 7 Writing a system of equations from an input-output table and solving with a graph: more party packages Skate World offers a Saturday afternoon party package: \$45 for the first 10 people and \$5 for each additional person. The Willamalane Wave Pool charges \$30 for a three-hour rental of its party room and \$3 per person. For how many people will the two locations cost the same? Make a table and graph to answer this question. What can we conclude about the costs?

Solution From Table 11 it appears that the cost of the two locations is the same for 5 people. We might quit there, but a graph suggests that there is more than one solution.

Number of People, x	Skate World	Wave Pool
5	45	$3(5) + 30 = 45$
10	45	$3(10) + 30 = 60$
15	$45 + 5(15 - 10) = 70$	$3(15) + 30 = 75$
20	$45 + 5(20 - 10) = 95$	$3(20) + 30 = 90$
For $x \leq 10$	45	$3x + 30$
For $x > 10$	$45 + 5(x - 10)$	$3x + 30$

Table II

The graph in Figure 9 shows two points of intersection: (5, 45) and about (17, 80). The wave pool is cheaper for fewer than 5 people and 18 or more people. Skate World is cheaper for $5 < x \leq 17$ people.

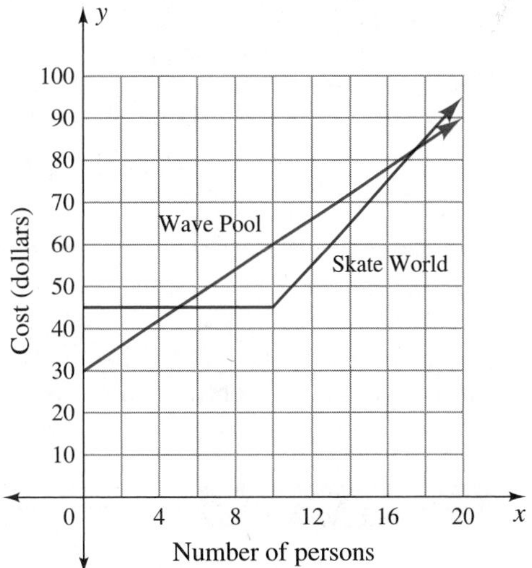

Figure 9

Graphing Calculator Solution For Skate World,

$$Y_1 = 45(0 < X)(X \leq 10) + (45 + 5(X - 10))(X > 10)$$

For Willamalane Wave Pool,

$$Y_2 = 3X + 30$$

Trace to find the two points of intersection: (5, 45) and (17.5, 82.5), or between 17 and 18 people. ●

Think about it 7: Suppose the party room rental at the wave pool is $10 per hour. How would the graph for the total cost at the wave pool change if the party room were rented for 2 hours instead of 3 hours? Change the graphing calculator equation and find the new solutions with a 2-hour rental.

The two solutions in Example 7 are due to the conditional nature of the Skate World prices. This realistic example illustrates why it is important to graph conditional equations.

Systems of Equations in Three Variables

In Example 8, we use guess and check to set up a system of equations to solve for three unknowns.

EXAMPLE  8

Writing a system of three equations in three variables: schedule planning Herman needs to include work, class, and study time in his schedule. He spends 12 more hours per week at work than he does in class. He studies 2 hours for each hour he is in class. He spends a total of 64 hours each week on the three activities. Set up a guess-and-check table to find the number of hours spent at each activity. In the last row of the table, show variables. Then write a system of equations to describe the problem.

Solution The guess-and-check table appears in Table 12.

The system is

$$c = w - 12$$
$$s = 2c$$
$$w + c + s = 64$$

Work Time (hr)	Class Time (hr)	Study Time (hr)	Total (hr)
20	$20 - 12 = 8$	$8 \times 2 = 16$	$20 + 8 + 16 = 44$
24	$24 - 12 = 12$	$12 \times 2 = 24$	$24 + 12 + 24 = 60$
w	c	s	64

Table 12

Think about it 8: Continue to guess and check until you find a solution.

ANSWER BOX

Warm-up: Quantities are 100 bags, 60 nickels, 30 hours, 20 quarters, 20 hours, and 400 bags. Values are 5 pounds per bag, \$0.25 per quarter, \$7.50 per hour, \$0.05 per nickel, 10 pounds per bag, and \$9.75 per hour.
Example 1:

Item	Quantity	Value	$Q \cdot V$
Nickels	60	\$0.05	\$3
Quarters	20	\$0.25	\$5
Total	80		\$8

The total value is too low, so we need more quarters and fewer nickels. Your own guess is probably correct, but here is a table for 40 nickels.

Item	Quantity	Value	$Q \cdot V$
Nickels	40	\$0.05	\$2
Quarters	40	\$0.25	\$10
Total	80		\$12

This time the total value is \$12 instead of \$10, so we need more nickels. **Think about it 1:** Our first guess of 60 nickels is too high, and our second guess of 40 nickels is too low. We might try the average: 50 nickels. A \$10 total comes from 50 nickels and 30 quarters. **Think about it 2:** x and $80 - x$ **Think about it 3:** 23 hours shipping and 27 hours filling orders **Think about it 4:** There are four times as many 5-pound bags as 10-pound bags, so we multiply the number of 10-pound bags by 4; 135 ten-pound bags and 540 five-pound bags **Think about it 5:** 130 hours and 113 hours **Think about it 6:** The cost for the first 10 people at each location is a flat fee; 34 people at each location costs \$199. **Think about it 7:** The vertical axis intercept would be \$20 instead of \$30. The new intersections are at between 8 and 9 people, $(8\frac{1}{3}, 45)$, and between 12 and 13 people, (12.5, 57.5). **Think about it 8:** $w = 25$ hr, $c = 13$ hr, $s = 26$ hr

EXERCISES 7.2

Set up quantity-value tables for Exercises 1 to 10. Complete one table with a guess as one input, and then complete a second table with variables as inputs. Write a system of equations to describe the problem situation. It is not necessary to solve the problem.

1. Celesta has \$2.90 in quarters and nickels. She has 22 coins altogether. How many of each does she have?

2. Larry has 24 nickels and quarters. He has \$2.60 altogether. How many nickels and how many quarters does he have?

3. Casey has 26 coins. He has \$5.45 altogether. If he has only dimes and quarters, how many of each does he have?

4. Nancy has \$3.90. She has only dimes and quarters. She has 27 coins altogether. How many of each does she have?

5. Lindsay has 65 nickels and dimes. She has \$5.40 altogether. How many nickels and how many dimes does she have?

6. Martina has 57 nickels and dimes. She has \$3.70 altogether. How many nickels and how many dimes does she have?

7. Se Ri earns \$316 from working 43 total hours at two jobs. She earns \$5.75 per hour at the first job and \$8.50 per hour at the second job. How many hours does she work at each job?

8. Paolo earns \$409 from working 54 total hours at two jobs. He earns \$6.25 per hour at the first job and \$9.50 per hour at the second job. How many hours does he work at each job?

9. The I. R. Rabbit Company sells carrots. It sells nine times as many pounds to grocery stores as to restaurants. It has 5250 pounds of carrots. How many pounds should be packaged for grocery stores and how many for restaurants?

10. Restaurants buy 10-pound and 25-pound bags of carrots. They want five times as many 10-pound bags as 25-pound bags. If the company estimates that restaurants will buy 525 pounds of carrots, how many bags of each size should be packaged?

In Exercises 11 to 18, set up a guess-and-check table. Make three guesses, and then label the columns in the table with variables. Write a system of equations that will solve the problem. It is not necessary to solve the problem.

11. Renee cuts a 10-meter hose into two lengths. One piece is 5 meters longer than the other. What is the length of each piece?

12. Jacques cuts a 12-decimeter submarine sandwich into two pieces. One piece is 6 decimeters longer than the other. What is the length of each piece?

13. Ed works on two projects during one month. He works 28 hours longer on one project than on the other. He works a total of 176 hours during the month. How many hours does he work on each project?

14. Alexandra works for two clients during the month. She works a total of 170 hours during the month. She works 62 hours more for one client than for the other. How many hours does she work for each client?

15. A party at Fun Base One costs \$30 for 8 children and \$3.50 for each additional child. A party at Papa's Pizza costs \$3.00 per child. For how many children will the costs of parties at the two locations be equal?

16. Lane County Ice costs \$85 for 10 children and \$4.75 for each additional child. Grand Slam U.S.A. charges \$5.95 per child for its basic party. For what number of children will the party costs be equal?

17. For its party package, the American Gymnastics Training Center charges \$70 for up to 15 children and \$3 for each additional child. Farrell's Ice Cream Parlour and Restaurant charges \$3.95 per child. For what number of children will parties at the two locations cost the same?

18. Fun Base One charges \$30 for 8 children and \$3.50 for each additional child. Grand Slam U.S.A. charges \$5.95 per child. For what number of children will parties at the two locations cost the same?

For Exercises 19 to 22, solve the given exercise by graphing.

19. Exercise 15

20. Exercise 16

21. Exercise 17

22. Exercise 18

For Exercises 23 to 26, write a system of three equations. Solve using guess and check.

23. Marielena spent \$620 on a watch, locket, and chain. She paid \$20 more for the locket than for the chain. She paid twice as much for the watch as for the locket. How much did she pay for each?

24. Katreen earns \$1375 one summer. She earns twice as much mowing lawns as shopping for the elderly. She earns \$500 more scraping old paint on houses than mowing lawns. How much does she earn at each job?

25. Andre buys notebooks, paperback books, and hardbound books for his classes. He buys 3 more paperback books than hardbound books. He buys 1 fewer notebooks than paperback books. He buys 20 items altogether. How many of each does he buy?

26. The I. R. Rabbit Company is packaging carrots. It has 2400 pounds of carrots to be packaged into bags for grocery stores. The bags are 1 pound, 2 pounds, and 5 pounds. The company sells twice as many 1-pound bags as 2-pound bags and ten times as many 1-pound bags as 5-pound bags. How many bags of each size should the company fill?

27. The number of equations we write to help solve a problem is related to the number of variables we use. Look for a pattern in the examples and exercises, and complete these statements:

a. When we use one variable, we need ____ equation(s).

b. When we use two variables, we need ____ equation(s).

c. When we use three variables, we need ____ equation(s).

Projects

28. ***Using Manipulatives to Solve Problems. Part I:*** One day Mr. McFadden decides to count his farm animals. He counts strangely and reports 10 heads and 28 feet. He has cows and ducks. (Note: It is permissible to have zero animals of one type.) How many of each does Mr. McFadden have? Use 10 coins, rubber bands, or buttons for heads. Use 28 toothpicks, paper clips, safety pins, or cotton swabs for feet. Model the animals.

a. Is it possible to have 10 heads and 24 feet?

b. Is it possible to have 10 heads and 30 feet?

c. What is the largest number of feet possible with the 10 heads?

d. What is the smallest number of feet possible with the 10 heads?

e. What patterns do you observe if you organize your data into a table? Use the following table, adding more rows or columns as needed.

Ducks	Cows	Total Heads	Total Feet

f. Write a system of equations to solve Mr. McFadden's problem.

Part II: Mr. McFadden's neighbor is Mr. Schaaf. Mr. Schaaf counts his animals the same way. He reports 12 heads and 28 feet. How many cows and ducks does Mr. Schaaf have?

g. What is the largest number of heads possible with 28 feet?

h. What is the smallest number of heads possible with 28 feet?

i. Organize your results in the table below to look for patterns. Add more rows or columns as needed.

Ducks	Cows	Total Heads	Total Feet

j. Write a system of equations to solve Mr. Schaaf's problem.

29. ***Equations in One Variable, I.*** This section focused on writing systems of equations in two (or more) variables. Review earlier work (page 324) by writing one equation in one variable for each of the given exercises. Solve the equation.

a. Exercise 1 (Hint: Let $x =$ the number of nickels, $22 - x =$ the number of quarters.)

b. Exercise 3

c. Exercise 5

d. Exercise 7 (Hint: Let $x =$ the number of hours at \$5.75, $43 - x =$ the number of hours at \$8.50.)

e. Exercise 9

f. Exercise 11

g. Exercise 13

30. ***Equations in One Variable, II.*** Write one equation in one variable for each of the given exercises. Solve the equation.

a. Exercise 2

b. Exercise 4

c. Exercise 6

d. Exercise 8

e. Exercise 10

f. Exercise 12

g. Exercise 14

7.3 Solving Systems of Equations by Substitution

OBJECTIVES

- Solve an equation for one variable in terms of a second variable.
- Solve a system of equations by substitution.
- Solve a system of equations when both equations are in $y = mx + b$ form.
- Use geometry facts in setting up a system of equations.
- Solve a system of three equations in three variables.

WARM-UP

Solve for the indicated variable.

1. $x = 4y$ for y

2. $x + y = 80$ for x

3. $2y = 5 - x$ for x

4. $2y = 5 - x$ for y

5. $x + y = 2000$ for x

6. $c = w - 12$ for w

7. $\dfrac{l}{w} = \dfrac{8}{5}$ for l

THERE ARE MANY WAYS of solving problems, and some work better than others in particular situations. We have used graphs and tables to solve equations. In this section, we solve equations in algebraic notation using a process called substitution. We will return to several examples from Sections 7.1 and 7.2 and work some new examples involving geometry facts.

Solving Systems of Equations by Substitution

Solving by Substitution

To solve by **substitution**, we replace variables with equivalent expressions or numbers.

Mathematical substitution resembles the substitution of players in a sports event or the substitution of ingredients in a recipe. In sports and cooking as well as in mathematics, there is a "taking out" and a "putting in" process. However, substitutions differ in that mathematics uses an equal replacement, whereas sports and cooking use a replacement that only approximates the player or ingredient removed.

EXAMPLE  Solving a system of equations by substitution We return to the onion bag problem from Section 7.2 (Example 4, page 417), where we had the system

$$x = 4y$$
$$5x + 10y = 4050$$

The point of intersection is the ordered pair that makes both equations true. Thus, at the point of intersection, the condition $x = 4y$ is true for the x and y in $5x + 10y = 4050$. Use the following substitution process to solve the system.

a. Replace x in the second equation with $4y$, and solve the new equation for y.

b. Substitute the y value from part a into $x = 4y$ to find x.

c. State the solution as an ordered pair, (x, y).

d. Check that the ordered pair makes both equations true.

Solution **a.**

$5x + 10y = 4050$	Substitute $4y$ for x.
$5(4y) + 10y = 4050$	Simplify.
$20y + 10y = 4050$	Add like terms.
$30y = 4050$	Divide both sides by 30.
$y = 135$	

See the Answer Box for parts b, c, and d. ●

Steps in Solving a System of Equations by Substitution

1. Solve one equation for a variable equal to an expression.
2. In the other equation, substitute the expression from step 1 for the variable, placing the expression in parentheses. Solve the second equation for the remaining variable.
3. Substitute the value from step 2 in the first equation.
4. State the solution as an ordered pair, (x, y), or as a system, $x =$, $y =$.
5. Check the solution in both equations.

In Example 2, we solve the equations for the coins in the ashtray problem from Section 7.2 (Example 2, page 416). The number of coins was described by $x + y = 80$ and the value of the coins by $0.05x + 0.25y = 10$.

EXAMPLE 2 Solving by substitution Using substitution, solve the system of equations for Robert's ashtray coins:

$$x + y = 80$$
$$0.05x + 0.25y = 10$$

Solution ***Step 1:*** We solve one equation for a variable equal to an expression.

$x + y = 80$	Subtract y from each side.
$x = 80 - y$	

Step 2: In the other equation, we substitute the expression from step 1 for the variable, placing the expression in parentheses. We then solve for the remaining variable.

$0.05x + 0.25y = 10$	Substitute $x = 80 - y$.
$0.05(80 - y) + 0.25y = 10$	Apply the distributive property.
$0.05(80) - 0.05y + 0.25y = 10$	Simplify.
$4 + 0.20y = 10$	Subtract 4 from each side.
$0.20y = 6$	Divide both sides by 0.20.
$y = 30$	

Step 3: We substitute the value from step 2 in the first equation.

$$x = 80 - y$$
$$x = 80 - 30$$
$$x = 50$$

Step 4: Stating the solution as an ordered pair and as a system, we have (50, 30) and $x = 50$, $y = 30$, respectively. Robert has 50 nickels and 30 quarters.

Step 5: We check the solution in both equations.

$$x + y = 80$$
$$50 + 30 \stackrel{?}{=} 80 \checkmark$$
$$0.05x + 0.25y = 10$$
$$0.05(50) + 0.25(30) \stackrel{?}{=} 10 \checkmark$$

"How do we know where to start?" is a common question. When one variable in the system has a coefficient of 1 or -1, the first step in substitution is easy to find: We start by isolating the variable with coefficient 1 or -1.

EXAMPLE 3 Solving a system of equations by substitution Name the equation that can be easily solved for one variable. Then solve the system by substitution.

$3x - 4y = -25$ Equation I

$2y = 5 - x$ Equation II

Solution The variable x in $2y = 5 - x$ has -1 as a coefficient, so we choose to isolate the variable in Equation II.

$2y = 5 - x$ Add x to each side.

$x + 2y = 5$ Subtract $2y$ from each side.

$x = 5 - 2y$

Next we substitute $5 - 2y$ for x in Equation I.

$3x - 4y = -25$ Substitute $x = 5 - 2y$.

$3(5 - 2y) - 4y = -25$ Apply the distributive property.

$15 - 6y - 4y = -25$ Add like terms.

$15 - 10y = -25$ Subtract 15 on each side.

$-10y = -25 - 15$ Add like terms.

$-10y = -40$ Divide each side by -10.

$y = 4$

Then we substitute $y = 4$ into Equation II.

$2y = 5 - x$ Substitute $y = 4$.

$2(4) = 5 - x$ Simplify.

$8 = 5 - x$ Subtract 5 on each side.

$3 = -x$ Multiply by -1 on each side.

$-3 = x$

Finally, we check that $x = -3$ and $y = 4$ make both equations true.

$$3(-3) - 4(4) \stackrel{?}{=} -25 \checkmark$$
$$2(4) \stackrel{?}{=} 5 - (-3) \checkmark$$

Think about it 1: What one step is needed to solve $2y = 5 - x$ for y? Why might substituting the resulting expression for y into $3x - 4y = -25$ be more difficult than substituting $5 - 2y$ for x?

Equations in $y = mx + b$ form are the most natural to solve by substitution. In Example 4, we replace y in one equation with $mx + b$ from the other equation.

EXAMPLE 4 Solving a system of equations in $y = mx + b$ form Using substitution, solve this system of equations from the wage options problem in Section 7.1 (Example 4, page 406):

$$y = 300 + 0.06x$$
$$y = 150 + 0.10x$$

Solution To solve this system of equations by substitution, we replace y in the second equation with $300 + 0.06x$ from the first equation and then solve for x.

$$y = 150 + 0.10x \quad \text{Let } y = 300 + 0.06x.$$
$$300 + 0.06x = 150 + 0.10x \quad \text{Subtract } 0.06x \text{ on each side.}$$
$$300 = 150 + 0.04x \quad \text{Subtract 150 on each side.}$$
$$150 = 0.04x \quad \text{Divide each side by 0.04.}$$
$$3750 = x$$

Then we solve for the second variable.

$$y = 300 + 0.06x \quad \text{Replace } x \text{ with 3750.}$$
$$y = 300 + 0.06(3750)$$
$$y = 525$$

The solution is $x = 3750$, $y = 525$.

Finally, we check the solution in both equations:

$$y = 300 + 0.06x$$
$$525 \stackrel{?}{=} 300 + 0.06(3750) \quad ✓$$
$$y = 150 + 0.10x$$
$$525 \stackrel{?}{=} 150 + 0.10(3750) \quad ✓$$

●

Think about it 2: What is it about the problem situation that allows us to set $300 + 0.06x$ equal to $150 + 0.10x$?

Algebraic Results for Special Systems

In Section 7.1, we saw that a system of equations that graphs as parallel lines has no solution. Also, a system that graphs as coincident lines has an infinite number of solutions. In the next two examples, we look at the algebraic results of solving these special systems. Example 5 is from More Wage Options in Section 7.1 (Example 6, page 409).

EXAMPLE 5 Finding the special algebraic results of solving a system that graphs as parallel lines Solve the system of equations

$$y = 300 + 0.06x$$
$$y = 150 + 0.06x$$

Solution We substitute $300 + 0.06x$ for y in the second equation.

$$y = 150 + 0.06x \quad \text{Let } y = 300 + 0.06x.$$
$$300 + 0.06x = 150 + 0.06x \quad \text{Subtract 150 from each side.}$$
$$150 + 0.06x = 0.06x \quad \text{Subtract } 0.06x \text{ from each side.}$$
$$150 = 0 \quad \text{The variables drop out, leaving a false statement.}$$

Because the statement $150 = 0$ is false, we say that the system of equations has no solution. ●

In Example 6, we return to the coincident lines system of Example 7 in Section 7.1 (page 409).

EXAMPLE 6 Finding the special algebraic results of solving a system that graphs as coincident lines Solve the following system of equations, which forms coincident lines:

$$y = \frac{1}{2}x - 2$$
$$2y + 4 = x$$

Solution We substitute $2y + 4$ for x in the first equation.

$y = \frac{1}{2}x - 2$ Let $x = 2y + 4$.

$y = \frac{1}{2}(2y + 4) - 2$ Simplify using the distributive property.

$y = y + 2 - 2$ Add 2 and -2.

$y = y$ Subtract y from each side.

$0 = 0$ The variables drop out, leaving a true statement.

The statement $0 = 0$ is always true. Thus, *any solution to the first equation is also a solution to the second equation. There are an infinite number of solutions.* ●

Think about it 3: In Example 6, we could have substituted $y = \frac{1}{2}x - 2$ into $2y + 4 = x$ as a starting step. Do so. Do you come to the same conclusion as in Example 6?

Table 13 summarizes the geometric and algebraic results for various types of systems of two linear equations.

Geometric Results	Intersecting lines	Parallel lines	Coincident lines
Algebraic Results	The equations can be solved for x and y: (x, y)	The variables drop out, and the remaining statement is false: $0 = 4$	The variables drop out, and the remaining statement is true: $0 = 0$
Solution	One ordered pair is the solution.	There is no real-number solution to the system.	An infinite number of ordered pairs satisfy the system.

Table 13 Geometric and Algebraic Results

Applications

QUANTITY AND VALUE In a quantity-value setting, one equation frequently expresses the sum of quantities: $x + y =$ a number. Both x and y have a coefficient of 1, and so the equation is easy to solve for either variable. We return to a quantity-value setting in Example 7.

EXAMPLE 7

Using substitution to solve a quantity-value problem: ration balancing You are the owner of a cattle-restoration ranch, saving endangered species from extinction. Suppose the local animal-science agent at the Extension Service advises you that your cattle need a 20% protein supplement. Ingredients for this supplement are soybean meal (44% protein) and corn (9% protein). Use Table 14 to find out how much of each ingredient is needed to produce 2000 pounds of supplement.

	Quantity (lb)	Value (% protein)	$Q \cdot V$ Total Protein (lb)
Soybean meal	x	0.44	$0.44x$
Corn	y	0.09	$0.09y$
Total	2000	0.20 (average)	400

Table 14 Protein Supplement

a. What equation describes the quantity column?
b. What equation describes the total protein column?
c. Solve the system of equations from parts a and b by substitution.
d. Summarize your results in terms of the problem situation.

Solution **a.** The two ingredients add to 2000 pounds:

$$x + y = 2000$$

b. The total amount of protein from each ingredient must add to 400 pounds:

$$0.44x + 0.09y = 400$$

c. We solve $x + y = 2000$ for x and substitute the result into the other equation.

$$0.44x + 0.09y = 400 \quad \text{Let } x = 2000 - y.$$
$$0.44(2000 - y) + 0.09y = 400 \quad \text{Apply the distributive property.}$$
$$880 - 0.44y + 0.09y = 400 \quad \text{Add like terms.}$$
$$880 - 0.35y = 400 \quad \text{Subtract 880 on each side.}$$
$$-0.35y = -480 \quad \text{Divide by } -0.35.$$
$$y \approx 1371 \text{ lb, rounded to the nearest pound}$$

We substitute $y \approx 1371$ to find x.

$$x = 2000 - y$$
$$x \approx 2000 - 1371$$
$$x \approx 629 \text{ lb}$$

Then we check the results in both equations:

$$x + y = 2000$$
$$629 + 1371 \stackrel{?}{=} 2000 \checkmark$$

$$0.44x + 0.09y = 400$$
$$0.44(629) + 0.09(1371) \stackrel{?}{=} 400 \checkmark$$

In the last equation, the left-hand side does not exactly equal 400 because the number of pounds was rounded.

d. A ration of 629 pounds of soybean meal at 44% protein and 1371 pounds of corn at 9% protein will provide a total of 400 pounds of protein. ●

SPECIAL ANGLES IN GEOMETRY A traditional protractor, like the one in Figure 10, shows 180° for a half circle. The numbers on a protractor permit measuring angles from either the left side or the right side.

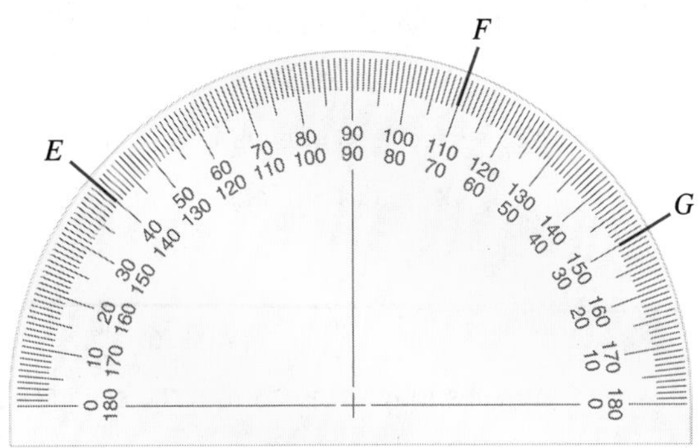

Figure 10

EXAMPLE 8 Finding patterns on a protractor What pattern can you find in the pairs of numbers labeled with the letters E, F, and G on the protractor in Figure 10?

Solution At position E, 40 matches with 140. At position F, 110 matches with 70. At position G, 150 matches with 30. Each pair of numbers adds to 180. ●

When we look at the two angles being described by each pair of numbers on a protractor, we find that *the angles share a side and their other sides form a straight line.* These angles are often called a **linear pair**. Figure 11(a) shows a linear pair.

The drawings of a parallelogram in Figure 11(b) and parallel lines in Figure 11(c) show angles that may be rearranged to form a linear pair. *Two angles that add up to a linear pair* (180°) are **supplementary angles**. The justification for the fact that the angles in Figure 11 are supplementary is left to a geometry course.

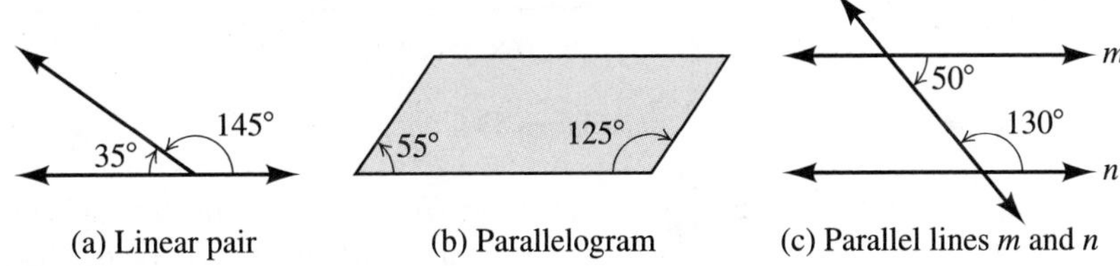

(a) Linear pair (b) Parallelogram (c) Parallel lines m and n

Figure 11

EXAMPLE 9 Solving systems by substitution: supplementary angles Angles M and N in Figure 12 are supplementary. Angle N is 24° more than five times angle M. What is the measure of each angle? Write a system of equations and solve by substitution.

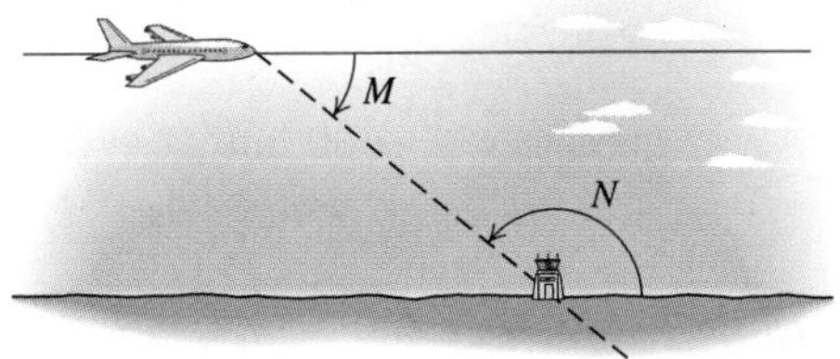

Figure 12

Solution The angles are supplementary, so they add to 180°:

$$M + N = 180$$

We translate the other sentence into a second equation,

$$N = 24 + 5M$$

We use the second equation to substitute for N in the first equation.

$$\begin{aligned} M + (24 + 5M) &= 180 \\ 6M + 24 &= 180 \\ 6M &= 156 \\ M &= 26 \end{aligned}$$

We substitute $M = 26$ into the second equation.

$$\begin{aligned} N &= 24 + 5(26) \\ N &= 154 \end{aligned}$$

The angles are $M = 26°$ and $N = 154°$.

Check:

$26 + 154 \stackrel{?}{=} 180$ ✓

$154 \stackrel{?}{=} 24 + 5(26)$ ✓

Complementary angles *add to 90°*. Any two angles forming a right angle are complementary; see Figure 13(a). In a right triangle, because one angle is 90°, the two smaller angles (called *acute angles*) add to 90° and are therefore complementary, as shown in Figure 13(b).

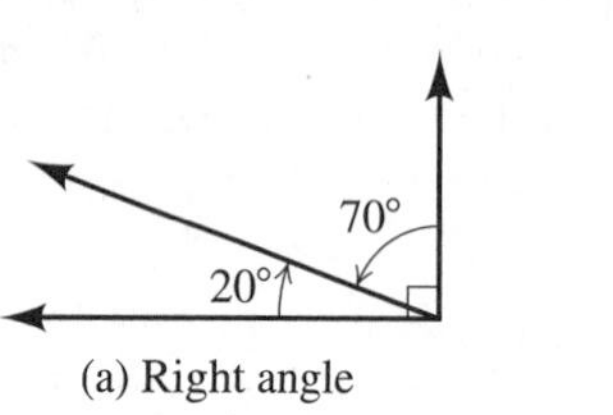

(a) Right angle

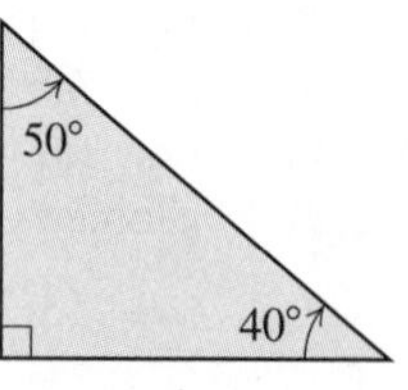

(b) Right triangle

Figure 13

EXAMPLE 10 Solving systems by substitution: complementary angles In Figure 14, the axes are perpendicular, and angle D is twice the size of angle C. What is the measure of each angle? Write a system of equations and solve by substitution.

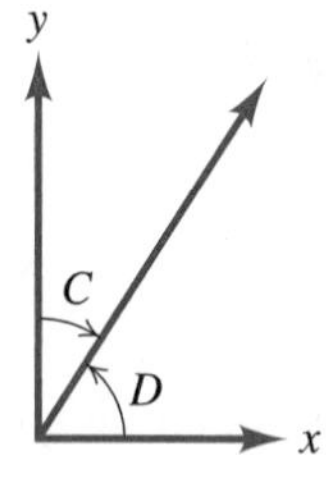

Figure 14

Solution Because the axes are perpendicular, $C + D = 90°$. From the second sentence, $D = 2C$. We substitute $D = 2C$ into the first equation.

$$\begin{aligned} C + D &= 90 && \text{Let } D = 2C. \\ C + 2C &= 90 && \text{Add like terms.} \\ 3C &= 90 && \text{Divide by 3.} \\ C &= 30 \end{aligned}$$

We substitute $C = 30$ into $D = 2C$.

$$D = 60$$

Check:

$C + D = 90$

$30 + 60 \stackrel{?}{=} 90$ ✓

$D = 2C$

$60 \stackrel{?}{=} 2(30)$ ✓

RATIOS AND PROPORTIONS This application refers back to concepts presented in Sections 5.1 and 5.2: ratios and proportions. When a word problem gives the ratio of two quantities, it may be that one equation in the system can be a proportion.

$$\text{If the ratio of } x \text{ to } y \text{ is } a \text{ to } b, \quad \text{then } \frac{x}{y} = \frac{a}{b}.$$

To solve such a proportion, we use cross-multiplication.

In Example 11, we build a system of two equations. One equation is a proportion, and the other is a geometry formula from Section 2.5.

EXAMPLE 11 Using substitution to solve a system of equations involving ratios and proportions
Pablo has 130 centimeters of expensive frame material. He wants to build a rectangular frame that has sides in the ratio of 8 to 5. Set up a system of equations to solve for the length and width of the frame, and then solve the system by substitution.

Solution Let l = length and w = width. If the ratio of length to width is 8 to 5, then the proportion is

$$\frac{l}{w} = \frac{8}{5}$$

The formula for the perimeter of a rectangle is $P = 2l + 2w$. Because the length has 1 as coefficient, we solve the proportion for the length, l.

$$\frac{l}{w} = \frac{8}{5} \qquad \text{Cross multiply.}$$
$$5 \cdot l = 8 \cdot w \qquad \text{Divide both sides by 5.}$$
$$l = \frac{8w}{5}$$

We then substitute $l = 8w/5$ into the perimeter formula, where $P = 130$ cm.

$$2l + 2w = 130 \qquad \text{Substitute } l = 8w/5.$$
$$2\left(\frac{8w}{5}\right) + 2w = 130 \qquad \text{Simplify.}$$
$$\frac{16w}{5} + 2w = 130 \qquad \text{Multiply both sides by 5.}$$
$$16w + 10w = 650 \qquad \text{Add like terms.}$$
$$26w = 650 \qquad \text{Divide both sides by 26.}$$
$$w = 25$$

Next we find the length.

$$l = \frac{8w}{5} \qquad \text{Substitute } w = 25.$$
$$l = \frac{8(25)}{5} \qquad \text{Simplify.}$$
$$l = 40$$

The length is 40 centimeters, and the width is 25 centimeters.

Check:

The ratio 40 to 25 equals 8 to 5. ✓

The perimeter is $2(40) + 2(25) = 130$ cm. ✓

Solving Systems of Three Equations by Substitution

Calculator technology now makes it easy to solve complicated systems of three or more linear equations in three or more unknowns. Thus, we will consider here only selected systems that can be solved reasonably quickly by hand. In the next two examples, not all the equations in the system contain all three variables. As a result, in both examples we will be able to arrange two of the equations so as to substitute for two variables in the third equation.

Solving a System of Three Equations in Three Variables

1. Look for a common variable in all three equations.
2. If possible, solve two of the equations for one variable *in terms of* the common variable.
3. Use the equations in step 2 to substitute for all but the common variable in the third equation.
4. Solve for the value of the common variable.
5. Substitute the value of the common variable into the first two equations to find the values of the other variables.

In Example 12, we return to the system of equations for Herman's work, study, and class time (Example 8, Section 7.2, page 421). We solve the system by substitution.

EXAMPLE 12 Solving a system of three equations in three unknowns Solve this system of equations by substitution:

$$c = w - 12$$
$$s = 2c$$
$$w + c + s = 64$$

Solution Each equation contains the variable c, so we solve the first two equations for the variables w and s in terms of c.

$$w = c + 12$$
$$s = 2c$$

We then substitute for w and s in the third equation, to build an equation containing only c. We place the substitutions in parentheses so that they show up clearly.

$$w + c + s = 64 \quad \text{Substitute for } w \text{ and } s.$$
$$(c + 12) + c + (2c) = 64 \quad \text{Simplify.}$$
$$4c + 12 = 64 \quad \text{Subtract 12 on each side.}$$
$$4c = 52 \quad \text{Divide by 4 on each side.}$$
$$c = 13$$

Next we substitute $c = 13$ into the other two equations and solve for w and s.

$$w = c + 12$$
$$w = 13 + 12$$
$$w = 25$$
$$s = 2c$$
$$s = 2(13)$$
$$s = 26$$

Finally, we check in all three equations.

$$c = w \cdot 12$$
$$13 \stackrel{?}{=} 25 - 12 \quad ✓$$
$$s = 2c$$
$$26 \stackrel{?}{=} 2(13) \quad ✓$$
$$w + c + s = 64$$
$$25 + 13 + 26 \stackrel{?}{=} 64 \quad ✓$$

The fact that *the angle measures of a triangle sum to 180 degrees* is an important geometric result. This fact is well enough known that it usually is assumed rather than stated in algebra problems. The project in Exercise 86 suggests a way to intuitively prove this result. In Example 13, we solve a puzzle problem based on the sum of the angle measures.

EXAMPLE  Using substitution to solve problems involving the sum of angle measures in a triangle If one angle of an isosceles triangle is 15° smaller than another angle, what are the angle measures? Use guess and check to build equations, and then solve the equations by substitution. (Geometry note: An **isosceles triangle** has *two equal sides with equal angles opposite these sides.*) Figure 15 shows an isosceles triangle with $AC = BC$ and angle A = angle B.

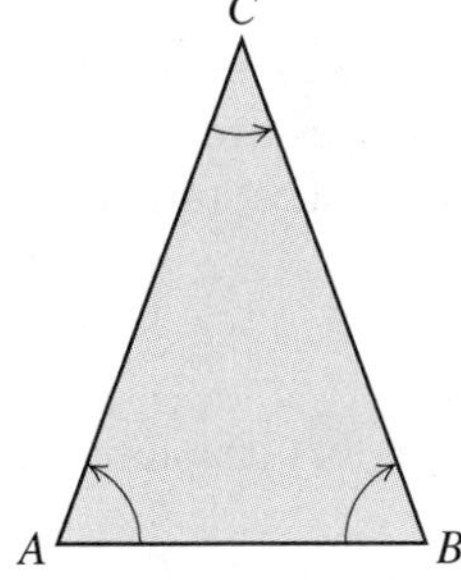

Figure 15

Solution We build the guess-and-check table shown in Table 15, with a column for each angle measure and a column for the angle sum. As indicated earlier, the sum of the measures of all three angles of the triangle must be 180°. We use 60° as our starting guess because that is the angle measure when all the angles are equal (180° divided by 3). We assume that angle C is 15° smaller than angles A and B and fill in the first row of the table.

Angle A	Angle B	Angle C	Total Degrees
60°	60°	45°	165°
A	B	C	180°

Table 15

The table helps us to see how the angles are related and to write equations.

Two angles are equal: $A = B$.

One angle is 15° smaller than another: $C = A - 15$.

The three angles add to 180°: $A + B + C = 180$.

Scanning the three equations, we see that each equation contains A, so we solve the system *in terms of* A. We substitute for B and C in the third equation, using $B = A$ and $C = A - 15$.

$$A + B + C = 180$$
$$A + A + (A - 15) = 180$$
$$3A - 15 = 180$$
$$3A = 195$$
$$A = 65$$

We substitute $A = 65$ into the first equation to solve for B.

$$B = A$$
$$B = 65$$

We substitute $A = 65$ into the second equation to solve for C.

$$C = A - 15$$
$$C = 65 - 15 = 50$$

We check by replacing the variables with their values in all three equations.

$$A = B$$
$$65 \stackrel{?}{=} 65 \quad ✓$$
$$C = A - 15$$
$$50 \stackrel{?}{=} 65 - 15 \quad ✓$$
$$A + B + C = 180$$
$$65 + 65 + 50 \stackrel{?}{=} 180 \quad ✓$$

●

Think about it 4: We made an assumption in Example 13: We assumed that angle C was the smallest angle. Assume that the two equal angles are smaller than angle C. Write the new system of equations, and solve it by substitution.

Summary of Strategies

Here is a guide to some of the strategies we have used so far in this chapter.

Strategies for Solving Systems of Equations

Use guess and check to help you work through the problem—reading, understanding, and writing equations. You may even want to guess the correct answer before writing the equations.

Solve by graphing if the two equations are easily written in $y = mx + b$ form.

Solve by substitution if one equation has a variable with a coefficient of 1 and the other equation is somewhat complicated.

ANSWER BOX

Warm-up: **1.** $y = x/4$ **2.** $x = 80 - y$ **3.** $x = 5 - 2y$ **4.** $y = \frac{5}{2} - \frac{1}{2}x$ **5.** $x = 2000 - y$ **6.** $w = c + 12$ **7.** $l = 8w/5$ **Example 1:** **b.** $x = 4y$, $x = 4(135)$, $x = 540$ **c.** $(540, 135)$ **d.** $540 = 4(135)$, $5(540) + 10(135) = 4050$ **Think about it 1:** We divide $2y = 5 - x$ by 2. The result, $y = \frac{5}{2} - \frac{1}{2}x$, contains fractions, and some students do not like to work with fractions in equation solutions. **Think about it 2:** We want the income levels, y, to be equal. **Think about it 3:** Upon substituting, we obtain $2\left(\frac{1}{2}x - 2\right) + 4 = x$. Applying the distributive property, we have $x - 4 + 4 = x$. Simplifying and subtracting x on both sides gives $0 = 0$. This always-true statement implies that the system is true for all x. **Think about it 4:** $A = B$, $A = C - 15$, $A + B + C = 180$; $A = B = 55°$, $C = 70°$

EXERCISES 

Exercises 1 to 22 provide practice with skills needed to solve equations by substitution. Solve each of the equations in Exercises 1 to 10 for the indicated variable.

1. $L = 2W$ for W

2. $L = 3W$ for W

3. $a + b = c$ for b

4. $x + 3 = y$ for x

5. $C = 2\pi r$ for r

6. $D = 2r$ for r

7. $x - y = 5$ for y

8. $a - b = c$ for b

9. $C = \pi d$ for d

10. $P = 4x$ for x

In Exercises 11 to 14, solve for x.

11. $2x + 3(2) = 12$

12. $9x - 2(6) = -3$

13. $-5x - 6(-3) = 3$

14. $7 = -2x - 13(-1)$

In Exercises 15 to 22, name the variable with a coefficient of 1 or −1. Solve for that variable in terms of the other variable.

15. $3x + y = 4$

16. $x + 3y = 26$

17. $x - 4y = 5$

18. $2y - x = 7$

19. $5y - x = 9$

20. $x - 3y = 3$

21. $3x - y = -2$

22. $2x - y = 13$

Use substitution to solve the systems of equations in Exercises 23 to 52.

23. $y = x - 8$
$3x + y = 4$

24. $x = 26 - 3y$
$x - 4y = 5$

25. $5x + 5 = y$
$y - 3x = 9$

26. $4x - 2y = 20$
$x = 2 - y$

27. $4x + 5y = 11$
$x = 9 + 5y$

28. $-x + 2y = 7$
$x = 3 + 3y$

29. $2x - y = 1$
$2y = 3x + 3$

30. $3y + x = -1$
$2x + 6 = -5y$

31. $2x + 3y = 0$
$3x + y = 7$

32. $y = 3x - 2$
$y = -2x + 13$

33. $y = \frac{4}{3}x$
$y = -\frac{8}{3}x + 8$

34. $y = -\frac{8}{3}x + 8$
$y = -\frac{2}{3}x + 4$

35. $y = 3x + 4$
$3x - y = 8$

36. $y = 4x - 3$
$4x - y = 3$

37. $y = 2x - 3$
$y - 2x = 5$

38. $y = -3x + 4$
$y + 3x = 4$

39. $x + y = 7$
$x = 7 - y$

40. $x - y = 5$
$x = y + 3$

41. $y = 3x + 2$
$y = -x - 2$

42. $y = 3x - 2$
$y = 3x + 2$

43. $y = -x - 2$
$y = 3x - 2$

44. $y = 3x + 2$
$y - 3x - 2 = 0$

45. $2x + 3y = 6$
$2y - x = -10$

46. $2x + y = 4$
$2y = 6 - 4x$

47. $y = \frac{1}{2}x - 2$
$2y + 4 = x$

48. $y = 0.10x + 40$
$y = 0.20x + 20$

49. $x + y = 50$
$7.50x + 9.75y = 435.75$

50. $y = 55 + 6(x - 10)$
$y = 85 + 4.75(x - 10)$

51. $y = 45 + 5(x - 10)$
$y = 30 + 3x$

52. $x + y = 60$
$0.10x + 0.25y = 9.75$

For Exercises 53 to 68, define variables, build equations, and then solve the resulting system.

53. Stephanie cuts a 20-yard ribbon. One piece is 3 yards longer than the other. What is the length of each piece?

54. Delores cuts a 16-inch salami. One piece is 4 inches longer than the other. What is the length of each piece?

55. Yoko has \$4.60 in quarters and nickels. She has 24 coins altogether. How many of each does she have?

56. Bart has 30 nickels and quarters. He has \$3.70 altogether. How many of each coin does he have?

57. Chen Chen has 28 coins. She has \$4.45 altogether. If she has only dimes and quarters, how many of each does she have?

58. Janice has \$5.05. She has only dimes and quarters. She has 25 coins altogether. How many of each does she have?

59. The perimeter of the front of a 15-ounce Cheerios® box is 40 inches. The height is 4 inches more than the width. What are the width and height of the front of the box?

60. The perimeter of the front of a 3-ounce Jello® box is 32 centimeters. The height is 2 centimeters less than the width. What are the width and height of the front of the box?

61. The perimeter of the front of a videotape box is 58 centimeters. The height of the front is 1 less than twice the width. What are the height and width of the box front?

62. The perimeter of the front of a 7-ounce Jiffy Muffin® Mix box is 44 centimeters. The height of the front is 2 less than twice the width. What are the height and width of the box front?

63. Peanuts are 27% protein (by weight), and cashews are 16% protein (by weight). How many grams of each need to be blended to make 270 grams of a mixture containing 66 grams of protein?

64. Shrimp is 24.7% protein (by weight), and cooked brown rice is 2.6% protein (by weight). How many grams of shrimp and brown rice need to be blended to make 1145 grams of a mixture containing 67 grams of protein?

65. Write one equation from the figure and the other from the sentence. Solve by substitution.

a. One angle is 24° more than the other.

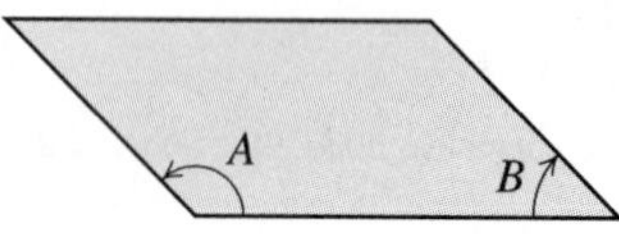

Parallelogram

b. One angle is 26° more than the other.

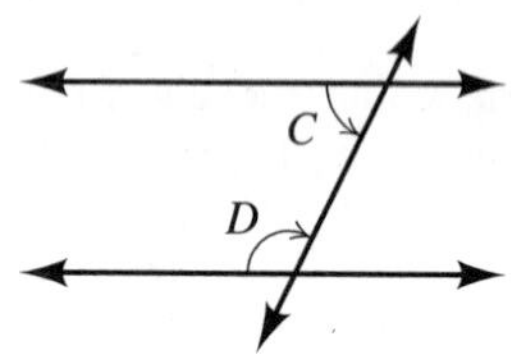

Parallel lines

c. One angle is 2° more than the other.

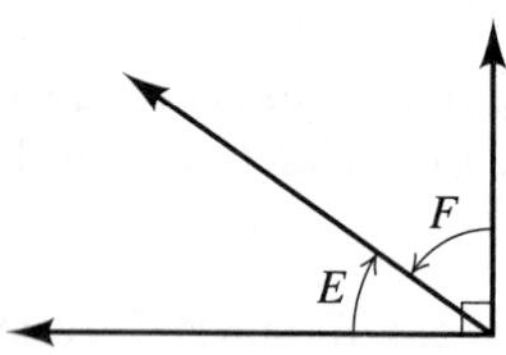

d. One angle is 3° more than twice the other.

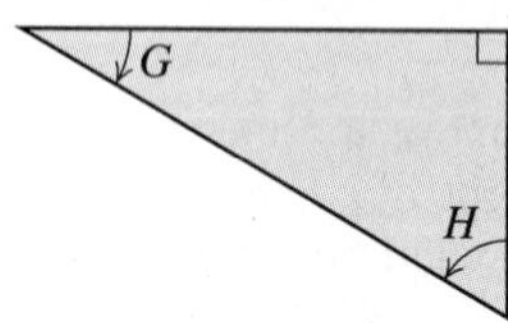

e. One angle is 45° less than twice the other.

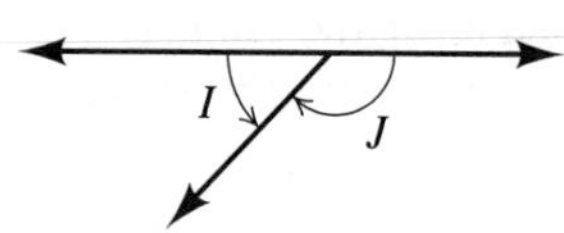

66. Write one equation from the figure and the other from the sentence. Solve by substitution.

a. One angle is 21° less than twice the other.

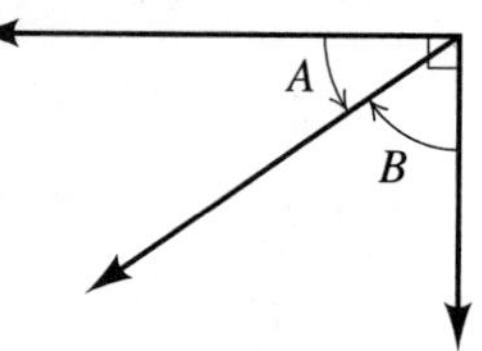

b. One angle is 40° less than three times the other.

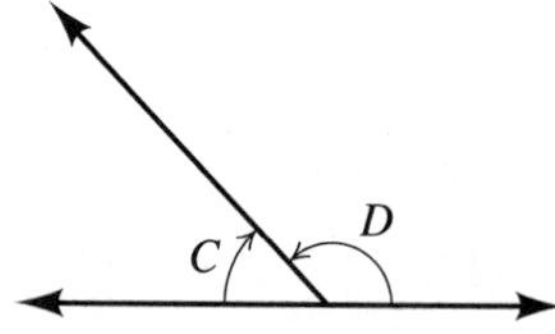

c. One angle is 18° less than three times the other.

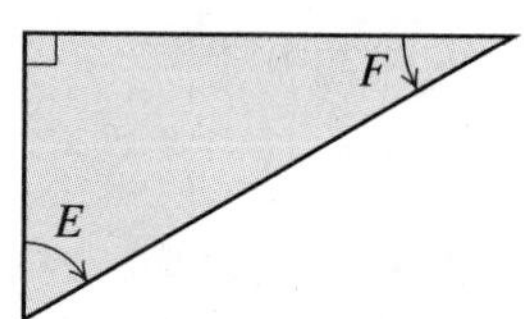

d. One angle is 3° less than twice the other.

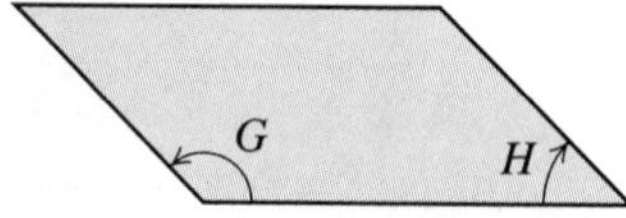

Parallelogram

e. One angle is 21° more than three times the other.

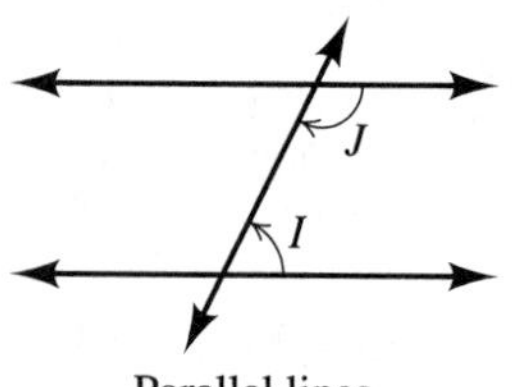

Parallel lines

67. a. The ratio of width to length of a fencing mat (the piste, shown in the figure) is 1 to 7. The perimeter of this rectangular surface is 32 meters. Find the width and length.

b. The ratio of length to width of a field hockey playing area is 5 to 3. The perimeter of the field is 320 yards. Find the length and width.

68. a. The ratio of width to length of a rectangle is 1 : 2. The length is 6 units less than three times the width. Find both dimensions. What is the area of the rectangle?

b. The base and height of a triangle are in the ratio 3 : 4. The height is 3 units more than the base. Find both dimensions. What is the area of the triangle?

For Exercises 69 to 72 (from Section 7.2), write and solve a system of equations.

69. Marielena spent \$620 on a watch, locket, and chain. She paid \$20 more for the locket than for the chain. She paid twice as much for the watch as for the locket. How much did she pay for each?

70. Katreen earns \$1375 one summer. She earns twice as much mowing lawns as shopping for the elderly. She earns \$500 more scraping old paint on houses than mowing lawns. How much does she earn at each job?

71. Andre buys notebooks, paperback books, and hardbound books for his classes. He buys 3 more paperback books than hardbound books. He buys 1 fewer notebooks than paperback books. He buys 20 items altogether. How many of each does he buy?

72. The I. R. Rabbit Company is packaging carrots. It has 2400 pounds of carrots to be packaged into bags for grocery stores. The bags are 1 pound, 2 pounds, and 5 pounds. The company sells twice as many 1-pound bags as 2-pound bags and ten times as many 1-pound bags as 5-pound bags. How many bags of each size should the company fill?

Solve the systems in Exercises 73 and 74.

73. $A + B + C = 180$
$A = 2B$
$B = 3C$

74. $A + B + C = 180$
$A = 3B$
$B = 2C$

In Exercises 75 to 80, write and solve a system of equations.

75. Angle A of a triangle is twice angle B. Angle C is 20° more than angle A. What is the measure of each angle?

76. The two equal angles of an isosceles triangle are each twice the size of the third angle. What is the measure of each angle?

77. The two equal angles of an isosceles triangle are each half the size of the third angle. What is the measure of each angle?

78. The two equal angles of an isosceles triangle are each four times the size of the third angle. What is the measure of each angle?

79. The length of a large gift box is 7 inches more than the width. The width is four times the height. The sum of the length, width, and height is 34 inches. Find the length, width, and height.

80. A gift box for a shirt is 5.5 inches longer than it is wide. The sum of the length, width, and height is 26.5 inches. The height of the box is 1 less than a fifth of the length. Find the length, width, and height.

81. Explain how to solve a system of two equations by substitution.

82. Explain how to choose a variable to use for the substitution.

83. Explain how, after looking at a system of equations, you would choose substitution, guess and check, or graphing as the solution process.

Projects

84. ***Matching Solutions with Results. Part I:*** Match each numerical, geometric, and algebraic result with the appropriate one of the following phrases:

a. The system has no solution.

b. The system has one solution, (x, y).

c. The system has an infinite number of solutions.

Numerical results:

$0 = 0$

$0 = 4$

$x = 3,\ y = 5$

Geometric results:

The lines intersect at a point.

The lines are coincident.

The lines are parallel.

Algebraic results:

The algebra always yields a false statement.

The algebra always yields a true statement.

The algebra yields a unique x and unique y value.

Part II: Write a system of two equations in two unknowns that fits each of the descriptions in parts a to c. Make a graph for each of the three systems to prove that your equations fit the requirements.

85. ***U.S. Currency.*** Who is pictured on each denomination of U.S. currency? Currency is printed in denominations of \$1, \$2, \$5, \$10, \$20, \$50, \$100, \$500, \$1000, \$5000, and \$10,000. Select an appropriate variable for each person pictured on the bills, as named in the following clues. Where two people's names start with the same letter, hints for variables are suggested. Write an equation for each relevant clue. Solve by substitution.

a. 10 Clevelands (C_l) equal 1 Chase (C_h).

b. Cleveland, Franklin, and Chase total \$11,100.

c. 10 Hamiltons make a Franklin.

d. 3 Washingtons plus a Jefferson (J_e) equal a Lincoln.

e. 4 Lincolns make a Jackson (J_a).

f. Hamilton, Franklin, and Chase were never president.

g. 2 Washingtons make a Jefferson.

h. 10 Grants make a McKinley (M_c).

i. 2 Grants make a Franklin.

j. Cleveland was born on March 18.

k. 2 Hamiltons make a Jackson.

l. Madison's picture is on the \$5000 bill.

m. Washington is on the \$1 bill.

86. ***Interior Angles of a Triangle.*** Using a ruler, carefully draw three different triangles. Make one triangle with an angle larger than 90°. Make another triangle close to a right triangle. Cut out the triangles. Taking one triangle at a time, tear the triangle in three pieces to separate the corners. Arrange the points of the corners together to show the total angle measure of all three corners. What do you observe about each triangle? Tape your corners to your homework paper.

87. ***Systems of Three Equations.*** Solve the systems of equations in parts a to d. Two of the three equations in each set can be solved in terms of one variable. The expressions for the two variables can then be substituted into the third equation. Write answers as ordered triples, (x, y, z).

a. $x + z = 2$
$x - y = 1$
$y - z = -2$

b. $x + y = 3$
$z - x = 2$
$y - z = -9$

c. $x + y = 2$
$z - x = 7$
$z - y = 2$

d. $x + z = 2$
$x - y = 5$
$z - y = 2$

e. There are three different ways to solve parts a to d. Find a second way to solve two of the exercises.

88. ***Formulas, Functions, and Substitution.*** In the formula $A = \pi r^2$, area is a function of radius. In $C = 2\pi r$, circumference is a function of radius. If we solve $C = 2\pi r$ for r and substitute the result into $A = \pi r^2$, we get $A = \pi(C/2\pi)^2$ or $A = C^2/4\pi$, which gives area as a function of circumference.

a. Write diameter as a function of circumference, using $d = 2r$ and $C = 2\pi r$.

b. Write the area of a square as a function of perimeter, using $A = x^2$ and $P = 4x$.

c. Write area as a function of height, using $A = bh/2$ and $b = h$.

d. Write area as a function of diameter, using $A = \pi r^2$ and $d = 2r$.

MID-CHAPTER 7 TEST

In Exercises 1 to 3, solve for the indicated variable.

1. $2x - y = 5000$ for y

2. $V = \dfrac{\pi r^2 h}{3}$ for h

3. $\dfrac{l}{w} = \dfrac{13}{6}$ for l

Solve the systems in Exercises 4 and 5 for the x and y that make both equations true.

4. $x + y = 5000$
$3x - 2y = -2500$

5. $x + y = 5000$
$3x - 2y = 2500$

6. a. The point (3, 4) is the solution to which two equations in the figure below?

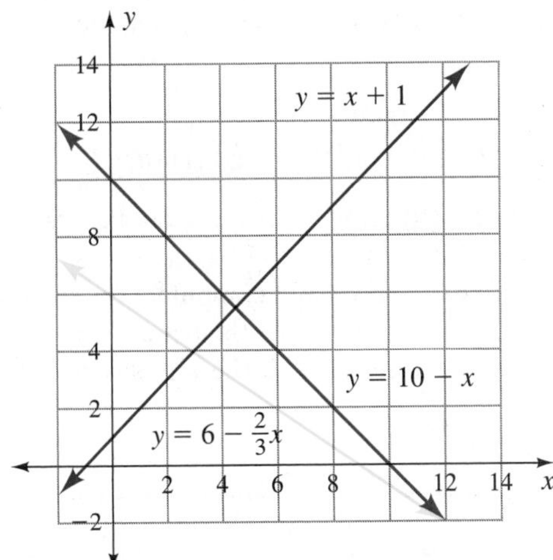

b. Estimate the point of intersection of $y = x + 1$ and $y = 10 - x$.

c. Use substitution to find the coordinates of the point of intersection of $y = 10 - x$ and $y = x + 1$.

d. Use substitution to find the coordinates of the point of intersection of $y = 10 - x$ and $y = 6 - \frac{2}{3}x$.

7. a. Graph the system

$$2x + y = 6$$
$$x - 2 = y$$

b. Estimate the point of intersection from the graph.

c. Solve the system by substitution.

8. Solve by substitution:

$$2x + y = 4$$
$$y = 4 - 2x$$

Comment on the result. What does the result imply about the graphs of the lines?

Use guess and check to start a solution to Exercises 9 to 11. Define your variables, and write a system of equations. Solve the system if your guess and check did not reach a solution.

9. A green turtle lays 12 times more eggs than an ostrich. If the sum of the eggs laid is 195, how many did each lay?

10. A produce distributor receives 10,000 pounds of potatoes and needs to bag them for retail sale. Local grocery stores sell three times as many 5-pound bags as 10-pound bags. How many bags of each weight should be prepared?

11. The ratio of length to width of a tennis court is 13 to 6. The perimeter of the court is 114 feet. Find the length and width.

12. Why does the intersection of two graphs solve the system of equations describing the graphs?

7.4 Solving Systems of Equations by Elimination

OBJECTIVES

- Solve a system of equations by eliminating like terms with addition or subtraction.
- Change a system of equations to make terms with opposite coefficients.
- Solve systems based on geometry facts.
- Solve systems related to quantity and value.
- Solve systems involving wind and current.

WARM-UP

The following exercises review the distributive property.

1. $3(2x + 4y) - 2(3x + 2y)$
2. $3(5x + y) - 5(3x - 2y)$
3. $1(2x + 4y) + 4(3x - y)$
4. $4(x - 3y) + 3(2x + 4y)$
5. $c(ax + by) - a(cx + dy)$
6. $d(ax + by) + b(cx - dy)$

THIS SECTION INTRODUCES a fourth method of solving systems of equations: elimination. We will apply elimination to equations with like and unlike numerical coefficients. Applications will include complementary and supplementary angles in geometry and quantity and value problems involving wind and current.

Principles Underlying Solving Systems by Elimination

Thus far, we have solved equations by graphing, guess and check, and substitution. We now consider a fourth method: elimination. Elimination is the foundation for techniques programmed into calculators and computers for solving large systems in applications. In business, economics, mathematics, and sociology, techniques using matrices, linear programming, and the simplex method build upon the elimination method.

Solving by Elimination

> **Elimination** is a process in which one variable is removed from a system of equations by adding (or subtracting) the respective sides of two equations.

It may be helpful to call the process *elimination of one variable* or *elimination of like terms.*

In the following Explorations, we look at graphical reasons why elimination works.

EXAMPLE  Exploring elimination*

a. Graph these four equations on one set of axes:

$$y = 2x + 3$$
$$y = -2x - 1$$
$$y = 1$$
$$x = -1$$

b. What is the point of intersection of the four graphs?

c. Add the terms on the left sides of the following equations, and then add the terms on the right sides. Solve the resulting equation.

$$y = 2x + 3$$
$$y = -2x - 1$$

d. Subtract the terms on the left sides of the following equations, and then subtract the terms on the right sides. Solve the resulting equation.

$$y = 2x + 3$$
$$y = -2x - 1$$

e. What do you observe in parts c and d?

Solution See the Answer Box.

*The structure of this and the next Exploration was suggested by my colleague, Jill McKenney, Lane Community College, Eugene, Oregon.

In Example 1, the equations in parts c and d were arranged so that like terms were lined up vertically. When we added the equations in part c, the terms $2x$ and $-2x$ added to zero and we were able to solve for the value of y. When we subtracted the equations in part d, the terms y and y subtracted to zero and we were able to solve for the value of x. The x and y make an ordered pair naming the point of intersection.

In Example 2, we will both add and subtract the equations. This time, although terms will not disappear, the results will still be surprising.

EXAMPLE 2 Exploring elimination

a. Graph these two equations on one set of axes:

$$3x - y = 4$$
$$2x + 3y = -1$$

b. What is the point of intersection of the two graphs?

c. Substitute the values of x and y from the point of intersection into each equation.

d. Add the left sides and right sides of the following equations:

$$3x - y = 4$$
$$2x + 3y = -1$$

Graph the resulting equation on your graph from part a.

e. Subtract the left sides and right sides of the following equations:

$$3x - y = 4$$
$$2x + 3y = -1$$

Graph the resulting equation on your graph from part a.

f. Compare the intersections of the graphs for parts d and e with those for part a. What do you observe?

Solution See the Answer Box. ●

The key idea in Examples 1 and 2 is that *adding or subtracting equations preserves the solution to the equations.* In the first example, addition helped us find the solution. In the second example, addition and subtraction did not help us find the solution, but the graphs of the resulting equations passed through the same point of intersection as the graphs of the original pair of equations (see the graph for part a in the Answer Box).

The addition property of equations is the principle underlying the elimination method.

Addition Property of Equations

> Adding the same number to both sides of an equation produces an equivalent equation. In algebraic notation,
>
> If $a = b$, then $a + c = b + c$.

The Explorations suggest this extension of the addition property of equations:

Extension to Addition Property of Equations

> Adding equal values to both sides of an equation produces an equivalent equation. In algebraic notation,
>
> If $a = b$ and $c = d$, then $a + c = b + d$.

The extended addition property says that if we add equal values to both sides of an equation, the solution set to the equation (or system of equations) remains the same.

Elimination with Like Coefficients

EXAMPLE 3 Solving by elimination Use the addition property of equations to eliminate a variable from the system of equations

$$2x - 3y = 16$$
$$x + 3y = -1$$

Solution We write one equation below the other, with like terms lined up, and then add.

$$\begin{aligned} 2x - 3y &= 16 \\ \underline{x + 3y} &\underline{= -1} && \text{Add the equations to eliminate } y. \\ 3x \qquad &= 15 && \text{Divide both sides by 3.} \\ x &= 5 \end{aligned}$$

We then substitute $x = 5$ into the first equation and solve for y:

$$\begin{aligned} 2x - 3y &= 16 && \text{Let } x = 5. \\ 2(5) - 3y &= 16 \\ 10 - 3y &= 16 && \text{Subtract 10 on each side.} \\ -3y &= 6 && \text{Divide both sides by } -3. \\ y &= -2 \end{aligned}$$

The solution is $x = 5$, $y = -2$ or, as a point of intersection of the graphs, $(5, -2)$.

Check:

$$2x - 3y = 16$$
$$2(5) - 3(-2) \stackrel{?}{=} 16 \checkmark$$
$$x + 3y = -1$$
$$5 + 3(-2) \stackrel{?}{=} -1 \checkmark$$

We now look more closely at features in the equations that make the elimination process possible.

In Section 1.3, we defined the **numerical coefficient** as *the sign and number multiplying the variable(s).* The reason addition eliminated $3y$ and $-3y$ in Example 3 was that the numerical coefficients were opposites and so the terms added to zero. *To identify like terms and terms with opposite numerical coefficients, we always line up the like terms in the equations.*

In Example 4, we set up a system of equations and solve by elimination.

EXAMPLE 4 Solving by elimination: a puzzle problem Two numbers add to 500. Their difference is 900. What are the two numbers?

Solution Each sentence in the problem suggests an equation:

$$x + y = 500 \quad \text{and} \quad x - y = 900$$

The left sides of the two equations contain y, but with opposite signs. If those left sides are added, the y variables will be eliminated. Because each equation is balanced over the equal sign, we are able to add the left sides and add the right sides.

$$\begin{aligned} x + y &= 500 \\ \underline{x - y} &\underline{= 900} && \text{Add the two equations to eliminate } y. \\ 2x \quad &= 1400 && \text{Divide by 2.} \\ x &= 700 \end{aligned}$$

To find y, we substitute $x = 700$ into one of the original equations:

$$\begin{aligned} x + y &= 500 \\ 700 + y &= 500 \\ y &= -200 \end{aligned}$$

Check: We substitute the values found for both x and y into the original equations:

$$\begin{aligned} x + y &= 500 \\ 700 + (-200) &\stackrel{?}{=} 500 \\ x - y &= 900 \\ 700 - (-200) &\stackrel{?}{=} 900 \quad ✓ \end{aligned}$$

●

In Example 5, the coefficients of like terms are not opposites. However, two of the terms have the same numerical coefficient. The extended addition property also applies to subtraction, because any subtraction may be written as addition of the opposite. In this case, we subtract the equations.

EXAMPLE 5 Solving by elimination Eliminate like terms in the following system by subtraction:

$$\begin{aligned} 2x + 3y &= 13 \\ 3y &= 5 - x \end{aligned}$$

Solution The second equation must be changed to $x + 3y = 5$ to line up like terms. We then subtract the second equation from the first equation.

$$\begin{aligned} 2x + 3y &= 13 \\ \underline{-(x + 3y)} &\underline{= -(5)} && \text{Subtraction eliminates } y. \\ x \quad &= 8 \end{aligned}$$

We substitute $x = 8$ into the first equation to find y:

$$\begin{aligned} 2x + 3y &= 13 \\ 2(8) + 3y &= 13 \\ 16 + 3y &= 13 && \text{Subtract 16 on both sides.} \\ 3y &= -3 && \text{Divide by 3.} \\ y &= -1 \end{aligned}$$

Check: We substitute $x = 8$ and $y = -1$ into each equation.

$$2x + 3y = 13$$
$$2(8) + 3(-1) \stackrel{?}{=} 13 \quad ✓$$
$$3y = 5 - x$$
$$3(-1) \stackrel{?}{=} 5 - 8 \quad ✓$$

●

Think about it 1: Multiply each term of the second equation in Example 5 by -1 and then add the equations. What does the result suggest about solving systems with like coefficients?

Example 6 returns to the subject of Cal and Joe's job sharing (Section 7.2, page 418).

EXAMPLE 6 Solving by elimination Cal and Joe's job sharing equations are $x + y = 243$ and $y = x - 17$. Arrange the equations to line up like terms, and solve by elimination.

Solution When the second equation is rearranged, the terms containing x have opposite signs, and we add the equations.

$$\begin{aligned} x + y &= 243 \\ -x + y &= -17 \quad \text{Add the equations to eliminate } x. \\ \hline 2y &= 226 \quad \text{Divide by 2 on each side.} \\ y &= 113 \end{aligned}$$

We substitute $y = 113$ into the first equation and solve for x:

$$x + y = 243$$
$$x + 113 = 243 \quad \text{Subtract 113 on each side.}$$
$$x = 130$$

The solution is $x = 130$, $y = 113$.

Check:

$$x + y = 243$$
$$130 + 113 \stackrel{?}{=} 243 \quad ✓$$
$$y = x - 17$$
$$113 \stackrel{?}{=} 130 - 17 \quad ✓$$

●

Think about it 2: Subtract the equations in Example 6: $x + y = 243$ and $-x + y = -17$. What do you observe?

Elimination with Unlike Coefficients

In the examples above, terms had like or opposite coefficients. The system

$$5x - y = 4$$
$$4x + 2y = -1$$

does not have like coefficients. In order to use elimination, we must change the equations to obtain like coefficients.

EQUATIONS IN WHICH ONE VARIABLE HAS A 1 OR −1 COEFFICIENT When a variable in one equation has a 1 or −1 coefficient, we may multiply that equation by the coefficient of the like term in the other equation and then add or subtract the equations.

EXAMPLE 7 Solving a system with a 1 or −1 coefficient Use multiplication on one equation in this system to obtain terms with opposite coefficients:

$$5x - y = 4$$
$$4x + 2y = -1$$

Solution The system can be written with a −1 coefficient on y in the first equation:

$5x - 1y = 4$	Multiply by 2.	$10x - 2y = 8$	
$4x + 2y = -1$		$\underline{4x + 2y = -1}$	Add the equations.
		$14x = 7$	Divide by 14.
		$x = \frac{1}{2}$, or 0.5	

We substitute $x = 0.5$ into the first equation and solve for y.

$5x - y = 4$	
$5(0.5) - y = 4$	Let $x = 0.5$.
$2.5 - y = 4$	Subtract 2.5 on each side.
$-y = 1.5$	Multiply both sides by −1.
$y = -1.5$	

The solution to the system is $x = 0.5$, $y = -1.5$.

Check:

$$5x - y = 4$$
$$5(0.5) - (-1.5) \stackrel{?}{=} 4 \quad ✓$$
$$4x + 2y = -1$$
$$4(0.5) + 2(-1.5) \stackrel{?}{=} -1 \quad ✓$$

●

EQUATIONS IN WHICH ALL VARIABLES HAVE UNLIKE COEFFICIENTS In most systems of equations, the variables have unlike coefficients. In this case, we use a process similar to that of finding a common denominator in order to add or subtract fractions.

Changing Unlike Coefficients to Like Coefficients

> When all variables have unlike coefficients, multiply both equations by numbers that make opposite coefficients on one variable.

The variables in Example 8 have unlike coefficients.

EXAMPLE 8 Solving a system with unlike coefficients Use multiplication on both equations to obtain like terms:

$$4x - 2y = 11$$
$$5x + 3y = 0$$

Solution The y terms have opposite signs. We multiply by numbers that make opposite coefficients.

$4x - 2y = 11$	Multiply by 3.	$12x - 6y = 33$	
$5x + 3y = 0$	Multiply by 2.	$10x + 6y = 0$	Add the equations.
		$22x = 33$	Divide by 22.
		$x = 1.5$	

We substitute $x = 1.5$ into the first equation and solve for y.

$4x - 2y = 11$	
$4(1.5) - 2y = 11$	Simplify.
$6 - 2y = 11$	Subtract 6 on both sides.
$-2y = 5$	Divide by -2 on both sides.
$y = -2.5$	

The solution is $x = 1.5$, $y = -2.5$.

Check:

$$4x - 2y = 11$$
$$4(1.5) - 2(-2.5) \stackrel{?}{=} 11 \checkmark$$
$$5x + 3y = 0$$
$$5(1.5) + 3(-2.5) \stackrel{?}{=} 0 \checkmark$$

●

Think about it 3: By what numbers could we multiply to give the x terms in Example 8 opposite coefficients? Do the multiplications and solve for y.

Steps in Solving Equations by Elimination

1. Arrange the equations so that like terms line up.
2. As needed, multiply one or both equations by numbers that make opposite coefficients on one variable.
3. Add the sides of the equations to eliminate one variable, and solve for the first variable.
4. Use substitution to find the second variable.
5. Check the solution in both equations.

SPECIAL ANGLES IN GEOMETRY, CONTINUED In Examples 9 and 10, we return to the examples on supplementary and complementary angles from Section 7.3 and solve them by elimination.

EXAMPLE  Solving by elimination: supplementary angles Angles M and N in Figure 16 are supplementary. Angle N is 24° more than five times angle M. What is the measure of each angle?

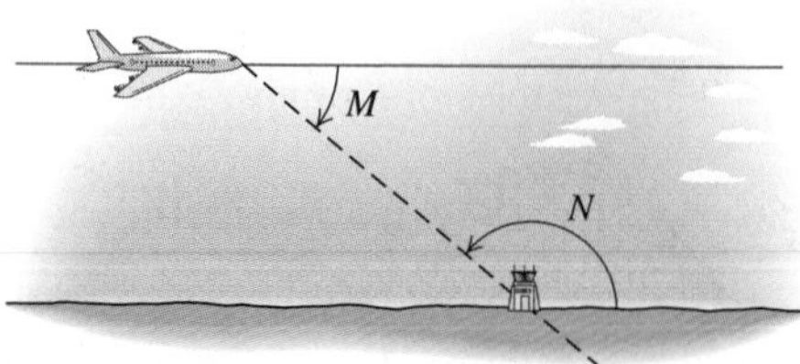

Figure 16

Solution The angles are supplementary, so they add to 180°:

$$M + N = 180$$

We translate the other sentence into a second equation,

$$N = 24 + 5M$$

which we can rearrange (by what steps?) to

$$5M - N = -24$$

To solve by elimination, we line up the variables.

$$\begin{aligned} M + N &= 180 \\ \underline{5M - N} &\underline{= -24} \quad \text{Add the equations to eliminate } N. \\ 6M &= 156 \quad \text{Divide by 6.} \\ M &= 26 \end{aligned}$$

By substitution, we have

$$N = 24 + 5(26)$$
$$N = 154$$

The angles are $M = 26°$ and $N = 154°$.

Check:

$$26 + 154 \stackrel{?}{=} 180 \checkmark$$
$$154 \stackrel{?}{=} 24 + 5(26) \checkmark$$

●

EXAMPLE 10

Solving by elimination: complementary angles In Figure 17, the axes are perpendicular, and angle D is twice the size of angle C. What is the measure of each angle?

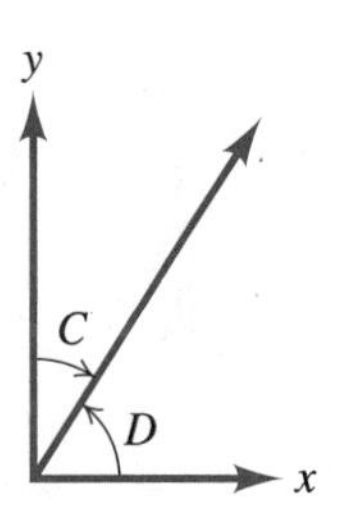

Figure 17

Solution As before, the equations are

$$C + D = 90$$
$$D = 2C$$

We line up like variables.

$$C + D = 90$$
$$2C - D = 0$$

We add the equations to eliminate D and then solve.

$$3C = 90$$
$$C = 30$$

We substitute $C = 30$ into the second equation and solve for D.

$$D = 2(30)$$
$$D = 60$$

The angle measures are $C = 30°$ and $D = 60°$.

Check:

$$30 + 60 \stackrel{?}{=} 90 \checkmark$$
$$60 \stackrel{?}{=} 2(30) \checkmark$$

●

QUANTITY AND VALUE In earlier quantity and value problems, the quantities were unknown and we were able to build two equations from a single quantity-value table. In the word problems in this section, the values are unknown.

Observe that each of the facts in Example 11 requires a separate quantity-value table and, furthermore, we can obtain only one equation from each table.

EXAMPLE 11 Solving by elimination: quantity-value tables Build a quantity-value table from each fact, and write an equation describing the total $Q \cdot V$.

a. 2 adult movie tickets and 4 student tickets cost \$31.

b. 3 adult tickets and 2 student tickets cost \$28.50.

Solution **a.** Table 16 is the quantity-value table. The equation is $2a + 4s = 31$.

	Quantity	**Value: Cost per Ticket**	$Q \cdot V$ **(\$)**
Adult tickets	2	a	$2a$
Student tickets	4	s	$4s$
			31

Table 16

b. Table 17 is the quantity-value table. The equation is $3a + 2s = 28.50$.

	Quantity	**Value: Cost per Ticket**	$Q \cdot V$ **(\$)**
Adult tickets	3	a	$3a$
Student tickets	2	s	$2s$
			28.50

Table 17

Think about it 4: Why can we not add a and s in Example 11 to obtain a second equation from each table? Solve the system formed by parts a and b.

The two equations in Example 11 describe the $Q \cdot V$ column and can be summarized by

$$\text{Quantity} \cdot \text{Value} + \text{Quantity} \cdot \text{Value} = \text{Total } Q \cdot V$$

We use the summary form of quantity and value in Example 12.

EXAMPLE 

Using elimination to solve quantity-value equations: calories Dietitians know that 3 grams of protein and 4 grams of fat make 48 calories. Also, 4 grams of protein and 3 grams of fat make 43 calories. Write each fact as an equation. Solve for the calories in a gram of protein and a gram of fat by the elimination method.

Solution The equations are $3p + 4f = 48$ and $4p + 3f = 43$. If we multiply the first equation by 3 and the second by -4, the terms containing f will be opposites.

$$\begin{array}{rcrl} 9p + 12f &=& 144 & \\ -16p - 12f &=& -172 & \text{Add the equations.} \\ \hline -7p &=& -28 & \text{Divide by } -7. \\ p &=& 4 & \end{array}$$

To find f, we substitute $p = 4$ into one of the original equations:

$$\begin{array}{rcll} 3p + 4f &=& 48 & \text{Let } p = 4. \\ 3(4) + 4f &=& 48 & \text{Simplify.} \\ 12 + 4f &=& 48 & \text{Subtract 12 on each side.} \\ 4f &=& 36 & \text{Divide by 4.} \\ f &=& 9 & \end{array}$$

Check: We substitute into both original equations:

$3(4) + 4(9) \stackrel{?}{=} 48$ ✓

$4(4) + 3(9) \stackrel{?}{=} 43$ ✓

In Example 12, we chose to multiply by 3 and -4 out of convenience, because they made the terms containing f opposites. There are other pairs of numbers that would work as well. We could multiply the first equation by -4 and the second equation by 3, or we could multiply the first equation by 4 and the second equation by -3. Confirming these results is left to the exercises.

TRAVEL PROBLEMS INVOLVING WIND AND CURRENT The formula $D = rt$ relates distance, rate, and time. A $D = rt$ problem is a special type of quantity and value problem. The time of travel is the quantity, and the speed of travel is the rate or value. The distance traveled is the product across the table: Time · Rate.

EXAMPLE 13

Relating quantity and value to time, rate, and distance Build a quantity-value table for the following setting:

Maxine drives 60 miles per hour (mph) on the toll road and 30 mph in town. If her total travel time for a 210-mile journey was 4 hours, how long did she drive at each speed?

Solve the resulting system of equations by elimination.

Solution We start by setting up the quantity-value table in Table 18.

	Quantity: Time (hr)	**Value: Rate (mph)**	**$Q \cdot V$: Distance (mi)**
On the toll road	x	60	$60x$
In town	y	30	$30y$
	4		210

Table 18

The system of equations is

$$x + y = 4$$
$$60x + 30y = 210$$

To get opposite coefficients on y, we multiply the first equation by -30. We then add the equations.

$$\begin{array}{r} -30x - 30y = -120 \\ 60x + 30y = 210 \\ \hline 30x = 90 \end{array}$$
$$x = 3$$

We substitute 3 for x in the first equation and solve for y.

$$3 + y = 4$$
$$y = 1$$

Check:

$3 + 1 \stackrel{?}{=} 4$ ✓

$60(3) + 30(1) \stackrel{?}{=} 210$ ✓

Maxine drove 3 hours on the toll road and 1 hour in town. ●

Think about it 5: How many miles did Maxine drive on the toll road and in town? What was her average rate for the trip?

In this section, *we assume that the direction of travel is parallel to the wind and river currents.* In Section 8.1, we will assume that the direction of travel is perpendicular to the wind and river currents. Other angles related to travel and wind (or currents) are addressed in trigonometry courses.

A strong wind called the *jet stream* flows eastward across the United States, as shown in Figure 18. The jet stream is at an altitude of about 12 kilometers (40,000 feet). Airplanes traveling eastbound take advantage of the jet stream in that their speed is increased by the wind speed. Those flying westbound must allow extra time and fuel for the flight because their speed is decreased by the wind speed.

Example 14 shows how wind can affect the speed of an airplane.

EXAMPLE 14 Finding rates given wind speed In winter, the jet stream blows at about 130 km/hr across the southern part of the country; in summer, the jet stream blows at about 65 km/hr across the northern part of the country. What is the speed of each aircraft?

a. An eastbound jet flying 500 km/hr in winter

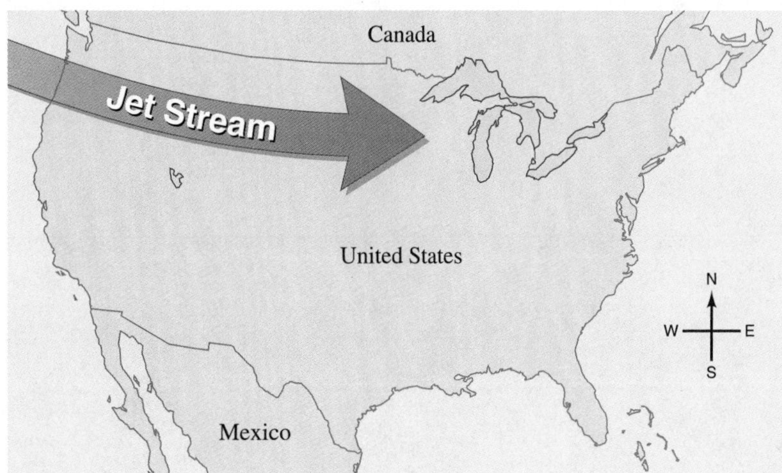

Figure 18

b. A westbound jet flying 500 km/hr in winter

c. An eastbound jet flying 500 km/hr in summer

d. A westbound jet flying 500 km/hr in summer

e. An eastbound jet flying 500 km/hr in still air

Solution We assume that the jet is flying parallel to the direction of the jet stream.

a. Net speed is $(500 + 130)$ km/hr $=$ 630 km/hr.

b. Net speed is $(500 - 130)$ km/hr $=$ 370 km/hr.

c. Net speed is $(500 + 65)$ km/hr $=$ 565 km/hr.

d. Net speed is $(500 - 65)$ km/hr $=$ 435 km/hr.

e. Net speed is 500 km/hr. In still air, there is no wind effect on the plane. ●

Finding Rates Given Wind Speed

> To find the speed of a plane flying directly into the wind (a head wind), we reduce the speed, r, by the wind, w:
>
> Net speed $= r - w$
>
> To find the speed of a plane flying with the wind (a tail wind), we increase the speed, r, by the wind, w:
>
> Net speed $= r + w$

When we add information about wind and current to travel problems, we modify the rate or value column in the quantity-value table. The north wind in Example 15 is due to an intense cold weather system that heads south from Canada during the winter.

EXAMPLE **15** Finding rates given distance and time: airplane travel Use a table to build a system of equations for this problem setting and then solve the system:

A plane leaves Fargo, North Dakota, and flies south to Dallas, Texas (see Figure 19). Another plane flies the opposite route. The planes fly at the same airspeed (speed in still air), r, but there is a wind, w, from the north that gives the

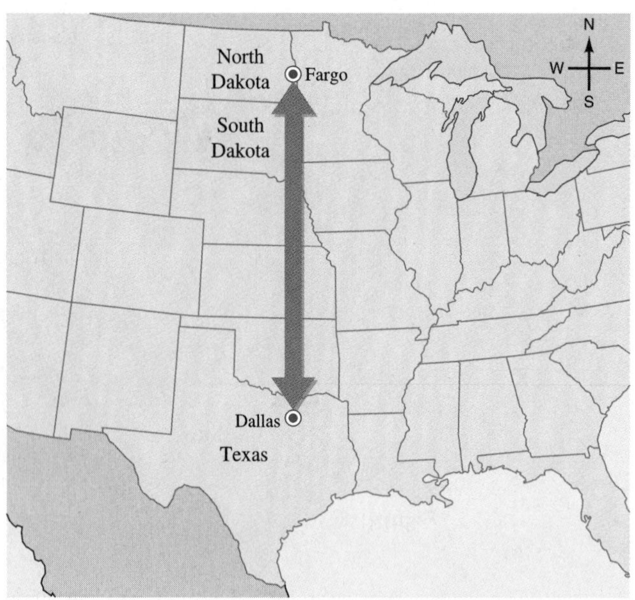

Figure 19

southbound plane a net speed of $r + w$ and slows the northbound plane to a net speed of $r - w$. The Fargo-to-Dallas plane takes 4 hours for the 1100-mile trip. The Dallas-to-Fargo plane takes 5 hours to travel the same distance. What is the airspeed of the two planes, and what is the wind speed?

Solution We enter the time, rate, and distance information in Table 19. Each row corresponds to the distance formula:

$$\text{Time} \cdot \text{Net rate relative to ground} = \text{Distance}$$

Quantity: Time (hr)	Value: Net rate relative to ground (mph)	Time · Net rate: Distance (mi)
4	$r + w$	1100
5	$r - w$	1100

Table 19

From each row ($t \cdot r = D$), the equations are

$$4(r + w) = 1100$$
$$5(r - w) = 1100$$

Rather than use the distributive property, we divide both sides of the equations by the times and obtain

$$r + w = 275$$
$$r - w = 220$$

Adding these equations gives

$$2r = 495$$
$$r = 247.5$$

Substituting r into $r + w = 275$ gives

$$247.5 + w = 275$$
$$w = 27.5$$

Check:

$$247.5 + 27.5 \stackrel{?}{=} 275 \checkmark$$
$$247.5 - 27.5 \stackrel{?}{=} 220 \checkmark$$

The speed of the airplane in still air (without wind) is 247.5 mph. The speed of the wind is 27.5 mph. ●

The current in a river affects boats in the same way as wind affects airplanes. Going downstream—*with* the current—adds to the speed at which a boat travels in still water. Going upstream—*against* the current—subtracts from the speed at which a boat travels in still water. (See Figure 20.)

If the wind exceeds the airspeed, the plane will move backwards; similarly, if the current exceeds the speed in still water, the boat will move backwards. This may sound like fantasy, but just imagine swimming 3 miles per hour against a flood traveling 20 miles per hour.

The passengers ought to be wearing life jackets!

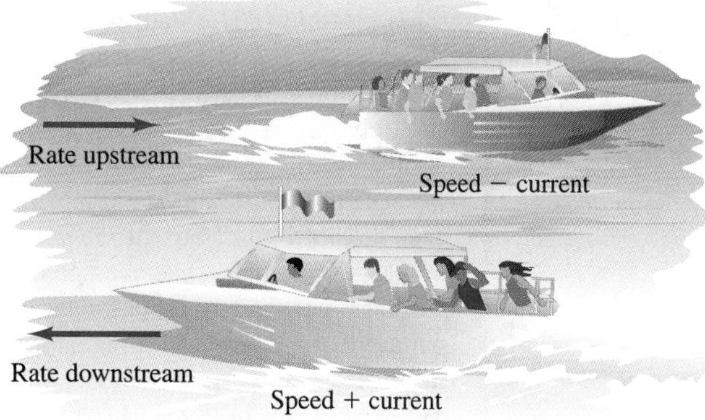

Figure 20

ANSWER BOX

Warm-up: 1. $8y$ **2.** $13y$ **3.** $14x$ **4.** $10x$ **5.** $bcy - ady$ **6.** $adx + bcx$
Example 1: a.

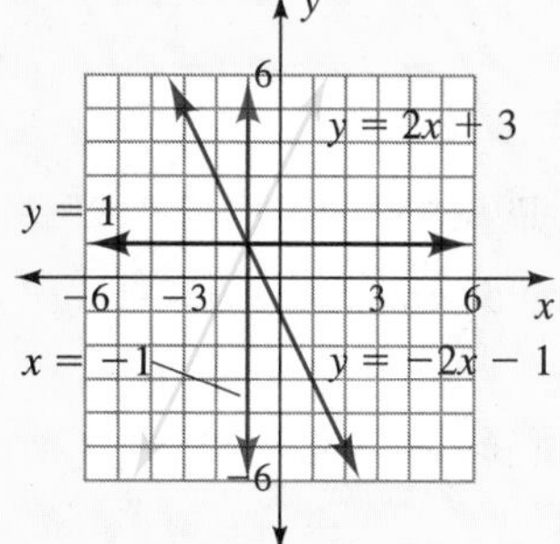

b. $(-1, 1)$ **c.** $2y = 2$, so $y = 1$. **d.** $0 = 4x + 4$, so $x = -1$. **e.** When we added (or subtracted) the equations, we eliminated one variable and were able to solve for the other variable. The values of the variables gave the point of intersection. The original two equations, as well as $x = -1$ and $y = 1$, all intersect at $(-1, 1)$.

Example 2: a.

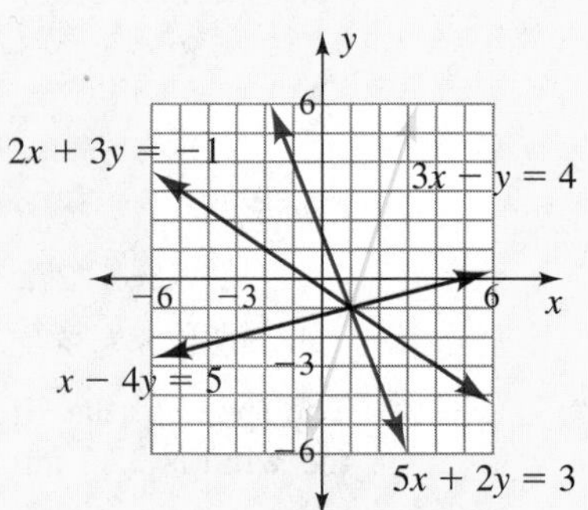

b. (1, −1) **c.** Substituting $x = 1$ and $y = -1$ makes each equation true. **d.** $5x + 2y = 3$ **e.** $x - 4y = 5$ **f.** The graphs of the equations in parts d and e pass through the same point of intersection as the original two lines. **Think about it 1:** Because subtraction is the same as adding the opposite, the results are the same:

$$\begin{aligned} 2x + 3y &= 13 \\ -x - 3y &= -5 \\ \hline x \quad\quad &= 8 \end{aligned}$$

Think about it 2: When we subtract the equations, we get the x value directly: $2x = 260$, so $x = 130$. This suggests an optional solution process for some elimination problems. **Think about it 3:** We multiply the first equation by 5 and the second equation by -4. The results are the same. **Think about it 4:** a and s cannot be added because they represent the prices for the tickets. If an entry were to be included in the last box in the value column, it would be the average value—in this case, the average cost for each group to attend the show. Because entries in a value (or rate) column are never added, this column cannot be used to build an equation. $a = \$6.50$, $s = \$4.50$. **Think about it 5:** She drove $3 \cdot 60 = 180$ mi on the toll road and $1 \cdot 30 = 30$ mi in town. Her average rate was $210/4 = 52.5$ mph.

EXERCISES 7.4

1. a. Solve by graphing:

$$y = 2x + 1$$
$$y = -x + 2$$

b. Add the equations and graph the resulting equation.

c. Subtract the equations and graph the resulting equation.

d. Comment on the results in parts b and c.

2. a. Solve by graphing:

$$y = 3x + 1$$
$$y = x + 3$$

b. Add the equations and graph the resulting equation.

c. Subtract the equations and graph the resulting equation.

d. Comment on the results in parts b and c.

Solve the systems of equations in Exercises 3 to 28 by the elimination method.

3. $x + y = -2$
$x - y = 8$

4. $x + y = -15$
$x - y = 3$

5. $m + n = 3$
$-m + n = -11$

6. $p - q = 9$
$p + q = -5$

7. $2x + y = -1$
$x + 2y = 4$

8. $3x + y = -6$
$x + 2y = -7$

9. $2a + b = -5$
$a + 3b = 35$

10. $3c + d = 28$
$c + 3d = -12$

11. $2x + 3y = 3$
$3x - 4y = -21$

12. $3m - 2n = 22$
$2m + 3n = -7$

13. $5p - 2q = -6$
$2p + 3q = 9$

14. $4x - 3y = -8$
$3x + 5y = -6$

15. $x + y = 6$
$x + y = 10$

16. $x - y = 5$
$x - y = 7$

17. $x - y = 7$
$2x - 2y = 14$

18. $x + y = 3$
$3x + 3y = 9$

19. $x + y = 5$
$y - x = -13$

20. $x + y = 3$
$y - x = 7$

21. $2x + 3y = 0$
$3x + 2y = 5$

22. $2x - 3y = -2$
$3x + 2y = -16$

23. $7 = 5m + b$
$3 = 3m + b$

24. $7 = -m + b$
$-1 = 3m + b$

25. $0.2x + 0.6y = 2.2$
$0.4x - 0.2y = 1.6$

26. $0.8x + 0.3y = 5.3$
$0.4x - 0.2y = 0.2$

27. $0.5x + 0.2y = 1.8$
$0.2x - 0.3y = -0.8$

28. $0.4x - 0.3y = 3.0$
$0.5x + 0.2y = 2.6$

In Exercises 29 to 32, identify two appropriate variables, and use the variables to write equations. Use either the elimination method or the substitution method to solve the equations.

29. The sum of two numbers is 25. Their difference is 8. Find the numbers.

30. The sum of two numbers is 35. Their difference is 10. Find the numbers.

31. The sum of two numbers is 20. Twice the second less twice the first is 21. Find the numbers.

32. The sum of two numbers is 12. Twice the larger less four times the smaller is 2. Find the numbers.

Use the alternative methods described in Exercises 33 and 34 to solve the system in Example 12. The equations are $3p + 4f = 48$ and $4p + 3f = 43$.

33. Multiply the first equation by -4 and the second equation by 3.

34. Multiply the first equation by 4 and the second equation by -3.

35. Solve the system from Example 6 by substitution:

$$x + y = 243$$
$$y = x - 17$$

36. Solve the system from Example 13 by substitution:

$$x + y = 4$$
$$60x + 30y = 210$$

In Exercises 37 to 40, define variables for the angles. Write equations based on the given facts as well as the complementary angles and supplementary angles. Solve the resulting system of equations by elimination.

37. a. Two angles are complementary. One angle is 50° larger than the other.

b. Two angles are supplementary. One angle is 50° larger than the other.

38. a. Two angles are supplementary. One angle is 60° less than the other.

b. Two angles are complementary. One angle is 60° less than the other.

39. Write one equation from the figure and the other from the sentence. Solve by elimination.

a. One angle is 24° more than the other.

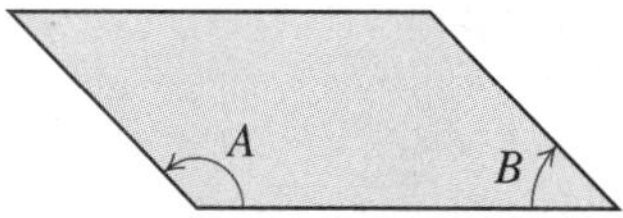

Parallelogram

b. One angle is 26° more than the other.

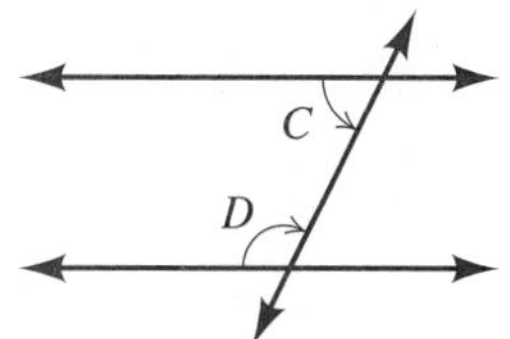

Parallel lines

c. One angle is 2° more than the other.

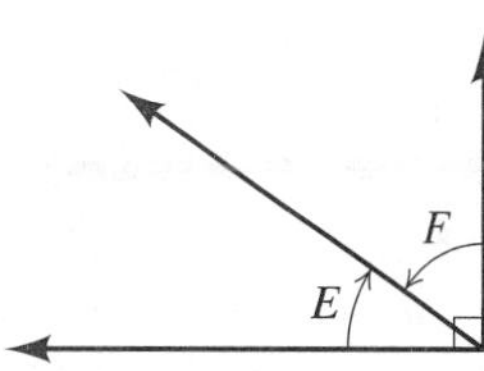

d. One angle is 3° more than twice the other.

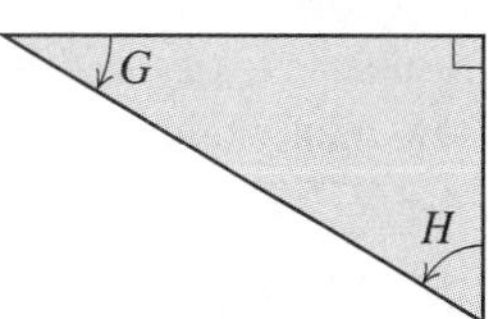

e. One angle is 45° less than twice the other.

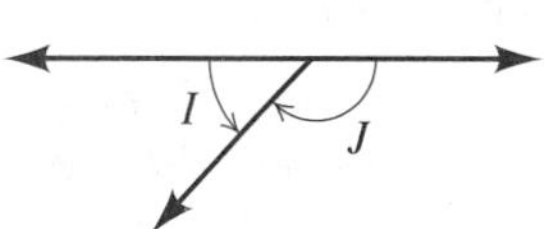

40. Write one equation from the figure and the other from the sentence. Solve by elimination.

a. One angle is 21° less than twice the other.

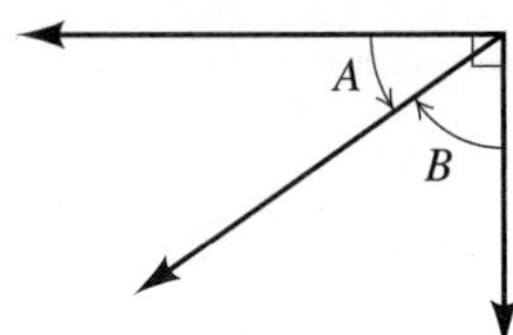

b. One angle is 40° less than three times the other.

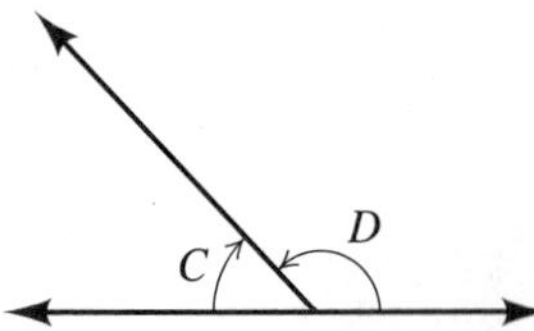

c. One angle is 18° less than three times the other.

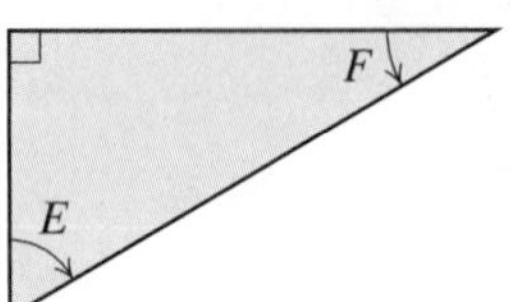

d. One angle is 3° less than twice the other.

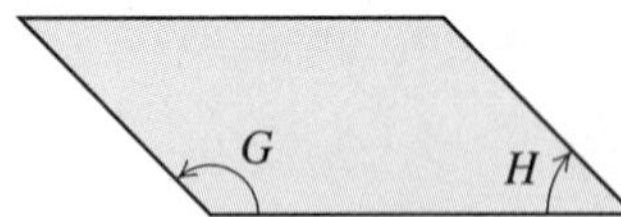

Parallelogram

e. One angle is 21° more than three times the other.

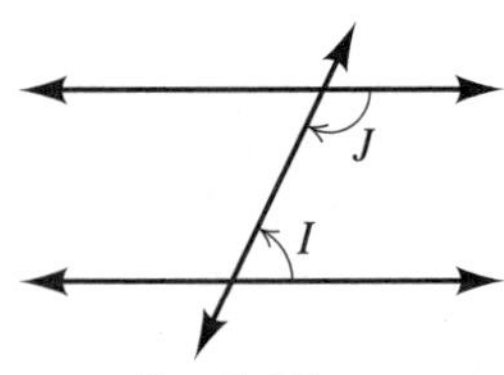

Parallel lines

For Exercises 41 to 60, define two variables. Then write equations. A quantity-value table may be helpful. Use either the elimination method or the substitution method to solve the equations.

41. A snack of 4 sugar cookies and 2 ginger snaps has 296 calories. Another of 3 sugar cookies and 10 ginger snaps has 329 calories. How many calories are in one of each kind of cookie?

42. A candy selection of 5 large gumdrops and 8 caramels has 511 calories. Another selection of 3 large gumdrops and 10 caramels has 525 calories. How many calories are in a piece of each kind of candy?

43. A fruit plate of 15 large cherries and 22 red grapes has 126 calories. Another assortment with 20 large cherries and 11 red grapes has 113 calories. How many calories are in one of each kind of fruit?

44. A bag of dried fruit, 10 dates and 3 figs, has 415 calories. Another bag of 6 dates and 5 figs has 425 calories. How many calories are in one of each fruit?

45. An appetizer of 8 green olives and 5 ripe olives has 285 calories. Another of 4 green olives and 10 ripe olives has 330 calories. How many calories are in one of each type of olive?

46. A vegetable snack of 6 small carrots and 4 stalks of celery has a total of 132 calories. Another snack of 3 small carrots and 10 stalks of celery has 90 calories. How many calories are in a small carrot? In a stalk of celery?

47. A set of 6 adult tickets and 3 student tickets to a basketball game costs \$58.50. Another set, consisting of 5 adult tickets and 4 student tickets, costs \$54. What is the cost of one of each type of ticket?

48. A group of 3 adult tickets and 8 student tickets to a football game costs \$67.50. For 4 adult tickets and 5 student tickets to the same game, the cost is \$64.50. What is the cost of one of each type of ticket?

49. Jordan buys 3 identical shirts and 2 identical ties for \$109.95. Gabe buys 4 of the same shirts and a tie for \$119.95. What is the cost of one of each?

50. At the bookstore, 2 notebooks and 5 pens cost \$14.91. At the same time, 6 notebooks and 3 pens cost \$32.85. What is the cost of one of each?

51. Two CDs and three game disks cost \$137.95. Four CDs and one game disk cost \$75.95. What does each cost?

52. Three sets of videotapes and four cassettes cost \$60.36. Two sets of videotapes and six cassettes cost \$61.24. What does each cost?

53. Ned drives a total of 7 hours and travels 405 miles. He drives 40 mph on gravel roads and 65 mph on paved roads. How long does he drive on each surface?

54. Nerine combines walking (at 6 km/hr) and running (at 18 km/hr) in her fitness program. Last week she exercised a total of 11 hours and traveled 126 kilometers. How many hours did she walk and how many hours did she run?

55. A plane makes a round trip between Dallas and Fargo, a one-way distance of 1100 miles. The flight with the wind takes 2 hours, and the flight against the wind takes 4 hours. What is the speed of the wind? What is the speed of the plane in still air?

56. A plane flies between San Francisco and Denver. Going east with the wind the plane travels 950 miles and takes 2 hours. Returning west (along a slightly different route) against the wind, the plane travels 975 miles and takes 3 hours. What speed in still air and what constant wind speed would give these results?

57. A flight from Cleveland, Ohio, to Washington, D.C., takes 4.5 hours against a wind, whereas the return flight on the same small airplane takes only 3 hours with the same wind. The one-way flight covers 360 miles. What is the airspeed (speed in still air) of the airplane? What is the speed of the wind?

58. A flight from Lincoln, Nebraska, to Dallas, Texas, takes 2.5 hours with the wind. The return flight takes 3.4 hours against the same wind. The distance between the cities is 612 miles. Find the airspeed (speed in still air) of the airplane and the speed of the wind.

59. A fishing boat goes 20 miles upstream, against a current, in 5 hours. The same boat goes 20 miles downstream, with the current, in 2 hours. What is the speed of the boat (in still water)? What is the speed of the current?

60. A jet boat goes 24 miles upstream, against a current, in 3 hours. It travels downstream the same distance in 2 hours. What is the speed of the boat (in still water)? What is the speed of the current?

61. Why is it possible to make forward progress when walking 3 miles per hour against a 40-mile-per-hour wind but not when swimming 3 miles per hour against a 6-mile-per-hour current?

62. Why is there a limit to acceptable wind levels in a short race (100 meters) but not in a long race that takes several laps of an oval track to complete?

63. Explain how to solve a system of equations by elimination.

64. In the example system

$$ax + by = c$$
$$dx + ey = f$$

explain how to eliminate x.

65. In the example system

$$ax + by = c$$
$$dx + ey = f$$

explain how to eliminate y.

66. Explain how, after looking at a system of equations, you would choose substitution, elimination, or graphing as the solution process.

Project

67. ***Changing Repeating Decimals to Fractions.*** We change repeating decimals into fractions with a process similar to the elimination method. Instead of using multiplication and subtraction to eliminate a variable, we use them to eliminate decimal portions of a number.

Example a: To change $0.3333\overline{3}$ to a fraction, we let f represent the fraction equivalent to the given repeating decimal:

$$f = 0.3333\overline{3}$$

Because only one digit is being repeated, we multiply both sides of the equation by 10. This shifts the decimal point one place to the right. Then we subtract the original equation.

$$\begin{aligned} 10f &= 3.3333\overline{3} \\ -f &= -0.3333\overline{3} \\ \hline 9f &= 3 \\ f &= \tfrac{3}{9}, \text{ or } \tfrac{1}{3} \end{aligned}$$

If the decimal repeated two digits, we would multiply both sides by 100 and repeat the process in Example a.

Example b: To change $0.454545\overline{45}$ to a fraction, we let $f = 0.45454545$. Because two digits are repeated, we multiply both sides by 100 to move the decimal point two places to the right.

$$\begin{aligned} 100f &= 45.4545\overline{45} \\ -f &= -0.4545\overline{45} \\ \hline 99f &= 45 \\ f &= \frac{45}{99} = \frac{9 \cdot 5}{9 \cdot 11} = \frac{5}{11} \end{aligned}$$

Use the process shown to change these repeating decimals to fractions. Simplify the fractions to lowest terms.

a. 0.4444... **b.** 0.7777... **c.** 0.151515...

d. 0.161616... **e.** 0.243243243...

f. 0.270270270... **g.** 0.567567...

68. ***Mr. Hall's Farm.****

a. Mr. Hall has a farm where he raises cows and ducks. There are 20 total heads among the animals and 50 total feet. How many cows and how many ducks are there?

b. Return to Exercise 28 in Section 7.2. Set up and solve systems of equations for Mr. McFadden's farm and Mr. Schaaf's farm.

c. Just for fun, now that you know how to do the problems mathematically, ask a child as young as kindergarten age to model cows and ducks with the equipment suggested in Exercise 28 in Section 7.2. Give the child exactly 10 heads (coins) and 28 feet (paper clips) with which to model cows and ducks, and tell him or her that no pieces should be left over. With the child, explore the other two problems you just solved in the text. Make up and explore your own problem.

d. Report on your observations of the problem-solving skills of young children.

*This problem is dedicated to the memory of three great problem solvers who were my mentors.

7.5 Solving Systems of Linear Inequalities by Graphing

OBJECTIVES

- Find out if an ordered pair is in the solution set to a system of inequalities.
- Solve a system of linear inequalities by graphing.

> WARM-UP
>
> Graph these inequalities in two variables. (If you need help, review Section 4.5.)
>
> **1.** $x + y \geq 5$ **2.** $x - y < -2$ **3.** $x \geq -1$ **4.** $y < 1$

IN THIS SECTION, we combine the concept of systems of equations, from this chapter, with the process of graphing inequalities in two variables, from Section 4.5.

Solutions to Systems of Inequalities

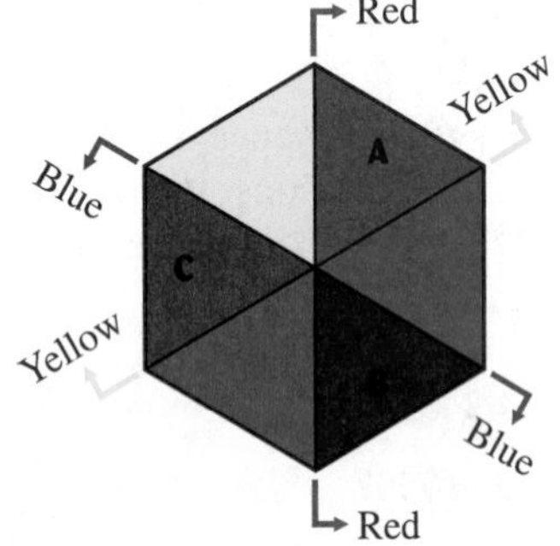

Figure 21

In Figure 21, which repeats the figure on the chapter opener, each of the three lines through the six-sided figure is a boundary line for one of the primary colors—red, yellow, or blue. Each color produces a shaded half-plane in the figure, just as the solution set to each inequality in a system forms a half-plane.

The overlapping of two primary colors creates another color; for example, the overlapping of yellow and red creates orange. One could say that orange is the solution set to the colors that contain both yellow and red.

A **system of inequalities** is *a set of two or more inequalities.* The **solution set to a system of inequalities** is *the set of ordered pairs that make all inequalities in the system true.* We can substitute an ordered pair into each inequality in a system to find out if the ordered pair is a solution to the system.

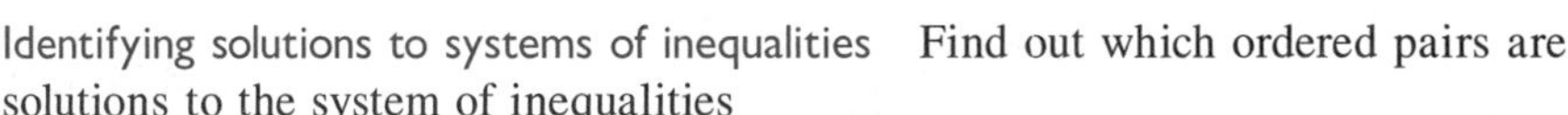

EXAMPLE 1 Identifying solutions to systems of inequalities Find out which ordered pairs are solutions to the system of inequalities

$$x + y \geq 5$$
$$x - y < -2$$

a. (0, 0) **b.** (5, 0) **c.** (0, 5) **d.** (−2, 8) **e.** (1, 3)

Solution We substitute the ordered pairs into the inequalities. A solution must make both inequalities true.

a. $0 + 0 \geq 5$ is false; $0 - 0 < -2$ is false; not a solution

b. $5 + 0 \geq 5$ is true; $5 - 0 < -2$ is false; not a solution

c. $0 + 5 \geq 5$ is true; $0 - 5 < -2$ is true; a solution

d. $-2 + 8 \geq 5$ is true; $-2 - 8 < -2$ is true; a solution

e. $1 + 3 \geq 5$ is false; $1 - 3 < -2$ is false; not a solution ●

We find solution sets to systems of inequalities by shading half-planes and finding the overlapping regions. Recall that a half-plane is the region on one side of a line in the coordinate plane. We graph each inequality in the system, as we did in Section 4.5.

Finding Solution Sets to Systems of Inequalities

To find the solution set to a system of inequalities:

1. **a.** Graph the first boundary line.
   b. Use a test point to find the half-plane that is the solution set.
   c. Shade that half-plane.
2. **a.** Graph the second boundary line.
   b. Use a test point to find the half-plane that is the solution set.
   c. Shade that half-plane.
3. The region that lies in both half-planes is the solution set to the system.

(The process can be extended to find solution sets to systems of three or more inequalities.)

Two-Variable Boundary Lines

In Example 2, we graph two inequalities that include boundary lines.

EXAMPLE 2 **Finding the solution set to a system of inequalities** Graph the inequalities and find the portions of their graphs that overlap:

$$x - y \leq 6$$
$$x + y \leq 3$$

Solution To graph $x - y \leq 6$, we graph the boundary line $x - y = 6$. Solving the equation for y gives $y = x - 6$. We use the slope and y-intercept to plot the line: $m = 1$ and $b = -6$. The boundary line is solid in Figure 22 because the inequality sign is $\leq$. We choose the test point (0, 0) and check:

$$0 - 0 \leq 6, \text{ true}$$

Thus, we shade the side of the line containing the origin, (0, 0).

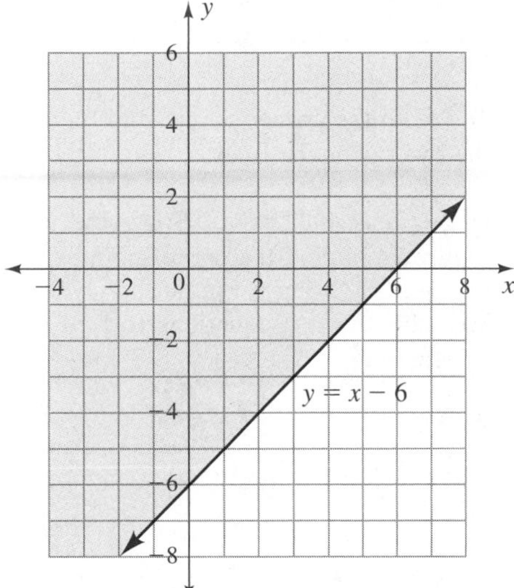

Figure 22

To graph $x + y \leq 3$, we graph the boundary line $x + y = 3$. Solving the equation for y gives $y = -x + 3$. We use the slope and y-intercept to plot the line: $m = -1$ and $b = 3$. The boundary line is solid in Figure 23 because the inequality sign is $\leq$. We choose the test point (0, 0) and check:

$$0 + 0 \leq 3, \text{ true}$$

Thus, we shade the side of the line containing the origin, (0, 0).

The two graphs are shown together in Figure 24. The overlapping region is the solution set.

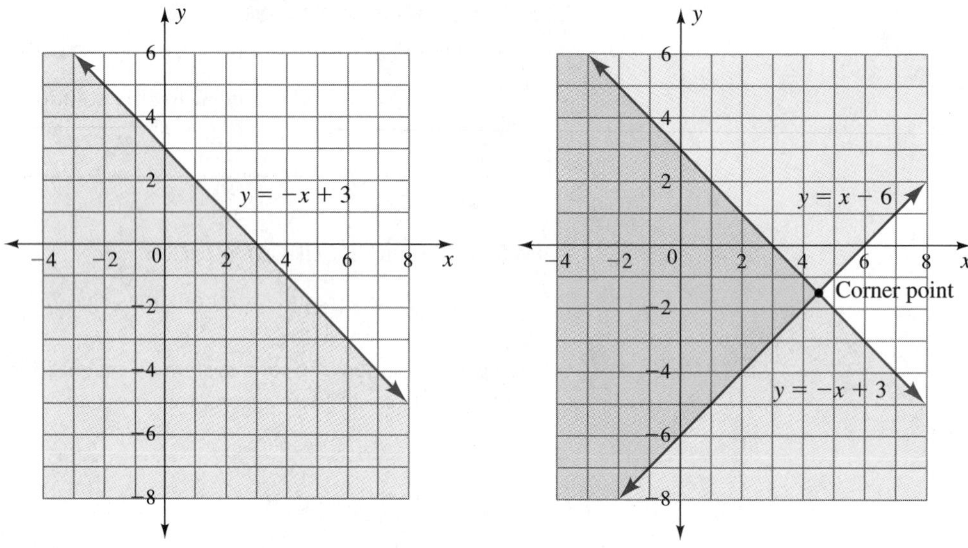

Figure 23

Figure 24

In linear programming and other applications in business and economics, the coordinates of the corner point are important.

Definition of Corner Point

> The **corner point** marks the point of intersection of the boundary lines of two inequalities.

To locate the corner point, find the ordered pair for the point of intersection of the two boundary lines.

EXAMPLE 3 Finding a corner point Solve the system consisting of the boundary line in Figure 24 to find the corner point.

Solution We solve the two equations to find the point of intersection of the two boundary lines.

$$\begin{aligned} x - y &= 6 \\ \underline{x + y} &\underline{= 3} && \text{Add the equations.} \\ 2x &= 9 && \text{Divide by 2 on each side.} \\ x &= 4.5 \end{aligned}$$

We substitute $x = 4.5$ into the first equation and solve for y.

$$\begin{aligned} x - y &= 6 \\ 4.5 - y &= 6 && \text{Subtract 4.5 on each side.} \\ -y &= 1.5 && \text{Multiply by } -1 \text{ on each side.} \\ y &= -1.5 \end{aligned}$$

The corner point is $(4.5, -1.5)$.

In Example 4, both boundary lines are dashed lines.

EXAMPLE 4 Finding the solution set to a system of inequalities Graph the inequalities and find the portions of their graphs that overlap:

$$y > x + 3$$
$$y < -x + 2$$

Solution To graph $y > x + 3$, we graph the boundary line $y = x + 3$, using the slope and y-intercept: $m = 1$ and $b = 3$. The boundary line is dashed in Figure 25 because the inequality sign is $>$. We choose the test point (0, 0) and check:

$$0 > 0 + 3, \text{ false}$$

Thus, we shade the side of the line opposite the origin, (0, 0).

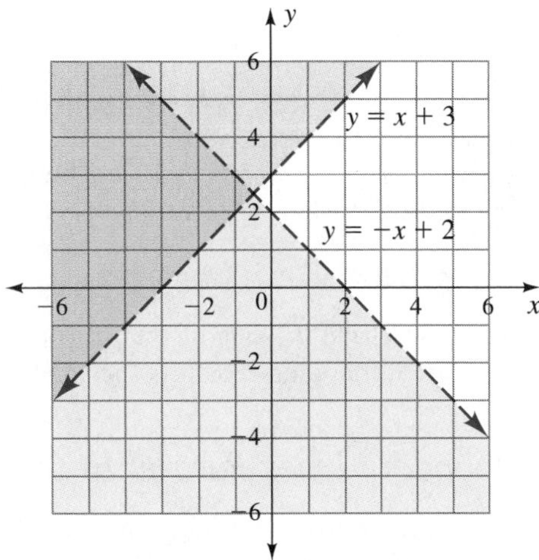

Figure 25

To graph $y < -x + 2$, we graph the boundary line $y = -x + 2$, using the slope and y-intercept: $m = -1$ and $b = 2$. The boundary line is dashed in Figure 25 because the inequality sign is $<$. We choose the test point (0, 0) and check:

$$0 < -0 + 2, \text{ true}$$

Thus, we shade the side of the line containing the origin, (0, 0).
The overlapping region is the solution set. ●

In Example 5, one of the boundary lines is solid and the other is dashed.

EXAMPLE 5 Finding the solution set to a system of inequalities Graph the inequalities and find the portions of their graphs that overlap:

$$x + y \le 5$$
$$2x + y > 5$$

Solution To graph $x + y \le 5$, we graph the boundary line $x + y = 5$. Solving the equation for y gives $y = -x + 5$. We use the slope and y-intercept to plot the line $m = -1$ and $b = 5$. The boundary line is solid in Figure 26 because the inequality sign is $\le$. We choose the test point (0, 0) and check:

$$0 + 0 \le 5, \text{ true}$$

Thus, we shade the side of the line containing the origin, (0, 0).

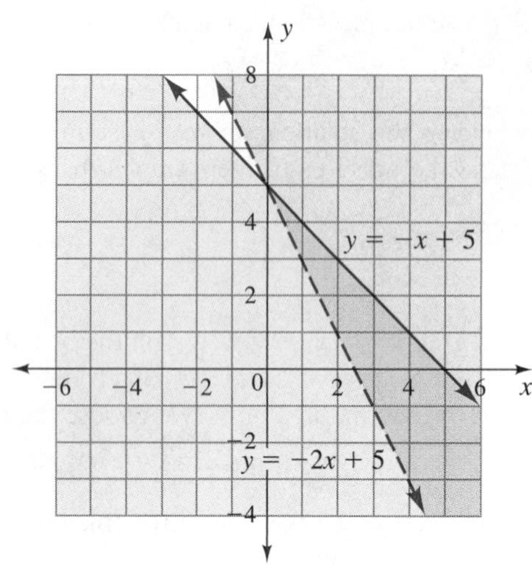

Figure 26

To graph $2x + y > 5$, we graph the boundary line $2x + y = 5$. Solving the equation for y gives $y = -2x + 5$. We use the slope and y-intercept to plot the line: $m = -2$ and $b = 5$. The boundary line is dashed in Figure 26 because the inequality sign is $>$. We choose the test point (0, 0) and check:

$$2(0) + 0 > 5, \text{ false}$$

Thus, we shade the side of the line opposite the origin, (0, 0).

The overlapping region is the solution set. ●

Boundary Lines Parallel to the Axes

Some inequalities have only one variable. Recall that equations parallel to the x-axis have equations $y = c$ and equations parallel to the y-axis have equations $x = c$. In Example 6, we graph four inequalities to create a region. Which inequalities have boundaries parallel to or on the axes?

EXAMPLE 6 Finding the solution set for a system of inequalities Graph the following system of inequalities and then describe the region containing the solution set:

$$x \geq 0$$
$$y \geq 0$$
$$y \leq 6$$
$$2y + 5x \leq 16$$

Solution The inequality $x \geq 0$ describes quadrants 1 and 4 and the y-axis. The inequality $y \geq 0$ describes quadrants 1 and 2 and the x-axis. The inequality $y \leq 6$ describes all points on or below the line $y = 6$.

To graph $2y + 5x \leq 16$, we graph the boundary line $2y + 5x = 16$. Solving the equation for y gives $y = -\frac{5}{2}x + 8$. We use the slope and y-intercept to plot the line: $m = -\frac{5}{2}$ and $b = 8$. The boundary is a solid line in Figure 27 because the inequality sign is $\leq$. We test (0, 0) in $2y + 5x \leq 16$:

$$2(0) + 5(0) \leq 16, \text{ true}$$

We shade the half-plane containing (0, 0).

The region containing the solution set is in the first quadrant, bounded above by $y = 6$ and on the right by the boundary line for $2y + 5x \leq 16$.

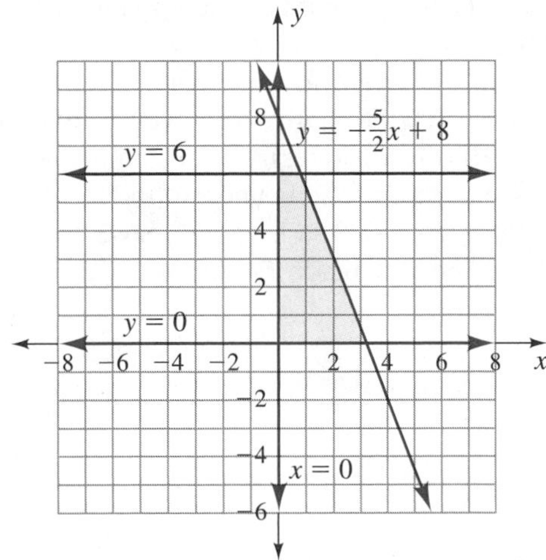

Figure 27

Graphing Calculator Technique: Shading Inequalities and Graphing Systems

Some calculators permit you to select a shading option in the [Y=] screen. Set the viewing window first. After entering the equation, move the cursor to the left of Y_1 and press ENTER until you see the correct shading option. Press GRAPH when all equations and shadings have been entered.

To graph the system of inequalities in Example 6, set the window for the first quadrant: $0 \le X \le 10$ and $0 \le Y \le 10$. With this setting, the conditions $x \ge 0$ and $y \ge 0$ will be satisfied by the viewing window itself.

Enter $Y_1 = 6$, and set the shading on Y_1 to below the line.

Solve $2y + 5x = 16$ for y:

$$2y = -5x + 16$$
$$y = -5x/2 + 8$$

Enter $Y_2 = -5x/2 + 8$.

Select a test point, such as (0, 0), and substitute it into the original inequality, $2y + 5x \le 16$. Since $2(0) + 5(0) \le 16$ is a true statement, set the shading on Y_2 to below the line.

Graph.

Applications

In Example 7, we return to the setting in Example 7 of Section 4.5 (page 253).

EXAMPLE  7 **Applying a system of inequalities: Smart-Mart cash** Smart-Mart keeps the currency in its overnight cash drawer limited to \$100, with a maximum of 10 five-dollar bills. Suppose the cash drawer has x one-dollar bills, y five-dollar bills, and no larger bills. Write a system of inequalities to describe the possible numbers of each type of currency. Graph and solve the system.

Solution It is a good idea to set up the boundary equations before graphing to make it easier to estimate the scales needed on the axes.

Because x is the number of one-dollar bills, $x \ge 0$, and because y is the number of five-dollar bills, $y \ge 0$. These conditions suggest a first-quadrant graph. The 10 five-dollar bills imply that $y \le 10$. We will shade below $y = 10$.

The $100 limit is a quantity and value statement: $1x + 5y \le 100$. Solving the boundary equation, $1x + 5y = 100$, for y gives $y = -\frac{1}{5}x + 20$. We use the slope and y-intercept to graph this solid line: $m = -\frac{1}{5}$ and $b = 20$. Because (0, 0) makes the inequality true, we shade below the boundary line.

If the cash register has only ones, then x can be as large as 100. The y-intercept of the boundary line for the total amount of cash is $y = 20$. We set the axes with $0 \le x \le 100$ and $0 \le y \le 20$.

The graph is in Figure 28.

The graph is shaded because the scale on the axes precludes using dots for a solution set.

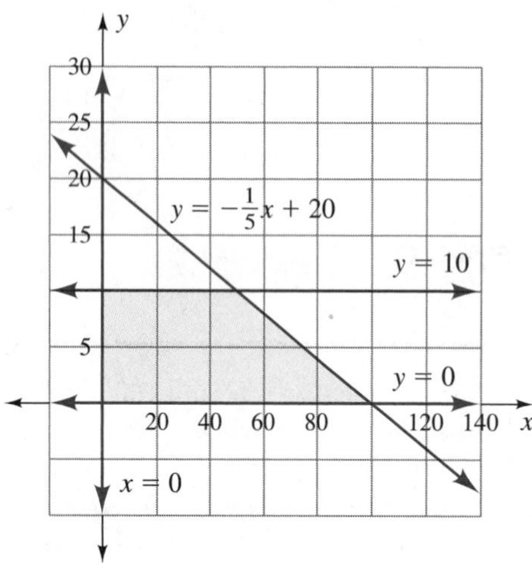

Figure 28

Graphing Calculator Solution

We solve the boundary equation, $1x + 5y = 100$, for y: $y = -x/5 + 20$. We enter the equation, and because (0, 0) makes the inequality true, we shade below the line. We enter the equation Y = 10 and set the shading below the line. A window with X on the interval [0, 100], scale of 10, and Y on the interval [0, 20], scale of 2, gives a first-quadrant graph and a complete view of the solution region. ●

In Example 8, we return to the setting in Example 8 of Section 4.5 (page 254).

EXAMPLE 

Applying a system of inequalities: calories in snacks You are allowed 160 calories in snacks. Caramel candies have 40 calories each, and ginger snaps have 20 calories each. You prefer no more than 5 ginger snaps each day. Let x be the number of caramels and y be the number of ginger snaps. Write a system of inequalities to describe the possible numbers of each type of snack. Graph and solve the system.

Solution Because the number of snacks must be zero or positive, this will be a first-quadrant graph with $x \ge 0$ and $y \ge 0$. The ginger snaps are limited, so $y \le 5$.

The calorie information is a quantity and value statement: $40x + 20y \le 160$. Solving the boundary equation, $40x + 20y = 160$, for y, gives $y = -2x + 8$. We use the slope and y-intercept to graph: $m = -2$ and $b = 8$.

To set the numbers on the axes in Figure 29, we look at the limits on the calories. If no caramels are eaten, then the limit is $160/20 = 8$ ginger snaps. If no ginger snaps are eaten, then the limit is $160/40 = 4$ caramels. We set $0 \le x \le 4$ and $0 \le y \le 8$.

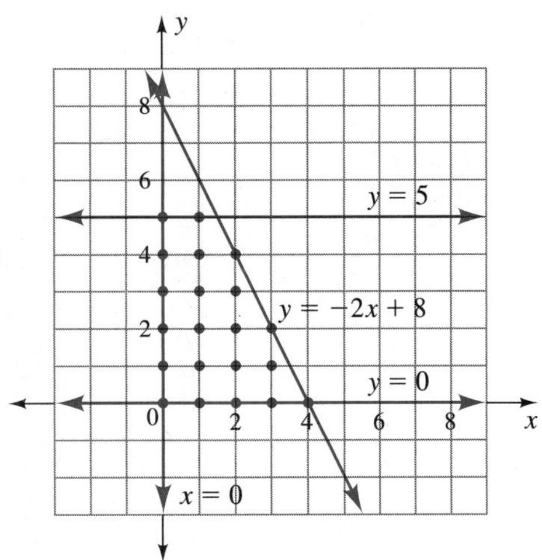

Figure 29

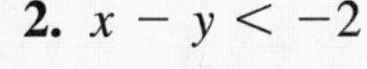

Warm-up:

1. $x + y \geq 5$

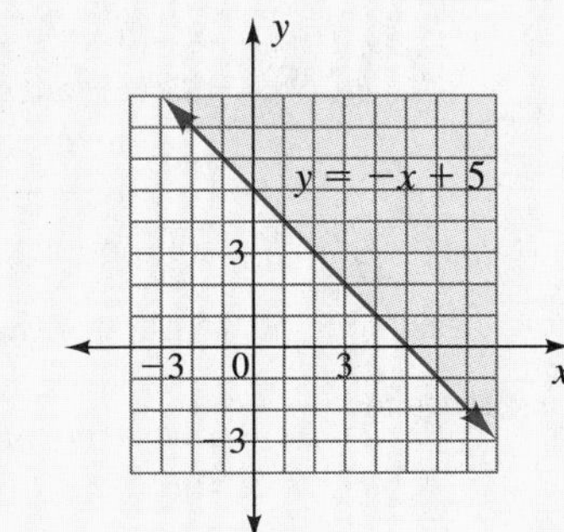

2. $x - y < -2$

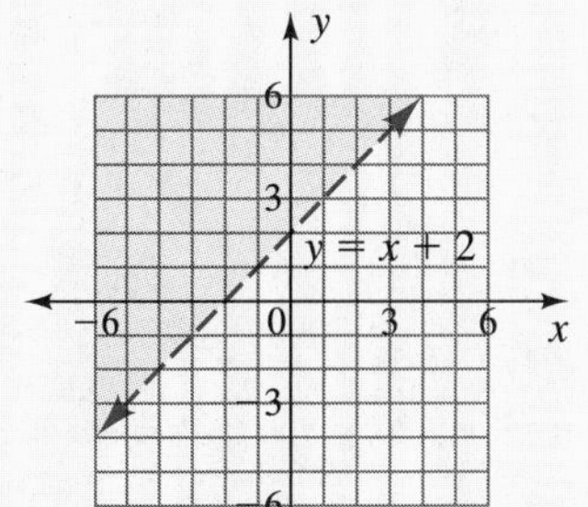

3. $x \geq -1$

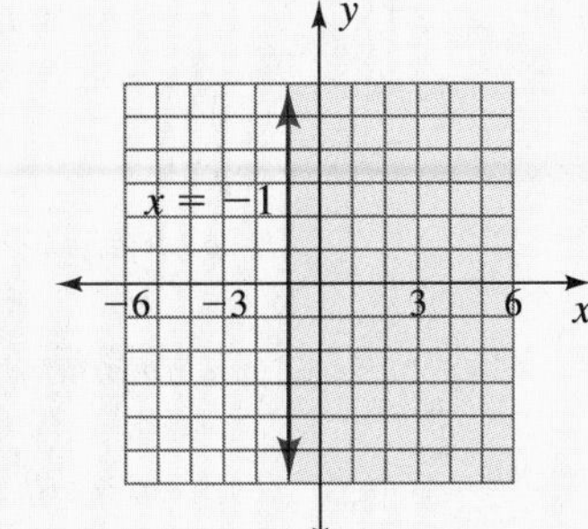

4. $y \leq 1$

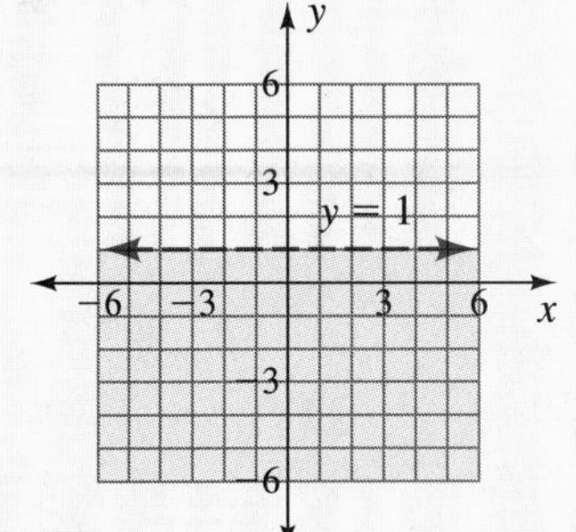

EXERCISES 7.5

In Exercises 1 to 4, which ordered pairs satisfy the system of inequalities given?

1. $y \leq x + 2$
$y \geq x - 2$
a. (0, 0) **b.** (1, 1) **c.** (1, −2) **d.** (−3, 3)

2. $x + y < 3$
$x + y > -3$
a. (0, 0) **b.** (2, 2) **c.** (−1, 4) **d.** (−2, −1)

3. $x + y > 4$
$x - y > 4$
a. (0, 0) **b.** (2, 2) **c.** (5, 0) **d.** (0, 5)

4. $y \leq -x - 2$
$y \geq x$
a. (0, 0) **b.** (−2, −2) **c.** (−2, 2) **d.** (0, −2)

In Exercises 5 to 12, graph the system of inequalities and show the solution set.

5. $y \leq 2x + 2$, $y \geq -x + 4$

6. $y \geq 3x + 3$, $y \leq -2x - 1$

7. $y \leq 2x + 2$, $y \geq 3x - 3$

8. $y \leq 2x + 5$, $y > -2x + 1$

9. $y \geq 2x - 3$, $y > -x + 3$

10. $y \geq 2x + 4$, $y > -2x - 3$

11. $y > 3x - 3$, $y < 2x - 2$

12. $y < -x + 1$, $y < -2x - 2$

In Exercises 13 to 18, graph the system of inequalities and show the solution set.

13. $x \geq 2$, $y \geq 3$

14. $x \leq -2$, $y \geq -2$

15. $x > -2$, $y \leq 3$

16. $x \leq 1$, $y < -2$

17. $x > -2$, $y < -1$

18. $x < 1$, $y > 2$

In Exercises 19 and 20, graph the system of inequalities and shade the solution set.

19. $x \geq 0$, $y \geq 0$, $y \leq 4$, $y \leq -2x + 5$

20. $x \geq 0$, $y \geq 0$, $x \leq 3$, $y \leq -x + 4$

21. $x \geq 0$, $y \geq 0$, $x + y \geq 3$, $x + y \leq 5$

22. $x \leq 0$, $y \leq 0$, $x - y \geq 2$, $x - y \leq 4$

23. $x \leq 0$, $y \geq 0$, $x \geq -3$, $y \leq -x + 2$

24. $x \geq 0$, $y \leq 0$, $y \geq -x - 1$, $x \leq 3$

25. Johanna has time for at most 12 workouts each week. She can jog at most 4 times each week. Let x = the number of times she jogs and y = the number of times she swims. Write a system of inequalities to describe the possible numbers of times she carries out each activity. Graph and show the solution set.

26. Jaime has 40 hours per week for doing math and reading. He can spend, at most, 20 hours reading. Let x = the number of hours spent doing math and y = the number of hours spent reading. Write a system of inequalities to describe the possible numbers of hours spent at each activity. Graph and show the solution set.

27. A birthday brunch caterer charges $15 for adults and $8 for young people between 6 and 15. Children under 6 are free. The total budget for a party is $240. Let x = the number of adults and y = the number of young people. There are a maximum of 12 young people who might attend. Write a system of inequalities to describe the possible numbers of people age 6 and over who can attend. Graph and show the solution set.

28. A dieter is allowed 140 calories for a snack. Apricots contain 20 calories each, and tangerines contain 35 calories. Sue wants no more than 2 tangerines each day. Let x = the number of apricots and y = the number of tangerines. Write a system of inequalities to describe the possible numbers of each item she can eat. Graph and show the solution set.

29. For each football game, the athletic department can sell 45,000 tickets. There are regular admission tickets and student tickets. The number of student tickets can be no more than 5000. Let x = the number of regular tickets and y = the number of student tickets. Write a system of inequalities to describe the possible numbers of each type of ticket sold. Graph and show the solution set.

30. Annette, the chief financial officer for a Fortune 500 company, has a $200,000 bonus to invest. She wants to invest no more than $40,000 in the stock market and the rest in municipal (tax-free) bonds. The bonds cost $1,000 each. Assume the stocks cost $100 each. Let x = the number of stocks and y = the number of bonds. Write a system of inequalities to describe the possible numbers of stocks and bonds. Graph and show the solution set.

Projects

These projects introduce new material and extend various ideas about inequalities.

31. ***Compound Inequalities.*** Inequalities containing two inequality symbols, such as $-3 < x < 4$, are called *compound inequalities.* To solve a compound inequality, do the same operation or operations to all three parts of the inequality, to isolate the variable between the two inequality signs.

$-1 < 2x + 3 < 5$ Subtract 3.
$-4 < 2x < 2$ Divide by 2.
$-2 < x < 1$

Number-line graph of solution set:

−3 −2 −1 0 1 2 x

Solve each compound inequality, and sketch the solution set on a number-line graph.

a. $-4 < 3x - 1 < 5$ **b.** $-1 \le 2x + 5 \le 7$

c. $-8 \le \frac{1}{2}x - 3 \le -4$ **d.** $-3 \le \frac{1}{4}x + 2 \le -1$

e. $-3 < 4 - x < 6$ **f.** $-5 \le 1 - 3x \le 7$

(*Hint:* If you multiply or divide by a negative number, reverse the direction of all inequality signs.)

32. ***Absolute Value Inequalities.*** The absolute value graph has a V-shape. Make a short input-output table to find where the V is located for each inequality in parts a to f. Graph the ordered pairs from the table and connect to form the V. Show the solution set to the inequality by shading above or below the graph, depending on the truth of the inequality for a test point. As examples, see the graphs of $y \le |x|$ and $y > -|x + 2|$ below.

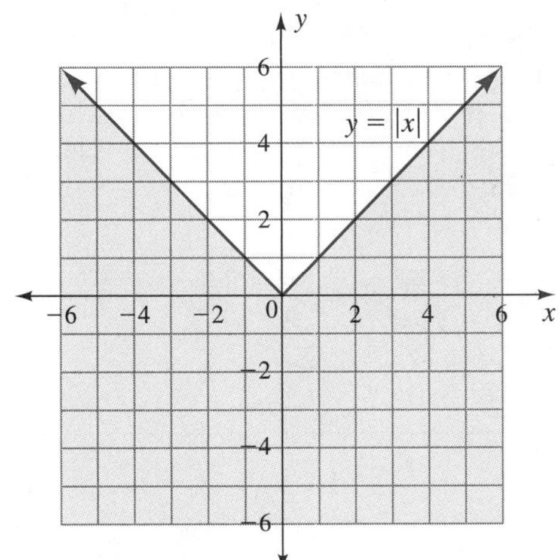

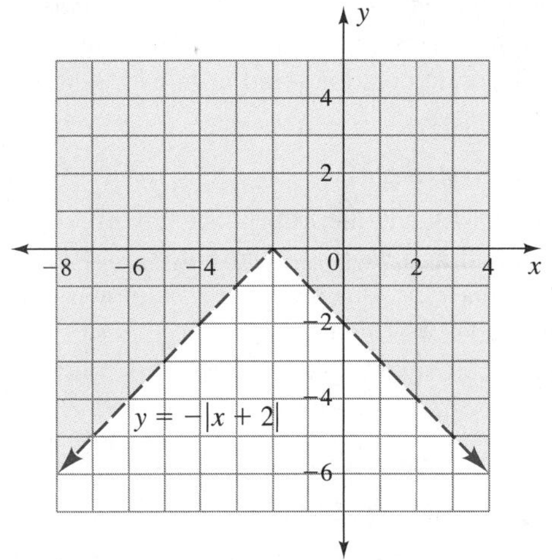

a. $y \le |x + 3|$ **b.** $y > -|x|$

c. $y \ge |x - 2|$ **d.** $y < -|x + 1|$

e. $y > |x|$ **f.** $y \le -|x - 3|$

CHAPTER 7 SUMMARY

Vocabulary

For definitions and pages references, see the Glossary/Index.

- coincident lines
- complementary angles
- corner point
- elimination
- isosceles triangle
- linear pair
- numerical coefficient
- point of intersection
- solution to a system
- solution set to a system of inequalities
- substitution
- supplementary angles
- system of equations
- system of inequalities

Concepts

7.1 Solving Systems with Graphs

A system of linear equations can have one, no, or an infinite number of solutions. (See the summary charts on page 404 and page 429.)

One solution means that the graphs of the two linear equations intersect in a single point; see Figure 30(a). A unique ordered pair describes the point of intersection.

No solution means that the graphs of the two linear equations are parallel; see Figure 30(b). No ordered pair makes both equations true. An algebraic solution gives a false statement.

An infinite number of solutions means that the graphs of the two linear equations are coincident; see Figure 30(c). The lines have every point in common. An algebraic solution gives a true statement.

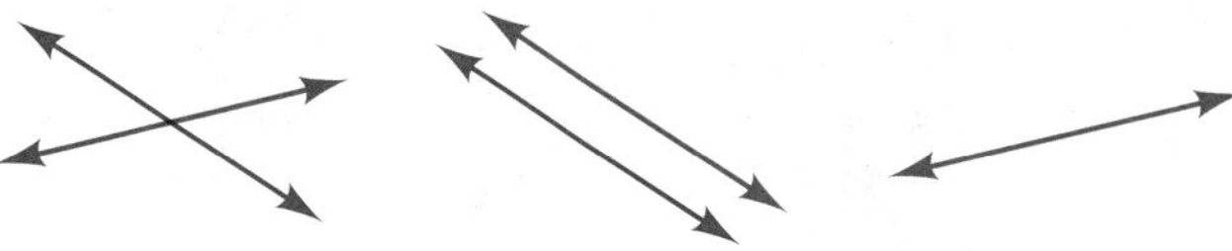

(a) Intersecting lines (b) Parallel lines (c) Coincident lines

Figure 30

There are four methods of solving a system of two linear equations: graphing (visual), guess and check (numerical), substitution (algebraic notation), and elimination (algebraic notation).

7.2 Setting Up Systems

Quantity-value tables may be helpful for organizing information or for building equations. Unknowns may be either quantities or values. If quantities are unknown, then a quantity-value table will contain two equations. If values are unknown, then a quantity-value table will contain only one equation. To create equations, use

$$\text{Quantity} \cdot \text{Value} + \text{Quantity} \cdot \text{Value} = \text{Total } Q \cdot V$$

Guess-and-check tables have a column for each unknown and a row for each guess. Showing the operations within the unknown column helps in writing algebraic expressions. Add other columns as needed for calculations.

The number of unknowns in a problem situation determines the number of equations needed to solve for the unknowns.

7.3 Solving Systems by Substitution

Solving by substitution is convenient when one variable has a coefficient of 1 or -1. For the steps in substitution, see page 426.

Supplementary angles add to 180°. Complementary angles add to 90°. The measures of the interior angles of a triangle add to 180°.

7.4 Solving Systems by Elimination

The elimination method is based on an extension to the addition property of equations: Adding equal values to both sides of an equation produces an equivalent equation. If $a = b$ and $c = d$, then $a + c = b + d$. For the steps in elimination, see page 448.

Distance, rate, and time are related by the formula $D = r \cdot t$.

The net speed against a wind (or current) is $r - w$ (or $r - c$).

The net speed with a wind (or current) is $r + w$ (or $r + c$).

7.5 Solving Systems of Inequalities

A system on the coordinate plane may contain any number of inequalities. Inequalities describe half-planes. Boundary lines may be any line, including the axes or lines parallel to the axes. Many problem settings assume first-quadrant graphs with $x \geq 0$ and $y \geq 0$. For the steps in solving systems of inequalities, see page 461.

CHAPTER 7 REVIEW EXERCISES

In Exercises 1 to 6, solve for the indicated variable.

1. $y - 2x = 5000$ for y

2. $P = 2b + 2h$ for h

3. $C = 2\pi r$ for r

4. $x - 3y = 3$ for y

5. $\dfrac{a}{b} = \dfrac{5}{8}$ for b

6. $\dfrac{x}{y} = \dfrac{4}{5}$ for x

In Exercises 7 to 10, solve the system of equations by graphing.

7. $y = x - 2$
$y = \dfrac{-2x}{3} + 3$

8. $y = -x$
$y = x - 3$

9. $x = 2$
$y = 4$

10. $y = -\dfrac{1}{2}x + 5$
$y = \dfrac{3x}{2} - 3$

In Exercises 11 to 14, solve the system of equations by substitution.

11. $y = x + 2$
$y = 3x + 3$

12. $x + y = 5$
$3x + 5y = 27$

13. $x + y = 5000$
$3x - 2y = 12{,}500$

14. $x + y = 5000$
$3x - 2y = -7500$

In Exercises 15 to 20, solve the system of equations by elimination.

15. $3 = -2m + b$
$2 = 3m + b$

16. $2x - 3y = 7$
$-2x + 3y = 4$

17. $2x - 3y = 7$
$3y + 4x = -1$

18. $x - 2y = 11$
$3x + 4y = -7$

19. $5x + 3y = -18$
$3x - 2y = -7$

20. $2x - 4 = 6y$
$3y - x = -2$

In Exercises 21 to 24, solve the system of equations by the method of your choice.

21. $x + y = 15$
$y = -6x$

22. $x + y = 12$
$y + 4x = -3$

23. $3y - 4x = 2$
$8x = 6y + 4$

24. $3y - 4x = 2$
$8x = 6y - 4$

25. What does an algebraic result of $0 = 0$ tell us about the graph of a system of two linear equations?

26. Write an equation of a line parallel to $x + y = 5$. Solve the system formed by $x + y = 5$ and your new equation, and tell at what point in the algebra you know that there are no solutions to the system.

For Exercises 27 to 36, define variables and write a system of equations. Use guess and check as needed. Solve the system.

27. The record centipede has 356 fewer legs than the record millipede. Together the two have 1064 legs. How many legs does each have?

28. A Boeing 747 holds 279 more people than a Boeing 707. If the two planes together carry 721 people, how many does each plane carry? (The total number of people includes crew.)

29. An English muffin and two fried eggs contain 330 calories. Three English muffins and one egg contain 515 calories. How many calories are in each item?

30. Janet has borrowed \$3500. Some of it is an 8% auto loan, and the rest is at 21% on her credit card. She pays \$345 per year interest on the loans. How much has she borrowed from each source?

31. Suppose 5 grams of carbohydrate and 2 grams of fat contain 38 calories. Furthermore, 2 grams of carbohydrate and 6 grams of fat contain 62 calories. How many calories are in 1 gram of carbohydrate? In 1 gram of fat?

32. Mr. McFadden has a farm where he raises cows and ducks. There are 20 total heads among the animals and 64 total feet. How many cows and how many ducks are there? State your assumptions.

33. Mr. Schaaf builds 3-legged stools and 4-legged tables. He has finished 19 pieces of furniture altogether and counts 64 legs. How many of each type does he have?

34. An airplane makes a round trip between Atlanta and New Orleans. The trip is 450 miles each way. The flight takes 1.5 hours with the wind and 2.25 hours against the wind. What is the speed of the plane in still air, and what is the speed of the wind?

35. A trout travels upstream 8 miles in 0.8 hour. Another trout travels downstream 16 miles in the same time. Assume that both trout travel at the same speed in still water. What is their speed in still water? What is the speed of the current?

36. The perimeter of a rectangular room is 60 feet. The ratio of width to length is 5 to 7. What are the width and length?

For Exercises 37 to 41, start with guess and check. Use your guesses to define variables and write a system of three equations to describe each problem situation.

37. The perimeter of an isosceles triangle is 32 inches. The two equal sides are each 2.5 inches longer than the third side. What is the length of each side?

38. One angle of a triangle is 16° less than a second angle. A third angle equals the sum of the first two angles. The total angle measure of all three angles is 180°. What is the measure of each angle?

39. One serving of 7 medium shrimp, fried in vegetable shortening, contains 198 calories. Protein and carbohydrates make 4 calories per gram; fat makes 9 calories per gram. The total weight of protein, fat, and carbohydrates in the shrimp is 37 grams. There are 5 more grams of protein than of carbohydrates. Find the number of grams of each (protein, fat, and carbohydrates) in this serving of shrimp.

40. The record number of children born to one mother is 69. There were 27 sets of births, including quadruplets, triplets, and twins. There were no single births. The woman bore four times as many sets of twins as sets of quadruplets. There were nine fewer sets of triplets than sets of twins. How many sets of twins, triplets, and quadruplets were delivered? (This problem may take three equations.)

41. A cup of peanuts roasted in oil contains 903 calories. The total weight in protein, fat, and carbohydrates is 137 grams. Fat contains 9 calories per gram; protein and carbohydrates each contain 4 calories per gram. There are 44 more grams of fat than of carbohydrates. How many grams of each (protein, fat, and carbohydrates) are there in the peanuts?

42. Why does the intersection of two graphs solve the system of equations describing the graphs?

43. For each of the situations described in parts a to e, explain which of the following methods would be most appropriate for solving the system of equations:

Graphing
Table and graph on a graphing calculator
Guess and check
Substitution
Elimination

a. One variable has a 1 or -1 coefficient.

b. We are trying to understand a problem.

c. We have no algebraic method for a solution.

d. The coefficients are integers or can be made into integers with multiplication.

e. The equations are easily placed into $y = mx + b$ form.

In Exercises 44 to 49, graph the system of inequalities. Shade the solution set.

44. $y > x + 3$
$y \le -x + 2$

45. $x + y < 4$
$y \ge x$

46. $x > 3$
$y \ge -2$

47. $x \le 2$
$y < 1$

48. $x \ge 0$
$y \ge 0$
$x \le 5$
$y \le -\frac{1}{2}x + 6$

49. $x \ge 0$
$y \ge 0$
$y \le 4$
$y \le -x + 6$

50. A wedding caterer charges \$18 for adults and \$10 for young people between 6 and 15. Children under 6 are free. The total budget for a wedding is \$3600. Let $x =$ the number of adults and $y =$ the number of young people. At most 100 young people will attend. Write a system of inequalities to describe the possible numbers of people over age 6 who can attend. Graph and show the solution set.

51. A dieter is allowed 715 calories for a snack. Popcorn (salted, popped in vegetable oil) contains 55 calories per cup, and large pretzels contain 65 calories. Jean wants no more than 2 pretzels each day. Let $x =$ the number of cups of popcorn and $y =$ the number of pretzels. Write a system of inequalities to describe the possible numbers of each item she can eat. Graph and show the solution set.

52. For each basketball game, the athletic department can sell 15,000 tickets. There are regular admission tickets and student tickets. The number of student tickets can be no more than 3000. Let $x =$ the number of regular tickets and $y =$ the number of student tickets. Write a system of inequalities to describe the possible numbers of each type of ticket sold. Graph and show the solution set.

CHAPTER  TEST

In Exercises 1 and 2, solve for the indicated variable.

1. $3x - y = 400$ for y **2.** $x - 2y = 3$ for y

In Exercises 3 to 8, solve the system of equations using any method.

3. $a + 2b = 6$
$3a - b = -17$

4. $2 = 5m + b$
$-1 = m + b$

5. $x + y = 5$
$y - 5 = -x$

6. $y = 2x - 1$
$y - 2x = 3$

7. $3x - 4y = 3$
$5x - 2y = 47$

8. $3x + 2y = 5700$
$x - y = 900$

9. What does "infinitely many solutions" tell us about the graph of a system of two linear equations?

10. What does an algebraic result of $0 = 1$ tell us about the graph of a system of two linear equations?

Use the figure for Exercises 11 to 13.

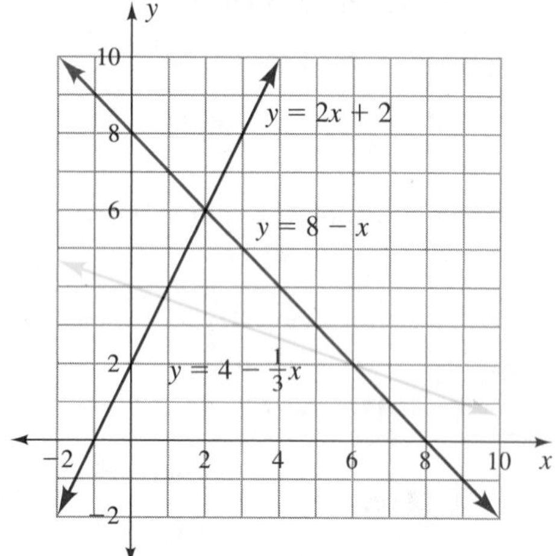

11. The point (6, 2) in the figure is the point of intersection of which two lines? What will be the result when you substitute the intersection coordinates into the two equations?

12. Shade the region in the figure that satisfies this system of inequalities:

$$y > 4 - \tfrac{1}{3}x$$

$$y \leq 8 - x$$

13. Is (7, 1) in the solution set to the system containing $y < 2x + 2$ and $y \leq 8 - x$? Explain your answer.

In Exercises 14 to 16, identify variables and write equations to describe the problem situation. Solve the problem either by guess and check or by using your equations.

14. A typical caterpillar has ten more legs than a butterfly. Eight butterflies and six caterpillars have a gross (144) of legs. How many does each have?

15. A snack of 16 peanuts and 5 cashews contains 135 calories. Another snack of 20 peanuts and 25 cashews contains 405 calories. How many calories are in each peanut? In each cashew?

16. A dolphin travels 135 miles in 3 hours with an ocean current. Another dolphin travels 100 miles in 4 hours against the same current. Assume that the dolphins travel at the same speed in still water. What is the speed of the dolphins in still water, and what is the speed of the current?

17. There are 216 coach seats on an airplane. The airline sells regular tickets and discount tickets. The airline limits the number of discount tickets on any flight to 50. Let $x =$ the number of regular tickets and $y =$ the number of discount tickets. Write a system of inequalities to describe the possible numbers of each type of ticket. Graph and show the solution set.

18. Describe how you decide whether to use elimination or substitution to solve a system of equations.

8

Squares and Square Roots: Expressions and Equations

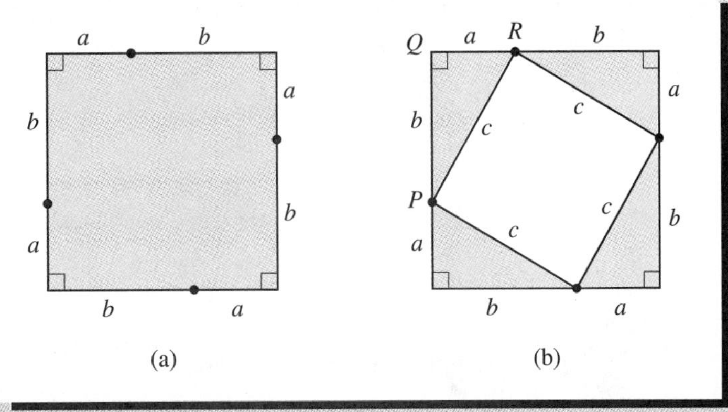

Figure 1

The squares in parts (a) and (b) of Figure 1 both have sides of length $a + b$. What is the area of the square in (a)? Write the area of each of the five parts of the square in (b). What is the sum of the areas?

Because both figures are squares with sides $a + b$, their areas are equal. Set the expressions for their areas equal and simplify. What do you observe? We will return to these questions in Example 2 of Section 8.1. This chapter applies the algebra of the Pythagorean theorem to radical expressions involving square roots and to quadratic expressions (expressions containing x^2). Our work will include forming ratios, factoring, and solving equations from tables and graphs. The chapter opens with an introduction to the Pythagorean theorem (Section 8.1), moves to formal work with radicals and radical equations (Sections 8.2 and 8.3), and then goes on to quadratic expressions and equations (Sections 8.4 to 8.6). It closes with an application of square roots in statistics (Section 8.7).

8.1 Pythagorean Theorem

OBJECTIVES

- Use the Pythagorean theorem to write equations and to solve for missing sides of right triangles.
- Use the converse of the Pythagorean theorem to find out whether triangles are right triangles.
- Apply the Pythagorean theorem in a variety of settings.

WARM-UP

1. List the values of these perfect squares:

1^2	6^2	11^2	16^2
2^2	7^2	12^2	20^2
3^2	8^2	13^2	25^2
4^2	9^2	14^2	30^2
5^2	10^2	15^2	40^2

2. Multiply out these binomials.
   a. $(x + 3)(x + 3)$ **b.** $(x - y)(x - y)$
   c. $(a + b)(a + b)$ **d.** $(x - 1)^2$
3. Your friend says $(x + y)^2 = x^2 + y^2$. Explain the error.

THE PYTHAGOREAN THEOREM introduced in this section will be applied to work with both square root functions (Sections 8.2 and 8.3) and quadratic functions (Sections 8.4, 8.5, and 8.6). The section provides practice in numerical work with squares and square roots. Familiarity with the numbers here will help you understand the concepts in the rest of the chapter.

Introduction to the Pythagorean Theorem

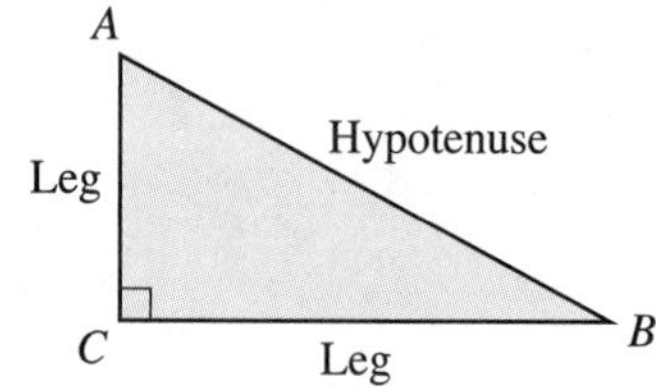

Figure 2

RIGHT TRIANGLES The **Pythagorean theorem** relates *the lengths of the perpendicular sides* (legs) *of a right triangle to the length of the longest side* (hypotenuse). The legs and hypotenuse are shown on the right triangle in Figure 2.

EXAMPLE 1 Exploring the right triangle Place the measures of the sides of the triangles in Table 1, and then complete the remaining columns. What patterns do you observe?

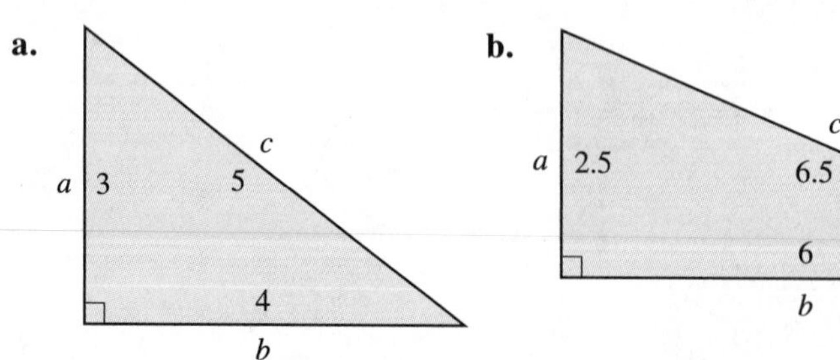

c.

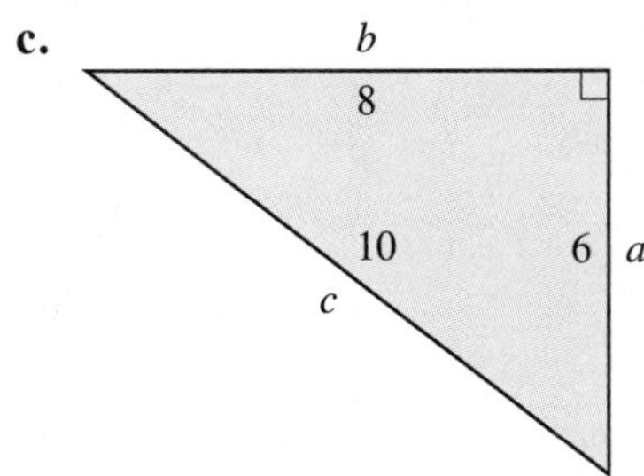

Triangle	Leg a	Leg b	Hypotenuse c	a^2	b^2	$a^2 + b^2$	c^2
a							
b							
c							

Table 1

Solution The Pythagorean theorem appears in the last columns of Table 1, where $a^2 + b^2 = c^2$. See the Answer Box for the completed table. ●

The Pythagorean Theorem

If a triangle is a right triangle, then the sum of the squares of the two shortest sides (legs) is equal to the square of the longest side (hypotenuse):

$$a^2 + b^2 = c^2$$

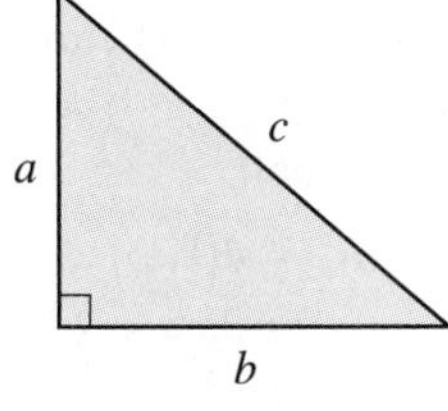

PROOF OF THE PYTHAGOREAN THEOREM A **proof** is *a logical argument that demonstrates the truth of a statement.* It is based on previously known facts or agreed-upon statements. The Greek mathematician Pythagoras is credited with being the first to record a proof of the relationship for right triangles given his name. However, documents from Chinese history indicate that the relationship was known long before Pythagoras's time.

Example 2 proves the Pythagorean theorem by showing that it is true for any a, b, and c that satisfy the conditions of the theorem: a and b are legs of a right triangle, and c is the hypotenuse.

EXAMPLE 2 Proving the Pythagorean theorem Using the following steps to prove the Pythagorean theorem, where $\triangle PQR$ is a right triangle.

a. What is the area of the square in part (a) of Figure 3, which has $a + b$ on each side?

b. Write the area of each of the five parts of the square in part (b) of Figure 3. Write the sum of the areas.

c. Because both figures are squares with sides $a + b$, their areas are equal. Set the expressions for their areas equal, and show that $a^2 + b^2 = c^2$.

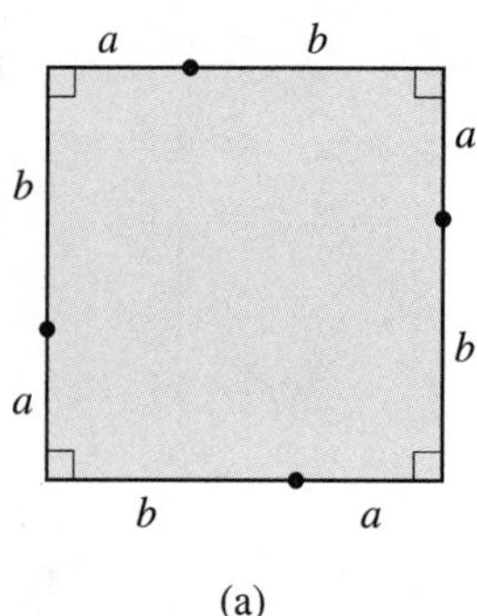

(a)

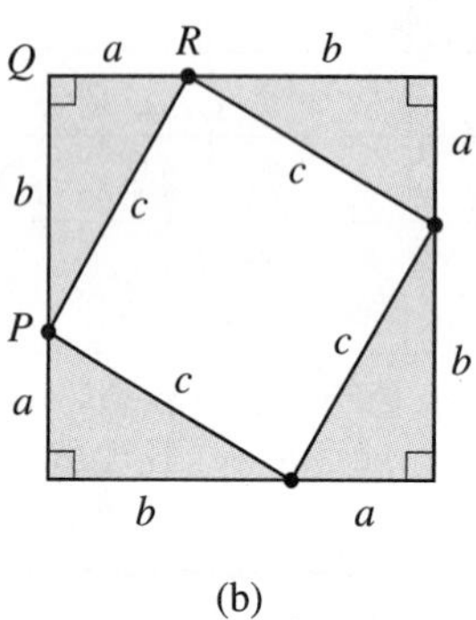

(b)

Figure 3

Solution **a.** The square in (a), with side $a + b$, has area $(a + b)^2$.

b. The square in (b) has four right triangles and an inner square of side c. The area of each right triangle is $\frac{1}{2}ab$. The area of the inner square is c^2. Thus, the total area is $4\left(\frac{1}{2}ab\right) + c^2$.

c. We set the areas equal:

$$(a + b)^2 = 4\left(\tfrac{1}{2}ab\right) + c^2$$

$$a^2 + 2ab + b^2 = 2ab + c^2 \qquad \text{Multiply to remove the parentheses.}$$

$$a^2 + b^2 = c^2 \qquad \text{Subtract } 2ab \text{ from both sides.}$$

Therefore, for right triangle PQR, $a^2 + b^2 = c^2$. ●

SOLVING FOR MISSING SIDES In Example 3, we solve for missing sides using the Pythagorean theorem. In the drawing in part c, there is a right angle on the circle and the hypotenuse is coincident with the diameter of the circle. The drawing has applications in navigation. In the drawing in part d, n is the length of the diagonal of the rectangle.

EXAMPLE 3 Finding missing sides What is the missing side, n, in each of the following drawings?

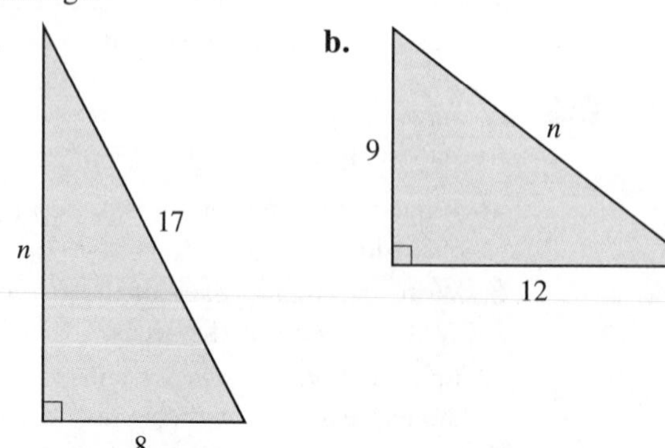

c.

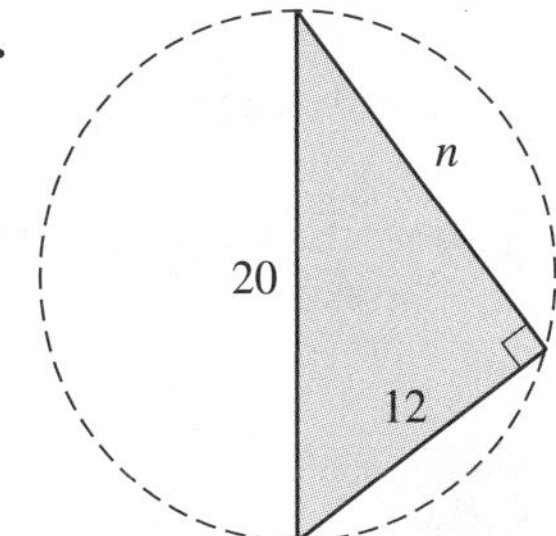

d.

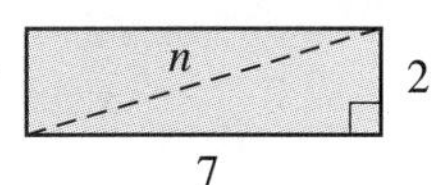

Solution **a.** The sides n and 8 are perpendicular and are the legs, a and b.

$$n^2 + 8^2 = 17^2$$
$$n^2 + 64 = 289$$
$$n^2 = 225$$
$$n = 15$$

Although -15 also makes $n^2 = 225$ true, length must be positive, so we disregard -15.

b. The sides 9 and 12 are perpendicular and are the legs.

$$9^2 + 12^2 = n^2$$
$$81 + 144 = n^2$$
$$225 = n^2$$
$$15 = n$$

c. The sides n and 12 are perpendicular.

$$n^2 + 12^2 = 20^2$$
$$n^2 + 144 = 400$$
$$n^2 = 256$$
$$n = 16$$

d. The sides 2 and 7 are perpendicular.

$$2^2 + 7^2 = n^2$$
$$4 + 49 = n^2$$
$$53 = n^2$$
$$\sqrt{53} = n$$

When a square root does not give an exact value, we generally leave it in square-root form or use the calculator to give a decimal approximation. In this case, $n \approx 7.28$ to the nearest hundredth. ●

When we solve for the missing side, we are using a new process: *taking the square root of both sides.* We will discuss this operation more formally in Example 2 of Section 8.5 (page 521).

CONVERSE OF THE PYTHAGOREAN THEOREM *Changing the position of the* if *and* then *statements* in the Pythagorean theorem gives us the **converse of the Pythagorean theorem**. The converse is also true.

Converse of the Pythagorean Theorem

> If the sum of the squares of the two shortest sides (legs) is equal to the square of the longest side (hypotenuse), then the triangle is a right triangle.

In Example 4, we apply the converse of the Pythagorean theorem to find out if a triangle is a right triangle.

EXAMPLE 4 Finding right triangles Each triangle below is drawn to look like a right triangle. Which *is* a right triangle?

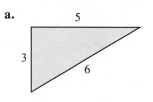

a. 5, 3, 6 **b.** 2, 1.5, 2.5 **c.** 9, 8, 12

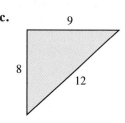

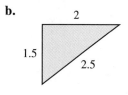

Solution **a.** $3^2 + 5^2 = 9 + 25 = 34$, which does not equal 36, the square of the third side. This triangle is not a right triangle.

b. $1.5^2 + 2^2 = 2.25 + 4 = 6.25 = 2.5^2$. This is a right triangle, with the right angle where the shortest sides meet.

c. $8^2 + 9^2 = 64 + 81 = 145$, which does not equal 144, the square of the third side. This triangle is not a right triangle. ●

Applications

LADDER SAFETY In Section 5.2, we used the 4-to-1 ladder-safety ratio (height of the ladder on the wall to the distance of the base of the ladder from the wall). We now use the Pythagorean theorem to find the length of the ladder needed for a particular job and safe positions for the ladder given its length.

In Example 5, we are given the height up the wall to be reached. After dividing by 4 to find the distance from the wall, we need to find the length of the ladder.

EXAMPLE 5 Finding ladder length A ladder needs to reach 21 feet up a wall (see Figure 4). For a safe ratio of 4 to 1, the base of the ladder must be 5.25 feet from the wall. To the nearest foot, what length ladder is needed?

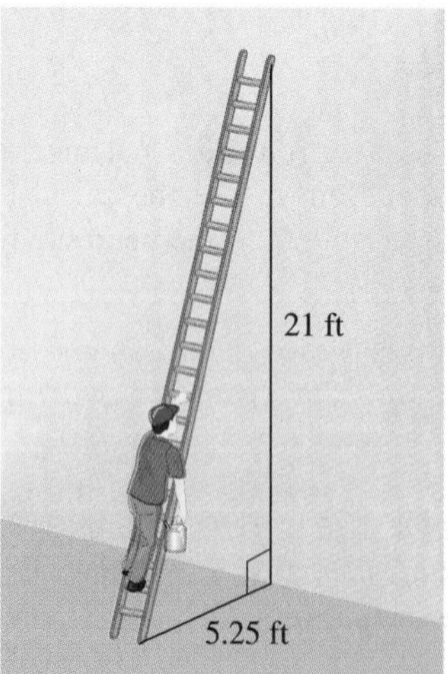

Figure 4

Solution The ground and the wall form the legs of a right triangle, and the ladder is the hypotenuse. The length of the ladder is c.

$$c^2 = (21 \text{ ft})^2 + (5.25 \text{ ft})^2$$
$$c^2 \approx 468.56 \text{ ft}^2$$
$$c \approx 21.6 \text{ ft}$$

Thus, a 22-foot ladder will be needed. ●

In Example 6, we are given the length of the ladder. Both the base and the height are unknown, but we know the ratio of base to height. Look carefully at how the example is solved in one variable and then try the "Think about it" with two variables.

EXAMPLE 6 Finding ladder positions given ladder length What is the safe ladder position for a 16-foot ladder (see Figure 5)?

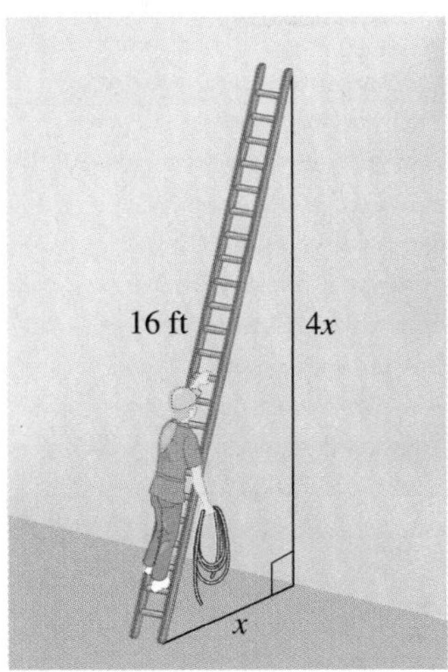

Figure 5

Solution The base-height ratio must be 1 to 4. Thus, the legs of the right triangle are x and $4x$. By the Pythagorean theorem,

$$a^2 + b^2 = c^2$$
$$x^2 + (4x)^2 = (16 \text{ ft})^2$$
$$17x^2 = 256 \text{ ft}^2$$
$$x^2 = \frac{256}{17}$$
$$x^2 = 15.059$$
$$x \approx 3.88 \text{ ft from the base of the wall}$$
$$4x \approx 15.52 \text{ ft up the wall}$$ ●

Think about it: Set up Example 6 in two variables. First, write a proportion for the 4 to 1 ratio, with l = length and w = width. Second, solve your proportion for l. Third, write the Pythagorean equation, using l and w for the wall height and base and 16 for the ladder length. Last, substitute for l in the Pythagorean equation.

FLIGHT AND WIND In the examples in Section 7.4, the wind and the direction of airplane flight were parallel. We now consider the result when the wind and the direction of flight are perpendicular. The figure in Example 7 shows a northbound plane and a cross wind from the west. The effect of the wind is to blow the plane off course to the right (or northeast).

The data in the example are given in miles per hour. To simplify, we will use a flying time of 1 hour. Here is how we change units to miles:

$$D = rt$$

$$D = \frac{x \text{ mi}}{\text{hr}} \cdot 1 \text{ hr} = x \text{ mi}$$

EXAMPLE 7 Finding distance traveled with a cross wind If the wind is blowing at 68 miles per hour from the west and the plane is traveling 100 miles per hour due north, calculate the distance flown after 1 hour (as shown in Figure 6).

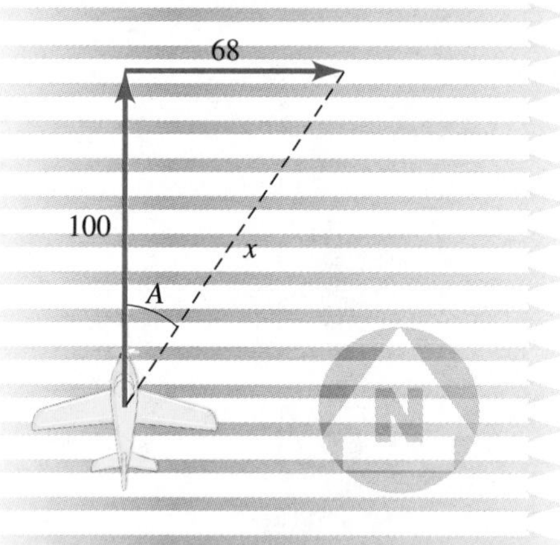

Figure 6

Solution In 1 hour, the plane will travel 100 miles north and 68 miles east. Because these directions are at right angles, the actual flight path will be the hypotenuse.

$$100^2 + 68^2 = x^2$$
$$14{,}624 = x^2$$
$$x = \sqrt{14{,}624} \approx 121 \text{ mi}$$

ANSWER BOX

Warm-up: 1. 1, 4, 9, 16, 25, 36, 49, 64, 81, 100, 121, 144, 169, 196, 225, 256, 400, 625, 900, 1600. A pattern can help you learn the squares: The last digits in 13^2 are reversed in 14^2. **2. a.** $x^2 + 6x + 9$ **b.** $x^2 - 2xy + y^2$ **c.** $a^2 + 2ab + b^2$ **d.** $x^2 - 2x + 1$ **3.** There is no middle term given; $(x + y)^2 = x^2 + 2xy + y^2$. Using a table to multiply $(x + y)(x + y)$ will show him the terms he missed. **Example 1:**

Triangle	Leg a	Leg b	Hypotenuse c	a^2	b^2	$a^2 + b^2$	c^2
a	3	4	5	9	16	25	25
b	2.5	6	6.5	6.25	36	42.25	42.25
c	6	8	10	36	64	100	100

Think about it: $\frac{l}{w} = \frac{4}{1}$, $l = 4w$, $l^2 + w^2 = 16^2$, $(4w)^2 + w^2 = 16^2$

EXERCISES 

In Exercises 1 to 6, find the length of the side marked with an x. Round to the nearest hundredth.

1.

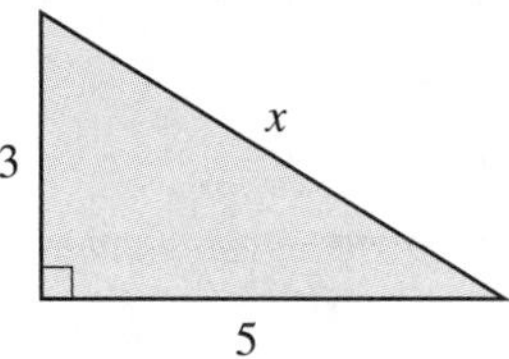

2.

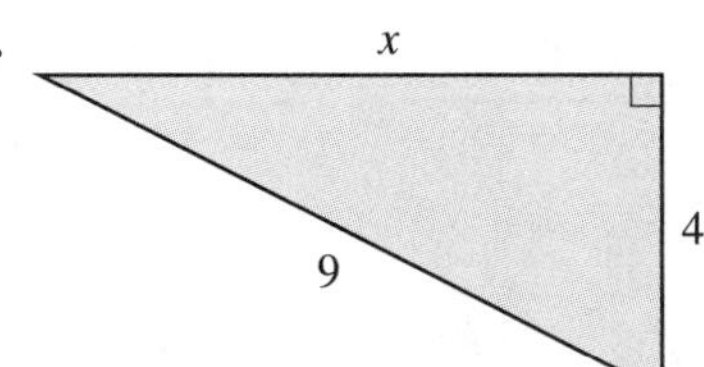

3.

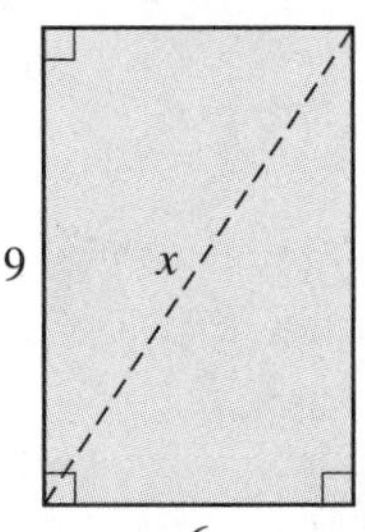

4.

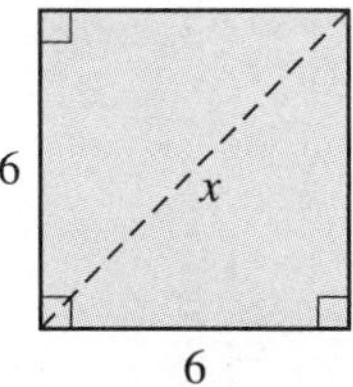

5.

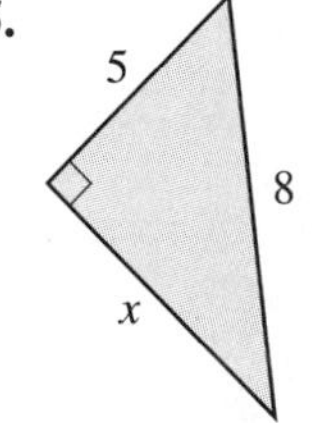

6.

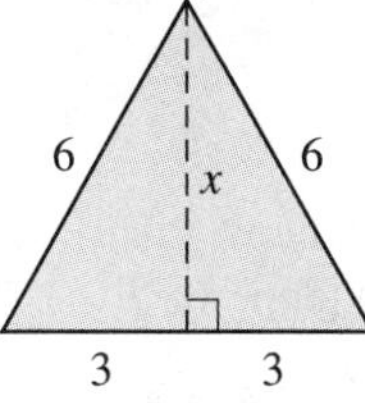

Which of the sets of three numbers in Exercises 7 to 12 could represent the sides of a right triangle?

7. {7, 8, 9} **8.** {12, 15, 18}

9. {12, 16, 20} **10.** {8, 15, 17}

11. {7, 24, 25} **12.** {9, 40, 41}

Use proportions to find the missing sides of the similar triangles shown in Exercises 13 and 14. Show, with the converse of the Pythagorean theorem, that they are all right triangles.

13.

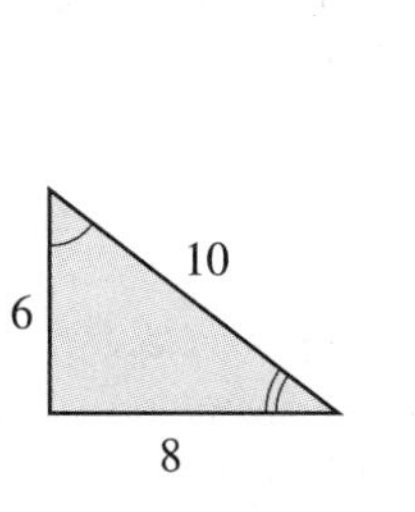

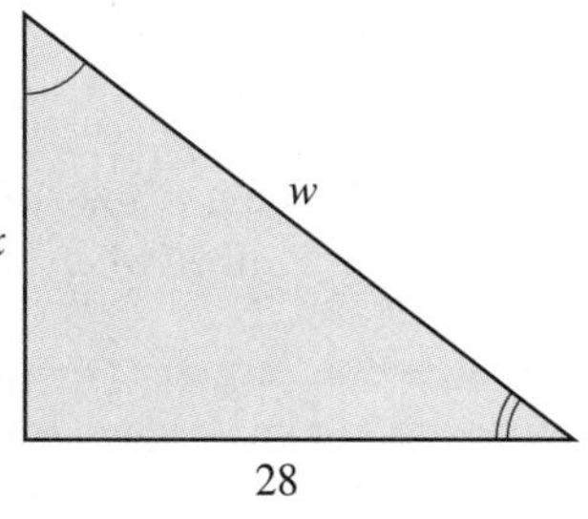

14.

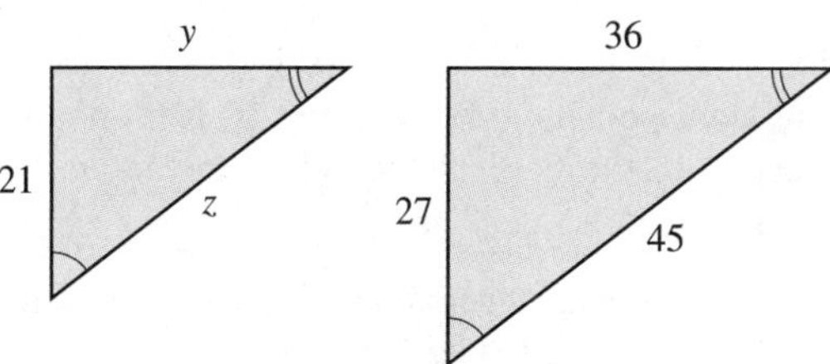

In Exercises 15 and 16, complete the tables, which show enlargements or reductions of {3, 4, 5} right triangles. Each row represents the sides of a different triangle.

15.

Leg	Leg	Hypotenuse
3	4	5
6		
		30
	12	
1		

16.

Leg	Leg	Hypotenuse
3	4	5
9		
		35
	20	
		1

17. What are the squares of these expressions?

a. $3x$ **b.** $4x$ **c.** $5x$ **d.** $6x$

18. In Example 6, the equation was $x^2 + (4x)^2 = 16^2$. Suppose we made a mistake and wrote $x^2 + 4x^2 = 16^2$. Solve the second equation, and explain how the resulting base (x) and wall height ($4x$) cannot match with a 16-foot ladder. Include a sketch with your explanation.

Use the variable expressions in Exercises 19 to 24 in the Pythagorean theorem and solve for x. Round answers to the nearest tenth. The sides are listed smallest to largest.

19. $\{x, x, 8\}$ **20.** $\{x, x, 12\}$

21. $\{x, 5x, 52\}$ **22.** $\{x, 3x, 20\}$

23. $\{x, 2x, 15\}$ **24.** $\{x, 7x, 50\}$

In Exercises 25 to 30, assume a safe-ladder ratio, height to base, of 4 to 1. Round to the nearest tenth.

25. A safe ladder position for reaching 12 feet up a wall is 3 feet from the base of the ladder to the wall. How long a ladder is needed?

26. A safe ladder position for reaching 8 feet up a wall is 2 feet from the base of the ladder to the wall. How long a ladder is needed?

27. A safe ladder position for reaching 9 feet up a wall is 2.25 feet from the base of the ladder to the wall. How long a ladder is needed?

28. A safe ladder position for reaching 10 feet up a wall is 2.5 feet from the base of the ladder to the wall. How long a ladder is needed?

29. What is the safe ladder position for reaching 14 feet up a wall? How long a ladder is needed?

30. What is the safe ladder position for reaching 19 feet up a wall? How long a ladder is needed?

Find the safe ladder positions for the ladder lengths in Exercises 31 to 34. Use a safe-ladder ratio, height to base, of 4 to 1. Give your answers first with decimal portions of a foot and then in feet and inches to the nearest inch.

31. 12-foot ladder

32. 22-foot extension ladder

33. 18-foot extension ladder

34. 25-foot extension ladder

In Exercises 35 to 38, draw triangles and show the direction actually flown in a cross wind.

35. A plane flies south for 1 hour at 250 miles per hour from Montreal, Canada. A 30-mile-per-hour wind is blowing from the east. How far will the plane have flown?

36. A plane flies east for 1 hour from Missoula, Montana, at 210 miles per hour. A 35-mile-per-hour wind from the north blows the plane off course. How far will the plane have traveled?

37. A plane flies for 3 hours on a heading due north from San Francisco. The plane is flying at 200 miles per hour while a 32-mile-per-hour wind is blowing from the west. What is the total distance flown?

38. A plane flies for 3 hours on a heading due north from Atlanta. The plane travels 225 miles per hour. A wind is blowing 20 miles per hour from the west during the trip. What is the total distance flown?

39. What is the height of the house in the figure?

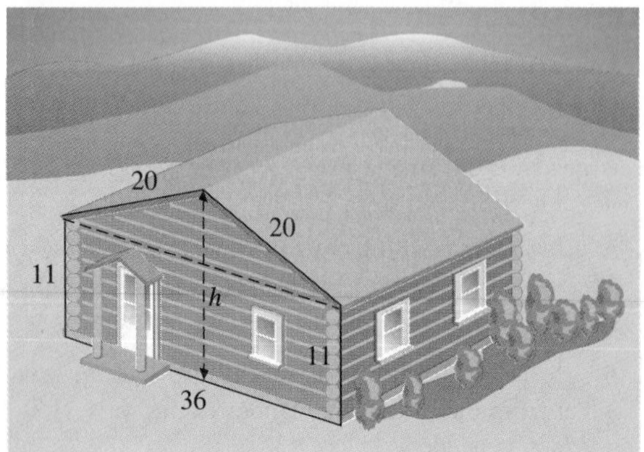

40. What are the lengths of the crossed pieces, AB and CD, needed to make the kite in the figure?

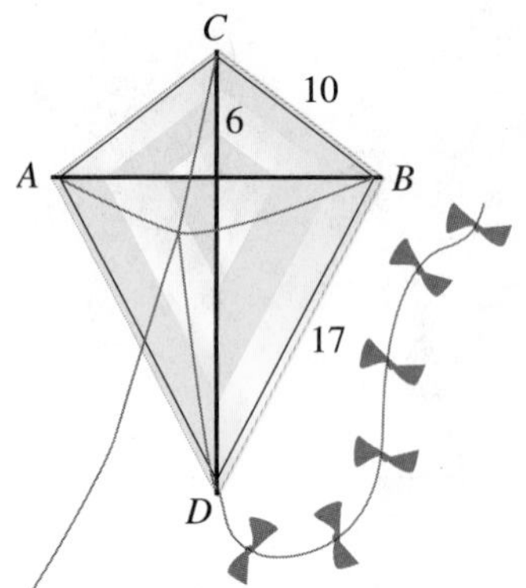

A rectangular house has a 28-foot by 40-foot floor plan (see the figure). For Exercises 41 to 44, assume that the roof ends at the edge of the wall. Assume both sides of the roof have the same area and slope.

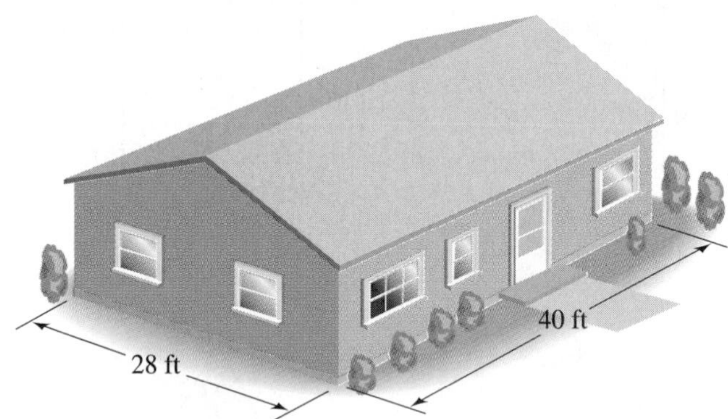

41. How many square feet of roofing material are needed for a slope of 3 to 14?

42. How many square feet of roofing material are needed for a slope of 5 to 14?

43. How many square feet of roofing material are needed for a slope of 9 to 14?

44. How many square feet of roofing material are needed for a slope of 10 to 14?

Projects

45. ***Calculator Investigations with Squaring.*** To answer the questions, use the $\boxed{x^2}$ key to guess and check. In parts a to d, write both the number and its square.

a. What is the largest three-digit number that gives a five-digit number when squared?

b. What is the largest four-digit number that gives a seven-digit number when squared?

c. What is the largest five-digit number that gives a nine-digit number when squared? Compare the digits with those in $\sqrt{10}$. Is this a coincidence?

d. What is the largest square possible with a five-digit number?

e. The author's grandmother was x years old in the year x^2. Grandmother was still alive in 1945. Find the age x and the year x^2. When was she born?

f. The author's great, great, great grandmother was x years old in the year x^2. She was alive during the American Civil War. Find the age x and the year x^2. When was she born?

g. Would it be possible for someone you know to be alive in 2010 and be x years old in the year x^2? Give the person's age, the year, and the person's year of birth.

46. ***Rounding Numbers.*** When we do several steps on a calculator, rounding at a middle step can cause us to overlook patterns or arrive at unacceptable answers. Simplify these calculator expressions and describe the differences in the answers.

a. Exact: $\sqrt{(5^2 + (\sqrt{11})^2)}$
Rounded: $\sqrt{(5^2 + 3.3^2)}$

b. Exact: $\sqrt{(6^2 + (\sqrt{13})^2)}$
Rounded: $\sqrt{(6^2 + 3.6^2)}$

c. Exact: $\sqrt{(7^2 + (\sqrt{15})^2)}$
Rounded: $\sqrt{(7^2 + 3.9^2)}$

d. What general number pattern is shown in parts a to c? *Hint*: $20^2 + (\sqrt{\_\_\_})^2 = 21^2$ and $21^2 + (\sqrt{\_\_\_})^2 = 22^2$.

The error from rounding gets worse as exponents get higher in compound interest problems. Describe the differences in the answers to parts e and f.

e. Exact: $1000(1 + 0.08/12)^{12 \cdot 10}$
Rounded: $1000(1.007)^{12 \cdot 10}$

f. Exact: $1000(1 + 0.05/12)^{12 \cdot 30}$
Rounded: $1000(1.004)^{12 \cdot 30}$

You can avoid working with rounded numbers by entering the entire expression into the calculator at once.

47. ***Garfield Proof.*** James A. Garfield, 20th president of the United States, is credited with a proof of the Pythagorean theorem based on the areas of a trapezoid (see the figure). The area of a trapezoid is given by

$$A = \tfrac{1}{2}(\text{height})(\text{top} + \text{bottom})$$

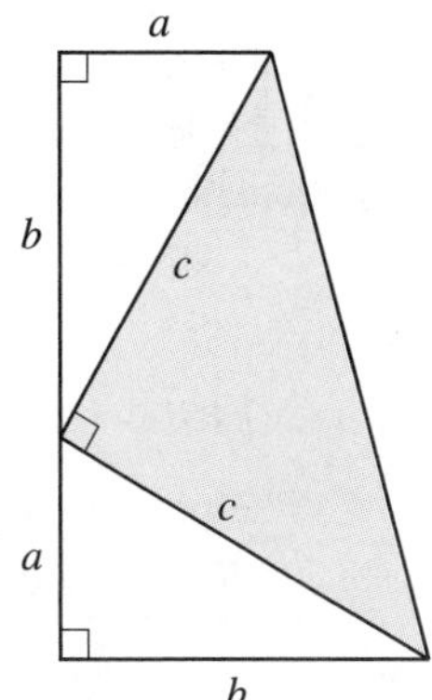

a. Find the area of the trapezoid using the formula.

b. Find the area of the trapezoid by adding the areas of the three triangles.

c. Because the areas in parts a and b both describe the trapezoid, we can set the areas equal. Show that the resulting equation simplifies to $a^2 + b^2 = c^2$.

8.2 Square Root Expressions and Their Applications, Including the Distance Formula

OBJECTIVES

- Give the principal square root of a number.
- Graph the square root curve.
- Simplify square root expressions.
- Multiply and divide square roots using the properties for products and quotients.
- Identify sides of right triangles.
- Apply the distance formula.

WARM-UP

1. Find these square roots without a calculator.

a. $\sqrt{64}$ b. $\sqrt{121}$ c. $\sqrt{49}$ d. $\sqrt{256}$

e. $\sqrt{400}$ f. $\sqrt{225}$ g. $\sqrt{144}$ h. $\sqrt{625}$

2. Practice mental estimation by giving the two consecutive integers above and below each square root. For example, $2 < \sqrt{6} < 3$.

a. $\sqrt{45}$ b. $\sqrt{75}$ c. $\sqrt{200}$ d. $\sqrt{20}$

e. $\sqrt{110}$ f. $\sqrt{12}$ g. $\sqrt{905}$ h. $\sqrt{395}$

THIS SECTION STARTS WITH the principal square root, irrational numbers, and the square root graph and introduces multiplication, division, and simplification of square root expressions. Square roots help us identify Pythagorean triples and find the length of line segments.

Square Roots

The **principal square root** of a number is *the positive number that, multiplied by itself, produces the given number.* In algebraic notation, $\sqrt{n} = x$ if $x^2 = n$ for $x \geq 0$.

EXAMPLE 1 Naming principal square roots What are the principal square roots?

a. 25 b. 256 c. 2 d. $x^2, x \geq 0$ e. $a, a \geq 0$

Solution See the Answer Box.

Student Note: For these short examples, solutions are listed in the Answer Box. You are encouraged to write your own answers and then look in the Answer Box to check your thinking.

Many numbers have exact square roots. **Perfect squares** are *those integers with exact square roots*: 1, 4, 9, 16, and so forth. Some decimals and fractions also have exact square roots.

EXAMPLE 2 Naming principal square roots Guess and check the square roots of these decimals and fractions.

a. 0.16 b. 2.25 c. 6.25 d. $\frac{1}{4}$ e. $\frac{9}{16}$ f. $\frac{144}{9}$

Solution See the Answer Box.

RATIONAL AND IRRATIONAL NUMBERS Any number with an exact square root can be written as a ratio of two integers (with the denominator not zero). This ratio is called a rational number. In algebraic notation, a *rational number* is a/b, where a and b are integers and $b \neq 0$. (This definition was first introduced in Section 1.2.)

Numbers without exact square roots are irrational numbers: $\sqrt{2}$, $\sqrt{3}$, $\sqrt{5}$, $\sqrt{6}$, and so forth. An **irrational number** is *a real number that cannot be written as the quotient of two integers*. Pi (π) is an irrational number. Together the rational and irrational numbers make up the set of *real numbers*.

Set of Real Numbers

Rational Numbers (can be written as the quotient of two integers)

Fractions: $\frac{1}{2}$, $\frac{1}{4}$
Repeating decimals: $0.33\overline{3}$, $0.1\overline{212}$
Terminating decimals: 0.5, 0.125
Integers: -2, -1, 365
Whole numbers: 3, 18
Square roots with exact decimal values: $\sqrt{6.25}$, $\sqrt{16}$, $\sqrt{256}$

Irrational Numbers (cannot be written as the quotient of two integers)

Pi: 3.14159265...
Square roots without exact decimal values: $\sqrt{2}$, $\sqrt{3}$, $\sqrt{5}$

EXAMPLE 3 Identifying rational and irrational numbers Complete Table 2. Round irrational numbers to three decimal places.

Square Root	Value	Rational or Irrational?
$\sqrt{9}$	3	rational
$\sqrt{19}$		
$\sqrt{29}$		
$\sqrt{39}$		
$\sqrt{49}$		
$\sqrt{1.21}$		
$\sqrt{.09}$		
$\sqrt{20.25}$		
$\sqrt{169}$		

Table 2

Solution The table contains only three irrational numbers. See the Answer Box. ●

SQUARE ROOT GRAPHS The square root equation is $y = \sqrt{x}$. We graph the equation in Example 4 and look at how important features of the square root graph are related to inputs and outputs.

EXAMPLE 4 Plotting a square root graph Table 3 shows ordered pairs from $y = \sqrt{x}$. Graph the ordered pairs. The irrational numbers in the table have been rounded to three decimal places.

Solution In graphing by hand, we can only estimate the positions of irrational numbers to about a tenth, so the other decimal places are excessive. The graph of $y = \sqrt{x}$ is shown in Figure 7.

Graphing Calculator Solution In the graphing calculator solution in Figure 8, we see the square root graph again, this time with all four quadrants showing. Try it on your calculator. Set $Y_1 = \sqrt{X}$. Set the window for X at $[-4, 10]$, scale 1, and Y at $[-4, 4]$. Graph.

Input x	Output $y = \sqrt{x}$	Coordinate Point (x, y)
0	0	(0, 0)
0.5	≈0.707	(0.5, 0.707)
1	1	(1, 1)
1.5	≈1.225	(1.5, 1.225)
2	≈1.414	(2, 1.414)
3	≈1.732	(3, 1.732)
4	2	(4, 2)
5	≈2.236	(5, 2.236)

Table 3

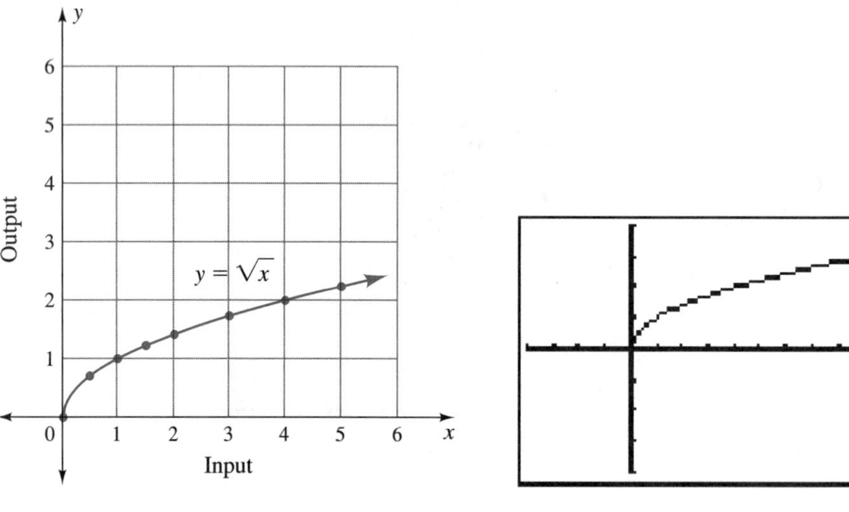

Figure 7 **Figure 8**

The square root graphs shown in Example 4 have two important features:

1. There is no graph to the left of the y-axis (in the second or third quadrant), where x is negative. Thus, *there are no outputs for inputs less than zero. We cannot take the square root of a negative number and obtain a real-number output.*
2. There is no graph below the x-axis (in the third or fourth quadrant), where y is negative. Thus, *there are no negative outputs to a square root; $\sqrt{x}$ is always positive.* That is why the principal square root is defined as the positive root of a number.

In Examples 5 and 6, we use numbers to look more closely at inputs and outputs of square roots. The results should agree with our observations from square root graphs.

EXAMPLE 5 Exploring square roots on a calculator Use a calculator to take the square root of -1, -4, and -9.

Solution Your calculator will probably give an error message, because *the square root of a negative number is undefined in the real numbers.* If your calculator is an advanced type, the answer may be in another notation. If so, see the Calculator Note. ●

Think about it: Predict what a calculator table for $y = \sqrt{x}$, with inputs -3 to 0, would look like. Then try it.

Square Root of a Negative Number

> The square root of a negative number is *undefined in the real numbers.* Thus, $y = \sqrt{x}$ for $x \geq 0$.

Calculator Note:
In response to $\sqrt{-1}$, $\sqrt{-4}$, and $\sqrt{-9}$, a few advanced calculators will give i, $2i$, and $3i$. These numbers are not real numbers. They contain the imaginary unit, $i = \sqrt{-1}$. Other calculators will give ordered pairs: (0, 1), (0, 2), and (0, 3). In this case, the ordered pair (a, b) describes the complex number $a + bi$. The term a in $a + bi$ is the real-number part and is zero for these square roots. The term bi is the imaginary part. Thus, (0, 1) indicates $0 + 1i$, (0, 2) is $0 + 2i$, and (0, 3) is $0 + 3i$. The ordered pair (0, 1) is not a pair of coordinates on a graph. A graph has real numbers on both axes, and there is no way to place an imaginary number on a graph.

MORE SQUARE ROOT GRAPHS Figures 7 and 8 showed the principal square root graph with no points to the left of the origin because negative numbers have no square roots. When graphing square roots containing expressions, though, we must look carefully for negative numbers. We graph two more square root expressions in Examples 6 and 7.

EXAMPLE 6 Graphing square root equations Make a table and graph for $y = \sqrt{x + 2}$. Use integer inputs on the interval $[-4, 7]$.

Solution The table is Table 4, and the graph appears in Figure 9. For $y = \sqrt{x + 2}$, there are defined values to the left of the y-axis. The expression $x + 2$ causes the inputs between -2 and 0 to be acceptable inputs.

x	$y = \sqrt{x + 2}$	x	$y = \sqrt{x + 2}$
-4	error	2	2
-3	error	3	2.236
-2	0	4	2.449
-1	1	5	2.646
0	1.414	6	2.828
1	1.732	7	3

Table 4

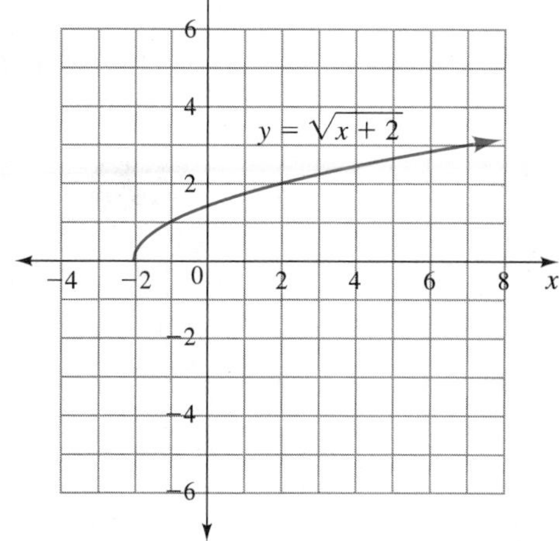

Figure 9

Graphing Calculator Solution

Enter $Y_1 = \sqrt{(X + 2)}$. Set the table feature, with -4 as the starting value and 1 as the change (Δ) in the table. Set the window for X at $[-4, 7]$, scale 1, and for Y at $[-4, 4]$, scale 1. Graph. The table and graph should agree with those in the prior solution. ●

EXAMPLE  Graphing square root equations Make a table and graph for $y = \sqrt{x-3}$.

Solution The table is Table 5, and the graph appears in Figure 10. For $y = \sqrt{x-3}$, there are no defined values to the left of $x = 3$. The expression $x - 3$ causes all the inputs from -4 to 2 to give square roots of negative numbers.

x	$y = \sqrt{x-3}$	x	$y = \sqrt{x-3}$
−4	error	2	error
−3	error	3	0
−2	error	4	1
−1	error	5	1.414
0	error	6	1.732
1	error	7	2

Table 5

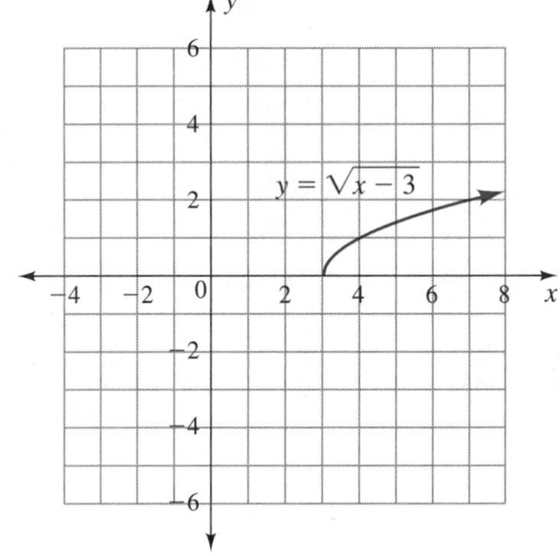

Figure 10

Graphing Calculator Solution  Enter $Y_1 = \sqrt{(X-3)}$. Set the table feature, with -4 as the starting value and 1 as the change (Δ) in the table. Set the window for X at $[-4, 8]$, scale 1, and for Y at $[-6, 6]$, scale 1. Graph. The table and graph should agree with those in the prior solution. ●

Negative Signs with Square Roots

Simplify can mean to evaluate a square root expression. To simplify, we need to interpret negative signs correctly. Note that a negative sign outside the square root sign means the opposite of a square root (which is defined), whereas a negative sign inside the square root sign means the square root of a negative (which is not defined in the real numbers). We use the *plus or minus sign* ($\pm$) outside the square root sign to indicate that the answer is both the positive root and its opposite.

EXAMPLE  Working with negatives in and on square root expressions Simplify these expressions.

a. $\sqrt{25}$ **b.** $-\sqrt{49}$ **c.** $\pm\sqrt{121}$

d. $\sqrt{-16}$ **e.** $\pm\sqrt{1.44}$ **f.** $-\sqrt{25}$

Solution **a.** $\sqrt{25} = 5$. We give the positive, or principal, root.

b. $-\sqrt{49} = -7$. The negative sign outside the square root sign means the opposite of the square root.

c. $\pm\sqrt{121} = \pm 11$. Both positive and negative numbers are given.

d. $\sqrt{-16}$. There is no real-number solution to the square root of a negative number.

e. $\pm\sqrt{1.44} = \pm 1.2$. Although a decimal, 1.44 has an exact square root.

f. $-\sqrt{25} = -5$. The negative sign outside the square root sign means the opposite of the square root. ●

> The principal square root of a real number is always zero or positive:
>
> $\sqrt{x} \geq 0$

Exponents and Square Roots

Example 9 shows a surprising result with exponents.

EXAMPLE 

Exploring the meaning of one-half as exponent Use a calculator exponent key, [^], [y^x], or [x^y], to find these values.

a. $25^{0.5}$ **b.** $16^{1/2}$ **c.** $225^{0.5}$ **d.** $1.44^{1/2}$

Solution The decimal 0.5 is equal to the fraction $\frac{1}{2}$. The meaning of the exponent is the same in both decimal and fraction notation: the square root. The answers are listed in the Answer Box.

One-half as Exponent

> When one-half is used as an exponent, it means to take the principal square root of the base. Thus, if a is zero or a positive real number,
>
> $a^{1/2} = a^{0.5} = \sqrt{a}$

Example 10 suggests why the $\frac{1}{2}$ power is the square root. Recall that one property of exponents is that $a^x a^y = a^{x+y}$.

EXAMPLE 

Simplifying exponential and square root expressions Simplify these expressions.

a. $36^{1/2} \cdot 36^{1/2}$ **b.** $\sqrt{36} \cdot \sqrt{36}$ **c.** $25^{1/2} \cdot 25^{1/2}$ **d.** $\sqrt{25} \cdot \sqrt{25}$

Solution In parts a and c, we add the exponents. In parts b and d, we take the square roots first and then multiply. See the Answer Box.

Multiplication and Division with Square Roots

PRODUCT PROPERTY In Example 11, we explore the order in which we do multiplication and square roots.

EXAMPLE 

Exploring the order of operations Do these without a calculator.

a. $\sqrt{4} \cdot \sqrt{9}$ **b.** $\sqrt{4 \cdot 9}$ **c.** $\sqrt{9 \cdot 25}$ **d.** $\sqrt{9} \cdot \sqrt{25}$

Solution In parts a and d, we take the square roots first. In parts b and c, we multiply first. See the Answer Box.

Example 11 suggests that the results are the same whether we (1) take the square root and then multiply or (2) multiply and then take the square root.

Product Property for Square Roots

> The square root of a product equals the product of the square roots. In algebraic notation,
>
> $\sqrt{a \cdot b} = \sqrt{a} \cdot \sqrt{b}$
>
> if a and b are positive numbers or zero.

We prove the product property of square roots by going back to the product property of exponents and applying the fact that $\frac{1}{2}$ as exponent means the square root.

EXAMPLE 12 Proving the product property of square roots Use the product property of exponents to show that

$$\sqrt{a \cdot b} = \sqrt{a} \cdot \sqrt{b}$$

Solution We assume a and b are zero or positive.

$$\sqrt{a \cdot b} = (a \cdot b)^{1/2} \quad \text{Definition of } \tfrac{1}{2} \text{ as exponent}$$
$$(a \cdot b)^{1/2} = a^{1/2} \cdot b^{1/2} \quad \text{Product property for exponents}$$
$$a^{1/2} \cdot b^{1/2} = \sqrt{a} \cdot \sqrt{b} \quad \text{Definition of } \tfrac{1}{2} \text{ as exponent}$$

Thus, $\sqrt{a \cdot b} = \sqrt{a} \cdot \sqrt{b}$. ●

In Example 13, we practice simplifying radical expressions without a calculator. These simplifications are useful when we want exact—not rounded—answers.

EXAMPLE 13 Using the product property of square roots Simplify without a calculator.

a. $\sqrt{8}\sqrt{8}$ **b.** $(\sqrt{2})^2$ **c.** $(2\sqrt{3})^2$ **d.** $\sqrt{3 \cdot 9}$
e. $\sqrt{25 \cdot 3}$ **f.** $\sqrt{9 \cdot 2}$ **g.** $\sqrt{50}$ **h.** $\sqrt{12}$

Solution

a. $\sqrt{8}\sqrt{8} = \sqrt{64} = 8$

b. To square a number, we can write it twice and multiply. Thus,

$$(\sqrt{2})^2 = \sqrt{2} \cdot \sqrt{2} = \sqrt{2 \cdot 2} = \sqrt{4} = 2$$

c. We can write $(2\sqrt{3})$ twice or use $2^2\sqrt{3^2}$:

$$(2\sqrt{3})^2 = 2\sqrt{3} \cdot 2\sqrt{3} = 4\sqrt{9} = 4 \cdot 3 = 12$$
$$\text{or} \quad (2\sqrt{3})^2 = 2^2 \cdot \sqrt{3^2} = 4 \cdot 3 = 12$$

d. $\sqrt{3 \cdot 9} = \sqrt{3} \cdot \sqrt{9} = \sqrt{3} \cdot 3 = 3\sqrt{3}$

e. $\sqrt{25 \cdot 3} = \sqrt{25} \cdot \sqrt{3} = 5\sqrt{3}$

f. $\sqrt{9 \cdot 2} = \sqrt{9} \cdot \sqrt{2} = 3\sqrt{2}$

g. We obtain an exact square root expression by factoring 50 into $25 \cdot 2$ and taking the square root of one factor:

$$\sqrt{50} = \sqrt{25 \cdot 2} = \sqrt{25} \cdot \sqrt{2} = 5\sqrt{2}$$

h. We factor 12 to obtain $4 \cdot 3$ and take the square root of the perfect square factor:

$$\sqrt{12} = \sqrt{4 \cdot 3} = \sqrt{4} \cdot \sqrt{3} = 2\sqrt{3}$$

Obtaining $2\sqrt{3}$ confirms our work in part c. ●

From the fact that we took the square root of certain factors in Example 13, we can conclude the following:

> A simplified square root expression contains no perfect square factors.

VARIABLES IN SQUARE ROOT EXPRESSIONS Because the square root of a negative number is not defined in the real numbers, we must state assumptions about the inputs when working with square roots and variables. In this section, we make the assumption that the variables have zero or positive value.

When the variables are positive,

$$\sqrt{a^2} = a, \qquad \sqrt{a^4} = a^2, \qquad \sqrt{a^6} = a^3$$

In Example 14, we work with roots containing variables.

EXAMPLE 14 Working with variables in square root expressions Simplify these expressions. Assume that all variables represent zero or a positive number.

a. $\sqrt{x}\sqrt{x}$ **b.** $\sqrt{y^4}$ **c.** $-\sqrt{4x^2}$

d. $\sqrt{25x^4y^2}$ **e.** $\sqrt{2}\sqrt{8a^2}$ **f.** $\sqrt{x}\sqrt{x^3}$

Solution If x and y are zero or positive, we have the following solutions.

a. $\sqrt{x}\sqrt{x} = \sqrt{x^2} = x$

b. $\sqrt{y^4} = y^2$

c. $-\sqrt{4x^2} = -2x$ (The negative sign here means "opposite," not the square root of a negative.)

d. $\sqrt{25x^4y^2} = 5x^2y$

e. $\sqrt{2}\sqrt{8a^2} = \sqrt{16a^2} = 4a$

f. $\sqrt{x}\sqrt{x^3} = \sqrt{x^4} = x^2$ ●

QUOTIENT PROPERTY In Example 15, we explore the order in which we do square roots and division.

EXAMPLE  Exploring the order of operations Simplify these expressions without a calculator.

a. $\dfrac{\sqrt{16}}{\sqrt{4}}$ **b.** $\sqrt{\dfrac{16}{4}}$ **c.** $\dfrac{\sqrt{144}}{\sqrt{16}}$ **d.** $\sqrt{\dfrac{144}{16}}$

Solution In parts a and c, we do the square root first. In parts b and d, we do the division first. See the Answer Box. ●

Example 15 suggests that the results are the same whether we (1) take the square roots first and then divide or (2) divide and then take the square root.

Quotient Property for Square Roots

> The square root of a quotient equals the quotient of the square roots. In algebraic notation,
>
> $$\sqrt{\frac{a}{b}} = \frac{\sqrt{a}}{\sqrt{b}}$$
>
> if a is zero or positive and b is positive.

We prove the quotient property of square roots by using the product property of exponents and the fact that $\frac{1}{2}$ as exponent means the square root. This proof is left as an exercise.

EXAMPLE 16 Using the quotient property of square roots Simplify these expressions. You may find it useful to simplify fractions before taking the square roots. Assume the variables represent only positive numbers.

a. $\sqrt{\dfrac{36}{81}}$ **b.** $\sqrt{\dfrac{64x^2}{16}}$ **c.** $\sqrt{\dfrac{50}{18}}$ **d.** $\sqrt{\dfrac{4x^3}{36x}}$

Solution **a.** $\sqrt{\dfrac{36}{81}} = \dfrac{6}{9} = \dfrac{2}{3}$

or $\sqrt{\dfrac{36}{81}} = \sqrt{\dfrac{4 \cdot 9}{9 \cdot 9}} = \sqrt{\dfrac{4}{9}} = \dfrac{2}{3}$

b. $\sqrt{\dfrac{64x^2}{16}} = \dfrac{8x}{4} = 2x$

or $\sqrt{\dfrac{64x^2}{16}} = \sqrt{4x^2} = 2x$

c. $\sqrt{\dfrac{50}{18}} = \sqrt{\dfrac{25 \cdot 2}{9 \cdot 2}} = \sqrt{\dfrac{25}{9}} = \dfrac{5}{3}$

d. $\sqrt{\dfrac{4x^3}{36x}} = \sqrt{\dfrac{4 \cdot x^2 \cdot x}{4 \cdot 9 \cdot x}} = \sqrt{\dfrac{x^2}{9}} = \dfrac{x}{3}$ ●

Applications

PYTHAGOREAN TRIPLES **Pythagorean triples** are *sets of three numbers that make the Pythagorean theorem true.* The set of numbers {3, 4, 5} satisfies the Pythagorean theorem: $3^2 + 4^2 = 5^2$.

Before calculators, students and instructors saved considerable time by learning selected sets of Pythagorean triples. Engineering textbooks still make frequent use of triples, as do nationally standardized exams for entry into college or graduate school. The triple {3, 4, 5} is the basis for a rope surveying instrument used by the early Egyptians (see the Historical Note on page 493).

EXAMPLE  Finding Pythagorean triples Show whether or not these sets of three numbers could be used as sides of right triangles. The numbers may not be in order from smallest to largest.

a. {4, 5, 6} **b.** $\{1, 1, \sqrt{2}\}$ **c.** $\{5, \sqrt{11}, 6\}$ **d.** $\{8, \sqrt{17}, 9\}$

e. $\left\{\frac{1}{3}, \frac{1}{4}, \frac{1}{5}\right\}$ **f.** $\left\{\frac{3}{7}, \frac{4}{7}, \frac{5}{7}\right\}$

Solution The two smaller sides of a right triangle can be placed into a^2 and b^2 in any order. The longest side (the hypotenuse) must be substituted into c^2. If $a^2 + b^2 = c^2$ is satisfied, the set of numbers is a Pythagorean triple.

a. $a^2 + b^2 = c^2$

$4^2 + 5^2 \stackrel{?}{=} 6^2$

$16 + 25 \neq 36$

The numbers do not satisfy $a^2 + b^2 = c^2$.

b. $1^2 + 1^2 \stackrel{?}{=} (\sqrt{2})^2$

$1 + 1 = 2$ ✓

c. $5^2 + (\sqrt{11})^2 \stackrel{?}{=} 6^2$

$25 + 11 = 36$ ✓

d. $8^2 + (\sqrt{17})^2 \stackrel{?}{=} 9^2$

$64 + 17 = 81$ ✓

HISTORICAL NOTE

To form a tool for making a right angle in construction and survey work, early Egyptians tied eleven equally spaced knots in a rope. They then tied the rope in a loop with a twelfth knot. The rope formed a {3, 4, 5} right triangle when staked to the ground.

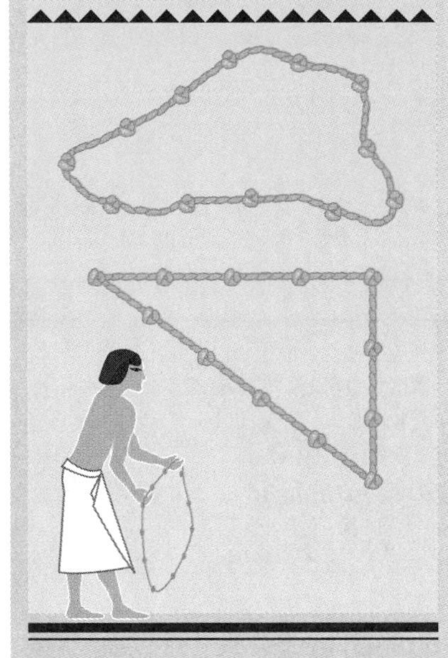

e. The side measuring $\frac{1}{3}$ is the longest, so it replaces c.

$$\left(\tfrac{1}{4}\right)^2 + \left(\tfrac{1}{5}\right)^2 \stackrel{?}{=} \left(\tfrac{1}{3}\right)^2$$

$$\tfrac{1}{16} + \tfrac{1}{25} \stackrel{?}{=} \tfrac{1}{9}$$

$$\tfrac{41}{400} \neq \tfrac{1}{9}$$

The numbers do not satisfy $a^2 + b^2 = c^2$.

f. $\left(\tfrac{3}{7}\right)^2 + \left(\tfrac{4}{7}\right)^2 \stackrel{?}{=} \left(\tfrac{5}{7}\right)^2$

$$\tfrac{9}{49} + \tfrac{16}{49} = \tfrac{25}{49} \checkmark$$

DISTANCE FORMULA When we did coordinate graphing, we found the slopes and equations of lines, the midpoints of lines, and the intersections of lines with the axes and other lines. We now find the length of lines with the distance formula.

The **distance formula** permits us to *find the length*, or distance, *between two coordinate points* (x_1, y_1) and (x_2, y_2), as shown in Figure 11. The distance formula is derived from the Pythagorean theorem. As in finding slope, we find the change in x, Δx, to be $(x_2 - x_1)$ and the change in y, Δy, to be $(y_2 - y_1)$. The angle at (x_2, y_1) (see Figure 11) is a right angle, so the Pythagorean theorem holds:

$$d^2 = (\Delta x)^2 + (\Delta y)^2$$

$$d^2 = (x_2 - x_1)^2 + (y_2 - y_1)^2$$

$$d = \sqrt{(x_2 - x_1)^2 + (y_2 - y_1)^2}$$

Distance Formula

$$d = \sqrt{(x_2 - x_1)^2 + (y_2 - y_1)^2}$$

As in calculating slope, the choice of which point is (x_1, y_1) and which is (x_2, y_2) is arbitrary.

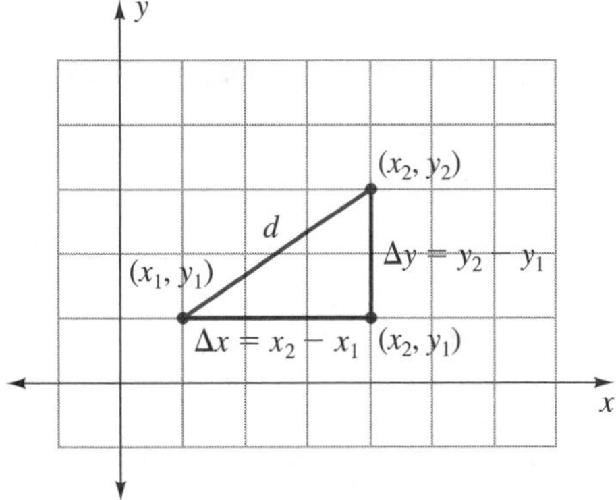

Figure 11

EXAMPLE 18 Finding distance Find the three distances between these coordinates: (1, 1), (3, 0), and (5, 4). Determine whether the three points form a right triangle. Make a sketch of the points and lines on coordinate axes. Identify the location of the right angle, if one exists.

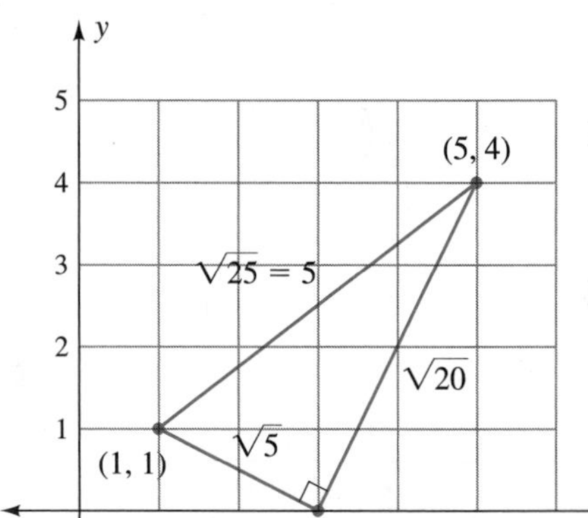

Figure 12

Solution Figure 12 shows the triangle. Between (1, 1) and (3, 0),

$$d = \sqrt{(3-1)^2 + (0-1)^2} = \sqrt{2^2 + 1^2} = \sqrt{5}$$

Between (3, 0) and (5, 4),

$$d = \sqrt{(5-3)^2 + (4-0)^2} = \sqrt{2^2 + 4^2} = \sqrt{20}$$

Between (1, 1) and (5, 4),

$$d = \sqrt{(5-1)^2 + (4-1)^2} = \sqrt{4^2 + 3^2} = \sqrt{25} = 5$$

By the Pythagorean theorem,

$$\sqrt{5}^2 + \sqrt{20}^2 = 25 = 5^2$$

The right angle is at (3, 0). ●

ANSWER BOX

Warm-up: 1. a. 8 **b.** 11 **c.** 7 **d.** 16 **e.** 20 **f.** 15 **g.** 12 **h.** 25 **2. a.** $6 < \sqrt{45} < 7$ **b.** $8 < \sqrt{75} < 9$ **c.** $14 < \sqrt{200} < 15$ **d.** $4 < \sqrt{20} < 5$ **e.** $10 < \sqrt{110} < 11$ **f.** $3 < \sqrt{12} < 4$ **g.** $30 < \sqrt{905} < 31$ **h.** $19 < \sqrt{395} < 20$ **Example 1: a.** 5 **b.** 16 **c.** $\sqrt{2}$ **d.** x **e.** $\sqrt{a}$ **Example 2: a.** 0.4 **b.** 1.5 **c.** 2.5 **d.** $\frac{1}{2}$ **e.** $\frac{3}{4}$ **f.** $\frac{12}{3}$, or 4

Example 3:

Square Root	Value	Rational or Irrational?
$\sqrt{9}$	3	rational
$\sqrt{19}$	4.359	irrational
$\sqrt{29}$	5.385	irrational
$\sqrt{39}$	6.245	irrational
$\sqrt{49}$	7	rational
$\sqrt{1.21}$	1.1	rational
$\sqrt{.09}$	0.3	rational
$\sqrt{20.25}$	4.5	rational
$\sqrt{169}$	13	rational

Think about it: The table will output "error" for each negative input and 0 for $\sqrt{0}$. **Example 9: a.** 5 **b.** 4 **c.** 15 **d.** 1.2 **Example 10: a.** $36^{\frac{1}{2}} \cdot 36^{\frac{1}{2}} = 36^{(\frac{1}{2}+\frac{1}{2})} = 36^1 = 36$ **b.** $\sqrt{36} \cdot \sqrt{36} = 6 \cdot 6 = 36$ **c.** $25^{\frac{1}{2}} \cdot 25^{\frac{1}{2}} = 25^{(\frac{1}{2}+\frac{1}{2})} = 25^1 = 25$ **d.** $\sqrt{25} \cdot \sqrt{25} = 5 \cdot 5 = 25$ **Example 11: a.** $\sqrt{4} \cdot \sqrt{9} = 2 \cdot 3 = 6$ **b.** $\sqrt{4 \cdot 9} = \sqrt{36} = 6$ **c.** $\sqrt{9 \cdot 25} = \sqrt{225} = 15$ **d.** $\sqrt{9} \cdot \sqrt{25} = 3 \cdot 5 = 15$ **Example 15: a.** $\frac{\sqrt{16}}{\sqrt{4}} = \frac{4}{2} = 2$ **b.** $\sqrt{\frac{16}{4}} = \sqrt{4} = 2$ **c.** $\frac{\sqrt{144}}{\sqrt{16}} = \frac{12}{4} = 3$ **d.** $\sqrt{\frac{144}{16}} = \sqrt{9} = 3$

EXERCISES 8.2

In Exercises 1 and 2, find the square roots without using a calculator.

1. a. $\sqrt{81}$ **b.** $\sqrt{1.96}$ **c.** $\sqrt{0.04}$ **d.** $\sqrt{3600}$

2. a. $\sqrt{169}$ **b.** $\sqrt{1.44}$ **c.** $\sqrt{0.64}$ **d.** $\sqrt{2500}$

Between which two consecutive integers is each square root in Exercises 3 and 4? Do not use a calculator.

3. a. $\sqrt{80}$ **b.** $\sqrt{54}$ **c.** $\sqrt{210}$ **d.** $\sqrt{18}$

4. a. $\sqrt{15}$ **b.** $\sqrt{115}$ **c.** $\sqrt{895}$ **d.** $\sqrt{405}$

Complete the tables in Exercises 5 and 6. Round numbers to three decimal places.

5.

Square Root	Value	Rational or Irrational?
$\sqrt{36}$	6	rational
$\sqrt{15}$		
$\sqrt{25}$		
$\sqrt{35}$		
$\sqrt{12.25}$		
$\sqrt{2.25}$		
$\sqrt{6}$		
$\sqrt{16}$		
$\sqrt{26}$		

6.

Square Root	Value	Rational or Irrational?
$\sqrt{4}$	2	rational
$\sqrt{14}$		
$\sqrt{24}$		
$\sqrt{34}$		
$\sqrt{64}$		
$\sqrt{144}$		
$\sqrt{.01}$		
$\sqrt{6.25}$		
$\sqrt{1.44}$		

Simplify the expressions in Exercises 7 to 10.

7. a. $\sqrt{-36}$ **b.** $-\sqrt{81}$ **c.** $\pm\sqrt{144}$

8. a. $\pm\sqrt{256}$ **b.** $\sqrt{-49}$ **c.** $-\sqrt{64}$

9. a. $\sqrt{49}$ **b.** $-\sqrt{225}$ **c.** $\pm\sqrt{400}$

10. a. $\sqrt{121}$ **b.** $\pm\sqrt{64}$ **c.** $-\sqrt{169}$

In Exercises 11 to 16, make a table and graph for inputs in [−4, 4], counting by 2s.

11. $y = \sqrt{x + 4}$ **12.** $y = \sqrt{x - 1}$

13. $y = \sqrt{x - 2}$ **14.** $y = \sqrt{x + 3}$

15. $y = \sqrt{2x}$ **16.** $y = \sqrt{\frac{1}{2}x}$

Simplify the expressions in Exercises 17 to 32. Assume the variables are positive.

17. a. $\sqrt{5} \cdot \sqrt{3}$ **b.** $(3\sqrt{5})^2$ **c.** $\sqrt{16 \cdot 3}$

18. a. $(5\sqrt{3})^2$ **b.** $\sqrt{11}\sqrt{11}$ **c.** $\sqrt{32}$

19. a. $(3\sqrt{7})^2$ **b.** $\sqrt{13}\sqrt{13}$ **c.** $\sqrt{36 \cdot 2}$

20. a. $\sqrt{3} \cdot \sqrt{7}$ **b.** $(4\sqrt{2})^2$ **c.** $\sqrt{75}$

21. a. $\sqrt{3} \cdot \sqrt{27}$ **b.** $(2\sqrt{5})^2$ **c.** $\sqrt{18}$

22. a. $\sqrt{8} \cdot \sqrt{18}$ **b.** $(5\sqrt{2})^2$ **c.** $\sqrt{98}$

23. a. $\sqrt{a}\sqrt{a}$ **b.** $\sqrt{b^2}$ **c.** $\sqrt{121a^2}$

24. a. $\sqrt{2x} \cdot \sqrt{18x}$ **b.** $\sqrt{3y}\sqrt{3y}$ **c.** $\sqrt{256x^2}$

25. a. $\sqrt{32a} \cdot \sqrt{2a}$ **b.** $\sqrt{2x}\sqrt{2x}$ **c.** $\sqrt{16b^2}$

26. a. $\sqrt{b}\sqrt{b}$ **b.** $\sqrt{c^2}$ **c.** $\sqrt{400x^2}$

27. a. $\sqrt{\frac{x^2}{9}}$ **b.** $\sqrt{\frac{4}{25}}$ **c.** $\sqrt{\frac{45}{5}}$

28. a. $\sqrt{\frac{y^2}{4}}$ **b.** $\sqrt{\frac{64}{9}}$ **c.** $\sqrt{\frac{48}{3}}$

29. a. $\sqrt{\frac{28x}{7x^3}}$ **b.** $\sqrt{\frac{3x^2}{27}}$ **c.** $\sqrt{\frac{8a^4}{32}}$

30. a. $\sqrt{\frac{6y^3}{24y}}$ **b.** $\sqrt{\frac{3x^2}{12}}$ **c.** $\sqrt{\frac{63b^4}{7}}$

31. a. $\sqrt{\frac{8x^2y}{2y}}$ **b.** $\sqrt{\frac{2x^4}{50y^2}}$

c. $-\sqrt{\frac{3xy^4}{27x}}$

32. a. $\sqrt{\frac{18x^2y}{2y}}$ **b.** $-\sqrt{\frac{32x^4y}{2x^2y}}$

c. $\sqrt{\frac{20xy^2}{5x}}$

Which of the sets of numbers in Exercises 33 to 36 are Pythagorean triples? The numbers might not be listed in the a, b, c order. Which sets satisfy $a^2 + b^2 = c^2$ and yet cannot satisfy the Pythagorean theorem? Explain why.

33. a. $\{2.1, 2.9, 2\}$ **b.** $\{2, \sqrt{3}, 1\}$

c. $\{1, 2, 3\}$

34. a. $\{-1, 0, 1\}$ **b.** $\{-5, -4, -3\}$

c. $\{7, \sqrt{15}, 8\}$

35. a. $\{3, \sqrt{7}, 4\}$ **b.** $\{-5, -12, -13\}$

c. $\{6, \sqrt{13}, 7\}$

36. a. $\{10, \sqrt{21}, 11\}$ **b.** $\{8, 15, 17\}$

c. $\{6, 6\sqrt{3}, 12\}$

For each pair of points in Exercises 37 to 44,

a. *find the distance between the two points. Simplify the square root.*

b. *find the slope of the line segment connecting the two points.*

c. *find the equation of the line passing through the points.*

37. (2, 3) and (4, 9) **38.** (3, 2) and (9, 4)

39. (2, 2) and (5, −1) **40.** (5, 3) and (6, −2)

41. (−3, 3) and (4, 2) **42.** (−3, 3) and (2, 4)

43. (−3, −1) and (3, −3) **44.** (−2, −1) and (2, −4)

In Exercises 45 to 50, the vertices of a triangle are given. Use the distance formula to find what kind of triangle each is. A sketch of the graph may be helpful. Note: An isosceles triangle has two equal sides. An equilateral triangle has three equal sides. A right triangle satisfies the Pythagorean theorem.

45. (3, 4), (0, 1), (6, 1) **46.** (3, 1), (0, 4), (−3, 1)

47. (6, 5), (4, 2), (8, 2) **48.** (3, 6), (6, 4), (3, 2)

49. (4, 8), (1, 6), (5, 0) **50.** (5, 8), (0, 7), (2, −3)

Evaluate the expressions in Exercises 51 to 56 without a calculator.

51. a. 25^0 **b.** 25^{-1} **c.** $25^{1/2}$ **d.** $25^{0.5}$

52. a. 4^{-1} **b.** 4^0 **c.** $4^{0.5}$ **d.** $4^{1/2}$

53. a. $9^{1/2}$ **b.** 9^0 **c.** $9^{0.5}$ **d.** 9^{-1}

54. a. $36^{1/2}$ **b.** $36^{0.5}$ **c.** 36^0 **d.** 36^{-1}

55. a. $\left(\frac{1}{4}\right)^{-1}$ **b.** $\left(\frac{1}{4}\right)^0$ **c.** $\left(\frac{1}{4}\right)^{1/2}$ **d.** $\left(\frac{1}{4}\right)^{0.5}$

56. a. $\left(\frac{1}{9}\right)^{1/2}$ **b.** $\left(\frac{1}{9}\right)^{0.5}$ **c.** $\left(\frac{1}{9}\right)^0$ **d.** $\left(\frac{1}{9}\right)^{-1}$

Evaluate the expressions in Exercises 57 and 58 without a calculator. Hint: Change decimals to fractions.

57. a. $(0.25)^{-1}$ **b.** $(0.01)^{0.5}$ **c.** $(6.25)^{0.5}$

d. $(0.25)^{0.5}$ **e.** $(0.02)^{-1}$ **f.** $(0.05)^{-1}$

58. a. $(2.25)^{0.5}$ **b.** $(0.36)^{0.5}$ **c.** $(0.01)^{-1}$

d. $(0.1)^{-1}$ **e.** $(0.125)^{-1}$ **f.** $(0.0001)^{0.5}$

59. Is $y = \sqrt{x + 3}$ defined for $x = -2$? Explain.

60. Is $y = \sqrt{x - 2}$ defined for $x = 1$? Explain.

61. Is $y = \sqrt{-2x}$ ever defined? Explain.

62. Is $y = \sqrt{|x|}$ ever defined? Explain.

63. True or false: The square root of x is smaller than x. Explain your reasoning and give an example.

64. True or false: Only whole numbers can have exact square roots. Explain your reasoning and give an example.

65. Prove the quotient property of square roots,

$$\sqrt{\frac{a}{b}} = \frac{\sqrt{a}}{\sqrt{b}}$$

where a is zero or positive and b is positive.

66. Is a triangle with sides {9, 16, 25} similar to one with sides {3, 4, 5}? Why or why not? Do sides {9, 16, 25} make a right triangle?

Project

67. ***Square Root Patterns***

a. Find a pattern in the values of these square roots. Describe your findings in words. (*Hint*: Your word description should relate to the number of zeros or decimal places in the radicand.)

$\sqrt{4}$	$\sqrt{9}$	$\sqrt{25}$
$\sqrt{40}$	$\sqrt{90}$	$\sqrt{250}$
$\sqrt{400}$	$\sqrt{900}$	$\sqrt{2500}$
$\sqrt{4000}$	$\sqrt{0.09}$	$\sqrt{0.25}$
$\sqrt{40,000}$	$\sqrt{0.0009}$	$\sqrt{0.025}$
$\sqrt{4,000,000}$	$\sqrt{0.000\,000\,9}$	$\sqrt{0.000\,25}$

b. Test your pattern statement on the following square roots. Check with a calculator.

$\sqrt{0.000\,000\,4}$	$\sqrt{9,000,000}$	$\sqrt{250,000,000}$
$\sqrt{0.000\,000\,000\,04}$	$\sqrt{0.000\,09}$	$\sqrt{0.000\,000\,25}$

c. Use your pattern on the following square roots. Suppose you know that $\sqrt{8} = 2.8284$ and $\sqrt{80} = 8.9443$. Do not use a calculator.

$\sqrt{800,000}$	$\sqrt{0.8}$	$\sqrt{8000}$
$\sqrt{8,000,000}$	$\sqrt{0.0008}$	$\sqrt{0.000\,000\,8}$

8.3 Simplifying Expressions and Solving Square Root Equations

OBJECTIVES

- Use absolute value in simplifying square root expressions.
- Find sets of inputs for square root functions.
- Solve square root equations by graphing.
- Solve equations and formulas by squaring both sides.

> WARM-UP
>
> **1.** For what values of x are these expressions undefined?
>
> **a.** x^0 **b.** $\dfrac{5}{x}$ **c.** $\dfrac{y-4}{x-5}$
>
> **2.** Think about the order of operations as you simplify these expressions.
>
> **a.** $\sqrt{(-1)^2}$ **b.** $(\sqrt{9})^2$ **c.** $\sqrt{(-3)^2}$ **d.** $(\sqrt{3})^2$

IN THIS SECTION, we look at square roots more formally: what vocabulary to use to describe parts of square root notation (radicals and radicands), how to guarantee positive outputs from the square root (with absolute values), and how to guarantee positive inputs to the square root (with inequalities for x).

We solve square root equations by graphing and by squaring both sides of the equation. We look at two applications in which square roots appear in formulas.

Radicals and Radicands

The more general name for square roots (and higher degree roots) is **radicals**. The *radical sign* is the symbol we use for square root. The **radicand** is *the number or expression under the radical sign.*

index → $\sqrt[n]{\text{radicand}}$ ← radical sign

The *index* is the small n on the radical sign. For square roots, the index is 2. We assume $n = 2$ and don't write it. See the project in Exercise 62 for examples of radicals with indices of 3, 4, and 5.

Radical vocabulary is useful because it is easier to say "radicand" than "the expression under the square root sign."

EXAMPLE 1 Identifying radicands What is the radicand in each of these expressions?

a. $5\sqrt{2}$ **b.** $b\sqrt{bc}$ **c.** $3\sqrt{x+2}$

d. $3\sqrt{x}+2$ **e.** $(\sqrt{x+2})^2$ **f.** $\sqrt{(x+2)^2}$

Solution **a.** 2 **b.** bc **c.** $x + 2$

d. x. The 2 is added after the square root is taken, so the 2 is not part of the radicand.

e. $x + 2$. The square applies to the entire expression, so the radicand is just the $x + 2$ under the radical sign.

f. $(x + 2)^2$. The square applies to the $x + 2$ and not the radical sign.

Square Root Function

The equation $y = \sqrt{x}$ is graphed in Figure 13. Note that for each input on the x-axis, there is only one output. Because its graph passes the vertical-line test, the equation $y = \sqrt{x}$ is a function.

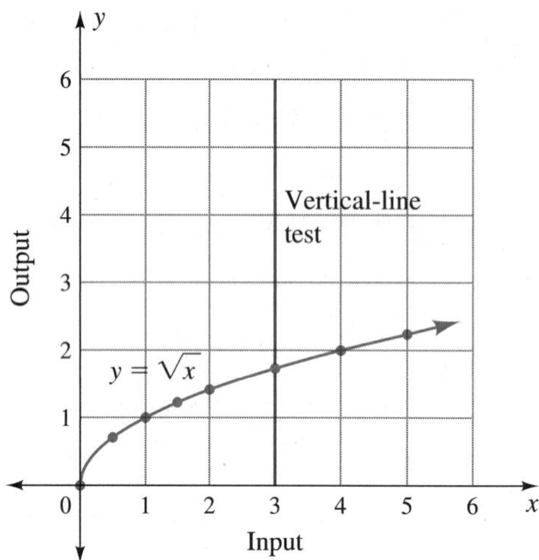

Figure 13

Square Root Function

> The square root function is $f(x) = \sqrt{x}$. The input set (or domain) is $x \geq 0$. The output set (or range) is $y \geq 0$ or $f(x) \geq 0$.

The square root has limitations on both inputs and outputs, as we saw in Section 8.2. Look again at the graph. The set of inputs is $x \geq 0$, and the graph is to the right of $x = 0$. The set of outputs is $f(x)$ or $y \geq 0$, and the graph is above the x-axis. Put simply, both the input and the output to a square root function must be zero or positive.

WHEN X IS ANY REAL NUMBER Because the square roots of negative numbers are undefined in the set of real numbers, we have to go to a lot of work in algebra to avoid having a negative number in the radicand (the expression under the square root sign). In Section 8.2, we avoided negative numbers by assuming that all variables represented positive numbers or zero. We wrote $x \geq 0$. There are times when we cannot make that assumption and have to let x be any real number.

The function $f(x) = \sqrt{x^2}$ is a square root function that is defined for all real numbers (yes, even negative numbers, $x < 0$). In Example 2, we evaluate $f(x) = \sqrt{x^2}$.

EXAMPLE 2 Evaluating a square root function For $f(x) = \sqrt{x^2}$, find

a. $f(3)$ **b.** $f(1)$ **c.** $f(0)$ **d.** $f(-1)$ **e.** $f(-3)$

Solution

a. $f(3) = \sqrt{3^2} = \sqrt{9} = 3$

b. $f(1) = \sqrt{1^2} = \sqrt{1} = 1$

c. $f(0) = \sqrt{0^2} = \sqrt{0} = 0$

d. $f(-1) = \sqrt{(-1)^2} = \sqrt{1} = 1$

e. $f(-3) = \sqrt{(-3)^2} = \sqrt{9} = 3$

Think about it 1: Without graphing Example 2, do you know what function takes every positive input and gives the same number as output and takes every negative input and gives its opposite (a positive number) as output?

In Example 3, we graph the data from Example 2.

EXAMPLE 3 Identifying the graph of $\sqrt{x^2}$

a. List the data from Example 2 as ordered pairs and graph them. Connect them from left to right.

b. What function describes the graph?

Solution **a.** The ordered pairs are $(-3, 3)$, $(-1, 1)$, $(0, 0)$, $(1, 1)$, and $(3, 3)$. The graph is in Figure 14.

b. The graph is the absolute value function, $y = |x|$.

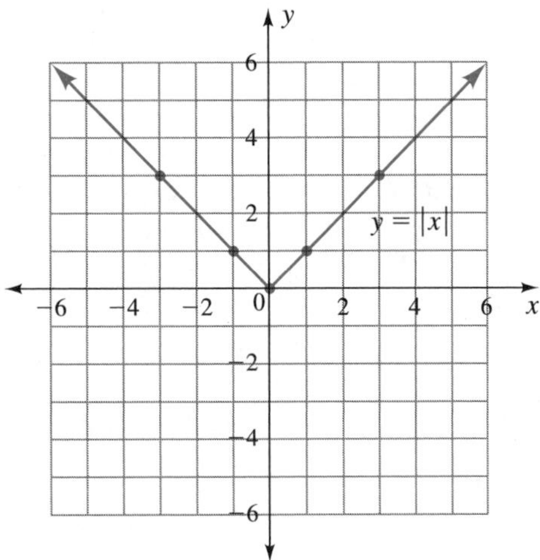

Figure 14

Graphing Calculator Solution  Graph $Y_1 = \sqrt{(X^2)}$ for X in $[-5, 5]$ and Y in $[-5, 5]$. Compare the graph with that of the absolute value, $Y_2 = \text{abs}(X)$. The absolute value key may be on the keyboard, [abs], or it may be under CATALOG. The two graphs should be identical.

> The function $f(x) = \sqrt{x^2}$ is the same as the absolute value function, $f(x) = |x|$.

The absolute value function gives us a way to simplify many square root functions. Here is a summary of square roots both for $x \geq 0$ and for x as any real number.

Square Roots and Absolute Values

- Roots of variables with exponents if x is zero or a positive number ($x \geq 0$):

$$\sqrt{x^2} = x$$
$$\sqrt{x^4} = x^2$$
$$\sqrt{x^6} = x^3$$
$$\sqrt{x^8} = x^4$$

- Roots of variables with exponents if x is any real number:

$\sqrt{x^2} = \lvert x \rvert$	Because x could be negative
$\sqrt{x^4} = x^2$	Because an even power is positive
$\sqrt{x^6} = \lvert x^3 \rvert$	Because x could be negative
$\sqrt{x^8} = x^4$	Because an even power is positive

In Example 4, we use the absolute value expressions from the chart to simplify expressions.

EXAMPLE 4 Simplifying radical expressions Let x be any real number, and simplify these expressions.

a. $\sqrt{4x^2}$ **b.** $\sqrt{x^6y^4}$ **c.** $\sqrt{a \cdot a^2}$ **d.** $\sqrt{c \cdot c^4}$

Solution No matter what the input expression, the square root function requires a positive output (note that the graph of the square root function is only in the first quadrant). Because x can be negative in this problem, every odd power will be negative and must have absolute value on it.

a. $\sqrt{4x^2} = 2\lvert x \rvert$

b. $\sqrt{x^6y^4} = \lvert x^3 \rvert y^2$; y^2 is positive and does not need absolute value.

c. $\sqrt{a \cdot b^2} = \lvert b \rvert \sqrt{a}$

d. $\sqrt{c \cdot c^4} = c^2\sqrt{c}$; c^2 is positive and does not need absolute value. ●

In Example 4, the absolute value keeps the output positive, satisfying $f(x)$ or $y \geq 0$. In the next example, we look at how to keep radicands positive when $x \geq 0$ doesn't work.

WHEN THE RADICAND IS AN EXPRESSION When solving square root equations, we keep the radicand positive by limiting the input set (domain) with an inequality.

> To keep the radicand positive, limit the input set by solving this inequality for the variable:
>
> radicand ≥ 0

EXAMPLE 5 Finding the input set (domain) For what inputs, x, is $\sqrt{x + 2}$ defined?

Solution $\sqrt{x+2}$ is defined if $x + 2$ is zero or positive.

$$x + 2 \geq 0 \quad \text{Find when the radicand is zero or positive.}$$
$$x + 2 - 2 \geq 0 - 2 \quad \text{Subtract 2 from each side.}$$
$$x \geq -2$$

Thus, $\sqrt{x+2}$ is defined when $x \geq -2$. ●

We will use this skill as we solve square root equations, beginning with Example 7.

Solving Square Root Equations

SOLVE BY GRAPHING We can solve a square root equation by graphing.

EXAMPLE 6 Solving square root equations Solve $\sqrt{2-x} = 1$ by graphing.

Solution The graphs of $y = \sqrt{2-x}$ and $y = 1$ are shown in Figure 15. The point of intersection is (1, 1), so $\sqrt{2-1} = 1$ at $x = 1$.

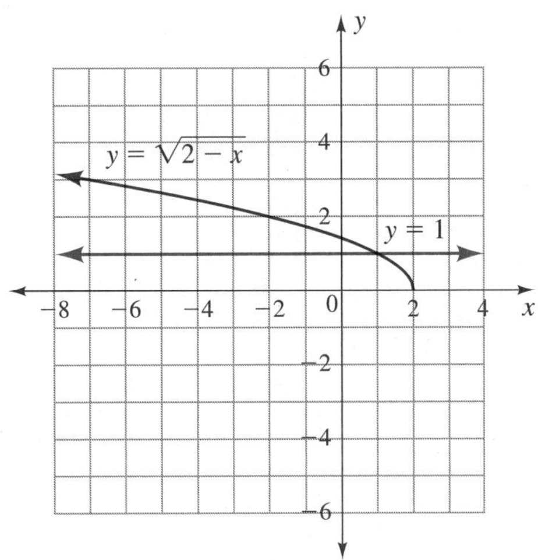

Figure 15

Check: $\sqrt{2-1} \stackrel{?}{=} 1$ ✓ ●

We look at a way to solve such equations with algebraic notation.

SOLVE BY SQUARING BOTH SIDES We now add another technique to our tools for solving equations with algebraic notation: squaring both sides.

> To solve an equation of the form $\sqrt{x} = a$, we square both sides.

Note that $(\sqrt{x})^2 = \sqrt{x}\sqrt{x} = x$ if $x \geq 0$. Again, the x represents any number or expression and, as a radicand, must be positive. In Example 7, the radicand is $2 - x$, and it must be zero or positive: $2 - x \geq 0$.

EXAMPLE 7 Solving by squaring both sides

a. For what x is $\sqrt{2 - x}$ defined?

b. Solve $\sqrt{2 - x} = 1$.

Solution **a.** To make the radicand zero or positive, $2 - x \geq 0$, we must have $2 \geq x$, or $x \leq 2$.

b.

$\sqrt{2 - x} = 1$	Square both sides.
$(\sqrt{2 - x})^2 = 1^2$	Simplify.
$2 - x = 1$	Add x to both sides.
$2 = x + 1$	Subtract 1 on each side.
$1 = x$	

Check: $x = 1$ satisfies the condition $x \leq 2$.

$\sqrt{2 - 1} \stackrel{?}{=} 1$ ✓

Think about it 2: How does the condition $x \leq 2$ show on the graph in Figure 15?

The following example shows why squaring both sides is a reasonable operation.

EXAMPLE 8 Squaring both sides of an equation In Example 5, we found that $\sqrt{x + 2}$ is defined when $x \geq -2$. Solve $\sqrt{x + 2} = 4$ by

a. using algebraic notation **b.** examining the graph

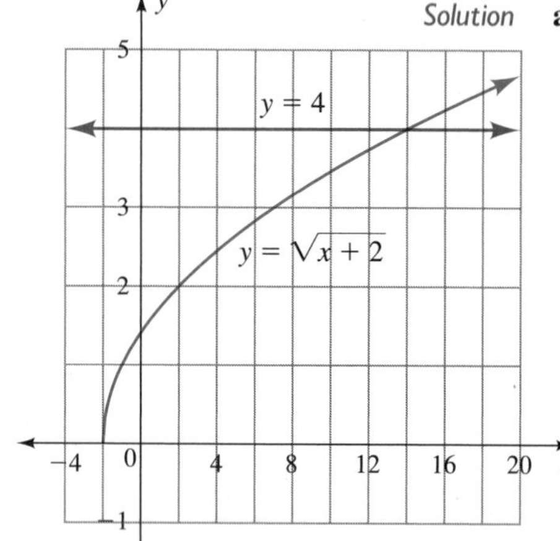

Figure 16

Solution **a.** Squaring both sides of an equation is a modification of multiplying both sides by the same number.

$\sqrt{x + 2} = 4$	
$4 \cdot \sqrt{x + 2} = 4 \cdot 4$	Multiply both sides by 4.
$\sqrt{x + 2} \cdot \sqrt{x + 2} = 4 \cdot 4$	Because $\sqrt{x + 2} = 4$, we can replace the 4 on the left by $\sqrt{x + 2}$; the result is the same as squaring both sides.
$x + 2 = 16$	Recall that $x + 2$ is positive.
$x = 14$	

Check: We must check the solution in the original equation to make sure that the answer does not make the equation false or undefined because of a negative radicand. $\sqrt{14 + 2} \stackrel{?}{=} 4$ ✓

b. The intersection of $y = \sqrt{x + 2}$ with $y = 4$ in Figure 16 gives the solution to $\sqrt{x + 2} = 4$. The point of intersection is (14, 4), so $x = 14$. The graph also shows the condition $x \geq -2$ in that there are no points on the square root graph to the left of $x = -2$.

EXAMPLE 9 Solving by squaring both sides

a. Find inputs for which the equation $y = \sqrt{x - 5}$ is defined.

b. Solve $\sqrt{x - 5} = 3$ by squaring both sides.

c. Solve with a graph.

Solution **a.** The expression $x - 5$ must be zero or positive: $x - 5 \geq 0$. By adding 5 to both sides, we have $x \geq 5$.

b. We solve the equation by squaring both sides.

$$\sqrt{x - 5} = 3 \qquad \text{Square both sides.}$$

$$(\sqrt{x - 5})^2 = 3^2$$

$$x - 5 = 9 \qquad (\sqrt{x - 5})^2 = x - 5 \text{ because } x - 5 \text{ is positive.}$$

$$x = 14$$

Check: $\sqrt{14 - 5} \stackrel{?}{=} 3$ ✓

c. The intersection of $y = 3$ with $y = \sqrt{x - 5}$ is at (14, 3) in Figure 17. Thus, $x = 14$ is the solution to $\sqrt{x - 5} = 3$. There are no points on the square root graph to the left of 5 because the expression $\sqrt{x - 5}$ is undefined for $x < 5$.

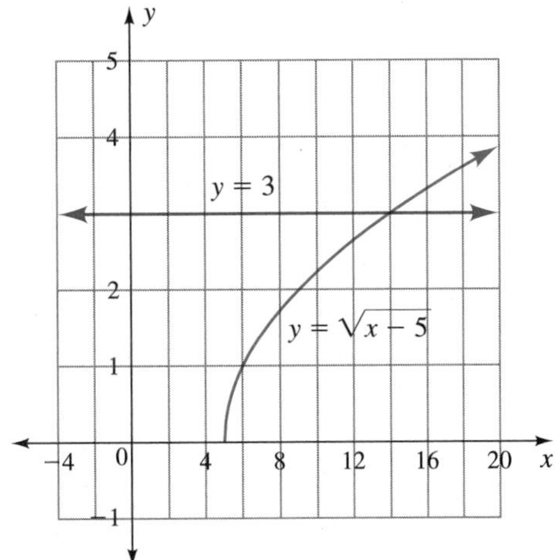

Figure 17

Applications

DISTANCE TO THE HORIZON

EXAMPLE **10** Using square root formulas: distance to the horizon at the seashore On Earth, the distance to the horizon is given by

$$d \approx \sqrt{\frac{3h}{2}}$$

where d is in miles to the horizon and h is the height in feet above sea level.

a. Find the distance seen from a height of 5 feet.

b. Find the distance seen from a height of 10 feet (Figure 18).

c. Find the distance seen from a height of 24 feet.

d. How high would we have to be to see 12 miles?

Solution **a.** $d \approx \sqrt{\frac{3h}{2}} \approx \sqrt{\frac{3 \cdot 5}{2}} \approx 2.7$ mi

b. $d \approx \sqrt{\dfrac{3 \cdot 10}{2}} \approx 3.9 \text{ mi}$

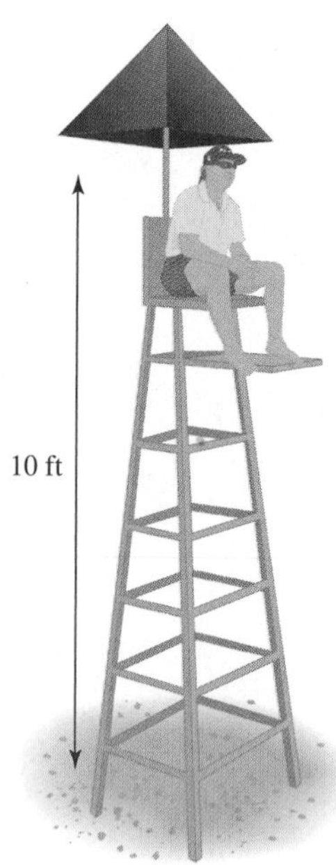

Figure 18

c. $d \approx \sqrt{\dfrac{3 \cdot 24}{2}} \approx 6 \text{ mi}$

d.
$$12 \approx \sqrt{\frac{3h}{2}}$$
$$144 \approx \frac{3h}{2}$$
$$288 \approx 3h$$
$$h \approx 96 \text{ ft}$$

The unit change from h feet to d miles is not shown in the formula. The units are included in the coefficient of $\frac{3}{2}$ on h. See the project in Exercise 63 for the steps in deriving this formula and the formula for the distance seen on the moon.

Squaring both sides can be used to solve formulas as well as equations.

EXAMPLE 11 Solving formulas Solve the formula $d = \sqrt{\dfrac{3h}{2}}$ for h.

Solution

$d = \sqrt{\dfrac{3h}{2}}$ Square both sides.

$d^2 = \left(\sqrt{\dfrac{3h}{2}}\right)^2$ Simplify.

$d^2 = \dfrac{3h}{2}$ Multiply both sides by 2.

$2d^2 = 3h$ Divide both sides by 3.

$\dfrac{2d^2}{3} = h$

FALLING TIME

EXAMPLE 12 Using square root formulas: Washington Monument As a child, I dropped a paper-wrapped sugar cube from the open window of the Washington Monument in our nation's capital. Fortunately, it was a cold winter day in the 1950s, and no one was at the base of the 555-foot tower.

The time required for a dropped object to hit the ground is given by

$$t = \sqrt{\frac{2h}{g}}$$

where h is the vertical distance traveled and g is the acceleration due to gravity, assumed to be 32.2 feet per second squared. The downward speed, or velocity v, such an object will be traveling is given by $v = gt$.

a. How long did it take for the cube to hit the ground?

b. What speed in feet per second was the cube traveling when it hit the ground?

c. Use unit analysis to change the speed to miles per hour.

Solution **a.** Assuming the distance from the window (just below the top of the tower) to the ground is 550 feet, we have

$$t = \sqrt{\frac{2h}{g}} = \sqrt{\frac{2 \cdot 550 \text{ ft}}{32.2 \dfrac{\text{ft}}{\text{sec}^2}}} \approx \sqrt{34.1615 \text{ sec}^2} \approx 5.8 \text{ sec}$$

b. Speed is $v = gt$.

$$v = gt = \frac{32.2 \text{ ft}}{\text{sec}^2} \cdot 5.8 \text{ sec} \approx 187 \text{ ft per sec}$$

c. Using unit analysis to change feet per second to miles per hour, we have

$$\frac{187 \text{ ft}}{\text{sec}} \cdot \frac{1 \text{ mi}}{5280 \text{ ft}} \cdot \frac{60 \text{ sec}}{1 \text{ min}} \cdot \frac{60 \text{ min}}{1 \text{ hr}} \approx 128 \text{ mi per hr}$$

P.S. My parents were furious! The Washington Monument windows are now sealed. ●

EXAMPLE **13** Solving formulas Solve the formula $t = \sqrt{\frac{2h}{g}}$ for h.

Solution

$$t = \sqrt{\frac{2h}{g}}$$ Square both sides.

$$t^2 = \left(\sqrt{\frac{2h}{g}}\right)^2$$ Simplify.

$$t^2 = \frac{2h}{g}$$ Multiply both sides by g.

$$gt^2 = 2h$$ Divide both sides by 2.

$$\frac{gt^2}{2} = h$$ ●

ANSWER BOX

Warm-up: 1. a. 0 **b.** 0 **c.** 5 **2. a.** $\sqrt{(-1)^2} = \sqrt{1} = 1$ **b.** 9 **c.** $\sqrt{(-3)^2} = \sqrt{9} = 3$ **d.** 3 **Think about it 1:** The absolute value function, $f(x)$ or $y = |x|$ **Think about it 2:** The graph is only to the left of $x = 2$.

EXERCISES 8.3

What are the radicands in the expressions in Exercises 1 and 2?

1. a. $\sqrt{-3}$ b. $a\sqrt{ab}$ c. $\sqrt{4x^2}$
   d. $5\sqrt{2}$ e. $\sqrt{x} + 2$ f. $\sqrt{x + 2}$
2. a. $\sqrt{-5}$ b. $\sqrt{9y^2}$ c. $x\sqrt{xy}$
   d. $3\sqrt{5}$ e. $\sqrt{x - 3}$ f. $\sqrt{x} - 3$

Evaluate the functions in Exercises 3 to 6 for f(−4), f(−1), f(0), f(4), and f(6).

3. $f(x) = \sqrt{4 - x}$
4. $f(x) = \sqrt{2 - x}$
5. $f(x) = \sqrt{3 - x}$
6. $f(x) = \sqrt{1 - x}$

Simplify each expression in Exercises 7 to 14. The variables may represent any real number, so absolute values may be needed.

7. a. $\sqrt{ab^2}$ b. $\sqrt{a^2b}$ c. $\sqrt{a^2b^2}$
8. a. $\sqrt{a^2b^4}$ b. $\sqrt{a^4b^2}$ c. $\sqrt{a^6b^4}$
9. a. $\sqrt{49x^2}$ b. $\sqrt{121y^2}$
10. a. $\sqrt{x^2y}$ b. $\sqrt{y^2x}$
11. a. $\sqrt{p^1p^2}$ b. $\sqrt{p^4}$
12. a. $\sqrt{p^8}$ b. $\sqrt{p^1p^8}$
13. a. $\sqrt{b \cdot b^4}$ b. $\sqrt{c \cdot c^2}$
14. a. $\sqrt{d \cdot d^2}$ b. $\sqrt{m \cdot m^6}$
15. Explain why $\sqrt{x^2y} = |x|\sqrt{y}$ needs absolute value, but $\sqrt{x^4y} = x^2\sqrt{y}$ does not.
16. Explain why $\sqrt{x^6y} = |x^3|\sqrt{y}$ needs absolute value, but $\sqrt{x^8y} = x^4\sqrt{y}$ does not.

In Exercises 17 to 20, for what inputs is each expression defined?

17. a. $\sqrt{x - 1}$ b. $\sqrt{x + 3}$
18. a. $\sqrt{x - 3}$ b. $\sqrt{x + 1}$
19. a. $\sqrt{4 - x}$ b. $\sqrt{3 - x}$
20. a. $\sqrt{2 - x}$ b. $\sqrt{1 - x}$

For Exercises 21 and 22, use the graphs to solve the equations.

21. $\sqrt{x - 2} = 2$

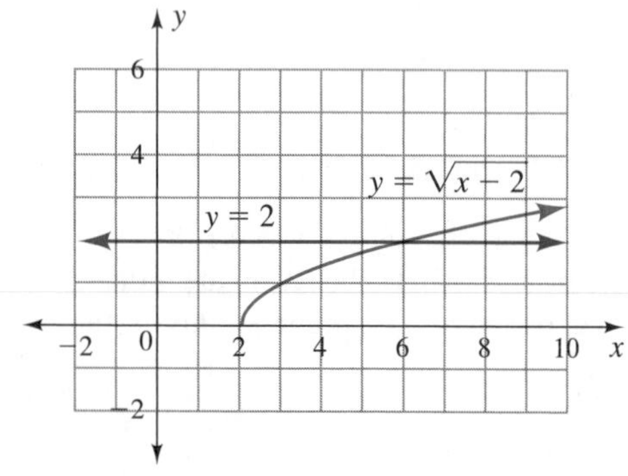

22. $\sqrt{x + 2} = 2$

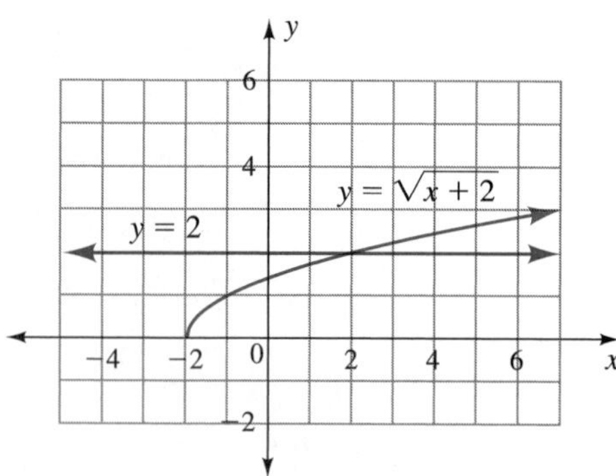

23. Graph $y = \sqrt{x + 3}$ and $y = 1$. For what x is $\sqrt{x + 3}$ defined? Solve $\sqrt{x + 3} = 1$ from your graph and with algebraic notation.
24. Graph $y = \sqrt{x - 3}$ and $y = 1$. For what x is $\sqrt{x - 3}$ defined? Solve $\sqrt{x - 3} = 1$ from your graph and with algebraic notation.
25. Graph $y = \sqrt{3 - x}$ and $y = 2$. For what x is $\sqrt{3 - x}$ defined? Solve $\sqrt{3 - x} = 2$ from your graph and with algebraic notation.
26. Graph $y = \sqrt{4 - x}$ and $y = 3$. For what x is $\sqrt{4 - x}$ defined? Solve $\sqrt{4 - x} = 3$ from your graph and with algebraic notation.

Solve the equations in Exercises 27 to 34. For what inputs, x, are the radical expressions defined?

27. $\sqrt{x - 2} = 3$
28. $\sqrt{x - 3} = 2$
29. $\sqrt{2x + 2} = 4$
30. $\sqrt{3x + 1} = 5$
31. $\sqrt{3x - 3} = 6$
32. $\sqrt{2x - 1} = 7$
33. $\sqrt{5 - x} = 3$
34. $\sqrt{6 - 2x} = 4$

Assume a clear day and flat terrain in the nearby region. How far could you see, in miles, from the top of each of the buildings whose heights are given in Exercises 35 to 38? See text examples for formulas.

35. Sears Tower, Chicago, 1454 feet
36. Bank of China, Hong Kong, 1209 feet
37. Texas Commerce Tower, Houston, 1002 feet
38. Transamerica Pyramid, San Francisco, 853 feet

On the moon, the distance seen in miles from a height of h feet is given by

$$d \approx \sqrt{\frac{3h}{8}}$$

For Exercises 39 to 42, calculate how far can be seen from the given heights. Why might the distance be different from that on Earth?

39. 24 feet
40. 5 feet

41. 96 feet **42.** 10 feet

43. a. How high do we need to be on Earth to see 30 miles?

b. How high do we need to be on the moon to see 30 miles?

c. How many times higher do we need to be in part b than in part a?

d. Extension: Explain why. A picture may be helpful.

44. a. How high do we need to be on Earth to see 5 miles?

b. How high do we need to be on the moon to see 5 miles?

c. How many times higher do we need to be in part b than in part a?

d. Extension: Explain why. A picture may be helpful.

How long would it take an object to fall from the top of each of the places whose heights are given in Exercises 45 to 48? To the nearest whole number, how fast would the object be traveling in feet per second when it hit the ground? See text examples for formulas.

45. C. N. Tower, Toronto, 1821 feet

46. World Trade Center, New York City, 1368 feet

47. C&C Plaza, Atlanta, 1063 feet

48. Eiffel Tower, Paris, 984 feet

49. How tall a building is required for an object to take 12 seconds to fall to the ground?

50. How tall a building is required for an object to take 10 seconds to fall to the ground?

51. How long would it take an object to fall from one mile up? (Assume no air resistance, wind, or other complicating factors.)

52. If a piece of ice falls off an airplane at 35,000 feet, how long until it will hit the ground? (Assume the ice does not melt.)

In Exercises 53 to 58, solve the formula for the indicated variable.

53. Electricity: $E = \sqrt{W \cdot R}$ for R

54. Electricity: $I = \sqrt{\dfrac{W}{R}}$ for W

55. Falling objects: $t = \sqrt{\dfrac{2d}{g}}$ for g

56. Distance seen on the moon: $d = \sqrt{\dfrac{3h}{8}}$ for h

57. Orbital velocity: $V_o = \sqrt{\dfrac{GM}{R}}$ for M

58. Escape velocity: $V_e = \sqrt{\dfrac{2GM}{R}}$ for M

59. Solve $\sqrt{\dfrac{3h_e}{2}} = \sqrt{\dfrac{3h_m}{8}}$ for the ratio h_e/h_m, the ratio of the height needed on Earth to the height needed on the moon to see the same distance.

60. How do we find the sets of inputs for which a radicand is defined?

61. How do we know when to use absolute value on an expression after taking the square root?

Projects

62. ***Radicals with Index Other than 2.*** If $2^3 = 8$, then 2 is a cube root of 8. If $3^4 = 81$, then the fourth root of 81 is 3. In radical notation, these facts are written $\sqrt[3]{8} = 2$ and $\sqrt[4]{81} = 3$, where 3 and 4 are the *indices* (plural of *index*). What are the values of these expressions? Do the problems mentally by thinking "What number raised to the [index] power gives [the radicand]?" For example, part a would be "What number raised to the 4th power gives 625?"

a. $\sqrt[4]{625}$ **b.** $\sqrt[4]{16}$ **c.** $\sqrt[3]{64}$

d. $\sqrt[5]{32}$ **e.** $\sqrt[3]{27}$ **f.** $\sqrt[3]{\frac{1}{8}}$

g. $\sqrt[5]{243}$ **h.** $\sqrt[4]{256}$ **i.** $\sqrt[3]{\frac{1}{27}}$

j. $\sqrt[3]{125}$ **k.** $\sqrt[3]{0.001}$ **l.** $\sqrt[3]{0.125}$

63. ***Distance to the Horizon***

a. We are given the formula $d = \sqrt{\dfrac{3h}{2}}$ to estimate the distance d seen, in miles, from a height h, in feet, on Earth. The Earth's radius at the equator is 3963 miles. There are 5280 feet in 1 mile. Follow the steps below to see where the formula came from and why the unit change from feet to miles doesn't show in the formula.

In the figure, the circle is the Earth at the equator. Let r be the radius in miles, $h/5280$ be the height above the surface in miles (so that h is in feet), and d be the distance from the given height to the horizon. (Geometry fact: There is a right angle between the line to the horizon and the radius at that point.)

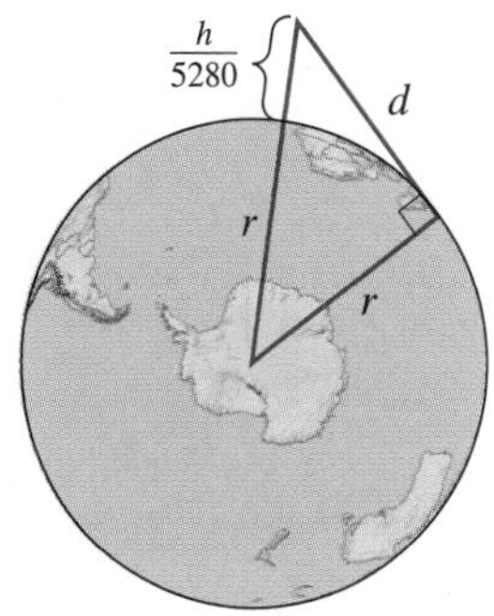

Step 1: Use the Pythagorean theorem to write the formula relating r, $h/5280$, and d. (*Hint*: The hypotenuse is $r + h/5280$ in miles.)

Step 2: Simplify the formula. (Don't forget that binomials like $(a + b)^2$ have three terms.) Note that r^2 drops from each side. Take the square root of both sides. Since the $h^2/5280^2$ term is virtually zero for heights of up to 1000 feet, throw that term away. (Scientists and engineers throw out squared terms all the time!)

Step 3: Substitute the radius of the Earth for r. Divide the resulting coefficient on the h term. Round to two decimal places. The result should be approximately $\frac{3}{2}$. Mission accomplished!

b. To verify the formula $d = \sqrt{\frac{3h}{8}}$ for the moon, repeat step 3 for the radius of the moon. The radius of the moon at the equator is 1090 miles.

MID–CHAPTER  TEST

1. Which of these sets of numbers represent the sides of a right triangle?

a. $\{4, 6, 8\}$ **b.** $\{10, 15, 20\}$ **c.** $\{15, 20, 25\}$

2. A plane flies due north at 120 miles per hour (mph). There is a 50-mph cross wind from the west. How many miles will the plane have flown in 1 hour?

3. The sides of a rectangle are 6 inches and 15 inches. What is the length of the diagonal of the rectangle?

4. Simplify these exponent and radical expressions.

a. $\sqrt{2} \cdot \sqrt{18}$ **b.** $(3\sqrt{2})^2$ **c.** $\sqrt{72}$

d. $(2\sqrt{3})^2$ **e.** $\sqrt{48}$ **f.** $-\sqrt{4}$

g. $\sqrt{-16}$ **h.** $\pm\sqrt{\frac{25}{16}}$ **i.** $\frac{\sqrt{2}}{\sqrt{32}}$

5. Simplify. Assume that x and y are any real numbers (except for zero in the denominator).

a. $\sqrt{3x}\sqrt{27x^3}$ **b.** $\sqrt{\frac{3x^2}{12y^4}}$

c. $\sqrt{\frac{196x}{x^3}}$ **d.** $\sqrt{a^2 + b^2}$

6. Find the distance between $(2, -3)$ and $(-4, 6)$.

7. Between what two consecutive integers is $\sqrt{135}$?

8. Solve $I = \sqrt{\frac{W}{R}}$ for R.

9. Solve by graphing and with algebraic notation: $\sqrt{x - 3} = 2$. For what inputs is the equation defined?

10. How does a square root graph show that the square root of a negative number is undefined in the real numbers?

8.4 Graphing and Solving Quadratic Equations

OBJECTIVES

- Make a table and graph for a quadratic function.
- Find the x-intercepts, y-intercept, vertex, and axis of symmetry of the graph of a quadratic function.
- Identify a, b, and c from a quadratic equation in standard form, and build a quadratic equation given a, b, and c.
- Solve a quadratic equation from a table and a graph.

WARM-UP

Make an input-output table for each equation, using the integers on the interval $[-2, 4]$ as inputs.

1. $y = x^2 - 3x - 4$ **2.** $y = 4 - x^2$

THIS SECTION INTRODUCES quadratic functions, special features of their graphs, and solving quadratic equations from tables and graphs.

Quadratic Functions

Linear functions are polynomials in which the highest power on the input variable is 1: $y = mx^1 + b$. When the highest power (called the *degree*) on a polynomial equation is 2, we have a quadratic function.

Definition of Quadratic Function

> A **quadratic function** may be written $f(x) = ax^2 + bx + c$, where a, b, and c are real numbers and a is not zero.

The equations in the Warm-up are quadratic equations because the highest power on x is 2, in the x^2 term. The set of inputs, or domain, for a quadratic function is all real numbers. The set of outputs, or range, for a quadratic function depends on the location of the vertex (more on that later this section).

Graphing Quadratic Functions

EXAMPLE 1 Exploring patterns in tables and graphs

a. Plot the points from Tables 6 and 7.

Input x	Output $y = x^2 - 3x - 4$
-2	6
-1	0
0	-4
1	-6
2	-6
3	-4
4	0

Table 6

Input x	Output $y = 4 - x^2$
-2	0
-1	3
0	4
1	3
2	0
3	-5
4	-12

Table 7

b. What is the name of the point on the graph where $x = 0$?

c. What is the name of the point(s) on the graph where $y = 0$?

d. Are any output numbers within a table the same?

e. What is the lowest or highest point on each graph?

f. What is the Δy pattern in each table? What is the same about the patterns in the two tables?

g. Use the patterns in the tables to predict the output for each equation when $x = -3$.

Solution See Examples 2 and 3 for the answers to parts a to e. See the Answer Box for parts f and g. ●

VERTEX, AXIS OF SYMMETRY, AND INTERCEPTS The *graph of a quadratic equation* is a curve called a **parabola**. *The highest or lowest point on the parabola* is called the **vertex**. The path of the bouncing ball in Figure 19 (repeated from Figure 1 in Chapter 2) is a series of parabolas. The vertex of each parabola is the highest point in the bounce.

Figure 19

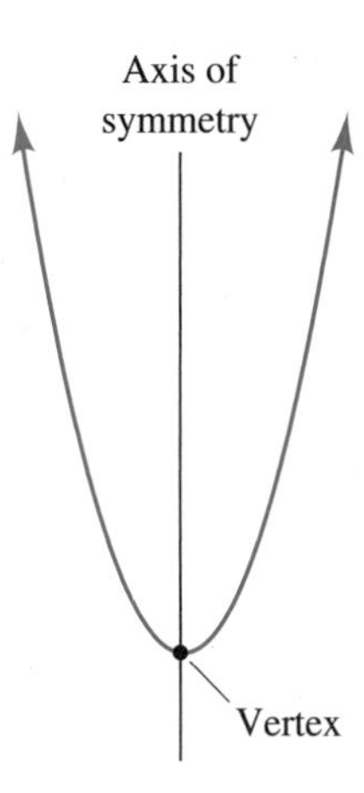

Figure 20 Parabola

An **axis**, or *line*, **of symmetry** is *a line across which the graph can be folded so that points on one side of the graph match up with points on the other side of the graph.* The parabolic graphs of quadratic functions have an axis of symmetry with equation $x = n$; see Figure 20. The axis of symmetry is halfway between any pair of inputs having equal outputs (such as the x-intercepts having output $y = 0$). The axis of symmetry passes through the vertex.

EXAMPLE 2 Graphing quadratic functions The graph of $y = x^2 - 3x - 4$ is shown in Figure 21.

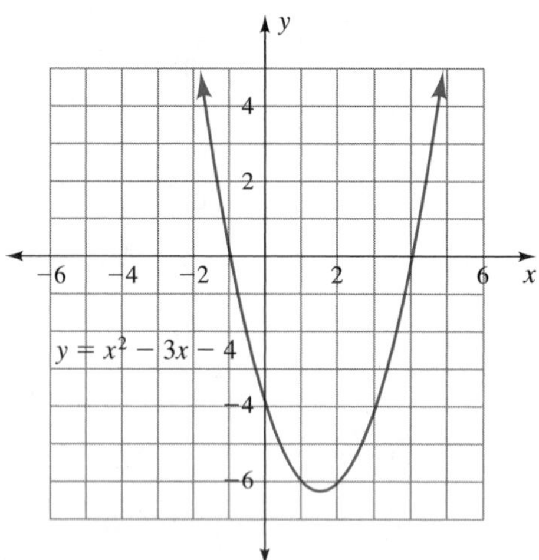

Figure 21

a. What is the y-intercept point?

b. What are the x-intercept points?

c. Give two or three sets of ordered pairs that have the same output. Will this pattern continue? What is the axis of symmetry?

d. What is the vertex?

Solution **a.** The y-intercept is -4. The y-intercept point is $(0, -4)$.

b. From the table at $y = 0$, the x-intercepts are $x = -1$ and $x = 4$. The x-intercept points are $(-1, 0)$ and $(4, 0)$.

c. Ordered pairs with the same output include $(0, -4)$ and $(3, -4)$, $(1, -6)$ and $(2, -6)$, as well as the x-intercept points. The axis of symmetry is halfway between the x-intercepts:

$$\frac{-1 + 4}{2} = \frac{3}{2} = 1\frac{1}{2}$$

or $x = 1.5$.

d. The vertex lies on the axis of symmetry, so $x = 1.5$. To find y, we have

$y = x^2 - 3x - 4$ Substitute $x = 1.5$.

$y = (1.5)^2 - 3(1.5) - 4$ Simplify.

$y = 2.25 - 4.5 - 4$

$y = -6.25$

The vertex is $(1.5, -6.25)$. ●

Think about it 1: When we substitute $x = 0$ or $x = 1.5$ into a function $f(x)$, what notation do we write to describe the substitution?

In many cases, the intercepts, vertex, and axis of symmetry are all you need to sketch the graph of a quadratic equation. If needed, one or two other points can be found by substituting x-values into the equation.

EXAMPLE  Graphing quadratic functions

a. What is the y-intercept point for the graph of $y = 4 - x^2$?

b. What are the x-intercept points?

c. What is the axis of symmetry?

d. What is the vertex?

e. Graph $y = 4 - x^2$.

Solution **a.** Let $x = 0$ in $y = 4 - x^2$. The y-intercept is $y = 4$. The y-intercept point is $(0, 4)$.

b. Let $y = 0$ in $y = 4 - x^2$.

$0 = 4 - x^2$

$x^2 = 4$

The x-intercepts are $x = -2$ and $x = 2$. The x-intercept points are $(-2, 0)$ and $(2, 0)$.

c. The axis of symmetry is halfway between the x-intercepts:

$$\frac{-2 + 2}{2} = 0$$

so $x = 0$, the y-axis.

d. The vertex lies on the axis of symmetry (y-axis), so $x = 0$.

$y = 4 - x^2$ Let $x = 0$.

$y = 4$

The vertex is $(0, 4)$.

e. The graph is in Figure 22.

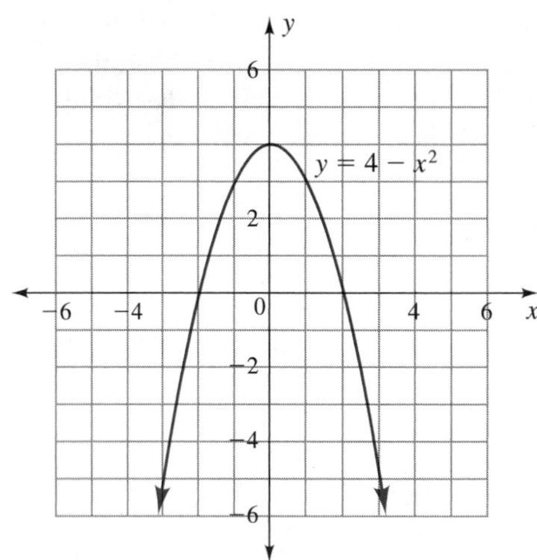

Figure 22

Think about it 2: What part of the equation in Example 3 causes the graph to open downward?

SUMMARY EXAMPLE OF GRAPHING In Example 4, we compare two graphs to observe what causes a graph to open upward or open downward.

EXAMPLE 4 Graphing quadratic equations

a. What is the y-intercept point for $y = x^2$ and for $y = -x^2$?

b. What are the x-intercept points?

c. For each equation, give two or three sets of ordered pairs that have the same output. Will this pattern continue? What is the axis of symmetry?

d. What is the vertex?

e. What part of the equation causes the parabola to open upward or open downward?

f. Graph $y = x^2$ and $y = -x^2$.

Solution **a.** At $x = 0$, $y = 0$. The y-intercept point is the origin, (0, 0).

b. The x-intercept point (where $x^2 = 0$ and $-x^2 = 0$) is also the origin.

c. Any pair of opposite inputs gives matching outputs; for example, −2 and 2 both give 4. The ordered pairs (−2, 4) and (2, 4) are in $y = x^2$. The ordered pairs (−2, −4) and (2, −4) are in $y = -x^2$. The axis of symmetry is the y-axis, $x = 0$.

d. The vertex for both graphs is the origin, (0, 0).

e. The −1 coefficient in $y = -x^2$ gives the negative values, causing the graph to open downward instead of upward, as $y = x^2$ does.

f. The graphs are shown in Figure 23.

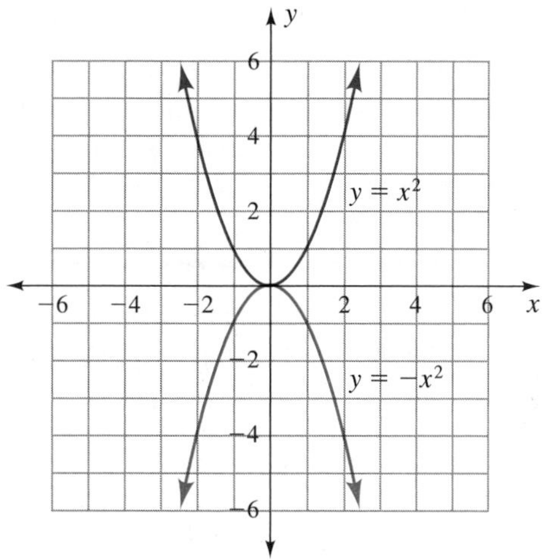

Figure 23

The special parts of the quadratic equation graph are summarized below, for reference in graphing.

Special Features of the Quadratic Equation Graph

1. The y-intercept is at $x = 0$. In function notation, the y-intercept is $f(0)$.
2. The x-intercepts are where $y = 0$. In function notation, the x-intercepts are where $f(x) = 0$.
3. The graph has an axis (or line) of symmetry, $x = n$. To find the axis of symmetry, average the inputs for two ordered pairs with the same output. Symmetry or the change in y can be used to extend the table once one side of the parabola is known.
4. The lowest (or highest) point on the parabolic graph is the vertex. The x-coordinate for the vertex is the same as x in the axis of symmetry.

RANGE The vertex of the parabola helps us to find the set of outputs, or range, for the quadratic function.

Finding the Range for a Quadratic Function

If the vertex is (x, n), then the range is

$y \geq n$ for a parabola that opens up

$y \leq n$ for a parabola that opens down

EXAMPLE 5 Finding the range Use the vertex to find the range for these two quadratic functions.

a. $f(x) = x^2 - 3x - 4$ **b.** $f(x) = 4 - x^2$

Solution

a. The vertex for $f(x) = x^2 - 3x - 4$ is $(1.5, -6.25)$. The parabola opens upward, so the range of the function includes all y values from a low of -6.25: $y \geq -6.25$.

b. The vertex for $f(x) = 4 - x^2$ is $(0, 4)$. The parabola opens downward, so the range includes all y values from a high of 4: $y \leq 4$.

Interpreting a, b, and c in a Quadratic Equation

With linear equations, we write $y = mx + b$ to identify the slope m and the y-intercept b. Similarly, with quadratic functions, we obtain certain information from the letters a, b, and c. In some applications, we are given an equation and need to identify a, b, and c.

> To identify a, b, and c, arrange a quadratic equation into standard form. For a one-variable equation, standard form is
>
> $$ax^2 + bx + c = 0$$
>
> For a two-variable equation, standard form is
>
> $$y = ax^2 + bx + c$$

EXAMPLE 6 *Identifying a, b, and c in the quadratic equation* What are the values of a, b, and c in each equation? In parts e and f, the input variable is t.

a. $3x^2 + 5x + 2 = 0$ **b.** $x^2 = 16$ **c.** $y = x^2 + 3x - 4$

d. $4 - y = x^2$ **e.** $h = -16t^2 + 48t$ **f.** $h = \frac{1}{2}gt^2 + v_0t + h_0$

Solution

a. $a = 3, b = 5, c = 2$

b. Rearranging to $x^2 - 16 = 0$ gives $a = 1, b = 0, c = -16$.

c. $a = 1, b = 3, c = -4$

d. Rearranging to $y = -x^2 + 4$ gives $a = -1, b = 0, c = 4$.

e. $a = -16, b = 48, c = 0$

f. $a = \frac{1}{2}g, b = v_0, c = h_0$ ●

In other applications, we are given a, b, and c and must place them into a quadratic equation. We practice this process in Example 7.

EXAMPLE 7 *Finding quadratic equations given a, b, and c* Write equations given the following information.

a. $a = 1, b = -2, c = 1$ **b.** $a = 1, b = 0, c = -9$

c. $a = 4, b = 0, c = 0$ **d.** $a = -1, b = 0, c = 0$

Solution **a.** $y = x^2 - 2x + 1$ **b.** $y = x^2 - 9$ **c.** $y = 4x^2$ **d.** $y = -x^2$ ●

Example 8 relates the graph of a quadratic equation to the letters a, b, and c.

EXAMPLE 8 *Using a, b, and c to describe key features of the graph of a quadratic equation*

a. Substitute $x = 0$ into $y = ax^2 + bx + c$ to find an equation for the y-intercept for all quadratic equations.

b. Which letter, a, b, or c, tells whether the parabola will turn up or turn down?

Solution

a. $y = ax^2 + bx + c$

$y = a(0)^2 + b(0) + c$

$y = c$

Since $y = c$ at $x = 0$, the letter c tells the y-intercept.

b. In $y = 4 - x^2$ and $y = -x^2$, the negative sign caused the graph to open downward. The letter a controls the direction of the parabola. ●

More on Features of the Graph of a Quadratic Equation

> A positive value for the letter a indicates that the graph opens upward.
>
> A negative value for the letter a indicates that the graph opens downward.
>
> The letter c describes the y-intercept.

You will learn more about a, b, and c later in this text and in higher mathematics courses.

Solving Quadratic Equations with Tables and Graphs

There are at least five different ways to solve a quadratic equation. In this section, we review solving equations from tables and graphs, first introduced in Section 3.2. The next two sections introduce three ways of using algebraic notation to solve quadratic equations.

Like other equations, quadratic equations can be solved with tables by looking for inputs that give the required output.

To solve with graphs, we look for inputs that correspond to the intersection of the graphs of the left and right sides of the equation.

EXAMPLE 9 Solving quadratic equations by table and graph Use a table and graph to solve these equations.

a. $x^2 - 3x - 4 = 0$ **b.** $x^2 - 3x - 4 = -6$ **c.** $x^2 - 3x - 4 = 6$

Solution The table is Table 8, and the graph is shown in Figure 24.

Input x	Output $y = x^2 - 3x - 4$
−2	6
−1	0
0	−4
1	−6
2	−6
3	−4
4	0

Table 8

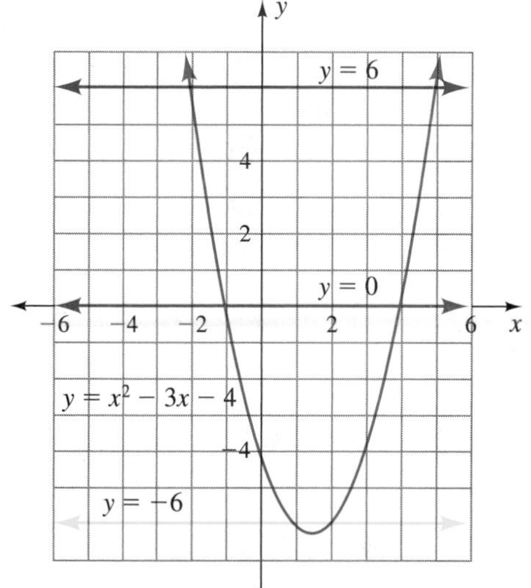

Figure 24

a. In Table 8, $y = 0$ when $x = -1$ and $x = 4$. On the graph in Figure 24, $y = 0$ is the x-axis, and the intersections with the parabola are at $x = -1$ and $x = 4$.

Check:

$(-1)^2 - 3(-1) - 4 \stackrel{?}{=} 0$ ✓

$(4)^2 - 3(4) - 4 \stackrel{?}{=} 0$ ✓

b. In Table 8, $y = -6$ when $x = 1$ and $x = 2$. On the graph in Figure 24, $y = -6$ intersects the parabola at $x = 1$ and $x = 2$.

Check:

$(1)^2 - 3(1) - 4 \stackrel{?}{=} -6$ ✓

$(2)^2 - 3(2) - 4 \stackrel{?}{=} -6$ ✓

c. Table 8 shows only one place, $x = -2$, where the output is 6. However, by symmetry, there will be another 6 at $x = 5$. On the graph in Figure 24, $y = 6$ intersects the parabola at $x = -2$ and $x = 5$.

Check:

$(-2)^2 - 3(-2) - 4 \stackrel{?}{=} 6$ ✓

$(5)^2 - 3(5) - 4 \stackrel{?}{=} 6$ ✓

Graphing Calculator Solution Enter $Y_1 = X^2 - 3X - 4$ and $Y_2 = 0$. To solve by table, set up the table with a starting value of $x = -2$ and a change in table (Δx) of 1. View the table.

To solve by graph, set the window for X at $[-6, 6]$, scale 1, and for Y at $[-7, 7]$, scale 1. Graph. Trace to the points of intersection. Zoom in as needed. Change Y_2 to -6 and then 6 to find the other solutions. ●

Applications

HEIGHT OF FALLING ITEMS The height h, at any given time t, of an object dropped from a starting height h_0 is described by the formula

$$h = h_0 - \tfrac{1}{2}gt^2$$

The letter g represents the acceleration due to gravity and may be approximated by 32.2 feet per second squared. (The square is on the second, not on the feet.)

EXAMPLE 10 Solving quadratic equations Write equations to find at what times a sugar cube dropped from a height of 550 feet will be at 400 feet, 300 feet, 200 feet, 100 feet, and zero feet. Solve the equations by table and graph.

Solution The table is Table 9, and the graph is shown in Figure 25.

Time (sec)	Height (ft)
0	550
1	533.9
2	485.6
3	405.1
4	292.4
5	147.5
6	−29.6

Table 9

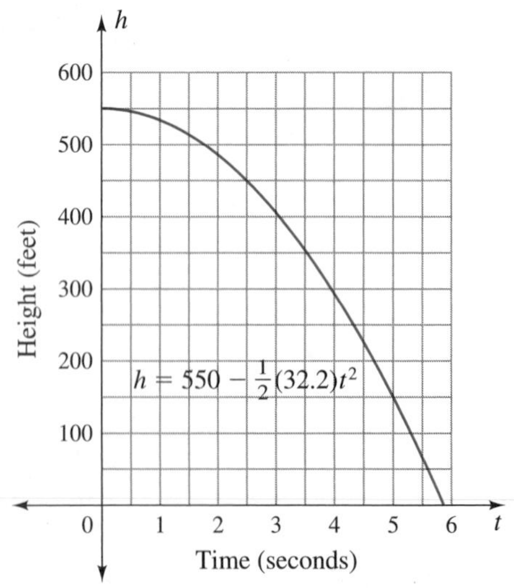

Figure 25

From $h = h_0 - \frac{1}{2}gt^2$, we get the following equations to be solved. Estimated solutions following each equation are from the table and graph.

	Corresponding input value:
$400 = 550 - \frac{1}{2}(32.2)t^2$	$t \approx 3$ sec
$300 = 550 - \frac{1}{2}(32.2)t^2$	$t \approx 4$ sec
$200 = 550 - \frac{1}{2}(32.2)t^2$	$t \approx 4\frac{3}{4}$ sec
$100 = 550 - \frac{1}{2}(32.2)t^2$	$t \approx 5\frac{1}{4}$ sec
$0 = 550 - \frac{1}{2}(32.2)t^2$	$t \approx 5\frac{3}{4}$ sec

For related information, see Examples 12 and 13 of Section 8.3 (page 504). ●

Think about it 3: Does the sugar cube drop the same amount each second? What part of the equation would explain any change in speed?

USING QUADRATIC EQUATIONS WITHIN ALGEBRA In this application, we return to the Pythagorean triples in Section 8.2. Because {3, 4, 5} is a Pythagorean triple, we might be tempted to think that {4, 5, 6} or {5, 6, 7} or some other set of three consecutive integers would also be a triple. We prove in Example 11 that the set {3, 4, 5} is the *only set of consecutive positive integers* that satisfies the Pythagorean theorem.

EXAMPLE 11 Proving that {3, 4, 5} is unique What sets of three consecutive integers satisfy $a^2 + b^2 = c^2$?

a. Let x, $x + 1$, and $x + 2$ be the three consecutive integers. Write the Pythagorean theorem using consecutive integer notation, and simplify.

b. Find the solutions to the resulting equation from a table.

c. Find the solutions to the resulting equation from a graph. Use a graphing calculator if one is available.

d. Substitute the solutions into the expressions in part a and state your conclusions.

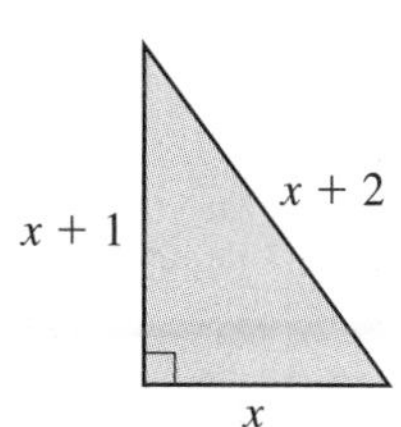

Figure 26

Solution **a.** The consecutive integers are placed on a right triangle in Figure 26. We substitute x, $x + 1$, and $x + 2$ into the Pythagorean theorem:

$$x^2 + (x + 1)^2 = (x + 2)^2$$

$$x^2 + x^2 + 2x + 1 = x^2 + 4x + 4 \quad \text{Square the binomials.}$$

$$2x^2 + 2x + 1 - x^2 - 4x - 4 = 0 \quad \text{Subtract } x^2 + 4x + 4 \text{ from both sides.}$$

$$x^2 - 2x - 3 = 0 \quad \text{Add like terms.}$$

b. We solve $x^2 - 2x - 3 = 0$ by completing Table 10 for $y = x^2 - 2x - 3$. The table shows $y = 0$ corresponding with the inputs $x = -1$ and $x = 3$.

c. In Figure 27, we graph the data together with $y = 0$, which is the x-axis. The x-intercepts, $x = -1$ and $x = 3$, are the solutions to $x^2 - 2x - 3 = 0$.

d. Substituting the solutions into the consecutive integer expressions gives

$x = -1$, $x + 1 = 0$, and $x + 2 = 1$ or $\{-1, 0, 1\}$

$x = 3$, $x + 1 = 4$, and $x + 2 = 5$ or $\{3, 4, 5\}$

Both sets of numbers, $\{-1, 0, 1\}$ and $\{3, 4, 5\}$, satisfy $a^2 + b^2 = c^2$. However, there is no triangle with sides of lengths -1, 0, and 1, so this solution must be disregarded. Thus, the only consecutive integer solution set to the Pythagorean theorem is $\{3, 4, 5\}$.

Input x	Output $y = x^2 - 2x - 3$
-2	5
-1	0
0	-3
1	-4
2	-3
3	0
4	5

Table 10

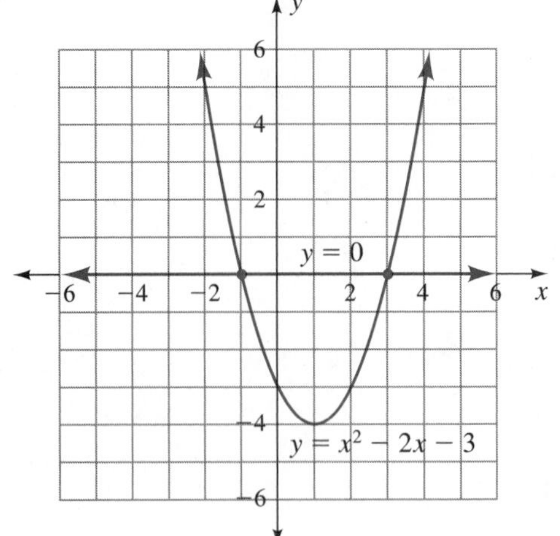

Figure 27

ANSWER BOX

Warm-up: See Tables 6 and 7. **Example 1: f.** The Δy pattern is consecutive even numbers for $y = x^2 - 3x - 4$ and consecutive odd numbers for $y = 4 - x^2$. Both patterns show consecutive numbers. **g.** The consecutive even number pattern for Δy suggests that $y = 14$ at $x = -3$ for $y = x^2 - 3x - 4$. The matching of outputs in the table suggests that $y = -5$ at $x = -3$ for $y = 4 - x^2$. **Think about it 1:** $f(0), f(1.5)$ **Think about it 2:** The subtraction of x^2 creates large negative values as x takes on numbers away from the origin. This causes the parabola to open downward. **Think about it 3:** The sugar cube drops 150 feet in the first 3 seconds. It drops the next 100 feet in 1 second and the next 100 feet in $\frac{3}{4}$ second. The cube is speeding up. The t^2 is causing this acceleration.

EXERCISES 

Find the x- and y-intercept points, axis of symmetry, and vertex for each of the parabolas shown in Exercises 1 to 4.

1. $y = x^2 + x - 6$

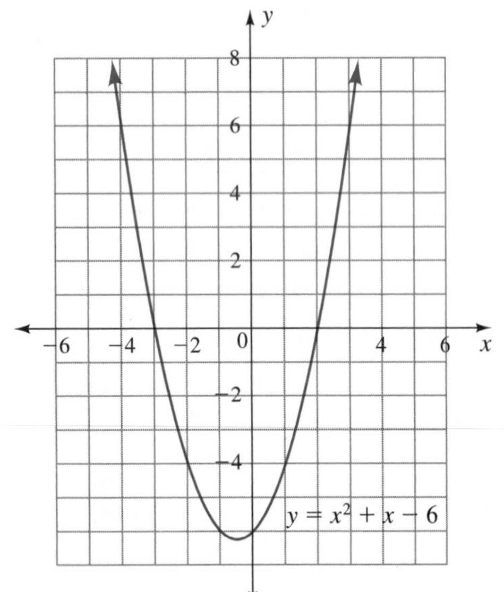

2. $y = x^2 + 2x - 3$

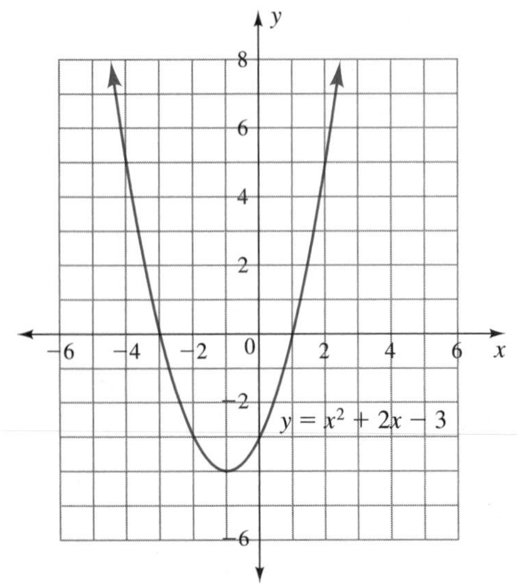

3. $y = 5x - x^2$

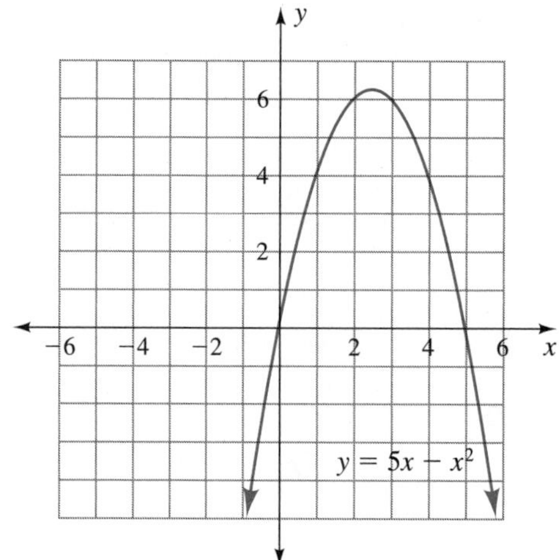

4. $y = 2 - x - x^2$

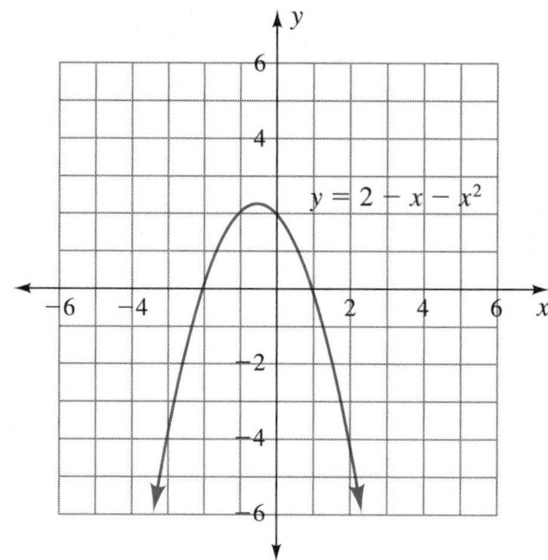

Make a table and graph for the equations in Exercises 5 to 12. Find the x- and y-intercept points, axis of symmetry, and vertex for each.

5. $y = x^2 - 6x + 7$

6. $y = x^2 - 3x - 10$

7. $y = x^2 + x - 12$

8. $y = x^2 - 2x - 8$

9. $y = x - x^2$

10. $y = 2x - x^2$

11. $y = 4x - 2x^2$

12. $y = 3x - x^2$

For Exercises 13 to 16, find the range of the function graphed in the given exercise.

13. Exercise 1

14. Exercise 2

15. Exercise 3

16. Exercise 4

In Exercises 17 to 32, name a, b, and c.

17. $y = 2x^2 + 3x + 1$

18. $y = 3x^2 + 2x - 1$

19. $r^2 - 4r + 4 = 0$

20. $r^2 - 6r + 9 = 0$

21. $y = x^2 - 4$

22. $y = x^2 - 25$

23. $4t^2 - 8t = 0$

24. $3t^2 - 27t = 0$

25. $4 = x - x^2$

26. $x = 6 - 2x^2$

27. $x^2 = x - 1$

28. $x^2 = 3 - x$

29. Height after time t: $h = -0.5gt^2 + vt + s$

30. Angle swept in time t for circular motion: $A = 0.5\alpha t^2 + wt$

31. Area of circle: $A = \pi r^2$

32. Surface area of cylinder of height 1: $SA = 2\pi r^2 + 2\pi r$

In Exercises 33 to 38, write the quadratic equation with these coefficients.

33. $a = 4, b = 4, c = 1$

34. $a = 2, b = -1, c = -6$

35. $a = 9, b = 0, c = -16$

36. $a = 25, b = 0, c = -1$

37. $a = 3, b = 6, c = 0$

38. $a = 5, b = 15, c = 0$

Solve the equations in Exercises 39 to 42 from the graphs in Exercises 1 to 4.

39. a. $x^2 + x - 6 = 6$ **b.** $x^2 + x - 6 = -4$

c. $x^2 + x - 6 = -8$ **d.** $x^2 + x - 6 = 0$

40. a. $x^2 + 2x - 3 = 5$ **b.** $x^2 + 2x - 3 = -3$

c. $x^2 + 2x - 3 = -4$ **d.** $x^2 + 2x - 3 = 0$

41. a. $5x - x^2 = 6$ **b.** $5x - x^2 = 8$

c. $5x - x^2 = 4$ **d.** $5x - x^2 = 0$

42. a. $2 - x - x^2 = 2$ **b.** $2 - x - x^2 = 4$

c. $2 - x - x^2 = -4$ **d.** $2 - x - x^2 = 0$

In Exercises 43 and 44, assume that the formula $h = h_0 - \frac{1}{2}gt^2$ holds true. Let g equal 32.2 feet per second squared.

43. Suppose a drop of water starts at the top of Ribbon Falls in Yosemite National Park and falls 1612 feet. How long will it take the drop to reach halfway? to reach the bottom?

44. Suppose a drop of water starts at the top of the American side of Niagara Falls and falls 182 feet. How long will it take the drop to reach halfway? to reach the bottom?

Projects

45. ***Largest Product***

a. List ten pairs of numbers that add to 12. Multiply the numbers in each pair to get a product, y. Describe any patterns.

b. Let x be the first number in each pair. Let y be the product of each pair. Graph the ordered pairs (x, y).

c. Let x be the first number in each pair. Write an expression for the second number in terms of x. Write an equation describing the products, y.

d. What is the largest product?

46. ***A Number Pattern***

a. What do bowling, pocket billiards, and stacks of paint cans at a paint store have in common? (*Hint*: How many bowling pins are in a full rack? How many balls are in a rack of pocket billiards? How many cans are pictured in the illustration?)

b. Complete the table, where the input represents the number of rows in the stack of cans and the output represents the total number of cans in the stack for that many rows.

Row of Cans x	Total Number of Cans y
1	
2	
3	
4	
5	
6	
7	

c. Find a rule for the table. (*Hint*: It is a quadratic equation and involves consecutive numbers.) To find out if your rule is correct, see if it gives 300 cans for 24 rows. Use your rule to find the number of cans in 15 rows.

8.5 Solving Quadratic Equations by Taking the Square Root and Factoring

OBJECTIVES

- Solve quadratic equations by taking the square root of both sides.
- Solve quadratic equations by factoring and applying the zero product rule.

WARM-UP

1. Simplify these absolute value expressions.

a. $|-2|$ **b.** $|2|$ **c.** $|-3|$

2. Use guess and check to solve these absolute value equations.

a. $|x| = 2$ **b.** $|x| = 3$

3. Factor these expressions.

a. $x^2 - 4$ **b.** $x^2 - 3x - 4$

c. $t^2 - 3t + 2$ **d.** $25x^2 - 16$

e. $3x^2 - 75$ **f.** $x^2 - 3x$

g. $-16t^2 + 48t$ **h.** $x^2 - x - 2$

THIS SECTION INTRODUCES two methods for solving quadratic equations with algebraic notation: taking the square root of both sides and factoring. We use the zero product rule and again apply the techniques to applications: a real-world application and an algebraic application from elsewhere in this chapter.

Solving Quadratic Equations by Taking the Square Roots

Taking the square root is a way to solve quadratic equations for which $b = 0$ in $ax^2 + bx + c = 0$. These quadratic equations are in the form $ax^2 + c = 0$. In the first example, we look at the graphical results from solving equations with $b = 0$. Observe that some solutions include both positive and negative numbers.

EXAMPLE  Exploring graphical solutions After completing Table 11, use the graph of $y = x^2$ in Figure 28 to solve these equations.

Input x	Output $y = x^2$
−3	
−2	
−1	
0	
1	
2	
3	

Table 11

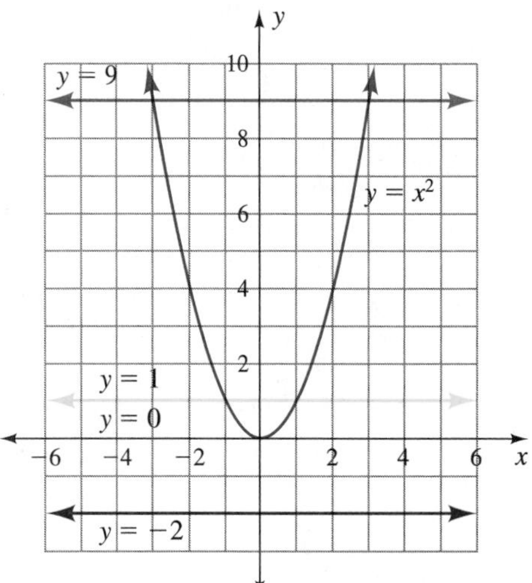

Figure 28

a. $x^2 = 9$ **b.** $x^2 = 1$ **c.** $x^2 = 0$ **d.** $x^2 = -2$

Solution See the Answer Box.

As noted in Sections 8.2 and 8.3, so long as we observe the restriction that the input to a square root is zero or positive and the output from taking the square root is zero or positive, we can take the square root of both sides of an equation.

> To solve an equation, we can take the square root of both sides.

Recall from page 499 that $\sqrt{x^2} = |x|$.

EXAMPLE  Solving by taking the square root of both sides Solve $x^2 - 4 = 0$. The variable x may be any real number.

Solution We start by taking the square root of both sides of $x^2 = 4$:

$$\sqrt{x^2} = \sqrt{4}$$

$$\sqrt{x^2} = 2$$

This time x may be either positive or negative. We still must have $\sqrt{x^2}$ be positive, and the only way to do this is with absolute value. Thus, $\sqrt{x^2} = |x|$.

$$|x| = 2$$

There are two numbers that make $|x| = 2$ true: $x = 2$ and $x = -2$. Thus, there are two solutions to $x^2 = 4$:

$$x = 2 \quad \text{and} \quad x = -2$$

EXAMPLE 3 Solving by taking the square root of each side Solve $3x^2 - 75 = 0$.

Solution

$3x^2 - 75 = 0$ Add 75 to both sides.

$3x^2 = 75$ Divide both sides by 3.

$x^2 = 25$

$\sqrt{x^2} = \sqrt{25}$ Take the square root of both sides.

$|x| = 5$ $\sqrt{x^2} = |x|$, the formal step.

$x = 5 \quad \text{or} \quad x = -5$ There are two solutions.

Check:

$$3(5)^2 - 75 \stackrel{?}{=} 0 \ ✓$$
$$3(-5)^2 - 75 \stackrel{?}{=} 0 \ ✓$$

In earlier sections of this chapter, we ignored the negative solution because we were working with triangles or other objects, which have only positive dimensions. In this section, we want both solutions. In Examples 2 and 3, the absolute value gives a formally correct answer. If it is too formal, just remember this:

> The equation $x^2 = n$ has two solutions:
>
> $$x = +\sqrt{n} \quad \text{and} \quad x = -\sqrt{n}$$

We will omit the absolute value step in the next example.

EXAMPLE 4 Solving by taking the square root of each side Solve $25x^2 - 16 = 0$.

Solution We solve first for x^2:

$$25x^2 - 16 = 0$$
$$25x^2 = 16$$
$$x^2 = \frac{16}{25}$$
$$\sqrt{x^2} = \sqrt{\frac{16}{25}}$$
$$x = \frac{4}{5} \quad \text{or} \quad x = -\frac{4}{5}$$

Check:

$$25\left(\frac{4}{5}\right)^2 - 16 \stackrel{?}{=} 0 \ ✓$$
$$25\left(-\frac{4}{5}\right)^2 - 16 \stackrel{?}{=} 0 \ ✓$$

Solving by Taking a Square Root

> To solve by taking a square root:
>
> **1.** Look for $b = 0$ in $ax^2 + bx + c = 0$.
>
> **2.** Solve for x^2.
>
> **3.** Take the square root of each side.
>
> **4.** Check.

Solving Quadratic Equations by Factoring

We start by looking at products that multiply to zero.

EXAMPLE 5 Exploring zero products Describe the role of zero in these equations and expressions.

a. $A \cdot 5 = 0$ **b.** $-5 \cdot B = 0$

c. $(x - 4)(x + 1)$ if $x = 4$ **d.** $(x - 4)(x + 1)$ if $x = -1$

Solution **a.** $A = 0$, because zero multiplied by 5 gives a zero product.

b. $B = 0$, because zero multiplied by -5 gives a zero product.

c. $(4 - 4)(4 + 1) = 0 \cdot 5 = 0$
The product $(x - 4)(x + 1)$ is zero if $x = 4$.

d. $(-1 - 4)(-1 + 1) = (-5) \cdot 0 = 0$
The product $(x - 4)(x + 1)$ is zero if $x = -1$.

Parts a and b show that if two numbers multiply to zero, then one of the numbers is zero. Parts c and d show that if one of two expressions in a product is zero, then the product is zero. ●

Example 5 suggests the **zero product rule**.

Zero Product Rule

> If $A \cdot B = 0$, then either $A = 0$ or $B = 0$.
>
> A and B usually represent monomial or binomial factors such as x, $(x + 2)$, $(x - 3)$, or $(2x - 5)$.

In Example 6, look for numbers in the factors that make the entire factor have a zero value.

EXAMPLE 6 Finding numbers that make a factor equal to zero For what values of x are these expressions true?

a. $(x + 2)(x - 2) = 0$ **b.** $(x - 4)(x + 7) = 0$

c. $(2x + 1)(3x - 2) = 0$

Solution **a.** Either $x + 2 = 0$ or $x - 2 = 0$
$x = -2$ or $x = 2$

b. Either $x - 4 = 0$ or $x + 7 = 0$
$x = 4$ or $x = -7$

c. Either $2x + 1 = 0$ or $3x - 2 = 0$
$2x = -1$ or $3x = 2$
$x = -\frac{1}{2}$ or $x = \frac{2}{3}$ ●

EXAMPLE  Solving by factoring Solve $x^2 + 3x - 28 = 0$.

Solution We will factor the left side and then set each of the factors equal to zero.

$$x^2 + 3x - 28 = 0$$
$$(x + 7)(x - 4) = 0$$

Either $x + 7 = 0$ or $x - 4 = 0$
$x = -7$ or $x = 4$

Check:

$(-7)^2 + 3(-7) - 28 \stackrel{?}{=} 0$ ✓

$4^2 + 3(4) - 28 \stackrel{?}{=} 0$ ✓

Solving by Factoring

To solve by factoring:

1. Write the equation in standard form, $ax^2 + bx + c = 0$.
2. Factor the expression on the left.
3. Use the zero product rule to write two factor equations.
4. Solve each factor equation.
5. Check.

EXAMPLE 8 Solving by factoring Solve $x^2 = 3x$ by factoring.

Solution

$x^2 = 3x$	Change to standard form.
$x^2 - 3x = 0$	Factor the left side.
$x(x - 3) = 0$	Use the zero product rule to set each factor equal to zero.
Either $x = 0$ or $x - 3 = 0$	Solve the factor equations.
$x = 0$ or $x = 3$	

Check:

$(0)^2 \stackrel{?}{=} 3 \cdot 0$ ✓

$3^2 \stackrel{?}{=} 3 \cdot 3$ ✓

In Examples 9 and 10, the equations from Examples 3 and 4 are solved by factoring. These examples show that many equations can be solved in several ways.

EXAMPLE 9 Solving by factoring Solve $3x^2 - 75 = 0$ by factoring.

Solution We factor the left side by first removing the greatest common factor.

$3(x^2 - 25) = 0$	
$3(x - 5)(x + 5) = 0$	Because $3 \neq 0$, set the other two factors equal to zero.
Either $x - 5 = 0$ or $x + 5 = 0$	Solve the factor equations.
$x = 5$ or $x = -5$	

Check:

$3(5)^2 - 75 \stackrel{?}{=} 0$ ✓

$3(-5)^2 - 75 \stackrel{?}{=} 0$ ✓

Example 10 shows that extra steps are needed to solve for x when the equation contains factors of the form $ax + b$.

EXAMPLE 10 Solving by factoring Solve $25x^2 = 16$ by factoring.

Solution

$$25x^2 = 16$$ First change to standard form.

$$25x^2 - 16 = 0$$ To factor, observe that $25x^2 - 16$ is the difference of squares.

$$(5x - 4)(5x + 4) = 0$$ Set the factors equal to zero.

Either $5x - 4 = 0$ or $5x + 4 = 0$ Solve the factor equations.

$$5x = 4 \qquad 5x = -4$$

$$x = \tfrac{4}{5} \qquad x = -\tfrac{4}{5}$$

Check:

$25\left(\tfrac{4}{5}\right)^2 \stackrel{?}{=} 16$ ✓

$25\left(-\tfrac{4}{5}\right)^2 \stackrel{?}{=} 16$ ✓

Applications

HEIGHT AND TIME The falling sugar cube in Example 10 of Section 8.4 (page 516) gave height in terms of time for an object already a given distance above the ground. In Example 11, the formula gives the height in terms of time for an object traveling up from the ground and back down again.

EXAMPLE 11

Finding the time required to reach given heights: spraying water Suppose a connector in a high-pressure water system has broken and a fine spray of water is shooting straight up into the air, according to the formula

$$h = -16t^2 + 48t$$

where h is height and t is time. In how many seconds after leaving the break will a droplet of water reach the 32-foot level? Solve from the graph in Figure 29 and by factoring.

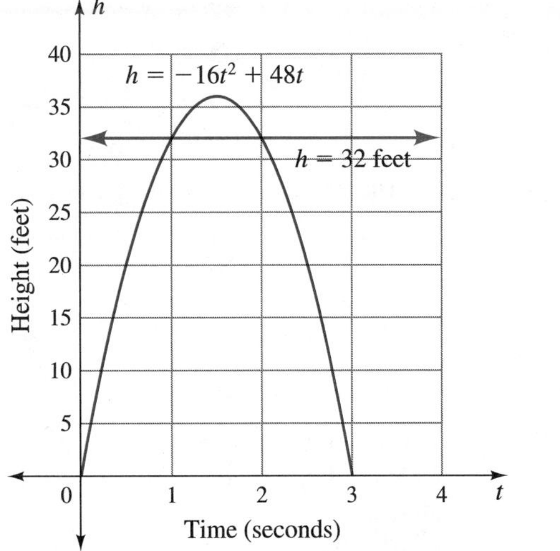

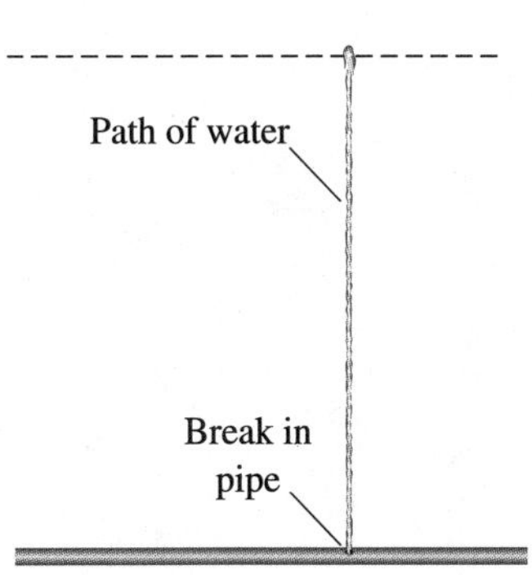

Figure 29

Solution According to the graph, a water droplet reaches 32 feet at $t = 1$ sec on the way up and again at $t = 2$ sec on the way down.

The equation to solve by factoring is $32 = -16t^2 + 48t$.

$32 = -16t^2 + 48t$	Change to equal zero.
$16t^2 - 48t + 32 = 0$	Factor out the greatest common factor.
$16(t^2 - 3t + 2) = 0$	Factor the remaining trinomial.
$16(t - 2)(t - 1) = 0$	Use the zero product rule.
Either $(t - 2) = 0$ or $(t - 1) = 0$	Solve the factor equations.
$t = 2$ or $t = 1$	

Check:

$$32 \stackrel{?}{=} -16(2)^2 + 48(2)$$
$$32 \stackrel{?}{=} -16(1)^2 + 48(1) \checkmark$$

The graph in Figure 28 does not show the path of the water. For the graph to show such a path, the x-axis would have to be labeled in feet instead of seconds. The path of the water in this example goes straight up and straight back down, as shown in the drawing to the right of the graph.

USING QUADRATIC EQUATIONS IN SOLVING RADICAL EQUATIONS In Example 12, we return to solving equations by squaring both sides. The example shows why it is important to check solutions to an equation.

EXAMPLE 12 Solving an equation containing square roots

a. Find the inputs for which $y = \sqrt{3 - x}$ is defined.

b. Solve $\sqrt{3 - x} = x - 1$ using symbols.

c. Solve $\sqrt{3 - x} = x - 1$ by graphing.

Solution **a.** The equation is defined for positive radicands: $3 - x \geq 0$. The solution to this inequality is $3 \geq x$, or $x \leq 3$.

b. Solve $\sqrt{3 - x} = x - 1$ by squaring both sides.

$$\sqrt{3 - x} = x - 1$$
$$(\sqrt{3 - x})^2 = (x - 1)^2$$
$$3 - x = x^2 - 2x + 1$$
$$0 = x^2 - x - 2$$
$$0 = (x + 1)(x - 2)$$
$$\text{Either } (x + 1) = 0 \quad \text{or} \quad (x - 2) = 0$$
$$x = -1 \quad \text{or} \quad x = 2$$

Check:

$$\sqrt{3 - (-1)} \stackrel{?}{=} (-1) - 1$$
$$\sqrt{3 - (2)} \stackrel{?}{=} 2 - 1 \checkmark$$

The input $x = 2$ gives a true statement. The input $x = -1$ gives $2 = -2$, which is false. The solution $x = -1$ must be discarded.

c. Figure 30 shows the intersection of $y = \sqrt{3 - x}$ and $y = x - 1$ at $x = 2$, but it shows no intersection at $x = -1$. This confirms the single solution found symbolically.

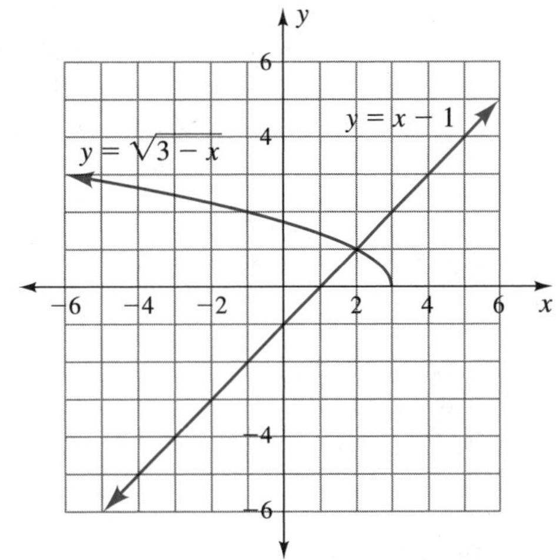

Figure 30

Solutions that give false statements are called **extraneous roots**.

ANSWER BOX

Warm-up: 1. a. 2 **b.** 2 **c.** 3 **2. a.** $x = 2, x = -2$ **b.** $x = 3$, $x = -3$ **3. a.** $(x - 2)(x + 2)$ **b.** $(x - 4)(x + 1)$ **c.** $(t - 1)(t - 2)$ **d.** $(5x - 4)(5x + 4)$ **e.** $3(x - 5)(x + 5)$ **f.** $x(x - 3)$ **g.** $-16t(t - 3)$ **h.** $(x - 2)(x + 1)$ **Example 1:** The missing table entries are 9, 4, 1, 0, 1, 4, and 9. **a.** $x = \pm 3$ **b.** $x = \pm 1$ **c.** $x = 0$ **d.** no real-number solutions

EXERCISES 8.5

Solve the equations in Exercises 1 to 18. The variable x is any real number (except zero in a denominator).

1. $x^2 = 5$

2. $x^2 = 10$

3. $2x^2 = 14$

4. $3x^2 = 33$

5. $x^2 = \frac{4}{25}$

6. $x^2 = \frac{36}{49}$

7. $100x^2 = 4$

8. $100x^2 = 25$

9. $49x^2 - 225 = 0$

10. $144x^2 = 169$

11. $36x^2 = 121$

12. $121x^2 - 64 = 0$

13. $75x^2 - 27 = 0$

14. $75x^2 - 12 = 0$

15. $\dfrac{3}{x} = \dfrac{x}{27}$

16. $\dfrac{2}{x} = \dfrac{x}{32}$

17. $\dfrac{x}{4} = \dfrac{9}{x}$

18. $\dfrac{x}{6} = \dfrac{24}{x}$

For what values of x are the equations in Exercises 19 to 24 true?

19. $(x - 4)(x + 4) = 0$

20. $(x + 3)(x - 5) = 0$

21. $(2x - 1)(3x + 2) = 0$

22. $(3x + 4)(2x - 1) = 0$

23. $(2x - 5)(x + 2) = 0$

24. $(2x + 1)(3x - 5) = 0$

Solve the equations in Exercises 25 to 52 by factoring.

25. $x^2 - 4 = 0$

26. $x^2 - 3x - 4 = 0$

27. $x^2 + x - 6 = 0$

28. $x^2 + 2x - 3 = 0$

29. $x^2 - 2x - 15 = 0$

30. $x^2 - 8x + 15 = 0$

31. $x^2 + 3x = 0$

32. $x^2 - 4x = 0$

33. $2x^2 = -x$

34. $3x^2 = 2x$

35. $x^2 - 4x = 12$

36. $x^2 + x = 12$

37. $2x^2 = x + 3$

38. $2x^2 = 5x + 3$

39. $x^2 = x + 12$

40. $x^2 = 15 - 2x$

41. $x^2 + 6 = 7x$

42. $x^2 = 4x - 3$

43. $2x^2 + 3x = 5$

44. $2x^2 + 3 = 7x$

45. $4x^2 - 25 = 0$

46. $9x^2 - 4 = 0$

47. $3x^2 - 12 = 0$

48. $5x^2 - 80 = 0$

49. $\dfrac{x+2}{2} = \dfrac{5}{x-1}$

50. $\dfrac{x-3}{5} = \dfrac{1}{x+1}$

51. $\dfrac{1}{x-3} = \dfrac{x-2}{2}$

52. $\dfrac{3}{x-4} = \dfrac{x+1}{2}$

53. In Example 11, the water is at ground level when the height, h, is zero. Use factoring to find the inputs, t, when $-16t^2 + 48t = 0$. Do the results agree with the graph?

54. Estimate the maximum height, h, reached by the water in Example 11, where $h = -16t^2 + 48t$. Estimate the input that gives maximum height. How could the input be used to find the maximum height?

55. An air-powered rocket is launched, and at time t it is at height $h = -16t^2 + 64t$. Use the graph in the figure to find the times when the rocket is at 48 feet. Use factoring to find the time. Do the results agree?

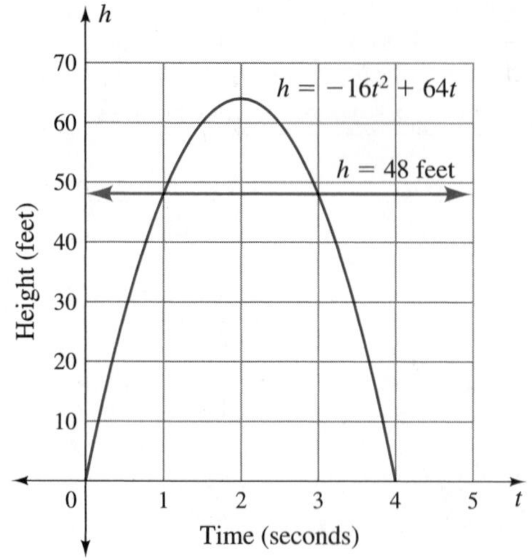

56. Estimate the maximum height reached by the rocket in Exercise 55 (see the figure). Estimate the input that gives maximum height. How could the input be used to find the maximum height?

57. Use factoring to find when the rocket in Exercise 55 is at ground level, $y = 0$. Does your result agree with the graph?

58. Use the graph to estimate how far the rocket in Exercise 55 travels in the first second and how far it travels in the second second. If rate, or speed, is distance divided by time, what is the average speed of the rocket during the first second? What is the average speed of the rocket in the second second?

Exercises 59 to 62 return to equations containing square roots (Section 8.3). Use factoring as needed to solve the quadratic equations. Show your check for each solution.

59. $\sqrt{4-x} = x - 2$

60. $\sqrt{3-x} = x - 1$

61. $\sqrt{x+5} = x + 3$

62. $\sqrt{3-x} = x - 3$

63. Refer to the figure below.

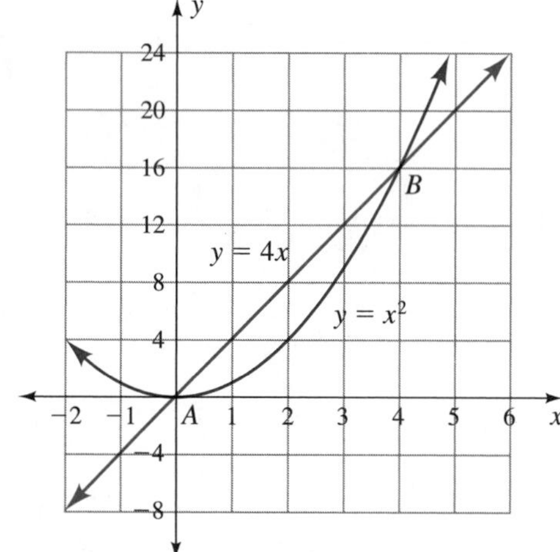

a. What are the ordered pairs at A and B?

b. The intersections at A and B solve what equation?

c. Solve your equation with algebra. Do you obtain the same results as with the graph?

64. Refer to the figure below.

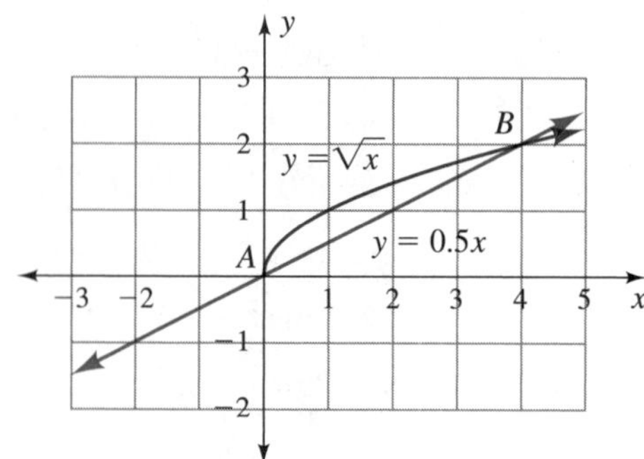

a. What are the ordered pairs at A and B?

b. The intersections at A and B solve what equation?

c. Solve your equation with algebra. Do you obtain the same results as with the graph?

65. a. Complete the table.

a	b	$a^2 + b^2$	$(a + b)^2$	$\sqrt{a + b}$	$\sqrt{a} + \sqrt{b}$
4	9				
1	3				
4	5				
3	6				

b. Explain why a student might think that $a^2 + b^2$ and $(a + b)^2$ were equal. How does the table show that they are not equal?

c. Explain why a student might think that $\sqrt{a + b}$ and $\sqrt{a} + \sqrt{b}$ were equal. How does the table show that they are not equal?

66. a. Complete the table.

a	b	$a^2 \cdot b^2$	$(a \cdot b)^2$	$\sqrt{a \cdot b}$	$\sqrt{a} \cdot \sqrt{b}$
4	9				
1	3				
4	5				
3	6				

b. What does the table indicate about $a^2 \cdot b^2$ and $(a \cdot b)^2$?

c. What does the table indicate about $\sqrt{a \cdot b}$ and $\sqrt{a} \cdot \sqrt{b}$?

Projects

67. ***From Solutions to Equations.*** The solutions to five different quadratic equations are given in parts a to d. Find an equation $y = x^2 + bx + c$ for each set of solutions.

a. $x = -3$ and $x = -2$

b. $x = -1$ and $x = 2$

c. $x = 4$ and $x = -3$

d. $x = 5$ and $x = 2$

e. $x = \frac{1}{2}$ and $x = 5$

68. ***Area Puzzle.*** Trace the figure shown. Cut it out, and rearrange the pieces to answer these questions.

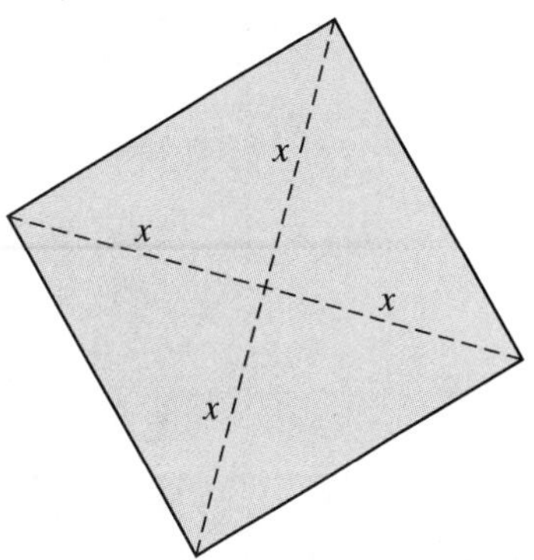

a. What is the area of the square in terms of x?

b. What is the length of a side of the square in terms of x?

69. ***Quadratic Formula.*** Not all quadratic equations factor easily. Some do not factor at all. When a quadratic equation is written in the form $ax^2 + bx + c = 0$, it may be solved for x with the quadratic formula. The quadratic formula is shown as two equations:

$$x = \frac{-b + \sqrt{b^2 - 4ac}}{2a} \quad \text{and}$$

$$x = \frac{-b - \sqrt{b^2 - 4ac}}{2a}$$

a. Use the quadratic formula on these equations to check that you are using it correctly. They are equations from earlier exercises.

$x^2 - 4x - 12 = 0$ Answer: $x = -2, x = 6$

$2x^2 - x - 3 = 0$ Answer: $x = 1.5, x = -1$

$4x^2 - 25 = 0$ Answer: $x = 2.5, x = -2.5$

b. Use the quadratic formula to solve these equations.

$2x^2 + 5x - 3 = 0$ $\qquad$ $3x^2 + 5x - 4 = 0$

$3x^2 - 2x - 4 = 0$ $\qquad$ $3x^2 - 2x - 3 = 0$

8.6 Solving Quadratic Equations with the Quadratic Formula

OBJECTIVES

- Simplify expressions based on the quadratic formula.
- Estimate the value of expressions containing radicals.
- Solve quadratic equations with the quadratic formula.
- Give reasons in a proof of the quadratic formula.

WARM-UP

Multiply these expressions.

1. $(2x - y)(2x - y)$
2. $(a + b)(a + b)$
3. $(ax + b)(ax + b)$
4. $(2ax + b)(2ax + b)$

Factor these expressions.

5. $x^2 + 2x + 1$
6. $x^2 - 2x + 1$
7. $x^2 - 1$
8. $x^2 - x - 1$
9. $x^2 + x + 1$
10. $x^2 - x - 2$

These radicals are between what two consecutive integers?

11. $\sqrt{5}$
12. $\sqrt{28}$
13. $\sqrt{84}$

THIS SECTION INTRODUCES the final method of solving quadratic equations: the quadratic formula. We review operations needed to simplify expressions obtained from the quadratic formula and solve equations with the quadratic formula.

The Quadratic Formula

Two expressions in the Warm-up could not be factored: $x^2 - x - 1$ and $x^2 + x + 1$. We explore $x^2 - x - 1$ further in Example 1 and $x^2 + x + 1$ in the Exercises.

EXAMPLE  Exploring solutions to quadratic equations Graph $y = x^2 - x - 1$. Solve the equation $x^2 - x - 1 = 0$ from your graph. Estimate solutions, if they exist.

Solution The graph is shown in Figure 31. The solutions are $x \approx -0.5$ and $x \approx 1.5$.

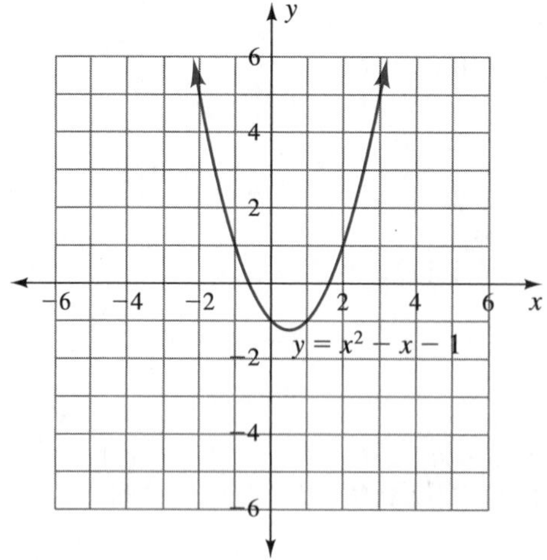

Figure 31

Our prior methods for solving quadratic equations don't work well—if at all—with the equation in Example 1. The expression does not factor. Graphing gives approximate results. The graph crosses the x-axis, but not at an integer. The expression has $b \neq 0$, so we cannot use the square root method.

To solve an equation like the one in Example 1, mathematicians developed *a more general procedure:* the **quadratic formula.** We may use the quadratic formula to solve for the variable x in any equation of the form $ax^2 + bx + c = 0$.

The quadratic formula may be shown as two equations:

Quadratic Formula

For $ax^2 + bx + c = 0$,

$$x = \frac{-b + \sqrt{b^2 - 4ac}}{2a} \quad \text{or} \quad x = \frac{-b - \sqrt{b^2 - 4ac}}{2a}$$

Two other useful forms of the quadratic formula are

$$x = \frac{-b \pm \sqrt{b^2 - 4ac}}{2a} \quad \text{and} \quad x = \frac{-b}{2a} \pm \frac{\sqrt{b^2 - 4ac}}{2a}$$

In both forms, the plus or minus sign ($\pm$) allows us to write two equations with one statement. In the second form, the fraction is separated into two parts over the common denominator, $2a$. The separation prevents many simplification errors. The second form also contains graphical information, which is explored in the project in Exercise 72.

Solving with the Quadratic Formula

To solve with the quadratic formula:

1. Place the equation in standard form, $ax^2 + bx + c = 0$.
2. Substitute a, b, and c into the quadratic formula.
3. Simplify the resulting expression, following the order of operations.
4. State and check the solutions.

In Example 2, we solve the equation from Example 1.

EXAMPLE 

Applying the quadratic formula Solve $x^2 - x - 1 = 0$ with the quadratic formula.

Solution We use the formula in standard form, with $a = 1$, $b = -1$, and $c = -1$:

$$x = \frac{-b \pm \sqrt{b^2 - 4ac}}{2a} \qquad \text{Substitute into the formula.}$$

$$x = \frac{-(-1) \pm \sqrt{(-1)^2 - 4(1)(-1)}}{2(1)} \qquad \text{Simplify.}$$

$$x = \frac{1 \pm \sqrt{1 + 4}}{2} \qquad \text{Write the two solutions.}$$

$$x = \frac{1 + \sqrt{5}}{2} \quad \text{or} \quad x = \frac{1 - \sqrt{5}}{2}$$

These solutions are irrational numbers. ●

Checking irrational-number solutions by substitution is beyond the level of this text, so we will simply estimate the values and see that the solutions agree with our estimates from the graph.

Estimation and Calculator Use

In Example 3, we will estimate and use a calculator to write the solutions without the radical notation.

EXAMPLE 3 Estimating and calculating irrational numbers Use an integer estimate for each radical, and estimate the value of the expression. Use a calculator to find the value of each expression to three decimal places.

a. $x = \dfrac{1 + \sqrt{5}}{2}$

b. $x = \dfrac{1 - \sqrt{5}}{2}$

Solution **a.** Since $\sqrt{5}$ is closest to 2, we replace the $\sqrt{5}$ with 2 and simplify:

$$x = \frac{1 + \sqrt{5}}{2}$$

$$x \approx \frac{1 + 2}{2} \approx 1.5$$

The answer is about 1.5, in agreement with the results in Figure 31, shown enlarged in Figure 32.

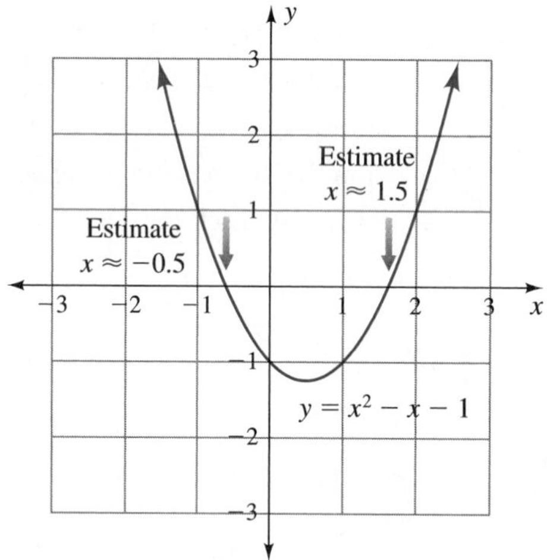

Figure 32

On a calculator, we place the numerator in parentheses and then divide by the denominator. It is important to do the entire expression at once:

$(1 + \sqrt{5}) \div 2$ [=]

The result is $x \approx 1.618$, which is close to our estimate of 1.5.
Warning: If your calculator automatically places an opening parenthesis, (, with the square root, you must have two closing parentheses after the 5:

$(1 + \sqrt{(5)}) \div 2$ [=]

Without this second closing parenthesis,

$(1 + \sqrt{(5)} \div 2 \approx 2.118.$

Our estimated result shows this value to be incorrect.

b. Taking $\sqrt{5}$ to be 2, we have

$$x = \frac{1 - \sqrt{5}}{2}$$

$$x \approx \frac{1 - 2}{2} \approx -0.5$$

The answer is about -0.5, in agreement with Figure 32.

On a calculator, $(1 - \sqrt{5}) \div 2 \approx -0.618$, which lies close to -0.5. Predicting the answer will again help us to reject the incorrect answer derived from omitting the second closing parenthesis after $\sqrt{5}$:

$$(1 - \sqrt{(5)} \div 2 \approx -0.118$$

●

Think about it 1: The calculator warning in Example 3 underscores the need to estimate results. If we omitted the parentheses around the numerator in the expressions in parts a and b of Example 3, incorrectly entering them as shown below, which of these incorrect calculator answers would pass and which would be rejected based on our estimate?

a. $1 + \sqrt{5} \div 2$ [=] **b.** $1 - \sqrt{5} \div 2$ [=]

Types of Solutions Found with the Quadratic Formula

In Examples 4 and 5, we return to solving equations with the quadratic formula. We examine the solutions to the equations and the graphs of the equations in more detail.

EXAMPLE 4 Solving equations by the quadratic formula Use the quadratic formula to solve these equations.

a. $2x^2 + 7x + 3 = 0$ **b.** $4x^2 = 5x + 2$ **c.** $2x^2 + 2x = -1$

Solution **a.** In $2x^2 + 7x + 3 = 0$, $a = 2$, $b = 7$, and $c = 3$.

$$x = \frac{-b \pm \sqrt{b^2 - 4ac}}{2a} = \frac{-7 \pm \sqrt{7^2 - 4(2)(3)}}{2(2)}$$

$$= \frac{-7 \pm \sqrt{49 - 24}}{4} = \frac{-7 \pm \sqrt{25}}{4}$$

The solution is

$$x = \frac{-7 + 5}{4} = \frac{-2}{4} = \frac{-1}{2} \quad \text{or} \quad x = \frac{-7 - 5}{4} = \frac{-12}{4} = -3$$

b. The equation $4x^2 = 5x + 2$ must be rearranged to equal zero. In $4x^2 - 5x - 2 = 0$, $a = 4$, $b = -5$, and $c = -2$.

$$x = \frac{-b \pm \sqrt{b^2 - 4ac}}{2a} = \frac{-(-5) \pm \sqrt{(-5)^2 - 4(4)(-2)}}{2(4)}$$

$$= \frac{-(-5) \pm \sqrt{25 + 32}}{8} = \frac{5 \pm \sqrt{57}}{8}$$

The solution is

$$x = \frac{5 + \sqrt{57}}{8} \approx 1.569 \quad \text{or} \quad x = \frac{5 - \sqrt{57}}{8} \approx -0.319$$

c. The equation $2x^2 + 2x = -1$ must be rearranged to equal zero. In $2x^2 + 2x + 1 = 0$, $a = 2$, $b = 2$, and $c = 1$.

$$x = \frac{-b \pm \sqrt{b^2 - 4ac}}{2a} = \frac{-2 \pm \sqrt{2^2 - 4(2)(1)}}{2(2)}$$

$$= \frac{-2 \pm \sqrt{4 - 8}}{4} = \frac{-2 \pm \sqrt{-4}}{4}$$

There are no real-number solutions because the square root of a negative number is not defined in the real numbers. Thus, we say "no real solution." ●

In each part of Example 4, the answer has a different form. Part a has rational solutions; we are able to find the exact square root of 25. The square root of 57 in part b cannot be expressed exactly by a decimal number, and so the solutions to the quadratic equation are irrational (not rational) and are approximated with decimals rounded to the nearest thousandth. As mentioned, the square root of -4 in part c is not a real number. In later courses, you will learn about other types of numbers that can be used to solve problems such as the one in part c.

We now examine the graphs related to the equations in Example 4.

EXAMPLE  Identifying solutions on graphs Graph these equations and identify the solutions to $y = 0$ on the graphs.

a. $y = 2x^2 + 7x + 3$ **b.** $y = 4x^2 - 5x - 2$ **c.** $y = 2x^2 + 2x + 1$

Solution The solutions to $y = 0$ are the x-intercepts.

a. The graph in Figure 33 confirms that the solutions to $2x^2 + 7x + 3 = 0$ are at $x = -\frac{1}{2}$ and $x = -3$.

b. The graph in Figure 34 confirms that one solution to $4x^2 - 5x - 2 = 0$ is near $x = 1\frac{1}{2}$ and the other is between $x = -\frac{1}{2}$ and $x = 0$.

c. The graph in Figure 35 does not cross the x-axis, so it has no x-intercepts. This confirms that the equation $2x^2 + 2x + 1 = 0$ has no real-number solutions.

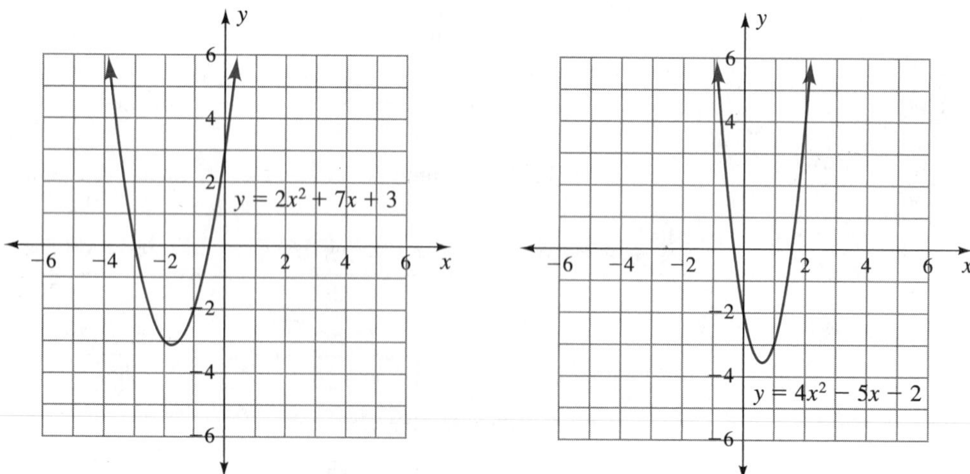

Figure 33

Figure 34

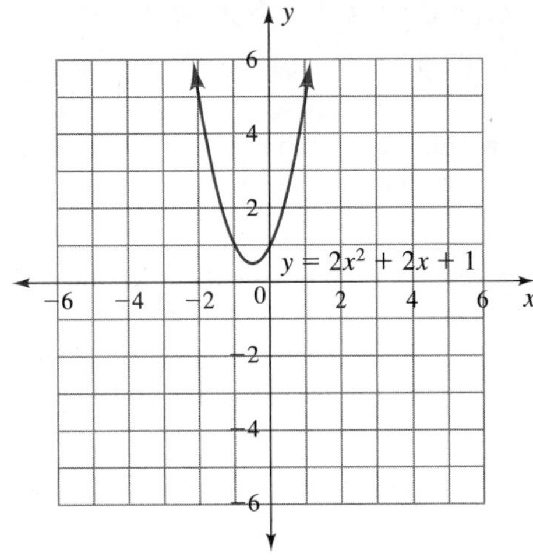

Figure 35

Think about it 2: What part of the quadratic formula shows whether there are no, one, or two solutions to a quadratic equation?

Proof of the Quadratic Formula

In the proof of the Pythagorean theorem, we squared a binomial, $(a + b)$, to obtain the perfect square trinomial, $a^2 + 2ab + b^2$. In proving the quadratic formula, we will build a perfect square trinomial and factor it back to the square of a binomial: $4a^2x^2 + 4abx + b^2 = (2ax + b)(2ax + b) = (2ax + b)^2$.

EXAMPLE 6 Proving the quadratic formula The steps below show how we obtain the quadratic formula from a quadratic equation in the form $ax^2 + bx + c = 0$. Explain what was done to obtain each step.

$$ax^2 + bx + c = 0$$

a. $4a^2x^2 + 4abx + 4ac = 0$

b. $4a^2x^2 + 4abx = -4ac$

c. $4a^2x^2 + 4abx + b^2 = b^2 - 4ac$

d. $(2ax + b)^2 = b^2 - 4ac$

e. Either $2ax + b = \sqrt{b^2 - 4ac}$ or $2ax + b = -\sqrt{b^2 - 4ac}$

f. Either $2ax = -b + \sqrt{b^2 - 4ac}$ or $2ax = -b - \sqrt{b^2 - 4ac}$

g. Either $x = \dfrac{-b + \sqrt{b^2 - 4ac}}{2a}$ or $x = \dfrac{-b - \sqrt{b^2 - 4ac}}{2a}$

Solution **a.** Multiply both sides of $ax^2 + bx + c = 0$ by $4a$.

b. Subtract $4ac$ from both sides of the equation.

c. Add b^2 to both sides. This creates a perfect square trinomial on the left.

d. Factor the left side.

e. Find the square root of both sides. Because $2ax + b$ may be either positive or negative, we obtain two equations.

f. Subtract b from both sides of each equation.

g. Divide by $2a$ in each equation.

Our purpose is not to justify why each step was done; rather, it is just to present a proof.

Deciding on a Solution Method

Regardless of which method you choose, you should either graph the equation or estimate answers so that you know your final results are sensible. Make a sketch showing the y-intercept and two or three other easy-to-find points, and use the sign on the coefficient of x^2 to decide whether the graph opens upward ($a > 0$) or downward ($a < 0$).

- Use a *graphing calculator graph* if the numbers in the equation are hard to work with or you need several digits in your answer. The graph makes it easy to find out if zero, one, or two answers are sensible.
- Use the *square root* if the equation is $ax^2 + c = 0$ (that is, if $b = 0$).
- Use *factoring* if it takes 30 seconds or less to factor the expression.
- Use the *quadratic formula* in all other cases.

ANSWER BOX

Warm-up: 1. $4x^2 - 4xy + y^2$ **2.** $a^2 + 2ab + b^2$ **3.** $a^2x^2 + 2abx + b^2$ **4.** $4a^2x^2 + 4abx + b^2$ **5.** $(x + 1)(x + 1)$ **6.** $(x - 1)(x - 1)$ **7.** $(x + 1)(x - 1)$ **8.** does not factor **9.** does not factor **10.** $(x + 1)(x - 2)$ **11.** $2 < \sqrt{5} < 3$ **12.** $5 < \sqrt{28} < 6$ **13.** $9 < \sqrt{84} < 10$ **Think about it 1: a.** $1 + \sqrt{5} \div 2 \approx 2.581$, which is not close to the estimate $x = 1.5$ and would be rejected **b.** $1 - \sqrt{5} \div 2 \approx -0.581$, which is close to the estimate $x = -0.5$ and might be accepted in error **Think about it 2:** The expression under the radical: $b^2 - 4ac$. See the project in Exercise 71.

EXERCISES 8.6

In Exercises 1 to 8, identify a, b, and c from the quadratic equations.

1. $9x^2 + 6x + 1 = 0$ **2.** $3x^2 - 7x + 2 = 0$

3. $3x^2 - 9x = 0$ **4.** $x^2 + 4 = 0$

5. $x^2 = 4x - 3$ **6.** $x^2 + 6 = 5x$

7. $x^2 = -9$ **8.** $2x^2 = 6x$

Simplify the expressions in Exercises 9 to 16 without a calculator.

9. $\dfrac{-(-4) - \sqrt{(-4)^2 - 4(1)(-12)}}{2 \cdot 1}$

10. $\dfrac{-(-4) + \sqrt{(-4)^2 - 4(1)(-12)}}{2 \cdot 1}$

11. $\dfrac{-(-5) + \sqrt{(-5)^2 - 4(6)(1)}}{2(6)}$

12. $\dfrac{-(-5) - \sqrt{(-5)^2 - 4(6)(1)}}{2(6)}$

13. $\dfrac{-1 - \sqrt{(1)^2 - 4(3)(-4)}}{2(3)}$

14. $\dfrac{-1 + \sqrt{(1)^2 - 4(3)(-4)}}{2(3)}$

15. $\dfrac{-6 + \sqrt{(6)^2 - 4(16)(-1)}}{2(16)}$

16. $\dfrac{-6 - \sqrt{(6)^2 - 4(16)(-1)}}{2(16)}$

Replace the radical with the nearest whole number and estimate the value of each expression in Exercises 17 to 24. Then use a calculator to approximate the value to three decimal places.

17. a. $\dfrac{2 - \sqrt{2}}{2}$ **b.** $\dfrac{3 + \sqrt{6}}{3}$ **c.** $\dfrac{3\sqrt{6}}{3}$

18. a. $\dfrac{3 - \sqrt{6}}{3}$ **b.** $\dfrac{2 + \sqrt{2}}{2}$ **c.** $\dfrac{2\sqrt{2}}{2}$

19. a. $\dfrac{3 - \sqrt{5}}{3}$ **b.** $\dfrac{2\sqrt{10}}{2}$ **c.** $\dfrac{2 + \sqrt{10}}{2}$

20. a. $\dfrac{3\sqrt{5}}{3}$ **b.** $\dfrac{2 - \sqrt{10}}{2}$ **c.** $\dfrac{3 + \sqrt{5}}{3}$

21. a. $\dfrac{-5 - \sqrt{13}}{2}$ **b.** $\dfrac{-9 + \sqrt{57}}{4}$

22. a. $\dfrac{-9 - \sqrt{57}}{4}$ **b.** $\dfrac{-5 + \sqrt{13}}{2}$

23. **a.** $\dfrac{-2 - \sqrt{28}}{6}$ **b.** $\dfrac{-2 + \sqrt{84}}{10}$

24. **a.** $\dfrac{7 + \sqrt{73}}{12}$ **b.** $\dfrac{7 - \sqrt{29}}{10}$

Solve the equations in Exercises 25 to 30 with the quadratic formula.

25. $4x^2 + 3x - 1 = 0$ **26.** $2x^2 + x - 6 = 0$

27. $7x^2 = 5x + 2$ **28.** $9x^2 = 2 - 3x$

29. $x = 2 - 10x^2$ **30.** $x = 2 - 6x^2$

The equations in Exercises 31 to 38 are in the form $h = at^2 + bt + c$. The input is the time t, in seconds. The output is the height h of an object tossed straight up from a height c, in feet, with an initial velocity b, in feet per second. Solve the equations for t, using the quadratic formula. Round to the nearest hundredth. Which answers are acceptable in the setting?

31. $-16t^2 + 30t + 150 = 0$

32. $-16t^2 + 20t + 50 = 0$

33. $-16t^2 + 15t + 150 = 0$

34. $-16t^2 + 25t + 100 = 0$

35. $-16t^2 + 30t + 150 = 200$

36. $-16t^2 + 15t + 150 = -50$

37. $-16t^2 + 20t + 50 = -40$

38. $-16t^2 + 25t + 100 = 150$

39. Solve the equation $x^2 + x + 1 = 0$ with the quadratic formula.

40. In Section 8.4 (page 517), we solved the equation $x^2 - 2x - 3 = 0$ and proved that the Pythagorean triple $\{3, 4, 5\}$ gave the only three consecutive numbers that satisfied the Pythagorean equation $a^2 + b^2 = c^2$ and made a right triangle. The triple $\{-1, 0, 1\}$ satisfied the equation but could not make a triangle. Solve the equation $x^2 - 2x - 3 = 0$ by factoring and by the quadratic formula.

Choose a reasonable method for solving each of the equations in Exercises 41 to 64. Round to three decimal places.

41. $2x^2 + x - 1 = 0$ **42.** $2x^2 + 3x - 2 = 0$

43. $3x^2 + x - 2 = 0$ **44.** $2x^2 + 2x + 1 = 0$

45. $x^2 + 2x + 2 = 0$ **46.** $3x^2 + 5x - 2 = 0$

47. $4x^2 = 10$ **48.** $9x^2 = 14$

49. $4x^2 - 9 = 0$ **50.** $9x^2 - 25 = 0$

51. $3x^2 + 2x = 6$ **52.** $2x^2 + 3x = 6$

53. $5x^2 + 4x + 1 = 0$ **54.** $7x^2 = 3 - 4x$

55. $5x^2 = 4 - 2x$ **56.** $5x^2 + 2x + 1 = 0$

57. $x^2 + 5 = -2x$ **58.** $x^2 + 4x = -5$

59. $x^2 + 9 = 6x$ **60.** $x^2 + 16 = 8x$

61. $4x^2 = 4x + 7$ **62.** $3x^2 + 4x = 7$

63. $4x^2 = 12x - 9$ **64.** $9x^2 + 4 = 12x$

65. Explain how to tell the number of solutions to $ax^2 + bx + c = 0$ from a graph of $y = ax^2 + bx + c$.

66. Explain how to estimate the square root of a number under 250.

67. Show the keystrokes for entering the quadratic formula for $2x^2 + 3x + 4$ into a calculator. Tell what calculator you use.

68. **a.** Compare the graphs for $y = x^2 - 4$ and $y = 4 - x^2$. Explain why the graphs are the same or different.

b. Solve $x^2 - 4 = 0$, and solve $4 - x^2 = 0$. Explain why the solutions are the same or different.

69. Write an equation that describes "The output is the product of two consecutive numbers." Solve your equation for $y = 0$. In what interval is the output negative?

70. Write an equation that describes "The output is the sum of the squares of two consecutive numbers." Solve your equation for $y = 0$. What does this say about the squares of real numbers?

Projects

71. ***The Discriminant.*** We can predict whether there will be no real solution, one real solution, or two real solutions to a quadratic equation from $b^2 - 4ac$, the discriminant.

a. Which part of $x = \dfrac{-b \pm \sqrt{b^2 - 4ac}}{2a}$ causes there to be two answers?

b. How many real numbers do we obtain from $\pm\sqrt{(\text{negative})}$?

c. From $\pm\sqrt{(\text{positive})}$?

d. From $\pm\sqrt{(\text{zero})}$?

e. How many real-number solutions will there be to a quadratic equation if $b^2 - 4ac = 0$?

f. If $b^2 - 4ac > 0$?

g. If $b^2 - 4ac < 0$?

h. If c is negative and a is positive?

i. If a and c are both positive and $4ac > b^2$?

Use the discriminant to find how many different real-number solutions there are to these equations:

j. $4x^2 + 4x + 1 = 0$ **k.** $5x^2 - 8x + 2 = 0$

l. $3x^2 - 9x = 0$ **m.** $x^2 + 4 = 0$

72. ***Quadratic Formula and Graphs.*** The highest or lowest point on the graph of a quadratic equation $y = ax^2 + bx + c$ is called the *vertex*. The quadratic formula in the form

$$x = \frac{-b}{2a} \pm \frac{\sqrt{b^2 - 4ac}}{2a}$$

shows why the vertex is symmetrically placed between the x-intercepts. The distance between the axis of symmetry and each intercept is

$$\frac{\sqrt{b^2 - 4ac}}{2a}, \quad a > 0$$

as shown in the figure. The axis of symmetry, which passes through the vertex V, has equation

$$x = \frac{-b}{2a}$$

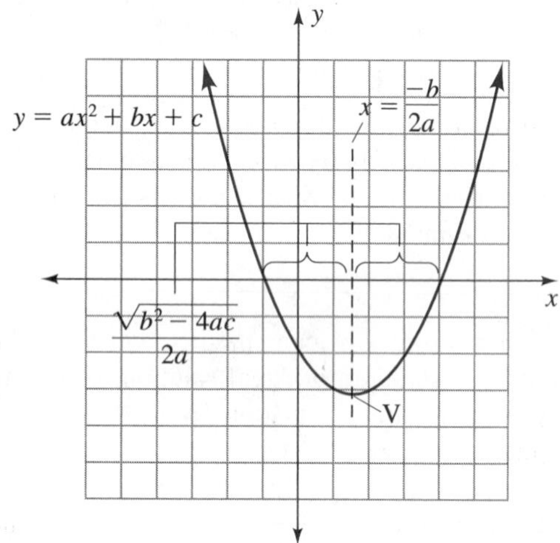

Using either the x-intercepts or the axis of symmetry, find the coordinates of the vertex for each quadratic equation:

a. $y = x^2 + 6x - 16$

b. $y = 2x^2 + 3x - 4$

c. $y = -16t^2 + 30t + 150$

d. $y = -16t^2 + 20t + 50$

e. If the graphs for parts c and d represent the height of a ball at time t, what is the meaning of the vertex?

f. Find the coordinates of the vertex for each part in Example 5. Check that each vertex agrees with the graphs.

73. ***Lines and Points***

a. Complete this table for the number of distinct straight lines that can pass through a given number of points.

Number of points, x	Number of lines, y
1	0
2	1
3	
4	6
5	
6	
7	
8	

b. What is the pattern in the outputs?

c. What is the input-output rule (in words if not an equation)? (*Hint:* The rule is quadratic.)

8.7 Range, Box and Whisker Plots, and Standard Deviation

OBJECTIVES

- Find the range from a set of data.
- Calculate the median and quartiles of a set of data, and summarize with a box and whisker plot.
- Calculate the standard deviation for a set of data.

WARM-UP

Exercises 1 to 4 each contain a list of measurements of students' drawings of a 3-inch line. Calculate the mean and median of each set. As needed, review the calculation of averages from Section 5.4.

1. 1.25, 1.25, 2.25, 2.75, 7.5

2. 6, 3.50, 0.75, 1.75, 2.50, 3.50

3. 4, 2, 3, 2.25, 3.75

4. 3.25, 3.25, 3.25, 2, 3.25

Calculate the distance between these points.

5. (3, 7) and (−2, −5)

6. (−3, 5) and (5, −10)

Without any rulers or coins, draw the following.

7. A line that you estimate to be 3 inches long

8. A line that you estimate to be 4 centimeters long

9. A circle that you estimate to be the size of a dime

10. A circle that you estimate to be the size of a quarter

When everyone has finished drawing, take out a ruler and coins and measure the lengths of the lines and the diameters of the circles. Measure lines to the nearest quarter inch and circles to the nearest eighth inch.

SECTION 5.4 INTRODUCED the mean, median, and mode. These calculations are called *measures of central tendency* because they indicate the average, middle, or center of a set of data.

In this section, we denote the mean with $\bar{x}$.

This section introduces ways you might find out how well you have drawn your lines and circles compared with others in the class. We will examine **measures of variation or dispersion**—that is, *measures that describe how close to the middle or how scattered a set of data is.* In studying measures of dispersion, we will consider range, box and whisker plots, and standard deviation.

Range

Exercises 1 to 4 in the Warm-up describe results when students drew 3-inch lines. The four sets have the same mean; that is, when we added the numbers and divided by the number of numbers, we obtained 3. (Again, the mean is denoted with $\bar{x}$.) The median, or middle number, in each set is also close to 3. If we looked only at the mean and median, we might conclude that the students did very well at drawing a 3-inch line. But the mean and median may be deceptive, so it is important to consider other facts as well in summarizing the data.

Finding the range is a first step in determining how sets of numbers differ. The range is calculated by subtracting the lowest number from the highest number.

Range

The **range** is the difference between the largest and the smallest number in a data set:

$$\text{Range} = \text{maximum number} - \text{minimum number}$$

EXAMPLE  Finding range As we saw in the Warm-up, the mean for each set of measurements is 3. Find the range of each set.

a. 1.25, 1.25, 2.25, 2.75, 7.50

b. 6, 3.50, 0.75, 1.75, 2.50, 3.50

c. 4, 2, 3, 2.25, 3.75

d. 3.25, 3.25, 3.25, 2, 3.25

Solution **a.** The range is $7.50 - 1.25 = 6.25$ in.

b. The range is $6 - 0.75 = 5.25$ in.

c. The range is $4 - 2 = 2$ in.

d. The range is $3.25 - 2 = 1.25$ in. ●

Although the sets all have the same mean, their ranges differ considerably. The range indicates the spread of the data, which is considerably greater in parts a and b than in parts c and d. The student measures in parts c and d are more accurate.

Box and Whisker Plots

The median, as found in the Warm-up, is used with the range to draw a **box and whisker plot**, which gives a visual summary of the data.

Drawing a Box and Whisker Plot

1. Find the median—the middle number when data are arranged in smallest to largest order. If there is no single middle number, average the two middle numbers. Locate the median on a horizontal line with a scale appropriate to the data set (see the figure).

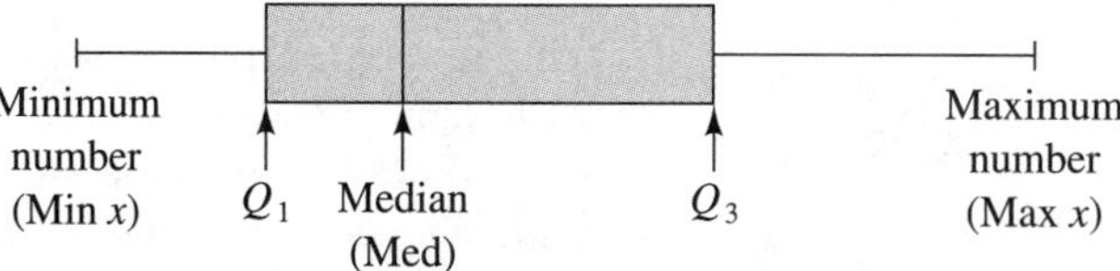

2. Find the **quartiles**, Q_1 and Q_3, as follows: Q_1 is *the middle number of the numbers below the median.* Q_3 is *the middle number of the numbers above the median.* If there is no single middle number, average the two middle numbers. Draw a box from Q_1 to Q_3, with length to scale.
3. Draw a line from the left end of the box to the minimum number. Draw another line from the right end of the box to the maximum number. These lines form the whiskers.

EXAMPLE 2 Building a box and whisker plot Calculate Q_1 and Q_3 for each data set in Example 1. Draw a box and whisker plot for each set.

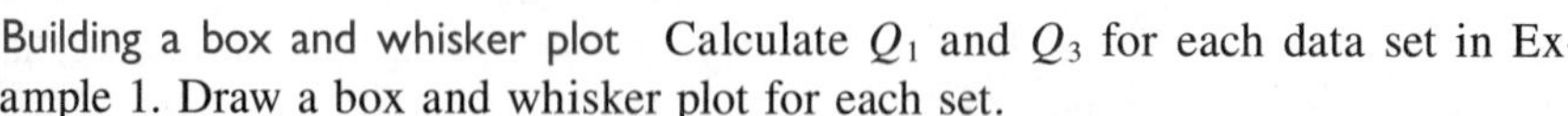

Solution **a.** In the set {1.25, 1.25, 2.25, 2.75, 7.50}, the median is 2.25. Q_1 is 1.25, the average of the two numbers smaller than the median: 1.25 and 1.25. Q_3 is 5.125, the average of the two numbers larger than the median: 2.75 and 7.5.

The box and whisker plot is in Figure 36. Q_1 and the minimum are at the same point, so there is no whisker on the left.

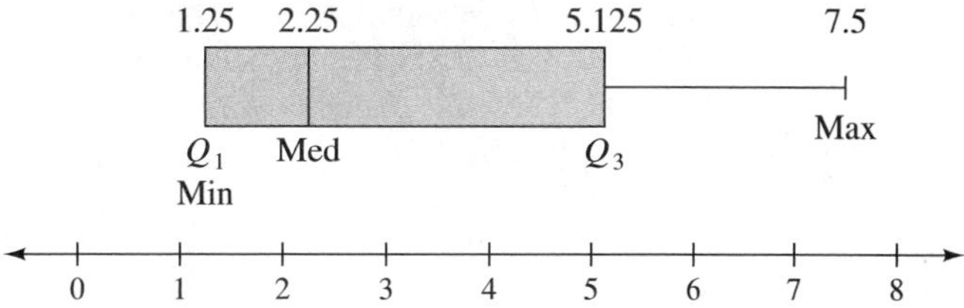

Figure 36

b. In the set {0.75, 1.75, 2.50, 3.50, 3.50, 6}, the median is 3, the average of 2.50 and 3.50. Q_1 is 1.75, the middle number of the three below the median. Q_3 is 3.50, the middle number of the three above the median. The box and whisker plot is in Figure 37.

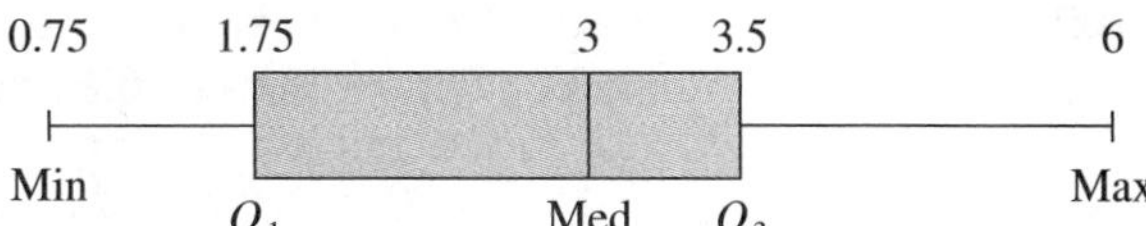

Figure 37

c. In the set {2, 2.25, 3, 3.75, 4}, the median is 3. Q_1 is 2.125, the average of the two numbers smaller than the median: 2 and 2.25. Q_3 is 3.875, the average of the two numbers larger than the median: 3.75 and 4. The box and whisker plot is in Figure 38.

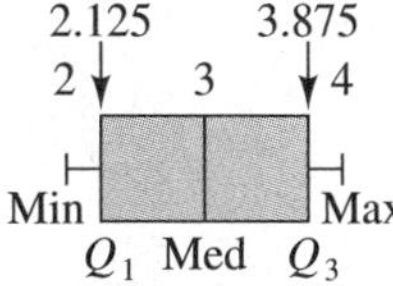

Figure 38

d. In the set {2, 3.25, 3.25, 3.25, 3.25}, the median is 3.25. Q_1 is 2.625, the average of the two numbers smaller than the median: 2 and 3.25. Q_3 is 3.25, the average of the two numbers larger than the median: 3.25 and 3.25. This time, the median, Q_3, and the maximum are all the same, which creates a short box plot with no whisker showing on the right. The box and whisker plot is in Figure 39.

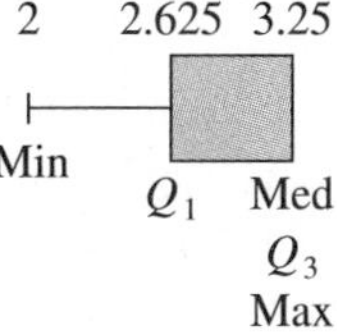

Figure 39

Standard Deviation

The box and whisker plot indicates visually where the data are located. A disadvantage of the box and whisker plot is that it does not give us a numerical description. One way to find a numerical measure of the spread of the data is to consider the differences between each data point, x_i, and the mean, $\bar{x}$. *The difference between each number and the mean* is called the **deviation**.

In developing the formula for what British mathematician Karl Pearson was to call "standard deviation" in 1894, statisticians may have thought about how we find the distance between two points. The distance between (x_1, y_1) and (x_2, y_2) is given by

$$d = \sqrt{(x_2 - x_1)^2 + (y_2 - y_1)^2}$$

The distance formula finds the deviation between the x-coordinates and between the y-coordinates. To assure that the distances are positive, we square the deviations and then add. To compensate for squaring the deviations, we then take the square root of the sum to obtain the distance. Observe the similarity between the distance formula and the formula for the **sample standard deviation.**

Sample Standard Deviation

For a set of n numbers $x_1, x_2, x_3, \ldots, x_n$ with mean $\bar{x}$, drawn randomly from a population, the **sample standard deviation** is

$$s_x = \sqrt{\frac{(x_1 - \bar{x})^2 + (x_2 - \bar{x})^2 + (x_3 - \bar{x})^2 + \cdots + (x_n - \bar{x})^2}{n - 1}}$$

The division by $n - 1$ in the sample standard deviation formula creates an "average" deviation.

EXAMPLE 3

Finding the standard deviation Find the sample standard deviation for the set of measures in part a of Example 1, {1.25, 1.25, 2.25, 2.75, 7.5}, with the mean $\bar{x} = 3$.

Solution Standard deviation, s_x

$$= \sqrt{\frac{(1.25 - 3)^2 + (1.25 - 3)^2 + (2.25 - 3)^2 + (2.75 - 3)^2 + (7.5 - 3)^2}{5 - 1}}$$

$$= \sqrt{\frac{(-1.75)^2 + (-1.75)^2 + (-0.75)^2 + (-0.25)^2 + (4.5)^2}{4}}$$

$$= \sqrt{\frac{27}{4}} \approx 2.6$$

There is *another form of the standard deviation used when the data represent the entire population,* such as the set of all ages of the presidents of the United States: the **population standard deviation,** σ_x. The Greek letter sigma (σ) is used in the population standard deviation. In the σ_x formula, n replaces the $n - 1$ used in the sample standard deviation formula. The sample standard deviation, s_x, is more commonly used, because our data set is usually not the *population* (the set of all possible outcomes available). Example 4 illustrates the population standard deviation, but with the data from Example 3 so that you can compare results.

EXAMPLE 4

Finding the population standard deviation Find the population standard deviation, σ_x, for the set of measures in Example 3.

Solution For the population standard deviation, we use n instead of $n - 1$ in the formula and divide by 5 instead of 4.

Population standard deviation, σ_x

$$= \sqrt{\frac{(x_1 - \overline{x})^2 + (x_2 - \overline{x})^2 + (x_3 - \overline{x})^2 + \cdots + (x_n - \overline{x})^2}{n}}$$

$$= \sqrt{\frac{(1.25 - 3)^2 + (1.25 - 3)^2 + (2.25 - 3)^2 + (2.75 - 3)^2 + (7.5 - 3)^2}{5}}$$

$$= \sqrt{\frac{(-1.75)^2 + (-1.75)^2 + (-0.75)^2 + (-0.25)^2 + (4.5)^2}{5}}$$

$$= \sqrt{\frac{27}{5}} \approx 2.32$$

Student Note: Unless σ_x is specifically requested in a problem, use the sample standard deviation, s_x.

Our purpose here is not to perform tedious calculations with numbers, but rather to understand how the results are found. In Example 5, we practice finding standard deviations by finding the standard deviations for Exercises 1 to 4 in the Warm-up. Graphing calculators and some scientific calculators will calculate the standard deviation.

Graphing Calculator Technique

The mean, the standard deviation, and the box and whisker inputs may be calculated with the statistical function on a graphing calculator.

Enter the set of data into the LIST function, usually in [STAT] 1. If there are already data in the list, clear them. Get the calculation of one-variable statistics. Q_1, the median, and Q_3 for the box and whisker plot may be listed after the standard deviations.

EXAMPLE 

Finding standard deviations Use a calculator to find the population standard deviation and the sample standard deviation for each of the four sets of measures in the Warm-up. The mean for each set is $\overline{x} = 3$.

Solution

a. The population standard deviation for $\{1.25, 1.25, 2.25, 2.75, 7.50\}$ is $\sigma_x \approx 2.32$. The sample standard deviation is $s_x \approx 2.60$, as described in Examples 3 and 4.

b. For $\{0.75, 1.75, 2.50, 3.50, 3.50, 6\}$, the population standard deviation is $\sigma_x \approx 1.65$. The sample standard deviation is $s_x \approx 1.81$.

c. For $\{2, 2.25, 3, 3.75, 4\}$, $\sigma_x \approx 0.79$ and $s_x \approx 0.88$.

d. For $\{2, 3.25, 3.25, 3.25, 3.25\}$, $\sigma_x = 0.5$ and $s_x \approx 0.56$.

Note that the standard deviation is larger for widely scattered data than for data that are close together.

Because we divide by a smaller number ($n - 1$ instead of n), the sample standard deviation gives a larger number than the population standard deviation. The subtle distinctions that determine when to use each form of the standard deviation are presented in full in a statistics course; they will not be discussed here.

Process Control

Student Note: In manufacturing, standard deviation is often abbreviated s.d.

The standard deviation from the mean is an important tool in manufacturing process control. Figure 40 shows a plot of the four sets of measures in the Warm-up. The data are indicated by dots above the line. The standard deviation for the combined data, s.d. ≈ 1.5, is shown below the line.

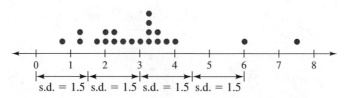

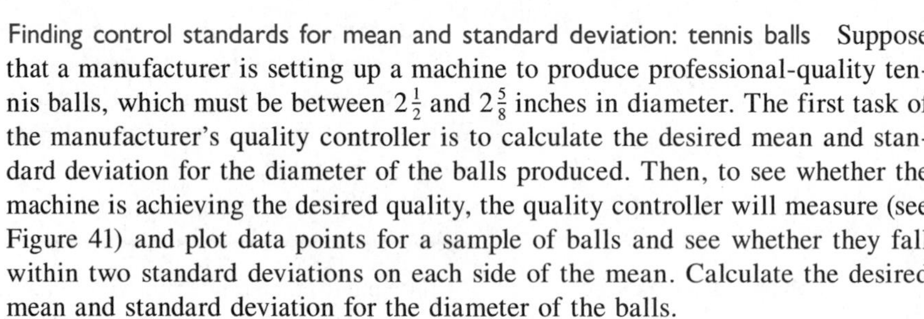

Figure 40

Observe that all data points but one are within two standard deviations of the mean, 3. It is common for data to fall within two standard deviations on each side of the mean; thus, control processes are often designed to find whether sample measurements fit within two standard deviations of the mean.

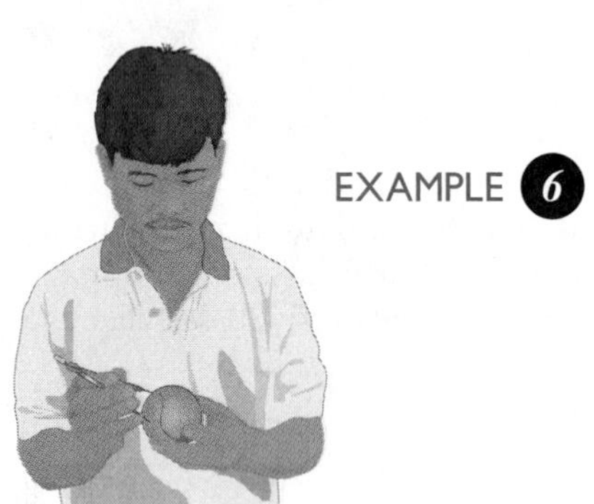

Figure 41

EXAMPLE 6 Finding control standards for mean and standard deviation: tennis balls Suppose that a manufacturer is setting up a machine to produce professional-quality tennis balls, which must be between $2\frac{1}{2}$ and $2\frac{5}{8}$ inches in diameter. The first task of the manufacturer's quality controller is to calculate the desired mean and standard deviation for the diameter of the balls produced. Then, to see whether the machine is achieving the desired quality, the quality controller will measure (see Figure 41) and plot data points for a sample of balls and see whether they fall within two standard deviations on each side of the mean. Calculate the desired mean and standard deviation for the diameter of the balls.

Solution The desired mean, halfway between $2\frac{1}{2}$ (2.5) and $2\frac{5}{8}$ (2.625), is

$$\frac{2.5 + 2.625}{2} = 2.5625$$

A distance of 0.0625 is available on each side of the mean to contain two standard deviations. Thus, half of 0.0625, or 0.03125, is the desired standard deviation. (See Figure 42.)

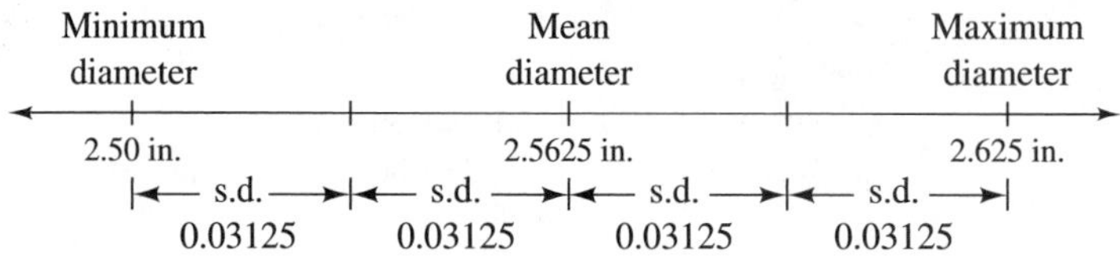

Figure 42

ANSWER BOX

Warm-up 1. 3, 2.25 **2.** 3, 3 **3.** 3, 3 **4.** 3, 3.25 **5.** 13 **6.** 17

EXERCISES 8.7

Calculate the mean and the median of each set of incomes in Exercises 1 to 4.

1. \$8000, \$10,000, \$12,000, \$13,000, \$13,000, \$100,000

2. \$4000, \$8000, \$9000, \$9000, \$100,000

3. \$20,000, \$27,500, \$27,500, \$27,500, \$27,500

4. \$23,000, \$25,000, \$26,000, \$27,000, \$29,000

In Exercises 5 to 8, find Q_1 and Q_3 for each set and draw a box and whisker plot.

5. Data from Exercise 1

6. Data from Exercise 2

7. Data from Exercise 3

8. Data from Exercise 4

In Exercises 9 and 10, the numbers represent thousands of dollars. Round to the nearest hundred dollars.

9. Find the standard deviation for the data in Exercise 1, using the following equation:

$$s_x = \sqrt{\frac{(8-26)^2+(10-26)^2+(12-26)^2+(13-26)^2+(13-26)^2+(100-26)^2}{6-1}}$$

10. Find the standard deviation for the data in Exercise 2, using the following equation:

$$s_x = \sqrt{\frac{(4-26)^2+(8-26)^2+(9-26)^2+(9-26)^2+(100-26)^2}{5-1}}$$

11. Using a calculator, find the mean, population standard deviation (σ_x), and sample standard deviation (s_x) for the data in Exercise 3. Round to the nearest dollar.

12. Using a calculator, find the mean, population standard deviation (σ_x), and sample standard deviation (s_x) for the data in Exercise 4. Round to the nearest dollar.

The average price of a home in one city in February 1995 was $107,761. This average was calculated from sales of 265 homes. The median home price was $96,500. Use this information in Exercises 13 to 16.

13. Why might the average be larger than the median?

14. What was the total value of homes sold?

15. Are we able to calculate the standard deviation from the given information?

16. How many homes sold for less than $96,500?

17. What might account for the variability in students' drawings of 13-inch lines in the Warm-up?

18. Why might some students have drawn a 4-centimeter line more accurately than a 3-inch line in the Warm-up?

19. Describe an experiment to find out whether drawings would improve if students were given a second opportunity to draw a 3-inch line in the Warm-up.

20. Describe an experiment to find out whether students are better able to draw a circle the size of a dime if they are given a nickel for reference. What results would you expect?

In Exercises 21 to 25, suppose the process is in normal operation if the current output or reading is within two standard deviations. For the mean and standard deviation given, calculate the outputs for which the process is normal.

21. Ceramic furnace temperature:
mean = 3500°F, s.d. = 15°F

22. Room heating control: mean = 20°C, s.d. = 0.75°C

23. Milling machine control: mean diameter = 6.075 cm, s.d. = 0.013 cm

24. Pressure control: mean = 84.9 psi, s.d. = 0.28 psi

25. Waiting line time: mean = 2.5 min, s.d. = 0.75 min

Some controls may be set for normal operation at other than two standard deviations from the mean. Exercises 26 to 29 give the mean, standard deviation, and condition for normal operation for several processes. Give the outputs for which the process is normal.

26. Steam flow: mean = 3.58 lb/sec, s.d. = 0.04 lb/sec, condition = ±1 s.d.

27. Air pollution: mean = 0.3 ppm (parts per million), s.d. = 0.06 ppm, condition = ±1.5 s.d.

28. Concrete strength: mean = 3000 psi, s.d. = 28 psi, condition = ±2.5 s.d.

29. Gasoline flow: mean = 48.4 gal/min, s.d. = 0.7 gal/min, condition = ±0.5 s.d.

Projects

30. ***Penny Plot***

a. Plot the data given below, with date on the horizontal axis. (The first number is the date; the second is the weight of the penny in grams.)

1983D, 2.501; 1994D, 2.510; 1969S, 3.161; 1982D, 2.518; 1972D, 3.107; 1964, 3.070; 1974, 3.130; 1967, 3.135; 1994D, 2.497; 1968D, 3.085; 1960D, 3.111; 1966, 3.100; 1977D, 3.084; 1963, 3.078; 1981D, 3.051; 1985D, 2.515; 1984D, 2.515; 1988D, 2.548; 1984D, 2.628; 1989D, 2.440; 1962, 3.037; 1973D, 3.134; 1979, 3.055; 1970, 3.140; 1991D, 2.538; 1978D, 3.100; 1980D, 3.119

b. What do you observe from your graph? Find a way to use the mean and standard deviation to justify your observation.

31. ***Drawing to Measure.*** Do the experiment you described in Exercise 19 or 20. See Exercise 32 for suggestions on writing up your experiment.

32. ***Learning a Skill.*** Design an experiment based on improving how someone learns a skill. Write out a plan for your experiment and discuss it with your instructor. In your plan, state your idea about the outcome and describe your experiment. Carry out the experiment. Calculate approximate measures. Summarize the results with graphs or charts. State any conclusions.

CHAPTER 8 SUMMARY

Vocabulary

For definitions and page references, see the Glossary/Index.

- axis of symmetry
- box and whisker plot
- converse of the Pythagorean theorem
- deviation
- distance formula
- extraneous roots
- irrational number
- measures of variation or dispersion
- one-half as exponent
- parabola
- perfect squares
- population standard deviation
- principal square root
- proof
- Pythagorean theorem
- Pythagorean triples
- quadratic formula
- quartiles
- radicals
- radicand
- range
- sample standard deviation
- vertex
- zero product rule

Concepts

8.1 Pythagorean Theorem

Pythagorean theorem: If a triangle is a right triangle, then the sum of the squares of the two shortest sides (legs) is equal to the square of the longest side (hypotenuse). If the legs are of lengths a and b and the hypotenuse is of length c, then $a^2 + b^2 = c^2$.

Converse of the Pythagorean theorem: If the squares of the sides of a triangle satisfy $a^2 + b^2 = c^2$, the triangle is a right triangle.

Two common Pythagorean triples are $\{3, 4, 5\}$ and $\{5, 12, 13\}$.

8.2 Square Root Expressions and Their Applications, Including the Distance Formula

The principal square root of a number is the positive root.

The square root of a negative is undefined in the set of real numbers.

A negative root or both positive and negative roots of a number are given only when specified with $-$ or $\pm$, respectively.

No portion of the graph of the square root function lies to the left of the vertical axis because the square root of a negative number is not a real number. No portion of the graph is below the x-axis because the output to the square root function is the principal, or positive, square root.

When one-half is used as an exponent, take the principal square root of the base:

$$a^{1/2} = \sqrt{a}, \quad a \geq 0$$

The square root property for products is given by

$$\sqrt{a \cdot b} = \sqrt{a} \cdot \sqrt{b}$$

The square root property for quotients is given by

$$\sqrt{\frac{a}{b}} = \frac{\sqrt{a}}{\sqrt{b}}$$

Use absolute value to assure a positive root:

$$\sqrt{a^2} = |a|, \quad \sqrt{a^6} = |a^3|$$

A simplified square root expression contains no perfect square factors inside the radical sign.

Distance formula: The distance between (x_1, y_1) and (x_2, y_2) is given by

$$d = \sqrt{(x_2 - x_1)^2 + (y_2 - y_1)^2}$$

8.3 Simplifying Expressions and Solving Square Root Equations

The function $f(x) = \sqrt{x^2}$ is the same as the absolute value function, $f(x) = |x|$.

If x is any real number,

$$\sqrt{x^2} = |x|$$
$$\sqrt{x^4} = x^2$$
$$\sqrt{x^6} = |x^3|$$
$$\sqrt{x^8} = x^4$$

If x is a zero or a positive number ($x \geq 0$),

$$\sqrt{x^2} = x$$
$$\sqrt{x^4} = x^2$$
$$\sqrt{x^6} = x^3$$
$$\sqrt{x^8} = x^4$$

To find the domain for a radical expression, limit the inputs to those which make the radicand zero or positive.

We can solve radical equations by table, graph, or squaring both sides.

8.4 Graphing and Solving Quadratic Equations

We can solve quadratic equations by table and graph (Section 8.4), taking the square root and factoring (Section 8.5), or the quadratic formula (Section 8.6).

The intercepts, vertex, and axis of symmetry are usually all you need to sketch the graph of a quadratic equation, $y = ax^2 + bx + c$. If a, the coefficient on x^2, is positive, the graph opens upward; if a is negative, the graph opens downward.

8.5 Solving Quadratic Equations by Taking the Square Root and Factoring

So long as we observe the restriction that the radicand is zero or positive and the output is zero or positive, we can take the square root of both sides of an equation.

The zero product rule permits us to set the factors on the left side of $ax^2 + bx + c = 0$ equal to zero. It may not always be possible to factor the left side.

8.6 Solving Quadratic Equations with the Quadratic Formula

The quadratic formula, $x = \dfrac{-b \pm \sqrt{b^2 - 4ac}}{2a}$, solves

$$ax^2 + bx + c = 0$$

To solve with the quadratic formula,

1. Place the equation in standard form, $ax^2 + bx + c = 0$.
2. Substitute a, b, and c into the quadratic formula.
3. Simplify the resulting expression, following the order of operations.
4. State and check the solutions.

See page 536 for hints on deciding on a solution method.

8.7 Range, Box and Whisker Plots, and Standard Deviation

The mean, median, and mode are measures of central tendency (Section 5.4). In this section, the mean is denoted by $\bar{x}$. The topics of this section are measures of variation or dispersion.

The range is the difference between the largest and smallest numbers in a data set.

To draw a box and whisker plot, find the median (after arranging the data in order from smallest to largest). Locate the median on a horizontal line with a scale appropriate to the data set. The quartiles are the midpoints of the first half and of the second half of the data. Locate the quartiles, Q_1 and Q_3, on the line. The box extends from Q_1 to Q_3. The whiskers extend out of the box to the minimum and maximum data points.

For a set of n numbers $x_1, x_2, x_3, \ldots, x_n$ with mean $\bar{x}$, drawn randomly from a population, the **sample standard deviation** is

$$s_x = \sqrt{\frac{(x_1 - \bar{x})^2 + (x_2 - \bar{x})^2 + (x_3 - \bar{x})^2 + \cdots + (x_n - \bar{x})^2}{n - 1}}$$

The sample standard deviation, s_x, contains division by $n - 1$. The population standard deviation, σ_x, contains division by n. The sample standard deviation is larger and considered more conservative. On calculators, the sample mean, $\bar{x}$, is used to calculate both deviations.

CHAPTER 8 REVIEW EXERCISES

1. Which of these sets of numbers represent the sides of a right triangle?

   a. {2, 3, 4} **b.** {5, 17, 18} **c.** {8, 15, 17}

2. Find the length of the side marked with an x.

   a. 7, x, 4

   b.

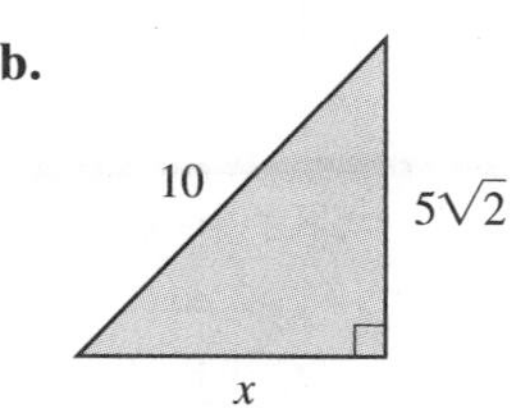

   c.

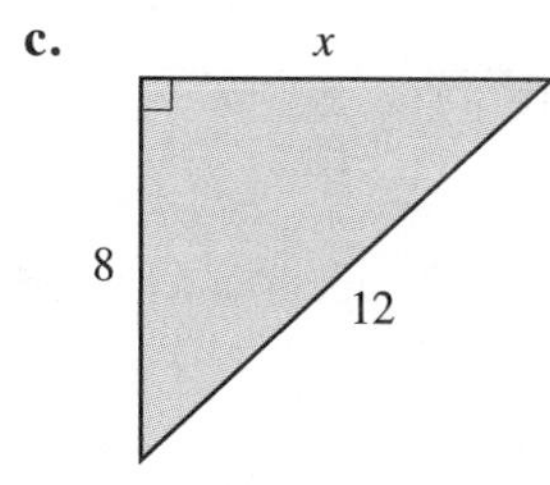

   d.

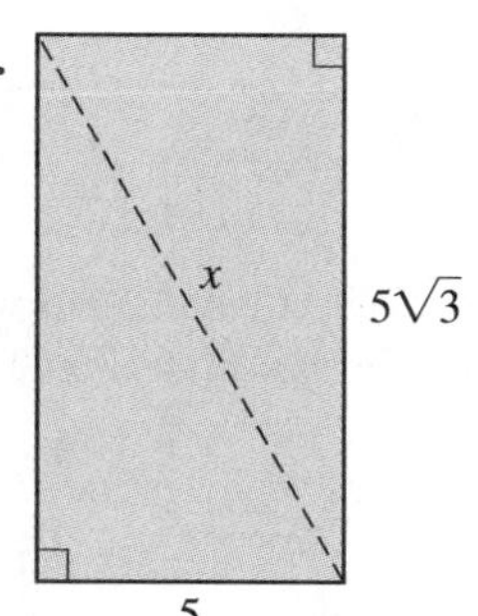

3. An extension ladder is to reach 24 feet up a wall. The safe ladder position for the base is 6 feet from the wall. How long a ladder is needed?

4. Simplify these exponent and radical expressions.

   a. $\sqrt{6} \cdot \sqrt{24}$ **b.** $(5\sqrt{2})^2$ **c.** $\sqrt{72} \cdot \sqrt{2}$

   d. $(2\sqrt{5})^2$ **e.** $\sqrt{3} \cdot \sqrt{12}$

5. Which of the three expressions in each set has the same value as the radical given first?

   a. $\sqrt{60}$ $\{6\sqrt{10}, 10\sqrt{6}, 2\sqrt{15}\}$

   b. $\sqrt{63}$ $\{7\sqrt{3}, 9\sqrt{7}, 3\sqrt{7}\}$

   c. $\sqrt{54}$ $\{6\sqrt{3}, 3\sqrt{6}, 9\sqrt{6}\}$

Use the definitions of exponents to simplify each expression in Exercises 6 to 10. Try them without a calculator first.

6. **a.** 49^{-1} **b.** $49^{1/2}$ **c.** $49^{0.5}$ **d.** 49^0
7. **a.** $144^{1/2}$ **b.** 144^{-1} **c.** 144^0 **d.** $144^{0.5}$
8. **a.** $\left(\frac{1}{25}\right)^{-1}$ **b.** $\left(\frac{1}{25}\right)^{1/2}$ **c.** $\left(\frac{1}{25}\right)^0$ **d.** $\left(\frac{1}{25}\right)^{0.5}$
9. **a.** $(0.36)^{1/2}$ **b.** $(0.36)^{-1}$ **c.** $(0.36)^{0.5}$ **d.** $(0.36)^0$
10. **a.** $(0.25)^{-1}$ **b.** $(0.25)^0$ **c.** $(0.25)^{1/2}$ **d.** $(0.25)^{0.5}$

11. Use the distance formula to find the lengths of the sides and the diagonals of the four-sided shapes below. (These figures are not drawn to scale.) If the diagonals are equal, then the shape is a rectangle (or square). Which are rectangles?

a.

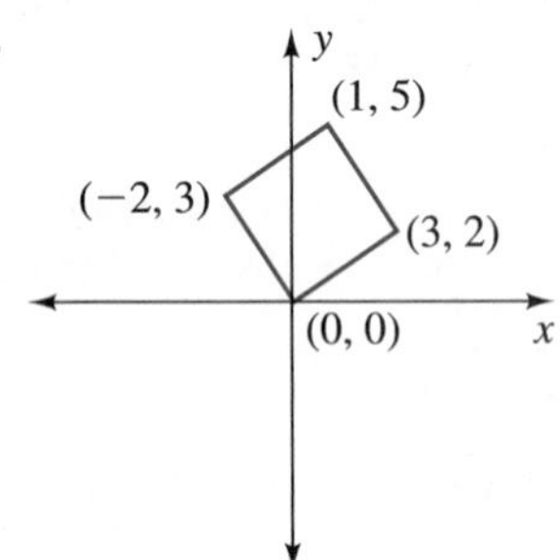

b.

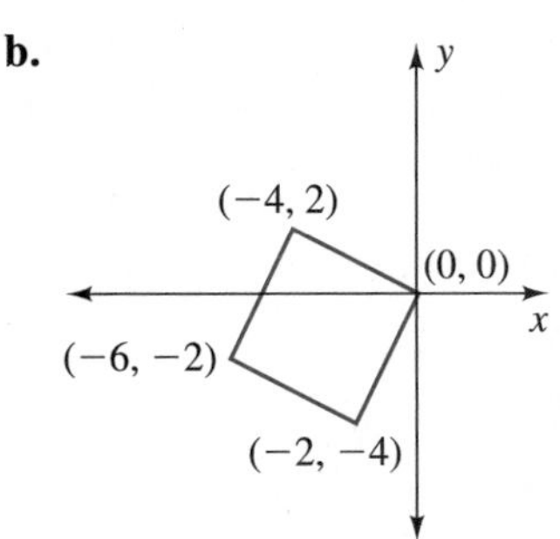

c.

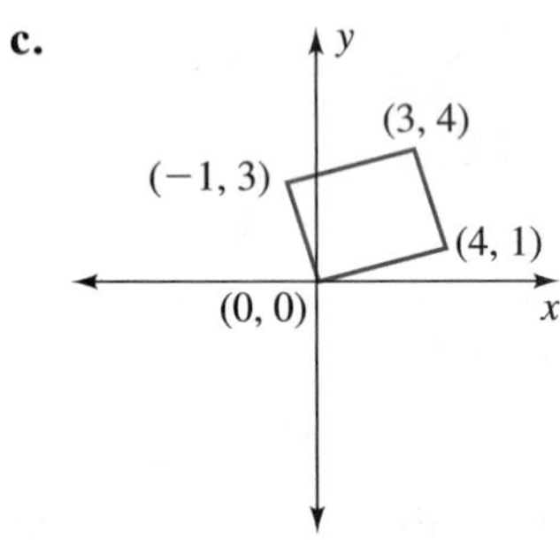

12. The sides of a square or rectangle are perpendicular. Find the slopes of two consecutive sides of each shape in Exercise 11. Verify that the two rectangles do have two consecutive perpendicular sides and the other shape does not.

13. Simplify these expressions. Assume that the variables represent only positive numbers.

a. $\sqrt{25x^2y^4}$ b. $\sqrt{169x^6y^2}$

c. $\sqrt{2.25a^3}$ d. $\sqrt{0.64b^5}$

e. $\sqrt{\dfrac{80x^3}{5x}}$ f. $\sqrt{\dfrac{3a^4}{27b^6}}$

g. $\sqrt{\dfrac{192a^6}{3}}$ h. $\sqrt{\dfrac{121}{49b^4}}$

14. Repeat Exercise 13 for variables representing any real number.

15. The f-stops on a camera lens are the numbers 1.4, 2, 2.8, 4, 5.6, 8, 11, and 16. Complete the table, and compare the results with the f-stops.

n	1	2	3	4	5	6	7	8
$(\sqrt{2})^n$								

16. Solve the equations from the graphs. How can we determine which inputs make the radical expressions undefined?

a. $\sqrt{5x+1} = 11$

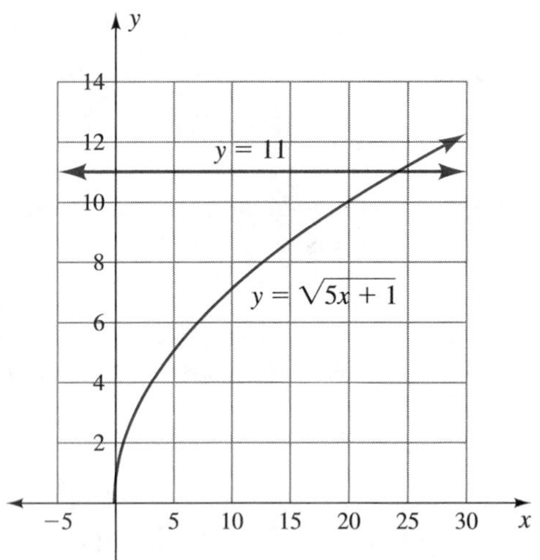

b. $\sqrt{5x-1} = 7$

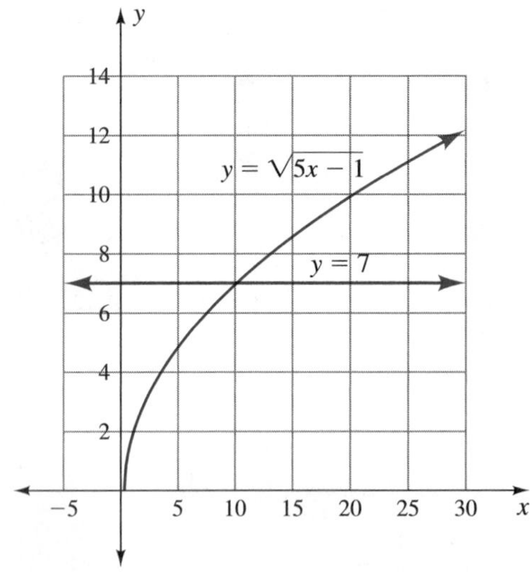

17. Solve these equations. For what inputs is each radical expression defined?

a. $\sqrt{7x-3} = 5$ b. $\sqrt{2-x} = x - 2$

c. $\sqrt{4x-3} = 7$

18. Solve these formulas for the indicated variable.

a. $E = \sqrt{W \cdot R}$ for W

b. $V_o = \sqrt{\dfrac{GM}{R}}$ for R

c. $n = \dfrac{1}{2l}\sqrt{\dfrac{T}{m}}$ for T

d. $n = \dfrac{1}{2rl}\sqrt{\dfrac{T}{\pi d}}$ for T

19. On the moon, the distance seen in miles from a height of h feet is given by

$$d \approx \sqrt{\frac{3h}{8}}$$

a. How far can be seen from a height of 20 feet?

b. How high would an astronaut need to climb to see 4 miles?

In Exercises 20 and 21, solve for x with algebraic notation. Describe how to find the solution on the graph.

20. $\sqrt{x+1} = 1$

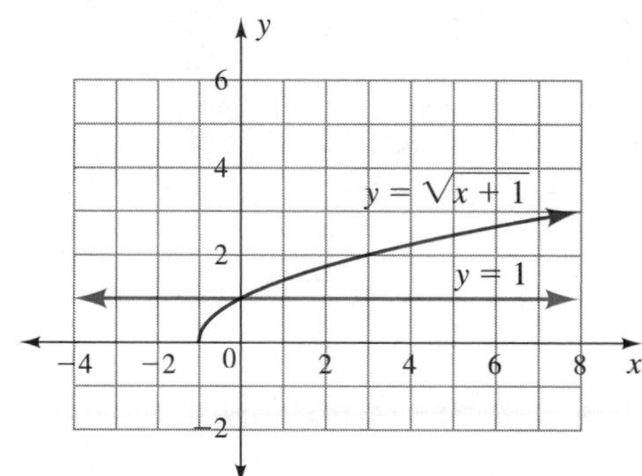

21. $\sqrt{x-1} = 2$

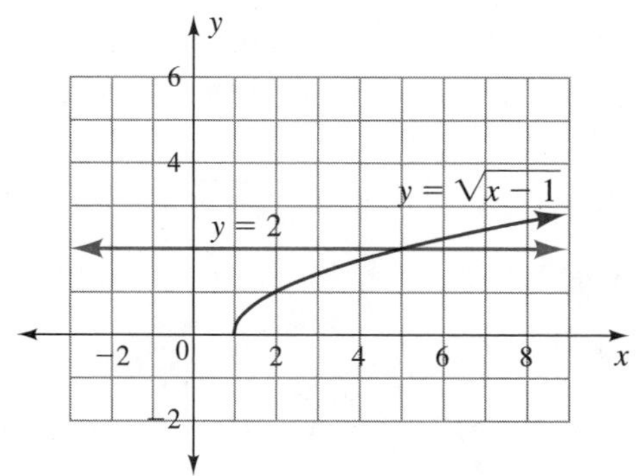

Use the graph of y = x² − 2x − 8 in the figure to solve the equations in Exercises 22 and 23.

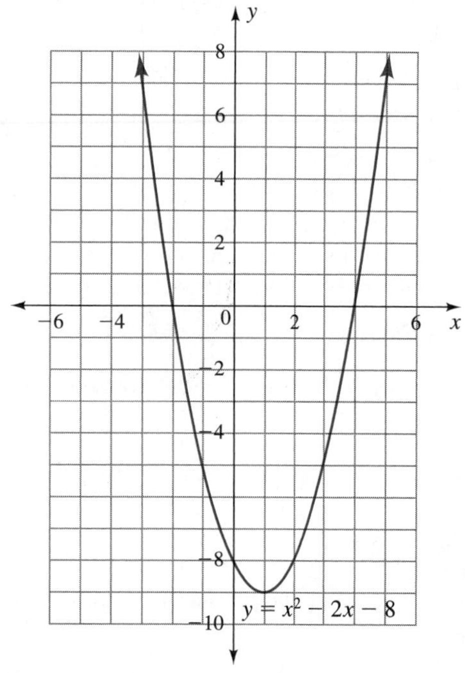

22. $x^2 - 2x - 8 = 8$

23. $x^2 - 2x - 8 = 0$

24. What are the vertex, axis of symmetry, y-intercept point, and x-intercept points of the graph for Exercises 22 and 23?

In Exercises 25 to 38, solve for x. Round irrational roots to the nearest thousandth.

25. $x^2 = \frac{16}{144}$

26. $2x^2 - 18 = 0$

27. $x^2 - 5x + 4 = 0$

28. $x^2 - 4x - 5 = 0$

29. $2x^2 + 5x - 6 = 0$

30. $2x^2 + 5x + 1 = 0$

31. $x^2 + 3x - 18 = 0$

32. $x^2 - 4x - 21 = 0$

33. $3x^2 = 4x + 7$

34. $4x - 3 + 4x^2 = 0$

35. $2x^2 + 3x + 5 = 0$

36. $8 - 4x + x^2 = 0$

37. $x^2 - 6x + 9 = 0$

38. $x^2 + 12x = -36$

39. Solve these formulas for the indicated variable. Assume variables and outputs take on positive values only.

a. $A = \pi r^2$ for r

b. $p = \frac{1}{2}dv^2$ for v

c. $S = 4\pi r^2$ for r

d. $h = \dfrac{v^2}{2g}$ for v

40. In traffic accident investigations, tire skid tests are used to find the coefficient of friction between tires and the road surface near an accident scene. An investigator measures the following tire skid marks, in feet, for a skid at 30 miles per hour:

Test 1: Left front, 50; right front, 49; left rear, 47; right rear, 48

Test 2: Left front, 47; right front, 50; left rear, 48; right rear, 51

a. Find the mean, range, and sample standard deviation (s_x) of the skid marks for each of the two tests. Round the standard deviation to two decimal places.

b. Find the coefficient of friction for each test with $f = \dfrac{S^2}{30D}$, where f is the coefficient of friction, S is the speed of the car making the tests, and D is the mean skid mark distance for the four tires.

41. The following data give the number of children of each of the presidents of the United States:

0, 5, 6, 0, 2, 4, 0, 4, 10, 15, 0, 6, 2, 0, 3, 4, 5, 4, 8, 5, 3, 5, 3, 2, 6, 3, 3, 0, 0, 2, 2, 5, 1, 1, 2, 2, 2, 4, 4, 2, 6, 1

a. Find the median of the data.

b. Make a box and whisker plot.

c. If a graphing calculator is available, find the mean and population standard deviation (σ_x) for the number of children.

CHAPTER 8 TEST

1. Which of these sets of numbers represent sides of a right triangle?

a. $\{3, 4, 5\}$ **b.** $\{1, 2, 3\}$ **c.** $\{1, \sqrt{3}, 2\}$

d. $\{\sqrt{2}, \sqrt{2}, 2\}$ **e.** $\{\sqrt{3}, \sqrt{4}, \sqrt{5}\}$

2. Which of the three expressions in each set has the same value as the radical given first?

a. $\sqrt{45}$ $\{9\sqrt{5}, 5\sqrt{9}, 3\sqrt{5}\}$

b. $\sqrt{44}$ $\{4\sqrt{11}, 2\sqrt{11}, 11\sqrt{4}\}$

3. Romeo plans to use an extension ladder to reach Juliet's balcony, 20 feet above the ground. He will set the base of the ladder 5 feet away from the wall below the edge of the balcony. To preserve this safe ladder position, how long a ladder does he need?

4. Simplify these expressions. Assume that the variables represent positive numbers.

a. $\sqrt{5} \cdot \sqrt{20}$ **b.** $\sqrt{3} \cdot \sqrt{27}$ **c.** $(3\sqrt{6})^2$

d. $(2\sqrt{7})^2$ **e.** $\sqrt{36x^2y}$ **f.** $\sqrt{0.81x^4y^3}$

g. $\sqrt{\dfrac{147}{3}}$ **h.** $\dfrac{\sqrt{18}}{\sqrt{32}}$ **i.** $\sqrt{\dfrac{a^5b^2}{a}}$

5. Find the missing sides in the similar right triangles below. Use the marks in the angles to determine which sides are proportional. Round decimals to the nearest tenth.

a.

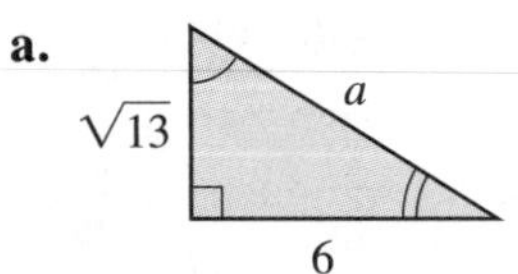

b.

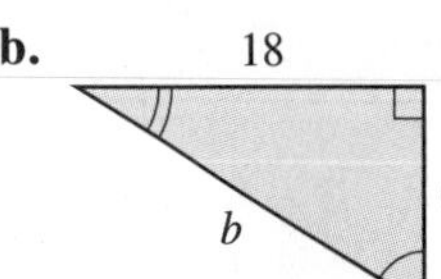

c.

d
10
e

6. Solve with the graph: $x^2 - 3 = 1$.

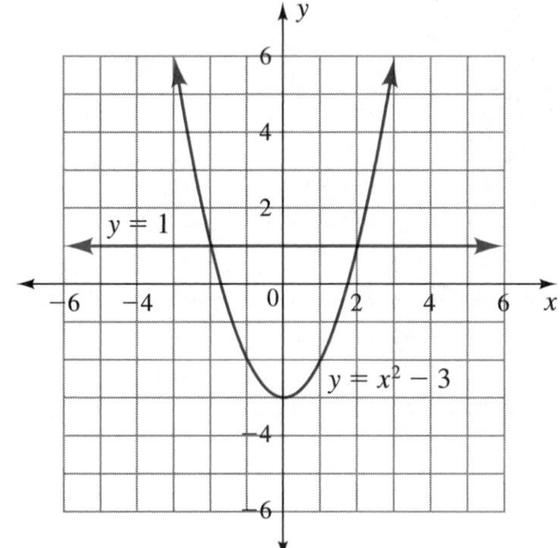

7. Solve for all inputs, x, that make these equations true. Use an inequality or an interval to indicate inputs that are defined.

a. $x^2 = \frac{36}{121}$ **b.** $\sqrt{3 - x} = x + 3$

c. $x^2 + x - 2 = 0$ **d.** $\sqrt{5x - 6} = 12$

e. $2x^2 = 8 - 15x$ **f.** $8x^2 + 5x = 4$

g. $(3x + 8)(3x - 8) = 0$ **h.** $4x^2 + 8 = 0$

i. $x^2 - 6x + 9 = 0$ **j.** $4x^2 - 25 = 0$

8. Solve these formulas for the indicated variable. Assume all variables and outputs represent positive numbers.

a. $V_e = \sqrt{\dfrac{2GM}{R}}$ for R **b.** $E = \frac{1}{2}mv^2$ for v

c. $E = \dfrac{kH^2}{8\pi}$ for H

9. Hydroplaning occurs when a tire slides on the surface of the water on a pavement instead of gripping the pavement's surface. The *Advanced Pilot's Flight Manual* gives the relationship between the minimum hydroplaning speed s, in miles per hour, and tire pressure t, in pounds per square inch, as

$$s = 8.6\sqrt{t}$$

The implication of this formula may not be obvious. Perhaps these thoughts will help: The softer the tire, the greater the surface area and the greater the tendency to slide along the surface, or hydroplane. A harder tire tends to cut through the water to the paved surface.

a. If the tire pressure is 36 pounds per square inch, what is the speed at which the tire will hydroplane?

b. If the tire pressure is 100 pounds per square inch, what is the speed at which the tire will hydroplane?

c. If a plane lands at 120 miles per hour on wet pavement, what is the tire pressure at which hydroplaning will occur?

10. Each shape has an area of 50 square feet. What is the length of x in each case? Round to the nearest thousandth. The area of an equilateral triangle is

$$A = \frac{x^2\sqrt{3}}{4}$$

a. Square, side x:

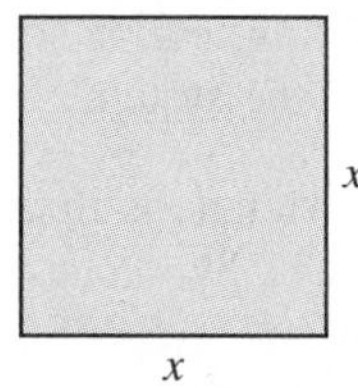

b. Circle, diameter x:

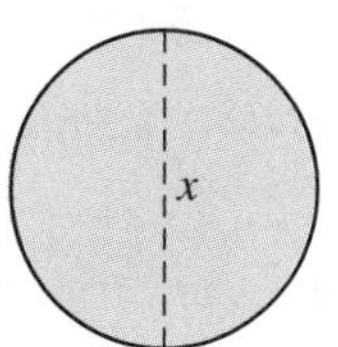

c. Equilateral triangle:

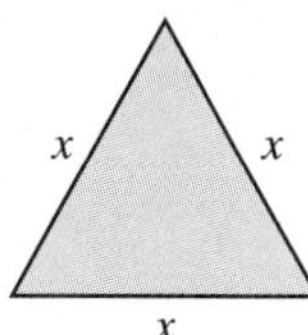

11. The coordinates (3, 5), (6, 8), and (3, 11) form a triangle.

a. Sketch them on coordinate axes.

b. What is the length of each side of the triangle?

c. Use slope to determine whether any two sides are perpendicular.

d. What kind of a triangle is this triangle? Explain why.

12. The table below gives food energy and sodium content for a variety of dry cereals.

Dry Cereal	Food Energy (calories)	Sodium Content (milligrams)
Cap'n Crunch®	120	145
Froot Loops®	110	145
Super Golden Crisp®	105	25
Sugar Frosted Flakes®	110	230
Sugar Smacks®	105	75
Trix®	110	181

a. Find the mean and sample standard deviation (s_x) for food energy.

b. Find the median and make a box and whisker plot for sodium content.

CUMULATIVE REVIEW OF CHAPTERS 1 TO 8

Describe each of the situations in Exercises 1 to 3 with an expression showing the appropriate operation. Then answer the question.

1. A surgical procedure takes $1\frac{1}{4}$ hours. How many procedures can be scheduled in a 12-hour day?

2. An attorney spends $3\frac{5}{6}$ hours on a client's business. What is the total bill to the client at \$180 per hour?

3. Simplify 3(\$15) + 2(−\$20) − (\$10) − (−\$25).

4. Simplify $-3(4) - 4(-5) + (-4)^2$.

5. Simplify $\dfrac{3x^3y^2}{27xy^3}$.

6. a. What are the perimeter and area of a square with sides of 90 feet (a baseball infield, or diamond)?

   b. What are the perimeter and area of a square with sides of 12 meters (a gymnastics mat)?

   c. Use unit analysis to write a ratio comparing the perimeter in part a to the perimeter in part b.

   d. Use unit analysis to write a ratio comparing the area in part a with the area in part b.

Solve the equations in Exercises 7 to 10.

7. $2 - 3x = 26$

8. $3(x - 5) = x + 9$

9. $3(x + 4) = 2(1 - x)$

10. $\frac{2}{3}x - 4 = 26$

11. Solve $2x - 1 \leq 2 - x$ with a graph and with symbols.

12. Solve $A = \frac{1}{2}bh$ for h.

13. If $f(x) = 3x - 2x^2$, what are $f(-1)$, $f(0)$, $f(1)$, and $f(2)$?

14. Sketch a line with slope $\frac{2}{3}$ that passes through $(2, -1)$.

15. Write a linear equation with slope $= \frac{4}{3}$ and y-intercept -2.

16. a. Make a table and graph for $y = \frac{1}{3}x - 2$. Circle the point that shows the solution for each of these equations. Solve the equation.

   b. $\frac{1}{3}x - 2 = 0$

   c. $\frac{1}{3}x - 2 = -1$

   d. $\frac{1}{3}x - 2 = -4$

17. Build an input-output table, graph, and equation for the value remaining on a \$20 prepaid transit card where each trip costs \$2.25. Let the input be the number of trips from 0 to 10, counting by 2.

18. You have a charge account at Ice Cream America. Each day after work, you have the same ice cream cone, which costs \$1.75. Let the input be the number of days. Build an input-output table, graph, and equation for the total on the charge account.

Write proportions for Exercises 19 to 22. Make an estimate of the answer before solving.

19. What is 125% of 52?

20. 35% of what number is 21?

21. 33 is what percent of 88?

22. Zee Paper Towels were two for a dollar in 1986. In 1995, they were \$0.69 each. What is the percent change in price?

23. A sneeze may reach 100 miles per hour. How far in feet could sneezed bacteria travel in 2 seconds? Solve with a unit analysis.

24. Solve for x: $\dfrac{3x + 2}{7} = \dfrac{5(x - 2)}{9}$.

25. Multiply.

   a. $(x + 3)(2x - 1)$ b. $x(x + 1)(3x - 1)$

26. Factor.

   a. $x^2 - 4x - 21$ b. $x^2 - 4x$

   c. $4x^2 - 10x$ d. $4x^2 - 11x - 3$

   e. $4x^2 - 4x + 1$ f. $4x^2 - 9$

27. a. The ratio of the distance the sun is from Earth to the diameter of the sun is 1.496×10^8 kilometers to 1.3914×10^6 kilometers. Divide the ratio.

   b. The ratio of the distance the moon is from Earth to the diameter of the moon is 384,000 kilometers to 3480 kilometers. Divide the ratio.

   c. Compare the quotients in parts a and b. Astronomers suggest that this is why the sun and moon appear to be of similar size in the sky.

Solve the systems of equations in Exercises 28 and 29.

28. $x + y = -13$
    $x - y = 23$

29. $7x - 8y = -25$
    $3x + 4y = -7$

30. Two angles are supplementary. One angle is 5° greater than four times the other. Write a system of equations to find the angles, and then solve the system.

31. The length of one leg of a right triangle is three more than the length of the other leg. The hypotenuse is 15. What are the lengths of the legs?

32. Solve $4x^2 - 20 = 0$.

33. a. Sketch $y = 2x^2 - 5x - 3$.

   b. Solve $2x^2 - 5x - 3 = 0$ from the graph.

   c. Solve $2x^2 - 5x - 3 = 0$ by factoring.

   d. Solve $2x^2 - 5x - 3 = 0$ with the quadratic formula.

34. For what values is $\sqrt{x + 24} = 3$ defined? Solve the equation.

35. The data below are for 6″ Subway sandwiches.

Sandwiches	**Fat (g)**	**Cholesterol (mg)**	**Calories**
Subway club®	5	26	312
Turkey breast & ham	5	24	295
Veggie delite™	3	0	237
Turkey breast	4	19	289
Ham	5	28	302
Roast beef	5	20	303
Roasted chicken breast	6	48	348

SUBWAY® regular 6″ subs include bread, veggies, and meat. Addition of condiments or cheese alters nutrition content.

a. What are the mean, median, and mode for the grams of fat?

b. Draw a box and whisker plot for the milligrams of cholesterol.

c. What are the range, mean, and sample standard deviation for the calories?

9

Rational Expressions

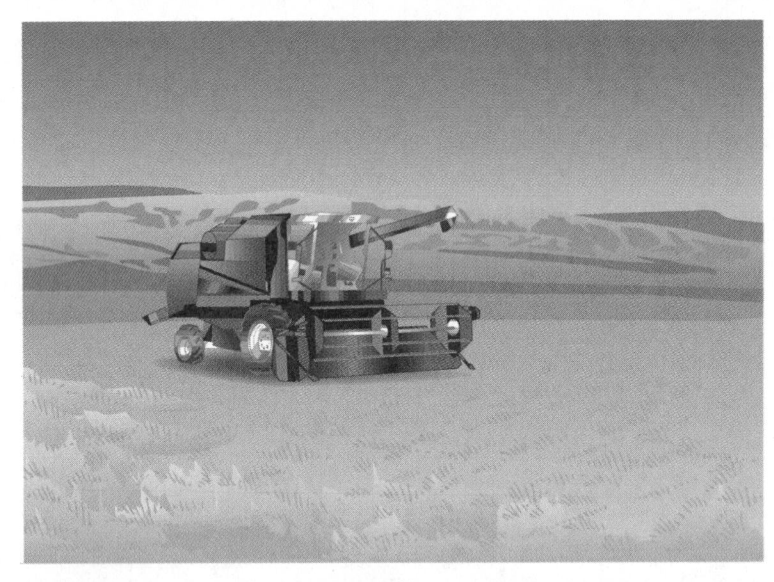

Figure 1

In Sherman County, Oregon, strong winds shatter the ripe wheat heads, and occasionally hail storms beat wheat crops to the ground. To get the ripened wheat harvested before storms can damage it, farmers from opposite ends of the county share harvest crews and equipment.

Farmer Lee's equipment could cut her north-end crop in 12 days. Farmer Terry of the south end of the county estimates he could cut the same acreage in 18 days. If the two farmers work together, how many days will it take to cut Lee's crop? We will examine this problem in Section 9.5.

In this chapter, we work with operations on expressions and equations containing fraction notation. We look at zero in the denominator in Section 9.1, simplifying expressions in Section 9.2, multiplication and division in Section 9.3, adding and subtracting in Section 9.4, and solving equations in Section 9.5.

9.1 Rational Functions: Graphs and Applications

OBJECTIVES

- Identify rational expressions.
- Identify inputs that create zero denominators.
- Explore graphical behavior near inputs that create undefined expressions.
- Use graphs to investigate applications of rational expressions.

> **WARM-UP**
>
> Suppose you have 10 miles to travel. How long will it take? The distance, rate, and time formula is $D = rt$. Make an input-output table showing the rate of travel as input and the time required to go 10 miles as output. Use inputs from 0 to 10 mph, counting by 2, and then include 50 mph and 100 mph.

THIS SECTION INTRODUCES rational functions. We look at places where rational functions are undefined, graphs of rational functions, and a selection of applications.

Rational Functions

Recall that **rational numbers** are *the set of numbers that may be written as the ratio of two integers a/b, with b not equal to zero.* The **rational function** $f(x)$ is *a function that may be written as the ratio of two polynomials* $\dfrac{p(x)}{q(x)}$, *where the denominator is not zero.*

Examples of rational functions are

$$f(x) = \frac{x}{x+1}, \qquad f(x) = \frac{x^2 + 2x + 1}{x - 3}, \qquad \text{and} \qquad f(x) = \frac{a+b}{a-b}$$

In less formal terms, **rational functions** contain *algebra in fraction notation.* Some formal definitions of rational functions exclude fractions that have only whole numbers in the denominator, such as $\dfrac{x+3}{2}$. We will include such expressions because the algebraic techniques are the same. Expect lots of other fraction work as well.

Division by Zero

EXAMPLE 1 Evaluating a rational function Complete Table 1 for $f(x) = \dfrac{x}{x-5}$. Complete Table 2 for $f(x) = \dfrac{x+3}{(x+2)(x-3)}$. Is there anything unusual about the answers?

x	$f(x) = \dfrac{x}{x - 5}$
−2	
−1	
0	
1	
2	
3	
4	
5	

Table 1

x	$f(x) = \dfrac{x + 3}{(x + 2)(x - 3)}$
−2	
−1	
0	
1	
2	
3	
4	
5	

Table 2

Look back at these tables as you do Example 2.

Solution See the Answer Box. ●

One of the simplest rational expressions is $\dfrac{1}{x}$. The variable x is in the denominator. A zero denominator would imply division by zero, which is an undefined operation. Thus, in rational expressions we must identify inputs that create a zero denominator and exclude them from our set of inputs.

EXAMPLE  Finding the domain (set of inputs) What inputs must be excluded for each rational expression to be defined? State the domain.

a. $\dfrac{1}{x}$ **b.** $\dfrac{x}{x - 5}$ **c.** $\dfrac{x + 3}{x^2 - x - 6}$

Solution Inputs that give zero denominators lead to undefined expressions.

a. When $x = 0$,

$$\frac{1}{x} = \frac{1}{0}$$

The domain, or set of inputs, is the set of all real numbers x, $x \neq 0$.

b. When $x = 5$.

$$\frac{x}{x - 5} = \frac{5}{5 - 5} = \frac{5}{0}$$

The domain is the set of all real numbers x, $x \neq 5$.

c. To see where $x^2 - x - 6 = 0$, we need to solve the quadratic equation. We factor the denominator:

$$x^2 - x - 6 = (x + 2)(x - 3)$$

When $x = -2$,

$$\frac{x + 3}{(x + 2)(x - 3)} = \frac{-2 + 3}{(-2 + 2)(-2 - 3)} = \frac{1}{(0)(-5)} = \frac{1}{0}$$

or when $x = 3$,

$$\frac{x+3}{(x+2)(x-3)} = \frac{3+3}{(3+2)(3-3)} = \frac{6}{(5)(0)} = \frac{6}{0}$$

The domain is the set of all real numbers x, $x \neq -2$, $x \neq 3$. ●

The notation for the set of all real numbers is $\mathbb{R}$. Use it in answers to the Exercises.

To save time and space, when we work with equations in this chapter, we will state numbers that must be excluded from the set of inputs. However, unless specifically requested to do so (as in Example 2), we will not state excluded numbers when working with expressions.

Graphs of Rational Expressions

If an expression is undefined for a certain input, there is no point on the graph of the expression for that input. Furthermore, the graph near such a point has unusual features, as shown in Example 3.

EXAMPLE 3 Exploring the graph as the denominator value approaches zero What happens to the graph of $y = \dfrac{2}{x}$ as x gets close to zero?

Solution Tables 3 and 4 show outputs for $y = \dfrac{2}{x}$ as x gets close to zero. The data in each table form a curve when graphed (see Figure 2). If we trace the third-quadrant curve from left to right, we find that as the inputs, x, approach zero, the curve turns downward. Although the curve gets close to the y-axis, it never touches the y-axis. We describe this behavior by saying *y approaches negative infinity as x approaches zero.*

Input x	Output $y = 2/x$
-5	$2/-5 = -0.4$
-4	$2/-4 = -0.5$
-2	$2/-2 = -1.0$
-1	$2/-1 = -2.0$
-0.5	$2/-0.5 = -4.0$
-0.25	$2/-0.25 = -8.0$
-0.1	$2/-0.1 = -20.0$

Table 3 $y = 2/x$ as x approaches zero from the left

Input x	Output $y = 2/x$
5	$2/5 = 0.4$
4	$2/4 = 0.5$
2	$2/2 = 1.0$
1	$2/1 = 2.0$
0.5	$2/0.5 = 4.0$
0.25	$2/0.25 = 8.0$
0.1	$2/0.1 = 20.0$

Table 4 $y = 2/x$ as x approaches zero from the right

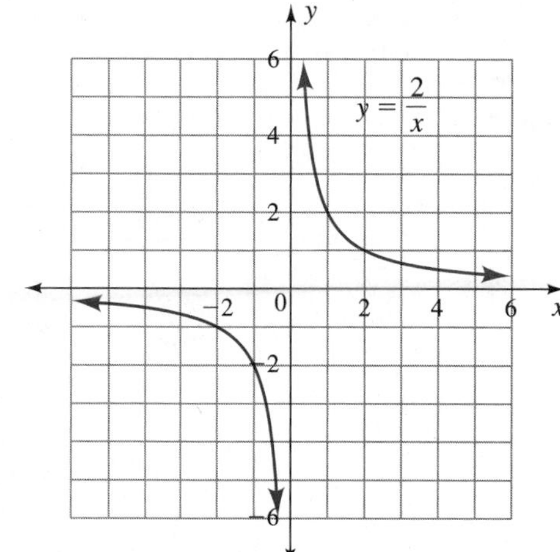

Figure 2

If we trace the first-quadrant graph from right to left, we see that as the inputs, x, approach zero, the curve rises. As the curve rises, it gets closer to the y-axis, but, like the third-quadrant curve, it never touches the y-axis. We describe this behavior by saying *y approaches positive infinity as x approaches zero.*

Because the equation $y = \dfrac{2}{x}$ has no output at $x = 0$, we say *the equation is defined for the set of real numbers x, $x \neq 0$, or $\mathbb{R}$, $x \neq 0$.* ●

> The graph of a rational expression (simplified to lowest terms) approaches infinity in a vertical direction whenever the denominator approaches zero.

Together the two curves in Figure 2 form a *rectangular hyperbola.* The name comes from the fact that the curve approaches but does not intersect the rectangular coordinate axes. Look for these curves and changes in their position in the remainder of this chapter.

EXAMPLE 

Exploring the graph as the denominator value approaches zero Make a table and graph for $y = \dfrac{-2}{x + 4}$. What inputs must be excluded? What do you observe about the behavior of the graph near that input?

Solution The expression has a zero denominator for $x = -4$; thus, $x \neq -4$. The graph in Figure 3 becomes nearly vertical as we approach $x = -4$ from either the left (Table 5) or the right (Table 6).

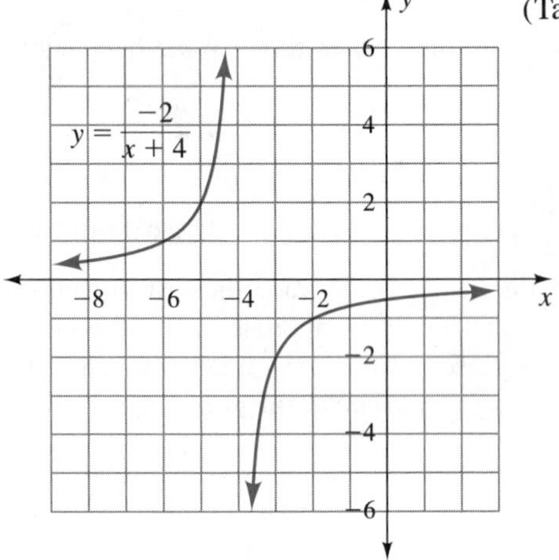

Figure 3

x	y
−9	0.4
−8	0.5
−7	0.67
−6	1.0
−5	2.0
−4	undefined

Table 5

x	y
−3	−2
−2	−1
−1	−0.67
0	−0.5
1	−0.4
2	−0.33

Table 6

Before we examine applications with rational expressions, two facts should be noted about graphing calculators and zero denominators. First, as we trace along the graph of a rational expression, the output will be blank each time the input creates a zero denominator. It often takes considerable adjustment of the viewing window to see the blank.

Second, in certain viewing-window settings, the calculator draws an almost vertical line on the graph at an undefined point. This line is an error made by the calculator. The calculator evaluates the functions to the left and right of the undefined point and connects the ordered pairs. The line will disappear if the calculator is set on dot mode rather than connected mode.

Applications of Rational Functions

In our applications of rational equations we return to two settings—distance and geometry (area and perimeter)—and introduce two new settings—resource management and light intensity.

DISTANCE, RATE, AND TIME The basic distance, rate, and time formula is $D = rt$. In Example 5, we let distance be a fixed number (or, as we usually say, *hold distance constant*) and examine the behavior of the two variables, rate and time.

EXAMPLE 5 Finding time of travel The rates in Table 7, of 1 to 10 miles per hour, range fr... a slow walk to a leisurely bicycling speed.

a. Complete Table 7.

Rate, r mph	Time, t hours	Distance, $D = r \cdot t$
1	10	
2	5	
5	2	
10	1	

Table 7 Variable rate and time

b. Set up a graph, with rate along the horizontal axis and time along the vertical axis. Label both axes from 0 to 12. Graph the points (r, t) from the table. Connect the points from left to right.

c. What do you observe about the graph?

d. Change $r \cdot t = 10$ to a rational equation that uses r as input and t as output.

Solution **a.** The third column is 10 miles for all entries.

b. The graph is shown in Figure 4.

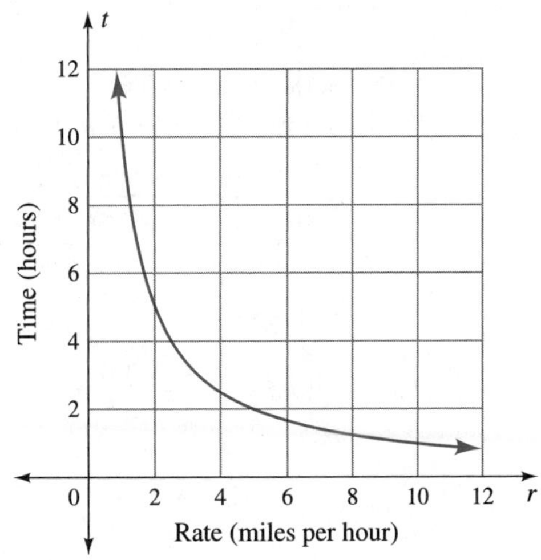

Figure 4

c. The graph of the points (r, t) in Figure 4 does not make a straight line. Thus, the relationship between rate and time is not linear when the product (distance) is constant.

d. The third column in Table 7 is $D = r \cdot t$. The relationship between r and t is $r \cdot t = 10$. If we solve for t as output with r as input, the equation becomes

$$t = \frac{10}{r}, \quad r \neq 0$$

Think about it 1: Calculate the "slope," $\frac{\Delta t}{\Delta r}$, in Table 7. How does this rate of change compare with that of a linear equation?

The graph of a rational function shows that for some inputs, change in output takes place rapidly and for other inputs, change in output takes place slowly. In the distance example, a change in speed at low speeds causes a large decrease in travel time. A change in speed at high speeds causes little change in travel time. What does this say about driving fast when you are late? (*Hint*: See Exercise 13.)

RESOURCE MANAGEMENT In the next example, we examine the application of rational expressions in resource management. Many resources, such as minerals, are limited in quantity. Strategic planners estimate the total quantity available and attempt to plan for future shortages.

EXAMPLE 6 Finding the number of years' supply of a resource: resource management Silver is used to produce electrical and electronic products and photographic film, as well as flatware and jewelry. The estimated world reserves of silver were 420,000 metric tons in 1990.

a. If the rate of use of silver reserves is x metric tons per day, write an equation for how many years the reserves will last.

b. If the 1990 rate of use of silver was 40 metric tons per day, how many years will the 1990 reserves last?

c. Graph the equation, letting x be use in metric tons per day and y be supply in years.

Solution **a.** The reserves will last y years, as shown by the equation

$$y = \frac{420{,}000}{365x}$$

b. If the world silver reserves are used up at the rate of 40 metric tons per day, the 1990 reserves will last y years:

$$y = \frac{420{,}000 \text{ metric tons}}{\dfrac{40 \text{ metric tons}}{1 \text{ day}} \cdot \dfrac{365 \text{ days}}{1 \text{ yr}}}$$

$$y = 28.8 \text{ yr}$$

c. The graph of $y = \dfrac{420{,}000}{365x}$ is shown in Figure 5.

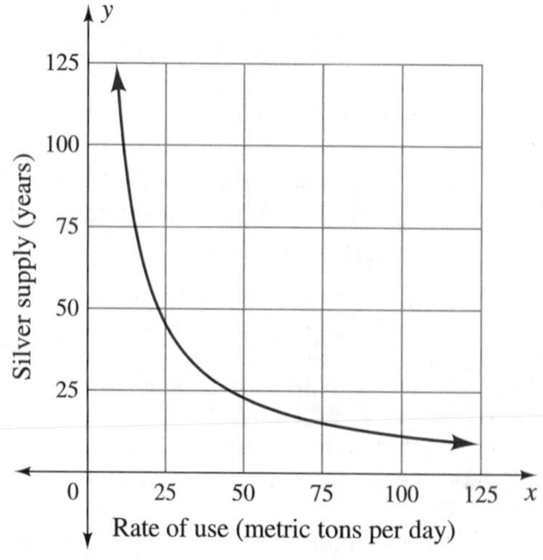

Figure 5

Think about it 2: If we reduce use to 15 metric tons per day, how many years will the silver reserves last? If we increase use to 100 metric tons per day, how many years will the silver reserves last? Finish this sentence: As the rate of use increases, the number of years' supply ____ .

In Sections 9.2 and 9.3, we will practice working with expressions containing units, such as those in part b of Example 6. For now, we observe that *metric tons* and *days* both cancel and only *years* remain.

GEOMETRY In Example 7, we solve a system of equations with a graph. One equation is a rational function.

EXAMPLE 7 Solving a system of equations from geometry Find the length, x, and width, y, of a rectangle with area 48 square centimeters and perimeter 38 centimeters. Write an equation for the area and another for the perimeter. Use the graphs of the equations to find the length and width that satisfy both equations.

Solution The area is $xy = 48$, or $y = \frac{48}{x}$, $x \neq 0$. The perimeter is $2x + 2y = 38$, or $y = 19 - x$. The graphs are shown in Figure 6. There are two intersections: (16, 3) and (3, 16).

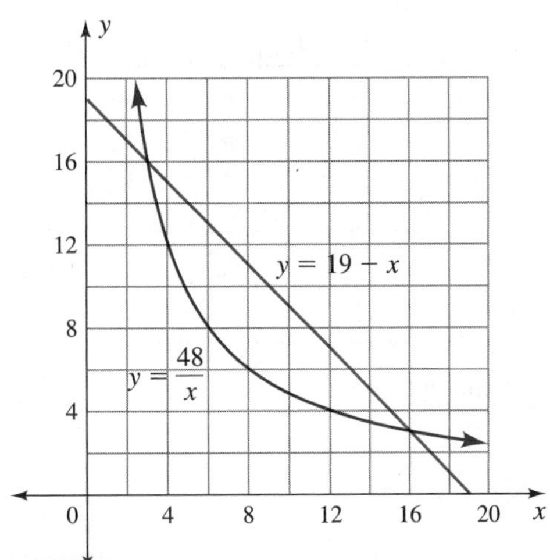

Figure 6

Check:

$$\text{Area} = 16(3) = 3(16) = 48 \text{ cm}^2$$

$$\text{Perimeter} = 2(16) + 2(3) = 38 \text{ cm}$$

Thus, the length and width are 16 centimeters and 3 centimeters, respectively. The mathematical results do not specify that length be the larger number, as is customary in everyday use. ✔ ●

INVERSE VARIATION Rational equations such as $y = \frac{a}{x}$ or $y = \frac{a}{x^2}$ illustrate **inverse variation**, or **inverse proportions**. *Two variables vary inversely if one gets smaller as the other gets larger.* In Example 5, for a fixed distance, the rate and time vary inversely. In Example 6, for a fixed amount of silver resources, the amount of daily use and the number of years the supply will last vary inversely.

In Example 7, for a fixed area of 48 square centimeters, the length and width of the rectangle vary inversely. The last three examples were of the form $y = \frac{a}{x}$. In the next example, the intensity of light and its distance from the source vary inversely according to $y = \frac{a}{x^2}$.

EXAMPLE 8 Finding a table and graph for the inverse square: light intensity Light intensity, I, varies inversely with the square of the distance, d, from the source of the light. The formula is

$$I = \frac{k}{d^2}, \quad d \neq 0$$

where k is a constant that changes the units from square feet to those needed for light intensity. For simplicity, we use $k = 1$ here. Make a table and graph for light intensity in terms of distance from the light source.

Solution The values for light intensity are shown in Table 8. The graph in Figure 7 is somewhat the same shape as earlier graphs, but because the distance variable is squared, this curve is not as symmetrically placed in the first quadrant.

Distance in Feet, d	Light Intensity, I
1	1
2	$\frac{1}{4}$
3	$\frac{1}{9}$
4	$\frac{1}{16}$
5	$\frac{1}{25}$

Table 8

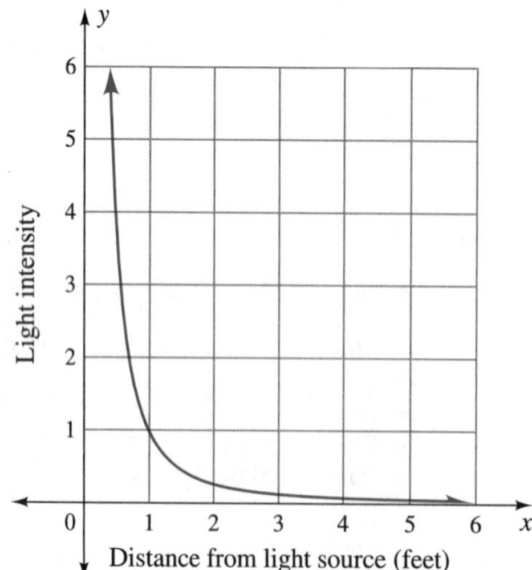

Figure 7

The project in Exercise 33 describes a light intensity activity related to Example 8. You can experiment with light intensity without any measuring device. In a darkened room with only one light source, hold a book 1 foot from the bulb and read. Then move to 2 feet away and read. Move to 3 feet away and read. Observe how quickly it becomes difficult to read the book.

ANSWER BOX

Warm-up: The ordered pairs for the table are (0, (undefined)), (2, 5), (4, 2.5), $\left(6, 1\frac{2}{3}\right)$, $\left(8, 1\frac{1}{4}\right)$, (10, 1), $\left(50, \frac{1}{5}\right)$, and $\left(100, \frac{1}{10}\right)$. **Example 1:**

x	$f(x) = \dfrac{x}{x - 5}$
−2	0.2857
−1	0.1667
0	0
1	−0.25
2	−0.6667
3	−1.5
4	−4
5	undefined

x	$f(x) = \dfrac{x + 3}{(x + 2)(x - 3)}$
−2	undefined
−1	−0.5
0	−0.5
1	−0.6667
2	−1.25
3	undefined
4	1.1667
5	0.5714

There are values of $y = f(x)$ that are undefined.
Think about it 1: Between (1, 10) and (2, 5), the "slope," $\Delta t/\Delta r$, is $-5/1$. Between (2, 5) and (5, 2), $\Delta t/\Delta r = -3/3$, or -1. Between (5, 2) and (10, 1), $\Delta t/\Delta r = -1/5$. The "slope" changes from $-5/1$ to $-1/5$. The slope of a linear equation is constant. **Think about it 2:** From the graph we can estimate an output of 75 years for an input near 15 metric tons per day. For 100 metric tons per day, the number of years drops to about $12\frac{1}{2}$. From the equation, 15 tons per day gives 76.7 years, and 100 tons per day gives 11.5 years. As the rate of use increases, the number of years' supply decreases.

EXERCISES 9.1

What inputs must be excluded for each expression in Exercises 1 to 8 to be defined?

1. $\dfrac{2}{x + 1}$

2. $\dfrac{3}{x - 1}$

3. $\dfrac{x}{2x - 1}$

4. $\dfrac{x}{2x + 3}$

5. $\dfrac{x}{3x + 1}$

6. $\dfrac{x}{3x - 2}$

7. $\dfrac{5}{x^2 + 4x + 3}$

8. $\dfrac{3x}{x^2 - 6x + 5}$

Using a calculator to build a table and a graph or using a graphing calculator, describe the behavior of the graph of each equation in Exercises 9 to 12 as x approaches the indicated number.

9. $y = \dfrac{x}{x - 5}$, as x approaches 5

10. $y = \dfrac{-x}{x + 3}$, as x approaches -3

11. $y = \dfrac{x + 3}{(x + 2)(x - 3)}$, as x approaches -2

12. $y = \dfrac{x + 3}{(x + 2)(x - 3)}$, as x approaches 3

13. Do you gain enough time by driving fast to risk getting a speeding ticket? Suppose you have 60 miles to travel. At what rate would you need to drive to travel the distance in these times? (*Hint*: First solve $D = rt$ for r.)

a. 2 hours **b.** 1 hour

c. $\frac{5}{6}$ hour **d.** $\frac{3}{4}$ hour

e. $\frac{2}{3}$ hour **f.** $\frac{1}{2}$ hour

14. Suppose you have 6 miles to travel on a sometimes congested freeway. How long would it take you to travel that distance at each of these speeds?

a. 3 mph (walk) **b.** 6 mph (jogging)

c. 10 mph **d.** 20 mph

e. 60 mph **f.** 0 mph

15. Estimated world crude oil reserves in 1990 were about 1000 billion barrels. If the average world production of crude oil is x barrels per day, what equation describes how many days the 1990 oil reserves will last? How many years?

16. A potential landfill site contains 1,000,000 cubic yards of space. If the average fill per day is x cubic yards, what equation describes how many days the landfill site may be used? How many years?

17. World production of crude oil uses up oil reserves. If crude oil production in 1990 was 60 million barrels per day, how many years will the oil reserves last? (See Exercise 15.)

18. If a city with a population of 100,000 produces 600 cubic yards of garbage each day, how many years will the landfill in Exercise 16 last?

19. A two-year college has financial aid available for a total of 90 credits. The input, x, is the number of credits taken each term.

a. What is the meaning of the output, y, for the equation $y = \dfrac{90}{x}$?

b. Describe the problem situation if $x = 12$ credits per term.

20. A student saves a total of \$2800 over the summer to spend during the school year. The input, x, is dollars spent per month.

a. What is the meaning of the output, y, for the equation $y = \dfrac{2800}{x}$?

b. Describe the problem situation if $x = \$400$.

21. In Example 5, $r = 0$ makes the equation $t = \dfrac{10}{r}$ undefined. How do we interpret $r = 0$ in the problem setting?

22. In Example 5, Figure 4 shows a first-quadrant graph only. Give five (r, t) coordinates from other quadrants that make the equation $t = \dfrac{10}{r}$ true. Why are these points most likely not relevant to the problem situation?

23. In the equation $t = \dfrac{10}{r}$, is there an input r that makes $t = 0$?

24. As time, t, changes from 0.1 to 0.01 hour, what happens to the rate, r, in the equation $t = \dfrac{10}{r}$?

25. a. List ten possible widths and lengths of rectangles with area of 30 square inches. Use the table headings shown. Calculate the perimeter for each rectangle.

Width	Length	Area	Perimeter
		30	
		30	

b. Graph the length (x-axis) and the perimeter (y-axis).

c. Where is the location of the point on the graph describing the smallest perimeter?

d. Find the equation (perimeter in terms of length, x) describing the graph.

26. Refer to the table completed in Exercise 25.

a. Graph the length and width pairs on coordinate axes, placing width on the x-axis and length on the y-axis. What shape is formed?

b. What length and width give a perimeter smaller than 22 for an area of 30?

c. Find the equation (length in terms of width) describing the graph.

The equations $y = \frac{1}{x}$ and $y = \frac{1}{x^2}$ are graphed in the figure below. Refer to this figure in Exercises 27 to 30.

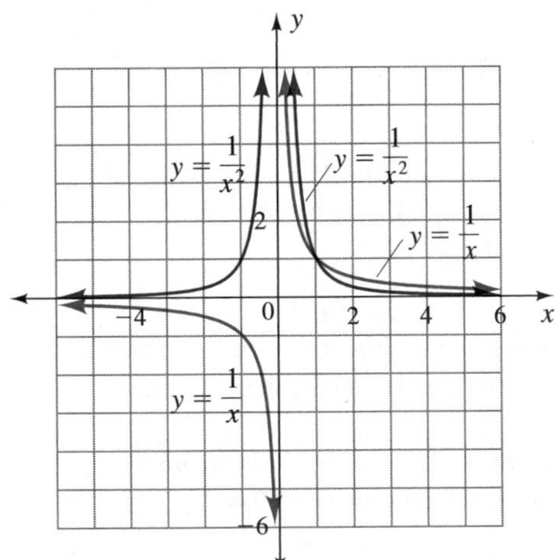

27. Draw the line of symmetry for each graph and write its equation.

28. Why does $y = \frac{1}{x^2}$ not have points in the third quadrant?

29. The coordinates $\left(2, \frac{1}{2}\right)$ and $\left(\frac{1}{2}, 2\right)$ both lie on $y = \frac{1}{x}$. Is there such a reversal for all points on the graph? Why may the coordinate be reversed?

30. For inputs $x > 1$, which graph is on top: $y = \frac{1}{x}$ or $y = \frac{1}{x^2}$? Why?

31. The force due to gravitational attraction between planets varies inversely with the square of the distance between the planets. Which equation would describe gravitational attraction?

a. $F = km_1m_2d^2$ **b.** $F = \frac{km_1m_2}{d}$

c. $F = \frac{km_1m_2}{d^2}$

32. Under certain circumstances, the intensity of a magnetic field varies inversely with the cube of the distance from the magnet. Which equation would describe the intensity?

a. $H = \frac{M}{r^2}$ **b.** $H = \frac{M}{r^3}$

c. $H = Mr^3$ **d.** $H = Mr^2$

Project

33. ***Measuring Light Intensity.*** Light intensity is measured by the formula $I = \frac{k}{d^2}$, where k is a constant and $d \neq 0$. Obtain a light meter or other device for measuring light intensity. In a darkened room, measure light intensity at several distances from a single light source (such as a bare light bulb).

a. Construct a three-column table. Record your distance, d, and intensity, I, data in the first two columns.

b. In the third column, calculate $k = I \cdot d^2$.

c. Plot your data on a graph, with distance on the horizontal axis and intensity on the vertical axis.

d. Comment on how accurately your experiment reflects the formula. In a perfect experiment, how should the k numbers in the table compare?

e. What factors might change the accuracy of your results?

9.2 Simplifying Rational Expressions

OBJECTIVES

- Simplify rational expressions.
- Simplify rational expressions containing units of measure.
- Find when a rational expression simplifies to 1 or to −1.

WARM-UP

Factor these binomials and trinomials by guess and check or the table method.

1. $x^2 - 4$ **2.** $x^2 + 4x - 5$ **3.** $2x^2 + 7x + 6$

4. $x^2 + 2x$ **5.** $x^2 - 5x + 6$ **6.** $2x^2 + 6x$

7. $x^2 + 3x + 2$ **8.** $x^2 + 6x + 9$ **9.** $4x^2 + 6x$

10. $x^2 - 9$ **11.** $6 - x - x^2$ **12.** $4 - 3x - x^2$

IN THIS SECTION, we simplify rational expressions containing variables and units of measure. We investigate the simplified result when the numerator is the opposite of the denominator.

As mentioned in Section 2.3, when we *simplify a fraction to an equivalent fraction*, we are using the **simplification property of fractions**.

Simplification Property of Fractions

For all real numbers, a not zero and c not zero,

$$\frac{ab}{ac} = \frac{a}{a} \cdot \frac{b}{c} = 1 \cdot \frac{b}{c} = \frac{b}{c}$$

When we simplify a fraction, we factor the numerator and denominator and remove the common factors, such as $\frac{a}{a}$. If there are no common factors, the fraction cannot be simplified and is said to be in lowest terms.

EXAMPLE  Simplifying rational expressions Assume there are no zero denominators and simplify the following:

a. $\frac{15}{35}$ **b.** $\frac{28}{18}$ **c.** $\frac{2a}{a^2}$ **d.** $\frac{6xy}{2y^2}$

Solution **a.** $\frac{15}{35} = \frac{3 \cdot 5}{5 \cdot 7} = \frac{3}{7}$ **b.** $\frac{28}{18} = \frac{2 \cdot 2 \cdot 7}{2 \cdot 3 \cdot 3} = \frac{14}{9}$

c. $\frac{2a}{a^2} = \frac{2 \cdot a}{a \cdot a} = \frac{2}{a}$ **d.** $\frac{6xy}{2y^2} = \frac{2 \cdot 3 \cdot x \cdot y}{2 \cdot y \cdot y} = \frac{3x}{y}$ ●

The expressions in Example 2 need to be factored before we can simplify to lowest terms. The factors may be a monomial and binomial or two binomials.

EXAMPLE  Simplifying rational expressions Assume there are no zero denominators and simplify these rational expressions.

a. $\frac{2x - 4}{3x - 6}$ **b.** $\frac{x^2 - 4x}{x^2 + 2x}$ **c.** $\frac{x - 3}{(x + 3)(x - 3)}$

d. $\frac{x + x^2}{1 + x}$ **e.** $\frac{x^2 - 4}{x^2 + 3x + 2}$ **f.** $\frac{x^2 + 4x - 5}{x^2 - 2x + 1}$

Solution **a.** $\frac{2x - 4}{3x - 6} = \frac{2(x - 2)}{3(x - 2)} = \frac{2}{3}$

b. $\dfrac{x^2 - 4x}{x^2 + 2x} = \dfrac{x(x - 4)}{x(x + 2)} = \dfrac{x - 4}{x + 2}$

c. $\dfrac{x - 3}{(x + 3)(x - 3)} = \dfrac{1(x - 3)}{(x + 3)(x - 3)} = \dfrac{1}{(x + 3)}$

d. $\dfrac{x + x^2}{1 + x} = \dfrac{x(1 + x)}{(1 + x)} = \dfrac{x}{1} = x$

e. $\dfrac{x^2 - 4}{x^2 + 3x + 2} = \dfrac{(x + 2)(x - 2)}{(x + 1)(x + 2)} = \dfrac{(x - 2)}{(x + 1)}$

f. $\dfrac{x^2 + 4x - 5}{x^2 - 2x + 1} = \dfrac{(x + 5)(x - 1)}{(x - 1)(x - 1)} = \dfrac{(x + 5)}{(x - 1)}$ ●

Equivalent Fractions

To add or subtract fractions (Section 9.4), we need fractions with common denominators. Once we know the common denominator, we change each fraction to an equivalent fraction. This change uses the **equivalent fraction property**, which is the *simplification property in reverse*. We multiply the numerator and denominator of the fraction by a common factor.

Equivalent Fraction Property

For all real numbers, a not zero and c not zero,

$$\frac{b}{c} = \frac{b}{c} \cdot \frac{a}{a} = \frac{ab}{ac}$$

EXAMPLE 3

Finding equivalent expressions Change each of the following to an equivalent expression with the indicated numerator or denominator. Assume there are no zero denominators.

a. $\dfrac{4}{5} = \dfrac{12}{\quad}$ **b.** $\dfrac{a}{2} = \dfrac{\quad}{2b}$ **c.** $\dfrac{3}{x} = \dfrac{3x}{\quad}$

d. $\dfrac{2}{x + 2} = \dfrac{\quad}{2x(x + 2)}$ **e.** $\dfrac{x}{x + 2} = \dfrac{\quad}{x^2 + 3x + 2}$

Solution **a.** $\dfrac{4}{5} = \dfrac{4 \cdot 3}{5 \cdot 3} = \dfrac{12}{15}$ **b.** $\dfrac{a}{2} = \dfrac{a \cdot b}{2 \cdot b} = \dfrac{ab}{2b}$ **c.** $\dfrac{3}{x} = \dfrac{3 \cdot x}{x \cdot x} = \dfrac{3x}{x^2}$

d. $\dfrac{2}{x + 2} = \dfrac{2x \cdot 2}{2x(x + 2)} = \dfrac{4x}{2x(x + 2)}$

e. $\dfrac{x}{x + 2} = \dfrac{x(x + 1)}{(x + 2)(x + 1)} = \dfrac{x^2 + x}{x^2 + 3x + 2}$ ●

We are not always told whether to simplify a fraction or to change it to an equivalent fraction in unsimplified form. Example 4 uses both operations.

EXAMPLE 4

Working with equivalent fractions Find the missing number in each pair of fractions. Assume there are no zero denominators.

a. $\dfrac{5}{8}, \dfrac{10}{\quad}$ **b.** $\dfrac{8}{6}, \dfrac{4}{\quad}$ **c.** $\dfrac{2}{5}, \dfrac{\quad}{15}$

d. $\dfrac{4}{3}, \dfrac{12}{\quad}$ **e.** $\dfrac{5x}{10x}, \dfrac{\quad}{2}$ **f.** $\dfrac{3a}{6a^2}, \dfrac{\quad}{12a^3}$

Solution **a.** $\frac{5}{8} = \frac{5 \cdot 2}{8 \cdot 2} = \frac{10}{16}$ **b.** $\frac{8}{6} = \frac{2 \cdot 4}{2 \cdot 3} = \frac{4}{3}$ **c.** $\frac{2}{5} = \frac{2 \cdot 3}{5 \cdot 3} = \frac{6}{15}$

d. $\frac{4}{3} = \frac{4 \cdot 3}{3 \cdot 3} = \frac{12}{9}$ **e.** $\frac{5x}{10x} = \frac{5 \cdot x}{2 \cdot 5 \cdot x} = \frac{1}{2}$ **f.** $\frac{3a}{6a^2} = \frac{3a \cdot 2a}{6a^2 \cdot 2a} = \frac{6a^2}{12a^3}$ ●

Expressions Containing Units

As noted earlier, many expressions involve units of measurement. The simplification property indicates that fractions containing units, such as

$$\frac{\text{meters}}{\text{meters}}, \quad \frac{\text{inches}}{\text{inches}}, \quad \frac{\text{gallons}}{\text{gallons}}, \quad \frac{\text{miles}}{\text{miles}}, \quad \text{or} \quad \frac{\text{hours}}{\text{hours}}$$

all simplify to 1.

EXAMPLE 5 Simplifying fractions containing units Simplify the following:

a. $\frac{48 \text{ cm}^3}{16 \text{ cm}}$ **b.** $\frac{5000 \text{ foot-pounds}}{10 \text{ feet}}$ **c.** $\frac{24 \text{ degree days}}{6 \text{ days}}$

Solution **a.** $\frac{48 \text{ cm}^3}{16 \text{ cm}} = \frac{16 \cdot 3 \text{ cm} \cdot \text{cm} \cdot \text{cm}}{16 \text{ cm}} = \frac{3 \text{ cm}^2}{1}$

b. $\frac{5000 \text{ foot-pounds}}{10 \text{ feet}} = \frac{500 \cdot 10 \text{ foot-pounds}}{10 \text{ feet}} = \frac{500 \text{ pounds}}{1}$

c. $\frac{24 \text{ degree days}}{6 \text{ days}} = \frac{6 \cdot 4 \text{ degree days}}{6 \text{ days}} = \frac{4 \text{ degrees}}{1}$ ●

Expressions Containing Opposites

The concept of opposites is central to the next two examples, so we restate the definition here.

Definition of Opposites

> Two expressions are **opposites** if they add to zero.

We use opposites as we look at some special forms of simplifying. The examples provide a shortcut and a caution.

EXAMPLE 6 Working with opposites In parts a to c, fill in the blank.

a. The opposite of 7 is ____. **b.** The opposite of x is ____.

c. The opposite of $7 + x$ is ____.

In parts d to k, add the two expressions in parentheses and note whether they are opposites.

d. $(7 + x) + (-7 - x)$ **e.** $(7 + x) + (7 - x)$ **f.** $(7 - x) + (7 + x)$

g. $(7 - x) + (x - 7)$ **h.** $(a - b) + (a + b)$ **i.** $(a - b) + (b - a)$

j. $(a - b) + (-a + b)$ **k.** $(a - b) + (a - b)$

Simplify the expressions in parts l and m.

l. $\frac{3 - 4}{4 - 3}$ **m.** $\frac{5 - (-2)}{-2 - 5}$

n. Why do parts l and m have the same answer?

Solution **a.** -7 **b.** $-x$ **c.** $-7 - x$

d. 0; $7 + x$ and $-7 - x$ are opposites

e. 14; not opposites **f.** 14; not opposites

g. 0; $7 - x$ and $x - 7$ are opposites

h. $2a$; not opposites **i.** 0; $a - b$ and $b - a$ are opposites

j. 0; $a - b$ and $-a + b$ are opposites

k. $2a - 2b$; not opposites **l.** -1 **m.** -1

n. The fractions have opposites in the numerator and denominator. Both simplify to -1. ●

Here are some things to remember when simplifying fractions (see Figure 8):

- When the numerator and denominator of a fraction are the same, the fraction simplifies to 1.
- When the numerator and denominator of a fraction are opposites, the fraction simplifies to -1.

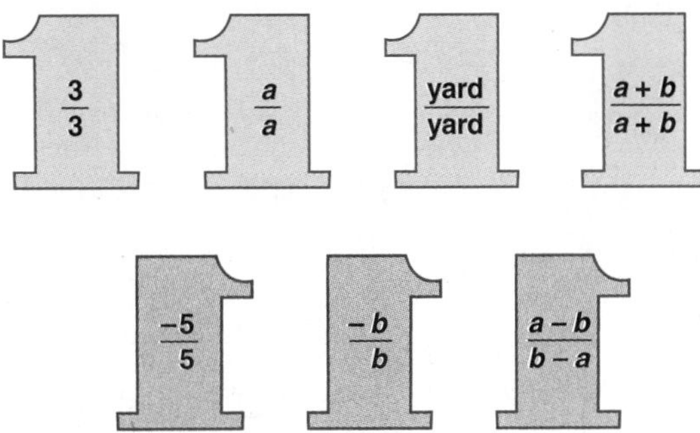

Figure 8

Example 7 shows why rational expressions containing opposite numerators and denominators simplify to -1.

EXAMPLE 7 Simplifying fractions containing opposites Show that $\dfrac{a-b}{b-a} = -1,\ a \neq b$.

Solution At least three methods are possible.

Method 1: Multiply the numerator and denominator of the fraction by -1:

$$\frac{a-b}{b-a} = \frac{(-1)(a-b)}{(-1)(b-a)} = \frac{(-1)(a-b)}{(a-b)} = -1$$

Method 2: Multiply either the numerator or the denominator by $(-1)(-1)$, which equals 1 and will not change the fraction:

$$\frac{a-b}{b-a} = \frac{(-1)(-1)(a-b)}{(b-a)} = \frac{(-1)(b-a)}{(b-a)} = -1$$

Method 3: Factor -1 from either the numerator or the denominator:

$$\frac{a-b}{b-a} = \frac{(-1)(b-a)}{(b-a)} = -1$$

With the first two methods, we multiplied out only one of the -1 factors. This changed one of the expressions to its opposite and thus permitted simplification. ●

It does not matter which method is used to simplify fractions containing opposites to -1. Choose a method that makes sense and consistently gives you the correct result.

ANSWER BOX

Warm-up: **1.** $(x-2)(x+2)$ **2.** $(x+5)(x-1)$ **3.** $(2x+3)(x+2)$ **4.** $x(x+2)$ **5.** $(x-2)(x-3)$ **6.** $2x(x+3)$ **7.** $(x+1)(x+2)$ **8.** $(x+3)(x+3)$ **9.** $2x(2x+3)$ **10.** $(x+3)(x-3)$ **11.** $(3+x)(2-x)$ **12.** $(4+x)(1-x)$

EXERCISES 9.2

Simplify the expressions in Exercises 1 to 20. Assume there are no zero denominators.

1. $\dfrac{2ab}{6a^2b}$ **2.** $\dfrac{3ac}{15ac^2}$

3. $\dfrac{12cd^2}{8c^2d}$ **4.** $\dfrac{15b^2c^3}{10b^3c}$

5. $\dfrac{(x-2)(x+3)}{x+3}$ **6.** $\dfrac{(x-2)(x+3)}{(x+3)(x+2)}$

7. $\dfrac{2-x}{(x+2)(x-2)}$ **8.** $\dfrac{1-x}{(x-1)(x+2)}$

9. $\dfrac{3ab+3ac}{5b^2+5bc}$ **10.** $\dfrac{2ac+4bc}{4ad+8bd}$

11. $\dfrac{4x^2+8x}{2x^2-4x}$ **12.** $\dfrac{3x^2-6x}{6x^2+12x}$

13. $\dfrac{x^2-4}{x^2+5x+6}$ **14.** $\dfrac{x^2-1}{x^2+4x+3}$

15. $\dfrac{x^2+x-6}{x^2-2x}$ **16.** $\dfrac{x^2+x-2}{2x+4}$

17. $\dfrac{x-3}{6-2x}$ **18.** $\dfrac{x-4}{12-3x}$

19. $\dfrac{x^2-5x+6}{x^2-9}$ **20.** $\dfrac{x^2-3x-4}{x^2-16}$

Find the missing number or expression in each pair of equivalent fractions in Exercises 21 to 30. Assume there are no zero denominators.

21. $\dfrac{6}{9}, \dfrac{\quad}{45}$ **22.** $\dfrac{15}{10}, \dfrac{3}{\quad}$

23. $\dfrac{24x}{3x^2}, \dfrac{8}{\quad}$ **24.** $\dfrac{24x}{3x^2}, \dfrac{72x^3}{\quad}$

25. $\dfrac{\quad}{10a^2}, \dfrac{b}{2a}$ **26.** $\dfrac{6cd}{9cd^2}, \dfrac{2}{\quad}$

27. $\dfrac{x+2}{\quad}, \dfrac{2x+4}{2x-6}$ **28.** $\dfrac{x+3}{x+5}, \dfrac{x^2+3x}{\quad}$

29. $\dfrac{a+b}{\quad}, \dfrac{a^2+2ab+b^2}{a^2-b^2}$ **30.** $\dfrac{\quad}{a^2+4a+4}, \dfrac{a-2}{a+2}$

In Exercises 31 to 36, simplify.

31. $\dfrac{108 \text{ m}^2}{6 \text{ m}}$ **32.** $\dfrac{125 \text{ in}^3}{5 \text{ in.}}$

33. $\dfrac{144 \text{ in}^2}{1728 \text{ in}^3}$ **34.** $\dfrac{27 \text{ yd}^3}{9 \text{ yd}^2}$

35. $\dfrac{2060 \text{ degree gallons}}{103 \text{ degrees}}$ **36.** $\dfrac{1200 \text{ foot-pounds}}{200 \text{ pounds}}$

Simplify the fractions in Exercises 37 to 40. What do you observe? Why?

37. $\dfrac{7-4}{4-7}$ **38.** $\dfrac{15-9}{9-15}$

39. $\dfrac{-3-4}{4-(-3)}$ **40.** $\dfrac{6-(-2)}{-2-6}$

What is the opposite of each expression in Exercises 41 to 46?

41. $a-b$ **42.** $a+b$ **43.** $-a+b$

44. $b-a$ **45.** $b+a$ **46.** $-a-b$

In Exercises 47 to 50, use both of the following statements to test whether the two expressions are opposites:

a. *Opposites add to zero.*

b. *If we multiply an expression by -1, we get its opposite.*

47. $n-m$ and $n+m$ **48.** $n-m$ and $m-n$

49. $x-2$ and $2-x$ **50.** $x-2$ and $x+2$

In Exercises 51 to 58, what numerator or denominator is needed in the equation to make a true statement? State any inputs that must be excluded.

51. $\dfrac{x+2}{\quad} = 1$ **52.** $\dfrac{x+2}{\quad} = -1$ **53.** $\dfrac{x-2}{\quad} = -1$

54. $\frac{x-2}{} = 1$ **55.** $\frac{}{a-b} = -1$ **56.** $\frac{b-a}{} = 1$

57. $\frac{3-x}{} = 1$ **58.** $\frac{3-x}{} = -1$

In Exercises 59 to 62, identify the equal expressions in each set. How are the other two expressions related?

59. $4, \frac{1}{4}, -4, 4^{-1}$ **60.** $-3, 3, \frac{1}{3}, 3^{-1}$

61. $\frac{1}{a}, a, a^{-1}, -a$ **62.** $b^{-1}, \frac{1}{b}, b, -b$

In Exercises 63 and 64, indicate whether the statement is true or false. If true, explain why; if false, give an example that shows why it is false.

63. A rational expression must be factorable in order to be simplified to lowest terms.

64. A rational expression can have the same variables in the numerator and denominator and still not be simplifiable.

65. If $\frac{a}{a} = 1$, list three fractions containing positive or negative a's that would simplify to -1.

66. What happens when we divide opposites?

67. Describe the role of factoring in simplifying fractions.

68. What may be concluded about $-\frac{a}{b}, \frac{-a}{b}$, and $\frac{a}{-b}$?

Projects

69. ***Fraction Pattern, I***

a. Simplify:

$$\frac{1+2+3}{4+5+6} \qquad \frac{7+8+9}{10+11+12} \qquad \frac{13+14+15}{16+17+18}$$

b. Predict the values of these fractions:

$$\frac{50+51+52}{53+54+55} \qquad \frac{100+101+102}{103+104+105}$$

c. Show that your pattern always works by building a rational expression (a fraction) with x as the first number, $x+1$ as the second number, and so on. Simplify the expression.

70. ***Wheat Harvest.*** Refer to the wheat harvest problem at the beginning of the chapter. Suppose each large rectangle in the following figure represents the entire wheat harvest. If Terry harvests the wheat in 18 days, then each day he harvests $\frac{1}{18}$ of the crop. The shading in the first rectangle represents the amount Terry harvests in one day. The shading in the second rectangle is the amount Lee harvests in one day: $\frac{1}{12}$. Together they harvest the amount shown in the third rectangle. The second day of harvest is shown by the second row of rectangles.

Day 1

\+ =

Day 2

\+ =

Day 3

\+ =

Day 4

\+ =

Day 5

\+ =

a. Shade the rectangles for each subsequent day's accumulative harvest. Add more rectangles as needed.

b. How will the rectangle look when the harvest is complete?

c. Estimate the total number of days needed to complete the harvest.

d. What is a better way to show the fraction parts $\frac{1}{12}$ and $\frac{1}{18}$ so that we can improve our estimate of the total number of days needed to complete the harvest?

9.3 Multiplication and Division of Rational Expressions

OBJECTIVES

- Factor, simplify, and multiply rational expressions.
- Change division problems to multiplication problems and complete the multiplication.
- Simplify complex rational expressions by changing to division.
- Apply multiplication and division principles to expressions containing units of measure.

WARM-UP

Translate each phrase into symbols, and perform the indicated operation.

1. The product of 3 and $\frac{1}{2}$
2. The quotient of 3 and $\frac{1}{2}$
3. The quotient of $\frac{1}{2}$ and 3
4. The quotient of $\frac{1}{2}$ and $\frac{1}{3}$
5. The product of $\frac{1}{3}$ and $\frac{3}{4}$
6. The quotient of $\frac{3}{4}$ and $\frac{3}{5}$

Factor these numbers or expressions.

7. 35
8. $cx + c$
9. $x^2 - 5x - 6$
10. $x^2 + 3x + 2$
11. $x^2 - 3x - 4$
12. $x^2 - 16$

IN THIS SECTION, we apply principles of multiplication and division to rational expressions, complex rational expressions, and expressions containing units of measure.

Multiplication of Rational Expressions

In the first three examples, observe the role of $\frac{a}{a} = 1$ in simplifying the multiplication and division of rational expressions. In each case, *the solution is found most easily by simplifying expressions before doing the final multiplication.*

EXAMPLE 1 Multiplying fractions Multiply $\frac{8}{35} \cdot \frac{25}{32} \cdot \frac{14}{9}$.

Solution

$$\frac{8}{35} \cdot \frac{25}{32} \cdot \frac{14}{9} = \frac{8 \cdot 25 \cdot 14}{35 \cdot 32 \cdot 9} = \frac{2 \cdot 2 \cdot 2 \cdot 5 \cdot 5 \cdot 2 \cdot 7}{5 \cdot 7 \cdot 2 \cdot 2 \cdot 2 \cdot 2 \cdot 2 \cdot 3 \cdot 3} = \frac{5}{2 \cdot 3 \cdot 3} = \frac{5}{18}$$

In this solution, we use the multiplication property to change the three fractions to a single fraction with one numerator and one denominator. We simplify and then multiply the remaining factors in the numerator and denominator. Changing to all primes may not be necessary. ●

EXAMPLE 2 Multiplying rational expressions Multiply $\frac{ax^2}{cx} \cdot \frac{cx + c}{a^2}$. Assume there are no zero denominators.

Solution

$$\frac{ax^2}{cx} \cdot \frac{cx + c}{a^2} = \frac{ax^2(cx + c)}{cx \cdot a^2} = \frac{a \cdot x \cdot x \cdot c(x + 1)}{c \cdot x \cdot a \cdot a} = \frac{x(x + 1)}{a}$$

In this solution, we again write the numerators and denominators as products and factor them completely. Simplification eliminates any further need for multiplication. ●

EXAMPLE 3 Multiplying rational expressions Multiply $\frac{x + 2}{x + 6} \cdot \frac{x^2 - 5x - 6}{x^2 + 3x + 2}$. Assume there are no zero denominators.

Solution

$$\frac{x+2}{x+6}\cdot\frac{x^2-5x-6}{x^2+3x+2}=\frac{(x+2)(x^2-5x-6)}{(x+6)(x^2+3x+2)}$$

$$=\frac{(x+2)(x+1)(x-6)}{(x+6)(x+1)(x+2)}=\frac{x-6}{x+6}$$

We apply the multiplication property of fractions by writing the two fractions as one fraction. We then factor the numerator and denominator expressions and simplify the fraction. ●

Caution: If we first multiply the fractions in Example 3, the resulting expression is not easily simplified. It requires factoring techniques beyond the level of this or the next mathematics course.

$$\frac{x+2}{x+6}\cdot\frac{x^2-5x-6}{x^2+3x+2}=\frac{(x+2)(x^2-5x-6)}{(x+6)(x^2+3x+2)}=\frac{x^3-3x^2-16x-12}{x^3+9x^2+20x+12}=?$$

If your homework solutions contain similar expressions, go back to the original problem, factor, and simplify before multiplying.

In all multiplication problems, keep in mind that we are eliminating factors, not terms. Any units are treated the same way as factors. We used this idea earlier in unit analysis.

EXAMPLE 4 Multiplying and simplifying units: water flow A shower head permits a flow of 5 gallons per minute. How many gallons of water are used in a $3\frac{1}{2}$-minute shower?

Solution

$$\frac{5\text{ gal}}{1\text{ min}}\cdot 3.5\text{ min}=\frac{5(3.5)}{1}\frac{\text{gal}\cdot\text{min}}{\text{min}}=17.5\text{ gal}$$ ●

Here are some things to remember when multiplying rational expressions:

- Factor, simplify to lowest terms, and then multiply, as needed.
- Simplify only like factors using $\frac{a}{a}=1$. The x's in $\frac{x+5}{x}$ are terms, not factors.
- Any units are treated the same way as factors.
- No common denominator is needed for multiplication.

Division of Rational Expressions

Division of rational expressions is based on the same property as division of fractions.

Division of Fractions

> To divide fractions, multiply the first fraction by the reciprocal of the second.

EXAMPLE 5 Dividing expressions in fraction notation Assume there are no zero denominators.

a. Divide $\frac{1}{18}$ and $\frac{1}{25}$. **b.** Divide $\frac{ax}{b}$ and $\frac{cx}{d}$.

Solution To divide, we change the division to multiplication by the reciprocal of the second fraction.

a. $\frac{1}{18} \div \frac{1}{25} = \frac{1}{18} \cdot \frac{25}{1} = \frac{25}{18}$

b. $\frac{ax}{b} \div \frac{cx}{d} = \frac{ax}{b} \cdot \frac{d}{cx} = \frac{adx}{bcx} = \frac{ad}{bc}$ ●

EXAMPLE 6 **Dividing rational expressions** Divide $\frac{x^2 - 3x - 4}{x - 3} \div \frac{x^2 - 16}{x^2 - 9}$. Assume there are no zero denominators.

Solution

$$\frac{x^2 - 3x - 4}{x - 3} \div \frac{x^2 - 16}{x^2 - 9} = \frac{x^2 - 3x - 4}{x - 3} \cdot \frac{x^2 - 9}{x^2 - 16}$$
$$= \frac{(x - 4)(x + 1)(x - 3)(x + 3)}{(x - 3)(x - 4)(x + 4)}$$
$$= \frac{(x + 1)(x + 3)}{x + 4}$$

After changing the problem to a multiplication problem, we factor the expression and simplify. No further simplification is possible because there are no common factors in the numerator and denominator. ●

EXAMPLE 7 **Dividing expressions containing units** Divide 450 miles by 60 miles per hour.

Solution

$$450 \text{ mi} \div \frac{60 \text{ mi}}{1 \text{ hr}} = 450 \text{ mi} \cdot \frac{1 \text{ hr}}{60 \text{ mi}} = 7.5 \text{ hr}$$ ●

The multiplication and division of rational expressions follow these rules:

Multiplication and Division of Rational Expressions

$$\frac{a}{b} \cdot \frac{c}{d} = \frac{ac}{bd}, \quad b \neq 0, d \neq 0$$

$$\frac{a}{b} \div \frac{c}{d} = \frac{a}{b} \cdot \frac{d}{c}, \quad b \neq 0, c \neq 0, d \neq 0$$

Here are some things to remember when dividing rational expressions:

- Change division to multiplication by the reciprocal.
- Factor, simplify to lowest terms, and then multiply.

Complex Rational Expressions

The technique of changing division to multiplication by a reciprocal may be applied to more complicated forms of rational expressions—the complex rational expressions. *Rational expressions that contain fractions in either the numerator or the denominator* are called **complex fractions**. When the numerator, the denominator, or both are single fractions, recall that the fraction bar means division and change the complex fraction to a division problem.

EXAMPLE 8 Dividing complex rational expressions Simplify $\dfrac{\frac{8}{15}}{\frac{4}{5}}$.

Solution The fraction bar means division, so we change the notation to two fractions separated by the ÷ sign.

$$\frac{\frac{8}{15}}{\frac{4}{5}} = \frac{8}{15} \div \frac{4}{5} = \frac{8}{15} \cdot \frac{5}{4} = \frac{2 \cdot 4 \cdot 5}{3 \cdot 5 \cdot 4} = \frac{2}{3}$$

EXAMPLE 9 Simplifying complex fractions Assume there are no zero denominators and simplify the complex fraction $\dfrac{\frac{a}{b}}{\frac{c}{d}}$.

Solution

$$\frac{\frac{a}{b}}{\frac{c}{d}} = \frac{a}{b} \div \frac{c}{d} = \frac{a}{b} \cdot \frac{d}{c} = \frac{ad}{bc}$$

We simplify the complex fraction by writing it as a division problem, with the longer fraction bar replaced by a division sign. The division is then changed to multiplication by a reciprocal.

We may include units in simplifying complex fractions.

EXAMPLE 10 Applying complex fractions Answer the question by writing a complex fraction and simplifying.

a. How many half-dollars are in \$5.00?

b. How many fourths are in $\frac{5}{8}$?

c. If a 19-passenger jet flying at 459 nautical miles (nm) per hour consumes 397 gallons of fuel per hour, how many nautical miles does it get per gallon?

Solution In parts a and b, *how many* implies division.

a. $\dfrac{5.00 \text{ dollars}}{\frac{1}{2} \text{ dollar}} = 5 \div \dfrac{1}{2} = 5 \cdot \dfrac{2}{1} = 10$

b. $\dfrac{\frac{5}{8}}{\frac{1}{4}} = \dfrac{5}{8} \div \dfrac{1}{4} = \dfrac{5}{8} \cdot \dfrac{4}{1} = \dfrac{20}{8} = \dfrac{5}{2} = 2\frac{1}{2}$

In part c, we are looking for nautical miles per gallon, so the expression containing nautical miles should be placed in the numerator.

c. $\dfrac{\frac{459 \text{ nm}}{\text{hr}}}{\frac{397 \text{ gal}}{\text{hr}}} = \dfrac{459 \text{ nm}}{\text{hr}} \div \dfrac{397 \text{ gal}}{\text{hr}} = \dfrac{459 \text{ nm}}{\text{hr}} \cdot \dfrac{\text{hr}}{397 \text{ gal}} \approx 1.16 \dfrac{\text{nm}}{\text{gal}}$

Unit analysis can also be used to solve this problem. We start with 459 nm per hour and multiply by an expression containing hours in the numerator in order to cancel the hours:

$$\frac{459 \text{ nm}}{\text{hr}} \cdot \frac{\text{hr}}{397 \text{ gal}} \approx 1.16 \frac{\text{nm}}{\text{gal}}$$

ANSWER BOX

Warm-up: **1.** 1.5 **2.** 6 **3.** $\frac{1}{6}$ **4.** $\frac{3}{2}$ **5.** $\frac{1}{4}$ **6.** $\frac{5}{4}$ **7.** $5 \cdot 7$ **8.** $c(x + 1)$ **9.** $(x - 6)(x + 1)$ **10.** $(x + 2)(x + 1)$ **11.** $(x - 4)(x + 1)$ **12.** $(x - 4)(x + 4)$

EXERCISES 9.3

Multiply and divide each pair of fractions in Exercises 1 to 4. Assume there are no zero denominators.

1. a. $\frac{1}{3}$ and $\frac{1}{4}$ **b.** $\frac{1}{2}$ and $\frac{1}{5}$

2. a. $\frac{1}{4}$ and $\frac{1}{12}$ **b.** $\frac{1}{8}$ and $\frac{1}{12}$

3. a. $\frac{3}{4}$ and $\frac{1}{6}$ **b.** $\frac{2}{3}$ and $\frac{1}{6}$

4. a. $\frac{a}{b}$ and $\frac{c}{d}$ **b.** $\frac{w}{x}$ and $\frac{y}{z}$

5. Calculate these fractional expressions. What may be observed about each pair of problems? What do they tell us about multiplication and division?

a. $100 \div \frac{4}{1}$ and $100 \cdot \frac{1}{4}$ **b.** $100 \div \frac{1}{5}$ and $100 \cdot \frac{5}{1}$

6. For $\frac{5}{8} \div \frac{1}{4}$, the answer is $2\frac{1}{2}$. The figure below shows $\frac{5}{8}$ of one rectangle shaded and $\frac{1}{4}$ of an identical rectangle shaded. Trace these rectangles and show why there are $2\frac{1}{2}$ fourths in $\frac{5}{8}$.

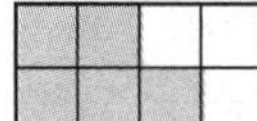 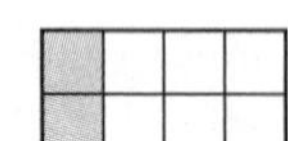

In Exercises 7 and 8, multiply or divide, as indicated. Assume there are no zero denominators.

7. a. $\frac{1}{x} \cdot \frac{x^2}{1}$ **b.** $\frac{1}{a} \div \frac{a^2b^2}{1}$

c. $\frac{a}{b} \cdot \frac{b^2}{a^2}$ **d.** $\frac{a}{b} \div \frac{a^2}{b^3}$

8. a. $\frac{1}{x} \cdot \frac{x^3}{1}$ **b.** $\frac{1}{b} \div \frac{a^2b^2}{1}$

c. $\frac{b}{a} \div \frac{a^2}{b^2}$ **d.** $\frac{a^2}{b^3} \div \frac{a}{b}$

In Exercises 9 to 12, multiply or divide, as indicated. Assume there are no zero denominators.

9. a. $\frac{x^2 + 2x + 1}{x + 1} \cdot \frac{x}{x^2 + x}$

b. $\frac{x^2 - 4}{x + 2} \cdot \frac{1}{x^2 - x}$

c. $\frac{x + 2}{x^2 - 4x + 4} \div \frac{x^2 + 2x}{x - 2}$

d. $\frac{x^2 - 5x}{x^2 + 5x} \div \frac{x^2 - 10x + 25}{x}$

e. $\frac{x^2 - 6x + 9}{x^2 + 3x} \div \frac{x^2 - 9}{x}$

10. a. $\frac{x^2 - 7x + 12}{x^2 - 4} \cdot \frac{x^2 + 2x}{x - 3}$

b. $\frac{x^2 - 2x}{x} \cdot \frac{x^2}{x^2 - 3x + 2}$

c. $\frac{x - 3}{x^2 - 4x + 3} \div \frac{x^2 + x}{x - 1}$

d. $\frac{x^2 - 6x + 9}{x^2 + 3x} \cdot \frac{x + 3}{x - 3}$

e. $\frac{x^2 + 3x}{x} \cdot \frac{x^2 - x - 6}{x^2 - 9}$

11. a. $\frac{x^2 + x}{x - 1} \cdot \frac{x^2 - 1}{x + 1}$

b. $\frac{x - 3}{x^2 + 6x + 9} \cdot \frac{x + 3}{x^2 - 9}$

c. $\frac{4 - 8x}{x + 1} \div \frac{1 - 2x}{x^2 - 1}$

d. $\frac{x - x^2}{x + 1} \cdot \frac{x - 1}{1 - x}$

e. $\frac{x^2 - x}{x^2 - 3x + 2} \div \frac{1 - x^2}{x^2 - 2x + 1}$

12. a. $\frac{3x + 3}{1 - x} \cdot \frac{1 - x^2}{x + 1}$

b. $\frac{2 - x}{x + 2} \div \frac{4 - 2x}{x^2 + 4x + 4}$

c. $\frac{x^2 - x}{x^2} \div \frac{x - 1}{x}$

d. $\frac{x + 3}{3x} \cdot \frac{9x^2}{x^2 - 9}$

e. $\frac{x^2 + 4x + 4}{x^2 - 4} \div \frac{x^2 + 2x}{2 - x}$

In Exercises 13 and 14, write each as a multiplication or division, and then perform the indicated operation. Assume there are no zero denominators.

13. **a.** Product, $\frac{1}{a}$ and $\frac{1}{b}$ **b.** Quotient, a and $\frac{1}{a}$

c. Product, $\frac{1}{b}$ and a **d.** Quotient, $\frac{1}{b}$ and $\frac{1}{a}$

e. Product, $\frac{a}{b}$ and $\frac{1}{b}$ **f.** Quotient, $\frac{a}{b}$ and $\frac{1}{a}$

14. **a.** Quotient, $\frac{1}{a}$ and $\frac{1}{b}$ **b.** Product, b and $\frac{1}{b}$

c. Quotient, $\frac{1}{b}$ and a **d.** Quotient, $\frac{1}{b}$ and b

e. Quotient, $\frac{a}{b}$ and $\frac{1}{b}$ **f.** Product, $\frac{1}{b}$ and $\frac{1}{a}$

In Exercises 15 to 26, use properties of fractions to simplify the expressions. The word per means division and may be replaced by a fraction bar.

15. $\dfrac{\text{miles}}{\dfrac{\text{miles}}{\text{hour}}}$

16. $\dfrac{\text{kilometers}}{\dfrac{\text{kilometers}}{\text{minute}}}$

17. $\dfrac{\text{93,000,000 miles}}{\text{186,000 miles per second}}$

18. $\dfrac{\text{5280 feet}}{\text{1130 feet per second}}$

19. $\dfrac{\text{300 miles per hour}}{\text{100 gallons per hour}}$

20. $\dfrac{\text{60 miles per hour}}{\text{25 miles per gallon}}$

21. $\dfrac{\text{12 cookies per dozen}}{\text{\$2.98 per dozen}}$

22. $\dfrac{\text{12 stitches per inch}}{\dfrac{\text{1 yd}}{\text{36 inches}}}$

23. $\dfrac{\text{85 words per minute}}{\text{300 words per page}}$

24. $\dfrac{\text{880 cycles per second}}{\text{344 meters per second}}$

25. $\dfrac{\text{40 moles}}{\text{12 moles per liter}}$

26. $\dfrac{\text{186 days}}{\text{5 days per week}}$

27. Choose one of the expressions from Exercises 15 to 26, and give a situation in which it would make sense.

28. Make up a division problem containing units of measure, and explain what the answer means.

29. The current, I, in an electrical circuit is found by dividing the voltage, V, by the resistance, R. Suppose the voltage and resistance vary with time as in the equations

$$V = \frac{t^2 - 4}{2t^2 - 3t - 2} \quad \text{and} \quad R = \frac{t + 2}{t^2}$$

Assume there are no zero denominators. Find a formula in terms of t that gives the current, I.

30. In economics, total quantity sold is the product of the price and the demand. Find the total quantity, Q.

$$\text{Price} = 3x + 6 \quad \text{and} \quad \text{Demand} = \frac{800}{x^2 + 2x}$$

Assume there are no zero denominators.

31. A student familiar with simplifying fractions such as $\frac{12}{15}$ is puzzled by the canceling of threes in $\frac{3}{5} \cdot \frac{4}{3}$. Explain why the product can be simplified when the threes are in different fractions.

32. Explain why

$$\frac{x(x+1)(x-3)}{x(x-3)} = x + 1$$

is correct and

$$\frac{x^2 + x + 1}{x^2} = x + 1$$

is not correct.

Project

33. ***Exiting a Theater, I.*** A movie theater has two exits. One door, by itself, can empty the theater in 5 minutes. The second door, by itself, can empty the theater in 8 minutes. Suppose we wish to answer this question: If both doors are available, how many minutes will it take to empty the theater?

Each large rectangle in the figure below represents a full theater. During each minute, the first door permits $\frac{1}{5}$ of the theater to empty. During each minute, the second door permits $\frac{1}{8}$ of the theater to empty.

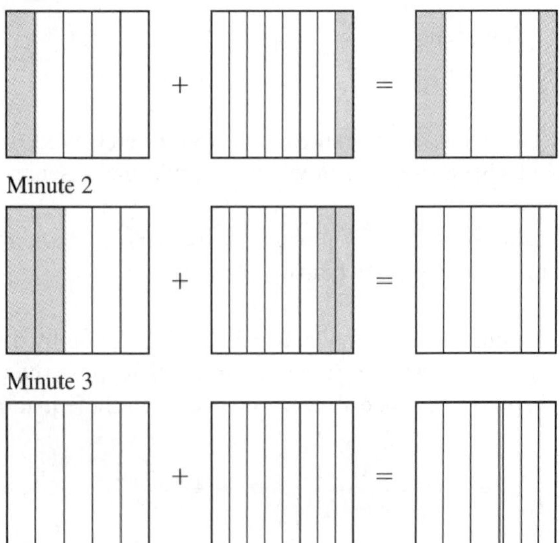

a. What does the shading in the first rectangle for Minute 1 represent?

b. What does the shading in the second rectangle for Minute 1 represent?

c. What does the shading in the third rectangle for Minute 1 represent?

d. Shade the rectangles for each subsequent minute's departures. Add more rectangles as needed.

e. How will the last rectangle look when the theater is empty?

f. Estimate the total number of minutes needed to empty the theater.

g. What is a better way to draw the $\frac{1}{5}$ and $\frac{1}{8}$ fractions to improve the addition?

MID–CHAPTER 9 TEST

1. For what inputs, x, will the expression $\dfrac{1}{(x+2)(x-1)}$ have a zero denominator?

For Exercises 2 and 3, assume that the budget for a credit union's annual meeting is \$4800. The budget covers food, chair set-up charges, and a souvenir gift for each member attending.

2. Make a table and graph for the possible spending per person for zero to 1200 members. Use number of members attending as input and spending per member as output.

3. What equation describes the relationship between the number of members attending the meeting and the spending per member?

Describe the behavior of the graph in the figure for the situations specified in Exercises 4 to 6.

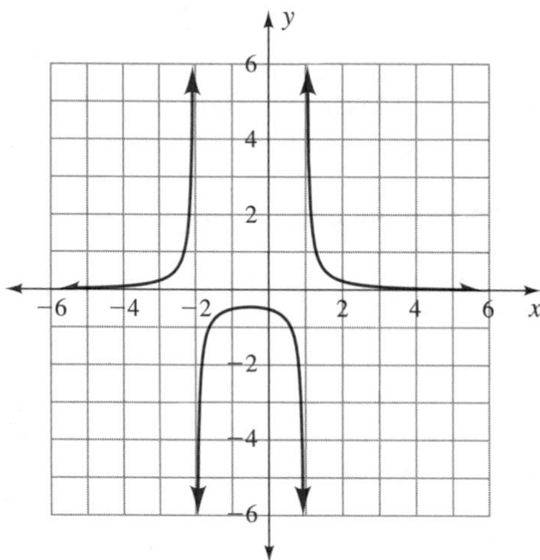

4. As x approaches -2 from the left

5. As x approaches 1 from the right

6. As x approaches 1 from the left

In Exercises 7 and 8, simplify. If the expression does not simplify, explain why. Assume there are no zero denominators.

7. a. $\dfrac{24ac}{28a^2}$ **b.** $\dfrac{a+2}{a-2}$

8. a. $\dfrac{x+2}{x^2-4}$ **b.** $\dfrac{x^2-2x+1}{x^2-3x+2}$

In Exercises 9 and 10, find the missing numerator or denominator. Assume there are no zero denominators.

9. a. $\dfrac{3x}{5y} = \dfrac{\quad}{10xy}$ **b.** $\dfrac{2a}{3b} = \dfrac{8a^2b}{\quad}$

10. a. $\dfrac{3}{x+5} = \dfrac{\quad}{5(x+5)}$ **b.** $\dfrac{2}{x-3} = \dfrac{\quad}{(x+2)(x-3)}$

Perform the indicated operations in Exercises 11 to 15. Simplify the answers. Assume there are no zero denominators.

11. $\dfrac{3x}{y^2} \div \dfrac{x^2}{y}$

12. $\dfrac{2x}{y} \div \dfrac{x^2}{y}$

13. $\dfrac{x^2-5x-6}{x+2} \cdot \dfrac{x^2-4}{x-1}$

14. $\dfrac{x^2-3x}{x^2-16} \div \dfrac{x-3}{x+4}$

15. $\dfrac{\frac{2}{3x}}{\frac{x^2}{6}}$

In Exercises 16 to 18, simplify these expressions containing units of measure.

16. $\dfrac{63 \text{ days}}{7 \text{ days per week}}$

17. $\dfrac{16 \text{ stitches per second}}{8 \text{ stitches per inch}}$

18. From dosage computation: $\dfrac{\frac{10 \text{ mL}}{100 \text{ mL}} \cdot 400 \text{ mL}}{\frac{50 \text{ mL}}{100 \text{ mL}}}$

9.4 Finding the Common Denominator and Addition and Subtraction of Rational Expressions

OBJECTIVES

- Determine the common denominator for two or more rational expressions.
- Add and subtract rational expressions with like denominators.
- Convert rational expressions to like denominators, and complete the addition or subtraction.
- Simplify complex fractions, using multiplication by a common denominator.

WARM-UP

Add or subtract, as indicated.

1. $\frac{3}{4} + \frac{5}{6}$ **2.** $\frac{1}{3} + \frac{1}{9}$ **3.** $\frac{1}{4} - \frac{1}{6}$ **4.** $\frac{1}{18} + \frac{1}{12}$ **5.** $\frac{2}{18} + \frac{2}{12}$ **6.** $\frac{3}{18} + \frac{3}{12}$

IN THIS SECTION, we find the common denominator and use it to add and subtract fractions and to simplify complex fractions.

Least Common Denominator

Factoring plays an important role in finding the **least common denominator** (**LCD**), *the smallest number into which both denominators divide evenly.*

EXAMPLE 1 Finding the least common denominator and adding fractions Add $\frac{3}{4} + \frac{5}{6}$.

Solution To find the least common denominator, we list the factors of each denominator:

$$4 = 2 \cdot 2$$
$$6 = 2 \cdot 3$$

The least common denominator needs to be divisible by both denominators and needs two factors of 2 and one factor of 3. (See Figure 9.)

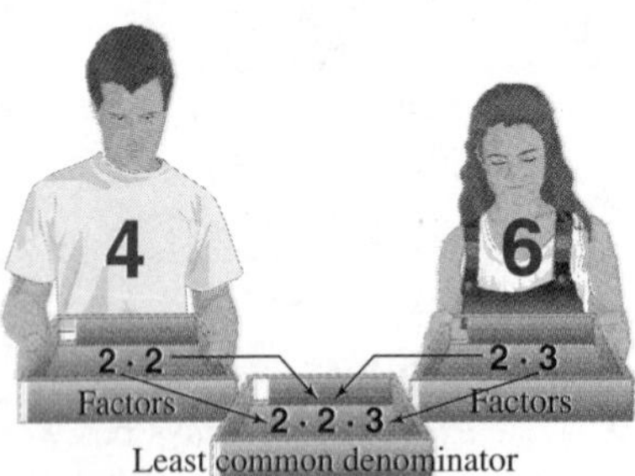

Figure 9

The product of these factors, $2 \cdot 2 \cdot 3$, gives the least common denominator: 12. To add, we change each fraction to an equivalent fraction with the common denominator.

$$\begin{array}{rl} \frac{3}{4} = \frac{3 \cdot 3}{4 \cdot 3} = & \frac{9}{12} \\ + \frac{5}{6} = \frac{5 \cdot 2}{6 \cdot 2} = & \frac{10}{12} \\ \hline & \frac{19}{12} \end{array}$$

Although any common denominator can be used to add or subtract fractions, there are two advantages of using the least common denominator. First, the fractions are simpler. Second, the answer is less likely to need simplifying. In the next example, we extend the common denominator to rational expressions.

Least Common Denominator (LCD)

To find the least common denominator:

1. List the prime factors of each denominator.
2. Compare the factored denominators.
3. **a.** If the denominators have no common factors, the LCD is the product of the denominators.
   b. If the denominators have common factors, list each factor the highest number of times it appears in any one denominator.
4. Write the LCD as the product of the listed factors.

EXAMPLE 

Finding the least common denominator Assume there are no zero denominators and find the LCD for these pairs of rational expressions.

a. $\dfrac{3}{ab^2}, \dfrac{5}{abc}$ **b.** $\dfrac{2}{x+1}, \dfrac{x}{(x+1)^2}$

Solution **a.** To find the least common denominator, we list the factors of each denominator:

$$ab^2 = a \cdot b \cdot b$$
$$abc = a \cdot b \cdot c$$

The least common denominator needs to be divisible by both denominators and needs two factors of b as well as one each of a and c.

$$\text{LCD} = a \cdot b \cdot b \cdot c = ab^2c$$

b. To find the least common denominator, we list the factors of each denominator:

$$x + 1 = (x + 1)$$
$$(x + 1)^2 = (x + 1)(x + 1)$$

The least common denominator needs to be divisible by both denominators and needs two factors of $(x + 1)$.

$$\text{LCD} = (x + 1)(x + 1)$$

●

Addition and Subtraction of Rational Expressions

Adding and Subtracting Rational Numbers

To add or subtract rational numbers:

1. Find a common denominator, if necessary.
2. Change each rational number to the common denominator by multiplying numerator and denominator by any factor missing from the common denominator.
3. Add (or subtract) the numerators and place over the common denominator.
4. Simplify the numerator.
5. Eliminate any *common factors* (not common terms) in the numerator and denominator to simplify the expression to lowest terms.

LIKE DENOMINATORS When rational expressions have the same denominator, they can be added or subtracted as written.

EXAMPLE 

Adding or subtracting rational expressions Add or subtract these rational expressions containing like denominators. Assume there are no zero denominators.

a. $\dfrac{8}{y} - \dfrac{1}{y}$ **b.** $\dfrac{2x}{x+2} + \dfrac{5}{x+2}$

Solution **a.** $\dfrac{8}{y} - \dfrac{1}{y} = \dfrac{8-1}{y} = \dfrac{7}{y}$ **b.** $\dfrac{2x}{x+2} + \dfrac{5}{x+2} = \dfrac{2x+5}{x+2}$

In both answers, the numerators and denominators contain no common factors, and therefore the expressions are in lowest terms. ●

UNLIKE DENOMINATORS When rational expressions have unlike denominators, first we must find a common denominator.

EXAMPLE 4 Adding rational expressions Add $\frac{3}{ab^2} + \frac{5}{abc}$. Assume there are no zero denominators.

Solution From Example 2, the least common denominator is ab^2c.

$$\frac{3}{ab^2} + \frac{5}{abc} = \frac{3 \cdot c}{ab^2 \cdot c} + \frac{5 \cdot b}{abc \cdot b}$$ Set up the common denominator.

$$= \frac{3c}{ab^2c} + \frac{5b}{ab^2c}$$ Add the numerators.

$$= \frac{3c + 5b}{ab^2c}$$

EXAMPLE 5 Subtracting rational expressions Subtract $\frac{2}{x + 1} - \frac{x}{(x + 1)^2}$. Assume there are no zero denominators.

Solution From Example 2, the least common denominator is $(x + 1)^2$.

$$\frac{2}{x + 1} - \frac{x}{(x + 1)^2} = \frac{2(x + 1)}{(x + 1)(x + 1)} - \frac{x}{(x + 1)^2}$$ Set up the common denominator and combine numerators.

$$= \frac{2(x + 1) - x}{(x + 1)^2}$$ Apply the distributive property.

$$= \frac{2x + 2 - x}{(x + 1)^2}$$ Add like terms.

$$= \frac{x + 2}{(x + 1)^2}$$

Subtraction problems must be worked carefully because there may be a sign change when numerators are subtracted. The next example illustrates both finding the common denominator and changing signs with subtraction.

EXAMPLE 6 Subtracting rational expressions Subtract $\frac{x}{x^2 + 3x + 2} - \frac{3}{2x + 2}$. Assume there are no zero denominators.

Solution To find the least common denominator, we factor the denominators:

$$x^2 + 3x + 2 = (x + 1)(x + 2)$$
$$2x + 2 = 2(x + 1)$$

The least common denominator will be the product of the three factors, $(x + 1)$, $(x + 2)$, and 2.

$$\text{LCD} = 2(x + 1)(x + 2)$$

$$\frac{x}{x^2 + 3x + 2} - \frac{3}{2x + 2} = \frac{x}{(x + 1)(x + 2)} - \frac{3}{2(x + 1)}$$ Factor the denominators and set up the LCD.

$$= \frac{2 \cdot x}{2(x + 1)(x + 2)} - \frac{3(x + 2)}{2(x + 1)(x + 2)}$$ Combine numerators.

$$= \frac{2x - 3(x + 2)}{2(x + 1)(x + 2)}$$ Apply the distributive property.

$$= \frac{2x - 3x - 6}{2(x + 1)(x + 2)}$$ Add like terms.

$$= \frac{-x - 6}{2(x + 1)(x + 2)}$$ Factor the numerator.

$$= \frac{-1(x + 6)}{2(x + 1)(x + 2)}$$

Note the sign change from the distributive property. Each time we subtract rational expressions, we must watch for such sign changes. ●

Applications

FORMULAS An important application for students is checking to see whether their answers match those in the back of the book. In Example 7, we work with two solutions to the problem "Solve $A = h(a + b)/2$ for b."

EXAMPLE  Working with formulas containing fractions Use common denominators and subtraction of fractions to show that $b = \frac{2A}{h} - a$ is equivalent to $b = \frac{2A - ha}{h}$, $h \neq 0$.

Solution

$$b = \frac{2A}{h} - a$$ Change a to fraction form.

$$b = \frac{2A}{h} - \frac{a}{1}$$ Build an equivalent fraction with a common denominator.

$$b = \frac{2A}{h} - \frac{h}{h} \cdot \frac{a}{1}$$ Combine numerators.

$$b = \frac{2A - ha}{h}$$ ●

Think about it: Solve the formula $A = h(a + b)/2$ for b. Do you obtain one of the formulas in Example 7 or yet another formula?

THEATER EXITS Example 8 introduces a formula for emptying a theater.

EXAMPLE  Working with formulas containing fractions: theater exits In architecture, the rate of flow of traffic through doors is important. A movie theater has two exit doors of slightly different sizes. The first can empty the theater in t_1 minutes, and the second can empty the theater in t_2 minutes. The formula to find the number of minutes, t, required to empty the theater if both doors are functioning is

$$\frac{1}{t_1} + \frac{1}{t_2} = \frac{1}{t}$$

Add the two fractions on the left side of the equation. Assume there are no zero denominators.

Solution

$$\frac{1}{t_1} + \frac{1}{t_2} = \frac{1}{t}$$

$$\frac{1 \cdot t_2}{t_1 \cdot t_2} + \frac{1 \cdot t_1}{t_2 \cdot t_1} = \frac{1}{t}$$

$$\frac{t_2 + t_1}{t_1 t_2} = \frac{1}{t}$$ ●

If we solve for t by cross multiplication, we obtain

$$t = \frac{t_1 t_2}{t_2 + t_1}$$

Thus, the time required to empty the theater is not a simple sum of the individual door times.

Simplifying Complex Fractions

We may use the least common denominator of two fractions within a fraction to simplify a complex fraction.

EXAMPLE 9 Simplifying complex fractions Simplify $\dfrac{\frac{3}{4}}{\frac{5}{8}}$.

a. Find the least common denominator of the fractions in the numerator and the denominator.

b. Simplify the complex fraction by multiplying the numerator and denominator by the least common denominator.

Solution **a.** The least common denominator of 4 and 8 is 8.

b. $$\frac{\frac{3}{4}}{\frac{5}{8}} = \frac{\frac{3}{4} \cdot 8}{\frac{5}{8} \cdot 8} = \frac{\frac{3}{1} \cdot 2}{\frac{5}{1} \cdot 1} = \frac{6}{5}$$

EXAMPLE 10 Simplifying complex fractions Find the slope of the line connecting $\left(\frac{a}{2}, \frac{b}{2}\right)$ with (a, b).

Solution A sketch of the coordinates is shown in Figure 10. We substitute the coordinates directly into the slope formula:

$$\text{Slope} = m = \frac{y_2 - y_2}{x_2 - x_1} = \frac{b - \frac{b}{2}}{a - \frac{a}{2}} = \frac{\frac{b}{2}}{\frac{a}{2}}$$

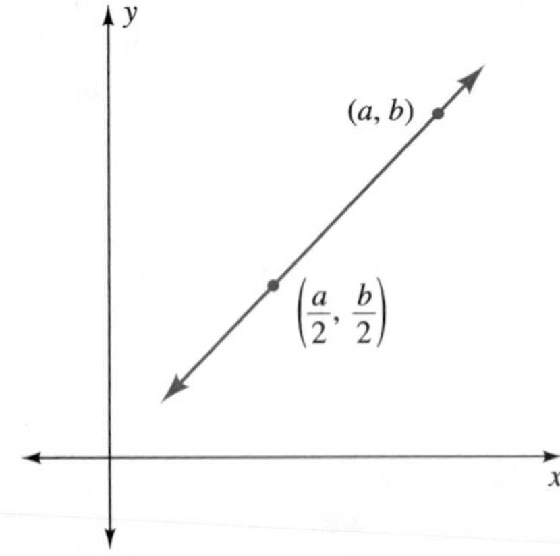

Figure 10

We have two ways to simplify the resulting complex fraction. We can change the fraction bar into division, as in Section 9.3, or we can multiply the numerator and denominator by the common denominator of the two fractions.

Method 1:

$$\frac{\frac{b}{2}}{\frac{a}{2}} = \frac{b}{2} \div \frac{a}{2} = \frac{b}{2} \cdot \frac{2}{a} = \frac{b}{a}$$

Method 2: The LCD is 2.

$$\frac{\frac{b}{2}}{\frac{a}{2}} = \frac{\frac{b}{2}\cdot\frac{2}{1}}{\frac{a}{2}\cdot\frac{2}{1}} = \frac{b}{a}$$

Thus, the slope of the line connecting $\left(\frac{a}{2}, \frac{b}{2}\right)$ with (a, b) is $\frac{b}{a}$.

ANSWER BOX

Warm-up: 1. $\frac{19}{12}$ **2.** $\frac{4}{9}$ **3.** $\frac{1}{12}$ **4.** $\frac{5}{36}$ **5.** $\frac{5}{18}$ **6.** $\frac{5}{12}$ **Think about it:** One solution method is to multiply both sides by 2, divide both sides by h, and then subtract a from both sides. These steps give the starting equation in Example 7.

EXERCISES 9.4

Add or subtract the fractions or rational expressions in Exercises 1 and 2. Assume there are no zero denominators.

1. a. $\frac{11}{7} - \frac{4}{7}$ **b.** $\frac{2}{3} + \frac{x}{3}$ **c.** $\frac{3}{2x} - \frac{5}{2x}$

d. $\frac{4}{x^2 + 1} - \frac{x^2}{x^2 + 1}$ **e.** $\frac{2}{x - 1} - \frac{x + 1}{x - 1}$

2. a. $\frac{13}{6} - \frac{5}{6}$ **b.** $\frac{2}{5} - \frac{x}{5}$ **c.** $\frac{5}{3x} - \frac{8}{3x}$

d. $\frac{4}{x^2 + 2} + \frac{x^2}{x^2 + 2}$ **e.** $\frac{2}{x + 1} - \frac{x - 1}{x + 1}$

What is the least common denominator for each set of rational expressions in Exercises 3 and 4? Assume there are no zero denominators.

3. a. $\frac{5}{12} + \frac{7}{20}$ **b.** $\frac{2}{x} + \frac{5}{2x}$ **c.** $\frac{8}{y} - \frac{1}{y^2}$

d. $\frac{3}{b} + \frac{2}{a} + \frac{5}{b} - \frac{3}{a}$ **e.** $\frac{4}{x - 3} - \frac{2}{x^2 - 9}$

f. $\frac{4}{x^2 + 5x + 6} - \frac{2}{x^2 - 9}$

4. a. $\frac{1}{2} + \frac{3}{8}$ **b.** $\frac{5}{8} + \frac{7}{18}$ **c.** $\frac{2}{3} - \frac{3}{a}$

d. $\frac{b}{a} - \frac{c}{a^2}$ **e.** $\frac{2}{x^2 + 2x} + \frac{5}{x^2 - 4}$

f. $\frac{4}{x^2 - 2x + 1} - \frac{3}{x^2 - 1}$

5. Add or subtract, as indicated, the expressions in Exercise 3.

6. Add or subtract, as indicated, the expressions in Exercise 4.

In Exercises 7 to 20, add or subtract the rational expressions, as indicated. Assume there are no zero denominators.

7. $\frac{3}{2b} + \frac{3}{4a}$

8. $\frac{5}{3a} + \frac{7}{6b}$

9. $\frac{1}{x - 1} + \frac{1}{x}$

10. $\frac{1}{x + 1} + \frac{1}{x - 1}$

11. $\frac{8}{x + 1} - \frac{3}{x}$

12. $\frac{5}{x + 1} - \frac{8}{x}$

13. $\frac{2}{x + 3} + \frac{3}{(x + 3)^2}$

14. $\frac{4}{x - 2} + \frac{2}{(x - 2)^2}$

15. $\frac{1}{a} - \frac{1}{a^2}$

16. $\frac{2}{b^2} - \frac{3}{b}$

17. $\frac{2}{ab} - \frac{3}{2b}$

18. $\frac{3}{ac} - \frac{5}{2a}$

19. $\frac{3}{x^2 - x} + \frac{x}{x^2 - 3x + 2}$

20. $\frac{x}{x^2 - 6x + 9} + \frac{5}{x^2 - 3x}$

Add the fractions on the right side of each question in Exercises 21 to 28 to obtain a single fraction for the application formula. Assume there are no zero denominators.

21. Temperature change: $\Delta T = \frac{T_0}{T} - 1$

22. Resistors in parallel in an electrical circuit: $\frac{1}{R} = \frac{1}{R_1} + \frac{1}{R_2}$

23. Condensers in series in an electrical circuit: $\frac{1}{C} = \frac{1}{C_1} + \frac{1}{C_2}$

24. Days to complete a wheat harvest with two machines: $\dfrac{1}{D} = \dfrac{1}{D_1} + \dfrac{1}{D_2}$

25. Total time for a round trip: $t = \dfrac{D}{r_1} + \dfrac{D}{r_2}$

26. Traffic accident analysis, preliminary to calculating vehicle speed: $R = \dfrac{C^2}{8M} + \dfrac{M}{2}$

27. Radius of curvature of a surface: $F = \dfrac{L^2}{6d} + \dfrac{d}{2}$

28. Approximating an exponential function:

$$e^x \approx 1 + x + \frac{x^2}{2} + \frac{x^3}{6} + \frac{x^4}{24}$$

For Exercises 29 to 34, find the slope of the line connecting the points. Simplify to eliminate fractions from the numerator and the denominator.

29.

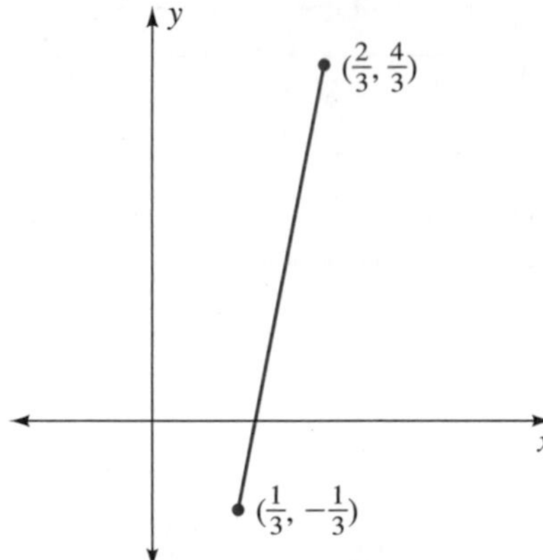

30.

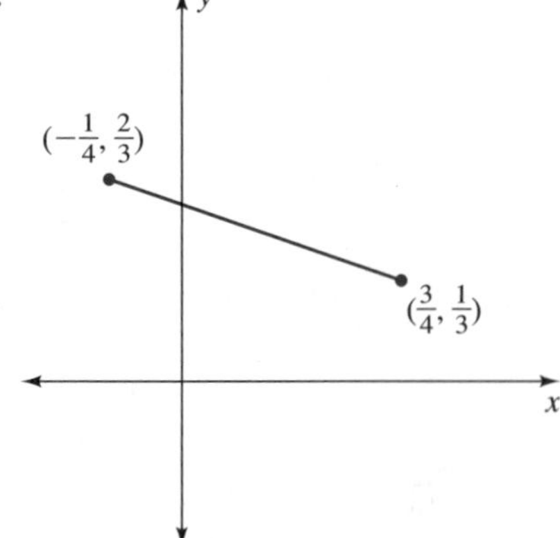

31. $\left(\frac{1}{5}, \frac{2}{5}\right), \left(\frac{-4}{5}, \frac{1}{5}\right)$ **32.** $\left(\frac{2}{3}, \frac{1}{4}\right), \left(\frac{-2}{3}, \frac{3}{4}\right)$

33.

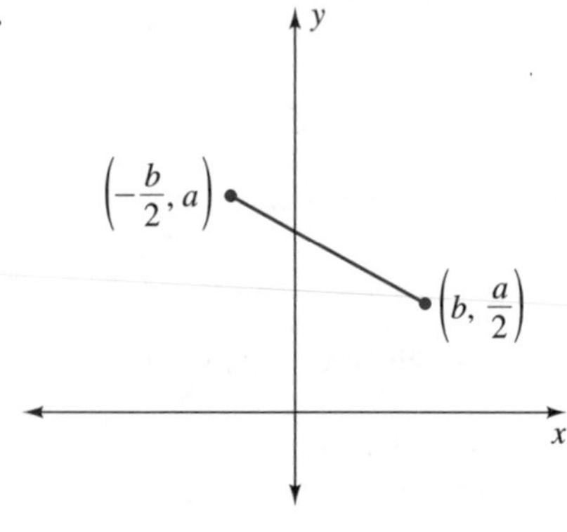

34.

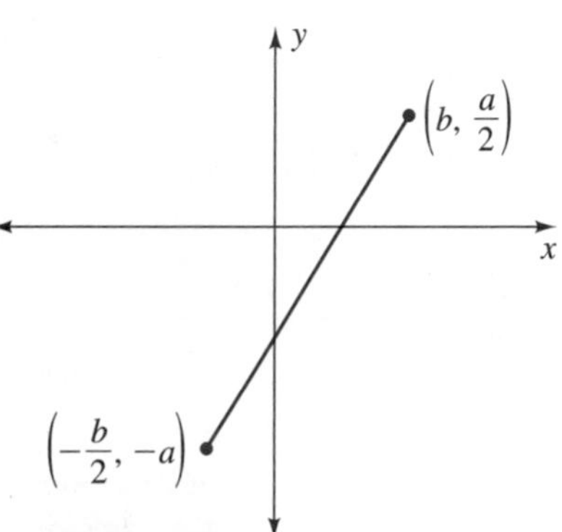

In Exercises 35 and 36, find the slopes of the line segments connecting the indicated points in each figure. What do you observe about the slopes?

35. Line segment (a, b) to $(c, 0)$
Line segment $\left(\dfrac{a}{2}, \dfrac{b}{2}\right)$ to $\left(\dfrac{c}{2}, 0\right)$

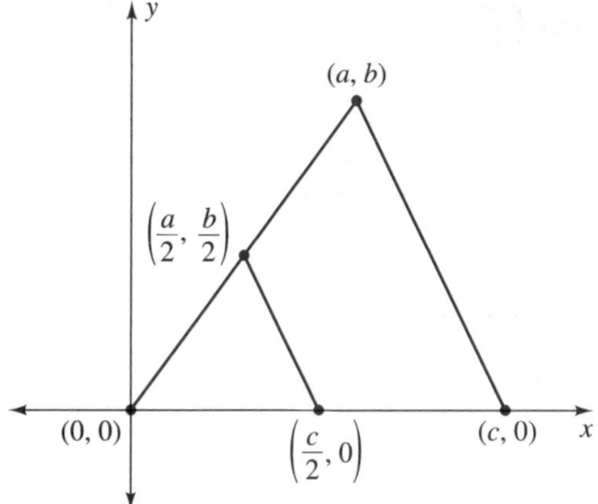

36. Line segment $(0, 0)$ to $(c, 0)$
Line segment $\left(\dfrac{a}{2}, \dfrac{b}{2}\right)$ to $\left(\dfrac{a + c}{2}, \dfrac{b}{2}\right)$

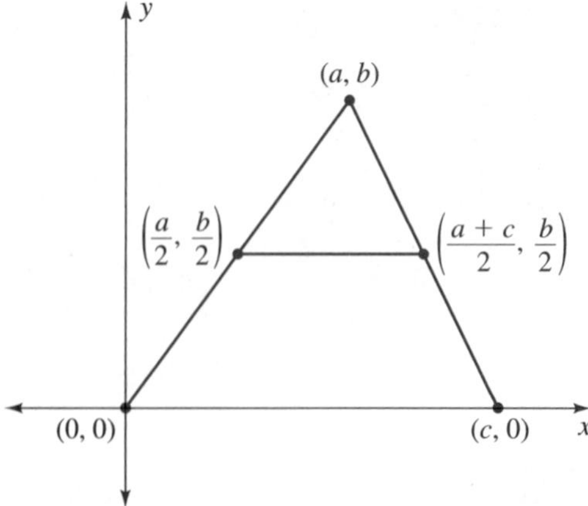

Simplify the complex fractions in Exercises 37 to 44 to eliminate fractions from the numerators and denominators. Assume there are no zero denominators.

37. $\dfrac{5}{\frac{1}{5} + 1}$ **38.** $\dfrac{3}{1 + \frac{1}{3}}$ **39.** $h = \dfrac{A}{\frac{1}{2}b}$

40. $h = \dfrac{V}{\frac{1}{3}\pi r^2}$ **41.** $\dfrac{x + \frac{x}{2}}{2 - \frac{x}{3}}$ **42.** $\dfrac{\frac{x}{3} + x}{3 - \frac{x}{2}}$

43. Refrigeration cycle: $\dfrac{1}{\dfrac{Q_H}{Q_L} - 1}$

44. Heat transfer: $\dfrac{1}{1 - \dfrac{Q_L}{Q_H}}$

Graphing Calculator Exploration

45. a. Graph the expression in Exercise 41 before and after simplifying.

b. Compare the graphs before and after simplifying.

c. Where is the graph nearly vertical? Why?

46. a. Graph the expression in Exercise 42 before and after simplifying.

b. Compare the graphs before and after simplifying.

c. Where is the graph nearly vertical? Why?

Problem Solving

Identify the missing operation symbol (+, −, ·, or ÷) in Exercises 47 to 50. Assume there are no zero denominators.

47. a. $\frac{3}{4} \square \frac{2}{5} = \frac{6}{20} = \frac{3}{10}$

b. $\frac{3}{4} \square \frac{2}{5} = \frac{15}{8}$

c. $\frac{3}{4} \square \frac{2}{5} = \frac{23}{20}$

48. a. $\frac{2}{5} \square \frac{3}{4} = \frac{8}{15}$

b. $\frac{3}{4} \square \frac{2}{5} = \frac{7}{20}$

c. $\frac{5}{4} \square \frac{2}{3} = \frac{23}{12}$

49. a. $\dfrac{1}{a} \square \dfrac{1}{b} = \dfrac{a + b}{ab}$

b. $\dfrac{1}{a} \square \dfrac{1}{b} = \dfrac{1}{ab}$

50. a. $\dfrac{1}{a} \square \dfrac{1}{b} = \dfrac{b - a}{ab}$

b. $\dfrac{1}{b} \square \dfrac{1}{a} = \dfrac{a}{b}$

Projects

51. *Fraction Pattern, II*

a. Add these two sets of fractions.

$\frac{1}{6} + \frac{1}{9} + \frac{1}{18}$ $\frac{1}{10} + \frac{1}{15} + \frac{1}{30}$

b. Look for a shortcut for finding the sum, and use it to add the next two sets of fractions.

$\frac{1}{22} + \frac{1}{33} + \frac{1}{66}$ $\frac{1}{18} + \frac{1}{27} + \frac{1}{54}$

c. Write a formula for the pattern, and show why the pattern always works.

52. *Exiting a Theater, II.* Return to the movie theater project in Exercise 33 of Section 9.3. The movie theater has two exists. One door, by itself, can empty the theater in 5 minutes. The second door, by itself, can empty the theater in 8 minutes. Suppose we wish to answer this question: If both doors are available, how many minutes will it take to empty the theater?

Minute 1

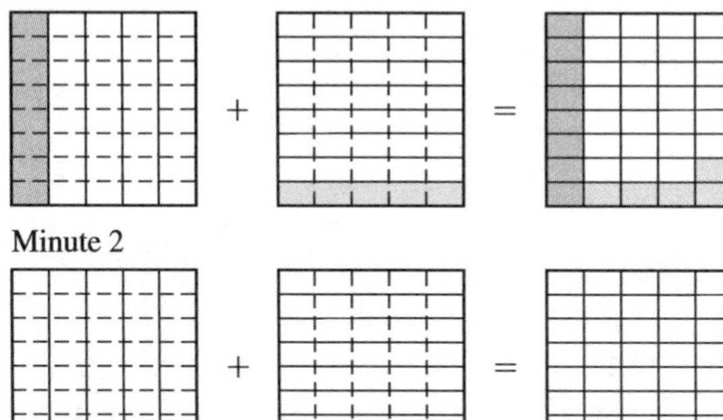

Minute 3

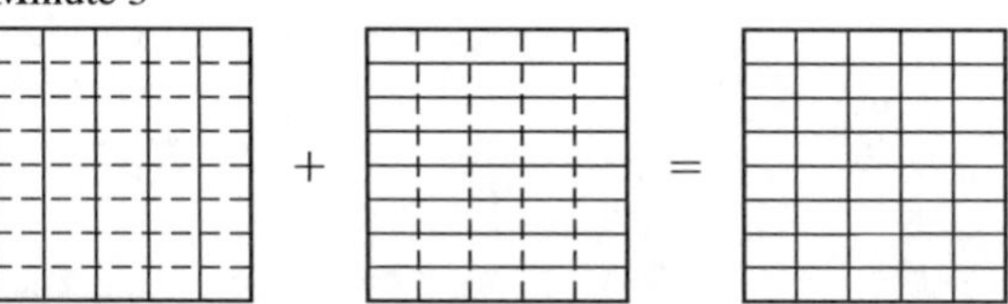

a. How are the rectangles in the figure the same as those in the figure in Exercise 33 of Section 9.3?

b. How are the rectangles different?

c. How many small pieces are there in each large rectangle? Why is this number important to adding the pieces together?

d. What does the shading in the first rectangle in the first row represent?

e. What does the shading in the second rectangle in the first row represent?

f. What does the shading in the third rectangle in the first row represent?

g. Shade the rectangles for each subsequent minute's departures. Add more rectangles as needed.

h. How will the rectangle look when the theater is empty?

i. Estimate the total number of minutes needed to empty the theater.

j. Why might the rectangles in this figure be a better way to represent the fractions $\frac{1}{5}$ and $\frac{1}{8}$ than those in the figure in Exercise 33 of Section 9.3?

k. Make up a problem of your own involving three doors.

9.5 Solving Rational Equations

OBJECTIVES

- Find the least common denominator of rational expressions in an equation.
- Multiply by the least common denominator to eliminate the denominators in an equation.
- Solve equations containing rational expressions.
- Solve application problems related to $\frac{1}{a} + \frac{1}{b} = \frac{1}{c}$.

> **WARM-UP**
>
> Multiply these expressions using the distributive property.
>
> **1.** $\frac{12}{1}\left(\frac{x}{4} + \frac{1}{6}\right)$ **2.** $\frac{2}{1}\left(\frac{2}{1} + \frac{3x}{2}\right)$ **3.** $36\left(\frac{1}{12} + \frac{d}{18}\right)$
>
> Find the least common denominator for each pair of fractions.
>
> **4.** $\frac{1}{12}, \frac{1}{18}$ **5.** $\frac{1}{9}, \frac{1}{15}$ **6.** $\frac{1}{8}, \frac{1}{18}$

IN THIS SECTION, we apply the distributive property to rational expressions. We solve equations containing rational expressions. We also examine one application of rational equations that has particular appeal to the mathematician: a common formula to describe patterns that occur in a variety of applications.

Rational Expressions and the Distributive Property

We will be solving rational equations in this section, and the first step in the solution process is to multiply by a number or expression in order to clear the denominators. We return to the table form of multiplication because tables remind us how important it is to multiply both terms in parentheses by the expression in front. Recall that the distributive property is

$$a(b + c) = ab + ac$$

EXAMPLE 1 Applying the distributive property to rational expressions Complete these tables. Summarize by writing the solution as a simplified product.

a.

Multiply	$\frac{1}{12}$	$+\frac{d}{18}$
36		

b.

Multiply	$\frac{1}{4}$	$+\frac{1}{3x}$
$12x$		

Solution **a.**

Multiply	$\frac{1}{12}$	$+\frac{d}{18}$
36	$\frac{36}{12}$	$+\frac{36d}{18}$

Summary:

$$36\left(\frac{1}{12} + \frac{d}{18}\right) = 36\left(\frac{1}{12}\right) + 36\left(\frac{d}{18}\right)$$
$$= \frac{36}{12} + \frac{36d}{18}$$
$$= 3 + 2d$$

b.

Multiply	$\frac{1}{4}$	$+\frac{1}{3x}$
$12x$	$\frac{12x}{4}$	$+\frac{12x}{3x}$

Summary:

$$12x\left(\frac{1}{4} + \frac{1}{3x}\right) = 12x\left(\frac{1}{4}\right) + 12x\left(\frac{1}{3x}\right)$$
$$= \frac{12x}{4} + \frac{12x}{3x}$$
$$= 3x + 4$$

●

Think about it: Look at the answers to Warm-up Exercises 1 to 3 and Example 1. Did multiplication eliminate all the denominators? How is the number being multiplied related to the expressions inside the parentheses?

The Harvest Problem

In Example 2, we return to the wheat harvest problem, page 554.

EXAMPLE **2** Exploring work done together: wheat harvest Farmers Lee and Terry can separately harvest Lee's wheat in 12 days and 18 days, respectively. How might we calculate the number of days needed to harvest the crop if they work together?

Consider the following proposed methods. Comment on the results.

Proposed method 1: Suppose we add the 12 days and 18 days.

Proposed method 2: Suppose we average the number of days.

Proposed method 3: Suppose we subtract 12 days from the 18 days.

Proposed method 4: Suppose we find how much the farmers harvest each day.

Solution ***Proposed method 1:*** Adding the 12 days and 18 days gives 30 days, which is not reasonable. If the farmers work together, the task should take fewer than the 12 days Lee requires alone.

Proposed method 2: If we average the number of days, we have

$$\text{Average} = \frac{12 \text{ days} + 18 \text{ days}}{2} = 15 \text{ days}$$

This result is not reasonable, as the two farmers should not take longer than Lee working by herself.

Proposed method 3: If we subtract 12 days from the 18 days, we get 6 days. Although subtraction might give a reasonable answer in this setting, consider subtraction in the case where both farmers harvest in 18 days. Subtraction would give zero days working together, whereas a reasonable answer might be 9 days.

Proposed method 4: Finding how much the farmers harvest each day is another approach. On the first day, Lee harvests $\frac{1}{12}$ of the total while Terry harvests $\frac{1}{18}$. Together they harvest $\frac{1}{12} + \frac{1}{18}$. Each day, they represent these fractions of the total job. The following summarizes the fraction of the total crop harvested by the end of each day, for 8 days:

Day 1: $\frac{1}{12} + \frac{1}{18} = \frac{3}{36} + \frac{2}{36} = \frac{5}{36}$

Day 2: $2\left(\frac{1}{12} + \frac{1}{18}\right) = 2\left(\frac{5}{36}\right) = \frac{10}{36}$

Day 3: $3\left(\frac{1}{12} + \frac{1}{18}\right) = 3\left(\frac{5}{36}\right) = \frac{15}{36}$

Day 4: $4\left(\frac{5}{36}\right) = \frac{20}{36}$

Day 5: $5\left(\frac{5}{36}\right) = \frac{25}{36}$

Day 6: $6\left(\frac{5}{36}\right) = \frac{30}{36}$

Day 7: $7\left(\frac{5}{36}\right) = \frac{35}{36}$

Day 8: $8\left(\frac{5}{36}\right) = \frac{40}{36}$

The harvest is complete when the fraction reaches $\frac{36}{36} = 1$. The harvest is nearly complete at the end of day 7. It is completely finished during day 8, when the fraction exceeds 1. The exact time to finish is d days, where

$$d\left(\frac{1}{12} + \frac{1}{18}\right) = 1$$

Traditionally, this equation is divided on both sides by d and written

$$\frac{1}{12} + \frac{1}{18} = \frac{1}{d}, \quad d \neq 0$$

●

Sum of Rates Applications

Our first step in solving the harvest equation will be to eliminate the denominators. (Yes, we just divided by d and put it in the denominator. Now we are going to take it out again. Keep in mind that the traditional form of the equation has the variable in the denominator.)

In the Warm-up and Example 1, we applied the distributive property to fractions. When we multiplied by the least common denominator, we eliminated the denominators. We use this approach to eliminate denominators in equations.

Eliminating Denominators

To eliminate denominators from an equation, multiply both sides of the equation by the least common denominator.

EXAMPLE 3 Solving a rational equation Solve $\frac{1}{12} + \frac{1}{18} = \frac{1}{d}$ for d, $d \neq 0$.

Solution The least common denominator for all fractions in the equation is $36d$.

$$\frac{1}{12} + \frac{1}{18} = \frac{1}{d}$$

$$36d\left(\frac{1}{12} + \frac{1}{18}\right) = 36d\left(\frac{1}{d}\right)$$

$$\frac{36d \cdot 1}{12} + \frac{36d \cdot 1}{18} = \frac{36d \cdot 1}{d}$$
$$3d + 2d = 36$$
$$5d = 36$$
$$d = \tfrac{36}{5} = 7.2 \text{ days}$$

Calculator check: $\frac{1}{12} + \frac{1}{18} \approx 0.0833 + 0.0556 \approx 0.1389 \approx \frac{1}{7.2}$ ✓

In Example 3, we could have added the fractions on the left side and created a proportion. The equation could then have been solved by cross multiplication. We leave this method as an exercise.

EXAMPLE 4

Building and solving a rational equation: wheat harvest When it is time to harvest Terry's wheat, Lee moves her equipment to the south end of the county. Terry normally harvests his wheat in 15 days. Lee's equipment could do the work in 9 days. If they work together, how long will it take to harvest Terry's wheat? Estimate an answer, and then write an equation and solve it.

Solution We have the same situation as before but with different numbers of days of work. The harvest should take fewer than 9 days (Lee's time) and more than $4\frac{1}{2}$ days (if they worked at equal rates).

The equation is

$$\frac{1}{9} + \frac{1}{15} = \frac{1}{d}, \quad d \neq 0$$

The least common denominator of the fractions within the equation is $45d$.

$$\frac{1}{9} + \frac{1}{15} = \frac{1}{d}$$
$$45d\left(\frac{1}{9} + \frac{1}{15}\right) = 45d\left(\frac{1}{d}\right)$$
$$\frac{45d \cdot 1}{9} + \frac{45d \cdot 1}{15} = \frac{45d \cdot 1}{d}$$
$$5d + 3d = 45$$
$$8d = 45$$
$$d = \tfrac{45}{8} = 5.625 \text{ days}$$

Calculator check: $\frac{1}{9} + \frac{1}{15} \approx 0.1111 + 0.0667 \approx 0.1778 \approx \frac{1}{5.625}$ ✓

The individual times to harvest the wheat may be described as d_1 and d_2. Working individually, the farmers could harvest the whole crop at a rate of $\dfrac{1}{d_1}$ and $\dfrac{1}{d_2}$, respectively, each day. Because both are working to finish 1 job, we add the rates for the two farmers to obtain the rate for working together, $\dfrac{1}{d}$:

$$\frac{1 \text{ wheat crop}}{d_1 \text{ days}} + \frac{1 \text{ wheat crop}}{d_2 \text{ days}} = \frac{1 \text{ wheat crop}}{d \text{ days together}}$$
$$\frac{1}{d_1} + \frac{1}{d_2} = \frac{1}{d}$$

We now return to an application from the projects in Sections 9.3 and 9.4. This application has a structure surprisingly similar to that of the harvest application.

EXAMPLE 5 Building an equation: exiting a theater Suppose a movie theater has two exits. One is a single door that, by itself, can empty the theater in 8 minutes. The other is a double-wide door that, by itself, can empty the theater in 5 minutes. Estimate the time required to empty the theater when both doors are open, and then build an equation.

Solution The time will be less than 5 minutes (double-wide door) and greater than $2\frac{1}{2}$ minutes (if both were double-wide doors).

We change exit times to rates for emptying the theater. During each minute, the doors permit $\frac{1}{8}$ and $\frac{1}{5}$ of the theater, respectively, to empty. The following expressions represent the portion of the theater that has been emptied by the end of each minute:

First minute: $\frac{1}{8} + \frac{1}{5} = \frac{5}{40} + \frac{8}{40} = \frac{13}{40}$

Second minute: $2\left(\frac{1}{8} + \frac{1}{5}\right) = 2\left(\frac{13}{40}\right) = \frac{26}{40}$

Third minute: $3\left(\frac{13}{40}\right) = \frac{39}{40}$

Fourth minute: $4\left(\frac{13}{40}\right) = \frac{52}{40}$

The theater will be entirely empty during the fourth minute, when the fraction $\frac{40}{40} = 1$ is reached. To find the exact time, x, needed to empty the theater, we solve the equation

$$x\left(\frac{1}{8} + \frac{1}{5}\right) = 1$$

Again, the traditional form of this equation is

$$\frac{1}{8} + \frac{1}{5} = \frac{1}{x}, \quad x \neq 0$$

●

EXAMPLE  Solving a rational equation Solve $\frac{1}{8} + \frac{1}{5} = \frac{1}{x}$ for x, where $x \neq 0$.

Solution The least common denominator for the equation is $40x$.

$$\frac{1}{8} + \frac{1}{5} = \frac{1}{x}$$

$$40x\left(\frac{1}{8} + \frac{1}{5}\right) = 40x\left(\frac{1}{x}\right)$$

$$5x + 8x = 40$$

$$13x = 40$$

$$x = \frac{40}{13} \approx 3.1 \text{ min}$$

Calculator check: $\frac{1}{8} + \frac{1}{5} = 0.125 + 0.200 = 0.325 \approx \frac{1}{3.1}$ ✓

The two-exit system permits the theater to be cleared rapidly. ●

Quadratic Equation Practice

Examples 7, 8, and 9 provide more illustrations of solving equations containing rational expressions. The first two use factoring to solve a quadratic equation.

EXAMPLE 7 Solving a rational equation List any inputs that must be excluded, and then solve $\frac{2}{x} + \frac{3x}{2} = 4$ for x.

Solution The expressions in the equation are undefined for a zero denominator; thus, $x \neq 0$.

$$\frac{2}{x} + \frac{3x}{2} = 4 \qquad \text{Multiply by the LCD.}$$

$$2x\left(\frac{2}{x} + \frac{3x}{2}\right) = 2x \cdot 4 \qquad \text{Distribute the LCD.}$$

$$\frac{2x \cdot 2}{x} + \frac{2x \cdot 3x}{2} = 2x \cdot 4 \qquad \text{Reduce the fractions.}$$

$$4 + 3x^2 = 8x \qquad \text{Solve for the zero form.}$$

$$3x^2 - 8x + 4 = 0 \qquad \text{Factor.}$$

$$(3x - 2)(x - 2) = 0 \qquad \text{Apply the zero product rule.}$$

$$\text{Either} \quad 3x - 2 = 0 \quad \text{or} \quad x - 2 = 0$$

$$x = \tfrac{2}{3} \quad \text{or} \quad x = 2$$

We check by substituting each x into the original equation. The check is left as an exercise. ●

EXAMPLE 8 Solving a rational equation List any inputs that must be excluded, and then solve $\frac{2}{x+1} + \frac{3}{x} = -2$ for x.

Solution The expressions in the equation are undefined for a zero denominator; thus, $x \neq -1$ and $x \neq 0$.

$$\frac{2}{x+1} + \frac{3}{x} = -2 \qquad \text{Multiply by the LCD.}$$

$$x(x+1)\left(\frac{2}{x+1} + \frac{3}{x}\right) = -2 \cdot x(x+1) \qquad \text{Distribute the LCD.}$$

$$\frac{x(x+1) \cdot 2}{x+1} + \frac{x(x+1) \cdot 3}{x} = -2x(x+1) \qquad \text{Reduce the fractions.}$$

$$2x + 3(x+1) = -2x^2 - 2x \qquad \text{Solve for the zero form.}$$

$$2x + 3x + 3 + 2x^2 + 2x = 0 \qquad \text{Combine like terms.}$$

$$2x^2 + 7x + 3 = 0 \qquad \text{Factor.}$$

$$(2x + 1)(x + 3) = 0 \qquad \text{Apply the zero product rule and solve.}$$

$$\text{Either} \quad x = -\tfrac{1}{2} \quad \text{or} \quad x = -3$$

The check by substitution is left as an exercise. ●

Example 9 is shown with two different solutions: multiplying both sides by the LCD and using cross multiplication (Section 5.2). Whenever a rational equation is a proportion, we have the option to cross multiply and eliminate denominators.

EXAMPLE 9 Solving a rational equation Solve $\frac{x+1}{x-3} = \frac{3}{x-3}$ for x, where $x \neq 3$, first by multiplying both sides by the common denominator and then by cross multiplication.

Solution When we multiply both sides by $x - 3$, the solution is $x = 2$.

$$\frac{x+1}{x-3} = \frac{3}{x-3}$$

$$(x-3) \cdot \frac{(x+1)}{(x-3)} = (x-3) \cdot \frac{3}{(x-3)}$$

$$x + 1 = 3$$

$$x = 2$$

When we cross multiply, we find a quadratic equation with two solutions.

$$\frac{x+1}{x-3} = \frac{3}{x-3}$$

$$(x-3)(x+1) = 3(x-3)$$

$$x^2 - 2x - 3 = 3x - 9$$

$$x^2 - 5x + 6 = 0$$

$$(x-2)(x-3) = 0$$

$$\text{Either} \quad x - 2 = 0 \quad \text{or} \quad x - 3 = 0$$

$$x = 2 \quad \text{or} \quad x = 3$$

Check: In cheching our solutions, we find that $x = 2$ satisfies the equation:

$$\frac{2+1}{2-3} \stackrel{?}{=} \frac{3}{2-3} \quad ✓$$

However, $x = 3$ gives a zero denominator:

$$\frac{3+1}{3-3} \stackrel{?}{=} \frac{3}{3-3}$$

As noted in the original problem, $x = 3$ has been excluded from the set of possible inputs. The solution $x = 3$ is an extraneous root. ●

Student Note:
Congratulations, this is the last section of the textbook! You are to be commended for finishing the course.

Example 9 emphasizes the importance of checking answers. The cross multiplication solution created an *extraneous root*—an answer that gave an undefined expression when we substituted it into the equation.

ANSWER BOX

Warm-up: 1. $3x + 2$ **2.** $4 + 3x$ **3.** $3 + 2d$ **4.** 36 **5.** 45 **6.** 72
Think about it: No denominators remain in any of the answers. The number being multiplied is the least common denominator of the expressions inside the parentheses.

EXERCISES 9.5

Multiply the expressions in Exercises 1 to 6, and simplify the results. Assume there are no zero denominators.

1. $12x\left(\frac{1}{12} + \frac{2}{3x}\right)$

2. $8x\left(\frac{1}{4x} + \frac{1}{2}\right)$

3. $4x^2\left(\frac{1}{2x} + \frac{3}{x^2}\right)$

4. $6x^2\left(\frac{2}{3x^2} + \frac{1}{6}\right)$

5. $x(x - 1)\left(\frac{1}{x - 1} + \frac{1}{x}\right)$

6. $x(x + 2)\left(\frac{1}{x} + \frac{3}{x + 2}\right)$

Find the least common denominator for each equation in Exercises 7 to 22.

7. $\frac{x}{4} + \frac{x}{6} = 28$

8. $\frac{x}{14} + \frac{x}{8} = 11$

9. $\frac{3}{4} + \frac{1}{5} = \frac{1}{x}$

10. $\frac{2}{3} + \frac{2}{5} = \frac{1}{x}$

11. $\frac{1}{8} + \frac{1}{x} = \frac{1}{2}$

12. $\frac{1}{10} + \frac{1}{x} = \frac{1}{6}$

13. $\frac{1}{x} = \frac{1}{3x} + \frac{1}{3}$

14. $\frac{4}{x} + \frac{2}{x} = \frac{3}{x}$

15. $\frac{3}{x} - \frac{2}{x} = \frac{4}{x}$

16. $\frac{1}{x} = \frac{1}{2} - \frac{1}{2x}$

17. $\frac{1}{x - 1} = \frac{2}{x + 3}$

18. $\frac{2}{x - 1} = \frac{1}{x - 4}$

19. $\frac{2}{x^2} - \frac{3}{x} + 1 = 0$

20. $1 + \frac{5}{x} - \frac{14}{x^2} = 0$

21. $\frac{1}{x - 3} - 3 = \frac{4 - x}{x - 3}$

22. $\frac{1}{x - 4} = \frac{5 - x}{x - 4} + 4$

For Exercises 23 to 38, solve the given exercise for x, using whichever method (cross multiplication, calculator, multiplication by the LCD) seems appropriate. Indicate any inputs that must be excluded.

23. Exercise 7

24. Exercise 8

25. Exercise 9

26. Exercise 10

27. Exercise 11

28. Exercise 12

29. Exercise 13

30. Exercise 14

31. Exercise 15

32. Exercise 16

33. Exercise 17

34. Exercise 18

35. Exercise 19

36. Exercise 20

37. Exercise 21

38. Exercise 22

Add the fractions on the left side of each equation in Exercises 39 to 42, and use cross multiplication on the resulting proportion to solve for the indicated variable. Assume d ≠ 0, x ≠ 0.

39. Solve $\frac{1}{12} + \frac{1}{18} = \frac{1}{d}$ for d.

40. Solve $\frac{1}{9} + \frac{1}{15} = \frac{1}{d}$ for d.

41. Solve $\frac{1}{8} + \frac{1}{5} = \frac{1}{x}$ for x.

42. Solve $\frac{1}{5} + \frac{1}{6} = \frac{1}{x}$ for x.

The reciprocal key, [x^{-1}] *or* [1/x], *gives a way to obtain decimals for fractions with 1 in the numerator. In Exercises 43 to 46, use the reciprocal key to solve the equation in the given exercise; list your keystrokes. Round to three decimal places.*

43. Exercise 39

44. Exercise 40

45. Exercise 41

46. Exercise 42

47. Substitute $x = 2$ into this equation from Example 7, and check that it is a solution:

$$\frac{2}{x} + \frac{3x}{2} = 4$$

48. Substitute $x = \frac{2}{3}$ into the equation in Example 47, and check that it is a solution. Use a calculator as needed.

49. Substitute $x = -\frac{1}{2}$ into this equation from Example 8, and check that it is a solution:

$$\frac{2}{x + 1} + \frac{3}{x} = -2$$

Use a calculator as needed.

50. Substitute $x = -3$ into the equation in Exercise 49, and check that it is a solution.

Set up and solve equations for Exercises 51 to 58. Round to the nearest tenth.

51. One farmer harvests his barley in 14 days. A second farmer harvests the same crop in 12 days. How many days will it take them working together?

52. One farmer bales her hay in 6 days. A second farmer does it in 8 days. How many days will it take them working together?

53. One ventilation fan changes the air in a house in 4 hours. A second fan vents the same volume of air in 5 hours. How long will it take to vent the house if both fans are working? If building code requires a complete change of air every 3 hours, will the two fans be sufficient?

54. A $\frac{5}{8}$-inch garden hose fills a child's pool in 1 hour. A $\frac{1}{2}$-inch garden hose fills a child's pool in 1.5 hours. How long will it take to fill the pool if both hoses are used at once?

55. One ventilation fan changes the air in a house in 5 hours. A second fan is to be installed. In order to satisfy code, both fans working together must change the air in 3 hours. Describe the fan needed to satisfy code.

56. A $1\frac{1}{2}$-inch pipe fills a swimming pool directly from a farm well in 5 days. A $\frac{5}{8}$-inch hose feeding water from the cistern is added, and the pool fills in 4 days. How long would the hose take to fill the pool by itself?

57. One door can clear a theater in 9 minutes. A second door can clear the theater in 6 minutes. How fast can the theater be emptied if both doors are available?

58. A large theater has three exits. Operating individually, each door can clear the theater in 6 minutes. How fast can the theater be emptied if all three doors are available?

In Exercises 59 to 62, solve for the indicated letter. Assume there are no zero denominators.

59. $\frac{1}{a} + \frac{1}{b} = \frac{1}{c}$ for b

60. $\frac{1}{a} + \frac{1}{b} = \frac{1}{c}$ for a

61. $\frac{1}{a} + \frac{1}{b} = \frac{1}{c}$ for c

62. $\frac{1}{R} = \frac{1}{R_1} + \frac{1}{R_2}$ for R_1

63. Compare the roles of multiplying by the least common denominator in simplifying complex fractions and in solving equations.

64. What was done wrong in the following simplification?

$$\overset{4}{12x}\left(\frac{2}{\underset{1}{3x}} + \frac{1}{4}\right) = \frac{4 \cdot 2}{1} + \frac{4}{4} = 8 + 1 = 9$$

In Exercises 65 to 72, indicate any inputs that must be excluded, and then solve the equation.

65. $5x + \frac{13}{2} = \frac{3}{2x}$

66. $\frac{1}{2} + 3x = \frac{1}{x}$

67. $\frac{1}{x} + 3 = \frac{5}{x(x + 1)}$

68. $4x + 7 = \frac{1}{x + 1}$

69. $x = 6 - \frac{6}{x - 1}$

70. $x + 1 = \frac{x + 1}{x}$

71. $\frac{x + 3}{x - 2} = \frac{1}{x - 2}$

72. $\frac{3 - x}{x - 1} = \frac{4}{x - 1}$

Projects

73. ***Exiting a Theater, III.*** A movie theater has three exit doors of slightly different sizes. The first can empty the theater in t_1 minutes; the second, in t_2 minutes; and the third, in t_3 minutes. The formula to determine the number of minutes, t, required to empty the theater if all three doors are functioning is

$$\frac{1}{t_1} + \frac{1}{t_2} + \frac{1}{t_3} = \frac{1}{t}$$

a. Add the three fractions on the left side of the equation.

b. Solve the equation in part a for t. Assume none of the times are zero.

c. Make up reasonable exit times for a theater with three doors, as described, and find the total exiting time.

74. ***Graphing Calculator***

a. Enter $y_1 = \frac{2}{x} + \frac{1}{x - 2}$ and $y_2 = \frac{2(x - 2) + x}{x(x - 2)}$.

b. Use the table function to compare the outputs. Try TblSet with TblMin $= -3$, ΔTbl $= 1$.

c. Why are there no outputs at $x = 0$ and $x = 2$?

d. Graph the equations.

e. What may be concluded about the two equations? Why?

CHAPTER 9 SUMMARY

Vocabulary

For definitions and page references, see the Glossary/Index.

complex fractions
equivalent fraction property
inverse proportions
inverse variation
least common denominator (LCD)
opposites
rational functions
rational numbers
simplification property of fractions

Concepts

9.1 Rational Functions: Graphs and Applications

A zero denominator means division by zero, an undefined operation.

The symbol $\mathbb{R}$ is the notation for the set of all real numbers.

The graph of a rational expression (simplified to lowest terms) approaches infinity in a vertical direction whenever the denominator approaches zero.

9.2 Simplifying Rational Expressions

A rational expression must be factored in order to simplify to lowest terms.

Simplify expressions containing units of measure, using facts such as feet/feet = 1.

When the numerator and denominator of a fraction are the same, the fraction equals 1.

When the numerator and denominator of a fraction are opposites, the fraction equals -1.

9.3 Multiplication and Division

Advice from Angie Cowles, student at Lane Community College: "Common factors will be more apparent if you leave the final expressions in factored form."

To multiply rational expressions, simplify first. Then multiply the numerators. Then multiply the denominators.

To divide rational expressions, multiply the first expression by the reciprocal of the second expression.

To simplify complex fractions or complex expressions, multiply the numerator by the reciprocal of the denominator.

9.4 LCD Addition and Subtraction

To find the least common denominator, form a product by including each factor the highest number of times it appears in any denominator.

To add or subtract rational expressions, rewrite the expressions with a least common denominator (if needed). Combine the numerators over the common denominator and then simplify by eliminating common factors.

To simplify a complex fraction or rational expression, find the LCD for the numerator and denominator fractions and multiply the LCD times both the numerator and the denominator.

9.5 Solving Rational Equations

To solve an equation containing rational expressions, multiply each side by the least common denominator.

When a rational equation is in the form of a proportion, $\frac{a}{b} = \frac{c}{d}$, the equation may be cross multiplied to yield $a \cdot d = b \cdot c$ and then solved.

Check answers to eliminate extraneous roots.

CHAPTER 9 REVIEW EXERCISES

1. For what input, x, is the expression $\frac{2 - x}{x + 3}$ undefined?

2. Find ten coordinate points that satisfy $y = \frac{6}{x + 3}$, and sketch a graph. Is there an input that gives $y = 0$? Why or why not?

3. World reserves of natural gas were estimated at 4000 trillion cubic feet in 1990. Write an equation that describes the number of years the gas reserves will last if x cubic feet are used per day.

In Exercises 4 and 5, describe the output behavior of the graph in the figure in the given situations.

4.

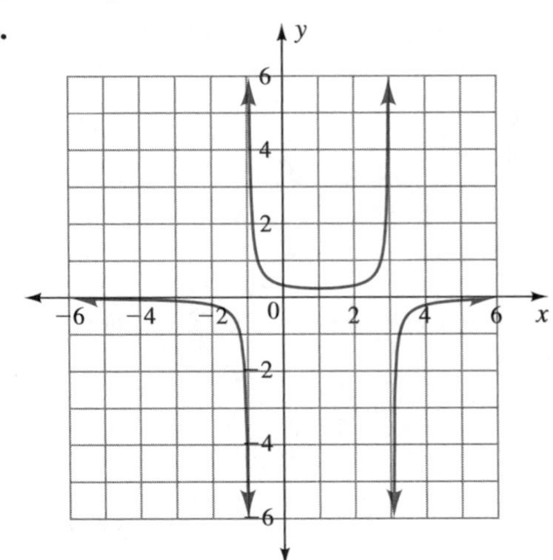

a. As x approaches -1 from the left

b. As x approaches 3 from the right

c. As x approaches -1 from the right

5.

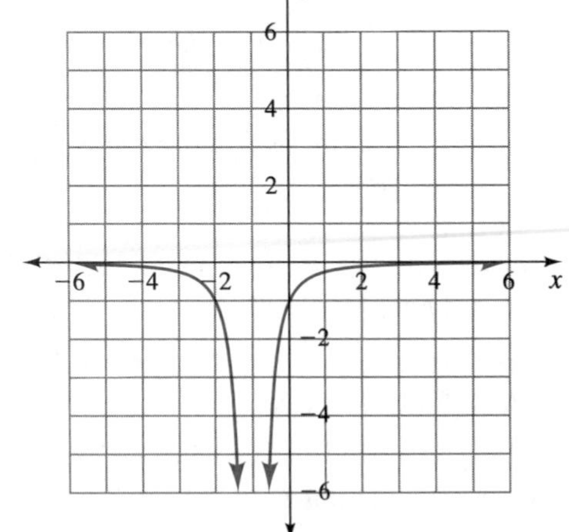

a. As x approaches -1 from the right

b. As x approaches -1 from the left

6. Simplify each expression. Which simplify to fractions that are equivalent to $\frac{6}{8}$? Why?

a. $\frac{6 \cdot 2}{8 \cdot 2}$ b. $\frac{6 \div 2}{8 \div 2}$ c. $\frac{6 - 2}{8 - 2}$ d. $\frac{6 + 2}{8 + 2}$

Simplify the rational expressions in Exercises 7 to 12, if possible. If not, explain why. Assume there are no zero denominators.

7. $\dfrac{2xy}{x^2}$ **8.** $\dfrac{x + y}{xy}$ **9.** $\dfrac{a^2 - b^2}{a + b}$

10. $\dfrac{ab}{a^2 + b}$ **11.** $\dfrac{ab}{a^2 + b^2}$ **12.** $\dfrac{ab}{a^2 + ab}$

Simplify the expressions in Exercises 13 and 14. Indicate any inputs that must be excluded.

13. $\dfrac{1 - a}{a - 1}$ **14.** $\dfrac{x^2 + 4x + 4}{x^2 + 3x + 2}$

In Exercises 15 to 20, what numerator or denominator is needed to make a true statement? Assume there are no zero denominators.

15. $\dfrac{x - 3}{\quad} = -1$ **16.** $\dfrac{\quad}{2 - x} = -1$

17. $\dfrac{\quad}{a - b} = 1$ **18.** $\dfrac{\quad}{a + b} = 1$

19. $\dfrac{\quad}{4 - x} = -1$ **20.** $\dfrac{b - 5}{\quad} = -1$

21. Is the pair of fractions $\dfrac{16}{9}$ and $\dfrac{\sqrt{16}}{\sqrt{9}}$ equal or not equal? Explain why.

22. Explain why $\dfrac{x - 6}{x + 6}$ will not simplify.

23. **a.** Calculate the fractional expressions $300 \div \frac{3}{1}$ and $300 \cdot \frac{1}{3}$.

b. What may be observed about the pair of problems?

c. What do they tell us about multiplication and division?

Identify the word clues that indicate the necessary operations (addition, subtraction, multiplication, or division), and then answer the questions in Exercises 24 to 29.

24. John ate $\frac{1}{3}$ of the pie; Sue ate $\frac{1}{4}$ of the pie. Altogether they ate what portion of the pie?

25. How many servings are there in 4 large pizzas if each person eats one-third of a pizza?

26. Sred stitches $\frac{2}{3}$ of the hem. Evelyn rips out $\frac{1}{2}$ of Sred's work. What fraction remains to be finished?

27. Sally ran $\frac{3}{4}$ mile. Jim ran half as far. How far did Jim run?

28. A box of corn flakes contains 18 ounces. A serving is $1\frac{1}{10}$ ounces. How many servings are in the box?

29. Half an animal shelter's funding comes from cat and dog licensing fees. A third comes from property taxes. What fraction remains to be raised from private donations?

In Exercises 30 to 35, multiply or divide, as indicated. Factor and simplify. Assume there are no zero denominators.

30. $\dfrac{x}{x - 1} \cdot \dfrac{x^2 - 1}{x^2}$

31. $\dfrac{1 - x}{x + 1} \div \dfrac{x^2 - 1}{x^2}$

32. $\dfrac{x^2 - 9}{3 - x} \div \dfrac{x^2}{x + 3}$

33. $\dfrac{n - 2}{n(n - 1)} \cdot \dfrac{(n + 1)n(n - 1)}{n - 2}$

34. $\dfrac{2x + 6}{x - 2} \cdot \dfrac{x^2 - 4}{x^2 + 3x}$

35. $\dfrac{3x - 9}{x + 3} \cdot \dfrac{x}{x^2 - 6x + 9}$

Add or subtract the expressions in Exercises 36 to 41, as indicated. Assume there are no zero denominators.

36. $\dfrac{1}{2a} + \dfrac{5}{6ab}$

37. $\dfrac{x}{x + 2} - \dfrac{2}{x^2 - 4}$

38. $\dfrac{x}{x + 1} - \dfrac{2}{(x + 1)^2}$

39. $\dfrac{a}{a + b} - \dfrac{b}{a - b}$

40. $\dfrac{4}{5} - \dfrac{3}{x} + \dfrac{2}{x^2}$

41. $1 - \dfrac{x^2}{2} + \dfrac{x^4}{24} - \dfrac{x^6}{720}$

42. Explain the role of factoring in the addition or subtraction of rational expressions.

Use the order of operations to simplify the expressions in Exercises 43 and 44. Assume $a \neq 0$ and $c \neq 0$.

43. $\dfrac{1}{a} + \dfrac{2a}{3} \cdot \dfrac{6}{a} \div \dfrac{1}{3} - \dfrac{a}{3}$

44. $\dfrac{3}{c} - \dfrac{4}{3c} \div \dfrac{2}{3} + \dfrac{1}{4} \cdot \dfrac{8}{c}$

In Exercises 45 and 46, simplify the expression or equation containing units of measurement.

45. $\dfrac{\text{4 buttons per card}}{\text{12 buttons per shirt}}$

46. $d = -\dfrac{1}{2}\left(\dfrac{9.81\text{ m}}{\text{sec}^2}\right)(5\text{ sec})^2 + \left(\dfrac{8\text{ m}}{\text{sec}}\right)(5\text{ sec}) + 50\text{ m}$

In Exercises 47 to 49, simplify the expressions from dosage computation. The abbreviation gr is for grain and precedes the number.

47. $\text{gr}\ \dfrac{1}{2} \cdot \dfrac{1\ \text{tab}}{\text{gr}\ \frac{1}{6}}$

48. $\dfrac{\dfrac{1\ \text{mL}}{25\ \text{mL}} \cdot 400\ \text{mL}}{\dfrac{1\ \text{mL}}{4\ \text{mL}}}$

49. $\text{gr}\ \dfrac{1}{150} \cdot \dfrac{1\ \text{mL}}{\text{gr}\ \frac{1}{750}}$

50. Find the slope of the line connecting the points $\left(\dfrac{a}{2}, b\right)$ and $\left(\dfrac{b}{2}, \dfrac{a}{3}\right)$.

51. Simplify the expression on the right side of the equation so that it contains no fractions in the numerator or denominator:

$$t^2 = \frac{d}{\frac{1}{2}g}, \quad g \neq 0$$

Multiply the expressions in Exercises 52 and 53, and simplify the result. Assume there are no zero denominators.

52. $x^2\left(\dfrac{1}{x} + \dfrac{2}{x^2}\right)$

53. $(x + 2)(x - 2)\left(\dfrac{2}{x - 2} + \dfrac{1}{x + 2}\right)$

Solve the equations in Exercises 54 to 57. What inputs must be excluded?

54. $\dfrac{1}{5} = \dfrac{1}{9} + \dfrac{1}{x}$

55. $\dfrac{1}{5} + \dfrac{1}{x} = \dfrac{8}{5x}$

56. $\dfrac{3}{x} + \dfrac{1}{x - 1} = \dfrac{17}{4x}$

57. $\dfrac{1}{2x} = \dfrac{2}{x} - \dfrac{x}{24}$

Add the expressions on the right side of each equation in Exercises 58 and 59.

58. Traffic flow through parallel doors: $\dfrac{1}{m} = \dfrac{1}{m_1} + \dfrac{1}{m_2}$

59. Focal distance for a lens, in optics: $\dfrac{1}{F} = \dfrac{1}{f_1} + \dfrac{1}{f_2}$

60. A can of paint covers 300 square feet. Mark four coordinates on the graph in the figure that show possible lengths and widths of a rectangular floor to be painted. Connect your points. Where on the graph will points be located that represent floor sizes requiring less than a full can of paint?

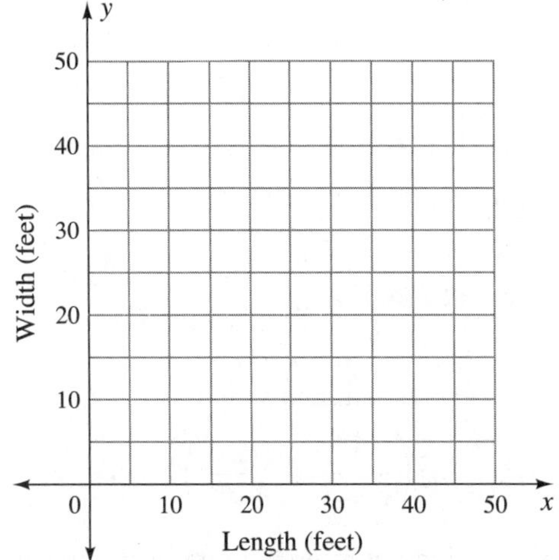

61. The hot water line fills the clothes washer in 3 minutes. The cold water line fills the washer in 2 minutes. How long will it take to fill the washer on the "warm" setting, using both water lines?

62. One clerk is able to issue all advance mail-order tickets for an event in 30 days. An additional clerk is hired. The two clerks together get the job finished in 20 days. In how many days could the second clerk have done the job alone?

CHAPTER 9 TEST

1. For what input is the expression $\dfrac{4 - x}{x - 4}$ undefined?

2. Explain why the expression $\dfrac{4 - x}{x - 4}$ may or may not be simplified?

3. Evaluate $y = \dfrac{4}{x - 2}$ for x as integers between -5 and 5. Sketch a graph.

4. Simplify each expression. Which simplify to fractions equivalent to $\frac{6}{9}$? Why?

a. $\dfrac{6 \div 3}{9 \div 3}$ **b.** $\dfrac{6 - 3}{9 - 3}$ **c.** $\dfrac{6 + 3}{9 + 3}$ **d.** $\dfrac{6 \cdot 3}{9 \cdot 3}$

5. Using factors, show whether the fractions $\dfrac{4}{25}$ and $\dfrac{\sqrt{4}}{\sqrt{25}}$ are or are not equal.

Simplify the expressions in Exercises 6 to 9, if possible. If not, explain why. Assume there are no zero denominators.

6. $\dfrac{b^2}{3ab}$

7. $\dfrac{a - b}{ab}$

8. $\dfrac{a^2 - b^2}{a - b}$

9. $\dfrac{xy}{xy + y}$

Multiply or divide the expressions in Exercises 10 to 13, as indicated. Factor and simplify. Assume there are no zero denominators.

10. $\dfrac{1 - x}{x + 1} \cdot \dfrac{x^2 - 1}{x^2}$

11. $\dfrac{x - 1}{1 - x^2} \div \dfrac{x^2 - 2x + 1}{1 + x}$

12. $\dfrac{n + 1}{n} \div \dfrac{n(n - 1)}{n^2}$

13. $\dfrac{x^2 - 2x}{x^2 - 4} \cdot \dfrac{x - 2}{x}$

14. Multiply this expression, and simplify:

$$(x + 1)(x - 1)\left(\frac{1}{x + 1} + \frac{2}{x - 1}\right)$$

15. One door can empty a meeting room in 16 minutes. A different width door is to be added. Both doors working together must be able to empty the room in 5 minutes. How fast must the new door empty the room?

Simplify each expression or equation in Exercises 16 to 18.

16. 4 yd per shirt · \$8.98 per yd

17. $\dfrac{\dfrac{\$2.50}{1 \text{ gal}}}{\dfrac{25 \text{ mi}}{1 \text{ gal}}}$

18. $3 \text{ mg} \cdot \dfrac{\text{gr } 1}{60 \text{ mg}} \cdot \dfrac{1 \text{ tab}}{\text{gr } \frac{1}{120}}$

19. Is $\dfrac{2\frac{1}{7}}{10} = \dfrac{3}{14}$ a true statement? Explain your answer.

20. Write the expression on the left side of this optics equation as a single fraction:

$$\frac{1}{p} + \frac{1}{q} = \frac{1}{f}$$

21. Simplify the right side:

$$h = \frac{A}{\frac{1}{2}b}$$

Add or subtract the expressions in Exercises 22 to 25. Assume there are no zero denominators.

22. $\dfrac{2}{9} + \dfrac{7}{15}$

23. $\dfrac{2}{ab^2} - \dfrac{a}{2b}$

24. $\dfrac{x}{x + 2} + \dfrac{2}{x^2 - 4}$

25. $\dfrac{2}{3} + \dfrac{3}{x} - \dfrac{4}{x^2}$

26. Simplify $\dfrac{a}{2} + \dfrac{2}{3} \cdot \dfrac{a}{2} - \dfrac{3a}{2} \div \dfrac{9}{2} + \dfrac{1}{4}$.

Solve the equations in Exercises 27 to 30. State any inputs that must be excluded.

27. $\dfrac{1}{3} + \dfrac{1}{6} = \dfrac{1}{x}$

28. $\dfrac{1}{x} = \dfrac{1}{2x} + \dfrac{1}{2}$

29. $\dfrac{3}{x} = \dfrac{2}{x - 3}$

30. $\dfrac{x - 2}{4} + \dfrac{1}{x} = \dfrac{19}{4x}$

31. Explain the role of factoring in multiplying and dividing rational expressions.

FINAL EXAM REVIEW

1. Simplify these expressions without a calculator.

a. $-7 + (-5)$ b. $-3 - (-8)$ c. $-6 + 11$

d. $(-5)(12)$ e. $(-4)(-16)$ f. $|4 + (-8)|$

g. $\left|\frac{-45}{9}\right|$ h. $\frac{3}{2} \cdot \frac{4}{15}$ i. $\frac{-2}{3} \div \frac{5}{6}$

2. Evaluate these expressions if $n = -3$.

a. $-4n$ b. $-n$ c. $-n^2$

d. $(2n)^2$ e. $(-n)^2$ f. $|n - 4|$

g. $2n - 3$ h. $3 - 2n$ i. $\frac{1}{n}$

3. Simplify these expressions. Leave answers without negative or zero exponents.

a. $\frac{bcd}{bdf}$ b. $\frac{-3rs}{12sx}$ c. $\frac{a^3}{a^2}$

d. $\frac{x - 2}{2 - x}$ e. $\frac{(4m^2n)^3}{6n^2}$ f. $\frac{ab}{c} \div \frac{ac}{b}$

g. n^3n^4 h. $(mn^2)^3$ i. $m^{-1}m^{-2}$

j. $(3x^3)^0$ k. $\frac{6x^{-3}}{2x^2y^{-2}}$ l. $(3x^{-3}y^2)^3$

4. Simplify these expressions.

a. $3 - 4(5 - 6) + 7(8)$ b. $|3 - 4| + |\sqrt{4} - 3|$

c. $\sqrt{9} + \sqrt{16} - \sqrt{49}$ d. $3x + 4y - 5x + 2y$

e. $3x^2 + 2x - 1 - (x^2 - 3x + 2)$

f. $3(x - 1) + 4(x + 2) - 5(x - 3)$

5. Solve for the indicated variable.

a. $2x + 3y = 12$ for y

b. $2x - 3y = 7$ for x

c. $ax + by = c$ for y

d. $\frac{x}{x + 1} = \frac{3}{8}$ for x

e. $\frac{P_1V_1}{T_1} = \frac{P_2V_2}{T_2}$ for P_2

f. $3(x + 2) = 7(x - 10)$ for x

g. $4^2 + x^2 = 8^2$ for x

h. $d^2 = [4 - (-3)]^2 + (5 - 2)^2$ for d

i. $6 = \sqrt{3x - 3}$ for x

6. Multiply these polynomials.

a. $3(x^2 - 6x + 4)$ b. $-4x(x^2 - 2x - 3)$

c. $(x - 4)(x + 4)$ d. $(x - 4)^2$

e. $(x - 4)(x^2 - 8x + 16)$ f. $(2x - 3)(3x - 2)$

7. Factor these expressions, using the table method as needed.

a. $12xy + 3xy^2 + 6x^2y^2$ b. $x^2 + x$

c. $x^2 + x - 20$ d. $4x^2 - 9$

e. $6x^2 + x - 2$ f. $6x^2 - 11x - 2$

g. $x^2 - 10x + 25$ h. $6x^2 + 3x - 9$

8. Multiply out each of these binomial squares.

a. $(n - 3)^2$

b. $(n - 1)^2$

c. $(n - 4)^2$

d. Simplify $(n - 1)^2 - (n - 4)^2$.

e. Show that $n^2 - (n - 3)^2 - [(n - 1)^2 - (n - 4)^2] = 6$.

9. Graph $x + y = 0$ and $x - y = 2$. Solve the system of equations from the graph.

Solve the systems in Exercises 10 to 15.

10. $x + y = 6$; $y = -\frac{2}{3}x$

11. $3x + 4y = 10$; $x = 6 - 2y$

12. $x + y = 8$; $x - 3y = 4$

13. $3x + 4y = -20$; $2x - 3y = -2$

14. $x + y = 8$; $x + y = 10$

15. $x + y = 8$; $x = 8 - y$

In Exercises 16 to 19, find the slope of the line through the points and the distance between the points.

16. (2, −3) and (3, 4)

17. (2, −3) and (2, 5)

18. (2, −3) and (−5, −3)

19. (2, −3) and (−5, −2)

20. Use the slopes found in Exercises 16 to 19 to identify pairs of perpendicular lines.

Solve the equations in Exercises 21 to 32. Indicate any inputs that must be excluded.

21. $\frac{4}{x} = \frac{6}{x}$

22. $\frac{x}{4} = \frac{x}{6}$

23. $\frac{2x + 26}{3} = \frac{x - 1}{2}$

24. $\frac{x - 1}{5} = \frac{3x + 3}{18}$

25. $\frac{x}{1} = \frac{14}{x + 5}$

26. $\frac{2x}{5} - \frac{x}{10} = -4.5$

27. $\frac{2x}{5} - \frac{x}{10} = \frac{3x}{10}$

28. $\frac{2x}{5} - \frac{x}{10} = \frac{2x}{10}$

29. $\sqrt{x + 13} = 5$

30. $\sqrt{x + 13} = x - 7$

31. $\sqrt{x + 13} = 0$

32. $\sqrt{x + 13} = -7$

33. The graph of $y = \sqrt{4 - x}$ is shown. Use it to answer the questions.

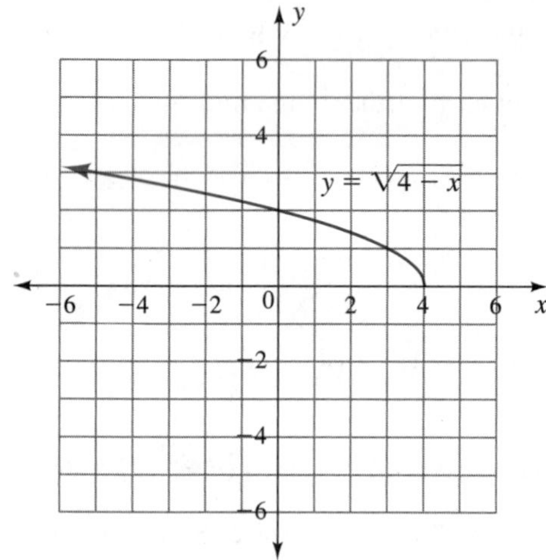

a. What is the x-intercept of the graph?

b. What is the y-intercept of the graph?

c. If $\sqrt{4 - x} = 1$, then $x =$ ____.

d. If $x = 13$, then $\sqrt{4 - x} =$ ________.

e. Why does the graph show no negative values for y?

34. Graph $y = \frac{-24}{x}$.

a. What is the y-intercept?

b. What is the x-intercept?

c. Does the graph of $y = x + 4$ intersect your graph?

d. How does solving $x + 4 = \frac{-24}{x}$, $x \neq 0$, with algebraic notation show whether or not the graphs in part c intersect?

35. The following graph shows a dieting experience. Describe what happened.

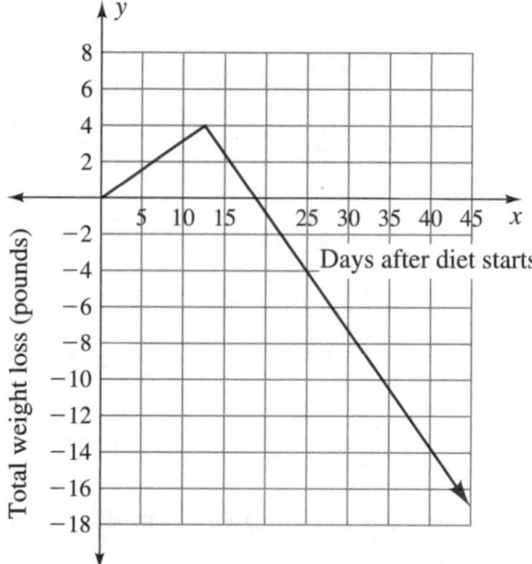

36. Solve these equations.

a. $x^2 + 3x - 4 = 0$

b. $x^2 - 4x + 3 = 0$

c. $2x^2 + 8x + 8 = 0$

d. $x(2x - 3) = 0$

37. Simplify these expressions using properties of exponents and radicals.

a. $3^0 + 4^2$

b. $5^2 + 5^{-1}$

c. $\sqrt{25} + 16^{1/2}$

d. $\sqrt{225} + 25^{1/2}$

e. $\sqrt{64x^2}$, assuming x is either positive or negative.

f. $\sqrt{64x^2}$, assuming x is only positive.

g. 0^0

h. $\sqrt{x}$, where x is a negative number

38. Draw an example that shows why calculating the slope of a vertical line results in division by zero.

39. Factor and simplify $\frac{4x^2 - 1}{2x^2 + 5x + 2}$.

40. For what inputs, x, will $\frac{-1}{(x - 3)(x + 1)}$ be undefined?

41. Simplify these expressions containing units of measure.

a. $\frac{\$6.00 \text{ per hour}}{3 \text{ rooms cleaned per hour}}$

b. $d = -\frac{1}{2}\left(\frac{32.2 \text{ ft}}{\text{sec}^2}\right)(3 \text{ sec})^2 + \left(\frac{22 \text{ ft}}{\text{sec}}\right)(3 \text{ sec}) + 100 \text{ ft}$

42. Simplify these expressions. Assume there are no zero denominators.

a. $\dfrac{2a + 2b}{a^2 - b^2}$ **b.** $\dfrac{4 - x}{6x - 24}$

43. Perform the indicated operations. Leave the answers in reduced form. Assume there are no zero denominators.

a. $\dfrac{2 - x}{xy^2} \cdot \dfrac{x^2}{x - 2}$

b. $\dfrac{x^2 - 3x - 4}{x^2 - 2x} \cdot \dfrac{x^2 - 4}{4 - x}$

c. $\dfrac{x - 4}{x + 2} + \dfrac{x^2 - 3x - 4}{x^2 - 4}$

d. $\dfrac{x^2 + 3x + 2}{x^2 + 6x + 9} \div \dfrac{x + 2}{x + 3}$

e. $\dfrac{1}{3x} - \dfrac{2}{5x}$

f. $8x\left(\dfrac{1}{2x} - \dfrac{3}{x}\right)$

g. $3(x - 2)\left(\dfrac{4}{x - 2} + \dfrac{x}{3}\right)$

44. a. Make an input-output table for $y = 3x - 4$ for inputs in the interval $[-1, 3]$.

b. Graph the equation.

c. Where do we find the slope in a linear equation? Explain how to find the slope from a table. Show the slope on the graph, with rise and run.

d. Where is the y-intercept in the equation? How do we find the y-intercept from the table? Indicate the y-intercept on the graph.

e. If we set $y = 0$ and solve $0 = 3x - 4$ for x, what point on the graph have we found?

f. Solve the equation $3x - 4 = -6.4$. Show the steps.

g. Describe how we might estimate the solution to $3x - 4 = -3$ from the graph.

45. A credit card payment schedule is shown in the table.

Charge Balance, x	**Payment Due, y**
(\$0, \$30]	Full amount
(\$30, \$100]	\$30 + 20% of amount in excess of \$30
(\$100, $+\infty$)	\$50 + 50% of amount in excess of \$100

a. How much is paid for charge balances of \$25, \$35, \$95, \$100, and \$105?

b. Write an equation that describes the payment due in terms of the charge balance for each of the three categories. The equations should use y in terms of x.

c. Explain the difference between the brackets, [ ], and the parentheses, ( ), in the charge balance column.

d. Does it matter which interval notation—brackets or parentheses—is used with the infinity symbol, ∞?

e. Write each charge balance interval using an inequality expression.

46. Describe each set of numbers. Choose from the following: real numbers, rational numbers, irrational numbers, integers, whole numbers, natural numbers.

a. $\{1, 2, 3, 4, \ldots\}$

b. $\{0, 1, 2, 3, 4, \ldots\}$

c. $\{-3, -2, -1, 0, 1, 2, 3, \ldots\}$

d. $\{\text{pi}, \sqrt{2}, \sqrt{3}\}$

47. Use the given facts and unit analysis, as needed, to answer the following questions.

The speed of light is 186,000 miles per second.

The minimum distance from Pluto to the sun is 2756.4×10^6 miles.

The maximum distance from Pluto to the sun is 4551.4×10^6 miles.

There are 60 seconds in a minute.

There are 60 minutes in an hour.

The speed limit on a freeway is 65 miles per hour.

There are 5280 feet per mile.

There are 1609 meters in a mile.

There are 1000 meters in a kilometer.

a. What is the minimum distance from Pluto to the sun, written in the standard form of scientific notation?

b. How long does it take light to travel the maximum distance from the sun to Pluto?

c. Mary Meagher set a world record in the butterfly in 1981. She swam 200 meters in 125.96 seconds. What was her average speed in miles per hour?

d. The orbit velocity of the space shuttle is 28,300 km/hr. To the nearest hundred, how many miles per hour is this?

48. Set up equations for each problem. If the problem describes a system of equations, solve using substitution or elimination or a graphing calculator.

a. Three strings of holiday lights and two packages of giftwrap cost \$18.90. Two strings of holiday lights and five packages of giftwrap cost \$19.86. What is the cost of each item?

b. Four cups of milk and a cup of cottage cheese contain 685 calories. Three cups of milk and two cups of cottage cheese contain 770 calories. How many calories are in each?

c. The three angles of a triangle have measures that add to 180°. The largest angle is four times the smallest. The middle angle is 9° less than the largest. Find the measure of each angle.

d. The equations $y = \frac{3}{5}x$ and $y = -\frac{3}{5}x + 6$ form the diagonals of a rectangle. What is the point of intersection of the two lines? If two sides of the rectangle lie on the coordinate axes, what is the area of the rectangle? What is the length of the diagonal of the rectangle?

e. A birthday party at McDonald's costs \$29.95 for a party of 10 and \$2.50 for each additional child. What equation describes the cost for x people?

f. A jumbo package contains 20 ounces of potato chips. If the input is the number of ounces eaten by each person, what equation describes the number of people served by the package?

49. The volume of a cube, or box with all edges equal to x, is found with the formula $V = x^3$. The surface area of the box is the amount of area covering all six faces. The surface area of a cube is $6x^2$.

a. Fill in this table.

Length of Edge of Box, x	Volume of Cube, x^3	Surface Area, $6x^2$
10		
20		
40		
n		
$2n$		

b. If we double the length of an edge of the box, what happens to the volume?

c. If we double the length of an edge of the box, what happens to the surface area?

d. Use the Pythagorean theorem to determine the length of the diagonal on the face of a cubical box if the length of the edge is 20.

50. In the early 1980s the federal government gave away surplus cheese, butter, and powdered milk. Eligibility guidelines for family monthly income were as follows: one person, \$507; two, \$682; three, \$857; four, \$1,032; five, \$1,207; six, \$1,382; seven, \$1,557; eight, \$1,732. Make a table and graph by following the four-step process. In step 4 (check and extend), how much does each additional person add to the income level? If the points were connected and continued to the left, where would the graph cross the y-axis? Write a rule in words and in symbols that describes the guidelines.

Calculator Objectives

Here is a checklist of important calculator objectives, to complement the mathematical objectives at the beginning of each section of the text.

Perform operations with integers using the change of sign key, [+/−] or [(−)].

Use the reciprocal key, [1/x] or [x^{-1}].

Use the parentheses keys, [( ] and [ )].

Find the values of expressions containing positive and negative exponents with the exponentiation key, [x^y], [y^x], or [^].

Use the correct keystrokes to obtain the required order of operations in evaluating expressions such as $\frac{5.3 + 7.8}{15.4 - 19.8}$.

Use the STORE TO MEMORY key, [STO], [M in], [x→M].

Use the RECALL FROM MEMORY key, [RCL], [M out], [M→x].

Use the pi key, [π].

Use the table feature to build a table when given an equation or expression or when evaluating formulas for a number of inputs.

Enter a function in [Y =].

Set a viewing window.

Graph a function entered in [Y =].

Trace a graph.

Use the statistical function to fit a straight line to two ordered pairs.

Explore the effect of increasing (or decreasing) slope on the graph of a straight line.

Explore the effect of changing the constant, b, in $y = mx + b$ on the graph of a straight line.

Explore the effect of replacing the slope with the negative reciprocal.

Check that a graph of $ax + by = c$ created with a table agrees with a calculator graph of $y = -\frac{a}{b}x + \frac{c}{b}$.

Select and deselect scientific notation mode.

Enter numbers in scientific notation with the EXP, EE, or EEX key.

Correctly read a calculator display showing numbers in scientific notation.

Use the square root key, √, and recognize the calculator's display for expressions undefined in the real numbers.

Answers to Selected Odd-Numbered Exercises and Tests

As you compare your answers to those listed here, keep these hints in mind:

- Don't give up too quickly.
- Have confidence that you worked the exercise correctly.
- Check that you looked up the right answer.
- Check that you copied the exercise correctly.
- See if you can use algebraic notation or simplification to change your answer to match the text's answer.

If the answers still don't match, try working the exercise again:

- Make sure you thoroughly understand the exercise. Read the exercise aloud. Shut the book and say it in your own words.
- On a separate piece of paper, copy the exercise from the text.
- Work the exercise without looking at your first attempt.
- Let the problem rest for an hour or two or overnight. Sometimes the solutions to problems become clear when you step away from them.
- Compare your work with that of another student. (Do this only after you have done the problem twice.)
- Review the text material and related examples.
- Go on to another problem. You can continue doing homework without having completed each and every exercise.
- At the next opportunity, ask your instructor to review your work. Note that this does not mean that you should ask him or her to show you how to do the exercise.

Only after trying the above steps should you assume either that your work is wrong or that the four to six human beings who worked every exercise for this book made an error. The latter is possible, and the author and publisher would appreciate corrections.

EXERCISES 1.1

1. You work 20 days a month.
3. Your teacher knows your last name and in which class you are enrolled.
5. 41 panels **7.** 21 panels **9.** 5, 8, 11, and 14

15. The surgeon is the boy's mother.

EXERCISES 1.2

1. 2 **3.** $\frac{5}{6}$, 0.83 **5.** $\frac{1}{3}$, 0.33 **7.** $\frac{4}{15}$, 0.27 **9.** $1\frac{5}{12}$, 1.42
11. $2\frac{2}{15}$, 2.13 **13.** $\frac{7}{15}$, 0.47 **15.** $3\frac{3}{4}$, 3.75 **17.** $\frac{2}{5}$, 0.4
19. $\frac{1}{2}$, 0.5 **21.** $576 **23.** 8 credit hours
25. a. $3\frac{1}{4}$ yd **b.** $8\frac{3}{4}$ yd **27. a.** $\frac{3}{8}$ yd **b.** 12 shirts

29.

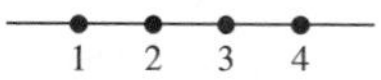

Number of Dots	Number of Segments
1	2
2	3
3	4
4	5
10	11

31.

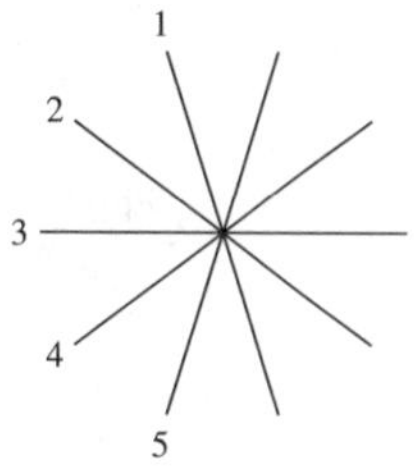

Number of Lines	Number of Regions
1	2
2	4
3	6
4	8
5	10
10	20

39.

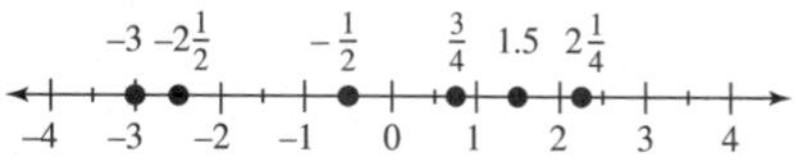

41.

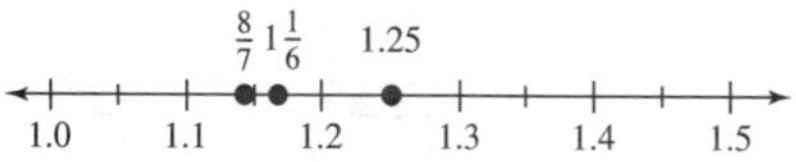

43. $37.50 **45.** $525

47.

Input	Output
0	5
1	2
2	5
3	6
4	5
5	10
6	5
7	14
8	5

49.

Input	Output
8	$440
9	$495
10	$550
11	$600
12	$600

53. a. natural numbers **b.** negative real numbers
c. rational numbers **d.** integers

55. a. 3, 9 **b.** 3, 7

57.

$a + b$	$a - b$	$a \cdot b$	$a \div b$
20	10	75	3
$1\frac{3}{20}$	$\frac{7}{20}$	$\frac{3}{10}$	$1\frac{7}{8}$
0.42	0.30	0.0216	6
6.3	4.9	3.92	8
3.75	0.75	3.375	1.5

59. Input a: 1500%, 75%, 36%, 560%, 225%;
Input b: 500%, 40%, 6%, 70%, 150%

MID–CHAPTER 1 TEST

1. A 20-sided figure will have 18 triangles.

2. a. True **b.** True

c. False; zero is added to the set of natural numbers to make the set of whole numbers.

d. False; for example, 3 divided by 4 is not an integer.

3. a. $6\frac{1}{12}$ **b.** $1\frac{7}{8}$ **c.** $1\frac{1}{2}$ **d.** 2

4. a. \$3.50 **b.** \$3.15 **c.** \$3.85 **d.** 7 min

EXERCISES 1.3

1. a. $3n$ **b.** $\frac{8}{n}$ **c.** $n - 4$ **d.** $\frac{n}{5}$ **e.** $15n$

3. a. $3 + 2n$ **b.** $4 - 3n$ **c.** $7n + 4$ **d.** n^2 **e.** $\$0.79n$

5. a. 2 and π, constants and numerical coefficients; r, variable

b. 1.5, constant and numerical coefficient; x, variable

c. −4, constant and numerical coefficient; 3, constant; n, variable

d. 1, constant and numerical coefficient; −9, constant; x, variable

11. a. $0.35n$ **b.** $0.10x$ **c.** $0.875n$ **d.** $0.375x$ **e.** $0.005n$

f. $1.08x$

13.

x	$3x + 2$
1	5
2	8
3	11
4	14
5	17

15.

Input	Output
4	9
20	41
50	101
100	201
n	$2n + 1$

c; twice the input plus 1

17.

Input	Output
4	15
20	79
50	199
100	399
n	$4n - 1$

b; one less than four times the input

19.

Number of Hours, t	Distance Traveled, D
1	55
2	110
3	165
t	$55t$

21.

Number of Sales	Income
1	\$325
2	\$400
3	\$475
n	$\$250 + \$75n$

25.

Input	Output
−6	0
−5	2
−4	2
−3	0
−2	2
−1	2
0	0
1	2
2	2
3	0
4	2
5	2
6	0

27.

Input	Output
−3	0
−2	0
−1	0
0	0
1	1
2	2
3	3

29.

Input	Output
8	9
50	51
101	202
Even n	$n + 1$
Odd n	$2n$

31.

Pairs of Pens	Panels
4	22
x	$5x + 2$
10	52
25	127

33.

Tables	Chairs
1	5
2	8
3	11
4	14
n	$3n + 2$
10	32

35.

x	y	xy	$x + y$
4	3	12	7
1	12	12	13
5	4	20	9
20	1	20	21
2	2	4	4
6	3	18	9

EXERCISES 1.4

1. $A(-2, 2)$; $B(-5, 0)$; $C(-3, -4)$; $D(0, -2)$; $E(4, -4)$; $F(2, -6)$; $G(4, 0)$; $H(4, 6)$; $I(0, 5)$

3. a. 2nd **b.** 4th **c.** 3rd **d.** 3rd

5. a. vertical axis **b.** vertical axis **c.** horizontal axis

7.

x	y
0	0
−2	1
−4	2
2	−1
4	−2
−6	3
5	−2.5

9.

x	$y = 2x + 5$
0	5
1	7
2	9
3	11
4	13
5	15

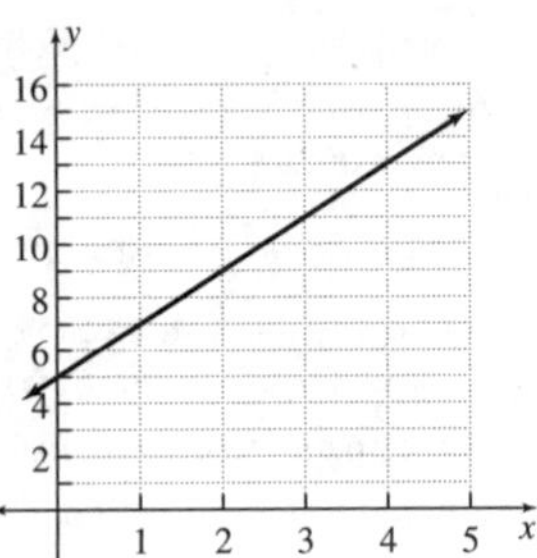

11.

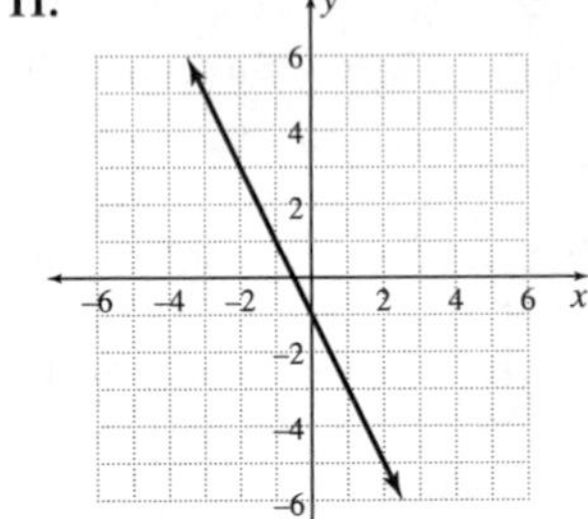

13.

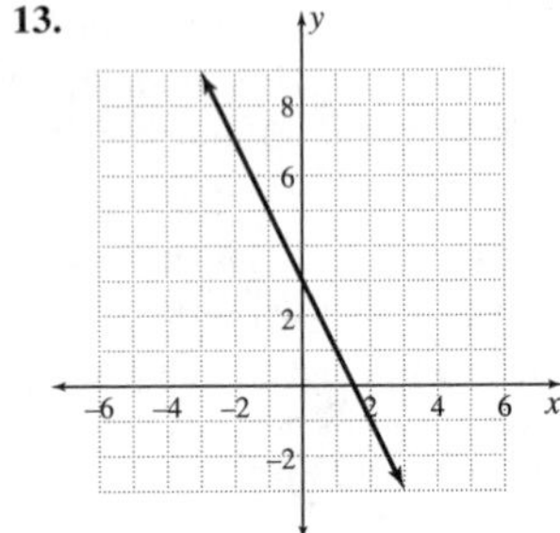

15.–21.

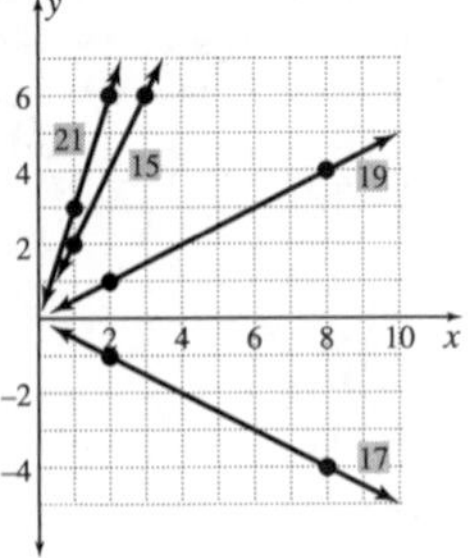

23. a.

Input	Output
0	\$ 0
1	\$ 6.50
2	\$13.00
3	\$19.50
4	\$26.00

b. Within \$0.50 of \$16.25, \$11.38, \$21.13

c. packaged nuts

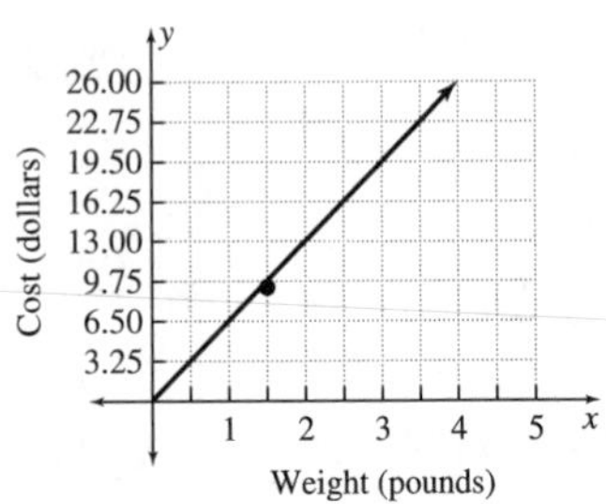

25. a.

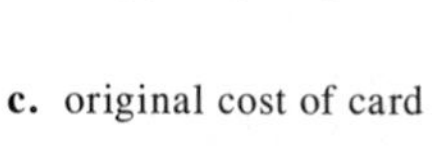

Minutes	Dollars
0	\$24.00
4	\$18.00
8	\$12.00
12	\$6.00
16	0

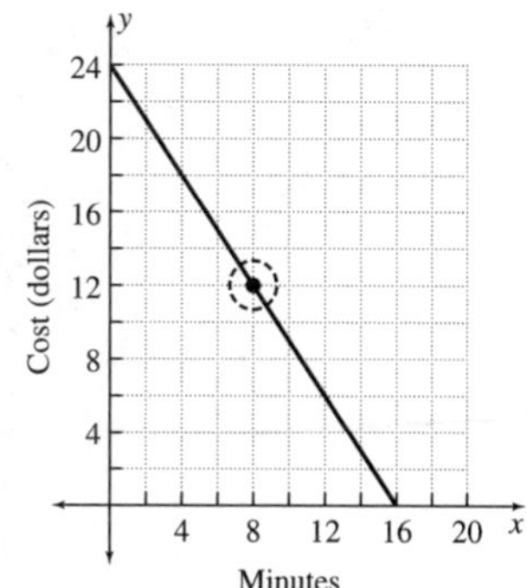

c. original cost of card

d. total minutes of time available on card

27. a.

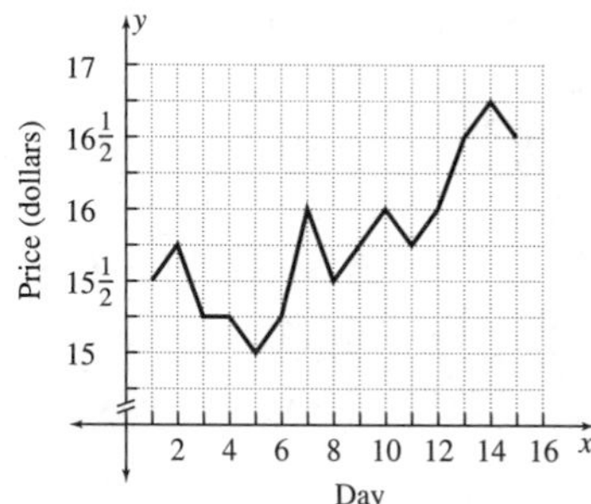

31. a. 9 **b.** ≈12 **c.** 16 **d.** ≈30

33. $2\frac{1}{2}$ yd

35. The scale is not the same as on the rest of that axis.

37. $\frac{3}{4}$ yd **39.** above the 44-inch-wide outputs

41. The order in which the numbers are written is important.

43. From the point, trace vertically to the horizontal axis for the x value, and trace horizontally to the vertical axis for the y value. The ordered pair is (x, y).

45. The second number in an ordered pair is zero on the horizontal axis, and the first number is zero on the vertical axis.

47. \$20 is greater than 10% of the input.

49. a. \$20 **b.** \$20 **c.** \$40 **d.** \$250 **51.** no

53. The graph becomes steeper.

55. $A(2, -5)$; $B(-2, -5)$

57. $A(3, 4)$; $B(5, 2)$

CHAPTER 1 REVIEW EXERCISES

3. a. 3 **b.** 6 **c. and d.**

Dots	Dominos
0	1
1	3
2	6
3	10
4	15
5	21
6	28

e. 55 dominos

5. a. $1\frac{5}{24}, \frac{11}{24}, \frac{5}{16}, 2\frac{2}{9}$ **b.** $3\frac{11}{12}, \frac{5}{12}, 3\frac{19}{24}, 1\frac{5}{21}$

7. a. real, rational **b.** real, rational, integer, whole number

c. real, rational **d.** real, rational **e.** real, rational

f. real, rational, integer

9.

Input	Output
1	\$0.79
2	\$1.68
5	\$4.35
n	\$0.89n − \$0.10

11.

n	$n + 4$
0	4
1	5
2	6
3	7
4	8
5	9
6	10

13.

Input	Output
0	5
1	−1
2	5
3	1
4	5
5	3
6	5

17. Addition:
increased by; sum; more than; longer than; farther; increases; altogether; plus; combined; faster than; a greater than b; b bigger than a; a exceeds b by 3; $a + b$

Subtraction:
decreased by; fewer than; less than; difference; slower than; loses; a less b; a diminished by b; b subtracted from a; a decreased by b; $a - b$

Multiplication:
product; half, $\frac{1}{2}(\ )$; twice; of; for each; times; one third, $\frac{1}{3}(\ )$; multiplied by; $a \cdot b$; $(a)(b)$; ab; $a(b)$; $a \times b$

Division:
per; half, $\frac{(\ )}{2}$; quotient; one third, $\frac{(\ )}{3}$; the fraction bar; a/b; $a \div b$; b/a

19. 1, 3

21. b.

Input	Output
50	54
100	104

c. The output is four more than the input: $n + 4$.

23. a.

Weight (lb)	Cost (\$)
x	$0.37x$
1	0.37
2	0.74
3	1.11
4	1.48

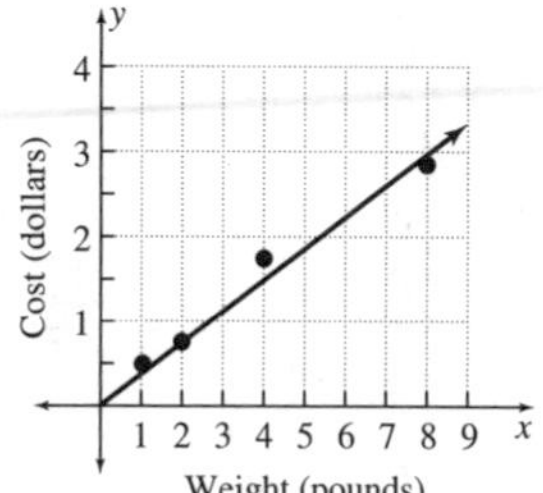

b. $y = \$0.37x$

25.

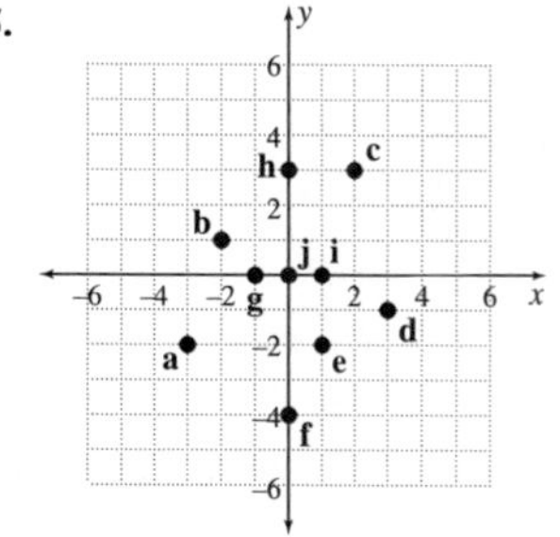

27. $A(7, 3)$; $B(2, 0)$

29.

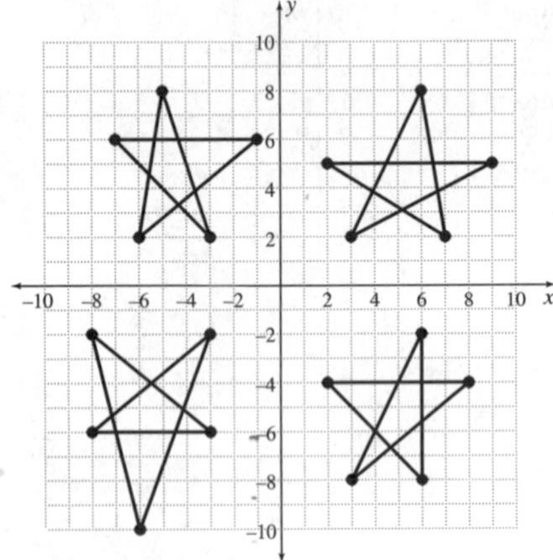

31.

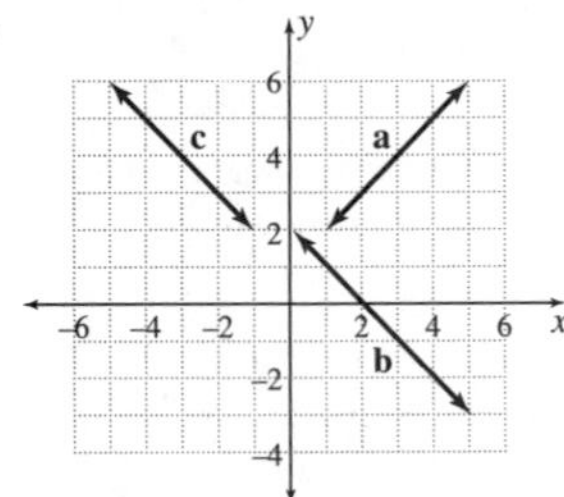

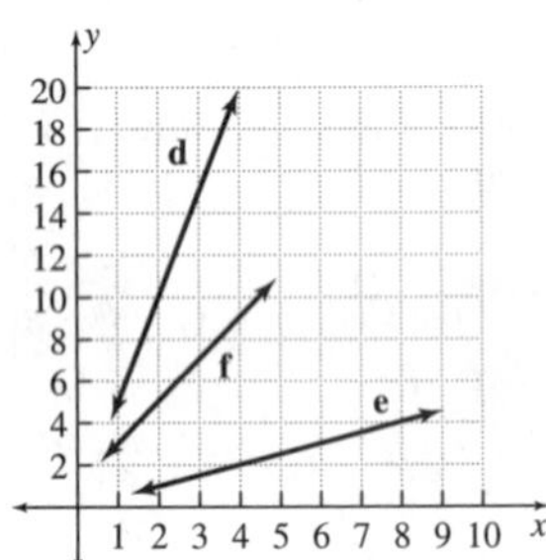

33. **a.** $y = x + 1$ **b.** $y = 2 - x$ **c.** $y = 1 - x$ **d.** $y = 5x$

e. $y = \dfrac{x}{2}$ **f.** $y = 2x + 1$

35.

Uses	Cost
0	0
1	0
2	0
3	0
4	0
5	0
6	0
7	0
8	0
9	\$0.75
10	\$1.50
11	\$2.25
12	\$3.00

37. Inputs are integers only.

CHAPTER 1 TEST

3. difference **4.** set **5.** integers

6. *a*, origin; *b*, vertical or *y*-axis; *c*, horizontal or *x*-axis; *d*, quadrant 3

7. *e*, $(-4, 4)$; *f*, $(2, -3)$; *g*, $(0, -5)$; *h*, $(-2, -7)$

8.

x	$2x + 3$
0	3
1	5
2	7
3	9
4	11
5	13

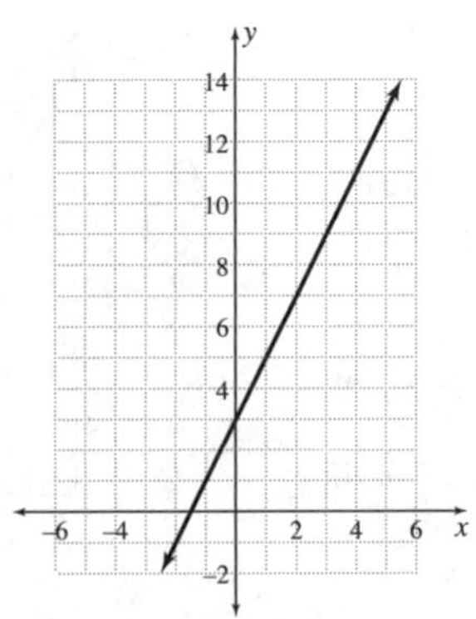

9. **a.** miles on x and cost on y, $\$0.10x + \$150$

c.

Miles, x	Cost, $\$0.10x + \$150$
100	\$160
200	\$170
300	\$180
400	\$190
500	\$200
600	\$210

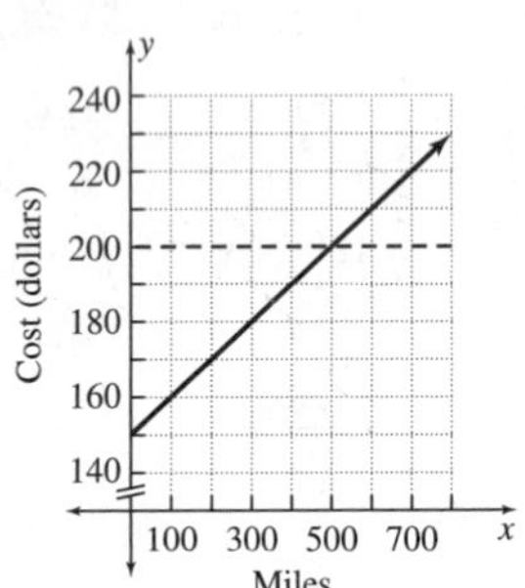

d. 500 mi

10.

Input	Output
0	0
1	2
2	1
3	6
4	2
5	10
6	3
7	14
8	4

11. $\frac{19}{24}$ **12.** $1\frac{31}{54}$ **13.** 3.15

14.

Input	Output
5	20
100	400
n	$4n$

Output is 4 times the input.

15. $A(-5, 4)$; $B(-2, 5)$

EXERCISES 2.1

1. **a.** -5 **b.** $\frac{1}{2}$ **c.** -0.4 **d.** $-x$ **e.** $2x$

3. **a.** 4 **b.** 6 **c.** 5 **d.** 2

5. **a.** -7 **b.** -8 **c.** 3 **d.** 7

7. **a.** 4 **b.** 5 **c.** 5 **d.** -7

9. **a.** 4 **b.** -4 **11.** **a.** -6 **b.** -6

13. The opposite of the absolute value of x

15. $-3 + 4 = 1$; net charge $= +1$

17. $5 + -5 = 0$; net charge $= 0$

19. $-7 + 5 = -2$; net charge $= -2$

21. $-3 + (+2) = -1$ **23.** $+3 + (-5) = -2$

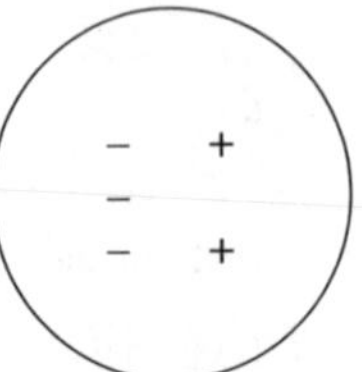

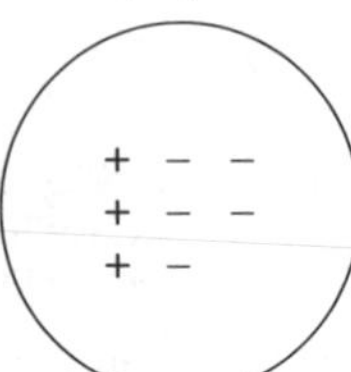

25. **a.** -5 **b.** -3 **c.** 0

27. **a.** 0 **b.** -11 **c.** -4 **29.** **a.** -2 **b.** -4

31. **a.** -7 **b.** -2 **33.** -14 **35.** -1

37. **a.** $1 - (-2) = 3$ **b.** $2 - (+3) = -1$ **39.** **a.** -1 **b.** 2

41. **a.** -3 **b.** 3 **c.** -19 **d.** -5 **e.** -21

43. **a.** 0 **b.** 4 **c.** 5 **d.** -1 **e.** -9

45. **a.** -5 **b.** -3 **c.** 7 **d.** 3 **e.** -5

47. **a.** 9 **b.** 1

49.

x	y	$x + y$	$x - y$
4	5	9	-1
5	-4	1	9
-4	5	1	-9
-4	-5	-9	1

51. 7000 m

53. 2244 m **55.** $|-15| + |-10| + 0 + |-5| + 10 = \$40$

57. $+5 + 12 + 0 + |-7| + |-10| = \$34$

59.

x	$y = \|x\|$
-3	3
-2	2
-1	1
0	0
1	1
2	2
3	3

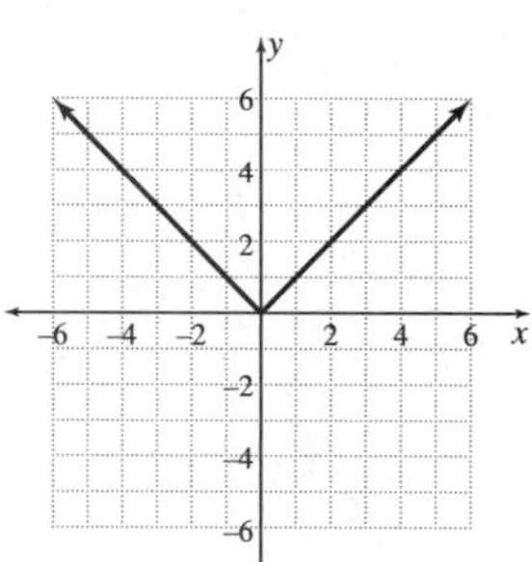

61.

x	$y = \|x + 1\|$
-3	2
-2	1
-1	0
0	1
1	2
2	3
3	4

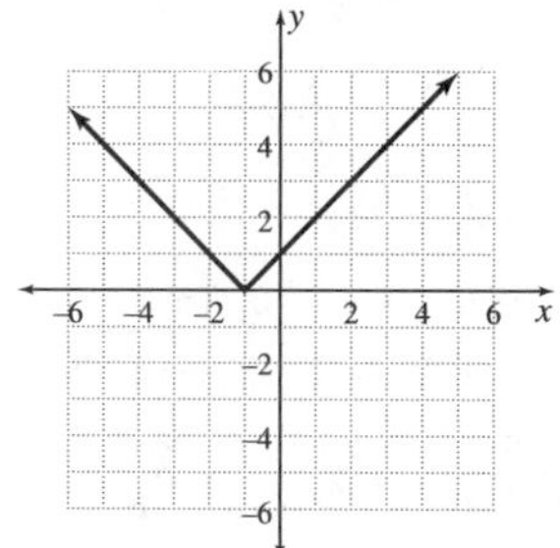

63.

x	$y = \|x\| - 2$
-3	1
-2	0
-1	-1
0	-2
1	-1
2	0
3	1

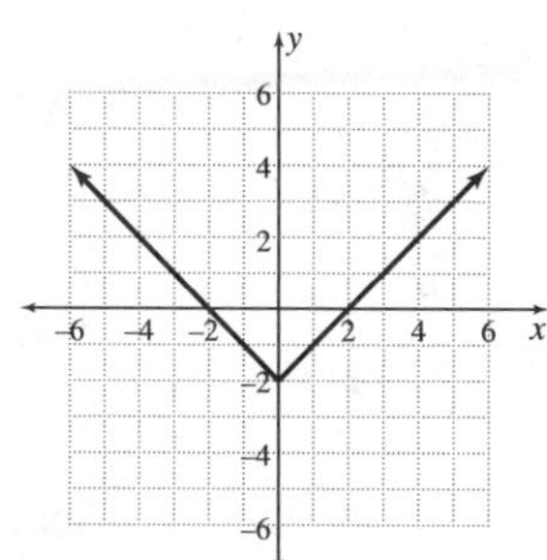

65. true **67.** distance from zero on a number line

69. Change subtraction to adding the opposite.

EXERCISES 2.2

1. $-2(+150) = -\$300$ **3.** $+2(+400) = +\$800$

5. $-8(-40) = +\$320$ **7.** $+3(-90) = -\$270$

9.

x	$3x$
2	6
1	3
0	0
-1	-3
-2	-6

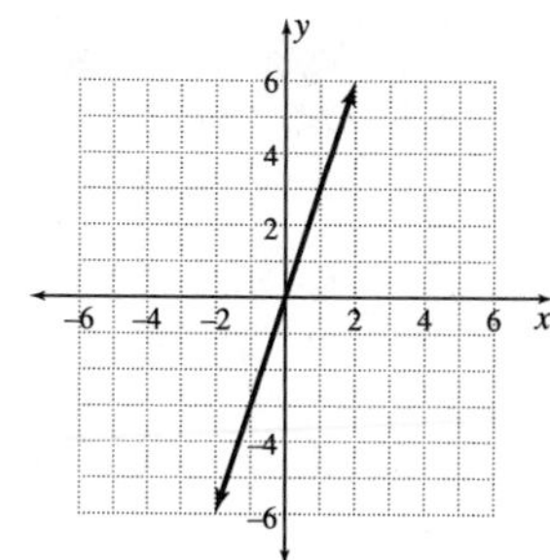

11.

x	$-2x$
2	-4
1	-2
0	0
-1	2
-2	4

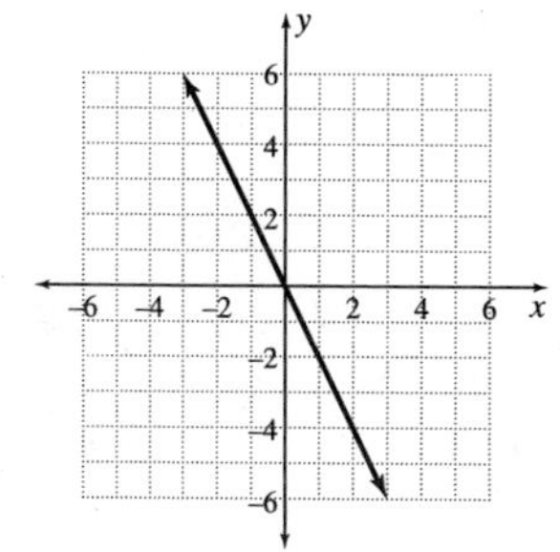

13. **a.** 42 **b.** -48 **c.** -45 **d.** 35

15. **a.** -48 **b.** -64 **c.** 60 **d.** 81

17. **a.** 48 **b.** 25 **c.** -24

19. **a.** -210 **b.** -108 **c.** 0

21. **a.** $\frac{1}{4}$ **b.** $-\frac{1}{2}$ **c.** 2 **d.** $-\frac{4}{3}$ **e.** 2

23. **a.** $\frac{3}{10}$ **b.** $\frac{2}{13}$ **c.** $\frac{1}{x}$ **d.** $\frac{b}{a}$ **e.** $-\frac{1}{x}$

25. **a.** -1 **b.** 1 **c.** -1 **d.** 1

27. **a.** -3 **b.** -5 **c.** -7 **29.** **a.** -7 **b.** 7 **c.** -7

31. **a.** 5 **b.** -7 **c.** -3

33.

a	b	$-\left(\frac{b}{a}\right)$	$\frac{-b}{a}$	$\frac{b}{-a}$
5	35	-7	-7	-7
-27	3	$\frac{1}{9}$	$\frac{1}{9}$	$\frac{1}{9}$

35. $\frac{-b}{a}$, $-\frac{b}{a}$, also $-\left(\frac{b}{a}\right)$

37.

Input	Output
-3	3
-2	2
-1	1
0	0
1	1
2	2
3	3

39. **a.** 45 **b.** $\frac{5}{6}$ **c.** $\frac{15}{16}$ **d.** 3

41.

x	y	$x \cdot y$	$x + y$
2	-2	-4	0
3	-2	-6	1
-4	3	-12	-1
-3	3	-9	0
2	-3	-6	-1

43. Multiply as if the numbers were positive, since the answer will be positive.

45. $\frac{1}{\frac{2}{3}} = 1 \div \frac{2}{3} = 1 \cdot \frac{3}{2} = \frac{3}{2}$

EXERCISES 2.3

1. a. 2 **b.** 3 **c.** 1 **d.** 2 **e.** 1 **f.** 2

3. a. 3 **b.** 2 **c.** 3 **d.** 4 **e.** 2 **5.** 13

7. 200 **9.** 8.98

11. 60 = 60, true, commutative property of multiplication

13. 6 = 6, true, commutative property of addition

15. 12 = 12, true, associative property of addition

17. $60 \neq 27$, false statement

19. a. 6 **b.** 0 **c.** 8 **d.** 2 **e.** no **f.** no

21. $4(5.00 - 0.03) = 20.00 - 0.12 = \$19.88$

23. $3(11.00 - 0.02) = 33.00 - 0.06 = \$32.94$

Hint: Watch signs carefully in Exercises 25 to 29.

25. a. $6x + 12$ **b.** $-3x + 9$ **c.** $-6x - 24$

27. a. $-3x - 3y + 15$ **b.** $-x + y + z$

29. a. $-x + 3$ **b.** $4y + y^2$ **c.** $-2 + y$

31. a. $\frac{1}{3}$ **b.** $-\frac{2}{3}$ **c.** $\dfrac{2}{5yz}$ **33. a.** $\dfrac{5x}{7}$ **b.** $\dfrac{3b}{d}$ **c.** $\dfrac{-y}{4z}$

35. a. $\dfrac{a}{c}$ **b.** $\dfrac{b}{c}$ **c.** $\dfrac{a}{b}$

37. a. $\frac{3}{4}x + 1$ **b.** $x + 2$ **c.** $x + y$ **d.** $a - c$

39. $3x^2 + 3x + 6$ **41.** $2a^2 + 4ab + 6b^2$

43. a. -4 **b.** 1 **c.** -1

45. a. $3x$ **b.** $-11y^2$ **c.** $5a + 15$ **d.** $x + 5$ **e.** $-x + 4$
f. $4x - 6y$

47. a. $\frac{1}{4}x + \frac{3}{4}y$ **b.** $2a + 0.25b$ **c.** $-12x + 3y$ **d.** $4c$

49. $x \cdot y = y \cdot x$

51. The product of factors $a(b + c)$ becomes the sum of terms $(ab + ac)$.

53. Variables and exponents are identical.

55. Factors are multiplied; terms are added.

57. Student forgot $\frac{2}{2} = 1$; the answer should be $x + y + 1$.

59. a. $3^2 \cdot 5$ **b.** prime **c.** $2^3 \cdot 3^2$ **d.** $3 \cdot 37$

MID-CHAPTER 2 TEST

1. a. -3 **b.** 2 **c.** 2 **2. a.** 7 **b.** -7 **c.** 3.5

3. a. 0.72 **b.** 3.72 **4. a.** -6 **b.** 15 **c.** 4

5. a. -9 **b.** 14 **c.** -8 **6. a.** $2x + 6y$ **b.** $-x$

7. a. $4z$ **b.** $\dfrac{3}{yz}$ **c.** $-\dfrac{1}{2x}$

8. a. $+6 - (+4) = +2$ **b.** $0 - (-3) = +3$ **9.** 6 **10.** 0

11. 780

12.

x	$y = x - 1$
−2	−3
−1	−2
0	−1
1	0
2	1
3	2

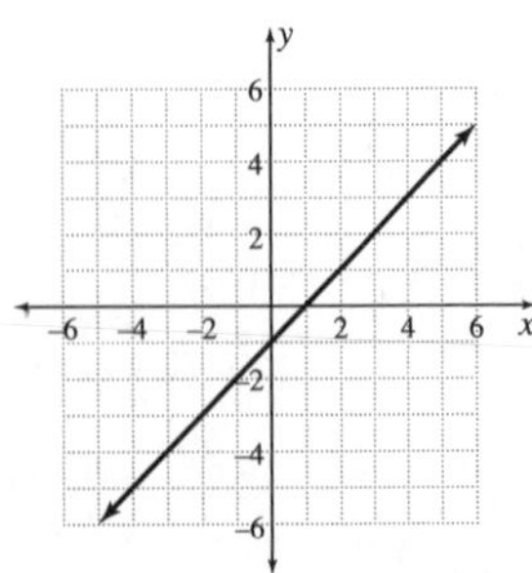

13.

x	$y = x + 1$
−2	−1
−1	0
0	1
1	2
2	3
3	4

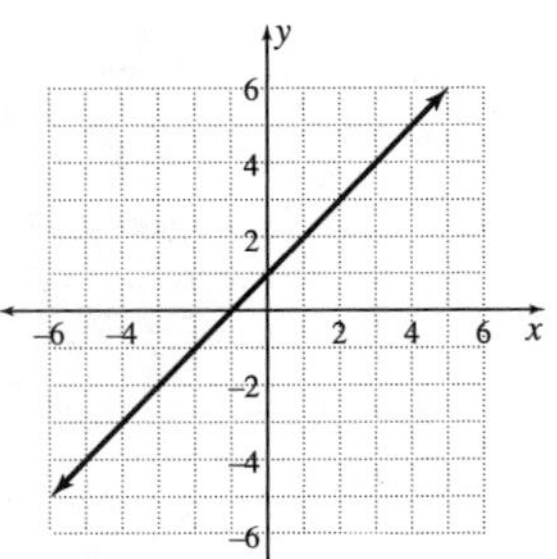

14.

x	$y = 2x - 3$
−2	−7
−1	−5
0	−3
1	−1
2	1
3	3

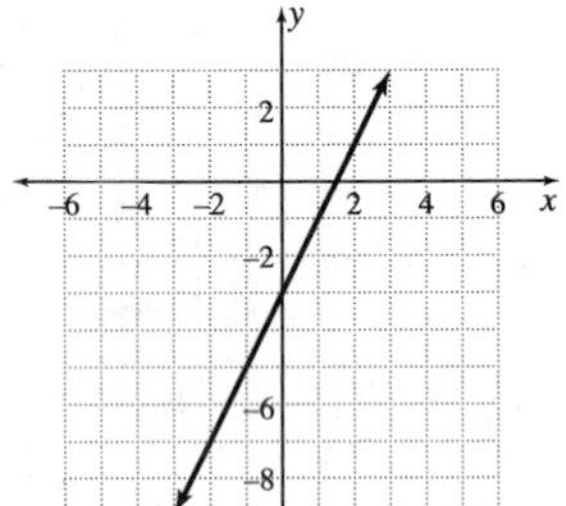

15.

x	$y = 3 - x$
−2	5
−1	4
0	3
1	2
2	1
3	0

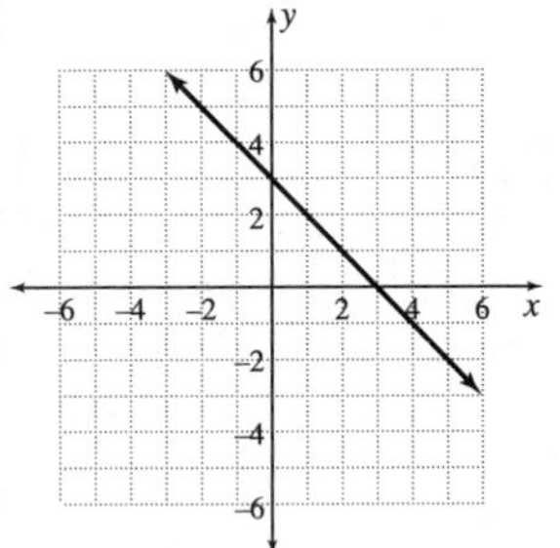

16. 9250 m **17.** 6280 m **18.** Mauna Kea, 1082 ft

19. $2x + y$ **20.** $a + c$

EXERCISES 2.4

1. a. base x; $3 \cdot x \cdot x$ **b.** base x; $-3 \cdot x \cdot x$
c. base $-3x$; $(-3x)(-3x)$ **d.** base x; $a \cdot x \cdot x$
e. base x; $-1 \cdot x \cdot x$ **f.** base $-x$; $(-x)(-x)$

3. a. 243 **b.** 64 **c.** 4 **d.** -27

5. a. $\frac{1}{27}$ **b.** $\frac{64}{125}$ **c.** $-\frac{8}{27}$ **d.** $-\frac{1}{9}$

7. a. 32 **b.** -4 **c.** 12 **d.** -36

9. a. 2^5, 32 **b.** 3^8, 6561 **c.** 4^5, 1024 **d.** 3^5, 243

11. a. m^8 **b.** n^8 **c.** a^8 **d.** a^8

13. a. x^3 **b.** a^3 **c.** $\dfrac{x^2}{y^2}$ **d.** $\dfrac{8x^3}{y^3}$ **15.** 3

17. 1, 11; 2, 10; 3, 9; 4, 8; 5, 7; 6, 6

19. a. x^6 **b.** x^2y^2 **c.** x^4y^6

21. a. $512a^9$ **b.** $108a^2b^3$ **c.** $-108a^3$

23. a. 1 **b.** $\dfrac{-2b}{3}$ **c.** a^2b **d.** $\frac{2}{5}x^2y^2$

25. a. x^4 **b.** $8x^6$ **c.** $\frac{1}{16}a^4$ **27. a.** y^6 **b.** $9y^6$ **c.** $\frac{1}{9}y^6$

29. a. 74 **b.** 20 **c.** 28 **31.** 94 **33.** 35 **35.** 100

37. 169 **39.** 1 **41.** 15 **43.** $-3x + 18$ **45.** $3x - 12$

47. 36 **49.** 4 **51.** 6 **53. a.** $-\frac{3}{5}$ **b.** -5 **c.** $\frac{3}{7}$

55. a. 2 **b.** $\frac{3}{2}$ **c.** 4 **57.** 15 **59.** 10 **61.** $\frac{8.5}{3} \approx 2.83$

63. 75.36 **65.** ≈ 33.49 **67.** 131.88 **69.** 18

73. **a.** When multiplying expressions with like bases, keep the base and add the exponents.

b. When applying an exponent to a power expression, multiply the exponents.

c. When applying an exponent to a product in parentheses, apply the exponent to all parts of the product.

d. When applying an exponent to a quotient in parentheses, apply the exponent to all parts of the quotient.

75. Student #3 is correct.

77. **a.** $-1 + 9 - 9 + 8 = 7$ **b.** $1 + (9 \div 9) + \sqrt{9} = 5$

79. $(-3) \cdot (-1)$ is 3, not -3 **81.** 5 days, Sunday **83.** 21 rounds

EXERCISES 2.5

1. $\frac{1}{2}$ pound **3.** 3600 inches **5.** feet^2 **7.** $\frac{1}{\text{inch}}$

9. $\approx$3281 feet **11.** $\approx$54,545 grams **13.** 86,400 seconds

15. 11.6 days **17.** 2,270,592,000 seconds **19.** 1728 in^3

21. 1296 in^2 **23.** 8.3 ft^2 **25.** 65,000 ft

27. **a.** $P = 9.9$ cm, $A = 4.8\ \text{cm}^2$ **b.** $P = 12.6$ cm, $A = 6.8\ \text{cm}^2$
c. $C \approx 12.6$ ft, $A \approx 12.6\ \text{ft}^2$ **d.** $P = 9.6$ cm, $A \approx 4.0\ \text{cm}^2$

29. **a.** $P = 60$ yd, $A = 120\ \text{yd}^2$ **b.** $P = 6.8$ cm, $A \approx 2.9\ \text{cm}^2$
c. $P = 85$ m, $A \approx 270.8\ \text{m}^2$ **d.** $C \approx 28.3$ in., $A \approx 63.6\ \text{in}^2$

31. **a.** $S = 280\ \text{ft}^2$, $V = 300\ \text{ft}^3$ **b.** $S = 256\ \text{in}^2$, $V = 240\ \text{in}^3$

33. **a.** $S \approx 527.5\ \text{cm}^2$, $V \approx 904.3\ \text{cm}^3$
b. $S \approx 131.9\ \text{cm}^2$, $V \approx 113.0\ \text{cm}^3$

35. **a.** $S \approx 803.8\ \text{cm}^2$, $V \approx 2143.6\ \text{cm}^3$
b. $S \approx 201.0\ \text{cm}^2$, $V \approx 267.9\ \text{cm}^3$

37. 2.25 times **39.** $V \approx 425.3\ \text{in}^3$; 425,250 in^3; $\approx$246.1 ft^3

41. 346.5 in^2 **43.** 19 gal **45.** $\approx$5.8 calories per cracker

47. $\approx$1392.5 ft **49.** $\approx$4,010,417 ft^2, $\approx$0.14 mi^2

51. **a.** $P = 8x$, $A = 4x^2$ **b.** $P = 16x$, $A = 16x^2$
The perimeter of a is $\frac{1}{2}$ the perimeter of b.
The area of a is $\frac{1}{4}$ the area of b.

53. $\frac{1}{2}$ base times height **55.** $\frac{4}{3}$ times π times radius cubed

59. You add 2 lengths and 2 widths. **61.** $\approx$1.9 cm^2

63. $\approx$11,483.3 m^2 **65.** 117 m^2

EXERCISES 2.6

1.
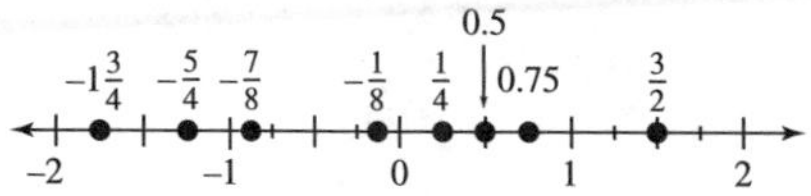

3. **a.** $-8 < -3$ **b.** $+4 > -9$ **c.** $(-3)^2 = 3^2$
d. $0.5 > 0.5^2$ **e.** $6 > -5$ **f.** $-2(6) < -2(-5)$
g. $-6 < -5$ **h.** $-2(-6) > -2(-5)$

5. **a.** $-3.75 < -3.25$ **b.** $3(-2) = -3(2)$ **c.** $\frac{1}{2} > -\frac{1}{2}$
d. $|-4| > |2|$ **e.** $(-2)(-3) > 2(-4)$ **f.** $(\frac{1}{2})^2 = (-\frac{1}{2})^2$
g. $-2.5 > -3$ **h.** $\frac{22}{7} > \pi$

7. **a.** $0 < x < 4$ **b.** $-5 < x \le -2$

9. **a.** $x > 3$ and $x < 8$ **b.** $x > -3$ and $x \le -1$
c. $x > -2$ and $x < 1$

11. f, s **13.** e, w **15.** b, r

17. **a.** $[-1, 3)$; x is between -1 and 3, including -1
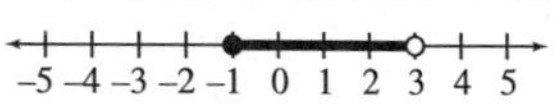

b. $-4 < x \le -1$; x is greater than -4 and less than or equal to -1
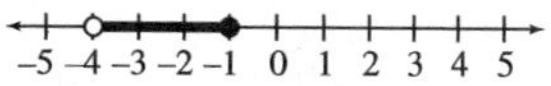

c. $-3 \le x < 5$; $[-3, 5)$

d. $(-\infty, -4)$; x is less than -4
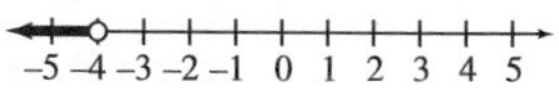

e. $-2 \le x < 5$; $[-2, 5)$; x is between -2 and 5, including -2

f. $(-2, +\infty)$; x is greater than -2
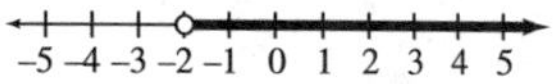

g. $-4 < x \le 2$; $(-4, 2]$
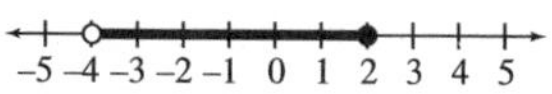

h. $x \ge -3$; x is greater than or equal to -3
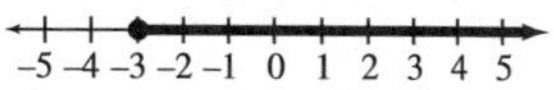

19. An inequality is a statement that one quantity is greater than or less than another quantity.

21. An interval is a set of numbers between endpoints that may also include one or both endpoints.

23. x cannot be less than 2 *and* greater than 4.

25. The $\le$ includes the possibility of equality.

27. **a.** $x \le 2000$ **b.** $2000 < x \le 5000$ **c.** $x > 5000$

29. **a.** $x \le -5$ **b.** $-5 < x < 5$ **c.** $x \ge 5$

31. **a.** $x < 5$ **b.** $5 \le x \le 50$ **c.** $x > 50$

33.

n percent	\$1.00	\$5.00	\$10.00	x
6%	\$0.06	\$0.30	\$0.60	$0.06x$
10%	\$0.10	\$0.50	\$1.00	$0.10x$
25%	\$0.25	\$1.25	\$2.50	$0.25x$
100%	\$1.00	\$5.00	\$10.00	$1x$
150%	\$1.50	\$7.50	\$15.00	$1.5x$

CHAPTER 2 REVIEW EXERCISES

1. **a.** -8 **b.** -5 **c.** -13 **d.** 13 **e.** -14 **f.** -7
g. 42 **h.** -24 **i.** -16 **j.** 12 **k.** 11 **l.** 13

3. **a.** -54 **b.** 54 **c.** 54 **d.** -54 **e.** 56 **f.** -56
g. -56 **h.** 56 **i.** 2 **j.** -4 **k.** -8 **l.** 8
m. 16 **n.** -16 **o.** -6

5. **a.** -4 **b.** 10 **c.** 12 **d.** -9 **e.** 5 **f.** 10
g. 15 **h.** 9 **i.** 2.5 **j.** 8 **k.** -4 **l.** 1

7. **a.** $2(x + y)$ **b.** $a(c + b)$ **c.** $4(x^2 - 2x + 3)$
d. $xy(3 + 4x)$ **e.** $3(2x + 4y - 5)$ **f.** $5ab(3ac + b + 2c)$

9. **a.** $\frac{a}{d}$ **b.** $\frac{2y}{3z}$ **c.** $-\frac{3c}{2a}$ **d.** $4x^2$ **e.** $-27y^3$ **f.** $16y^4$
g. a^2b^2 **h.** a^2b^2 **i.** m^9 **j.** m^9 **k.** m^3 **l.** m^3
m. $x + 2y$ **n.** $m + n$ **o.** $\frac{1}{2}a + b$

11. **a.** 10 **b.** -2 **c.** 11 **d.** $-\frac{3}{4}$ **e.** -1 **f.** 34
g. $-\frac{1}{5}$ **h.** 66

13. **a.** $A \approx 19.6\ \text{ft}^2$ **b.** $A = 10\ \text{yd}^2$ **c.** $V \approx 113.1\ \text{m}^3$
d. $V \approx 3.4\ \text{cm}^3$

15. **a.** $P = 9.5$ cm, $A \approx 4.1\ \text{cm}^2$ **b.** $P = 9$ ft, $A \approx 5.9\ \text{ft}^2$
c. $P = 18$ in., $A = 15\ \text{in}^2$ **d.** $P \approx 20.6$ m, $A \approx 28.6\ \text{m}^2$

17. One possible example: $(3 + 4) + 5 = 7 + 5 = 12$;
$3 + (4 + 5) = 3 + 9 = 12$

19. The small circle excludes the point and is used with < and >. The dot includes the point and is used with ≤ and ≥.

21. $-x^2$ is the opposite of $x \cdot x$, while $(-x)^2$ is $(-x)(-x)$.

23. Change $2\frac{1}{4}$ to $\frac{9}{4}$ and write $\frac{4}{9}$.

29. **a.** $4 > -3$ **b.** $2(-3) < (-2)(-3)$ **c.** $(-2)^2 > -2^2$

d. $|-4| = |4|$ **e.** $-2^3 = (-2)^3$ **f.** $|-5| > -|5|$

g. $-\frac{1}{4} > -\frac{1}{2}$ **h.** $-1.3 > -1.5$

31. $0 \le x < 50$; $[0, 50)$
$50 \le x \le 500$; $[50, 500]$
$x > 500$; $(500, +\infty)$

CHAPTER 2 TEST

1.

x	$y = 3 - x$
−2	5
0	3
2	1
4	−1

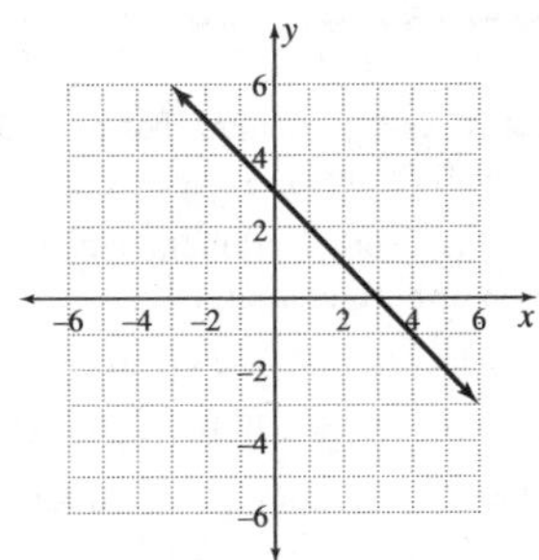

2. **a.** $\frac{19}{12}$ **b.** $-\frac{1}{12}$ **c.** $-\frac{5}{8}$ **d.** $-\frac{9}{10}$

3. commutative property of addition

4. associative property of addition

5. **a.** 4 **b.** −2.5 **c.** −1 **d.** 60 **e.** −1 **f.** 10

g. m^{11} **h.** m^4 **i.** 37 **j.** $\frac{ce}{ft}$ **k.** $\frac{2x}{3}$ **l.** $-x$

m. a^6b^9 **n.** $\frac{a^9}{8b^6}$ **o.** $x - 3$ **p.** $x + 2$

6. **a.** $5x - 5y$ **b.** $x^3 - 5x^2 + 7x - 2$ **c.** $5x - 7$

d. $7x - 7$ **e.** $3a^2 + ab + 6b^2$

7. **a.** $6x + 27y$ **b.** $6a + 27b$ **c.** $6x^2 + 8x - 4$

d. $ab^2 - a^2b^2 + a^3b$

8. No; $(-x)^2$ is $(-x)(-x)$, while $-x^2$ is the opposite of $x \cdot x$.

9. **a.** ≈12.56 ft² **b.** 1256 ft² **c.** 100 times larger; 100

10. 65 days **11.** **a.** 128,100 ft² **b.** 1220 ft **c.** $883\frac{1}{3}$ yd

12. **a.** ≈6.14 ft **b.** 130 m

13. **a.** $V = 864$ ft³, $S = 600$ ft² **b.** $V = 1130$ in³, $S = 603$ in²

14. $0 \le x \le 20$; $[0, 20]$

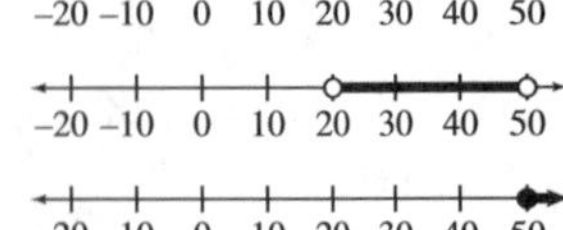

$x \ge 50$; $[50, +\infty)$

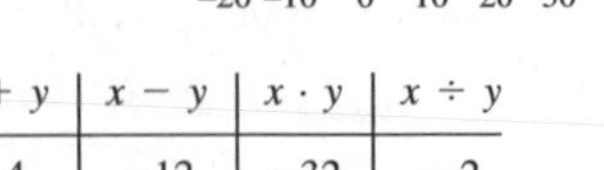

15.

x	y	$x + y$	$x - y$	$x \cdot y$	$x \div y$
−8	4	−4	−12	−32	−2
−6	−2	−8	−4	12	3
6	−3	3	9	−18	−2

CUMULATIVE REVIEW OF CHAPTERS 1 AND 2

1. **a.**

x	$y = x^2 - 1$
−2	3
−1	0
0	−1
1	0
2	3
3	8

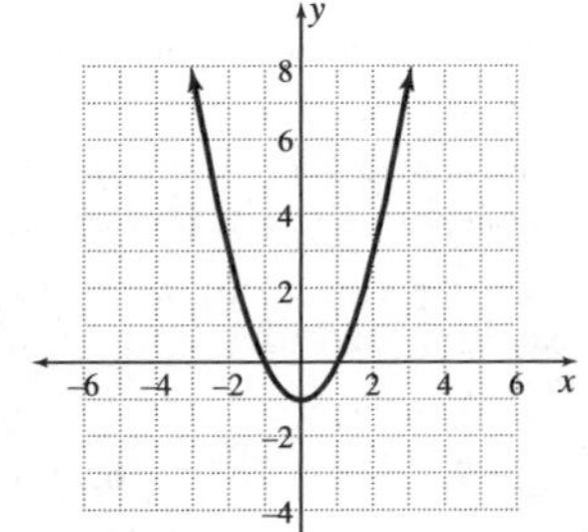

b.

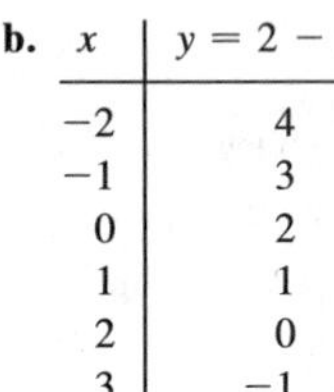

x	$y = 2 - x$
−2	4
−1	3
0	2
1	1
2	0
3	−1

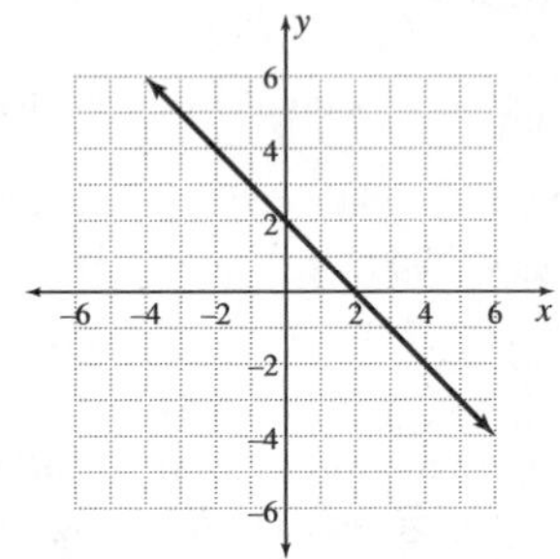

c.

x	$y = 2x + 3$
−2	−1
−1	1
0	3
1	5
2	7
3	9

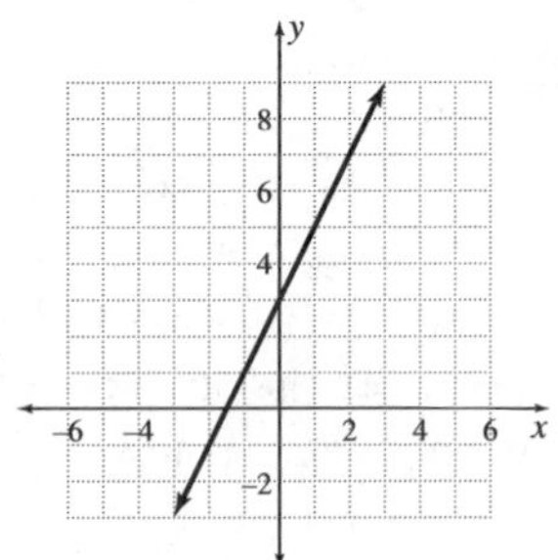

d.

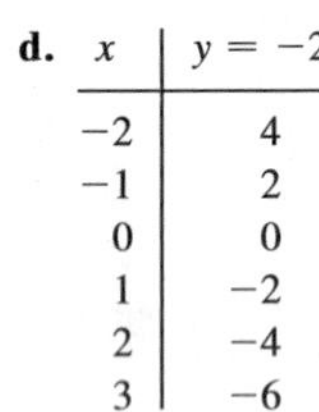

x	$y = -2x$
−2	4
−1	2
0	0
1	−2
2	−4
3	−6

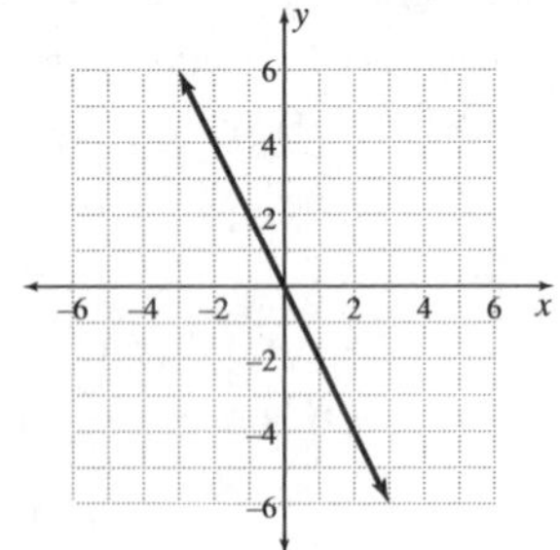

e.

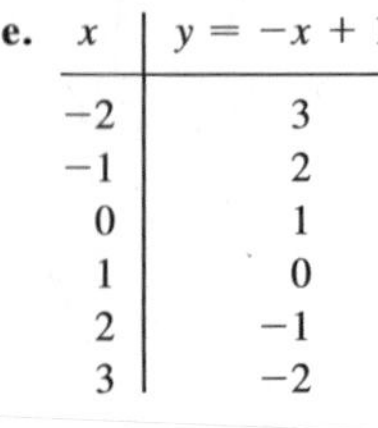

x	$y = -x + 1$
−2	3
−1	2
0	1
1	0
2	−1
3	−2

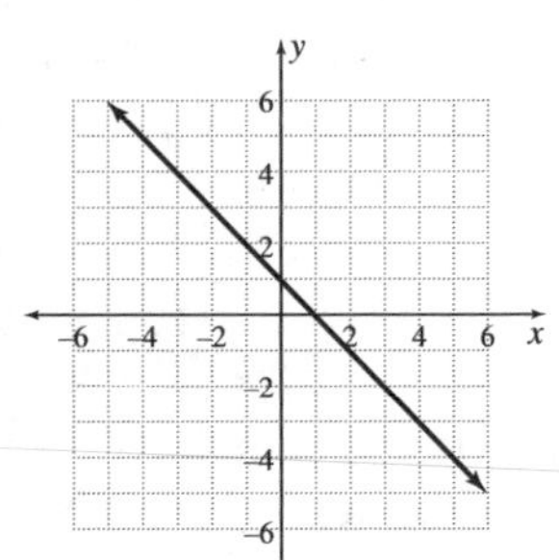

f.

x	$y = \|x - 1\|$
−2	3
−1	2
0	1
1	0
2	1
3	2

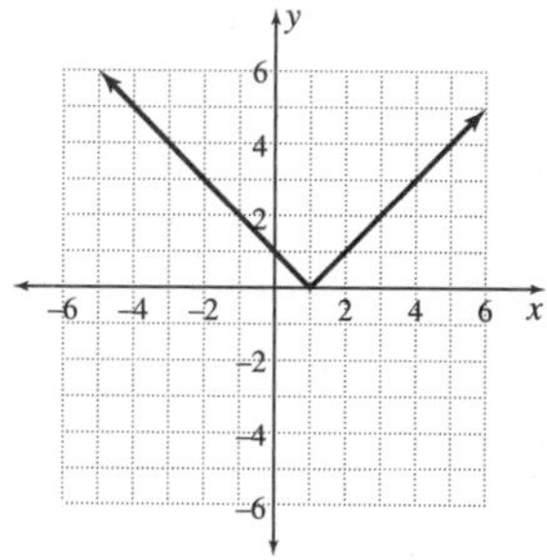

3. a. 4 is less than x.

b. The product of 3 and x is subtracted from 5.

c. The opposite of the opposite of x

d. The absolute value of the difference between 3 and x

5. a. $8x - 7$ **b.** $-2x + 10$ **c.** $5x + 22$ **d.** $3x - 1$

7.

x	y	$x + y$	$x - y$	$x \cdot y$	$x \div y$
−4	6	2	−10	−24	$-\frac{2}{3}$
5	−3	2	8	−15	$-\frac{5}{3}$
$-\frac{2}{3}$	$\frac{3}{4}$	$\frac{1}{12}$	$-\frac{17}{12}$	$-\frac{1}{2}$	$-\frac{8}{9}$
$\frac{7}{8}$	$-\frac{7}{10}$	$\frac{7}{40}$	$\frac{63}{40}$	$-\frac{49}{80}$	$-\frac{5}{4}$
1.44	1.8	3.24	−0.36	2.592	0.8
0.25	−0.5	−0.25	0.75	−0.125	−0.5

9. a. 14 **b.** −60 **c.** 30 **d.** 28 **e.** 1 **f.** −1

g. $2a + \frac{2}{3}$ **h.** $y - x$

11. $\approx 1.65 \text{ yd}^3$

13. a. One example: The distance from Erie to Pittsburgh equals the distance from Pittsburgh to Erie.

b.

	Erie	Pittsburgh	Reading	Scranton
Erie	—	135	325	300
Pittsburgh	135	—	255	272
Reading	325	255	—	99
Scranton	300	272	99	—

c. The commutative properties of addition and multiplication

EXERCISES 3.1

1. $\frac{4}{5}$ **3.** $7\frac{1}{2}$ **5.** 30% **7.** 26

9. a. conditional, 2 variables **b.** conditional, 1 variable

c. conditional, 3 variables **d.** identity **e.** identity

f. identity

11. $\frac{x}{2} = 8$ **13.** $6 + x = 4$ **15.** $2x - 15 = -9$

17. $19 = 2x + 3$ **19.** $3x - 4 = 17$ **21.** $26 - 4x = 2$

23. Four less than a number is 6.

25. Five less than the product of 3 and a number is 16.

27. Two-thirds of a number is 24.

29. Six less twice a number is 10.

31. a. $2(4) \stackrel{?}{=} 8$ correct **b.** $(-6) + 1 \stackrel{?}{=} -5$ correct

c. $(12) - 3 \stackrel{?}{=} 15$; no—add 3

33. a. $55(20) \stackrel{?}{=} 440$; no—divide by 55

b. $\frac{1}{2}(5) \stackrel{?}{=} 10$; no—multiply by 2 **c.** $1.05(40) \stackrel{?}{=} 42$ correct

35. Add 5; $x = 13$ **37.** Subtract 12; $x = -3$

39. Add 6; $x = -4$ **41.** Divide by 2; $x = 13$

43. Divide by 8; $x = \frac{3}{8}$ **45.** Multiply by 4; $x = 64$

47. Multiply by −12; $x = -48$ **49.** Multiply by 2; $x = -8$

51. Multiply by $-\frac{4}{3}$; $x = -16$ **53.** f; $x = \frac{1}{2}$ **55.** c; $x = 14$

57. b; $x = 20$ **59.** $x = 5$ **61.** $x = 15$ **63.** $x = 12$

65. $x = 7$ **67.** $x = 1$ **69.** $x = -2$ **71.** $x = -10$

73. $x = -2$ **75.** $x = -3$ **77.** $x = -1\frac{1}{2}$ **79.** $x = 2.5$

81. Divide by 2. **83.** Subtract 5. **85.** Add 4.

87. Multiply by 2. **89.** $t = 2$ **91.** $C = 100$ **93.** $t = 3\frac{7}{11}$

95. $r = 66\frac{2}{3}$ **97.** $p \approx \$1.67$ **99.** $x \approx 1884.25$ kwh

101. $h = 10$ **103.** Subtraction is the same as adding the opposite.

105. Add a to each side.

107. An identity is true for all values of the variable; a conditional equation is true for only certain values.

EXERCISES 3.2

1. a. linear **b.** linear **c.** nonlinear **d.** linear

e. nonlinear **f.** linear

3. $y = 2x + 5$ **5.** $y = 3x - 6$ **7.** $y = \frac{1}{2}x - 5$

9. $y = 0.15x$ **11.** $y = 0.0145x$ **13.** $y = 1.49x$ **15.** $y = 3x$

17. $y = 75x + 32$ **19.** $y = -3.25x + 26$

21. a. $x = 2$ **b.** $x = 4$ **c.** $x = 7$ **d.** $x = 12$

23. a. $x = 1$ **b.** $x = 2$ **c.** $x = -1$ **d.** $x = 2\frac{1}{2}$

25.

x	$y = 2x + 1$
3	7
$3\frac{1}{2}$	8
4	9
$4\frac{1}{2}$	10
5	11
$5\frac{1}{2}$	12

27.

x	$y = 5x - 4$
3	11
$3\frac{1}{5}$	12
$3\frac{2}{5}$	13
$3\frac{3}{5}$	14
$3\frac{4}{5}$	15
4	16

29. a. $x = -4$ **b.** $x = -6$ **c.** $x = -10$

31. a. $x = 1$ **b.** $x = -1$ **c.** $x = 2$

33. a. $\{-2, 2\}$ **b.** $\{-1, 1\}$ **c.** $\{0\}$ **d.** $\{\ \}$

35. a.

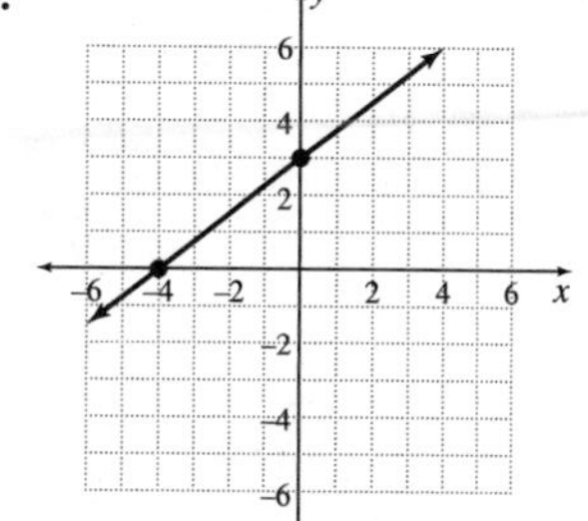

(−4, 0); (0, 3)

b.

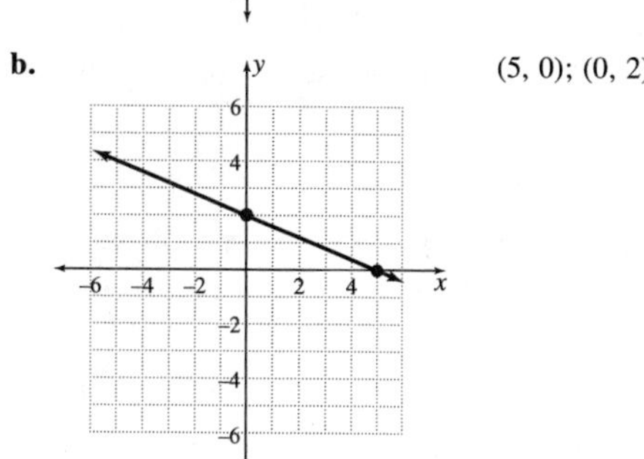

(5, 0); (0, 2)

37. a. (6, 0); (0, −4)

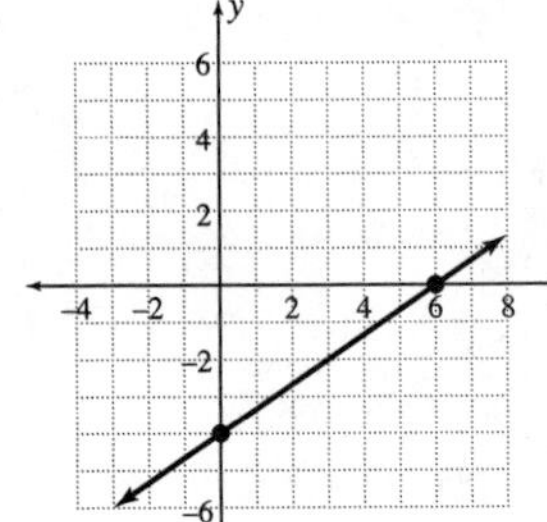

b. (−4, 0); (0, 3)

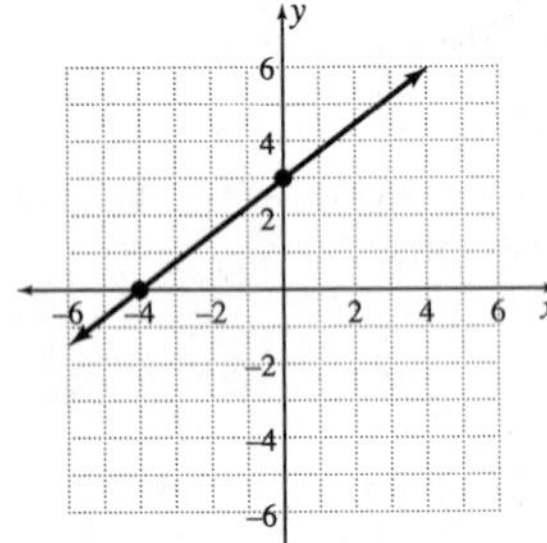

39. a. $-2 \stackrel{?}{=} 3(0) - 2$ correct **b.** $2 + 5(-2) \stackrel{?}{=} -8$ correct **c.** $4 \stackrel{?}{=} \frac{1}{2}(2)$; no **d.** $3 + 5 \stackrel{?}{=} 8$ correct **e.** $\frac{1}{2}(8)(9) \stackrel{?}{=} 36$ correct

41. A linear equation can be written in the form $y = mx + b$ (2 variables) or $a = mx + b$ (1 variable).

43. Find the input that gives n as the output.

45. Find where the graph crosses the y-axis or where $x = 0$.

47. { } is the set with nothing in it; {0} contains the number 0.

49. True **51.** False

53. a. $x = 30$ **b.** $y = 5.00$ **c.** $x = 100$

55. a. 55 copies **b.** c **c.** b

57. a.

x	y
0	12.00
10	9.00
20	6.00
30	3.00
40	0
50	0

b. $3 **c.** 10 min **d.** $33\frac{1}{3}$ min **e.** 40 **f.** Total minutes on the card **g.** $12 **h.** Original cost of the card **i.** $y = -0.30x + 12$

59. a.

r	D
0	0
10	30
20	60
30	90
40	120

b. 5 mph **c.** 25 mph **d.** 0; distance traveled at 0 mph

MID–CHAPTER 3 TEST

1. a. conditional **b.** identity **c.** identity **d.** conditional

2. a. equivalent **b.** not equivalent **c.** not equivalent **d.** not equivalent

3. yes **4.** no **5.** yes **6.** yes **7.** $x = 7$ **8.** $x = 36$

9. $x = -5$ **10.** $x = 14$ **11.** $x = \frac{1}{6}$ **12.** $x = -2.5$

13. $2x + 5 = 10$ **14.** $5 = \frac{1}{2}x - 6$ **15.** $\$545 = \$85x + \$35$

16. $\$7.20 = -\$0.40x + \$20$

17. Five is 4 less than 3 times a number.

18. Four more than twice a number is −3.

19. a. $x = 2$ **b.** $x = 8$ **c.** $x = 0$ **d.** $x = 5$

20. a. $x = 2$ **b.** $x = -4$ **c.** $x = -1$

21. a. {−4, 3} **b.** {−3, 2} **c.** {−1, 0}

22. It says that 10 = 5. There should be no equal sign between 10 and x; a comma would be correct.

23. a.

x	y
5	70
10	70
15	70
20	85

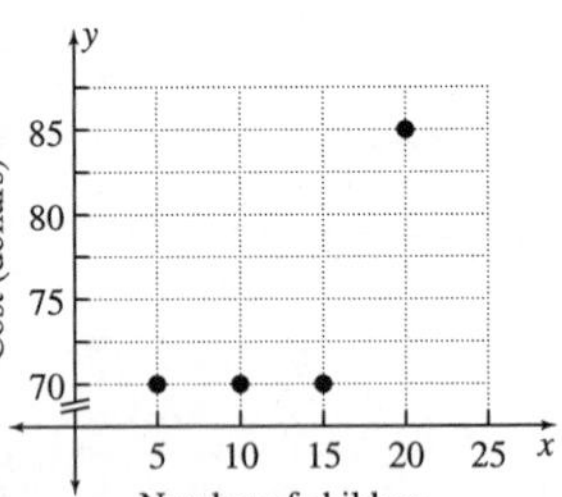

b. $70

c. $85

d. $70; cost for 0 children may not be meaningful, unless it is the charge if the party is canceled at the last minute.

e. $\$100 = \$3(x - 15) + \$70$

EXERCISES 3.3

1.

x	$5x - 8$
−1	−13
0	−8
1	−3
2	2
3	7
4	12

x	$2(x + 2)$
−1	2
0	4
1	6
2	8
3	10
4	12

(4, 12); $x = 4$

3.

x	$3(x - 3)$
−1	−12
0	−9
1	−6
2	−3
3	0
4	3

x	$6(x - 2)$
−1	−18
0	−12
1	−6
2	0
3	6
4	12

(1, −6); $x = 1$

5. a. $x = 1$ **b.** $x = 0$ **c.** $x = -2$ **d.** $x = 2$ **e.** $x = 2$

7. a. $x = -1$ **b.** $x = -3$ **c.** $x = 2$ **d.** $x = 0$ **e.** $x = -2$

9. a.

x	$y = 8 - 3(x + 2)$
0	2
1	−1
2	−4
−1	5
−2	8
−3	11

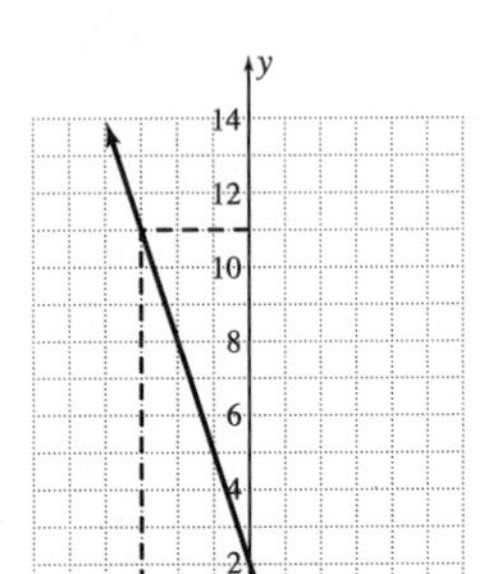

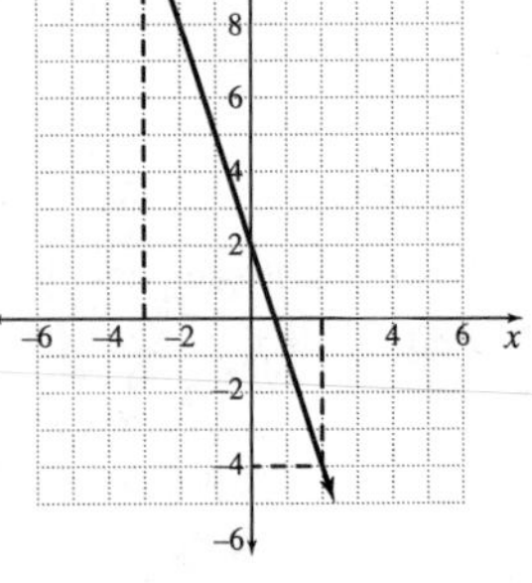

b. $x = -3$

c. $x = \frac{2}{3}$

d. $x = 2$

11. $x = 3$ **13.** $x = -3\frac{1}{2}$ **15.** $x = \frac{1}{3}$ **17.** $x = 1\frac{1}{2}$

19. $x = \frac{2}{3}$ **21.** $x = 8$ **23.** $x = 25$ **25.** $x = 18$

27. $x = -12$ **29.** $x = 13$ **31.** $x = -2$ **33.** $x = 2\frac{1}{2}$

35. $x = -3\frac{1}{2}$ **37.** $x = -4$ **39.** $x = 4$

41. distribute 2; distributive property

43. add -7 to each side; addition property

45. add x to each side; addition property

47. $2(x + 5) = 14; x = 2$ **49.** $-2(x - 4) = 6; x = 1$

51. $\frac{1}{2}(x - 5) = y$ **53.** $y = 15(3 - x)$ **55.** $y = 0.80(200 - x)$

57. $100 = 0.15(x - 100) + 65; x \approx 333$ mi

59. $6.50 = 1.50(x - 1) + 2.00; x = 4$ copies

61.

	4	5	6
$x = 4$	x	$x + 1$	$x + 2$
$x = 5$	$x - 1$	x	$x + 1$
$x = 6$	$x - 2$	$x - 1$	x

63.

	5	7	9
$x = 5$	x	$x + 2$	$x + 4$
$x = 7$	$x - 2$	x	$x + 2$
$x = 9$	$x - 4$	$x - 2$	x

65. $x + (x + 1) + (x + 2) = 42; x = 13$; integers are 13, 14, 15

67. $x + (x + 2) + (x + 4) = 177; x = 57$; integers are 57, 59, 61

69. The numbers circled or boxed are 2 apart.

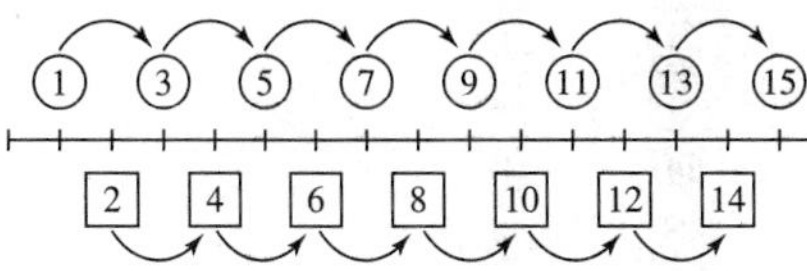

71. 2012; $x, x + 4, x + 8, x + 12$ **73.** 1942 **75.** $x = 400$

77. $x = 350$ **79.** $x = 1000$

81. Find the common ordered pairs in the tables built from each side of the table.

83. Add or subtract to move variable terms to one side and constant terms to the other. Add like terms. Divide by the numerical coefficient on x.

85. Subtract c and multiply by $\frac{b}{a}$.

87. $x = 1$: 1, 2, 4; $x = 2$: 2, 3, 5 **89.** x is positive.

EXERCISES 3.4

1. −3 −2 −1 0 1 2 3 4 5 6 7 $(5, +\infty)$

3. −5 −4 −3 −2 −1 0 1 2 3 4 5 $(-\infty, -2]$

5. −5 −4 −3 −2 −1 0 1 2 3 4 5 $[0, +\infty)$

7. −5 −4 −3 −2 −1 0 1 2 3 4 5 $(-\infty, -1)$

9. −5 −4 −3 −2 −1 0 1 2 3 4 5 $x > 2$

11. −5 −4 −3 −2 −1 0 1 2 3 4 5 $x < 1$

13. −5 −4 −3 −2 −1 0 1 2 3 4 5 $x > 2$

15. −5 −4 −3 −2 −1 0 1 2 3 4 5 $x < 4$

17.

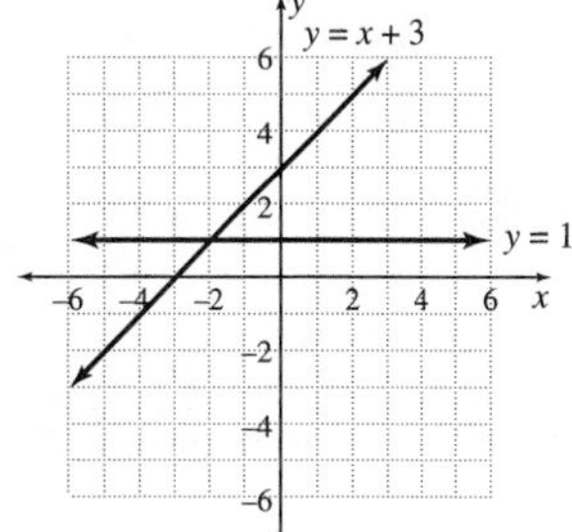

−6 −4 −2 0 2 4 6 $x > -2$

19.

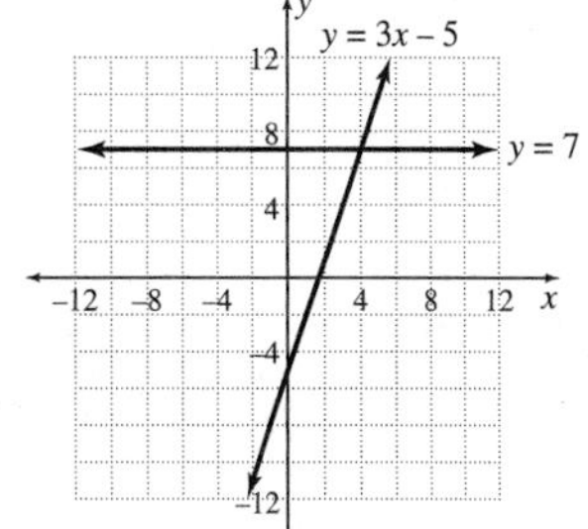

−12 −8 −4 0 4 8 12 $x \geq 4$

21.

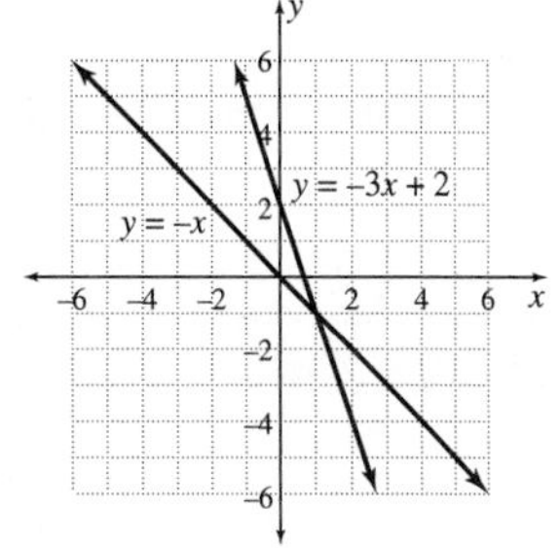

−6 −4 −2 0 2 4 6 $x < 1$

23.

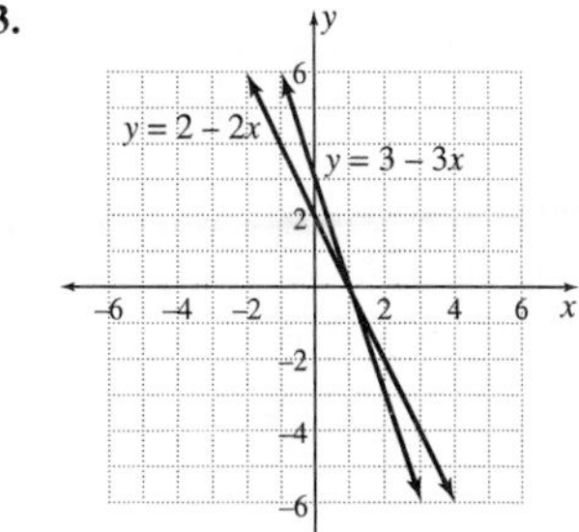

−6 −4 −2 0 2 4 6 $x < 1$

25. $x > 2, (2, +\infty)$ **27.** $x < 1, (-\infty, 1)$ **29.** $x > 2, (2, +\infty)$

31. $x < 4, (-\infty, 4)$ **33.** $x < 1, (-\infty, 1)$ **35.** $x \leq 2, (-\infty, 2]$

37. $x \geq 4, [4, +\infty)$ **39.** $x < -9, (-\infty, -9)$

41. $x > 1\frac{1}{4}, \left(1\frac{1}{4}, +\infty\right)$ **43.** $x < -1\frac{1}{2}, \left(-\infty, -1\frac{1}{2}\right)$

45. $x < 5, (-\infty, 5)$ **47.** $x \leq 2, (-\infty, 2]$ **49.** $x < 1, (-\infty, 1)$

51. $x > 2, (2, +\infty)$ **53.** $x < -8, (-\infty, -8)$

55. $x < -2, (-\infty, -2)$

57. $\dfrac{78 + 84 + 72 + 5 + x}{520} \geq 0.80$; $x \geq 177$. The student needs more points than the final is worth; the student cannot earn a B.

59. $\dfrac{x + 70 + 135}{520} \geq 0.90$; $x \geq 263$ points

61. $\dfrac{88 + 84 + 89 + 70 + x}{100 + 100 + 100 + 70 + 200} \geq 0.90$; $x \geq 182$

63. $5.50x + 17 \leq 160$; $x \leq 26$ 65. $8.50x + 350 \leq 1030$; $x \leq 80$

67. $8.50x + 350 \leq 17.50x + 100$

69. If $a < b$, then $\dfrac{a}{c} > \dfrac{b}{c}$, where a and b are real numbers and $c < 0$.

71. Inequality sign reverses: $+4 > +3$.

CHAPTER 3 REVIEW EXERCISES

1. equivalent equations
3. two-variable equation, conditional equation
5. conditional equation 7. yes 9. yes 11. $x = -7$
13. $x = 9$ 15. $x = 24$ 17. $x = 6$ 19. $x = -3$
21. $x = -5$ 23. $x = 7$ 25. $x = 1$ 27. $x = 1\frac{1}{5}$
29. $3x - 6 = -15$ 31. $-7x = 21$ 33. $3x = 2x + 4$
35. $6(2 + x) = -6$ 37. The quotient of a number and 5 is 15.
39. Three times the difference between a number and 4 is -18.
41. 11, 13 43. 18, 21 45. $x - 1, x + 1$ 47. $x - 2, x + 2$
49. $x + 2(x + 1) = 17$; $x = 5$
51. a. $x = 2$ b. $x = 5$ c. $x = 3\frac{1}{3}$ d. $x = 6$
53. a. $\{2\}$ b. $\{1, 3\}$ c. $\{0, 4\}$
55. a. $x = 2$ b. $x = -3$ c. $x = -1$
57. $y = 4x$; intercept $(0, 0)$ 59. $y = 0.075x$; intercept $(0, 0)$
61. $y = 300x + 150$; intercepts $(-0.5, 0)$ and $(0, 150)$
63. $y = -10x + 520$; intercepts $(52, 0)$ and $(0, 520)$
65. $440 = 20(x + 3)$; $x = 19$ seats
67. $x + (x + 5) + (x + 10) = 60$; 15, 20, 25
69. a, b, c does not indicate that the numbers are 1 unit apart.
71. $\dfrac{82 + 72 + 20 + 0 + 20 + 20 + 18 + 12 + x}{100 + 100 + 20 + 20 + 20 + 20 + 20 + 70 + 150} \geq 0.80$; $x \geq 172$; not possible
73. $x > 5$

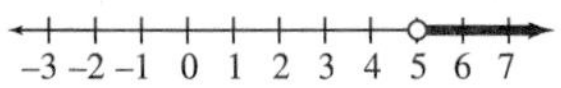

75. $5 < x$, $(5, +\infty)$ 77. $x < -16$, $(-\infty, -16)$
79. $x < -2\frac{2}{3}$, $\left(-\infty, -2\frac{2}{3}\right)$ 81. $x \leq -18$, $(-\infty, -18]$
83. Blue line is $y = 15 - 2x$; red line is $y = x - 6$; $x > 7$.
85. $x \leq 2$

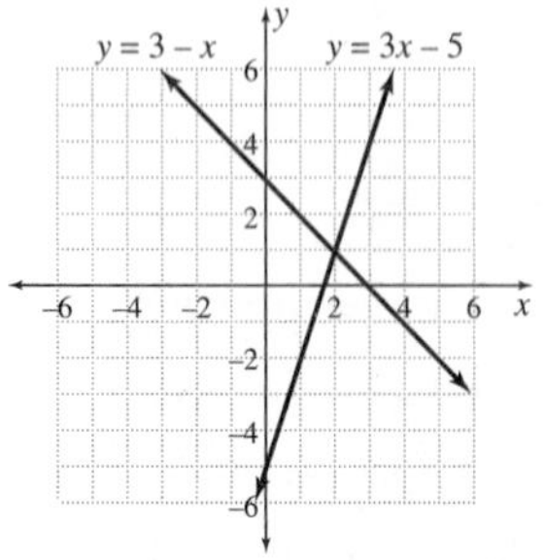

87. $x > -1$, $(-1, +\infty)$ 89. $-3 > x$, $(-\infty, -3)$
91. $x \leq 1$, $(-\infty, 1]$ 93. $4 > x$, $(-\infty, 4)$ 95. $x \geq \frac{1}{5}$, $\left[\frac{1}{5}, +\infty\right)$
97. $3500 \geq 25x + 550$; $x \leq 118$ people
99. $12x + 250 < 8x + 320$; $x < 17.5$ ft

CHAPTER 3 TEST

1. $x = -11$ 2. $x = -1$ 3. $x = 75$ 4. $x = -\frac{1}{2}$
5. $x = 1$ 6. $x = -16$ 7. $x = 1$ 8. $x = 4\frac{1}{2}$
9. $x = 3$ 10. $x = 2$ 11. $(1, 3)$; $x = 1$
12. Zero is halfway between -1 and 1, so x is halfway between 2 and 3.
13. Find where the graph crosses the x-axis.
14. a. $(1, -2)$
b. $-2 \overset{?}{=} 2(1) - 4$ correct, $-2 \overset{?}{=} -(1) - 1$ correct
c. $x = 1$
d. The solution is the x value of the intersection, $x = 1$.
15. $\frac{1}{2}x + 6 = 15$; $x = 18$ 16. $3x - 7 = -31$; $x = -8$
17. $y = \frac{1}{3}x$ 18. $y = 2x - 2$
19. $x + (x + 1) + (x + 2) + (x + 3) = -74$; $x = -20$
20.

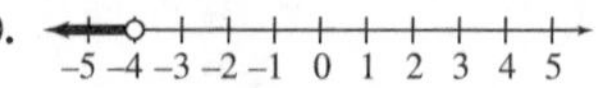

$x < -4$, $(-\infty, -4)$

21.

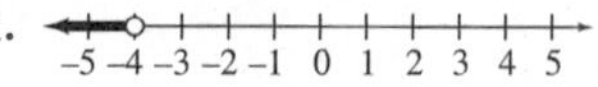

$-4 > x$, $(-\infty, -4)$

22.

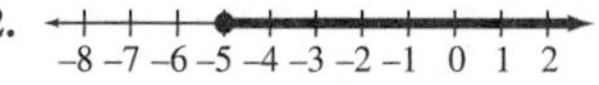

$x \geq -5$, $[-5, +\infty)$

23. The x value of the point of intersection gives the starting number for the solution set, $x < 7$.

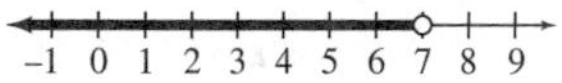

24. a.

x	y
0	4
20	16.30
40	28.60
60	40.90
80	53.20
100	65.50

b.

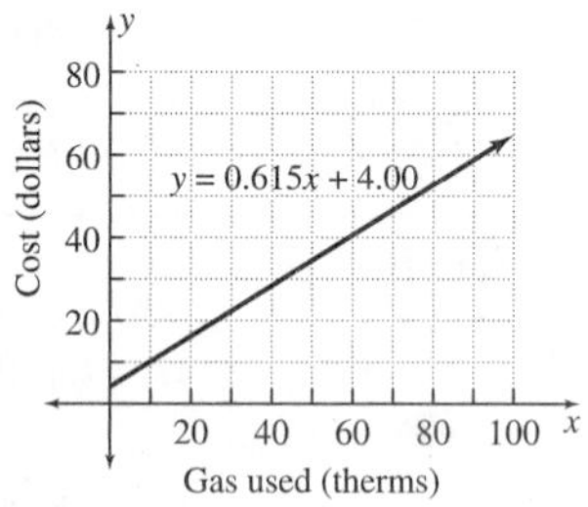

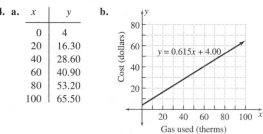

c. $y = 0.615x + 4.00$ d. 70 therms
e. $(-6.50, 0)$; no meaning f. $(0, 4)$; monthly fee

EXERCISES 4.1

1. Interest in terms of principal, rate, and time
3. Rate in terms of distance and time
5. Grade percent in terms of tests, homework, final exam, and total points possible
7. $A = lw$ 9. $A = \pi r^2$ 11. $A = \frac{1}{2}bh$ 13. $P = 2l + 2w$
15. $C = fm$ 17. $n = \dfrac{p}{5}$ 19. $A = H + P$ 21. $r = \dfrac{C}{2\pi}$
23. $h = \dfrac{A}{b}$ 25. $t = \dfrac{I}{pr}$ 27. $d = \dfrac{C}{\pi}$ 29. $C = R - P$
31. $n = \dfrac{PV}{RT}$ 33. $V_1 = \dfrac{C_2V_2}{C_1}$ 35. $c = P - a - b$
37. $h = \dfrac{2A}{a + b}$ 39. $r^2 = \dfrac{3V}{\pi h}$ 41. $b = -2ax$
43. $b = y - mx$ 45. $h = \dfrac{2d^2}{3}$ 47. $g = \dfrac{2d}{t^2}$
49. $y = \dfrac{-4}{x}$ 51. $y = 3x - 10$ 53. $y = \dfrac{x + 5}{2}$ or $y = \frac{1}{2}x + \frac{5}{2}$
55. $y = \frac{3}{2}x - 3$ 57. a. $r = \dfrac{A - P}{Pt}$ b. $r = 0.0525$ or $5\frac{1}{4}\%$
59. a. $F = \frac{9}{5}C + 32$ b. $F = 98.6°$

61. **a.** 48 **b.** $A = 200 - H/0.8$ **c.** 170

63. **a.** $b = y - mx$ **b.** $b = -2$ **c.** $b = 10$
d. $b = 2.5$ **e.** $b = 5.5$

EXERCISES 4.2

1. **a.** function; inputs $\{-6, -5, 5, 6\}$; outputs $\{5, 6\}$
b. not a function

3. **a.** function; inputs $\{2, 3, 4\}$; outputs $\{\frac{1}{2}, \frac{1}{3}, \frac{1}{4}\}$
b. not a function

5. function; inputs {Eden, Tuckman, McClintock}; output {Barbara}

7. not a function

9. input {units of time}; output {cost of call}

11. input {distance from equator}; output {hours of sunlight}

19. C = auto registration cost, v = value of car, $C = f(v)$

21. C = circumference, d = diameter, $C = f(d)$

23.

x	$f(x) = 2x - 1$
−2	−5
−1	−3
0	−1
1	1
2	3

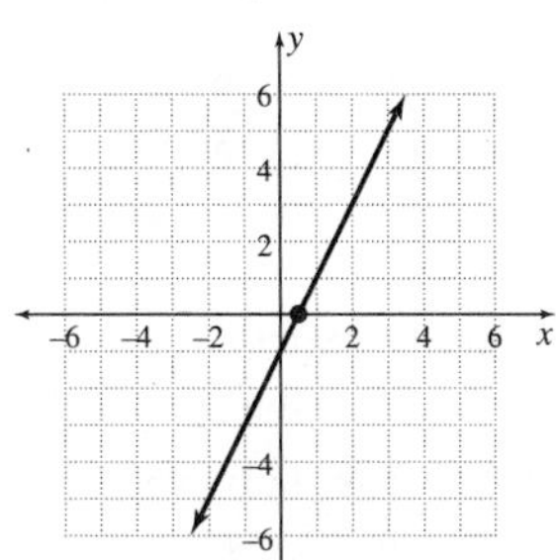

25.

x	$f(x) = 2 - 3x$
−2	8
−1	5
0	2
1	−1
2	−4

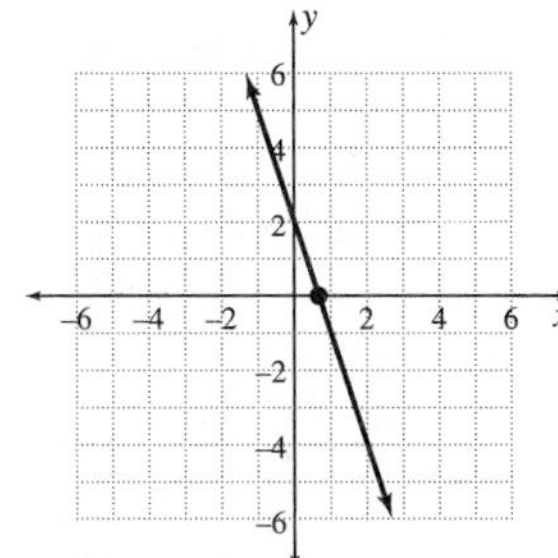

27.

x	$f(x) = \frac{1}{4}x + 1$
−2	$\frac{1}{2}$
−1	$\frac{3}{4}$
0	1
1	$1\frac{1}{4}$
2	$1\frac{1}{2}$

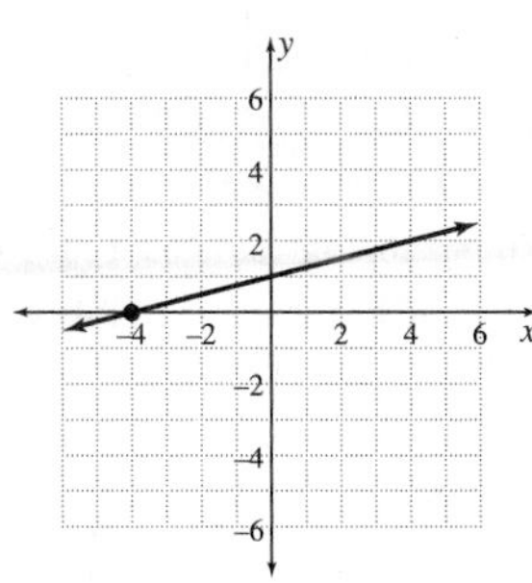

29. $x = \frac{1}{2}$ **31.** $x = \frac{2}{3}$ **33.** $x = -4$ **35.** function of x

37. function of x **39.** function of x **41.** not a function of x

43. not a function of x **45.** $h(4) = 2$, $h(-4) = 2$

47. $H(4) = -12$, $H(-4) = -20$ **49.** $g(-2) = 5$, $g(1) = 2$

51. $G(-2) = -2$, $G(1) = 1$

53. for 45, 0; for 47, 0; for 49, 1; for 51, 2 **55.** $x = 3$

57. $x = 4$ **59.** $x = 3$

61. $f(0) = 48$, the fees included in tuition; $f(x) = 0$, $x = -\frac{48}{115}$, no meaning

63. $f(0) = \$19.50$, original value of phone card; $f(x) = 0$, $x = 26$, total minutes the card will buy

65. not a function **67.** function **69.** not a function

71. Substitute a for every x in the function $f(x)$. **73.** Find $f(0)$.

75. $f(x)$ is special notation reserved for functions; usually we write the product of two variables a and b as $a \cdot b$ or ab.

77. d = distance, r = rate, t = time; $d = f(r, t)$

79. V = volume, r = radius, h = height; $V = f(r, h)$

MID-CHAPTER 4 TEST

1. Temperature in Celsius in terms of temperature in Fahrenheit

2. Volume in terms of radius **3.** $b = 11$ **4.** $b = 0$

5. $K = C + 273$ **6.** $h = \dfrac{8d^2}{3}$ **7.** $d = \dfrac{l - a}{n - 1}$

8. **a.** 18 **b.** 5 **c.** 4 **d.** 34

9.

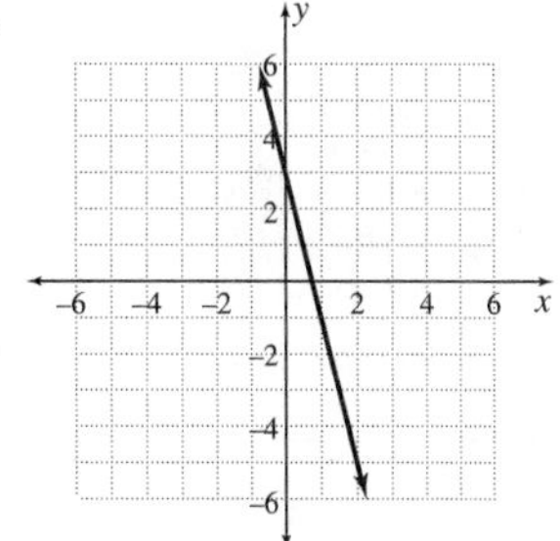

10. **a.** yes **b.** $\{2, 3, 4, 5, 6, 7\}$ **c.** $\{3, 4\}$

11. **a.**

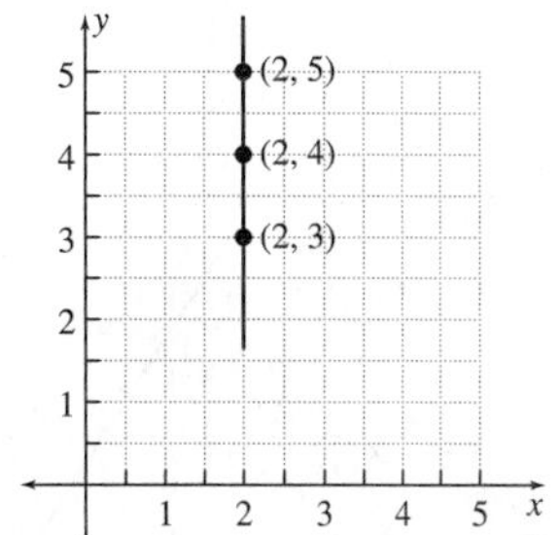

b. one **c.** One input has 3 outputs. **d.** does not pass

12. $f(0) = -5$ **13.** $x = -4$ **14.** vertical axis intercept

EXERCISES 4.3

1. **a.** positive, 3 **b.** positive, $\frac{2}{3}$ **c.** zero, 0

3. **a.** negative, −2 **b.** positive, $\frac{3}{2}$ **c.** undefined

5. **a.** negative, −50 **b.** positive, 40 **c.** negative, −20

7. positive, $\frac{1}{4}$ **9.** negative, $-\frac{7}{2}$ **11.** undefined

13. zero, 0 **15.** negative, $-\frac{2}{3}$ **17.** zero, 0 **19.** negative, $-\frac{4}{5}$

21. undefined **23.** negative, linear, slope = −3, no units

25. positive, linear, slope = 7, no units

27. positive, linear, slope = 9, earnings per hour

29. positive, linear, slope = 0.50, cost per kg

31. positive, linear, slope = 0.32, cost per pound

33. positive, nonlinear, feet per second

35. negative, linear, slope = −0.25, dollar value per copy

37. positive, nonlinear, miles per hour **39.** $\dfrac{b - d}{a - c}$ **41.** $\dfrac{-b}{a}$

43. undefined

45.

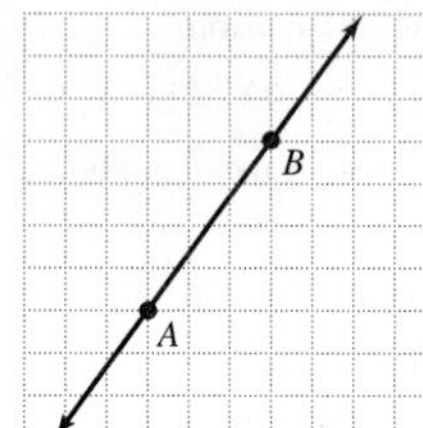

47.

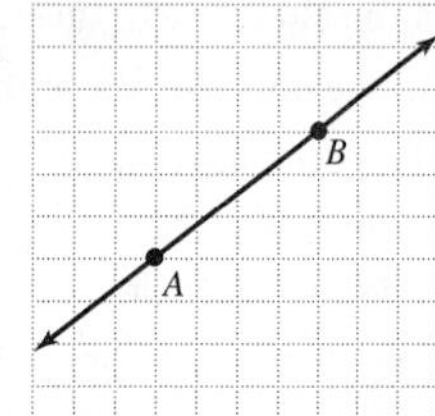

49.

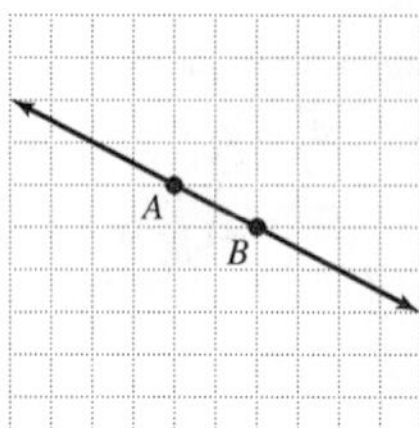

51.

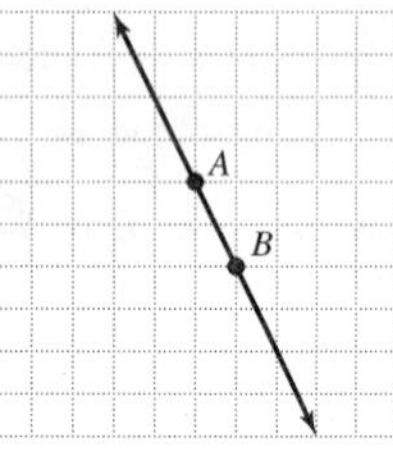

53.

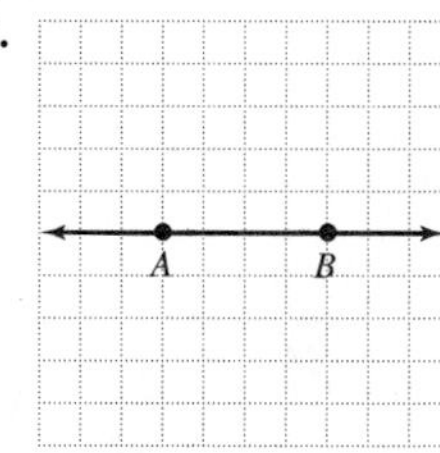

55.

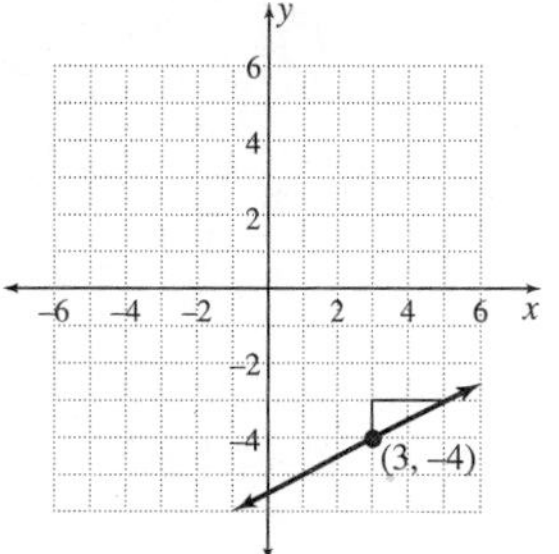

57.

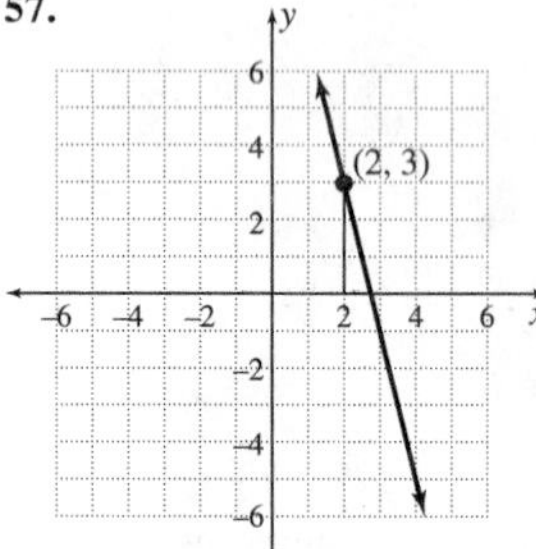

59.

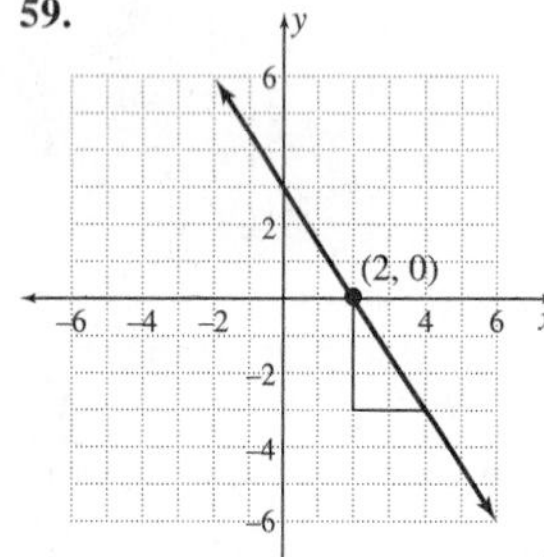

61.

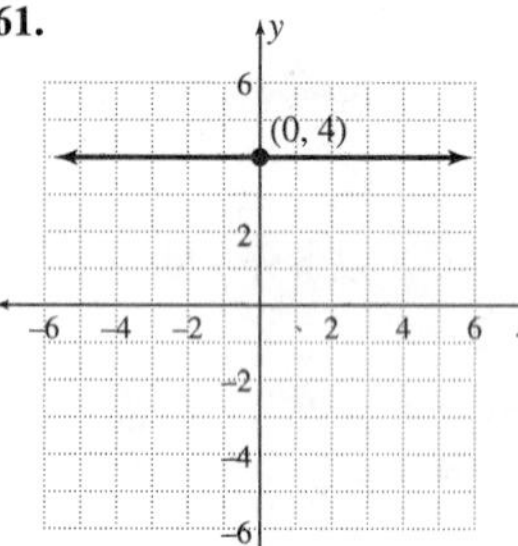

63.

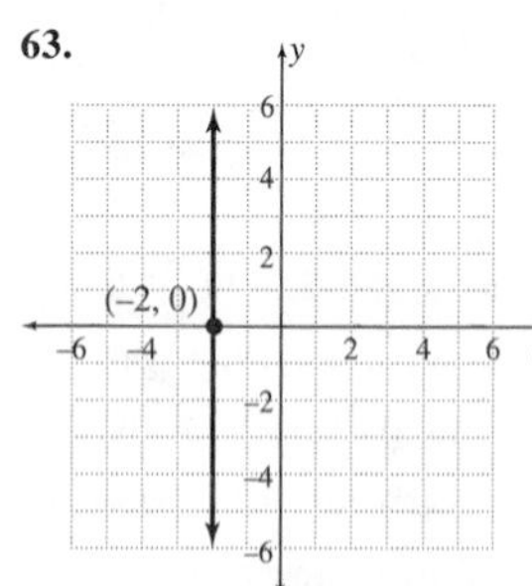

65. $\frac{3}{4}, 1, \frac{4}{3}$ **67.** $0, -\frac{1}{2}, -\frac{2}{1}$ **69.** $\frac{3}{5}, \frac{3}{4}, 1, \frac{3}{2}$ **71.** $\frac{1}{2}$ **73.** $\frac{1}{2}$

75. a.

Gallons, g	Cost, c
0	0
1	\$1.55
2	\$3.10

b. slope = 1.55, dollars per gallon **c.** $c = \$1.55g$

77. a.

Hours, h	Earnings, e
0	0
1	\$6.25
2	\$12.50

b. slope = 6.25, dollars per hour **c.** $e = \$6.25h$

79. a.

Hours, h	Distance, d (in km)
0	0
1	80
2	160

b. slope = 80, kilometers per hour **c.** $d = 80h$

81. a.

Hours, x	Cost, c
0	\$3
1	\$4
2	\$5

b. slope = 1, dollars per hour **c.** $c = \$1x + \$3$

83. a.

Meal Cost, x	Total Cost, c
0	0
1	\$1.15
2	\$2.30

b. slope = 1.15, total cost dollars per meal cost dollar

c. $c = \$1.15x$

85. The fee in each exercise is constant.

87. freezing (0, 32), boiling (100, 212), slope = $\frac{9}{5}$, °F/°C

89. relative direction and steepness

91. Output gets smaller as input gets larger.

93. Select two points on a line. Going left to right, count the change in y and divide by the change in x.

95. Starting at any point, count a units vertically (up or down, as the sign dictates) and b units to the right horizontally.

97. The correct formula is the reciprocal of the student's answer.

99. a.

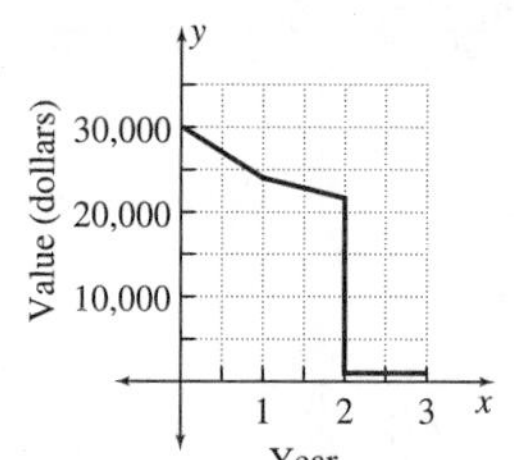

b. −6,000 **c.** −2,400

d. Extreme loss of value happened suddenly. **e.** 0

f. One possible answer: After driving it for two years, she had a wreck in the third year.

EXERCISES 4.4

1. $y = -x + 1$ **3.** $y = 2x + 2$ **5.** $y = -1$

7. $C = \frac{1}{30}n + 1$ **9.** $C = -4t + 160$ **11.** $m = 2, b = -\frac{1}{2}$

13. $m = -4, b = 15$ **15.** $m = -\frac{3}{4}, b = 0$ **17.** $m = 2, b = -4$

19. $m = -\frac{2}{3}, b = 4$ **21.** $m = \frac{2}{5}, b = 2$ **23.** $m = \frac{1}{4}, b = -1$

25. $m = -0.30, b = 12$ **27.** $m = 55, b = 0$

29. $m = 2\pi, b = 8$ **31.** $m = 2.98, b = 0.50$

33. $m = -0.29, b = 50$ **35.** $m = -0.8, b = 160$

37. $m = 0.15, b = 50$ **39.** $y = \frac{1}{2}x + 3$ **41.** $y = \frac{2}{3}x - 2$

43. $y = 5x + \frac{1}{4}$ **45.** $y = -\frac{3}{2}x + 1$ **47.** $y = 4x - 13$

49. $y = -x + 2$ **51.** $y = \frac{1}{2}x + 3$ **53.** $y = \frac{4}{5}x + 11$

55. $y = \frac{5}{3}x + 6$ **57.** $y = -2x + 6$ **59.** $y = 4x - 3$

61. $y = -2x + 2$ **63.** $y = -\frac{2}{3}x - 1$ **65.** $y = 2x - 20$

67. $y = -8x - 34$ **69.** $y = -\frac{1}{2}x + \frac{9}{2}$

71. a and c are perpendicular, b and d are parallel

73. a and c are parallel, b and d are perpendicular

75.

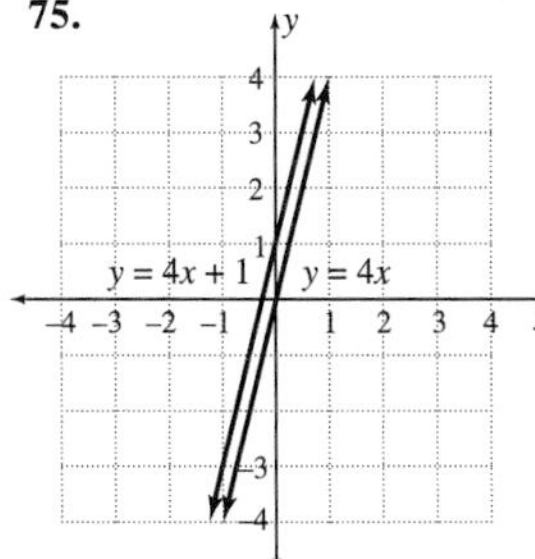

77.

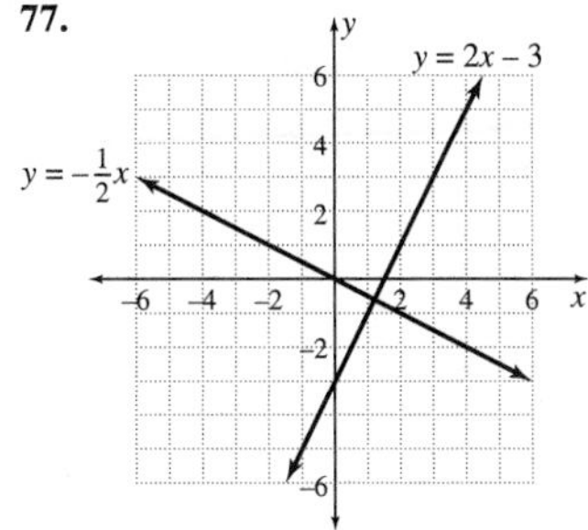

79.

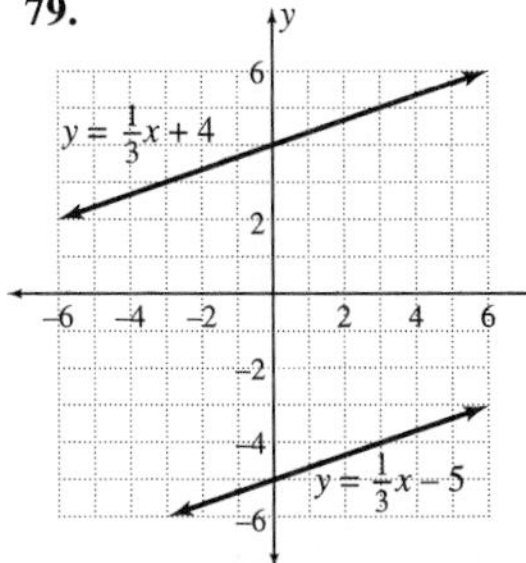

81.

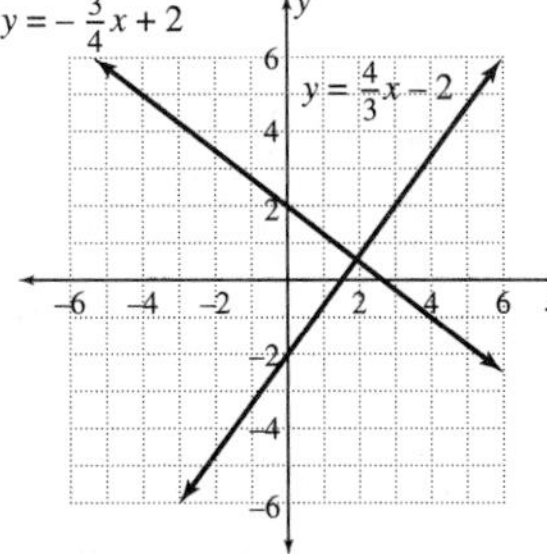

83. $y = 4$

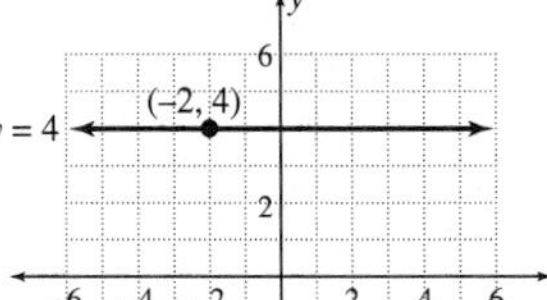

85. $x = 4$

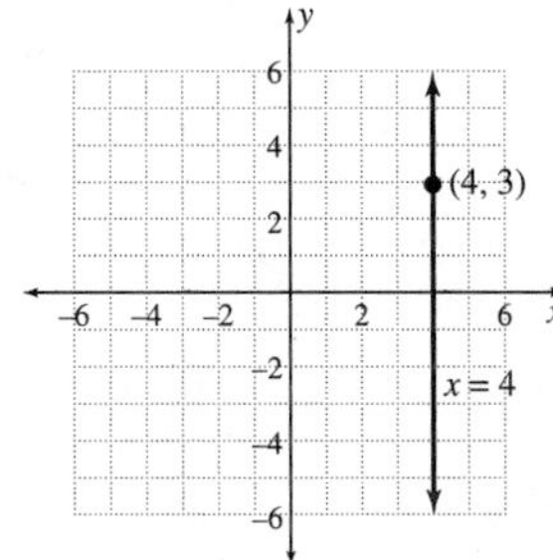

87. $x = 3$

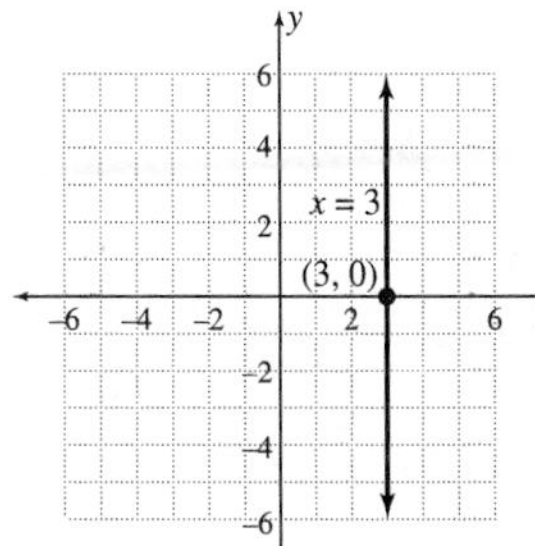

89. $y = 0$

91. a. \$2 rental per hour **b.** \$50 prepaid amount

c. $y = -\$2x + \$50$

93. a. \$42 rental per hour **b.** \$28 insurance fee

c. $y = \$42x + \$28$

95. per hour, percent of **97.** parallel line, $C = \$0.25n + \$45$

99. steeper line, $C = \$1.60g$ **101.** parallel line, $C = 0.01x + \$5$

103. $y = \$0.78x + \$0.55$

105. a. AB, 1; BC, 0; CD, $\frac{1}{10}$; DE, 1

b. AB, 0; BC, 20; CD, 0; DE, -450

c. AB, $y = x$; BC, $y = 20$; CD, $y = \frac{1}{10}x$; DE, $y = x - 450$

d. Assume we include the right-hand endpoint in each inequality. AB, $0 \le x \le 20$; BC, $20 < x \le 200$; CD, $200 < x \le 500$; DE, $500 \le x \le 600$

e. Assume we include the right-hand endpoint in each inequality. AB, $0 \le y \le 20$; BC, $y = 20$; CD, $20 < y \le 50$; DE, $50 \le y \le 150$; AB and DE are parallel, as they have the same slope

107. Find the slope from $m = \dfrac{y_2 - y_1}{x_2 - x_1}$. Then substitute m and one ordered pair into $b = y - mx$ to find b. Substitute m and b in $y = mx + b$.

109. c = slope, d = y-intercept

111. slope $= -\dfrac{b}{a}$

113. a. $-\dfrac{b}{a}$ **b.** opposite of b over a **c.** $-\frac{4}{5}$

EXERCISES 4.5

1. Multiply both sides by -1; reverse the inequality sign.

3. Multiply both sides by -2; reverse the inequality sign.

5. Subtract $2x$ from both sides; divide both sides by -3; reverse the inequality sign.

7. (4, 0), (1, 4) **9.** (−2, 3), (6, −1) **11.** (1, −1), (3, 0)

13. (−2, 0), (0, −2)

15.

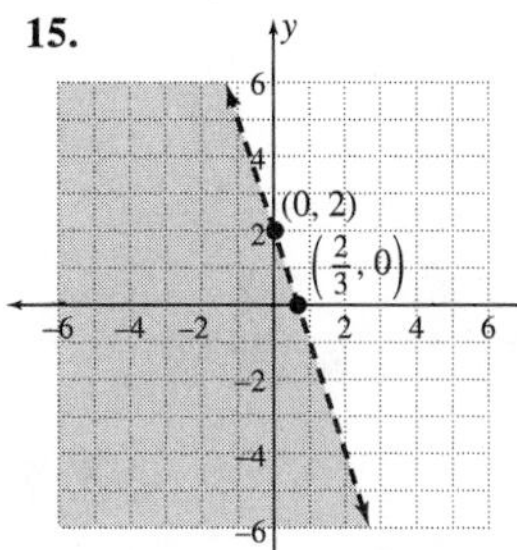

17.

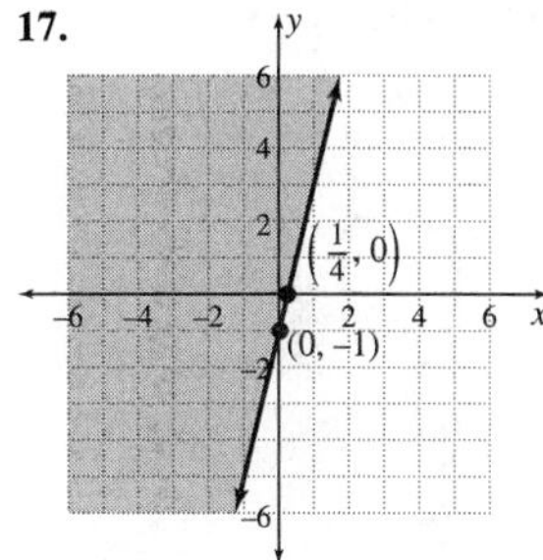

19.

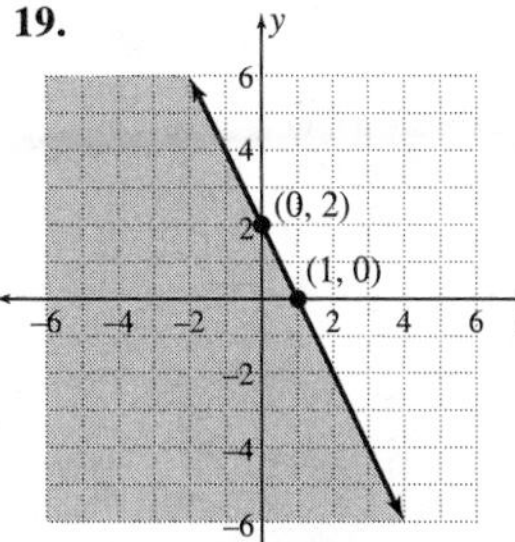

21.

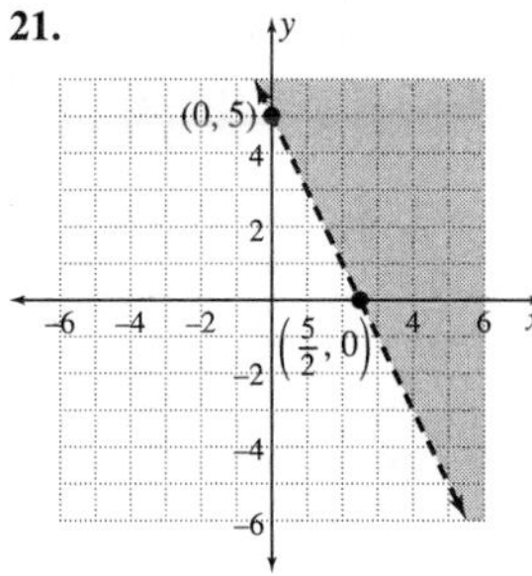

23.

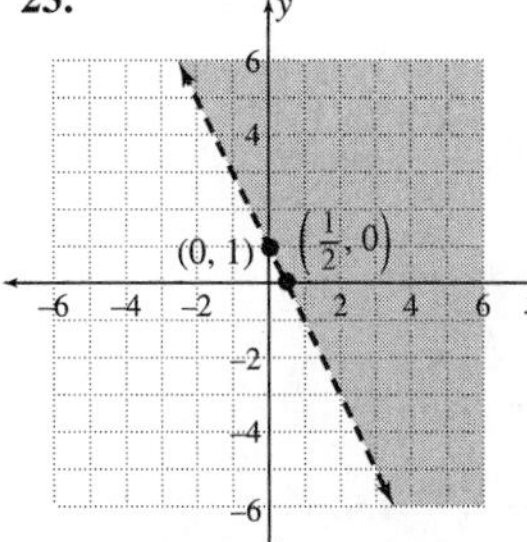

25.

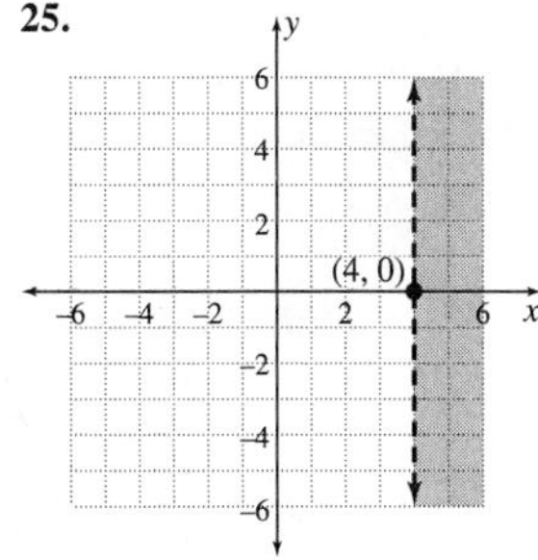

27.

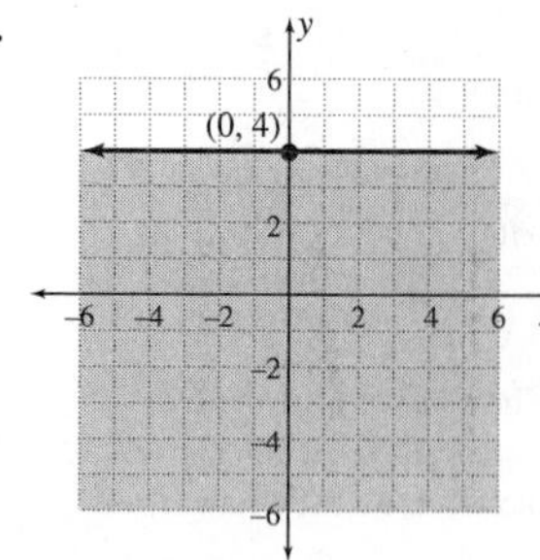

29. $y = 3, y < 3$

31. $x = 3, x \leq 3$ **33.** $y = x + 4, y \leq x + 4$

35. $y = 2x + 1, y > 2x + 1$

37. Test a point in the equation. If it is true, shade the side containing the point; if it is false, shade the other side.

39. The first is a line graph with a dot at $x = 3$ and arrow to the right. The second is a vertical boundary line, $x = 3$, with shading to the right.

41. $16x + 12y \geq 2400$

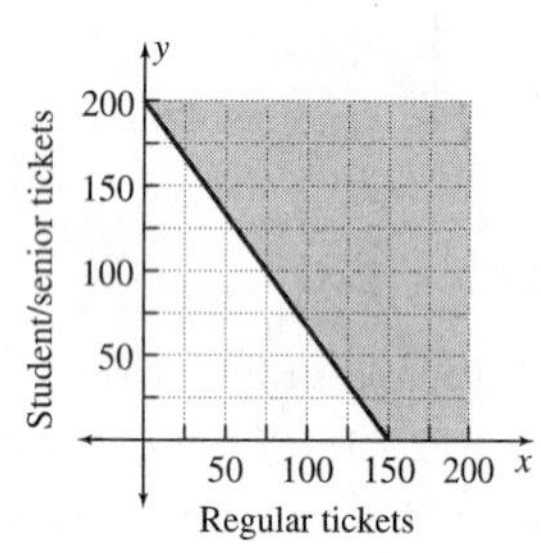

43. a. Possibilities include 0 apricots and 4 tangerines, 7 apricots and 0 tangerines, and 3 apricots and 2 tangerines.

b. 1st quadrant below the line $20a + 35t = 140$, including the line and both axes

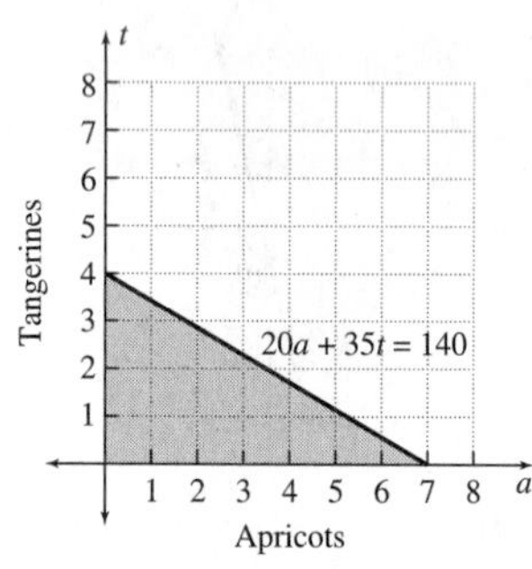

c. Inputs are positive numbers; $20a + 35t \leq 140$, $a \leq 7$, $t \leq 4$

45. a. (10, 4); (20, 0) **b.** (15, 2); (0, 8) **c.** $0.10x$ **d.** $0.25y$

e. $0.10x + 0.25y \leq \$2$

f. $0 \leq x \leq 20$, $0 \leq y \leq 8$

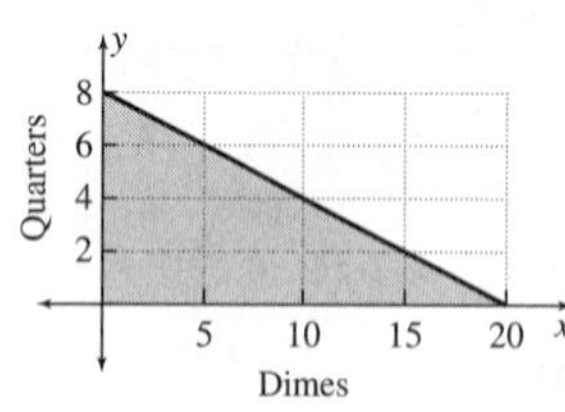

47. 4th quadrant including positive x-axis

49. 3rd quadrant including negative x-axis

51. $x > 0, y < 0$

CHAPTER 4 REVIEW EXERCISES

1. $b = 10$ **3.** $F = 98.6$ **5.** Area in terms of base and height

7. $h = \dfrac{W}{p}$ **9.** $b = \dfrac{2A}{h}$ **11.** $T = \dfrac{AH}{I}$ **13.** $T = \dfrac{PV}{nR}$

15. $x = \dfrac{c - by}{a}$ **17.** $k = \dfrac{C}{5} + 93$ **19.** $P_2 = \dfrac{P_1V_1}{V_2}$

21. $b_1 = \dfrac{2A}{h} - b_2$ **23.** $b = 9$

25. $f(0) = 7, f(3) = 1, f(-5) = 17, f(a) = 7 - 2a$

27. $f(0) = 0, f(3) = 12, f(-5) = 20, f(a) = a^2 + a$

29. $x = \frac{5}{2}$ **31.** $x = -\frac{4}{3}$ **33.** $x = 12$ **35.** y-intercept

37. a. {2, 4, 6, 8} **b.** {25, 50, 75, 100}

c. value of bits in cents

39. a. {1, 2, 3, 4, 5, 6} **b.** {2, 6, 10, 20, 40, 60}

c. the name of the nail in "pennies," usually abbreviated d

41. function **43.** function **45.** not a function **47.** b

49. a **51. a.** $\frac{3}{1}$ **b.** $-\frac{7}{10}$ **c.** $\frac{1}{5}$ **d.** $\frac{2}{15}$

53.

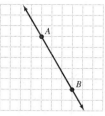

55.

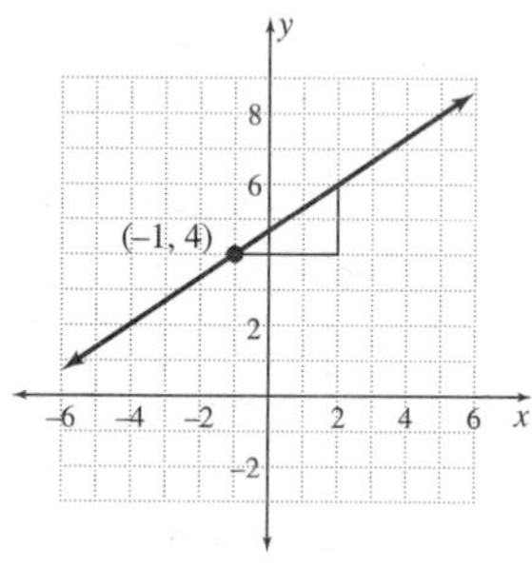

57. Line goes down from left to right.

59. Locate (x, y) and count a spaces up (or down) and b spaces to the right.

61. $m = \frac{3}{1}, b = 2, y = 3x + 2$ **63.** $m = -\frac{1}{2}, b = 1, y = -\frac{1}{2}x + 1$

65. linear, $m = -\frac{3}{2}$ **67.** nonlinear

69. a. (14, 4.44), (5, 2.10); x = minutes, y = cost

b. $m = 0.26$ **c.** cost per minute

71. a. (12, 12), (17, 32); x = hours, y = feet **b.** $m = 4$

c. feet per hour

73. $y + b = mx$ **75.** $y = -\frac{3}{5}x + 3, m = -\frac{3}{5}, b = 3$

77. $y = \frac{5}{2}x - 5, m = \frac{5}{2}, b = -5$ **79.** $y = \frac{3}{4}x + 2, m = \frac{3}{4}, b = 2$

81. $y = -3, m = 0, b = -3$ **83.** undefined slope, no y-intercept

85. $m = 1.5, b = 0.5$

87. $y = 2x + 4$

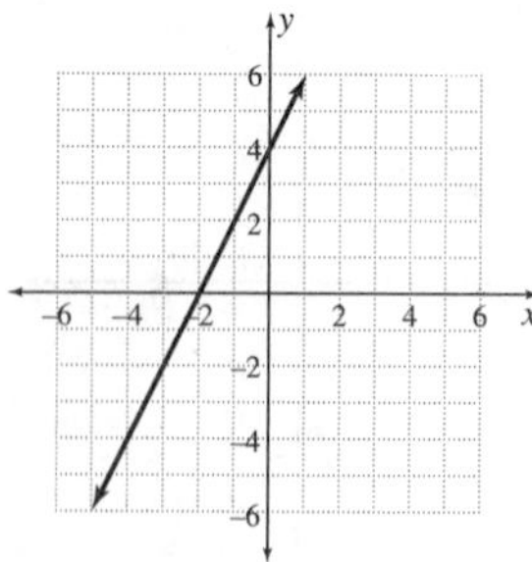

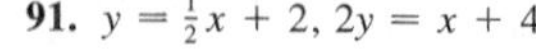

89. $y = 3x$

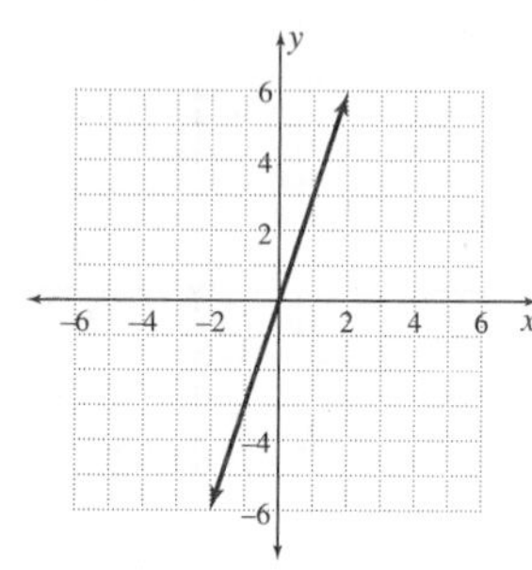

91. $y = \frac{1}{2}x + 2, 2y = x + 4$

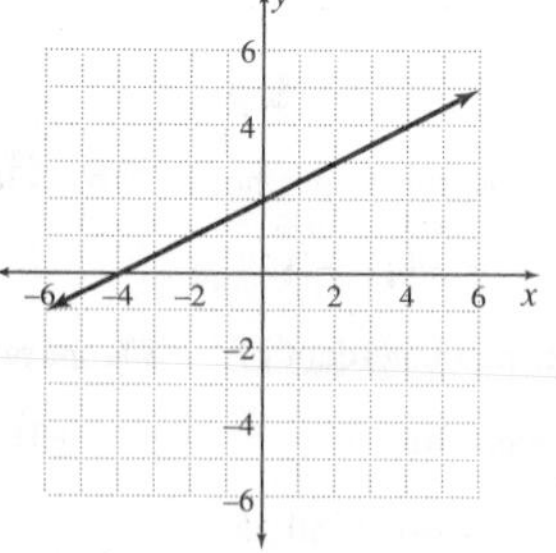

93. $y = -3x + \frac{1}{4}$, $4y = -12x + 1$

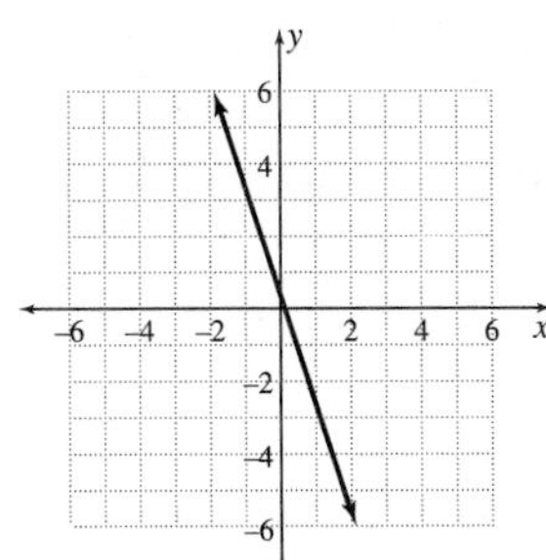

95. $x = 4$

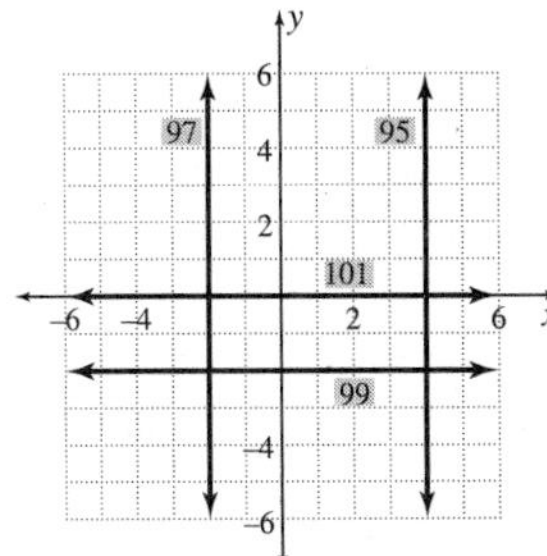

97. $x = -2$ **99.** $y = -2$ **101.** $y = 0$

103. $y = 3x$ **105.** $y = \frac{3}{2}x - 2$

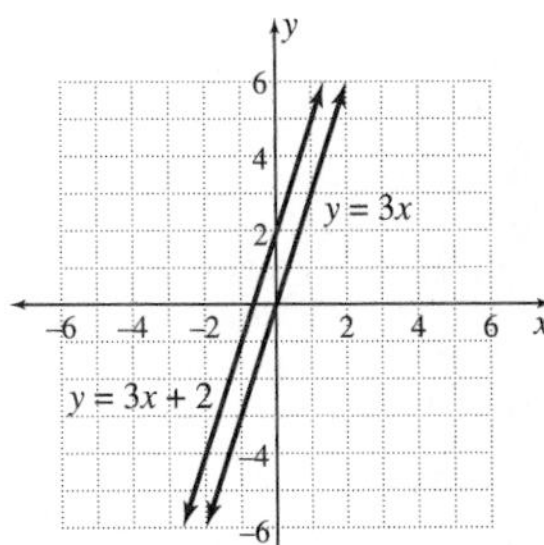

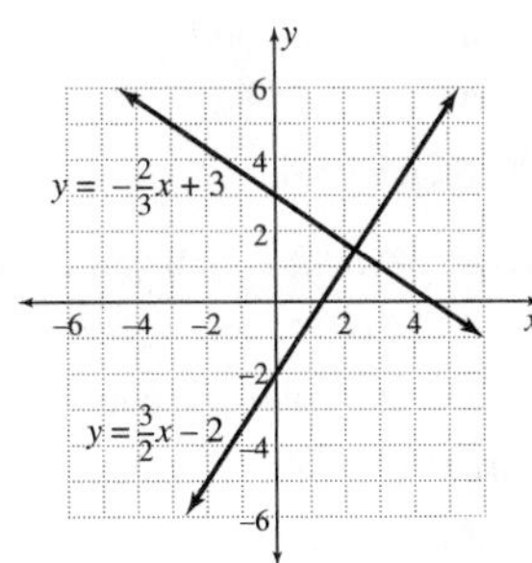

107. $y = 3 - 0.5x$ **109.** $C = 10w$

111. linear; y = income, x = number of people; $y = \$175x + \$332$

113. $(-1, -6)$

115. **a.** $y \geq 0$ **b.** $x \geq 0$ **c.** $x < 0$ **d.** $y < 0$ **e.** $x > 0$ **f.** $y \leq 0$

117.

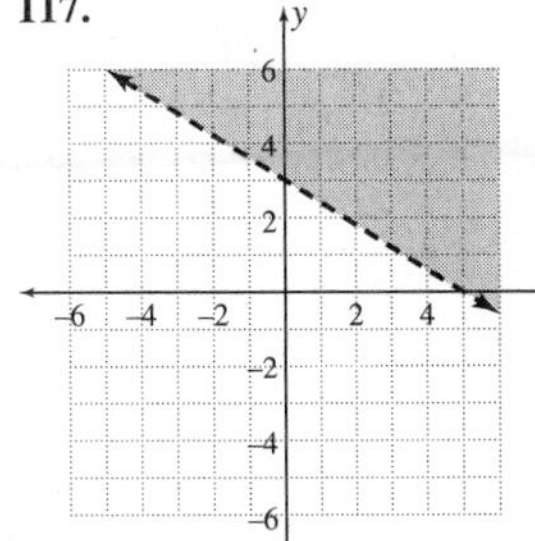

119.

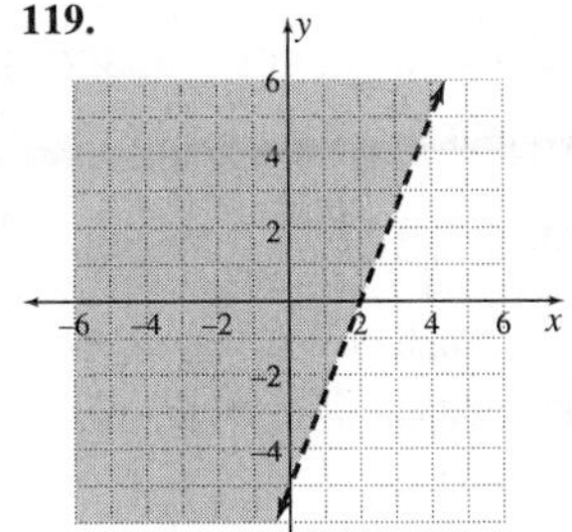

121.

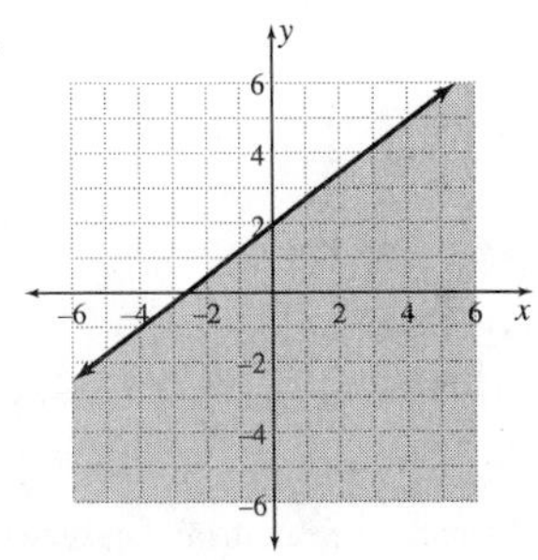

123. x = green, y = ripe; $20x + 25y \leq 400$

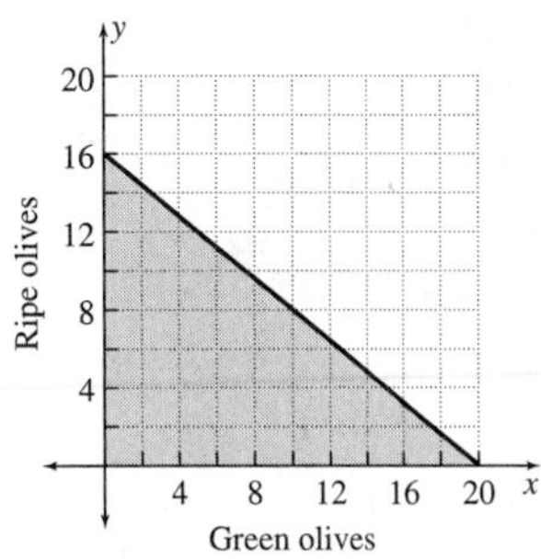

CHAPTER 4 TEST

1. $b = 9$ **2.** Possible answer: $G = \dfrac{T_1 + T_2 + H + F}{P}$

3. **a.** $d = \dfrac{C}{\pi}$ **b.** $h = \dfrac{2A}{b}$ **c.** $b = y - mx$ **d.** $V_2 = \dfrac{P_1V_1}{P_2}$

4. **a.** function **b.** not a function **c.** function **d.** function **e.** not a function **f.** function

5. **a.** $f(x) = 0$ **b.** $f(0)$ **c.** $f(a)$

6. **a.** graph 1, $m = 0$; graph 2, $m = 10$; graph 3, $m = 50$; graph 4, undefined

b. graph 2

7. **a.** $-\frac{2}{3}$ **b.** 0 **c.** -7 **d.** undefined

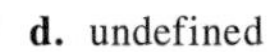

8.

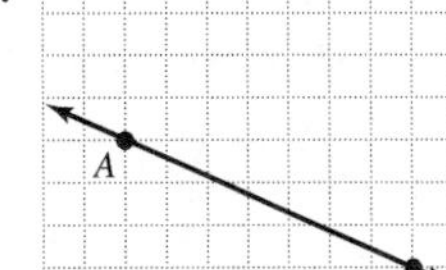

9.

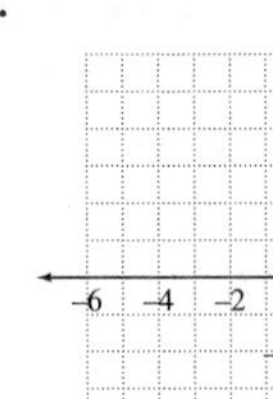

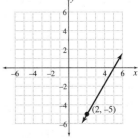

10. **a.** $m = -2, b = 5$ **b.** $m = \frac{1}{3}, b = -\frac{1}{2}$

11. **a.** neither **b.** parallel **c.** perpendicular **d.** neither

12. Substitute slope and one ordered pair into $b = y - mx$. Solve for b. Write $y = mx + b$ using slope and b.

13. slope $= a$, $b = k$, $y = ax + k$ **14.** $y = k$

15. $y = 5x - 1$ **16.** $y = \frac{1}{3}x + 2$ or $3y = x + 6$

17.

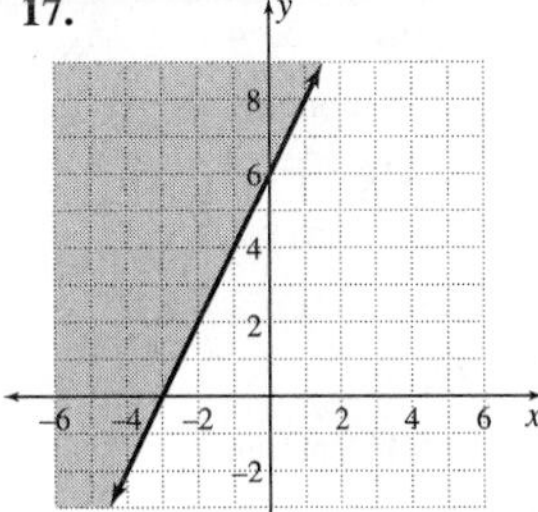

18.

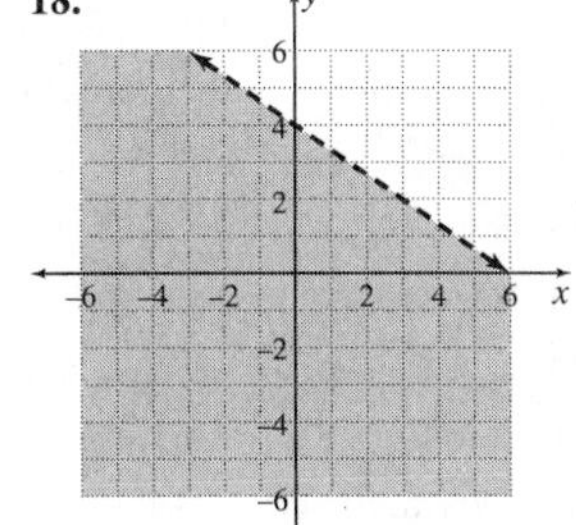

19. **a.** x, miles **b.** y, cost **c.** \$42.50 **d.** 270 miles **e.** \$30, base price for rental **f.** \$0.10, cost per mile **g.** $y = \$0.10x + \$30$ **h.** -300; no meaning

CUMULATIVE REVIEW OF CHAPTERS 1 TO 4

1. $\frac{16}{21}, -\frac{2}{21}, \frac{1}{7}, \frac{7}{9}$ **3.** $-1.6, 11.2, -30.72, -0.75$

5. $4x^2 + 6x$, $4x^2 - 6x$, $24x^3$, $\frac{2}{3}x$ **7.** $\approx$694.4 days

9. $h = 14$ inches **11.** $x = -7$ **13.** $x = -3$

15. a. $-\frac{3}{2}$ **b.** 9 **17.**

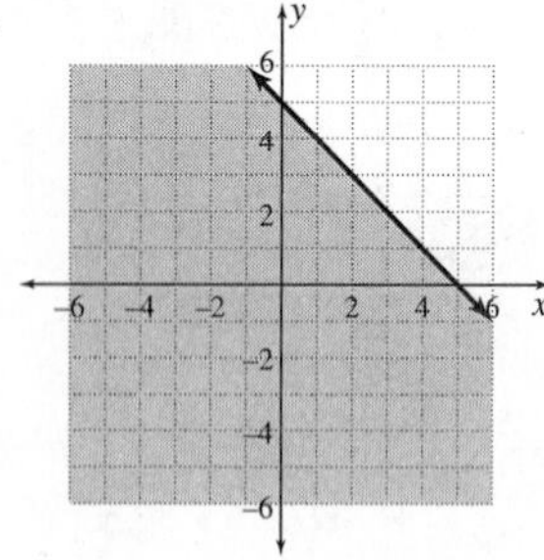

19. $f(-2) = 4$ **21.** $m = -1, b = -4, y = -x - 4$

EXERCISES 5.1

1. a. $\frac{5}{3}$, 5:3 **b.** 3:2, 3 to 2 **c.** $\frac{4}{9}$, 4 to 9

3. a. 5, $\frac{3}{7}$ **b.** 16, $\frac{3}{1}$ **c.** 240, $\frac{22}{15}$

5. a. $4x$, $3:x$ **b.** $2xy$, 1 to $3x$ **c.** x, $1:3x^3$

7. a. ac, b to e **b.** a, $(b + c)$ to c **c.** $a + b$, $1:a$

9. $\frac{3}{1}$ **11.** $-\frac{4}{1}$ **13.** $\frac{1}{2}$ **15.** $\frac{2.98}{1}$ **17.** $-\frac{1}{20}$

19. a. $\frac{1}{5}$ **b.** $\frac{3}{4}$ **c.** $\frac{9}{14}$ **d.** $\frac{29}{39}$

21. a. .357 **b.** .260

23. $4x + 3x + 1x = 400$, $x = \$50$; $A = \$200$, $B = \$150$, $C = \$50$

25. Cosmo, 1000 crowns; Timo and Sven, 750 crowns each

27. 9 carbon, 8 hydrogen, 4 oxygen

29. 1 calcium, 1 carbon, 3 oxygen **31.** $\frac{5}{8}$ **33.** $8\frac{1}{3}\%$ **35.** $12\frac{1}{2}\%$

37. 30%, 11%, 20%, 7%, 8%, 4%, 5%, 2%, and 13%, respectively

39. 33.3%, 133.3% **41.** −40.2%, 59.8%

43. a. \$540 **b.** \$375

45. a. 6 to 1 **b.** 3 to 1 **c.** 2 to 5 **d.** 4 to 1 **e.** 1 to 100 **f.** 4 to 25 **g.** 4 to 3

47. ≈0.6% **51.** 3 to 1 **53.** \$40 to 1 credit **55.** 1 to 16

57. 15 cc per hour **59.** 44 ft/sec **61.** $117\frac{1}{3}$ ft/sec

63. 60 mph **65.** 150 mph **67.** 352 ft, more than a city block

69. ≈56.8 miles **71.** 3 miles **73.** 20.8 microdrops/min

75. 8 minutes **77.** 17.8 grams NaCl

EXERCISES 5.2

1. 3 to 1, unsafe, slip **3.** 4.5 to 1, unsafe, tip **5.** 4 to 1, safe

7. proportion **9.** proportion **11.** false statement

15. $x = 11\frac{1}{4}$ **17.** $x = 3\frac{1}{3}$ **19.** $x = 2\frac{1}{7}$ **21.** $x = 9\frac{1}{3}$

23. 80 **25.** 66.7% **27.** 25.2 **29.** 140% **31.** 4.5

33. 25 **35.** 130% **37.** 70% **39.** $x = 9$ **41.** $x = 2$

43. $x = 5$ **45.** $x = 2$ **47.** $x = 26$ **49.** $x = -3$

51. $x = 8$ **53.** ≈1.65 meters **55.** ≈24.15 kilometers

57. 27,878,400 ft^2 **59.** 26 ft **61.** 36 ft **63.** ≈12.57 ft

65. $\frac{1}{6864}$ **67.** 130,000 units **69.** 1875 ft **71.** 950 ft

73. ≈488 birds

75. Units are the same on both sides after cross multiplication.

EXERCISES 5.3

1. 5 cm **3.** height = $1\frac{1}{4}$ in. **5.** height ≈ $\frac{3}{16}$ in.

7. $\frac{16}{25}$; similar triangles; sides AB and ND, AT and NE, BT and DE

9. $\frac{29}{34} \neq \frac{10}{29}$; not similar

11. $\frac{6}{12} = \frac{16}{32}$; similar figures; sides FG and HK, GO and KW, and others

13. $\frac{11}{8} = \frac{22}{16}$; similar figures; radii RT and OE, diameters EN and VD

15. $n = 31.5$ **17.** $n = 48$ **19.** ≈4.7 ft **21.** 20 ft

23. ≈40.4 ft **25.** $x = 6\frac{2}{3}$ **27.** $x = 9$ **29.** $x = 3$

31. $A(6, 0)$, $B(3, 2)$ **33.** $A(0, 2)$, $B(0, 4.8)$ **35.** 8 ft

37. $AB = 10 - x$ **39.** $BC = x - 10$

41. circles, squares, spheres

MID-CHAPTER 5 TEST

1. $4x$ to $5yz$ **2.** $\dfrac{x-2}{x}$ **3.** 1:4 **4.** 4 to 1 **5.** 70

6. 144 **7.** 25% **8.** ≈12.5% **11.** $x = \frac{1}{100{,}000}$

12. $x = 26\frac{2}{3}$ **13.** $x = 15$ **14.** $x = 17$ **15.** $x = 2$

16. $b = \dfrac{ad}{c}$ **17.** $x = 11.25$ **18.** $A(5, 0)$, $B(7, 5.6)$

19. 300, 450, 750 **20.** 45°, 45°, 90° **21.** 48 ft

22. 400 native fish, assuming all hatchery fish survive

23. a. \$295,000,000 **b.** \$9.35/sec

EXERCISES 5.4

1. a. 58, 57, 57 **b.** 58.2, 57, none **c.** 58.4, 61, none **d.** 54.6, 54, none

3. a. 385.4, 390, 3 modes (365, 395, 415)

b. 437.1, 422.5, 4 modes (410, 435, 515, 550)

5. no; for example, mean of 6 and 8 = 7 = mean of 5 and 9

7. Move the mean away from the median. **9.** is not

11. When we want to reduce the effect of large or small data

13. No; the median means half the data are above the score.

15. Some data are comparatively very small.

17. There is usually not time to test for everything.

19. It is in the middle of the traffic lanes.

21. GRCC, 31; LCC, 29; BCC, 30 **23.** 0.85 **25.** 0.70

27. ≈0.86 **29.** not possible **31. a.** (2.5, 3) **b.** (1, 0)

33. a. $\left(\dfrac{a}{2}, \dfrac{b}{2}\right)$ **b.** $\left(\dfrac{a}{2}, \dfrac{a}{2}\right)$ **c.** $\left(0, \dfrac{b}{2}\right)$

35. midpoints (0, 3), (4.5, 3), (4.5, 0); centroid (3, 2)

37. midpoints (0, 3), (4, 3), (4, 0); centroid $\left(2\frac{2}{3}, 2\right)$

EXERCISES 5.5

1. value **3.** quantity **5.** value **7.** value **9.** value

11.

	Quantity	Value	$Q \cdot V$
Peanuts	5 kg	\$8.80/kg	\$44.00
Cashews	2 kg	\$24.20/kg	\$48.40
	7 kg	\$13.20/kg	\$92.40

b. Sum of quantity column is total kg of peanuts and cashews.

c. \$13.20/kg

d. Sum of $Q \cdot V$ column is total worth of peanuts and cashews.

13.

	Quantity	Value	$Q \cdot V$
Grapes	3 lb	\$0.98	\$2.94
Potatoes	5 lb	\$0.49	\$2.45
Broccoli	2 lb	\$0.89	\$1.78
	10 lb	\$0.717	\$7.17

b. Sum of quantity column is total lb of vegetables and fruit purchased.

c. average price \$0.72

d. Sum of $Q \cdot V$ column is total purchase amount.

15.

	Quantity	Value	$Q \cdot V$
Dimes	15	\$0.10	\$1.50
Quarters	20	\$0.25	\$5.00
	35	\$0.1857	\$6.50

b. Sum of quantity column is total coins.

c. average coin value \$0.19

d. Sum of $Q \cdot V$ column is total amount of money.

17.

Quantity	Value	$Q \cdot V$
\$1500	0.09	\$135
\$1500	0.06	\$90
\$3000	0.075	\$225

b. Sum of quantity column is total invested.

c. average rate 7.5%

d. Sum of $Q \cdot V$ column is total return on investment.

19.

Quantity	Value	$Q \cdot V$
100 lb	0.12	12.0
50 lb	0.15	7.5
150 lb	0.13	19.5

b. Sum of quantity column is total lb of dog food.

c. average value 0.13

d. Sum of $Q \cdot V$ column is total lb of protein.

21.

Quantity	Value	$Q \cdot V$
150 lb	0.10	15
25 lb	0	0
175 lb	≈0.086	15

b. Sum of quantity column is total weight.

c. average value ≈0.086

d. Sum of $Q \cdot V$ column is total lb of protein.

23.

	Quantity	Value	$Q \cdot V$
Ds	5 hr	1	5
Cs	4 hr	2	8
Bs	3 hr	3	9
	12 hr	1.83	22

b. Sum of quantity column is total hr. **c.** average ≈1.83

d. Sum of $Q \cdot V$ column is total points.

25.

Quantity	Value	$Q \cdot V$
3 hr	80 kph	240 km
2 hr	30 kph	60 km
5 hr	60 kph	300 km

b. Sum of quantity column is total hr. **c.** average 60 kph

d. Sum of $Q \cdot V$ column is total km.

27.

Quantity	Value	$Q \cdot V$
150 mL	18	2700
100 mL	3	300
250 mL	12	3000

b. Sum of quantity column is total mL. **c.** average 12

d. Sum of $Q \cdot V$ column, divided by 1000, is total moles of sulfuric acid. (The division is needed because molarity is in terms of liters, not milliliters.)

29. $300 - x$ **31.** $\$15{,}000 - x$ **33.** $16 - x$

35.

	Quantity	Value	$Q \cdot V$
Boeing	200	\$54	\$10,800
Nike	x	\$71	$\$71x$
	$200 + x$		\$25,000

$10{,}800 + 71x = 25{,}000$
$x = 200$ shares

37.

Quantity	Value	$Q \cdot V$
30	\$5.80	\$174.00
x	\$7.20	$\$7.20x$
$30 + x$	\$6.36	2 equations

$(30 + x)6.36 = 174 + 7.20x$
$x = 20$ hours

39. $(30 + x)7 = 174 + 7.20x$
$x = 180$; not reasonable

41.

	Quantity	Value	$Q \cdot V$
Colombian	$300 - x$	\$7.25	$7.25(300 - x)$
Sumatran	x	\$10.00	$10x$
	300	\$8.35	\$2505

$7.25(300 - x) + 10x = 2505$
$x = 120$ lb
120 lb Sumatran
180 lb Colombian

43. ≈7.1 kg cashews, ≈42.9 kg peanuts

45. ≈28.6 kg cashews, ≈21.4 kg peanuts

47. a. earns \$1200, average rate is 8%
b. earns \$975, average rate is 6.5%

49. a. $A \approx \$4666.67$, $B \approx \$10{,}333.33$
b. $A = \$12{,}500$, $B = \$2500$

51. 3 hr **53.** ≈10.3 gal **55.** ≈27.6 gal

57. 87.5°; yes **59.** 200 mL

61. In Example 5, money invested is multiplied by the interest rate to get earnings. In Example 7, the number of shares is multiplied by price per share to get total money invested.

63. The average value is closer to the value of the item with the larger quantity.

65. If $Q \cdot V$ is larger for one item than for the others, that item will have a larger effect.

67. The greatest percent decrease is from \$0.20 to \$0.05. The smallest percent decrease is from \$0.30 to \$0.20. 50%, 40%, 33.3%, 75%

CHAPTER 5 REVIEW EXERCISES

1. 4 to x^2 **3.** 10:3 **5.** 144 **9.** 55,000, 55,000, 11,000

11. 450 cal, 50 g **13.** ≈15% **15.** ≈\$1.59 U.S./gal

17. 20 ft **19.** 4.5 gal **21.** $x = 9$ **23.** $x = 4$ **25.** $d = \dfrac{bc}{a}$

27. a. $x = 8.4$ **b.** $x = 4$ **c.** $x = \dfrac{3h}{b}$ **d.** $x = \dfrac{w(l-4)}{l}$

29. $A(0, 1.5)$, $B(-3, 0)$

31.

Quantity	Value	$Q \cdot V$
\$10,000	0.08	\$800
5,000	0.035	\$175
\$15,000	0.065	\$975

33.

Quantity	Value	$Q \cdot V$
8 L	90°C	720 L · °C
50 L	5°C	250 L · °C
58 L	≈16.7°C	970 L · °C

35.

	Quantity	Value (Assumed)	$Q \cdot V$
A	12 credits	4 points	48 credit points
B	4 credits	3 points	12 credit points
	16 credits	3.75 points	60 credit points

37. \$15,000 **39.** 31.73, 31.75, 31.8 **41.** 44.8, 45.0, 45.0

43. a. EG, (1, 7); EF, (1, 2); HE, (−3, 7); HF, (1, 7)
b. HF, $-\frac{5}{4}$; EG, $\frac{5}{4}$ **c.** Division by zero is undefined.
d. (1, 7)

45. (1, 1) **47.** 90

CHAPTER 5 TEST

3. $\dfrac{b}{4a}$ **4.** $\dfrac{1}{a-b}$ **5.** $\frac{5}{1}$ **6.** 84 **7.** \$1277.50 per year

8. 1.4 ft **9.** $x = 32.5$ **10.** $x = 15$ **11.** 94%

12. \$0.29, \$0.59 **13.** \$0.69, \$0.59

15. Clockwise from (3, 2): (5, 5), (7.5, 5), (5.5, 2) **16.** (6, 4)

17.

Quantity	Value	$Q \cdot V$
\$17,000	0.058	\$986
\$2,000	0.149	\$298
\$19,000 (total debt)	≈0.068 (average interest rate on total debt)	\$1284 (total interest on debt)

18. a. $x = 16$ **b.** $x = 4.8$ **19.** 12 credits

20. $I_1 = 10$ mA; no; the products from cross multiplication are different.

EXERCISES 6.1

1. a. −5 **b.** −5 **c.** −13 **d.** −10 **e.** −4 **f.** 7

3. a. $a + 12b - 8c$; trinomial **b.** $3m - n$; binomial
c. $21y$; monomial **d.** $x^2 - x - 12$; trinomial
e. $-x^2 + 2x - 3$; trinomial

5. $-x^3 - 3x^2 + 5x + 5$ **7.** $-x^4 + x^2 - x + 1$

9. a. $x^2 + 5x + 6$; trinomial **b.** $3x^2 + 7x + 2$; trinomial
c. $-4x + 17$; binomial **d.** $6x^2 + 5x + 1$; trinomial
e. $a^2 - b^2$; binomial

11. a. $x^2 - 4xy + 4y^2$; trinomial **b.** $x^2 - 4y^2$; binomial
c. $x^3 + 3x^2 + 3x + 1$; four-term polynomial
d. $x^3 + 2x^2 - 4x - 8$; four-term polynomial
e. $a^3 + b^3$; binomial

13. length $= 2b$, width $= a + b$, $P = 2a + 6b$, $A = 2ab + 2b^2$

15. length $= x + 1$, width $= x + 2$, $P = 4x + 6$, $A = x^2 + 3x + 2$

17. a. $5a$ **b.** $4x$ **c.** $5x - 3$ **d.** $2.5x$ **e.** $3 + \pi$

19. a. side $= 3c$, $A = 15ac$ **b.** side $= 2.5y$, $A = 3.75xy$
c. side $= 5y$, $A = 25y^2$

21. a. $2a^2 + 4ab$ **b.** $4ab + 2b^2$ **c.** $x^2 + 3x$ **d.** $2x^2 + x$

23. a. $2x^3 + 6x^2$ **b.** $x^3 - x^2$ **c.** $x^4 + 2x^3 + x^2$
d. $ab^3 - ab$ **e.** $ab^2 - b^3$ **f.** $a^2 + a^2b - a^2b^2$

25. $3x^2 + 2x$ **27.** $2b^2 + 10b$ **29.** $2b^2 - 6b$

31. x 1 1 / x 1 **33.** a b / a b b b

35. a. 3 · 37 **b.** 7 · 13

37. a. gcf $= 9$; $\frac{4}{11}$ **b.** gcf $= 66$; $\frac{1}{15}$ **c.** gcf $= 37$; $\frac{5}{27}$
d. gcf $= m$; $\dfrac{n}{p}$ **e.** gcf $= 4n$; $\dfrac{p}{6m}$

39. $y^2 + 2xy + x^2$ **41.** $a - 2b + 3b^2$

43. gfc $= x$; $x(x^2 + 4x + 4)$ **45.** gcf $= b$; $b(a^2 + ab + b^2)$

47. gcf $= 2x$; $2x(3x + 1)$ **49.** gcf $= 3y$; $3y(5y - 1)$

51. gcf $= 5xy$; $5xy(3x + 2y)$ **53.** $4(-x - 3)$, $-4(x + 3)$

55. $2y(-x + 2y)$, $-2y(x - 2y)$

57. $4(-3x^2 - 2x - 2)$, $-4(3x^2 + 2x + 2)$

59. $y^2(-1 + 4y - 8y^2)$, $-y^2(1 - 4y + 8y^2)$

61. a. $2a(1 - 2b)$ **b.** $xy(x + y)$ **63.** 2 **65.** 1 **67.** 3

69. 3 **71.** 4 **73.** 4 **75.** distributive property

77. factors, sum, terms

79. Variables and exponents are identical.

81. Find the factors that are the same in each term, and then multiply.

EXERCISES 6.2

1. $(x + 1)(x + 1) = x^2 + 2x + 1$

3. $(a + 2b)(2a + b) = 2a^2 + 5ab + 2b^2$

5. $(a + 2b)(a + 2b) = a^2 + 4ab + 4b^2$ **7.** $2x^2 - 3x - 20$

9. $2x^2 - x - 6$ **11.** $2x^2 - 5x + 2$ **13.** $5x^2 - 21x + 4$

15. a. $x^2 - 4x + 4$ **b.** $x^2 - 4$

17. a. $a^2 + 10a + 25$ **b.** $b^2 - 25$

19. a. $a^2 - b^2$ **b.** $a^2 - 2ab + b^2$

21. a. $x^2 + 8x + 7$ **b.** $x^2 - 6x - 7$

23. a. $b^2 + 14b + 49$ **b.** $a^2 - 49$

25. a. $x^2 + 2xy + y^2$ **b.** $x^2 - 2xy + y^2$

27. 15a, 16a, 17a, 18b, 19b, 20b, 23a, 24b, 25a, 25b

29. pst; $4x^2 + 12x + 9$ **31.** ds; $4x^2 - 9$

33. neither; $-4x^2 + 12x - 9$ **35.** pst; $9x^2 - 12x + 4$

37. ds; $9x^2 - 4$ **39.** $x^2 + 10x + 25$ **41.** $a^2 - 12a + 36$

43. $2x^2 + 12x + 18$ **45.** $3x^2 - 30x + 75$

47. The exponent was applied incorrectly:
$(x - a)^2 = (x - a)(x - a) = x^2 - 2ax + a^2$

49. 11 **51.** 1 **53.** 4 **55. a.** $x^2 + 9x + 8$ **b.** $x^2 - 7x - 8$

57. a. $x^2 + 6x + 8$ **b.** $x^2 - 2x - 8$

59. a. $6x^2 - 13x + 6$ **b.** $6x^2 + 5x - 6$

61. a. $6x^2 + 37x + 6$ **b.** $6x^2 + 35x - 6$

63. a. $2x^2 + 7x + 5$ **b.** $2x^2 + 3x - 5$

65. a. $2x^2 + 9x - 5$ **b.** $2x^2 + 11x + 5$

67. $(x \pm 1)(x \pm 12)$, $(x \pm 2)(x \pm 6)$, $(x \pm 3)(x \pm 4)$

69. $(x \pm 1)(x \pm 20)$, $(x \pm 2)(x \pm 10)$, $(x \pm 4)(x \pm 5)$

71. 25, 14, 11, 10

73. The coefficient is from the sum of the terms obtained by multiplying $3 \cdot x$ and multiplying $-5 \cdot x$.

MID–CHAPTER 6 TEST

1. a. $-3a + 11b - 13c$; trinomial **b.** $2x^2 - 4x - 6$; trinomial
c. $2x^3y^2 - 3x^2y + 2xy^2$; trinomial
d. $2a - 7b + 4c$; trinomial **e.** $x^3 - 125$; binomial
f. $-3x^2 - 9x + 9$; trinomial **g.** $4x - 8$; binomial

2. a. gcf $= 2$; $2(3x^2 - x + 4)$ **b.** gcf $= a$; $a(2bc - 3c + 4b)$

3. a. length $= 2a + b$, width $= a + 2b$, $P = 6a + 6b$, $A = 2a^2 + 5ab + 2b^2$
b. length $= x + 2$, width $= x + 1$, $P = 4x + 6$, $A = x^2 + 3x + 2$

4. Terms are added (or subtracted), as in $a + b$; factors are multiplied, as in $a \cdot b$.

5. $(a + 2b)(3a + 2b)$ **6.** $6x^2 - 11x - 10$ **7.** $-60x^2$

8. $(x - 2)(3x^2 - 2x + 1) = 3x^3 - 8x^2 + 5x - 2$

9. $-6x^2$ and $-2x^2$, $4x$ and x **10.** $x^2 + 8x + 15$

11. $3x^2 - 7x - 20$ **12.** $6x^2 + 19x - 7$ **13.** $x^2 - 9$

14. $4x^2 - 20x + 25$

15. a. neither **b.** neither **c.** ds **d.** pst

16. 1 and 10, 2 and 5; $\pm 11x$, $\pm 9x$, $\pm 7x$, $\pm 3x$

EXERCISES 6.3

1.

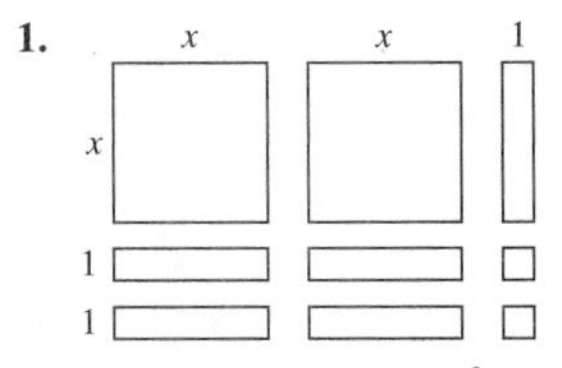

$(2x + 1)(x + 2) = 2x^2 + 5x + 2$

3.

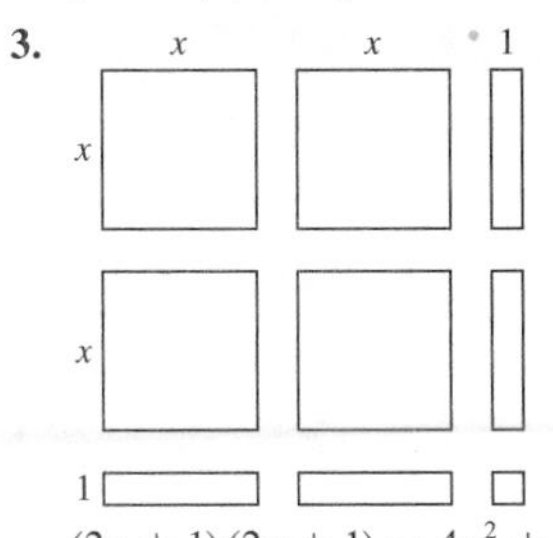

$(2x + 1)(2x + 1) = 4x^2 + 4x + 1$

5. $(x + 5)(x + 4) = x^2 + 9x + 20$

7. $(x + 2)(x - 10) = x^2 - 8x - 20$

9. $(6x + 1)(x - 3) = 6x^2 - 17x - 3$

11. The product gives $x^2 + 3x - 5x - 15$, and $3x - 5x = -2x$.

13. $(x + 2)(x + 6)$ **15.** $(x - 12)(x - 1)$

17. cannot be factored **19.** $(x + 3)(x + 4)$ **21.** $(x + 4)(x - 3)$

23. $(x - 12)(x + 1)$ **25.** $(x + 3)^2$ **27.** $(x + 6)(x + 5)$

29. $(x + 15)(x - 2)$ **31.** $(x + 2)(x - 8)$ **33.** $(x + 16)(x - 1)$

35. $(x + 5)(x - 5)$ **37.** $(x + 2)(x - 2)$ **39.** $4(x + 2)(x - 2)$

41. $(x + 6)^2$ **43.** $4(x + 1)^2$ **45.** cannot be factored

47. $(x - 3)^2$ **49.** $(2x + 3)(x + 4)$ **51.** $(2x + 3)(x - 3)$

53. $(2n + 3)(n - 1)$ **55.** $(3x - 1)(x + 2)$ **57.** $(3a + 1)(a - 4)$

59. cannot be factored **61.** cannot be factored

63. $(3x + 7)(3x - 7)$ **65.** $(4x + 3)(4x - 3)$

67. $(3x + 2)(2x - 1)$ **69.** $(3x - 2)(2x + 3)$

71. $(2n - 1)(n + 5)$ **73.** $(5x - 6)(5x + 6)$

75. $3(x + 3)(x + 1)$ **77.** $3(x + 3)(x - 3)$ **79.** $5(x - 1)^2$

81. $2(3x + 5)(3x - 5)$ **83.** $3(x - 5)^2$ **85.** $3(x^2 + 2x + 4)$

EXERCISES 6.4

1. Output: 3^x
9
3
1
$\frac{1}{3}$
$\frac{1}{9}$
$\frac{1}{27}$

3. Output: 4^x
16
4
1
$\frac{1}{4}$
$\frac{1}{16}$
$\frac{1}{64}$

5. a. $\frac{1}{4}$ **b.** $\frac{1}{2}$ **c.** 1 **7. a.** 1 **b.** 16 **c.** 4

9. a. 2 **b.** 1 **c.** 4 **11.** -1 **13.** 0

15. One example: $\frac{1}{4}$ **17. a.** 5^{-4} **b.** 6^3 **c.** 10^{-3} **d.** 10^{-5}

19. a. 10^{-30} **b.** 10^{-9} **c.** 2^8 **d.** 2^8

21. a. 3^7 **b.** 10^{-3} **c.** 10^{-16} **d.** 10^{-2}

23. a. 2^{12} **b.** 2^{-12} **c.** 10^{10} **d.** 10^{-18}

25. a. a^{-7} **b.** x^8 **c.** n^{-4} **27. a.** x^{-2} **b.** a **c.** b^3

29. a. a^9 **b.** a^{-8} **c.** x^6 **31. a.** x^{-8} **b.** x^{12} **c.** b^{-6}

33. a. $\dfrac{1}{x}$ **b.** $\dfrac{y}{x}$ **c.** $\dfrac{x}{y}$ **d.** 1 **e.** 1 **f.** $\dfrac{bc}{a}$

35. a. $\dfrac{1}{y^3}$ **b.** $\dfrac{x^2}{y^2}$ **c.** b^3 **d.** $\dfrac{b^3}{a^3}$ **e.** $\dfrac{c^2}{16a^4}$ **f.** $\dfrac{b^6}{a^3}$

37. a. $\dfrac{1}{xy^5}$ **b.** y^3 **c.** $\dfrac{b^3}{a^5}$

39. a. $4x^2 - 12x + 9$ **b.** $8x^3 + 36x^2 + 54x + 27$
c. $27x^3 - 54x^2 + 36x - 8$

41. a. x **b.** x **c.** x

43.

x	10^x as fraction	10^x as decimal
0	1	1
-1	$\frac{1}{10}$	0.1
-2	$\frac{1}{100}$	0.01
-3	$\frac{1}{1000}$	0.001
-4	$\frac{1}{10{,}000}$	0.0001
-5	$\frac{1}{100{,}000}$	0.00001

45. a. 1.47×10^{11} **b.** 5.12×10^{-13}

47. a. 0.0234 **b.** 3140 **c.** 62,800,000 **49.** 1.39×10^6 km

51. 2.76×10^9 miles **53.** 1.8×10^3 g

55. a. 3.4×10^4 **b.** 5.6×10^0

57. a. 4.32×10^3 **b.** 5.67×10^{-6}

59. 1,990,000,000,000,000,000,000,000,000,000 kg

61. $-0.000\,000\,000\,000\,000\,000\,160\,2$ coulomb

63. 200,000,000 years

65. $0.000\,000\,000\,000\,000\,000\,000\,000\,001\,675$ kg

67. a. $>$ **b.** $<$ **69. a.** $<$ **b.** $>$ **71. a.** $>$ **b.** $<$

73. electron **75.** 1.58×10^{17} miles **77.** 1.80×10^7 quarts

79. 3.48×10^{12} calories; ≈ 622 calories **81.** 4.90×10^{-9} mph

83. a. 6×10^{27} **b.** 1.2×10^{32}

85. a. 3×10^{-26} **b.** 4.8×10^{-24}

87. a. 4×10^3 **b.** 2×10^{-13} **89.** 1×10^3

CHAPTER 6 REVIEW EXERCISES

1. a. $7a^2 - 3ab - 10b^2$; trinomial **b.** $4x^2 + 3x + 5$; trinomial
c. $x^3 - 64$; binomial **d.** $4x^2 + 7x$; binomial
e. $-4x + 21$; binomial

3. a. $P = 10x + 2, A = 6x^2 + 2x$
b. $P = 8x + 4, A = 3x^2 + 6x$ **c.** $P = 7x - 3, A = 2x^2 + x$

5. $(3x + 1)(x + 1) = 3x^2 + 4x + 1$ **7.** $x^2 + 7x + 12$

9. $4x^2 - 20x + 25$ **11.** $9x^2 + 9x - 10$ **13.** $6x^2 - 5x - 6$

15. $a^2 - 2ab + b^2$ **17.** 9, 15

19. a. $(x + 7)(x - 2) = x^2 + 5x - 14$
b. $(3x + 5)(2x + 3) = 6x^2 + 19x + 15$

21. a. $(x - 2)(x - 7) = x^2 - 9x + 14$
b. $(x + 7)(2x - 3) = 2x^2 + 11x - 21$
c. $(3x - 1)(4x + 1) = 12x^2 - x - 1$
d. $(2x + 3)(10x - 3) = 20x^2 + 24x - 9$

23. $(x - 2)(x - 1)$; neither **25.** $(3x + 4)(3x - 4)$; ds

27. $(2x - 3)(x + 2)$; neither **29.** $(3x + 2)(3x - 1)$; neither

31. $(x + 2)(x + 4)$; neither **33.** $(x - 1)(x - 10)$; neither

35. $(5 + 3x)(5 - 3x)$; ds **37.** $(x + 2)(x + 6)$; neither

39. $(2x + 7)(x - 5)$; neither **41.** $4(x - 1)^2$; pst

43. $x(x + 2)^2$; contains pst **45.** $3(x + 3)(x - 3)$; contains ds

47. $x(x - 2)(x - 5)$; neither **49. a.** $\frac{1}{3}$ **b.** 1 **c.** $\frac{1}{9}$

51. a. 1 **b.** $\frac{3}{2}$ **c.** $\frac{9}{4}$ **53. a.** x^5 **b.** 1 **c.** $\dfrac{1}{b^{10}}$

55. a. n^9 **b.** n **57. a.** $\dfrac{1}{b^8}$ **b.** x^6 **c.** 1

59. a. $\dfrac{b^2}{a^2}$ **b.** $\dfrac{b^6c^3}{8a^3}$ **61. a.** $\dfrac{1}{x^8y}$ **b.** $\dfrac{1}{a^3b^2}$

63. a. x^3 **b.** 1

65.

^{40}K	1,400,000,000
^{41}Ca	120,000
^{219}Rn	0.000 000 124 3
^{212}Po	0.000 000 30

67. $\approx 9.5 \times 10^{-15}$ year **69. a.** 4.2×10^7 **b.** 3.2×10^5

71. a. 9×10^{16} **b.** 6×10^{14}

73. a. 2.34×10^7 **b.** 4.36×10^6 **75.** $\approx 1.26 \times 10^{-4}$ mph

CHAPTER 6 TEST

1. terms **2.** greatest common factor **3.** factoring

4. -1 **5.** $x = 0$ **6.** $x = -2$ **7.** 0.000 348 2

8. 4.5×10^{10} **9.** $4a - 2b + 5c - 2d$; four-term polynomial

10. $x^3 - 27$; binomial **11.** $-4x^2 + 20x$; binomial

12. $2y(3x^2 + 7x - 9y)$ **13.** $(3x + 4)(2x - 5) = 6x^2 - 7x - 20$

14. $P = 12x + 10, A = 8x^2 + 20x$ **15.** $3x$ **16.** $x^2 + 3x - 28$

17. $x^2 - 14x + 49$ **18.** $4x^2 - 49$ **19.** $2x^2 - 16x + 32$

20. $(x - 4)(x - 5)$ **21.** $(2x + 1)(x - 2)$ **22.** $2(x + 2)(x - 2)$

23. $(x - 4)^2$ **24.** $(3x + 1)^2$ **25.** 18, 22

26. a. b **b.** $\dfrac{1}{x^6}$ **27. a.** b^5 **b.** $\dfrac{b}{a}$

28. a. 1 **b.** $\dfrac{625y^4}{81x^4}$ **29.** 0.000 000 000 000 001

30. 5×10^{-5}

31. factors: $(x \pm 1)(x \pm 21)$, $(x \pm 3)(x \pm 7)$; trinomials:
$x^2 \pm 22x + 21$; $x^2 \pm 10x + 21$, $x^2 \pm 20x - 21$, $x^2 \pm 4x - 21$;
2 and 21, 3 and 7 are the only factors of 21

32. There are two places where a and b multiply each other:

	a	$+b$
a	a^2	($+ab$)
$+b$	($+ab$)	b^2

33. 10 [EE] 3 is $10 \times 10^3 = 10^4 = 10{,}000$.
10 [y^x] 3 is $10^3 = 1000$.

CUMULATIVE REVIEW OF CHAPTERS 1 TO 6

1.

Input x	Output $y = 0.25x + 0.25$
0	0.25
5	1.50
10	2.75
15	4.00
20	5.25
25	6.50
30	7.75

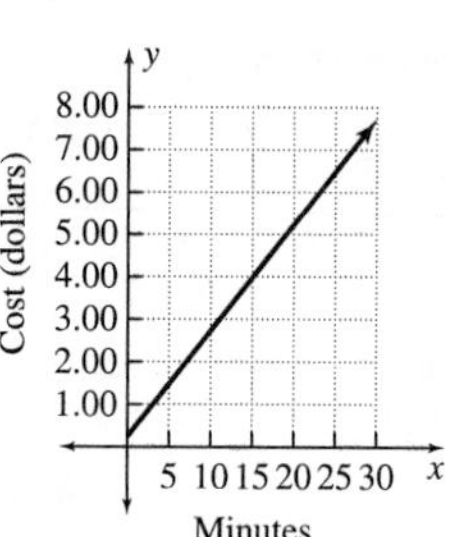

3. $2x - 3$ **5.** x^{11} **7.** $R = 12$

9. $x = 5.2$

11. $x \geq 2$

−5 −4 −3 −2 −1 0 1 2 3 4 5

13. $y = -\frac{3}{5}x + 3$

15. $a = 3m - b - c$

17.

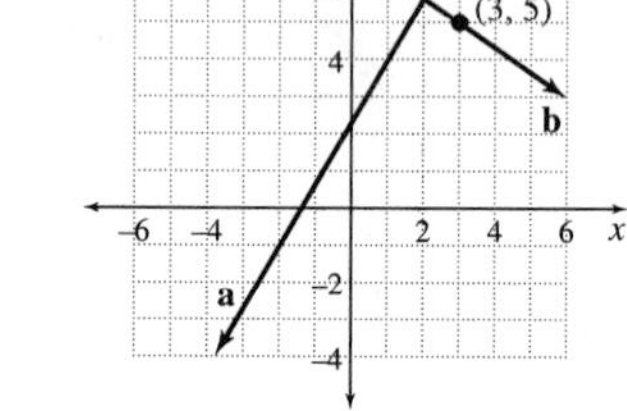

19. $y = -\frac{3}{5}x$ **21.** $\approx 29.3\%$

23. $x = 9$

25. $x^3 - 3x^2 + 3x - 1$

27. $4x^2 + 8x + 3$

29. $(2x + 5)(2x - 5)$; difference of squares

EXERCISES 7.1

1.

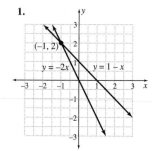

(−1, 2) $y = -2x$ $y = 1 - x$

3.

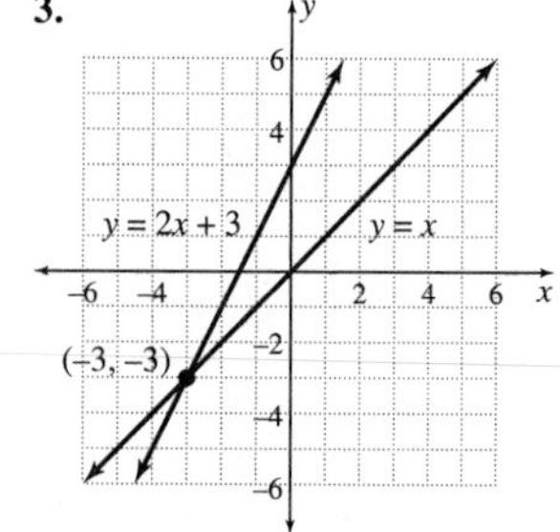

5.

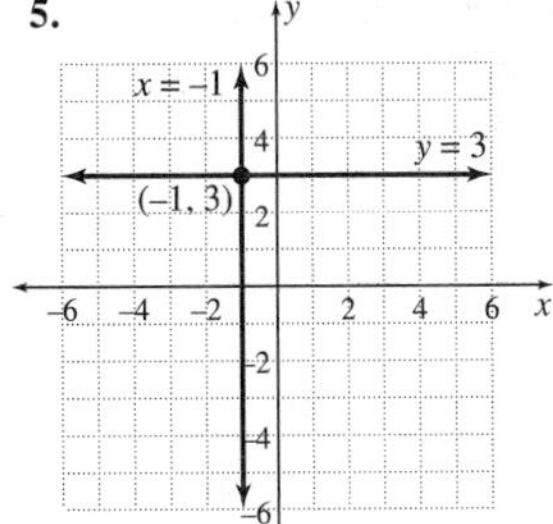

7.

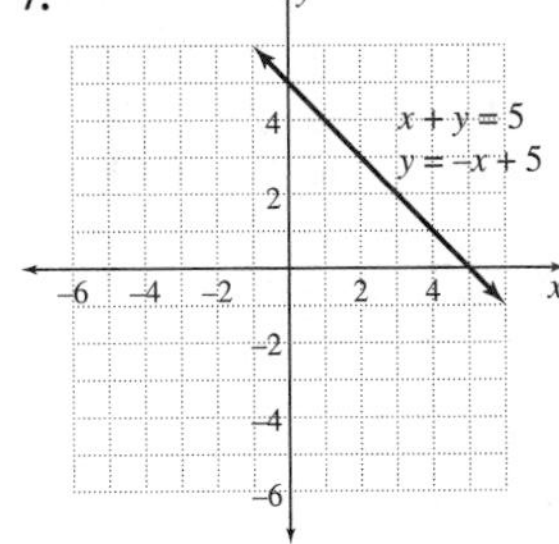

coincident lines; infinite number of solutions

9.

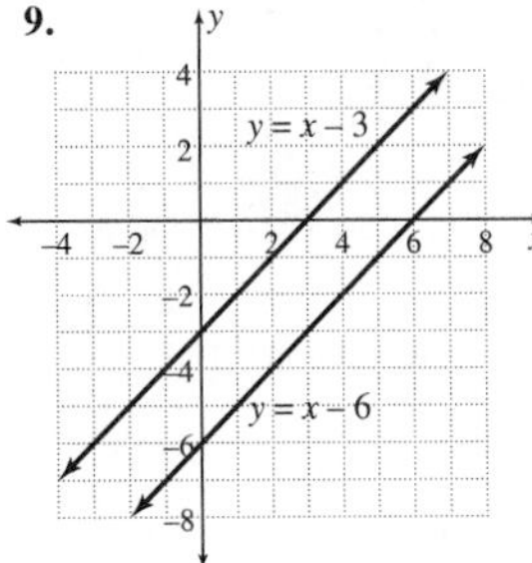

parallel lines; no solution

11.

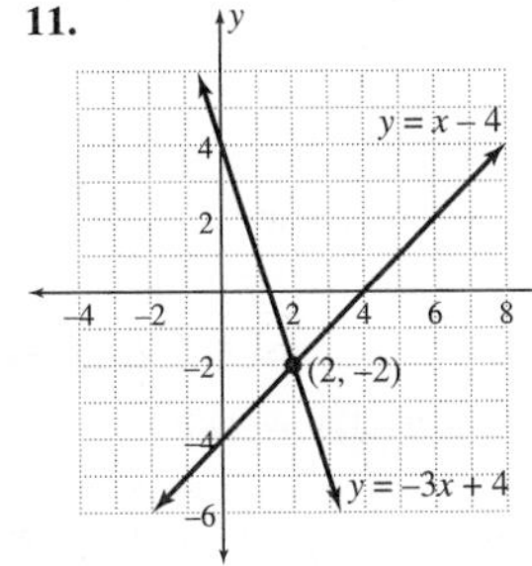

13.

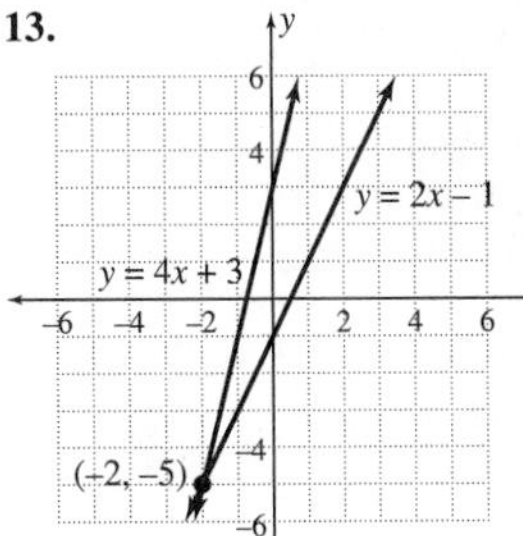

15.

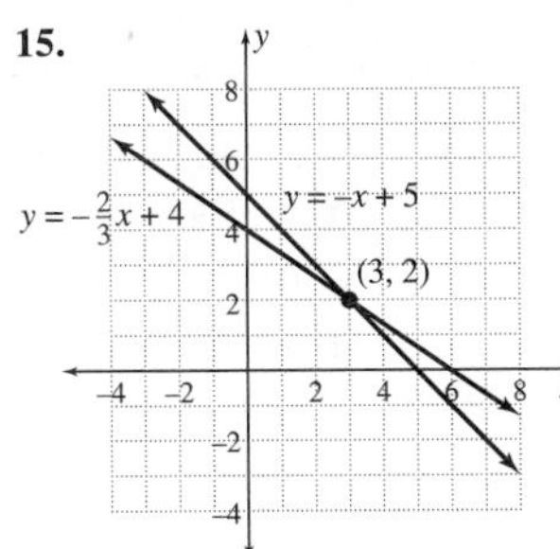

17. parallel **19.** not parallel **21.** parallel

23. not coincident **25.** coincident **27.** not coincident

29. $y = -2x + 5$, $y = -x + 3$; intersection (2, 1)

31. $y = \frac{3}{2}x + 1$, $y = \frac{5}{2}x - 1$; intersection (2, 4)

33. **a.** Sense, $y = 0.05x + 50$; Herr's, $y = 0.20x + 20$

b.

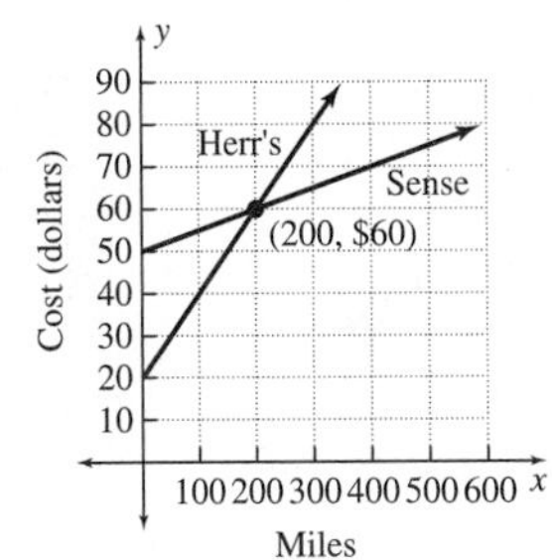

c. (200, \$60); mileage where costs are equal

d. Herr's; greater slope (cost per mile)

e. rental fee **f.** $x > 200$ mi

35. **a.** $y = 50 + 0.12x$ rewards large sales

b.

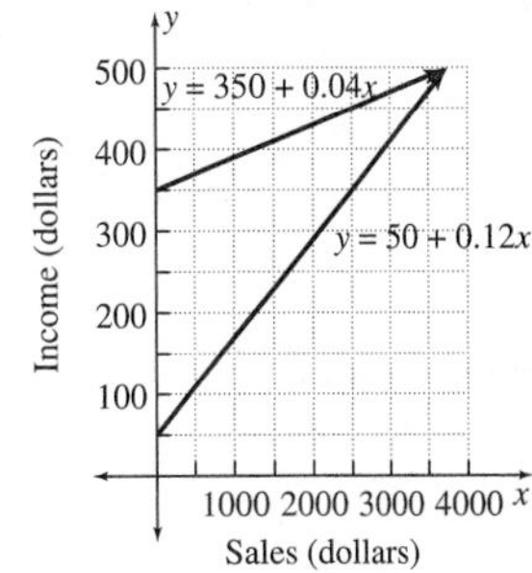

c. (3750, 500) **d.** sales less than \$3750

e. $y = 350 + 0.08x$, $y = 50 + 0.24x$

37. **a.** \$525 **b.** \$500 **c.** −\$25 **d.** cost

39. **a.** $C = 250 + 8.50x$ **b.** $R = 10x + 200$

Other forms of the equations are possible.

c.

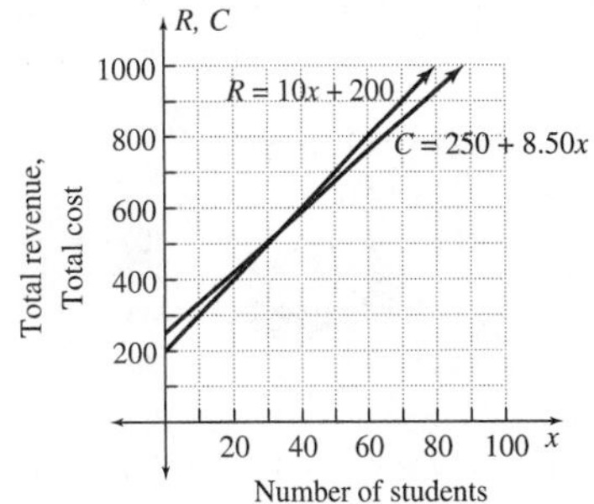

d. ≈(34, 540) (rounded to nearest whole person)

41. **a.**

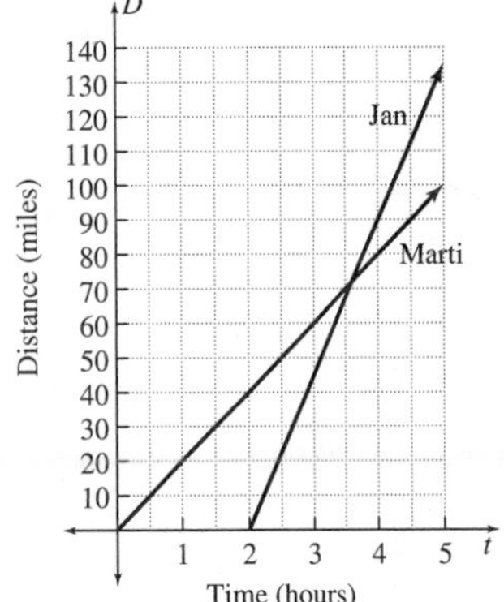

b. $D = 20t$ **c.** $D = 45t - 90$ **d.** (3.6, 72)

e. elapsed time and distance when Jan catches up with Marti

f. Jan; greater slope (higher mph)

g. Jan starts 2 hours after Marti.

EXERCISES 7.2

1.

Item	Quantity	Value (\$)	$Q \cdot V$ (\$)
Nickels	x	0.05	$0.05x$
Quarters	y	0.25	$0.25y$
Total	22		2.90

$x + y = 22$, $0.05x + 0.25y = 2.90$

3.

Item	Quantity	Value ($)	$Q \cdot V$ ($)
Dimes	x	0.10	$0.10x$
Quarters	y	0.25	$0.25y$
Total	26		5.45

$x + y = 26$, $0.10x + 0.25y = 5.45$

5.

Item	Quantity	Value ($)	$Q \cdot V$ ($)
Nickels	x	0.05	$0.05x$
Dimes	y	0.10	$0.10y$
Total	65		5.40

$x + y = 65$, $0.05x + 0.10y = 5.40$

7.

Job	Hours	Wage ($)	Earnings ($)
A	x	5.75	$5.75x$
B	y	8.50	$8.50y$
Total	43		316

$x + y = 43$, $5.75x + 8.50y = 316$

9.

Grocery Stores (lb)	Restaurants (lb)	Total (lb)
x	y	5250

$x + y = 5250$, $x = 9y$

For Exercises 11 through 17, guesses will vary.

11. x = long segment, y = short segment; $x + y = 10$, $x = y + 5$

13. x = first project hours, y = second project hours; $x + y = 176$, $y = x + 28$

15. x = number of persons, y = total cost; for Fun Base, $y = 30$ for $x \le 8$. $y = 30 + 3.50(x - 8)$ for $x > 8$; for Papa's, $y = 3x$

17. x = number of persons, y = total cost; for American Gymnastics, $y = 70$ for $x \le 15$, $y = 70 + 3(x - 15)$ for $x > 15$; for Farrell's, $y = 3.95x$

19.

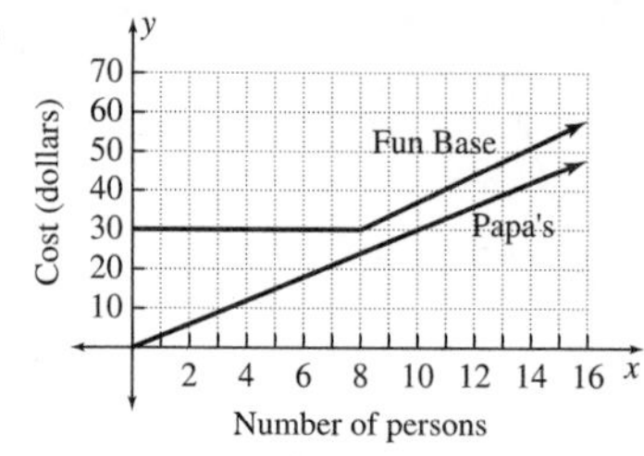

Papa's will always be less.

21.

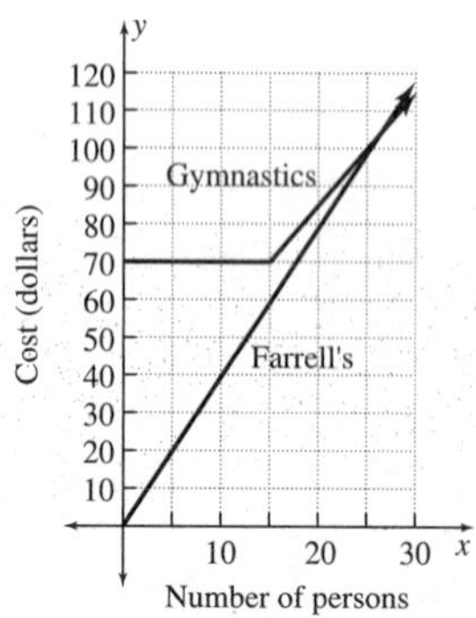

Cost is equal for $\approx$26 children.

23. $w + l + c = \$620$ watch $= w = \$320$
$l = c + 20$ locket $= l = \$160$
$w = 2l$ chain $= c = \$140$

25. $n + p + b = 20$ notebooks $= n = 7$
$p = b + 3$ paperbacks $= p = 8$
$n = p - 1$ hardbound $= b = 5$

27. **a.** 1 **b.** 2 **c.** 3

EXERCISES 7.3

1. $W = \frac{L}{2}$ **3.** $b = c - a$ **5.** $r = \frac{C}{2\pi}$ **7.** $y = x - 5$

9. $d = \frac{C}{\pi}$ **11.** $x = 3$ **13.** $x = 3$ **15.** $y = 4 - 3x$

17. $x = 4y + 5$ **19.** $x = 5y - 9$ **21.** $y = 3x + 2$

23. $x = 3, y = -5$ **25.** $x = 2, y = 15$ **27.** $x = 4, y = -1$

29. $x = 5, y = 9$ **31.** $x = 3, y = -2$ **33.** $x = 2, y = \frac{8}{3}$

35. no solution **37.** no solution

39. infinite number of solutions **41.** $x = -1, y = -1$

43. $x = 0, y = -2$ **45.** $x = 6, y = -2$

47. infinite number of solutions **49.** $x = 23, y = 27$

51. $x = 17.5, y = 82.5$

For Exercises 53 to 71, the answers are listed in the order in which items appear in the exercises.

53. 8.5 yd, 11.5 yd **55.** 17, 7 **57.** 17, 11 **59.** 8 in., 12 in.

61. 19 cm, 10 cm **63.** $\approx$207 g, $\approx$63 g

65. **a.** $A = 102°$, $B = 78°$ **b.** $C = 77°$, $D = 103°$
c. $E = 46°$, $F = 44°$ **d.** $G = 29°$, $H = 61°$
e. $I = 75°$, $J = 105°$

67. **a.** 2 m, 14 m **b.** 100 yd, 60 yd **69.** $320, $160, $140

71. 7, 8, 5 **73.** $A = 108$, $B = 54$, $C = 18$

75. $A = 64°$, $B = 32°$, $C = 84°$ **77.** $A = 45°$, $B = 45°$, $C = 90°$

79. length = 19 in., width = 12 in., height = 3 in.

MID–CHAPTER 7 TEST

1. $y = 2x - 5000$ **2.** $h = \frac{3V}{\pi r^2}$ **3.** $l = \frac{13w}{6}$

4. $x = 1500, y = 3500$ **5.** $x = 2500, y = 2500$

6. **a.** $y = 6 - \frac{2}{3}x$, $y = x + 1$ **b.** (4.5, 5.5) **c.** (4.5, 5.5)
d. (12, −2)

7. **a.**

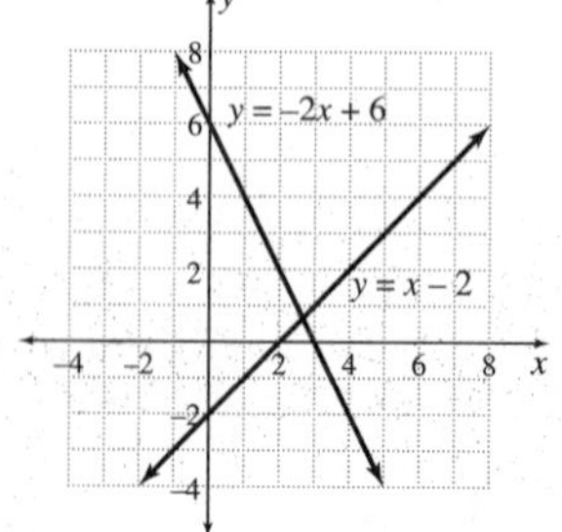

b. (2.5, 1) **c.** $\approx$(2.7, 0.7) or $(\frac{8}{3}, \frac{2}{3})$

8. A true statement is obtained. There are an infinite number of solutions. The lines are coincident.

9. 180, 15 **10.** 1200, 400 **11.** 39 ft, 18 ft

12. The ordered pair for the point of intersection is where the same input gives the same output in each equation.

EXERCISES 7.4

1. a.

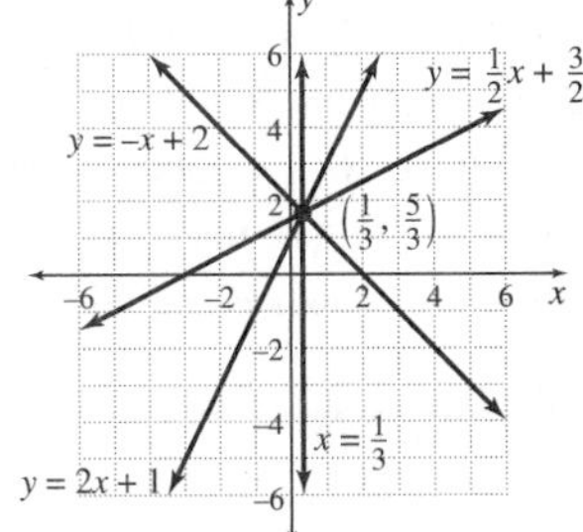

b. $y = \frac{1}{2}x + \frac{3}{2}$ c. $x = \frac{1}{3}$

d. All graphs intersect at the same point, $\left(\frac{1}{3}, 1\frac{2}{3}\right)$.

3. $x = 3, y = -5$ 5. $m = 7, n = -4$

7. $x = -2, y = 3$ 9. $a = -10, b = 15$

11. $x = -3, y = 3$ 13. $p = 0, q = 3$

15. no solution 17. infinite number of solutions

19. $x = 9, y = -4$ 21. $x = 3, y = -2$

23. $b = -3, m = 2$ 25. $x = 5, y = 2$

27. $x = 2, y = 4$ 29. 8.5, 16.5

31. 4.75, 15.25 33. $f = 9, p = 4$

35. $x = 130, y = 113$

37. a. 20°, 70° b. 65°, 115°

39. a. $A = 102°, B = 78°$ b. $C = 77°, D = 103°$

c. $E = 46°, F = 44°$ d. $G = 29°, H = 61°$

e. $I = 75°, J = 105°$

For Exercises 41 to 59, answers are listed in the order in which the items are listed in the exercise.

41. ≈67.7 cal, ≈12.6 cal 43. 4 cal, 3 cal

45. 20 cal, 25 cal 47. \$8.00, \$3.50

49. \$25.99, \$15.99 51. \$8.99, \$39.99

53. 2 hr, 5 hr 55. 137.5 mph, 412.5 mph

57. 100 mph, 20 mph 59. 7 mph, 3 mph

61. friction with the ground

EXERCISES 7.5

1. a, b 3. c

5.

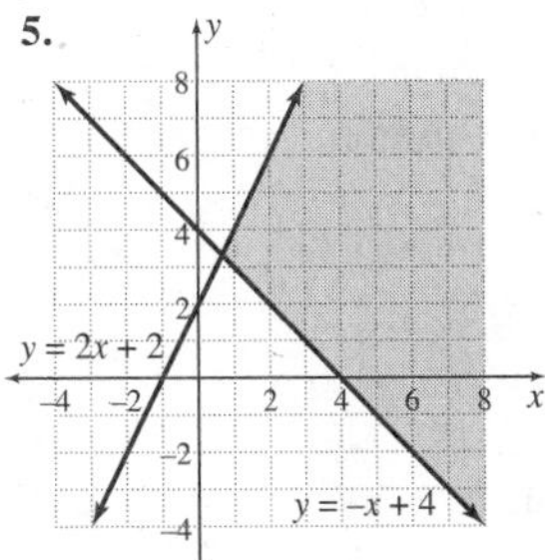

7.

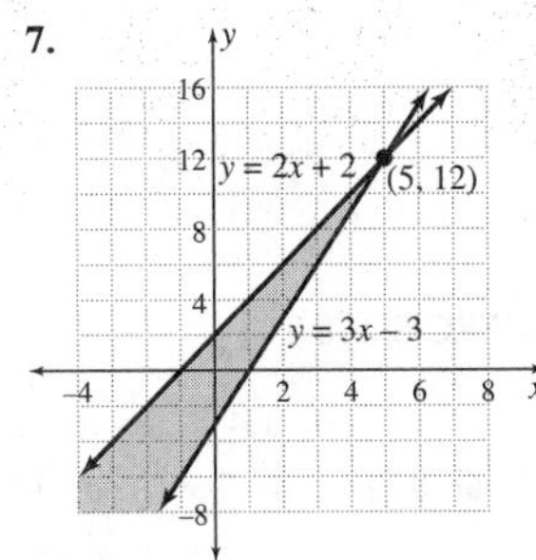

9.

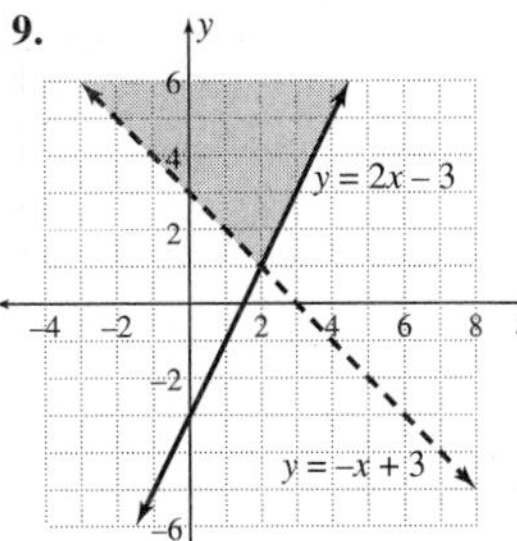

11.

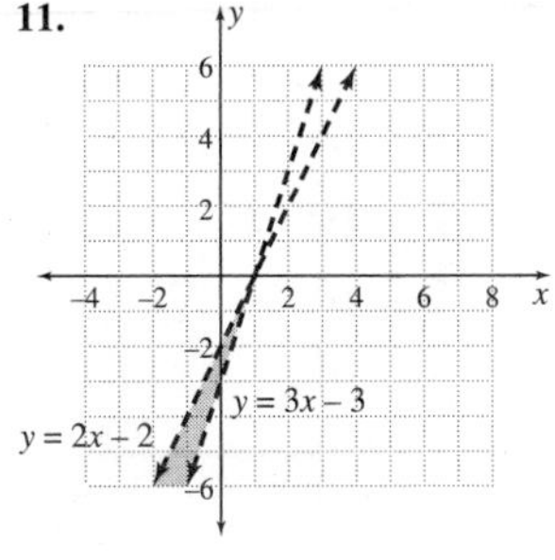

13.

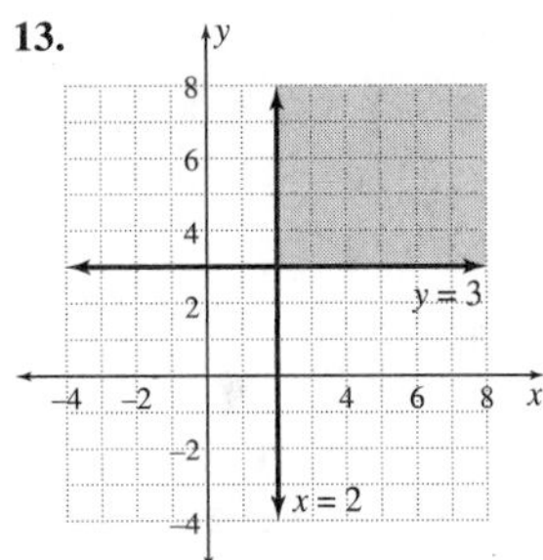

15.

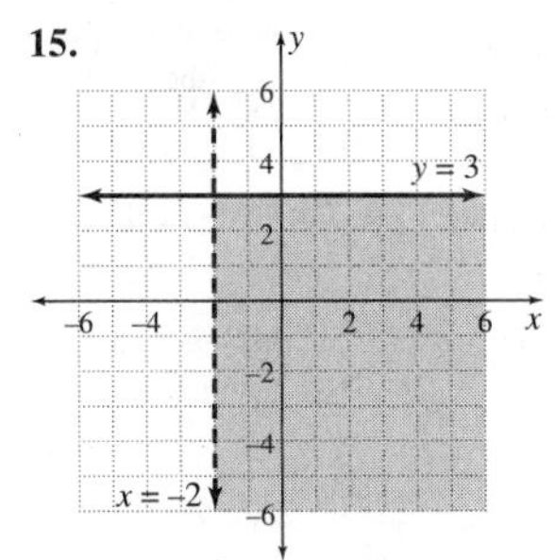

17.

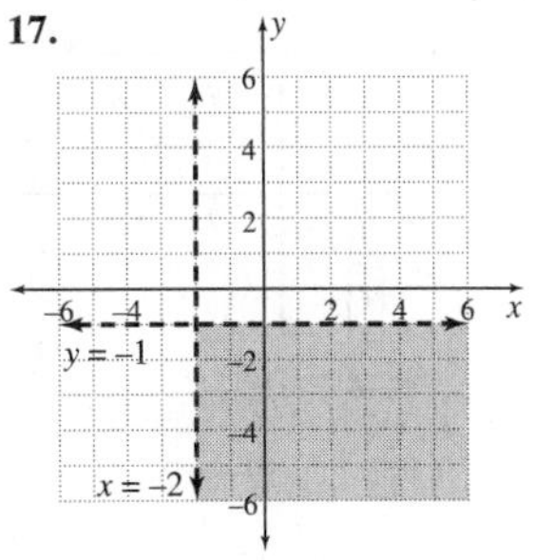

19.

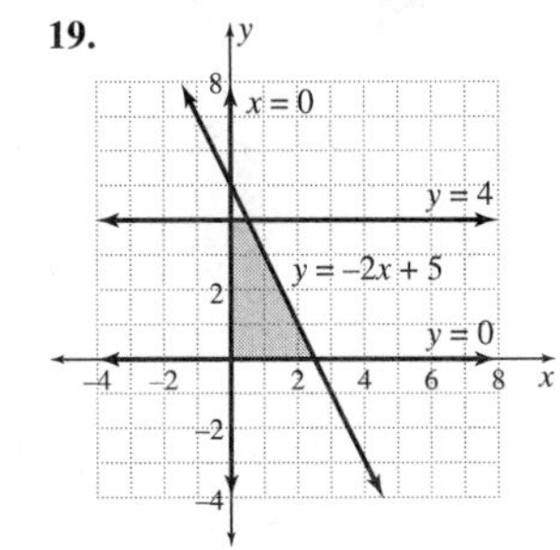

21.

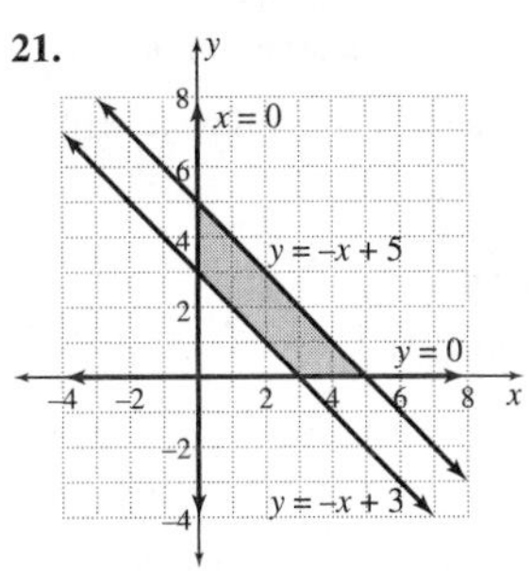

23.

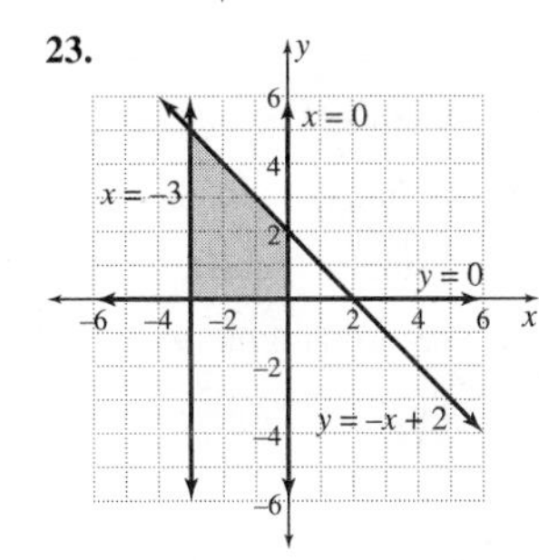

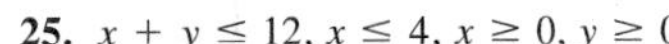

25. $x + y \leq 12, x \leq 4, x \geq 0, y \geq 0$

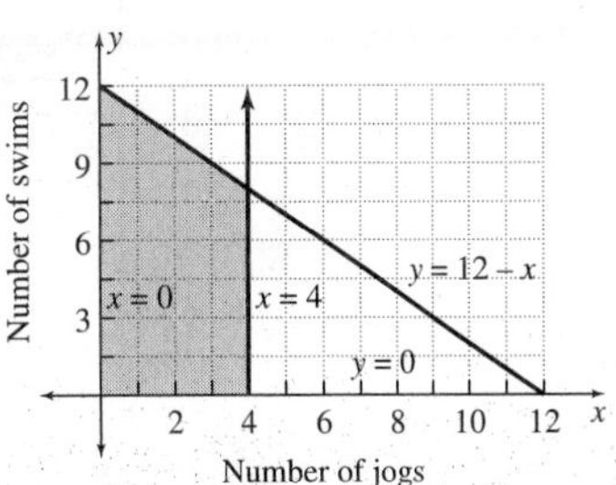

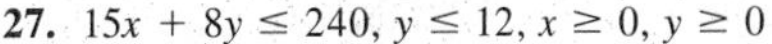

27. $15x + 8y \leq 240, y \leq 12, x \geq 0, y \geq 0$

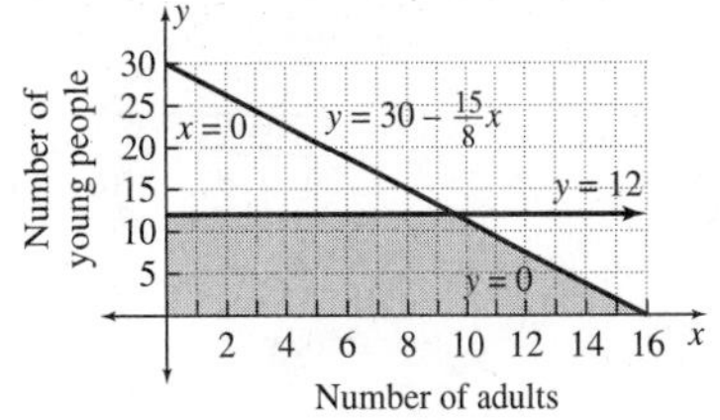

29. $x + y \leq 45{,}000$, $y \leq 5000$, $x \geq 0$, $y \geq 0$

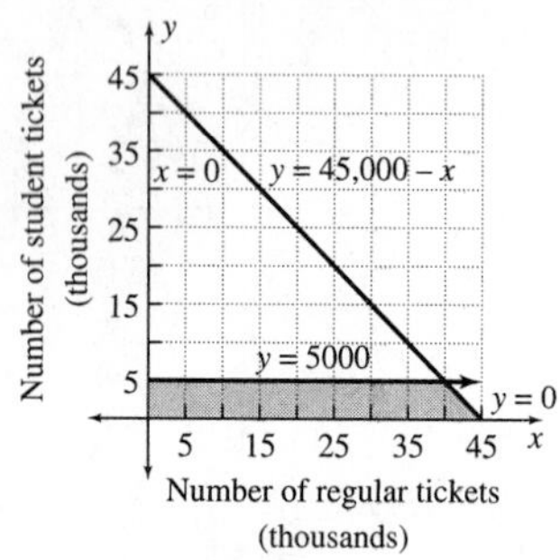

CHAPTER 7 REVIEW EXERCISES

1. $y = 2x + 5000$ **3.** $r = \dfrac{C}{2\pi}$ **5.** $b = \dfrac{8a}{5}$

7.

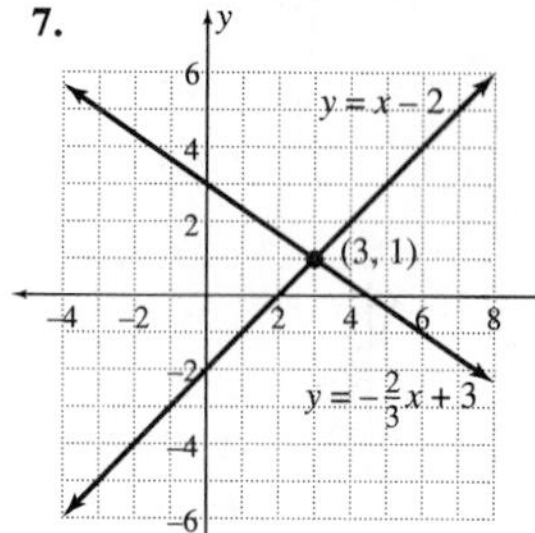

9.

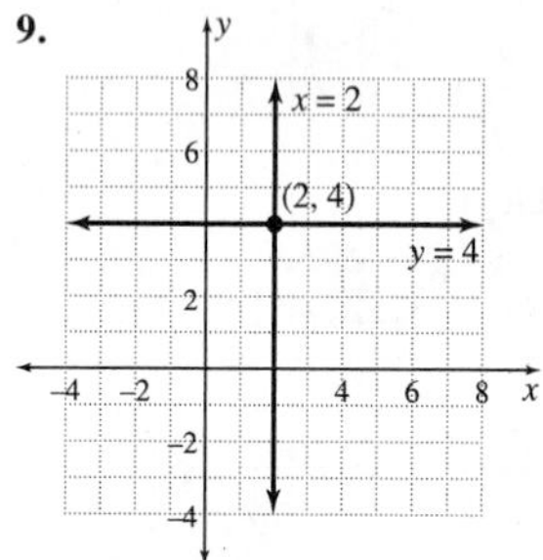

11. $x = -\frac{1}{2}, y = \frac{3}{2}$ **13.** $x = 4500, y = 500$

15. $m = -\frac{1}{5}, b = \frac{13}{5}$ **17.** $x = 1, y = -\frac{5}{3}$

19. $x = -3, y = -1$ **21.** $x = -3, y = 18$ **23.** no solution

25. The lines are coincident. **27.** centipede, 354; millipede, 710

29. muffin, 140 cal; egg, 95 cal **31.** 4 cal, 9 cal

33. 12 stools, 7 tables **35.** 15 mph, 5 mph

37. 11.5 in., 11.5 in., 9 in. **39.** 16 g, 10 g, 11 g

41. 39 g, 71 g, 27 g

45.

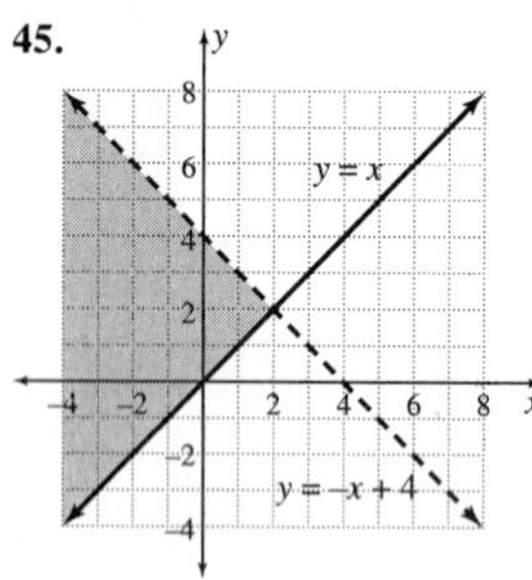

47.

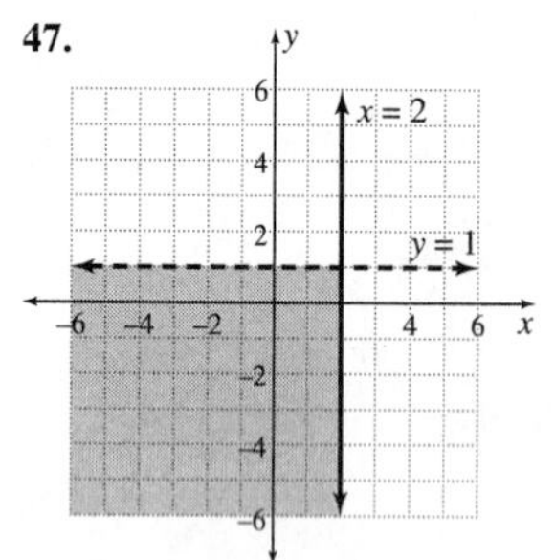

49.

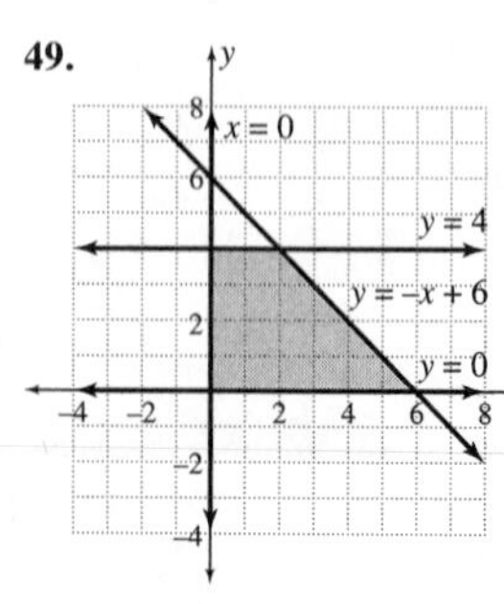

51.

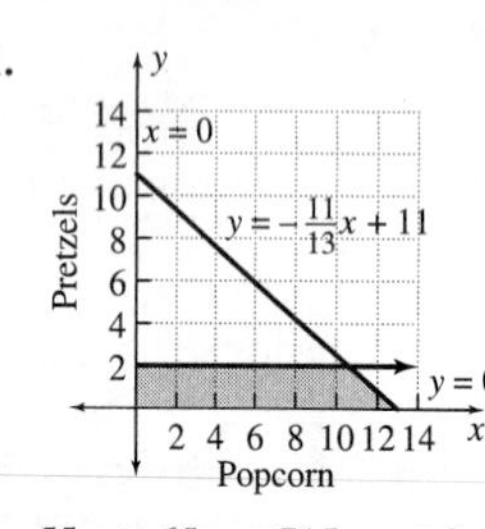

$55x + 65y \leq 715$, $y \leq 2$, $x \geq 0$, $y \geq 0$

CHAPTER 7 TEST

1. $y = 3x - 400$ **2.** $y = \frac{1}{2}x - \frac{3}{2}$ **3.** $a = -4, b = 5$

4. $m = 0.75, b = -1.75$ **5.** infinite number of solutions

6. no solution **7.** $x = 13, y = 9$ **8.** $x = 1500, y = 600$

9. The graphs are coincident lines. **10.** The lines are parallel.

11. $y = 4 - \frac{1}{3}x$ and $y = 8 - x$; the ordered pair makes both equations true.

12.

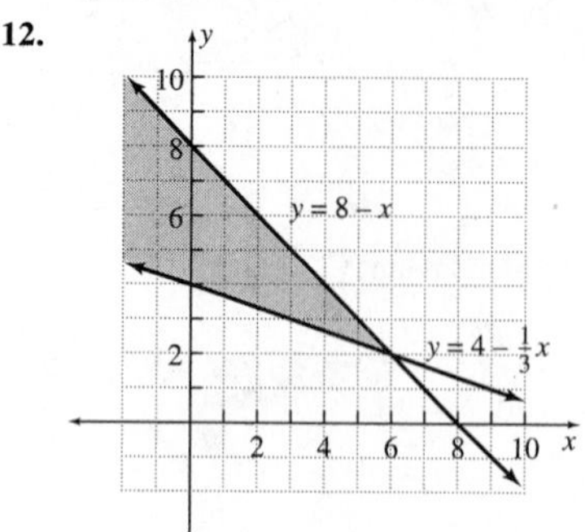

13. (7, 1) is a solution; the ordered pair makes both inequalities true.

14. 16, 6 **15.** 4.5 cal, 12.6 cal

16. dolphins, 35 mph; current, 10 mph

17.

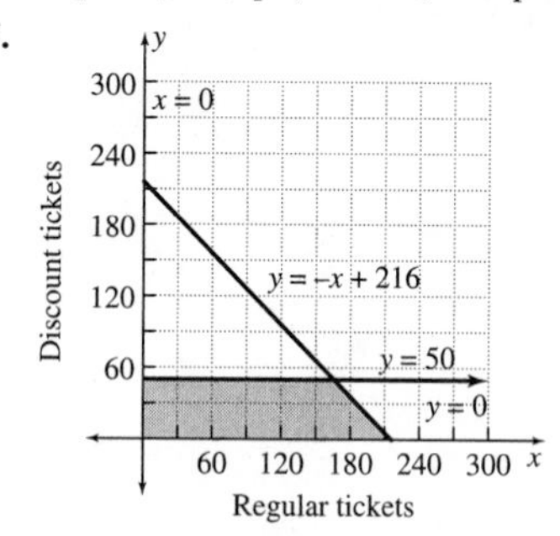

$x + y \leq 216$, $y \leq 50$, $x \geq 0$, $y \geq 0$

EXERCISES 8.1

1. $x \approx 5.83$ **3.** $x \approx 10.82$ **5.** $x \approx 6.24$ **7.** no

9. yes **11.** yes **13.** $x = 21$; $w = 35$

15.

Leg	Leg	Hypotenuse
3	4	5
6	8	10
18	24	30
9	12	15
1	$\frac{4}{3}$	$\frac{5}{3}$

17. **a.** $9x^2$ **b.** $16x^2$ **c.** $25x^2$ **d.** $36x^2$

19. $x \approx 5.7$ **21.** $x \approx 10.2$ **23.** $x \approx 6.7$ **25.** $\approx$12.4 ft

27. $\approx$9.3 ft **29.** $\approx$3.5 ft; $\approx$14.4 ft

31. $\approx$2.91 ft $\approx$ 2 ft 11 in.; $\approx$11.64 ft $\approx$ 11 ft. 8 in.

33. $\approx$4.37 ft $\approx$ 4 ft 4 in.; $\approx$17.46 ft $\approx$ 17 ft 6 in.

35. $\approx$251.8 mi **37.** $\approx$607.6 mi **39.** $\approx$19.72 ft

41. ≈ 1146 ft^2 **43.** ≈ 1332 ft^2

EXERCISES 8.2

1. **a.** 9 **b.** 1.4 **c.** 0.2 **d.** 60

3. **a.** 8, 9 **b.** 7, 8 **c.** 14, 15 **d.** 4, 5

5.

3.873	irrational
5	rational
5.916	irrational
3.5	rational
1.5	rational
2.449	irrational
4	rational
5.099	irrational

7. a. no real-number solution **b.** −9 **c.** ±12

9. a. 7 **b.** −15 **c.** ±20

11.

x	$y = \sqrt{x+4}$
−4	0
−2	1.414
0	2
2	2.449
4	2.828

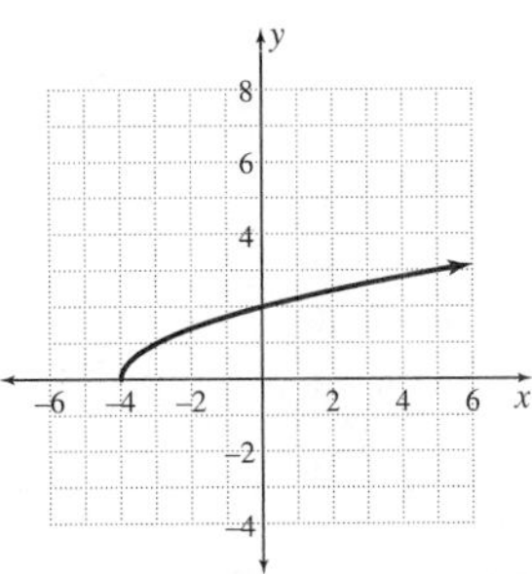

13.

x	$y = \sqrt{x-2}$
−4	not a real number
−2	not a real number
0	not a real number
2	0
4	1.414

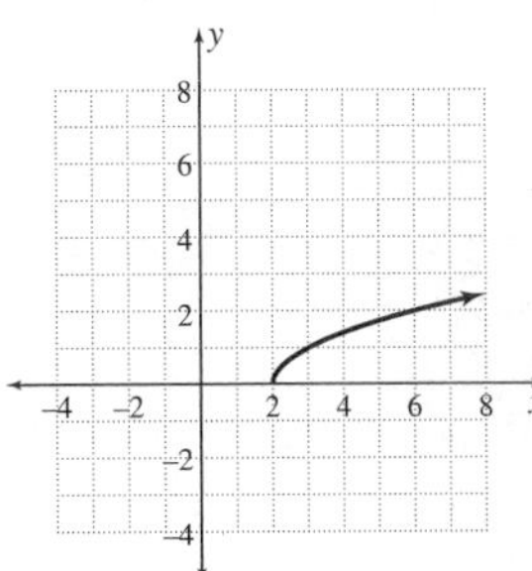

15.

x	$y = \sqrt{2x}$
−4	not a real number
−2	not a real number
0	0
2	2
4	2.828

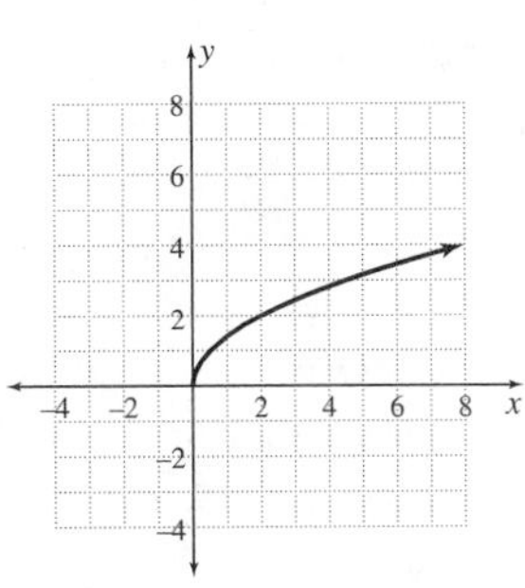

17. a. $\sqrt{15}$ **b.** 45 **c.** $4\sqrt{3}$ **19. a.** 63 **b.** 13 **c.** $6\sqrt{2}$

21. a. 9 **b.** 20 **c.** $3\sqrt{2}$ **23. a.** a **b.** b **c.** $11a$

25. a. $8a$ **b.** $2x$ **c.** $4b$ **27. a.** $\frac{x}{3}$ **b.** $\frac{2}{5}$ **c.** 3

29. a. $\frac{2}{x}$ **b.** $\frac{x}{3}$ **c.** $\frac{a^2}{2}$ **31. a.** $2x$ **b.** $\frac{x^2}{5y}$ **c.** $-\frac{y^2}{3}$

33. a. yes **b.** yes **c.** no

35. a. yes **b.** yes; not a triangle **c.** yes

37. a. $2\sqrt{10}$ **b.** 3 **c.** $y = 3x - 3$

39. a. $3\sqrt{2}$ **b.** −1 **c.** $y = -x + 4$

41. a. $5\sqrt{2}$ **b.** $-\frac{1}{7}$ **c.** $y = -\frac{1}{7}x + 2\frac{4}{7}$

43. a. $2\sqrt{10}$ **b.** $-\frac{1}{3}$ **c.** $y = -\frac{1}{3}x - 2$

45. isosceles right **47.** isosceles **49.** right

51. a. 1 **b.** $\frac{1}{25}$ **c.** 5 **d.** 5

53. a. 3 **b.** 1 **c.** 3 **d.** $\frac{1}{9}$

55. a. 4 **b.** 1 **c.** $\frac{1}{2}$ **d.** $\frac{1}{2}$

57. a. 4 **b.** 0.1 **c.** 2.5 **d.** 0.5 **e.** 50 **f.** 20

59. yes; $y = 1$ **61.** yes; if $x \le 0$ **63.** False if $0 < x < 1$

65. $\sqrt{\frac{a}{b}} = \left(\frac{a}{b}\right)^{1/2}$ Definition of $\frac{1}{2}$ as exponent

$\left(\frac{a}{b}\right)^{1/2} = \frac{a^{1/2}}{b^{1/2}}$ Quotient property of exponents

$\frac{a^{1/2}}{b^{1/2}} = \frac{\sqrt{a}}{\sqrt{b}}$ Definition of $\frac{1}{2}$ as exponent

Thus, $\sqrt{\frac{a}{b}} = \frac{\sqrt{a}}{\sqrt{b}}$

EXERCISES 8.3

1. a. −3 **b.** ab **c.** $4x^2$ **d.** 2 **e.** x **f.** $x + 2$

3. $f(-4) = 2\sqrt{2}, f(-1) = \sqrt{5}, f(0) = 2, f(4) = 0, f(6)$ is not a real number

5. $f(-4) = \sqrt{7}, f(-1) = 2, f(0) = \sqrt{3}, f(4)$ is not a real number, $f(6)$ is not a real number

7. a. $|b|\sqrt{a}$ **b.** $|a|\sqrt{b}$ **c.** $|ab|$ **9. a.** $7|x|$ **b.** $11|y|$

11. a. $|p|\sqrt{p}$ **b.** p^2 **13. a.** $b^2\sqrt{b}$ **b.** $|c|\sqrt{c}$

15. x^2 is always positive. **17. a.** $x \ge 1$ **b.** $x \ge -3$

19. a. $x \le 4$ **b.** $x \le 3$ **21.** $x = 6$ **23.** $x \ge -3; x = -2$

25. $x \le 3; x = -1$ **27.** $x \ge 2; x = 11$ **29.** $x \ge -1; x = 7$

31. $x \ge 1, x = 13$ **33.** $x \le 5; x = -4$ **35.** ≈46.7 mi

37. ≈38.8 mi **39.** 3 mi **41.** 6 mi

43. a. 600 ft **b.** 2400 ft **c.** 4 **45.** ≈10.6 sec, ≈341 ft/sec

47. ≈8.1 sec, ≈261 ft/sec **49.** ≈2318.4 ft **51.** ≈18.1 sec

53. $R = \frac{E^2}{W}$ **55.** $g = \frac{2d}{t^2}$ **57.** $M = \frac{RV_0^2}{G}$ **59.** $\frac{h_e}{h_m} = \frac{1}{4}$

MID-CHAPTER 8 TEST

1. a. no **b.** no **c.** yes **2.** 130 mi **3.** $\sqrt{261}$ in.

4. a. 6 **b.** 18 **c.** $6\sqrt{2}$ **d.** 12 **e.** $4\sqrt{3}$ **f.** −2

g. not a real number **h.** $\pm\frac{5}{4}$ **i.** $\frac{1}{4}$

5. a. $9x^2$ **b.** $\frac{|x|}{2y^2}$ **c.** $\frac{14}{|x|}$ **d.** already simplified

6. $3\sqrt{13} \approx 10.8$ **7.** 11, 12 **8.** $R = \frac{W}{I^2}$ **9.** $x = 7, x \ge 3$

10. There are no values for y when the radicand is negative.

EXERCISES 8.4

1. (−3, 0), (2, 0); (0, −6); axis of symmetry $x = -\frac{1}{2}$, vertex (−0.5, −6.25)

3. (0, 0), (5, 0); (0, 0); axis of symmetry $x = 2.5$; vertex (2.5, 6.25)

5.

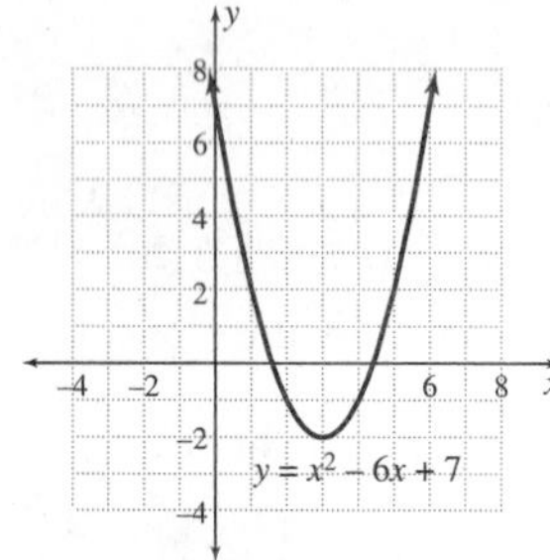

x	$y = x^2 - 6x + 7$
-4	47
-2	23
0	7
2	-1
4	-1

(4.4, 0), (1.6, 0); (0, 7); $x = 3$; (3, -2)

7.

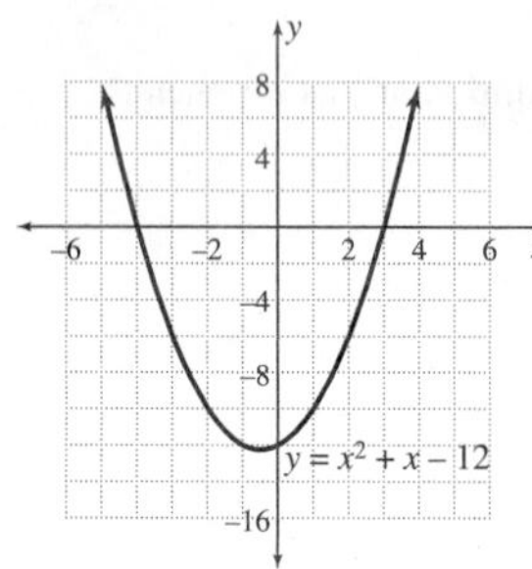

x	$y = x^2 + x - 12$
-3	-6
-1	-12
0	-12
1	-10
3	0

(-4, 0), (3, 0); (0, -12); $x = -0.5$; (-0.5, -12.25)

9.

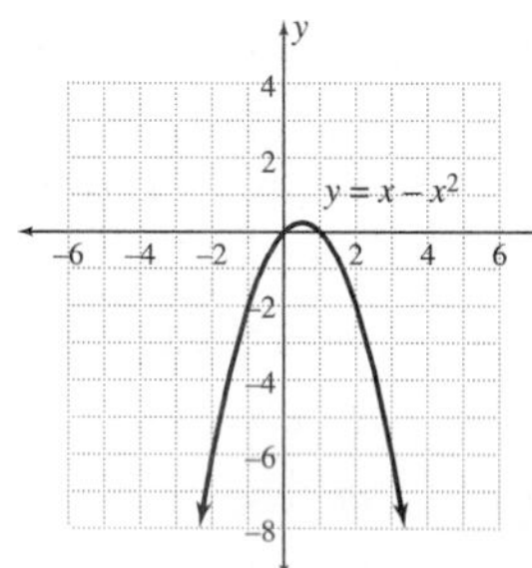

x	$y = x - x^2$
-3	-12
-1	-2
0	0
1	0
3	-6

(0, 0), (1, 0); (0, 0); $x = 0.5$; (0.5, 0.25)

11.

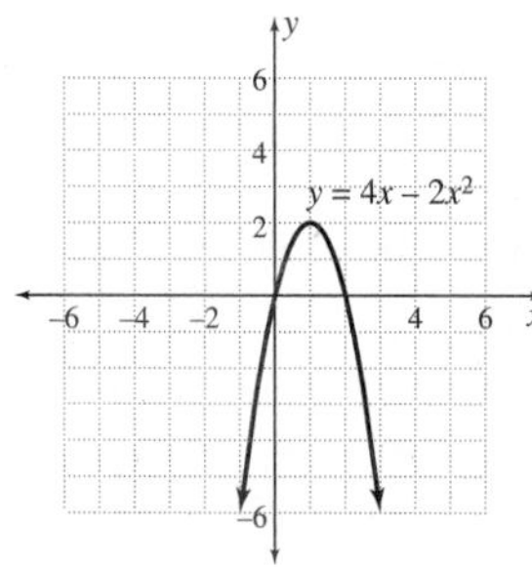

x	$y = 4x - 2x^2$
-2	-16
1	-6
0	0
1	2
2	0

(0, 0), (2, 0); (0, 0); $x = 1$; (1, 2)

13. $y \geq -6.25$ **15.** $y \leq 6.25$ **17.** $a = 2, b = 3, c = 1$

19. $a = 1, b = -4, c = 4$ **21.** $a = 1, b = 0, c = -4$

23. $a = 4, b = -8, c = 0$ **25.** $a = 1, b = -1, c = 4$

27. $a = 1, b = -1, c = 1$ **29.** $a = -0.5g, b = v, c = s$

31. $a = \pi, b = 0, c = 0$ **33.** $y = 4x^2 + 4x + 1$

35. $y = 9x^2 - 16$ **37.** $y = 3x^2 + 6x$

39. a. $x = -4, x = 3$ **b.** $x = -2, x = 1$

c. no real-number solution **d.** $x = -3, x = 2$

41. a. $x = 2, x = 3$ **b.** no real-number solution

c. $x = 1, x = 4$ **d.** $x = 0, x = 5$

43. $\approx$7.1 sec; $\approx$10 sec

EXERCISES 8.5

1. $x = \pm\sqrt{5}$ **3.** $x = \pm\sqrt{7}$ **5.** $x = \pm\frac{2}{5}$ **7.** $x = \pm\frac{1}{5}$

9. $x = \pm\frac{15}{7}$ **11.** $x = \pm\frac{11}{6}$ **13.** $x = \pm\frac{3}{5}$ **15.** $x = \pm 9$

17. $x = \pm 6$ **19.** $x = 4$ or $x = -4$ **21.** $x = \frac{1}{2}$ or $x = -\frac{2}{3}$

23. $x = \frac{5}{2}$ or $x = -2$ **25.** $x = 2$ or $x = -2$

27. $x = 2$ or $x = -3$ **29.** $x = 5$ or $x = -3$

31. $x = 0$ or $x = -3$ **33.** $x = 0$ or $x = -\frac{1}{2}$

35. $x = -2$ or $x = 6$ **37.** $x = \frac{3}{2}$ or $x = -1$

39. $x = 4$ or $x = -3$ **41.** $x = 1$ or $x = 6$

43. $x = -\frac{5}{2}$ or $x = 1$ **45.** $x = \pm\frac{5}{2}$ **47.** $x = \pm 2$

49. $x = -4$ or $x = 3$ **51.** $x = 4$ or $x = 1$

53. $t = 0$ sec or $t = 3$ sec **55.** $t = 1$ sec or $t = 3$ sec; yes

57. $t = 0$ sec or $t = 4$ sec; yes **59.** $x = 3$ **61.** $x = -1$

63. a. $A(0, 0)$, $B(4, 16)$ **b.** $4x = x^2$ **c.** $x = 0$ or $x = 4$; yes

65. a.

$a^2 + b^2$	$(a + b)^2$	$\sqrt{a + b}$	$\sqrt{a} + \sqrt{b}$
97	169	$\sqrt{13} \approx 3.6$	5
10	16	2	$1 + \sqrt{3} \approx 2.73$
41	81	3	$2 + \sqrt{5} \approx 4.24$
45	81	3	$\sqrt{3} + \sqrt{6} \approx 4.18$

EXERCISES 8.6

1. $a = 9, b = 6, c = 1$ **3.** $a = 3, b = -9, c = 0$

5. $a = 1, b = -4, c = 3$ **7.** $a = 1, b = 0, c = 9$

9. -2 **11.** $\frac{1}{2}$ **13.** $-\frac{4}{3}$ **15.** $\frac{1}{8}$

17. a. $\frac{1}{2}$, 0.293 **b.** $\frac{5}{3}$, 1.816 **c.** 2, 2.449

19. a. $\frac{1}{3}$, 0.255 **b.** 3, 3.162 **c.** $\frac{5}{2}$, 2.581

21. a. $-\frac{9}{2}$, -4.303 **b.** $-\frac{1}{4}$, -0.363

23. a. $-\frac{7}{6}$, -1.215 **b.** $\frac{7}{10}$, 0.717 **25.** $x = -1, x = \frac{1}{4}$

27. $x = -\frac{2}{7}, x = 1$ **29.** $x = -\frac{1}{2}, x = \frac{2}{5}$

31. $t \approx 4.14$ or $t \approx -2.26$ (negative time is not acceptable)

33. $t \approx 3.57$ or $t \approx -2.63$ (negative time is not acceptable)

35. no real-number solution

37. $t \approx 3.08$ or $t \approx -1.83$ (negative time is not acceptable)

39. no real-number solution **41.** $x = -1, x = \frac{1}{2}$

43. $x = -1, x = \frac{2}{3}$ **45.** no real-number solution

47. $x = \pm\dfrac{\sqrt{10}}{2} \approx \pm 1.581$ **49.** $x = \pm\frac{3}{2}$

51. $x \approx -1.786, x \approx 1.120$ **53.** no real-number solution

55. $x \approx -1.117, x \approx 0.717$ **57.** no real-number solution

59. $x = 3$ **61.** $x \approx -0.914, x \approx 1.914$ **63.** $x = 1.5$

65. Number of solutions is number of x-intercepts.

69. $y = x(x + 1)$; $x = 0$ or $x = -1$; interval is $(-1, 0)$

EXERCISES 8.7

1. \$26,000; \$12,500 **3.** \$26,000; \$27,500

5. \$10,000; \$13,000

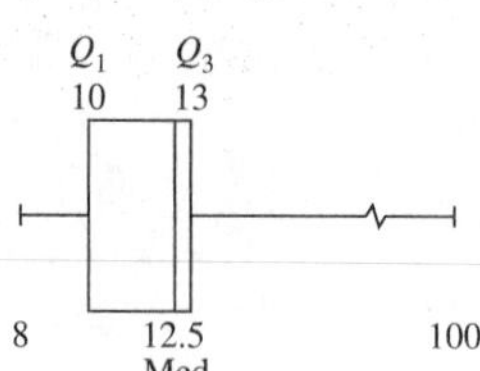

Numbers in thousands

7. \$23,750; \$27,500

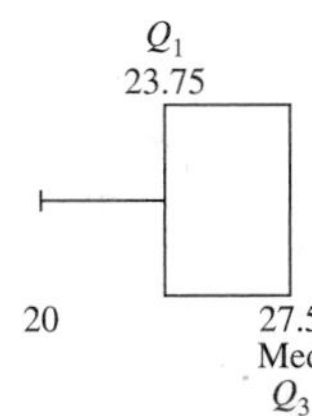

Numbers in thousands

9. ≈\$36,300 **11.** \$26,000; \$3000; \$3354

13. A few high-priced sales would raise the average. **15.** no

17. varying experience with measurement **21.** 3470°F to 3530°F

23. 6.049 cm to 6.101 cm **25.** 1 min to 4 min

27. 0.21 ppm to 0.39 ppm **29.** 48.05 gal/min to 48.75 gal/min

CHAPTER 8 REVIEW EXERCISES

1. a. no **b.** no **c.** yes **3.** ≈24.7 ft

5. a. $2\sqrt{15}$ **b.** $3\sqrt{7}$ **c.** $3\sqrt{6}$

7. a. 12 **b.** $\frac{1}{144}$ **c.** 1 **d.** 12

9. a. 0.6 **b.** $\frac{25}{9}$ **c.** 0.6 **d.** 1

11. a. sides all $\sqrt{13}$; diagonals $\sqrt{26}$; square

b. sides all $2\sqrt{5}$; diagonals $2\sqrt{10}$; square

c. sides $\sqrt{10}$ and $\sqrt{17}$; diagonals $3\sqrt{3}$ and 5; not a rectangle

13. a. $5xy^2$ **b.** $13x^3y$ **c.** $1.5a\sqrt{a}$ **d.** $0.8b^2\sqrt{b}$ **e.** $4x$

f. $\dfrac{a^2}{3b^3}$ **g.** $8a^3$ **h.** $\dfrac{11}{7b^2}$

15. ≈1.41, 2, ≈2.83, 4, ≈5.66, 8, ≈11.3, 16

17. a. $x = 4, x \geq \frac{3}{7}$ **b.** $x = 2, x \leq 2$ **c.** $x = 13, x \geq \frac{3}{4}$

19. a. ≈2.7 min **b.** $42\frac{2}{3}$ ft **21.** $x = 5$ **23.** $x = -2, x = 4$

25. $x = \pm\frac{1}{3}$ **27.** $x = 1, x = 4$ **29.** $x \approx 0.886, x \approx -3.386$

31. $x = -6, x = 3$ **33.** $x = 2\frac{1}{3}, x = -1$

35. no real-number solution **37.** $x = 3$

39. a. $r = \sqrt{\dfrac{A}{\pi}}$ **b.** $v = \sqrt{\dfrac{2p}{d}}$ **c.** $r = \dfrac{1}{2}\sqrt{\dfrac{S}{\pi}}$

d. $v = \sqrt{2gh}$

41. a. 3

b.

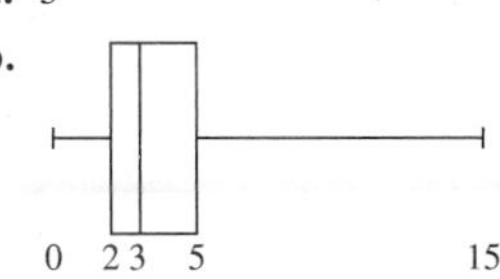

c. ≈3.38; ≈2.92

CHAPTER 8 TEST

1. a. yes **b.** no **c.** yes **d.** yes **e.** no

2. a. $3\sqrt{5}$ **b.** $2\sqrt{11}$ **3.** ≈20.6 ft

4. a. 10 **b.** 9 **c.** 54 **d.** 28 **e.** $6x\sqrt{y}$ **f.** $0.9x^2y\sqrt{y}$

g. 7 **h.** $\frac{3}{4}$ **i.** a^2b

5. a. $a = 7$ **b.** $b = 21, c \approx 10.8$ or $3\sqrt{13}$

c. $d \approx 16.6, e \approx 19.41$

6. $x = -2, x = 2$

7. a. $x = \pm\frac{6}{11}$ **b.** $x = -1, x \leq 3$ **c.** $x = -2, x = 1$

d. $x = 30, x \geq \frac{6}{5}$ **e.** $x = \frac{1}{2}, x = -8$

f. $x \approx 0.46, x \approx -1.09$ **g.** $x = \pm\frac{8}{3}$

h. no real-number solution **i.** $x = 3$ **j.** $x = \pm\frac{5}{2}$

8. a. $R = \dfrac{2GM}{V_e^2}$ **b.** $v = \sqrt{\dfrac{2E}{m}}$ **c.** $H = 2\sqrt{\dfrac{2\pi E}{k}}$

9. a. 51.6 mph **b.** 86 mph **c.** ≈194.7 psi

10. a. $x = 5\sqrt{2} \approx 7.071$ **b.** $x = 10\sqrt{\dfrac{2}{\pi}} \approx 7.979$

c. $x \approx 10.746$

11. a.

(3, 11), (6, 8), (3, 5)

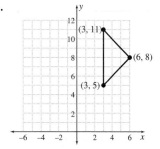

b. 6, 4.24, 4.24 **c.** yes; slopes are 1, −1, and undefined.

d. isosceles right triangle; 2 equal sides and 1 right angle

12. a. 110, ≈5.5 **b.** 145

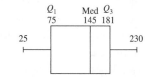

Q_1 75, Med 145, Q_3 181, 25, 230

CUMULATIVE REVIEW OF CHAPTERS 1 TO 8

1. 9 **3.** \$20 **5.** $\dfrac{x^2}{9y}$ **7.** $x = -8$ **9.** $x = -2$

11. $x \leq 1$ **13.** $f(-1) = 5; f(0) = 0; f(1) = 1; f(2) = -2$

15. $y = \frac{4}{3}x - 2$

17.

Trips	Value (\$)
0	20.00
2	15.50
4	11.00
6	6.50
8	2.00
10	−2.50

$y = 20 - 2.25x$

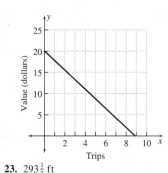

19. 65 **21.** 37.5% **23.** $293\frac{1}{3}$ ft

25. a. $2x^2 + 5x - 3$ **b.** $3x^3 + 2x^2 - x$

27. a. $\approx 1.075 \times 10^2$ **b.** $\approx 1.103 \times 10^2$

29. $x = -3, y = \frac{1}{2}$ **31.** 9, 12

33. a.

b. $x = -\frac{1}{2}, x = 3$

35. a. 4.7, 5, 5

b. median = 24; Q_1 = 19; Q_3 = 28

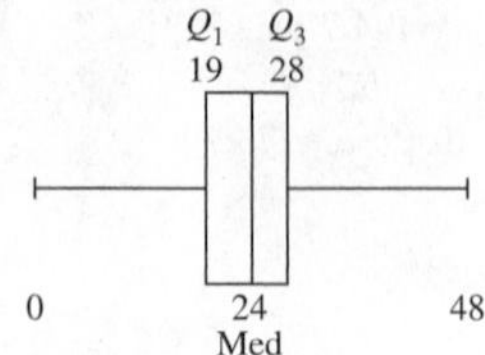

c. 111, 298; ≈33

EXERCISES 9.1

1. $\mathbb{R}, x \neq -1$ **3.** $\mathbb{R}, x \neq \frac{1}{2}$ **5.** $\mathbb{R}, x \neq -\frac{1}{3}$

7. $\mathbb{R}, x \neq -1, -3$

9. y approaches $-\infty$ from the left and ∞ from the right.

11. y approaches ∞ from the left and $-\infty$ from the right.

13. a. 30 mph **b.** 60 mph **c.** 72 mph **d.** 80 mph

e. 90 mph **f.** 120 mph

15. $y = \dfrac{1000 \cdot 10^9}{x}$ days; $y = \dfrac{1000 \cdot 10^9}{365x}$ years **17.** ≈46 yr

19. a. number of terms the aid will last

b. Available aid will last $7\frac{1}{2}$ terms.

21. You are not moving. **23.** no

25. a.

Width	Length	Area	Perimeter
1	30	30	62
2	15	30	34
3	10	30	26
4	7.5	30	23
5	6	30	22
6	5	30	22
7.5	4	30	23
10	3	30	26
15	2	30	34
30	1	30	62

b.

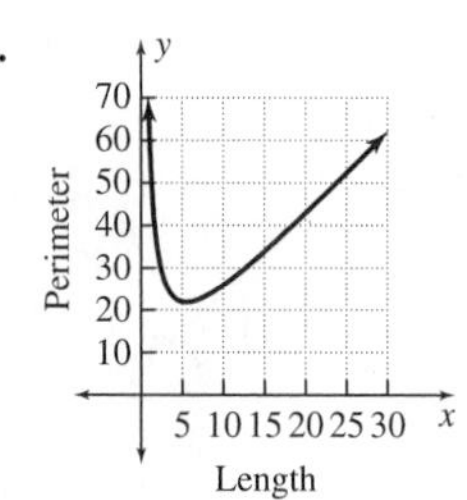

c. ≈(5.5, 21.9) **d.** $y = 2x + 2\left(\dfrac{30}{x}\right)$

27. $y = x, y = -x, x = 0$

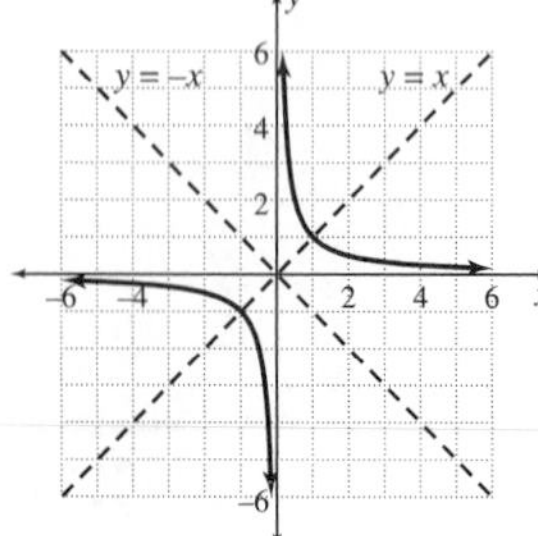

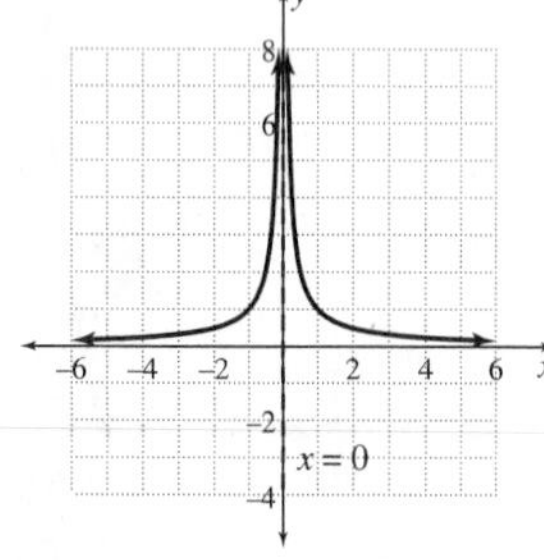

29. yes; $y = \dfrac{1}{x}$ is equivalent to $x = \dfrac{1}{y}$. **31.** c

EXERCISES 9.2

1. $\dfrac{1}{3a}$ **3.** $\dfrac{3d}{2c}$ **5.** $x - 2$ **7.** $\dfrac{-1}{x + 2}$ **9.** $\dfrac{3a}{5b}$

11. $\dfrac{2(x + 2)}{x - 2}$ **13.** $\dfrac{x - 2}{x + 3}$ **15.** $\dfrac{x + 3}{x}$ **17.** $-\frac{1}{2}$

19. $\dfrac{x - 2}{x + 3}$ **21.** 30 **23.** x **25.** $5ab$ **27.** $x - 3$

29. $a - b$ **31.** 18 m **33.** $\dfrac{1}{12 \text{ in.}}$ **35.** 20 gal

37. −1; opposite numerator and denominator simplify to −1.

39. −1 **41.** $-a + b$ **43.** $a - b$ **45.** $-a - b$

47. not opposite **49.** opposites **51.** $x + 2; x \neq -2$

53. $-x + 2; x \neq 2$ **55.** $b - a; a \neq b$ **57.** $3 - x; x \neq 3$

59. $\frac{1}{4} = 4^{-1}$; 4 is the opposite of −4

61. $\dfrac{1}{a} = a^{-1}$; a is the opposite of $-a$

63. True; only factors may be simplified to lowest terms.

65. $\dfrac{a}{-a}, \dfrac{-a}{a}, -\dfrac{a}{a}$

67. Fractions simplify only if the numerator and denominator contain common factors.

EXERCISES 9.3

1. a. $\frac{1}{12}, \frac{4}{3}$ **b.** $\frac{1}{10}, \frac{5}{2}$ **3. a.** $\frac{1}{8}, \frac{9}{2}$ **b.** $\frac{1}{9}, 4$

5. a. 25 **b.** 500

The expressions within each part are equal; division by a number is the same as multiplication by its reciprocal.

7. a. x **b.** $\dfrac{1}{a^3b^2}$ **c.** $\dfrac{b}{a}$ **d.** $\dfrac{b^2}{a}$

9. a. 1 **b.** $\dfrac{x - 2}{x(x - 1)}$ **c.** $\dfrac{1}{x(x - 2)}$ **d.** $\dfrac{x}{(x + 5)(x - 5)}$

e. $\dfrac{x - 3}{(x + 3)^2}$

11. a. $x(x + 1)$ **b.** $\dfrac{1}{(x + 3)^2}$ **c.** $4(x - 1)$ **d.** $\dfrac{x(x - 1)}{x + 1}$

e. $-\dfrac{x(x - 1)}{(x - 2)(x + 1)}$

13. a. $\dfrac{1}{ab}$ **b.** a^2 **c.** $\dfrac{a}{b}$ **d.** $\dfrac{a}{b}$ **e.** $\dfrac{a}{b^2}$ **f.** $\dfrac{a^2}{b}$

15. hour **17.** 500 sec **19.** 3 mpg **21.** ≈4 cookies/dollar

23. ≈0.28 pages/min **25.** $3\frac{1}{3}$ L **29.** $I = \dfrac{t^2}{2t + 1}$

MID-CHAPTER 9 TEST

1. $x = -2, x = 1$

2.

Input	Output	Budget
0	undefined	4800
10	480	4800
20	240	4800
100	48	4800
1000	4.80	4800
1200	4.00	4800

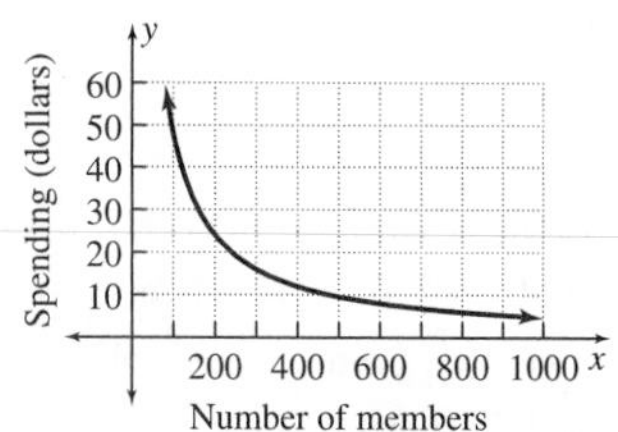

3. $y = \frac{4800}{x}, x \neq 0$ 4. $+\infty$ 5. $+\infty$ 6. $-\infty$

7. a. $\frac{6c}{7a}$ b. no common factors 8. a. $\frac{1}{x-2}$ b. $\frac{x-1}{x-2}$

9. a. $6x^2$ b. $12ab^2$ 10. a. 15 b. $2(x+2)$

11. $\frac{3}{xy}$ 12. $\frac{2}{x}$ 13. $\frac{(x-6)(x+1)(x-2)}{x-1}$ 14. $\frac{x}{x-4}$

15. $\frac{4}{x^3}$ 16. 9 weeks 17. 2 in. per sec 18. 80 mL

EXERCISES 9.4

1. a. 1 b. $\frac{2+x}{3}$ c. $-\frac{1}{x}$ d. $\frac{4-x^2}{x^2+1}$ e. -1

3. a. 60 b. $2x$ c. y^2 d. ab e. $(x+3)(x-3)$
f. $(x+3)(x+2)(x-3)$

5. a. $\frac{23}{30}$ b. $\frac{9}{2x}$ c. $\frac{8y-1}{y^2}$ d. $\frac{8a-b}{ab}$

e. $\frac{4x+10}{(x+3)(x-3)}$ f. $\frac{2x-16}{(x+3)(x+2)(x-3)}$

7. $\frac{6a+3b}{4ab}$ 9. $\frac{2x-1}{x(x-1)}$ 11. $\frac{5x-3}{x(x+1)}$

13. $\frac{2x+9}{(x+3)(x+3)}$ 15. $\frac{a-1}{a^2}$ 17. $\frac{4-3a}{2ab}$

19. $\frac{x^2+3x-6}{x(x-1)(x-2)}$ 21. $\frac{T_0-T}{T}$ 23. $\frac{C_2+C_1}{C_1C_2}$

25. $\frac{D(r_2+r_1)}{r_1r_2}$ 27. $\frac{L^2+3d^2}{6d}$ 29. 5 31. $\frac{1}{5}$ 33. $-\frac{a}{3b}$

35. $\frac{b}{a-c}$; slopes are equal 37. $\frac{25}{6}$ 39. $\frac{2A}{b}$ 41. $\frac{9x}{2(6-x)}$

43. $\frac{Q_L}{Q_H - Q_L}$

45. a. and b. Both graphs are the same:

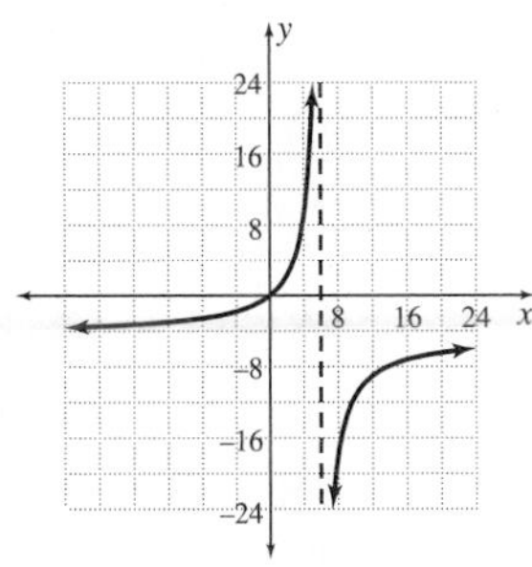

c. near $x = 6$; undefined when $x = 6$

47. a. · (mult.) b. ÷ c. + 49. a. + b. · (mult.)

EXERCISES 9.5

1. $x+8$ 3. $2x+12$ 5. $2x-1$ 7. 12 9. $20x$

11. $8x$ 13. $3x$ 15. x 17. $(x-1)(x+3)$ 19. x^2

21. $x-3$ 23. $x = 67.2$ 25. $x = \frac{20}{19}; x \neq 0$

27. $x = \frac{8}{3}; x \neq 0$ 29. $x = 2; x \neq 0$ 31. no solution; $x \neq 0$

33. $x = 5; x \neq 1, x \neq -3$ 35. $x = 1$ or $x = 2; x \neq 0$

37. no solution; $x \neq 3$ 39. $d = \frac{36}{5}$ 41. $x = \frac{40}{13}$ 43. $d = 7.2$

45. $x \approx 3.077$ 51. ≈6.5 days 53. ≈2.2 hr; yes

55. one that vents in $7\frac{1}{2}$ hr 57. 3.6 min 59. $b = \frac{ac}{a-c}$

61. $c = \frac{ab}{a+b}$ 65. $x \neq 0; x = \frac{1}{5}, x = -\frac{3}{2}$

67. $x \neq 0, x \neq -1; x = \frac{2}{3}, x = -2$ 69. $x \neq 1; x = 3, x = 4$

71. $x \neq 2, x = -2$

CHAPTER 9 REVIEW EXERCISES

1. $x = -3$ 3. $y = \frac{4000 \text{ trillion}}{365x}$ 5. a. $-\infty$ b. $-\infty$

7. $\frac{2y}{x}$ 9. $a - b$ 11. no common factors 13. $-1; a \neq 1$

15. $3 - x$ 17. $a - b$ 19. $x - 4$

21. $\frac{16}{9} \neq \frac{4}{3}$; no common factors

23. a. 100, 100 b. Both equal 100.

c. Division by a number is the same as multiplication by its reciprocal.

25. each, one-third; 12 27. half as far; $\frac{3}{8}$ mi 29. remains; $\frac{1}{6}$

31. $\frac{-x^2}{(x+1)^2}$ 33. $n+1$ 35. $\frac{3x}{(x+3)(x-3)}$

37. $\frac{x^2-2x-2}{(x+2)(x-2)}$ 39. $\frac{a^2-2ab-b^2}{(a+b)(a-b)}$

41. $\frac{720 - 360x^2 + 30x^4 - x^6}{720}$ 43. $\frac{3+36a-a^2}{3a}$

45. $\frac{1}{3}$ shirt per card 47. 3 tab 49. 5 mL 51. $\frac{2d}{g}$

53. $3x + 2$ 55. $x = 3; x \neq 0$ 57. $x = -6$ or $x = 6; x \neq 0$

59. $\frac{f_2 + f_1}{f_1 f_2}$ 61. 1.2 min

CHAPTER 9 TEST

1. $x = 4$

2. Because $4 - x$ is the opposite of $x - 4$, the expression simplifies to -1.

3.

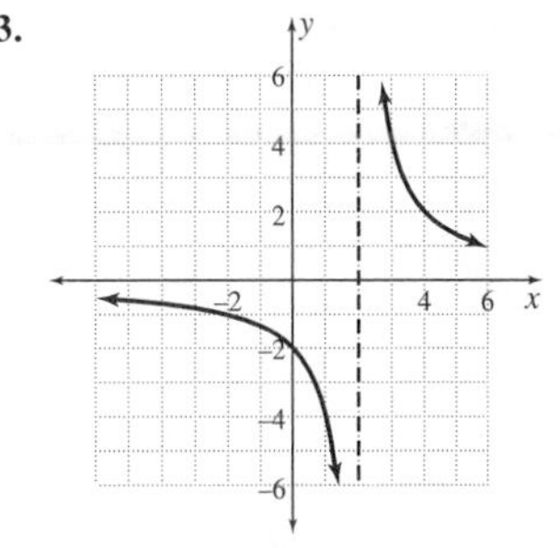

4. a. $\frac{2}{3}$; equivalent b. $\frac{1}{2}$ c. $\frac{3}{4}$ d. $\frac{2}{3}$; equivalent
Only a and d satisfy the simplification property of fractions.

5. $\frac{4}{25} \neq \frac{2}{5}$; no common factors 6. $\frac{b}{3a}$ 7. no common factors

8. $a + b$ 9. $\frac{x}{x+1}$ 10. $\frac{(1-x)(x-1)}{x^2}$ or $\frac{-(x-1)^2}{x^2}$

11. $\frac{1}{(1-x)(x-1)}$ or $\frac{-1}{(x-1)^2}$ 12. $\frac{n+1}{n-1}$ 13. $\frac{x-2}{x+2}$

14. $3x + 1$ 15. ≈7.3 min 16. \$35.92/shirt 17. \$0.10/mi

18. 6 tab **19.** true, by cross multiplication **20.** $\dfrac{q+p}{pq}$

21. $\dfrac{2A}{b}$ **22.** $\frac{31}{45}$ **23.** $\dfrac{4-a^2b}{2ab^2}$ **24.** $\dfrac{x^2-2x+2}{(x+2)(x-2)}$

25. $\dfrac{2x^2+9x-12}{3x^2}$ **26.** $\dfrac{2a+1}{4}$ **27.** $x=2;\ x\neq 0$

28. $x=1;\ x\neq 0$ **29.** $x=9;\ x\neq 0,\ x\neq 3$

30. $x=5$ or $x=-3;\ x\neq 0$

31. It permits us to simplify before multiplying.

FINAL EXAM REVIEW

1. a. -12 **b.** 5 **c.** 5 **d.** -60 **e.** 64 **f.** 4 **g.** 5
h. $\frac{2}{5}$ **i.** $-\frac{4}{5}$

3. a. $\dfrac{c}{f}$ **b.** $\dfrac{-r}{4x}$ **c.** a **d.** -1 **e.** $\dfrac{32m^6n}{3}$ **f.** $\dfrac{b^2}{c^2}$

g. n^7 **h.** m^3n^6 **i.** $\dfrac{1}{m^3}$ **j.** 1 **k.** $\dfrac{3y^2}{x^5}$ **l.** $\dfrac{27y^6}{x^9}$

5. a. $y=4-\frac{2}{3}x$ **b.** $x=\frac{7}{2}+\frac{3}{2}y$ **c.** $y=\dfrac{c}{b}-\dfrac{a}{b}x$

d. $x=\frac{3}{5}$ **e.** $P_2=\dfrac{P_1V_1T_2}{T_1V_2}$ **f.** $x=19$ **g.** $x\approx\pm 6.9$

h. $d\approx\pm 7.6$ **i.** $x=13$

7. a. $3xy(4+y+2xy)$ **b.** $x(x+1)$ **c.** $(x+5)(x-4)$
d. $(2x-3)(2x+3)$ **e.** $(2x-1)(3x+2)$
f. $(x-2)(6x+1)$ **g.** $(x-5)^2$ **h.** $3(x-1)(2x+3)$

9.

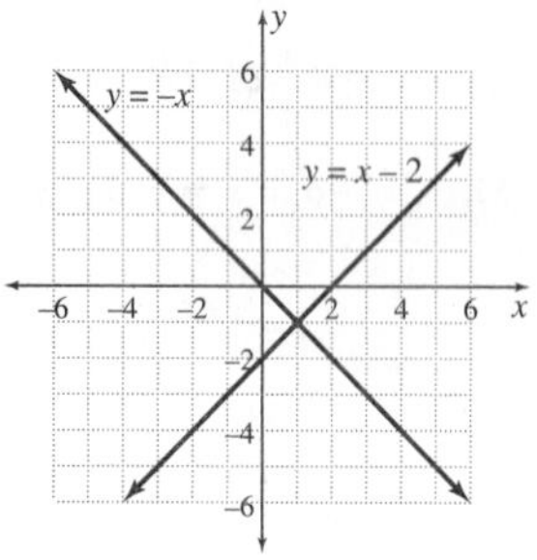

(1, −1)

11. $x=-2,\ y=4$ **13.** $x=-4,\ y=-2$

15. infinite number of solutions

17. slope is undefined; distance = 8

19. slope $=-\frac{1}{7}$; distance $=5\sqrt{2}$ **21.** no solution; $x\neq 0$

23. $x=-55$ **25.** $x=2$ or $x=-7;\ x\neq -5$

27. infinite number of solutions **29.** $x=12;\ x\geq -13$

31. $x=-13;\ x\geq -13$

33. a. (4, 0) **b.** (0, 2) **c.** 3 **d.** undefined
e. because $\sqrt{4-x}$ is always positive

35. Person went off the diet after about 13 days and gained weight.

37. a. 17 **b.** 25.2 **c.** 9 **d.** 20 **e.** $8|x|$ **f.** $8x$
g. undefined **h.** undefined in the real numbers

39. $\dfrac{2x-1}{x+2}$ **41. a.** \$2/room cleaned **b.** 21.1 ft

43. a. $-\dfrac{x}{y^2}$ **b.** $-\dfrac{(x+1)(x+2)}{x}$ **c.** $\dfrac{(2x-1)(x-4)}{(x-2)(x+2)}$

d. $\dfrac{x+1}{x+3}$ **e.** $-\frac{1}{15}x$ **f.** -20 **g.** $x^2-2x+12$

45. a. \$25, \$31, \$43, \$44, \$52.50
b. $y=x$; $y=30+0.20(x-30)$; $y=50+0.50(x-100)$
c. [ ] includes endpoints; ( ) excludes endpoints
d. yes; use parentheses
e. $0<x\leq 30$; $30<x\leq 100$; $x>100$

47. a. 2.76×10^9 mi **b.** ≈6.8 hr **c.** ≈3.55 mph
d. 17,600 mph

49. a.

x^3	$6x^2$
1000	600
8000	2400
64,000	9600
n^3	$6n^2$
$8n^3$	$24n^2$

b. 8 times larger **c.** 4 times larger **d.** ≈28.28 in.

Index of Projects

Following are the section numbers, exercise numbers, and titles of the projects throughout this text. Projects marked with an asterisk are particularly suited for small groups, inside or outside class.

Glossary/Index

Absolute value The distance a number is from zero. 53
addition using, 54
Adding like terms Counting like terms and writing a single term to describe the total count. 76
Addition
associative property of, 71
commutative property of, 72
of like terms, 77
using absolute value, 54
Addition property of equations The property that says that adding the same number to both sides of an equation produces an equivalent equation. 130, 443, 444
Addition property of inequalities, 176
Additive inverse The opposite of a number. 50
Algebra, 8
Algebraic notation, 20
solving equations with, 158–160
solving inequalities with, 174–178
Area The measure of the surface enclosed within a flat object; also, the amount of surface a figure covers. 96, 97, 98
operations involving, 346
Ascending order For a polynomial, arrangement so that exponents on one variable are listed from lowest to highest. 340
Associative property The property that says that numbers can be grouped in any way that is convenient for adding or multiplying. 71
Assumption Something not stated but taken as a fact. 3
Average value In a quantity-value table, the number derived by dividing the sum of the $Q \cdot V$ column by the sum of the quantity column. 321
Axes Number lines that divide a coordinate plane into four quadrants. 30
Axis of symmetry A line across which a graph can be folded so that points on one side of the graph match up with points on the other side of the graph. 510

Base The number repeated in multiplication; b in b^n. 81
division property of, 84, 85
multiplication property of, 84, 85
negative sign and, 82
power property of, 85
Base property The property that says that an exponent outside of parentheses applies to all parts of a product or quotient inside the parentheses. 83, 85
Bimodal Having two modes. 310
Binomial A polynomial expression containing two terms. 330
Boundary line The line between two half-planes. 251
of inequalities, 461–464
parallel to axes, 464–465
Box and whisker plot A visual summary of data. 540

Calculators, graphing
operations on, 99, 236, 255, 294, 421, 466, 499, 516, 558
techniques for, 133, 142, 143, 157, 172, 256, 277, 384, 391, 406, 407, 420, 465, 487, 488, 543
Capacity The amount (especially of liquids) that a container holds. 100
Cartesian coordinates, 31
Center of gravity, 313
Center of mass, 313
Centroid The center of a flat or solid geometric shape; the average of the coordinates of the vertices of a rectangle, square, or triangle. 313–314
Circle, formulas for, 97
Circumference The perimeter of a circle. 96
Coefficients
positive, 177
unlike, changing to like, 447
Coincident lines Two lines that have the same equation. 410
Common factor A factor that appears in each term. 344
Commutative property The property that lets us change the order in which actions or operations are done. 72
of addition, 72
of multiplication, 72
Complementary angles Two angles that form a right angle. 432
Complex fractions Rational expressions that contain fractions in either the numerator or the denominator. 574
simplifying, 575, 584–585
Composite number A number with factors other than 1 and itself. 79
Compound inequality Two inequalities in one statement. 109, 468–469
Condition A requirement or restriction stated within a problem setting. 3
Conditional equation An equation that is true for only certain values of the variable(s) in the equation. 126
Conditional inequality, 135
Conditional rule A rule that changes for inputs in different intervals. 14

Multiplication property of equations The property that says that multiplying both sides of an equation by the same nonzero number produces an equivalent equation. 130
Multiplication (by a negative number) property of inequalities The property that says that the direction of the inequality sign must be changed when both sides of an equality are multiplied by the same negative number. 175
Multiplication (by a positive number) property of inequalities The property that says that the direction of the inequality sign is not changed when both sides of an inequality are multiplied by the same positive number. 175
Multiplicative inverse The number that, when multiplied by n, gives 1; the reciprocal. 65

Natural numbers The numbers in the set $\{1, 2, 3, 4, \ldots\}$. 10
Nonlinear equation An equation whose graph does not form a straight line. 137
Notation, 9–11. *See also* Algebraic notation.
Numerator The top number in fraction notation. 10
Numerical coefficient The sign and number multiplying a variable or variables. 23, 444

Odd numbers The integers not divisible by two. 16, 161
One-half as exponent Alternative notation for the principal square root. 489
Opposites Numbers on opposite sides of zero and the same distance from zero on a number line; two expressions that add to zero. 10, 50, 568
Order of operations An agreed-upon order in which mathematical operations are performed. 86
Ordered pair The pair of numbers that identify a position on a coordinate plane. 30
  finding linear equations from, 235–239
  finding slope from, 215
Origin The point where the number lines on a coordinate plane cross. 30

Parabola The name given to the graph of a quadratic equation. 510
Parallel lines Lines with the same slope but different y-intercepts. 240–241
Parallelogram, formulas for, 97
Pearson, Karl, 542
Percent Ratio of a number to the total, expressed per hundred. 273
Percent change An expression of change found by subtracting the original number from the new number and dividing the difference by the original number. 274
Percent grade An expression of the slope of a highway. 273
Perfect square A whole number generated by placing a positive integer into x^2. 484
Perfect square trinomial An expression of the form $a^2 + 2ab + b^2$ that can be factored to $(a + b)^2$, the square of a binomial. 356–357
Perimeter The distance around the outside of a flat object. 96, 97
Perpendicular lines Two lines that cross at a right angle. 96, 241–242
Pi The number found by dividing the circumference of any circle by its diameter. 98
Point of intersection The point where two graphs coincide, which identifies a solution to the two equations graphed. 402
Polya, George, 2
Polynomial An expression, with only positive integer exponents, containing one or more terms being added or subtracted. 338–340
  adding and subtracting, 341–342
  arranging terms in, 340
  multiplying, 342–344
Population standard deviation An alternative form of the standard deviation used when the data represent the entire population being considered. 542–543
Power The base and exponent together; the name given to b^n. 81
Power property The property that says that to apply an exponent to an expression in exponential form, multiply the exponents. 84, 85
Prime number A number with no integer factors except 1 and itself. 79
Principal square root The positive number that, multiplied by itself, produces a given number. 484
Problem-solving
  four steps in, 2–5, 414–415
  strategy for, 128–129
Product The answer to a multiplication problem. 8
Proof A logical argument that demonstrates the truth of a statement. 475
Proportion An equation formed by two equal ratios. 283
  solving application problems as, 288–289
  solving equations as, 287–288
  solving percent problems as, 286–287
  and statistics, 289–290
Pythagoras, 475
Pythagorean theorem If a triangle is a right triangle, then the sum of the squares of the two shortest sides (legs) is equal to the square of the longest side (hypotenuse). 473, 474–478
Pythagorean triples Sets of three numbers that make the Pythagorean theorem true. 492–493, 517

Quadrants The four sections of a coordinate plane. 30
Quadratic equation
  solving by factoring, 523–525
  solving with tables and graphs, 515–516
  solving by taking square roots, 521–522
  special features of graph of, 513, 515
Quadratic formula For $ax^2 + bx + c = 0$,

$$x = \frac{-b + \sqrt{b^2 - 4ac}}{2a} \quad \text{or} \quad x = \frac{-b - \sqrt{b^2 - 4ac}}{2a}.$$

530–531
  proof of, 535
Quadratic function A function that may be written as $f(x) = ax^2 + bx + c$, where a, b, and c are real numbers and a is not zero. 509
  graphing, 509–513
  range of, 513
Quantity A number that answers the question "How many?" or "How much?" 318
Quantity-value table A table that displays the quantities and values (or rates) from a particular problem setting, along with their product. 319
  building equations from, 325, 415–418
  solving equations using, 322–328
Quartiles Q_1 is the middle number of the numbers below the median; Q_2 is the median; Q_3 is the middle number of the numbers above the median. 540
Quotient The answer to a division problem. 8

Radical The square root symbol; more generally, the name for square roots and higher-degree roots. 88, 497
Radicand The number or expression under the radical sign. 497

Range The set of outputs from a function; also, the difference between the largest and the smallest number in a data set. 198, 539–540
Rate A special ratio for comparing quantities with different units. 276–277
Ratio A comparison of two (or more) like or unlike quantities. 268
simplifying, 269–270
Rational expressions
addition and subtraction of, 581–583
and the distributive property, 588–589
division of, 573–574
graphs of, 557–558
multiplication of, 572–573
Rational function A function that may be written as the ratio of two polynomials $p(x)/q(x)$, where the denominator is not zero; a function that contains algebra in fraction notation. 555
Rational numbers The set of numbers that may be written as a quotient of integers a/b, $b \neq 0$. 11, 484, 555
Real numbers Rational or irrational numbers. 11, 484, 485
Reciprocal The number that, when multiplied by n, gives 1; the multiplicative inverse. 65
Rectangle, formulas for, 97
Rise The vertical change in a line. 214
Run The horizontal change in a line. 214

Sample standard deviation The square root of the average of the squares of a set of deviations from the mean. 542
Scale The value assigned to the spacing between numbers shown on a coordinate plane. 34
Scientific notation A short way of writing large and small numbers. 385–387
converting to decimal notation, 389
operations with, 390–391
unit analysis and, 390
Set A collection of objects or numbers. 10
Similar figures Geometric figures with corresponding sides that are proportional and corresponding angles that are equal. 295, 296
Simplification property of fractions The property that says that if the numerator and denominator of a fraction contain a common factor, those factors can be eliminated:

$$\frac{ab}{ac} = \frac{a}{a} \cdot \frac{b}{c} = \frac{b}{c}. \qquad 74, 75, 566$$

Simplify To use the simplification property of fractions to eliminate common factors as well as to do the given operations; to combine expressions with like bases and use the properties of exponents to do the indicated operations; to use the order of operations to calculate the value of an expression, to apply the properties of exponents, to add like terms, to change fractions to lowest terms, or to use the number properties. 75, 84, 87
Slope The steepness of a line, expressed as the vertical change relative to the horizontal change. 213, 224, 269–270
finding, 213–216, 234–235
formula for, 215
of horizontal and vertical lines, 216–217
and linear functions, 217–218, 224, 232
negative, 215
use of in graphing, 218–219
Slope formula For ordered pairs (x_1, y_1) and (x_2, y_2),

$$\text{slope} = \frac{y_2 - y_1}{x_2 - x_1}. \qquad 215$$

Solution A value of the variable that makes the equation true. 128
Solution set The set of all solutions to an equation. 128
Solution set of a one-variable inequality The set of values of the input variable that make the inequality a true statement. 169
Solution set to a system of inequalities The set of ordered pairs that make all inequalities in a system true. 460
Solution to a system An ordered pair that makes all of the equations in the system true. 402
Solving an equation The process of finding the values of the variable for the solution set. 128
Special products, 356–359
factoring, 373–374
Square, formulas for, 97
Square root, 12, 484–488
exponents and, 489
multiplication and division with, 489–492
negative signs with, 488–489
product property for, 489–490
quotient property for, 491–492
solving quadratic equations with, 521–522
Square root equations
solving, 501–503
Square root function, 498–501
Squared Raised to the power of 2. 82
Standard deviation, 541–543
Statistics, 289
Subscripts Small letters or numbers placed to the right of variables to distinguish a particular item from a group of similar items. 189
Substitution The replacement of variables with equivalent expressions or numbers. 425
solving systems of equations by, 425–428, 434–436
Subtraction of integers, 57
Sum The answer to an addition problem. 8
Supplementary angles Two angles that add up to a linear pair (180°). 431
Surface area The area needed to cover a three-dimensional object. 100
System of equations A set of two or more equations that are to be solved for the values of the variables making all the equations true. 402
building with guess-and-check tables, 418–421
building with quantity-value tables, 415–418
with an infinite number of solutions, 409–410
with no solution, 408–409
with one solution, 404–408
results of, 429
solving by substitution, 425–428
strategies for solving, 436
in three variables, 421–422, 434–436
System of inequalities A set of two or more inequalities. 460
finding solution sets to, 461

Tables
finding slope from, 215–216
solving equations from, 140–142, 155–158, 163–164, 515–516
Tangent, 294
Terms Numbers, variables, or expressions being added or subtracted. 69, 346
dividing, 75
Tile model, 76
Trapezoid, formulas for, 97
Triangles
formulas for, 97
similar, 298–302

Symbols and Formulas

Symbols

Symbol	Meaning
$\lvert\ \rvert$	absolute value
$a + b$	addition of a and b
b^n	base b with exponent n
{ }	braces
[ ]	brackets
○	circle on a graph: the point is excluded from the graph
^{3}	cube (exponent 3)
$^\circ$	degree (temperature)
Δ	delta: change
$\frac{a}{b}$	division of a by b
●	dot or filled-in circle on a graph: the point is included in the graph
//	double slash on a graph: the spacing between the origin and the first number on the axis is different from the spacing between the other numbers
$a \stackrel{?}{=} b$	is a equal to b?
$\approx$	is approximately equal to
$=$	is equal to
$>$	is greater than
$\geq$	is greater than or equal to
$<$	is less than
$\leq$	is less than or equal to
$\neq$	is unequal to
$a \cdot b$, $a(b)$, ab, $(a)(b)$	multiplication of a and b
$-\infty$	negative infinity
-3	negative 3
$-b$	opposite of b
( )	parentheses
%	percent
$\perp$	perpendicular
π	pi, approximately 3.14
$\pm$	plus or minus (add or subtract)
$+3$	positive three
$+\infty$	positive infinity
. . .	repeats or continues, as in a pattern of numbers
$\mathbb{R}$	set of real numbers
∟	square corner at perpendicular lines
^{2}	square (exponent 2)
$\sqrt{\ }$	square root, radical sign
$a - b$	subtraction of a and b
x_1	variable x with subscript 1

Formulas

(See pages 97 and 100 for perimeter, area, volume, and surface area formulas.)

Distance formula (when traveling): Distance = rate · time, $d = rt$

Distance formula (length of a line segment on a graph): $d = \sqrt{(x_2 - x_1)^2 + (y_2 - y_1)^2}$

Interest formula: Interest = principal · rate · time, $i = prt$

Linear equation: $ax + by = c$ (standard), $y = mx + b$ (slope-intercept)

Pythagorean theorem, where c is always the hypotenuse: $a^2 + b^2 = c^2$

Quadratic equation: $y = ax^2 + bx + c$

Quadratic formula: If $ax^2 + bx + c = 0$, then $x = \dfrac{-b + \sqrt{b^2 - 4ac}}{2a}$ or $x = \dfrac{-b - \sqrt{b^2 - 4ac}}{2a}$

Slope formula: $m = \dfrac{y_2 - y_1}{x_2 - x_1}$

y-intercept: $b = y - mx$